SEQUENCE STRATIGRAPHY AND FACIES ASSOCIATIONS

Sequence Stratigraphy and Facies Associations

Edited by Henry W. Posamentier,
Colin P. Summerhayes,
Bilal U. Haq and George P. Allen

SPECIAL PUBLICATION NUMBER 18 OF THE
INTERNATIONAL ASSOCIATION OF SEDIMENTOLOGISTS
PUBLISHED BY BLACKWELL SCIENTIFIC PUBLICATIONS
OXFORD LONDON EDINBURGH BOSTON
MELBOURNE PARIS BERLIN VIENNA

and published for them by
Blackwell Scientific Publications
Editorial Offices:
Osney Mead, Oxford OX2 0EL
25 John Street, London WC1N 2BL
23 Ainslie Place, Edinburgh EH3 6AJ
238 Main Street, Cambridge
Massachusetts 02142, USA
54 University Street, Carlton
Victoria 3053, Australia

Other Editorial Offices:
Librairie Arnette SA
2, rue Casimir-Delavigne
75006 Paris
France

Blackwell Wissenschafts-Verlag GmbH
Düsseldorfer Str. 38
D-10707 Berlin
Germany

Blackwell MZV
Feldgasse 13
A-1238 Wien
Austria

First published 1993

Set by Semantic Graphics, Singapore
Printed and bound in Great Britain
at the University Press, Cambridge

DISTRIBUTORS

Marston Book Services Ltd
PO Box 87
Oxford OX2 0DT
(*Orders*: Tel: 0865 791155
Fax: 0865 791927
Telex: 837515)

USA
Blackwell Scientific Publications, Inc.
238 Main Street
Cambridge, MA 02142
(*Orders*: Tel: 800 759-6102
617 876-7000)

Canada
Oxford University Press
70 Wynford Drive
Don Mills
Ontario M3C 1J9
(*Orders*: Tel: 416 441-2941)

Australia
Blackwell Scientific Publications Pty Ltd
54 University Street
Carlton, Victoria 3053
(*Orders*: Tel: 03 347-5552)

A catalogue record for this title
is available from the British Library

ISBN 0-632-03548-X

Library of Congress
Cataloging-in-Publication Data

Sequence stratigraphy and facies associations/
edited by Henry W. Posamentier . . . [et al.].
p. cm.
(Special publication number 18
of the International Association of Sedimentologists,
ISSN 0141-3600)
ISBN 0-632-03548-X
1. Geology, Stratigraphic.
I. Posamentier, Henry W.
II. Series: Special publication . . . of the
International Association of Sedimentologists; no. 18.
QE651.S46 1993
551.7—dc20

Contents

Preface

Since the publication of SEPM Special Publication 42 (Wilgus *et al.*, 1988) there has been a virtual explosion of stratigraphic studies utilizing principles of sequence stratigraphy (Eleventh Annual Research Conference, SEPM, 1990; MacDonald, 1991). Although the concept of time stratigraphy is not new (Barrell, 1917; Wheeler, 1958), the packaging of depositional units into systems tracts and sequences is. This new approach (Vail *et al.*, 1977; Vail, 1987; Haq *et al.*, 1988; Posamentier *et al.*, 1988; Posamentier & Vail, 1988; Van Wagoner *et al.*, 1990) has led to the reassessment of the geology of areas that in some cases have been the subject of intense geological scrutiny for decades (e.g. Elliott & Pulham, 1991). Sequence-stratigraphic principles have been applied to data bases ranging from regional seismic (Erskine & Vail, 1988; Haq *et al.*, 1992; Boyd *et al.*, this volume, pp. 581–603), to high-resolution seismic (Tesson *et al.*, 1990 and this volume, pp. 183–196; Okamura and Blum, this volume, pp. 213–232), to outcrop and subsurface (Van Wagoner *et al.*, 1990, Devlin *et al.*, this volume, pp. 501–520; Hadley & Elliott, this volume, pp. 521–535; Posamentier & Chamberlain, this volume, pp. 469–485), to flume-scale data (Posamentier *et al.*, 1992; Wood *et al.*, this volume, pp. 43–53), and to modern systems (Allen & Posamentier, 1991, 1993). The principles have been applied to carbonate as well as clastic facies (Sarg, 1988; Hunt & Tucker, this volume, pp. 307–341; Tucker *et al.*, this volume, pp. 397–415). Clearly, the fundamental principles upon which sequence stratigraphy is based are applicable at a broad range of temporal and physical scales (Posamentier *et al.*, 1992; Posamentier & James, this volume, pp. 3–18).

The modern incarnation of time stratigraphy, i.e. *sequence stratigraphy*, has its roots in the development of seismic stratigraphy in the 1960s by workers at Exxon Production Research Company (Vail *et al.*, 1977). With the advent of high-quality seismic data, stratigraphic information rather than merely structural information became available from these data. Stratal geometries were inferred from seismic-reflection patterns, and a predictive tool for relative age and lithology was developed (Vail *et al.*, 1977).

The observation of apparent global synchroneity of coastal-onlap geometry suggested that a global forcing mechanism was responsible (Vail *et al.*, 1977). The forcing parameter that was thought to be responsible at least at third-order scales and less (i.e. 1–3 million years and less) was global sea level or eustasy rather than local tectonics. The issue of global synchroneity of unconformities and coastal-onlap pattern is still a contentious one, however (Miall, 1986, 1991; Summerhayes, 1986; Hubbard, 1988), in spite of publication of more comprehensive sequence-stratigraphic based, sea-level information recently (Haq *et al.*, 1987, 1988).

With the publication of Jervey (1988), Posamentier *et al.* (1988), Posamentier and Vail (1988), Sarg (1988) and Van Wagoner *et al.* (1990), the emphasis of sequence stratigraphy shifted from age-model prediction, as based on the sea-level cycle charts (Haq *et al.*, 1987, 1988), to lithologic prediction. The sequence-stratigraphic approach initially was strongly embraced by petroleum explorationists and more recently by academicians. The inherent utility of the sequence-stratigraphic approach is the predictive aspect of what appear commonly to be cyclic rock successions. Consequently, it would, perhaps, be more appropriate to refer to this approach as *sequential stratigraphy*, thus placing emphasis on the repetitive nature of rock successions and de-emphasizing the occurrence of sequences and sequence-bounding surfaces. Exactly what constitutes a sequence and what is the nature of the bounding surfaces has become somewhat of an issue in light of the proposal by Galloway (1989a,b) to bound sequences not by unconformities as proposed by Vail *et al.* (1977) and Mitchum (1977) but rather by maximum flooding surfaces. The contrast between these two approaches is addressed by Posamentier and James (this volume, pp. 3–18). However a sequence boundary is chosen, it is important nonetheless to note that prior to the publication of AAPG Memoir 26 the formation of unconformities was commonly attributed to tectonic events. With the publication of Vail *et al.* (1977) came the suggestion that the sedimentary rock record consisted of a succession of responses to 1 to 3 Ma sea-level (third-order) cycles superimposed on lower frequency (first and second) cycles. The observation of high-frequency unconformities (Erskine & Vail, 1988) suggested that sea-level change may occur at even higher frequencies (fourth-, fifth- and sixth-order). The rock record was thus observed to be punctuated by sea-level related unconformities at a variety of scales. It has been suggested that the principles of sequence stratigraphy are applicable at a broad range of temporal and spatial scales (Posamentier *et al.*, 1992; Wood *et al.*, this volume, pp. 43–53). In any event, no matter which type of surface is chosen to bound a sequence, there commonly will be a stratigraphic (in contrast with tectonic) unconformity surface either bounding or within the unit.

This volume is an outgrowth of several sessions on sequence stratigraphy held at the Thirteenth International Sedimentological Congress in Nottingham, England, in August, 1990. The emphasis at these sessions was on facies associations within a sequence-stratigraphic framework. The papers included in this volume comprise most of the presentations at the Nottingham meeting. The emphasis here, as it was at the conference, is facies associations within a sequence-stratigraphic framework.

This volume includes five papers on concepts and six papers on methodology, followed by applications of sequence concepts in a number of different tectonic and stratigraphic settings. With the exception of those papers based on high-resolution seismic data, which are grouped together, the papers are grouped by geographic locality, including examples from North America, Europe, Greenland and Australia. This volume represents the first comprehensive compilation of sequence-stratigraphic applications with an emphasis on the integration of facies associations. We believe that the papers included here are merely the forerunners of what will eventually be an extensive body of literature in the not too distant future.

Henry W. Posamentier
Plano, Texas

Colin P. Summerhayes
Godalming, UK

Bilal U. Haq
Washington, DC

G.P. Allen
St Remy les Chevreuses, France

REFERENCES

Allen, G.P. & Posamentier, H.W. (1991) Facies and stratal patterns in incised valley complexes: examples from the Recent Gironde Estuary (France), and the Cretaceous Viking Formation (Canada). *Am. Assoc. Petrol. Geol., Annual Meeting, Dallas TX, April 7–10, 1991, Programs with Abstracts*, p. 70.

Allen, G.P. & Posamentier, H.W. (1993) Sequence stratigraphy and facies model of an incised valley fill: the Gironde Estuary, France. *J. Sedement. Petrol.* (*in press*).

Barrell, J. (1917) Rhythms and the measurement of geological time. *Bull. Geol. Soc. Am.* **44**, 745–904.

Eleventh Annual Research Conference, SEPM (1990) Sequence stratigraphy as an exploration tool. *SEPM Gulf Coast Section, December 2–5, 1990, Program and Extended and Illustrated Abstracts*.

Elliott, T. & Pulham, A. (1991) The sequence stratigraphy of Upper Carboniferous deltas, western Ireland. *Am. Assoc. Petrol. Geol., Annual Meeting, Dallas TX, April 7–10, 1991. Programs with Abstracts*, p. 104.

Erskine, R.D. & Vail, P.R. (1988) Seismic stratigraphy of the Exmouth Plateau. In: *Atlas of Seismic Stratigraphy*, Vol. 2 (Ed. Bally, A.W.) Am. Assoc. Petrol. Geol., Stud. Geol. 27, 163–173.

Galloway, W.E. (1989a) Genetic stratigraphic sequences in basin analysis I: architecture and genesis of flooding-surface bounded depositional units. *Bull. Am. Assoc. Petrol. Geol.* **73**, 125–142.

Galloway, W.E. (1989b) Genetic stratigraphic sequences in basin analysis II: application to northwest Gulf of Mexico Cenozoic Basin. *Bull. Am. Assoc. Petrol. Geol.* **73**, 143–154.

Haq, B.U., Boyd, R.L., Exon, N.F. & von Rad, U. (1992) Evolution of central Exmouth Plateau: a post drilling perspective. In: *Proceedings of ODP Scientific Results* (Eds von Rad, U., Haq, B.U., *et al.*), 122, pp. 801–816.

Haq, B.U., Hardenbol, J. & Vail, P.R. (1987) Chronology of fluctuating sea levels since the Triassic. *Science* **235**, 1156–1166.

Haq, B.U., Hardenbol, J. & Vail, P.R. (1988) Mesozoic and Cenozoic chronostratigraphy and cycles of sea level change. In: *Sea-level Changes: An Integrated Approach* (Eds Wilgus, C.K., Hastings, B.S., Kendall, C.G.St.C., Posamentier, H.W., Ross, C.A. & Van Wagoner, J.C.), Spec. Publ. Soc. Econ. Paleontol. Mineral. 42, 71–108.

Hubbard, R.J. (1988) Age and significance of sequence boundaries on Jurassic and early Cretaceous rifted continental margins. *Bull. Am. Assoc. Petrol. Geol.* **72**, 49–72.

Jervey, M.T. (1988) Quantitative geological modeling of siliciclastic rock sequences and their seismic expression. In: *Sea-level Changes: An Integrated Approach* (Eds Wilgus, C.K., Hastings, B.S., Kendall, C.G.St.C., Posamentier, H.W., Ross, C.A. & Van Wagoner, J.C.), Spec. Publ. Soc. Econ. Paleontol. Mineral. 42, 47–69.

MacDonald, I.M. (1991) *Sedimentation, Tectonics and Eustasy*. IAS Special Publication Number 12, Blackwell Scientific Publications, Oxford.

Miall, A.D. (1986) Eustatic sea level changes interpreted from seismic stratigraphy: a critique of the methodology with particular reference to the North Sea Jurassic record. *Bull. Am. Assoc. Petrol. Geol.* **70**, 131–137.

Miall, A.D. (1991) Stratigraphic sequences and their chronostratigraphic correlation. *J. Sediment. Petrol.* **61**, 497–505.

Mitchum, R.M., Jr. (1977) Seismic stratigraphy and global changes of sea level, part 1: glossary of terms used in seismic stratigraphy. In: *Seismic Stratigraphy—Applications to Hydrocarbon Exploration* (Ed. Payton, C.E.), Mem. Am. Assoc. Petrol. Geol. 26, 205–212.

Posamentier, H.W. & Vail, P.R. (1988) Eustatic controls on clastic deposition II—sequence and systems tract models. In: *Sea-level Changes: An Integrated Approach* (Eds Wilgus, C.K., Hastings, B.S., Kendall, C.G.St.C., Posamentier, H.W., Ross, C.A., & Van Wagoner, J.C.), Spec. Publ. Soc. Econ. Paleontol. Mineral. 42, 125–154.

Posamentier, H.W., Allen, G.P. & James, D.P. (1992) High resolution sequence stratigraphy—The East Coulee delta. *J. Sediment. Petrol.* **62** 310–317.

Posamentier, H.W., Jervey, M.T. & Vail, P.R. (1988) Eustatic controls on clastic deposition I—conceptual framework. In: *Sea-level Changes: An Integrated Approach* (Eds Wilgus, C.K., Hastings, B.S., Kendall, C.G.St.C., Posamentier, H.W., Ross, C.A. & Van Wagoner, J.C.), Spec. Publ. Soc. Econ. Paleontol. Mineral. 42, 109–124.

Sarg, J.F. (1988) Carbonate sequence stratigraphy. In: *Sea-level Changes: An Integrated Approach* (Eds Wilgus, C.K., Hastings, B.S., Kendall, C.G.St.C., Posamentier, H.W., Ross, C.A. & Van Wagoner, J.C.) Spec. Publ. Soc. Econ. Paleontol. Mineral. 42, 155–181.

Summerhayes, C.P. (1986) Sea level curves based on seismic stratigraphy: their chronostratigraphic significance. *Palaeogeogr., Palaeoclim., Palaeoecol.* **57**, 27–42.

Tesson, M., Gensous, B., Allen, G.P. & Ravenne, C. (1990) Late Quaternary deltaic lowstand wedges on the Rhône continental shelf, France. *Mar. Geol.* **91**, 325–332.

Vail, P.R. (1987) Seismic stratigraphy interpretation procedure. In: *Atlas of Seismic Stratigraphy* (Ed. Bally, A.W.), Am. Assoc. Petrol. Geol. Stud. Geol. 27, 1–10.

Vail, P.R., Mitchum, R.M., Jr. & Thompson, S., III (1977). Seismic stratigraphy and global changes of sea level, part 3: relative changes of sea level from coastal onlap. In: *Seismic Stratigraphy—Applications to Hydrocarbon Exploration* (Ed. Payton, C.E.), Mem. Am. Assoc. Petrol. Geol. 26, 63–81.

Van Wagoner, J.C., Mitchum, R.M., Jr, Campion, K.M. & Rahmanian,V.D. (1990) Siliciclastic sequence stratigraphy in well logs, cores, and outcrops: concepts for high-resolution correlation of time and facies. *Am. Assoc. Petrol. Geol. Meth. Expl. Ser.* **7**, 55 pp.

Wheeler, H.E. (1958) Time stratigraphy. *Bull. Am. Assoc. Petrol. Geol.*, **42**, 1047–1063.

Wilgus, C.K., Hastings, B.S., Kendall, C.G.St.C., Posamentier, H.W., Ross, C.A. & Van Wagoner, J.C. (Eds) (1988) *Sea-level Changes: An Integrated Approach.* Spec. Publ. Soc. Econ. Paleontol. Mineral. 42, 407 pp.

Sequence-stratigraphic Concepts and Principles

Spec. Publs Int. Ass. Sediment. (1993) **18**, 3–18

An overview of sequence-stratigraphic concepts: uses and abuses

H.W. POSAMENTIER* *and* D.P. JAMES†

**ARCO Exploration and Production Technology, 2300 West Plano Parkway, Plano, TX 75075, USA; and*
†Saskoil, 140 4th Avenue SW, Calgary, Alberta T2P 3S3, Canada

ABSTRACT

There has been a proliferation in recent years of applications of sequence-stratigraphic concepts since the publications by Vail *et al.* (1977a,b), Posamentier *et al.* (1988), Posamentier and Vail (1988a,b) and Van Wagoner *et al.* (1990). The use of this approach to enhance understanding of geological relationships within a time-stratigraphic framework clearly has achieved widespread acceptance. Along with the rapid acceptance of these concepts, however, has come a number of problems that are addressed in this paper.

The inevitable introduction of a morass of terminology as well as occasional misunderstanding and misapplication of the stratigraphic model has led to considerable confusion. Moreover, certain key contentious issues within sequence stratigraphy remain that need to be resolved. Most importantly, however, sequence stratigraphy must be viewed as a *tool* or *approach* rather than a *rigid template*. The conceptual stratigraphic models are based on *first principles* and it is the understanding of these first principles rather than memorization of the model that is the key to successful application of sequence stratigraphy.

INTRODUCTION

The publications by Vail *et al.* (1977a), Vail (1987), Posamentier and Vail (1988a), Posamentier *et al.* (1988) and Van Wagoner *et al.* (1990), as well as many of the studies included in SEPM Special Publication 42 (Wilgus *et al.*, 1988) have stimulated much discussion about the application of sequence-stratigraphic concepts (Miall, 1986, 1991; Summerhayes, 1986; Hubbard, 1988; Walker, 1990). Critical comments range from questioning the validity of interbasinal stratigraphic correlations upon which the Haq *et al.* (1987) global eustatic curve is based (Hubbard, 1988), to questioning the validity of certain aspects of the stratigraphic models (Miall, 1991) as presented by Jervey (1988), Posamentier *et al.* (1988) and Posamentier and Vail (1988a). These discussions and the ensuing studies are and will be useful and healthy as sequence concepts continue to evolve.

Sequence stratigraphy — tool versus template

All too often sequence-stratigraphic concepts are applied without being fully understood. The key to correct application lies in the understanding of key first principles. Armed with these fundamental concepts the user can then appropriately utilize the sequence concepts to build models suitable to each unique geologic setting. As Posamentier and Vail (1988b) pointed out, the conceptual block diagrams presented in Posamentier and Vail (1985, 1988a) and Posamentier *et al.* (1988) are just one possible stratigraphic response to fluctuating sea level. Posamentier and Vail (1988a) specifically state that before applying sequence-stratigraphic concepts, local conditions regarding tectonics, sediment flux and physiography *must* be taken into account. Consequently, given the unlimited variations possible,* the sequence-stratigraphic concepts should be applied as a *tool* or *approach* rather than as a *template*. It is extremely unlikely that any geologic setting perfectly conforms to the simplified block diagrams presented in Posamentier *et al.* (1988). As with any model, the caution stated by Anderton (1985) must be kept in mind so as to minimize misuse and abuse: 'whether it is the intention of their creators or not, there is a danger that ['ideal' or 'generalized'] models are taken too literally and applied out of context'. *Thus, sequence stratigraphy should be considered as a way of looking at and ordering geologic data, rather than an end in itself.*

As recent work has shown, sequence-stratigraphic principles apply at all scales, both temporal and spatial. These principles have been applied at the flume scale (Posamentier *et al.*, 1992a; Wood *et al.*, this volume, pp. 43–53), at outcrop and core scale (Posamentier & Chamberlain, 1989; Van Wagoner *et al.*, 1990), at the scale of high-resolution seismic data (Suter & Berryhill, 1985; Suter *et al.*, 1987; Okamura, 1989 and this volume, pp. 213–232; Tesson *et al.*, 1990), as well as at the scale of conventional multi-channel seismic data (Erskine & Vail, 1988). It is important to re-emphasize that sequence-stratigraphic concepts deal with the stratigraphic response to the interaction between sediment flux on the one hand and the space that is made available on the shelf for those sediments to fill on the other. The shelf space (or accommodation) that is made available is a function of eustatic change and total sea-floor subsidence (due to tectonics, loading, sediment compaction, etc.). *These parameters are essentially space and time independent.*

Definitions

Sequence stratigraphy has been defined as 'the study of rock relationships within a chronostratigraphic framework of repetitive, genetically related strata bounded by surfaces of erosion or nondeposition, or their correlative conformities' (Van Wagoner *et al.*, 1988). Posamentier *et al.* (1988) provide a similar definition, though they take a somewhat more generic approach by not specifying the nature of the bounding surfaces in their definition. Implicitly, in both views the *key* attributes of the sequence-stratigraphic approach are the *cyclic nature* of stratigraphic successions and the use of the *chronostratigraphic framework* to enhance lithologic prediction.

Sequence stratigraphy has its roots in earlier works by Sloss (1962, 1963), Wheeler (1958, 1959), Weller (1960) and others (Wheeler & Murray, 1957; Moore, 1964). Their works also were based on the establishment of chronostratigraphic frameworks and the interpretation of time lines across rock sections. They, too, recognized the importance of relative sea-level change in the evolution of basin fill. Sequence stratigraphy has experienced a major period of development since the mid-1970s (Vail *et al.*, 1977a) due largely to the advent and availability of high-quality seismic data and the recognition that chronostratigraphic information could be inferred directly and reasonably unambiguously from these data (Ross, pers. com.).

Similar approaches to what has been referred to as *sequence stratigraphy* are embodied within *event stratigraphy* (Einsele & Seilacher, 1982), *cyclostratigraphy* (Perlmutter & Matthews, 1989), *allostratigraphy* (North American Commission on Stratigraphic Nomenclature, 1983, p. 865; Bergman & Walker, 1988) and *genetic stratigraphy* (Galloway, 1989a,b). The common thread linking these approaches is the recognition of cyclicity in the rock succession and the recognition of the utility of establishing a time-stratigraphic framework (Fig. 1). The key differences between these approaches lie primarily in the determination of what should be the sequence-bounding surfaces and what should be their recognition criteria.

Sequence stratigraphy versus genetic stratigraphy

The issue of where sequences begin and end, or what should be the bounding surfaces, constitutes the fundamental difference between *sequence stratigraphy* and *genetic stratigraphy*. The approach taken by Galloway (1989a,b) in his discussion of *genetic stratigraphy* is to subdivide the stratigraphic succession at *condensed sections* (CS) or *maximum flooding surfaces* (MFS) (in the terminology of Posamentier & Vail, 1988a). In contrast, the *sequence-stratigraphic* approach employed by the Exxon group (Mitchum, 1977; Vail *et al.*, 1977a; Posamentier & Vail, 1985, 1988a; Posamentier *et al.*, 1988; Van Wagoner *et al.*, 1990) is to subdivide the stratigraphic succession at *unconformity surfaces* (Fig. 2) and their correlative conformities.

*Posamentier and Vail (1988b) conservatively identify 1152 possible combinations of variables leading to an equal number of possible block diagram scenarios.

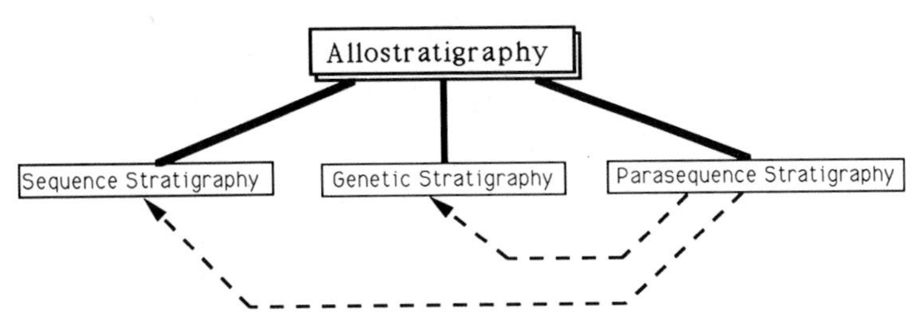

Fig. 1. The relationship of allostratigraphy to sequence stratigraphy, genetic stratigraphy and parasequence stratigraphy. Each involves the recognition of disconformities though the relative significance of each of the disconformity types varies from one approach to the other. The allostratigrapic approach involves the identification of all disconformities without necessarily assigning increased significance to any particular types. In contrast, the sequence-stratigraphic approach emphasizes the importance of unconformities whereas the genetic-stratigraphic approach emphasizes the importance of flooding surfaces. Parasequence stratigraphy involves the recognition of shallowing-upward successions bounded by surfaces associated with abrupt increases of water depth. Note that parasequences form the building blocks of the units identified in both sequence and genetic stratigraphy.

In many situations, it appears easier at first pass to recognize condensed sections than it is to recognize sequence boundaries. Condensed sections are distinctively thin, marine stratigraphic units consisting of pelagic to hemipelagic sediments (Loutit *et al.*, 1988) and have widespread nearly uniform distribution. Unconformities, on the other hand, can have a variety of expressions. They can be characterized by a sharp erosional contact where incised valleys occur (Harms, 1966; Weimer, 1983, 1984; Posamentier & Vail, 1988a; Van Wagoner *et al.*, 1990). In areas between incised valleys (i.e. interfluves) the unconformity may be more difficult to identify because in this setting this unconformable surface* may be characterized by a shale-on-shale contact, possibly separated by a thin, transgressive lag deposit. On basin margins, where gentle ramp-margin physiography precludes the formation and/or preservation of incised valleys, unconformities may be expressed exclusively as a succession of E/T surfaces overlying in some cases shallow marine sandy units and in others overlying offshore shaley units (Plint *et al.*, 1987; Posamentier & Chamberlain, 1989 and this volume, pp. 469–485). In contrast, the recognition of the condensed sections, with their blanket-like distribution, appears to be more straightforward, punctuating a succession of regressive events.

The rationale for bounding the sequence with unconformities rather than with condensed sections in spite of the greater difficulties involved is both *philosophical* as well as *economic*. The term *sequence* is defined as 'the following of one thing after another in chronological, causal, or logical order' or 'a continuous or related series' (*Webster's New World Dictionary*, 1982). The key concept relevant here is that internally a sequence should be characterized by a continuum of 'things' (in geology, this would refer to beds, bedsets, etc.).

In a stratigraphic succession punctuated by unconformities and condensed sections the most significant and recognizable breaks (e.g. temporal, stratigraphic, depositional, etc.) occur at unconformities, not at condensed sections. The condensed section is merely a very fine-grained lithofacies that is part of the *continuum of deposition* with the sediments immediately above and below, albeit deposited at varying rates and deposited by varying processes. In contrast, a distinct break occurs at an unconformity. At this surface, variable amounts of section may be lost by erosion associated with subaerial processes such as fluvial incision as well as nearshore marine processes such as beach erosion during transgression†. During intervals of subaerial exposure, fluvial incision or at least sedimentary bypass also may result in intervals of time characterized by non-deposition in certain areas. For

*This surface has been referred to as an E/T (erosion/truncation) surface by Plint *et al.* (1987); Posamentier and Chamberlain (1991, in press) provide an example of this type of surface.

†Transgression across an eroded surface may produce a ravinement surface that is superimposed on an unconformity. The result of the transgressive erosion may be to eliminate all evidence of subaerial exposure. Nonetheless, implicit in the interpretation of a ravinement surface is the conclusion that the shoreline must have been seaward of and subsequently passed or transgressed the location of the study area, thus providing indirect evidence for subaerial exposure there.

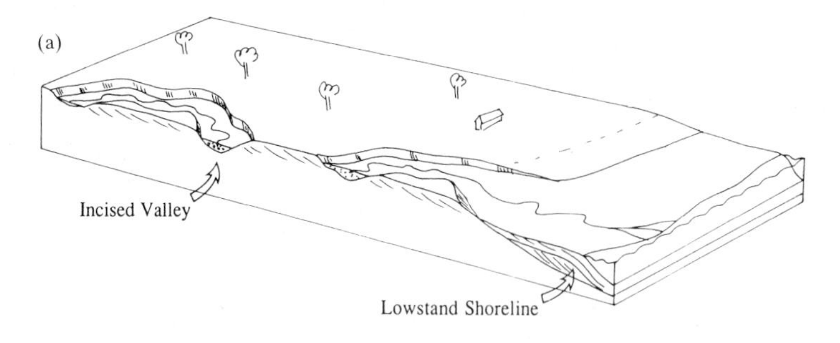

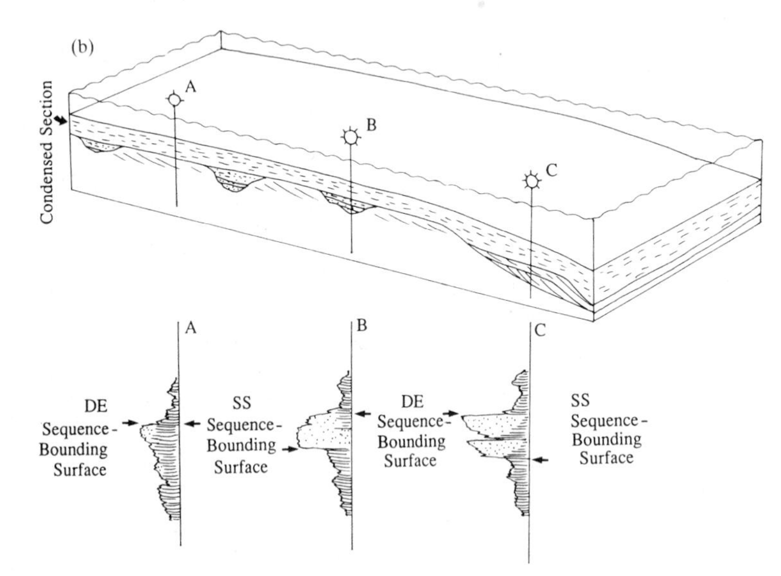

Fig. 2. Schematic illustration of the difference in packaging of sequences: sequence stratigraphy versus depositional episodes. (a) Illustrates a lowstand of relative sea level characterized by fluvial incision and deposition of lowstand shorelines. (b) Illustrates a later situation that post-dates a transgressive event. The hypothetical well logs at locations A, B and C are annotated with the position of the sequence-bounding surface according to both the sequence-stratigraphic approach after Posamentier *et al.* (1988) and the depositional episodes (i.e. genetic stratigraphy) approach after Galloway (1989a,b).

example, during falls and subsequent lowstands of relative sea level passive margins may be characterized by incised valleys and sedimentary bypass on the shelf and, in contrast, submarine fan deposition in the deeper waters of the slope and basin. Consequently, the time gap across unconformity surfaces marks the break in the continuum of depositional processes and can be very pronounced.

From an *economic* perspective (i.e. *vis-à-vis* oil exploration), the choice of unconformities as sequence-bounding surfaces is preferred because commonly these surfaces immediately underlie a variety of siliciclastic reservoir facies in a variety of environmental settings. These facies range from fluvial/estuarine incised valley fills to lowstand shorelines to deep-sea submarine fans. This facies association is in marked contrast with the lack of reservoir facies associated with the condensed section. Moreover, because of the seaward shift of facies assemblages associated with unconformities, these reservoir deposits are often encased in shales. Thus, reservoir, source and seal are in close proximity to each other at these surfaces. It therefore behoves the petroleum explorationist to identify these key surfaces and subdivide the stratigraphic succession using unconformities as sequence-bounding surfaces.

Within the condensed section, the maximum flooding surface, representing the time of maximum flooding, is nonetheless an important surface in the sequence-stratigraphic approach. It separates the transgressive systems tract from the subsequent highstand systems tract. Under certain circumstances, it can immediately overlie the unconformity where the transgressive systems tract is very thin or absent, or it can immediately underlie the unconformity where the highstand systems tract is very thin or absent.

When working a data set from a new area, it is common sequence-stratigraphic practice initially to identify the condensed sections or maximum flooding surfaces. These are characteristically easier to identify and their identification represents a useful starting point in tackling a new data set. On well logs condensed sections may occur as highly radioactive shales, in core and outcrop as organic-rich shales and on seismic data as downlap surfaces*. Once these surfaces are identified the next and potentially more difficult but arguably more important step is to identify the sequence boundaries inferred to occur between successive condensed sections or maximum flooding surfaces.

APPLICATION OF SEQUENCE-STRATIGRAPHIC CONCEPTS

Sequence-stratigraphic concepts may be applied in two fundamentally different ways. One application involves the *construction of age models* for given stratigraphic successions based on correlation of local stratigraphy with the global cycle chart published most recently by Haq *et al.* (1987) (Fig. 3). The other application involves *lithology prediction* based on the interpretation of cyclicity in the rock record (Figs 1 to 6 in Posamentier *et al.*, 1988).

This use of sequence-stratigraphic principles in the construction of age models is based on the assumption that stratigraphic successions preserved in a basin are a function primarily of eustatic rather than local tectonic fluctuations†. In order for this assumption to be valid, eustatic change must vary with a higher frequency and amplitude than tectonics and consequently be the dominant signal to which the sediments respond. Criticisms have been levelled at this application based on the assertion that local tectonics can vary at a frequencies at least equivalent to proposed third-order cycles (1 to 3 Ma) of sea-level change (Cloetingh, 1986). If this assertion is correct, then interbasinal correlations would be invalid in so far as tectonic effects may vary from basin to basin. Nonetheless, Cloetingh *et al.* (1989) suggest that local tectonics of the North Atlantic could have influenced stratigraphic evolution in a similar fashion in a number of basins in this area and therefore could be characterized by a similar stratigraphy. Furthermore, they contend that the Haq *et al.* (1987) sea-level curves, which are based largely on Atlantic-margin stratigraphy, are consequently relevant only to these basins rather than globally. Further criticisms regarding this application are based on the imprecision of palaeomagnetic, palaeontologic and radiometric age-dating upon which the Haq *et al.* (1987) curves are based, as well as the inadequacy of documentation

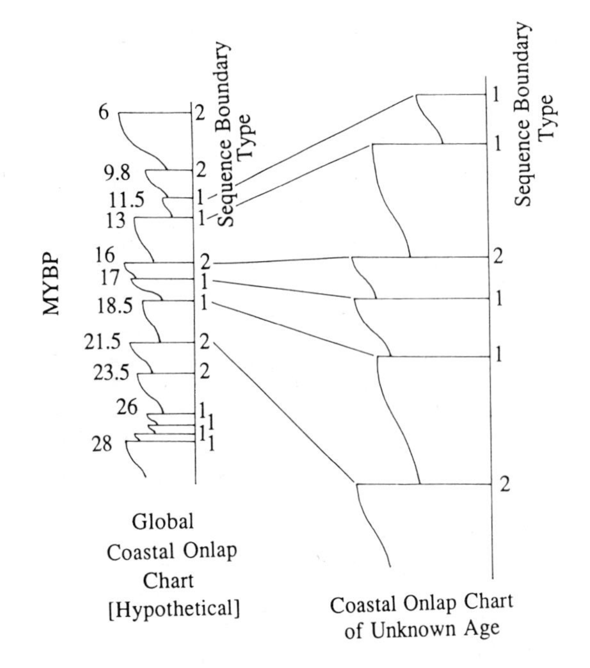

Fig. 3. Construction of age model based on comparisons of local coastal onlap data with global coastal onlap chart. Hypothetical data are shown. The procedure involves first establishing coastal onlap geometry in the area of interest and then correlating this with the global coastal onlap chart by sliding it up or down until a best fit is achieved.

*Note that not all hot shales are within the condensed section facies. For example, hot shales also can be associated with tectonically induced reduced oceanic circulation (e.g. Kimmeridgian shales in the North Sea). The potential pitfall to the stratigrapher is that if two successive hot shales are interpreted as condensed sections, the occurrence of a sequence boundary between them may be erroneously inferred.

†Depositional sequences develop in response to changes in the amount of space available on the shelf for sediment to fill (i.e. accommodation). The measure of this changing accommodation is relative sea-level change. Relative sea level is a function of both sea surface (i.e. eustatic) and sea floor (including tectonics and sediment compaction) movements (see discussion in Posamentier & Vail, 1988a).

provided (Summerhayes, 1986; Hubbard, 1988; Miall, 1991). Consequently, confidence in the validity of these curves is not uniform.

Construction of age models using sequence concepts has its most common application in data-poor areas where correlation with global sea-level charts is a useful starting point. In some cases, where only seismic data are available, this application serves as *the only* basis for age-model construction. This approach is useful also in addressing questions of local tectonic activity if one assumes that eustatic change is real and is known. There are nonetheless those who would argue that global sea-level charts have no relevance since globally synchronous sea-level change is a fallacy (e.g. Mörner, 1976; Miall, 1991). If this assertion were correct, then, of course, these types of analyses would be rendered baseless and meaningless.

The second application of sequence-stratigraphic concepts involves the *recognition of a predictable lithologic succession* occurring in response to variations of *relative* sea level (a function of *both* eustasy and tectonics) as well as sediment flux and physiography. This application does not depend on whether eustatic change or tectonics is the dominant factor, or what is (are) the cause(s) of sea-level cyclicity, or, indeed, whether the interbasinal correlations and hence the compilation of a 'global' sea-level curve are valid. The principal consideration here is that *relative sea-level changes* (regardless of the varying relative contributions of eustasy and tectonics) operating in concert with *sediment flux* and *physiography* are the primary control on the stratigraphic succession.

The application of sequence-stratigraphic concepts for lithologic prediction is most useful in mature basins where lithologic prediction rather than age prediction has clearly greater exploration significance. In these types of basins, sequence concepts can be invaluable in the prediction of source and seal, as well as reservoir lithologies. Sequence-stratigraphic concepts can also be useful in the development of new play concepts such as incised valley fills (e.g. Van Wagoner *et al.*, 1990; Allen & Posamentier, 1991) and lowstand shorelines (e.g. Plint, 1988; Posamentier & Chamberlain, 1989 and this volume, pp. 469–485; Posamentier *et al.*, 1992b).

Parasequence versus sequence

The terms *parasequence* and *sequence* have been used in inconsistent and somewhat confusing ways by certain authors who have used these terms interchangeably. As discussed below, the parasequence is a descriptive term devoid of issues regarding origin, nature and significance of their bounding surfaces and spatial and temporal relationships. In contrast, current usage of the term *sequence* does include consideration of the aforementioned issues. Consequently, the parasequence *should not be considered a sequence* at a small scale as it has sometimes been misused.

In order to address this concern it is useful to examine the origins of these terms. The term *parasequence* was loosely derived from Fig. 1 in Vail *et al.* (1977b) to refer to the sediments deposited during a *paracycle* (Van Wagoner, pers. com.). This usage differs from that of Vail *et al.* (1977b). The *paracycle* was proposed by Vail *et al.* (1977b) to be part of the 'hierarchy of supercycle, cycle and paracycle [which is meant to] reflect relative changes of sea level of different orders of magnitude'. From their Fig. 1, it appears that the duration of the paracycle is approximately 2 million years and is described as a fourth-order cycle of sea-level change. They observe that unless sediment flux is high, the sediments deposited during this interval typically are below seismic resolution. Nonetheless, Vail *et al.* (1977b) argue that depositional sequences are deposited during each paracycle of sea-level change. Herein lies the crux of the problem: Vail *et al.* (1977b) assert that paracycles are no different from cycles and supercycles in the sense that each is associated with a sea-level cycle (albeit of different duration). *The common thread that links the hierarchy is that each scale of cycle is manifested by a depositional sequence.* On the other hand, the parasequence, as it has been defined subsequently*, is not in and of itself a depositional sequence. Consequently, the parasequence is not part of this hierarchy although it may form the building blocks that ultimately comprise the sequence (Van Wagoner, 1985).

The term 'parasequence' introduced by Van Wagoner (1985) was intended to be a rock-based descriptive term and as such *without significance regarding temporal and spatial considerations.* A shoaling-upward succession may range in thickness

*The term *parasequence* has been defined as a relatively conformable succession of genetically related beds or bedsets bounded by marine flooding surfaces and their correlative surfaces (Van Wagoner, 1985).

from decimetres to tens of metres and likewise may be deposited anywhere from a period of days to millions of years. This usage is clearly at odds with any usage related to paracycles in the sense of Vail *et al.* (1977b). Consequently, any reference to a succession being at *parasequence scale* (or for that matter *sequence scale*) is therefore unclear and misleading.

The term 'sequence' has been defined by Mitchum (1977) as 'a relatively conformable succession of genetically related strata bounded at its top and base by unconformities or their correlative conformities'. Subsequently, this definition was refined to include the concept that it is 'composed of a succession of systems tracts and is interpreted to be deposited between eustatic fall inflection points' (Posamentier, *et al.*, 1988). Temporal and spatial considerations clearly are not part of these definitions and, consequently, *sequence-stratigraphic principles may be said to be spatially and temporally scale independent.*

This leads to the question regarding the relationship between the two terms sequence and parasequence. In one sense they are like apples and oranges in so far as the latter describes strictly a shoaling-upward succession devoid of inferences regarding sea-level change, systems tracts, or other such interpretive aspects, whereas the former explicitly involves inferences regarding sea-level change and component systems tracts. Nonetheless, a parasequence can be analysed or interpreted in sequence-stratigraphic terms. The first step is the recognition of the shoaling-upward nature of the section (i.e. identification of parasequences), followed by the recognition and interpretation of key bounding surfaces (e.g. ravinement surfaces, maximum flooding surfaces, unconformities, etc.), condensed sections, etc., and then recognition of stratigraphic relationships and identification of systems tracts and sequences. In this way, parasequences serve as the *building blocks* of sequences (Van Wagoner, 1985).

The following illustration (Fig. 4) based on data from Posamentier and Chamberlain (1989 and this volume, pp. 469–485) may be the best way to illustrate how these terms may be used in conjunction with one another. In a limited *window to the world* (i.e. a data set with restricted areal distribution) a shoaling-upward succession bounded by flooding surfaces has been observed and described (well 10-21-47-22W4, Line 2, Fig. 4) (and can be referred to as a *parasequence*). Farther basinward, a similar shoaling-upward succession bounded by flooding surfaces has been observed and described (well 5-5-48-20W4, Line 2, Fig. 4) (and also can be described or referred to as a *parasequence*). Upon closer inspection, it is observed that the two units are not coeval; in fact the parasequence in the landward setting was interpreted as part of the highstand systems tract (in the sense of Posamentier & Vail, 1988a), whereas the parasequence in the basinal setting can be interpreted as part of the lowstand systems tract. This implies that the flooding surface observed at the top of the parasequence in the landward setting is, in fact, both a flooding surface as well as an unconformity (implying a significant hiatus) and therefore a *sequence boundary*. The flooding surface at the top of the parasequence observed in the basinal setting marks the contact between the lowstand and subsequent transgressive systems tracts. It is merely a ravinement surface and not an unconformity (thus implying no significant hiatus). The sequence boundary here occurs at the base of the shoaling-upward succession and may be characterized by a sharp base to the shoreface succession (e.g. wells 7-35-48-22W4 and 5-36-48-22W4, Line 1, Fig. 4) (see discussion in Plint, 1988 and Posamentier *et al.*, 1992b) or may be simply a correlative conformity not readily distinguishable (e.g. wells 5-5-48-20W4 and 7-6-48-20W4, Line 2, Fig. 4).

The *sequence*, then, comprises the parasequences observed at each locality and places them, as well as their bounding surfaces, into a sequence framework. Consequently, the initial interpretation or recognition of parasequences is only the first step towards ultimate interpretation of the sections in a sequence-stratigraphic framework. The parasequences are the more observational building blocks used in construction of the more interpretive sequence-stratigraphic framework.

In this example, though parasequences are observed at each location, their sequence-stratigraphic significance with regard to temporal and spatial relationships with each other and interpretation of bounding surfaces is radically different. The value of going the additional step from interpretation of parasequences to interpretation within a sequence-stratigraphic framework lies in the enhanced predictive ability regarding reservoir occurrence, stratigraphic trapping potential and source and seal considerations.

Figure 5 illustrates the relative effects of eustasy and subsidence on sequence and parasequence

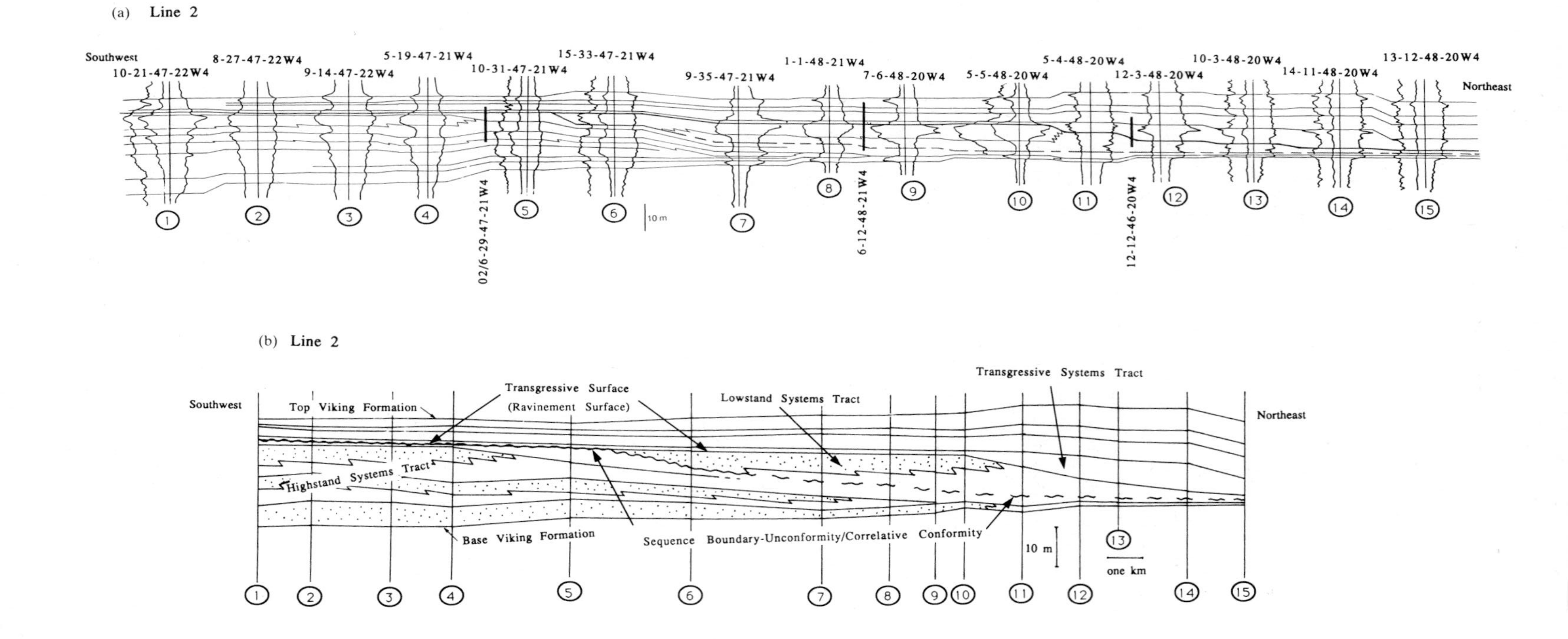
(a) Line 2
Southwest
10-21-47-22W4
8-27-47-22W4
9-14-47-22W4
5-19-47-21W4
10-31-47-21W4
15-33-47-21W4
9-35-47-21W4
1-1-48-21W4
7-6-48-20W4
5-5-48-20W4
5-4-48-20W4
12-3-48-20W4
10-3-48-20W4
14-11-48-20W4
13-12-48-20W4
Northeast
02/6-29-47-21W4
6-12-48-21W4
12-12-46-20W4
10 m
1
2
3
4
5
6
7
8
9
10
11
12
13
14
15
(b) Line 2
Southwest
Northeast
Top Viking Formation
Transgressive Surface
(Ravinement Surface)
Lowstand Systems Tract
Transgressive Systems Tract
Highstand Systems Tract
Base Viking Formation
Sequence Boundary-Unconformity/Correlative Conformity
10 m
one km

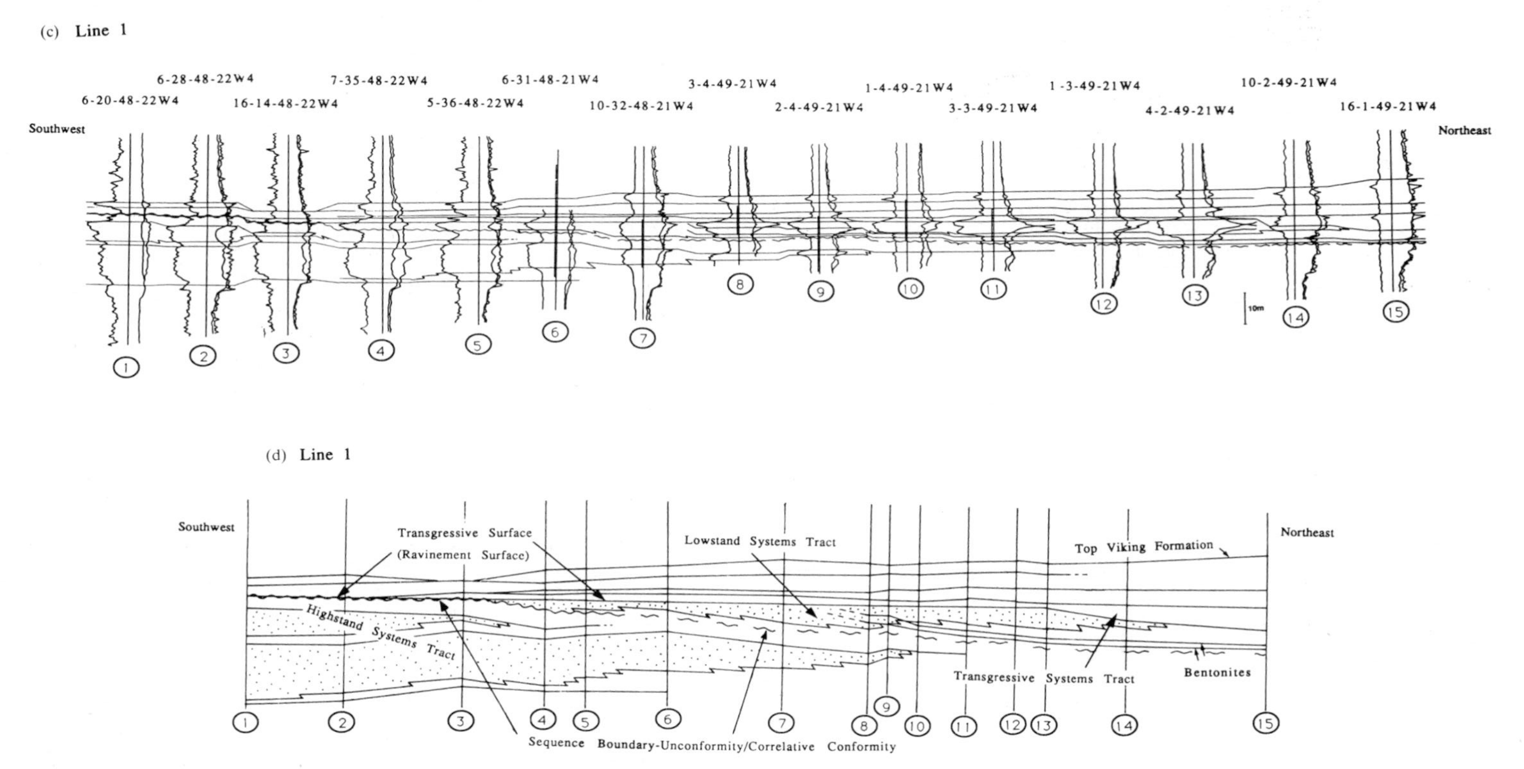

Fig. 4. Two dip-oriented well-log cross-sections across Joarcam Field, Alberta, Canada, illustrate shoreface progradation from the southwest to the northeast. Initial deposition of highstand systems tract deposits is separated from subsequent lowstand systems tract deposits by an unconformity and correlative conformity. The transgressive systems tract follows deposition of the lowstand systems tract and is separated from it in the proximal areas by a flooding surface (ravinement surface) and in the distal areas by a drowning surface. Landward of where the lowstand systems tract pinches out the transgressive systems tract directly overlies the highstand systems tract and is separated from it by a flooding surface (ravinement surface). (From Posamentier & Chamberlain, this volume, pp. 469–485.)

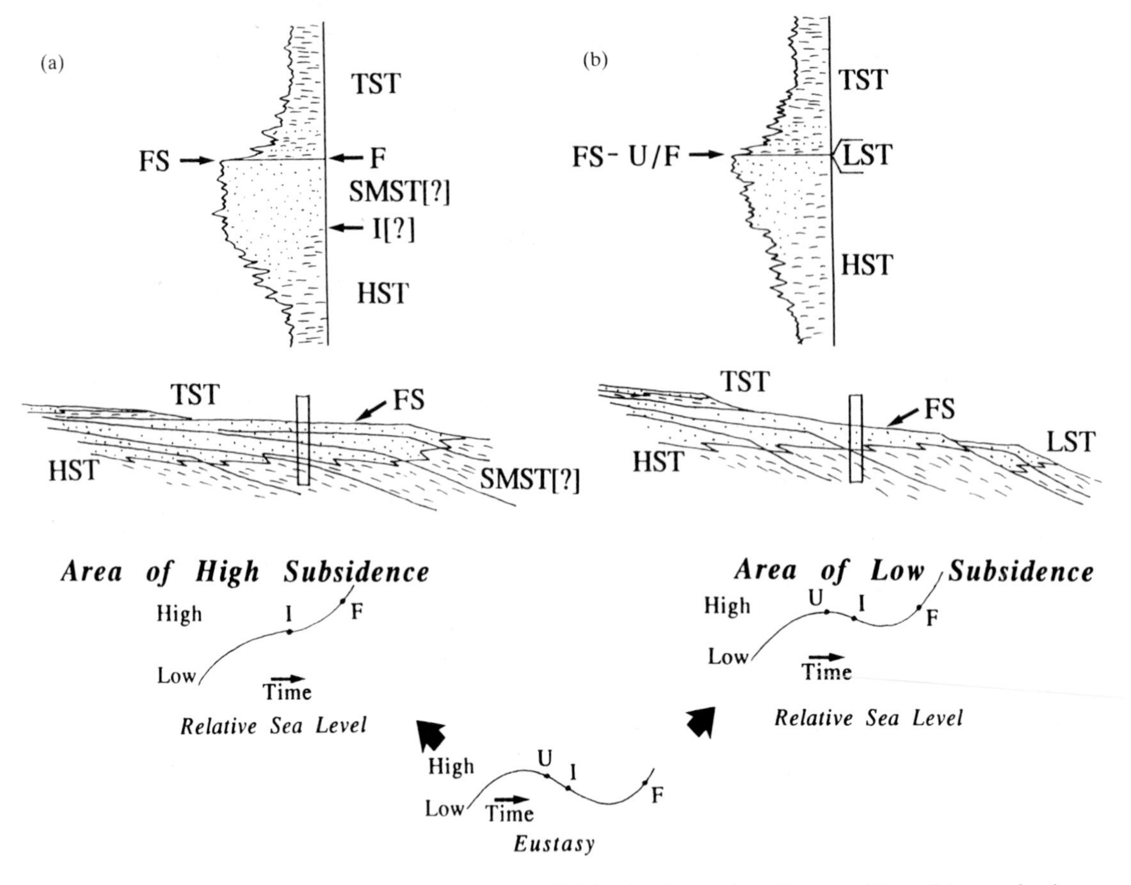

Fig. 5. Parasequence and sequence development in areas of high subsidence (a) and low subsidence (b) occurring in response to the same eustatic event. A relative sea-level fall beginning at time U occurs in the area of low subsidence, whereas the same interval is characterized by a slowdown and then an acceleration of relative sea-level rise. Note that in (a) a flooding surface bounds the top of the parasequence and in (b) a flooding surface merged with an unconformity (and sequence boundary) occurs at the top of the parasequence.

development. Stratal geometries from two areas are shown: one area is characterized by high rates of subsidence (Fig. 5(a)) and one area is characterized by low rates of subsidence. In the area affected by high rates of subsidence no relative sea-level fall occurs and a type 1 unconformity develops. The stratigraphic succession is virtually indistinguishable from that which occurs in the area of low subsidence (Fig. 5(b)). Only, in the latter case the same flooding surface that caps the shoaling-upward succession is also a sequence-bounding unconformity surface. Thus, although a parasequence is described at both localities, time I (the inflection point on the sea-level curve) occurs *within* the succession shown in Fig. 5(a) and *at the top* of the succession shown in Fig. 5(b). The stratal geometry shown in Fig. 5(b) is the response to the process of *forced regression* as discussed in Posamentier *et al.* (1990, 1992b). This further interpretive step from parasequence identification to sequence identification highlights the difference between the two terms.

Once again it is important to re-affirm that both 'sequence' and 'parasequences' are *generic terms that are time and space independent*. Consequently, it is somewhat misleading to refer to a particular depositional unit as being of *parasequence scale* or even *sequence scale*. Sequences deposited in response to varying-duration sea-level cycles can be referred to as first-order sequences, second-order sequences, third-order sequences, etc., in the sense of Vail *et al.* (1977b).

Eustasy versus relative sea-level versus water depth

There has been some confusion regarding the terms *eustasy*, *relative sea-level* and *water depth*. There is a distinction between these terms that is significant but is often obscured. Posamentier *et al.* (1988) point out that whereas eustasy is a function of sea-surface movement alone with reference to some fixed point such as the centre of the Earth, relative sea level is a function of sea-surface movement *in addition to* sea-floor movement. This latter parameter can be a function of tectonics, thermal cooling, loading by sediments or by water, or compaction. Relative sea-level change can therefore be quite variable from location to location as these factors vary from location to location. Most significantly, therefore, *relative* sea-level change describes how sediment accommodation varies (Posamentier *et al.*, 1988).

The term base level has been preferred by some to describe this interaction between sea-surface and sea-floor movement (Shanley & McCabe, 1991). Base level may be defined as the elevation of the point to which fluvial systems will be graded. Thus, position in vertical space of the graded or steady-state or equilibrium profile is determined by the elevation of base level. Clearly, then, base level also has application in the non-marine or continental setting (see discussion of graded stream profiles in Mackin, 1948). In this environment, however, steady-state profile position and changes of the shape of the steady-state profile can be a function of factors other than relative sea-level change. For example, increasing stream discharge or decreasing stream sediment load will tend to cause streams to incise and therefore seek a lower equilibrium profile. Consequently, the term relative sea level rather than base level is preferred though the usage can be quite similar.

In general, the effect of relative sea-level change on fluvial systems decreases or is dampened upstream largely because of the increased relative importance of climatic change as well as autocyclic changes such as sediment flux variations, local tectonics and changes in fluvial discharge in the upstream direction (Fig. 6). Thus, the climatic and autocyclic 'noise' eventually may drown out the relative sea-level 'signal' upstream (Fig. 7). It is possible, however, that major relative sea-level changes can nonetheless have a significant effect great distances upstream, thereby allowing the application of sequence-stratigraphic concepts in this setting.

Transgression and regression versus relative sea-level rise and fall

It is important to recognize that the terms *transgression* and *regression* describe the direction of movement of the shoreline landward and seaward, respectively. The direction of shoreline movement is a function of the balance between sediment flux and space on the shelf available for sediment to fill (i.e. accommodation). When relative sea level is rising, areas of low sediment flux may be characterized by transgressive shorelines whereas areas of high sediment flux may be characterized by regressive shorelines. Consequently, a one-to-one connection between relative sea-level rise and transgression is incorrect. When relative sea level falls,

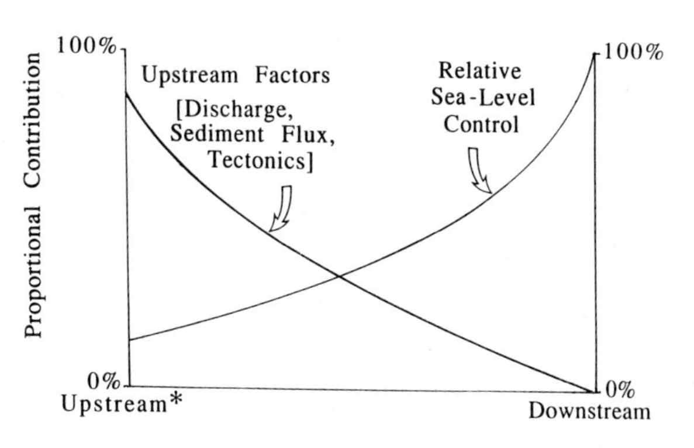

Fig. 6. Schematic depiction of the proportional importance of upstream factors (e.g. discharge, sediment flux and tectonics) and downstream factors (i.e. relative sea-level change) to changes in fluvial systems. Note that the upstream influence of relative sea-level change is effective only as far as the first significant knickpoint. Towards the downstream end of fluvial systems, the effect of changes in upstream factors becomes damped and approaches zero.

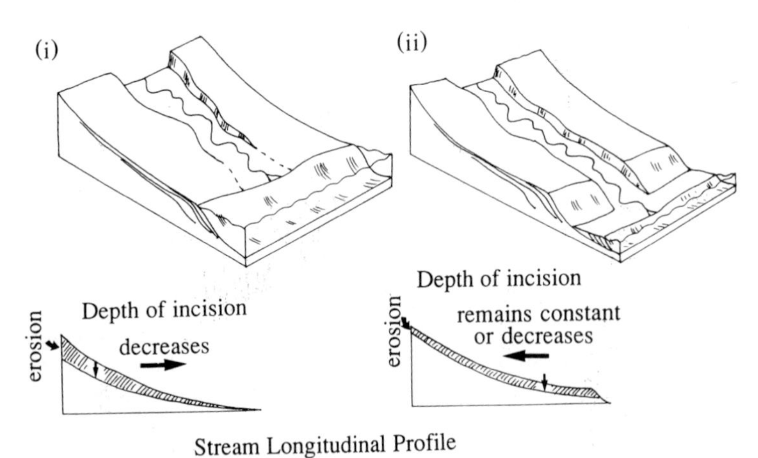

Fig. 7. Fluvial incision as a function of (i) changes of fluvial discharge, changes in sediment load and tectonic uplift in the non-marine areas contrasted with (ii) changes of relative sea level. Note that in situation (i) the depth of incision decreases and approaches zero in the downstream direction and in situation (ii) the depth of incision is maximum at the mouth of the stream and either remains the same or decreases in the upstream direction.

shoreline *regression* invariably occurs. This shoreline regression occurs irrespective of variations of sediment flux. Regression in response to relative sea-level fall alone has been referred to as a *forced regression* (Posamentier *et al.*, 1992b).

Incised valleys and relative sea-level fall

Not all instances of relative sea-level fall are characterized by fluvial rejuvenation and valley incision and conversely not all instances of fluvial rejuvenation and valley incision are associated with relative sea-level fall. In general, valley incision will occur if (i) relative sea-level fall exposes a land surface steeper than the graded or equilibrium profile (see discussion in Posamentier *et al.*, 1992b); or (ii) relative sea level remains constant or slowly rises but stream discharge increases significantly; or (iii) relative sea level remains constant or slowly rises, stream discharge remains unchanged, but stream sediment load decreases significantly; or (iv) relative sea level remains constant or slowly rises but tectonic uplift occurs in the non-marine areas. In each case, fluvial downcutting results in sedimentary bypassing and development of at least localized unconformities. Only cases (i) and possibly (iv), however, are functions of allocyclic processes and may therefore have greater regional significance.

Consequently, because incised valleys can occur in response to both autocyclic as well as allocyclic processes, the user of sequence-stratigraphic concepts is cautioned not to immediately interpret incised valleys as direct indicators of sequence boundaries. A useful, if not essential, associated line of evidence to confirm that relative sea-level fall (i.e. case (i)) was responsible for the observed stream incision and hence related to a sequence boundary is the occurrence of a lowstand shoreline or forced regression coeval with the incised valley.

As Fig. 7 illustrates, valley incision occurring as a function of change in upstream parameters (e.g. changes in sediment load, changes in stream discharge, etc.) will inevitably be damped towards the mouth of the stream if the position of relative sea level has remained unchanged.

Type 1 versus type 2 unconformities

Type 1 and type 2 unconformities were defined originally by Vail and Todd (1981) and were based on observations from seismic data. The definitions were subsequently modified and refined by Posamentier and Vail (1988a). The type 1 unconformity develops in response to *relative sea-level fall* and is associated with an abrupt basinward shift of coastal onlap characterized by forced regressions and in some cases fluvial incision. The type 2 unconformity develops in response to *decelerating* and then *accelerating relative sea-level rise*. In this case no relative sea-level fall occurs because the maximum rate of eustatic fall (at the inflection point) never quite attains the rate of subsidence. This unconformity is characterized by an abrupt basinward shift of coastal onlap without forced regressions and significant fluvial incision.

Type 1 unconformities are readily observed in outcrop (see Van Wagoner *et al.*, 1990). Conversely, type 2 unconformities are much less clear if not impossible to recognize in outcrop. This is because in the marginal marine setting and seaward, no relative sea-level fall occurs to punctuate the succession. Rather, the manifestation of a deceleration and then an acceleration of relative sea-level rise will be a change from an increasingly progradational to decreasingly progradational and subsequently aggradational stacking pattern. Landward of the marginal marine setting (i.e. the non-marine environment), the stratal geometry may be characterized by onlap onto the sequence boundary. The precise expression of the type 2 unconformity in the non-marine environment remains unclear, however, but may be manifested as extensive palaeosol development.

Systems tracts versus parasequence sets

The terms *systems tract* and *parasequence set* have often been used interchangeably. However, it is critical to note there is an important distinction between these terms. Systems tracts refer to a linkage of coeval depositional systems (Brown & Fisher, 1977) and combine to form sequences* (Posamentier *et al.*, 1988). Parasequence sets are defined as a succession of genetically related parasequences that form a distinctive stacking pattern (Van Wagoner, 1985). Systems tracts are composed of either progradationally stacked parasequences or a *progradational parasequence set* (characterizing the late highstand and early lowstand systems tracts), aggradationally stacked parasequences or an *aggradational parasequence set* (characterizing the early highstand and late lowstand systems tracts) and retrogradationally stacked parasequences or a *retrogradational parasequence set* (characterizing the transgressive systems tract). The distinction between *systems tracts* and *parasequence sets* is analogous to the distinction between sequences and parasequences. Parasequence sets and parasequences constitute a description of a succession that is devoid of inferences regarding sea-level change, stratal geometries, or other such interpretive aspects. In contrast, systems tracts *explicitly* involve interpretations regarding sea-level change, temporal and spatial relationships between facies tracts and the nature and significance of bounding surfaces.

*Type 1 sequences comprise lowstand, transgressive and highstand systems tracts, whereas type 2 sequences comprise shelf margin, transgressive and highstand systems tracts.

COMMON ABUSES OF SEQUENCE-STRATIGRAPHIC CONCEPTS

There are several general ways that sequence-stratigraphic concepts have been misapplied and misunderstood that have led to general confusion in the geologic community. Until recently, there has been a tendency to address the issue regarding the validity of 'global' sea-level curves as a measure of the validity of *all* aspects of the sequence-stratigraphic concept. Thus, the strength of the sequence-stratigraphic tool for lithologic prediction has been overshadowed and in some cases discredited due to a process of 'guilt by association' with the eustatic question. Once the utility of the sequence-stratigraphic concept had taken root in the geologic community (Wilgus *et al.*, 1988) this became less of an issue.

An example of misunderstanding occurs in Miall (1991) wherein it is claimed that Posamentier *et al.* (1988) argue that *significant fluvial deposition* occurs during intervals of relative sea-level fall. This is not what Posamentier *et al.* (1988, p. 139) had said; in fact they clearly state that intervals of relative sea-level fall are characterized by 'sedimentary bypass of the shelf through actively incising valleys'.

A common misunderstanding is that all sea-level cycles, no more — no less, as shown on the Haq *et al.* (1987) sea-level curve can have a stratigraphic expression in every study area. Depending on the rate of sediment flux, local tectonics and physiography, these cycles and more ... or fewer, can be observed locally. For example, in areas of high sediment flux far more cycles may be observed than are shown on the Haq *et al.* (1987) curve (e.g. Erskine & Vail, 1988). Moreover, data bases of relatively high resolution such as well logs, outcrops and high-resolution seismic, are likewise likely to allow recognition of additional cycles (e.g. Posamentier & Chamberlain, 1989 and this volume, pp. 469–485; Tesson *et al.*, 1990). The Haq *et al.* (1987) curve is not meant to show every sea-level cycle regardless of how short the duration. Rather it depicts primarily third-, second- and first-order cycles (in the sense of Vail *et al.*, 1977b). In contrast, in areas of low sediment flux it is likely that not all of the sea-level cycles will have clear stratigraphic expression. Consequently, it would be extremely fortuitous (but, more to the point, highly unlikely!) for any given area to be characterized by exactly the number of sea-level cycles depicted on the Haq *et al.* (1987) curve.

A parallel misconception is that at any given location all systems tracts of all sequences will be present. Clearly, this is extremely unlikely. For example, at the initiation of the lowstand systems tract, lowstand fans may be forming while at the same time sedimentary bypass of the shelf is occurring. Thus, the early lowstand systems tract is not present on the shelf. Likewise, if highstand progradation is aborted due to a relative sea-level fall, then the highstand systems tract may occur only as a thin veneer of condensed section over broad shelfal areas and may not be distinguishable from either the underlying or overlying systems tracts.

A common thread to these misapplications of the sequence-stratigraphic concepts is that models are being applied without taking local factors into consideration. Posamentier *et al.* (1988) specifically point out 'that the models are *generally* applicable. The effects of local factors ... must be incorporated into the models before these models can be applied to a particular basin'. In other words the geologist is cautioned not to force-fit his observations into someone else's model without first making appropriate modifications to it. The block diagrams shown in Posamentier *et al.* (1988, their Figs 1 to 6) are not necessarily valid worldwide and in all basins; *but the first principles upon which the models are based are.*

Distinguishing between *autocyclic* and *allocyclic* events has also led to some confusion. For example, every sharp-based sand does not imply the occurrence of a sequence boundary; the occurrence of some fluvial/estuarine 'blocky' channelized sands may merely be the result of avulsion. Moreover, as discussed earlier, valley incision alone may have little or no allocyclic or regional significance.

Another potential misapplication of the concepts is to maintain that all hot (i.e. high gamma-ray wireline log response) shales are condensed sections. This may or may not be the case. In some situations the 'hottest' shales occur just above the transgressive surface, significantly below the condensed section. The Kimmeridgian hot shales of the North Sea may be 'hot' but are probably not a condensed section in their entirety. Likewise, not all condensed sections are 'hot'. This, of course, will be a function of preservation of organic matter within these sections.

SUMMARY AND CONCLUSIONS

Sequence-stratigraphic concepts have recently been experiencing more widespread application. As is typical of most new approaches, however, some confusion regarding terminology and concepts persists. Of primary importance, however, is the caveat that sequence stratigraphy constitutes an approach and should therefore be viewed as a tool rather than a template. There is a tendency for some to apply the sequence concepts with blinders, erroneously force-fitting their data to fit someone else's model (e.g. the block diagrams in Posamentier *et al.*, 1988). The key to proper application of these concepts is to understand the first principles involved and then tailor the sequence model to account for local factors such as tectonics, sediment flux and physiography. When properly applied, sequence-stratigraphic concepts can be a powerful tool in moving beyond the small windows to the world offered by many data sets.

ACKNOWLEDGEMENTS

The authors wish to acknowledge the feedback received from various colleagues, in particular George Allen, V. Kolla, Dale Leckie and Pete Vail. This feedback and subsequent discussions have clarified where confusion lay regarding understanding and correct application of sequence-stratigraphic concepts. These interactions have in some cases opened the way to possibilities that had not been anticipated.

REFERENCES

Allen, G.P. & Posamentier, H.W. (1991) Facies and stratal patterns in incised valley complexes: examples from the recent Gironde Estuary (France) and the Cretaceous Viking Formation (Canada). *Annual Meeting, Dallas, Texas, April 7–10, 1991, Abstracts with Programs*, p. 70.

Anderton, R. (1985) Clastic facies models and facies analysis. In: *Sedimentology: Recent Developments and Applied Aspects* (Eds Brenchley, P.J. & Williams, B.J.P.) Geol. Soc. London, Spec. Publ. 18, Blackwell Scientific Publications, Oxford, pp. 31–47.

Bergman, K.M. & Walker, R.G. (1988) Formation of Cardium erosion surface E5, and associated deposition of conglomerate: Carrot Creek Field, Cretaceous western interior seaway, Alberta. In: *Sequences, Stratigraphy, Sedimentology: Surface and Subsurface* (Eds James, D.P. & Leckie, D.A.) Mem. Can. Soc. Petrol. Geol. 15, 15–24.

Brown, L.F., Jr. & Fisher, W.L. (1977) Seismic-stratigraphic interpretation of depositional systems: examples from Brazilian rift and pull-apart basins. In: *Seismic Stratigraphy – Applications to Hydrocarbon Exploration* (Ed. Payton, C.E.) Mem. Am. Assoc. Petrol. Geol. 26, 213–248.

Cloetingh, S. (1986) Intraplate stresses: a new mechanism for relative fluctuations of sea level. *Geology* **14**, 617–620.

Cloetingh, S., Tankard, A.J., Welsink, H.J. & Jenkins, W.A.M. (1989) Vail's coastal onlap curves and their correlation with tectonic events, offshore eastern Canada. In: *Extensional Tectonics and Stratigraphy of the North Atlantic Margins* (Eds Tankard, A.J. & Balkwill, H.R.) Mem. Am. Assoc. Petrol. Geol. 46, 283–293.

Einsele, G. & Seilacher, A. (1982) *Cyclic and Event Stratification; Symposium.* Springer-Verlag, Berlin, 550 pp.

Erskine, R.D. & Vail, P.R. (1988) Seismic stratigraphy of the Exmouth Plateau. In: *Atlas of Seismic Stratigraphy* (Ed. Bally, A.W.) Am. Assoc. Petrol. Geol. Stud. Geol. 27, 163–173.

Galloway, W.E. (1989a) Genetic stratigraphic sequences in basin analysis I: architecture and genesis of flooding-surface bounded depositional units. *Bull. Am. Assoc. Petrol. Geol.* **73**, 125–142.

Galloway, W.E. (1989b) Genetic stratigraphic sequences in basin analysis II: application to northwest Gulf of Mexico Cenozoic Basin. *Bull. Am. Assoc. Petrol. Geol.* **73**, 143–154.

Haq, B.U., Hardenbol, J. & Vail, P.R. (1987) Chronology of fluctuating sea levels since the Triassic. *Science* **235**, 1156–1167.

Harms, J.C. (1966) Stratigraphic traps in a valley-fill, western Nebraska. *Bull. Am. Assoc. Petrol. Geol.* **50**, 2119–2149.

Hubbard, R.J. (1988) Age and significance of sequence boundaries on Jurassic and early Cretaceous rifted continental margins. *Bull. Am. Assoc. Petrol. Geol.* **72**, 49–72.

Jervey, M.T. (1988) Quantitative geologic modeling of siliciclastic rock sequences and their seismic expression. In: *Sea-level Changes – An Integrated Approach* (Eds Wilgus, C.K., Hastings, B.S., Kendall, C.G.St.C., Posamentier, H.W., Ross, C.A. & Van Wagoner, J.C.) Spec. Publ. Soc. Econ. Paleontol. Mineral. Tulsa, 42, 47–69.

Loutit, T.S., Hardenbol, J., Vail, P.R. & Baum, G.R. (1988) Condensed sections: the key to age dating and correlation of continental margin sequences. In: *Sea-level Changes – An Integrated Approach* (Eds Wilgus, C.K., Hastings, B.S., Kendall, C.G.St.C., Posamentier, H.W., Ross, C.A. & Van Wagoner, J.C.) Spec. Publ. Soc. Econ. Paleontol. Mineral. Tulsa, 42, 183–213.

Mackin, J.H. (1948) Concept of the graded river. *Bull. Geol. Soc. Am.* **59**, 463–512.

Miall, A.D. (1986) Eustatic sea-level changes interpreted from seismic stratigraphy: a critique of the methodology with particular reference to the North Sea Jurassic record. *Bull. Am. Assoc. Petrol. Geol.* **70**, 131–137.

Miall, A.D. (1991) Stratigraphic sequences and their chronostratigraphic correlation. *J. Sediment. Petrol.* **61**, 505.

Mitchum, R.M., Jr. (1977) Glossary of terms used in seismic stratigraphy. In: *Seismic Stratigraphy – Applications to Hydrocarbon Exploration* (Ed. Payton, C.E.) Mem. Am. Assoc. Petrol. Geol. 26, 205–212.

Moore, R.C. (1964) Paleoecological aspects of Kansas Pennsylvanian and Permian cyclothems. In: *Symposium on Cyclic Sedimentation* (Ed. Merriam, D.F.) Bull. Kansas Geol. Survey, 169, 449–460.

Mörner, N.-A. (1976) Eustasy and geoid changes. *J. Geol.* **84**, 123–151.

North American Commission on Stratigraphic Nomenclature (1983) North American stratigraphic code. *Bull. Am. Assoc. Petrol. Geol.* **67**, 841–875.

Okamura, Y. (1989) Multi-layered progradational sequences in the shelf and shelf-slope of the southwest Japan forearc. In: *Sedimentary Facies in the Active Plate Margin* (Eds Taira, A. & Masuda, F.) Terra Publications, pp. 295–317.

Perlmutter, M.A. & Matthews, M.D. (1989) Global cyclostratigraphy — a model. In: *Quantitative Dynamic Stratigraphy* (Ed. Cross, T.A.) Prentice Hall, New Jersey, pp. 233–260.

Plint, A.G. (1988) Sharp-based shoreface sequences and 'offshore bars' in the Cardium Formation of Alberta; their relationship to relative changes in sea level. In: *Sea-level Changes – An Integrated Approach* (Eds Wilgus, C.K., Hastings, B.S., Kendall, C.G.St.C., Posamentier, H.W., Ross, C.A. & Van Wagoner, J.C.) Spec. Publ. Soc. Econ. Paleontol. Mineral. Tulsa, 42, 357–370.

Plint, A.G., Walker, R.G. & Bergman, K.M. (1987) Cardium Formation 6. Stratigraphic framework of the Cardium in subsurface. *Bull. Can. Petrol. Geol.* **34**, 213–225.

Posamentier, H.W. & Chamberlain, C.J. (1989) Viking lowstand beach deposition at Joarcam Field, Alberta. *Can. Soc. Explor. Geophys./Can. Soc. Explor. Geol. Annual Convention – Exploration Update*, June 11–15, 1989, pp. 96–97.

Posamentier, H.W. & Vail, P.R. (1985) Eustatic controls on depositional stratal patterns. *Soc. Econ. Paleontol. and Mineral. Res. Conf. No. 6, Sea Level Changes – An Integrated Approach*, October 20–23, 1985.

Posamentier, H.W. & Vail, P.R. (1988a) Eustatic controls on clastic deposition II — sequence and systems tract models. In: *Sea-level Changes – An Integrated Approach* (Eds Wilgus, C.K., Hastings, B.S., Kendall, C.G.St.C., Posamentier, H.W., Ross, C.A. & Van Wagoner, J.C.) Spec. Publ. Soc. Econ. Paleontol. Mineral. Tulsa, 42, 125–154.

Posamentier, H.W. & Vail, P.R. (1988b) Sequences, systems tracts, and eustatic cycles. *Am. Assoc. Petrol. Geol, Annual Meeting, Houston, TX, March 20–23, 1988.*

Posamentier, H.W., Allen, G.P. & James, D.P. (1992a) High resolution sequence stratigraphy — the East Coulee Delta. *J. Sediment. Petrol.* **62**, 310–317.

Posamentier, H.W., Allen, G.P. James, D.P. & Tesson, M. (1992b) Forced regressions in a sequence stratigraphic framework: concepts, examples, and exploration significance. *Am. Assoc. Petrol. Geol.* **76**, 1687–1709.

Posamentier, H.W., James, D.P. & Allen, G.P. (1990)

Aspects of sequence stratigraphy: recent and ancient examples of forced regressions. *Bull. Am. Assoc. Petrol. Geol.* **74**, 742.

Posamentier, H.W., Jervey, M.T. & Vail, P.R. (1988) Eustatic controls on clastic deposition I — conceptual framework. In: *Sea-level Changes — An Integrated Approach* (Eds Wilgus, C.K., Hastings, B.S., Kendall, C.G.St.C., Posamentier, H.W., Ross, C.A. & Van Wagoner, J.C.) Spec. Publ. Soc. Econ. Paleontol. Mineral. Tulsa, 42, 110–124.

Shanley, K.W. & McCabe, P.J. (1991) Predicting facies architecture through sequence stratigraphy — an example from the Kaiparowits Plateau, Utah. *Geology* **19**, 742–745.

Sloss, L.L. (1962) Stratigraphic models in exploration. *Bull. Am. Assoc. Petrol. Geol.* **46**, 1050–1057.

Sloss, L.L. (1963) Sequences in the cratonic interior of North America. *Bull. Geol. Soc. Am.* **74**, 93–114.

Summerhayes, C.P. (1986) Sealevel curves based on seismic stratigraphy: their chronostratigraphic significance. *Palaeogeogr., Palaeoclim., Palaeoecol.* **57**, 27–42.

Suter, J.R. & Berryhill, H.L. (1985) Late Quaternary shelf-margin deltas, northwest Gulf of Mexico. *Bull. Am. Assoc. Petrol. Geol.* **69**, 77–91.

Suter, J.R., Berryhill, H.L. & Penland, S. (1987) Late Quaternary sea-level fluctuations and depositional sequences, southwest Louisiana continental shelf. In: *Sea Level Fluctuations and Coastal Evolution* (Eds Nummedal, D., Pilkey, O.H. & Howard, J.P.) Spec. Publ. Soc. Econ. Paleontol. Mineral. Tulsa, 41, 199–219.

Tesson, M., Gensous, B., Allen, G.P. & Ravenne, C. (1990) Late Quaternary deltaic lowstand wedges on the Rhône continental shelf, France. *Mar. Geol.* **91**, 325–332.

Vail, P.R. (1987) Seismic stratigraphy interpretation procedure. In: *Seismic Stratigraphy Atlas* (Ed. Balley, B.) Am. Assoc. Petrol. Geol. Stud. Geol. 27, 1–10.

Vail, P.R. & Todd, R.G. (1981) North Sea Jurassic unconformities, chronostratigraphy and sea-level changes from seismic stratigraphy. *Petroleum Geology of the Continental Shelf of Northwest Europe, Proceedings*, pp. 216–235.

Vail, P.R., Mitchum, R.M., Jr. & Thompson, S., III. (1977a) Seismic stratigraphy and global changes of sea level from coastal onlap. In: *Seismic Stratigraphy — Applications to Hydrocarbon Exploration* (Ed. Payton, C.E.) Mem. Am. Assoc. Petrol. Geol. 26, 63–81.

Vail, P.R., Mitchum, R.M., Jr, Todd, R.G., *et al.* (1977b) Seismic stratigraphy and global changes of sea level. In: *Seismic Stratigraphy — Applications to Hydrocarbon Exploration* (Ed. Payton, C.E.) Mem. Am. Assoc. Petrol. Geol. 26, 49–212.

Van Wagoner, J.C. (1985) Reservoir facies distribution as controlled by sea-level change. *Soc. Econ. Paleontol. Mineral., Annual Midyear Meeting*, Golden, Colorado, **2**, 91.

Van Wagoner, J.C., Campion, K.M. & Rahmanian, V.D. (1990) Siliciclastic sequence stratigraphy in well logs, core, and outcrops: concepts for high-resolution correlation of time and facies. *Am. Assoc. Petrol. Geol., Meth. Expl. Ser.* **7**, 55 pp.

Van Wagoner, J.C., Posamentier, H.W., Mitchum, R.M., Jr. *et al.* (1988) An overview of the fundamentals of sequence stratigraphy and key definitions. In: *Sea-level Changes — An Integrated Approach* (Eds Wilgus, C.K., Hastings, B.S., Kendall, C.G.St.C., Posamentier, H.W., Ross, C.A. & Van Wagoner, J.C.) Spec. Publ. Soc. Econ. Paleontol. Mineral. Tulsa, 42, 39–45.

Walker, R.G. (1990) Facies modeling and sequence stratigraphy. *J. Sediment. Petrol.* **60**, 777–786.

Weimer, R.J. (1983) Relation of unconformities, tectonics, and sea level changes, Cretaceous of the Denver Basin and adjacent areas. In: *Mesozoic Paleogeography of the West-Central United States* (Eds Reynolds, M.W. & Dolly, E.D.) Soc. Econ. Paleontol. Mineral., Rocky Mountain Section — Rocky Mountain Paleogeography Symposium 2, 359–376.

Weimer, R.J. (1984) Relation of unconformities, tectonics, and sea level changes, Cretaceous of the Western Interior, USA. In: *Interregional Unconformities and Hydrocarbon Accumulations* (Ed. Schlee, J.S.) Mem. Am. Assoc. Petrol. Geol. 36, 7–36.

Weller, J.M. (1960) *Stratigraphic Principles and Practices*. Harper, New York, 725 pp.

Wheeler, H.E. (1958) Time stratigraphy. *Bull. Am. Assoc. Petrol. Geol.* **42**, 1047–1063.

Wheeler, H.E. (1959) Unconformity bounded units in stratigraphy. *Bull. Am. Assoc. Petrol. Geol.* **43**, 1975–1977.

Wheeler, H.E. & Murray, H.H. (1957) Baselevel control patterns in cyclothemic sedimentation. *Bull. Am. Assoc. Petrol. Geol.* **41**, 1985–2011.

Wilgus, C.K., Hastings, B.S., Kendall, C.G.St.C., Posamentier, H.W., Ross, C.A. & Van Wagoner, J.C. (Eds) (1988) *Sea-level Changes — An Integrated Approach*. Spec. Publ. Soc. Econ. Paleontol. Mineral. Tulsa, 42, 407 pp.

Spec. Publs Int. Ass. Sediment. (1993) **18**, 19–41

Modelling passive margin sequence stratigraphy

M.S. STECKLER, D.J. REYNOLDS*, B.J. COAKLEY*†,
B.A. SWIFT‡ *and* R. JARRARD§

Lamont-Doherty Earth Observatory of Columbia University, Palisades, New York, USA;
**also at Department of Geological Sciences, Columbia University, New York, NY, USA;*
†now at Department of Geology and Geophysics, University of Wisconsin-Madison, Madison, WI, USA;
‡United States Geological Survey, Office of Marine Geology, Woods Hole, MA, USA; and
§now at Department of Geology and Geophysics, University of Utah, Salt Lake City, UT, USA

ABSTRACT

We have modelled stratigraphic sequences to aid in deciphering the sedimentary response to sea-level change. Sequence geometry is found to be most sensitive to sea level, but other factors, including subsidence rate and sediment supply, can produce similar changes. Sediment loading and compaction also play a major role in generating accommodation, a factor often neglected in sequence-stratigraphic models. All of these parameters can control whether a type 1 or type 2 sequence boundary is produced. The models indicate that variations in margin characteristics produce systematic shifts in sequence boundary timing and systems tract distribution. The timing of the sequence boundary formation and systems tracts may differ by up to one-half of a sea-level cycle. Thus correlative sequence boundaries will not be synchronous. While rates of sea-level change may exceed the rate of thermal subsidence, isostasy and compaction may amplify the rate of total subsidence to several times greater than the thermal subsidence. Thus, total subsidence does not vary uniformly across the margin since it is modified by the sediment load. The amplitude of sea-level changes cannot be determined accurately without accounting for the major processes that affect sediment accumulation. Backstripping of a seismic line on the New Jersey margin is used to reconstruct continental margin geometry. The reconstructions show that the pre-existing ramp-margin geometry, rather than sea level, controls clinoform heights and slopes and sedimentary bypass. Backstripping also reveals progressive deformation of sequences due to compaction. Further work is still needed to understand quantitatively the role of sea level and the tectonic and sedimentary processes controlling sequence formation and influencing sequence architecture.

INTRODUCTION

Sequence stratigraphy, a methodology for identifying relatively conformable packages of genetically related strata, has become an important analytical tool for geology and petroleum exploration. The recognition of the chronological significance of seismic reflection surfaces and the identification of packages of strata based on geometric relationships allows the subdivision of sedimentary successions into their natural depositional units called sequences. Within sequences, the depositional units show stacking patterns that consist of lateral shifts from seaward to landward and back again. Facies belts similarly exhibit systematic shifts through the stratigraphic succession within sequences.

Most conceptual models of sequence development assume that eustatic fluctuations are responsible for the production of unconformity-bound sequences (Vail *et al.*, 1977, 1984; Pitman, 1978; Van Wagoner *et al.*, 1988), although some authors champion the role of tectonics (Watts, 1982; Summerhayes, 1986) or sediment supply (Galloway, 1988). Cyclic sea-level changes are the simplest means of explaining globally correlative sequences. However, the cyclic pattern within stratigraphic sequences can be generated from a variety of different models of the relationship of eustasy to sedimentation (Pitman & Golovchenko, 1983, 1988; Watts & Thorne, 1984; Galloway, 1988; Jervey, 1988; Reynolds *et al.*, 1991). Predominant among the models of the stratigraphic response to the eustatic cycle is the model developed by Exxon (Jervey, 1988; Posamentier *et al.*, 1988; Posamentier & Vail, 1988). Their model provides a basis for estimating the amplitude of sea-level fluctuations from the sequence geometry. Application of this model to global stratigraphic data has resulted in the first detailed global, eustatic sea-level curves for the Mesozoic and Cenozoic (Vail & Hardenbol, 1979; Vail *et al.*, 1984; Haq *et al.*, 1987, 1988).

Even if eustatic change is the most significant factor in generating sequence boundaries, it is but one of the several important variables in the formation of sedimentary sequences at passive margins. Figure 1 shows a block diagram of the New Jersey continental margin labelled with some of the factors that influence the geometry of sedimentary accumulations. Tectonics, eustasy and sediment supply have been considered to be the three primary factors in models of sequence formation. Tectonics determines the space created (or destroyed) for sediment accumulation. Sea level modifies the size of that space. Sediments transported to or created within the basin fill the available space. Indeed, first-order models, utilizing these three variables (Vail *et al.*, 1984; Jervey, 1988; Posamentier *et al.*, 1988; Posamentier & Vail, 1988), produce qualitatively satisfying sequence geometries, reproducing many of the observed characteristics of sedimentary sequences. However, they neglect many additional processes (Fig. 1) that are also expected to play major roles in sequence development and thus alter the stratigraphic response to sea level. Sediment transport and geomorphic processes control where available sediment is deposited or eroded (Pitman and Golovchenko, 1991; Swift & Thorne, 1991). Climate affects both the volume and type of sediment delivered to the margin and the type of depositional environments present at the margin (Perlmutter & Matthews, 1990). Compaction and the isostatic compensation of the sediments significantly modify the space made available (Reynolds *et al.*, 1991) and provide feedbacks between the tectonic and depositional processes. These other processes will certainly affect sequence geometry, physical dimension and depositional environments. Indeed, the very large, rapid sea-level changes interpreted based on these simplified models (Haq *et al.*, 1987, 1988; Greenlee *et al.*, 1988) have generated considerable scepticism (see Miall, 1986; Christie-Blick *et al.*, 1990).

If past fluctuations of sea level are to be determined, they must be inverted from the stratigraphic record. The strata within sequences are the physical record of deposition caused by continuous changes of the sedimentary regime. However, the complexity of geologic processes hinders prediction of the stratigraphic consequences of sea-level change. In order to determine sea-level fluctuations we must attain a better understanding of what controls the response of a margin to sea-level change. Which parameters are significant in sequence generation and which are not?

The aim of this paper is to examine the role of basic physical parameters on stratigraphic-sequence development. We have developed a model to test the effects of what are commonly recognized as the three key processes controlling relative sea-level change and hence sequence development: (i) tectonic (thermal subsidence); (ii) sediment flux; and (iii) eustasy. In addition, we have included effects due to two important physical properties known to occur at passive margins, but often neglected or minimized in stratigraphic models: (i) flexural isostatic compensation; and (ii) sediment compaction. These enable us to explore the role of feedbacks in the stratigraphic system, in this case, how sediment loading modifies the accommodation space into which it is deposited. Our objective in this research has been to produce simple yet flexible models which, unlike complex simulations, are capable of testing the fundamental effects of important parameters and evaluating a variety of interactions of these parameters. We examine several cases with our model to determine the modifications these processes cause in sequence formation. These results are compared to the results of other sequence models to evaluate the critical parameters, as well as those which are insensitive, for sequence geometry, sequence timing and our ability to estimate the magnitude of sea-level changes.

The importance of the sedimentary processes that control the location of deposition and erosion on the shelf has been acknowledged but it is difficult to quantify simply for analysis. To investigate controls on sequence geometry that are not yet adequately covered by numerical models, we reconstruct the physiography and stratigraphy of the Tertiary New Jersey continental shelf. This is accomplished by

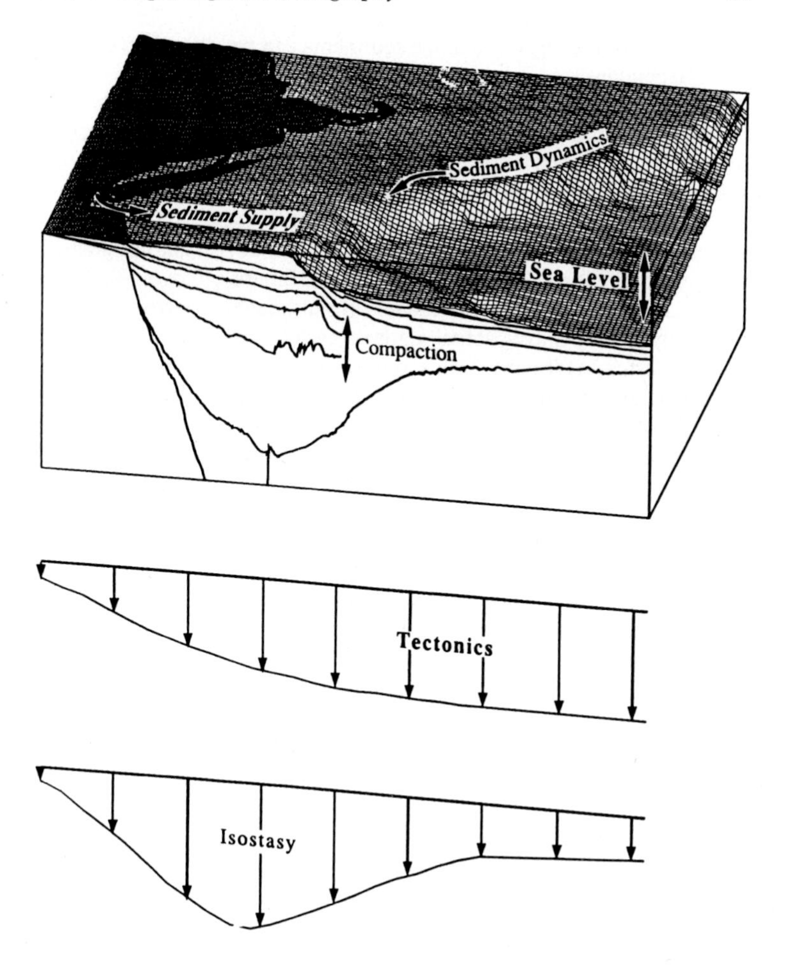

Fig. 1. Three-dimensional illustration showing margin geometry off New Jersey. The top of the cube is at sea level with the mesh showing topography and bathymetry. The front face of the cube shows some of the important stratigraphic horizons which have been imaged by seismic reflection data. The labels indicate some of the major controls on sedimentary sequences. In boldface are the three factors that are generally acknowledged as the primary components interacting at sedimentary basins: sea level, sediment supply and tectonic subsidence. However, other factors such as compaction, sediment dynamics and isostasy also significantly modify the creation and filling of accommodation.

backstripping the sedimentary layers to restore the margin geometry at times of sequence boundary formation. This enables us to investigate the response of a margin to sea-level changes by analysing the stepwise evolution of sequence geometry of the margin.

SEQUENCE-STRATIGRAPHIC MODEL

We have developed a model for sequence stratigraphy at Atlantic-type continental margins. This model incorporates the three basic parameters of sea level, tectonics and sediment supply (Fig. 1). The subsidence caused by thermal contraction is simulated by linear tilting the shelf platform about a hinge point 50 km from the left-hand side of the model. In the runs presented, a constant subsidence rate and sinusoidally varying sea level create and modify the accommodation. A proscribed sediment flux infills the accommodation made available by subsidence and sea-level change. Sediment deposition is controlled by a fixed equilibrium profile. For the computations presented here, the equilibrium surface consists of a horizontal shelf surface and a more steeply inclined slope face. Isostasy and compaction of the sediment column, whose importance in the overall sediment accumulation record of passive margins has long been recognized (Watts & Ryan, 1976; Steckler & Watts, 1978; Van Hinte, 1978), have generally been awarded a minor role in sequence generation. We include flexural isostasy in which the lithospheric response is modelled by an elastic beam of constant thickness, T_e. We assume that compaction of the sediment column follows an exponential decrease in porosity with depth (Sclater & Christie, 1980). The lithology and compaction parameters of the shelf and slope sediments can be set independently.

At each timestep, first the tectonic subsidence is incremented and then sea level is adjusted. The load of water added or removed is compensated using the flexural rigidity of the lithosphere below the margin. A specified volume of sediment is then added from the left-hand side and distributed according to the equilibrium profile. The sediment flux was kept constant for the model runs although it is likely to vary during actual sea-level cycles. For each timestep, sediment builds out to an extent delimited by the equilibrium profile until the volume of sediment deposited matches the input sediment flux. The weight of the sediments constitutes a load on the surface of the lithosphere. This load causes the lithosphere to subside until it is isostatically supported. Compaction of the sediment column is then calculated and the positions of the horizons adjusted. Although time is not explicitly used in the calculation, the length of the sea-level cycles in the model experiments is 1.6 million years. This is the average length of cycles as determined by the study of Haq *et al.* (1988). We have calibrated the rates and fluxes for 100,000-year timesteps. Thus our model takes 16 timesteps per sea-level cycle and generates 16 layers.

Sediment infill (or erosion) takes place to a base level that is controlled by sea level. Base level in our model runs is set to sea level across the entire depositional shelf surface. Thus a shoreline is not defined and we determine our sequence components using geometric criteria. We consider suspect the tie between the geometrical model of stratal relationships within a sequence and the location of facies belts, particularly the shoreline, and their stacking pattern. The depositional system across passive margins passes from an alluvial wedge through a coastal plain and shoreline facies to deeper marine deposits on the sloping front of the depositional system and the deep basin on the right-hand side. The break-in-slope that is analogous to the one in our model is referred to as the 'depositional shoreline break' and separates the region where the sea floor is at depositional base level from the basinward region where the sea floor is well below base level (Van Wagoner *et al.*, 1988). This is distinguished from the classical continental shelf edge which is only sometimes coincident with the shoreline break. However, the depositional shoreline break in the conceptual model appears to be more closely analogous to the leading edge of a prograding clinoform wedge, whose upper surface is at base level but not necessarily near the shoreline. The scale of this break as observed on seismic lines and outcrop is of the order of hundreds of metres and therefore distinct from the relatively minor break in slope (5–10 m) at the shoreface. Thus there are three distinct physiographic breaks at continental margins while the Exxon model only recognizes two. We therefore use the term 'depositional shelf break' for the physiographic break-in-slope in our model. Recognition of the distinction between the shoreface and the depositional shoreline break has led Vail *et al.* (1991) to propose renaming the latter the 'offlap break'.

The interpretation of the model output in a sequence-stratigraphic framework is shown in Fig. 2 where the model results and the widely reproduced archetypical Exxon sequence (Baum & Vail, 1988) are separated into their component systems tracts. The stratigraphic sections in Fig. 2 show two sequence boundaries, between which lie a sequence of relatively conformable sediments. In both our numerical and the Exxon conceptual model the shape and position of the depositional wedge shift as accommodation varies with sea level. This is most clearly seen in the time-stratigraphic, or Wheeler, diagrams. For our model results, the zone of deposition during each timestep is indicated by a thick black bar. The depositional shelf break is marked by an open square. The relationship of the shifting deposition to sea level is shown by the curves at the left of the Wheeler diagrams.

A type 1 unconformity (SB1) bounds the bottom of the sequence (Fig. 2). A type 1 sequence boundary is interpreted to form when deposition is restricted seaward of the previous depositional shelf break, and thus no deposition occurs on the shelf. In the conceptual model the depositional shelf (and upper slope of the clinoform package) is exposed and dissected by incised valleys (Posamentier *et al.*, 1988; Posamentier & Vail, 1988). Sediments derived from the shelf, the incised valleys, fluvial transport that bypasses the shelf and slumping are then deposited seaward to form the lowstand fan (lsf). Our model contains no provision for such a change in depositional style. We interpret the oblique clinoforms deposited above the sequence boundary, for which deposition remains restricted seaward of the preceding depositional shelf break, as the equivalent of lowstand fan deposits of the lowstand systems tract (LST). The overlying lowstand wedge (lsw) is a progradational unit which

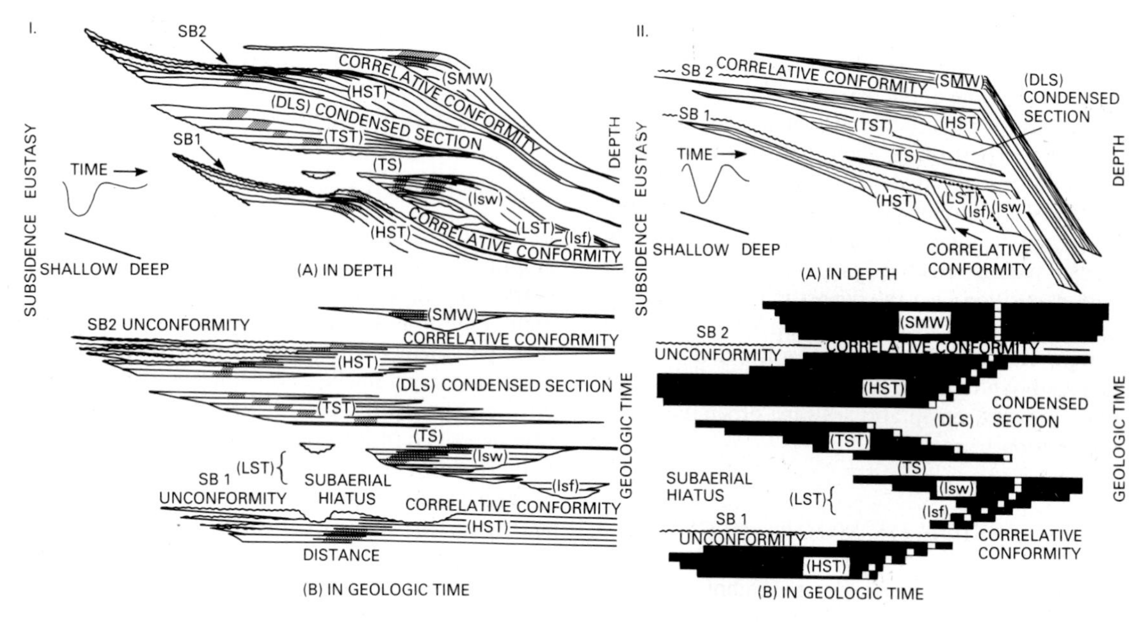

Fig. 2. Comparison of sequence-stratigraphic framework with our model. Both sets of diagrams have been 'exploded' to enable easier comparison of the systems tracts and stacking patterns. (IA) Archetypical stratigraphic cross-section of a sequence bounded on top and bottom by type 2 and type 1 sequence boundaries, respectively. The section includes a complete type 1 sequence with its depositional packages overlying a 'highstand' depositional package from the previous sequence. The top of the sequence is marked by a type 2 sequence boundary which is overlain by a shelf margin wedge systems tract. The dotted pattern represents shoreline deposits at the depositional shoreline break. (IB) Chronostratigraphic diagram or Wheeler plot of the stratigraphic section in (A) illustrating the temporal relationships and lateral shifting of deposition in sequences. The individual components of the sequences are labelled to show their relationships. (Modified from Baum & Vail, 1988.) (IIA) Our model contains many of the same elements as the conceptual stratigraphic model. In analogy to IA, the section shows a sequence bounded on top and bottom by type 2 and type 1 sequence boundaries, respectively. (IIB) A chronostratigraphic diagram, or Wheeler plot, of the same. The thick bars indicate the zone of deposition for each timestep and the open squares indicate the positions of the depositional shelf breaks that are equivalent to the shaded zones of IA and IB.

onlaps the slope of the preceding sequence and fills the incised valleys (Van Wagoner *et al.*, 1990). We similarly define the lowstand wedge by the occurrence of onlap back over the oblique clinoforms and distal edge of the underlying sequence while the depositional shelf break is progradational.

The lowstand systems tract (lowstand fan overlain by the lowstand wedge) is followed by the transgressive systems tract (TST) as relative sea level rises and migrates back over the former depositional shelf break and sediments reflood the entire shelf surface. The base of this systems tract is the transgressive surface (TS). The transgressive systems tract corresponds to the period when the depositional shelf break shifts landward in response to increased availability of space on the shelf.

When rate of accommodation development on the shelf decreases to less than the sediment supply, progradation of the depositional shelf break resumes and the highstand systems tract (HST) is established. The base of this systems tract is a downlap surface (DLS). Onlap continues during the early highstand but offlap develops during the late highstand in our model. In the model of Posamentier *et al.* (1988), the seaward shift in the bayline and river profile induces alluvial accumulation that enables coastal onlap to continue throughout the entire highstand systems tract until the time of the next sequence boundary. This alluvial deposition helps create the sharp asymmetric coastal onlap curve as seen in Fig. 2(B). That coastal onlap extends to the top of the sequence followed by an instantaneous downward shift has been disputed (Christie-Blick *et al.*, 1990). Indeed, the Posamentier *et al.* (1988) model is only partially consistent with geomorphic processes.

A type 2 sequence boundary is illustrated by the upper boundary of the sequence in Fig. 2 (SB2). A type 2 sequence boundary is developed when some deposition is maintained landward of the shelf break at the time of the maximum offlap. There is a downward shift in coastal onlap and subaerial exposure of part of the shelf, but stream rejuvenation and its associated erosion and a seaward shift in facies are absent (Van Wagoner *et al.*, 1988). Overlying the type 2 unconformity is a shelf margin wedge systems tract (SMW) which is associated with deposition near the depositional shelf edge on the outer part of the depositional shelf. The shelf margin wedge systems tract overlies a type 2 sequence boundary and is characterized by concurrent landward onlap and seaward progradation to aggradation.

DISCUSSION OF MODEL RESULTS

Our model, as described above, exhibits the basic components of sequence architecture. As the parameters that we have incorporated into the model are varied, systematic changes develop in the geometry and stacking patterns of the interpreted systems tracts and their relationship to sea level. In this section we present model results relative to three facets of sequence stratigraphy: (i) the features controlling the type of sequence boundary developed; (ii) the timing relative to the input sea-level curve of the sequence boundaries; and (iii) the implications of our model for estimating the amplitude of eustatic fluctuations.

Sequence boundary type

Initially, we vary the subsidence rate, sea-level amplitude and sediment supply, three of the primary controls on sedimentary sequence development, to test the sensitivity of sequence geometry to these parameters. The effects of varying each of the three major parameters are shown in Fig. 3. The cross-section in the centre represents a reference case described in Reynolds *et al.* (1991). Because it is the ratio of the parameters that determines the sequence architecture, the results are not dependent on the absolute levels chosen. Each of the cross-sections shown in Fig. 3 contains 100 timesteps or 10 million years by our scaling. Each axis represents variation of one parameter. For example, the vertical axis represents departures from the reference case by varying sea level only. The models plotted on the widened axes ends (b, c, e) represent a threefold increase in value of the indicated parameter. The models at the tapered end (a, d, f) of the axes indicate a reduction of the given parameter by a factor of three.

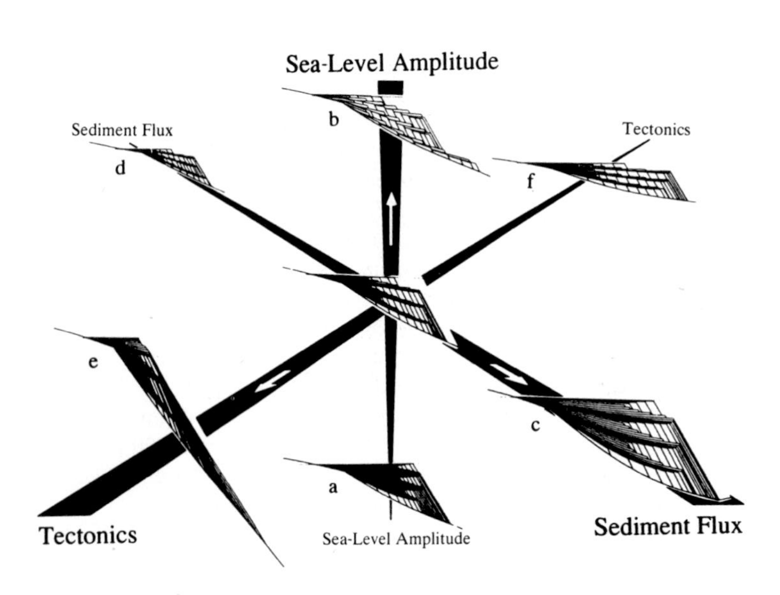

Fig. 3. Perspective view of the variation in sequence architecture as a function of the three major parameters: sea level, tectonics and sediment supply. All models represent 10 Ma (100 timesteps). The reference case is shown at the intersection of the three axes. Direction of increasing influence of parameter is marked by arrows, broadening of axes and enlarged type. The plots shown along each axis were generated by increasing or decreasing the magnitude of that parameter by a factor of three from the reference case. All of the plots are shown at the same scale. The lower three plots (Figs 3(a), 3(c), and 3(e)) exhibit type 2 sequence boundaries. The three upper plots (Figs 3(b), 3(d) and 3(f)) and the reference case exhibit type 1 sequence boundaries. The vertical exaggeration in this and all model runs is 25:1.

Major differences in the geometry may be seen along the sea-level axis. The magnitude of the horizontal shifts in deposition increases with sea-level amplitude. As a result, an increase in the amplitude of the sea-level fluctuations (b) favours exposure of the depositional shelf during sea-level falls and thus the formation of type 1 sequence boundaries. In the case where the sea-level variation is reduced (a), deposition is never restricted solely to positions seaward of underlying shelf breaks, indicating the formation of type 2 sequence boundaries.

Similar variations in sequence architecture are seen along the other axes, although these are not as pronounced as those in models where sea-level amplitude is changed. For example, variations in tectonic subsidence affect the rate at which space is made available. Where subsidence is fast (e), more space is made available and sea-level effects are minimized. This leads to the production of type 2 sequence boundaries. However, if the subsidence rate is low (f), sea-level effects become increasingly important and sediment is distributed over a broader area. This favours the formation of type 1 sequence boundaries. Finally, sediment supply alone can influence sequence geometry simply by the extent to which it fills accommodation. Increasing sediment supply shifts the shelf break from a position on the margin of relatively low tectonic subsidence to one of relatively high tectonic subsidence. Thus, low sediment supply (d) will favour type 1 sequence formation and high sediment supply (c) will favour type 2 sequence formation. Examples of this are seen along the 'sedimentation' axis. However, the magnitude of this effect will depend upon the horizontal gradient in tectonic subsidence.

In general, models plotting in the upper portion of Fig. 3 are composed of sequences defined by type 1 boundaries. Sequences plotting in the lower portion are delimited by type 2 boundaries. Variation in any one parameter can alter the sequence architecture and type of sequence boundary formed. In general, large sea-level amplitude, low tilt rate and low sediment supply favour the development of type 1 sequence boundaries. Production of type 2 sequence boundaries is favoured by low sea-level amplitude, high tilt rate and high sediment supply. However, tilt rate and especially sediment supply may vary appreciably not only from one margin to another but along strike within any given margin. In this fashion, various sequence geometries and boundaries may be created both regionally and globally from a single sea-level change.

The influence of the main parameters on sequence formation may be viewed as a tripartite system. Figure 4 shows how variations in these parameters may be plotted in a triangular or ternary diagram. Shaded circles indicate the scaling of the parameters for the runs shown in Fig. 3. For example, the cases that represent amplified or attenuated tectonic effects are plotted in Fig. 4 at (e) and (f), respectively. The fields which contain type 1 and type 2 sequences have been defined by 50 permutations of the reference model. The type 1 field has been contoured based on the percentage of time during sequence-boundary formation that the depositional shelf edge is exposed and oblique

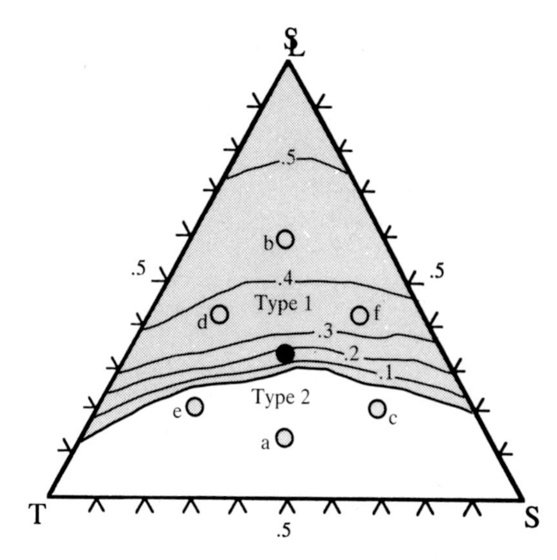

Fig. 4. Ternary diagram showing sequence type as a function of relative strength of three parameters with respect to the reference case. Values of each parameter are calculated as the normalized strength divided by the sum of the three parameters. Thus the reference case plots in the centre of the ternary diagram and the three axes in Fig. 3 correspond to lines bisecting the apexes of the triangle. The positions of other cases from Fig. 3 are also noted. Decreasing sea level ($L), increasing the tectonic subsidence rate (T) and/or the sediment flux (S) creates conditions favourable for type 2 sequence boundaries. However, the system is most sensitive to sea level. Type 1 conditions correspond to the shaded field, and the relative strength of a type 1 sequence boundary is denoted by the contours of the fraction of time during a sea-level cycle in which deposition is restricted to oblique clinoforms. ●, reference case; ○, Other cases shown in Fig. 3.

clinoforms are deposited. These values increase in a generally vertical sense in the diagram reflecting the greater effect of sea-level amplitude in modifying sequence architecture than the other parameters.

These results concur with ones obtained by other models. Jervey (1988) varied sediment flux in his stratigraphic models and obtained stratal patterns that correspond to the shift from type 1 to type 2 sequence boundaries we observed with increasing sediment flux (Figs 3(c), (d)). Thorne and Swift (1991) created a model that included a curvilinear subsidence profile and different depositional slopes for the shelf, slope and lowstand wedge. They determined their shoreline position using a method similar to Pitman (1978). Thorne and Swift (1991) also found that the type of sequence boundary formed was controlled by the sediment supply. In addition, they found the two depositional slopes influenced the sequence boundary type through their effect on the depth of the depositional break in slope. High values for these parameters caused an increase in the modelled depth to the depositional shelf break. In these cases, sea level did not drop below the depositional shelf break and resulted in type 2 sequence boundary formation.

While tectonic subsidence and sea-level changes may be the most important factors in generating accommodation, sediments have an effect which has been mostly disregarded in sequence-stratigraphic models. The weight of the sediments deposited during a time interval has two important effects. One is the depression of the lithosphere that occurs as a result of isostatic compensation. The other is the compaction of underlying sediments. Both of these processes cause additional subsidence of the sea floor and thus contribute to the creation of accommodation. Thus these parameters can significantly modify the available space and provide feedback between sedimentation and the apparent tectonic subsidence.

The effect of isostatic compensation is examined in Fig. 5. We hold all parameters constant except for elastic thickness (T_e) which we set at 0 km (Airy compensation) and 10 km (Figs 5(a), (b)). While the total accommodation space created by sediment loading is dependent only on the total weight of the sediments, the manner in which flexure distributes the accommodation space is a function of lithospheric rigidities. The margin with no flexural strength ($T_e = 0$ km) has deposition restricted to a narrow shelf and no space is created landward of the hinge. Accommodation is created vertically above the sediment load and the sediment load 'sinks into itself'. This additional accommodation shifts the model results toward type 2 sequence boundaries (Fig. 5(c)). For higher flexural rigidities ($T_e = 10$ km), accommodation space created by isostasy is broadly distributed. The margin is much wider with onlap occurring landward of the hinge. However, less vertical space is made available at any point and thus the formation of type 1 sequences is favoured (Fig. 5(d)). While the result may appear superficially similar to models in which isostasy is neglected, the additional accommodation induced by loading is identical to the Airy case. This accommodation results in the generation of the flexural wedge landward of the hinge point ($x = 50$ km) and the longer downdip progradation at the foot of the clinoforms. Thus, while the total accommodation made available by isostatic processes in each case is identical, the manner in which this space is partitioned (i.e. horizontally or vertically) influences the sequence architecture and boundary type.

The weight of deposited sediments causes the compaction and consolidation of the underlying sediments. While compaction has been considered a secondary process, in that it is a reaction to external forcing, the magnitude of the effect is not of secondary importance. Compaction of a thick sediment pile can induce substantial additional subsidence directly below a sediment load and can exceed 50% of the load thicknesses. The sediments deposited on a compactable margin literally sink into the substratum. Compaction is computed with an exponential decrease in porosity with depth. Figures 6(a) and (b) present the evolution of the model with compaction following sandstone and shale parameters (Sclater & Christie, 1980). The model assumes a 10 km sediment column underlying the modelled sequences. Despite the inclusion of a 10 km elastic thickness, the additional accommodation generated by compaction is concentrated directly below the sediments and strongly shifts the model results towards the formation of type 2 sequence boundaries. This additional space restricts deposition on the shelf and delays offlap. The end result is a narrower shelf compared to models where compaction has been ignored. Only subtle differences are observed between the all-sand (Fig. 6(a)) and all-shale models (Fig. 6(b)). The width of the all-sand margin is slightly greater than that of the all-shale margin. Also offlap occurs slightly sooner in the all-sand margin. The all-shale margin, however, has

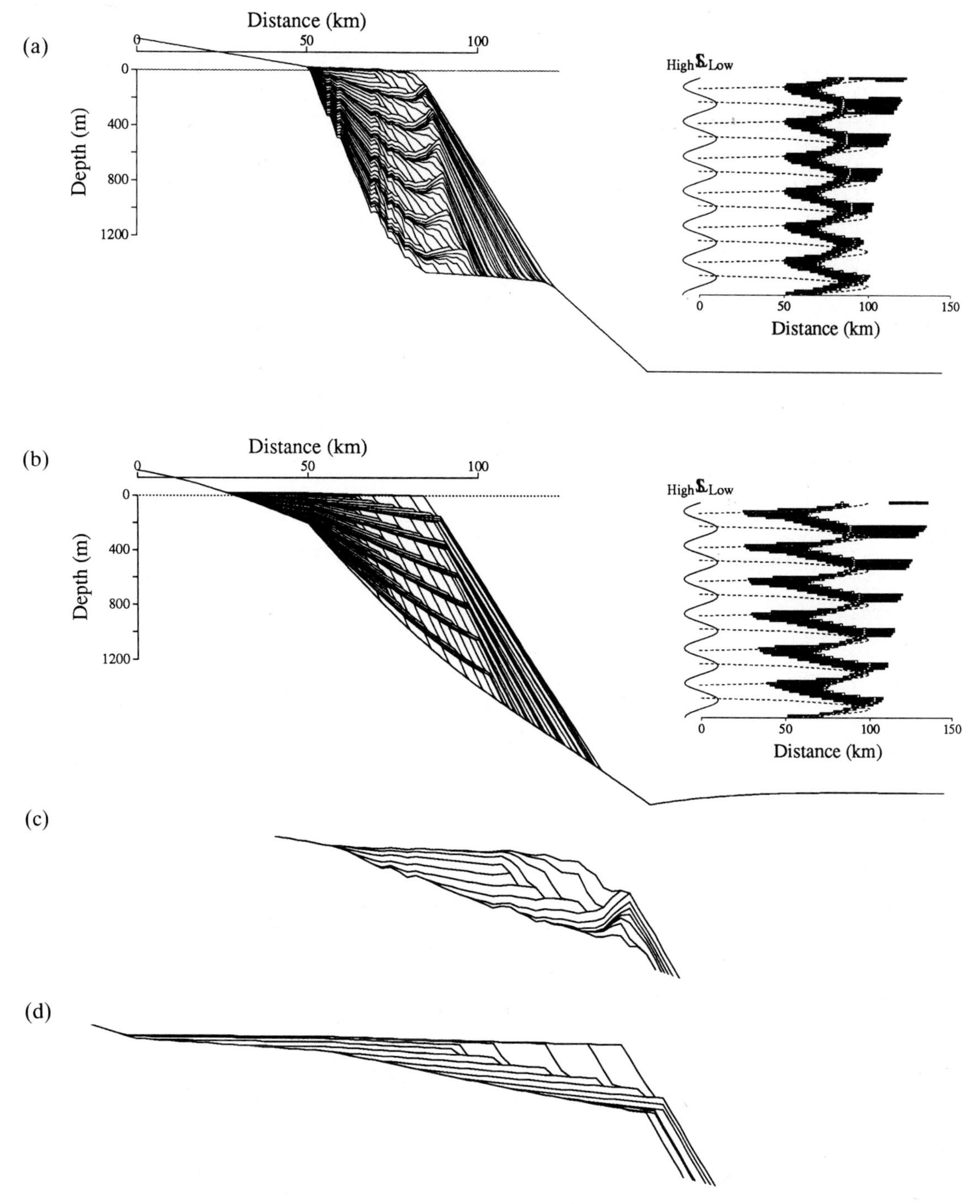

Fig. 5. (a,b) Cross-sections of two model runs illustrating the modifications in sequence architecture due to variations in flexural rigidity. Models were run for 100 timesteps (~10.0 Ma). Wheeler plots for both cases are shown at right. (a) For the Airy case ($T_e = 0$ km), deposition is restricted to a narrow shelf and little or no accommodation space is created landward of the hinge. Significant vertical space is created beneath the sediment load and type 2 sequence boundaries are formed. (b) For $T_e = 10$ km, space is made available laterally by flexural isostasy and deposition occurs over a wider area and a broad zone of onlap occurs landward of the hinge. However, less vertical space is available and type 1 sequence boundaries are created. (c) Close up of the youngest layers for the Airy case. (d) Close up of the youngest layers for the case with $T_e = 10$ km.

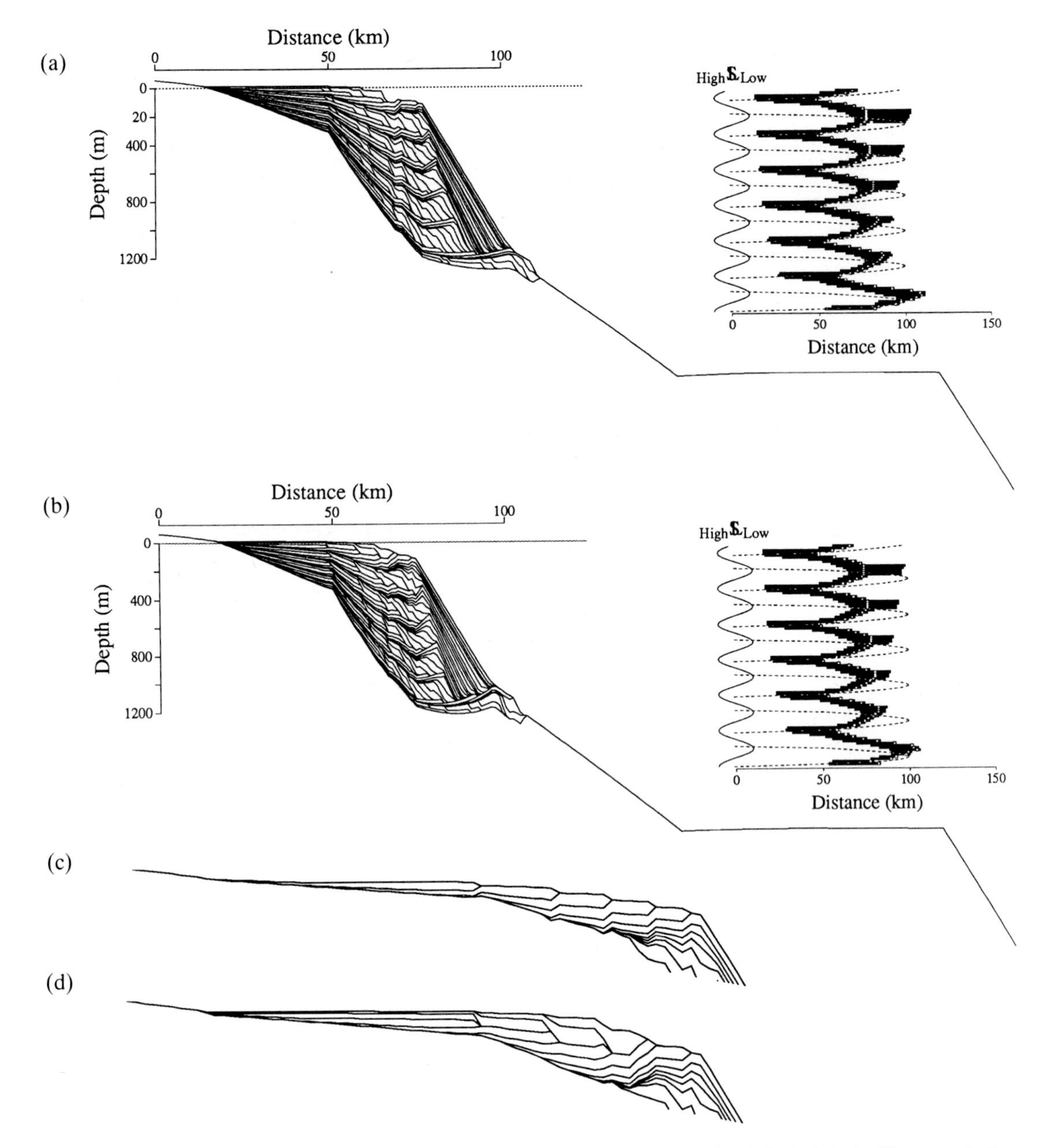

Fig. 6. For the case where compaction of a 10-km thick pre-existing sediment section is included, significant vertical space is created wherever sediments are deposited including the flexural wedge. An all-sand lithology is shown in (a) whereas (b) models an all-shale lithology. Both models include a $T_e = 10$ km. The all-sand model shows a slightly wider depositional extent and earlier offlap than the all-shale model. However, the all-shale model has a slightly thicker stratigraphic section directly beneath the sediment load. Note the restriction and shifting of the lateral extent of the zone of deposition due to compaction, the rotations of the section and the variations in sediment thicknesses, particularly near the shelf break. The timesteps shown are identical to those shown in Fig. 5. Note the differences in the relative position of the last deposits for the models in Figs 5 and 6 showing the various lags in deposition. The lags in sequence timing can also be seen in the Wheeler plots. (c) The all-shale model after 96 timesteps. (d) The all-shale model after 100 timesteps. Note the compaction and rotation of the underlying strata due to progradation of the overlying section.

a slightly thicker stratigraphic section directly beneath the sediment load. One effect of compaction is to produce rotation of strata underlying prograding sediments. Figures 6(c) and (d) show the process where the prograding clinoforms expel fluids from the underlying strata, producing anticlinal closure. This closure is a transient feature as it is lost or reduced when the shelf break progrades past the anticline.

The results of these models therefore suggest that compaction and compensation can also control variations in sequence-stratigraphic architecture and the type of sequence boundary produced (i.e. type 1 or type 2).

Sequence timing

In conjunction with the differences in the type of sequence boundary, the model runs also exhibit systematic differences in the timing of the sequence boundary formation and the systems tract composition of the sequences. This may be roughly seen in Fig. 3. Although all the models shown contain the same number of timesteps the distance that sedimentation has prograded seaward in the last timestep is greater in the type 2 case (Fig. 3(a), (c), (e)) than in the type 1 case (Fig. 3(b), (d), (f)) for each axis. In the reference case, sedimentation has extended to an intermediate distance. These differences in lateral position of the progradational clinoforms correspond to differences in the timing of sequence boundaries. Analysis of the sequence architectures displayed in Fig. 3 together with their Wheeler diagrams demonstrates systematic shifts in sequence boundary timing and systems tract distribution.

Figure 7 illustrates the systematic changes obtained in models generated as a function of sea level, subsidence rate and sediment supply but ignoring the effects of isostasy and compaction. The type 2 sequences consist of predominantly highstand and shelf margin wedge systems tracts with a transgressive systems tract of short duration. The shelf margin wedge and highstand systems tracts occupy positions nearly from inflection point to inflection point on the sea-level curve. The type 2 sequence boundary occurs at the falling inflection point of the sea-level curve and the transgressive systems tract occupies only a short interval at the inflection point associated with the maximum rate of sea-level rise. This implies that systems tracts in the model are responding to the rate of sea-level change. The type 1 sequences vary depending on the amplitude of sea-level fluctuation relative to the subsidence rate and sediment supply. As the magnitude of sea-level fluctuations increases, type 1 sequences become dominated by lowstand and transgressive systems tracts with a highstand of short duration. In the type 1 sequence, the timing of the lowstand and transgressive systems tracts progressively shifts toward the peak and trough of the sea-level curve, respectively. This occurs because as

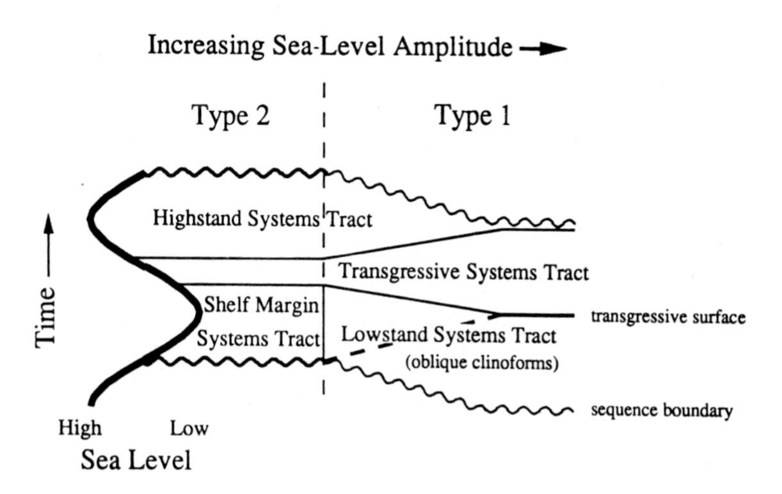

Fig. 7. Diagram showing the relationship of the timing of systems tracts and unconformity surfaces to sea level as a function of increasing sea-level amplitude (or equivalently, decreasing sediment supply or subsidence rate). These relationships are only valid when isostasy and compaction are neglected as these also affect the timing. While the sequence boundary occurs at the falling inflection point for type 2 sequences, it shifts earlier toward the sea-level peak as sea-level changes increase in amplitude. The transgressive surface marking the beginning of onlap also shifts by a one-quarter cycle, but in the opposite sense. Deposition shifts from dominantly highstand and shelf margin systems tracts in a type 2 to lowstand and transgressive systems tract in a type 1.

the magnitude of the sinusoidal sea-level curve increases, the rate of sea-level fall exceeds the subsidence rate at the depositional shelf break progressively earlier in the cycle. Thus, it is a combination of the amplitude of the sea-level fall as well as rate of sea-level fall relative to the subsidence rate that exerts primary control on the type of sequence boundary produced. Depending on the amplitude of sea level (or, alternatively, sediment supply or tectonics) the timing of the sequence boundary formation and systems tracts may shift by up to one-quarter of a cycle.

An example of the phase shift that can occur as the result of an extreme sea-level signal is seen in the Pleistocene of the Rhône continental shelf (Tesson *et al.*, 1990). Interpretation of high-resolution seismic profiles shows that the glacial cycles are composed entirely of transgressive and lowstand systems tracts. Further, dating of the sediments suggests that the sequence boundaries correspond to the times of sea-level highstands. This corresponds to the response of a margin when the rate and amplitude of eustatic change greatly exceed those of subsidence.

Differences in sequence architecture seen in Fig. 7 can also be produced by variations in sediment supply or subsidence rates. Contrasts within a single eustatic event can be seen by comparing the current systems tracts of the Louisiana Gulf coast, which is undergoing early highstand progradation, to the Atlantic margin of eastern North America, which is still experiencing a transgression (Boyd *et al.*, 1989). This discrepancy in systems tracts is possible during the late phase of a sea-level rise (Fig. 7). Since these two margins have similar ages and low thermal subsidence rates, the difference is due to the large sediment influx at the Gulf coast from the Mississippi River. The Louisiana Gulf coast has the attributes of a type 2 or 'weak' type 1 sequence, whereas the systems tracts present along the eastern North American margin appear to be characteristic of a type 1 sequence boundary (see Fig. 7). The timing shifts in the systems tracts are due to the direct effect of a large sediment supply together with the subsidence induced by compaction and other deformation of the underlying sediment column resulting from the sediment loading.

Compaction and isostasy affect the timing of the sequence development because of the changes in accommodation caused by the sediment. The space created by the sediments differs from linear tilting of the margin, a factor usually neglected in sea-level models (Jervey, 1988; Pitman & Golovchenko, 1988; Posamentier *et al.*, 1988; Thorne & Swift, 1991). Both compaction and isostasy generate additional subsidence in the vicinity of the sediment load creating a tendency for deposition to remain centred at its current location. This induces a lag in the downdip migration of deposition and therefore the timing of sequence boundaries. These lags can be seen in the Wheeler diagrams at the right of Figs 5 and 6. A lower flexural rigidity produces more of a lag because the isostatic compensation is more concentrated beneath the sediment load. Compaction tends to have a larger effect with the lags slightly more pronounced with shale parameters. The maximum flooding surface at the top of the transgressive systems tract is the most delayed in the example using shale compaction parameters (Fig. 6(b)), where it occurs at the sea-level maximum. The subsidence induced landward of the depocentre from flexural isostasy also shifts the lateral position of the entire sediment package landward. Compaction further shifts deposition landward because more sediment is retained in the flexural wedge.

While lags and leads in the response of margins to eustatic fluctuations are acknowledged in the Exxon model, in practice they are minimized. In the model of Posamentier *et al.* (1988), the sediment response to sea-level fall enables absolute synchroneity of type 2 sequence boundaries. Their model is based on assumptions about the relationship of the bayline, the point to which river profiles are adjusted, to the equilibrium point, the position where the rates of sea-level fall and subsidence are equal. Aside from the geomorphic problems with their assumptions, inclusion of isostasy and compaction would make the equilibrium point position a function of the sediments, as well as sea level and tectonic subsidence. This and the lags induced by isostasy and compaction thus disrupt the timing of type 2 sequence boundaries by changing the movement of the balance point between relative sea-level fall and rise. Since two margins rarely undergo identical sediment-load feedback, relative sea-level changes will vary between margins even though eustatic changes are, by definition, identical.

Phase shifts in the timing of sequence boundaries have been obtained by other stratigraphic models as well. The modelling developed by Pitman (1978) and by Pitman and Golovchenko (1983, 1988, 1991) has emphasized the role of the geomorphology of the coastal plain and shelf complex on timing. They assume that geomorphic processes maintain graded slopes on both the coastal plain and the shelf and specify rules governing shoreline migration. The introduction of these features into the model yields a number of major stratigraphic effects. One major effect is the development of important phase lags in the response of margins as a function of the graded slope and rate of subsidence. The shoreline migration rate is limited by the slope of the shelf. This contrasts with Posamentier and Vail (1988) where these phase shifts are avoided by forcing bayline migration to keep up with the equilibrium point during sea-level falls. Angevine (1989) quantified the shifts in timing of the shoreline migration and amplitude of its motion for the graded-shelf model as a function of the subsidence rate of the shelf and sea-level oscillation period and as a function of sedimentation response time. He concluded that phase shifts can reach up to 90° and will differ for different cycle frequencies.

Helland-Hansen *et al.* (1988) used their stratigraphic model to simulate the unscaled archetypical diagrams from Vail *et al.* (1984). To do this with sea-level curves similar to Vail *et al.* (1984), they needed to vary significantly the sediment flux and subsidence rate during the sea-level cycle. They found that their sequence boundaries fell near the maxima on their sea-level curve for type 1 and near the sea-level minima for type 2 sequence boundaries, both close to 90° off from the interpretation in Vail *et al.* (1984).

While the above models all proscribed the shapes of sediment packages or water depths, Jordan and Flemmings (1991) developed a stratigraphic model that uses diffusion of topography to simulate fluvial and marine sediment transport. Jordan and Flemmings (1991) stratigraphic sections replicate type 1 and type 2 sequence boundaries created during sea-level fall. However, they show that the timing of sequence boundary formation varies by up to one-quarter cycle from the maximum rate of sea-level fall to the sea-level minima. In their model, phase relationship depends on subsidence, sediment flux, efficiency of sediment transport and the period and amplitude of sea-level change. In addition, the timing of transgressive and regressive peaks also varies.

Thus, all numerical stratigraphic models exhibit significant variability in the timing relationship between sea level and sequence boundaries. The modifications in timing of sequence boundary development suggested by the models imply that correlative sequence boundaries are not necessarily synchronous. The magnitude of the phase shift is dependent on a number of parameters and can vary by ±90°. Thus, while the identification and correlation of equivalent sequence boundaries (i.e. produced by the same eustatic fall) can be performed, these surfaces may differ in age by up to one-half of a sea-level cycle. Attempts at temporal correlations between margins must therefore take into account leads and lags that are introduced into the sequence-stratigraphic record by differing magnitudes of tectonic subsidence and sediment supply and by the sediment-load effects. All sediments above a particular sequence boundary are not necessarily younger than all sediments below that sequence boundary on a global basis.

Sea-level amplitude

At this phase of development we are reluctant to estimate magnitudes of sea-level fluctuations from observed stratigraphic sections based on our model. However, the results do indicate several factors that will interfere with direct measurements of sea level from stratigraphic sections. Determinations of sea level from margins make use of the lateral shifts in deposition due to a vertical motion of sea level and/or vertical facies changes on a vertically moving platform. The basic pattern of sedimentation within sequences is a shifting of deposition from seaward to landward and back again with the eustatic cycles. This simple pattern can be generated for a variety of models of the relationship between eustasy and sedimentation. The amount and rate of the shift are affected by the slope of the shelf (Pitman, 1978), for example. Thus, the stratigraphy of actual shelves depends on a complex interaction of multiple factors in addition to sea level. Similarly, the vertical stacking pattern of facies belts depends on all of the significant forces that affect accommodation. It is therefore necessary to analyse the factors controlling the response of a margin to sea level and determine their effects and relative importance in order to invert the stratigraphic record for sea level.

Interpretation of the large third-order cycles in the Haq *et al.* (1987, 1988) sea-level curve was guided by Jervey's (1988) stratigraphic model. While the vertical scales in the Jervey (1988) model are normalized, calibrating them to realistic continental margin subsidence rates would yield sea-level changes (<10 m), considerably smaller than in the Haq *et al.* (1987, 1988) sea-level curve. We believe that the low amplitude for sea level is primarily a consequence of the flat equilibrium profile which allows sea level to drop below the depositional shelf break with minor falls in sea level. This limitation afflicts all models with flat shelf profiles (Burton *et al.*, 1987; Reynolds *et al.*, 1991) and clearly indicates the importance of realistic bathymetric slopes to sea-level models. If the break-in-slope at sequences actually corresponded to a shoreline break (Van Wagoner *et al.*, 1990), then only 10 m scale sea-level fluctuation would be required to generate type 1 unconformities.

Sequence-stratigraphic models generally assume that the rate of change of sea level is much greater than the rate of change of subsidence (Posamentier *et al.*, 1988). These arguments are made because the primary source of subsidence at continental margins is due to smoothly varying thermal contraction of the lithosphere (McKenzie, 1978). The model runs presented in this paper also incorporate this assumption. Since the rate of sea-level drop is presumed to be much greater than the local rate of subsidence, subsidence of the margin is commonly neglected in short-term sea-level calculations (Greenlee *et al.*, 1988).

However, the total subsidence is much greater than the thermal subsidence. The accommodation created by compaction and isostasy can amplify the total subsidence rates to greater than several times the thermal subsidence (Steckler *et al.*, 1988). However, this total subsidence, because it includes factors that shift with the location of sediment deposition, will not be smoothly varying. The assumption that all margin subsidence can be grouped into a simple tilt rate (Pitman, 1983; Posamentier *et al.*, 1988) is incorrect. The rate of sea-level change for third-order cycles is not necessarily much greater than fluctuations in the total subsidence rate.

The total subsidence rate at a margin reaches values similar to the rate of sea-level change. This can be shown by considering the Haq *et al.* (1988) sea-level curve. According to Haq (1991), the proportion of type 1 and type 2 sequence boundaries since the late Jurassic is about equal (type 2 boundaries dominate in the Triassic to late Jurassic). Although the sequence boundary types on the global chart necessarily average the variation from margin to margin, this result suggests that on average the maximum rate of eustatic fall is about equal to the total rate of subsidence at the depositional shelf break. Since the total subsidence rate can decrease to values closer to the thermal subsidence rate from depositional shifts, the fluctuations in the total subsidence rate can be of the same order as eustatic fluctuations. Considerable caution should be applied when using equilibrium-point calculations (Posamentier *et al.*, 1988) based on a constant rate of total subsidence to determine the depositional pattern.

It is necessary to understand better the feedback in the depositional processes on continental shelves before sea-level amplitudes can be confidently estimated. Stratigraphy is an indirect record of sea-level changes. On a perfectly graded shelf in equilibrium with transport processes, no sediment deposition occurs. Deposition occurs as the result of a deviation from the equilibrium state. This can result from either a change in the overall geometry of the shelf, perhaps from subsidence or sea-level change, or a modification of the equilibrium state, perhaps due to climate change. The coupling between processes prevents a simple prediction of the adjustments in the system. Forward models of shoreline migration within a sequence framework are still based on simple assumptions of the relationship between sea level and the shoreline (Pitman, 1978; Posamentier & Vail, 1988).

The coupling between processes prevents a simple prediction of the adjustments in the system. For example, a rise in sea level will not only increase the water depth on the shelf, but also cause adjustments in the shape of the equilibrium profile. Shelf morphology on prograding, regressive shelves (for example, the Louisiana Gulf coast) differs from the shape of the transgressive continental shelves (for example, the US Atlantic coast) (Swift & Thorne, in press). The sediments are coarser and the shelf is steeper on transgressive shelves with sediment supplied by shoreface erosion. Thus it seems likely that as transgressions and regressions occur throughout sea-level cycles, the relative gradients of the shelf and coastal plain will also change with time. These changing gradients constitute systematic changes in the margin morphology that co-vary with sea level. Previous attempts to model numerically the stratigraphic response to sea-level fluctuations, whether utilizing proscribed bathymetric profiles or diffusion coefficients, have neglected the cyclic changes that occur in conjunction with sea-level changes. The effect of a dynamically varying shelf morphology on sequence geometry and methodologies for measuring sea-level amplitude is unknown.

CONTINENTAL MARGIN RECONSTRUCTION

The amount of sea-level fall during the formation of a type 1 unconformity is estimated from reflector geometry by measuring the stratal thickness from the depositional shelf break to the point of first coastal onlap. To measure the subsequent sea-level rise the stratal thickness from the first true coastal onlap to the top of the sequence is used. These measurements must be corrected for thermal subsidence, sediment loading and compaction (Greenlee & Moore, 1988; Greenlee *et al.*, 1988). The water depth on the shelf before the sea-level fall must also be added. Critical to the calculation of the amplitude of sea-level fall and subsequent rise is determination of whether the lowstand wedge onlap is truly coastal or deeper marine. Facies analysis from nearby well data, when available, makes such a determination possible (Posamentier *et al.*, 1988). Direct determination by appropriately placed wells that sample this position is limited (Vail, pers. comm., 1987) and none have been publicly documented. One way to address these problems is to try to reconstruct the geometry of continental margins through time. Detailed reconstruction of the margin allows us to examine the geometry of the sequences and bounding surfaces as they appeared. We can then investigate the response of a margin to sea-level changes by analysing the geometry of the margin through time.

The Baltimore Canyon Trough, along the New Jersey coast of the Atlantic continental margin (Fig. 1), has been widely investigated for both its lithospheric and sedimentary structure. Available seismic reflection and refraction and gravity sample the deep structure (Diebold *et al.*, 1988; Grow *et al.*, 1988; Sheridan *et al.*, 1988). In addition, extensive thermal modelling of the subsidence (Watts & Thorne, 1984; Steckler *et al.*, 1988; Keen & Beaumont, 1991) has been undertaken. The depositional sequences across the Baltimore Canyon Trough have been investigated by a number of authors (Poag & Schlee, 1984; Poag, 1985; Poag *et al.*, 1987; Greenlee *et al.*, 1988; Greenlee & Moore, 1988). Currently available data include the USGS grid of seismic lines and over 30 commercial and COST stratigraphic wells on the shelf and upper slope. The Baltimore Canyon Trough also contains a series of thick prograding sequences of Miocene age. These sequences are thick enough to have their structure resolved on seismic lines. Moore *et al.* (1987) and Greenlee *et al.* (1988) used a sequence-stratigraphic framework to estimate sea level at the New Jersey margin. We have therefore chosen the Baltimore Canyon Trough for a detailed stratigraphic reconstruction.

In backstripping, individual wells or cross-sections are used to isolate each component of basin subsidence. Generally, palaeontological estimates of water depth are used as an input to calculate the tectonic subsidence. However, in an old margin such as the US Atlantic margin, the tectonic subsidence curve during the Tertiary is a smooth exponential one that is insensitive to the details of the rifting history. The overall tectonic subsidence of this margin has been computed by a number of authors based on well data and deep-penetration seismic lines (Steckler *et al.*, 1988; Keen & Beaumont, 1991). Thus the tectonic subsidence together with the observed sediment thicknesses and ages can be used to estimate the palaeowater depth through time. This is similar to the backtracking of sediments on oceanic crust in the deep sea. Backstripping of the New Jersey shelf enables us to reconstruct the morphology of the Miocene sequences at intervals throughout their formation.

Two-dimensional backstripping routines have been used to perform flexural backstripping. This process entails several steps (Fig. 8). The sediment load corresponding to each horizon is flexurally unloaded, allowing isostatic rebound of the underlying layers. These layers are then decompacted to their earlier porosities. Then the thermal subsidence that has occurred since deposition of the horizon of interest is removed. Finally, differential water loading by sea level is also removed. The result is a reconstruction of the palaeobathymetry and underlying horizons across the seismic line for the time period of the reconstruction. A simplified version of this procedure was used by Steckler *et al.* (1988) to reconstruct the Hauterivian to Eocene palaeobathymetric evolution of the Baltimore Canyon Trough. They showed that in the earliest Cretaceous, the New Jersey shelf consisted of a shallow water carbonate platform with a steep continental slope. Following the extinction of the shelf margin 'reef', the margin became sediment starved and thermal subsidence transformed the outer shelf into a gently seaward-dipping ramp by Eocene time.

Our initial analysis has concentrated on USGS line 6 off southern New Jersey (Fig. 9) which has been reprocessed to improve its resolution (Swift *et al.*, 1988). Our study of this profile has concen-

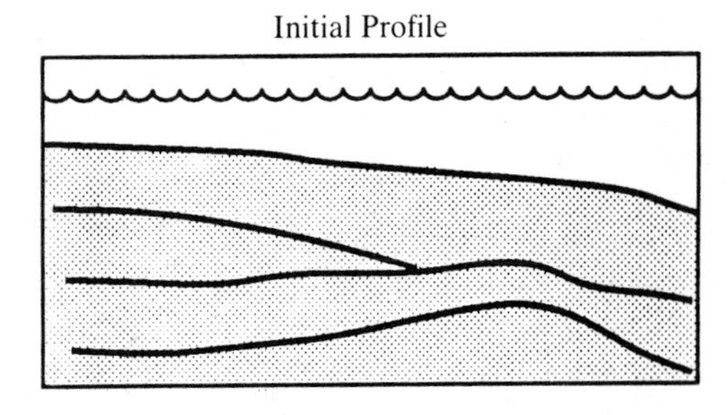

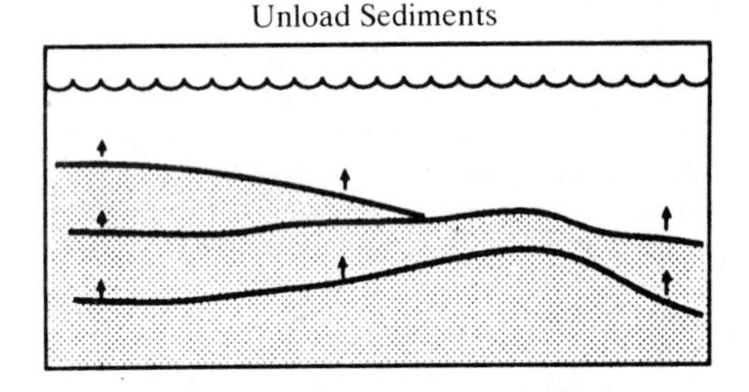

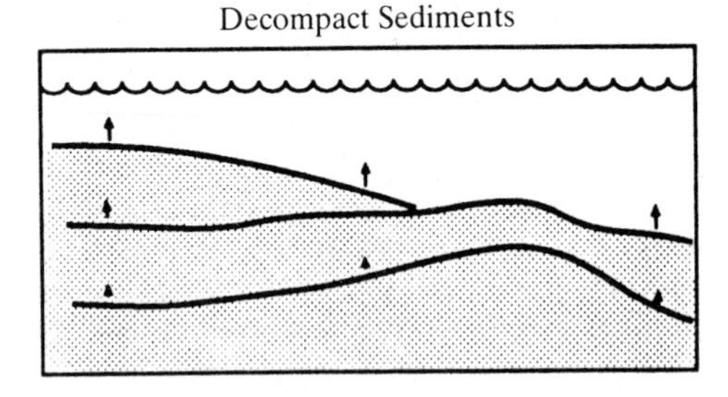

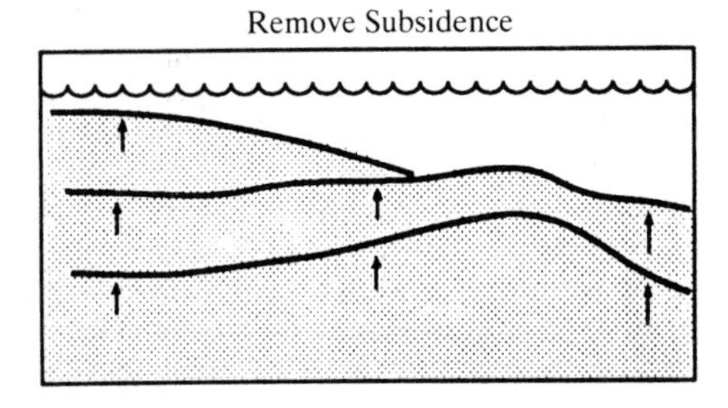

Fig. 8. Diagram illustrating the methodology for reconstructing the bathymetry and underlying stratal surfaces. At the top is the initial profile with digitized horizons. The sediments younger than the horizon of interest are flexurally unloaded and the remaining layers are isostatically adjusted. Next, the amount of compaction that was due to the removed sediment is computed and the thicknesses are restored. Finally, the thermal subsidence and eustatic change that occurred in the time interval are estimated and removed to restore the earlier attitude of the horizons.

trated on the thick Miocene progradational sequences where seismic data yield excellent resolution within the sequences and image the critical shoreline regions. Using a Landmark seismic interpretation workstation, we have identified a number of Tertiary sequence boundaries shown in Fig. 9(b). The sequence boundaries identified correlate well to those analysed by Greenlee *et al.* (1988) further north. Figure 10 shows a closeup of the well-developed clinoforms and interpreted sequence boundaries for the inner part of USGS line 6.

At this stage we have only reconstructed the geometry of the sequence boundaries that we have identified on the shelf. Due to erosion at the base of the slope, the shelf sequences cannot be directly tied to the reflectors on the continental rise. Therefore the reconstructions are based solely on data from the continental shelf. However, to account for the dramatic changes in thicknesses seaward of the shelf edge (Fig. 1), we linear tapered the thicknesses to zero at the position of the middle Eocene outcrop. For the continental rise we linearly increased the sediment thickness to the values at the shelf edge. While this is clearly in error, these sediments are too far from the region of interest to have significant effect on the reconstruction. As a result, we still consider the results to be subject to revision. However, modifications, while they may affect the absolute values of the depths in the reconstructions and modify some of the slopes, should not affect the conclusions of this study.

In order accurately to backstrip the sediments and reconstruct the bathymetry, a considerable amount of data on the lithology, porosity and density of the sediments are required. These data, obtained from three nearby wells, have been tied to the seismic lines using synthetic seismograms. Velocities from the well ties were used to convert the seismic line to depth. The well logs have been processed to estimate both lithology and porosity. Because the shallow depths of greatest interest are commonly the least reliably logged, the mineralogy was separated into only three percentages — porosity, clay content and other minerals — in the interval of greatest interest to this study. The sands are the dominant non-clay mineral in the sections analysed. We found that the porosity–depth curves for sandstone and shale determined by Sclater and Christie (1980) provided a good fit to the porosity data. For backstripping, the lithology was averaged for each sequence.

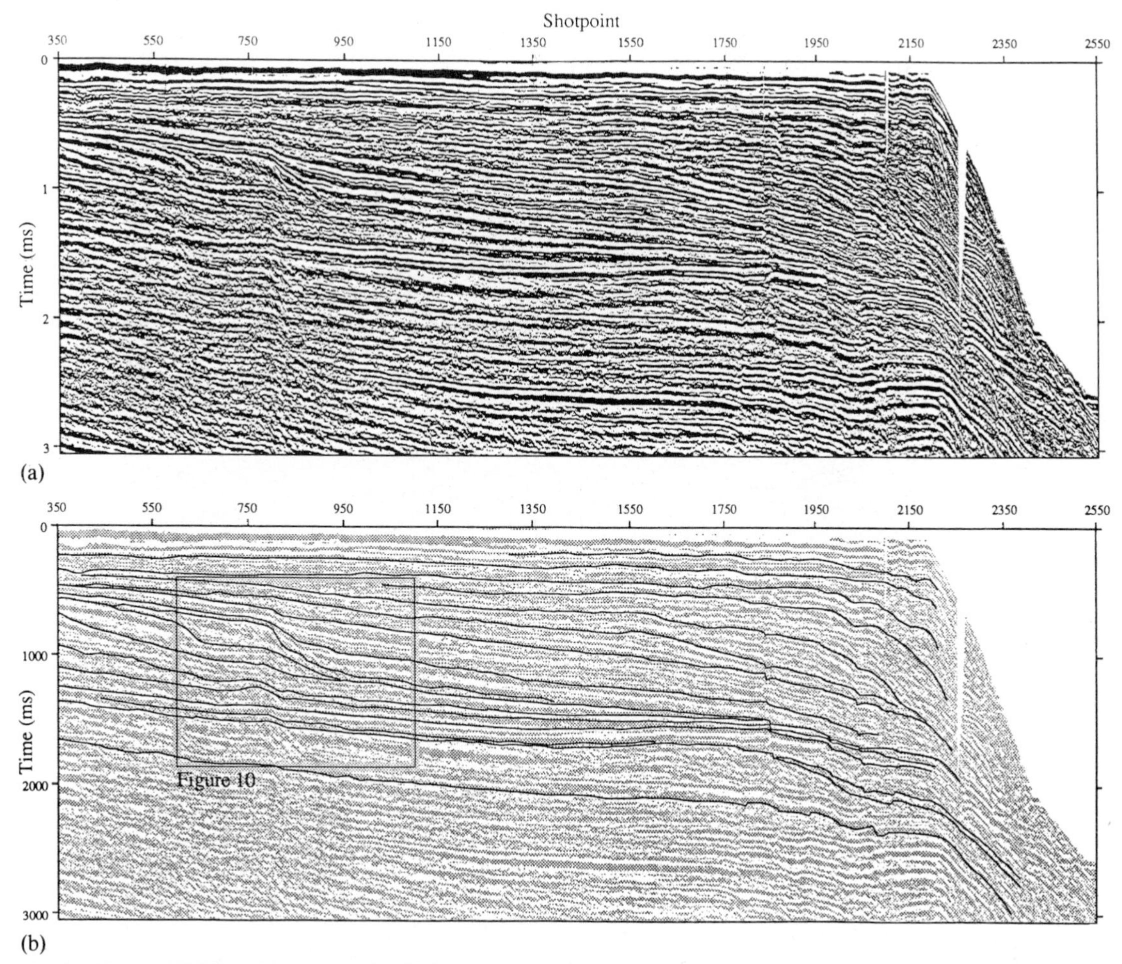

Fig. 9. Display USGS multichannel seismic line 6 across the Baltimore Canyon Trough, offshore New Jersey, that was used for the reconstructions. The horizons used in the reconstructions are shown in the lower display. On all seismic displays, 100 shotpoints equal 5 km (2000 ms of two-way travel time is approximately equal to 2 km).

The thermal subsidence for the Eocene to present was computed using a two-dimensional thermal model for the Baltimore Canyon similar to that of Steckler *et al.* (1988). The width of the zone of highly thinned crust was decreased slightly to account for the differences in the margin between USGS lines 25 and 6. The sea-level curve of Kominz (1984) was used to correct for long-term eustatic change.

The preliminary reconstruction of the palaeobathymetry and stratigraphy of the New Jersey shelf for the past 50 Ma is shown in Fig. 11(a–f). As in Moore *et al.* (1987), we have adopted the ages of the sequence boundaries from Haq *et al.* (1988) for convenience and do not have the age resolution to date them exactly. During the Eocene (Fig. 11(a)), the margin geometry initially corresponded to a gentle ramp, as was indicated by the reconstructions of Farre (1985) and Steckler *et al.* (1988). The slight uplift and more rapid increase in water depth at the seaward end of the profile is an artifact of growth faults at shotpoints 1800–2050. However, their impact on the reconstructions decreases through the series of time slices. In the Eocene, upper bathyal conditions existed on most of the present-day shelf and the shoreline was located well landward of the profile. The margin sloped seaward at an angle of approximately 0.5°. Throughout the Miocene a large increase in the sediment supply resulted in a massive progradation that filled in most of this

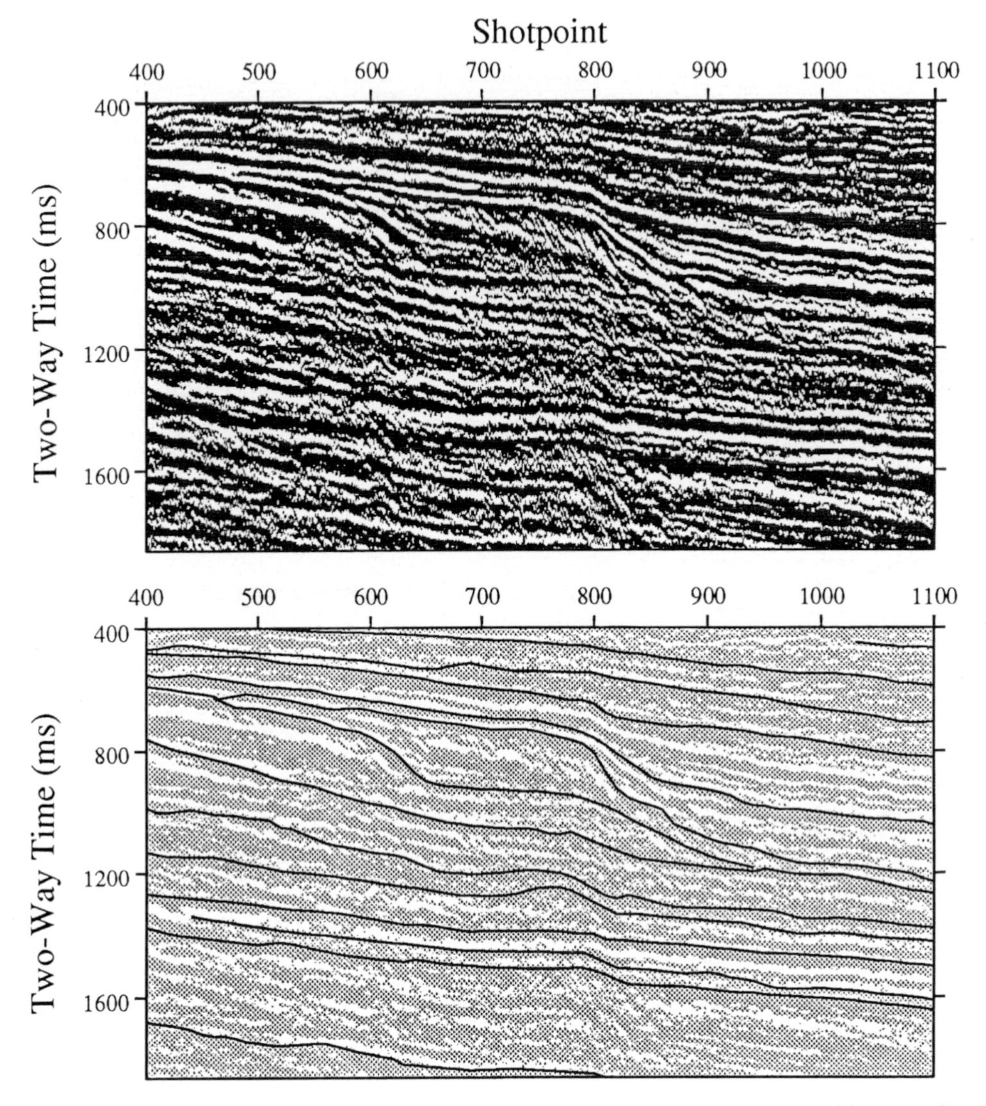

Fig. 10. Closeup of part of the seismic section USGS line 6 and interpretation showing the depositional shelf breaks of the lower Miocene sequence boundaries.

available space to produce the present-day broad continental shelf. The beginning of this progradation is observable on the mid-Oligocene reconstruction (Fig. 11(b)) as a thickening of the sediments at the landward end of line 6. The depositional shelf break for this time has been identified a short distance landward of this profile (Benson *et al.*, 1986).

In the following reconstruction near the base of the Miocene (Fig. 11(c)), a clear break in slope is visible near shotpoint 400. Seaward of this depositional shelf break, the deposits form a gently seaward tapering wedge. With the deposition of two subsequent sequences (Fig. 11(d)), the progradation continued to shift the depositional shelf break seaward. Backtilting of the underlying horizons due to sediment loading and compaction below the prograding sediments is now evident. This feature is a transient that disappears as the prograding sediments cover the entire present-day shelf removing the differential compaction effects. However, this tilting and untilting of horizons may have influenced fluid migration pathways as the margin developed. Reconstructions such as those presented here are required to reveal these transient features.

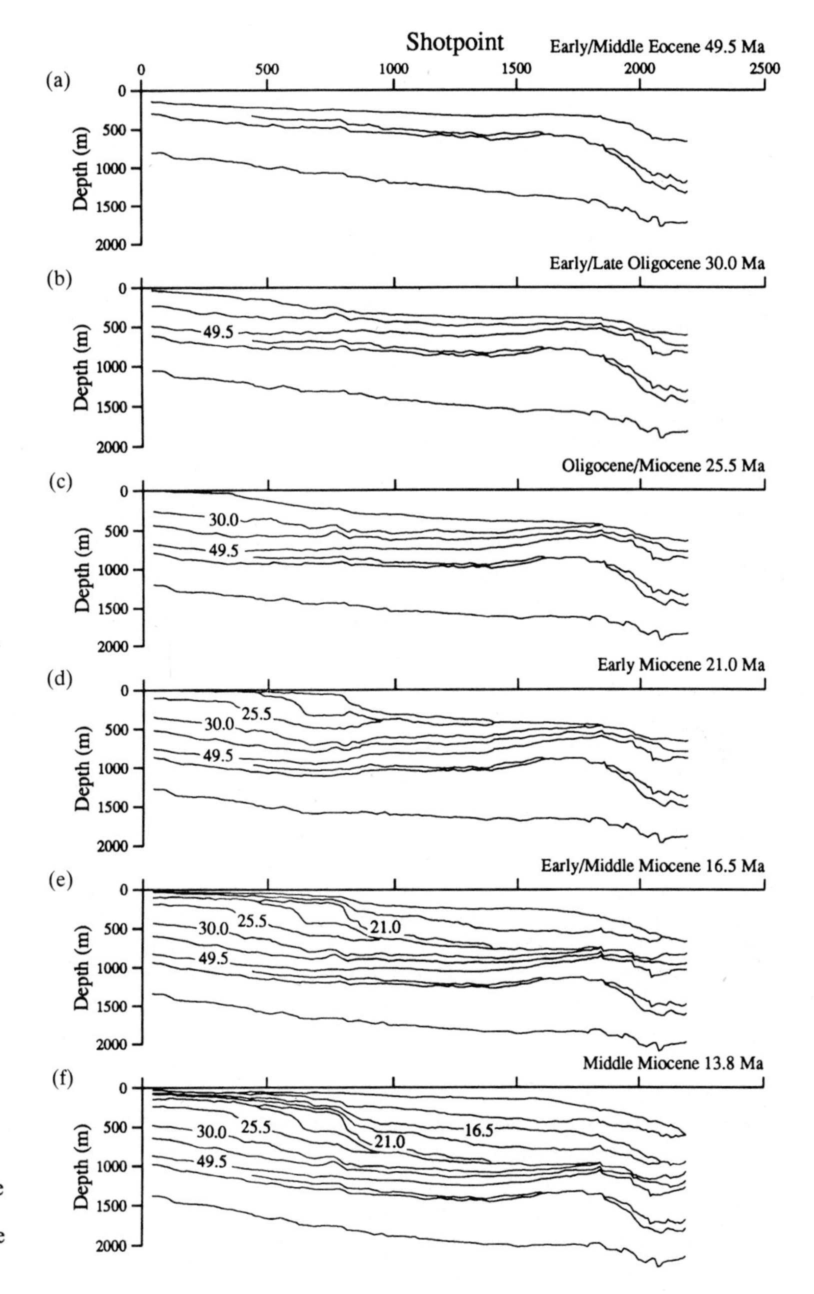

Fig. 11. Reconstructions of the margin geometry for six of the sequence boundaries in the Baltimore Canyon. The approximate geologic age and the timing of the sequence boundary according to Haq *et al.* (1988) are given for each restoration. Note the changing shape of the sequence boundary surfaces and underlying horizons as the sediments prograde across the ramp margin.

As the flatter shelf progrades over the earlier ramp margin, the height of the front of the clinoform increases. In conjunction with the increase in clinoform height is an increase in the slope of the concave-upward clinoform front. The reconstructions indicate that the changing slope at the clinoform front is depositional and is not the result of rotation due to a change in flexural rigidity across the margin as interpreted by Watts (1989).

In the next reconstruction (Fig. 11(e)), the pattern of continuous progradation of the depositional shelf break is altered. Instead the depositional shelf break remains in virtually the same position near shotpoint 800. Most of the sediments bypass the developing shelf and are deposited on the ramp. This has the effect of building up the ramp to a shallower

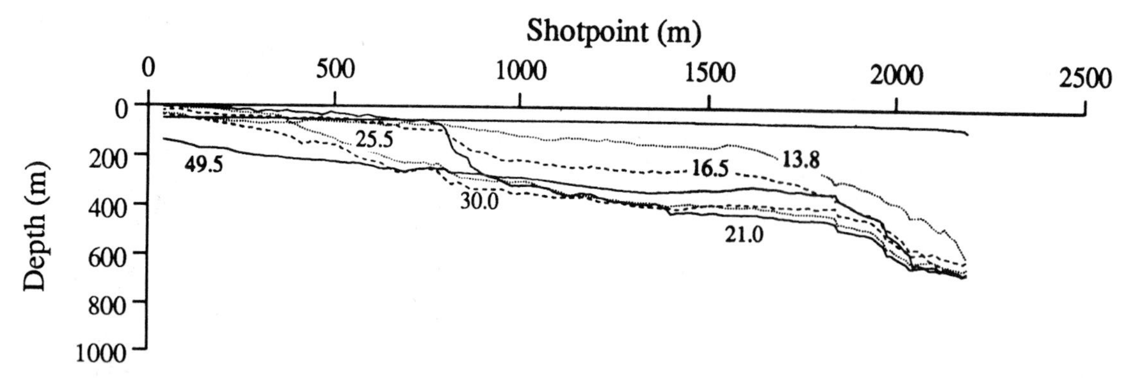

Fig. 12. Stacked bathymetric profiles for the six reconstructions shown in Fig. 11. Note the changes in the height and slope of the clinoforms as they progress across the initial ramp margin.

level. As a result both the height and slope of the clinoform front decrease. In the final reconstruction (Fig. 11(f)) the shelf once again progrades, rapidly outbuilding over the now shallower ramp to a new depositional shelf break position near shotpoint 1600. This break-in-slope is not well imaged by USGS line 6 which crosses this middle Miocene shelf break obliquely. It is better seen in the seismic lines shown by Greenlee *et al.* (1988) further to the north. Examination of line 6 and the line in Greenlee *et al.* (1988) shows that this pattern of stagnation of the shelf break coupled with bypass followed by rapid progradation seaward is repeated three times in the Miocene of the New Jersey shelf. This punctuated advance produces the lower Miocene (shown here), middle Miocene and upper Miocene shelf break positions along the New Jersey margin.

We interpret this pattern of progradation to be controlled by the pre-existing margin geometry interacting with sediment stability criteria rather than sea level and sediment supply. As the depositional shelf break progrades across the ramp margin the clinoform heights and slopes increase. The evolution of the topography of the margin is shown in Fig. 12. We note that the depositional shelf break depth for all of the reconstructions is $\simeq$50 m. The clinoform heights rise from $\simeq$200 m at 25.5 Ma to $\simeq$400 m at 21 Ma. At this point the progradation of the depositional shelf break was limited by the pre-existing ramp geometry of the margin. As the height of the clinoforms increases, the maximum slope of the front of the clinoforms, which follows an exponential curve, increases. The depositional shelf break cannot prograde further because the clinoforms have reached their maximum stable slope for this environment. Subsequently, most of the sediment that reached the shelf bypassed or slumped off the front of the oversteepened clinoforms thus building up the level of the downslope ramp. The buildup of the ramp occurs over the next several sequences. The major progradation resumed after the ramp margin built up to a shallower level.

Thus, in addition to the sea-level cyclicity that generates the sequence boundaries, the progradation of the margin is strongly modulated by the margin geometry. This controls the punctuated nature of the progradation, not a second-order eustatic cycle (Greenlee & Moore, 1988). The non-eustatic control of the clinoforms also has implications for the estimation of the magnitude of sea-level fall. The amplitude of sea-level fall is measured from the vertical shift from the last highstand below the sequence boundary to the first coastal onlap above the sequence boundary. Differentiating between marine onlap and coastal onlap against the clinoform fronts is difficult. We caution against equating the basal onlap with coastal onlap and therefore using the clinoform heights as a measure of the amplitude of sea-level falls. We interpret the maximum clinoform height to reflect sediment dynamics and not the amplitude of sea-level falls. Most of the reflectors lapping out against the base of the clinoforms dip seaward with no clear depositional shelf break. Flat-lying deposits with a clear break in slope during lowstand wedge deposition can be seen only when the wedge has built up virtually to the height of the earlier highstand. These are probable shallow-water sediments, but most of the lowstand wedge was probably deposited in deeper water. Thus the maximum height of clinoforms does not provide a measure of the amplitude of sea-level cycles.

SUMMARY AND CONCLUSIONS

Conceptual and numerical models of stratigraphic sequence formation are beginning to aid the deciphering of sedimentary response to sea level. The shifting of deposition in response to sea-level change inherent to sequences is probably so basic that a variety of approaches can replicate many first-order features of sequences. However, the detailed aspects of the sedimentary response of a margin to sea-level change, the understanding of which are necessary for correctly unravelling the stratigraphic record, are complex. Sequence geometry is most sensitive to sea level, but other factors, such as subsidence rate and sediment supply, can produce similar geometric variations. This is accomplished in part by changing the effective ratio of the rates of sea-level change to accommodation creation at the depositional shelf break. The role the sediments play in generating accommodation has mostly been neglected in sequence-stratigraphic models, but at old passive margins this space is larger in magnitude than that created by thermal subsidence. Thus, the same sea-level change can produce a type 1 or type 2 sequence boundary depending on proximity to sediment supply. Similarly, variations in shelf morphology and sedimentary processes between different margins can result in changes in sequence type.

Numerical models indicate that variations in margin characteristics produce systematic shifts in sequence boundary timing and systems tract distribution. The timing of the sequence boundary formation and systems tracts may vary by up to one-half of a sea-level cycle. These shifts are a function of both sea-level amplitude and rates. Different parameters can generate either lags (compaction and isostasy) or leads (tectonics and sediment supply). Since two margins will rarely undergo identical sediment-load feedback, relative sea-level changes will vary between margins even though eustatic change will, by definition, be identical. Sequence boundaries that are correlative (i.e. produced by the same eustatic event) will rarely be synchronous. Furthermore, because of the phase shifts, all sediments above a sequence boundary will not necessarily be younger than all sediments below it on a global basis.

The amplitude of sea-level changes cannot be determined accurately without accounting for the major processes that affect sediment accumulation. Stratigraphy records the generally horizontal depositional response to vertical sea-level fluctuations. This response and the feedback between different processes on continental shelves must be better understood before sea-level amplitudes can be confidently estimated. If the shape of the shelf is modified and not just translated during sea-level cycles then simple geometric estimates of sea-level falls may be incorrect. Furthermore, the similar frequency existence of type 1 and type 2 sequence boundaries implies similar rates for sea-level fall and total subsidence. This is also intrinsically assumed by conceptual models utilizing the equilibrium point concept. Thus, assuming subsidence is negligible during rapid third-order sea-level falls to calculate amplitudes is inconsistent.

Backstripping of sequences can be used to reconstruct past continental margin geometry. Analysis of a seismic line on the New Jersey margin reveals the role of pre-existing margin geometry and sediment stability in producing and modifying sequence geometries. As the depositional shelf break progrades across a ramp margin, the clinoform heights and slopes increase. The interaction of the ramp geometry with the clinoform stability modulates the progradational style of the margin. This is as important as the eustatic cyclicity in shaping the sequence geometry. The clinoform heights are not a proxy for sea level. Backstripping also reveals transient features such as the progressive deformation of sequences due to compaction. More importantly, the original shape of margin and sequence geometries can be restored enabling improvements in the analysis of the stratigraphic evolution.

In order to be able to rely on either conceptual or numerical models of sea-level controls on stratigraphy and the interpretations based on them, we need to understand the influence of the individual components. Sequences differ greatly in shape from cycle to cycle or from margin to margin. This variability contains important information on processes and parameters that control sequence formation and the depositional environments at continental margins. We need to understand the interactions between the processes involved in generating sequences in order to deal quantitatively with differences in sequence architecture and invert them for the causal eustatic, tectonic and sedimentary signals.

ACKNOWLEDGEMENTS

We would like to thank Bil Haq for inviting us to present this material in Nottingham. The manuscript benefited from reviews by Nick Christie-Blick, Bill Devlin, Bil Haq and Dennis O'Leary. Myung Lee and Warren Agena helped with reprocessing of USGS line 6. This work was supported by National Science Foundation grants OCE-88-00703 and OCE-90-13111. Lamont-Doherty Earth Observatory publication no. 5048.

REFERENCES

ANGEVINE, C.L. (1989) Relationship of eustatic oscillations to regressions and transgressions on passive continental margins. In: *Origin and Evolution of Sedimentary Basins and their Energy and Mineral Resources* (Ed. Price, R.A.) Am. Geophys. Union Geophys. Monogr. Ser., 48, 29–35.

BAUM, G.R. & VAIL, P.R. (1988) Sequence stratigraphic concepts applied to Paleogene outcrops, Gulf and Atlantic basins. In: *Sea-level Changes: An Integrated Approach* (Eds Wilgus, C.K., Hastings, B.S., Kendall, C.G.St.C., Posamentier, H.W., Ross, C.A. & Van Wagoner, J.C.) Spec. Publ. Soc. Econ. Paleontol. Mineral. **42**, 309–328.

BENSON, R.N., ANDRES, A.S., ROBERTS, J.H. & WOODRUFF, K.D. (1986) Seismic stratigraphy along three multichannel seismic reflection profiles off Delaware's Coast, Delaware Geological Survey, *Miscellaneous Map Series*, no. 4.

BOYD, R., SUTTER J. & PENLAND S. (1989) Relation of sequence stratigraphy to modern sedimentary environments, *Geology* **17**, 926–929.

BURTON, R., KENDALL, C.G.St.C. & LERCHE, I. (1987) Out of our depth: on the impossibility of fathoming eustatic sea level from the stratigraphic record. *Earth Sci. Rev.* **24**, 237–277.

CHRISTIE-BLICK, N., MOUNTAIN, G.S. & MILLER, K.G. (1990) Stratigraphic and seismic stratigraphic record of sea-level change. In: *Sea-level Change*. National Academy Press, Washington, DC, pp. 116–140.

DIEBOLD, J.B. & STOFFA, P.L. *et al.* (1988) A large aperture seismic experiment in the Baltimore Canyon Trough. In: *The Geology of North America, Vol. I-2, The Atlantic Continental Margin* (Eds Sheridan, R.E. & Grow, J.A.) Geological Society of America, Boulder, Colo., pp. 387–398.

FARRE, J.A. (1985) The importance of mass-wasting processes on the continental slope. Ph.D thesis, New York, Columbia University, 200 pp.

GALLOWAY, W.E. (1988) Genetic stratigraphic sequences in basin analysis, part II. Application to northwest Gulf of Mexico Cenozoic Basin. *Bull. Am. Assoc. Petrol. Geol.* **73**, 143–154.

GREENLEE, S.M. & MOORE T.C. (1988) Recognition and interpretation of depositional sequences and calculation of sea-level changes from stratigraphic data — offshore New Jersey and Alabama Tertiary. In: *Sea-level Changes: An Integrated Approach* (Eds Wilgus, C.K., Hastings, B.S., Kendall, C.G.St.C., Posamentier, H.W., Ross, C.A. & Van Wagoner, J.C.) Spec. Publ. Soc. Econ. Paleontol. Mineral. **42**, 329–353.

GREENLEE, S.M., SCHROEDER, F.W. & VAIL, P.R. (1988) Seismic stratigraphy and geohistory analysis of Tertiary strata from the continental shelf off New Jersey — calculation of eustatic fluctuations. In: *The Geology of North America, Vol. I-2, The Atlantic Continental Margin* (Eds Sheridan, R.E. & Grow, J.A.) Geological Society of America, Boulder, Colo. pp. 437–444.

GROW, J.A., KLITGORD, K.D. & SCHLEE, J.S. (1988) Structure and evolution of Baltimore Canyon Trough. In: *The Geology of North America, Vol. I-2, The Atlantic Continental Margin* (Eds Sheridan, R.E. & Grow, J.A.) Geological Society of America, Boulder, Colo. pp. 269–290.

HAQ, B.U. (1991) Sequence stratigraphy, sea-level change, and significance for the deep sea. *Publ. Int. Assoc. Sediment* **12**, 1–36.

HAQ, B.U., HARDENBOL, J. & VAIL, P.R. (1987) Chronology of fluctuating sea levels since the Triassic. *Science* **235**, 1156–1167.

HAQ, B.U., HARDENBOL, J. & VAIL, P.R. (1988) Mesozoic and Cenozoic chronostratigraphy and cycles in sea level change. In: *Sea-level Changes: An Integrated Approach* (Eds Wilgus, C.K., Hastings, B.S. Kendall, C.G.St.C., Posamentier, H.W., Ross, C.A. & Van Wagoner J.C.) Spec. Publ. Soc. Econ. Paleontol. Mineral. **42**, 71–108.

HELLAND-HANSEN, W., KENDALL, C.G.St.C., LERCHE, I. & NAKAYAMA, K. (1988) A simulation of continental basin margin sedimentation in response to crustal movements, eustatic sea level change, and sediment accumulation rates. *Math. Geol.* **20**, 777–802.

JERVEY, M.T. (1988) Quantitative geological modeling of siliciclastic rock sequences and their seismic expression. In: *Sea-level Changes: An Integrated Approach* (Eds Wilgus, C.K., Hastings, B.S., Kendall, C.G.St.C. Posamentier, H.W., Ross, C.A. & Van Wagoner, J.C.) Spec. Publ. Soc. Econ. Paleontol. Mineral, **42**, 47–69.

JORDAN, T.E. & FLEMMINGS, P.B. (1991) Large scale stratigraphic architecture, eustatic variation, and unsteady tectonism: a theoretical evaluation. *J. Geophys. Res.* **96**, 6681–6699.

KEEN, C.E. & BEAUMONT, C. (1990) Geodynamics of rifted continental margins. In: *Geology of the Continental Margin of Eastern Canada* (Eds Keen, M.J. & Williams, G.L.) Geol. Surv. Can., Geology of Canada, no. 2, 393–472.

KOMINZ, M.A. (1984) Oceanic ridge volumes and sea-level change — an error analysis. In: *Interregional Unconformities and Hydrocarbon Accumulation* (Ed. Schlee, J.B.). Mem. Am. Assoc. Petrol. Geol. **36**, 109–127.

MCKENZIE, D. (1978) Some remarks on the development of sedimentary basins. *Earth Planet. Sci. Lett.* **40**, 25–32.

MIALL, A.D. (1986) Eustatic sea level changes interpreted from seismic stratigraphy: a critique of the methodology with particular reference to the North Sea Jurassic record. *Bull. Am. Assoc. Petrol. Geol.* **70**, 131–137.

MOORE, T.C., LOUTIT T.S. & GREENLEE, S.M. (1987) Estimating short-term changes in sea level. *Paleoceanography* **2**, 625–637.

PERLMUTTER, M.A. & MATTHEWS, M.D. (1990) Global cyclostratigraphy — a model. In: *Quantitative Dynamic Stratigraphy* (Ed. Cross, T.A.) Prentice Hall, Englewood Cliffs, N.J. pp. 233–260.

PITMAN, W.C. III (1978) The relationship between eustasy and stratigraphic sequences of passive margins. *Geol. Soc. Am. Bull.* **89**, 1389–1403.

PITMAN, W.C. III & GOLOVCHENKO, X. (1983) The effect of sea-level change on the shelf edge and slope of passive margins. In: *The Shelfbreak: Critical Interface on Continental Margins* (Eds Stanley, D.J. & Moore, G.T.) Spec. Publ. Soc. Econ. Paleontol. Mineral. **33**, 41–60.

PITMAN, W.C. III & GOLOVCHENKO, X. (1988) Sea level changes and their effect on the stratigraphy of Atlantic-type margins. In: *The Geology of North America, Vol. I-2, The Atlantic Continental Margin* (Eds Sheridan, R.E. & Grow, J.A.) Geological Society of America, Boulder, Colo., pp. 429–436.

PITMAN, W.C. III & GOLOVCHENKO, X. (1991) Modeling sedimentary sequences. In: *Special Symposium on Controversies in Modern Geology* (Ed. Muller, J.A., McKenzie, J.A. & Weissert, H.) Academic Press, London, 279–309.

POAG, C.W. (1985) *Geologic Evolution of the United States Atlantic Margin*. Van Nostrand Reinhold, New York, 383 pp.

POAG, C.W. & SCHLEE, J.S. (1984) Depositional sequences and stratigraphic gaps on submerged U.S. Atlantic margin. In: *Interregional Unconformities and Hydrocarbon Accumulation* (Ed. Schlee, J.S.) Am. Assoc. Petrol. Geol. Mem. **36**, 165–182.

POAG, C.W., WATTS, A.B. *et al.* (1987) *Initial Reports of the Deep Sea Drilling Project*, Vol. XCV. US Government Printing Office, Washington.

POSAMENTIER, H.W. & VAIL P.R. (1988) Eustatic controls on clastic deposition, II. Sequence and systems tract models. In: *Sea-level Changes: An Integrated Approach* (Eds Wilgus, C.K., Hastings, B.S., Kendall, C.G.St.C., Posamentier, H.W., Ross, C.A. & Van Wagoner, J.C.) Spec. Publ. Soc. Econ. Paleontol. Mineral. **42**, 125–154.

POSAMENTIER, H.W., JERVEY M.T. & VAIL, P.R. (1988) Eustatic controls on clastic deposition, I. Conceptual framework. In: *Sea-level Changes: An Integrated Approach* (Eds Wilgus, C.K., Hastings, B.S., Kendall, C.G.St.C., Posamentier, H.W., Ross, C.A. & Van Wagoner, J.C.) Spec. Publ. Soc. Econ. Paleontol. Mineral. **42**, 109–124.

REYNOLDS, D.J., STECKLER, M.S. & COAKLEY, B.J. (1991) The role of the sediment load in sequence stratigraphy: the influence of flexural isostasy and compaction. *J. Geophys. Res.* **96**, 6931–6949.

SCLATER, J.G. & CHRISTIE, P.A.F. (1980) Continental stretching; an explanation of the post mid-Cretaceous subsidence of the central North Sea. *J. Geophys. Res.* **85**, 3711–3739.

SHERIDAN, R.E., GROW, J.A. & KLITGORD, K.C. (1988) Geophysical data. In: *The Geology of North America, Vol. I-2, The Atlantic Continental Margin* (Eds Sheridan, R.E. & Grow, J.A.) Geological Society of America, Boulder, Colo., pp. 177–195.

STECKLER, M.S. & WATTS, A.B. (1978) Subsidence of the Atlantic-type continental margin off New York. *Earth Planet. Sci. Lett.* **41**, 1–13.

STECKLER, M.S., WATTS, A.B. & THORNE, J.A. (1988) Subsidence and basin modeling at the U.S. Atlantic passive margin. In: *The Geology of North America, Vol. I-2, The Atlantic Continental Margin* (Eds Sheridan, R.E. & Grow, J.A.) Geological Society of America, Boulder, Colo., pp. 399–416.

SUMMERHAYES, C.P. (1986) Sealevel curves based on seismic stratigraphy: their chronostratigraphic significance. *Palaeogeogr., Palaeoclimatol., Palaeoecol.* **57**, 27–42.

SWIFT, B.A., LEE, M.W. & AGENA, W.F. (1988) Reprocessing of U.S.G.S. Multichannel Line, 6 (off Cape May, New Jersey). *U.S. Geol. Surv. Open File Report* 88–551, 27 pp.

SWIFT, D.J.P. & THORNE, J.A. (1991) Sedimentation on continental shelves, part I. A general model for shelf sedimentation. In: *Shelf Sands and Sandstone Bodies: Geometry, Facies and Distribution.* (Eds Swift, D.J.P., Oertel, G.F., Tillman, R.W. & Thorne, J.A.) Spec. Publ. Int. Assoc. Sedimentol. **14**, 3–32.

TESSON, M., GENSOUS, B., ALLEN, G.P. & RAVENNE, C. (1990) Late Quaternary deltaic lowstand wedges on the Rhône continental shelf, France. *Mar. Geol.* **91**, 325–332.

THORNE, J.A. & SWIFT, D.J.P. (1991) Sedimentation on continental shelves, part II. Application of the regime concept. In: *Shelf Sands and Sandstone Bodies: Geometry, Facies and Distribution* (Eds Swift, D.J.P., Oertel, G.F., Tillman, R.W. & Thorne, J.A.) Spec. Publ. Int. Assoc. Sedimentol. **14**, 33–58.

VAIL, P.R. & HARDENBOL, J. (1979) Sea-level changes during the Tertiary. *Oceanus* **22**, 71–79.

VAIL, P.R., AUDEMARD, F., BOWMAN, S.A., EINSELE, G. & PEREZ-CRUZ, G. (1991) The stratigraphic signatures of tectonics, eustacy and sedimentation. In: *Cycles and Events in Stratigraphy* (Eds Einsele, G., Ricken, W. & Seilacher, A.) Springer-Verlag, Berlin, 617–659.

VAIL, P.R., HARDENBOL, J. & TODD, R.G. (1984) Jurassic unconformities, chronostratigraphy, and sea-level changes from seismic stratigraphy and biostratigraphy. In: *The Jurassic of the Gulf Rim* (Eds Ventress, W.P.S., Bebout, D.G., Perkins, B.F. & Moore, C.H.) Soc. Econ. Paleontol. Mineral. 347–364.

VAIL, P.R., MITCHUM, R.M., THOMPSON, S., III, SANGREE, J.R., DUBB, J.N. & HASLID, W.G. (1977) Sequence stratigraphy and global changes of sea-level. In: *Seismic Stratigraphy — Applications to Hydrocarbon Exploration* (Ed. Payton, C.E.) Mem. Am. Assoc. Petrol. Geol. **26**, 49–205.

VAN HINTE, J.E. (1978) Geohistory analysis — application of micropaleontology in exploration geology. *Am. Assoc. Petrol. Geol. Bull.* **62**, 201–222.

VAN WAGONER, J.C., MITCHUM R.M., CAMPION, H.W. & RAHMANIAN, V.D. (1990) Siliciclastic sequence stratigraphy in well logs, cores, and outcrops. *Am. Assoc. Petrol. Geol. Meth. Expl. Ser.* **7**, 55 pp.

VAN WAGONER, J.C., POSAMENTIER, H.W., MITCHUM, R.M., JR, *et al.* (1988) An overview of the fundamentals of sequence stratigraphy and key definitions. In: *Sea-level Changes: An Integrated Approach* (Eds Wilgus, C.K., Hastings, B.S., Kendall, C.G.St.C., Posamentier, H.W., Ross, C.A. & Van Wagoner, J.C.) Spec. Publ. Soc. Econ. Paleontol. Mineral. **42**, 39–46.

WATTS, A.B. (1982) Tectonic subsidence, flexure and global changes of sea level. *Nature* **297**, 469–474.

WATTS, A.B. (1989) Lithospheric flexure due to prograding sediment loads: implications for the origin of offlap/onlap patterns in sedimentary basins. *Basin Res.* **2**, 133–144.

WATTS, A.B. & RYAN, W.B.F. (1976) Flexure of the lithosphere and continental margin basins. *Tectonophysics* **36**, 25–44.

WATTS, A.B. & THORNE, J.A. (1984) Tectonics, global changes in sea-level and their relationship to stratigraphic sequences at the U.S. Atlantic continental margin. *Mar. Petrol. Geol.* **1**, 319–339.

Spec. Publs Int. Ass. Sediment. (1993) **18**, 43–53

The effects of rate of base-level fluctuation on coastal-plain, shelf and slope depositional systems: an experimental approach

L.J. WOOD*, F.G. ETHRIDGE *and* S.A. SCHUMM

Department of Earth Resources, Colorado State University, Fort Collins, CO 80523, USA

ABSTRACT

Developments in sequence stratigraphy have suggested many assumptions about the manner in which coastal-plain, shelf and slope systems along passive margins react to base-level changes. A series of experiments were performed in a 7 m long by 4.5 m wide by 0.9-m deep non-recirculating flume to test these assumptions.

Results suggest that the rate of base-level change has a strong influence on the number of incised valleys that develop on the shelf and slope, on the thickness, character and geometry of the valley-fill and delta deposits and on the preservation potential of fluvial, deltaic and slope deposits. A rapid fall of base level produced thin, widespread shelf delta deposits, multiple, thin, fine-grained, time-synchronous shelf margin delta deposits and multiple, straight fluvial channels. One valley eventually captured all of the flow. During subsequent base-level rise, this valley filled with coarse-grained sediments from the hinterland while the others filled with dominantly fine-grained sediment. A rapid rise resulted in thin, backstepping, transgressive deltas that were much coarser grained than previous lowstand deposits.

In contrast, a slow fall in base level produced thicker, less widespread shelf deltas than those that formed during the rapid fall of base level and a single, large sinuous valley formed at the shelf break and on the shelf. A series of large, diachronous shelf margin deltas formed during slow fall. During subsequent base-level rise, the single valley was filled with coarse-grained sediments from the fluvial channel. It contained thicker transgressive delta deposits than those formed during a fast rise of base level. As in the rapid base-level rise, these transgressive deposits were coarser grained than the deposits of the previous lowstand time.

INTRODUCTION

The development of sequence-stratigraphic concepts at Exxon Production Research (EPR) (Vail *et al.*, 1977; Haq *et al.*, 1987; Posamentier & Vail, 1988) and elsewhere (Brown & Fisher, 1980) has revolutionized the manner in which one views stratigraphic relationships. These concepts suggest a temporal relationship between stratal architectures and base-level fluctuations. Some of these models, however, are based on an oversimplified view of how a fluvial system reacts to base-level fluctuations.

For example, Schumm (1974, 1991) discussed complex response and episodic behaviour of fluvial systems, because of extrinsic influences, such as base-level lowering. When there is a significantly large fall in base level, incision of an affected stream may be episodic (Fig. 1(b)). That is, periods of incision are followed by deposition until a new condition of relative stability is achieved. When base-level lowering is of lesser magnitude, modest incision is followed by sediment storage and flushing (Fig. 1(a)). In both cases sediment yields fluctuate about an exponential decay curve and sediment delivery to the depositional site will vary.

Because the actions of the fluvial system have a significant effect on deposition, erosion, distribution and character of sediments in coastal-plain, shelf and slope setting, it is important to under-

*Present Address: Amoco Corporation, PO Box 3092, Houston, TX 77253, USA.

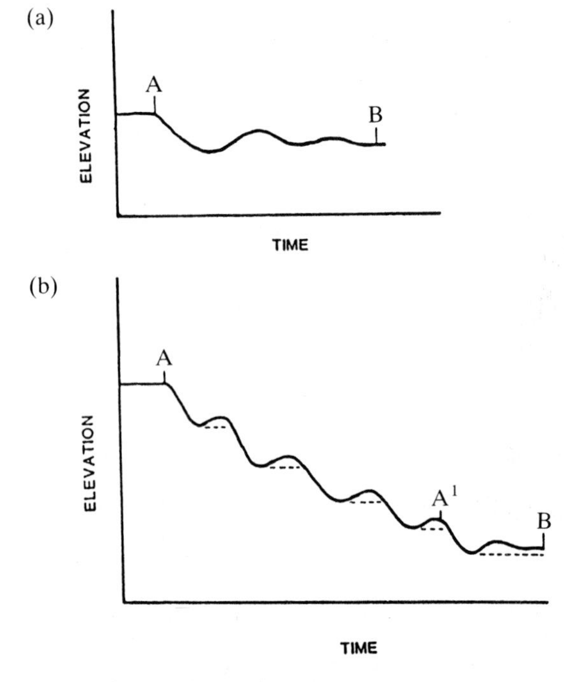

Fig. 1. Diagrams illustrating complex response (a) and episodic erosion (b) when a stream is affected at time (A) by an extrinsic impulse. When the impact is small (a), the stream degrades, aggrades and degrades until a new condition of relative stability is achieved at time (B). When the change is large (b), the major degradation is episodic being interrupted by periods of aggradation until relative stability is achieved at time (B). Area above dashed lines indicates the extent of aggradation. The complex response of (a) occupies the space between A^1 and B on the diagram of episodic erosion (b). (From Schumm, 1977.)

stand the complexities of the fluvial system and how it responds to fluctuations in base level. Although effects of base-level lowering have been studied, effects of both rise and fall of base level on the fluvial system have not. Also, researchers have not examined the effects of varying rates of base-level rise and fall on coastal-plain, shelf and slope depositional systems. It is the intent of this research to evaluate the effects of both a rise and fall of base level and the effects of changing rates of base-level fluctuations on coastal-plain, shelf and slope depositional systems.

In this study, sequence-stratigraphic concepts are scrutinized by comparing the development of stratigraphic sequences in a large flume with the conceptual stratigraphic sequences as proposed by the EPR group (Posamentier & Vail, 1988) and with sequence-stratigraphic interpretations of other authors (Suter & Berryhill, 1985; Suter *et al.*, 1987). The research discussed in this paper involves physical experiments that examine the effects of base-level change on depositional systems active within the coastal-plain, shelf and slope areas of a basin. Flume experiments allow the researcher to hold subsidence rates, discharge and sediment supply constant and to isolate effects of differing rates of base-level fluctuations on the geometry, vertical and lateral extent and sedimentological character of developed sequences. In addition, experimental studies allow observation and documentation of the effects of base-level change.

SCALING PROBLEMS

Questions always have existed concerning scaling problems, when attempting to compare the results of physical experiments with the natural systems they are intended to represent. These miniature depositional systems are not scale models of any specific natural river or coastal system; therefore, they should be regarded as analogous experiments rather than as scale models. Albertson *et al.* (1960, p. 489) points out that: 'complete similarity in the model is sometimes difficult or impossible to obtain'. However, hydraulic conditions as defined by Reynolds and Froude numbers are realistic. Therefore, results should not be extrapolated quantitatively to any modern or palaeo-environment. None the less, they should provide qualitative insight into the processes acting and the evolutionary development of coastal-plain, shelf and slope sedimentary deposits. The results of the work of Schumm and Khan (1972), Shepard and Schumm (1974) and Schumm *et al.* (1987) have provided valuable insight into the morphology and dynamics of the fluvial system.

TERMINOLOGY

The EPR sequence-stratigraphic approach is described elsewhere, but a brief summary of terminology is presented here. A short review of terminology related to fluvial geomorphology is also included to clarify some observations discussed in this paper.

A sequence is a relatively conformable succession of genetically related strata, that are bounded by unconformities and their correlative conformities (Mitchum, 1977). Sequence stratigraphy is the study of rock relations within a chronostratigraphic framework of repetitive, genetically related strata, that are bounded by surfaces of erosion or non-deposition, or their correlative conformities (Van Wagoner *et al.*, 1987). Base level is: 'The theoretical limit or lowest level toward which erosion of the Earth's surface constantly progresses but seldom, if ever, reaches; especially the level below which a stream cannot erode its bed. The general or ultimate base level for the land surface is sea level, but temporary base levels may exist locally' (Gary *et al.*, 1972). This definition is in error in part because streams do erode below base level (Shepard & Schumm, 1974), but the general concept is reasonable.

The 'equilibrium profile' is the smooth, concave-upward, stable longitudinal profile that rivers develop as a function of discharge and sediment load. One of the mechanisms by which this profile is established is the migration of knickpoints along the channel. A knickpoint is a break-in-slope that separates upper and lower reaches of a river. In this case the knickpoint results from rejuvenation by base-level lowering.

Shelf deltas are defined by Suter and Berryhill (1985) as widespread, thin, prograding depositional systems located on the shelf, and are characterized by low-angle clinoforms and numerous buried channels. Shelf-margin deltas are defined by Suter and Berryhill (1985) as localized, wedge-shaped depositional sequences of increased thickness with steepened well-developed clinoforms that are located on the upper slope.

EXPERIMENTAL DESIGN

Equipment

Experiments were conducted in a large-scale flume approximately 7 m long by 4.5 m wide by 0.9 m deep (Fig. 2). This non-tilting, non-recirculating flume was constructed from cement cinder blocks set on a poured concrete base. Water enters the upper end of the flume through a V-shaped weir cut into the headbox and exits through a series of adjustable valves in the lower end of the flume. The lower 1.5 m of the flume acts as a depositional basin and its water level (position of the shoreline and associated fluvial base level) is controlled by an adjustable overflow pipe. A point gauge with millimetre accuracy is mounted on a movable carriage that spans the width of the flume. X, Y and Z coordinates of points within the flume are measured with this point gauge and are used to construct contour maps and profiles of the flume surface at different times during base-level fluctuations.

Sediment

Sediment was selected that would simulate properties of the coastal plain–shelf–slope environment, and that would respond to imposed hydraulic con-

Fig. 2. Photograph of the non-recirculating flume used in these experiments. A vibrating sediment feeder is located near the head of the flume which is at the top of the picture. The upper ribbon marks the break between the coastal plain (with a slope of 2°) and the shelf (with a slope of 8°). The lower ribbon marks the break between the shelf and the continental slope (with a slope of 20°).

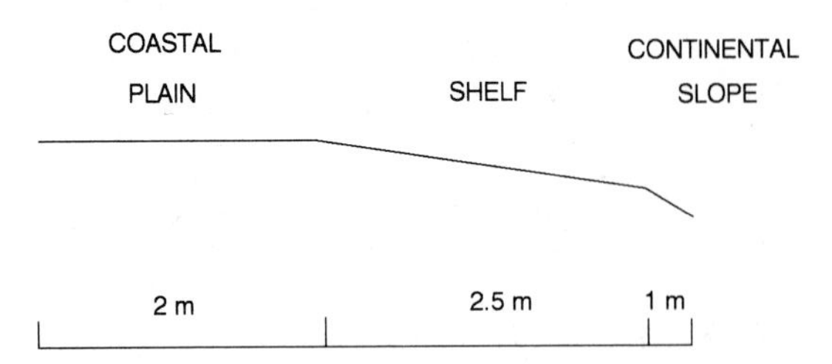

Fig. 3. Profile (in degrees) of the initial sediment surface used prior to the start of each experiment. Actual length of the flume corresponding to each physiographic region is shown, in metres, beneath the profile.

ditions within a realistic period. Preliminary experiments with different mixtures of sand, silt and clay indicated that a fine to very fine sand with 10% silt–clay would provide a uniform, cohesive and isotropic material that was used to construct the coastal-plain, shelf and slope surface.

Two mixtures of sediment were added to the flow at the entrance of the flume using an electric sand feeder, which simulated the drainage-basin, or hinterland sediment source (Fig. 2). The first was an equal mixture of coarse sand, fine sand and the material that was used to construct the coastal-plain, shelf and slope surface. This mixture was fed into the flow during base-level lowering and standstill conditions. The second mixture was identical except that the coarse sand was coloured blue, the fine sand was coloured red, and the remaining material was uncoloured. This mixture was fed into the flow during base-level rise. Water discharge was held constant at $0.1 \, l \, s^{-1}$.

Coastal-plain, shelf and slope angles

Coastal-plain, shelf and continental-slope physiography was simplified for the purposes of these experiments. The coastal-plain surface, which comprised the upper 2.0 m of the flume, was graded to a slope of 2°. The next 2.5 m of the flume simulated a shelf, which was graded to a slope of 8°. The continental slope, which occupied the final metre of the flume, was graded to a slope of 20° (Figs 2 & 3). Surfaces were carefully constructed and smoothed to minimize the influence of surface irregularities. Although coastal-plain, shelf and slope angles can vary significantly from location to location, the coastal-plain and continental-slope angles chosen for these experiments represent a fivefold exaggeration of the norm. The shelf angle represents a fivefold exaggeration of a very steep shelf (Vanney & Stanley, 1983; Thompson & Turk, 1991).

Fluctuations of base level

Base level was raised and lowered 106 cm during 4-h, 6-h and 8-h periods. This procedure produced different rates of base-level change for the same area (Table 1). Water level in the reservoir basin was measured every 12 min, and this base level was changed, during the three periods, to simulate a sinusoidal rise and fall in base level (Fig. 4).

Each experiment began with the shoreline held constant half way up the shelf for 6 h. After 6 h a stable fluvial channel had developed, bed forms were actively migrating in the channel and the amount of sediment fed into the channel was approximately equal to that sediment deposited at the mouth of the fluvial channel. After this initial 6-h period, the outlet pipe of the flume was adjusted so that the shoreline rose to the coastal-plain/shelf break. This level simulated maximum highstand (Fig. 4 (MHS)). For each period (2, 3 and 4 h), the water level was lowered progressively, as shown in Fig. 4, to a point half way down the continental slope, which simulated maximum lowstand (Fig. 4 (MLS)). A second 6-h lowstand equilibrium run followed. Water level was then raised back to the initial highstand level half way up the shelf, which resulted in transgression of the continental slope, shelf and lowstand coastal plain, to a final highstand (Fig. 4 (FHS)). This procedure, which completed a single run, was repeated for each of the three different periods. Following completion of a

Table 1. Duration, magnitude and amplitude of base-level fluctuations for each run

Run	Run time to complete one cycle of fluctuation (h)	Amplitude of base-level cycle (cm)	Average rate of rise and fall ($cm \, h^{-1}$)
1	4	106	26.5
2	6	106	17.66
3	8	106	13.25

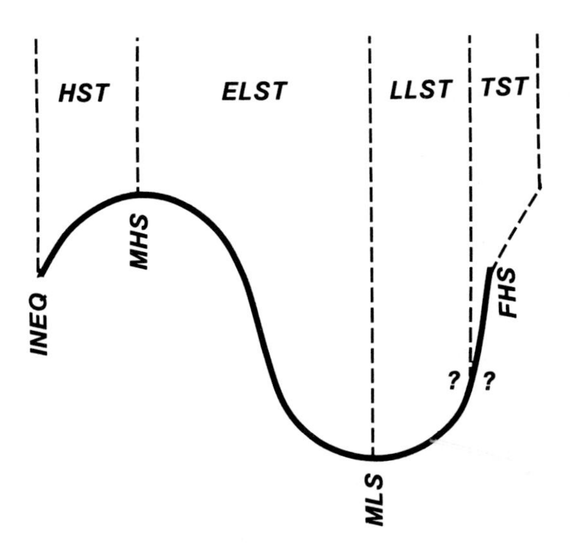

Fig. 4. Diagram showing the sinusoidal base-level curve during each run. Points of initial standstill (INEQ), maximum highstand (MHS), maximum lowstand (MLS), final highstand (FHS) and systems tract type during each interval are labelled. HST, highstand systems tract; ELST, early lowstand systems tract; LLST, late lowstand systems tract; and TST, transgressive systems tract. The boundary between LLST and TST (?) is a function of sediment flux versus base-level rise, and will vary according to the specific setting.

Fig. 5. Photograph showing a single incised valley that formed during slow fall of base level. The U-shaped cross-section was formed by slumping during base-level rise. Note multiple, backstepping deltas, which formed during final base-level rise.

run, the surface was regraded to the initial surface conditions before a new experiment was begun.

EFFECTS OF RATE OF BASE-LEVEL CHANGE AND LAG TIME ON EROSION AND DEPOSITION

Results of these experiments suggest that the rate at which base level fluctuates has a significant effect on the erosional and depositional features that develop on the coastal-plain, shelf and continental slope. During a slow fall in base level, the lower portion of the experimental channel was able to extend basinward at the same rate as the migrating shoreline. Thus a single, continuous valley was incised into the shelf. In contrast, during a rapid fall in base level the shoreline moved rapidly basinward and the experimental channel was unable to keep pace. Consequently, the initial highstand channel dispersed water and sediment onto the shelf as a broad shelf delta apron. As base level fell below the shelf, several valleys formed at the shelf edge and grew landward by headward erosion.

A slow rate of base-level fall produced thick, areally restricted shelf-delta deposits, one well-defined, sinuous, incised valley and a few, thick, diachronous shelf margin deltas (Figs 5 & 6; Table 2). In contrast a rapid base-level fall produced thin, widespread shelf delta deposits, multiple, poorly defined, straight incised valleys and thin, time-synchronous shelf margin deltas (Figs 7 & 8; Table 2).

During both rapid and slow base-level fall, aggradation of drainage basin sediment continued within the coastal-plain and upper shelf regions (0 and 3 m in the flume (Figs 6 & 8)), while incision was occurring into the lower shelf and upper slope (3–5 m in the flume (Figs 6 & 8)). A change from aggradational to erosional regimes will not occur in the upstream reaches of a fluvial system until knick-

Table 2. Number of channel incisions, sinuosity of active, late, lowstand channels and thickness of shelf margin and shelf phase delta deposits for each run (see Table 1 for duration of each run)

Run	Number of major incised channels	Sinuosity* last active channel	Shelf margin delta thickness (cm)	Shelf delta thickness (cm)
1	4	1.042	1.64	0.87
2	4	1.048	2.97	1.56
3	1	1.090	6.18	1.26

*Sinuosity = channel length/straight-line valley length

points generated by falling base level migrate headward into this region. In other flume experiments, channels have been noted to regrade their longitudinal profile from their mouths towards the upper reaches of the flume (Shepard & Schumm, 1974; Koss *et al.*, 1990). This phenomenon was also noted in the experiments discussed in this paper. As the shoreline fell over the shelf and shelf break, it generated a sudden increase in the slope of the lower channel reaches, which increased the capacity of these lower reaches to transport sediment. These factors combined to produce channel incision and a large volume of fine-grained sediment that was derived from slope and shelf deposits. Coarse sediment delivered by the channel was diluted by this large volume of finer grained sediment. Eventually, channel incision progressed into the upper reaches of the shelf and coastal-plain area. This process remobilized coarse-grained sediments that had previously been deposited here and transported them to the lower shelf and upper slope during late lowstand and transgression time (Table 3).

The lag time between maximum lowstand (Fig. 4) and the transport of large volumes of coarse-grained sediment from the upper reaches of the fluvial system is an important factor affecting the character

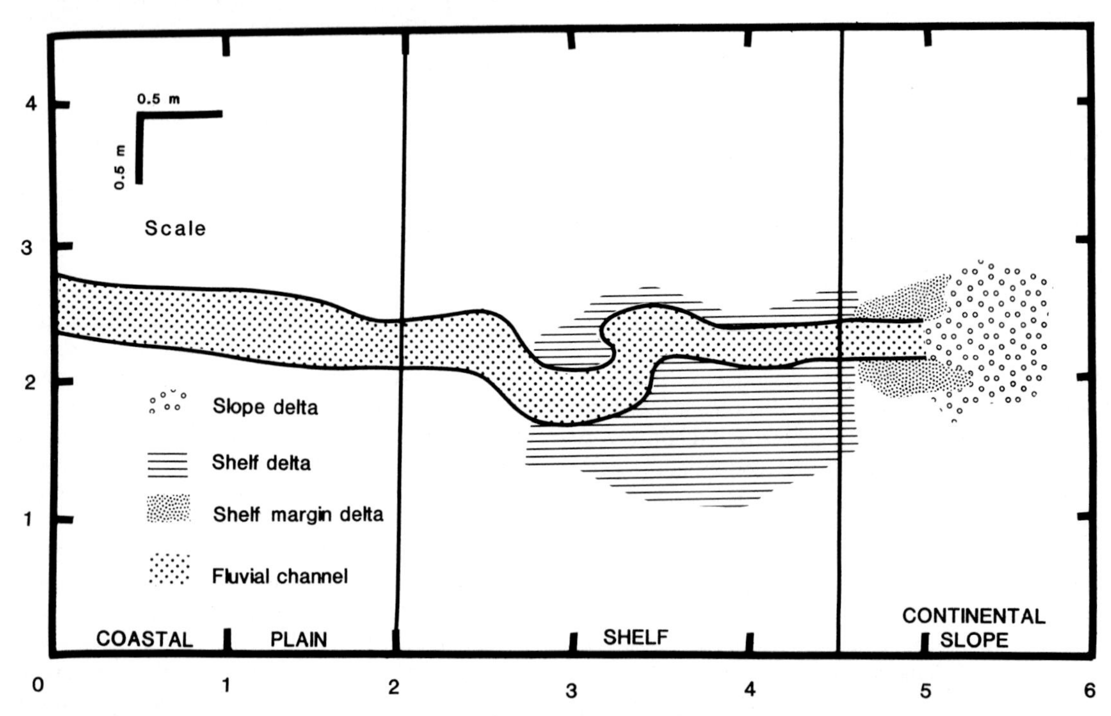

Fig. 6. Lithofacies map of the maximum lowstand surface for the 8-h run. Map shows a single incised valley, a limited number of shelf margin deltas and a narrow band of shelf delta sediments formed during a slow base-level fall and rise.

Fig. 7. Photograph showing numerous incised valleys and shelf margin delta deposits that formed during a rapid fall of base level. Note that at this time the incised valleys lack a distinct link to the upper drainage network. Eventually a single incised valley will capture all of the flow. The string running from the bottom to the top of the picture marks the break of slope between the shelf on the right and the continental slope on the left.

of the lowstand and transgressive deposits. These experiments suggest that late lowstand and transgressive deposits will have a higher percentage of coarse-grained sediment than early lowstand deposits (Fig. 4; Table 3). Any coarse-grained sediments that are present in early lowstand deposits are probably reworked from local sources and have not been transported directly from the hinterland. Without this reworked element, early lowstand shelf margin deltas will consist mainly of fine-grained, previously deposited upper-slope and outer-shelf sediments that are cannibalized by the

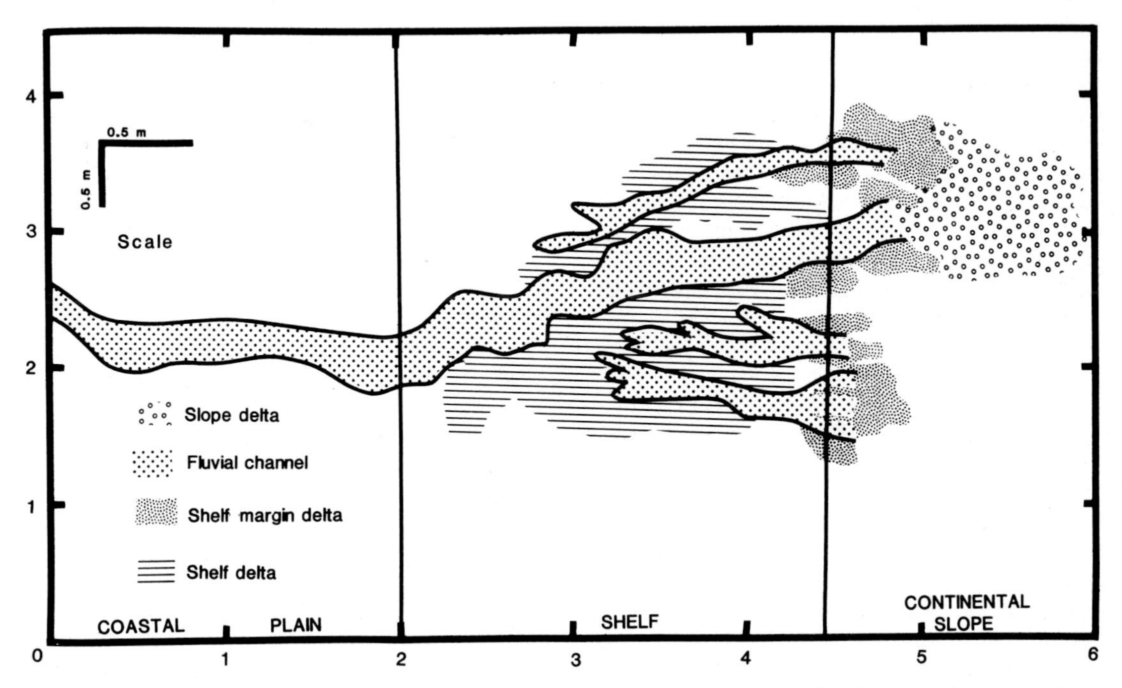

Fig. 8. Lithofacies map of the maximum lowstand surface for the 4-h run. Map shows four incised valleys, multiple shelf margin deltas and a wide band of shelf delta sediments formed during a rapid base-level fall and rise.

Table 3. Mean grain-size differences between shelf deltas, shelf margin deltas, lowstand and transgressive deposits are shown for each run. (See Table 1 for duration of each run.) Average mean grain size of depositional facies is shown at the bottom of the table. Grain sizes are in ϕ (mm)

Run	Shelf delta grain size ϕ (mm)	Slope delta grain size ϕ (mm)	Shelf margin delta grain size ϕ (mm)	Transgressive delta grain size ϕ (mm)
1	–	2.62 (0.1627)	2.96 (0.1285)	2.77 (0.1466)
2	–	3.11 (0.1158)	2.74 (0.1497)	2.38 (0.1921)
3	3.14 (0.1134)	3.04 (0.1216)	3.04 (0.1216)	2.89 (0.1349)
Av. mean	3.14 (0.1134)	2.92 (0.1321)	2.91 (0.1331)	2.68 (0.1560)

incised valleys eroding headward across the shelf (Fig. 7). The tie between lowstand incised valleys and the hinterland drainage network will not occur until well into lowstand time. This lag time between maximum lowstand and the transport of large volumes of coarse sediment by a valley also has important implications for the sedimentological character of basin-floor fans and submarine mass-wasting features. Deep-water deposits sourced from the slope may be finer grained in early lowstand time, but they become coarser grained during late lowstand time.

Base-level rise does not immediately affect the upper reaches of the incised valley (Fig. 6; 0–3 m). Knickpoints which originate in the incised valley during base-level fall and standstill continue to migrate up the incised valley during base-level rise. These knickpoints remobilize the transport of previously deposited, coarse-grained sediments in the upper reaches of the incised valley (Fig. 6; 0–3 m), and deposition of this sediment at the mouth of the incised valley continues well into the transgression. In the experiments, this factor led to the deposition of backstepping transgressive shelf deltas that contained significant amounts of coarse-grained sediment (Fig. 5; Table 3).

INCISED VALLEY MORPHOLOGY AND SEDIMENTOLOGY

Multiple valleys developed at the shelf edge with a rapid fall in base level (Figs 7 & 8). As they developed by headward erosion of the shelf, they captured different amounts of discharge. One or several of these valleys were active or inactive at any one time during base-level fall. Eventually, one valley captured all of the discharge, and the other valleys became inactive. Capture occurred before they received significant amounts of coarse-grained sediment from the hinterland. Therefore, inactive valleys may contain small amounts of reworked coarse-grained sediments from a previous highstand systems tract deposit, but they were dominantly filled with fine-grained muds and fine to coarse grained mass failure sediments, the latter of which were deposited as the valleys were flooded by rising sea level. Timing of the incised valley fill has important implications for the nature of sediment associated with incised valley features. An incised valley may not be linked directly upstream to a hinterland sediment source and, therefore, deltas at the valley mouth may be significantly finer grained than those at an adjacent locality where the valley was linked directly to a hinterland sediment source. Submarine fan deposits derived from these inactive valleys will not contain appreciable amounts of coarse-grained sediment derived from the hinterland. Any coarse-grained sediment contained within such submarine deposits will be derived from a local, reworked coarse-grained sediment source.

With rising base level, both inactive and active incised valleys were flooded. Flooding resulted in a rising water table, increased pore pressure in the valley walls and saturation of valley-wall sediments already weakened from undermining by shoreline and fluvial processes. The effect of these changes on the valleys was to induce mass failure of the valley walls. The result was a U-shaped valley cross-section with slump deposits partially filling the valley and overlying any previously deposited fluvial sand (Fig. 5).

TRANSGRESSIVE DEPOSITION

The rate at which the shelf and incised valleys are flooded has a pronounced effect on the thickness and preservation of transgressive deposits and previously deposited regressive units. A slow transgression of the shelf means that sediments have a longer period to accumulate at specific sites of deposition, resulting in thicker transgressive deposits at any one depositional site. However, a slow transgression also exposes lowstand and transgressive deposits to a prolonged reworking by fluvial and shoreline processes, resulting in a decreased preservation potential for these deposits.

On the other hand, a rapid rise of base level causes rapid transgression of the shelf, and hence less time to accumulate sediment at specific depositional sites, resulting in thinner transgressive deposits. A rapid transgression also means that lowstand and transgressive deposits are exposed to reworking by fluvial and shoreline processes for a shorter period. The result is an increased preservation potential for these deposits.

THE FLUVIAL SYSTEM AND COMPLEX RESPONSE

Fluvial systems are dynamic and complex. Episodes of deposition and erosion within these systems may have little to do with the extrinsic impulse of base-level change, the effects of which may be overwhelmed by the intrinsic variables of discharge and sediment supply.

The initial response of a fluvial channel to a lowering of base level is one of vertical incision. This incision starts at the mouth of the channel and progresses upstream. After vertical stability is achieved, a period of lateral erosion occurs. With lateral erosion, the channel transports successively more sediment, and lower portions of the channel begin to aggrade (Schumm *et al.*, 1984). Aggradation within the channel oversteepens the longitudinal profile of these reaches to a point where secondary incision occurs (Schumm, 1974). Upper portions of the channel also may be storing sediment while incision is taking place downstream, but as incision progresses upstream the previously stored coarse-grained sediment will be transported downvalley to sites of deposition on shelf margin or slope deltas. This may not happen until late in base-level lowstand, after effects of rejuvenation of the drainage system have ceased.

COMPARISON OF PUBLISHED INTERPRETATIONS AND FLUME RESULTS

Similarities, as well as differences, were noted between the experiments, published theoretical sequence-stratigraphic models and subsurface data from the Gulf of Mexico.

Other workers have noted the relationship between rapid rates of transgression and a higher preservation potential for shelf and coastal-plain deposits (Belknap & Kraft, 1981; Suter & Berryhill, 1985; Siringan *et al.*, 1990). Experimental results suggest a higher preservation potential for deposits that are being rapidly transgressed. Suter and Berryhill (1985) and Suter *et al.* (1987) relate thin, widespread shelf-deltaic deposits to rapid falls in base level, and the experimental results corroborate this observation.

Sequence-stratigraphic models suggest that incised valleys become U-shaped due to mass wasting and slumping of the valley walls, as these valleys are flooded by rising base level (Posamentier *et al.*, 1991). Similar processes occurred in the flume experiments. Rising water tables increased pore-water pressure in the valley walls, and flooding of the sediments saturated areas already weakened by fluvial undercutting. These factors contributed to mass failure of the valley walls, which decreased valley-wall slopes, widened the valley and produced chaotic sedimentation in the valley itself. Modified experimental valleys were characterized by a U-shaped cross-sectional geometry (Fig. 5).

Differences between sequence-stratigraphic models and the experimental results exist most obviously in the timing of transfer of coarse-grained hinterland sediment to the downvalley reaches of incised valleys. Posamentier and Vail (1988) suggest that the initiation of base-level fall marks the onset of a wave of coarse-grained fluvial sediment that progrades down the channel and is deposited at the channel mouth as coarse-grained, early lowstand deltaic deposits. Conversely, experimental results show that there is a significant lag between downstream impulses and upstream responses in a fluvial channel. Base-level lowering does not immediately result in a wave of upstream sediment prograding down the system. During the initial stages of base-level fall, any coarse-grained sediment contained in early lowstand deltaic deposits is derived from the reworking of highstand deposits.

Posamentier and Vail (1988) stated that the onset of base-level rise initiates a cessation of coarse-grained fluvial deposition at the mouth of the fluvial channel. Experimental results suggest that this is not true. Late lowstand time and early transgressive time are periods when the highest volume of coarse-grained sediment is delivered to the downstream portions of the fluvial channel.

Based on the same concept, sequence-stratigraphic models suggest that early and middle lowstand fan deposits have the greatest potential to contain significant amounts of coarse-grained sediment (Posamentier & Vail, 1988). However, the lag times documented by these experiments suggest that the greatest volume of coarse-grained sediment is delivered to the depositional basin during late lowstand and transgressive time. This delay in the transfer of large amounts of drainage basin sediments to the lower reaches of the incised valley ensures a continual supply of coarse sediment to the lower reaches of the incised valley into transgressive time. Therefore, fan deposits of the late lowstand may have a higher potential for containing coarse-grained sediment than those of early and middle lowstand time.

Suter and Berryhill (1985) stated that 'shelf-margin deltas ... mark ancient fluvial depocenters that probably contain increased sand content in proximity to fine-grained sediments of relatively high organic content'. They cite Berg (1982) who stated that shelf-margin deltas 'can also be used as markers in the search for turbidite sands carried downslope'. However, during the experiment, headward erosion cannibalized the shelf and limited the initial sediment supply to shelf-margin deltas to sediments making up the shelf and slope. For this reason, shelf-margin deltas may contain minor amounts of coarse-grained sediments reworked from previous highstand deposits, but may be composed dominantly of fine-grained sediments. Experimental results suggest that caution must be used in implying that shelf-margin deltas and associated downslope deposits contain significant amounts of coarse sediment.

CONCLUSIONS

The fluvial system is complex and it responds to internal and external impulses in a manner that restores dynamic equilibrium. The manner in which the fluvial channel responds has a significant effect on the coastal-plain, shelf and slope deposits associated with a rise and fall in base level. In addition, it is not enough to consider only the effects of a rise or fall in base level because the effect of changing rates of rise and fall must also be taken into account.

Sequence-stratigraphic models are based on an oversimplified understanding of how a fluvial system responds to base-level change. Complex response and lag of system response must be taken into account when attempting to predict the character of sediments eroded, transported and deposited during base-level fluctuations. A rapid fall in base level produces multiple fluvial incisions on the shelf and slope, thin, widespread shelf deltaic deposits and multiple, time-synchronous, dominantly fine-grained shelf-margin deltas. A slow fall in base level produces one large, incised valley on the shelf and slope, thicker, less widespread shelf deltaic deposits and multiple slightly coarser-grained, thicker, time-transgressive shelf-margin deltas.

A rapid rise in base level produces thinner transgressive deposits, but these deposits have a lower potential to be reworked. A slow rise in base level produces thicker transgressive deposits, but these deposits have a high potential to be reworked.

In all situations, a single incised valley will eventually capture all of the flow, and it will be the only valley to receive coarse-grained, hinterland sediment during late lowstand and transgression. The remaining incised valleys may become inactive, and thereafter receive only fine-grained sediment and possibly minor amounts of reworked, local, coarse-grained sediment. Thus, only those lowstand fans linked to the active cross-shelf valley during late lowstand time will contain significant amounts of coarse-grained sediment.

Base-level rise results in flooding of incised valleys, a rise in the water table and a weakening of unstable valley walls. These conditions lead to valley-wall slumping and development of a U-shaped cross-section in incised valleys. Marine muds will overlie these deposits, completing the stratigraphic sequence.

Experimental studies can add a new dimension to the problem of predicting the reaction of fluvial systems to base-level change. Added insight into this problem should help to refine sequence-stratigraphic models, stratigraphic correlation and reservoir-quality predictions.

ACKNOWLEDGEMENTS

Research on base-level experiments was funded by the National Science Foundation (CES 8802213). Special thanks are extended to Jeff Ware, John Koss and Dru Germanoski for their expert advice and co-operation and to John Kilcullen, Gregg

Macke, Bill Schmoker and Sharon Schmidt for their able assistance. This paper has benefited greatly from 'incisive' reviews of an earlier manuscript by Henry Posamentier and Bill Arnott.

REFERENCES

Albertson, M.L., Barton, J.R. & Simons, D.B. (1960) *Fluid Mechanics for Engineers*. Prentice-Hall, Hempstead, New York. 561 pp.

Belknap, D.F. & Kraft, J.C. (1981) Preservation potential of transgressive coastal lithosomes on the U.S. Atlantic shelf. In: *Sedimentary Dynamics of Continental Shelves* (Ed. Nitrouer, C.A.), Elsevier, Amsterdam, pp. 429–442.

Berg, O.R. (1982) Seismic detection and evaluation of delta and turbidite sequences: their application to exploration for the subtle trap. *Bull. Am. Assoc. Petrol. Geol.* **66**, 1271–1288.

Brown, L.F. & Fisher, W.L. (1980) *Seismic Stratigraphic Interpretation and Petroleum Exploration*. Am. Assoc. Petrol. Geol. Continuing Education Course Note Series 16, 181 pp.

Gary, M., McAffce, R. Jr. & Wolf, C.L. (1972) *Glossary of Geology*. American Geological Institute, Washington, DC, 805 pp.

Haq, B.U., Hardenbol, J. & Vail, P.R. (1987) The chronology of fluctuating sea level since the Triassic. *Science* **235**, 1156–1167.

Hubbert, M.K. (1937) Theory of scale models as applied to the study of geologic structures. *Bull. Geol. Soc. Am.* **48**, 1459–1520.

Koss, J., Ethridge, F.G. & Schumm S.A. (1990) The effects of base-level change on coastal plain and shelf systems — an experimental approach. *Bull. Am. Assoc. Petrol. Geol.* (Abstr.) **74**, 697.

Mitchum, R.M. (1977) Seismic stratigraphy and global changes of sea level. Part I: glossary of terms used in seismic stratigraphy. In: *Seismic Stratigraphy — Applications to Hydrocarbon Exploration*. Mem. Am. Assoc. Petrol. Geol., 26, 49–212.

Posamentier, H.W. & Vail, P.R. (1988) Eustatic controls on clastic deposition II — sequence and systems tract models. In: *Sea-level Changes: An Integrated Approach* (Eds Wilgus, C.K., Hastings, B.S., Kendall, C.G.St.C., Posamentier, H.W., Ross, C.A. & Van Wagoner, J.C.), Spec. Publ. Soc. Econ. Paleontol. Mineral. Tulsa, 42, 109–124.

Posamentier, H.W., Erskine, R.D. & Mitchum, R.M., Jr. (1991) Models for submarine fan deposition within a sequence stratigraphic framework. In: *Seismic Facies and Sedimentary Processes of Submarine Fans and Turbidite Systems*. Springer-Verlag, New York.

Schumm, S.A. (1974) Geomorphic thresholds and complex response of drainage systems. In: *Fluvial Geomorphology* (Ed. Morisawa M.), SUNY-Binghamton, Binghamton, New York, pp. 299–310.

Schumm, S.A. (1977) *The Fluvial System*. Wiley, New York, 338 pp.

Schumm, S.A. (1991) *To Interpret the Earth: Ten Ways to be Wrong*. Cambridge University Press, Cambridge, 133 pp.

Schumm, S.A. & Kahn, H.R. (1972) Experimental study of channel patterns. *Bull. Geol. Soc. Am.* **83**, 1755–1770.

Schumm, S.A., Harvey, M.D. & Watson, C.C. (1984) *Incised Channels. Morphology, Dynamics and Control*. Water Resources Publications, Littleton, Colorado, 200 pp.

Schumm, S.A., Mosley, P.M. & Weaver, W.E. (1987) *Experimental Fluvial Geomorphology*. Wiley, New York, 413 pp.

Shepard, R.G. & Schumm, S.A. (1974) Experimental study of river incision. *Bull. Geol. Soc. Am.* **85**, 257–268.

Siringan, F.P., Abdulah, K.C., Anderson, J.B., Bartek, L.R. & Thomas, M.A. (1990) Variability and preservation of coastal lithosomes during a transgression. A case study of north Texas Gulf Coast and Shelf. (Abstr.) *Ann. Meet. Geol. Soc. Am. Dallas, Texas*, A91.

Suter, J.R. & Berryhill, H.L. (1985) Late Quaternary shelf-margin deltas, northwest Gulf of Mexico. *Bull. Am. Assoc. Petrol. Geol.* **69**, 77–91.

Suter, J.R., Berryhill, H.L. & Penland, S. (1987) Late Quaternary sea-level fluctuations and depositional sequences southwest Louisiana continental shelf. In: *Sea-level Changes: An Integrated Approach* (Eds Wilgus, C.K., Hastings, B.S., Kendall, C.G.St.C., Posamentier, H.W., Ross, C.A. & Van Wagoner, J.C.) Spec. Publ. Soc. Econ. Paleontol. Mineral. Tulsa, 42, 199–219.

Thompson, G.R. & Turk, J. (1991) *Modern Physical Geology*. Saunders, Philadelphia, pp. 307–309.

Vail, P.R., Mitchum, R.M., Jr, Todd, R.G., *et al.* (1977) Seismic stratigraphy and global changes of sea level. In: *Seismic Stratigraphy — Applications to Hydrocarbon Exploration* (Ed. Payton, C.E.) Mem. Am. Assoc. Petrol. Geol. 26, 49–212.

Vanney, J.-R. & Stanley, D.J. (1983) Shelfbreak physiography: an overview. In: *The Shelfbreak: Critical Interface on Continental Margins* (Eds Stanley, D.J. & Moore G.T.) Spec. Publ. Soc. Econ. Paleontol. Mineral. 33, 1–24.

Van Wagoner, J.C., Mitchum, R.M., Jr., Posamentier, H.W. & Vail, P.R. (1987) Key definitions of sequence stratigraphy. In: *Atlas of Seismic Stratigraphy* (Ed. Bally, A.W.) Am. Assoc. Petrol. Geol. Stud. Geol. 27, 11–14.

Spec. Publs Int. Ass. Sediment. (1993) **18**, 55–68

High-resolution sequence architecture: a chronostratigraphic model based on equilibrium profile studies

D. NUMMEDAL, G.W. RILEY*, *and* P.L. TEMPLET†

Department of Geology and Geophysics, Louisiana State University, Baton Rouge, LA 70803, USA

ABSTRACT

The sequence-stratigraphic signature of a shoreline excursion is controlled by sediment accumulation rates relative to the distribution and rates of generation of accommodation space. Accommodation space, in turn, is controlled by slopes and rates of migration of the local equilibrium surfaces ('base levels'). Based on forward modelling of the migrations of these equilibrium profiles, we demonstrate that the sediment package produced by a single cycle of relative sea-level change may contain as many as five separate diastems.

Regressions associated with sea-level fall produce two separate surfaces: exposure and bypass generate a subaerial erosion surface (first diastem) and decreasing water depth on the inner shelf creates a submarine erosion surface (second diastem). Regressions associated with sea-level stillstand or slow rise, in contrast, deposit conformable successions. During transgressions the flooding of the coastal plain produces a bypass surface (initial transgressive surface and third diastem), the erosion of the shoreface cuts a ravinement surface (fourth diastem) and outer-shelf environments which are depleted in clastic input form a starvation surface on top of the transgressive shelf strata (fifth diastem).

This model predicts the stratigraphy of a deposit resulting from a single regression and transgression of a shoreline. If no relative sea-level fall is associated with such a cycle, the resulting deposit is a *parasequence*, bounded by transgressive surfaces. If the cycle does include a relative fall, the resulting deposit becomes two *half-sequences*, separated by a regressive surface of erosion, which is the sequence boundary. The models highlight the observation that *parasequences* and *sequences* are simply two different ways of subdividing the stratigraphic record; parasequences are bounded by surfaces of deepening (transgressive surfaces) and sequences are bounded by surfaces of shallowing (sequence boundaries). Both may, and often do, represent the same thickness of rocks and increments of time.

INTRODUCTION

Contemporary concepts of sequence stratigraphy have primarily evolved from interpretation of seismic-reflection data (Vail *et al.*, 1984) and secondarily from high-resolution studies of well logs, cores and outcrops (Wilgus *et al.*, 1988). During the past few years, emphasis has been placed increasingly on 'forward modelling' of sequence architecture based on physical, yet qualitative, models of shifts in depositional systems in response to sea-level changes (Posamentier *et al.*, 1988) and quantitative, geometric modelling of the equilibrium profiles of different systems (Cross, 1990; Helland-Hansen *et al.*, 1988, Jervey, 1988; Lawrence *et al.*, 1990; Ross, 1990; Reynolds *et al.*, 1991).

Forward modelling predicts the temporal change of a system from its initial to some future state. The accuracy of model predictions depends on how well we understand the processes which drive the change, and how well the model algorithms are

*Present address: Amoco Production Corporation, PO Box 3092, Houston, TX 77253, USA.
†Present address: Exxon Company USA, PO Box 4279, Houston, TX 77210, USA.

formulated. Modelling, therefore, drives sequence stratigraphy well beyond its initial emphasis on recognition of seismic-reflection patterns and into the realm of depositional systems analysis.

Our objective in this paper is to apply a simple forward model to examine the distribution of the erosional and bypass surfaces that form within a coastal and shallow marine (deltaic) sedimentary package in response to one single cycle of sea-level change. In addition, we will address the implications of these surfaces regarding the internal architecture of sequences and parasequences. It will become clear that sequences and parasequences are simply two different ways of subdividing the stratigraphic record. In delineating sequences, the investigator chooses boundaries associated with demonstrable, sudden shallowing (sequence boundaries). In delineating parasequences, he chooses boundaries associated with deepening (transgressive surfaces).

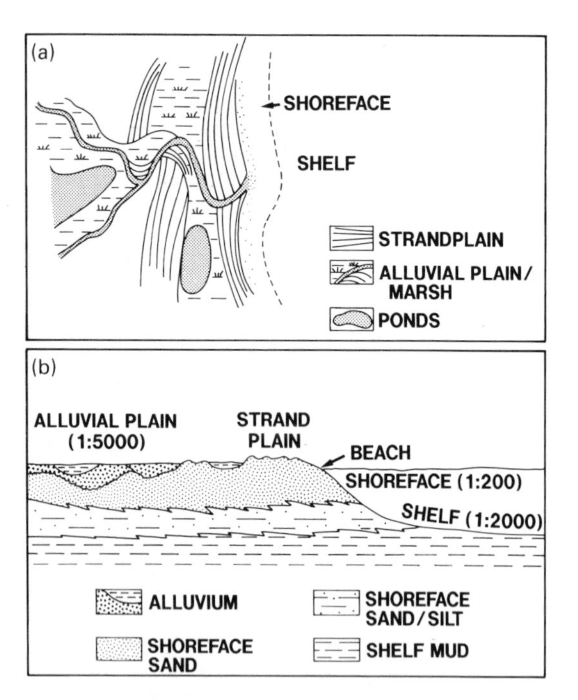

Fig. 1. Generic model of wave-dominated strandplain and associated fluvial, deltaic and shelf systems. The chosen slopes are typical values for the US Gulf coast. (a) Map view. (b) Schematic cross-section.

MODEL SYSTEMS

Geometry

The model addresses a wave-dominated strandplain and its associated fluvial, deltaic and shelf systems (Fig. 1). We assume that sediment deposition and erosion in these environments are controlled by migration of equilibrium profiles that are adjusting to shifts in shoreline position. Along modern coasts, the site-specific profiles vary a great deal in response to sediment yields, textures and wave climate. The values used in this paper characterize 'typical' conditions along the coast of the Gulf of Mexico.

Fluvial

Slopes for several Gulf coast streams were calculated from topographic maps and from data in Thomas (1990). Distal fluvial slopes depend on river size; a large stream like the Mississippi has an average slope of 0.00002 (1:50,000) whereas the Brazos and the Trinity Rivers average about 0.0002 (1:5000). We have chosen to use a value of 1:5000 for the fluvial systems in our model.

Shoreface

Shoreface profiles along the Gulf coast are well documented in a series of different settings. The gentler shoreface profiles occur along the relatively low-energy Louisiana coast; the steeper ones are found in south Texas. As a typical model value, we chose a slope of 1:200, which is characteristic of the south-central Texas coast (Snedden, 1985). Along most of this coast, the break-in-slope between the inner shelf and shoreface lies at a water depth of 10 to 15 m.

Continental shelf/ravinement

The slope of the northwest Gulf shelf ranges from 1:3500 off the Sabine River (Thomas, 1990) to 1:700 off south Texas (Snedden, 1985). Our model operates with a slope of 1:2000. The regional slope of the ravinement surface equals that of the presently transgressed shelves. In general, such shelves are steeper than those associated with present regressions (e.g. central Louisiana versus Texas shelf) and we assign a value of 1:1000 for the ravinement in our models.

Transgressive coastal plain

Environments such as bays, lagoons, tidal flats, marshes and estuaries characterize transgressive coasts, although similar environments also occur where strong longshore drift maintains regressive 'barrier islands' far away from the deltaic headlands. The point to stress in the model is that regressions require direct supply of fluvial sediments to the shoreline and, therefore, the river mouth. Such delivery, in turn, requires a continuous fluvial slope (energy gradient) to the sea. In contrast, transgressive shorelines erode because there is an insufficient supply of sediment to the shoreline to keep up with the rate of generation of accommodation space. For modelling purposes, therefore, the transgressive coastal plain is considered to be horizontal; there is no hydraulic energy gradient moving fluvial sediments to the sea.

Sensitivity

The stratigraphic models are not very sensitive to the exact values of the chosen slopes; they are valid in all cases where the relative slopes, from the most gentle to the steepest, are: (i) transgressive coastal plain; (ii) distal fluvial; (iii) shelf; (iv) ravinement; and (v) shoreface.

PROFILE ADJUSTMENTS TO BASE-LEVEL CHANGE

The shoreface profile

Concept

Bruun (1962) formulated the concept that marine shorefaces maintain equilibrium profiles (Fig. 2), the slope and depth of which depend on the local wave and current regime. These profiles vary alongshore but are assumed to remain invariant along depositional dip during regressions and transgressions associated with a single cycle of sea-level change. Field and laboratory testing of the 'Bruun rule' have confirmed its basic validity (Schwartz, 1967; Dean *et al.*, 1987), although contemporary coastal engineering studies have expanded its mathematical formulation to incorporate situations where there are longshore gradients in sediment transport. In this study, we apply the Bruun rule in its original two-dimensional form.

Sea-level rise

Because of the power of waves and currents along most shorelines, sediment dispersal is always rapid enough to maintain the equilibrium profile. Features such as 'overstepped barriers' in response to very 'rapid transgressions' have not been documented (Swift & Moslow, 1982). Bedding planes, seismic reflectors and time lines of shoreface deposits follow the equilibrium profile. This profile also controls the changes in accommodation space through vertical and lateral shifts in response to sea-level fluctuations.

When a shoreline erodes during rising sea level, the shoreface profile moves upward and landward (Fig. 3(a)). The pre-existing shoreface is truncated and new accommodation space is created beneath the rising seaward limb of the equilibrium profile. Sediment removed from the eroding shoreface will accumulate as an onlapping wedge of transgressive shelf strata within this growing accommodation space (Fig. 3(a)).

Erosion of the shoreface leaves a sharp surface at the base of this onlapping wedge; this is the *ravinement surface* (Swift, 1968; Nummedal & Swift, 1987). Inner-shelf sediments above the ravinement surface equilibrate with the local wave and current regime. Fine-grained sediments may accumulate in the form of landward-tapering wedges, probably onlapping the ravinement; whereas sand and gravel generally form shelf sand ridges (Stride, 1982; Swift *et al.*, 1984, 1991, a, b). Because of textural segregation of the material eroded from the shoreface and long-distance transport of the fines, the volume of the transgressive wedge would be expected to be less than the volume of the truncated shoreface.

Sea-level fall

The shoreface equilibrium concept is much less tested for conditions of relative sea-level fall. The few studies that do exist, however, suggest that it is valid (Dominguez & Wanless, 1991). During falling sea level, the zones of deposition and erosion would be reversed. In response to profile lowering, increased bottom shear stress during storms would cause shelf erosion (Fig. 3(b)). The resulting erosional surface (diastem) is referred to as the *regressive surface of marine erosion*. New accommodation space is created behind the seaward-moving upper limb of the equilibrium profile. Sediment transported alongshore from updrift deltas or other sources rapidly fills this available space and the shoreface progrades. Continued sea-level fall will

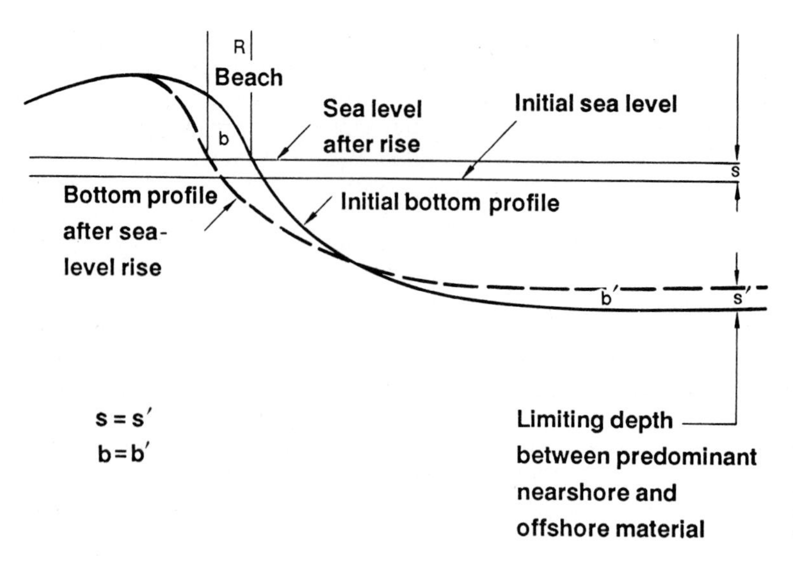

Fig. 2. Theoretical shoreface profile movement in response to a relative sea-level rise of magnitude S. As the beach erodes (distance R), the material removed from the shoreface (b) is moved offshore to fill part of the newly generated accommodation space on the inner shelf (b′). This is the 'Bruun rule'. (From Dean *et al.*, 1987.)

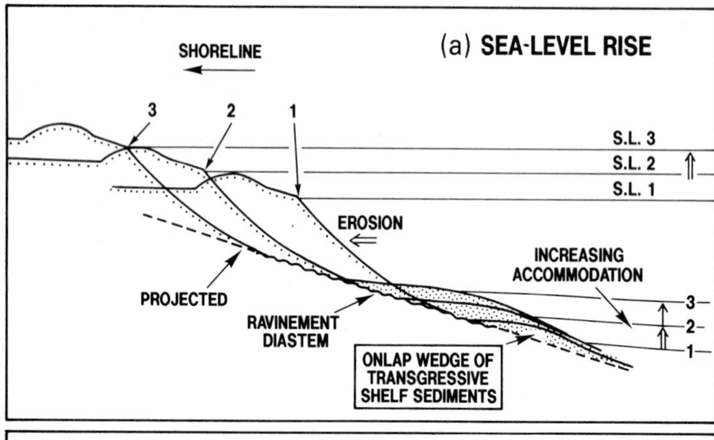

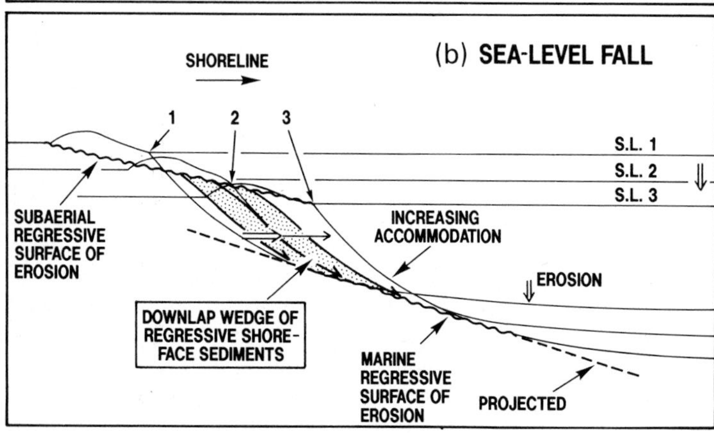

Fig. 3. Schematic inner shelf-to-shoreface stratigraphy formed in response to sea-level rise (a) and fall (b) assuming profile adjustments in accordance with the Bruun rule. (a) Rising sea level yields a wedge of transgressive shelf strata onlapping a sharp surface formed by shoreface erosion (the ravinement diastem). (b) Falling sea level produces a wedge of shoreface strata downlapping onto a regressive surface of marine erosion formed by truncation of shelf deposits.

induce subaerial erosion at the top of the older shoreface prism and remobilize some of the sand to help fill the moving accommodation space. Sandy shoreface progradation is a continuous process during sea-level fall. Bedding planes in the shoreface prism downlap onto the regressive surface of marine erosion (Fig. 3(b)).

Shoreface regression in response to relative sea-level fall has been defined by Posamentier *et al.* (1992) as a *forced regression.* Shorefaces formed by forced regressions can be differentiated from those related simply to high sediment supply by the presence of an erosional surface at their base (Plint, 1988). The sharp base may only be expressed at the proximal locations, near the updip stratigraphic pinch-out of these shoreface sandbodies, because this is where the relative change in water depth is the greatest.

The great volumes of beach and shoreface sand left behind by retreating shorelines along the late Pleistocene pluvial lakes in the Great Basin (e.g. Lake Bonneville; Smith *et al.*, 1989), demonstrate that shoreface sand accumulation during periods of water-level fall is quite effective.

The regressive fluvial/deltaic profile

Concept

Sequence-stratigraphic studies of fluvial accommodation space deal with the concept of the graded stream. To paraphrase Mackin (1948, p. 471), 'the graded stream is delicately adjusted to carry the available sediment by the discharge provided'. A river at grade is one that will adjust back to its original form in response to a 'perturbance', such as added sediment load, river-mouth progradation and dam building. Graded streams, however, can and do aggrade or erode at geological time scales, as expressed by Schumm and Lichty (1965) through the concept of 'graded time'.

Much debate has recently centred on the patterns of change in the graded-stream profile in response to base-level shifts (Miall, 1991; Paola *et al.*, 1991). The issue is complex because rivers aggrade and degrade in response to changes in discharge and sediment load as well as base level (Schumm, 1969). We have found nothing, however, in recent literature that invalidates the findings of an early and elegant study of rivers at grade published by Green (1936). In his study of small coastal-plain rivers in southern England, Green documented that the river profiles changed over time by 'stretching'; the river delta prograded concurrent with headwater retreat (Fig. 4). Green's (1936) profiles are logarithmic, they approach base level at an angle (they are not asymptotic) and become infinitely steep at their upstream termination. Our solutions of Green's (1936) equations (not included in this paper) demonstrate that aggradation characterizes the lower alluvial valley in response to river-mouth progradation (in agreement with suggestions in Posamentier & Vail, 1988), and that the upper valley reaches degrade (Fig. 4, profile 3). The calculations demonstrate further that during continued delta growth fluvial aggradation extends progressively further up the valley. As a consequence, fluvial strata onlap an erosion surface formed internally in the upper valley reaches (Fig. 5(a)).

The profile shifts illustrated in Fig. 5(a) are supported by numerical models that indicate that: 'the distance upstream over which the effects of variation in sea level can be felt is proportional to the

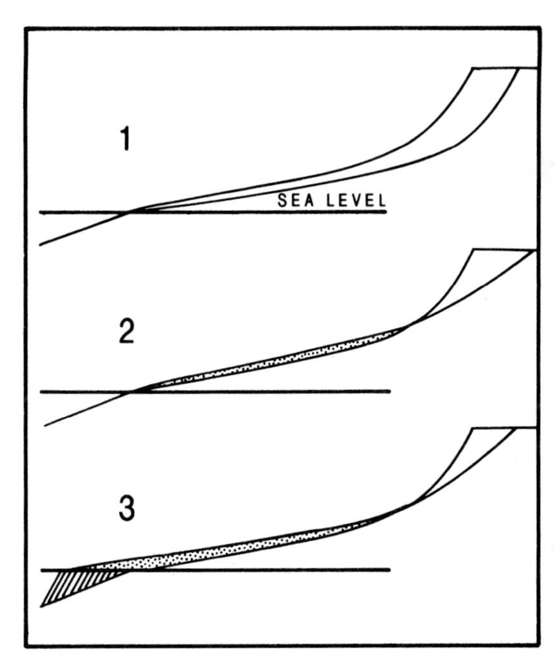

Fig. 4. Idealized river profiles, based on equations for coastal-plain streams in southern England. Cases 1 and 2 illustrate different profiles graded to the same position of the mouth. Case 3 illustrates that the river responds to progradation of the mouth by aggrading the lower alluvial valley and degrading the headwater region. (From Green, 1936.)

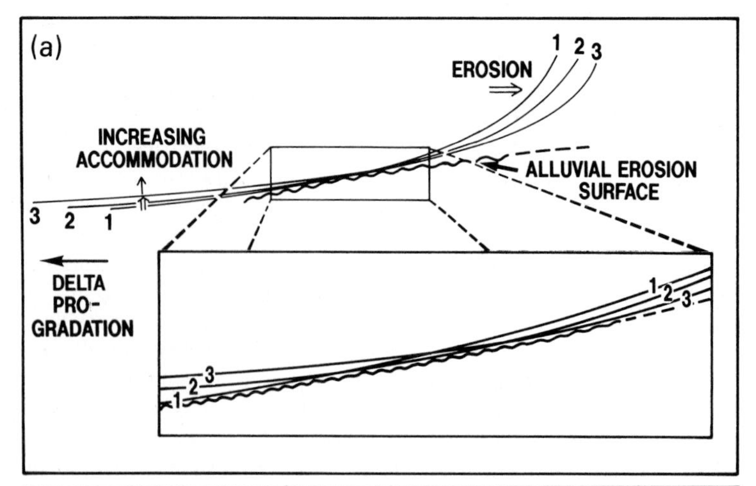

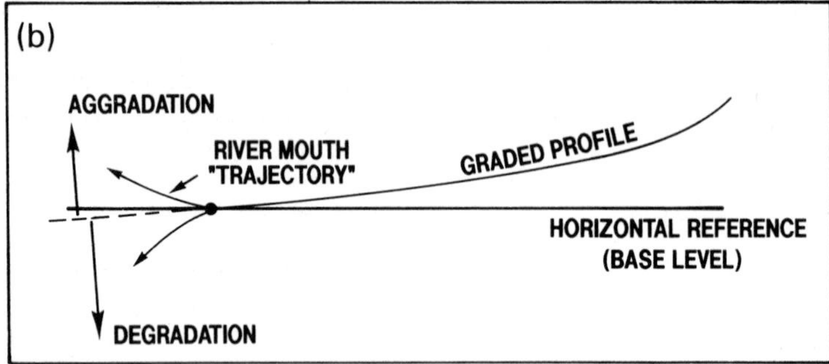

Fig. 5. (a) Generation of an internal alluvial erosion surface in response to temporal evolution of the river basin. Solution of Green's (1936) equations demonstrates that aggradational river profiles will onlap an internal erosion surface in the alluvial valley. 1, 2 and 3 refer to successively younger, and increasingly stretched, profiles. The generation of onlap by profile shifting is illustrated in the enlarged window. (b) Fluvial aggradation occurs whenever the river mouth follows a 'trajectory' above the continued projection of the graded river profile. Degradation occurs if the river mouth tracks below this line. Note that graded streams are not asymptotic to sea level; they approach their mouths at slopes slightly steeper than horizontal.

square root of the period of the variation' (Paola *et al.*, 1991). As typical values, Paola *et al.* (1991) suggest that the effects of 100,000-year sea-level cycles may be felt 100 km up the valley.

Several field studies, starting in the early part of this century, documented the fluvial onlap wedge that typically forms behind dams (Gilbert, 1917; Mackin, 1948; Leopold & Bull, 1979), but the debate continues to the present regarding the 'ultimate' equilibrium profile once fully graded conditions are re-established. We believe that the data and analyses presented above provide adequate information to predict the behaviour of graded streams at 'geological' time scales.

The behaviour of streams in their upper valley reaches is of little interest in sedimentary basin analysis because of the low preservation potential. We need to identify which patterns of base-level change are associated with aggradation or degradation in the lower alluvial valley and delta plain. The discussion will consider only one variable — base level; we allow no changes in sediment load or discharge in the fluvial system.

Profile shifts

Aggradation is assumed to occur when accommodation space increases. Increase in fluvial accommodation space occurs any time the river mouth follows a 'trajectory' along or above the seaward continuation of the graded profile (Fig. 5(b)). Because the river mouth follows the top of the regressive delta package, this trajectory has only an indirect relationship to the slope of the continental shelf (contrast: Miall, 1991). The distal alluvial valley will aggrade during delta progradation for conditions of relative sea-level rise, stillstand, or fall, as long as the rate of fall is sufficiently slow to keep the delta top at or above the projected graded stream profile. The river-mouth trajectory will fall below the graded-stream profile when the ratio between sea-level fall and sediment supply is such that the delta top can no longer keep up. During such conditions the lower alluvial valley becomes incised. Rapid relative sea-level falls, therefore, are more conducive to valley incisions than are slow ones.

The transgressive coastal-plain profile

Regarding fluvial response to transgressions, we also concur with Green (1936, p. LXIV) that 'there is no reason to suppose that such a (sea level) rise, changing the lower part of a river valley into an estuary, can have any direct effect on the processes of erosion and aggradation in the rest of the course until aggradation begins at the new mouth; and then only in the lowest graded segment'. Fluvial aggradation certainly may occur within the transgressive *systems tract* (Shanley & McCabe, 1991), but it occurs during the progradational phases of individual backstepping parasequences. If the concept of grade has any validity, there can be no fluvial aggradation during periods of river-mouth flooding.

In our model, therefore, which deals with a single cycle of shoreline regression and transgression, there is no fluvial aggradation during transgression. Of course, the fluvial system transmits sediment from the headwaters to its mouth regardless of the direction of movement of the mouth. During transgression, however, that sediment is delivered into environments that we classify as 'transgressive coastal plain', including bays, estuaries, marshes, lagoons and tidal-channel systems. The resulting horizontal strata onlap the underlying graded stream profile formed during the preceding regression.

The shelf profile

Continental shelf equilibrium profiles are inadequately understood and alternative large-scale sequence geometries have been predicted from different model assumptions. Our models assume that subsidence rates are spatially invariant across the shelf because shoreline migration distances during individual sea-level cycles are relatively small compared to overall basin dimensions. For the same reason, the 'equilibrium point' of Posamentier *et al.* (1988) does not enter into the models.

The *equilibrium slope* of a transgressive shelf (slope of seaward limb of the profile in Fig. 2) is generally less than the slope of the ravinement (Everts, 1987). Sedimentation rates during a transgression, however, are generally too low to fill all the available shelf accommodation space such that the average profile of a transgressed shelf will be very near the slope of the ravinement. This concept is consistent with the observation that Holocene transgressive sediments on most of the world's modern shelves form a thin sheet, rather than significant seaward-thickening wedges.

MODEL STRATIGRAPHY

Case 1. Sea-level stillstand and rise

Lithostratigraphy

The first model is constructed for the simple case where shoreline progradation occurs during relative sea-level stillstand and the subsequent transgression is a response to a constant rate of sea-level rise. The regressive component of this deposit consists of a conformable succession of shelf, shoreface and fluvial strata that formed between initiation of progradation at time T_0 and seaward shoreline turnaround at T_{50} (Fig. 6). During the subsequent transgression (T_{50} to T_{100}), the ravinement surface truncates regressive shoreface strata seaward of point C (Fig. 6), fluvial strata between C and D and only transgressive coastal-plain strata landward of D. In accordance with the discussion presented above, that there is no new accommodation space being generated within the fluvial system during transgression, our model predicts that the horizontal, *transgressive*, coastal-plain strata onlap a fluvial bypass surface. Transgressive shelf sediments accumulate within the shelf accommodation space (far right of Fig. 6). To avoid clutter in this diagram, they are not depicted.

Chronostratigraphy

The format of the chronostratigraphic diagrams and the nomenclature used to express intervals of time not represented by rocks follow Wheeler (1958). We agree with Sloss (1984) that the terms lacuna, hiatus and vacuity, as used by Wheeler, are 'essential linguistic symbols in the discussion and comprehension of unconformities'.

When depicted in a 'Wheeler diagram', the regressive strata (T_0 to T_{50}) form a simple, conformable succession of shelf, shoreface and fluvial strata (Fig. 7). The shoreface is strongly diachronous, becoming younger seaward. The chronostratigraphic signature of the transgression, in contrast, is quite complex. Shoreline transgression starts at T_{50}. This time line and the corresponding stratigraphic surface are defined as the *initial transgressive surface* (ITS). During transgression the eroding

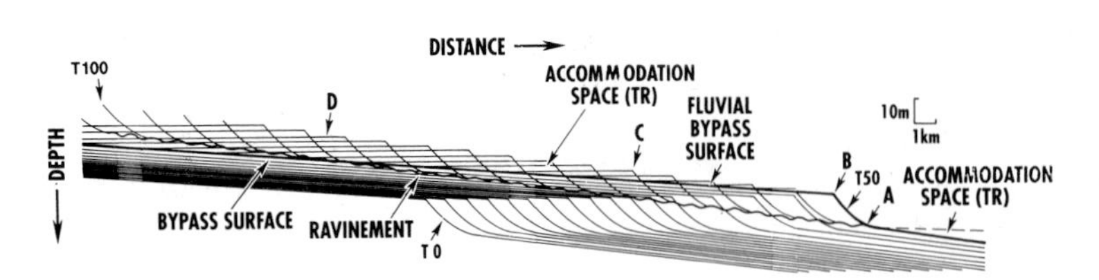

Fig. 6. Lithostratigraphic model of a deltaic deposit formed in response to progradation during sea-level stillstand (time T_0 to T_{50}) followed by transgression during sea-level rise (time T_{50} onwards). The ravinement surface truncates shoreface (between points A and C), fluvial (between points C and D) and transgressive coastal-plain strata (landward of point D) in a landward succession.

shoreface truncates successively older strata (the regressive shoreface) between points A and C (Fig. 6), creating a corresponding erosional vacuity as part of the lacuna on the ravinement surface (Fig. 7). Between points C and D the retreating shoreface truncates only fluvial strata. Because the slope of the equilibrium fluvial profile is less than that of the ravinement, the surface truncates successively younger strata landward. Landward of point D the ravinement no longer cuts deep enough to affect any of the regressive strata (deposits predating T_{50}) and truncation is limited to the upper part of the transgressive coastal plain. The bypass surface on top of the fluvial deposits represents a non-depositional hiatus expanding in duration landward (Fig. 7). Where the ravinement surface cuts across the bypass surface (D, Figs 6 & 7), its chronostratigraphic representation undergoes a vertical shift because the bypass surface has zero thickness but does represent a finite interval of time. From point D landward, the ravinement lacuna remains constant in duration but it becomes progressively younger in age. The ravinement surface, which commonly is a prominent lithostratigraphic marker (Swift, 1968; Nummedal & Swift, 1987), is strongly time transgressive.

Sediments deposited on the inner shelf, seaward of the eroding shoreface, form transgressive shelf strata (Fig. 7). As water depth increases during continued transgression, terrigenous shelf deposition becomes inactive and replaced by 'condensed' draping of hemipelagic mud. The physical surface on top of this section is referred to as a *marine starvation surface.* This surface forms during an interval of time that generally increases offshore. There will be one time line, however, that corresponds to the moment of landward shoreline turnaround. This instant is defined as the *time of final transgression.* The surface formed at that precise time is the *final transgressive surface* (FTS).

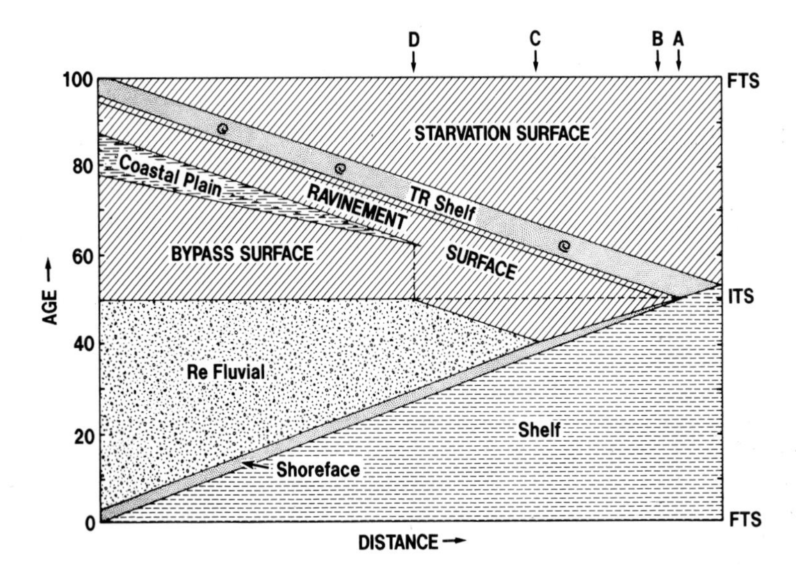

Fig. 7. Wheeler diagram of the deltaic succession illustrated in Fig. 6. The regressive component (T_0 to T_{50}) is conformable but the transgressive one contains three lacunae (cross-hatched). The lacuna on the bypass surface is entirely a non-depositional hiatus, the one on the ravinement is an erosional vacuity and the lacuna on the starvation surface is a complex combination of the two. This succession is a parasequence, as currently defined. The parasequence is bounded by final transgressive surfaces (FTS), and contains both transgressive and regressive components.

Discussion

The sedimentary package just modelled is a parasequence (Van Wagoner *et al.*, 1988, 1990). It is a package of genetically related strata bounded by surfaces of water deepening. The parasequence can be divided into two temporally equal halves: a conformable succession of shelf, shoreface and fluvial strata formed during the first half of the cycle, followed by a disconformable succession of coastal-plain and shelf strata deposited during the transgressive half of the cycle.

Relatively little is known about the duration of regressions and transgressions during formation of 'real world' parasequences. Based on known rates of growth and decay of crevasse-splays in the Mississippi River delta (Wells & Coleman, 1987), it seems reasonable to assume that the two processes operate at comparable rates. In terms of thickness the regressive part of the parasequence is generally dominant.

The term *flooding surface* has not yet been used in this discussion because the word 'flooding' does not describe adequately the physical actions during a transgression. As defined by Van Wagoner *et al.* (1988), the term 'flooding surface' is an adequate descriptor of the parasequence boundary in situations where the three surfaces discussed above all merge (the ITS, ravinement and FTS). Such will be the case in areas of slow rates of transgressive sedimentation or where the investigator is limited to low-resolution data sets.

In the terminology recommended here, a parasequence is bounded by final transgressive surfaces, and the regressive and transgressive parts are separated by the initial transgressive surface. The ravinement surface is a highly diachronous surface connecting the initial and final transgressive surfaces. Because it has no chronostratigraphic significance (i.e. all strata above are not younger than all strata below), the ravinement is not suitable as a surface for temporal subdivision of a depositional sequence.

Arguments can be made for referring to the final transgressive surface as a 'maximum flooding surface' (MFS) because it is associated with landward shoreline turnaround. However, we hold the opinion that the MFS terminology should be restricted to the boundary between the transgressive and highstand *systems tracts* (as in Posamentier & Vail, 1988), and not used at the boundaries of individual parasequences.

Vail *et al.* (1991) recommend that the surface separating the lowstand and transgressive systems tracts be referred to as the 'top of lowstand surface'. We support this change because the term 'transgressive surface' is thus removed from the 'third-order' depositional sequence. It can then be applied without confusion to the surfaces that actually form during periods of shoreline retreat, such as those discussed above.

Case 2. Sea-level cycle including a fall

Lithostratigraphy

The second model relates to the more general situation of shoreline migration in response to a relative sea-level cycle including both rise and fall. The model sea-level cycle consists of five segments (Fig. 8). The lithostratigraphic model for this case (Fig. 9) depicts only those strata that were preserved following regressive or transgressive erosion. The principles of equilibrium profile migration are those addressed in the first half of this paper.

Early highstand (T_0 to T_{20}; Figs 8 & 9) is characterized by progradation and aggradation in response to sea-level rise at a rate of 1 m per thousand years. The shoreline progrades at a moderate rate (0.5 km per thousand years) because much of the supplied clastics are accommodated in the rapidly aggrading fluvial system. During late highstand (stillstand: T_{20} to T_{30}) the shoreline progrades faster (1.0 km per thousand years) because of a decrease in the rate of generation of fluvial accommodation

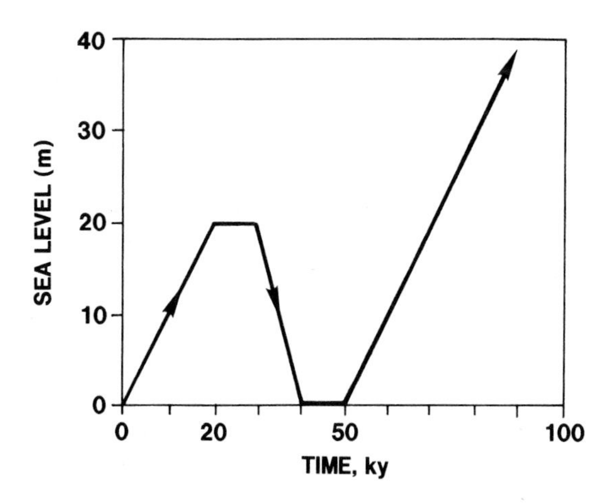

Fig. 8. Sea-level curve for model case 2.

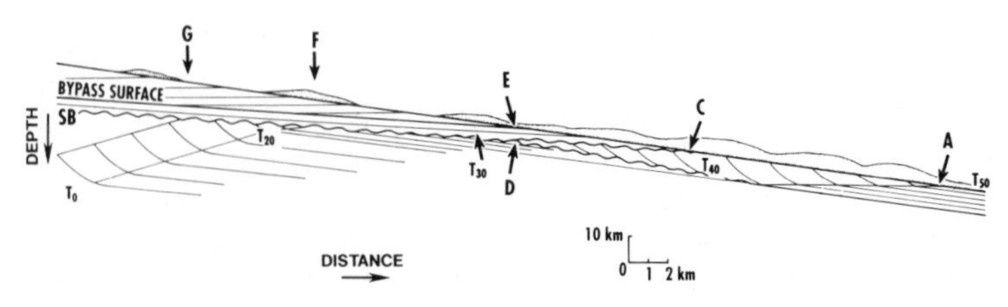

Fig. 9. Lithostratigraphic model of a deltaic deposit formed in response to progradation during sea-level rise, high stillstand and fall (time T_0 to T_{40}) followed by low stillstand (T_{40} to T_{50}) and transgression during sea-level rise (time T_{50} onwards). Surface SB forms during the regression in response to sea-level fall. At its seaward end, this surface splits into two; there is a subaerial surface on top of the regressive shoreface and a marine erosion surface at its base. Surface SB truncates much of the highstand shoreface. The ravinement surface truncates the shoreface (between points A and C) lowstand fluvial (between points C and E) and transgressive coastal-plain strata (landward of point E) in a landward succession.

space. Falling sea level (T_{30} to T_{40}) is characterized by rapid shoreline progradation (1.5 km per thousand years) because of upland fluvial erosion. The associated subaerial erosion surface truncates previous highstand strata and extends seaward to the point of shoreline location at the end of relative-sea-level fall (C, Fig. 9). Between points F and D (Fig. 9), the regressive (highstand) shoreface is entirely removed (in this two-dimensional model case; in a real world, three-dimensional case shoreface deposits of this phase may be preserved on interfluves). Lowstand (T_{40} to T_{50}) is characterized by renewed sea-level stillstand, shoreline progradation at a rate of 1.0 km per thousand years and renewed fluvial deposition beneath the rising, graded river profile.

In this model, we kept the fluvial profile invariant throughout the experiments. Therefore, time lines within the lowstand fluvial wedge (between SB and bypass surface in Fig. 9) are parallel to the erosion surface (SB). In reality, as discussed in the first part of this paper, fluvial fill in response to river-mouth progradation will be recorded upstream as a landward-accreting wedge. Time lines in the fluvial fill, therefore, would onlap the underlying erosion surface (SB).

The parameters controlling transgression (T_{50} onwards) are a sea-level rise of 1.0 m per thousand years and a shoreline migration rate of 2.0 km per thousand years. The transgression in this case proceeds exactly as in case 1 (Fig. 6), the only difference being the greater complexity of the underlying truncated units. In a landward succession the ravinement truncates lowstand shoreface (Fig. 9, segment A–C), lowstand fluvial strata (C–E) and transgressive coastal-plain deposits (landward of E).

Chronostratigraphy

The additional surface developed within this model succession is the *regressive surface of erosion*, which is associated with the phase of relative sea-level fall (T_{30} to T_{40}). This term refers to any surface formed by erosion while the shoreline moves seaward and downward. Landward of the lowstand shoreface wedge (left of D in Figs 9 & 10), the only preserved regressive surface of erosion is the subaerial one. Its lacuna consists of: (i) an erosional vacuity formed by truncation of much of the highstand fluvial strata deposited before time T_{30}, plus (ii) a hiatus associated with non-deposition between T_{30} and T_{40}, which is the period of relative sea-level fall.

Two regressive surfaces of erosion bracket the landward part of the preserved shoreface (between locations C–D, Figs 9 & 10): a subaerial erosion surface at the top and a surface of marine erosion at the base. The descending shoreface truncates earlier shelf deposits (landward of point C) in accordance with the 'inverse' Bruun rule discussed above. Renewed (inner) shelf deposition is possible only after sea level has stabilized at the new, lower level after time T_{40}. At that time, lowstand fluvial aggradation also resumes and a conformable succession of shelf, shoreface and distal fluvial facies forms during lowstand (T_{40} to T_{50}). The lowstand shelf strata at the base of this succession are separated from the underlying highstand by a *regressive surface of marine erosion.* The lacuna represented by the regressive surface of marine erosion covers essentially the same age interval as the equivalent subaerial surface (Fig. 10). This lacuna is a sequence boundary.

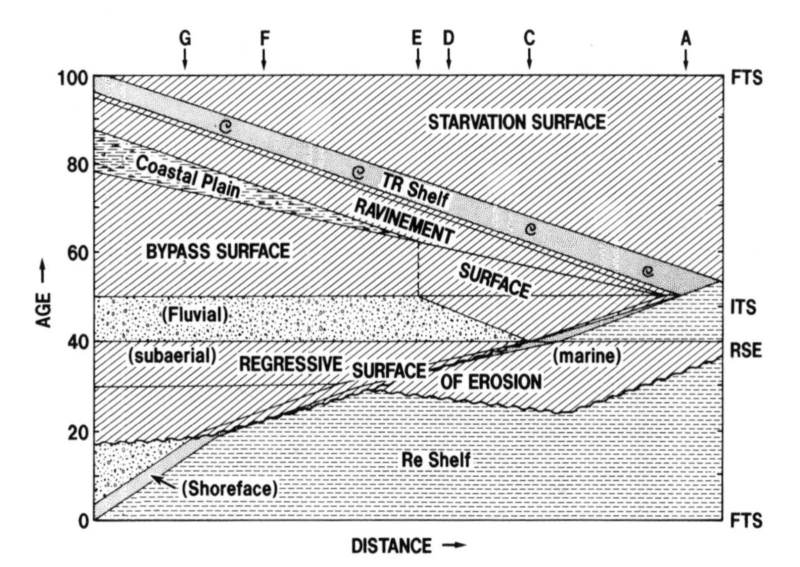

Fig. 10. Wheeler diagram of the deltaic succession illustrated in Fig. 9. The lacuna on the subaerial regressive surface of erosion (RSE) consists of an erosional vacuity (below T_{30}) and a hiatus (T_{30} to T_{40}). The lacuna on the marine part of this surface (below T_{40}) is entirely a vacuity. The transgressive component is identical to that in Fig. 7. This model represents a stack of two half-sequences, in which the combined temporal and spatial extent is equal to the parasequence in Fig. 7.

Discussion

Both of these model cases consider a sea-level cycle lasting 100,000 years (T_0 to T_{100}). The same models apply, however, at all time scales. In case 1, where the entire sea-level cycle consisted of a stillstand and rise, we modelled the chronostratigraphy of a parasequence. In case 2, where the sea-level cycle encompassed a phase of relative fall, we modelled a sequence (or better: two half-sequences, with a sequence boundary between them). In both cases, the entire succession reflects one regression followed by one transgression. The models make it very clear that parasequences and sequences are simply two different ways of subdividing the stratigraphic record; parasequences are bounded by surfaces of deepening (transgressive surfaces) and sequences are bounded by surfaces of shallowing (sequence boundaries). Both may, and often do, represent the same thickness of rocks and increments of time.

In a natural succession of deltaic rocks, stacks of sequences often are followed by stacks of parasequences of similar scale (typically 10 to 30 m). Excellent examples are found in Upper Cretaceous deposits of the Book Cliffs in Utah (Van Wagoner *et al.*, 1991). Such alternations between parasequences and sequences may reflect a eustatic signal consisting of superimposed sea-level curves of different frequencies, as proposed by Van Wagoner *et al.* (1990), or temporal variability in rates of tectonic subsidence superimposed on a steady, sinusoidal sea-level signal. Figure 11, redrafted from Van Wagoner *et al.* (1990), illustrates this concept. Fourth-order (relative sea-level) cycle 'A' would produce a stratigraphy as in our model in Figs 9 and 10, whereas fourth-order cycle 'B' would produce a parasequence as modelled in Figs 6 and 7.

CONCLUSIONS

This paper models the litho- and chronostratigraphy of a (fluvio-deltaic) deposit that results from a single regression and transgression of the shoreline. If there is no relative sea-level fall associated with such a cycle, the resulting deposit is a parasequence, bounded by transgressive surfaces. If the cycle does include a relative fall the resulting deposit becomes a stack of two half-sequences, separated by a regressive surface of erosion, which is the sequence boundary. The models highlight the observation that parasequences and sequences are simply two different ways of subdividing the stratigraphic record; parasequences are bounded by surfaces of deepening (transgressive surfaces) and sequences are bounded by surfaces of shallowing (sequence boundaries). Both may, and often do, represent the same thickness of rocks and increments of time.

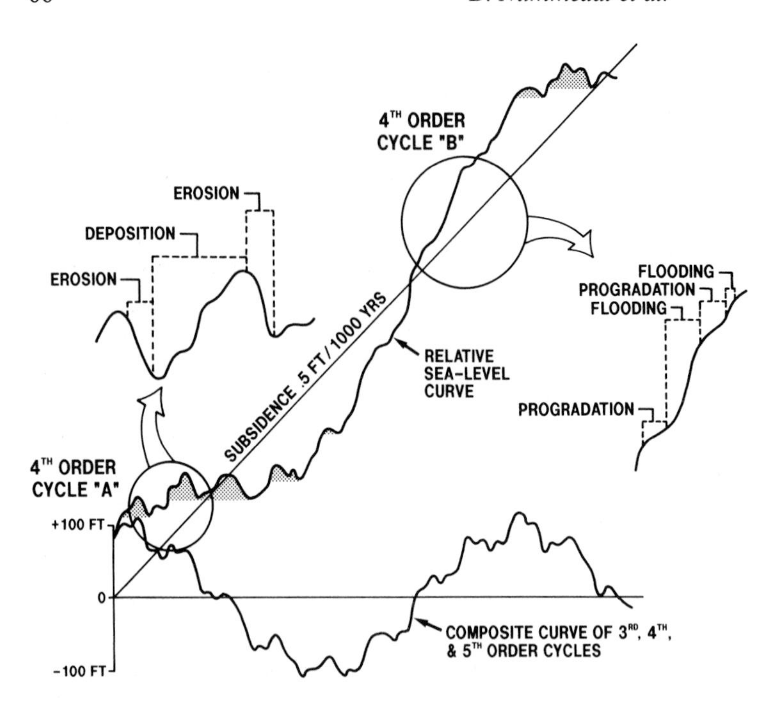

Fig. 11. Sea-level cycles of 'third, fourth and fifth order' superimposed on a linear basin subsidence trend. The resultant relative sea-level curve consists of cycles of rise and fall during the declining phase of the long-term (third-order) trend (cycle 'A') and alternating periods of slow and rapid rise during the ascending phase of the long-term trend (cycle 'B'). Sequences, as modelled in Fig. 10, form during cycle 'A'; parasequences, as modelled in Fig. 7, form during cycle 'B'. (Modified from Van Wagoner *et al.*, 1990.)

In response to a complete cycle of relative sea-level change, the models yield the following stratigraphy. During stillstand the regressive fluvio-deltaic system forms a conformable succession of shelf, shoreface (delta front) and fluvial deposits. As relative sea level begins to fall, a regressive surface of erosion will commence truncation of stillstand (highstand) fluvial, shoreface and inner-shelf strata. This regressive surface of erosion contains both subaerial and submarine parts. During the subsequent stillstand (lowstand), this surface is buried by additional regressive fluvial strata and a downdip, landward-thinning wedge of lowstand shoreface and shelf deposits. In these specific models, sea-level rise is always associated with transgression. Transgressive coastal-plain strata onlap lowstand fluvial deposits at a bypass surface and are, in turn, truncated by the ravinement surface. Transgressive shelf strata onlap the ravinement and are capped by a marine starvation surface. A complete relative sea-level cycle will produce five separate diastems (bypass or erosional surfaces).

The development of multiple erosional surfaces within parasequences and sequences has been recognized well before this study (Bergman & Walker, 1987; Plint, 1988; Walker, 1990; Dominguez & Wanless, 1991; Posamentier *et al.*, 1992; Swift *et al.*, 1991, a, b; Wright Dunbar *et al.*, 1992). A new element of this paper is the explicit modelling of generation and loss of accommodation space in response to migrations of fluvial and shoreface equilibrium profiles. By driving this forward model with equilibrium profile migrations which are well supported with field data, we have constructed realistic chrono- and lithostratigraphic diagrams for wave-dominated fluvio-deltaic systems in ramp settings.

ACKNOWLEDGEMENTS

The ideas presented in this paper have been stimulated by lively discussions with many colleagues. In particular we wish to express our appreciation to Don Swift, Peter Vail, John Van Wagoner, Robyn Wright Dunbar, Marilyn Huff and Doug Levin. Critical reviews of the paper by K. Molenaar, Dale Leckie, Henry Posamentier and Don Owen were very helpful. Research in part supported by NSF grant EAR–9205811.

REFERENCES

Bergman, K.M. & Walker, R.G. (1987) The importance of sea level fluctuations in the formation of linear conglomerate bodies: Carrot Creek Member, Cretaceous

Western Interior Seaway, Alberta, Canada. *J. Sediment Petrol.* **57**, 651–665.

Bruun, P. (1962) Sea-level rise as a cause of shore erosion. *J. Waterways and Harbors Div. Am. Soc. Civ. Eng.* **88**, 117–130.

Cross, T.A. (Ed.) (1990) *Quantitative Dynamic Stratigraphy.* Prentice-Hall, Englewood, NJ.

Dean, R.G., Dalrymple, R.A., Fairbridge, R.W., *et al.* (1987) *Responding to Changes in Sea Level – Engineering Implications.* National Academy Press, Washington, 148 pp.

Dominguez, J.M.L. & Wanless, H.R. (1991) Facies architecture of a falling sea level strandplain, Doce River coast, Brazil. In: *Shelf Sand and Sandstone Bodies; Geometry, Facies and Sequence Stratigraphy* (Eds Swift D.J.P. *et al.*) Int. Assoc. Sediment. Spec. Publ. **14**, 259–281.

Everts, C.H. (1987) Continental shelf evolution in response to a rise in sea level. In: *Sea-Level Change and Coastal Evolution* (Eds Nummedal, D., Pilkey, O.H. & Howard, J.D.) Soc. Econ. Paleontol. Mineral. Spec. Publ. 41, 49–58.

Gilbert, G.K. (1917) Hydraulic-mining debris in the Sierra Nevada. *U.S. Geol. Surv. Prof. Pap.* **105**, 154 pp.

Green, J.F.N. (1936) The terraces of southernmost England: *Geol. Soc. London, Quart. Jour.*, **92**, LVIII–LXXXVIII (Presidential address).

Helland-Hansen, W., Kendall, C.G.St.C., Lerche, I. & Nakayama, K. (1988) A simulation of continental margin sedimentation in response to crustal movements, eustatic sea level change and sediment accumulation rates. *Math. Geol.* **20**, 777–802.

Jervey, M.T. (1988) Quantitative geological modeling of siliciclastic rock sequences and their seismic expression. In: *Sea-level Changes: An Integrated Approach* (Eds Wilgus, C.K., Hastings, B.S., Kendall, C.G.St.C., Posamentier, H.W., Ross, C.A. & Van Wagoner, J.C.) Soc. Econ. Paleontol. Mineral. Spec. Publ. 42, 47–69.

Lawrence, D.T., Doyle, M. & Aigner, T. (1990) Stratigraphic simulation of sedimentary basins: concepts and calibrations. *Bull. Am. Assoc. Petrol. Geol.* **74**, 273–295.

Leopold, L.B. & Bull, W.B. (1979) Base level, aggradation and grade. *Am. Philos. Soc. Proc.* **123**, 168–202.

Mackin, J.H. (1948) Concept of the graded river. *Bull. Geol. Soc. Am.* **59**, 463–512.

Miall, A.D. (1991) Stratigraphic sequences and their chronostratigraphic correlation. *J. Sediment. Petrol.* **61**, 497–505.

Nummedal, D. & Swift, D.J.P. (1987) Transgressive stratigraphy at sequence-bounding unconformities: some principles derived from Holocene and Cretaceous examples. In: *Sea-Level Fluctuation and Coastal Evolution* (Eds Nummedal, D., Pilkey, O.H. & Howard, J.D.) Soc. Econ. Paleontol. Mineral. Spec. Publ. 41, 241–260.

Paola, C., Heller, P.L. & Angevine, C.L. (1991) The response distance of river systems to variations in sea level. *Geol. Soc. Am. Annual Meeting (Abstr.)*, San Diego, A170–A171.

Plint, A.G. (1988) Sharp-based shoreface sequences and 'offshore bars' in the Cardium Formation of Alberta: their relationship to relative changes in sea level. In: *Sea-level Changes: An Integrated Approach* (Eds Wilgus, C.K., Hastings, B.S., Kendall, C.G.St.C., Posamentier H.W., Ross, C.A. & Van Wagoner, J.C.) Spec. Publ. Soc. Econ. Paleontol. Mineral. 42, 357–370.

Posamentier, H.W. & Vail, P.R. (1988) Eustatic controls on clastic deposition II — sequence and systems tract models. In: *Sea-level Changes: An Integrated Approach* (Eds Wilgus, C.K., Hastings, B.S., Kendall, C.G.St.C., Posamentier, H.W., Ross, C.A. & Van Wagoner, J.C.) Soc. Econ. Paleontol. Mineral. Spec. Publ. 42, 125–154.

Posamentier, H.W., Allen, G.P., James, D.P. & Tesson, M. (1992) Forced regressions in a sequence stratigraphic framework: concepts, examples, and exploration significance. *Bull Am. Assoc. Petrol. Geol.* **76**, 1687–1709.

Posamentier, H.W., Jervey, M.T. & Vail, P.R. (1988) Eustatic controls on clastic deposition I — conceptual framework. In: *Sea-level Changes: An Integrated Approach* (Eds Wilgus, C.K., Hastings, B.S., Kendall, C.G.St.C., Posamentier H.W., Ross, C.A. & Van Wagoner, J.C.) Soc. Econ. Paleontol. Mineral. Spec. Publ. 42, 109–124.

Reynolds, D.J., Steckler, M.S. & Coakley, B.J. (1991) The role of the sediment load in sequence stratigraphy: the influence of flexural isostasy and compaction. *J. Geophys. Res.* **96**, 6931–6949.

Ross, W.C. (1990) Modeling base-level dynamics as a control on basin fill geometries and facies distribution: a conceptual framework. In: *Quantitative Dynamic Stratigraphy* (Ed. Cross, T.A.) Prentice-Hall, Engelwood Cliffs, NJ, 387–399.

Schwartz, M.L. (1967) The Bruun theory of sea-level rise as a cause of shore erosion. *J. Geol.* **73**, 528–532.

Schumm, S.A. (1969) River metamorphosis. *J. Hydraulics Div. Am. Soc. Civil Eng.* **95**, 225–273.

Schumm, S.A. & Lichty, R.W. (1965) Time, space and causality in geomorphology. *Am. J. Sci.* **263**, 110–119.

Shanley K.W. & McCabe, P.J. (1991) Predicting facies architecture through sequence stratigraphy — an example from the Kaiparowits Plateau, Utah. *Geology* **19**, 742–745.

Sloss, L.L. (1984) Comparative anatomy of cratonic unconformities. In: *Interregional Unconformities and Hydrocarbon Accumulation* (Ed. Schlee, J.S.) Am. Assoc. Petrol. Geol. Mem. 36, 1–6.

Smith, G.I., Benson, L.V. & Currey, D.R. (1989) *Quaternary Geology of the Great Basin.* 28th International Geological Congress Guidebook, T117; Am. Geophys. Union, Washington, DC, 78 pp.

Snedden, J.W. (1985) Origin and sedimentary characteristics of discrete sand beds in modern sediments of the central Texas continental shelf. PhD thesis, Louisiana State University, Baton Rouge, LA, 247 pp.

Stride, A.H. (Ed.) (1982) *Offshore Tidal Sands.* Chapman & Hall, London, 222 pp.

Swift, D.J.P. (1968) Coastal erosion and transgressive stratigraphy. *J. Geol.* **76**, 444–456.

Swift, D.J.P. & Moslow, T.F. (1982) Holocene transgression in south-central Long Island, New York — Discussion. *J. Sediment. Petrol.* **52**, 1014–1019.

Swift, D.J.P., McKinney, T.F. & Stahl, L. (1984) Recognition of transgressive and post-transgressive sand ridges on the New Jersey continental shelf: discussion. In: *Siliciclastic Shelf Sediments* (Eds Tillman, R.W. & Siemers, C.T.) Soc. Econ. Paleontol. Mineral. Spec. Publ. 34, 25–36.

Swift, D.J.P., Philips, S. & Thorne, J.A. (1991a) Sedimentation on continental margins; Part IV: lithofacies and depositional systems. In: *Shelf Sand and Sandstone Bodies; Geometry, Facies and Sequence Stratigraphy*. (Ed. Swift, D.J.P., *et al.*) Int. Assoc. Sediment. Spec. Publ. **14**, 89–152.

Swift, D.J.P., Philips, S. & Thorne, J.A. (1991b) Sedimentation on continental margins, part V: parasequences. In: *Shelf Sands and Sandstone Bodies: Geometry, Facies and Sequence Stratigraphy* (Eds Swift, D.J.P., *et al.*) Int. Assoc. Sediment. Spec. Publ. **14**, 153–187.

Thomas, M.A. (1990) The impact of long-term and short-term sea level changes on the evolution of the Wisconsinan-Holocene Trinity/Sabine incised valley system, Texas continental shelf. PhD thesis, Rice University, 247 pp.

Vail, P.R., Audemard, F., Bowman, S.A., Eisner, P.N. & Perez-Cruz, G. (1991) The stratigraphic signature of tectonics, eustacy and sedimentation. In: *Cycles and Events in Stratigraphy*. (Eds Einsele, G., Ricken, W. & Seilacher, A.) Springer-Verlag, Berlin, 617–659.

Vail, P.R., Hardenbol, J. & Todd, R.N. (1984) Jurassic unconformities, chronostratigraphy, and sea-level changes from seismic stratigraphy and biostratigraphy. In: *Interregional Unconformities and Hydrocarbon Accumulation (Ed. Schlee, J.S.)* Am. Assoc. Petrol. Geol. Mem. 36, 129–144.

Van Wagoner, J.C., Mitchum, R.M., Campion, K.M. & Rahmanian, V.D. (1990) *Siliciclastic Sequence Stratigraphy in Well Logs, Cores and Outcrops*. Am. Assoc. Petrol. Geol. Meth. Expl. Ser. **7**, 55 pp.

Van Wagoner, J.C., Nummedal, D., Jones, C.R., Taylor, D.R., Jennette, D.C. & Riley, G.W. (1991) *Sequence Stratigraphy Applications to Shelf Sandstone Reservoirs*. Am. Assoc. Petrol. Geol. Guidebook Ser.

Van Wagoner, J.C., Posamentier, H.W., Mitchum, R.M., Jr., Vail, P.R., Sarg, J.F., Loutit, T.S. & Hardenbol, J. (1988) An overview of the fundamentals of sequence stratigraphy and key definitions. In: *Sea-level Changes: An Integrated Approach* (Eds Wilgus, C.K., Hastings, B.S., Kendall, C.G.St.C., Posamentier, H.W., Ross, C.A. & Van Wagoner, J.C.) Spec. Publ. Soc. Econ. Paleontol. Mineral. 42, 39–46.

Walker, R.G. (1990) Facies modelling and sequence stratigraphy. *J. Sediment. Petrol.* **60**, 777–786.

Wells, J.T. & Coleman, J.M. (1987) Wetland loss and the subdelta life cycle. *Estuarine, Coastal, Shelf Sci.* **25**, 111–125.

Wheeler, H.E. (1958) Time stratigraphy. *Am. Assoc. Petrol. Geol. Bull.* **42**, 1047–1063.

Wilgus, C.K., Hastings, B.S., Kendall, C.G.St.C. Posamentier, H.W., Ross, C.A. & Van Wagoner, J.C. (Eds) (1988) *Sea-level Changes: An Integrated Approach*. Spec. Publ. Soc. Econ. Paleontol. Mineral. 42, 407 pp.

Wright Dunbar, R., Devine, P.E. & Katzman, D. (1992) The transgressive depositional record at parasequence scale: a model from the Upper Cretaceous Point Lookout Sandstone, New Mexico and Colorado. *Ann. Mtg. Am. Assoc. Petrol. Geol.* (Abstr.) 1992, Calgary, 146.

Spec. Publs Int. Ass. Sediment. (1993) **18**, 69–92

Incised valleys in the Pleistocene Tenryugawa and Oigawa coastal-fan systems, central Japan: the concept of the fan-valley interval

T. MUTO*

Department of Earth and Planetary Sciences, Kyusha University, Hakozaki 6-10-1, Kukuoka 812, Japan

ABSTRACT

The *fan-valley interval* is a time interval between two successive periods of alluvial open-fan development corresponding with highstands of relative sea level. The 0.2 Ma–Recent Tenryugawa and 0.9–0.7 Ma Oigawa systems, both located on the Pacific coast of central Japan, experienced long and short fan-valley intervals, respectively, and show significant variations of coastal-fan sequences. Comparative analyses of the two systems, and of two fans of different ages in the Tenryugawa system, suggest that incised valleys form a significant portion of coastal-fan systems situated adjacent to steep submarine slopes. The sea-level control model of coastal fans, which has been improved so as to contain the concept of the fan-valley interval, gives a synthetic explanation to a considerable part of the observed variations.

During a fan-valley interval, a fan valley exists in a coastal-fan system and becomes wider owing to lateral erosion of the walls by the accommodating rivers. A long fan-valley interval, generally accompanying tectonic uplift, leads to: (i) large lateral dimensions of the fan valley and dissection channels; (ii) low frequency of overflow at the end of the interval; (iii) low productivity of the overflow sediments; and (iv) accumulation of the overflow sediments, if any, within the dissection channels rather than on the existing open-fan surface. The complete set of open-fan deposits tends to have bipartite organization and low-relief geometry after a long fan-valley interval and tripartite organization and high-relief geometry after a short fan-valley interval.

INTRODUCTION

This paper proposes a new concept, that of the *fan-valley interval*. The concept has been developed from Muto's (1987a, 1988) sea-level control model and is expected to provide a useful framework for the sequence-stratigraphic analysis of coastal alluvial fans. A fan-valley interval indicates a time interval during which a fan valley exists in the system; the length and frequency of the fan-valley interval depend on relative sea-level changes and thus are partly related to vertical tectonism. The *fan valley* (Muto, 1987a) is the fan-dissecting valley connected to the feeder canyon for the fan (Fig. 1).

Nemec and Steel (1988) defined a *coastal fan* as 'the alluvial feeder system of a fan delta', and a *fan delta* as 'a prism of sediments delivered by an alluvial fan and deposited, mainly or entirely subaqueously, at the interface between the active fan and a standing body of water'. According to these definitions, subaqueous sediments are completely excluded from the coastal fan. Such terminology would be useful to describe sedimentation under steady state without geomorphic modification. However, geomorphic changes including dissection and subsequent marine filling are one of the essential properties of coastal fans (Muto, 1987a). So the term coastal fan becomes inappropriate and the valleys are regarded as part of a *coastal-fan system* (Muto, 1989), a term that encompasses both coastal-fan and fan-delta deposits.

The sea-level control model is a geomorphologic framework for coastal fans located adjacent to steep submarine slopes, i.e. for 'slope-type fan deltas' of Ethridge and Wescott (1984); its essence is

*Present address: Department of Geology, Nagasaki University, Bunkyomachi 1-14, Nagasaki 852, Japan.

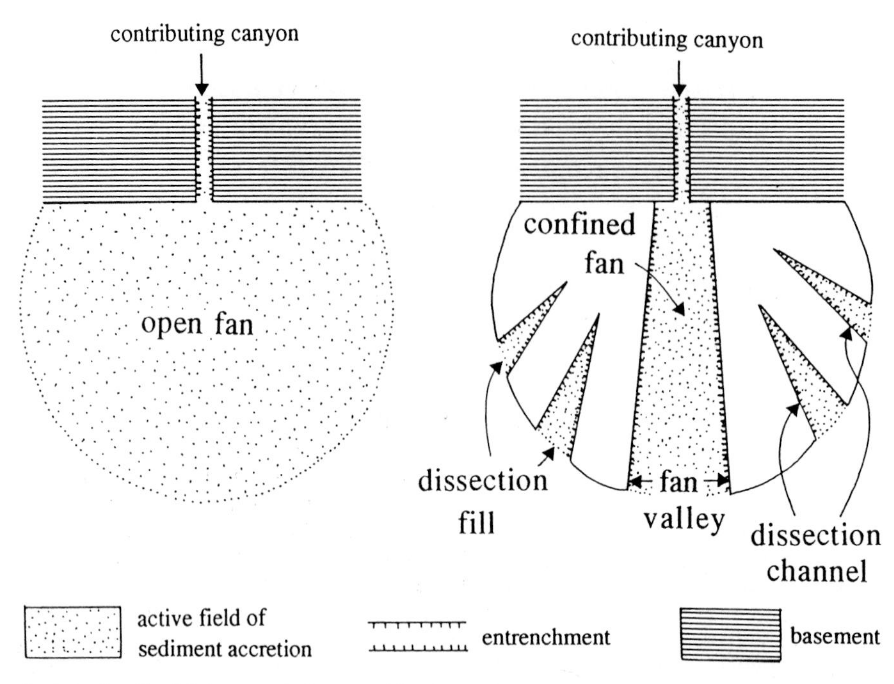

Fig. 1. Terminology for alluvial-fan morphology used in the sea-level control model. (After Muto, 1987a.)

summarized below. Relative sea-level changes directly control aggradation and dissection of coastal alluvial environments because the sea level works as the critical base level for the coastal rivers. The river bed aggrades with rise of sea level, while lowering of sea level causes degradation of the river. Even if a larger volume of sediment is episodically supplied into the coastal system, excess sediment is mostly preserved below sea level but does not cause significant coastal aggradation and progradation because the system faces a steep submarine slope. Two types of fan-dissecting valleys are distinguished: one is a fan valley, and the other includes *dissection channels* (Fig. 1), which are not connected with the feeder canyon and thus develop particularly in the distal area. Evolution of coastal-fan morphology consists of three cyclic phases (Fig. 2): (i) fan dissection with sea-level lowering (*fan-valley phase*); (ii) active aggradation confined within the fan valley with sea level rising (*confined-fan phase*); and (iii) open-fan aggradation over the entire system with further sea level rising beyond the previous highest levels (*open-fan phase*). The open-fan phase may commence another cycle, but sea-level rise which does not reach the maximum of the previous one results in the alternation of confined-fan and fan-valley phases. *Overflows* occur intermittently from the fan valley just before and after the open-fan phase, i.e. at the end of a confined-fan phase and at the beginning of a fan-valley phase. The competence and frequency of overflows increase with decreasing capacity of the fan valley (confined-fan phase) and decrease with its increasing capacity (fan-valley phase). Therefore, overflows occurring in a confined-fan phase tend to construct a coarsening-upward sequence, whereas those in a fan-valley phase cause a fining-upward sequence. Likely sites for the accumulation of such sequences are the surface of the existing open fan and/or within dissection channels. The term *geomorphic sequence* (Muto, 1988) is used to designate a sequence of strata formed through a complete cycle of relative sea-level rise and subsequent fall. For a given fan, such an idealized geomorphic sequence comprises a single set of open-fan deposits, a single fan-valley fill and multiple dissection fills (see Fig. 1). As a general trend, tectonic subsidence combined with eustatic sea-level changes will lead to predominance of the open-fan phases in the sedimentary record, whereas, under tectonic uplift, the fan-valley phase will be predominant.

Recently, an increasing number of publications

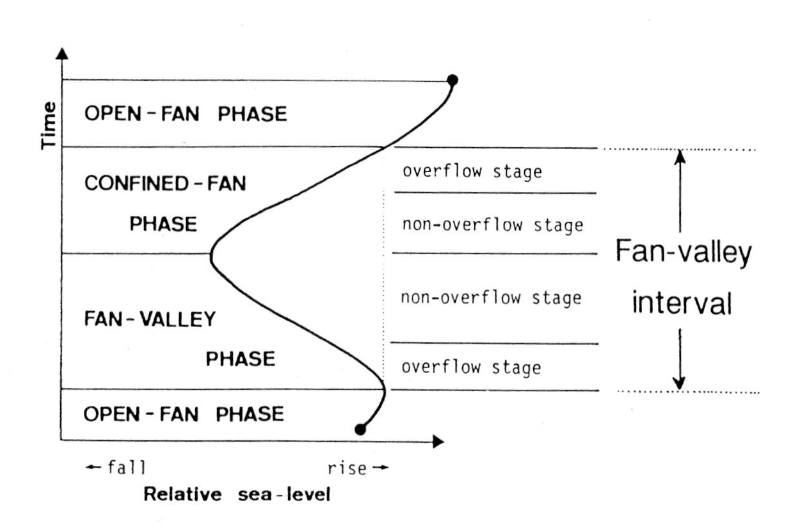

Fig. 2. The sea-level control model of coastal alluvial fans proposed by Muto (1987a, 1988). Sea level operates as the critical base level for coastal rivers. In coastal-fan systems situated adjacent to steep submarine slopes, evolution of fan morphology consists of three cyclic phases; (i) dissection of the fan, with sea-level lowering (fan-valley phase); (ii) active aggradation confined within the dissecting valleys, with sea level rising (confined-fan phase); (iii) open-fan aggradation over the entire system, with further sea level rising beyond the previous highest levels (open-fan phase). Overflows occur intermittently from the fan valley when it has an insufficient capacity to convey a high discharge flow from the feeder canyon for the system (overflow stage).

have reported the crucial importance of relative sea-level changes on coastal alluvial environments (e.g. McGowen, 1970; Muto, 1987a, 1988; Colella, 1988; Postma & Cruickshank, 1988; Muto & Blum, 1989; Bardaji *et al.*, 1990; see also Kazanci, 1988 for lake-level control). Yet it has been insufficiently understood what influences sea-level changes exert on coastal-fan architectures and stratigraphy. Sequence stratigraphy of coastal fans should be examined in terms of both: (i) *relative sea level* and its vertical ranges (e.g. amplitude); and (ii) *time* for relative sea-level fluctuations (e.g. period, frequency), which are usually two independent variables in sea-level curve diagrams. The sea-level control model, outlined above, mainly explains the amplitude influence. The time influence is not yet understood properly, so that the model allows a great variety of coastal-fan architectures. In fact, the 0.2 Ma–Recent Tenryugawa (*gawa* = river) and 0.9–0.7 Ma Oigawa fan systems on the Pacific coast of central Japan (Fig. 3) illustrate the model (Muto, 1987a; Muto & Blum, 1989), but show quite different dimensions, geometries and sediment types from each other. Even in the Tenryugawa system, two fans of different ages present significant variations. Muto and Blum (1989) suggested the possibility that the different patterns of relative sea-level changes, dependent largely on different tectonic regimes, were responsible for a considerable part of the observed variations. In this paper I aim to refine the idea into a new concept, to provide a framework which predicts sequence stratigraphy of coastal fans, and to improve the sea-level control model so as to include an element of time control.

FAN-VALLEY INTERVAL

The fan-valley interval is defined as the time interval between two successive open-fan phases (Fig. 2), during which a fan valley exists in the coastal-fan system. A fan-valley interval consists of a single or multiple couples of fan-valley and subsequent confined-fan phases. Tectonic subsidence tends to make the fan-valley interval shortened and increased in frequency, whereas, by tectonic uplift, the interval will be lengthened and decreased in frequency (Fig. 4).

The duration of a fan-valley interval equals the life span of a fan valley. As long as the fan valley exists, it progressively becomes wider because the accommodating rivers are attacking its walls. The downcutting of a fan valley is limited to a fan-valley phase, whereas the lateral erosion can take place during the entire fan-valley interval, either fan-valley or confined-fan phases. Dissection channels also exist only during a fan-valley interval, growing partly by headward erosion. A longer fan-valley interval will lead both to a wider fan valley and to wider dissection channels. However, lateral erosion of dissection channels may not be so distinct as of

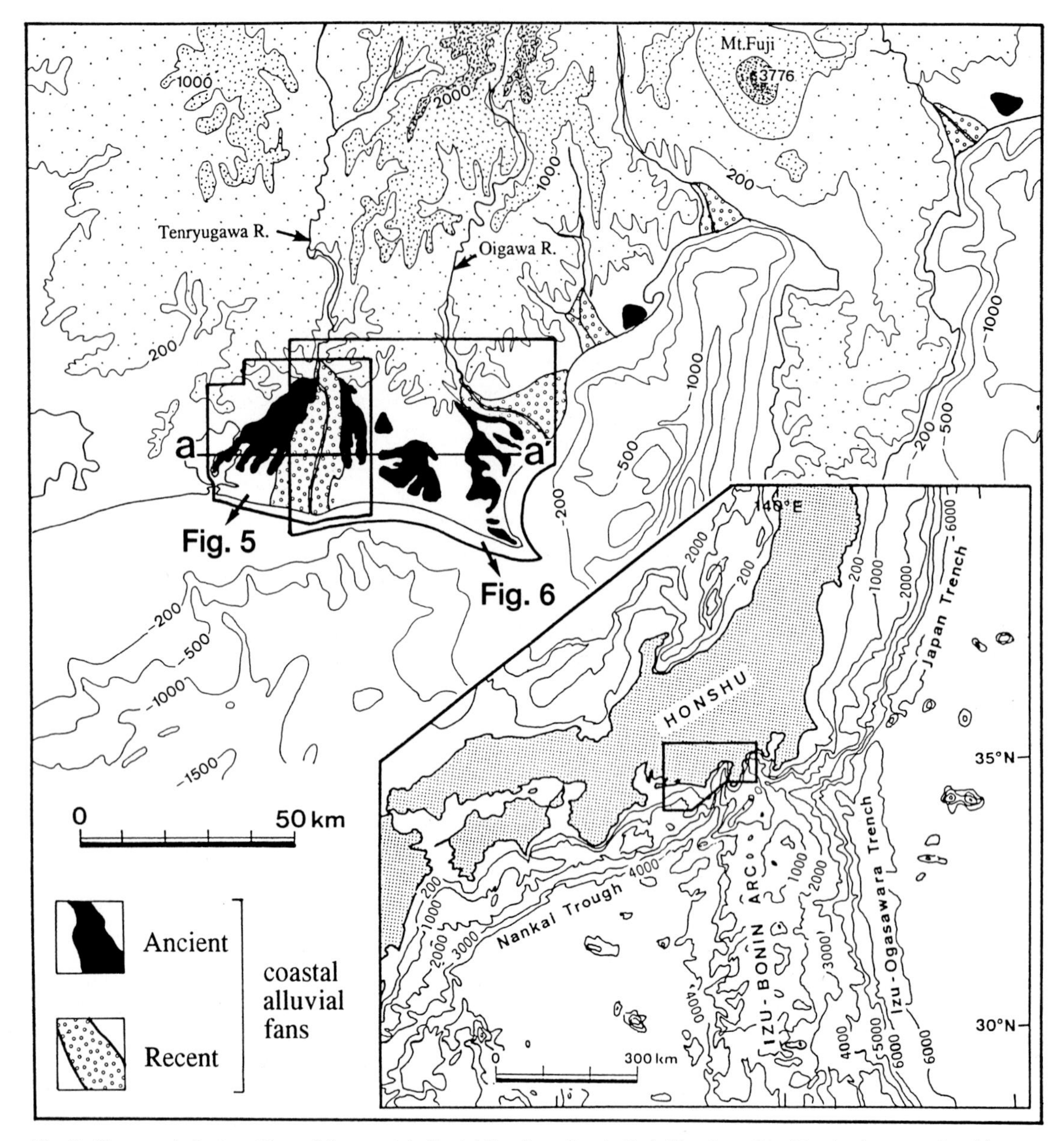

Fig. 3. Geomorphologic setting of the coastal alluvial-fan deposits studied. Numbers for altitude above sea level in metres. Line a–a′ is the index for Fig. 19(b).

the fan valley, because they are separated from the main fan drainage system.

The frequency of overflows depends on the capacity of the fan valley and the discharge of the accommodating rivers. With a given threshold capacity and a given range of river discharge, the wider fan valley is less likely to be flooded and supplies less overflow sediments beyond itself.

TENRYUGAWA AND OIGAWA SYSTEMS

The 0.2 Ma–Recent Tenryugawa and 0.9–0.7 Ma Oigawa systems are here compared in terms of relative sea-level changes experienced and fan architectures. The Tenryugawa system is located on the narrow, tectonically uplifted coast; the fan

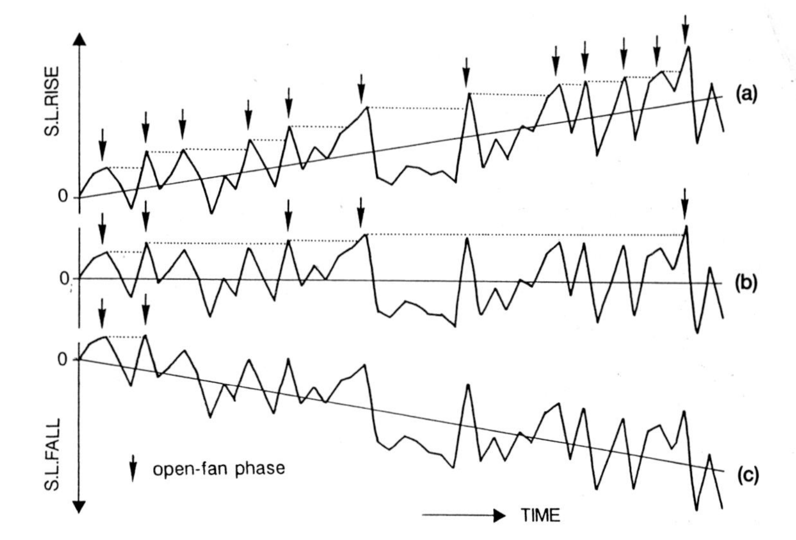

Fig. 4. Model for tectonic control of fan-valley intervals. (a) Curve modified from (b) by tectonic subsidence. (b) Hypothetical eustatic curve (tectonism-free situation). (c) Curve modified from (b) by tectonic uplift. Lengths of fan-valley intervals are indicated by dotted lines. Note that the intervals tend to decrease and increase in number with tectonic uplift and subsidence, respectively.

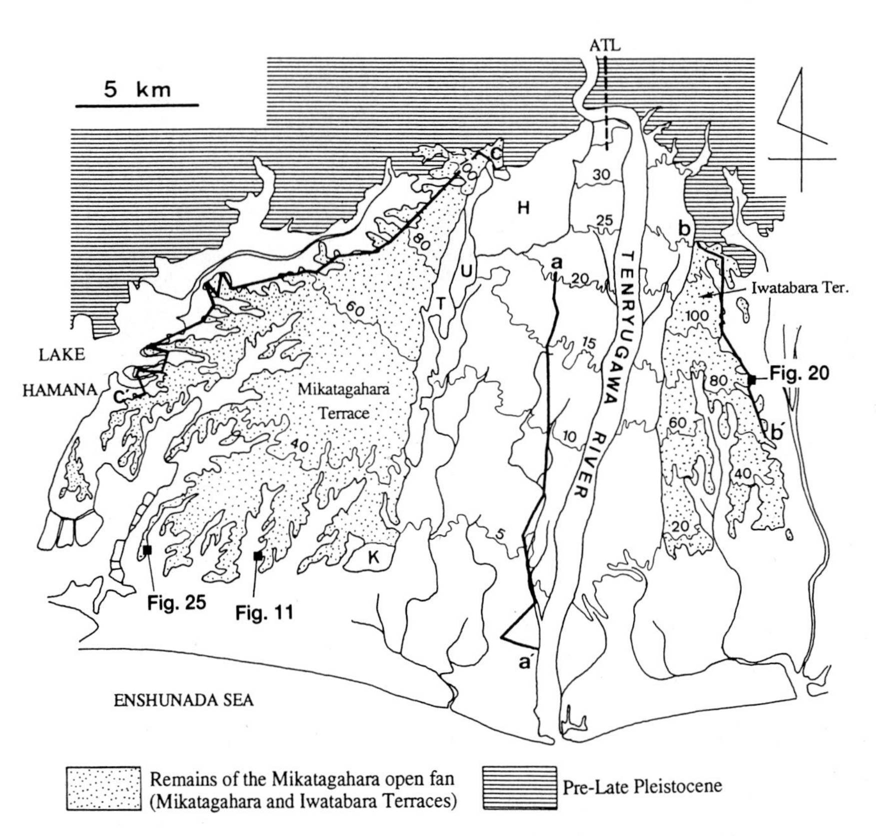

Fig. 5. Present dissected fan morphology in the Tenryugawa system. The Mikatagahara and Iwatabara Terraces are remains of the 'Mikatagahara open fan' (see text), which formed at about 0.125 Ma. K, Kamoe Terrace; T, Tomioka Terrace; U, Ubagaya Terrace; H, Hamakita Terrace. Numbers for elevation in metres. ATL indicates the Akaishi Tectonic Line. Lines a–a′, b–b′ and c–c′ are the indices for Figs 18(a), 18(b) and 10, respectively. (Modified from Muto, 1987a.)

deposits have been preserved as terraces gently sloping southward (Fig. 5). The Oigawa system is recorded in the early to middle Pleistocene Ogasayama Formation, which is exposed in the Ogasayama Hills (264 m in maximum height) near the Iwatabara Terrace to the west (Fig. 6). This formation is derived from the ancient Oigawa and thus independent of the Tenryugawa. The Oigawa

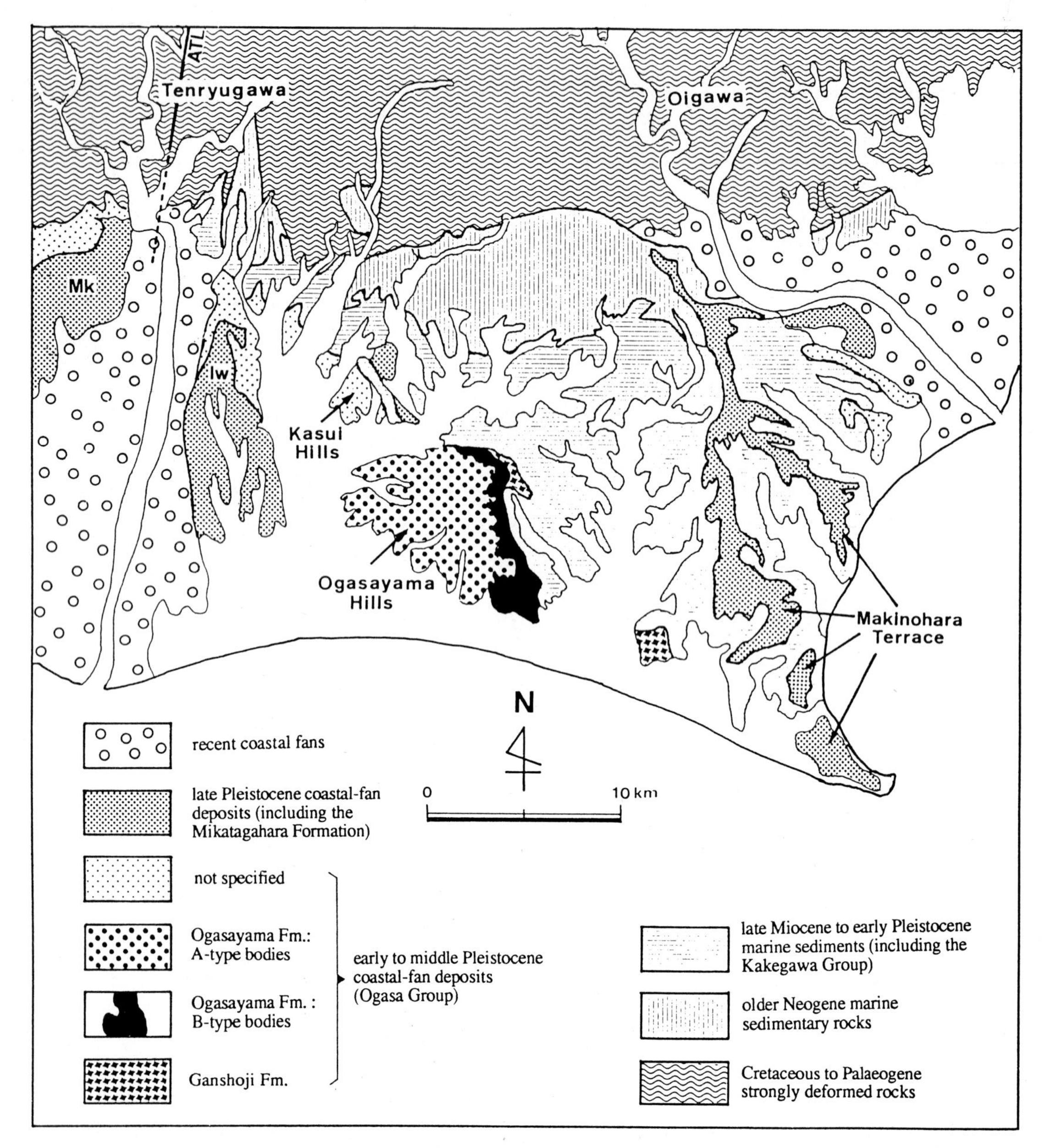

Fig. 6. Geologic setting of the Pleistocene Ogasayama Formation, which consists entirely of coastal-fan and related deposits derived from the ancient Oigawa River. The Ogasa Group consists of early to middle Pleistocene coastal-fan deposits derived from either the Tenryugawa or Oigawa; the Ogasayama Formation comprises a part of the Ogasa Group.

system is bounded by a structural high made of the late Miocene to early Pleistocene Kakegawa Group to the east but is free to expand toward the west (Ujiié, 1962).

Relative sea-level changes

Tenryugawa

The Tenryugawa system experienced three major highstands of sea level during the past 0.2 Ma, at 0.19 Ma, 0.125 Ma and 0.005 Ma (Fig. 7). The sedimentary records of the earlier two have been preserved as two sets of open-fan deposits, indicating the occurrences of two open-fan phases. The latest highstand, lower than the highstand at 0.125 Ma, did not lead to construction of an open fan because of an insufficient sea-level rise. The last fan-valley interval (0.125 Ma to Recent), still continuing, is twice as long as the previous one (0.19 to 0.13 Ma).

Major eustatic sea-level rises occurred just before 0.125 Ma with the Shimosueyoshi (last interglacial) transgression and at 0.02 to 0.005 Ma with the Yurakucho (postglacial) transgression (Fig. 7). The average rates of the sea-level rises have been estimated approximately from the oxygen-isotope data by Shackleton and Opdyke (1973, 1976), using the rough equivalence of 0.1 per mm in δ^{18}O-value to a 10 m change in sea level (Shackleton & Opdyke, 1973). The calculated rates are about 8 to 9 mm year^{-1} for each case.

The two broadest terraces — Mikatagahara and Iwatabara — represent remains of the younger open fan, which underwent tectonic uplift after the deposition (Nakagawa, 1961; Tsuchi, 1970). Stratigraphically, this fan body occupies the upper part of the Mikatagahara Formation, and thus is called the Mikatagahara open fan; whereas the older one, included in the Ogasa Group ('not specified' in Fig. 6), was called the 'Kamoe open fan' (both named by Muto, 1987a; stratigraphic nomenclature after Muto, 1985, 1987b).

The uplift rate of the Mikatagahara open fan can be calculated from the altitude of the distal end of the Mikatagahara Terrace, 30 m, and the eustatic sea level at 0.125 Ma, +6 m relative to the present sea level (Mesolella *et al.*, 1969; Sterns, 1976). Assuming that the altitude of the terrace coincides with the eustatic sea level, the rate is *c.* 0.2 mm year^{-1} (24 m/0.125 Ma). Such slow tectonism was insufficient to modify the fan-valley intervals

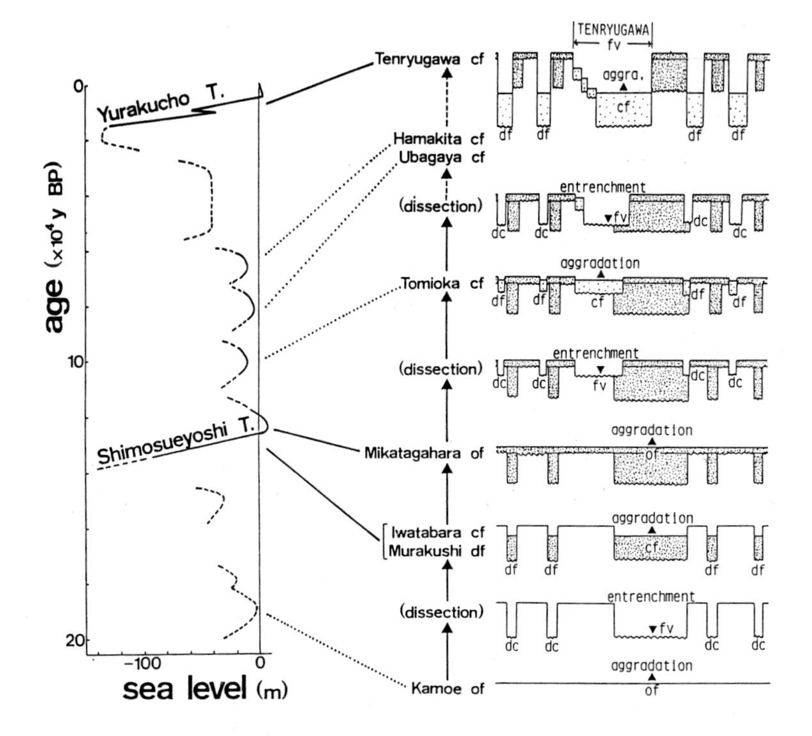

Fig. 7. Correlation of the fan deposits in the Tenryugawa system with the independently known eustatic sea-level changes in Japan during the past 0.2 Ma. The eustatic curve is compiled from Machida (1975), Kaizuka *et al.* (1977) and Naruse (1983). The morphologic development of the Tenryugawa system is schematically drawn. The three solid lines indicate that the correlation is supported by previous studies, while the dotted lines indicate correlation based on the sea-level control model. For simplification, the aggradation and dissection of the Ubagaya and Hamakita confined fans are not shown here. of, open fan; fv, fan valley; cf, confined fan; dc, dissection channel; df, dissection fill. (After Muto, 1987a.)

determined by the eustatic sea-level changes. However, the difference of sloping angle between the Mikatagahara (0.29°) and Iwatabara (0.57°) Terraces indicate tectonic deformation of the Mikatagahara open fan, probably related to the active Akaishi Tectonic Line (Figs 5 & 6).

Oigawa

The 450-m thick lower Ogasayama Formation of entirely coastal strata is the result of deposition during the total sea-level rise. Because the net eustatic sea-level rise during the time interval between 0.9 and 0.7 Ma is approximately zero (Fig. 8(a)), the general sea-level rise can be ascribed to tectonic subsidence, the rate of which was 2.3 mm year^{-1} (450 m/0.2 Ma) on average. The number of open-fan phases predicted from the relative sea-level curve (Fig. 8(b)) is 11; the field observation by Muto and Blum (1989) indicated 13 complete sets of open-fan deposits and thus the occurrence of 13 open-fan phases. The average length of the fan-valley intervals is less than 0.015 Ma, only one-sixth as long as in the Tenryugawa system (0.09 Ma on average; Fig. 9). The short fan-valley intervals resulted largely from the tectonic subsidence; only four open-fan phases would have occurred without tectonic subsidence (Fig. 8(a); see also Fig. 4).

Open fans

Tenryugawa

The Mikatagahara open fan is characterized by thin and broad geometry (3–15 m thick, 25+ km wide, 15+ km long) (Figs 5, 10 & 11). The thickness does not significantly change from the fan-apex area towards the distal margin. The altitude of the depositional surface ranges from 120 m in the north to 30 m in the south. The remains of the fan include a single fan valley and numerous dissection channels; the dissectional morphology is the result of

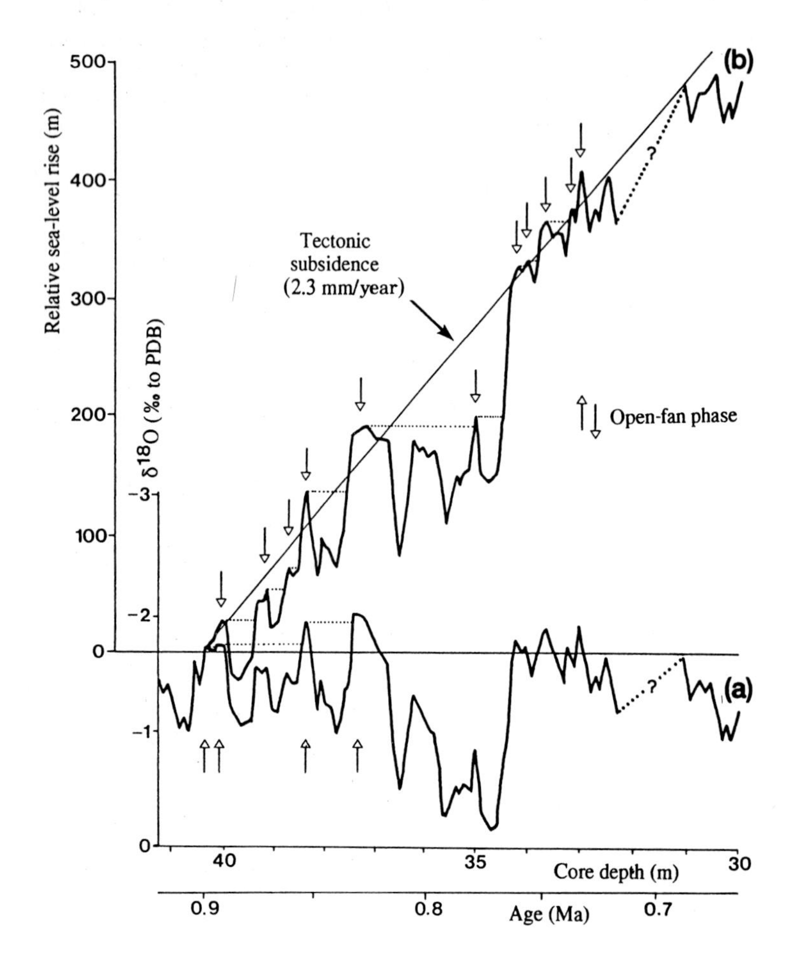

Fig. 8. (a) Oxygen-isotope curve after Shackleton and Hall (1983); core depth and age are correlated, assuming uniform depositional rate of 4.36 mm/1000 years. (b) Relative sea-level curve as arithmetic sum of curve (a) and the estimated local, uniform tectonic subsidence of 2.3 mm year^{-1}. (Simplified from Muto & Blum, 1989.)

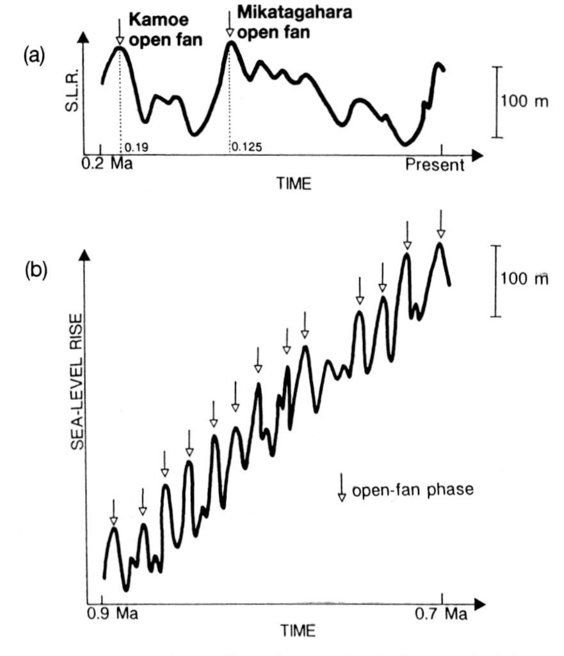

Fig. 9. Comparison of relative sea-level changes in (a) the Tenryugawa system and (b) the Oigawa system. Curve (a) is modified from the eustatic curve in Fig. 7 with the regional tectonic uplift of 0.2 mm year^{-1}. Curve (b), in consideration of Fig. 8(b), is a schematic drawing, where: (i) the number of open-fan phases and the total sea-level rise are reliable; and (ii) sea-level amplitudes are interpreted from the thicknesses of open-fan deposits and the depths of fan valleys.

dissection for 0.125 Ma, i.e. during the last fan-valley interval.

The Mikatagahara open fan is vertically divided into the upper and lower units, which are separated by a sharp to gradational boundary (Fig. 12). The lower unit rests on the Kamoe open fan with an undulating, erosional contact, so that a considerable portion of the latter may have been denuded. The lower unit shows clast-supported texture and consists mainly of imbricated pebble to cobble gravels with loosely packed sandy matrix; no general upward change in clast size is observed in this unit. The upper unit is finer grained than the lower and presents a fining-upward trend, from cobble gravels to a weathered mud layer (Fig. 12). The top mud layer (up to 3 m thick), included in the upper unit, is composed of poorly sorted, massive silt or clay with granules and scattered fine-grained pebbles, mostly near the bottom. The top part of the Kamoe open fan also shows a weathered mud layer (Kobayashi, 1964).

The sea-level control model explains that: (i) the lower unit was formed in the open-fan phase; (ii) the erosional base was caused by a preceding denudation of the Kamoe open fan by overflows in the preceding confined-fan phase; and (iii) the upper fining-upward unit was constructed by overflows of the subsequent fan-valley phase. The top mud layers were probably accumulated directly from suspension in the final overflow event. The overflows during the preceding fan-valley phase are not recorded in the Mikatagahara open fan.

Oigawa

The Ogasayama Formation consists of multiple open-fan sequences (A-type bodies), multiple Gilbert-type deltas prograding into fan valleys (B-type bodies) and a single dissection fill (C-type body) (Muto & Blum, 1989). The open-fan bodies occupy a western, major part of the Ogasayama Formation (Figs 6 & 13). Individual open-fan bodies are less than 20 km in possible width (in NW–SW directions) and range from 20 to 190 m, most less than 90 m thick. These have sheet-like geometry, but are thick and narrow relative to the Mikatagahara and Kamoe open fans. The high-relief geometry of the Oigawa open fans is abundantly accompanied by 4 to 20 m-thick, fine-grained interfan deposits (facies M, see below) yielding a freshwater molluscan fauna (Makiyama & Sakamoto, 1957), erect stumps and vertical rootlets particularly, which are absent in the 0.2 Ma–Recent Tenryugawa system.

The open-fan and associated deposits were described with five facies distinguished on the basis of predominant grain size and of sequence: M (mud, no general upward change), Sc (sand, coarsening-upward), FGf (pebble, fining-upward), FGc (pebble, coarsening-upward) and CG (cobble to boulder, no general upward change). Each facies has lateral dimensions of several to a few tens of metres in thickness and is traceable over a distance of several kilometres (Fig. 13). In most cases the vertical facies sequences exhibit a tripartite organization consisting of a lower, coarsening-upward unit (M/Sc, M/FGc, Sc/FGc, M/Sc → FGc), a middle, coarsest-grained unit with no general upward trend (/CG) and an upper fining-upward unit (→ FGf, → FGf → M), where '/' and '→' mean an abrupt or erosional upward change and a gradational upward change, respectively (Figs 14, 15 & 16). Because individual sequences are commonly truncated by overlying strata, vertical amalgamation of CG

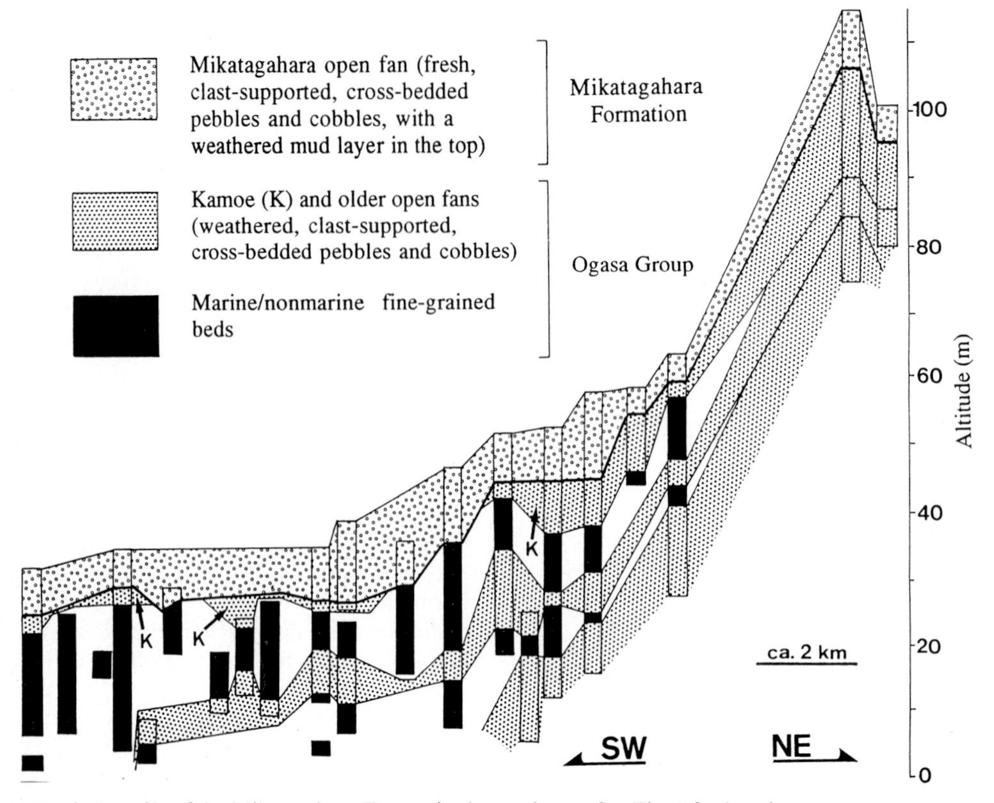

Fig. 10. Vertical profile of the Mikatagahara Terrace in the northwest. See Fig. 5 for location.

bodies can occur locally. Incomplete sequences without CG, such as M/Sc → FGc → FGf → M, were also observed.

The sea-level control model, as summarized in Fig. 17, suggests that: (i) the middle coarsest-grained unit (CG) represents an open-fan phase; (ii) the upper fining-upward unit was formed by overflows in a fan-valley phase; and (iii) the lower coarsening-upward unit was formed by overflows in a confined-fan phase. The overflow sediments in confined-fan phases comprise the lower units of the open-fan sequences.

Fig. 11. Thin and flat geometry of the Mikatagahara open fan, the base of which is indicated with arrows. The substrata represent the Kamoe open fan. Motorcycle for scale is 2 m long. See Fig. 5 for location.

Fig. 12. Bipartite organization of the Mikatagahara open fan. The lower unit (L) is coarser grained than the upper unit (U) and does not conspicuously change upward in predominant grain size, whereas the upper unit fines upward and is topped by a weathered mud layer. The boundary between the lower and upper units is gradational in this outcrop. K, Kamoe open fan. Bar for scale is 1 m. This photograph represents a close-up of Fig. 11.

Fan valleys

Tenryugawa

The Recent Tenryugawa fan valley is the large rectangular depression (11 km wide, 15 km long) between the Mikatagahara and Iwatabara Terraces and drained by the trunk river of the fan system. The fan valley is 10 to 50 times as wide as the contributing canyon, so that the alluvial infill shows a fan-like morphology spreading out southward and confined laterally in the fan valley ('confined fan' of Muto, 1987a; see Figs 1 & 5). A gravel-dominant, braided alluvial system, commonly seen in alluvial fans (e.g. Rust, 1978), is identified from the present micro-landforms of the alluvial-body surface (Kadomura, 1971). Inside the fan valley are three lower and smaller terraces than the Mikatagahara and Iwatabara. These are the Tomioka, Ubagaya and Hamakita Terraces, in descending order, preserving remains of confined-fan bodies which were formed and subsequently dissected in the last fan-valley interval (Fig. 5).

The infilling body of the Recent fan valley thickens seaward to be 90 m thick at the present river mouth (Fig. 18(a)), the vertical dimension reflecting

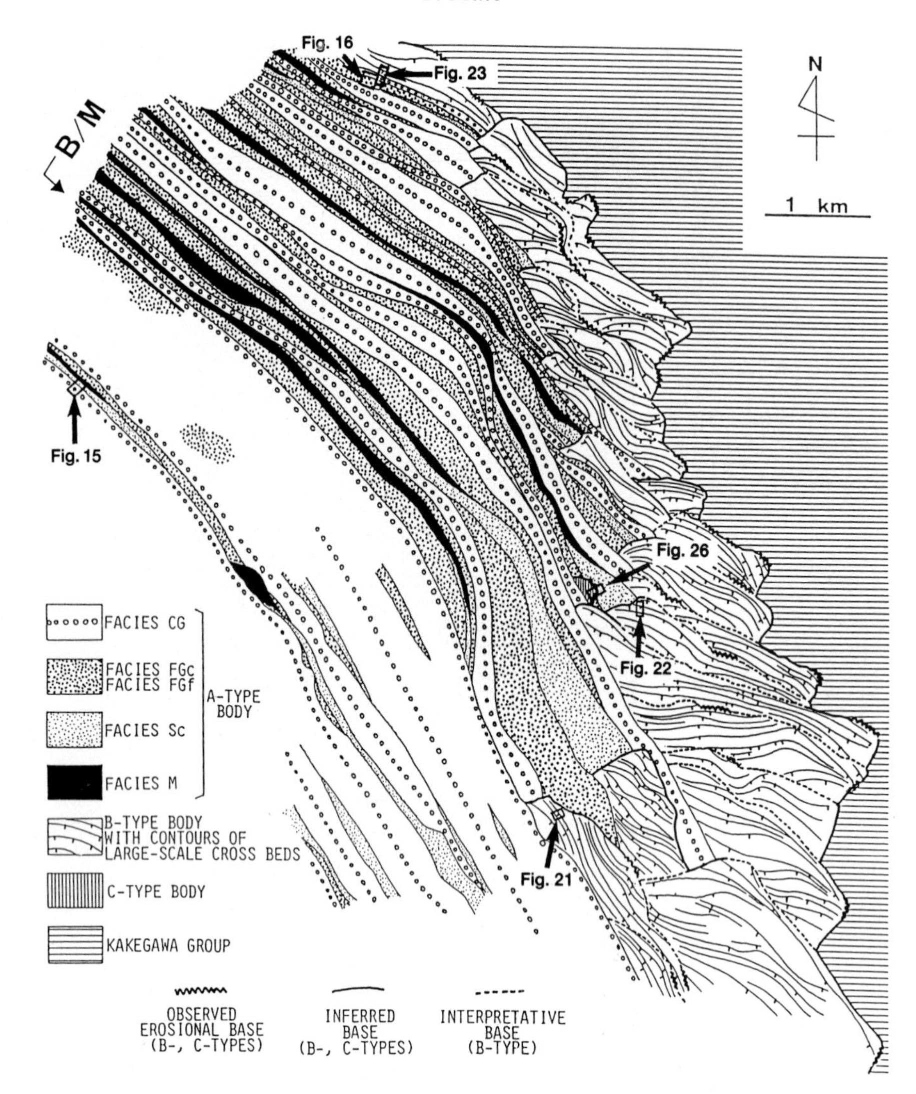

Fig. 13. Lithofacies map of the lower Ogasayama Formation. The A-type bodies include open-fan deposits (CG, FGc, FGf, Sc) and interfan deposits (M); the B-type bodies consist of Gilbert-type deltas prograding into fan valleys; the C-type body is a single dissection fill. Facies boundaries and geological structures are projected into the horizontal plane at sea level at the angle of structural dip (10° SW), in order to correct for topographic distortion. B/M indicates the Brunhes/Matuyama boundary. (After Muto & Blum, 1989.)

the sea-level rise of more than 100 m during the Yurakucho transgression. The body consists of coarse-grained, braided alluvial sediments in the upper and lower parts and a seaward-thickening wedge (over 40 m in maximum thickness) of silt and sandy silt with molluscs (Kobayashi, 1963) in the middle part. Such fine-grained wedges have been recorded from numerous coastal plains of latest Pleistocene to Recent age in Japan, and ascribed to the quick eustatic sea-level rise during the transgression (Kaizuka *et al.*, 1977).

The dissected morphology of the Kamoe open fan, as well as of the Mikatagahara open fan, includes a single fan valley and multiple dissection channels (Muto, 1987a, b; Fig. 19). The ancient fan valley was developed in the Iwatabara Terrace area and vicinity. Its infilling body thickens seaward (Fig. 18(b)) and is inferred to have been 100 m thick

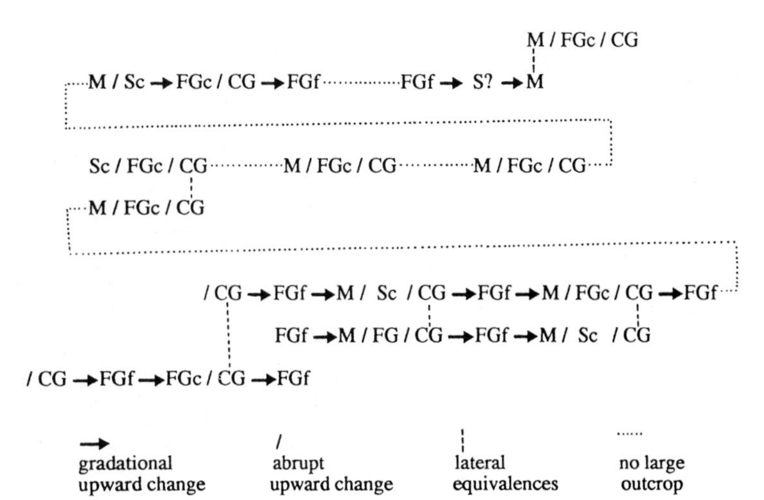

Fig. 14. Facies sequences in the A-type bodies of the lower Ogasayama Formation observed in large outcrops. The upward trend in this figure is left to right. 'S?' indicates sand-rich facies with unclear upward trend in predominant grain size.

at the present shoreline, reflecting the sea-level rise during the Shimosueyoshi transgression.

The precise width of the ancient fan valley is unknown because the walls are not preserved. However, it is clear that the western and eastern walls were, respectively, to the east of the Mikatagahara Terrace and to the west of the Ogasayama Hills (Fig. 19). The inferred lateral dimensions were 11+ km in length and 5 to 15 km in possible width. The main lithology of the ancient fan-valley fill is poorly sorted, clast-supported cobbles and boulders up to 40 cm in diameter, within a loosely packed

Fig. 15. A-type bodies of the Ogasayama Formation showing a facies sequence of M/Sc/CG. Arrows indicate the facies boundaries. Undulations in Sc are convoluted lamina. Bar for scale is 3 m. See Fig. 13 for location.

Fig. 16. A-type bodies of the Ogasayama Formation showing a facies sequence of CG → FGf → FGc/CG. Bar for scale is 5 m. See Fig. 13 for location.

sandy matrix. The gravel beds predominantly show horizontal stratification and trough and planar cross-bedding (Fig. 20), suggesting that major depositional processes took place in braided-channel/bar systems. Based on the lithological, geometrical and geomorphologic evidence, Muto (1987a) interpreted that the ancient fan valley was entirely filled with alluvial confined-fan deposits and underwent no marine invasion such as the Recent fan valley experienced.

Oigawa

Twenty-one fan valleys have been found in the Oigawa system (Fig. 13). The entrenchment position of the fan valleys progressively shifted to the

OBSERVED SEQUENCE OF A-TYPE DEPOSITS	INTERPRETATION	
	DEPOSITIONAL ENVIRONMENT	TIME OF DEPOSITION
FGf	OPEN FAN	OVERFLOW STAGE OF FAN-VALLEY PHASE
CG		OPEN-FAN PHASE
FGc		OVERFLOW STAGE OF CONFINED-FAN PHASE
Sc		
M	INTER-FAN LOWLAND	NON-OVERFLOW STAGE OF CONFINED-FAN PHASE

Fig. 17. An observed facies sequence of the A-type bodies of the Ogasayama Formation, presenting all the facies types, and their interpretation. (After Muto & Blum, 1989.)

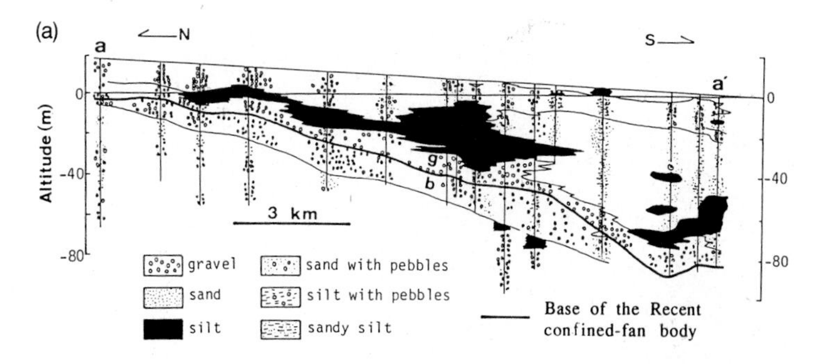

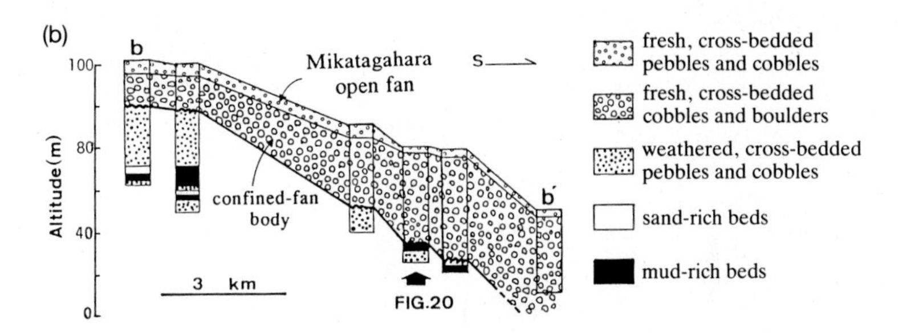

Fig. 18. Longitudinal profiles of the fan-valley fills in the Tenryugawa system. (a) Recent fan-valley fill. (After Kobayashi, 1963.) g = grey or bluish gravels, b = yellowish grey gravels. The strata beneath the base of the confined-fan body is included into the Ogasa Group. (b) Ancient fan-valley fill. See Fig. 5 for each location.

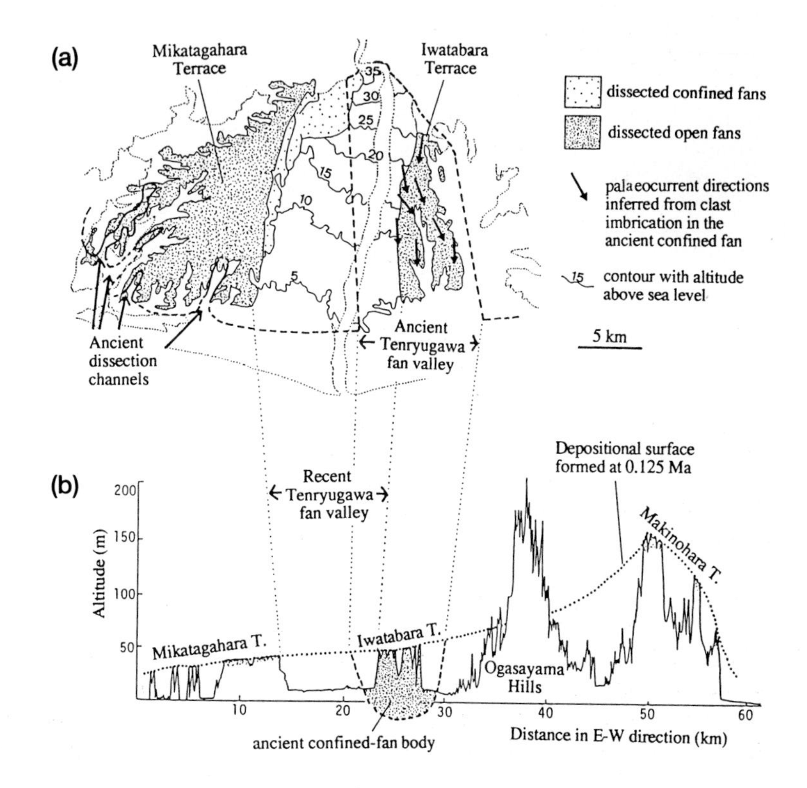

Fig. 19. (a) Present and ancient dissected-fan morphologies in the Tenryugawa system.
(b) Topographic section of the studied coastal region in an E–W direction. See Fig. 3 for location. (Simplified from Muto, 1989.)

Fig. 20. Section of the ancient confined-fan body in the Tenryugawa system. The body consists entirely of cross- and horizontally bedded pebble-to-boulder gravels. Bar for scale is 2 m. See Figs 5 and 18(b) for location.

ESE for adistance of more than 5 km, due to tectonic tilting events in the coastal region (Muto, 1989), so that their development was limited to the eastern narrow portion of the depositional system. The depth of the fan valleys ranges from 40 to 100 m, but in one case attains 190 m. The width, in the NW–SE direction, ranges from 1 to 3 km (mostly up to 2 km), one-tenth to possibly one-half the width of the Tenryugawa fan valleys. Such a comparison may not be valid, however, because the two drainage areas of the systems were independent of each other. Considering a ratio of fan-valley width to open-fan width, its possible range is between 0.2–0.5 for the Tenryugawa system and between 0.07–0.2 for the Oigawa system. The shorter fan-valley intervals certainly correspond with the narrower morphology of the fan valleys.

All the Oigawa fan valleys were filled with coarse-grained Gilbert-type deltas consisting mainly of inclined, pebble to cobble beds with less commonly horizontal beds at top and base (Figs 21, 22 & 23). The dimensions of the foresets correspond with those of fan valleys accommodating the deltas; no complete set has therefore been observed in one exposure. Because of the lateral confinement, the directions of delta progradation which are indicated by dip directions of the foreset beds range from SW to S with a small deviation (Fig. 13).

The sea-level control model predicts that sedimentary infill of a fan valley is the coarsest grained

Fig. 21. Sand-rich topset and foreset beds of the B-type body of the Ogasayama Formation. Arrows indicate a gravel bed traceable from the topset into the foreset. Bar for scale is 2 m. See Fig. 13 for location.

accumulation because the fan valley has the highest river-flow energy in the system. However, the observed fan valleys were filled with finer grained sediments than the open-fan deposits (CG). Muto and Blum (1989) suggested that alternate occurrence of humid and dry climatic conditions associated with the eustatic cycle was responsible for the finer grained infill. Assuming that high and low eustatic sea levels coincided with pluvial and interpluvial periods, respectively, in the study area during the early Pleistocene (Kaizuka, 1962; Suzuki, 1971; Saito, 1985), the pluvial periods corresponded to high river-flow discharge and competence, and vice versa. The fan valleys, relative to the open fans (CG), may have been supplied with a lower volume of sediments of smaller grain size.

The progradation of Gilbert-type deltas within the fan valleys is not explained explicitly by the sea-level control model, too. The progradation requires (i) $Ra/Rs < 1$ at an early stage (to inundate the fan valleys with sea water); and (ii) $Ra/Rs > 1$ or $Rs = 0$ at a later stage, where Ra and Rs are the rates of deposition and sea-level rise, respectively (see Curtis, 1970). The former stage has to be regularly

Fig. 22. Coarse-grained foreset beds of B-type body of the Ogasayama Formation. Bar for scale is 3 m. See Fig. 13 for location.

Fig. 23. Foreset beds of B-type and overlying A-type bodies (CG→FGf) of the Ogasayama Formation. Bright part in the left is a silt-dominated foreset bed. The open-fan sequence represents a lateral equivalence of the lower part of the outcrop shown in Fig. 16. Bar for scale is 2 m. See Fig. 13 for location.

followed by the latter for every confined-fan phase. Muto and Blum (1989) interpreted that, for a constant value of Rs, Ra was small at the beginning of the confined-fan phase and later increased relative to Rs because of the palaeoclimatic control (see Muto & Blum, 1989, for detailed discussion).

Dissection channels

Tenryugawa

The dissection channels in the Recent Tenryugawa system mostly range up to 1.3 km in width, 6 km in length and 30 m in depth; the five largest are 2 to 3 km wide and 6 to 13 km long and have their upstream ends at distances of 10 to 17 km from the mouth of the feeder canyon (Fig. 5). The infilling deposits do not present clear coarsening-upward sequences (Kobayashi, 1963), probably because the large dimensions of the Recent fan valley disfavoured occurrence of overflows.

During the previous fan-valley interval there were four ancient dissection channels at least in the west of the Mikatagahara Terrace area (Fig. 24).

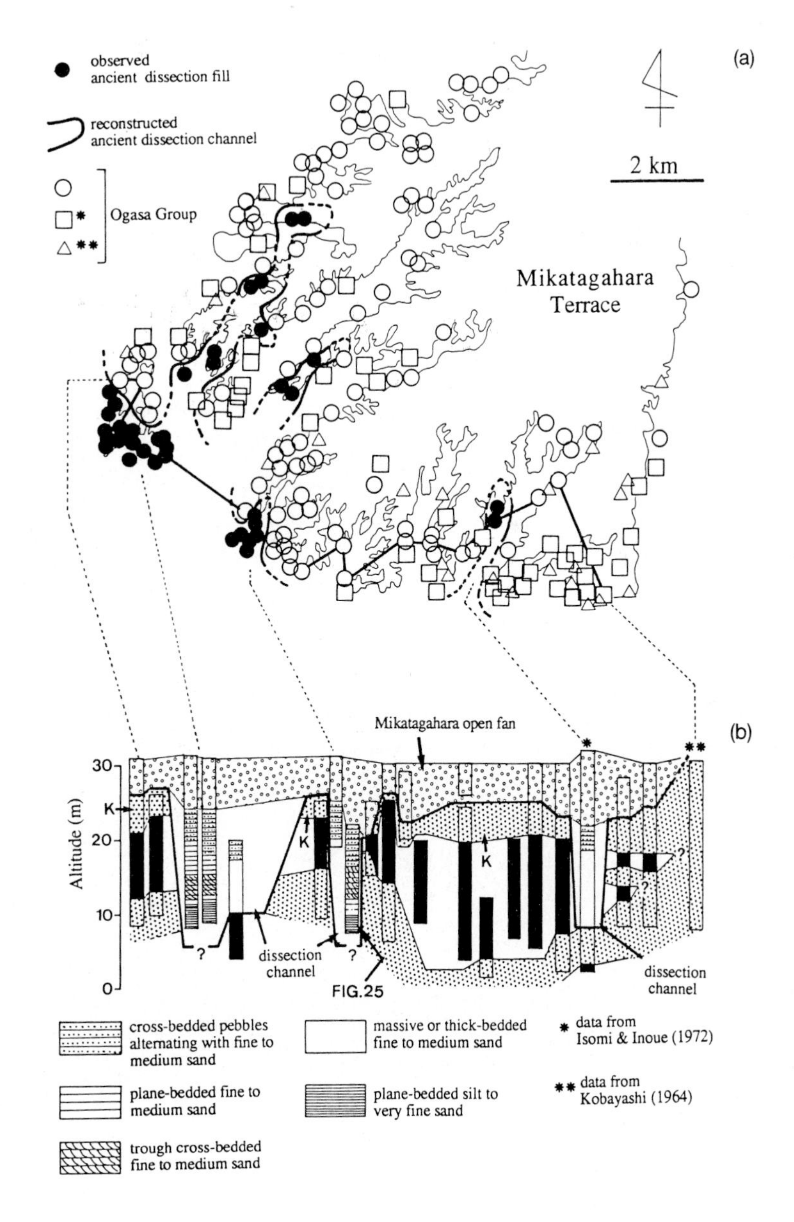

Fig. 24. (a) Spatial lithofacies distribution just below the Mikatagahara open fan.
(b) Vertical profile of the southern part of the Mikatagahara Terrace. See Fig. 10 for symbols which are not explained here. (Modified from Muto, 1987b.)

They were up to 1 km wide, 3 to 11 km long and 20 m deep and had their upstream end at distances of 17 to 22 km from the mouth of the feeder canyon. These were clearly less developed than the Recent dissection channels, reflecting the shorter fan-valley interval. The infilling deposits show a clear coarsening-upward sequence from uniformly fine-grained units yielding marine molluscan shells (Isomi & Inoue, 1972) in the lower unit, to medium to coarse sand beds alternating with granule to cobble beds showing framework-supported texture in the upper unit (Fig. 25).

The sea-level control model explains that: (i) the upper unit was formed by overflows in the confined-fan phase preceding the open-fan phase at 0.125 Ma; and (ii) the lower unit was also accumulated in the confined-fan phase but before the overflow events started.

Oigawa

Only a single dissection channel, 100 m wide (NW–SE) and 20 m deep, has been found from the entire stratigraphic record of the Oigawa system (Figs 13 & 26). The minor development relative to the Tenryugawa system corresponds with the shorter fan-valley intervals. However, the infilling deposits display a coarsening-upward sequence basically the same with the ancient dissection fills in the Tenryugawa system, consisting of: (i) alternating beds of fine-grained sand and mud in the lower unit (Fig. 26); and (ii) horizontally stratified, sandy beds in which pebble beds become thicker and coarser-grained and increase in number upward, in the upper part.

IMPROVED SEA-LEVEL CONTROL MODEL

The Tenryugawa system experienced longer fan-valley intervals than the Oigawa system; in the Tenryugawa system, the Mikatagahara open fan experienced a longer fan-valley interval than the

Fig. 25. Vertical section of the ancient dissection fill in the Tenryugawa system. L, lower, silt-dominated unit; U, upper, coarsening-upward unit (sand- and gravel-dominated); Mk, Mikatagahara open fan. Bar for scale is 2 m. See Figs 5 and 24 for location.

Fig. 26. C-type body of the Ogasayama Formation. Arrows indicate the erosional base of the dissection fill (D), which is underlain by an A-type body (Sc). Hammer 55 cm long. See Fig. 13 for location.

Kamoe open fan. The longer fan-valley interval led to larger lateral dimensions of the fan valley and dissection channels, as already noted. Most of the observed variations in the geomorphic sequences can be explained with the sea-level control model improved so as to contain the concept of the fan-valley interval (Fig. 27).

With a given sediment supply from the feeder canyon and a given amount of incision, a wide fan valley, relative to a narrow one, tends to be associated with a low rate of aggradation (sediment thickness per unit area per unit time) because the sediments are dispersed to cover a relatively large area (cf. Curtis, 1970). In such a situation, the fan-valley infill is more likely to contain subaqueous or marine deposits as the sediment supply is insufficient to suppress marine invasion. This is the case of the Recent Tenryugawa fan valley. The lack of the marine beds in the ancient fan valley resulted from the narrow geometry due to the short fan-valley interval.

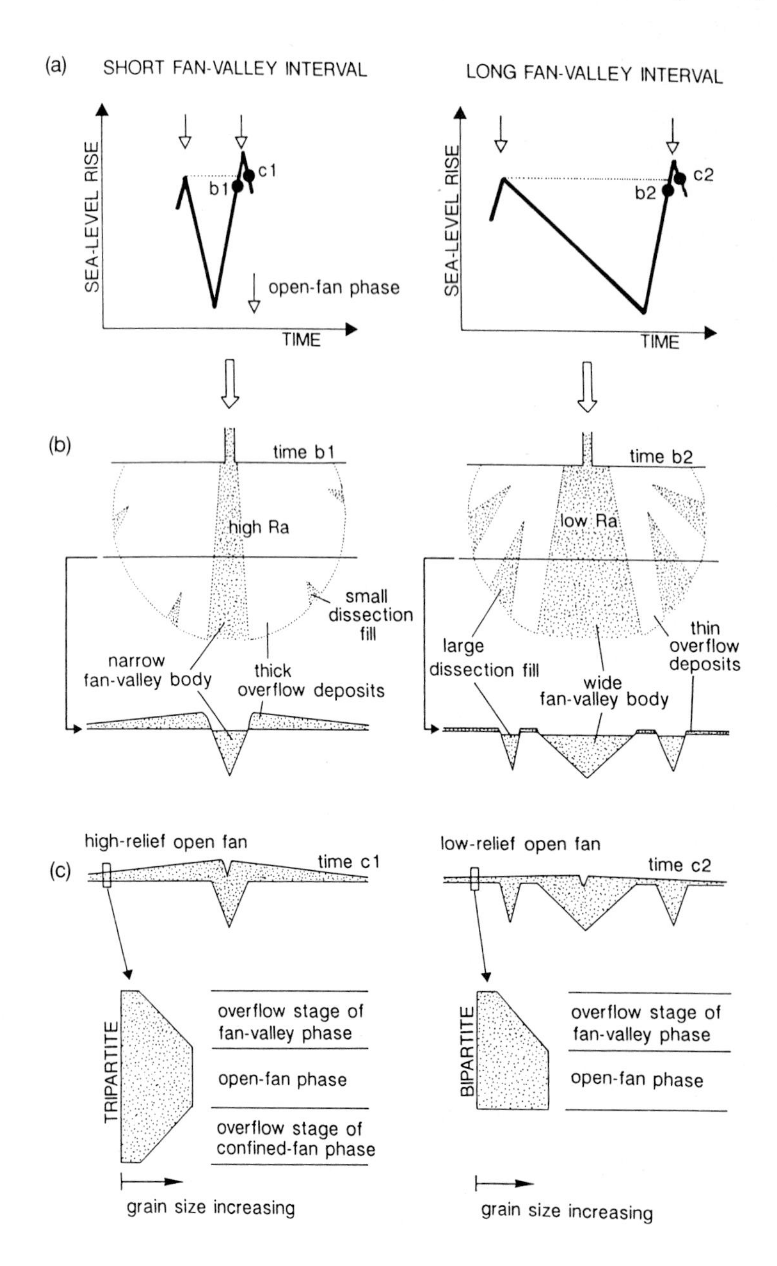

Fig. 27. Improved sea-level control model. (a) Hypothetical sea-level curves with short and long fan-valley intervals. (b) Expected dissectional morphology of the fans, at times b1 and b2. A longer fan-valley interval leads to larger lateral dimensions of a fan valley and dissection channels and minor development of overflow sediments. Ra indicates aggradation rate. (c) Expected transverse profiles of the geomorphic sequences, at times c1 and c2. The open fan has tripartite organization and high-relief geometry after a short fan-valley interval, and bipartite organization and low-relief geometry after a long fan-valley interval.

The dimensions of the dissectional morphology exert a great influence on accumulation sites of overflow sediment. At initiation, the lateral dimensions of the nascent fan valley and dissection channels are inevitably very small, so that overflow sediments are mostly accreted onto the open fan to construct a fining-upward sequence; whereas, at the end of the fan-valley interval, the overflow sediments may be accumulated within enlarged dissection channels. If the interval is long: (i) the sediment is more likely to be trapped in the fan valley; (ii) the valley is less likely to be flooded; and (iii) overflow sediments, if any, will be of small volume and fill only dissection channels. Therefore, the open fans will have a bipartite organization consisting of a lower 'ordinary' open-fan unit (open-fan phase) and an upper, fining-upward overflow unit (fan-valley phase). In general, alluvial fans show a high-relief, semi-conical landform in response to a rapid decrease in transport efficiency as the stream emerges from its confined valley. However, if the fan-valley interval is long relative to the preceding and/or following open-fan phases, most of the sediment goes to bury the fan valley and dissection channels, and cannot contribute to the direct construction of open fans. Therefore, the open fans will have a thin and low-relief geometry, well illustrated by the Mikatagahara open fan.

Conversely, a short fan-valley interval favours thick stacking of overflow sediments just above the existing open fan because of the minor development of dissection channels. Most of the overflow sediments are likely to be preserved as part of the open-fan deposits. As indicated by the Oigawa system, the entire set of the open-fan deposits will present a tripartite organization consisting of a lower, coarsening-upward overflow unit (confined-fan phase), a middle, coarsest grained 'ordinary' open fan (open-fan phase) and upper, fining-upward overflow sediments (fan-valley phase). The supplied sediments will mostly contribute to build up high-relief open fans, because the fan valley and dissection channels have small dimensions.

CONCLUSIONS

From detailed field analyses of the 0.2 Ma–Recent Tenryugawa and 0.9–0.7 Ma Oigawa systems it is suggested that incised valleys form a significant portion of coastal-fan systems situated adjacent to steep submarine slopes. The fan-valley interval, defined as a time interval between two successive open-fan phases, accounts for a considerable part of the observed variations of fan architectures. Such intervals are influenced by relative sea-level changes, i.e. partly by vertical tectonism, and characterized by: (i) lateral dimensions of a fan valley and dissection channels; (ii) occurrence frequency and sediment volume of overflows; (iii) accumulation sites of the overflow sediments; and (iv) internal organization and geometry of open-fan deposits.

The improved sea-level control model, including the concept of the fan-valley interval, is expected to provide a deeper understanding of sequence stratigraphy of coastal fans, when considered together with palaeoclimatic and tectonic controls of the fan sedimentation.

ACKNOWLEDGEMENTS

I very much appreciate discussions with P. Blum and K. Chinzei in the early stage of the study. K. Shefield kindly suggested usage of the term that was newly proposed in this paper. I am grateful to H. Okada for critical reading of an early version of the manuscript. This work was supported in part by a grant-in-aid for scientific research from the Ministry of Education, Science and Culture, Japan (no. 01790343).

REFERENCES

Bardaji, T., Dabrio, C.J., Goy J.L., Somoza, L. & Zazo, C. (1990) Pleistocene fan deltas in southeastern Iberian peninsula: sedimentary controls and sea level changes. In: *Coarse-grained Deltas* (Eds Colella, A. & Prior, D.). Spec. Publ. Int. Assoc. Sediment. **10**, 129–151.

Colella, A. (1988) Fault-controlled marine Gilbert-type fan deltas. *Geology* **16**, 1031–1034.

Curtis, D.M. (1970) Miocene deltaic sedimentation, Louisiana Gulf Coast. In: *Deltaic Sedimentation, Modern and Ancient* (Ed. Morgan, J.P.) Soc. Econ. Mineral. Paleontol. Spec. Publ. **15**, 293–308.

Ethridge, F.G. & Wescott, W.A. (1984) Tectonic setting, recognition, and hydrocarbon reservoir potential of fan-delta deposits. In: *Sedimentology of Gravels and Conglomerates* (Eds Koster, E.H. & Steel, R.J.) Can. Soc. Petrol. Geol. Mem. **10**, 217–235.

Isomi, H. & Inoue, M. (1972) *Geology of the Hamamatsu District. Geol. Surv. Japan* 35 pp. (in Japanese with English abstract).

Kadomura, H. (1971) Micromorphology of low-dip alluvial fans. In: *Alluvial Fans, Regional Variety* (Eds Yazawa, H., Toya, H. & Kaizuka, S.), Kokin Shoin, Tokyo, pp. 85–96. 318 pp. (in Japanese).

Kaizuka, S. (1962) Vegetation of Würm Glacial Age and some climatic terraces in Japan. *Quat. Res. (Tokyo)* **2**, 159–160 (in Japanese with English abstract).

Kaizuka, S., Naruse, Y. & Matsuda, I. (1977) Recent Formations and their basal topography in and around Tokyo Bay, central Japan. *Quat. Res.* **8**, 32–50.

Kazanci, N. (1988) Repetitive deposition of alluvial fan and fan-delta wedges at a fault-controlled margin of the Pleistocene–Holocene Burdur lake graben, southwestern Anatolia, Turkey. In: *Fan Deltas: Sedimentology and Tectonic Settings* (Eds Nemec, W. & Steel, R.J.). Blackie & Son, London, pp. 186–196, 444 pp.

Kobayashi, K. (1963) Epitome of Quaternary history of Hamamatsu and its environs in central Japan. *Fac. Liberal Arts and Sci. Shinshu Univ. part II, Nat. Sci.* **13**, 21–46.

Kobayashi, K. (1964) *Geology of Hamamatsu City*. The Hamamatsu Municipal Office, 165 pp. (in Japanese).

McGowen, J.H. (1970) Gum Hollow fan delta, Nueces Bay, Texas. *Texas Univ. Bur. Econ. Geol. Rept. Inv.* **72**, 57 pp.

Machida, H. (1975) Pleistocene sea level of south Kanto, Japan, analysed by tephrochronology. *Roy. Soc. New Zealand Bull.* **13**, 215–222.

Makiyama, J. & Sakamoto, T. (1957) *Mitsuke and Kakezuka: an Explanatory Text of the Geological Map of Japan* (in scale 1 : 50000). Geol. Surv. Japan, 50 pp. (in Japanese).

Mesolella, J.K., Matthews, R.K., Broecker, W.S. & Thurber, D.L. (1969) The astronomical theory of climatic change, Barbados data. *J. Geol.* **77**, 250–274.

Muto, T. (1985) Fossil submarine channels found from the Pleistocene strata in the Kakegawa district, Shizuoka Prefecture, central Japan. *J. Geol. Soc. Japan* **91**, 439–452 (in Japanese with English abstract).

Muto, T. (1987a) Coastal fan processes controlled by sea level changes: a Quaternary example from the Tenryugawa fan system, Pacific coast of central Japan. *J. Geol.* **95**, 716–724.

Muto, T. (1987b) Geology of the Mikatagahara and Iwatabara Terraces, in the lower Tenryugawa River area, Japan—an interpretation from the present dissected alluvial fan. *J. Geol. Soc. Japan* **93**, 259–273 (in Japanese with English abstract).

Muto T. (1988) Stratigraphical patterns of coastal-fan sedimentation adjacent to high-gradient submarine slopes affected by sea-level changes. In: *Fan Deltas: Sedimentology and Tectonic Settings* (Eds Nemec, W. & Steel, R.J.) Blackie & Son, London, pp. 84–90, 444 pp.

Muto T. (1989) A method of detecting tectonic tilting events from geologic records of coastal alluvial fans. *J. Geol.* **97**, 640–645.

Muto, T. & Blum, P. (1989) An illustration of a sea level control model from a subsiding coastal fan system: Pleistocene Ogasayama Formation, central Japan. *J. Geol.* **97**, 451–463.

Nakagawa, H. (1961) Pleistocene eustasy and glacial chronology along the Pacific coastal region of Japan. *Inst. Geosci. Paleontol. Tohoku Univ.* **54**, 1–61 (in Japanese with English abstract).

Naruse, Y. (1983) Position of the high stands of sea level during the late Pleistocene in Japan. *Osaka Keidai Ronshu,* **152**, 382–412 (in Japanese with English abstract).

Nemec, W. & Steel, R.J. (1988) What is a fan delta and how do we recognize it?. In: *Fan Deltas: Sedimentology and Tectonic Settings* (Eds Nemec, W. & Steel, R.J.), Blackie & Son, London, pp. 3–13, 444 pp.

Postma, G. & Cruickshank, C. (1988) Sedimentology of a late Weichselian to Holocene terraced fan delta, Varangerfjord, northern Norway. In: *Fan Deltas: Sedimentology and Tectonic Settings* (Eds Nemec, W. & Steel, R.J.) Blackie & Son, London, pp. 144–157, 444 pp.

Rust, B. (1978) Depositional models for braided alluvium. In: *Fluvial Sedimentology* (Ed. Miall, A.D.). Can. Soc. Petrol. Geol. Mem. 5, 605–625.

Saito, K. (1985) Comparison between dynamic equilibrium model and climatic linked model for alluvial fans in Japan. *J. Hokkai Gakuen Univ.* **52**, 35–81.

Shackleton, N. & Hall, M.A. (1983) Stable isotope record of Hole 504 sediments: high resolution record of the Pleistocene. *Init. Rep. Deep Sea Drilling Project,* **69**, 431–441.

Shackleton, N. & Opdyke, N.D. (1973) Oxygen isotope and paleomagnetic stratigraphy of Equatorial Pacific core V28-238: oxygen isotope temperatures and ice volume on a 10^5 year and 10^6 year scale. *Quat. Res.* **3**, 39–55.

Shackleton, N. & Opdyke, N.D. (1976) Oxygen-isotope and paleomagnetic stratigraphy of Equatorial Pacific core V28-239 late Pliocene to latest Pleistocene. *Geol. Soc. Am. Mem.* **145**, 449–464.

Sterns, C.E. (1976) Estimates of the position of sea level between 140,000 and 75,000 years ago. *Quat. Res.* **6**, 445–449.

Suzuki, H. (1971) Climatic zones of the Würm Glacial Age. *Bull. Dept. Geog. Univ. Tokyo,* **3**, 35–46.

Tsuchi, R. (1970) Quaternary tectonic map of the Tokai region, the Pacific coast of central Japan. *Rep. Fac. Sci. Shizuoka Univ.* **5**, 103–114.

Ujiié, H. (1962) Geology of the Sagara-Kakegawa sedimentary basin in central Japan. *Sci. Rep. Tokyo Kyoiku Univ. sect. C,* **8**, 1–66.

Spec. Publs Int. Ass. Sediment. (1993) **18**, 93–106

Deep-sea response to eustatic change and significance of gas hydrates for continental margin stratigraphy

B.U. HAQ

National Science Foundation, Washington, DC 20550, USA

ABSTRACT

Eustatic fluctuations on the continental shelf cause the familiar sequence-stratigraphic depositional patterns that result from shifting depocentres as sea level rises and falls. Sedimentary patterns on the basin floor are also influenced, albeit indirectly, by changing sea level. The erosive–corrosive model of deep-sea response to these changes maintains that erosion on the sea floor takes place largely during the sea-level fall and the ensuing lowstand. On the margins, steepened stream gradients induce shelfal incision, enhanced erosion and increased turbidity. Intensified thermal gradients also may lead to climatic deterioration that contributes to enhanced weathering, adding to the sediment load flux. In the deep sea, increased bottom-water activity reinforces its erosive capability. Moreover, while erosion increases during the lowstand phase, dissolution is reduced on the sea floor due to increased influx of carbonate and siliciclastics to the basins which suppresses the CCD (calcite compensation depth) and enhances productivity offshore due to increased nutrient supply. The corrosive phase occurs largely during the transgressive and early highstand phase of the eustatic cycle when accommodation on the shelves is at an optimum. The sequestering of carbonate and clastics nearshore and depletion offshore elevate the CCD and may cause widespread dissolution on the sea floor. Weakened thermal gradients and climatic amelioration induce reduced bottom-water activity, diminishing its erosive capability. In the late highstand phase when accommodation on the shelves and banks is reduced considerably, carbonate and clastics may once again prograde out or be transferred to the deeper basin (late highstand shedding). The suppressed CCD and increased productivity offshore in the late highstand time translate into high preservation potential due to lack of both erosive and corrosive forces.

These depositional patterns may be altered periodically by another process that is forced by fluctuating sea level. The growth and decay of methane hydrate on the margins may significantly influence the long-term preservability of depositional patterns and complicate the sequence-stratigraphic models, especially during the lowstand phase. A major sea-level fall and reduced hydrostatic pressure on the shelf/slope and rise can effect the breakdown of hydrates, substituting a zone with weakened sediment strength where solid hydrate converts to free gas and water. These zones of weakness would be more prone to faulting and/or slumping. Major slumps can release large amounts of methane into the atmosphere. If the preceding sea-level fall is glacially forced, addition of methane from large natural gas pools in the low latitudes could provide a negative feedback to the cooling trend, eventually reversing the course of glaciation. As high-latitude temperatures ameliorate, additional methane may be added rapidly from near-surface sources of these regions, providing a positive feedback to the warming trend, eventually terminating the glacial cycle. Thus, gas hydrates may be an important factor in climatic change, and as agents of tectonic activity along the margins. At present few empirical data are available on hydrates as the field has been largely ignored. A need for considerable research effort is obvious if we are to learn more about this important, perhaps first-order, forcing mechanism for continental margin stratigraphy and tectonics.

INTRODUCTION

Rising and falling seas generate predictable sediment accumulation patterns at the edge of the continents. The combined effects of eustasy (changes in global sea level) and margin subsidence produce relative changes in the base level that controls the location of depocentres and determines the nature of resultant stratal geometries. Climate and orography of sediment source area regulate the rate of weathering and sediment supply, and along with topography of the shelf and slope depocentres, they exert additional controls over the resultant depositional geometries along continental margins. These stratal geometries form the basis of sequence-stratigraphic interpretations.

Eustasy also plays an important, though circuitous, role in determining the pattern of sediment accumulation on the deep-sea floor. Sea-level change can exercise control over deep-sea sedimentation through sequestering or releasing siliciclastics and carbonates along the margins which in turn promotes changes in ocean chemistry, nutrient levels and productivity. Eustasy can also affect climatic change and fluctuations in the vertical and horizontal thermal gradients of the ocean. These in turn cause changes in both surface- and deep-water circulation that regulate erosive action along the margins and in the deep sea.

Sea-level oscillations can directly influence the sedimentary patterns in the deep sea when the basin floor is close to a terrigenous sediment source, and is separated only by a narrow shelf and/or supplied by a point source through incised valleys. Clastic sediments may be preferentially delivered to the basin during the lowstand times, especially if the sediment supply rates are high relative to increasing accommodation space made available by subsidence. Slope failure due to steepened topography or due to destabilization of the gas-hydrate concentration zone following major sea-level falls also may deliver large slumps to the foot of the slope, leaving behind slump-scar unconformities. To a lesser degree, slumping also may occur during highstand times along distal and especially steep parts of the slope. However, when a nearby terrigenous source is lacking, the deeper parts of the basin may not be directly influenced by eustasy. Under these circumstances the basin floor is covered by a relatively uniform blanket of biogenic sediments, that are punctuated by stratigraphic gaps or by dissolution intervals where no fossils can be identified. Many of the dissolutional, as well as non-depositional or erosional, gaps have been documented to be of ocean-wide extent.

Much can be learned about the direct and indirect relationship between the deep-sea depositional processes and the rise and fall of sea level by comparing eustatic cycles with the sedimentary and biotic patterns. Times of significant gaps in sedimentation, changes in the thermal structure of the oceanic hydrographic regimes indicated by stable-isotopic records, as well as biogeographic and evolutionary patterns of marine biota preserved in the palaeontological record, are all important clues to the deep-sea response to eustasy.

DEEP-SEA RECORD VERSUS EUSTATIC CHANGE

We can attempt to discern the relationship between eustasy and deep-sea sedimentation by comparing the age of eustatic cycles with the age of events that control deposition in the deep sea. To accomplish this, four sets of Coenozoic sedimentary, biotic and isotopic data have been compared with eustatic cycles in Fig. 1. These data include the ages of continental margin and deep-sea seismic reflectors inferred to have been caused by widespread dissolution or erosion, known deep-sea hiatuses, periods of active canyon cutting on the shelves, latitudinal migrations of oceanic microplankton through time and palaeoclimatic trends based on benthic foraminiferal oxygen-isotopic data from the Coenozoic. These have been plotted against the Coenozoic sea-level curve of Haq *et al.* (1988). The timing of all events has been converted to the Haq *et al.* (1988) time scale to allow comparison with the eustatic cycles.

The left half of Fig. 1 includes chronostratigraphic and biostratigraphic data as well as sequence stratigraphically derived eustatic curves. The middle column, showing major oceanic seismic reflectors, also includes five important canyon-cutting events along continental shelves. The Neogene reflectors (marked eM-O to IP-G) have been documented by Mayer *et al.* (1986) and occur over large areas in the equatorial Pacific. These reflectors have been linked to carbonate dissolution or diagenesis caused by reorganization of surface- and bottom-water circulation and changes in ocean chemistry. Other plotted reflectors (see Haq, 1991, for sources) include several regionally identifiable seismic horizons in the latest Cretaceous and Palaeogene that have been documented in the western North Atlantic (A*, A^c, A^u/R4 and R3). Major episodes of canyon incision on the continental margins include canyon-cutting events (CCE) in the mid-Palaeocene (PCCE) corresponding to the major fall at 58.5 Ma, in the early Eocene (ECCE) at the major drop at 49.5 Ma, in the mid-Oligocene (MOCCE) at the major fall at 30 Ma, at the Burdigalian sea-level drop at 16.5 Ma (BCCE) and in the late middle Miocene (MCCE) corresponding to the major drop at 10.5 Ma.

The adjacent column shows the Palaeogene (PHa–PHe) and Neogene (NH1A–NH8) gaps in the deep-sea sedimentary record that have been documented on an ocean-wide basis (Barron & Keller, 1982; Keller & Barron, 1983, 1987; Keller *et al.*, 1987). Sedimentary gaps are most often caused by active removal of sediments by erosive bottom currents. They may also be caused by depositional starvation, such as on the outer shelf and slope during rapid shoreline transgression following a sea-level rise and landward movement of depocentres, or by dissolution and diagenesis of predominantly calcareous sediments by shoaling of the calcite compensation depth (CCD). The gaps caused by dissolution are often pervasive, but generally seem to be of shorter duration than those caused by sediment removal. Erosion may conceivably erase the record of several earlier sedimentary sequences, which makes dating the inception of the erosive events uncertain. On Fig. 1 the deep-sea hiatuses have been placed into two categories: (i) those that are dominantly erosional; and (ii) those that are caused by carbonate dissolution.

Figure 1 also delineates calcareous microplankton migrationary events (interpreted as a response to cooling (C) or warming (W) of surface waters) and documented from the Coenozoic Atlantic Ocean (Haq *et al.*, 1977; Haq, 1980). High-latitude assemblages invade the low latitudes during cooling, while tropical and temperate assemblages migrate to higher-latitude areas during warm intervals.

The right-hand column of Fig. 1 includes climatic trends depicted by composite benthic foraminiferal $\delta^{18}O$ curves from the low-latitude Pacific and low- to mid-latitude Atlantic. The decline in palaeotemperature near the Eocene/Oligocene boundary from the relative highs of early Palaeogene is now generally ascribed to the development of a major ice cap on Antarctica. The pre-Oligocene earth, on the other hand, is inferred to have been largely ice free.

Figure 1 reveals several temporal coincidences that suggest causal relationship between eustasy and sedimentary patterns on continental margins and ocean basins. For example, all major cycle (supercycle of second-order cycle) boundaries, at 68.0, 58.5, 49.5, 39.5, 30.0, 21.0, 16.5 and 10.5 Ma, are either associated with major erosional seismic reflectors (unconformities) or with periods of canyon incision on the margins, or both. Where documentation is available, these events also occur near climatic cooling events, indicated by microplankton migrationary patterns and oxygen-isotopic data.

The comparison also reveals that carbonate dissolution/diagenesis-linked seismic reflectors from the equatorial Pacific (Mayer *et al.*, 1986) are associated with times of sea-level rise, transgression of shorelines and the formation of condensed (sediment-starved) sections on the continental shelves. The deep-sea erosional hiatuses caused by bottom-current activity, on the other hand, seem to be centred around times of lowered sea level. Erosional hiatuses commonly are also associated with climatic cooling events, while corrosional hiatuses occur during periods of climatic amelioration.

These comparisons indicate that carbonate dissolution in the deep sea is favoured during times of eustatic rise and the ensuing early highstands. Conversely, removal of sediments along the margins and from the deeper basins is intensified during sea-level fall and the ensuing lowstands. Major falls (supercycle boundaries) also may be associated with widespread canyon incision on the continental shelves.

Models of deep-sea sedimentation and their eustatic connection

Several models have attempted to explain the observed relationships between eustatic high- and lowstands and the deep-sea sedimentary patterns. The basin-shelf carbonate fractionation model of Berger (1970) and Berger and Winterer (1974) attempted to explain increased dissolution and hiatuses during highstands in terms of sequestering carbonate on the continental shelves. Milliman (1974) considered the Pleistocene record and suggested that shallow-water carbonate accumulation during interglacials leads to a deficiency of carbonate in ocean waters, raising the CCD and increasing deep-sea dissolution. During the glacial times when the sea level was lowered and shallow areas were subaerially exposed, the transfer of carbonate to the oceans lowered the CCD and reduced deep-sea dissolution.

Subaerial exposure of the shelf and the basinward shift in depositional patterns accompanying a fall of sea level were believed to be responsible for the development of unconformities on the shelf, slope and the ocean floor by Vail *et al.* (1980). Based on an analysis of palaeontological data, Poag and Ward

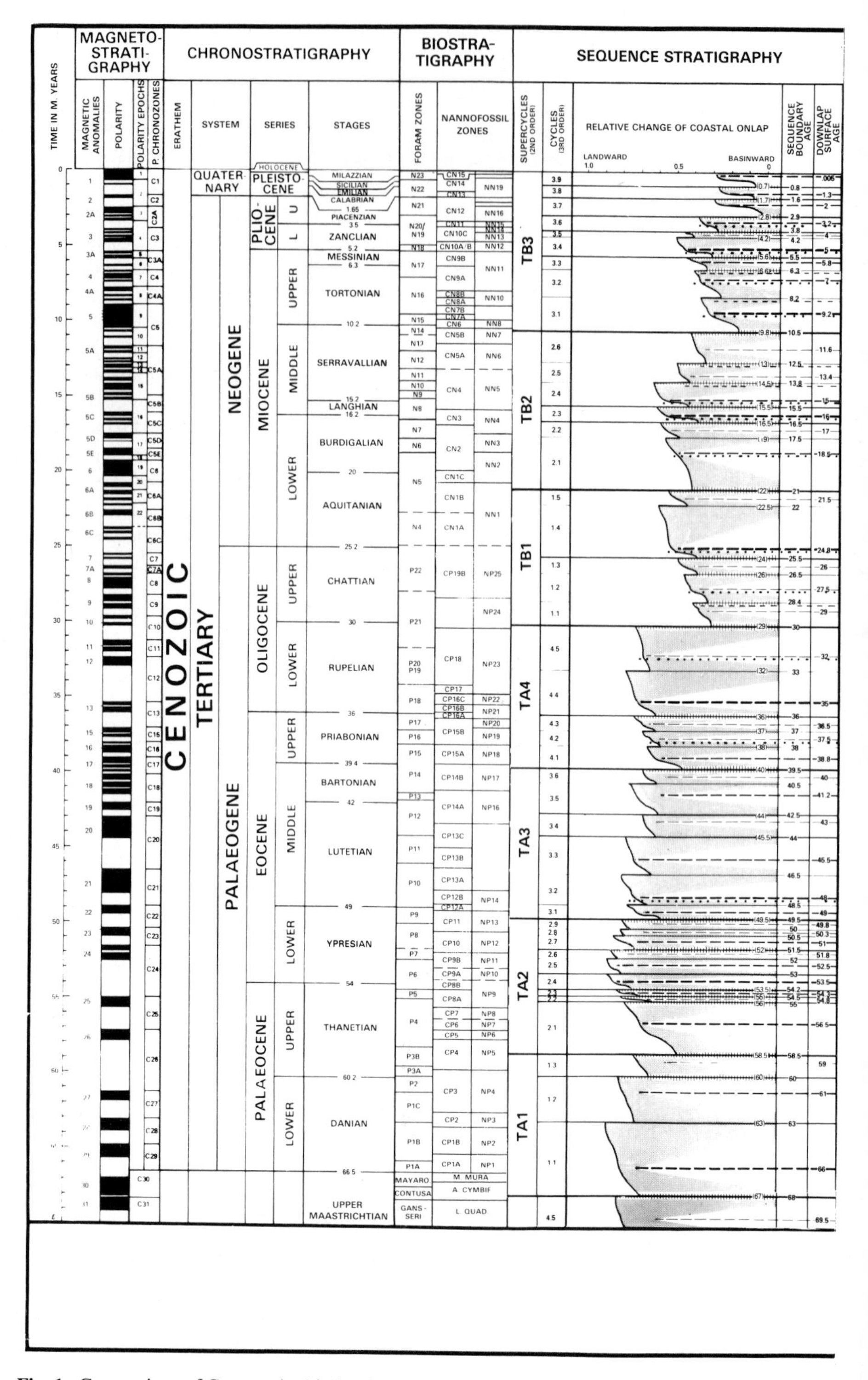

Fig. 1. Comparison of Coenozoic third-order sea-level changes with deep-sea reflectors (diagenetic or erosional horizons), major canyon-cutting events (PCCE, ECCE, MOCCE, BCCE and MCCE), times of known ocean-wide erosional or dissolution events (NH8–NH1A and PHa–PHe),

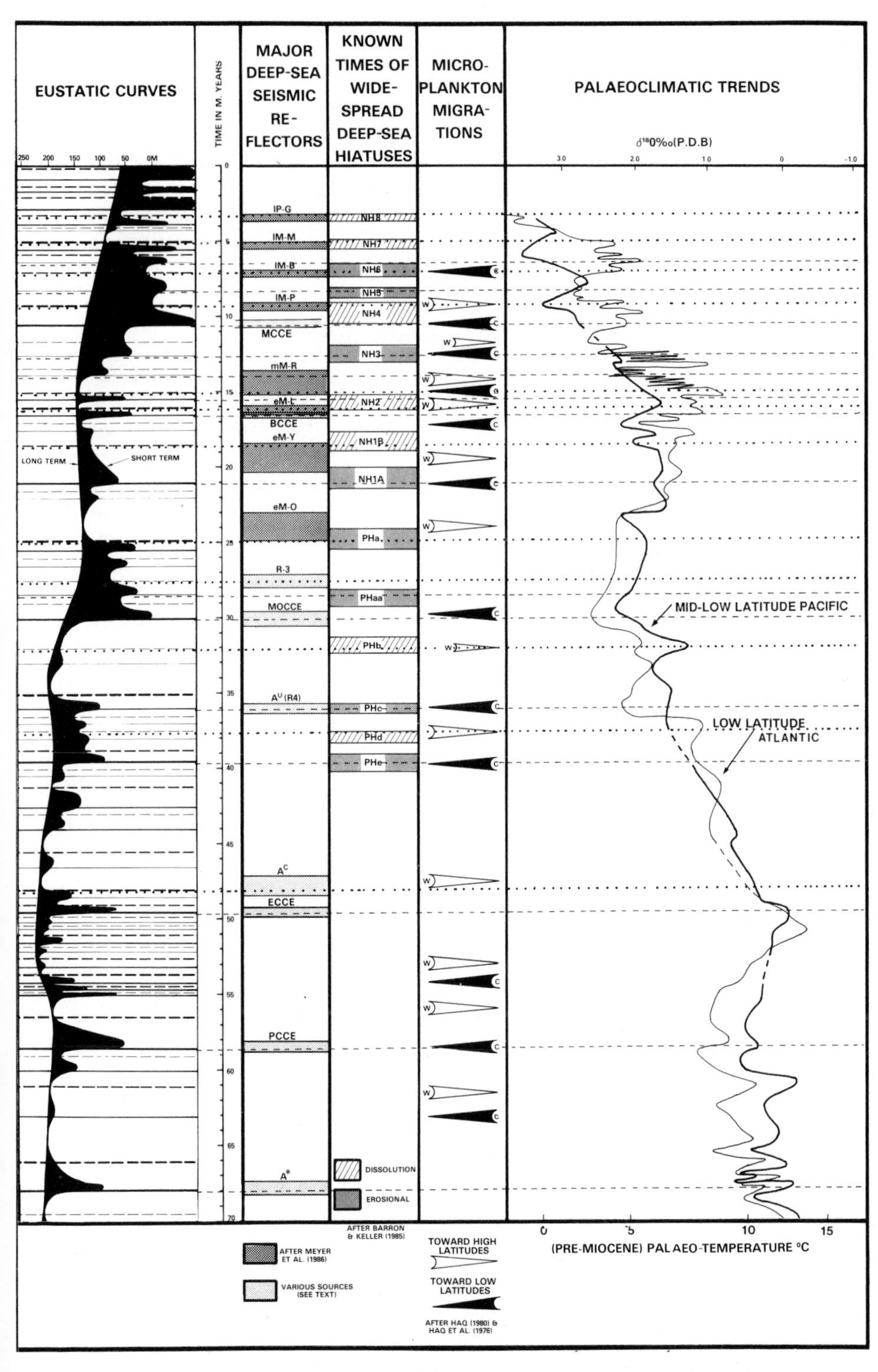

calcareous microplankton latitudinal migrations (interpreted as a response to cooling or warming of surface waters) and benthic oxygen-isotopic curves showing palaeoclimatic trends from low-latitude Pacific and mid to low latitude Atlantic. (After Haq, 1991. For sources of data see op. cit.)

(1987) showed a general agreement in the timing of major sea-level fall events and the development of unconformities on the US and Irish margins of the North Atlantic Basin. Haq *et al.* (1987) explained the observed patterns of sediment removal from the sea floor during lowstands and dissolution during highstands partially through a climatic feedback mechanism. They discussed earlier models, i.e. that highstands lead to sequestering of clastics and carbonates on the inner shelves, shoaling of the CCD and increased dissolution in the deep basins. To this they added that elevated sea levels also cause climatic equitability, weakening of latitudinal and vertical thermal gradients and reduced bottom-water current activity, all of which lead to decreased potential for erosional hiatuses on the sea floor. During sea-level falls, on the other hand, terrigenous as well as carbonate detritus bypass the shelves into the basins, leading to increased carbonate in the oceans and a depressed CCD. In addition, inequitable climates and strengthened thermal gradients during lowstands enhance bottom-water production and circulation causing greater erosion on the sea floor.

Open-ocean microplankton show an apparent empirical relationship to eustatic changes. Haq (1973) offered quantitative data from calcareous nannoplankton that demonstrated that plankton diversity increased during highstands and subsequent expansion of epicontinental seas, while widespread orogenic events and falls of sea level were followed by reductions in the number of taxa. Smaller temperature contrasts and equitable global climates during the highstands and the expansion of ecospace were seen to favour growth and speciation. Major eustatic falls, on the other hand, were argued to result in greater thermal contrast, extreme climate and loss of ecospace, all of which are conditions less favourable for the growth and speciation of marine plankton.

An empirical relationship between times of high seastands, higher diversities amongst pelagic organisms, and increased community complexity was also seen by Fischer and Arthur (1977). They also ascribed these *polytaxic* times to more equitable climates, reduced oceanic convection and greater niche availability. The opposite relationship was seen for periods of lowstands when sharper temperature gradients and increased current activity lead to lower diversity *oligotaxic* assemblages. These authors argued that productivity increased in the open ocean during lowstands and on the epicontinental seas during highstands.

A more inclusive, climatic feedback model of the influence of eustatic changes on deep-sea sedimentary patterns was outlined by the author (Haq, 1991) and is also discussed below. This discussion draws upon many of the earlier models, but also takes into consideration the observations offered by sequence-stratigraphic analyses of continental-margin sections in carbonate, siliciclastic and mixed systems.

The erosive–corrosive cycle model

In its simplest scenario three phases can be recognized within the erosive–corrosive cycle: (i) a largely erosive phase that dominates the period of sea-level fall and lowstand; (ii) a corrosive phase that is activated after the first major transgression of the shelf and continues into the early highstand time; and (iii) a preservational phase that encompasses the late highstand time when accommodation on the shelves and bank tops is at a minimum. However, this cycle may be altered periodically by large-scale slumps along the margins during lowstand times due to the destabilization of the gas hydrates that occur within continental margin sediments (see discussion on pp. 102–105).

Erosive phase

Erosion along the continental margins and on the basin floor occurs predominantly during the fall of sea level and the ensuing lowstand. Erosion is enhanced during these times in two ways (Fig. 2):

1 The steepened stream gradients and the increased sediment load of the incised valley system during falling sea level leads to greater erosive potential and increased frequency of turbidity currents. Major sea-level falls also may be associated with canyon cutting on the margins or reactivation of existing canyons.

Canyon cutting into the shelf and slope is more likely to be initiated during major sea-level falls. For example, seven of the largest submarine canyons in the world have been documented to have been cut into the continental margins of the Bering Sea during the late-Coenozoic glacial lowstand times (Carlson & Karl, 1988). The heavy sediment load during falling sea level induces turbidity currents with greater erosive power to cut and deepen canyons. During lowstand time the canyons can channel turbidity currents and debris flows to the basin floor. Canyon cutting also can be initiated through localized slope failure of unconsolidated sediments and subsequent headward erosion until the depression breaches the shelf break gaining access to the shelf sediments, which may further cut the depression (Farre *et al.*, 1983). The modern Mississippi canyon is inferred to have formed following retrogressive slope failure and mass slumping of sediments (Coleman *et al.*, 1983), most likely during the glacial lowstand time.

During lowstands, siliciclastics bypass the shelves to the basins through the incised valleys. If the shelf is exposed subaerially, erosion of the shelf top adds to the sediment load being transferred to the basin. Thus, potentially large amounts of sediment can be removed from the shelf/slope and redistributed during lowstands. Slope erosion due to sediment creep, gravity-driven slope failure and mass movement of sediments downslope also contributes to the redistribution process.

Lowered sea level and reduced hydrostatic pressure on the shelf and slope also are thought to be responsible for the degradation of gas hydrates within the continental margin sediments. The loss of mechanical strength in sediments due to hydrate destabilization may cause major slumps along the margins during lowstand times which can complicate the overall picture during the erosive phase of the cycle (see more detailed discussion of gas hydrates on pp. 102–106).

2 Erosion in the deep sea is also enhanced through climatic deterioration and the intensification of thermal gradients (Fig. 2). Increased albedo during lowstands favours extreme climates, and this in turn leads to enhanced thermal contrast between land and sea, between high and low latitudes and between surface and bottom water. Such extreme climates also increase weathering and erosion on land. The coincidence of lowstands and cooling climates (glacials, since late Neogene) would also enhance the production of cold, potentially erosive bottom waters in polar regions, which can remove

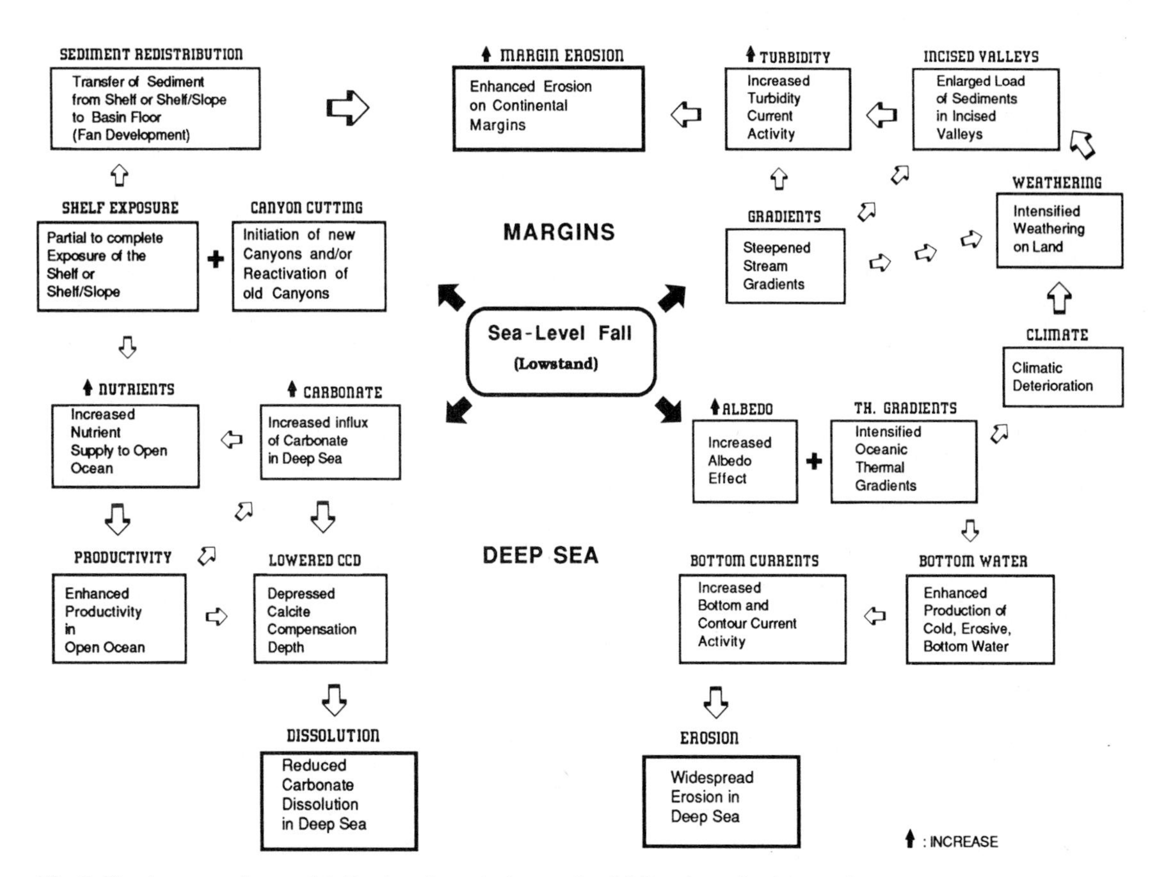

Fig. 2. Erosive–corrosive model. Erosive phase during sea-level fall and ensuing lowstand.

the extensive volume of sediments in their pathways.

Increased influx of carbonate in the deep sea during lowstands would depress the CCD (Fig. 2) and favour decreased dissolution. The influx of terrigenous material into the basins results in increased nutrient supply and thus increased productivity over the open ocean which can help to suppress the CCD even further.

This pattern changes abruptly when the eustatic sea level begins to rise rapidly and the shelves are flooded, cutting off the terrigenous sediment supply and reducing the carbonate influx into the basin. This reverses the climate feedback loop at the same time.

The erosive phase of the erosive–corrosive cycle explains the observation that erosional hiatuses are centred around major, type 1 sequence boundaries (see Haq, 1991) and canyon incision in the shelf is initiated during major falls of sea level, i.e. around second-order cycle (supercycle) boundaries.

Corrosive phase

Following the first major transgression and flooding of the shelf, the sediment-supply spigot is turned off as the depocentres move landward. The transgressive deposits consist of landward shifting, retrogradational sediment packages, representing consecutive flooding events of the shelf. As the transgression proceeds, terrigenous sediments are sequestered increasingly in more landward positions, and hemipelagic sediments begin to blanket the shelf at the trailing (seaward) edge. Whilst the shelf is sediment starved during maximum flooding, a condensed section is created on the outer shelf and slope (Fig. 3). During this time pelagic or hemipelagic sediments may be deposited over broad areas of the shelf. The relative duration of the sediment starvation within the condensed section increases basinward, until beyond the area of terrigenous influence, the biogenic deposition in the deeper basins can be thought of as a series of

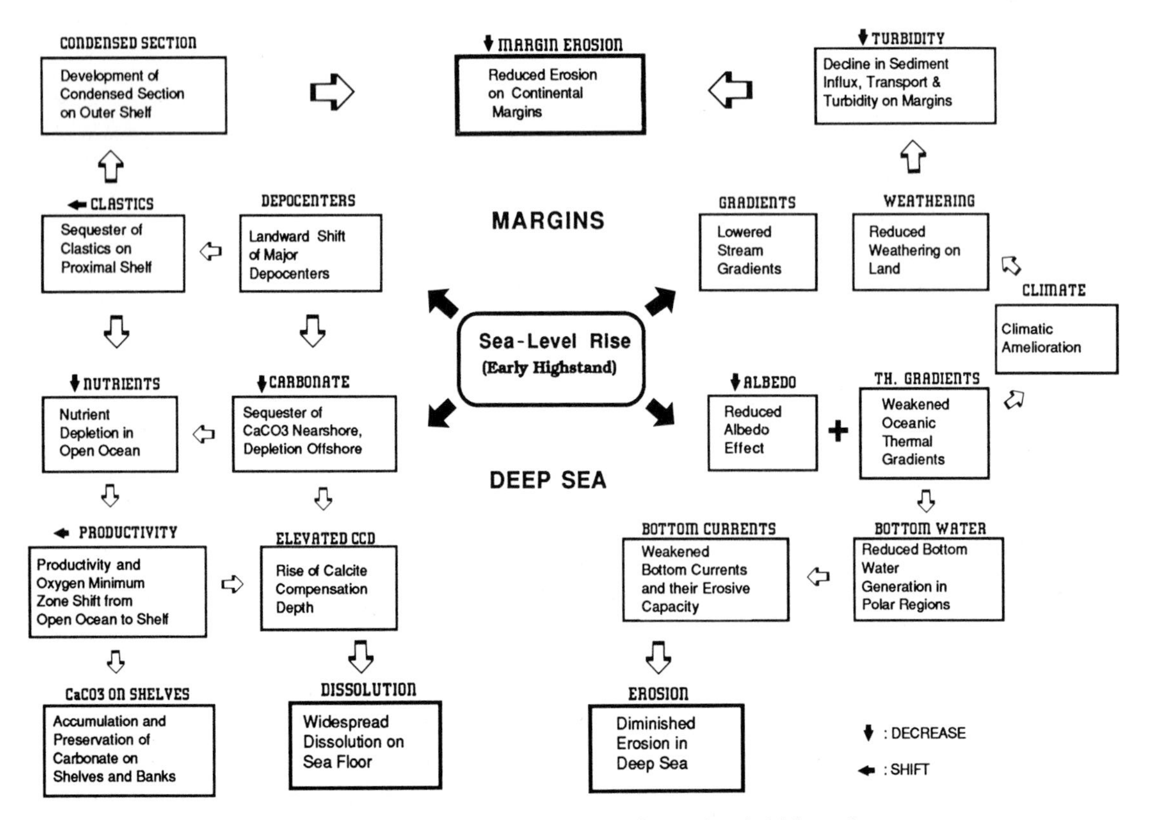

Fig. 3. Erosive–corrosive model. Corrosive phase during transgression and early highstand.

stacked condensed sections (Haq *et al.*, 1988). Sediment starvation on the shelf also may concentrate authigenic minerals (glauconite, phosphorite, siderite), pelagic fossils (diverse planktonic and benthic assemblages), organic matter and trace elements (Loutit *et al.*, 1988). In carbonate systems the condensed section is often a shaly micrite package with submarine hardgrounds. In well-oxygenated conditions over the shelf, a slower sea-level rise allows the carbonate production to keep pace with the increasing accommodation space and skeletal, grainstone-rich 'keep-up' carbonates are deposited. In conditions that are adverse to productivity, or if the sea level rises more rapidly, 'catch-up'-type carbonates may be deposited that cannot keep pace with the rate of sea-level rise. When the sea level rises too rapidly, the carbonate system may 'give up' (Neumann & Macintyre, 1985), often leaving behind well-cemented hardgrounds.

The sequestering of siliciclastics and carbonate on the proximal shelves depletes both the carbonate and nutrient budgets of the ocean (Fig. 3). The locus of productivity shifts from the open ocean to the shelves where carbonate and organic matter accumulate in large quantities. The decreased carbonate in the open ocean elevates the CCD, favouring widespread carbonate dissolution on the sea floor. The deposition in the early highstand phase is initially aggradational, but as accommodation space nearshore is filled, the aggradational depositional pattern changes to progradational as the shoreline begins to regress in the late highstand time. The highstand deposits are widespread on the shelf, often prograding out beyond the shelf break.

Preservational phase

The open-ocean carbonate-deficiency phase lasts only through the period of transgression and early phase of high seastand. In the late highstand time, when the sediments prograde out over the distal part of the shelf, beyond the shelf break or shed off the banks and platforms, the carbonate flux to the deep sea increases once again, and the locus of productivity moves offshore (Fig. 4). Thus, the early to late highstand is the transition period from predominant dissolution in the deep sea to enhanced preservation and relatively high rates of sediment accumulation.

The sea-level rise and decreased albedo also lead to weakening of vertical and latitudinal thermal gradients during the highstand. The reduced thermal contrast and coincidence of highstands with periods of global warming lower both the potential for bottom-water generation in the polar regions and the intensity of bottom currents. The lack of terrigenous influx along the margins and the cut-back in bottom-current activity favours reduced erosion along the margins and the deep sea.

During sea-level rise and maximum flooding of the shelves the oceanic oxygen-minimum zone may rise enough to abut the shelf in some areas. The increased productivity of the shelves, combined

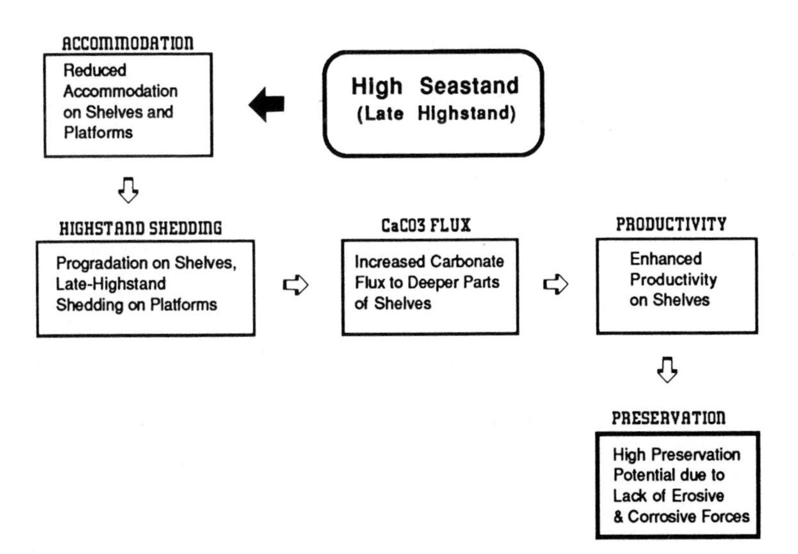

Fig. 4. Preservational phase during late highstand of sea level.

with the greater preservational potential for organic matter in the expanded oxygen-minimum zone, makes the transgressive and the early highstand times the most likely phases of the sea-level cycle for the accumulation of hydrocarbon source sediments.

The above scenario explains the general coincidence of dissolution-related reflectors and hiatuses in the deep sea with the periods of transgression and early highstand. For the reasons argued above, late highstands are perhaps the most favourable times for deep-sea sedimentation adjacent to continental margins. The record of these intervals holds the highest potential for being preserved relatively intact when both erosive and corrosive forces are considerably weakened (Fig. 4).

The erosive–corrosive model conceivably can operate at the third- as well as higher order cycle levels, and should be testable, particularly for the last glacio-eustatic event where the sequence-stratigraphic data as well as chronostratigraphic resolution are adequate to differentiate between the various phases of the sea-level cycle.

The above discussion contains the implicit assumption that relative sea-level change and responding shifts of depocentres nearshore and in the deep sea are the main cause and effect processes that determine the ultimate stratigraphic patterns preserved on continental margins. As mentioned earlier, these depositional patterns may be significantly altered periodically by another mechanism which is also forced by fluctuating sea level and the resultant changes in hydrostatic pressure on the shelf and rise, namely, the decay of gas hydrates during lowstand time (discussed below). This process may affect the long-term preservability of the depositional patterns of accumulated sediments on continental margins, and complicate sequence-stratigraphic models, especially during the lowstand phase of the depositional cycle.

HOW GAS HYDRATES MODIFY CONTINENTAL MARGIN SEDIMENTARY RECORD

Gas hydrates

Gas hydrate (clathrate) is the solid ice-like crystalline phase of natural gas (largely methane) and water which is stable under moderately low-temperature and high-pressure conditions and occurs within sea-floor sediments. The low-temperature–high-pressure conditions necessary for stability of gas hydrates are met in a high percentage of the sea floor where water depth exceeds 300 m. The gas-hydrate zone within the sediment column extends from the sea floor down to several hundred metres to 1000 m sub-bottom. At especially high latitudes of the Arctic (and presumably the Antarctic) where water temperatures are sufficiently low, hydrates may form in marine sediments at shallower depth. On land at these latitudes they are known to exist near the surface in association with permafrost. Rather high gas content is required to form hydrates and sediments with high organic content are more amenable to the production of large quantities of methane by bacterial degradation of buried organic matter. Thus, the limiting factor for gas-hydrate occurrence may be the generation of gas needed in large enough quantities to stabilize the clathrate structure (Kvenvolden & Barnard, 1983).

Gas hydrates can be detected on continental margins through the appearance of the so-called bottom-simulating reflections (BSR) on seismic-reflection profiles, which delineate the base of the hydrate concentration zone. However, BSRs may not be observed in a large number of cases where hydrates may be present. BSRs represent diagenetic boundaries and often cut across local depositional surfaces (Tucholke *et al.*, 1977). The acoustic velocity change at the BSR is inferred to mark the transition between gas-hydrate cemented sediments above and water-filled uncemented or gas-charged sediments below (Dillon & Paull, 1983). BSRs have been mapped over wide areas of the margins around the world. However, clathrates have only been sampled rarely and the estimates of methane trapped within gas-hydrate zones or the free gas below them remain speculative. Unanswered questions about BSRs include: Do BSRs always mark the transition between solid phase and free gas? How thick are the gas-hydrate zones above the BSR and do they extend (as is often assumed) to the sea floor? Are large quantities of free methane trapped below BSRs (which would make gas hydrates function as a stratigraphic seal for significant natural gas reservoirs)? These and other questions need to be answered before BSRs can be used for a more meaningful survey of the size of the methane reservoir in gas hydrates.

The lack of direct observations and sampling makes estimates of total methane trapped in gas hydrates somewhat conjectural. These estimates range from 1.7×10^3 to 41.1×10^4 Gt (1 gigaton $= 10^{15}$ g) of methane in gas hydrates guesstimated by various authorities. Kvenvolden (1988) cites widely differing estimates of total methane carbon in gas hydrates (between 2×10^3 to 4×10^6 Gt) but favours a conservative estimate of 1×10^4 Gt of carbon sequestered in methane hydrates on continental margins. This estimate of gas-hydrate carbon exceeds estimates of carbon from other sources and is therefore potentially a significant portion of total carbon within the shallow geosphere (Kvenvolden, 1988).

Gas-hydrate instability and implications for climate and sedimentary patterns

The special temperature–pressure relationship necessary for the stability of gas hydrates implies that any major change in either of these controlling factors will tend to alter the zone of hydrate stability. For example, a significant drop in eustatic sea level will reduce the hydrostatic pressure on the shelf/slope and rise, altering the temperature–pressure regime, leading to destabilization of the gas hydrates. Recently, Paull *et al.* (1991) have offered a model of sea-level fall associated with the Pleistocene glaciation, leading to gas-hydrate instability and major slumping on the continental margins. They envision a sea-level drop of *c.* 120 m during the Pleistocene, reducing hydrostatic pressure sufficiently to raise the lower limit of gas hydrates by an estimated 20 m (Dillon & Paull, 1983). The ensuing destabilization created a zone of weakness where sedimentary failure could take place, leading to major slumps worldwide. They ascribe the occurrence of common Pleistocene slumps on the sea floor to this mechanism. The slumping also could be accompanied by the liberation of a large volume of methane trapped below the level of the slump, injecting a significant amount of this greenhouse gas into the atmosphere. The volume of methane released would increase with the frequency of slumps as glaciation progresses. This would eventually trigger a negative feedback to advancing glaciation, eventually terminating the glacial cycle. Paull *et al.* (1991) suggested that the abrupt nature of Pleistocene glacial terminations may be due to this process.

The glacially forced sea-level lowering leading to slumping and release of methane providing negative feedback to glaciation can at first function effectively only in tropical to temperate latitudes. At higher latitudes glacially induced freezing would tend to delay the negative-feedback effect, but once deglaciation begins, even a relatively small increase in atmospheric temperature of the higher latitudes could cause release of methane from near-surface sources, providing a positive feedback to warming. Nisbet (1990) suggested that a small triggering event and liberation of one or more Arctic gas pools could initiate massive release of methane from the permafrost. The strong positive feedback would provide increased emissions of methane and accelerated warming. Nisbet ascribes the abrupt nature of the younger Dryas termination to such an event and contends that gas hydrates may play a dominant role, more important than ocean degassing, in recharging the biosphere with CO_2 at the end of glaciation.

The negative–positive feedback loop

The initially delayed effect of sea-level fall in the high versus low latitudes constitutes a negative–positive feedback loop that could be an effective mechanism for terminating ice ages. Glaciation is believed to be initiated by Milankovitch orbital forcing, a mechanism that also can explain the broad variations in glacial cycles, but not the relatively abrupt terminations. Oceanic degassing of CO_2 alone cannot explain the relatively rapid switch from glacials to interglacials (Nisbet, 1990). It may be that a combined effect of lowstand-induced slumping and methane emissions in low latitudes triggers a negative feedback to glaciation as suggested by Paull *et al.* (1991) and the ensuing positive feedback to warming in higher latitudes and further release of methane from near-surface sources as envisioned by Nisbet (1990). The two mechanisms would reinforce a relatively rapid termination of the glacial cycle.

For the optimal functioning of the negative–positive feedback model, methane would have to be constantly replenished during the switchover from new and larger sources. Although methane is nearly ten times as effective as CO_2 (by weight) as a greenhouse gas, its residence time in the atmosphere is relatively short (of the order of a decade and a half), after which it oxidizes to CO_2 and water (Lashof & Ahuja, 1990). CO_2 accounts for up to 80% of the contribution to greenhouse warming in the atmosphere. The atmospheric retention of CO_2 is somewhat more complex because it is readily transferred to other reservoirs, such as oceans and the biota, from which it can re-enter the atmosphere. Lashof and Ahuja (1990) estimate an effective average residence time of *c.* 230 years for CO_2. These retention times are short enough such that, for cumulative impact of methane and CO_2 through the negative–positive feedback loop to be effective, methane would have to be continuously replenished from gas-hydrate and permafrost sources.

Sea-level fall and gas hydrate related massive slumping

These observations open an intriguing possibility that past sea-level lowerings, especially major eustatic falls, may have been accompanied by massive slumping along the margins. Slumping need not be forced by gravity-driven slope failure alone, but development of zones of weakness within the sedimentary column after the breakdown of gas hydrates may also contribute significantly to tectonic rearrangement of sediment wedges on the margins.

In a simple scenario, the breakdown of gas hydrate in lower latitudes occurs after a major sea-level drop and in response to a drastic reduction in hydrostatic pressure on the shelf/slope and rise. This causes the lower limit of the hydrate zone to migrate upward by an amount that depends on the overall change in the temperature–pressure regime that determines the clathrate stability. The base of the gas hydrate is the first to destabilize because it is at the limit of stability, below which the geothermal gradient increases more rapidly. Where the solid clathrate turns into free gas and water it generates a zone of greatly decreased sediment strength which can act as a lubricated horizon that is more prone to faulting and block slumping. Weakening of mechanical strength of sediments leading to megaslumps may be an important first-order mechanism for tectonic activity on continental margins. This may be evidenced in the Carolina Trough area by the association of slump features and numerous faults that sole out at or above the BSRs (Paull *et al.*, 1989).

Timing of gas-hydrate development

When did the gas hydrates first develop? The special low-temperature–high-pressure requirement for the stability of gas hydrates suggests that they have existed at least since the latest Eocene, the timing of the first development of the modern oceanic psychrosphere and cold bottom waters. Prior to that, bottom waters in the world ocean are inferred to have been relatively warm even in the higher latitudes. This raises questions about the presence of hydrates in pre-psychrospheric times. Does this imply that gas hydrates are a relatively recent, post-psychrospheric, phenomenon? Or could hydrostatic pressure alone have maintained the clathrate stability? According to the gas hydrate stability window (Kvenvolden & Barnard, 1983) it is apparent that bottom-water temperature need not be very low, but instead the geothermal gradient within the sediments and the hydrostatic pressure above would be more critical for clathrate stability. Clathrates could exist on the slope and rise when bottom waters reach temperatures estimated for late Cretaceous and Palaeogene (*c.* 7 to 10°C), though they would occur deeper within the sedimentary column and would be relatively thinner.

In the pre-Oligocene there were no large ice caps, and the mechanism for short-term sea-level changes also remains uncertain. However, the Mesozoic–early Cainozoic eustatic history is replete with major sea-level falls of 100 m or more that are comparable in magnitude, if not in frequency, to glacially induced eustatic changes (see Haq *et al.*, 1988). If gas hydrates existed in the pre-Oligocene, major sea-level falls would imply that hydrate destabilization may have contributed significantly to shallow-seated tectonics along continental margins.

Is there geological evidence of increased frequency of slumping associated with major falls in the pre-Oligocene sea level that can be ascribed to gas-hydrate breakdown? As a test case we could look for such evidence in a careful study of buried canyons, slumps and erosion along the New Jersey margin of the US based on seismic data (Mountain, 1987). It has been obvious that sediment accumulation and preservation patterns along this margin (as along any other continental margin) resulted from a complex interaction of sea-level changes, fluctuating sediment supply rates and subsidence and along-margin and abyssal current flow. However, perhaps destabilization and movement of sediment wedges caused by gas-hydrate breakdown during lowstand times were equally important.

Mountain (1987) documents four periods of Palaeogene slope failure, slumping and infilling along the slope of the New Jersey margin, i.e. near the Cretaceous/Tertiary boundary, at the Palaeocene/Eocene boundary, at the top of the Lower

Eocene and in the Middle Eocene. In addition, Mountain and Tucholke (1985) have shown a widespread unconformity near the Eocene/Oligocene boundary which wiped out much of the Oligocene stratigraphic record. These unconformities have little or no shallow-water debris resting on them and all these events occur very close to major sea-level lowstands. The events near the Palaeocene/Eocene boundary and at the top of the lower Eocene also are associated with major slumps. The latter is associated with clear evidence of megaslump, which is compositionally similar to enclosing sediments and may have travelled a few kilometres downslope to its present position. Mountain (1987) ascribed the slope detachment and slumping to diagenesis and/or local faulting. However, this downward movement of a sediment wedge with original bedding still largely intact is more readily explained by gas-hydrate destabilization following lowered sea level and reduced hydrostatic pressure. Mountain (1987) wonders what process could be responsible for unconformities that appear to develop simultaneously on shelf and rise. Slump-scar unconformities caused by downslope movement of sediment blocks over lubricated horizons of destabilized gas hydrates would produce just such an effect.

In the non-glacial world, for methane emissions related to sea-level fall to have been effective agents of global climate change over the long term (of the order of millions of years), methane must be replenished continuously over a long period. This implies that the total duration of lowstand would be an important factor in determining the long-term effect on climatic change — long, sustained lowstands would cause continued and increasingly frequent slumps and release of methane, leading to climatic change.

As examples, two Palaeogene sea-level falls could be examined. A major sea-level drop of estimated 110 to 130 m occurred near the termination of early Eocene (the 49.5 Ma event of Haq *et al.*, 1987) but was relatively short lived, lasting *c.* 0.5 Ma. In contrast, the major sea-level fall in the middle Oligocene (the 30 Ma event of Haq *et al.*, 1987) is estimated to be *c.* 150 to 180 m and the sea level then fluctuated for several Ma, remaining lower than the levels in early Oligocene, before reaching a high in the Aquitanian. The first event already has been shown to be associated with megaslump features on the New Jersey margin (the top of the lower Eocene event of Mountain, 1987). The mid-Oligocene event should be associated with a significantly larger number of slumps along ancient continental margins and should have had a more lasting effect on the climate. Mountain and Tucholke (1985) have suggested that extensive slumping they see resting over the regional A^u unconformity beneath Blake Outer Ridge is lower-to-middle Oligocene in age, implying an association with the mid-Oligocene event.

A re-examination of seismic and stratigraphic data along continental margins for evidence of normal and growth faults and major slumping and sliding within gas-hydrate field depths could reveal causal connections between hydrates and sedimentary tectonic processes hitherto unaccounted for.

CONCLUSIONS

It is likely that gas hydrates have been an important factor in continental margin stratigraphy. As agents of periodic slumping and block sliding they may have played a very significant role in modifying sequence-stratigraphic patterns, particularly those of the lowstand phase of the sequence cycle.

The role of gas-hydrate breakdown in the rearrangement of sediment wedges is obviously a complex issue. Some of the ideas concerning this role have been touched upon here, but many major questions remain unanswered. Yet there has been little research on this issue of fundamental importance to sedimentary geology. A better understanding of gas hydrates may well show their considerable role in modifying continental margin stratigraphy and tectonics, as well as in global climatic change, and through it, as agents of biotic evolution. Greater research efforts are warranted to unravel the enigma of gas hydrates which may also prove to be an enormous untapped energy resource for future utilization.

ACKNOWLEDGEMENTS

The author benefited from discussions about gas hydrates with many colleagues, especially Bill Dillon, Art Green and Charlie Paull. The paper was reviewed by Bill Dillon, John Milliman, Greg Mountain and Charlie Paull which improved the quality of the text. The input of all of these colleagues is gratefully acknowledged.

REFERENCES

Barron J.A. & Keller, G. (1982) Widespread Miocene deep-sea hiatuses: coincidence with periods of global cooling. *Geology* **10**, 577–581.

Berger, W. (1970) Biogenous deep-sea sediments: fractionation by deep sea circulation. *Bull. Geol. Soc. Am.* **81**, 1385–1402.

Berger, W. & Winterer, E.L. (1974) Plate stratigraphy and the fluctuating carbonate line. *Int. Assoc. Sediment.* **1**, 11–48.

Carlson, P.R. & Karl, H.A. (1988) Development of large submarine canyons in the Bering Sea, indicated by morphologic, seismic, and sedimentologic characteristics. *Bull. Geol. Soc. Am.* **100**, 1594–1615.

Coleman, J.M., Prior, D.B. & Lindsay, J.F. (1983) Deltaic influences on shelf-edge instability processes. *Spec. Publ. Soc. Econ. Paleontol. Mineral.* **33**, 121–137.

Dillon, W.P. & Paull, C.K. (1983) Marine gas hydrates II: geophysical evidence. In: *Natural Gas Hydrates, Properties, Occurrence, Recovery* (Ed. Cox, L.L.) Butterworth, Woburn, MA, pp. 73–90.

Farre, J.A., McGregor, B.A., Ryan, W.B.F. & Robb, J.M. (1983) Breaching the shelfbreak: passage from youthful to mature phase in submarine canyon evolution. *Spec. Publ. Soc. Econ. Paleontol. Mineral.* **33**, 25–39.

Fischer, A.G. & Arthur, M.A. (1977) Secular variations in the pelagic realm. *Spec. Publ. Soc. Econ. Paleontol. Mineral.* **25**, 19–50.

Haq, B.U. (1973) Transgressions, climatic change and the diversity of calcareous nannoplankton. *Mar. Geol.* **15**, M25–M30.

Haq, B.U. (1980) Biogeographic history of Miocene calcareous nannoplankton and paleoceanography of the Atlantic Ocean. *Micropaleontology* **26**, 414–443.

Haq, B.U. (1991) Sequence stratigraphy, sea-level change, and significance for the deep sea. *Spec. Publ. Int. Assoc. Sediment.* **12**, 3–39.

Haq, B.U., Hardenbol, J. & Vail, P.R. (1987) Chronology of fluctuating sea levels since the Triassic. *Science,* **235**, 1156–1167.

Haq, B.U., Hardenbol, J. & Vail, P.R. (1988) Mesozoic and Cenozoic chronostratigraphy and eustatic cycles. *Spec. Publ. Soc. Econ. Paleontol. Mineral.* **42**, 71–108.

Haq, B.U., Premoli-Silva, I. & Lohmann, G.P. (1977) Calcareous plankton paleobiogeographic evidence for major climatic fluctuations in the Early Cenozoic Atlantic Ocean. *J. Geophys. Res.* **82**, 3861–3876.

Keller, G. & Barron, J.A. (1983) Paleoceanographic implications of Miocene hiatuses. *Bull Geol. Soc. Am.* **94**, 590–613.

Keller, G. & Barron, J.A. (1987) Paleodepth distribution of Neogene deep-sea hiatuses. *Paleo-oceanography* **2** (6), 697–713.

Keller, G., Herbert, T., Dorsey, R., D'Hondt, S., Johnson, M. & Chi, W.R. (1987) Global distribution of Paleogene hiatuses. *Geology* **15**, 199–203.

Kvenvolden, K.A. (1988) Methane hydrates — a major reservoir of carbon in shallow geosphere. *Chem. Geol.* **71**, 41–51.

Kvenvolden, K.A. & Barnard, L.A. (1983) Hydrates of natural gas in continental margins. *Mem. Am. Assoc. Petrol. Geol.* **34**, 631–640.

Lashof, D.A. & Ahuja, D.R. (1990) Relative contribution of greenhouse gas emissions to global warming. *Nature* **344**, 529–531.

Loutit, T., Hardenbol, J., Vail, P.R. & Baum, G.R. (1988) Condensed sections: the key to age dating and correlation of continental margin sequences. *Spec. Publ. Soc. Econ. Paleontol. Mineral.* **42**, 183–215.

Mayer, L.A., Shipley, T.H. & Winterer, E.L. (1986) Equatorial Pacific seismic reflectors as indicators of global oceanographic events. *Science* **233**, 761–764.

Milliman, J.D. (1974) *Marine Carbonates.* Springer Verlag, Heidelberg, pp. 242–249.

Mountain, G.S. (1987) Cenozoic margin construction and destruction offshore New Jersey. Cushman Found. *Foraminiferal Res. Spec. Publ.* **24**, 57–83.

Mountain, G.S. & Tucholke, B.E. (1985) Mesozoic and Cenozoic geology of the U.S. Atlantic continental slope and rise. In: *Geologic Evolution of the United States Atlantic Margin* (Ed. Poag, W.C.) Van Nostrand Reinhold, New York, pp. 293–341.

Neumann, A.C. & Macintyre, I.G. (1985) Reef response to sea level rise: keep-up, catch-up or give-up. *Fifth Int. Coral Reef Congress Proc.* **3**, 105–110.

Nisbet, E.G. (1990) The end of ice age. *Can. J. Earth Sci.* **27**, 148–157.

Paull, C.K., Schmuck, E.A., Chanton, J., Mannheim, F.T. & Bralower, T.J. (1989) Carolina Trough diapirs: salt or shale? *EOS Trans. Am. Geophys. U.* (Abstr.) **70**, 370.

Paull, C.K., Ussler, W. & Dillon, W.P. (1991) Is the extent of glaciation limited by marine gas hydrates. *Geophys. Res. Lett.* **18**, 432–434.

Poag, C.W. & Ward, L.W. (1987) Cenozoic unconformities and depositional supersequences of North Atlantic continental margins: testing of the Vail model. *Geology,* **15**, 159–162.

Tucholke, B.E., Bryan, G.M. & Ewing, J.I. (1977) Gas hydrate horizon detected in seismic reflection-profiler data from western North Atlantic. *Bull. Am. Assoc. Petrol. Geol.* **61**, 698–707.

Vail, P.R., Mitchum, R.M., Shipley, T.H. & Buffler, R.T. (1980) Unconformities of the North Atlantic. *Phil. Trans. Roy. Soc. London* **A294**, 137–155.

Sequence-stratigraphic Methods and Tools

Spec. Publs Int. Ass. Sediment. (1993) **18**, 109–123

Foraminifera indicators of systems tracts and global unconformities

J. REY*, R. CUBAYNES*, A. QAJOUN* *and* C. RUGET†

**Laboratoire de Géologie sédimentaire et Paléontologie et UA 1405 du CNRS 'Stratigraphie séquentielle et Micropaléontologie', Université Paul-Sabatier, 39 allées Jules Guesde, F-31062 Toulouse Cedex, France; and*
†Centre de Paléontologie stratigraphique et Paléoécologie, UA 11 du CNRS et Laboratoire de Micropaléontologie, Université Catholique, 25 rue du Plat, F-69288 Lyon Cedex 02, France

ABSTRACT

Successive systems tracts and global unconformities of third-order depositional sequences can be determined by foraminiferal analysis. Several examples of benthic assemblages studied in the middle and upper Liassic sediments in the southeast Aquitaine basin show that various criteria can be used, including abundance of different species, the presence and development of various genera having uniserial (*Citharina*) or uncoiling shape (*Marginulinopsis*, *Falsopalmula*) and small-size coiled *Lenticulina*. These groups monitor signs of a stable environment occurring during a high sea-level phase (transgressive systems tracts and the lower part of highstand systems tracts). Reduced abundance of species, predominant coiled morphotypes, large-size Lenticulines and robust or primitive species usually mean a disturbed environment occurring during a phase of relatively low sea level (shelf margin wedge systems tracts and the upper part of highstand systems tracts).

INTRODUCTION

In marine-shelf environments it is sometimes difficult to distinguish the different systems tracts of a depositional sequence (Vail *et al.*, 1984, 1987; Posamentier & Vail, 1988; Sarg, 1988). This is particularly true when dealing with homogeneous lithology, intermittent data and irregular dating elements as no real isochronous dividing line can be drawn. Therefore it is most helpful to use the various markers of high and low sea-level phases contained within the strata to confirm interpretations. In this context benthic foraminifera seem to provide particularly reliable data as they are very sensitive to any change in environment. The validity of this concept has been checked by studying the distribution of benthic foraminiferal associations in deposits where the cycles of eustatic rise and fall of sea level were already well known (Cubaynes *et al.*, 1989a). After encouraging preliminary results (Cubaynes *et al.*, 1989b, 1990a, 1990b) comparing well with parallel studies in more open (Olsson, 1988) or deeper environments (Magniez-Jannin & Jacquin, 1990), this paper aims to develop this theory with more complete micropalaeontological data obtained from benthic populations in the Domerian and Toarcian of the Quercy area (southeast Aquitaine).

THE STUDIED DATA

This study is based on the inventory and analysis of more than 60,000 specimens extracted by sieving and washing 69 different samples from marls (16 in the Domerian, 53 in the Toarcian). The benthic foraminifera assemblage consists essentially of *Rotaliina* belonging to the Nodosariid family. Three morphologically different groups can be distinguished (Fig. 1):

1 Planispiral, completely coiled forms: genus *Lenticulina sensu stricto.*

(a) Planispiral coiled Forms
Lenticulina sensu-stricto
(b) Uncoiling Forms
Morphogenera of Lenticulina sensu-lato
Astacolus
Marginulinopsis
Falsopalmula
Planularia

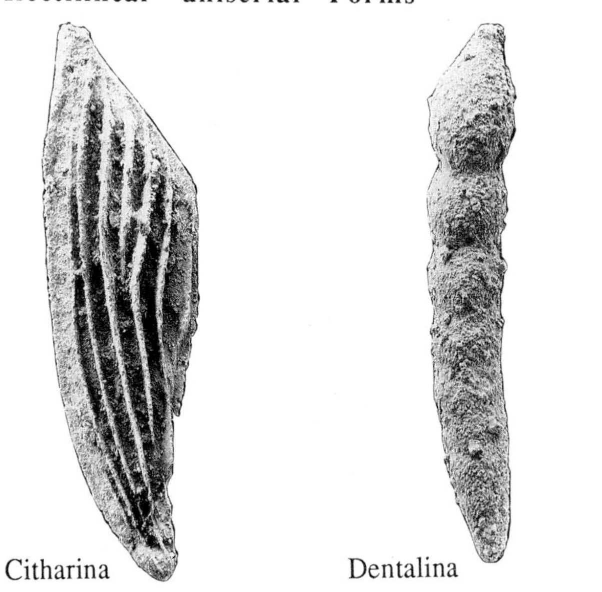
(c) Rectilinear uniserial Forms
Citharina
Dentalina

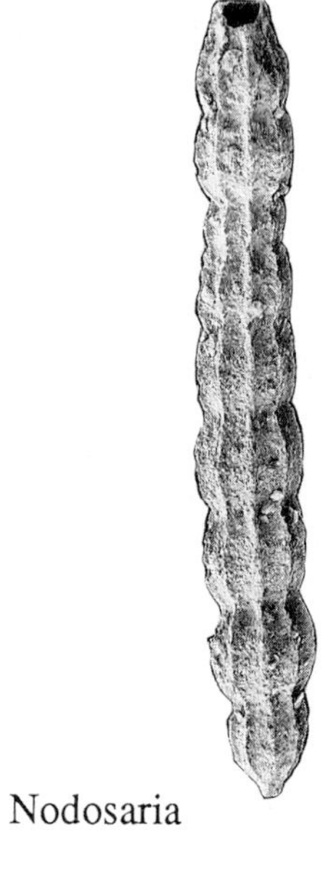
Nodosaria

2 Uncoiling forms: morphogenera *Marginulinopsis, Falsopalmula, Astacolus, Planularia*, within the genus *Lenticulina sensu lato* (Ruget, 1985).
3 Rectilinear forms, uniserial (genera *Citharina, Dentalina, Nodosaria, Pseudonodosaria, Lingulina, Ichthyolaria*) or biserial (*Bolivina*).
The species was determined for each form. After a systematic count of the specimens belonging to each species, we were able to evaluate:
1 The abundance of different species within each sampling level indicating the faunal diversity. The lack of information regarding the rate of sedimentation for each level does not allow the faunal abundance to be determined accurately.
2 The ratio of different genera and morphogenera to each other.
3 The limits of appearance and disappearance of each species.
4 The renewal rate of species (ratio of new species to the total number of species).

A biometric study was also carried out on the *Lenticulina sensu stricto* in the Toarcian, by mathematical analysis of form and size as in the 'Eigenshape-analysis' method. The analysis (Lohman, 1983; Granlund, 1986) consisted of the application of an algorithm to obtain pairs of coordinates at regular intervals around a contour beginning at a homologous point. In our study, this point was the contact of the opening with the previous coil. A model of the average composition of populations for each sampling point was obtained using this type of analysis. The usual norms (De Wever *et al.*, 1989) were respected and statistics were based on no fewer than 30 individual forms per sample group.

EXAMPLE: THE QUERCY DOMERIAN

Stratigraphy, systems tracts and depositional sequences

In the southeast Aquitaine basin (Fig. 2), lower and middle Domerian (Stokesi and Margaritatus zones) consists mainly of a clayey formation, the Valeyre Formation (Fig. 3), which is 40 m thick in the representative Boulbène cross-section (Cubaynes, 1986). This formation overlays alternate levels of limestone and marls (Brian de Vere Formation), and can be divided into two members:
1 The 'grey argillites member' at the base (Stokesi zone; the lower part of the Margaritatus zone).
2 The 'marls with slope taphosequence member' at the top (upper part of the Margaritatus zone and the base of the Spinatum zone).

At the boundary between the Stokesi and Margaritatus zones in the upper part of the grey argillites there is an ammonite-rich ferruginous crust of regional extent (unconformity Db, Cubaynes *et al.*, 1989a) indicating an interruption in sedimentation. The vertical progression of crinoiid associations up to this ferruginous crust (*Chladocrinus*, then *Chladocrinus* and *Balanocrinus*, then *Balanocrinus* alone) suggests, by comparison with recent data, a phase of gradually deepening deposits, during which mudflats developed around the coast. 2.5 m above the Db discontinuity, the 'marls with slope taphosequences member' indicate disturbed sediments and shallower deposits (tempestites?).

In the Grésigne area to the southwest (Fig. 2), the Valeyre Formation has a normal thickness of 35 m which, however, thins out towards the north–northeast (25 m at Figeac, 15 m at Capdenac), at the edge of the emerged Hercynian basement, as a result of the disappearance of the lower and middle 'marls with slope taphosequences'.

The palaeoecological indications and the stratal patterns point to an interpretation of the 'grey argillites' as a superposed transgressive systems tract and highstand systems tract. The Db discontinuity marks the condensed surface. The 'marls with slope taphosequences' indicate a shelf margin wedge systems tract (the lower and middle part of the member) and a transgressive systems tract (the upper part). The boundary between the 'grey argillites' member and the 'marls with slope taphosequences' member would coincide with the sequence boundary between the UAB 4-1 and UAB 4-2 cycles (Haq *et al.*, 1988).

Foraminifera in the Domerian systems tracts

The studied samples were taken from the representative cross-section of Boulbène in the Grésigne area (Fig. 3) where the sedimentary environment is deepest. The populations consist essentially of Nodosariidae associated with a few *Textulariina*.

Fig. 1. (*opposite*) The main forms of Nodosariidae in the Domerian and Toarcian deposits.

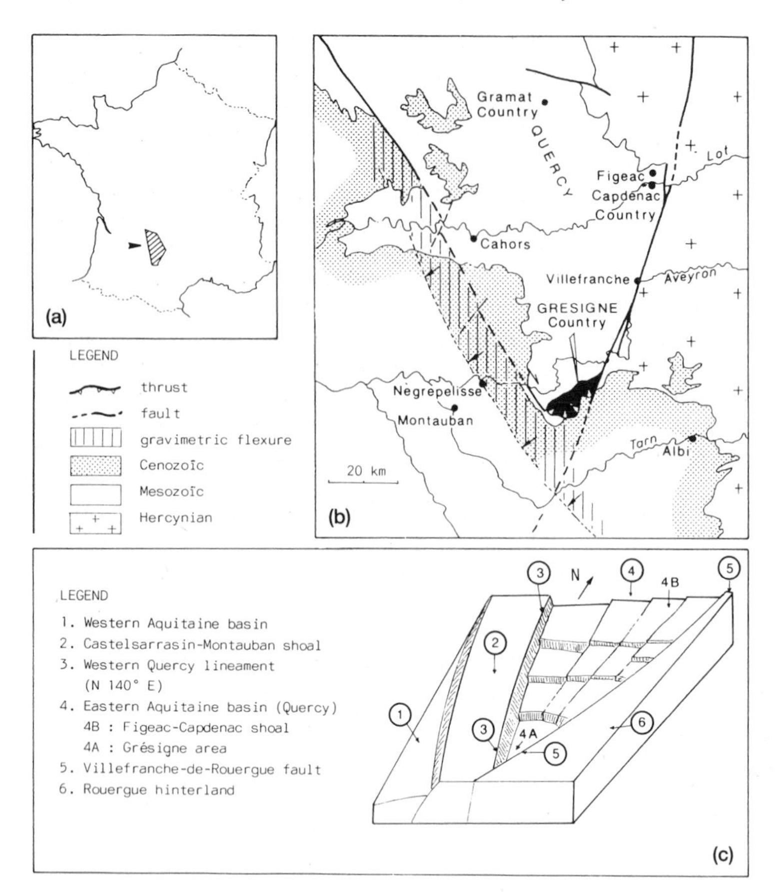

Fig. 2. (a) Localization of the Aquitaine basin, (b) geological sketch of the Quercy area localization of sampling sites and (c) morphology of the Liassic basin (Grésigne country and Capdenac–Figeac country).

Variety of species

This varies greatly from one systems tract to another (Figs 4 & 10):

1 Thirty-four species of foraminifera gradually colonized the transgressive systems tract, at first very quickly, then more slowly; 32 species are Nodosariidae (mostly *Lenticulina sensu lato*). These were joined by *Epistommina* in the middle of the deposit and *Ammodiscus* at the top. Eighteen species are inherited from the previous systems tract. The species renewal rate is 0.528.

2 The highstand systems tract contains 19 species, 18 from the preceding tract and one new one. Sixteen species present in the underlying deposits have disappeared. The renewal rate is 0.058.

3 The shelf margin wedge systems tract provides 14 species: seven are inherited and seven are new (principally belonging to the genera *Dentalina* and *Nodosaria*). Twelve species of the previous systems tract have disappeared. The renewal rate is 0.50.

Link between foraminiferal morphology and systems tracts

The transgressive systems tract contains (Figs 5 & 10) a high percentage (71 to 82%) of uncoiling morphogenera (*Astacolus, Marginulinopsis, Falsopalmula, Planularia*) and rectilinear genera (*Ichthyolaria, Marginulina, Bolivina*). The maximum concentration of these forms is found in the middle of the deposit.

Nine out of the 16 species which disappear in the highstand systems tract are uncoiling and rectilinear forms, resulting in an increase in coiled *Lenticulina* (22.5 to 48%).

The shelf margin wedge systems tract is marked by another significant decrease in the percentage of uncoiling and rectilinear forms (11 to 29%), linked to a decrease in species (four which disappear belong to this group). In the same layer there is an increase in robust forms (large-size *Lenticulina sensu stricto, Nodosaria metensis, Dentalina clavata*), which could be considered as opportunist forms (Cubaynes & Ruget, 1988).

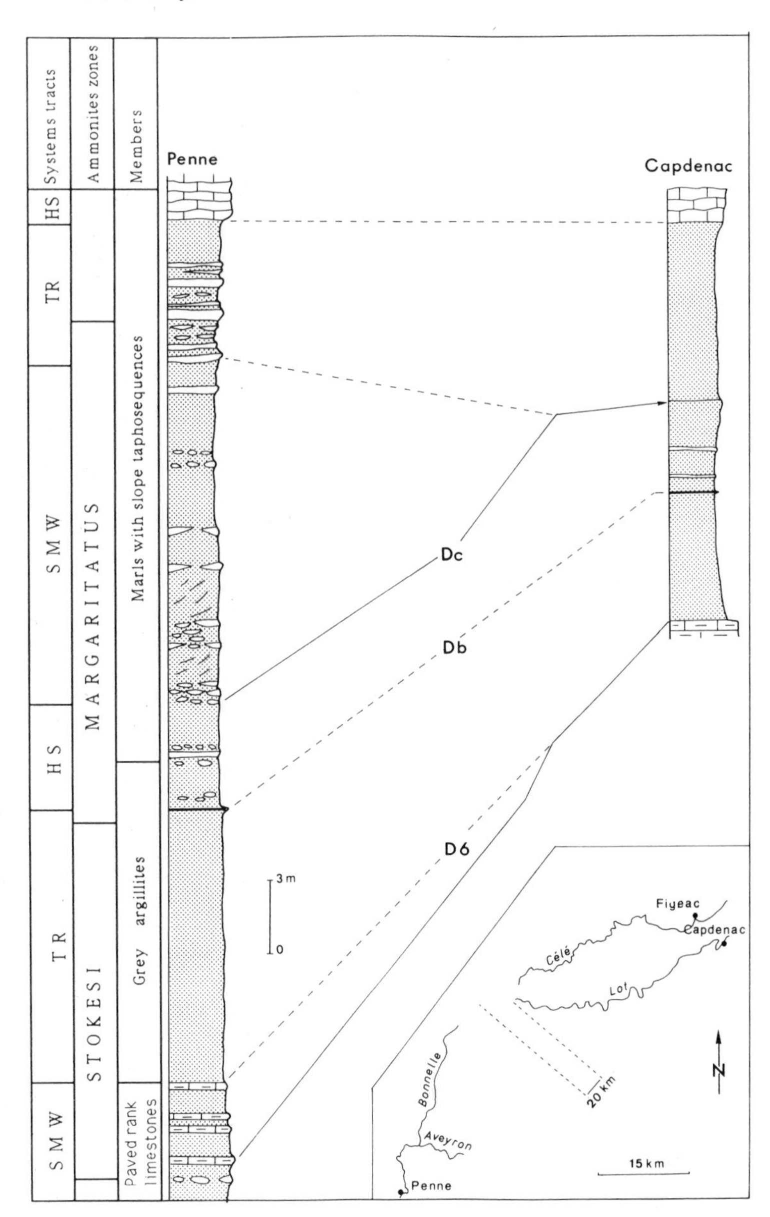

Fig. 3. The various systems tracts in the Quercy Domerian. SMW, shelf margin wedge deposits; TR, transgressive deposits; HS, highstand deposits.

EXAMPLE: LOWER AND MIDDLE TOARCIAN IN QUERCY

Stratigraphy, systems tracts and depositional sequences

In the Grésigne region (Fig. 2) the lower and middle Toarcian (Tenuicostatum, Serpentinus, Bifrons and Variabilis zones) are represented (Fig. 6) by the 50 m thick Penne Formation (Cubaynes, 1986) which is subdivided into three members, from top to base:

1 'Black marls with Pseudogrammoceras' (21 m).
2 'Hildoceratid marls and limestones' (25 m).
3 'Schistes carton' (4 m).

The micropalaeontological data were drawn from the first 18 m of 'Hildoceratid marls and limestones', based on 32 samples taken from alternating marls and biomicrites (a sample in each layer of marls).

The Gramat region, 80 km to the north, shows the same lithostratigraphical series but with a sub-

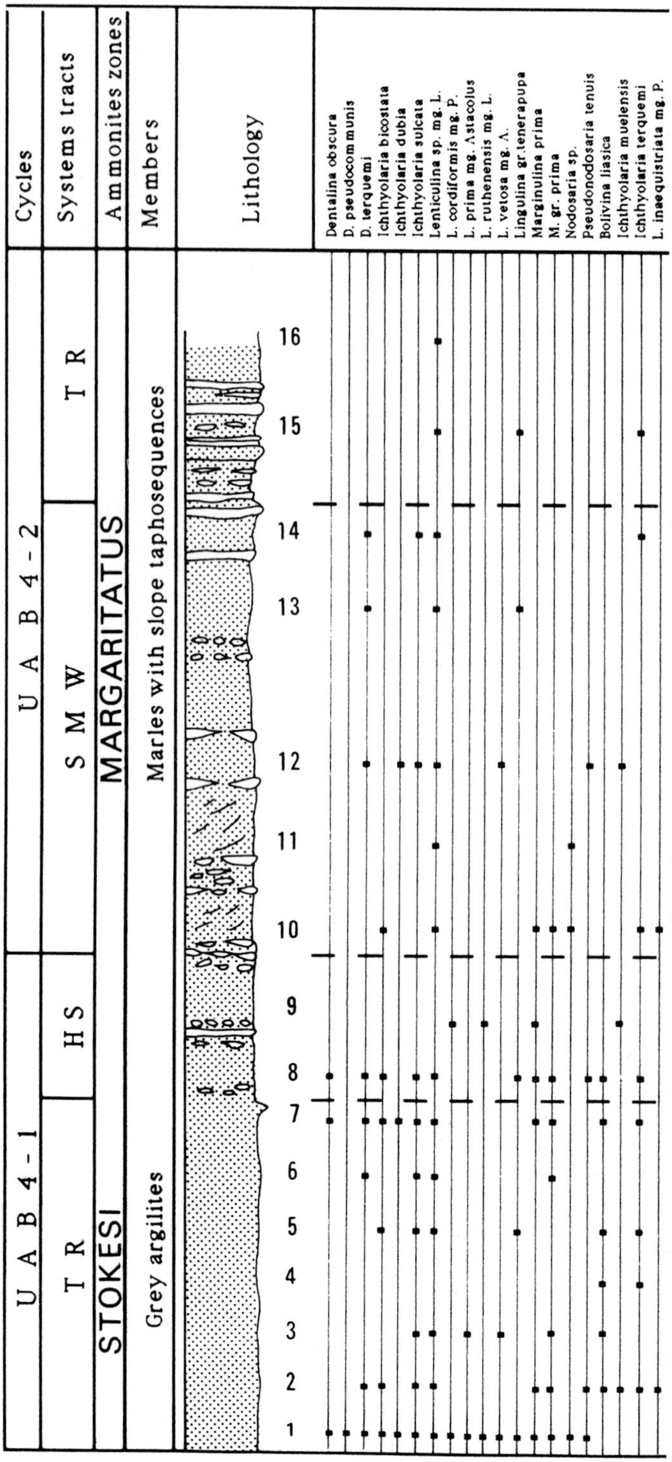

Cycles	UAB 4-1		UAB 4-2	
Systems tracts	TR	HS	SMW	TR
Ammonites zones	STOKESI		MARGARITATUS	
Members	Grey argilites		Marles with slope taphosequences	

Lithology

1 2 3 4 5 6 7 8 9 10 11 12 13 14 15 16

Dentalina obscura
D. pseudocommunis
D. terquemi
Ichthyolaria bicostata
Ichthyolaria dubia
Ichthyolaria sulcata
Lenticulina sp. mg. L.
L. cordiformis mg. P.
L. prima mg. Astacolus
L. ruthenensis mg. L.
L. vetosa mg. A.
Lingulina gr.tenerapupa
Marginulina prima
M. gr. prima
Nodosaria sp.
Pseudonodosaria tenuis
Bolivina liasica
Ichthyolaria muelensis
Ichthyolaria terquemi
L. inaequistriata mg. P.

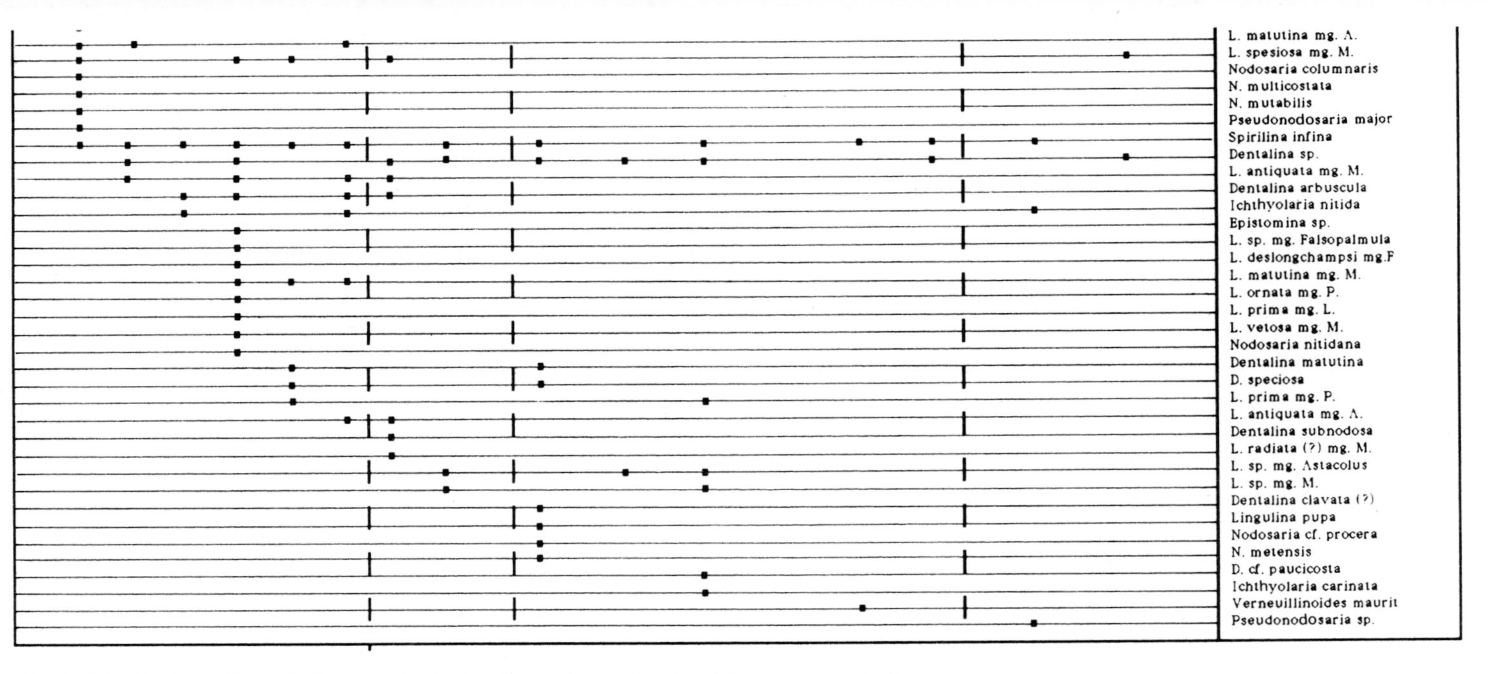

Fig. 4. Distribution of foraminifera species in the Quercy Domerian (Boulbène cross-section).

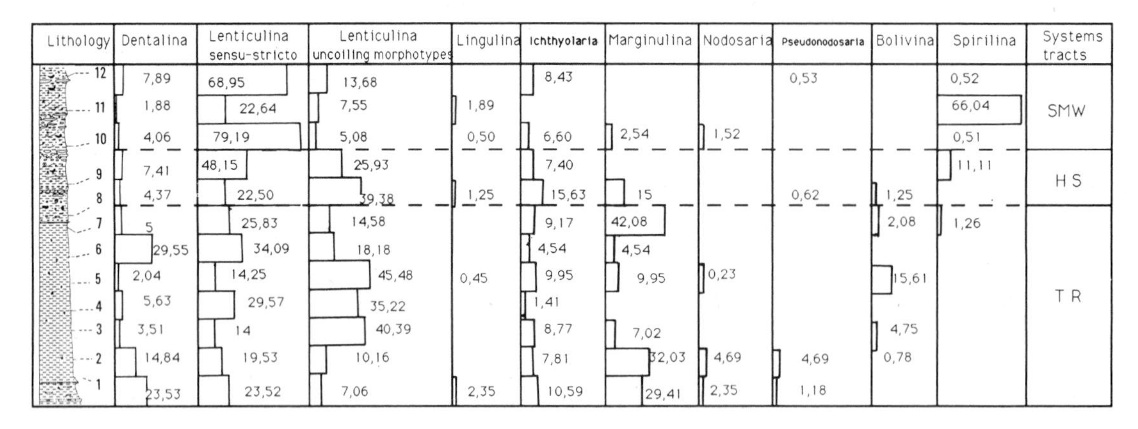

Fig. 5. Distribution and ratios of the main forms of benthic foraminifera in Quercy Domerian (Boulbène cross-section).

stantial increase in the thickness of the sandy clays below the 'Schistes carton' and in the middle part of the 'Hildoceratid marls and limestones'; in the lower part, 21 samples were taken from the 'Hildoceratid marls and limestones' (a sample in each layer of marls). The sequence interpretation is based on five criteria:

1 A detailed biostratigraphy permitting recognition of Toarcian levels II to XII (Cubaynes & Fauré, 1981; Cubaynes, 1986) in the 'Schistes carton' (levels II to IV) and 'Hildoceratid marls and limestones' (levels V to XII).

2 The succession of crinoiid fauna suggesting shallower palaeodepths in levels IX and X (with *Ectacrinus*) than in V to VIII and XI to XII (with *Isocrinus*).

3 The comparison between the stratal patterns in the Grésigne area, the Gramat area 80 km to the north and the Figeac–Capdenac area 80 km to the north–northeast (Fig. 6). The layers of level IX increase in thickness towards the Gramat area and wedge out rapidly to the northeast between Caylus and Puylagarde; likewise, levels VI and VII are condensed in the Figeac region, whereas levels II to III and X to XII extend to Capdenac.

4 The recognition of sedimentary unconformities: D7 at the summit of Pecten rock bar, Dd which interrupts the 'Schistes carton' and De at the top of the two Hildoceratid layers.

5 The concordance of our interpretation and the global cycle chart drawn up by Haq *et al.* (1988).

Using these criteria, the Dd unconformity can be interpreted as a condensed interval of levels IV and V containing abundant ammonites which was caused by the rapid rise in sea level at — 183.5 Ma. This unconformity separates the transgressive systems tracts from the highstand deposits (levels V to VIII of the 'Hildoceratid marls and limestones') of the UAB 4-3 cycle of Haq *et al.* (1988). The De surface is seen as an unconformity caused by the rapid fall in sea level that terminated cycle UAB 4-3. Above this, cycle UAB 4-4 begins with a shelf-margin wedge (levels IX and part of X), which is followed by a transgressive systems tract (levels X to XII) at the top of the 'Hildoceratid marls and limestones' and the base of the 'black marls with Pseudogrammoceras'.

Foraminifera in the lower and middle Toarcian systems tracts

This is a rather special case as the transgressive systems tract of the UAB 4-3 cycle ('Schistes carton') contains no benthic organisms due to anoxic conditions. The following systems tracts (the highstand systems tract of the UAB 4-3 cycle, the shelf margin wedge systems tract of the UAB 4-4 cycle) have been deposited in an infralittoral environment that was much shallower, more stable and less turbid than that of the Domerian (Cubaynes, 1986). The transgressive systems tract of the UAB 4-4 cycle indicates a gradual acceleration in the sedimentation rate and an increase in terrigenous deposits.

Variety of species

Above the transgressive systems tract having no foraminifera, the highstand systems tract of the

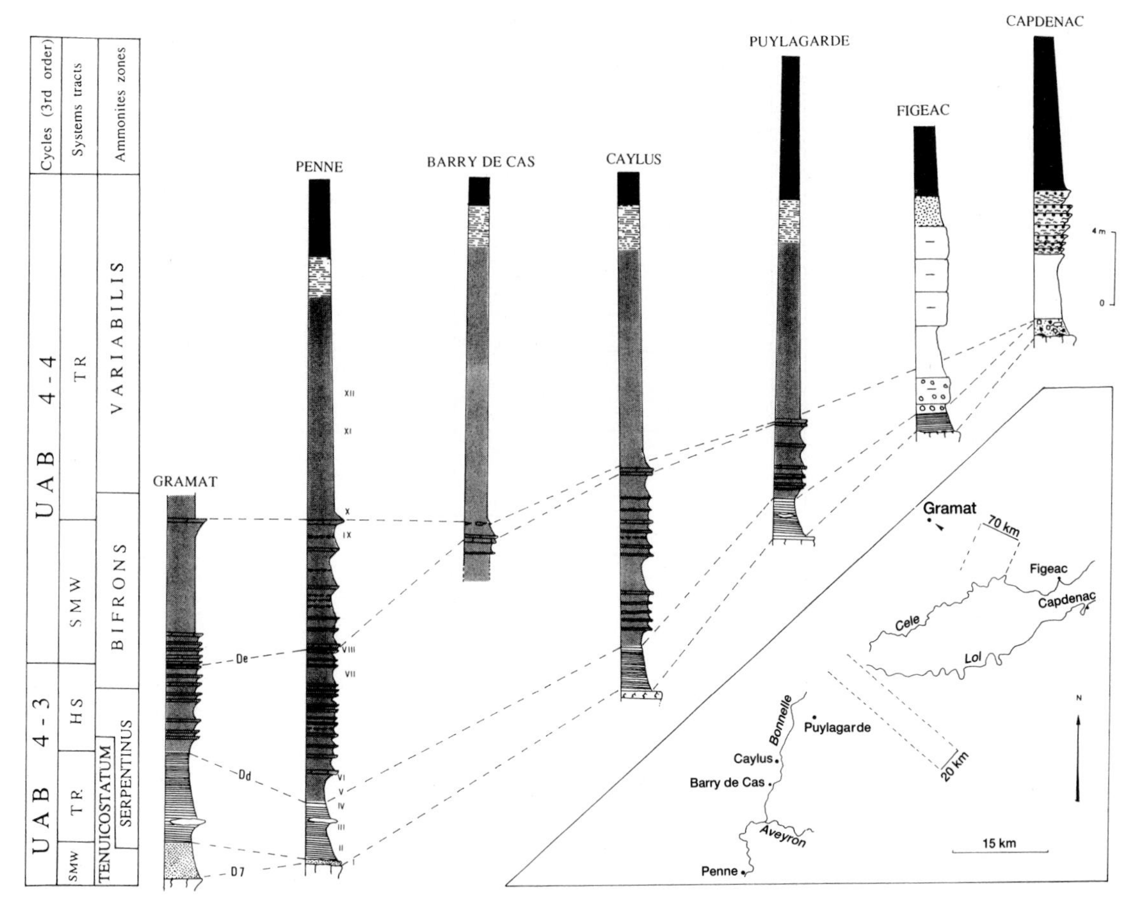

Fig. 6. The various systems tracts of the lower and middle Toarcian in Quercy. SMW, shelf margin wedge deposits; TR, transgressive deposits; HS, highstand deposits.

UAB 4-3 cycle contains 39 species. The rate of appearance is first rapid, then gradually decelerates (Figs 7 & 10). Only Nodosariidae are present in the lower and middle part of the deposit, but in the upper part they are associated with a few *Textularina* (*Ammobaculites*). Thus benthic foraminifera colonized the Toarcian in exactly the same way as in the Domerian transgressive systems tract, but at a later stage of sedimentary evolution when the anoxic conditions disappeared and the first sea-floor conditions favourable to benthic life occurred.

The shelf margin wedge systems tract of the UAB 4-4 cycle contains only 28 species; 17 of the underlying deposit have disappeared. There are six new species (belonging mostly to the genera *Nodosaria* and *Dentalina*), and 22 inherited from the previous systems tract. The renewal rate is 0.214.

The transgressive systems tract of the UAB 4-4 cycle indicates another decrease in the number of species (19 species in all, 11 inherited species, 17 which have disappeared and eight new ones) and a relatively high renewal rate (0.421); the latter might be linked to regional conditions unfavourable to benthic organisms due to an increasing sedimentation rate and more terrigenous deposits.

Relationship between foraminiferal morphology and systems tracts

In the highstand systems tract of the UAB 4-3 cycle, there is a high percentage of rectilinear Nodosariidae (*Citharina, Dentalina, Marginulinopsis, Falsopalmula, Planularia, Astacolus*); 21% in the Grésigne area (Figs 8 & 10) and 29% in the Gramat

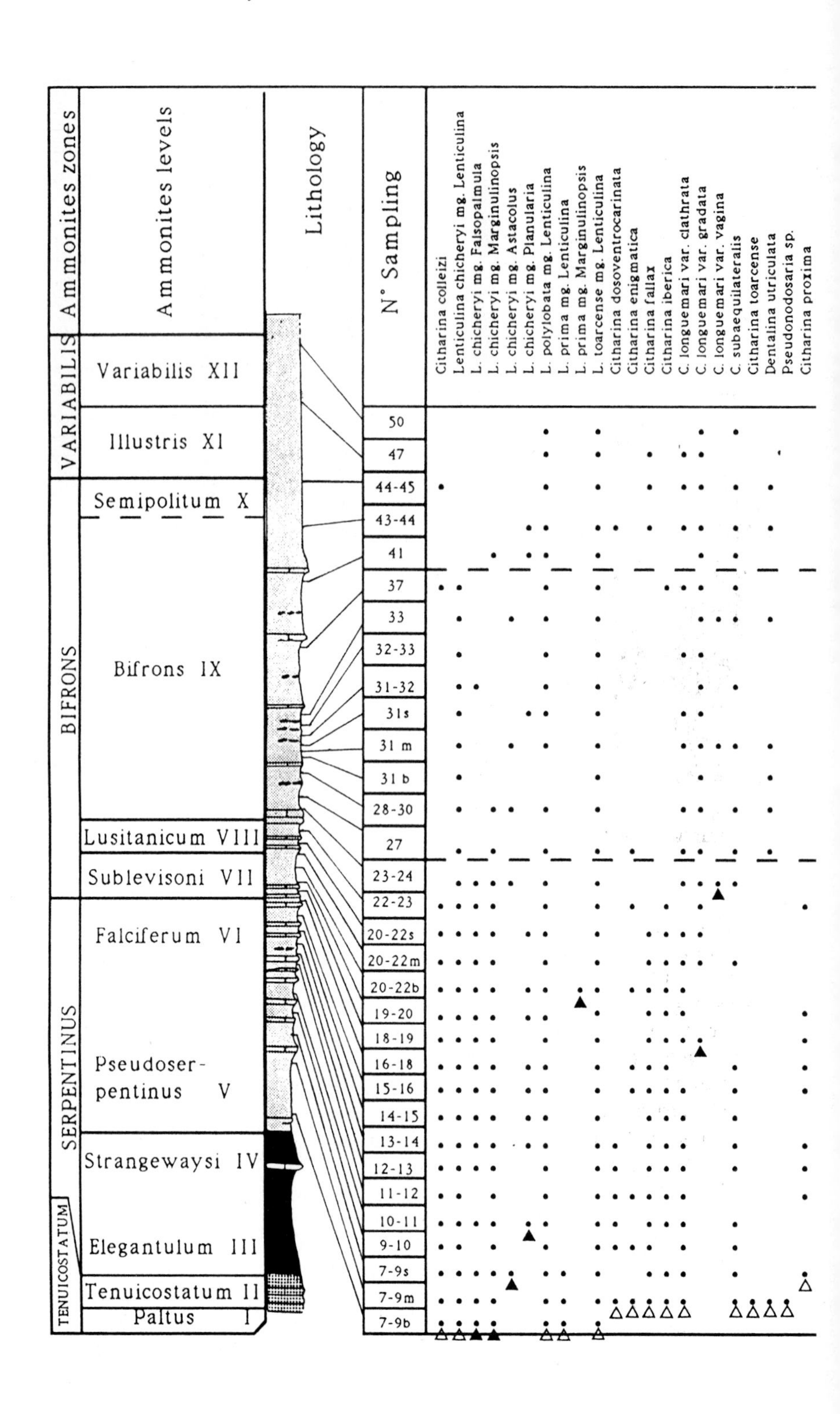
Ammonites zones
Ammonites levels
Lithology
N° Sampling
Citharina colleizi
Lenticulina chicheryi mg. Lenticulina
L. chicheryi mg. Falsopalmula
L. chicheryi mg. Marginulinopsis
L. chicheryi mg. Astacolus
L. chicheryi mg. Planularia
L. polylobata mg. Lenticulina
L. prima mg. Lenticulina
L. prima mg. Marginulinopsis
L. toarcense mg. Lenticulina
Citharina dosoventrocarinata
Citharina enigmatica
Citharina fallax
Citharina iberica
C. longuemari var. clathrata
C. longuemari var. gradata
C. longuemari var. vagina
C. subaequilateralis
Citharina toarcense
Dentalina utriculata
Pseudonodosaria sp.
Citharina proxima
VARIABILIS
BIFRONS
SERPENTINUS
TENUICOSTATUM
Variabilis XII
Illustris XI
Semipolitum X
Bifrons IX
Lusitanicum VIII
Sublevisoni VII
Falciferum VI
Pseudoser-pentinus V
Strangewaysi IV
Elegantulum III
Tenuicostatum II
Paltus I
50
47
44-45
43-44
41
37
33
32-33
31-32
31s
31 m
31 b
28-30
27
23-24
22-23
20-22s
20-22m
20-22b
19-20
18-19
16-18
15-16
14-15
13-14
12-13
11-12
10-11
9-10
7-9s
7-9m
7-9b

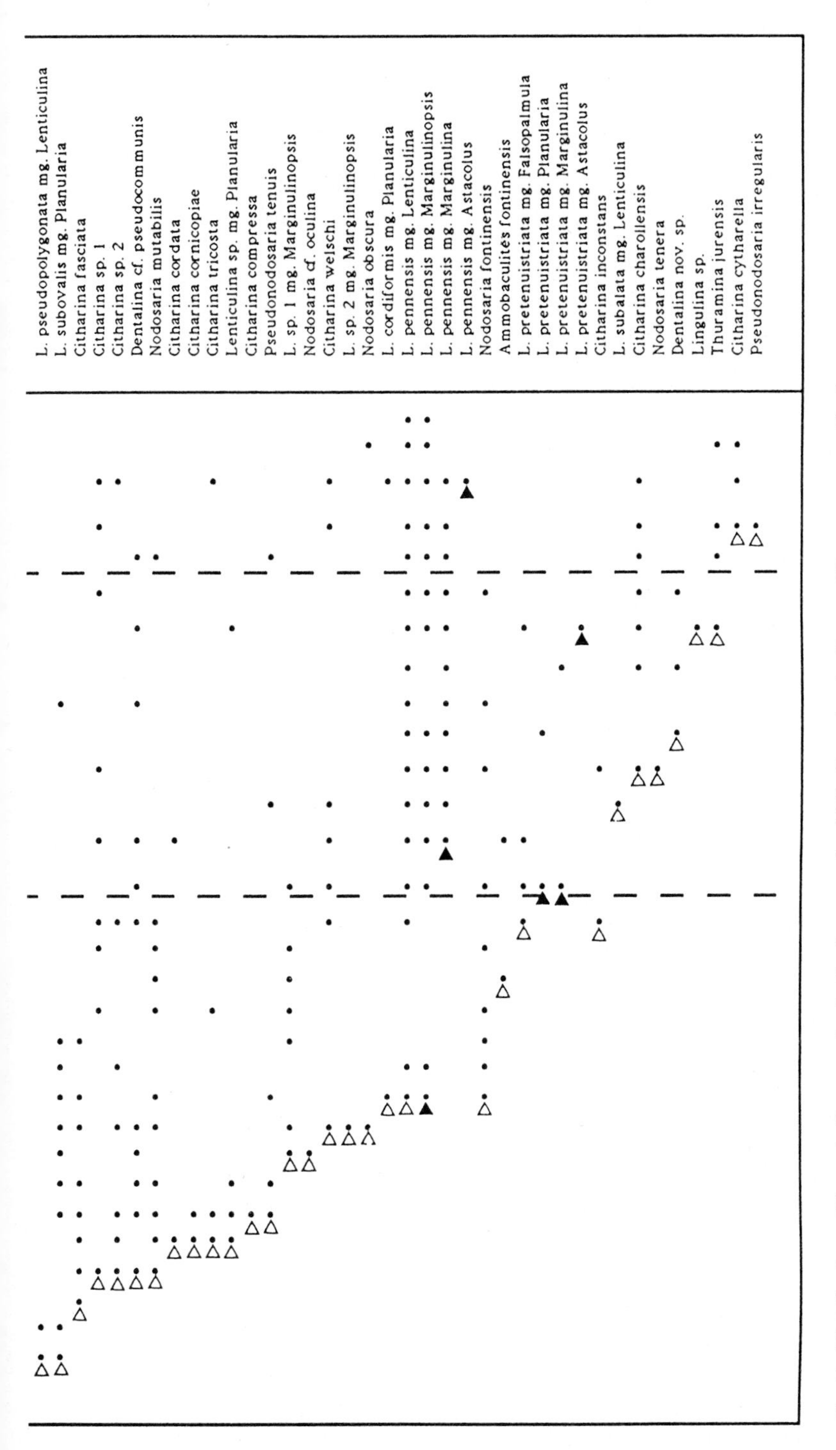

Fig. 7. Distribution of foraminifera species in the lower and middle Toarcian in Quercy (Penne cross-section).

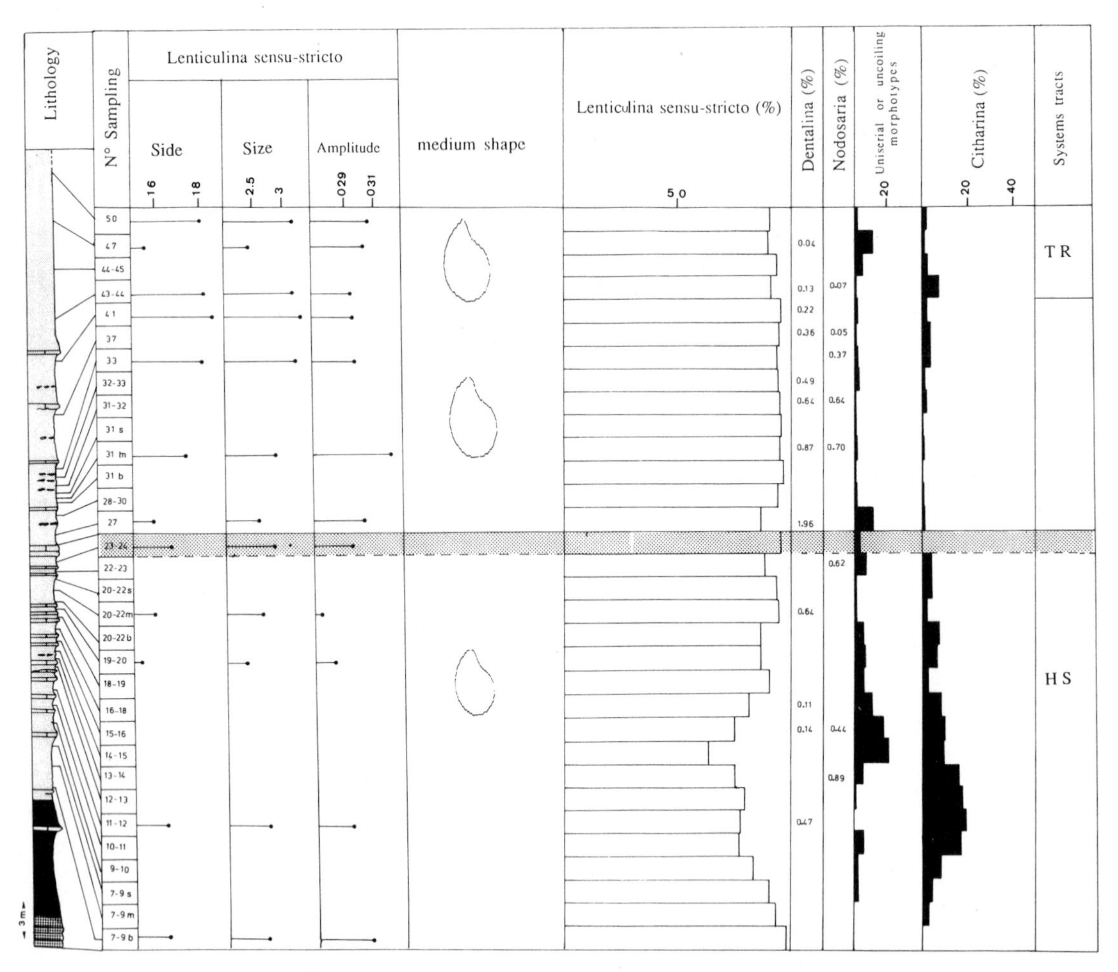

Fig. 8. Distribution and ratios of the main forms of benthic foraminifera in the lower and middle Toarcian in Quercy (Penne cross-section, Grésigne area).

area (Fig. 9), with maximum concentration in the middle of the deposit. Moreover, the coiled *Lenticulina* are small in size.

The shelf margin wedge systems tract of the UAB 4-4 cycle is marked by the disappearance of citharines (10 species) and of uncoiling morphogenera (three species). Large-size *Lenticulina* predominate (96% on average in the Grésigne area and 93% in the Gramat area).

In the following transgressive systems tract the *Citharina* and uncoiling morphogenera of *Lenticulina* reappear, making up 10% of the population in the Grésigne area and 20% in the Gramat area. Coiled *Lenticulina* are still of large size: this can be interpreted as a result of unfavourable environmental conditions.

Localization of primitive forms

Dentalina are more abundant in the shelf-margin wedge systems tract of the UAB 4-4 cycle than in the underlying highstand systems tract. The most widespread species (*Dentalina utriculata*) resembles the couple *Dentalina terquemi-obscura* (prevalent in the middle Lias) in form, ornamentation and robustness.

In the same way, under the 'Schistes carton' member, the shelf margin wedge systems tract of the UAB 4-3 cycle (the Tenuicostatum zone of the lower Toarcian mainly represented in the Gramat area) contains typically Domerian foraminiferal assemblages, in particular *Saracenaria sublaevis.*

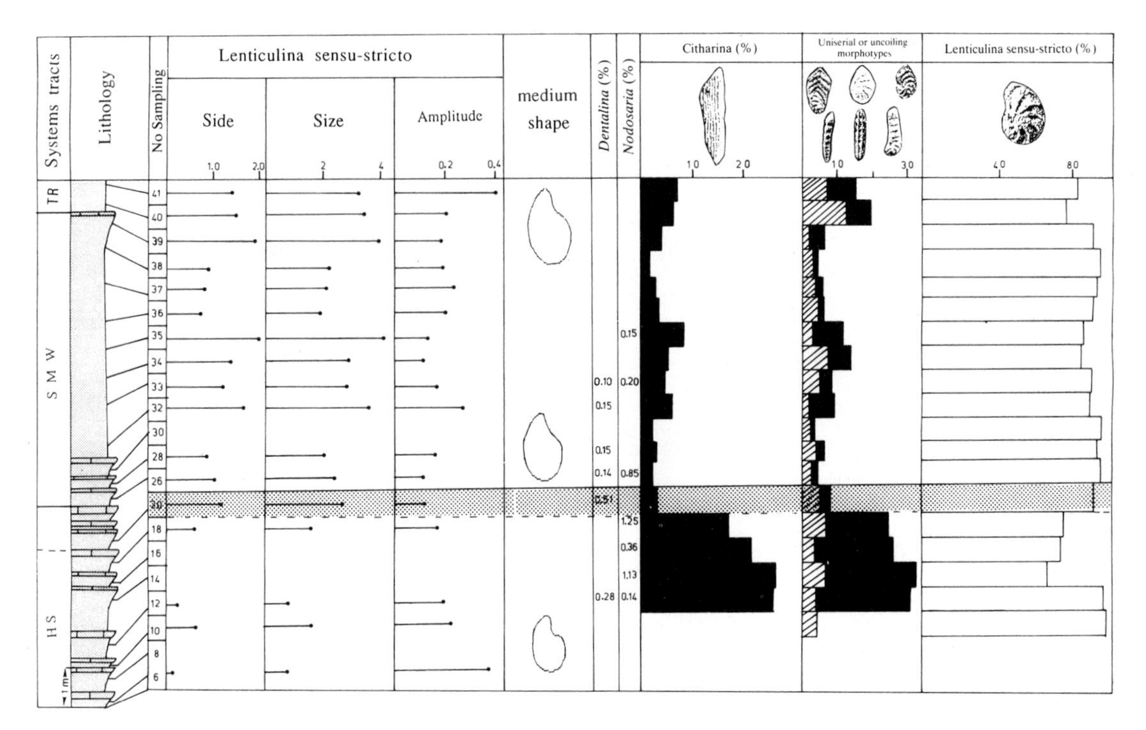

Fig. 9. Distribution and ratios of the main forms of benthic foraminifera in lower and middle Toarcian in Quercy (Cérède cross-section, Gramat area).

SYNTHESIS: FORAMINIFERAL POPULATIONS IN SYSTEMS TRACTS

The increase in abundance of species suggests colonization of the sea floor during a phase of sea-level rise (Fig. 10) during sedimentation of a transgressive systems tract (Domerian) or a lower highstand systems tract (Toarcian) if the preceding transgressive systems tract was barren. There are fewer different species at the top of the highstand systems tract and this tendency is more marked in the shelf margin wedge systems tract. The species renewal rate is fairly high in the shelf margin wedge systems tract (21 to 50%) and even more so in the transgressive systems tract (42 to 53%). It is very low in the Domerian highstand systems tract (about 5%). Of course the renewal rate of 100% found in the highstand systems tract of the lower and middle Toarcian is meaningless because the previous environmental conditions were unfavourable to benthic organisms.

The uncoiling or rectilinear shapes apparently indicate a stable environment (the Domerian transgressive systems tract; the lower part of the Toarcian highstand systems tract). They are more abundant in the distal, highstand systems tracts (Domerian), than in the more proximal highstand systems tracts (Toarcian) where more turbulent hydrodynamic conditions and larger continental deposits are prevalent. For any given systems tract in the Toarcian, the above shapes are more abundant in the distal regions of the basin (e.g. the Gramat area) than in the more proximal regions (e.g. the Grésigne area). On the contrary, the robust (large-coiled *Lenticulina*) or primitive forms (some *Dentalina, Saracenaria*) are more characteristic of the shelf margin wedge systems tract.

MICROPALAEONTOLOGICAL PATTERNS AND SYSTEMS TRACT BOUNDARIES

In many cases there is a discrepancy between the distributional modifications in foraminiferal associations suggesting environmental changes and the actual geometrical and sedimentological boundaries of the systems tracts.

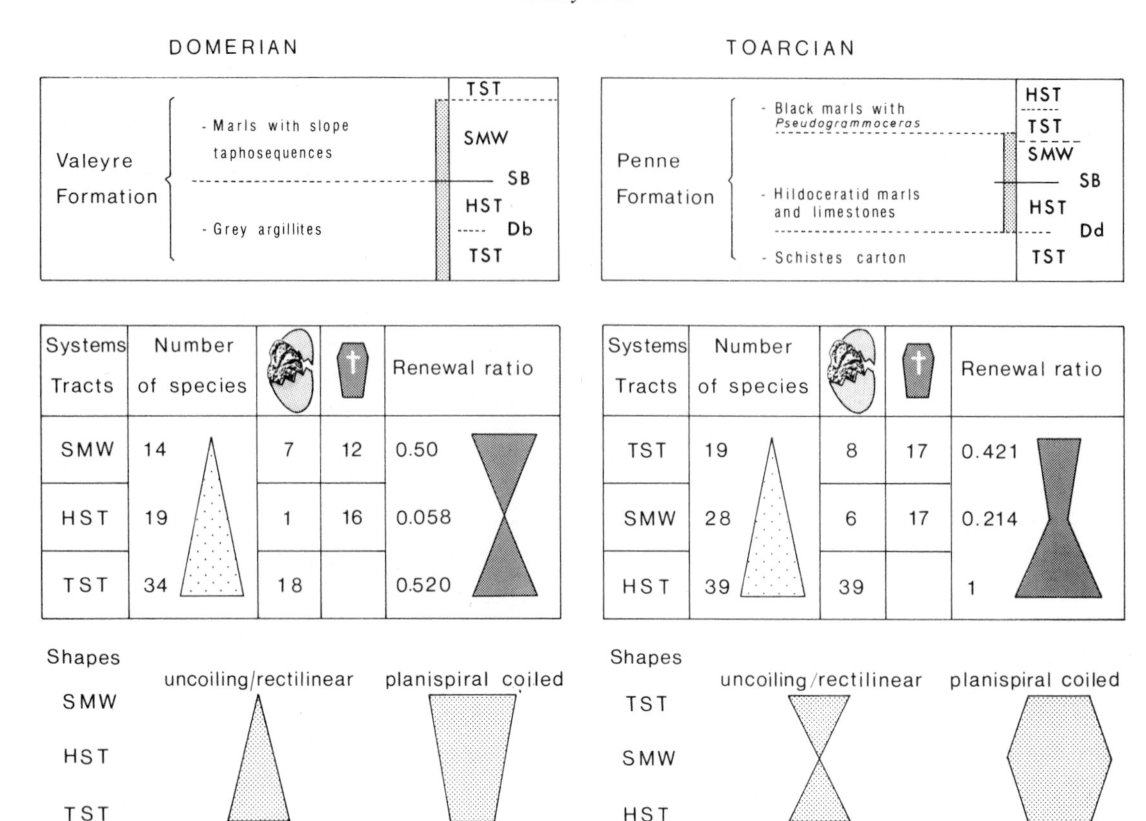

Fig. 10. Sketch summarizing the species diversities and the shapes in the Domerian and Toarcian deposits.

For example, in the Toarcian, eight species disappear 40 to 50 cm below the depositional sequence boundary (Fig. 7). *Lenticulina sensu stricto* start to increase in size, whereas they had previously decreased (Figs 8 & 9). In the same way a change in microfauna appears 1.50 m below the flooding surface which separates the shelf margin wedge systems tract from the transgressive systems tract in the UAB 4-4 cycle. This change especially concerns new species such as *Thurammina jurensis* considered to be a cold-water foraminifera. Moreover in the Domerian most species disappear well below the sequence boundary (Fig. 4).

In the studied cases it would almost seem that the microfauna perceived variations in the physicochemical parameters which anticipated major environmental changes. So (for reasons yet to be established) the gradual modifications in the composition and distribution of foraminiferal assemblages preceded the shift in stratal patterns.

CONCLUSIONS

Various modifications which can be distinguished easily in foraminiferal populations suggest directly or indirectly environmental changes due to relative sea-level variations. Such changes in fauna appear to precede the systems tract boundaries in lithologically homogeneous strata at least in the cases studied in this paper. Some features of these populations indicate a stable environment usually provided during a phase of sea-level rise; for example, an abundance of different species, small-size coiled *Lenticulina* and the presence and development of certain rectilinear (*Citharina*) or uncoiling (morphogenera of *Lenticulina sensu lato*) forms. Other conditions being equal, the transgressive systems tracts can be distinguished from the highstand systems tracts (in our examples) by a faster species renewal rate. Fewer different species, robust or primitive forms and predominant large-size coiled

Lenticulina indicate a disturbed environment generally occurring during a phase of sea-level fall (the shelf margin wedge systems tracts; the top of the highstand systems tracts). When sea-level rise is too rapid or too strong, anoxic conditions are created on the sea floor and the benthic foraminiferal assemblages then do not appear in the transgressive systems tract but in the lower part of the following highstand systems tract.

REFERENCES

Cubaynes, R. (1986) Le Lias du Quercy méridional: étude lithologique, biostratigraphique, paléoécologique et sédimentologique. *Strata, Toulouse,* **2** (6), 502 pp.

Cubaynes, R. & Faure, P.H. (1981) Première analyse biostratigraphique du Lias supérieur du Sud Quercy (bordure Bord-Est Aquitaine). *C.R. Acad. Sci. Paris* **292** (2), 1031–1034.

Cubaynes, R. & Ruget, C. (1988) Contrôle de l'environnement dans la composition des faunes pionnières de Nodosariidés. Exemple du Carixien inférieur du Sud-Quercy. *Rev. Paléobiologie, Genève,* sp. vol. **2** (1), 177–182.

Cubaynes, R., Faure, P., Hantzpergue, P., Pelissie, T. & Rey, J. (1989a) Le Jurassique du Quercy: unités lithostratigraphiques, stratigraphie et organisation séquentielle, évolution sédimentaire. *Géol. de la France, Orléans* **3**, 33–62.

Cubaynes, R., Ruget, C. & Rey, J. (1989b) Essai de caractérisation des prismes de dépôt d'origine eustatique par les associations de Foraminifères benthiques: exemple du Lias moyen et supérieur sur la bordure Est du Bassin Aquitain. *C.R. Acad. Sci. Paris* **308** (2), 1517–1522.

Cubaynes, R., Rey, J., Ruget, C., Courtinat, B. & Bodergat, A.M. (1990a) Relations between system tracts and micropaleontological assemblages on a Toarcian carbonate shelf (Quercy, southwest France). *Bull. Soc. Géol. France, Paris,* **VI** (8), 6, 989–993.

Cubaynes, R., Rey, J. & Ruget, C. (1990b) Renouvellement des espèces de Foraminifères benthiques et variations globales du niveau des mers. Exemples du Lias du Quercy et de l'Eocène des Corbières. *Rev. Micropaléontol. Paris* **33**, 3–4, 233–240.

De Wever, P., Granlund, A.H. & Cordey, F. (1989) Icule, un système d'analyse de contour d'image pour micropaléontologie, une étape vers un système paléontologique intégré. *Rev. Micropaléontol. Paris* **32** (3), 215–225.

Granlund, A.H. (1986) Quantitative analysis of microfossils. A methodological study with applications to Radiolaria. *Meddelanden f. Stockholm Univ. Geol. Inst.* **268**, 99 pp.

Haq, B.U., Hardenbol, J. & Vail, P.R. (1988) Mesozoic and Cenozoic chronostratigraphy and cycles of sea-level change. In: *Sea-level Changes: An Integrated Approach.* (Eds Wilgus, C.K., Hastings, B.S., Kendall, C.G.St.C., Posamentier, H.W., Ross, C.A. & Van Wagoner, J.C.) Spec. Publ. Soc. Econ. Paleontol. Mineral. Tulsa, 42, 71–108.

Lohman, P. (1983) Eigenshape analysis of microfossils: a general morphometric procedure for describing changes in shape. *J. Int. Assoc. Math. Geol.* **15**, 659–672.

Magniez-Jannin, F. & Jacquin, T. (1990) Validité du découpage séquentiel *sensu* Vail en domaine de bassin: arguments fournis par les Foraminifères dans le Crétacé inférieur du Sud-Est de la France. *C.R. Acad. Sci. Paris,* **310** (2), 263–269.

Olsson, R.K. (1988) Foraminiferal modeling of sea-level change in the late Cretaceous of New Jersey. In: *Sea-level Changes: An Integrated Approach* (Eds Wilgus, C.K., Hastings, B.S., Kendall, C.G.St.C., Posamentier, H.W., Ross, C.A. & Van Wagoner, J.C.) Spec. Publ. Soc. Econ. Paleontol. Mineral. Tulsa, 42, 289–297.

Posamentier, H.W. & Vail, P.R. (1988) Eustatic controls on clastic deposition. II — Sequences and systems tract models. In: *Sea-level Changes: An Integrated Approach* (Eds Wilgus, C.K., Hastings, B.S., Kendall, C.G.St.C., Posamentier, H.W., Ross, C.A. & Van Wagoner, J.C.) Spec. Publ. Soc. Econ. Paleontol. Mineral. Tulsa. 42, 125–154.

Ruget, C. (1985) Les Foraminifères (Nodosariidés) du Lias de l'Europe occidentale. *Docum. Lab. Géol. Lyon* **94**, 273 pp.

Sarg, J.F. (1988) Carbonate sequence stratigraphy. In: *Sea-level Changes: An Integrated Approach* (Eds Wilgus, C.K., Hastings, B.S., Kendall, C.G.St.C., Posamentier, H.W., Ross, C.A. & Van Wagoner, J.C.) Spec. Publ. Soc. Econ. Paleontol. Mineral. Tulsa, 42, 155–181.

Vail, P.R., Colin, J.P., Du Chene, R.J., Kuchly, J., Mediavilla, F. & Trifilieff, V. (1987) La stratigraphie séquentielle et son application aux corrélations chronostratigraphiques dans le Jurassique du Bassin de Paris. *Bull. Soc. Géol. France, Paris,* **III** (8), 7, 1301–1321.

Vail, P.R., Hardenbol, J. & Todd, R.G. (1984) Jurassic unconformities, chronostratigraphy, and sea-level changes from seismic stratigraphy and biostratigraphy. *Mem. Am. Assoc. Petrol. Geol. Tulsa,* **36**, 129–144.

Spec. Publs Int. Ass. Sediment. (1993) **18**, 125–160

The expression and interpretation of marine flooding surfaces and erosional surfaces in core; examples from the Upper Cretaceous Dunvegan Formation, Alberta foreland basin, Canada

J.P. BHATTACHARYA

ARCO Exploration and Production Technology, 2300 West Plano Parkway, Plano, TX 75075, USA

ABSTRACT

This paper presents examples of marine flooding surfaces and erosional surfaces in core and well logs from the Upper Cretaceous Dunvegan Formation, Alberta, Canada. Two categories of surface are defined. The first category comprises those surfaces associated with relative sea-level rise and includes the transgressive surface of erosion and several types of flooding surfaces, including minor, major and maximum marine flooding surfaces. The second category of surfaces comprises those associated with relative sea-level fall and includes the sequence boundary and its correlative surfaces. The sequence boundary is most readily recognized as an erosional surface at the base of incised valleys but correlative surfaces can include subaerial exposure surfaces in interfluve areas, marine erosional surfaces at the base of incised sharp-based shorefaces and correlative conformities in the seaward realm. A transgressive surface of erosion may be coincident with or may enhance the unconformity associated with the sequence boundary in a landward position and may pass seaward into a flooding surface.

The Dunvegan Formation is characterized by seven major marine flooding surfaces, each with a sharp contact between shallow-water sediments and overlying deeper water mudstones. These easily recognized surfaces could be correlated over hundreds of kilometres and were chosen as the bounding discontinuities for seven defined allomembers. Less extensive, minor marine flooding surfaces were used to recognize shingled offlapping units within each allomember. Sequence boundaries were also identified and used to separate highstand and lowstand systems tracts within several of the allomembers. Where the transgressive surface of erosion can be recognized, it is used to distinguish lowstand from transgressive systems tracts. The identification of systems tracts allows interpretation of the Dunvegan allomembers in terms of fourth-order, relative sea-level changes.

INTRODUCTION

Allostratigraphy is defined as the packaging of rocks bounded by discontinuities within a time-stratigraphic framework (North American Commission of Stratigraphic Nomenclature, 1983). It is a formally recognized way of naming discontinuity-bounded rock successions but places no particular emphasis on which type of discontinuity should be used as the 'fundamental' stratigraphic break. Allostratigraphic units may therefore include both unconformity-bounded *depositional sequences*, as defined by Mitchum (1977), and the *genetic stratigraphic sequences* proposed by Galloway (1989), which are based on marine flooding surfaces. Allostratigraphy represents a relatively generic way of *naming* discontinuity-bounded rock successions and emphasizes mappability. Sequence stratigraphy represents a more powerful way of *interpreting* rock successions in the context of relative sea-level change (Posamentier & Vail, 1988; Posamentier *et al.*, 1988). Sequence stratigraphy places bounding discontinuities into a distinct hierarchy with major emphasis on the unconformity or sequence boundary as the fundamental stratigraphic break (Van Wagoner *et al.*, 1990). Sequences may also be developed at different scales reflecting varying rates and frequency of sea-level change (e.g. first-, second-, third- and fourth-order).

This paper illustrates in detail the different types of physical surfaces that have been recognized within the Dunvegan Formation of the Canadian Alberta foreland basin. Examples of the surfaces in individual cores and well logs illustrate how the surfaces and the facies successions they bound change regionally. The paper also shows how these surfaces have been used to package the rocks into genetic units and how these units represent a basis for interpretation in terms of fourth-order relative sea-level changes. The allostratigraphic breakdown of Bhattacharya and Walker (1991a) is thus interpreted here in the context of sequence stratigraphy.

DUNVEGAN ALLOSTRATIGRAPHY AND FACIES

The Dunvegan Formation of northwest Alberta and northeast British Columbia comprises an Upper Cretaceous (mid-Cenomanian) clastic wedge deposited into the Western Interior epeiric seaway and shed from the actively rising Western Cordillera (Fig. 1). Previous lithostratigraphic representations of the Dunvegan Formation (Fig. 2) showed an undifferentiated eastward-tapering sand-prone wedge interfingering with shales of the Shaftesbury, Kaskapau and La Biche Formations (Singh, 1983). In a more detailed subsurface study, Bhattacharya and Walker (1991a) elucidated the nature of this interfingering and showed that the Dunvegan could be subdivided into seven regionally extensive allomembers (A to G in Fig. 3). Although broadly regressive in character, the allomembers include a

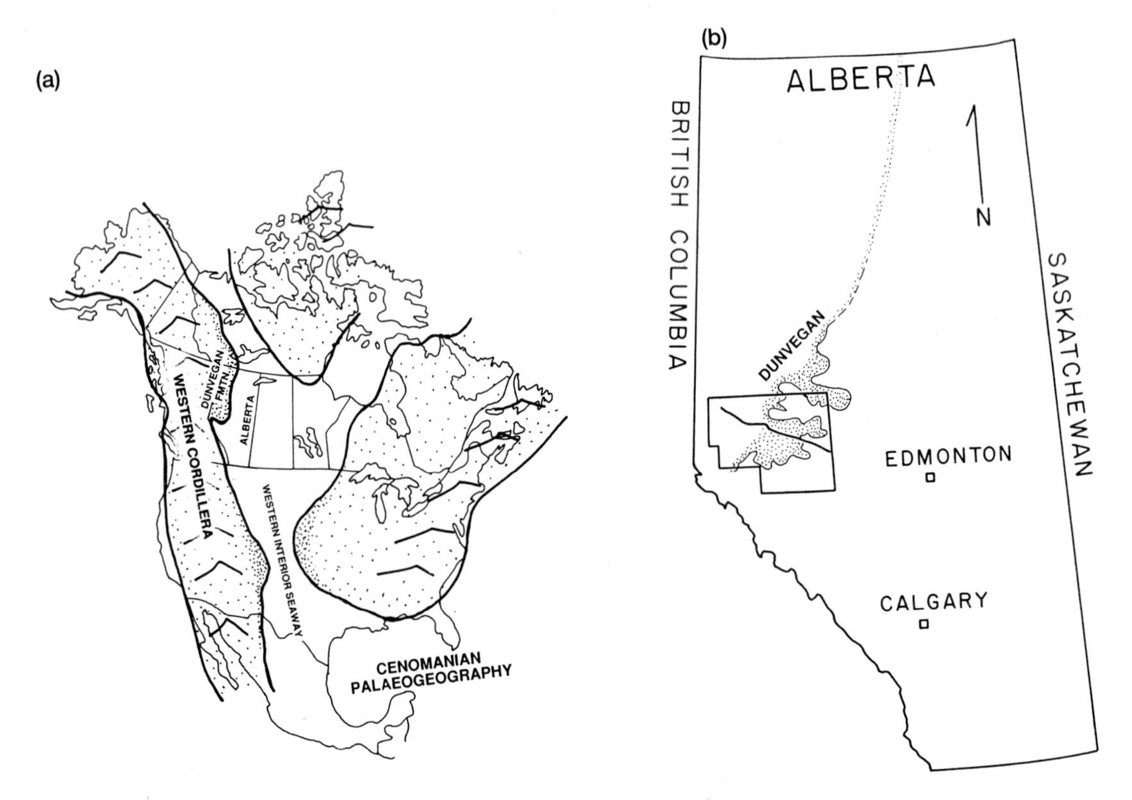

Fig. 1. Location maps. (a) Shows the palaeogeography of the Western Interior seaway during the late Cretaceous (mid-Cenomanian). (b) Shows the location of the study area in western Alberta with the approximate maximum limit of shoreline progradation in the Dunvegan indicated with the stippled pattern. The location of the regional cross-section, shown in Fig. 3, is indicated in the inset of the study area. The study area covers about 30,000 km^2.

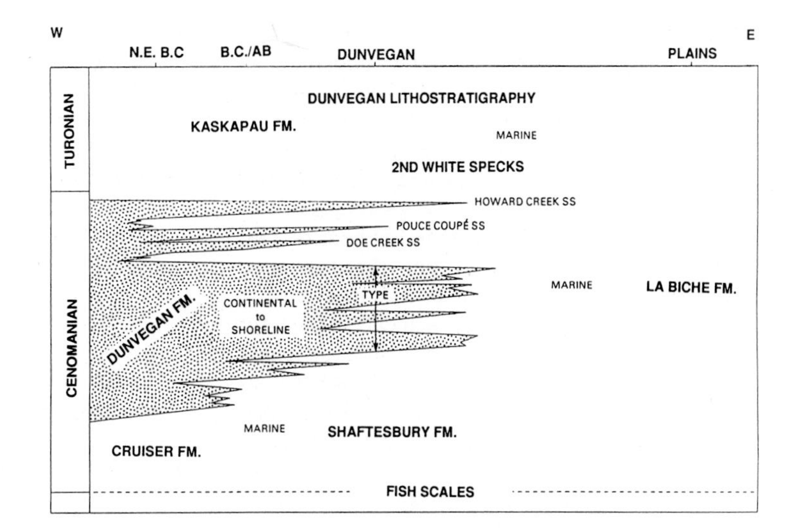

Fig. 2. Lithostratigraphic cross-section of the Dunvegan Formation shows an eastward-tapering wedge, intertonguing with shales of the Shaftesbury, La Biche and Kaskapau Formations. (Modified after Singh, 1983.)

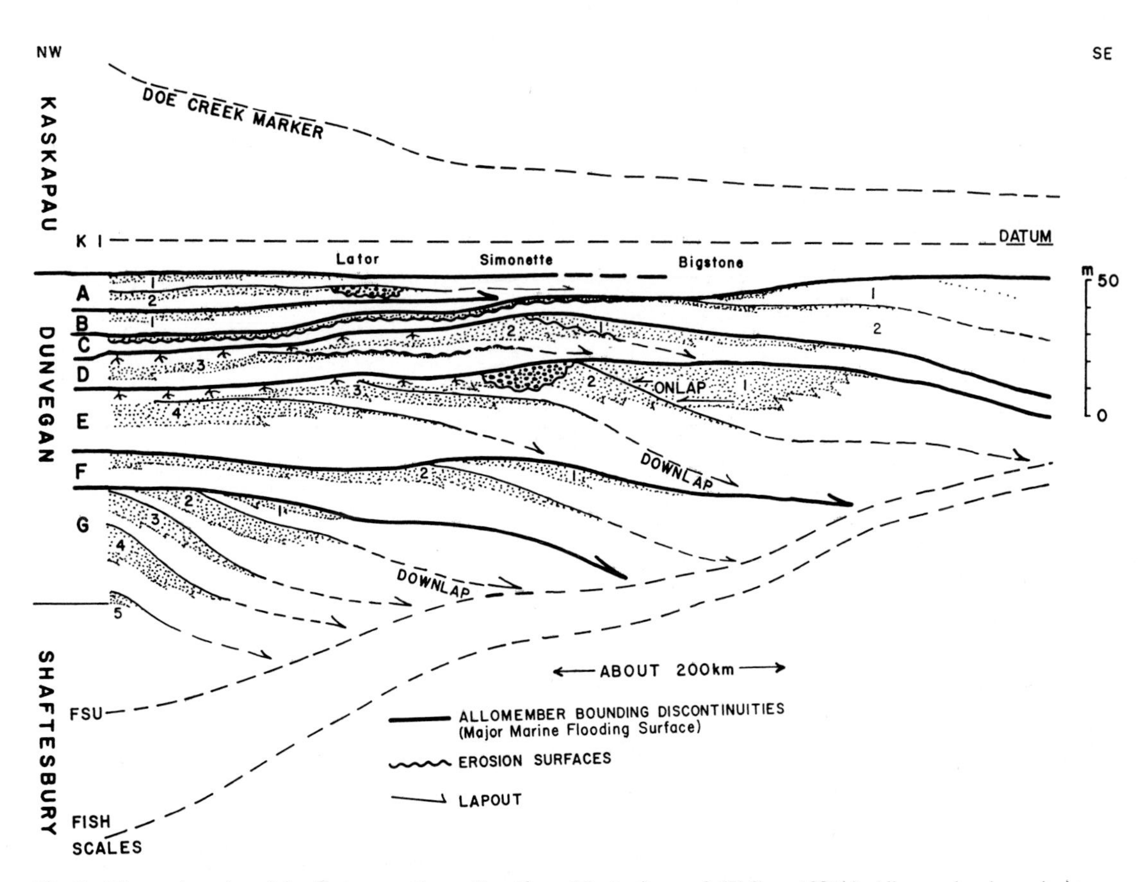

Fig. 3. Allostratigraphy of the Dunvegan Formation (from Bhattacharya & Walker, 1991b). Allomember boundaries cross-cut the lithostratigraphic boundaries. Letters refer to the allomembers and numbers refer to the shingles. Location shown in Fig. 1(b).

transgressive unit at the top that ranges from 0–2 m thick. The transgressive units are underlain in places by a transgressive surface of erosion (TSE) but are capped by regionally extensive marine flooding surfaces that could be correlated laterally for 100 km or more (Bhattacharya, 1988; Bhattacharya & Walker, 1991a). These flooding surfaces were designated as allomember boundaries and were shown to pass laterally into the associated shaly formations implying a genetic relationship between lithofacies previously included in different formations (Bhattacharya & Walker, 1991a). These allomembers are easy to recognize and map regionally, and are descriptively similar to the depositional episodes of Galloway (1989). The allomembers are in turn composed of more localized offlapping subunits (numbered in Fig. 3) which Bhattacharya and Walker (1991a) termed *shingles*. These shingles are similar to parasequences as defined by Van Wagoner *et al.* (1990).

Within the Dunvegan, 19 facies were grouped into seven facies successions (Tables 1 & 2 in Bhattacharya & Walker, 1991b). The seven facies successions represent the deposits of prograding wave-, river- and mixed-influence deltas (successions 1, 2 & 3), fluvial, estuarine and tidal channel fills (successions 4, 5 & 6) and associated lagoonal and delta-plain facies (succession 7). The various surfaces associated with these facies successions allowed them to be linked into distinctive depositional systems (Bhattacharya & Walker, 1991b). The depositional systems were separated from each other by a variety of key surfaces as defined below and were used to identify specific systems tracts within the allomembers.

KEY SURFACES

Figure 4 shows the types of surfaces associated with a typical Dunvegan allomember. Two major categories of surfaces can be recognized, those associated with relative sea-level rise and those associated with relative sea-level fall.

Sea-level rise may be accompanied by marine transgression and associated shoreface erosion which results in a *transgressive surface of erosion* (TSE), also called a *ravinement surface* (Swift, 1968). The ravinement process may result in planation of up to 20 m of sediment (e.g. Demarest & Kraft, 1987; Nummedal & Swift, 1987; Plint, 1988; Trincardi & Field, 1991). Where there is no sig-

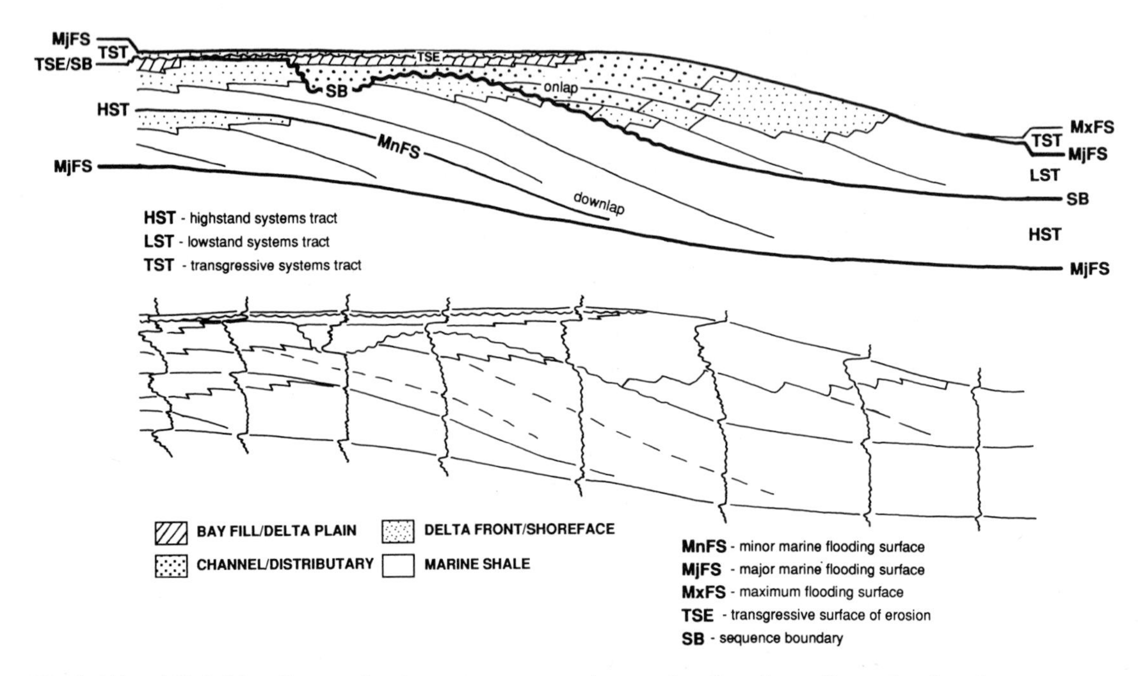

Fig. 4. (Above) Definition diagram showing systems tracts and types of surfaces in an allomember, based on allomember E. (Below) Hypothetical well-log cross-section based on above example.

nificant erosion associated with transgression, the more passive process of flooding may occur resulting in a *marine flooding surface*. A marine flooding surface is defined as a surface which separates younger from older strata across which there is evidence of an abrupt increase in water depth but without significant accompanying erosion (Van Wagoner *et al.*, 1988, 1990). Van Wagoner *et al.* (1988) suggest that this deepening is commonly accompanied by minor submarine erosion. In this paper, however, where erosion can be documented, the term TSE is favoured. In the examples shown below, it is usually possible to distinguish the TSE from the overlying flooding surface. Although the word flooding implies inundation of previously dry areas, the term is used here to encompass correlative areas always under water across which there is simply evidence of deepening. Flooding surfaces may also pass laterally into ravinement surfaces.

Within the Dunvegan, it was possible to develop a hierarchy of different types of flooding surfaces (Bhattacharya & Walker, 1991a). The flooding surfaces that cap individual shingles could be correlated only over a few tens of kilometres or less and were designated as *minor marine flooding surfaces* (MnFS). The flooding surfaces that cap the allomembers, in contrast, were more extensive and could be correlated over distances of several hundred kilometres. These were therefore designated as *major marine flooding surfaces* (MjFS). In places it was also possible to recognize *maximum flooding surfaces* (MxFS) formed at the time of peak transgression and commonly associated with condensed sections. Van Wagoner *et al.* (1990) recognize the same hierarchy of flooding surfaces and suggest that minor marine flooding surfaces cap parasequences whereas major marine flooding surfaces bound parasequence sets. It is important to note, however, that aside from areal extent, the appearance of major versus minor flooding surfaces may be virtually identical in any given core or well log.

The second category of surfaces are those associated with relative sea-level fall and have been referred to as sequence boundaries (Posamentier *et al.*, 1988; Van Wagoner *et al.*, 1988, 1990). A *sequence boundary* (SB), including its correlative non-erosional surfaces, is defined as a surface that separates younger from older strata across which there is evidence of a basinward shift in shoreline position accompanied by a significant hiatus and which is usually erosional (Van Wagoner *et al.*, 1990). Bhattacharya and Walker (1991a) referred to this as a *regressive surface of erosion* (RSE) also termed a *lowstand surface of erosion* by Weimer (1988). In areas of channelling, these surfaces may be expressed as *subaerial erosional surfaces* (Nummedal & Swift, 1987) whereas in interfluves they may be expressed as *subaerial exposure surfaces* (SES). The seaward expression of the sequence boundary may reflect marine erosion resulting from the impingement of fairweather processes on a previously quiescent shelf (Plint, 1988, 1991). Where no erosion occurs, the sequence boundary passes into its *correlative conformity*.

In some places, the erosional surface produced by relative sea-level fall will be subsequently removed by the ravinement process (Plint *et al.*, 1986; Kraft *et al.*, 1987; Plint, 1988; Trincardi & Field, 1991; Posamentier & Chamberlain, this volume, pp. 469–485). The ravinement process has the effect of enhancing the unconformity resulting in a more profound stratigraphic break. The sequence-bounding unconformity, in this case, is a composite surface produced by a combination of regressive and transgressive processes.

RECOGNITION AND CORRELATION OF SURFACES IN THE DUNVEGAN

This section illustrates the nature of surfaces and facies associations in allomembers E and D of the Dunvegan (Fig. 3). Allomember E comprises the deposits of river-dominated depositional systems whereas allomember D comprises the deposits of more wave-dominated systems as described in more detail in Bhattacharya (1989a,b, 1991) and Bhattacharya and Walker (1991b). Despite the differences in the nature of the depositional systems, the stratigraphic geometries and nature of the various key surfaces associated with each allomember are very similar (Fig. 3).

Allomember E

Allomember E comprises four offlapping shingles (Fig. 3). Bhattacharya (1991) showed that the sandbody geometries associated with each shingle are characteristic of river-dominated deltas (Fig. 5). The facies within the coarsening-upward successions indicate high sedimentation rates and a lack of wave reworking (Bhattacharya & Walker, 1991b; Bhattacharya, 1991). The shingles appear

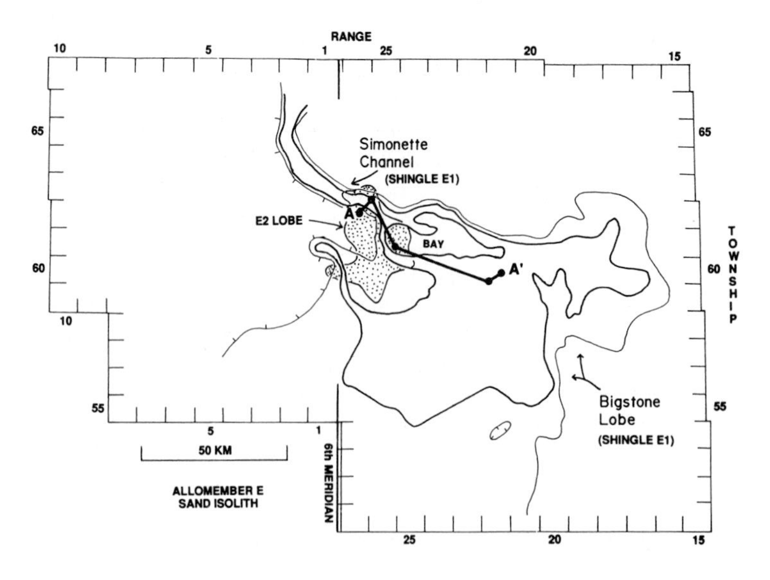

Fig. 5. Map showing position of cross-section AA′ (Fig. 6) and outline of sand bodies in shingle E1 and one sublobe in shingle E2 (stippled). The Simonette channel breaks up into a distributary network which in turn is interpreted to feed the Bigstone delta lobes. Shingle E1 represents the lowstand systems tract. (Modified after Bhattacharya & Walker 1991b.)

to downlap to the southeast onto the top of the underlying allomember F (Fig. 3). Calculations of the average clinoform dip indicate very low values of the order of 0.03 to 0.06° and suggest a gentle ramp-type shelf margin. These calculations, however, are dependent on the datum used and dips may have been higher in the originally uncompacted sediments.

A map view of shingles E2 and E1 (Fig. 5) shows a major channel (Simonette channel) incising into the older E2 lobe. The main channel splits into a distributary network which feeds a lowstand delta in the Bigstone area to the southeast.

The detailed stratigraphic relationship of the units shown on this map is illustrated in Fig. 6. The wells on this cross-section were chosen from more detailed cross-sections shown in Bhattacharya (1991). The cored wells (Fig. 6(b)) show facies successions and associated surfaces at the interfluve, channel, interdistributary bay and delta-front setting. The erosional surface at the base of the incised channel in the Simonette area (well 2-1-63-26W5, Fig. 6(b)) is interpreted as a sequence boundary which cuts into older (highstand) deltas of shingle E2. In the interfluve (well 15-31-62-26W5, Fig. 6(b)) these older highstand deltas are preserved and the sequence boundary is interpreted to correlate with a subaerial exposure surface characterized by *in situ* root traces. The sequence boundary can be traced seaward through an interdistributary bay (well 16-10-61-25W5, Fig. 6(b)) and below the associated lowstand delta (well 10-22-60-22W5, Fig. 6(b)) where it becomes more difficult to recognize, eventually passing into a correlative conformity.

There are several surfaces associated with the thin transgressive unit that lies at the top of allomember E. In wells 15-31-62-26W5, 2-1-63-26W5 and 16-10-61-25W5 the base is erosional and thus forms a TSE (see Fig. 6(b)). The TSE is interpreted to mark the base of the transgressive systems tract. Seaward, in well 6-19-60-21W5 (Fig. 6(b)), the degree of erosion is less and the TSE passes into a correlative marine flooding surface. In most of the wells, the thin transgressive deposits are sharply overlain by stratified shales and mudstones. This sharp contact is taken as a marine flooding surface which marks an abrupt deepening. This flooding surface was correlatable over hundreds of kilometres and was taken as the allomember boundary (MjFS in Fig. 6(b)). The maximum flooding surface is interpreted to coincide with the allomember bounding the major marine flooding surface over most of the cross-section. In well 6-19-60-21W5 (Fig. 6(b)) it occurs in the overlying mudstones about 1 m above the major marine flooding surface.

The following sections describe in detail how these various surfaces change in appearance and explain how the facies successions bounded by the surfaces change in a progressively more seaward position.

Well 15-31-62-26W5

Core description and photographs from 15-31-62-26W5 (Figs 7 & 8) illustrate the nature of the surfaces associated with the older preserved highstand systems tract and the associated overlying thin, lowstand and transgressive systems tracts. The core contains two coarsening-upward successions correlating with shingles E3 and E2 (Fig. 7). E3 is represented by distal delta-front deposits and is sharply overlain by deeper marine mudstones of shingle E2 at about 6.8 m in Fig. 7. The contact is marked by a thin burrowed zone interpreted as a marine flooding surface (MnFS in Fig. 8(a)). Correlation of this surface suggests that it is a minor marine flooding surface (Bhattacharya & Walker, 1991a; Bhattacharya, 1991). The facies in E2 also coarsen upward (Fig. 7) and have been interpreted as representing the deposits of a prograding river-dominated delta front (Bhattacharya, 1988; Bhattacharya & Walker, 1991b; Bhattacharya, 1991). E2 is interpreted to represent a highstand systems tract *relative* to shingle E1. The top of E2 is penetrated by root traces (R in Fig. 8(b)) interpreted as indicating subaerial exposure. This subaerial exposure surface (SES in Fig. 8(b)) is interpreted to correlate with the sequence boundary at the base of the Simonette channel which also marks the lower boundary of shingle E1 (Fig. 6(b)). The exposure surface is sharply overlain by a thin unit of black carbonaceous and sideritic mudstone, interpreted as representing deposition in a shallow-water swamp, marsh, or lake. The aggradation of these shallow-water deposits may be the first indication of deepening and subsequent transgression in this area, in which case they would be assigned to the late lowstand or early transgressive systems tract. Alternatively, these deposits may be interpreted in terms of the normal progradation of non-marine facies over the rooted horizon in which case they would belong to the underlying highstand systems tract. Thus the assignment of these facies to a specific systems tract is equivocal.

The first unequivocal indication of transgression is indicated by about 20 cm of bioturbated to rippled muddy sandstone (17.7 m in Fig. 7). The trace fauna are large and represent a diverse assemblage characteristic of the *Skolithos* ichnofacies (Pemberton *et al.*, 1992) and including *Bergaueria*. The sharp base of this sandstone is interpreted to represent a TSE (Fig. 8(b)). If the non-marine mudstones discussed above belong to the highstand systems tract then the TSE would be interpreted to be coincident with the sequence boundary rather than overlying it. The muddy sandstone is sharply overlain by about 10 cm of pervasively bioturbated sandy mudstone containing *Zoophycos* burrows. The sharp contact between sandstone and mudstone is interpreted as indicative of an abrupt deepening and, based on regional correlation, is taken as the major marine flooding surface that terminates allomember E (MjFS, Fig. 8(b)). The contact between the bioturbated sandy mudstone and the overlying stratified silty marine mudstone is also sharp and indicates further deepening. This is interpreted to represent the maximum flooding surface (MxFS, Fig. 8(b)). The better stratified mudstones above are interpreted to belong to the progradational deposits of allomember G, which are interpreted to downlap onto the maximum flooding surface (Fig. 3). The 20 cm or so of sediment between the TSE and the maximum flooding surfaces are interpreted to comprise the transgressive systems tract (TST). Here the TST is highly condensed and, as a result, too thin to distinguish on the well-log cross-section (well 15-31-62-26W5, Fig. 6(a)).

Well 2-1-63-26W5

Figures 9, 10 and 11 illustrate the facies successions and surfaces associated with the Simonette channel in well 2-1-63-26W5. The basal, erosional sequence boundary incises at least 10 m into underlying stratified prodelta mudstones of shingle E2 (Fig. 6(b)), although in places it erodes up to 20 m (Bhattacharya, 1991). The contact is sharp (Fig. 10) and in other wells, sideritized mud clasts and shell fragments form a distinct intraformational lag up to 30 cm thick (Fig. 12). The overlying lowstand channel fill is sandy at the base and grades upward into muddier facies (Fig. 9). The total lack of burrowing and wave rippling, and the presence of unidirectional current ripples suggest a non-marine fill. The channel-fill succession is truncated at 13.6 m (Fig. 9) by a surface overlain by about 20 cm of bioturbated sandy mudstone (Facies 3A in Fig. 9) that contains well-developed *Zoophycos* burrows indicating a marine environment (Fig. 11). The sandy mudstone is coarser than the immediately underlying non-marine facies suggesting an origin as a transgressive lag. The truncational contact beneath this thin but coarser grained facies is interpreted to represent a TSE (Fig. 11). A sharp contact occurs at 13.75 m in Fig. 9 between the sandy mudstone and overlying stratified marine mudstone. This contact is interpreted to represent the MjFS (Fig. 11) that caps allomember E. The sediments between the TSE and major flooding surface are very thin and are practically indistinguishable on the well-log cross-section (Fig. 6(a)). The stratified mudstones are again interpreted to belong to

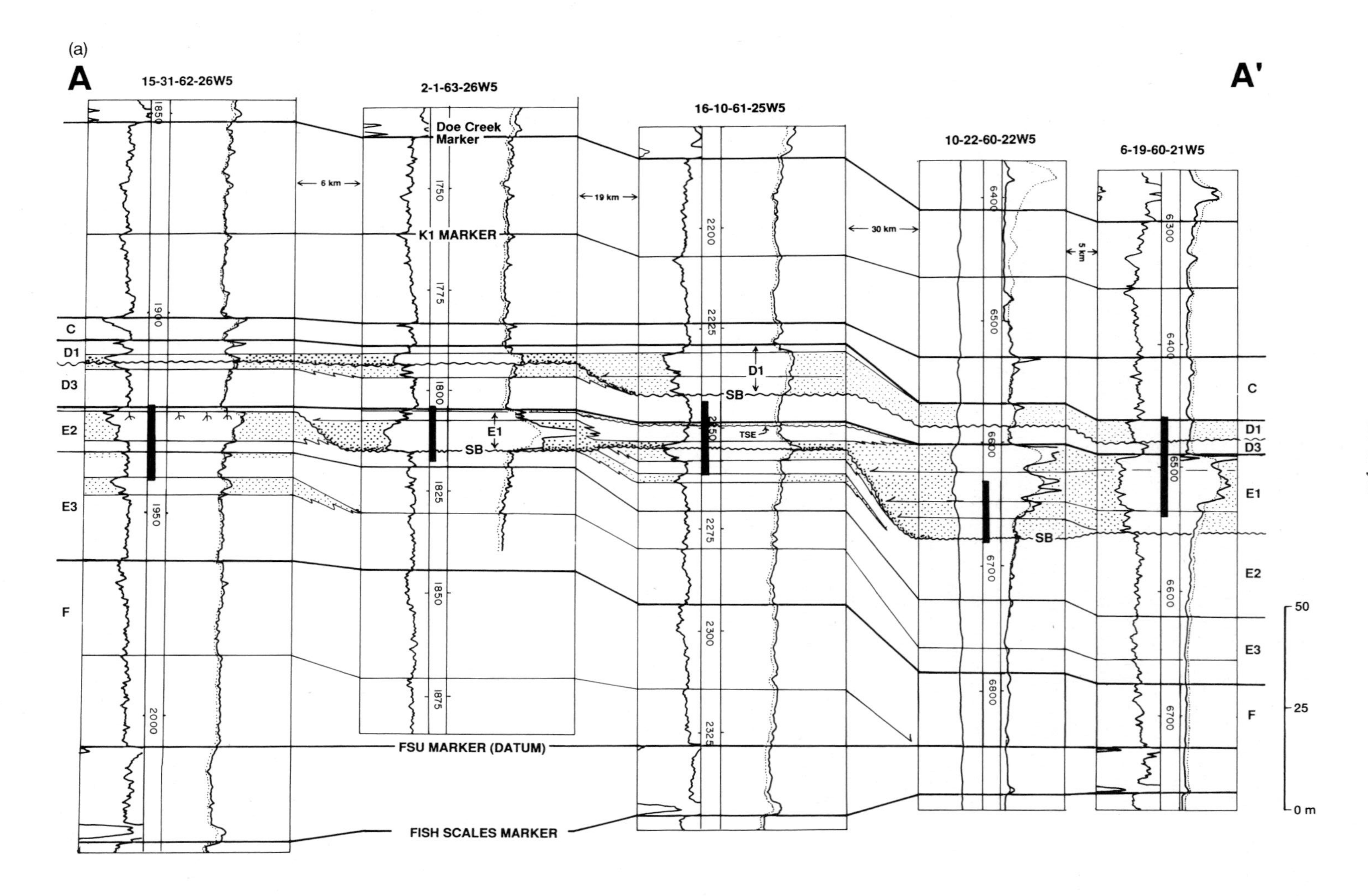
(a)
A
A'
15-31-62-26W5
2-1-63-26W5
16-10-61-25W5
10-22-60-22W5
6-19-60-21W5
Doe Creek Marker
K1 MARKER
FSU MARKER (DATUM)
FISH SCALES MARKER
6 km
19 km
30 km
5 km
SB
TSE
C
D1
D3
E1
E2
E3
F
50
25
0 m

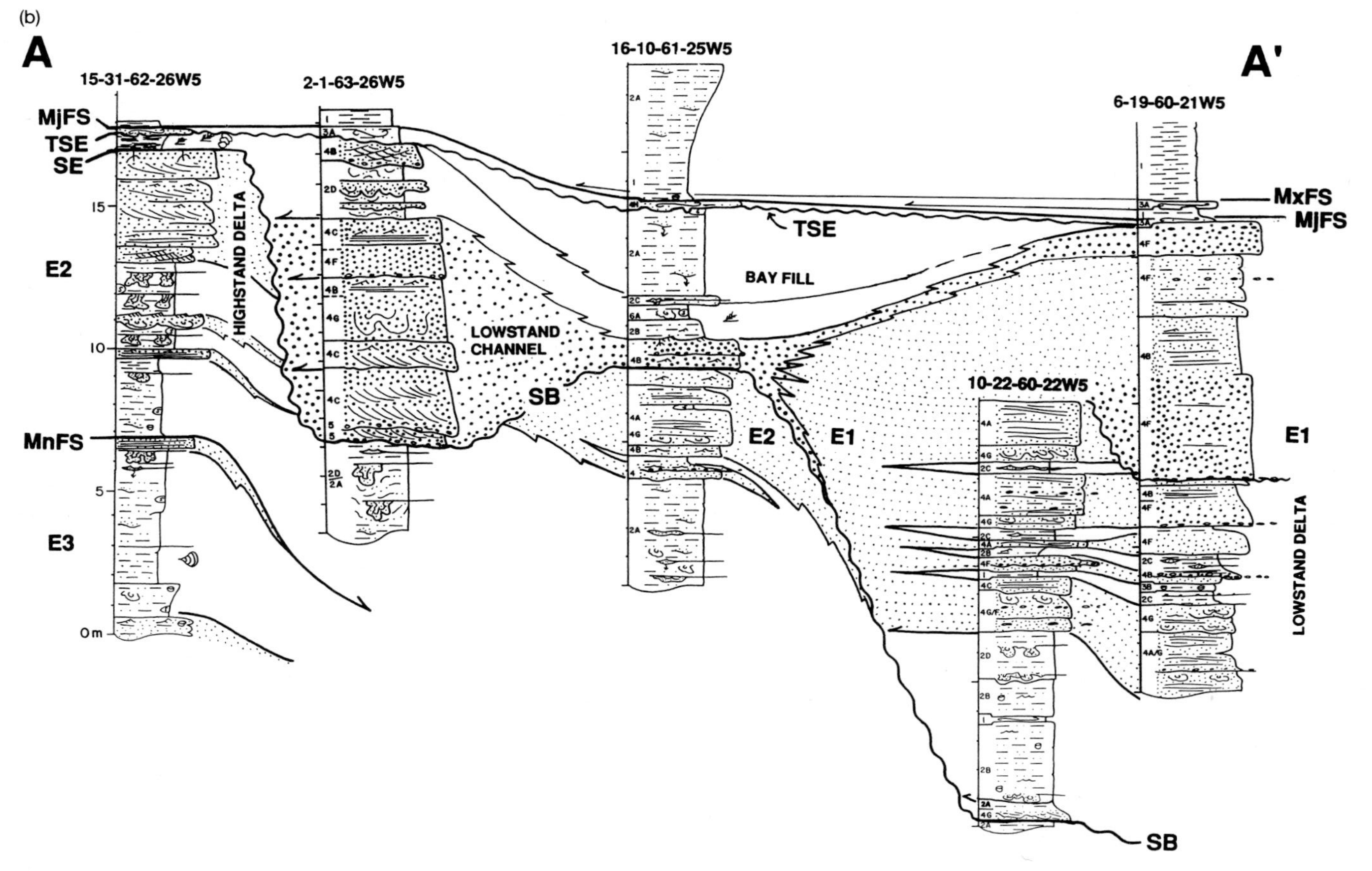

Fig. 6. Cross-section AA′ through the Dunvegan Formation highlighting allomember E (sands are stippled). The wells on this cross-section have been selected for illustrative purposes from more detailed well-log cross-sections published in Bhattacharya (1989a, 1991). Cross-sections are located in Fig. 5. (a) The well-log cross-section highlights the upper and lower markers used to constrain the correlations (FSU, Fish Scales Upper marker; SB, sequence boundary; TSE, transgressive surface of erosion). The lowstand delta in allomember D is thickest in well 16-10-61-25W5 (D1). Numbers in between well-log traces refer to depths in metres or feet. Cored intervals indicated with the black bars. Gamma log traces are on the left, except in well 10-22-60-22W5 which is an SP log. Resistivity or induction logs are on the right, with the shallow (solid line) and deep traces (dotted line) shown. The FSU marker has been used as the datum except in well 2-1-63-26W5 where the K1 marker has been used. (See Bhattacharya and Walker (1991a) for further discussion of the marker horizons.) (b) The core cross-section highlights the geometries associated with allomember E. Note the thin, onlapping, transgressive marine mudstones at the top of allomember E in well 6-19-60-21W5 (TST). SB, sequence boundary; MnFS, minor marine flooding surface; MjFS, major marine flooding surface; MxFS, maximum marine flooding surface; SE, subaerial exposure surface. (See text for discussion.)

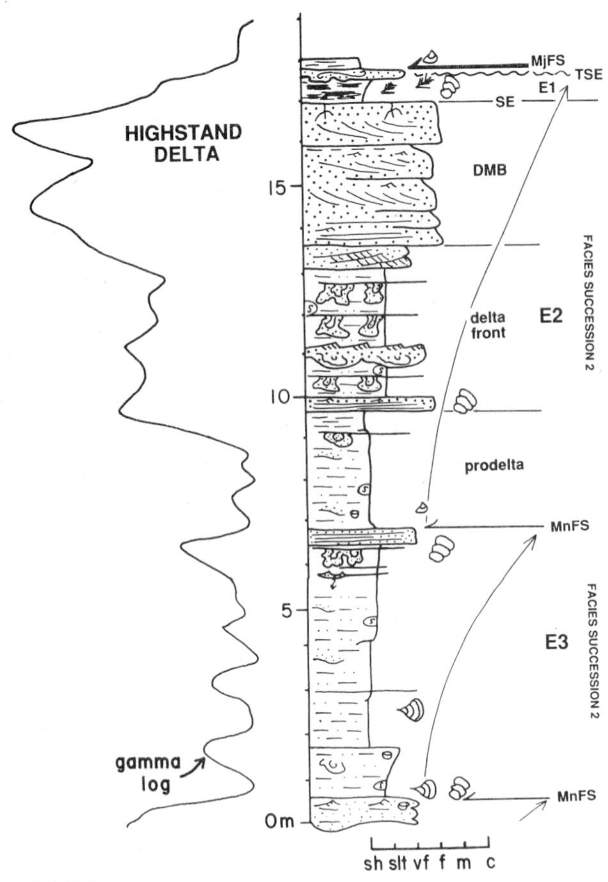

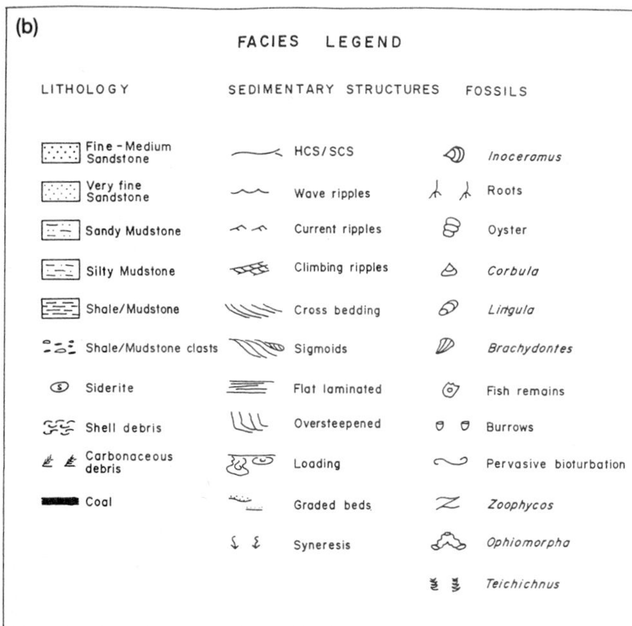

Fig. 7. (a) Detailed vertical core section from 'highstand' (E2) delta front succession in well 15-31-62-26W5 showing significant surfaces. (Modified after Bhattacharya, 1988, 1991). Photographs of the middle and upper parts of the cores are shown in Fig. 8. (Details of the facies successions are given in Bhattacharya & Walker, 1991b.) MjFS, MnFS, TSE and SE as in other figures. (b) Facies legend.

Fig. 8. Core photographs of well 15-31-62-26W5. (a) Bioturbated contact between deformed interbedded HCS sandstones and mudstones below into silty stratified mudstones above is interpreted as a minor marine flooding surface (MnFS) capping shingle E3 (also shown at about 6.8 m in Fig. 7(a)). (b) Top of allomember E showing rooted surface at the top of shingle E2. The roots (R) are interpreted to indicate a subaerial exposure surface (SES), also shown at about 17 m in Fig. 7(a). This is the expression of the sequence boundary on top of the highstand delta in the interfluve position. The SES is overlain by non-marine mudstones and is interpreted to indicate delta-plain aggradation during late lowstand time. The bioturbated sandstone indicates the first significant transgression and is underlain by a TSE. The transgressive systems tract (facies between the TSE and MxFS) is only about 0.5 m thick. It contains a major marine flooding surface (MjFS) and is capped by a maximum marine flooding surface (MxFS). Core diameter is about 4 in, length of core sleeves is about 60 cm.

the progradational lower portion of allomember G suggesting that the maximum and major flooding surfaces are coincident in this well. The lowstand systems tract (LST) is interpreted to be about 9 m thick whereas the TST is only about 15 cm.

Well 16-10-61-25W5

Well 16-10-61-25W5 lies within an interdistributary bay. Description of the core indicates an older, highstand, deltaic coarsening-upward succession (E2, Fig. 13) overlain by an irregular bay-fill succession (E1, Fig. 13).

In this well, recognition of the various surfaces discussed so far is more difficult. The sharp contact between very fine and fine-grained sandstone at 7.7 m (Fig. 13) is interpreted as the sequence boundary although it is not as obvious on the well log (Fig. 6(a)). An intraformational siderite lag, at 8 m, was considered as another candidate for the sequence boundary, although the facies above and below the lag are the same (fine-grained current-rippled sandstone), suggesting no basinward shift in facies. Vertical structures at the top of the sandstone (8.5 m, Fig. 13) were interpreted as possible root traces and considered a third possibility for the

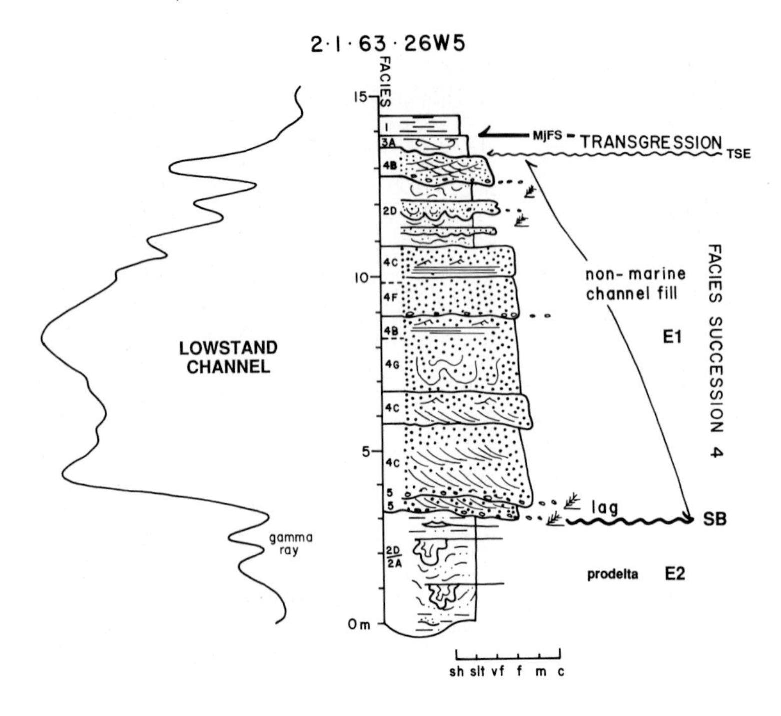

Fig. 9. Fining-upward fluvial channel fill through the E1 lowstand channel (Simonette channel, Fig. 5) in well 2-1-63-26W5 showing significant surfaces. (Modified after Bhattacharya & Walker, 1991b.) Photographs of the upper 4 m are shown in Fig. 11. SB, TSE and MjFS as in other figures.

location of the sequence boundary although the doubtfulness of the root structures and the lack of a grain-size increase suggest that the sequence boundary is lower down.

The sandstone is overlain by heterolithic sandstones and mudstones, with syneresis cracks and a very restricted trace fauna, indicating a shallow brackish water setting. These are interpreted to represent a late lowstand facies which fills the interdistributary bay, similar to the muddy delta-plain facies overlying the rooted surface in the 15-31-62-26W5 well (compare Figs 8(b) & 14(a)). These heterolithic bay-fill sediments of the lowstand systems tract are truncated by a surface (TSE, Fig. 14(a)) overlain by a thin bioturbated sandstone at 13 m in Fig. 13. This sandstone is coarser grained than in the immediately underlying facies suggesting an origin as a transgressive lag. The basal contact is thus interpreted as a TSE and represents the first significant transgressive surface. The TSE is marked by a well-developed burrowed firmground (Fig. 14(b)) indicated by the presence of the *Glossifungites* ichnofacies (McEachern *et al.*, 1992). The overlying muddy sandstone, in contrast, contains *Zoophycos* burrows. The transition between *Glossifungites* ichnofacies (McEachern *et al.*, 1990). The overlying muddy sandstone, in contrast, contains *Zoophycos* burrows. The transition between *Glossifungites* and *Zoophycos* may indicate progressive deepening within the 20 cm of sediment. The bioturbated sandstone is in turn sharply overlain by about 5 cm of bioturbated sandy mudstone. This sharp contact is interpreted as representing the MjFS (Fig. 14(a)) that caps allomember E. The bioturbated mudstones are in turn sharply overlain by stratified silty mudstones interpreted as belonging to the early progradational phase of allomember D. This uppermost contact, at 13.5 m in Fig. 13, is interpreted as the MxFS (in Fig. 14(a)). The sediments between the TSE and MxFS are interpreted to comprise a very thin, condensed, transgressive systems tract. The various surfaces associated with this TST are difficult to distinguish on the well-log cross-section (Fig. 6(a)).

Wells 10-22-60-22W5 and 6-19-60-21W5

Wells 10-22-60-22W5 and 6-19-60-21W5 occur in the position of the lowstand delta (Fig. 5). Cores described from both wells show a coarsening-upward facies succession (Figs 15 & 16) interpreted

Fig. 10. Sequence boundary from well 2-1-63-26W5. Medium-grained sandstone containing small sideritized mud clasts and shell debris sharply overlies stratified silty marine mudstones. Scale is 3 cm.

as representing progradation of a river-dominated delta front.

The facies succession in well 10-22-60-22W5 (Figs 15 & 17) begins with silty marine mudstone. This marine mudstone is sharply overlain by about 50 cm of stratified to deformed sandstone. Correlation (Fig. 6(b)) suggests that the correlative surface to the sequence boundary should be in about this position. This sharp contact between mudstone and sandstone is thus interpreted as the distal expression of the sequence boundary in the marine realm (SB, Fig. 17). The sandstone directly above the sequence boundary is in turn overlain by silty stratified mudstones that are moderately burrowed and contain wave-rippled sandy beds (Fig. 17). The overlying facies are interpreted as indicating deposition in shallower water than the underlying mudstones.

The overlying succession coarsens upward into the deposits of the overlying lowstand delta. In the 6-19-60-21W5 well the upper part of this lowstand succession is cored (Fig. 16). A 5.6 m thick graded sandstone unit, between 7.4 and 13 m in Fig. 16, exhibits a scoured erosional base indicated by ripped-up mud clasts. Unlike the lowstand channel in well 2-1-63-26W5, this sandstone is not interpreted to be underlain by a sequence boundary. It does not appear to erode into prodelta mudstones or into the underlying shingle E2 and is not characterized by an overall increase in grain size. In addition, it was not possible to correlate this erosion surface between wells (Bhattacharya, 1991). This erosional surface is thus interpreted to represent the extension of the distributary channel into its own delta front within the lowstand, rather than marking the base of a new sequence.

The sandy, lowstand delta succession is sharply capped by a thin pervasively bioturbated sandy mudstone (Fig. 16) which contains *Zoophycos* burrows. There does not seem to be significant erosion associated with the base and the TSE is interpreted to have passed seaward into its correlative marine flooding surface. A second bioturbated zone above the initial flooding surface is interpreted as part of the onlapping transgressive mudstone facies within the transgressive systems tract. The MxFS (in Figs 16 & 18) is interpreted to lie immediately above this second bioturbated zone. An alternative interpretation is that the maximum flooding surface lies a little higher at the slight γ-log increase at 19 m in Fig. 16. A third interpretation suggests that the MxFS may have been even higher but was eroded by the overlying sequence boundary in allomember D. Given the three alternatives, there is therefore some uncertainty as to whether the mudstones above allomember E in these wells should be assigned to the transgressive or highstand systems tract.

Allomember D

Allomember D contains three offlapping shingles (1, 2 and 3 in Fig. 3). The plan-view geometry of sandstones within allomember D (Fig. 19) has been interpreted as representing the deposits of wave-dominated prograding barrier and deltaic systems

Fig. 11. (a) Upper 4 m of the fluvial channel fill in well 2-1-63-26W5 showing graded, rippled sandstone beds with basal mud-chip layers and interbedded convolute-laminated mudstones. This interpreted non-marine facies is truncated by a TSE. Scale in lower right is 20 cm. (b) Detail of the thin transgressive sediments containing *Zoophycos* burrows, which overlie the TSE. The sharp contact between the bioturbated sandy mudstone and the stratified silty shale at the top of the photograph marks the MjFS. Scale is 3 cm.

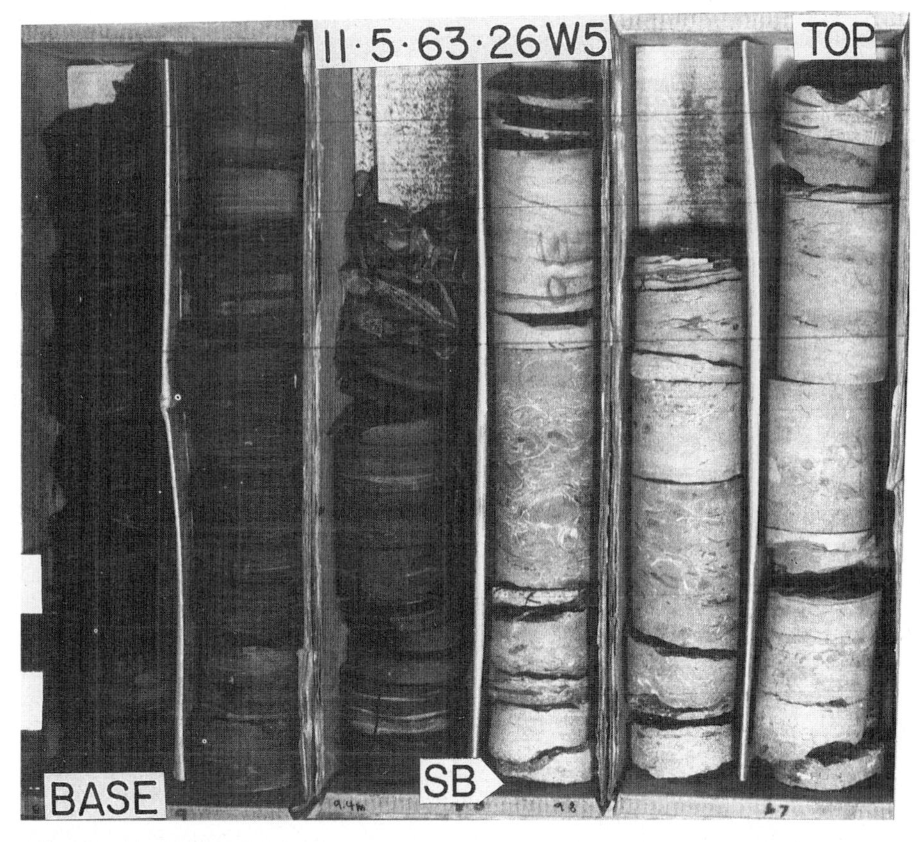

Fig. 12. Sequence boundary (SB) at the base of the lowstand channel in allomember E, in well 11-5-63-26W5. The sequence boundary overlies deeper marine stratified, silty mudstones deposited in a prodeltaic environment. The sequence boundary is overlain by a well-developed intraformational lag comprising mud clasts and shell debris. It passes up into cross-bedded sandstones. Scale at lower left is 20 cm.

(Bhattacharya & Walker, 1991b). Shingle D3 is poorly preserved and is truncated over much of the study area by an erosional surface (Bhattacharya, 1989a). Shingle D2 has been interpreted as a wave-dominated highstand barrier (D2 barrier, Fig. 19). The highstand-barrier shoreface succession is characterized by wave-formed sedimentary features including wave ripples and abundant hummocky cross-stratification and is cut in places by tidal inlets (Bhattacharya, 1989b; Bhattacharya & Walker, 1991b). Shingle D1 is characterized by an abrupt seaward shift in shoreline position accompanied by channelling (Waskahigan channel, Fig. 19) and is interpreted as a lowstand systems tract. The Waskahigan channel is interpreted as feeding a lowstand delta (D1 lobe, Fig. 19). The shape of the lobe and the predominance of wave-formed sedimentary structures suggest a wave-dominated delta (Bhattacharya & Walker, 1991b).

The detailed stratigraphic relationship of the units shown on this map is illustrated in Fig. 20. Like Fig. 6, the wells on this cross-section were chosen from more detailed cross-sections shown in Bhattacharya (1989a,b). The cored wells (Fig. 20(b)) show facies successions and associated surfaces at the highstand barrier (well 7-10-63-1W6), lowstand channel (well 7-21-64-23W5) and lowstand delta (wells 3-28-61-24W5, 6-29-62-23W5 and 6-19-60-21W5). The erosional surface at the base of the Waskahigan channel (well 7-21-64-23W5, Fig. 20(b)) is interpreted as a sequence boundary and cuts out all of the older highstand barrier deposits of shingle D2. Adjacent to the channel these older highstand deltas are preserved

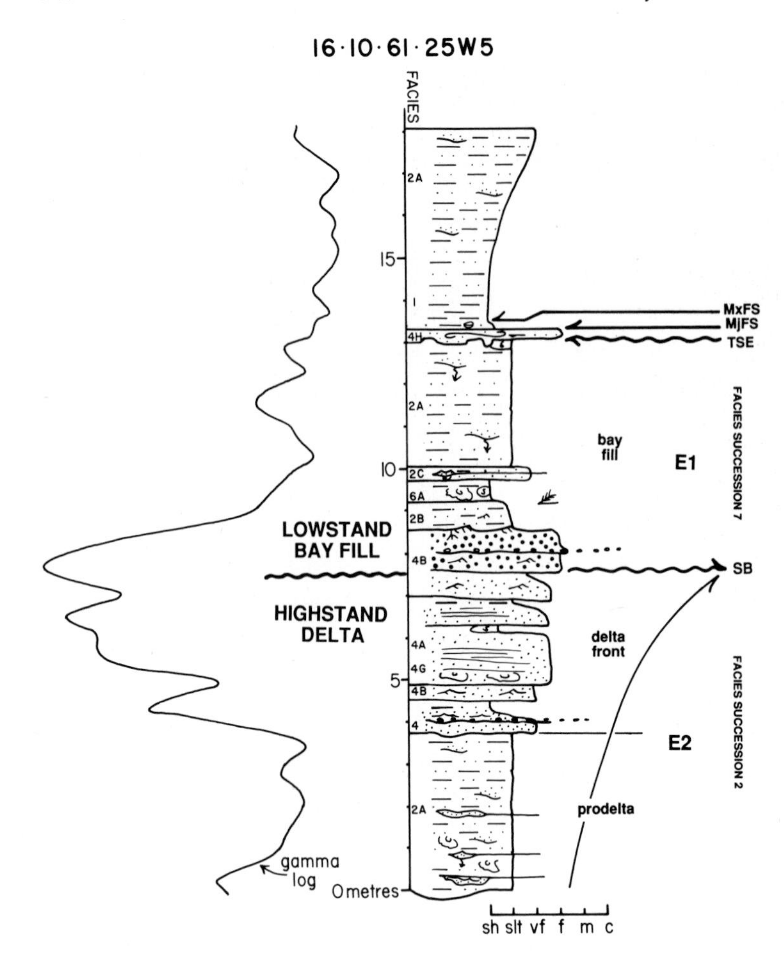

Fig. 13. Core description from well 16-10-61-25W5. The sequence boundary (SB) separates underlying highstand delta-front deposits from overlying lowstand bay-fill deposits. The bay-fill succession is terminated by a TSE. The thin, transgressive bioturbated sandstones (Facies 4H) are in turn sharply overlain by bioturbated mudstones. This contact is interpreted as a MjFS. The MxFS occurs just above the MjFS. Photographs of these surfaces are shown in Fig. 14.

(well 7-10-63-1W6, Fig. 20(b)) and the sequence boundary is interpreted to correlate with a subaerial exposure surface characterized by *in situ* root traces. The sequence boundary can be traced seaward below the associated lowstand delta (well 3-28-61-24W5, Fig. 20(b)) where it becomes more difficult to recognize. In a shelfal position, the sequence boundary is expressed as a sharp contact between deeper water mudstones below and shallow-water mudstones above in wells 6-29-62-23W5 and 6-19-60-21W5 (Fig. 20(b)).

The allomember-bounding the major marine flooding surface sharply overlies lowstand sandstones in wells 6-19, 6-29 and 3-28 (MjFS, Fig. 20(b)). Further landwards, in wells 7-21 and 7-10 (Fig. 20(b)) the MFS overlies a thin, transgressive sandy mudstone underlain by a transgressive surface of erosion. The MxFS is interpreted to coincide with the allomember bounding MjFS over the entire cross-section.

The following sections describe in detail the changes in appearance of the various surfaces outlined above. They focus on the wells shown in Fig. 20 and explain how the facies successions bounded by the surfaces change in a progressively more seaward position.

Well 7-10-63-1W6

Figures 21 and 22 illustrate the nature of the surfaces and facies successions associated with the highstand systems tract in allomember D. Description of the core from the 7-10-63-1W6 well shows two coarsening-upward facies successions (Fig. 21). The lower succession (shingle D3) is capped by a thin, siderite pebble lag, and some erosion and

(a)

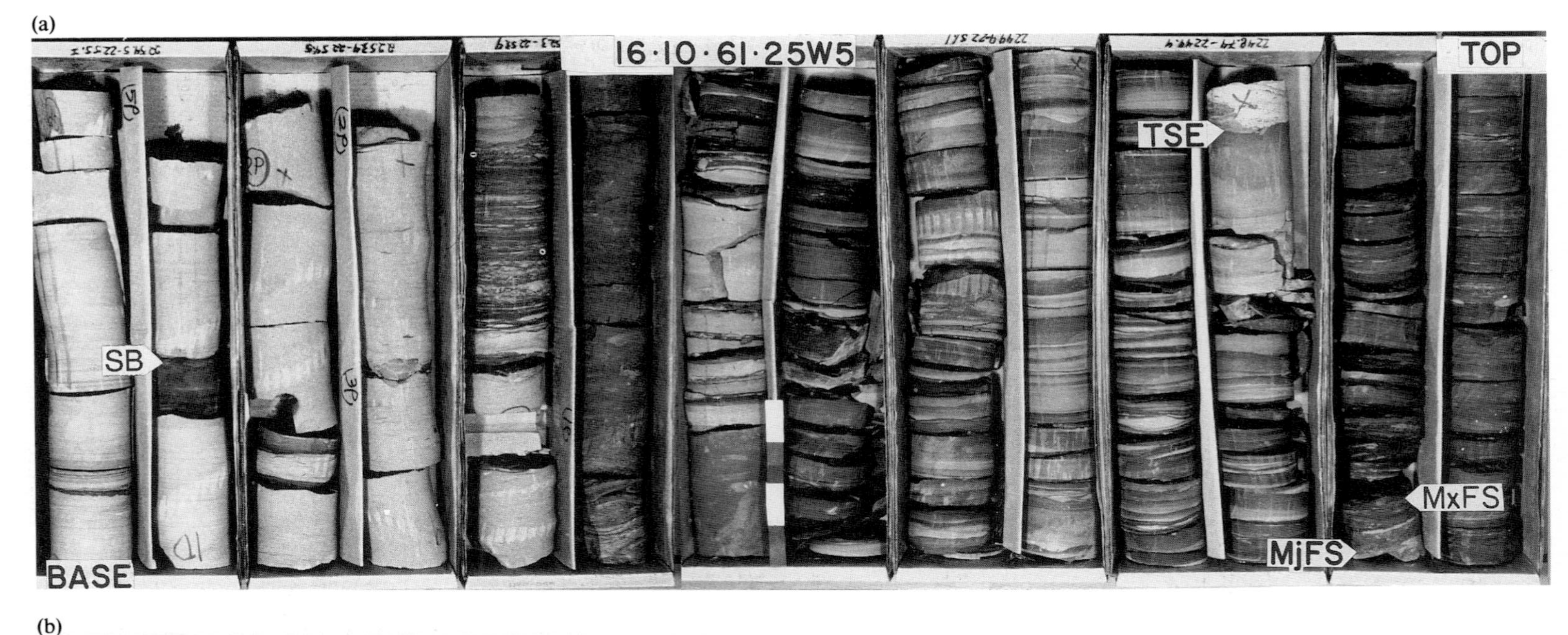

(b)

Fig. 14. Core photographs of well 16-10-61-25W5. (a) Box photographs of interval between 7 and 14 m in Fig. 13. Sequence boundary (SB) marks contact between very fine and fine-grained sandstone and separates underlying highstand delta sandstone from overlying lowstand sandstone. Heterolithic facies above the lowstand sandstone and below the TSE represent lowstand bay-fill deposits. Sediments between the MxFS and the TSE represent the transgressive systems tract. Scale in middle of photograph is 20 cm. (b) Detail of TSE in well 16-10-61-25W5 (left: bottom; right: top) showing firmground burrowing characteristic of the *Glossifungites* ichnofacies. Core is about 3.25 in in diameter.

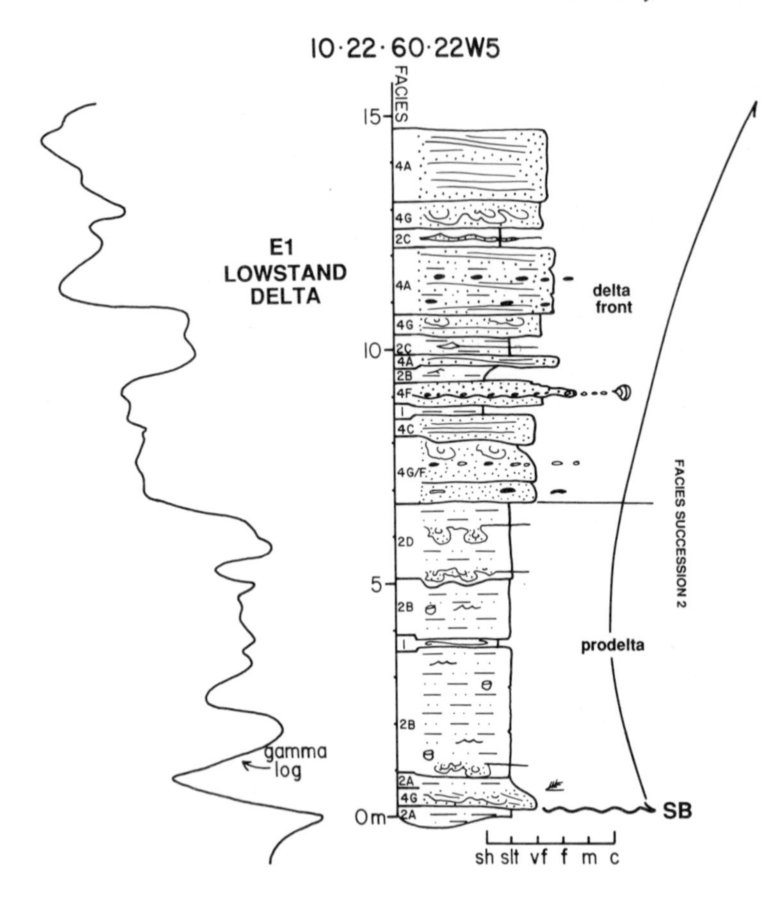

Fig. 15. Core description through the lower portion of the E1 lowstand delta in well 10-22-60-22W5. Sequence boundary (SB) is marked by a thin sandstone bed sharply overlying black deep-water mudstones (Facies 2A). Shallower water rippled to burrowed silty mudstones (Facies 2B) overlie the sandstone.

truncation have been documented (Bhattacharya, 1989a). Regional correlations do not indicate that this erosion is accompanied by channelling or a basinward shift in shoreline position, so it may represent a transgressive surface of erosion within the highstand systems tract (Bhattacharya, 1989a). The next succession has been interpreted as a prograding storm- and wave-dominated shoreface. This is indicated by its shore-parallel, linear geometry (Fig. 19) and by the nature of the shoreface sandstones which are dominated by wave ripples and hummocky cross-stratification (Bhattacharya & Walker, 1991b). It is interpreted as a highstand shoreface *relative* to shingle D1. The uppermost, flat stratified beach sandstones are penetrated by abundant root traces (R in Fig. 22) overlain by a thin muddy coal (C, Fig. 22) followed by a few centimetres of rippled sandy mudstones (Figs 21 & 22). The rooted surface is interpreted as a subaerial exposure surface preserved at the top of the highstand shoreline (shingle D2). This exposure surface is interpreted to represent the local expression of the SB away from the incised valley (Fig. 20(b)).

There are two ways of interpreting the thin coaly unit above the exposure surface. It may represent the preserved remnants of material responsible for producing the root traces during normal progradation of the highstand, in which case the SB would be above the coal. Alternatively, the coal may represent aggradation of the coastal plain as a result of deepening and may herald the beginning of transgression in that area. In this case, the coaly unit would belong to the late lowstand or early transgressive systems tract. Because roots were not observed in the coaly facies and because of the presence of mud and silt throughout this thin interval, the latter interpretation is favoured.

The non-marine coal and mudstone facies are sharply overlain by about 30 cm of pervasively bioturbated sandy mudstone containing a *Zoophycos*

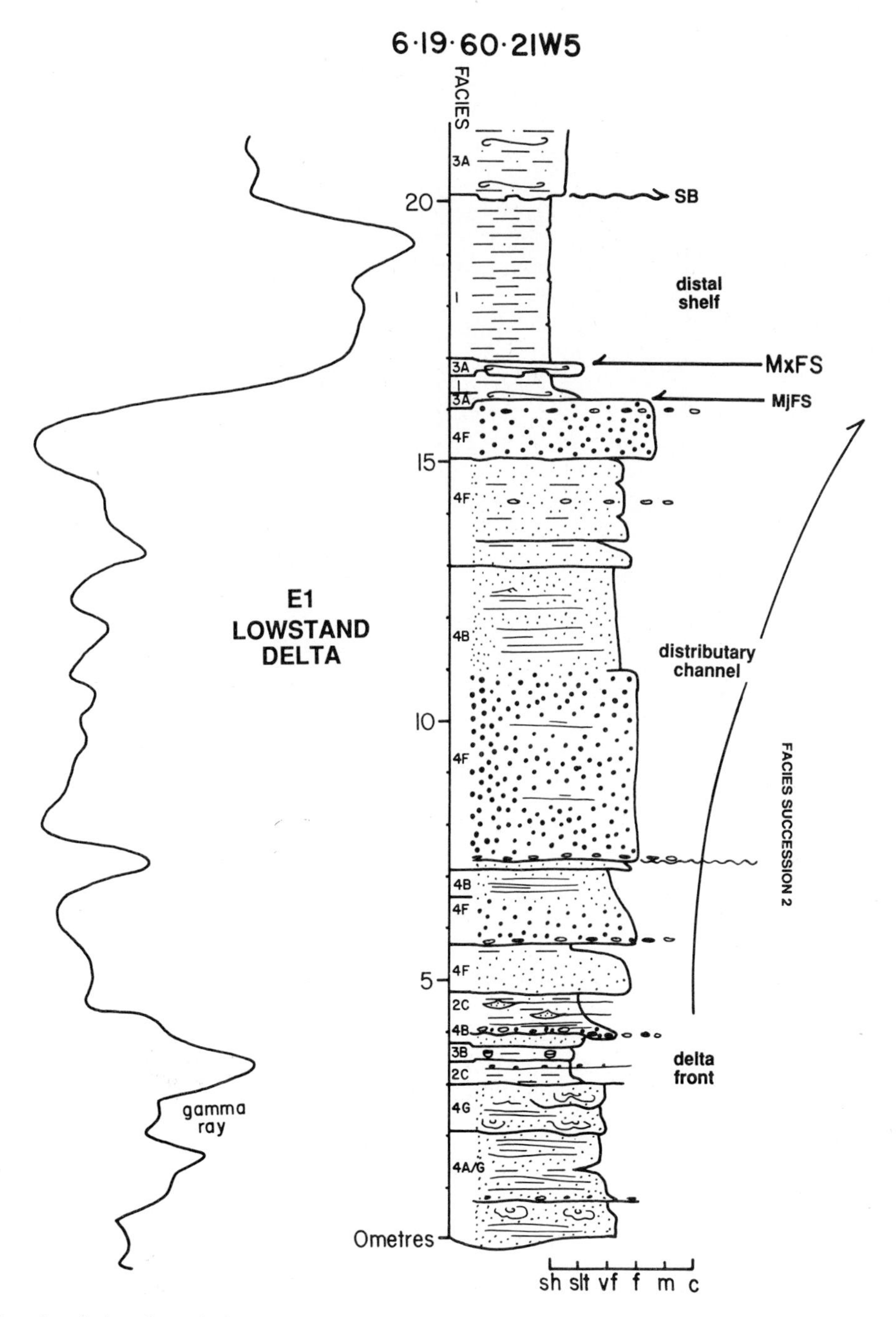

Fig. 16. Core description through the upper portion of the E1 lowstand delta in well 6-19-60-21W5. Massive sandstone between 7.4 m and 13 m is interpreted as the distributary channel cutting into its own mouth-bar sandstone. The prograding delta-front succession is capped by transgressive mudstones at 16 m. MjFS and MxFS as in other figures.

Fig. 17. Photograph of lower portion of core from well 10-22-60-22W5. Deep-water laminated mudstones underlie the sequence boundary (SB). SB is marked by a sharp contact between the mudstones and the overlying stratified to convoluted sandstone. Interbedded wave-rippled and moderately burrowed sandstones and mudstones above are interpreted to be deposited in shallower water than the black mudstones at the base of the core. Scale below the SB label in the lower left is 20 cm. (See text for further explanation.)

ichnofacies (Pemberton *et al.*, 1992). The base of the sandy mudstone is interpreted to represent the first significant transgressive surface (TSE in Fig. 22). The contact with the sandy mudstone and the overlying marine shale represents the major marine flooding surface that caps allomember D (MjFS in Fig. 22). The maximum flooding surface is interpreted to coincide with the allomember boundary.

Well 7-21-64-23W5

Description of the core from well 7-21-64-23W5 (Fig. 23) suggests that it represents an estuarine fill (Bhattacharya, 1989a; Bhattacharya & Walker, 1991b). The base of the estuary fill is interpreted as the sequence boundary. The occurrence of a lowstand delta further seaward suggests that the erosion surface was cut by fluvial processes. In this well, the estuary-fill thickness indicates about 10 m of erosion, although in other parts of the valley this erosion surface indicates incision of up to 35 m (Bhattacharya, 1989b; Bhattacharya & Walker, 1991a). The coarser grained sandy facies in the lower portions of the estuary fill (Figs 23 & 24) were probably derived from upstream fluvial systems, although the presence of double mud drapes and marine trace fauna indicate that marine processes affected and reworked these sediments. Tidal processes may therefore have resulted in modification and additional scouring of the originally fluvially formed erosion surface. The sandy portion of the channel fill is gradationally overlain by about 5 m of bioturbated interbedded mudstones and sandstones (Fig. 23). The interbedded sandstones become hummocky cross-stratified upwards. The transgressive estuarine fill is ultimately capped by a thin, bioturbated sandy mudstone at 11.4 m in Fig. 23. The base of the mudstone may represent an initial transgressive surface and is probably correlative with the TSE in well 7-10-63-1W6 (Fig. 20(b)). The sharp contact between the sandy

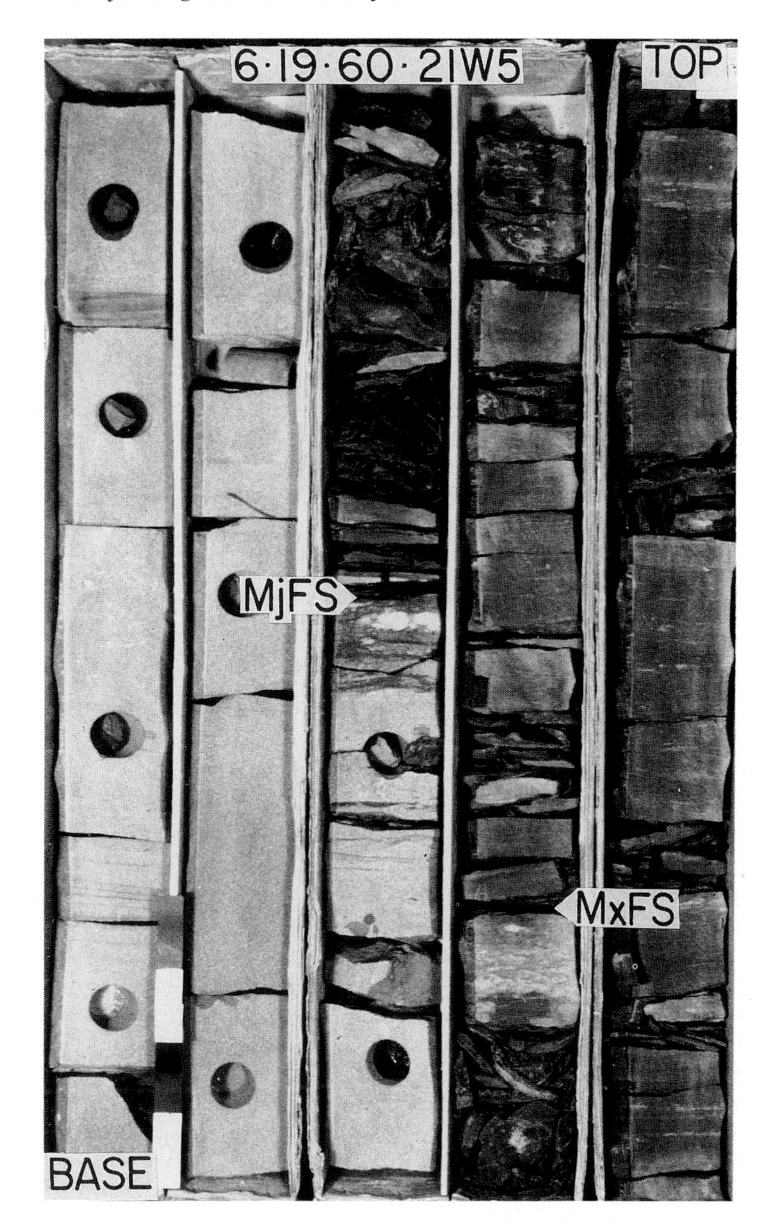

Fig. 18. Photographs of interval from 14–18 m in well 6-19-60-21W5 (Fig. 16). Bioturbated sandy mudstone below the MjFS caps massive poorly stratified sandstones belonging to the E1 lowstand delta. The TST comprises laminated to bioturbated silty mudstone. The top of the TST is indicated by the MxFS and is overlain by laminated shales. (See text for further explanation.)

mudstone and the overlying shale is interpreted to mark the marine flooding surface that caps allomember D (MjFS, Fig. 23). The overlying stratified shales and silty mudstones are interpreted as belonging to the progradational phase of the next highstand (allomember C) and the major and maximum flooding surfaces are interpreted to coincide in this well.

Wells 3-28-61-24W5, 6-29-62-23W5 and 6-19-60-21W5

Wells 3-28-61-24W5, 6-29-62-23W5 and 6-19-60-21W5 contain core from the lowstand delta in shingle D1. Although the top of allomember D is not cored in the 3-28-61-24W5 well, the well-log response indicates a sharp contact between sand-

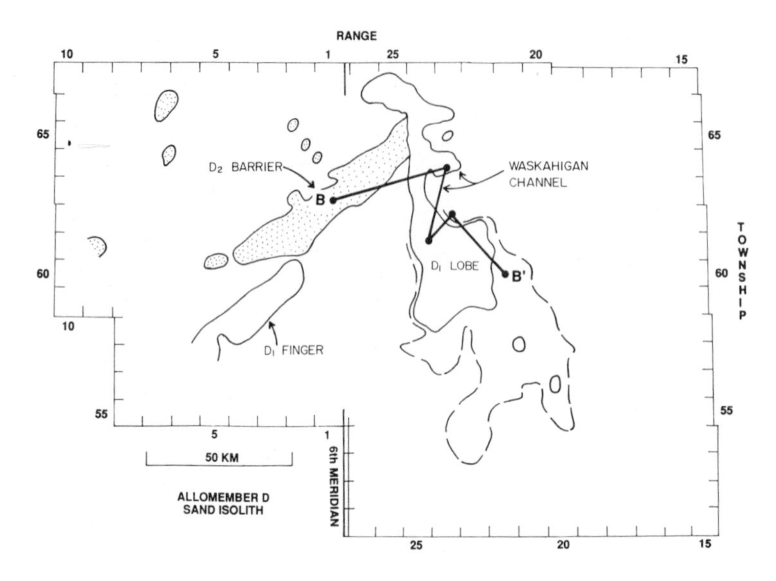

Fig. 19. Map showing position of cross-section BB′ (Fig. 20) and outline of sand bodies in all of the shingles within allomember D. The Waskahigan channel (D1) cuts into the northern end of the older D2 highstand barrier (stippled) and feeds a lowstand wave-dominated delta to the southeast (D1 lobe). The D1 finger is another channelized unit relating to the D1 incision. (Based on sandstone isolith map in Bhattacharya & Walker, 1991b.)

stone and mudstone interpreted to represent the marine flooding surface that caps allomember D (MjFS, Fig. 25). A TSE, if present, is not distinguishable on the well log alone. The high γ-ray log response at 13.4 m (Fig. 25) may indicate the presence of a condensed section, implying coincidence of the major and maximum flooding surface (Bhattacharya & Walker, 1991a). Two coarsening-upward facies successions are observed in well 3-28-61-24W5 (Figs 25 & 26). The lower succession coarsens from a shale into stratified siltstones and mudstones. A minor flooding surface at 3.8 m in Fig. 25 is indicated by a burrowed top and a sharp transition back into a bioturbated shale (MnFS, Fig. 26). The next succession coarsens into interbedded sandstones and mudstones. At 5.7 m there appears to be a relatively abrupt contact with the interbedded facies below and the more massive sandstones above. The overlying massively bedded sandstone is hummocky cross-stratified and is interpreted as a sharp-based shoreface. The sharp base is interpreted as the marine expression of the SB. These sharp-based shorefaces are common in the Western Interior and in some cases are thought to represent shoreface incision related to relative drops of sea level termed 'forced regressions' (Plint, 1988, 1991; Posamentier *et al.*, 1992; Posamentier & Chamberlain, this volume, pp. 469–485).

Well 6-29-62-23W5 lies laterally adjacent to well 3-28-61-24W5 on the inner fringes of the D1 lowstand delta (Fig. 19). There appears to be a relatively sharp transition between deeper water stratified silty mudstones below and shallow-water burrowed sandy mudstones above at 1.2 m in Fig. 27. This contact is interpreted to represent the distal expression of the SB, although its expression is subtle and the SB looks like a correlative conformity. The facies succession above the SB indicates a relatively gradual coarsening upward from stratified burrowed mudstones and sandstones into hummocky cross-stratified sandstones with thin, interbedded bioturbated mudstone (Facies 4A, Fig. 27). The uppermost metre of sandstone (Facies 4H, Fig. 27) is rather more bioturbated than the sandstones immediately below, and may indicate biological reworking during the last phases of the lowstand progradation. The sharp contact between the sandstone and overlying marine mudstones at 9.9 m in Fig. 27 is interpreted as the marine flooding surface that caps allomember D. There is no evidence of significant marine erosion here.

Well 6-19-60-21W5 contains the most distal core from the D1 lowstand (Figs 19, 28 & 29). There is a sharp contact at 20 m in Fig. 28 between stratified, unburrowed silty mudstones below, interpreted to have been deposited in deeper water, and the burrowed to rippled sandier mudstones above, interpreted to have been deposited in shallower water. Like well 6-29-62-23W5, this sharp contact is interpreted as the seaward expression of the SB (Fig. 29). The overlying sandy mudstones gradually coarsen upward into bioturbated to wave-rippled muddy sandstones (Fig. 29). These sandstones are interpreted to represent the shelf equivalents of the lowstand shoreface observed in the 3-28-61-24W5 well (Fig. 20). There is a sharp, burrowed and sideritized contact at 25.8 m in Fig. 28, between the sandy mudstones below and the overlying thin bioturbated mudstone. The bioturbated mudstone (Facies 3A in Fig. 28) is about 10 cm thick and passes abruptly into deeper marine shale (Facies 1 in Fig. 28). The contact between the sandstone and mudstone is interpreted as a marine flooding surface (MjFS in Fig. 29) and marks the top of allomember D. The contact between mudstone and shale is interpreted as the maximum flooding surface (MxFS in Fig. 29), although the γ-ray log response (Fig. 28) indicates that it might be higher, at about 28 m.

INTERPRETATION OF DUNVEGAN SURFACES

Development of an allomember

The stratigraphic geometries in allomembers E and D are similar and are interpreted to result from deposition during a fourth-order cycle of relative sea-level change. Figure 30 shows the sequence of events during a cycle of sea-level change which yields stratigraphic geometries similar to those described above in allomembers E and D.

Time 1 is characterized by normal shoreline progradation during highstand. At time 2, a relative fall of sea level produces a sequence boundary. In places the SB is expressed as a fluvially incised surface cut into the older highstand systems tract (e.g. well 2-1-63-26W5, Fig. 6(b) & well 7-21-64-23W5, Fig. 20(b)) whereas in interfluves it is expressed as a subaerial exposure surface capping the older highstand shoreline deposits and characterized by rooting such as seen in wells 15-31-62-26W5 (Fig. 6(b)) and 7-10-63-1W6 (Fig. 20(b)). The fall is accompanied by an abrupt seaward shift in facies and the deposition of a lowstand delta or shoreface further seaward (time 2). In places (e.g. well 3-28-61-24W5, Fig. 20(b)), the lowstand shoreface may incise into underlying deeper marine facies. In this case, the SB is expressed as a sharp-based shoreface indicating a forced regression, different from the normal regression during highstand (Plint, 1988, 1991; Posamentier *et al.*, 1992). On the shelf, the SB may be expressed by a sharp contact between shallower water burrowed to bioturbated sandy mudstones and underlying deeper water stratified mudstones or marine shales (e.g. well 10-22-60-22W5, Fig. 6(b) & well 6-19-60-21W5, Fig. 20(b)). Some scouring may be caused by impingement of fairweather processes onto a previously deeper water shelf (Plint, 1988, 1991). Farther seaward, the SB is expressed as a correlative conformity.

During late lowstand (time 3), characterized by a slow relative rise of sea level, the upper portion of the lowstand systems tract may become more highly affected by marine processes as deposition slows. For example, the upper part of the D1 lowstand in well 6-29-62-23W5 (Fig. 27) was more bioturbated. Backfilling of lowstand channels and valleys and infilling of interdistributary bays may also occur at this time as a result of the inability of the sediments in the channel to be transported seaward. Some of the originally fluvial channels may be transformed into estuaries, especially in cases where the energy of fluvial systems is low, such as was interpreted for allomember D. Aggradation of the delta plain may also occur.

In time 4, rapid relative sea-level rise causes the shoreface to migrate landward forming an extensive transgressive surface of erosion (TSE). The valley has been filled completely and the TSE has migrated across the now abandoned channel and valley fills. As the TSE migrates, it may overlie the aggradational delta-plain facies deposited during the late lowstand. Because the TSE is diachronous (in contrast to the SB) some of these aggradational delta-plain deposits underneath the TSE may actually be younger than the bioturbated sandy mudstones that overlie the TSE in wells further seaward. This diachroneity of facies associated with the TSE has also been shown by Nummedal and Swift (1987). The TSE may also coincide with the SB forming a composite surface.

Hierarchy of sequences and systems tracts

The concept of systems tracts, as applied to the Dunvegan, has been used at two different scales. At the scale of the Alberta basin the Dunvegan allomembers have been grouped into three systems tracts which are defined as components of a third-order cycle of eustatic sea-level change (Bhattacharya, 1988, in press; Bhattacharya & Walker, 1991a). The allomembers are interpreted as fourth-order cycles and contain a variety of systems tracts defined on the basis of the various surfaces described in detail above. The sequence boundaries in allomembers E and D are here interpreted to sepa-

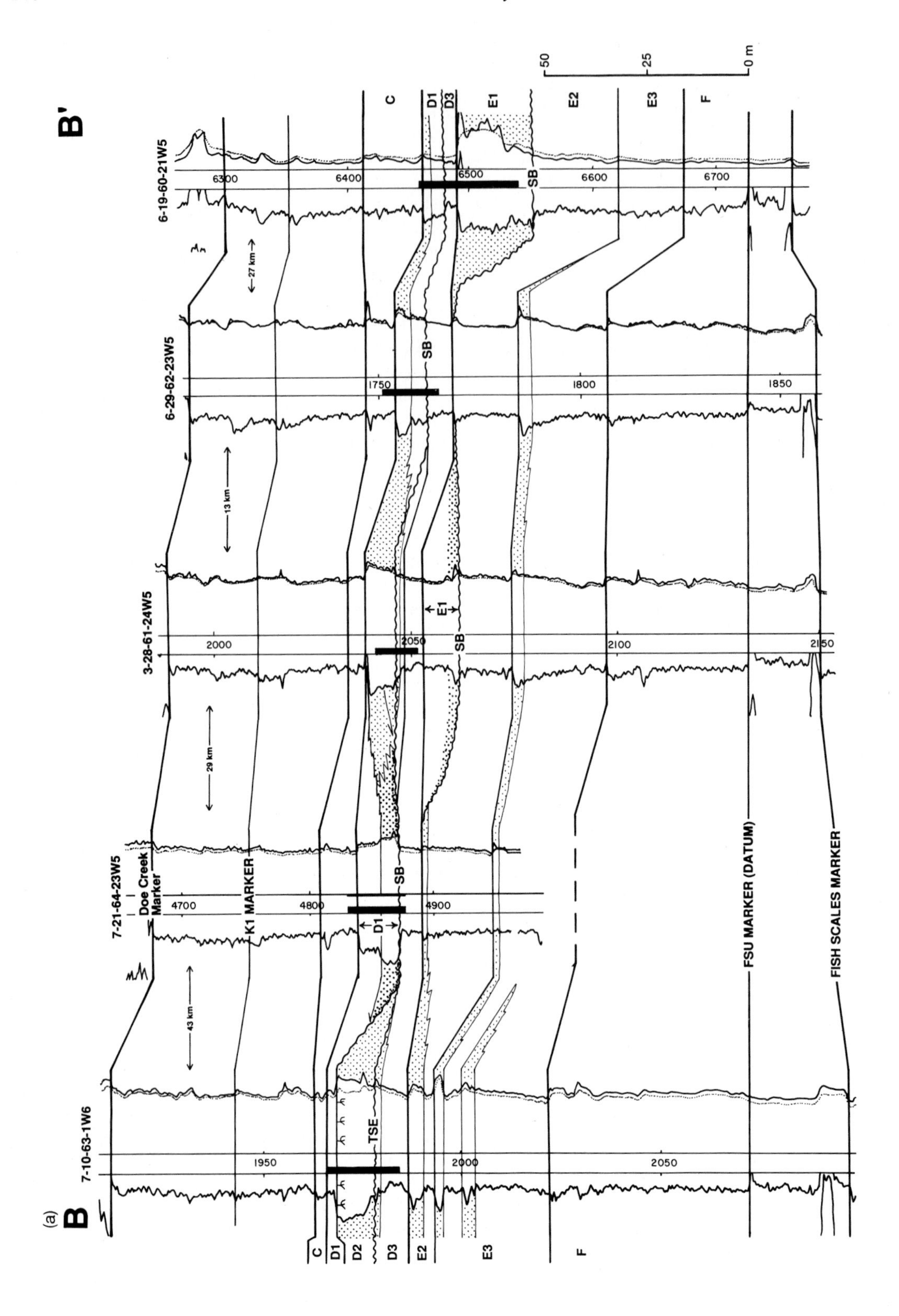
(a)
B
B'
7-10-63-1W6
7-21-64-23W5
3-28-61-24W5
6-29-62-23W5
6-19-60-21W5
Doe Creek Marker
K1 MARKER
FSU MARKER (DATUM)
FISH SCALES MARKER
43 km
29 km
13 km
27 km
1950
2000
2050
4700
4800
4900
2000
2050
2100
2150
1750
1800
1850
6300
6400
6500
6600
6700
C
D1
D2
D3
E1
E2
E3
F
SB
TSE
50
25
0 m

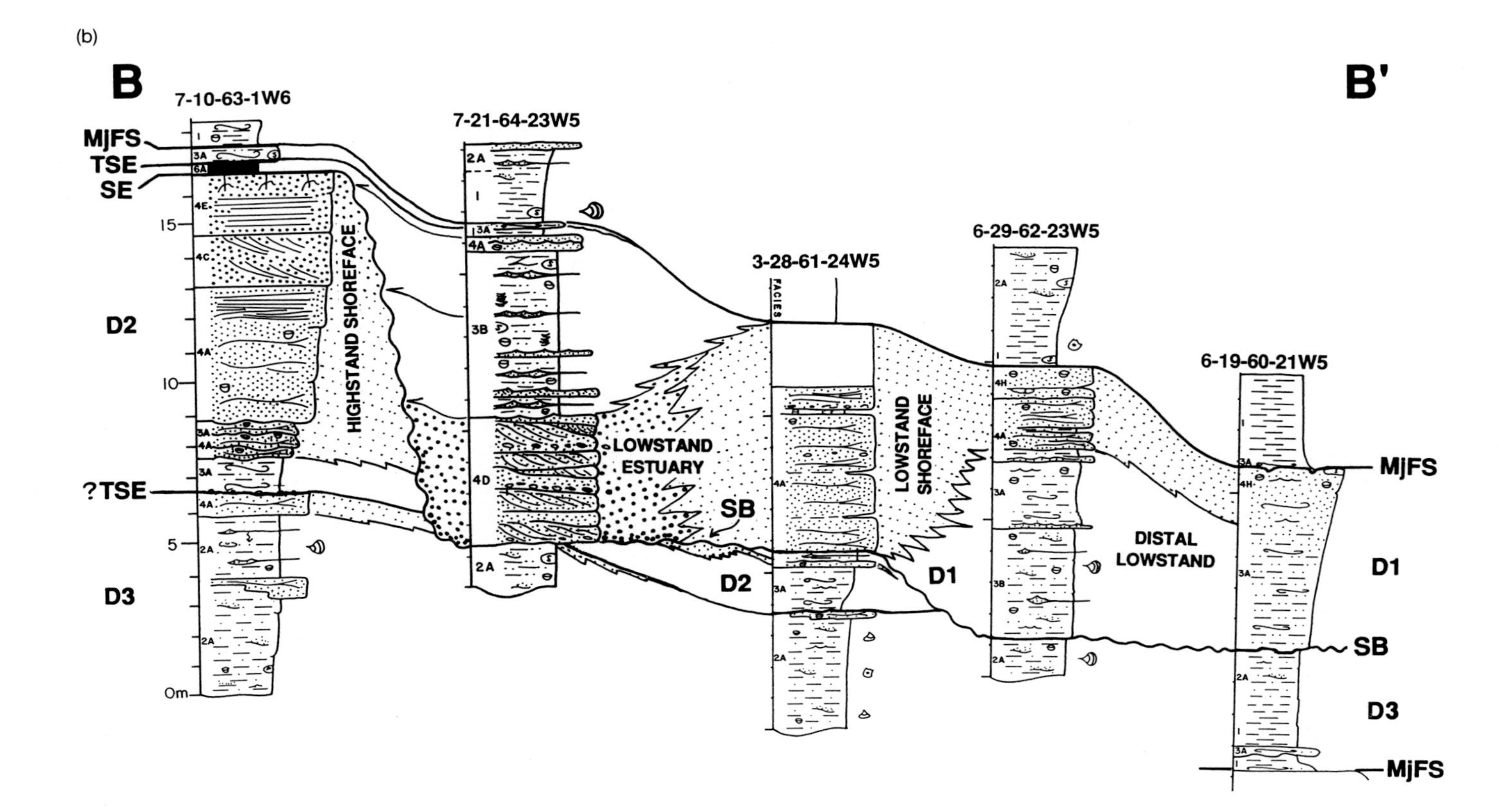

Fig. 20. Cross-section BB′ through the Dunvegan Formation highlighting allomember D. Location of cross-sections shown in Fig. 19. The wells on this cross-section have been selected for illustrative purposes from more detailed well-log cross-sections published in Bhattacharya (1989a,b). (a) The well-log cross-section highlights the upper and lower markers used to constrain the correlations. The letters and symbols as in Fig. 6(a). Well 6-19-60-21W5 is common to both cross-sections AA′ and BB′. Numbers in between well-log traces refer to depths in metres or in feet. The vertical bars indicate the cored intervals shown in Fig. 20(b). Gamma-log traces are on the left, resistivity or induction logs are on the right, with the shallow trace (solid line) and deep trace (dotted line) shown. The FSU marker has been used as the datum which has been inferred for well 7-21-64-23W5. (See Bhattacharya and Walker (1991a) for further discussion of the marker horizons.) (b) The core cross-section highlights the geometries and facies associated with the highstand (D2) and lowstand (D1) systems tracts within allomember D. Sequence boundary (SB) separates the highstand and lowstand systems tracts. It correlates with a subaerial exposure surface (SE) at the top of the highstand shoreface in well 7-10-63-1W6. The transgressive systems tract is interpreted to be very thin and is underlain by a transgressive surface of erosion (TSE) that overlies the sequence boundary in well 7-10-63-1W6. The major marine flooding surface (MjFS) caps allomember D. Portion of core in well 3-28-61-24W5 without a lithology pattern is uncored and vertical profile is based on well log. Details of the facies and surfaces in each well are explained in the text.

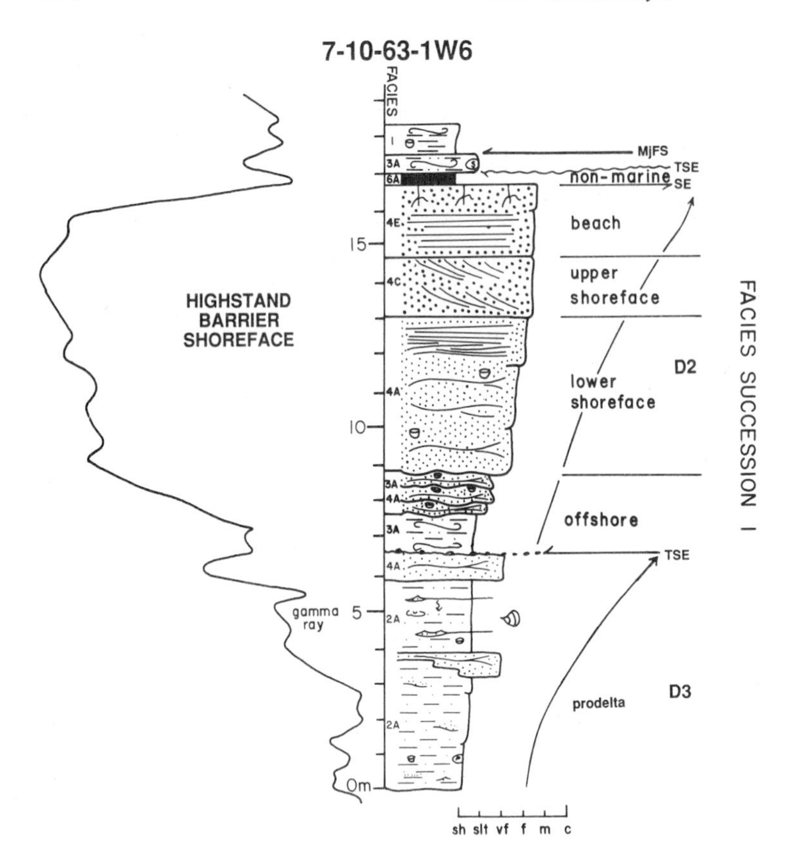

Fig. 21. Core description through the highstand barrier shoreface in well 7-10-63-1W6. SE, subaerial exposure; TSE & MjFS, as in other figures. Photographs of upper portion of well are shown in Fig. 22 (Figure modified from Bhattacharya, 1988.)

rate highstand and lowstand systems tracts within the allomembers. Bhattacharya and Posamentier (in press) suggest that at a larger scale of observation, these internal sequence boundaries may in fact separate successive stacked lowstands, similar to the stacked lowstand wedges described by Tesson *et al.* (1990). This suggests that recognition of systems tracts may be dependent on the scale of observation.

In several places, there is some uncertainty about which facies belongs to which systems tract. As shown above (Fig. 30), regression of the shoreline can occur both within the highstand and lowstand systems tract. The lowstand, however, is characterized by a regression during a fall of sea level, and is thus said to be forced. The lowstand systems tract is defined by its position in the sequence and is deposited during times of relative sea-level fall and subsequent slow rise (Posamentier & Vail, 1988; Van Wagoner *et al.*, 1990). During the last stages of deposition within the lowstand, however, valleys and channels will begin to backfill and may be transformed into estuaries which exhibit a transgressive fill. Distinguishing sediments belonging to the late lowstand versus early transgressive systems tract may be difficult in these cases, especially at the scale of allomembers where distinct parasequences are harder to distinguish and entire systems tracts may be represented by only a few metres or less of sediment.

Posamentier and James (this volume, pp. 3–18) suggest that sequence stratigraphy may be applied at very different scales. This study concurs with this point, although, as suggested above, distinction of systems tracts may become more difficult when represented by very thin intervals.

Significance of key surfaces

Much debate has centred on which surface is the most appropriate to pick as the fundamental stratigraphic break (Galloway, 1989; Walker, 1990; Posamentier & James, this volume, pp. 3–18). In this study, the choice of a marine flooding surface as the allomember boundary is a compromise be-

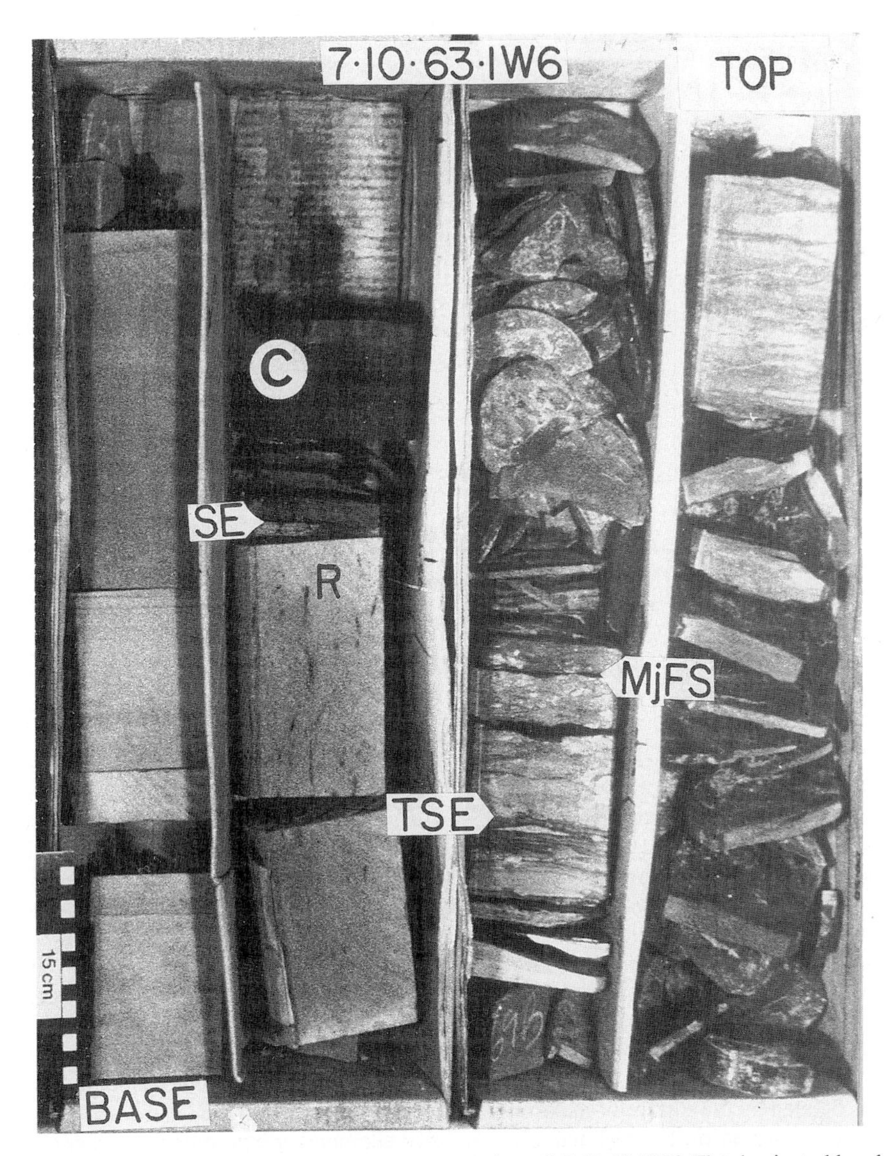

Fig. 22. Photographs of upper part of core from allomember D in well 7-10-63-1W6. Flat, laminated beach sandstones are rooted (R) and overlain by coal (C). The sequence boundary is expressed as a subaerial exposure surface (SE) in the interfluve. The overlying non-marine coal and mudstones are truncated by a transgressive surface of erosion (TSE), which marks the base of the thin transgressive systems tract. The top of the bioturbated sandy mudstone is marked by a major marine flooding surface (MjFS), which is taken as the allomember boundary.

tween the desire to provide meaningful and interpretable stratigraphic subdivisions and the desire to define mappable subdivisions in areas where core data are lacking. The allostratigraphic subdivision presented by Bhattacharya and Walker (1991a) represents the first breakdown of the Dunvegan into mappable subunits which are interpreted to have chronostratigraphic significance (compare Figs 2 & 3). The allomembers and shingles provide an objective basis, a stepping stone, toward the more detailed sequence-stratigraphic interpretation presented here.

In the Dunvegan, the most obvious lithostratigraphic break, interpreted to have chronostrati-

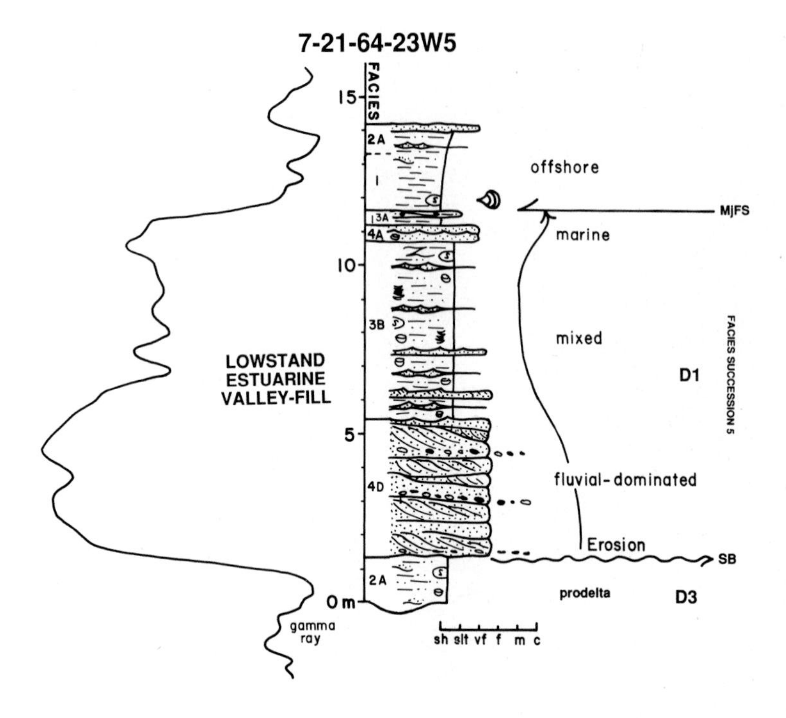

Fig. 23. Core description through D1 lowstand channel fill from well 7-21-64-23W5. The channel fill shows a transgressive facies succession and is interpreted as estuarine. (Modified after Bhattacharya & Walker, 1991b.) SB, sequence boundary; MjFS, major marine flooding surface. (Compare with fluvial channel fill in Fig. 9.)

graphic significance, is the allomember-bounding major marine flooding surface. This surface is distinct from the less extensive minor marine flooding surfaces which cap many of the shingles. The major marine flooding surface is contained within a thin, fourth-order transgressive systems tract, in places underlain by a transgressive surface of erosion (TSE). The TSE may pass seaward into a correlative flooding surface, but in many places is difficult to pick without core, especially where there is a sand on sand contact (e.g. well 3-28-61-24W5, Fig. 25). The top of the fourth-order transgressive systems tract is defined by the surface of maximum flooding which is expressed on the regional cross-section (Fig. 3) as a downlap surface. On the detailed cross-sections (Figs 6(b) & 20(b)) the amount of onlapping transgressive mudstones between the allomember-bounding major flooding surfaces and the maximum flooding surfaces is interpreted to be less than 1 m, although in several instances there was uncertainty in picking the maximum flooding surface.

The sequence boundaries in the Dunvegan have a more variable expression and are more difficult to pick in areas away from the incised channels and valleys, especially where core is unavailable. In several of the examples discussed in this paper, there was more than one possibility for the position of the sequence boundary. Donovan (pers. comm.), for example, suggested an alternative interpretation of cross-section B–B′ (Fig. 20) in which the sequence boundary underlying the estuarine channel fill in the 7-21-64-23W5 well correlated with the top of the sandstone in the 3-28-61-24W5 well rather than underneath it as shown here. Bhattacharya (1988, 1991) and Bhattacharya and Walker (1991a) also expressed some uncertainty as to distinguishing channelized erosional surfaces of purely autocyclic origin from those of more regional significance.

In well 15-31-62-26W5, the sequence boundary and allomember boundary were very close together and in many places the sequence boundary may be enhanced by a TSE. Unfortunately, the temporal significance of all of the surfaces described in this paper cannot be quantitatively determined owing to limited biostratigraphic resolution.

CONCLUSIONS

The nature and expression of flooding surfaces and erosional surfaces are similar in the two allomembers despite the differences in the nature of the depositional systems (river dominated versus wave

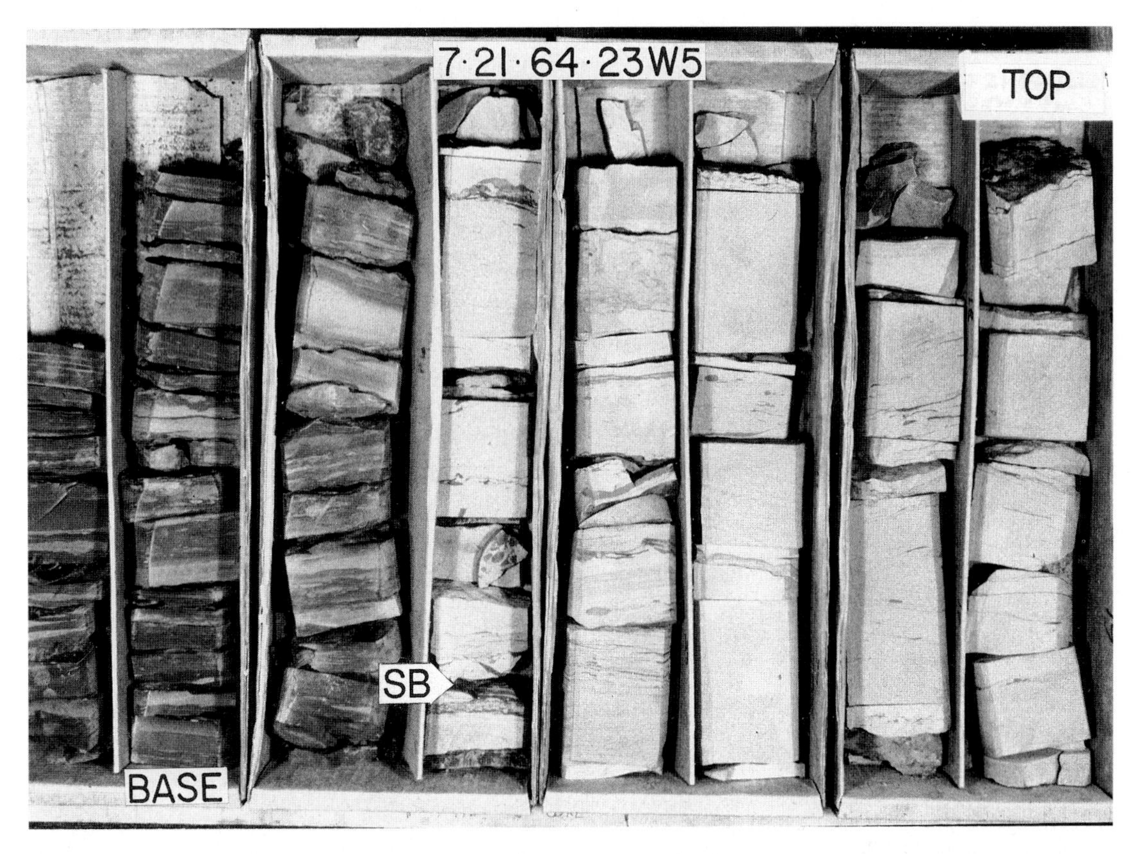

Fig. 24. Photographs of lower 4.7 m of core from well 7-21-64-23W5 (Fig. 23). Sequence boundary (SB) separates deeper water stratified prodelta mudstones from overlying cross-bedded sandstones. The cross-bedding is defined by muddy laminae and the sandstone is burrowed suggesting a marine (and probably tidal) influence. These sandstones are interpreted to represent the lower, fluvially derived sediments in an estuarine fill. The SB may have been modified by tidal scour.

dominated). Each allomember represents deposition during a fourth-order cycle of relative sea-level change and is bounded by a major marine flooding surface. The allomembers comprise thicker progradational deposits (regressive phase) bounded by thinner, transgressive facies tracts.

The sequence boundary allows subdivision of the regressive portion of each allomember into highstand and lowstand systems tracts (HST and LST). This reflects the difference between 'normal' and 'forced' regression. The sequence boundary is most readily recognized as an erosional surface at the base of incised channels and valleys, but correlative surfaces can include subaerial exposure surfaces in interfluve areas (e.g. rooted surfaces), marine erosional surfaces at the base of incised sharp-based shorefaces and correlative conformities. Away from the incised valleys (e.g. at correlative conformities) and in areas where core control is lacking, recognition of the sequence boundary may be more difficult and may have to be based on regional correlation.

In the two allomembers, the transgressive systems tract (TST) is interpreted to be very thin, less than 2 m, and may be underlain by a transgressive surface of erosion (TSE). Landward, and in areas where the LST deposits are absent, the TSE may coincide with and enhance the sequence boundary. Seaward, the TSE may pass into a conformable flooding surface. Like the sequence boundary, the TSE may also be difficult to pick in areas without core control. Within the thin TST, major marine flooding surfaces (MjFS) are marked in core by a sharp contact between coarser grained sandy facies

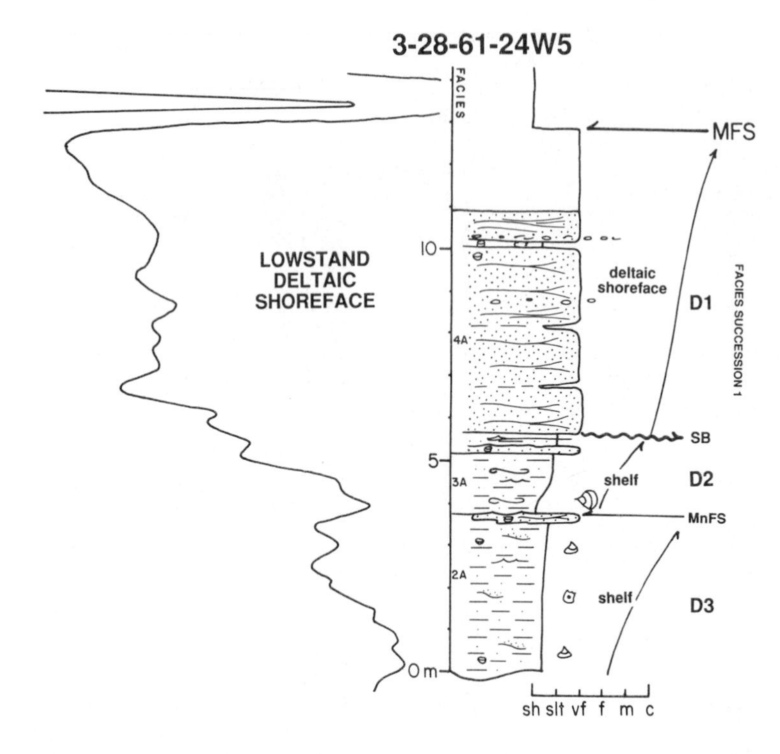

Fig. 25. Core description through the D1 lowstand delta from well 3-28-61-24W5. Lithology between 11 and 14 m is interpreted from the well log. A sharp contact between sandstone and mudstone is interpreted as the allomember-bounding major marine flooding surface (MjFS). The base of the shoreface is sharp and is taken as the seaward expression of the sequence boundary (SB) interpreted as typical of a lowstand shoreline produced by a 'forced regression' (explained in text). The minor marine flooding surface (MnFS) that caps shingle D3 correlates with a probable TSE surface further landward (see Fig. 20(b)).

and overlying deeper water mudstones and could be recognized in areas without core. These surfaces could be correlated over hundreds of kilometres and thus were used as the basis of an allostratigraphic subdivision.

The base of the HST is marked by the maximum flooding surface and in most of the examples presented here is concluded to coincide with the MjFS. This is indicated by downlap onto this surface and in core is seen as a transition from bioturbated reworked transgressive facies below the contact into silty stratified facies above. The HST also contains shingled offlapping subunits in each of the allomembers which are capped by less extensive minor marine flooding surfaces.

At the scale of fourth- and fifth-order sequences, assignment of some of the facies to distinct systems tracts was equivocal, especially in finer grained deposits on the delta plain, floodplain and in the estuaries. In the more river-dominated allomember E, lowstand channels are fluvial, whereas in the wave- and tide-dominated deposits of allomember D the lowstand valleys are estuarine in nature and are characterized by a transgressive fill. Distinct parasequences were not observed in either the TST or LST at this scale.

ACKNOWLEDGEMENTS

This paper stems from work begun as part of the author's PhD, supervised by R.G. Walker at McMaster University, and continued as a Postdoctoral Fellowship at the Alberta Geological Survey (AGS). Funding was provided by the Natural Science and Engineering Research Council (NSERC). The examples presented here formed part of a course in sequence stratigraphy taught by the author. The comments of attenders from the AGS, Norcen Energy Resources Ltd. and Petro Canada Resources helped tremendously in clarifying these examples for publication. The penetrating comments of reviewers Art Donovan and David James were also much appreciated and resulted in significant changes to the original manuscript. Later discussion with the editor, H.W. Posamentier, also helped, although any shortfalls in the interpretations presented rest solely with the author.

REFERENCES

Bhattacharya, J. (1988) Autocyclic and allocyclic sequences in river- and wave-dominated deltaic sedi-

Fig. 26. Photograph of entire core from well 3-28-61-24W5. Distal prodelta mudstones of shingle D3 are capped by a minor marine flooding surface (MnFS). The next succession also coarsens up into interbedded bioturbated mudstones and thin, hummocky cross-stratified sandstones. These heterolithic facies are truncated by a sequence boundary (SB) overlain by more massively bedded hummocky cross-stratified sandstones interpreted to be deposited in a shoreface. Sharp-based shorefaces such as illustrated here are thought to characterize progradation of the shoreface during a relative fall of sea level ('forced regression'). Horizontal scale immediately to the left of the 'TOP' label in the upper right is 5 cm.

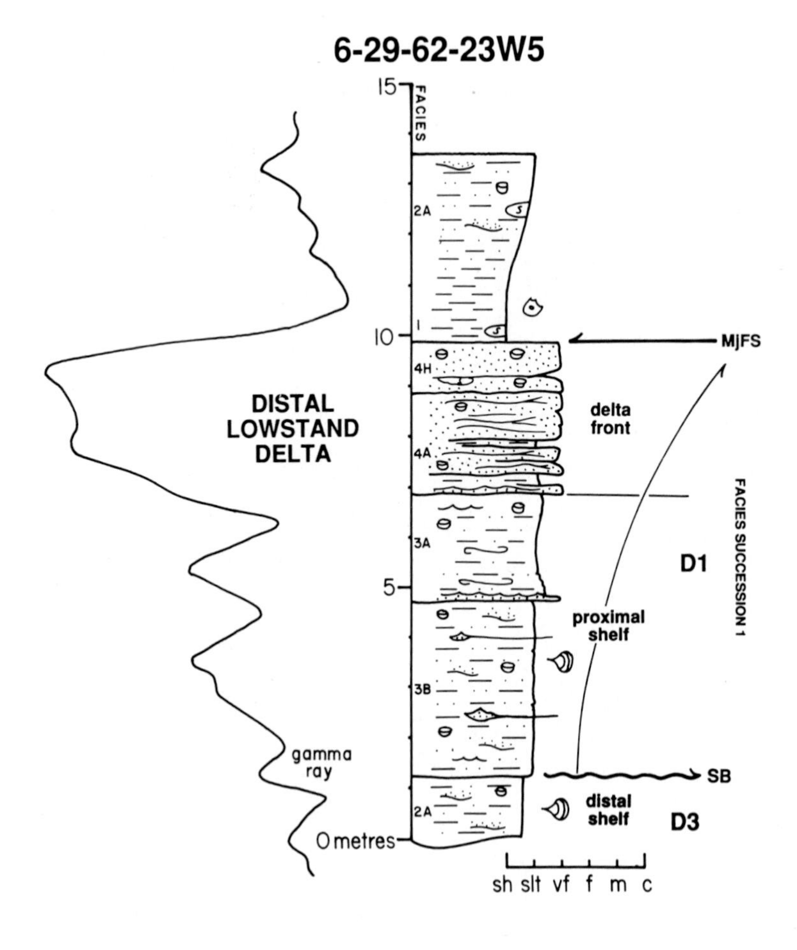

Fig. 27. Core description from well 6-29-62-23W5 through the lateral edge of the D1 lowstand delta. The core shows a coarsening-upward facies succession interpreted as being deposited in a distal wave-dominated delta front. The marine expression of the sequence boundary (SB) is here marked by a sudden increase in the proportion of interbedded sandstone and an increase in burrowing. The uppermost lowstand sandstones (Facies 4H) are more highly burrowed than the underlying hummocky cross-stratified sandstones (Facies 4A). The sharp contact between the lowstand sands (Facies 4H) and the overlying marine shale (Facies 1) is taken as the allomember-bounding major marine flooding surface (MjFS). (See text for further explanation.)

ments of the Upper Cretaceous, Dunvegan Formation, Alberta: core examples. In: *Sequences Stratigraphy, Sedimentology; Surface and Subsurface* (Eds James, D.P. & Leckie, D.A.) Mem. Can. Soc. Petrol. Geol. 15, 25–32.

Bhattacharya, J. (1989a) Allostratigraphy and river and wave-dominated depositional systems of the Upper Cretaceous (Cenomanian) Dunvegan Formation, Alberta. PhD thesis, McMaster University, Hamilton, Ontario, 588 pp.

Bhattacharya, J. (1989b) Estuarine channel fills in the Upper Cretaceous Dunvegan Formation: core example. In: *Modern and Ancient Examples of Clastic Tidal Deposits, A Core and Peel Workshop* (Ed. Reinson, G.E.). Can. Soc. Petrol. Geol. Calgary, Alberta, pp. 37–49.

Bhattacharya, J.P. (1991) Regional to subregional facies architecture of river-dominated deltas, Upper Cretaceous Dunvegan Formation, Alberta Subsurface. In: *The Three Dimensional Facies Architecture of Terrigenous Clastic Sediments, and its Implications for Hydrocarbon Discovery and Recovery* (Eds Miall, A.D. & Tyler, N.) SEPM (Society For Sedimentary Geology) Concepts and Models in Sedimentology and Paleontology 3, pp. 189–206.

Bhattacharya, J.P. (in press) Dunvegan Cretaceous Formation strata of the Western Canada Sedimentary Basin. In: *Geological Atlas of the Western Canada Sedimentary Basin* (Eds Mossop, G.D. & Shetsen, I.) Can. Soc. Petrol. Geol./Alberta Res. Council, Calgary, Alberta.

Bhattacharya, J.P. & Posamentier, H.W. (in press) Sequence stratigraphic and allostratigraphic applications in the Alberta Foreland basin. In: *Geological Atlas of the Western Canada Sedimentary Basin* (Eds Mossop, G.D. & Shetsen, I.) Can. Soc. Petrol. Geol./Alberta Res. Council, Calgary, Alberta.

Bhattacharya, J. & Walker, R.G. (1991a) Allostratigraphic subdivision of the Upper Cretaceous, Dunvegan, Shaftesbury, and Kaskapau Formations in the subsurface of northwestern Alberta. *Bull. Can. Petrol. Geol.* **39**, 145–164.

Bhattacharya, J. & Walker, R.G. (1991b) Facies and facies successions in river- and wave-dominated depositional systems of the Upper Cretaceous, Dunvegan Formation, northwestern Alberta. *Bull. Can. Petrol. Geol.* **39**, 165–191.

Demarest, J.M. & Kraft, J.C. (1987) Stratigraphic record of Quaternary sea levels: implications for more ancient

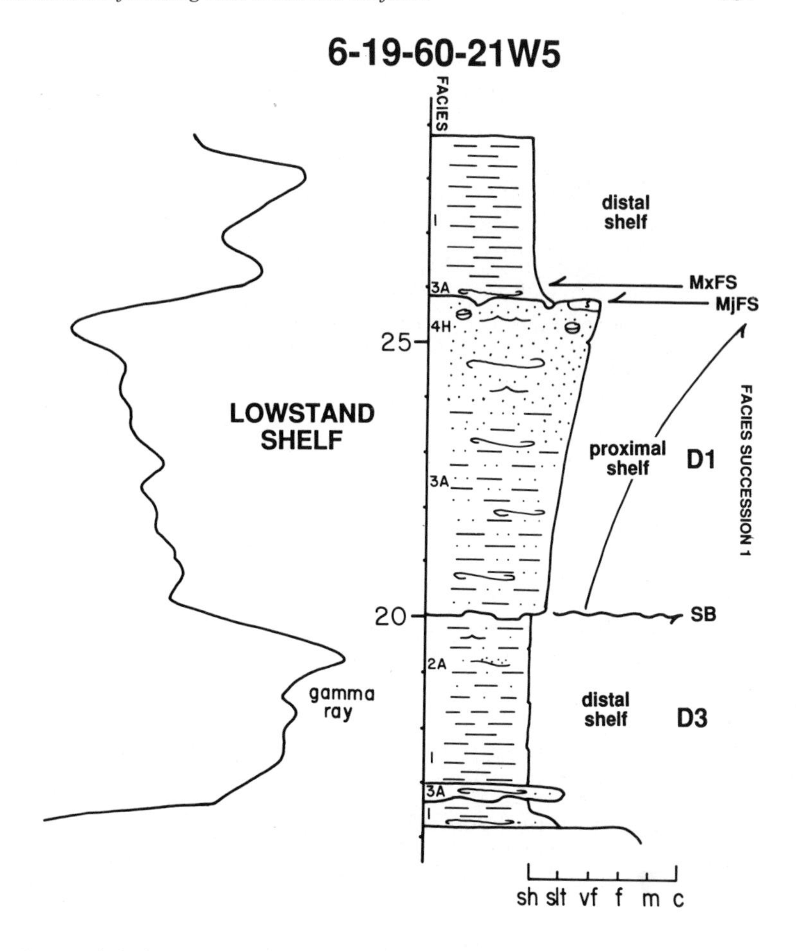

Fig. 28. Core description through the distal end of the D1 lowstand delta from the upper part of well 6-19-60-21W5. These facies were probably entirely deposited in a shelf environment. The sequence boundary (SB) is indicated by a sharp contact between stratified silty mudstones (Facies 2A) and bioturbated silty mudstones (Facies 3A). The allomember boundary is taken at the major marine flooding surface (MjFS) at the top of the lowstand sandstones. The top of the transgressive systems tract is taken at the maximum flooding surface (MxFS). (Photographs of core shown in Fig. 29.)

strata. In: *Sea-level Fluctuations and Coastal Evolution* (Eds Nummedal, D., Pilkey, O.H. & Howard, J.D.). Spec. Publ. Soc. Econ. Paleontol. Mineral. **41**, 223–239.

Galloway, W.E. (1989) Genetic stratigraphic sequences in basin analysis I: architecture and genesis of flooding-surface bounded depositional units. *Bull. Am. Assoc. Petrol. Geol.* **73**, 125–142.

Kraft, J.C., Chrzastowski, M.J. & Belknap, D.F. (1987) The transgressive barrier-lagoon coast of Delaware: morphostratigraphy, sedimentary sequences and responses to relative rise in sea level. In: *Sea-level Fluctuations and Coastal Evolution* (Eds Nummedal, D., Pilkey, O.H. & Howard, J.D.) Spec. Publ. Soc. Econ. Paleontol. Mineral. **41**, 129–143.

MacEachern, J.A., Raychaudhuri, I. & Pemberton, S.G. (1992) Stratigraphic applications of the Glossifungites Ichnofacies: Delineating Discontinuities in the Rock Record. In: *Applications of Ichnology to Petroleum Exploration* (Ed. Pemberton, S.G.) SEPM Core Workshop No. 17, pp. 169–198.

Mitchum, R.M., Jr. (1977) Seismic stratigraphy and global changes of sea level, part 1: glossary of terms used in seismic stratigraphy. In: *Seismic Stratigraphy Applications to Hydrocarbon Exploration* (Ed. Payton, C.E.). Mem. Am. Assoc. Petrol. Geol. **26**, 205–212.

North American Commission on Stratigraphic Nomenclature (1983) North American stratigraphic code. *Bull. Am. Assoc. Petrol. Geol.* **67**, 841–875.

Nummedal, D. & Swift, D.J.P. (1987) Transgressive stratigraphy at sequence-bounding unconformities: some principles derived from Holocene and Cretaceous examples. In: *Sea-level Fluctuations and Coastal Evolution* (Eds Nummedal, D., Pilkey, O.H. & Howard, J.D.) Spec. Publ. Soc. Econ. Paleontol. Mineral. **41**, 241–260.

Pemberton, S.G., MacEachern, J.A. & Frey, R.W. (1992) Trace fossil/facies models: Environmental & Allostratigraphic significance. In: *Facies Models: response to sea level change* (Eds Walker, R.S. & James, W.P.) pp. 47–72. Geological Assoc. of Canada.

Plint, A.G. (1988) Sharp-based shoreface sequences and 'offshore bars' in the Cardium Formation of Alberta: their relationship to relative changes in sea level. In: *Sea-level Changes: An Integrated Approach* (Eds Wilgus, C.K., Hastings, B.S., Kendall, C.G.St.C., Posamentier, H.W., Ross, C.A. & Van Wagoner, J.C.) Spec. Publ. Soc. Econ. Mineral. Paleontol. Tulsa, 42, 357–370.

Plint, A.G. (1991) High-frequency relative sea-level oscillations in Upper Cretaceous shelf clastics of the Alberta foreland basin: possible evidence for a glacio-eustatic control? In: *Sedimentation, Tectonics and Eustasy; Sea-*

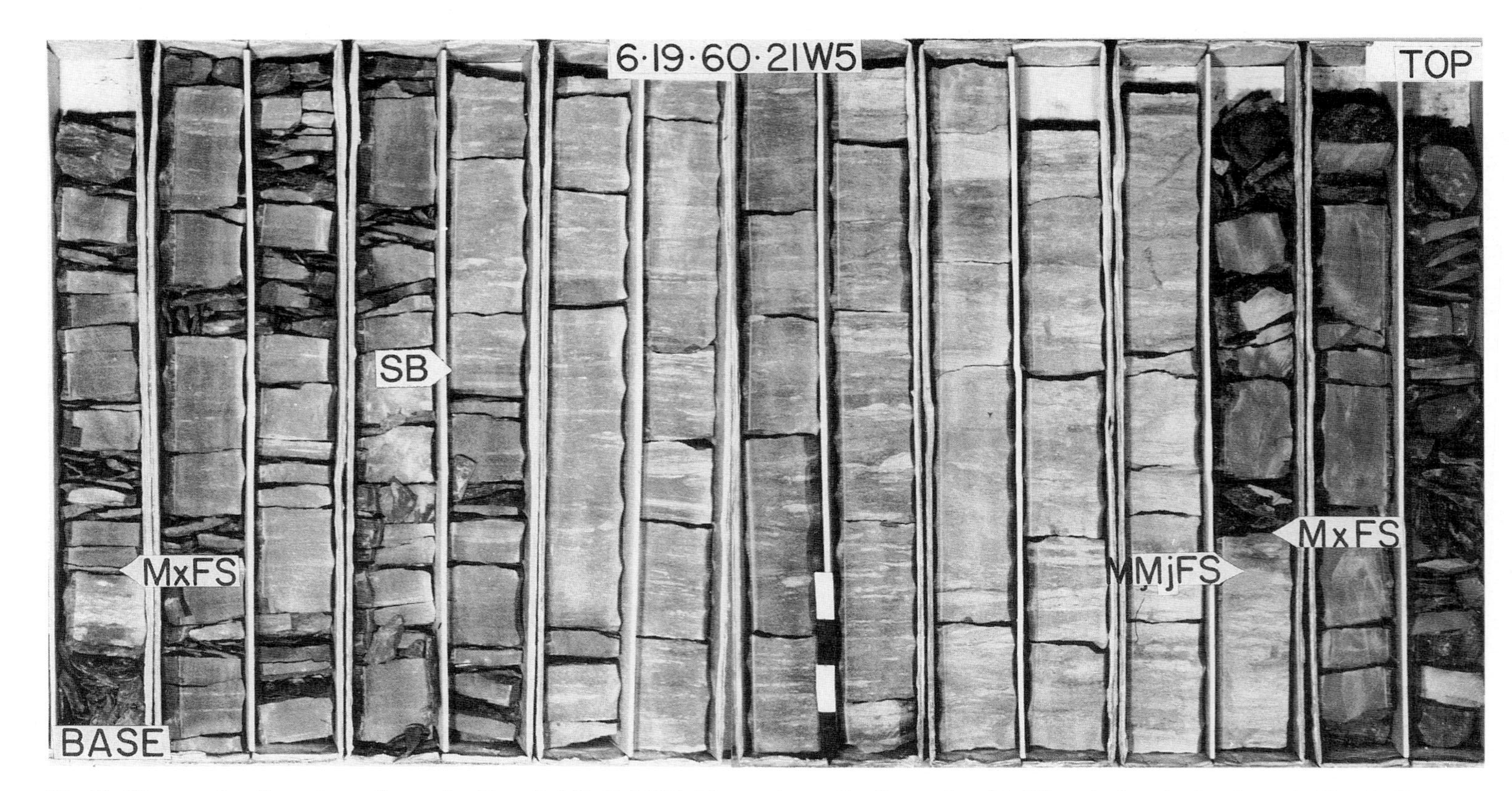

Fig. 29. Photographs of core from allomember D, well 6-19-60-21W5. The maximum flooding surface (MxFS) at the base is also shown in Fig. 18. The sequence boundary (SB) marks a sharp contact between deeper marine laminated mudstones below from the more bioturbated and siltier mudstones above. These bioturbated lowstand mudstones become sandier upwards. The lowstand muddy sandstones are sharply overlain by a thin layer of bioturbated sandy mudstone interpreted as the major marine flooding surface (MjFS) that caps allomember D. The bioturbated sandy mudstone is interpreted to represent a very thin, transgressive systems tract and is overlain by a maximum flooding surface (MxFS). Scale in lower middle is 20 cm.

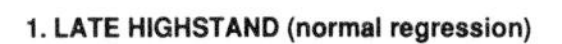

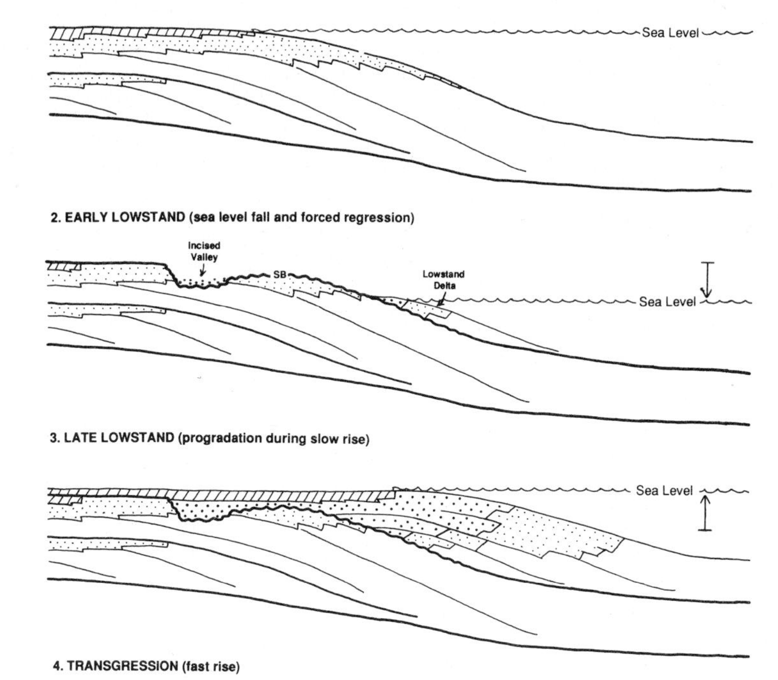

Fig. 30. Sequence of events in the development of a Dunvegan allomember during a single cycle of relative sea-level change. See Fig. 4 for definition of systems tracts and surfaces. Highstand (time 1) is characterized by progradation (normal regression). Sequence boundary (SB) is associated with a 'forced regression' during time 2 and results from a fall in sea level. The late lowstand (time 3) is characterized by backfilling of incised valleys, aggradation on the delta plain and continued progradation of the delta front. The transgressive surface of erosion (TSE) forms during subsequent transgression (time 4) and may enhance the sequence-bounding unconformity in interfluves. (See text for more details.)

level Changes at Active Margins (Ed. Macdonald, D.I.M.) Spec. Publ. Int. Assoc. Sedimentol. 12, 409–428.

Plint, A.G., Walker, R.G. & Bergman, K.M. (1986) Cardium Formation 6. Stratigraphic framework of the Cardium in subsurface. *Bull. Can. Petrol. Geol.* **34**, 213–225.

Posamentier, H.W. & Vail, P.R. (1988) Eustatic controls on clastic deposition II — sequence and systems tract models. In: *Sea-level Changes: An Integrated Approach* (Eds Wilgus, C.K., Hastings, B.S., Kendall, C.G.St.C., Posamentier, H.W., Ross, C.A. & Van Wagoner, J.C.) Spec. Publ. Soc. Econ. Mineral. Paleontol. Tulsa, 42, 125–154.

Posamentier, H.W., Allen, G.P. & James, D.P. (1992) High resolution sequence stratigraphy. The East Coulee delta *J. Sediment. Petrol.* **62**, 310–317.

Posamentier, H.W., Allen, G.P., James, D.P. & Tesson, M. (1992) Forced Regression in a sequence stratigraphic framework: concepts, examples, and exploration significance. *Bull. Am. Assoc. Petrol. Geol.* **76**, 1687–1709.

Posamentier, H.W., Jervey, M.T. & Vail, P.R. (1988) Eustatic controls on clastic deposition I — conceptual framework. In: *Sea-level Changes: An Integrated Approach* (Eds Wilgus, C.K., Hastings, B.S., Kendall, C.-G.St.C., Posamentier, H.W., Ross, C.A. & Van Wagoner, J.C.) Spec. Publ. Soc. Econ. Mineral. Paleontol. Tulsa, 42, 109–124.

Singh, C. (1983) Cenomanian microfloras of the Peace River area, northwestern Alberta. *Alberta Res. Council Bull.* **44**, 322 pp., Edmonton, Alberta.

Swift, D.J.P. (1968) Coastal erosion and transgressive stratigraphy. *J. Geol.* **76**, 444–456.

Tesson, M., Gensous, B., Allen, G.P. & Ravenne, C. (1990) Late Quaternary deltaic lowstand wedges on the Rhone continental shelf, France. *Mar. Geol.* **91**, 325–332.

Trincardi, F. & Field, M.E. (1991) Geometry, lateral variation, and preservation of downlapping regressive shelf deposits: Eastern Tyrrhenian Sea margin, Italy. *J. Sediment. Petrol.* **61**, 775–790.

Van Wagoner, J.C., Mitchum, R.M., Campion, K.M. &

Rhamanian, V.D. (1990) Siliciclastic sequence stratigraphy in well logs, cores, and outcrops. *Am. Assoc. Petrol. Geol. Meth. Expl. Ser.* **7**, 55 pp.

Van Wagoner, J.C., Posamentier, H.W., Mitchum, R.M., Jr., *et al.* (1988) An overview of the fundamentals of sequence stratigraphy and key definitions. In: *Sea-level Changes: An Integrated Approach* (Eds Wilgus, C.K., Hastings, B.S., Kendall, C.G.St.C., Posamentier, H.W., Ross, C.A. & Van Wagoner J.C.) Spec. Publ. Soc. Econ. Paleontol. Mineral. Tulsa, 42, 39–45.

Walker, R.G. (1990) Perspective, facies modelling and sequence stratigraphy. *J. Sediment. Petrol.* **60**, 777–786.

Weimer, R.J. (1988) Record of relative sea-level changes, Cretaceous of Western Interior, USA. In: *Sea-level Changes: An Integrated Approach* (Eds Wilgus, C.K., Hastings, B.S., Kendall, C.G.St.C., Posamentier, H.W., Ross, C.A. & Van Wagoner, J.C.) Spec. Publ. Soc. Econ. Paleontol. Mineral. Tulsa, 42, 285–288.

Spec. Publs Int. Ass. Sediment. (1993) **18**, 161–179

Sequence stratigraphy in coastal environments: sedimentology and palynofacies of the Miocene in central Tunisia

T.J.A. BLONDEL*, G.E. GORIN*
and R. JAN DU CHENE†

**Department of Geology and Palaeontology, University of Geneva, 13 rue des Maraîchers, 1211 Geneva 4, Switzerland; and*
†Geological Consultant, Square du Pomerol, Domaine de Beausoleil, 33170 Gradignan, France

ABSTRACT

Continental, coastal and shelf near-shore sediments of the Hajeb el Aïoun Group were deposited on the innermost Tunisian shelf during lower to middle Miocene transgressive flooding maxima (Burdigalian to early Serravallian). Only the most marine unit can be precisely dated by nannofossils and foraminifera, whereas the age of the other units is indirectly attributed by vertical and lateral correlations. A thorough sedimentological study of numerous field sections, locally complemented by palynofacies analyses, led to the establishment of a detailed sedimentological and palaeoenvironmental model, which can be interpreted in terms of sequence stratigraphy.

On the Tunisian shelf, the stratigraphic signature of these deposits is comparable to that recorded on other contemporaneous continental shelves. In coastal environments of the lower to middle Miocene of Tunisia, sequence boundaries and transgressive surfaces are well marked by macroscopical sedimentological features. In such environments, transgressive surfaces often overprint the underlying sequence boundary. Maximum flooding surfaces in tidal flat, estuarine and lagoonal environments are more difficult to identify in the field, but can often be singled out by palynofacies analysis because they coincide with a significant increase in abundance and diversity of dinoflagellate cysts. In the field, transgressive systems tracts are characterized by fining-up and deepening-up parasequences, while highstand systems tracts are characterized by coarsening-up and shallowing-up parasequences. Increasingly shallower water depths during late highstand deposits are often emphasized by a decrease in dinoflagellate cyst abundance and diversity.

INTRODUCTION

In central Tunisia, coastal to shelf near-shore sediments of Burdigalian to Langhian age were deposited during a period of transgressive flooding maxima, alternating with short-lived relative falls in sea level. The southern limit of these sediments lies to the west of the city of Sfax and is oriented NW–SE (Fig. 1). In this area, outcropping Burdigalian and Langhian sediments form a clearly identifiable lithological unit defined as the Hajeb el Aïoun Group by Beseme and Blondel (1989). This group is subdivided into three formations from base to top: the Oued el Hajel Formation, the Aïn Grab Formation *sensu lato* and the Mahmoud Formation *sensu stricto* (Fig. 2).

Detailed lithological, structural, sedimentological and micropalaeontological studies of these siliciclastic to mixed siliciclastic and calcareous deposits have been published by Beseme and Blondel (1989) and Blondel (1991). Blondel (1991) integrated all available data and attempted a sequence-stratigraphic interpretation of the Hajeb el Aïoun Group. In coastal environments sequence stratigraphy is often difficult to use because sediments are often fairly poor in biostratigraphically significant

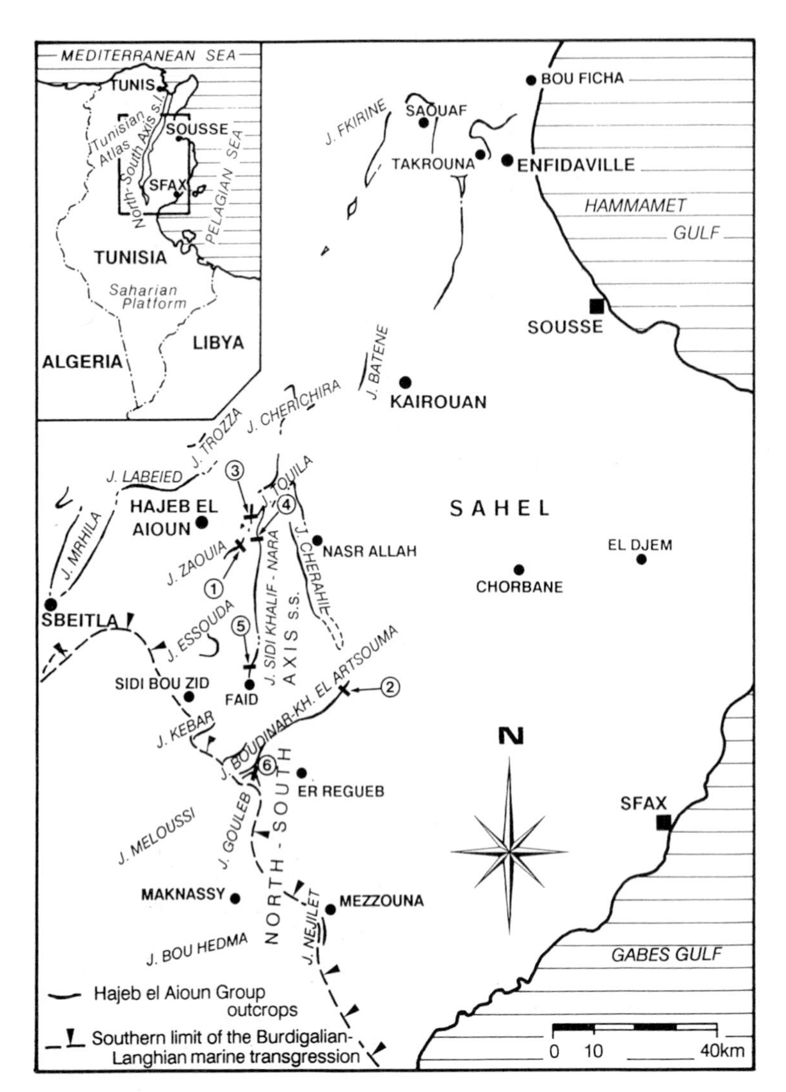

Fig. 1. Location map of central Tunisia. Numbers indicate field sections of the Miocene Hajeb el Aïoun Group investigated for their organic content: (1) Jebel Zaouia; (2) Jebel Khechem el Artsouma; (3) Jebel Touila SW; (4) Jebel Bou Gobrine NW; (5) Jebel Gatrana; (6) Jebel Boudinar SE. Sections 1 and 2 are detailed in Fig. 3 and were used to compile the schematic sedimentological log shown in Figs 5, 6 and 8. The area lying SW of the southern limit of the Burdigalian–Langhian marine transgression is referred to in the text as the northeastern termination of the 'Kasserine Island'.

fossils and they show major lateral and vertical facies and thickness variations. Nevertheless, despite the influence of tectonics and diapirism on the partly mobile inner shelf of central Tunisia, Blondel (1991) tentatively recognized the main systems tracts and major surfaces of eustatic origin which characterize Neogene deposits worldwide (Bartek *et al.*, 1991). Some of the most representative sections in central Tunisia were also analysed for their palynofacies content. These data were combined with standard sedimentological observations to develop an integrated palaeoenvironmental interpretation.

This study shows that the whole sedimentary record of eustatic variations can be identified quite far onto the inner shelf, essentially in incised valleys and tectonically induced subsiding depressions. In central Tunisia, within a partly well-defined biostratigraphic framework, this interpretation is based mainly on comprehensive detailed sedimentology supported locally by palynofacies.

GEOLOGICAL FRAMEWORK AND BIOSTRATIGRAPHY

Tunisia, together with Sicily and the Pelagian Platform, constitutes a structural and geographical tran-

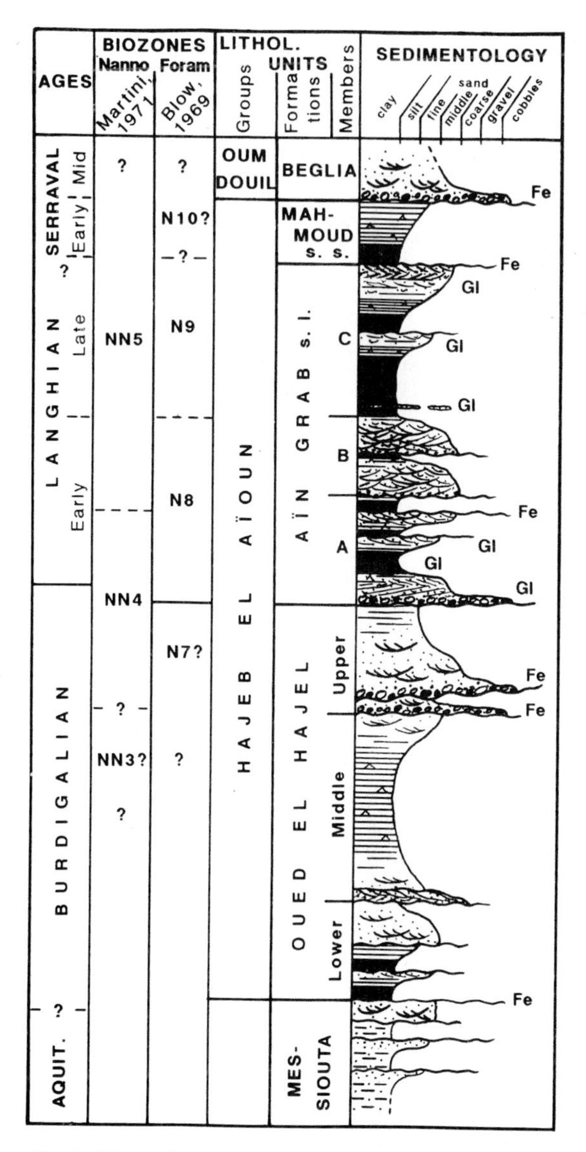

Fig. 2. Biostratigraphic framework and lithological units of the Hajeb el Aïoun Group. The schematic sedimentological log (not to scale) is derived from the study of numerous outcrops in central Tunisia. (After Blondel, 1991.) A similar log is used in Fig. 9.

sition zone separating the Western and Eastern Mediterranean basins. Mountain ranges in central Tunisia correspond to the eastern termination of Atlas lineaments on the northern edge of the African continent. These ranges form a large part of the Tunisian Atlas, whose structural lineaments are oriented NE–SW. Central Tunisia is bounded to the east by an alignment of more or less N–S trending ridges referred to as the 'North–South Axis' (Burollet, 1956; Fig. 1).

The studied Miocene sedimentary series belong to the Hajeb el Aïoun Group (Beseme & Blondel, 1989; Blondel, 1991; Fig. 2). Its most representative outcrops in central Tunisia are located in incised valleys and tectonically induced subsiding depressions of the inner Tunisian shelf, close to the 'North–South Axis' and to the palaeo-highs forming the northern and eastern margin of the 'Kasserine Island' (Flandrin, 1948; Burollet, 1956). The north-eastern edge of this palaeo-island marks the southern limit of the Burdigalian–Langhian transgressive floodings (Fig. 1). On the inner Tunisian platform, the Neogene deposits are characterized by their continental, coastal and shelf-near-shore environments. Sediments of the Hajeb el Aïoun Group are siliciclastic or mixed siliciclastic and calcareous (Fig. 2). In outcrop, they show well-marked variations in facies and thickness. In the vicinity of palaeo-highs, the total thickness of lower Neogene deposits may not exceed a few tens of centimetres, whereas in incised valleys and tectonically induced subsiding depressions, it may reach a few hundreds of metres (Blondel, 1991).

Deposits of the Hajeb el Aïoun Group are subdivided into three formations: the Oued el Hajel Formation, the Aïn Grab Formation *sensu lato* and the Mahmoud Formation *sensu stricto* (Burollet, 1956; Biely *et al.*, 1972; Beseme & Blondel, 1989; Blondel, 1991; Fig. 2). The first two are further subdivided into three members: lower, middle and upper members of the Oued el Hajel Formation, and A, B and C members of the Aïn Grab Formation *sensu lato* (Beseme & Blondel, 1989; Blondel, 1991; Figs 2 & 5).

Planktonic foraminifera and nannofossils assemblages date the Aïn Grab Formation *sensu lato* as latest Burdigalian to late Langhian, with possibly the Langhian–Serravallian boundary marking the base of the overlying Mahmoud Formation *sensu stricto* (Fig. 2; Blondel, 1991). These formations correspond to the sedimentary record of transgressive flooding maxima on the lower to middle Miocene Tunisian platform, spanning the foraminiferal zones N8 and N9, and the nannofossils zones NN4 and NN5 *pro parte* (Fig. 2; Beseme & Blondel, 1989; Blondel, 1991).

The continental to coastal Messiouta Formation underlying the Hajeb el Aïoun Group is attributed to the Aquitanian (Schoeller, 1933; Blondel *et al.*,

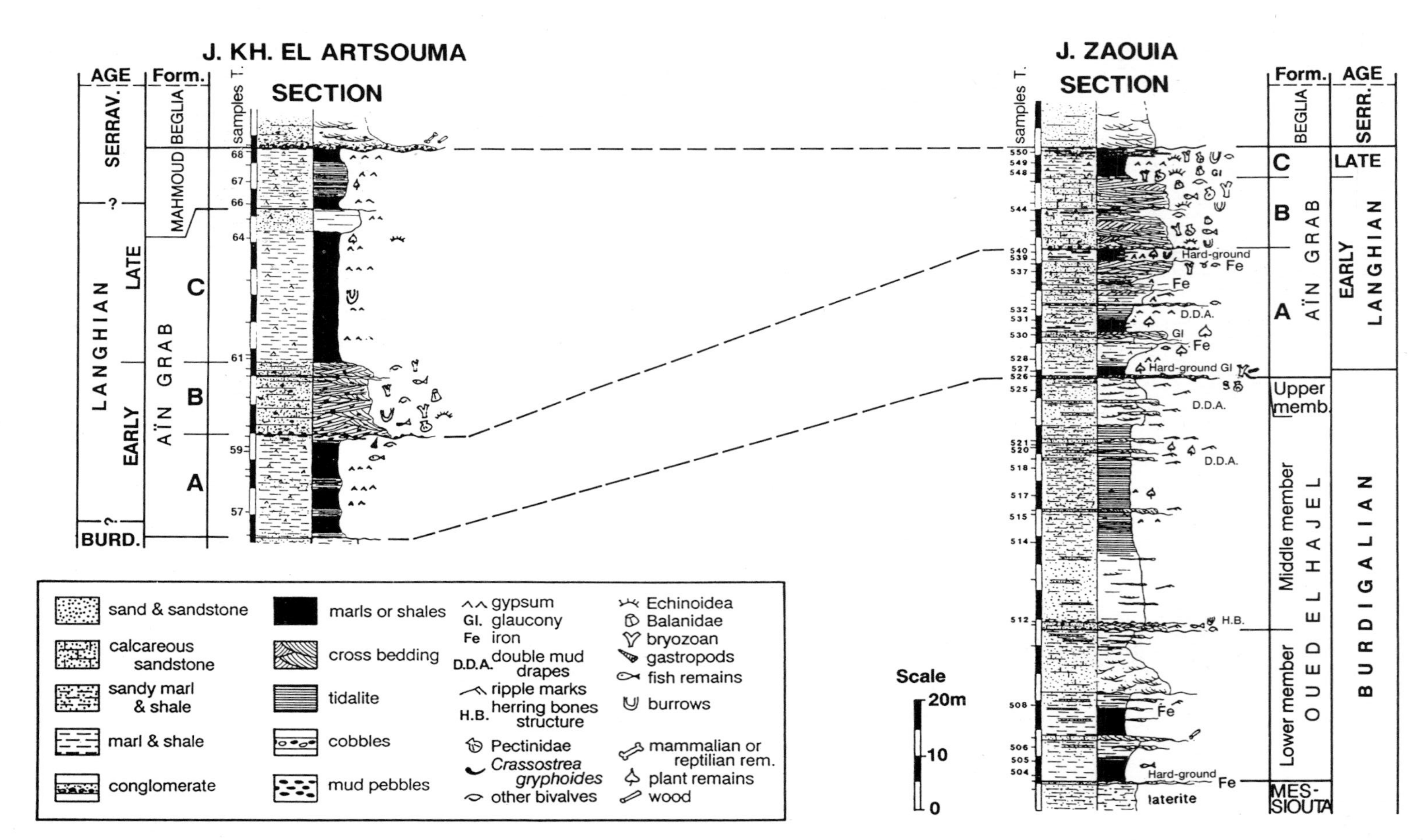

Fig. 3. Sedimentological analysis of two representative field sections of the Hajeb el Aïoun Group: the Jebel Zaouia and the Jebel Khechem el Artsouma sections. Numbered samples are those analysed for their organic content. These two sections are used to produce the schematic sedimentological log shown in Figs 5, 6 and 8.

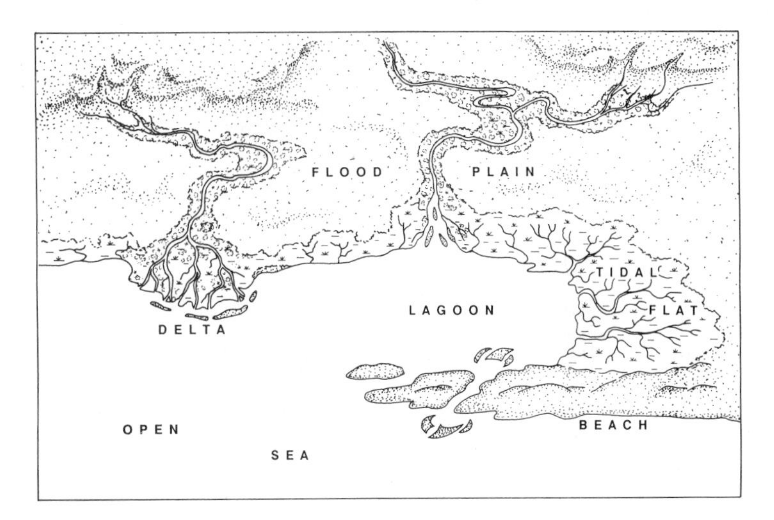

Fig. 4. Synthetic sedimentological framework for the deposits of the Hajeb el Aïoun Group.

1988). The continental to innermost shelf deposits constituting the lower and upper members of the Oued el Hajel Formation and the upper part of the Mahmoud Formation *sensu stricto* lack biostratigraphic markers. Their age can be approximated by vertical and lateral correlations with outcrops or onshore and offshore wells in central, eastern and northeastern Tunisia (Castany, 1951; Burollet, 1956; Biely *et al.*, 1972; Hooyberghs, 1977, 1987; Fournié, 1978; Ben Ismaïl-Lattrache, 1981; Touati, 1985; Beseme & Blondel, 1989), as well as by reworked microfossils in the Aïn Grab Formation *sensu lato* (Blondel, 1991). The Oued el Hajel Formation can be attributed to the Burdigalian and the Mahmoud Formation *sensu stricto* to the latest Langhian to early Serravallian. The fluvio-deltaic to estuarine Beglia Formation overlying the Hajeb el Aïoun Group is dated as Serravallian on the basis of vertebrate fossils (Biely *et al.*, 1972; Black, 1972; Robinson & Wiman, 1976; Robinson & Black, 1981).

SEDIMENTOLOGY AND PALAEOGEOGRAPHICAL INTERPRETATION

Detailed sedimentological analyses of the Hajeb el Aïoun Group deposits in central Tunisia were carried out on numerous field sections (Blondel, 1991). Those numbered and located in Fig. 1 were also analysed palynologically. The schematic sedimentology of the complete Hajeb el Aïoun Group is presented in Fig. 2. The detailed lithology and sedimentology of the group are illustrated in Fig. 3 by two representative field sections: the Jebel Khechem el Artsouma and the Jebel Zaouia sections (see Fig. 1 for location), and summarized in a simplified log (Figs 5, 6 & 8). A schematic sedimentological framework for the Hajeb el Aïoun Group is shown in Fig. 4. The sections palynologically analysed are numbered and located in Fig. 1. The palaeoenvironmental interpretation of Fig. 5 is derived from palaeontological and sedimentological data acquired by Blondel (1991).

When present, the Messiouta Formation deposits underlying the Oued el Hajel Formation are composed of reworked lateritic material. These red clays and silts originate from the erosion, reworking and resedimentation in tectonically induced subsiding depressions of lateritic soils overlying adjacent palaeoreliefs (Blondel *et al.*, 1988; Blondel, 1991). Up to a hundred metres thick in some outcrops, these deposits are arranged mainly in coarsening-up parasequences often terminated by palaeosoils or palustrine deposits with root traces. In the most subsiding areas, the upper part of the Messiouta Formation consists of lacustrine or palustrine to tidal flat limestones, fluviatile to estuarine sands (Fig. 2) or sabkha to lagoonal gypsum. An erosional surface separates this thin interval from the underlying lateritic sediments.

A ferruginous and sometimes bored hardground often characterizes the boundary between the top of the Messiouta Formation and the base of the Oued el Hajel Formation (Fig. 2).

In outcrop, the thickness of the lower member of the Oued el Hajel Formation is highly variable, with

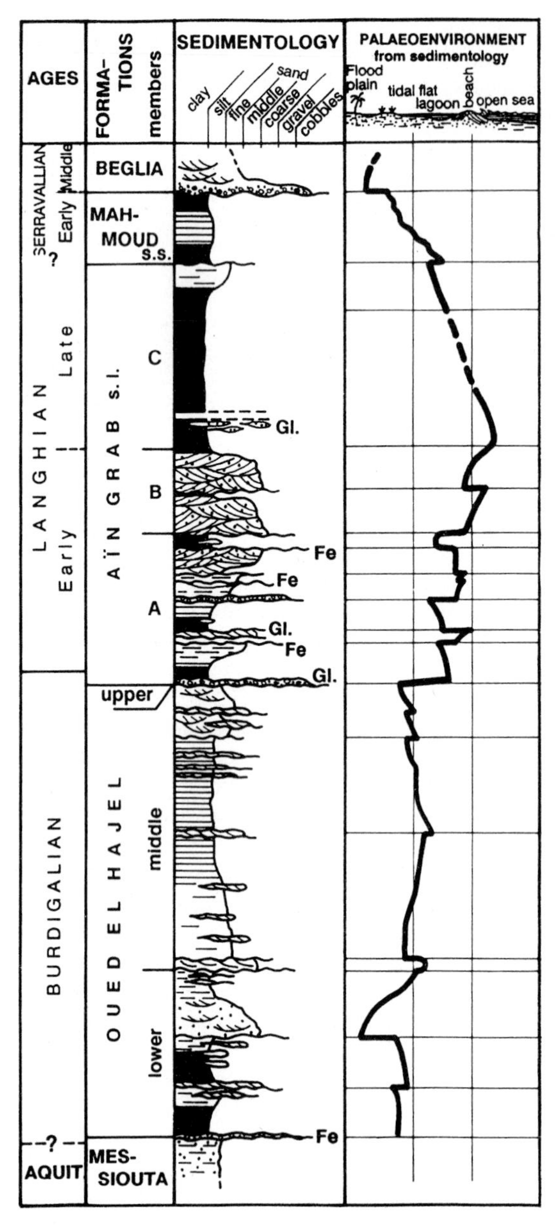

Fig. 5. Palaeoenvironmental interpretation of the Hajeb el Aïoun Group based on standard sedimentology. The sedimentological log is derived from the two field sections detailed in Fig. 3.

a maximum of some 30 m. The lower part of this member consists of greenish to varicoloured marls with intercalated sandy horizons, whereas the upper part is made up of cross-bedded medium to coarse sandstones with an erosive base. Marls were deposited in more or less restricted lagoonal to littoral mud flats or coastal swamps, the coarse sandstones in the upper part characterizing a more continental environment, i.e. fluvial to fluvio-deltaic becoming estuarine at the top (Fig. 5). Parasequences forming the lower part of the lower member are coarsening-up and shallowing-up, while the upper part, when preserved, shows generally one single fining-up and deepening-up parasequence.

The middle member of the Oued el Hajel Formation has a thickness varying from a few to tens of metres. It consists of fine sands, sandy clays and clays with intercalated, lenticular horizons (a few to tens of centimetres thick) of calcareous and bioclastic sands, sometimes with an erosive and conglomeratic base. It displays numerous sedimentary structures: ripple-marks, trough and sigmoid cross-bedding, flaser and lenticular bedding, single to double clay draping of tidalites which can form bundle sequences interpreted as the sedimentary record of lunar cycles (Blondel, 1991). These structures characterize coastal environments such as tidal flats and lagoons (Fig. 5), which are locally crossed by some tidal channels containing lagoonal-to-littoral macrofossils remains like ostreids, bryozoans and shark teeth. There are also local indications of estuarine deposits. The middle member of the Oued el Hajel Formation represents the sedimentary record, during the Burdigalian, of the most important phase of lower Miocene transgression and flooding onto the inner part of the Tunisian shelf. Parasequences forming the middle member of the Oued el Hajel Formation are generally fining-up and deepening-up in the lower part, and coarsening-up and shallowing-up in the upper part.

In central Tunisia, the upper member of the Oued el Hajel Formation is only locally developed inside incised valleys and tectonically induced synclinal axes oriented NE–SW and NW–SE (Blondel, 1991). It consists of cross-bedded coarse sandstones with an erosive and conglomeratic base grading progressively upwards into fine sands and sometimes into an alternation of sands and clays at the top (Fig. 2). This unit is composed of fining-up and deepening-up parasequences. In areas where this member is the thickest, its base shows two polygenic, ferruginous, conglomeratic horizons on top of each other which contain cobbles affected by pedogenesis and wood fragments and vertebrate bones indicating an earlier continental environment. The upper conglomeratic horizon is especially marked by its strongly erosive base (northern part of the 'North–South Axis', north of the Jebel Touila; see Figs 1 & 2). These deposits characterize a fluvial to fluvio-deltaic environment grading towards the top into transgressive estuarine conditions. This member is attributed to the latest Burdigalian by Blondel (1991). It is absent in the Jebel Zaouia section shown in Fig. 3.

The overlying Aïn Grab Formation *sensu lato* displays a more pronounced littoral to marine tendency than the underlying lower Miocene formations. Some horizons of the Aïn Grab Formation *sensu lato* contain an abundant and diversified marine macrofauna composed mainly of pectinids, bryozoans, echinids and balanids, indicating a coastal to shelf near-shore environment (Blondel *et al.*, 1990; Blondel & Demarcq, 1990; Blondel, 1991). The foraminiferal and nannofossil content gives an age ranging from the Burdigalian–Langhian boundary to the late Langhian (foraminiferal biozones N8 and N9, nannofossil biozones NN4 and NN5; Beseme & Blondel, 1989; Blondel, 1991).

When preserved, the member A of the Aïn Grab Formation *sensu lato* begins with one or more polygenic, bioclastic and conglomeratic horizons, sometimes very well cemented. In a few outcrops (see schematic section in Fig. 2), conglomerates are overlain by a few centimetres to 5 m of calcareous to marly glauconitic sandstones, sometimes bioclastic, showing well-marked low-angle to sigmoid cross-stratifications. This unit forms a fining-up and deepening-up parasequence. From base to top, it characterizes a highly hydrodynamic coastal environment such as pebble beach, retrograding to coastal to shelf near-shore deposits such as foreshore to shoreface bioclastic sand bars with tidal influence. In most of the thickest outcrops (e.g. Jebel Zaouia section; Fig. 3), conglomerates forming the base of member A are directly overlain by some 10 to 20 m of gypsiferous and glauconitic clays, marls, sands and calcareous sands, the last two being sometimes bioclastic and cross-bedded, with sigmoid to low-angle cross-stratifications. Muds characterize lagoonal to littoral low-energy environments, while sands belong to tidal nearshore bars or barrier beaches (Fig. 5). These sediments are organized in stacked, thickening-up and coarsening-up parasequences, indicative of prograding and shallowing-up coastal to shelf near-shore environments. Near the top of member A, the latter sands, when preserved (e.g. Jebel Zaouia section; Fig. 3), are terminated by a striking ferruginous, bored hardground. This surface indicates a hiatus in the sedimentation, probably accompanied by subaerial exposure. This sedimentary event may correspond to a period of lower relative sea level (see below). In some sections (e.g. Jebel Zaouia section; Fig. 3), this surface might be locally overlain by a lenticular thin level of greenish gypsiferous clays interpreted as swamp to lagoonal low-energy deposits (Fig. 5).

The member B of the Aïn Grab Formation *sensu lato* stands out in outcrops as one or more stacked bars of cross-bedded calcareous sandstones, sometimes glauconitic and highly bioclastic, with coarse, conglomeratic basal parts containing quartz gravels and mud pebbles. Each of these calcareous fining-up sand bars forms a deepening-up parasequence (Fig. 3). These sandy deposits correspond to a transgressive coastal to shelf near-shore environment with the strong hydrodynamic conditions encountered in barrier beaches and foreshore to nearshore sand bars (Fig. 5).

The member C of the Aïn Grab Formation *sensu lato* consists of one, sometimes two, coarsening-up and shallowing-up parasequences. The lower part of these parasequences is argillaceous and gypsiferous. The base of member C often displays very thin, marly to sandy glauconitic and bioclastic horizons. The upper part becomes marly to sandy, calcareous and often highly bioclastic with well-preserved fossils. Parasequences are locally terminated by calcareous sandstones exhibiting oblique, low-angle to sigmoid cross-stratifications or herringbone structures indicating a major tidal influence. Clays were deposited in low-energy lagoonal–littoral tidal flats, grading progressively into more hydrodynamic coastal deposits such as near shore to foreshore tidal-bar or barrier beach sands. Parasequences in member C are progradational. In the tectonically most subsiding depressions, the total thickness of this member may reach up to 50 m.

A ferruginous and sometimes bored hardground often marks the top of the Aïn Grab Formation *sensu lato* (not present in the Jebel Khechem el Artsouma section of Fig. 3; see Fig. 2; Blondel, 1991). This surface indicates a hiatus, probably accompanied by subaerial exposure of the top of member C. Like the previous ferruginous and bored surface recognized in some sections near the top of member A, this sedimentary event may correspond to a short-lived period of lower relative sea level (see below).

The Mahmoud Formation *sensu stricto* was deposited or preserved only in the most rapidly subsiding areas of central Tunisia. It consists of greenish to grey-blue clays, grading towards the top into centimetres to tens of centimetres thick alternations of clays and fine sands interpreted as tidalites, and sometimes into silts and fine white sands. This coarsening-up and shallowing-up parasequence shows a relatively low-energy lagoonal–littoral environment in the lower part, overlain by prograding silty to sandy tidal flat or estuarine deposits (Fig. 5). It forms another progradational parasequence on top of the one or two evidenced in member C of the underlying Aïn Grab Formation *sensu lato*.

The top of the Hajeb el Aïoun Group is cut by a major unconformity: the base of the overlying Beglia Formation is usually strongly erosive and its basal unconformity is often marked by a polygenic and ferruginous conglomerate which contains cobbles affected by pedogenesis, wood fragments and vertebrate bones indicating a former continental environment. This conglomerate is overlain by fine to coarse cross-bedded sandstones, with rare intercalations of greenish gypsiferous clays. These deposits are indicative of continental fluviatile conditions grading locally to fluvio-deltaic or estuarine conditions. The lowermost strata of the Beglia Formation are formed by stacked fining-up parasequences.

ORGANIC MATTER AND PALAEOENVIRONMENT

Methods

Organic matter (OM) in sediments of the Hajeb el Aïoun Group was analysed both by pyrolysis and palynological methods.

Rock-Eval pyrolysis of total rock (Espitalié *et al.*, 1985–6) provides quick determination of type, richness and thermal maturity of OM, using parameters such as TOC (total organic carbon), HI (hydrogen index), OI (oxygen index) and T_{max} (pyrolysis temperature).

Palynofacies preparations are performed using standard palynological techniques. All mineral constituents are destroyed by hydrochloric and hydrofluoric acids prior to heavy-liquid separation. Remaining OM is sieved at 10 μm prior to slide mounting. No oxidation by nitric acid is carried out, because it affects the fluorescence of hydrogen-rich particles. Slides are systematically examined in normal transmitted and incident ultraviolet light in order to determine the organic facies (or palynofacies; Combaz, 1980). Because sedimentary OM behaves as other sedimentary particles, observation of its various constituents (amorphous or figured) and determination of their relative abundance, together with other sedimentological data, allow the identification of the OM source and of the local palaeoenvironment (Denison & Fowler, 1980; Batten, 1982; Habib, 1983; Boulter & Riddick, 1986; Hart, 1986; Tyson, 1987; Bustin, 1988; Lister & Batten, 1988; Van Pelt & Habib, 1988; Habib & Miller, 1989; Van der Zwan, 1990).

The following figured organic constituents are distinguished in this study (Figs 6 & 7):

1 Relatively allochthonous terrestrial components: spores and pollens and woody debris defined as inertinite (= palynomaceral 4 of Bryant *et al.*, 1988 and Van der Zwan, 1990); vitrinite (= palynomaceral 1 of Van der Zwan, 1990); land plant tissues (= palynomaceral 2 of Van der Zwan, 1990); and cutinite (= palynomaceral 3 of Van der Zwan, 1990).

2 Relatively autochthonous aquatic components: fresh to brackish water palynomorphs — mainly the algae *Pediastrum*, and secondarily *Incertae sedis* Ovoidites — and marine palynomorphs, mainly dinoflagellate cysts (= dinocysts) and some foraminiferal linings.

In addition to this distinction based on biological origin, biological and physical degradation of OM constituents may provide another clue to the depositional environment. In particular, the summed-up relative abundance of inertinite and vitrinite may be used as an indicator of the energy/oxidation level (Figs 6, 7 & 8), because these constituents are the less buoyant and the most resistant to degradation.

Results and palaeoenvironmental interpretation

Rock-Eval pyrolysis results (Table 1) show that the overall organic content of the Hajeb el Aïoun Group is low, most of the analysed samples having TOCs lower than 0.25%, if not lower than 0.1%. In the middle member of the Oued el Hajel Formation, green clays rich in plant remains may have a significant amount of type III OM (Espitalié *et al.*, 1985–6): samples T 520 and T 521 of the Jebel

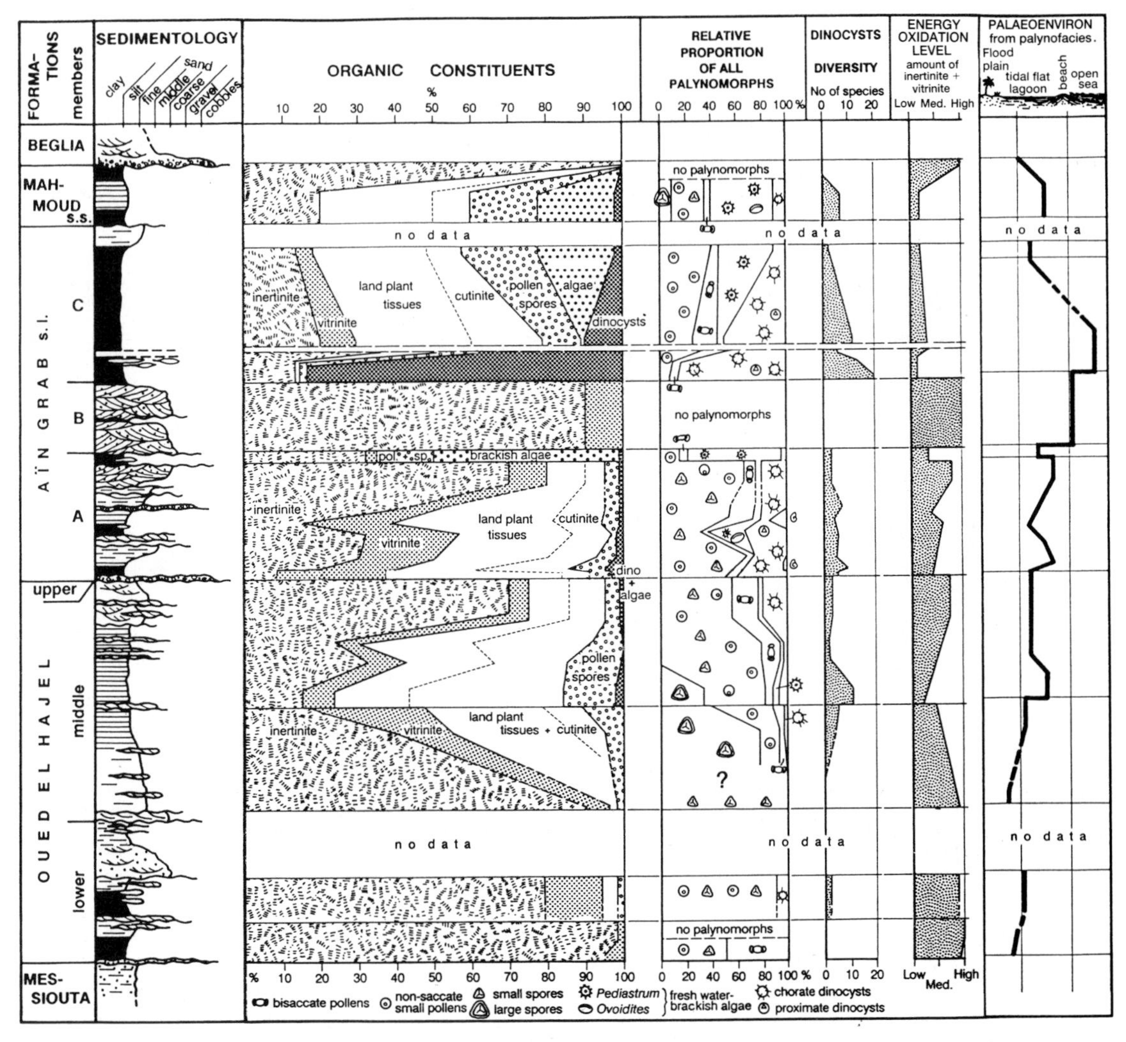

Fig. 6. Results of palynofacies analyses in the Hajeb el Aïoun Group and palaeoenvironmental interpretation based on palynofacies. The sedimentological log is derived from the two field sections detailed in Fig. 3.

Zaouia section (Fig. 3) were also analysed palynologically and this confirmed the dominance of humic continental material (Fig. 6). The T_{max} values of around 420°C clearly show that these sediments are thermally immature, these values corresponding to a vitrinite reflectance of *c.* 0.4% (Espitalié *et al.*, 1985–6). In the Aïn Grab Formation *sensu lato*, some green clays of the member A may have some marginal amounts of essentially type III to mixed type III/II OM (sample T 531 of Jebel Zaouia section). The member C of the Aïn Grab Formation *sensu lato* is extremely poor in OM, whereas the Mahmoud Formation may have marginal TOCs. A lignite from the overlying Beglia Formation confirms the thermal immaturity of the Miocene sediments in the studied area (T_{max} of 426°C; cf. Espitalié *et al.*, 1985–6).

Forty-five field samples were prepared for palynofacies studies, most of them coming from the representative sections of Jebel Khechem el Artsouma and Jebel Zaouia (numbered samples in Fig. 3), the others being mainly spot checks carried out in four other field sections located in Fig. 1. Palynofacies results and the palaeoenvironmental interpretation of the Hajeb el Aïoun Group are summarized in Fig. 6.

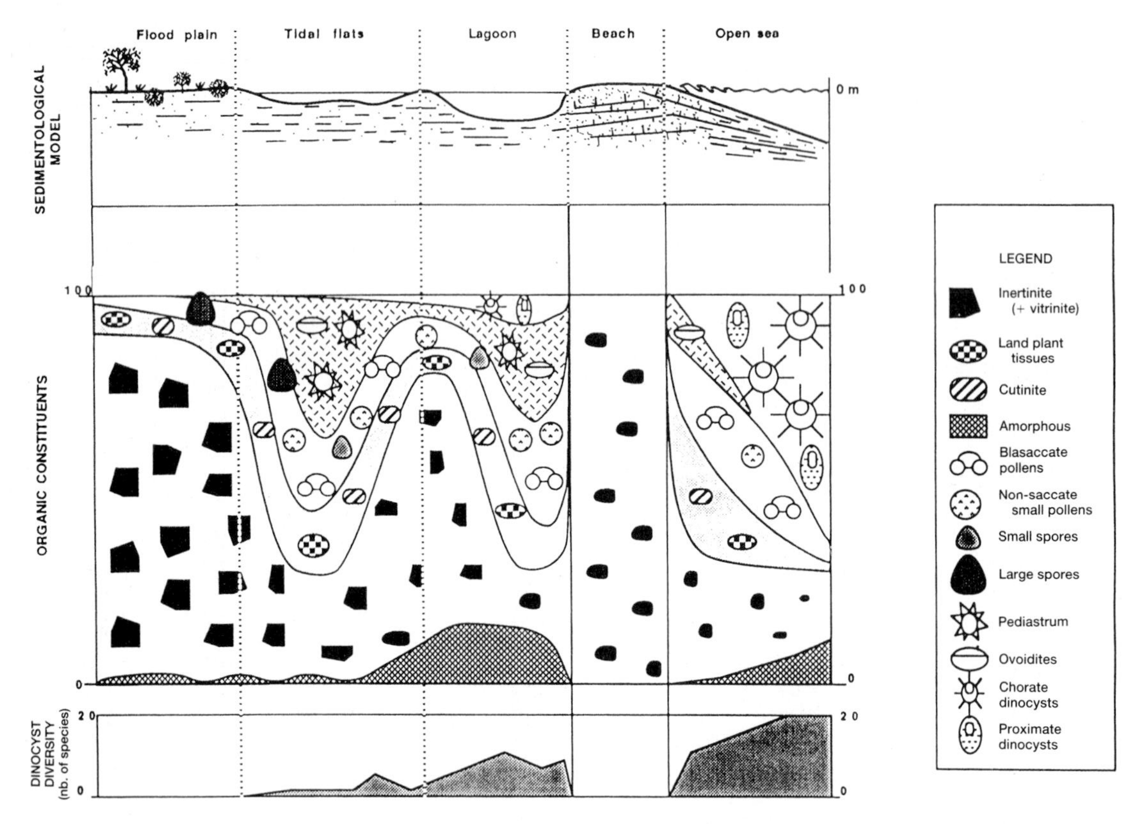

Fig. 7. Schematic distribution of palynofacies constituents across flood-plain to coastal-plain environments (see the sedimentological framework of Fig. 4).

In the Oued el Hajel Formation, the lower argillaceous part of the lower member is dominated by inertinite and to a lesser extent by vitrinite. Palynomorphs are not abundant and consist essentially of continentally derived pollen and spores. A few dinocysts may be present, but diversity is very low. Within the sedimentological framework of Fig. 4, this type of palynofacies, marked by a high oxidation level, is characteristic of flood-plain deposits becoming locally estuarine with a tidal influence. Apart from a few corroded inertinite grains, the medium to coarse sandstones forming the upper part of the lower member are devoid of OM, because of strongly oxidizing conditions encountered in such fluvial to fluvio-deltaic deposits.

On the basis of its organic content, the middle member of the Oued el Hajel Formation can be subdivided into two intervals. The lower interval is dominated by continental constituents, and the large proportion of land plant tissues and cutinite versus inertinite and vitrinite indicates lower oxidizing conditions than in the lower member. Palynomorphs increase up to 10% in the top of the interval and are dominated by large spores. A small amount of poorly diversified dinocysts appears at the top of this lower interval. These palynofacies show palaeoenvironmental conditions similar to those in the lower member of the Oued el Hajel Formation, i.e. flood plain to estuarine at the base, becoming tidal flat at the top. The upper interval is also dominated by continental fragments: land plant tissues and cutinite are dominant at the base, but decrease towards the top where inertinite is the most abundant constituent and shows the return to higher oxidizing conditions. Pollen and spores are the dominant palynomorphs throughout this upper interval. Dinocysts are also present throughout, but their absolute abundance and diversity are higher at the base of the interval. Fresh-water to brackish algae are present throughout in small amounts. The relatively high diversity of dinocysts at the base of this interval marks an influx of more open marine

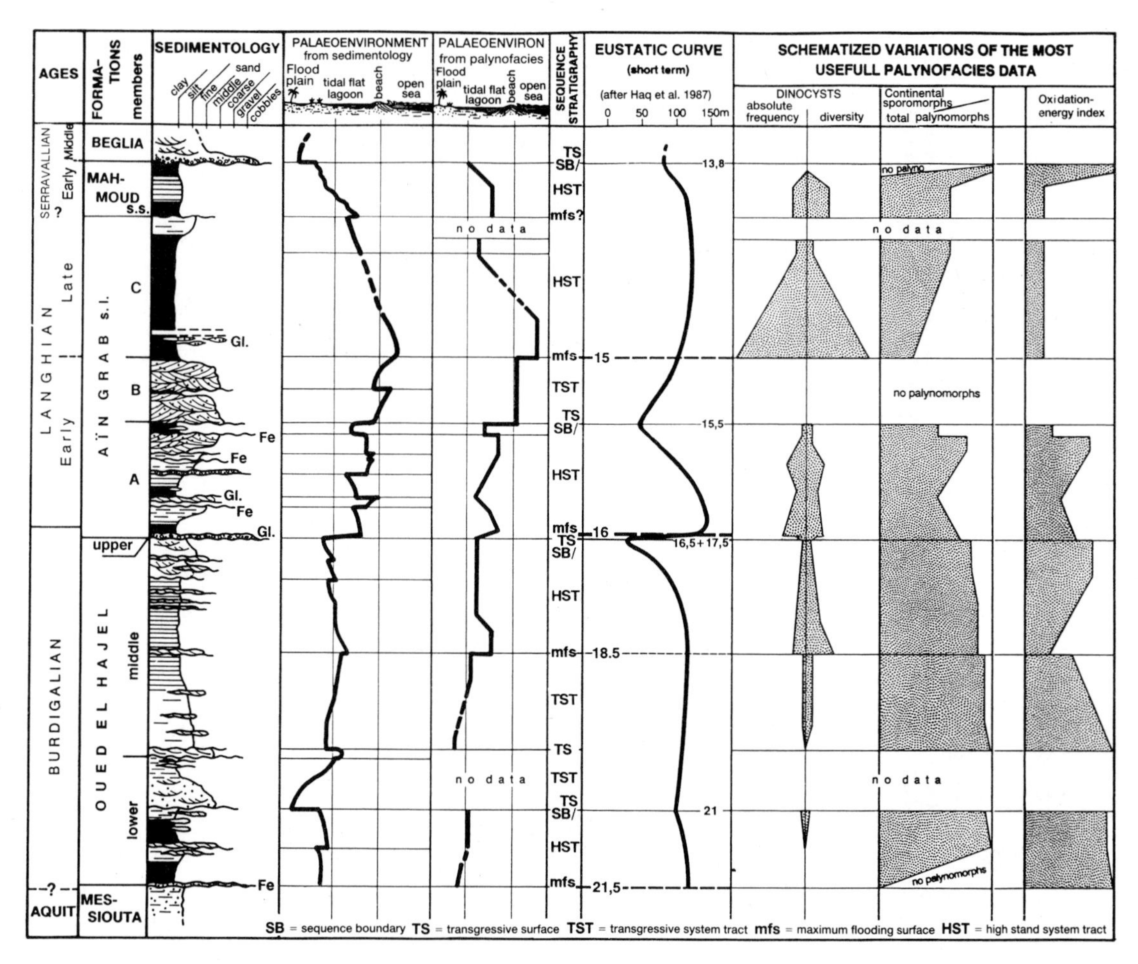

Fig. 8. Correlation between palaeoenvironmental interpretation derived from combined sedimentological and palynofacies analyses and the stratigraphic signature of eustasy in the Hajeb el Aïoun Group (lower to middle Miocene of central Tunisia). Note how the maximum flooding surfaces are remarkably emphasized by the increased dinocyst frequency and diversity. Dinoflagellate cyst frequency and diversity decrease towards the top of the highstand systems tract as a result of shallower water depths. The sedimentary log is derived from the two field sections detailed in Fig. 3.

waters in a relatively low-energy lagoonal to tidal-flat environment where continental constituents are still the most abundant. This diversity decreases rapidly towards the top where all organic parameters point to a more restricted tidal-flat to estuarine environment.

In the member A of the Aïn Grab Formation *sensu lato*, two intervals can be distinguished in terms of OM. The lower interval is the thickest and extends up to the major ferruginous hardground near the top of the member. Continental fragments of humic origin are the main organic constituents, dominated by land plant tissues and cutinite in the lower part and by inertinite in the upper part. The absolute frequency and diversity of dinocysts are highest just above the basal conglomerate and decrease towards the top of the interval. This decrease in dinocyst diversity is accompanied by the appearance of fresh-water to brackish algae. Pollen and spores are abundant throughout the interval. Some foraminiferal linings are present in the lower half of the interval. This palynofacies characterizes low to medium energy lagoonal conditions with a stronger marine influence in the basal part, and indications of restriction and of a slightly higher oxidation level at the top. The upper interval

Table 1. Results of Rock-Eval pyrolysis in the lower-to-middle Miocene Hajeb el Aïoun Group. TOC, total organic carbon in % weight; samples with TOC ≤ 0.25% give totally unreliable HI, OI and T_{max} values; samples with TOC ≤ 0.5% give HI, OI and T_{max} values that may be unreliable because of mineral matrix effect (see Espitalié *et al.*, 1985–6); HI, hydrogen index in μg HC/g TOC; OI, oxygen index in μg CO_2/g TOC; T_{max}, pyrolysis temperature in °C.

Sample no.	Field section (see Fig. 1 for location)	Lithol. unit	Age	Lithology	Rock-Eval pyrolysis			
					TOC (%)	HI	OI	T_{max}
T 173	Jebel Gatrana W	Beglia Fm	Serravallian	Lignite	3.73	7	93	426
T 701	Jebel Touila	Mahmoud Fm	Langh.–Serr.	Grey clay	0.34	876?	61	?
T 66	Jebel Khechem	Mahmoud Fm	Langh.–Serr.	Grey-blue clay	0.17			
T 699	Jebel Touila	Aïn Grab Fm, C Mb	Late Langh.	Green clay	< 0.1			
T 549	Jebel Zaouia	Aïn Grab Fm, C Mb	Late Langh.	Dark green clay	< 0.1			
T 64	J. Khechem	Aïn Grab Fm, C Mb	Late Langh.	Brown clay	< 0.1			
T 468	Jebel Boudinar	Aïn Grab Fm, C Mb	Late Langh.	Green clay	< 0.1			
T 527	Jebel Zaouia	Aïn Grab Fm, A Mb	Burdig.–Langh.	Dark green clay	0.13			
T 531	Jebel Zaouia	Aïn Grab Fm, A Mb	Early Langh.	Dark green clay	0.31	309	190	?
T 539	Jebel Zaouia	Aïn Grab Fm, A Mb	Early Langh.	Grey clay	0.16			
T 57	J. Khechem	Aïn Grab Fm, A Mb	Burdig.–Langh.	Dark green sandy clay	< 0.1			
T 59	J. Khechem	Aïn Grab Fm, A Mb	Burdig.–Langh.	Dark green clay	< 0.1			
T 526A	Jebel Zaouia	Aïn Grab Fm, A Mb	Late Burdig.	Greyish carbonates (cobble matrix)	0.15			
T 526B	Jebel Zaouia	Aïn Grab Fm, A Mb	Late Burdig.	Wood fragment	3.39	67	247	411
T 521	Jebel Zaouia	Oued el Hajel, mid. Mb	Burdig.	Green clay + plant rem.	2.20	53	178	418
T 520	Jebel Zaouia	Oued el Hajel, mid. Mb	Burdig.	Green sandy clay + plant remains	0.28	150	164	423
T 517	Jebel Zaouia	Oued el Hajel, mid. Mb	Burdig.	Green sandy clay	0.13			
T 459	J. Boudinar	Oued el Hajel, mid. Mb	Burdig.	Brown clay	< 0.1			
T 508	Jebel Zaouia	Oued el Hajel, lwr. Mb	Early Burdig.	Grey sandy clay	< 0.1			

is characterized by some 50% of fresh-water to brackish algae, dominated by the genus *Pediastrum*. Pollen and spores are well represented and dinocyst frequency and diversity are very low. This palynofacies is indicative of low-energy lagoonal conditions with a much stronger fresh-water influx than in the upper part of the lower interval.

The medium to coarse beach and tidal-bar sands in member B of the Aïn Grab Formation *sensu lato* contain only inertinite and accessorily vitrinite, indicative of high-energy oxidizing conditions.

In the Aïn Grab Formation *sensu lato*, the organic content of the lower part of member C shows the most open marine conditions of the entire Hajeb el Aïoun Group: in the Jebel Zaouia section, the palynofacies is dominated by a highly diversified and abundant (up to 80%) dinocyst population, associated with small-size inertinite fragments, a facies clearly indicative of open marine conditions (Gorin & Steffen, 1991). In its upper part (Jebel Khechem el Artsouma section), member C displays evidence of an increased continental influx: larger-sized continental fragments predominate, mainly land plant tissues and cutinite associated with abundant pollen and spores. Dinocyst absolute frequency and diversity decrease considerably, whereas fresh-water to brackish algae become abundant. This indicates the return to low–medium energy lagoonal–littoral environment.

The organic content of the Mahmoud Formation *sensu stricto* is dominated by continental fragments of humic origin, pollen and spores and fresh-water to brackish algae. Dinocysts are present but their frequency and diversity are low. This indicates low-energy, lagoonal–littoral to tidal-flat conditions. In the studied section, the uppermost part shows a strongly oxidizing environment compatible with sand-flat to estuarine deposits.

The palaeoenvironmental curve derived from palynofacies (Fig. 6), although not as detailed as that obtained from standard sedimentology (Fig. 5), shows a very similar trend reinforcing the results from the two approaches. One can therefore derive with confidence a schematic palynofacies profile (Fig. 7) that is valid for flood-plain to coastal-plain environments similar to that of the Hajeb el Aïoun Group proposed in Fig. 4.

Finally, two palynofacies observations are worth mentioning for their possible palaeoenvironmental connotation. Firstly, the absence or relatively low percentage throughout the analysed samples of hydrogen-rich amorphous organic matter, generally indicative of low-energy, stagnant and oxygen-depleted environments (Boulter & Riddick, 1986; Tyson, 1987; Bryant *et al.*, 1988; Van der Zwan, 1990; Fig. 7). This confirms the low TOCs obtained by pyrolysis and the relatively well-aerated depositional conditions of the Hajeb el Aïoun Group. Secondly, it is interesting to note the overwhelming predominance of chorate over proximate dinocysts in these coastal deposits. At present, it is difficult to evaluate the significance of the latter observation, but it may indicate that locally cyst morphology does have a palaeoenvironmental value, as already noted by other authors (Wall, 1965; Denison & Fowler, 1980; Monteil & Cornu, 1987; Tyson, 1987; Lister & Batten, 1988; Honigstein *et al.*, 1989).

CORRELATIONS BETWEEN SEDIMENTOLOGICAL AND PALAEOENVIRONMENTAL INTERPRETATIONS AND THE STRATIGRAPHIC SIGNATURE OF EUSTASY

Sedimentological analyses in the field and laboratory, together with localized palynofacies studies, led to a sequence-stratigraphic interpretation of the Hajeb el Aïoun Group (Blondel, 1991). This interpretation was performed using the stratigraphic framework described above. Each formation and member of the group was thoroughly investigated in order to recognize the main systems tracts (LST, TST and HST are lowstand, transgressive and highstand systems tracts — Haq *et al.*, 1987, 1988; Van Wagoner *et al.*, 1990; Vail *et al.*, 1991) and the major surfaces of 'discontinuity' (SB, sequence boundary; TS, transgressive surface) and 'relative continuity' (mfs, maximum flooding surface) observed at a regional scale on the inner Tunisian platform.

In continental to coastal areas constituting the innermost part of a platform, sequence boundaries are most often overprinted by transgressive surfaces. Except inside incised valleys or tectonically induced subsiding axes, the erosive base of the TST backstepping parasequences erodes and reworks whole or part of the underlying continental-to-coastal sediments constituting the LST, if deposited, as well as the non-deposition or erosive surfaces corresponding to sequence boundaries (Van Wagoner *et al.*, 1990; Vail *et al.*, 1991).

Figure 8 shows the detailed correlation between

the sedimentology- and palynofacies-derived palaeoenvironmental curves, the sequence-stratigraphic interpretation and the eustatic curve for the two representative field sections illustrated in Fig. 3. Figure 9 shows the synthetic sequence-stratigraphic interpretation for the ideal, complete schematic sedimentological log of the Hajeb el Aïoun Group shown in Fig. 2.

The sequence-stratigraphic interpretation of the Hajeb el Aïoun Group is described below from base to top (Figs 8 & 9; Blondel, 1991), integrating all data described above under sedimentology and palynofacies.

The partly reworked continental red clays and silts of the Messiouta Formation (which underlies the Hajeb el Aïoun Group), arranged in coarsening-up parasequences, are interpreted as an innermost shelf equivalent to late highstand deposits (HST). The interval of lacustrine, palustrine to tidal-flat limestones, fluviatile to estuarine sands or lagoonal to sabkha gypsum terminating the Messiouta Formation is interpreted as the sedimentary

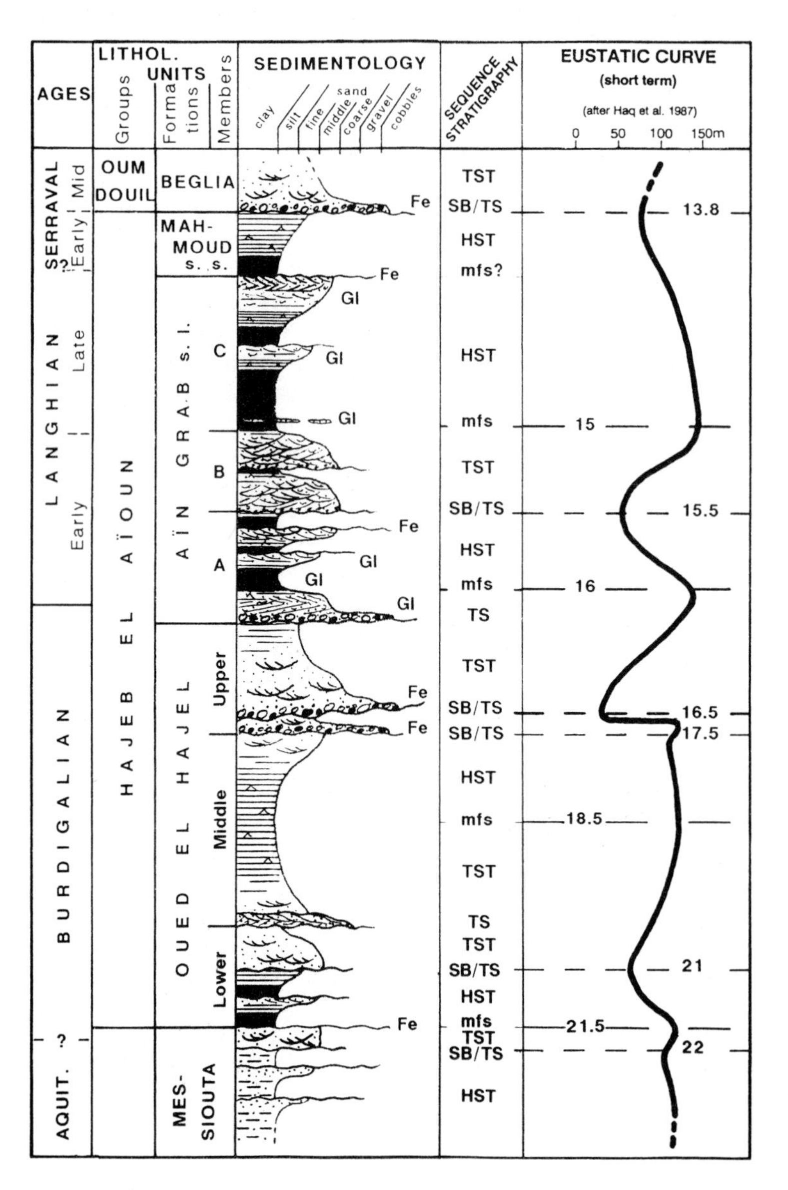

Fig. 9. Correlation between the schematic sedimentological log of the Hajeb el Aïoun Group (same as that of Fig. 2), the sequence-stratigraphic interpretation and the global eustatic curve of Haq *et al.* (1987, 1988).

record on the inner shelf of a TST (Fig. 9). By vertical and lateral correlations, this interval seems to be contemporaneous to the Aquitanian–Burdigalian boundary (Blondel, 1991). Thus, the erosional surface which separates this thin interval and the underlying lateritic sediments has to be considered as a SB overprinted by the flood-plain to inner-coastal-plain equivalent of a TS. This SB is tentatively dated as 22 Ma after the global cycle chart of Haq *et al.* (1987, 1988).

The often ferruginous and sometimes bored hardground marking the top of the Messiouta Formation – which indicates a condensed interval considered as an innermost coastal equivalent of a maximum flooding surface (Blondel, 1991) – may be tentatively correlated with the maximum flooding surfaces dated 21.5 Ma in Haq *et al.* (1987, 1988).

During the Burdigalian, in the subsiding depressions of the inner Tunisian shelf, close to the northeastern edge of the 'Kasserine Island' (Fig. 1), a two-stage marine inundation is well marked by lagoonal–littoral to tidal-flat deposits belonging to the Oued el Hajel Formation.

The first unequivocal marine indicators (dinocysts, ostracods, echinids remains) appear in the argillaceous to silty and sandy basis of the lower member of the Oued el Hajel Formation. This lower part consists of a few prograding coarsening-up and shallowing-up parasequences. These deposits are interpreted as equivalent to a HST in a coastal environment (Figs 8 & 9). Biostratigraphic data do not permit a precise dating of this sequence: teeth of cyprinids (fresh to brackish water fish) found in this unit indicate only an early Miocene age (Blondel, 1991). On the basis of vertical and lateral correlations, Blondel (1991) considers these deposits as lower Burdigalian. The fluvio-deltaic to estuarine sands forming the upper part of the lower member represent a fining-up and deepening-up parasequence. It may correspond to a new TST in the inner-shelf domain. The erosive base of this unit would then be a TS overprinting a previous SB.

In central Tunisia, the boundary between the lower and middle members of the Oued el Hajel Formation has been defined as equivalent to the first true marine transgressive surface (Beseme & Blondel, 1989; Blondel, 1991). We consider this surface as a lithostratigraphic boundary that might be regionally diachronous. In outcrops, it often corresponds to the erosive base of a tidal channel (e.g. Jebel Zaouia section; Figs 3 & 8) followed by sand-flat or estuarine deposits. The lower part of the middle member contains one or more fining-up and deepening-up parasequences interpreted as terminating the TST which began in the upper part of the lower member. A maximum flooding surface is tentatively defined within the middle member, on the basis of a sudden frequency and diversity increase of dinocysts in coastal sandy clays. It may be tentatively correlated with the maximum flooding surface dated as 18.5 Ma in the global cycle chart of Haq *et al.* (1987, 1988).

Where present and thickly developed (e.g. north of the Jebel Touila; see Fig. 1), the coarse-to-medium sands of the upper member of the Oued el Hajel Formation display at their base two stacked conglomeratic horizons, the upper one being better marked and strongly erosive. Overall, the upper member siliciclastic deposits form a retrograding fining-up and deepening-up parasequence. The two stacked conglomeratic horizons at the base are interpreted as erosional surfaces corresponding to sequence boundaries. The lower one is tentatively correlated with SB 17.5 Ma, and the upper one with SB 16.5 Ma on the global cycle chart of Haq *et al.* (1987, 1988; Fig. 9). They characterize relative sea-level drops marked locally by the downcutting of incised valleys on the inner Tunisian shelf (Blondel, 1991). These valleys were filled up by fluvio-deltaic to estuarine siliciclastics constituting the overlying upper member deposits and considered as the proximal equivalent of a TST (Fig. 9). SB 16.5 Ma is a Neogene surface well marked worldwide (Bartek *et al.*, 1991).

Sediments of the Aïn Grab Formation *sensu lato* – dated as latest Burdigalian to late Langhian by foraminifera and nannofossils (Fig. 2) – correspond to a major two-stage marine transgression and flooding onto a large part of the Tunisian shelf but for a few highs bordering the north and northeastern edge of the emerged 'Kasserine Island' (Fig. 1).

The base of member A of the Aïn Grab Formation *sensu lato* is defined by the first clearly marine facies underlined by erosive basal conglomerates (transgressive lag) which characterize the boundary between the Oued el Hajel and Aïn Grab Formations. In some sections (e.g. Jebel Touila section, see Fig. 1 for location and the schematic log of Figs 2 & 9), a 1–5 m thick marine, transgressive backstepping unit overlies the conglomerates, whereas in other sections (e.g. Jebel Zaouia section; Figs 3 & 8) the transgressive unit is very thin. In this case, the TST corresponds to a very thin 'condensed interval' (10 to 20 cm thick) characterized by polygenic glauconitic and bioclastic conglomerates sometimes very well indurated. Palynofacies show that the basal clay of member A overlying these conglomerates contains a higher frequency and diversity of dinocysts (Figs 6 & 8). Together with sedimentological and micropalaeontological data, these observations indicate that the mud layer directly overlying the basal transgressive deposits of member A of the Aïn Grab Formation *sensu lato* corresponds to the sedimentary record of a maximum flooding surface on the inner Tunisian shelf. This maximum flooding surface is very well dated as latest Burdigalian to earliest Langhian (base of biozone N8 and biozone NN4; Beseme & Blondel, 1989; Blondel, 1991; Fig. 2). It can be correlated with the maximum flooding surface dated 16 Ma in the global cycle chart of Haq *et al.* (1987, 1988). Above this maximum flooding surface, member A consists of a few stacked, prograding, coarsening-up, and shallowing-up parasequences, each one grading from marine clays to tidal–littoral bar sands. In some sections (e.g. Jebel Zaouia section; Figs 3 & 8), the last of these sandy parasequences is terminated by a ferruginous and bored hardground, which is overlain by gypsiferous to sandy lenticular clays interpreted as fresh to brackish water, low-energy swamp to lagoon deposits. This argillaceous parasequence is truncated by another well-marked transgressive surface forming the base of member B (Figs 3 & 8). Defining a SB within the top of the highstand downstepping unit of member A is difficult in outcrop. In the absence of an obvious candidate, we tentatively consider the TS forming the base of member B as overprinting the previous SB. The ferruginous hardground near the top of member A could represent the boundary between early and late highstand deposits. The overprinted SB may correspond to SB 15.5 Ma of the global cycle chart (Haq *et al.*, 1987, 1988).

The fining-up and deepening-up parasequences constituting the member B of the Aïn Grab Formation *sensu lato* correspond to transgressive backstepping deposits. They most probably belong to the base of the depositional sequence TB 2.4 reported on the global cycle chart (15.5 to 13.8 Ma; Haq *et al.*, 1987, 1988).

At the base of member C of the Aïn Grab Formation *sensu lato*, thin, bioclastic and glauconitic marly to sandy horizons are interpreted as a condensed interval associated with the most important phase of flooding on the Tunisian shelf during the Miocene. The base of this interval is correlated with the maximum flooding surface 15 Ma of the global cycle chart (Haq *et al.*, 1987, 1988), which clearly represents worldwide the maximum flooding during Neogene times (Bartek *et al.*, 1991). This surface is especially characterized by the high frequency and diversity of dinocysts (Figs 6 & 8). The one or two coarsening-up and shallowing-up prograding parasequences forming member C characterize a highstand downstepping unit in the inner-shelf domain of central Tunisia. Decrease in accommodation space from the base to the top of the HST — which corresponds to shallower water depths in terms of palaeoenvironment — is remarkably highlighted here by the significant decrease in absolute frequency and diversity of dinocysts (Fig. 8).

The top of the Aïn Grab Formation *sensu lato* is normally indicated by a ferruginous and bored hardground, overlain by clays, silts and sands of the Mahmoud Formation *sensu stricto*. It is tentatively interpreted as the boundary between the early and late HST, although it could also correspond to a (?) type II SB (Blondel, 1991). This surface can be correlated throughout the inner shelf of central Tunisia and probably corresponds to a low-amplitude drop of relative sea level marked mainly in the shallow inner zone of the shelf. This drop does not seem to be significant worldwide because it is not reported on the global cycle chart of Haq *et al.* (1987, 1988). It could also be the result of local tectonics. Further studies in Tunisia and on other shallow Miocene shelves may identify with certainty the global or local origin of this eustatic event marking the top of the Aïn Grab Formation *sensu lato*. Stratigraphically, this surface has been dated as early late Langhian by Beseme and Blondel (1989) or latest Langhian by Blondel (1991). The latter author tentatively considers that this hardground coincides with the Langhian–Serravallian boundary in central Tunisia. Additional studies of microfossil associations will hopefully permit a more precise dating.

The age of the base of the Mahmoud Formation *sensu stricto* is given as early late Langhian by Beseme and Blondel (1989), but it is based on a poorly preserved and sparse microfauna studied in only one field section (Jebel Touila SW, section 3 in Fig. 1). Blondel (1991) considers the Mahmoud Formation *sensu stricto* as early Serravallian by equating approximately the base of this formation with the Langhian–Serravallian boundary. The argillaceous to gypsiferous base of the Mahmoud

Formation *sensu stricto* corresponds to another period of 'major' flooding on the inner Tunisian shelf, but with a lower amplitude than those observed in the underlying Aïn Grab Formation *sensu lato*. The base of the Mahmoud Formation *sensu stricto* is characterized by a fairly abundant and diversified dinocyst population, but a rare microfauna. Overall, the Mahmoud Formation *sensu stricto* consists of a coarsening-up and shallowing-up parasequence corresponding to highstand downstepping deposits. Depending on the interpretation of the hardground at the top of the underlying Aïn Grab Formation *sensu lato*, the Mahmoud Formation *sensu stricto* could either be correlated with the late HST deposits of the sequence TB 2.4 beginning at the base of member B of the Aïn Grab Formation *sensu lato*, or interpreted as another new small sequence.

The strongly erosive conglomeratic base of the Beglia Formation — approximately dated as close to the early–middle Serravallian boundary using vertebrate fossils — is observed in ancient incised valleys or synclinal axes and is also very well marked near palaeo-highs (Blondel, 1991). It represents the stratigraphic signature of a major drop in relative sea level, locally amplified by tectonics or diapirism, and recognized over the entire Tunisian shelf. It is considered as an important SB correlated with SB 13.8 Ma on the global cycle chart (Haq *et al.*, 1987, 1988). This worldwide eustatic drop originates from the global cooling of oceanic and atmospheric masses during the Serravallian (Shackleton & Kennett, 1975; Bartek *et al.*, 1991). The siliciclastic sediments overlying this SB represent (?) LST and TST deposits on the innermost shelf (Blondel, 1991).

CONCLUSIONS

The continental to coastal deposits developed on the innermost shelf of central Tunisia during early to middle Miocene present important lateral and vertical facies variations and are often fairly poor in biostratigraphically significant fossils. Despite these negative factors, it has been possible to identify the stratigraphic signature of global eustatic variations. This has been achieved only after acquiring a minimum of reliable datings and establishing a palaeoenvironmental model for the inner-shelf areas. This model was constructed from a thorough sedimentological study of numerous field sections, leading to the lateral and vertical recognition of the main continental to coastal types of environments. These sedimentological observations were complemented locally by palynofacies analyses, a less accurate palaeoenvironmental tool that should always be used in conjunction with sedimentology in coastal deposits. The vertical and lateral variations of palaeoenvironments derived from the sedimentary record lead to the interpretation of the different lithological units in terms of sequence stratigraphy, i.e. sedimentary systems tracts, 'discontinuities' (sequence boundaries and transgressive surfaces) and 'relative continuities' (maximum flooding surfaces).

In inner coastal environments (such as tidal flats, estuaries or lagoons), sequence boundaries and transgressive surfaces are normally well defined by macroscopic sedimentological features, especially when TSs overprint SBs. In such environments, maximum flooding surfaces are more difficult to identify in the field, but they become more obvious with palynofacies studies because they usually coincide with a significant increase in abundance and diversity of dinocysts. In the field, transgressive backstepping units are characterized by fining-up and deepening-up parasequences, whereas highstand downstepping units are marked by coarsening-up and shallowing-up parasequences (Van Wagoner *et al.*, 1990; Vail *et al.*, 1991). Increasingly shallower water depths during late highstands are particularly well corroborated in palynofacies studies by the decrease in dinocysts abundance and diversity.

The stratigraphic signature of Burdigalian-to-early Serravallian deposits on the Tunisian shelf is comparable to that recorded on other contemporaneous continental margins, especially those of the Antarctic continent (Bartek *et al.*, 1991). The Miocene Tunisian shelf, which was located on the North African margin facing the Mesogean sea, has recorded the global, relative sea-level oscillations, even in its innermost areas. These global variations were mainly the consequence of successive coolings and warmings of oceanic and atmospheric masses accompanying the progressive onset of the Antarctic ice sheet (Bartek *et al.*, 1991). Since the end of the Palaeogene, glacio-eustasy has been one of the main governing factors of sedimentation and sedimentary facies distribution on all continental shelves.

ACKNOWLEDGEMENTS

This work was partly supported by the Swiss National Science Foundation (grant No. 20-5648.88). We are especially indebted to J. Charollais from the Geological Department of the University of Geneva for his permanent help and encouragement, to H. Ben Dhia and K. Medhioub from the Geological Department of the National Engineering School of Sfax, Tunisia, as well as to all the scientists and collaborators of the latter Department for their scientific help and their logistical support in the field.

We are also indebted to the Institut Français du Pétrole (G. Pichaud and J.P. Herbin) for kindly performing Rock-Eval analyses, to E. Monteil and M. Floquet for their precious advice on palynological preparations, to J. Metzger for his invaluable help in drafting and critically commenting on the figures and to J. Fellmann for preparing Table 1.

Finally, we thank D.K. Goodman and B.U. Haq for critically reviewing this paper.

REFERENCES

Bartek, L., Vail, P.R., Anderson, J.B., Emmet, P.A. & Wu, S. (1991) The effect of Cenozoic ice sheet fluctuations in Antarctica on the stratigraphic signature of the Neogene. *J. Geophys. Res.* (in press).

Batten, D.J. (1982) Palynofacies, palaeoenvironments and petroleum. *J. Micropalaeontol.* **1**, 107–114.

Ben Ismail-Lattrache, K. (1981) Etude micropaléontologique et biostratigraphie des séries paléogènes de l'anticlinal du Jebel Abderahman (Cap Bon, Tunisie nord-orientale). PhD Thesis, Faculty of Science, University of Tunis, 229 pp.

Beseme, P. & Blondel, T. (1989) Les séries à tendance transgressive marine du Miocène inférieur à moyen en Tunisie centrale: données sédimentologiques, biostratigraphiques et paléoécologiques. *Rev. Paléobiol. Genève* **8**(1), 187–207.

Biely, A., Rakus, M., Robinson, P. & Salaj, J. (1972) Essai de corrélation des formations miocènes du sud de la Dorsale tunisienne. *Notes Serv. Géol. Trav. Géol. Tunis* **7**, 73–92.

Black, C.C. (1972) Introduction aux travaux sur la Formation Beglia. *Notes Serv. Géol. Tunis* **35**, 5–6.

Blondel, T. (1991) Les séries à tendance transgressive marine du Miocène inférieur à moyen de Tunisie centrale: biostratigraphie, paléoenvironnements et analyse séquentielle des formations du Groupe Hajeb el Aïoun intégrés au contexte géodynamique, climatique et eustatique régional et global. DSc thesis, University of Geneva, No 2469, *Publ. Dépt. Géol. et Paléont.* **9**, 587 pp.

Blondel, T. & Demarcq, G. (1990) Les Pectinidae du Burdigalien terminal-Langhien de Tunisie. Données paléobiologiques et paléogéographiques. *Rev. Paléobiol. Genève* **9**(2), 243–255.

Blondel, T., Pouyet, S. & David, L. (1990) Les bryozoaires du Burdigalien terminal-Langhien de Tunisie. Données préliminaires. *Bull. Soc. Géol. France*, **VI**(4), 667–672.

Blondel, T., Yaich, C. & Decrouez, D. (1988) La Formation Messiouta de Tunisie centrale (Miocène inférieur continental): lithologie, sédimentologie et mise en place de cette Formation. *Géol. Méditerr. Marseille* **XII–XIII**/(3–4), (1985–86), 155–165.

Boulter, M.C. & Riddick, A. (1986) Classification and analysis of palynodebris from the Palaeocene sediments of the Forties Field. *Sedimentology* **33**, 871–886.

Bryant, I.D., Kantorowicz, J.D. & Love, C.F. (1988) The origin and recognition of laterally continuous carbonate-cemented horizons in the Upper Lias Sands of southern England. *Mar. Petrol. Geol.* **5**, 108–133.

Burollet, P.F. (1956) Contribution à l'étude stratigraphique de la Tunisie centrale. *Ann. Mines et Géol. Tunis* **18**, 350 pp.

Bustin, R.M. (1988) Sedimentology and characteristics of dispersed organic matter in Tertiary Niger delta: origin of source rocks in a deltaic environment. *Bull. Am. Assoc. Petrol. Geol.* **72/3**, 277–298.

Castany, G. (1951) Etude géologique de l'Atlas tunisien oriental. DSc. thesis. *Ann. Mines et Géol. Tunis* **8**, 632 pp.

Combaz, A. (1980) Les kérogènes vus au microscope. In: *Kerogen, Insoluble Organic Matter in Sedimentary Rocks* (Ed. Durand, B.), Technip, Paris, pp. 55–112.

Denison, C.N. & Fowler, R.M. (1980) The sedimentation of the North Sea reservoir rocks. *Norw. Petrol. Soc.* **12**, 1–22.

Espitalié, J., Deroo, G. & Marquis, F. (1985–86) La pyrolyse Rock-Eval et ses applications. *Rev. Inst. Franç. Pétrole* **40**(5), 563–579; **40**(6), 755–784; **41**(1), 73–89.

Flandrin, J. (1948) Contribution à l'étude stratigraphique du Nummulitique algérien. *Bull. Serv. Carte Géol. Algérie, (2): Stratigr. Alger* **4**(19), 334 pp.

Fournie, D. (1978) Nomenclature lithostratigraphique des séries du Crétacé supérieur au Tertiaire de Tunisie. *Bull. CREP Elf-Aquitaine* **2**(1), 97–148.

Gorin, G. & Steffen, D. (1991) Organic facies as a tool for recording eustatic variations in marine fine-grained carbonates — example of the Berriasian stratotype at Berrias (Ardèche, SE France). *Palaeogeogr, Palaeoclimatol., Palaeoecol.* **85**, 303–320.

Habib, D. (1983) Sedimentation-rate-dependent distribution of organic matter in the North Atlantic Jurassic–Cretaceous. *Init. Rep. Deep Sea Drill. Proj.* **76**, 781–794.

Habib, D. & Miller, J.A. (1989) Dinoflagellate species and organic facies evidence of marine transgression and regression in the Atlantic Coastal Plain. *Palaeogeogr., Palaeoclimatol., Palaeoecol.* **74**, 23–47.

Haq, B.U., Hardenbol, J. & Vail, P.R. (1987) Chronology of fluctuating sea levels since the Triassic. *Science*, **235**, 1156–1167.

Haq, B.U., Hardenbol, J. & Vail, P.R. (1988) Mesozoic and Cenozoic chronostratigraphy and cycles of sea-level change. In: *Sea-level Changes: An Integrated Approach* (Eds Wilgus, C.K., Hastings, B.S., Kendall, C.G.St.C., Posamentier, H.W., Ross, C.A. & Van Wagoner, J.C.) *Soc. Econ. Paleontol. Mineral. Spec. Publ.* **42**, 71–108.

Hart, G.F. (1986) Origin and classification of organic matter in clastic systems. *Palynology* **10**, 1–23.

Honigstein, A., Lipson-Benitah, S., Conway, B., Flexer, A. & Rosenfeld, A. (1989) Mid-Turonian anoxic event in Israel — a multidisciplinary approach. *Palaeogeogr., Palaeoclimatol., Palaeoecol.* **69**, 103–112.

Hooybergh, H.J.F. (1977) Stratigrafie van de Oligo-, Mio- en Pliocene afzettingen in het NE van Tunisië, met een bijzondere studie van de planktonische foraminiferen. PhD Thesis, Catholic University Leuven, Belgium, 148 pp.

Hooybergh, H.J.F. (1987) Foraminifères planctoniques d'âge langhien (Miocène) dans la Formation Aïn Grab au Cap Bon (Tunisie). *Notes Serv. Géol., Tunis* **55**, 5–17.

Lister, J.K. & Batten, D.J. (1988) Stratigraphic and palaeoenvironmental distribution of early Cretaceous dinoflagellate cysts in the Hurlands Farm borehole, West Sussex, England. *Palaeontographica Abstr. B* **210** (1–3), 9–89.

Monteil, E. & Cornu, P. (1987) Systématique et apport stratigraphique d'études phylogéniques chez les cératioides. Le rapport *Muderongia/Phoberocysta*: apport paléogéographique. Abstract, Xth Symposium of Assoc. Palynol. Langue Française, Bordeaux, Sept. 1987. *Trav. Doc. Géogr. Trop. CEGET, Bordeaux-Talence*, **59**, 90.

Robinson, P. & Black, C. (1981) Vertebrate localities and faunas from the Neogene of Tunisia. *Actes 1er Congr. Nat. Sci. Terre* (Tunis, September 1981), 253–261.

Robinson, P. & Wiman, S.K. (1976) A revision of the stratigraphic subdivision of the Miocene rock of sub-Dorsale Tunisia. *Notes Serv. Géol., Tunis* **42**, 71–86.

Schoeller, H. (1933) Présence de l'Aquitanien en Tunisie. *C. R. Somm. Seances Soc. Géol. France*, **11**, 158–159.

Shackleton, N.J. & Kennett, J.P. (1975) Palaeotemperate history of the Cenozoic and the initiation of Antarctic glaciation: oxygen and carbon analyses in D.S.D.P. Sites 277, 279 and 281. *Init. Rep. Deep Sea Drill. Proj.* **29**, 743–755.

Touati, M. (1985) Etude géologique et géophysique de la concession Sidi-el-Itayem en Tunisie orientale, Sahel de Sfax. DSc Thesis, University of Paris VI, 266 pp.

Tyson, R.V. (1987) The genesis and palynofacies characteristics of marine petroleum source rocks. In: *Marine Petroleum Source Rocks* (Eds Brooks, J. & Fleet, A.J.) Geol. Soc. Spec. Publ., 26, 47–67.

Vail, P.R., Audemard, F., Bowman, S.A., Eisner, P.N. & Perez-Cruz, G. (1991) *The Stratigraphic Signatures of Tectonics, Eustasy and Sedimentation — An Overview.* Elsevier, Amsterdam (in press).

Van der Zwan, C.J. (1990) Palynostratigraphy and palynofacies reconstruction of the Upper Jurassic to lowermost Cretaceous of the Draugen Field, offshore Mid Norway. *Rev. Palaeobot. Palynol.* **62**, 157–186.

Van Pelt, R.S. & Habib, D. (1988) Dinoflagellate species abundance and organic facies in Jurassic Twin Creek Limestone signal episodes of transgression and regression. *Abstr. 7th Int. Palynol. Congress, Brisbane*, p. 168.

Van Wagoner, J.C., Mitchum, R.M., Campion, K.M. & Rahmanian, V.D. (1990) Siliciclastic sequence stratigraphy in well logs, cores and outcrops: concepts for high-resolution correlation of time and facies. *Am. Assoc. Petrol. Geol. Meth. Expl. Ser. Tulsa* **7**, 1–55.

Wall, D. (1965) Microplankton, pollen and spores from the Lower Jurassic in Britain. *Micropalaeontology* **11**(2), 151–190.

Quaternary Applications of Sequence-stratigraphic Concepts

Spec. Publs Int. Ass. Sediment. (1993) **18**, 183–196

Late Pleistocene shelf-perched lowstand wedges on the Rhône continental shelf

M. TESSON*, G.P. ALLEN† *and* C. RAVENNE‡

**Université de Perpignan, 66100 Perpignan, France;*
†Total, Centre Scientifique et Technique, 78470 St Remy les Chevreuses, France; and
‡Institut Français du Pétrole, 92500 Rueil-Malmaison, France

ABSTRACT

A grid of high-resolution seismic profiles on the shelf of the Rhône delta has shown the existence of a complex of superimposed prograding sediment wedges on the outer shelf. These wedges represent alternating episodes of coastal progradation and transgression, and constitute a shelf-perched lowstand wedge system. Although only the uppermost wedge has been dated (Würm), it is believed that the entire system was probably deposited during the latest Pleistocene glacial lowstand. The geometry and stratal patterns of the wedges resulted from a combination of high-frequency eustatic cycles and episodic shelf tilting which coincided with periods of transgression. This shelf tilting increased accommodation on the distal extremity of the wedges and controlled the pattern of onlap terminations of the individual wedges.

INTRODUCTION

Sedimentation patterns on siliciclastic sedimentary shelves affected by relative sea-level fluctuations can be interpreted on the basis of sequence-stratigraphic concepts as outlined by Vail *et al.* (1977), Haq *et al.* (1987) and Posamentier and Vail (1988). One of the most significant features to have been described by these papers is the coastal lowstand wedge deposits which accumulate on the outer shelf and upper slope during lowstand. Such late-Pleistocene shelf-edge deposits have been observed on the Louisiana continental shelf (Suter & Berryhill, 1985; Berryhill, 1986; Suter *et al.*, 1987), the Mississippi–Alabama shelf (Kindinger, 1988, 1989) and concerned Wisconsinan depositional sequences. More recently Boyd *et al.* (1989) and Tesson *et al.* (1990a) have shown that shelf sediments deposited during Pleistocene glacio-eustatic cycles exhibit stratal patterns identical to those predicted by sequence-stratigraphic concepts. These deposits accumulated on the outer shelf and upper slope when the shelf was exposed, and form thick sediment wedges which have been identified by Tesson *et al.* (1990b) as representing the 'shelf-perched lowstand wedges' defined by Posamentier and Vail (1988). The geometry and stratal patterns of these shelfal lowstand deposits have been the focus of a regional study on the shelf adjacent to the Rhône delta, in SE France. In July 1990, a shelf-wide grid of single-channel high-resolution seismic-reflection profiles was established. This provided for the first time a detailed insight into the regional stratal geometry of deltaic shelfal lowstand wedges which underlie the post-glacial Holocene deposits (18,000 BP to Present). This paper presents the preliminary conclusions of this study.

REGIONAL SETTING

The study area is located off the French Mediterranean coast, on a broad arcuate shelf stretching between the Pyrenees and the Alps (Fig. 1). The shelf is about 250 km long between Pyrenean and Alpine orogenic belts, with a width varying between 73 and 36 km. The coast is wave dominated and composed of sediment furnished mainly by the

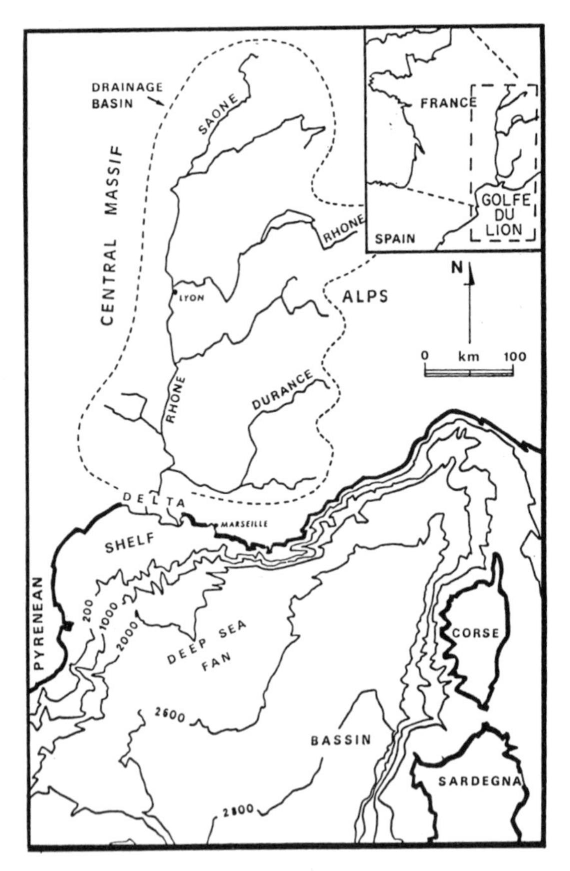

Fig. 1. Study area. Golfe du Lion is part of the northwestern Mediterranean sea.

Rhône river which accumulates as coast-parallel barrier and beach sands alternating with lagoonal muds. At the Rhône river mouth has formed an extensive wave-dominated delta system which has prograded about 15 km since the post-Holocene highstand (Kruit, 1955; Bertrand & L'Homer, 1975). The shelf is relatively flat and slopes gently seaward to the shelf break located at a depth of 120 to 130 m. The shelf-slope break is very irregular and incised by numerous canyons and shelf-edge slumps (Got *et al.*, 1979; Coutellier, 1985; Canals & Got, 1986).

Oil-company seismic lines and wells have shown that, since the Miocene, a thick wedge of clastic sediments (about 1500 to 3000 m thick) has been accumulating on the shelf (Cravatte *et al.*, 1974). Although subsidence of the shelf has been continuous throughout the Tertiary, these deposits do not appear to have been affected by major Alpine faulting and folding during the Pleistocene (Lefebvre, 1980), except in the southwestern and eastern parts of the shelf adjacent to the Pyrenean and Alpine orogens. The study area is restricted to the eastern extremity of the present shelf which is bounded to the east by the westward thrusting Alpine massif (Ellenberger & Gottis, 1967).

The surficial shelf sediments have been described by several authors (Kruit, 1955; Van Straaten, 1959; Monaco, 1971; Got *et al.*, 1985; Aloïsi, 1986; Gensous *et al.*, 1989, this volume, pp. 197–211). The existence of prograding Quaternary sediment wedges on the mid and outer shelf was first established by Alla *et al.* (1969), Leenhardt *et al.* (1969) and Alla *et al.* (1973), who carried out the first seismic grid on the entire shelf. In a later study concerning the southwestern part of the shelf (Monaco, 1971), these wedges were associated with glacial eustatic lowstands and the intervening discontinuities between the wedges with interglacial highstands (Labeyrie *et al.*, 1976). In 1986, Aloïsi suggested that these wedges resulted from prodelta silt and mud progradation during falling sea level, but more recently Aloïsi and Mougenot (1989) modified this interpretation and proposed that coastal progradation occurred during interglacial high sea-level stillstand.

Recently, on the basis of a new extensive grid of high-resolution seismic data covering the entire shelf, Tesson *et al.* (1990a, b) suggested that the sediment wedges represented lowstand deposits formed by coastal progradation during periods of low sea level and were separated by transgressive ravinement surfaces formed during periods of relative sea-level rise. Although only superficial piston cores (< 5 m) have been taken in the uppermost wedge, dated as >30 000 years, it is believed that the entire wedge system was deposited during the late Pleistocene.

STRATAL PATTERNS

A regional grid of 2000 km of high-resolution seismic profiles was completed over the entire Rhône shelf during the summer of 1990 (Fig. 2). This grid has confirmed the existence of a major incised Quaternary valley system cutting across the entire shelf to the west of the present Rhône mouth (Fig. 3). The internal seismic reflections within this valley are extremely discontinuous and no mappable markers can be distinguished. Older conven-

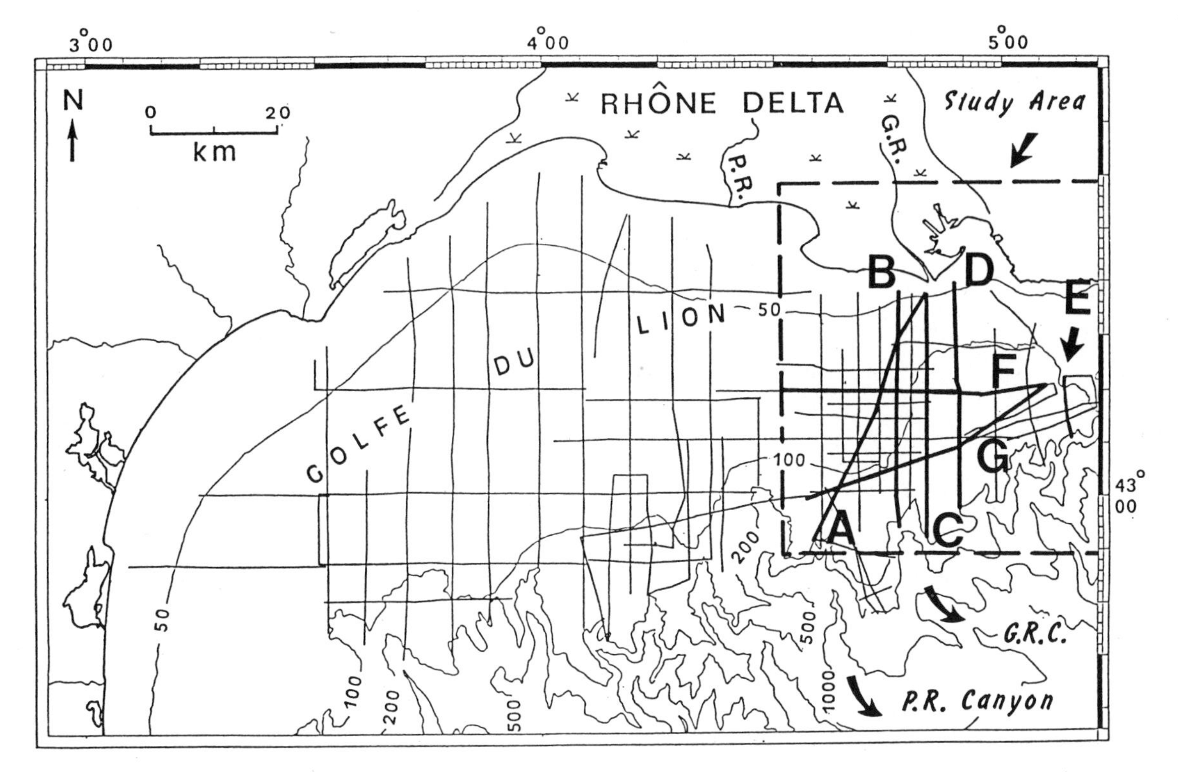

Fig. 2. High-resolution seismic trackline (1987–90). Acoustic systems were: (a) 1000 J, surf-boom; (b) 300 J, Uniboom; (c) 50 J, mini-sparker (mainly). Present publication is concerned with the area bounded by the dashed line.

tional multichannel seismic profiles have shown the existence of deeper (up to 1000 m) erosional features indicating that Rhône fluvial incision has occurred on this part of the shelf since the Miocene (Lefebvre, 1980). The same shelf valley appears to have been utilized throughout the successive late-Quaternary glacio-eustatic cycles, but its depth remains conjectural. Its width, estimated from the capping chaotic reflections, reaches 15 km.

In its seaward extremity the valley abuts onto a major shelf-edge canyon which incises the outer shelf and slope. This canyon, formed by incision and slumping during periods of lowstand, served as a sediment conduit for a large basin-floor fan developed during successive Quaternary lowstands (Droz, 1983). On the shelf, to either side of the valley, have accumulated superimposed prograding sediment wedges.

Geometry of the wedges

A high-resolution seismic transect across the shelf to the east of the incised valley is illustrated in Fig. 4. The sediments are organized into a system of stacked seaward prograding wedges (Fig. 5). Each wedge tapers landward, onlapping onto a mid-shelf unconformity (Fig. 6), and thickens seaward, attaining a thickness up to 50 m. at the shelf edge. All the wedges exhibit seaward prograding clinoforms. These clinoforms have a maximum dip of 1.5° and tangent the downlap surface. Episodically, the clinoforms are truncated by steeper erosion surfaces (Fig. 7) the dip of which decreases downward and becomes parallel to the clinoforms. The clinoforms immediately seaward of an erosion surface downlap onto this surface, and at times the toplap surface steps downward and onlaps onto the erosion surface. These erosional surfaces divide each wedge into a succession of lozenge-shaped units.

Each prograding wedge is bounded by a basal downlap surface and an upper toplap surface. Both surfaces are frequently erosional and usually coincide. Locally, however, a downlap surface can be separated from the underlying toplap surface by

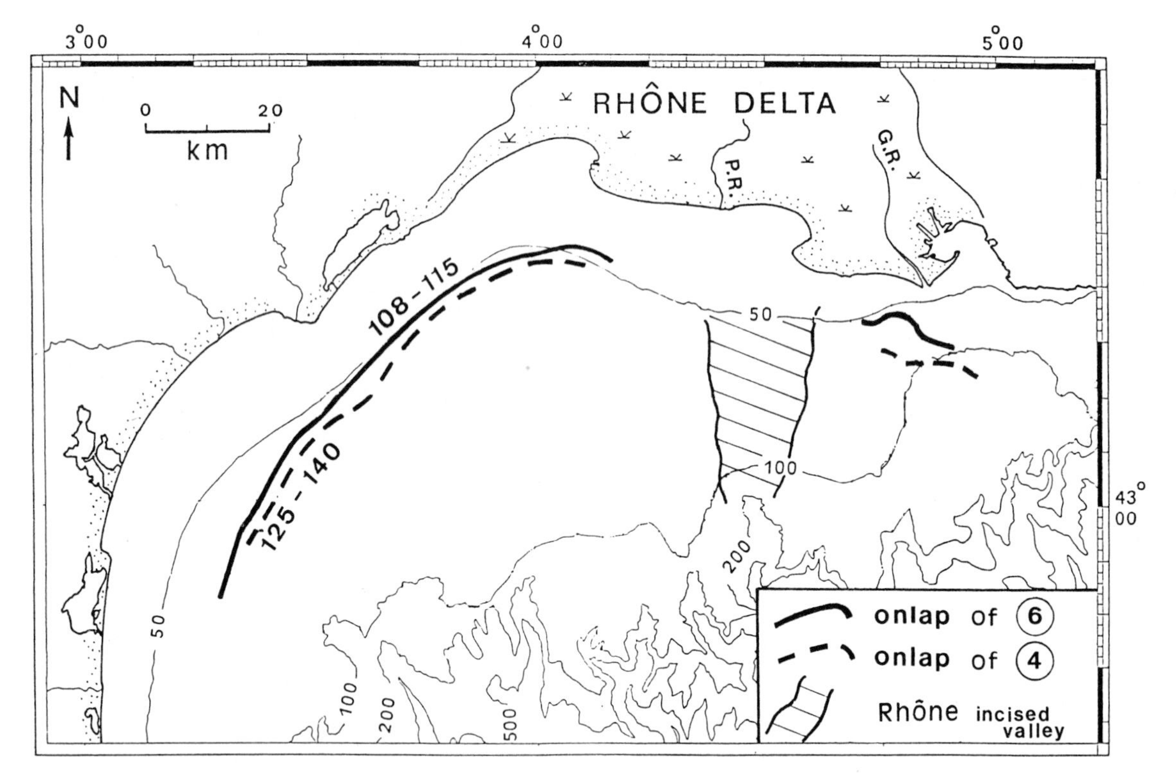

Fig. 3. Coastal-onlap location of discontinuities 6 and 4 (upper prograding wedges f and d). Palaeodepths are in metres per second (two-way travel time) and 100 m s^{-1} approximates 75 m (for 1500 m s^{-1} sound velocity).

thin, horizontal bedded or chaotic deposits which frequently form a landward-thickening unit intercalated between two prograding wedges (Figs 5, 8(a) & 8(b)). Also observed in some transects are small channel-like features incised into the upper part of the clinoforms (Fig. 5, kp. 20). These channels are up to 500 m wide, 10 to 15 m deep and filled with discontinuous horizontal reflectors which onlap onto the channel walls.

As the wedges prograde seaward and reach the edge of the shelf, they are affected by synsedimentary faulting. These faults are due to gravity mass movements which occur on the slope (Coutellier, 1985). Shelf-edge rotational slumps are also common. They result in large trough-like erosional features up to 5 km wide, 60–70 m thick, containing chaotic slumped and faulted sediments (Fig. 5, kp. 25) which can extend updip onto the shelf for distances up to 25 km from the physiographic shelf break (Fig. 10).

Regional correlations

Several regional shelf-normal and shelf-parallel transects are illustrated in Fig. 8. The individual wedges and the discontinuities which separate them are numbered and correlated over the entire study area (Fig. 9). On the basis of geometrical relations it appears that the wedges are continuous over the entire shelf. Thus, an individual wedge represents a major shelf-wide regressive event.

Figure 8 also shows that the wedges are organized into a complex stacking pattern. The toplap surface of each wedge onlaps landward onto a shelf-wide unconformity. This onlap pattern is quite complex, as the onlaps of successive wedges can migrate either landward or seaward. For example, wedge b in Fig. 8(b) exhibits a seaward shift of onlap with respect to the underlying wedge a, whereas the overlying wedges (c and d) onlap successively landward.

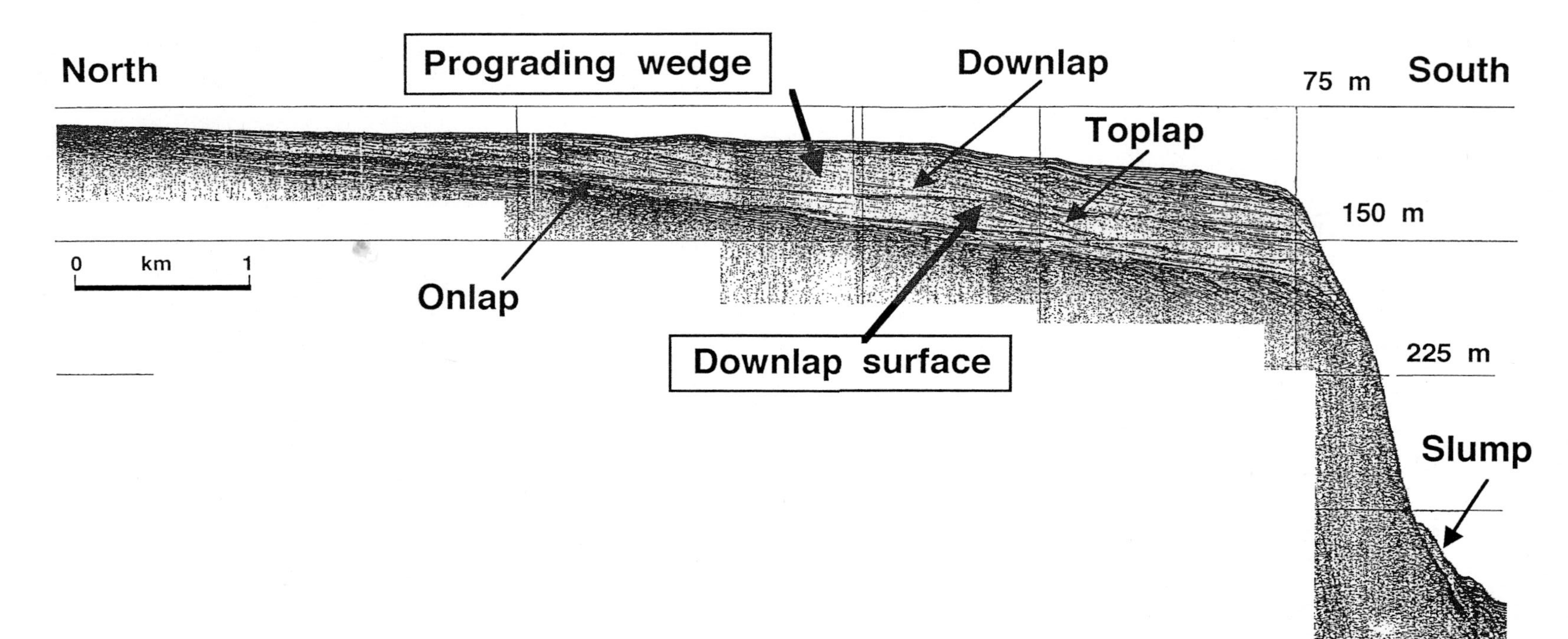

Fig. 4. Quaternary prograding sedimentary wedges on Golfe du Lion shelf. Boomer seismic system with about 2 m resolution. Reduction rate is 0.18. (For location see Fig. 2.)

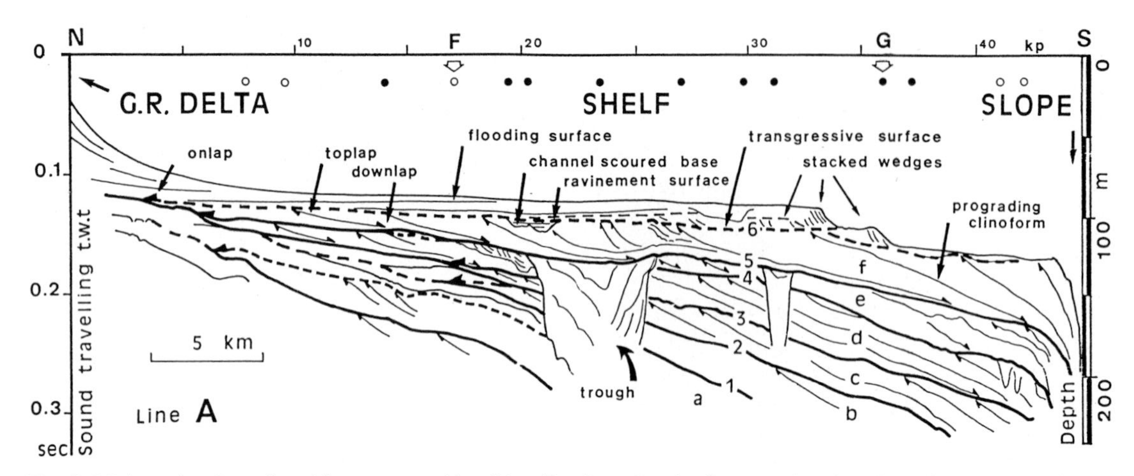

Fig. 5. Main regional stratigraphic pattern and key identification of major features. (For location of the line drawing A, see Fig. 2.)

The large-scale stacking pattern of the wedges is both aggradational and progradational. A rapid progradation of the slope occurred during wedges a, b and c, whereas wedges d, e and f exhibit a more aggradational pattern with a lower rate of slope progradation. This overall regressive evolution was terminated after wedge f, since the uppermost wedge (g) is only observed in some sections and exhibits a backstepping, i.e. transgressive stacking pattern. The present-day Rhône delta is prograding over this uppermost wedge.

Together, these wedges form a large regressive–transgressive cycle during which the shelf break and slope have prograded basinward more than 15 km,

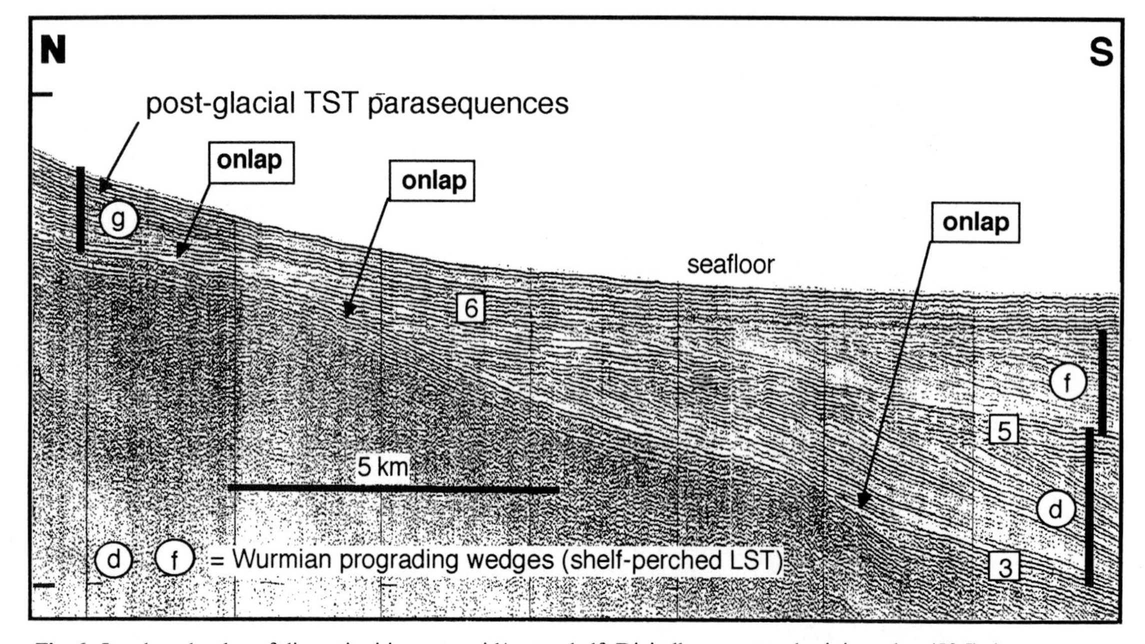

Fig. 6. Landward onlap of discontinuities onto mid/outer shelf. Digitally remastered mini-sparker (50 J) data.

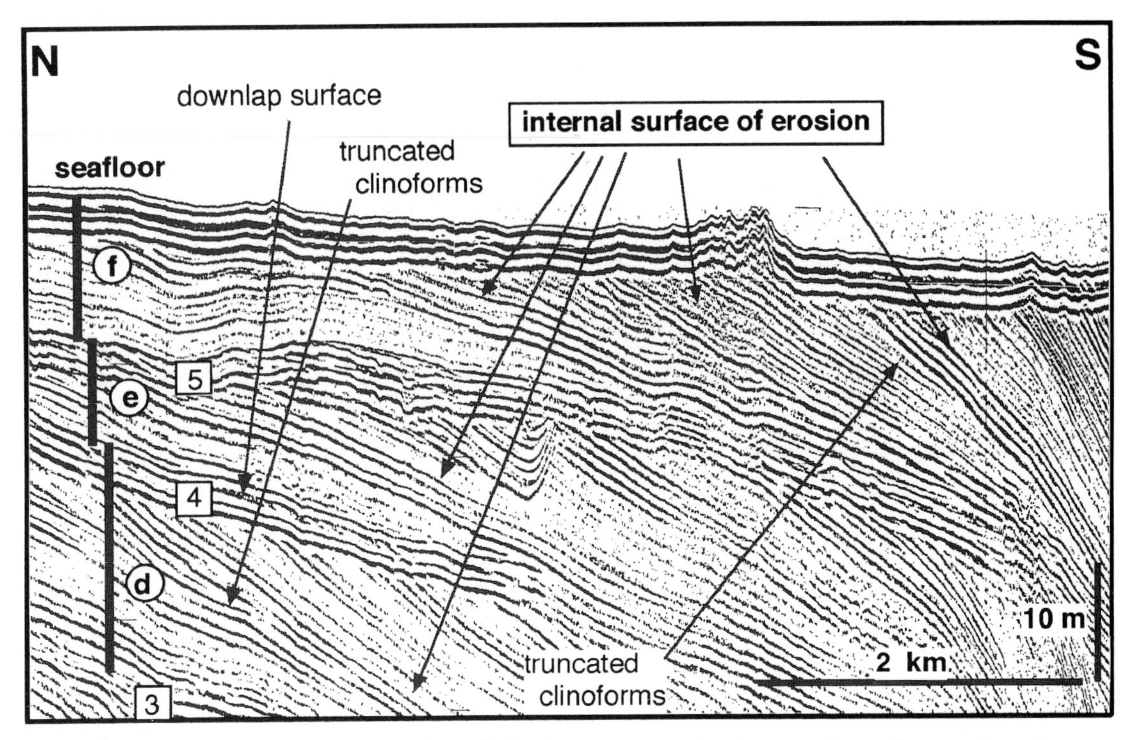

Fig. 7. Internal erosional surfaces and associated facies changes punctuating the prograding wedges, as observed on high-resolution seismic line D. Similar features occur on line drawing C, at kp. 25 (Fig. 8(b)). Digitally remastered mini-sparker (50 J) data.

and the outer shelf has aggraded 150 m. This large-scale cycle, as well as the individual wedges, can be correlated over the entire shelf, indicating that the stratigraphic patterns were controlled by allocyclic mechanisms. These occurred on at least two time scales, i.e. associated with the regressive–transgressive cycles of the individual wedges and the longer period regressive–transgressive pattern of the entire wedge complex; moreover the downward shifts observed within the prograding wedges, in some cases, could represent a much higher-frequency scale of allocyclic event.

In addition to these cycles on different scales, it is obvious from the geometry of the wedges that the shelf has been subjected to a progressive seaward and westward tilting during sedimentation. The seaward component of tilting provided most of the accommodation on the outer shelf for the wedges to accumulate (Figs 4, 5, 8(a) & 8(b)). On east–west sections through the outer shelf (Figs 8(c), (d)), a westward and shelf-parallel tilt clearly occurred, resulting mainly in a westward dipping of the wedges. The hinge line of this westward tilting was to the east of the study area, i.e. in the zone of the uplifting Alpine arc. The hinge of the seaward-tilting component (Figs 8(a), (b)) is located on the inner shelf, in the vicinity of the present-day coastline.

In addition to this regional shelf tilting, sediment was locally deformed, as indicated by the folding of wedges c and d (Fig. 8(d)) along N–S axes perpendicular to the shelf break. The amplitude of the folds reaches a maximum at the shelf edge and diminishes landward (Fig. 8(c)). The overall geometry of this early synsedimentary deformation is presently being studied in detail; however, it appears to be an episodic effect related either to ongoing compressive deformation related to the westward-thrusting Alpine arc, or to landward-attenuating collapses and accompanying shelf subsidence related to large-scale slumping and sediment failure on the slope.

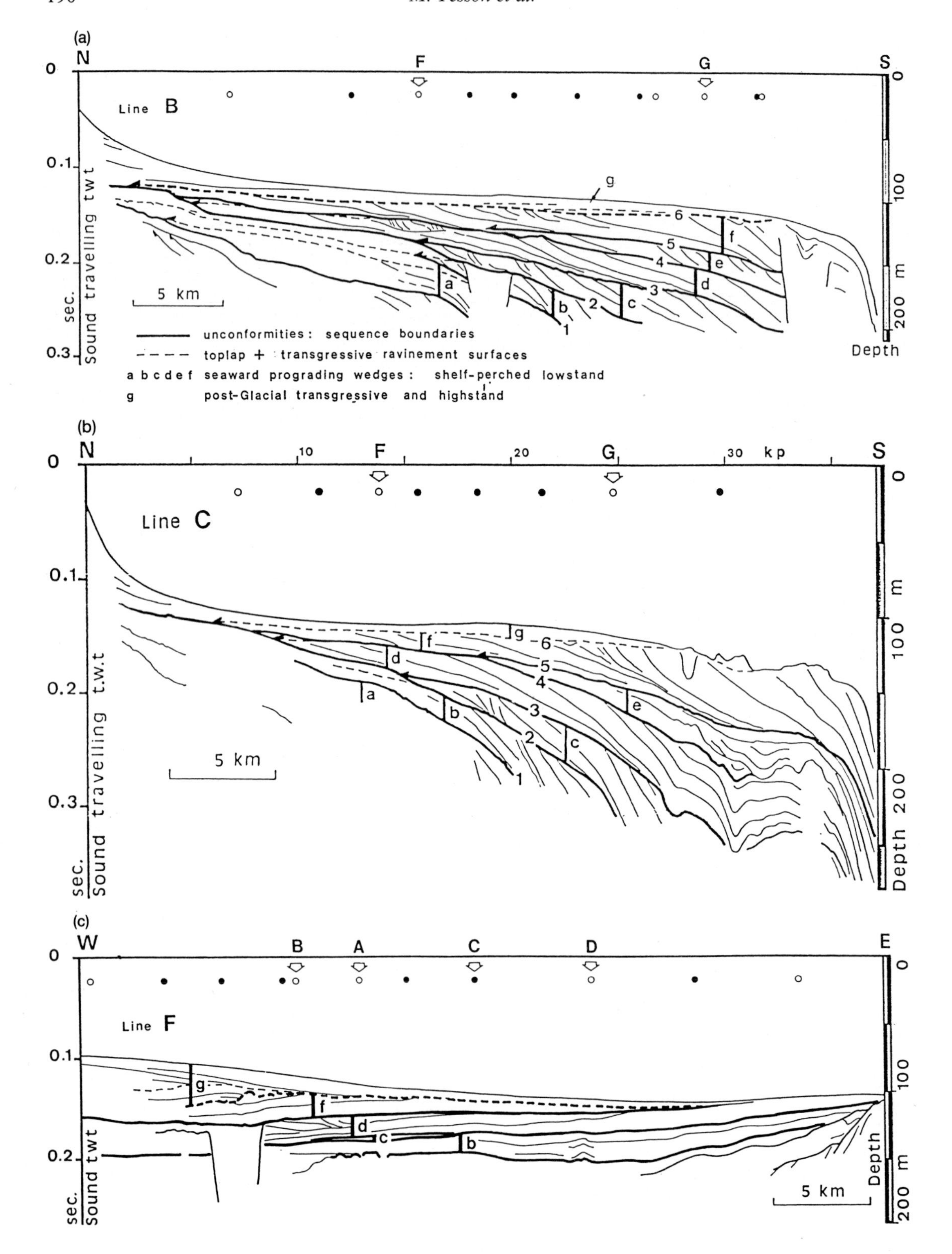
(a)
N
F
G
S
Line B
0
0.1
0.2
0.3
sec. Sound travelling twt
g
6
5
4
3
2
1
a
b
c
d
e
f
5 km
0
100
200 m
Depth
unconformities: sequence boundaries
toplap + transgressive ravinement surfaces
a b c d e f seaward prograding wedges: shelf-perched lowstand
g post-Glacial transgressive and highstand
(b)
N
10
F
20
G
30 kp
S
Line C
0
0.1
0.2
0.3
sec. Sound travelling t.w.t
g
6
5
4
3
2
1
a
b
c
d
e
f
5 km
0
100
200
Depth m
(c)
W
B
A
C
D
E
Line F
0
0.1
0.2
sec. Sound twt
g
f
d
c
b
5 km
0
100
200
Depth m

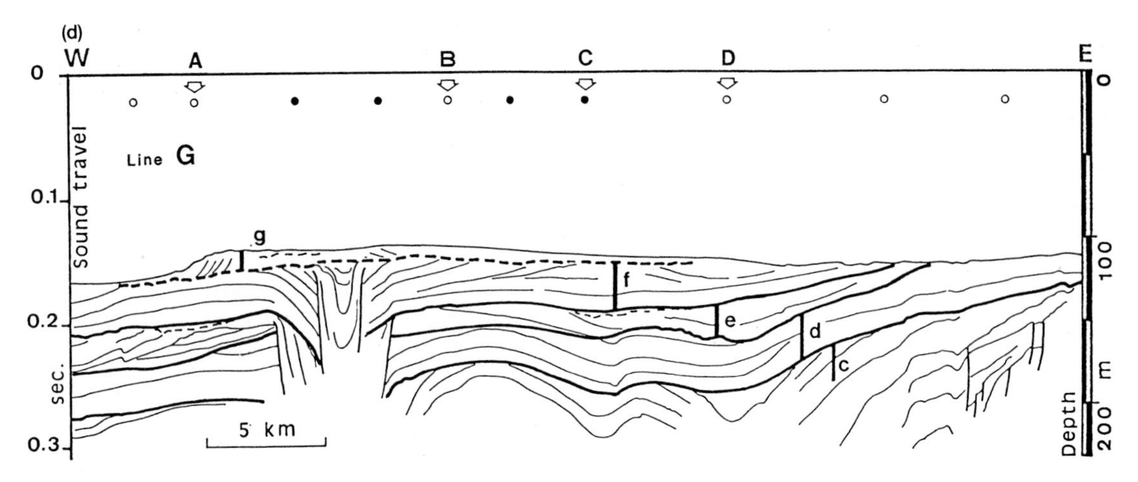

Fig. 8. (*oppposite and above*) Seismic line drawings used for regional correlations in the eastern part of the Golfe du Lion shelf. (For location see Fig. 2.)

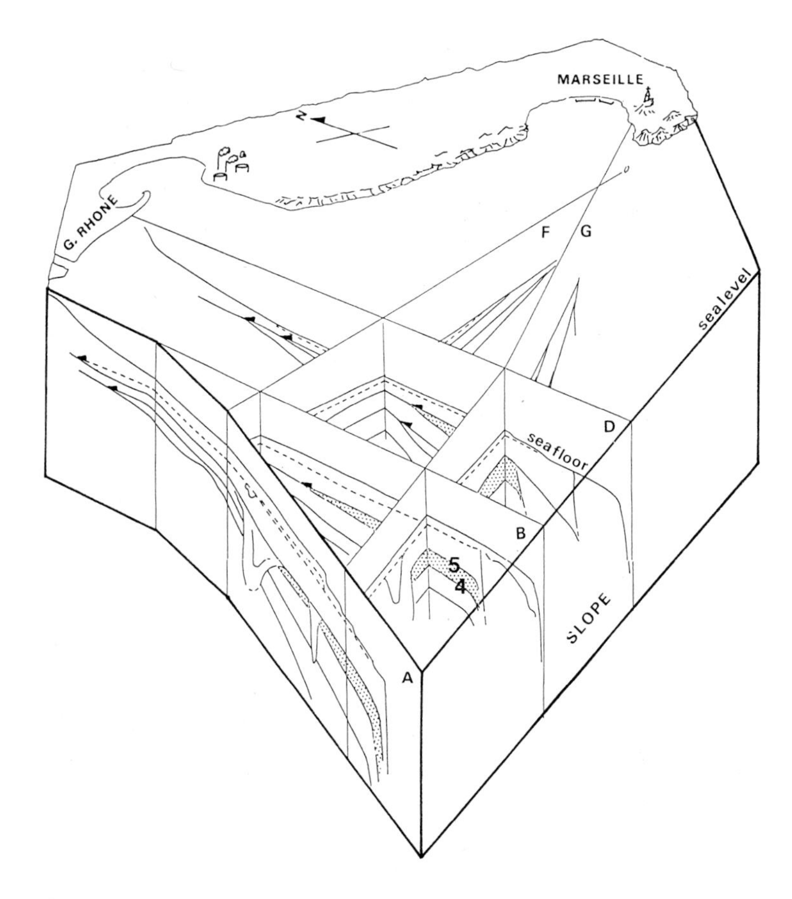

Fig. 9. Three-dimensional pattern of the late-Quaternary sedimentary wedges on the eastern Golfe du Lion shelf. The upper boundary (5) of the wedge d exhibits a seaward shift of coastal onlap onto the lower discontinuity (4).

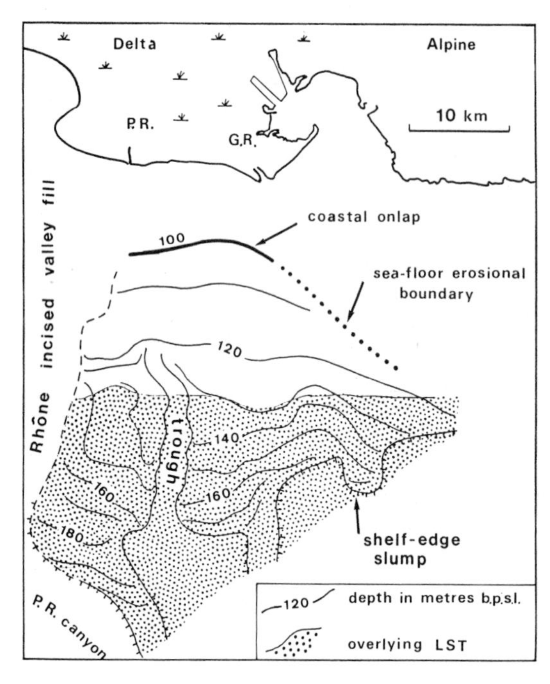

Fig. 10. Bathymetric map of the discontinuity 4 and superimposed location of the overlying prograding wedge (dots). Note the seaward shift of the onlapping discontinuity 5.

SEDIMENTOLOGICAL INTERPRETATION

Such shelf prograding sediment wedges could be formed either by coastal progradation or by a system of shelf sand bars. The widespread regional continuity of these wedges and the lack of any present-day shelf bars imply that the wedges represent coastal progradation. High-resolution seismic data of the inner shelf and nearshore (3.5 kHz mud penetrator) show that the geometry and dip of the clinoforms are similar to those of the modern shoreface. The present-day coast is wave dominated with beach and shoreface sand belts deposited by longshore drift from the Rhône river mouth. The present Rhône delta is forming a prograding sediment wedge on the inner shelf, accumulating as a regressive highstand deposit (Oomkens, 1967, 1970; Aloïsi *et al.*, 1978; L'Homer *et al.*, 1981; Gensous *et al.*, 1989, 1991).

Surface cores taken on the mid to outer shelf at the top of the clinoforms of wedge f (Fig. 5) indicate (B. Gensous, pers. com.) the presence of sand and mud with littoral fauna, dated from the Würm period. The cored sediments represent upper-shoreface deposits, and comprise parallel bedded-to-massive medium sand. These are truncated by a sharp erosion surface and a pebble and shell lag, overlain by thin silts and muds. No obvious beach deposits appear to have been preserved, and it is probable that the upper shoreface was subjected to wave erosion during transgression, so that the toplap surface represents a transgressive ravinement surface (Swift, 1975). Therefore, at least several metres of sediment were removed.

This ravinement surface is overlain by horizontally bedded transgressive deposits, which are usually thin (< 3 m) on the mid-shelf but can thicken landward to several metres in thickness. Similar transgressive deposits have also been observed on the Ebro shelf (Farran & Maldonado, 1990). These transgressive sediments are inferred to be storm deposits.

The outer-shelf wedges are interpreted as prograding shoreface deposits accumulated during a period of lower sea level. During glacio-eustatic lowstands, the Rhône river was located in the axis of the incised valley. Longshore drift would have dispersed sediment away from the river mouth to accumulate prograding beach and shoreface deposits to either side. The vertical stacking of individual wedges would then represent repeated cycles of transgression and renewed coastal progradation. The channelized features which sporadically erode the top of the wedges are interpreted as being minor distributary channels, such as the actual 'Petit Rhône' channel in the present Rhône delta.

DISCUSSION

Each wedge represents a single episode of coastal progradation which onlaps onto the mid shelf at a constant bathymetry (Fig. 3). This onlap represents a coastal onlap in the sense of Vail *et al.* (1977). Therefore the shelf landward of the onlap constituted a subaerial unconformity surface during progradation, and the point of onlap coincided with the bayline, as defined by Posamentier and Vail (1988). As indicated previously (Tesson *et al.*, 1990a,b), the present bathymetry and location of the onlap on the present-day mid-shelf region indicate that the wedges accumulated during a period of relative sea-level lowstand, corresponding to periods of glaciation, rather than during maximum highstand

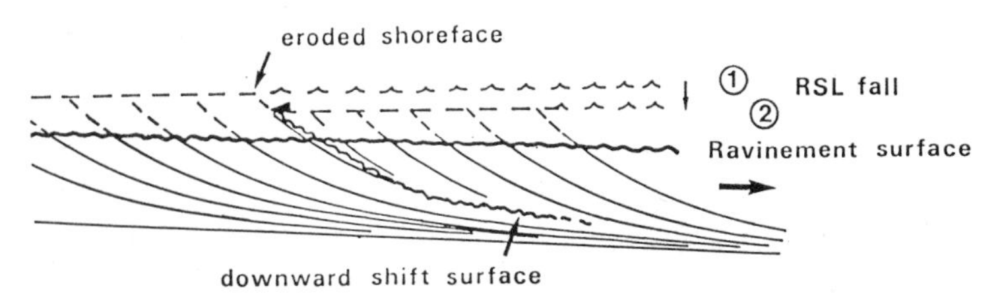

Fig. 11. Internal structure of a prograding wedge and inferred relationship to short-term sea-level falls: 'forced regressions'. (Compare to seismic section in Fig. 7.)

stillstand as proposed by Savelli *et al.* (1990), who suggested that the present bathymetry of coastal onlap (>70 m) could be explained by widespread and deep fluvial erosion at the time of the maximum sea-level lowstands. This would imply the preservation of only the most distal marine mud and silt facies which is not the case inferred from seismic lines and core analysis. Alternatively, a large amount of subsidence could account for a deep present-day bathymetry of highstand deposits but there is no indication that such an amount of subsidence occurred during the late Pleistocene (Lefebvre, 1980). Therefore the wedges constitute a 'shelf-perched' lowstand prograding wedge in the sense of Posamentier and Vail (1988). Each individual wedge forms a single progradational event, bounded by marine flooding surfaces, and therefore constitutes a parasequence (Van Wagoner *et al.*, 1988, 1990), or a high-frequency sequence. The stacking pattern of these sequences indicates that, during the period of overall lowstand, relative sea level was subjected to small-scale periodic rises or falls, followed by stillstand and coastal progradation. Most of the wedges were deposited during conditions of stepwise relative sea-level rise. Thus, wedges b, c and d aggrade, with a landward-stepping coastal onlap. Wedge e, however, is marked by a seaward shift of the coastal onlap (Figs 8(a), (b)). These different types of onlap patterns will be discussed later.

The internal erosion surfaces which punctuate the progradation of each wedge represent small-scale (a few metres) episodic sea-level falls (Fig. 11). These events would suddenly lower the wave base, resulting in the erosion of the underlying offshore muds and formation of sharp-based shoreface deposits analogous to those described in the Cardium formation by Plint (1988). Seaward, this sharp contact becomes conformable and would represent a higher order type 1 sequence boundary. This occurrence of progradation associated with relative sea-level fall represents episodes of 'forced regression' as described by Posamentier *et al.* (1988, 1992). Although the wedges have not been cored and dated (except for the uppermost wedge, f, which is pre-Holocene), it appears plausible that the lowstand wedge complex formed during the latest (Würm) glacio-eustatic cycle. This cycle lasted about 100,000 years, i.e. constitutes a fourth-order eustatic cycle in the terminology of Vail *et al.* (1977). During this lowstand, several authors (Fairbanks & Mathews, 1978; Moore, 1982; Aharon, 1983; Chappell & Shackleton, 1986) have documented the existence of higher frequency and smaller amplitude sea-level fluctuations (fifth-order cycles of Vail *et al.*, 1987). These could have formed the individual wedges observed on the sections.

In addition to these glacio-eustatic cycles, sedimentation and stratal geometry were affected by differential subsidence, as shown by the seaward tilting of the shelf and the wedges. This seaward tilting increased shelf accommodation, and probably enhanced the transgression between wedges. Even under conditions of constant eustatic sea level (Figs 12(a), (b)), a sudden shelf tilt would result in transgression and a seaward increase in accommodation, enabling another wedge to prograde when tilting stopped. If the hinge point of the tilting occurred landward of the coastal onlap of the initial wedge, the onlap of the succeeding wedge would step landward, in an aggradational pattern (Fig. 12(a)). This occurred for the wedges b, c, d, e and f. If the hinge point were located seaward of the initial coastal onlap, the succeeding wedge would onlap onto the top of the initial wedge and a seaward shift of onlap would occur (Fig. 12(b)), as shown by wedges d and e. In the case of a eustatic-sea-level fall simultaneous with the tilting (Fig. 12(c)), a seaward shift of the coastal onlap would occur if the hinge point were located landward of the initial onlap (this appears the most probable in view of the regional sections). Before wedge e prograded, it seems reasonable to suppose that these conditions prevailed.

The possible mechanisms of this shelf tilting, as well as the sequence-stratigraphic implications, are discussed in more detail in a forthcoming paper.

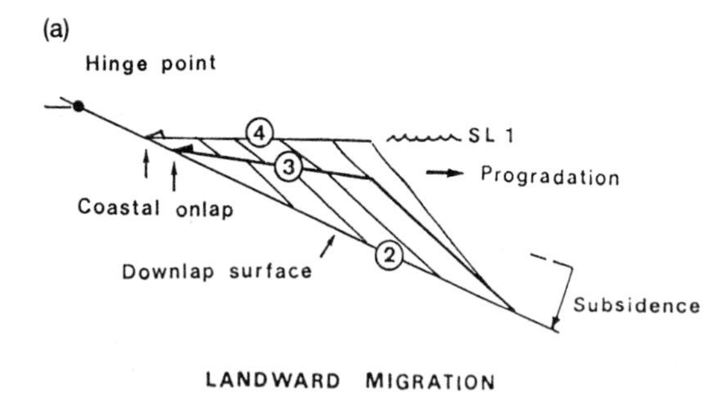

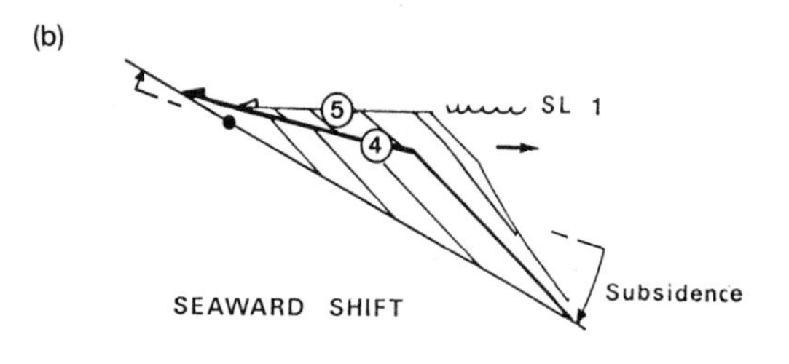

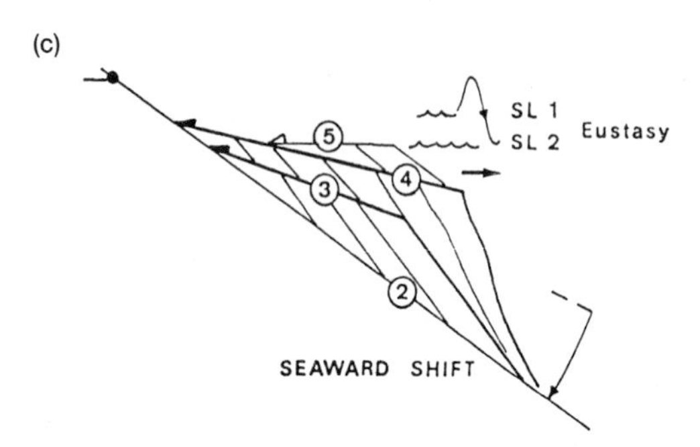

Fig. 12. Relative effects of subsidence and glacio-eustasy on Rhône shelf stratigraphic pattern. Numbers refer to discontinuities labelled on previous seismic line drawing. (a) Constant sea level (SL 1). Hinge line landward of the coastal onlap of 3. The coastal onlap of 4 steps landward. (b) Constant sea level (SL 1). Hinge line seaward of the coastal onlap of 4. The coastal onlap of 5 steps seaward. (c) Fluctuations of sea level. Hinge line landward of the coastal onlap of 4. The coastal onlap of 5 shifts seaward.

CONCLUSIONS

1 The late-Quaternary deposits of the Golfe du Lion form an aggrading and prograding lowstand wedge complex containing regionally correlatable unconformities, which converge and onlap landward onto a mid-shelf unconformity.
2 This complex is composed of a number of individual prograding wedges, which are bounded by a basal downlap surface and an upper toplap surface locally eroded by distributary channels. In the absence of transgressive deposits between two individual wedges both surfaces coincide.
3 Surface coring and the internal geometry of the wedges indicate that they represent wave-built prograding shoreface deposits originating from the Rhône river. Their progradation was probably sporadically punctuated by small-scale relative sea-level falls, which created internal downward shift surfaces.
4 The upper toplap surface of the wedges represents a transgressive ravinement surface overlain either by transgressive deposits or by the clinoforms of the overlying prograding wedge.
5 Analysis of onlap terminations of the individual wedges indicates both landward and seaward stepping onlap patterns. In both cases, a transgression occurs between two successive wedges. However, a seaward stepping onlap indicates that a shelf tilting occurred during or following the transgression, and prior to renewed progradation. Geometrical considerations imply that a seaward shift of coastal onlap will only occur if the axis of the tilting is located seaward of the onlap of the underlying older wedge.

These episodes of shelf tilting interacted with high-frequency glacio-eustatic cycles to control stratal geometries and appear related to a number of factors such as regional tectonism, sediment loading and compaction, mass movement on the slope and isostasy.
6 Although more chronostratigraphic data are needed, it is proposed that this lowstand prograding wedge complex formed during the latest (Würm/Wisconsinan) fourth-order glacio-eustatic cycle (less than 100,000 years). The individual wedges represent high-frequency parasequences and depositional sequences formed during the intervening smaller-scale (fifth-order) sea-level fluctuations.

ACKNOWLEDGEMENTS

We would like to thank the captain and crew of N/O *Catherine-Laurence* for their help at sea. Special appreciation is expressed to B. Gensous for contribution in collecting seismic data and for cored-deposits analysis. Critical comments were made by H. Posamentier and the two reviewers, R. Boyd and J.R. Suter.

This work is a CNRS/INSU contribution n° 377 to the national programme 'Dynamique et Bilan de la Terre', thème II, supported by grants from TOTAL and Institut Français du Pétrole.

REFERENCES

Aharon, P. (1983) 140.000 yr isotope climatic record from raised coral reef in New Guinea. *Nature* **304**, 720–723.

Alla, G., Dessolin, D., Got, H., Leenhardt, O., Rebuffatti, A. & Sabatier, R. (1969) Résultats préliminaires de la mission 'François Blanc' en sondage sismique continu. *Vie et Milieu* **XX** (fasc. 2-B), 211–220.

Alla, G., Byramjee, R., Didier, J., *et al.* (1973) Structure géologique de la marge continentale du Golfe du Lion. *Rapp. CIESMM, Monaco* **22**, 38–40.

Aloïsi, J.C. (1986) Sur un modèle de sédimentation deltaïque. Contribution à l'étude des marges passives. PhD thesis, University of Perpignan, 162 pp.

Aloïsi, J.C. & Mougenot, D. (1989) Stratigraphie séquentielle du Quaternaire des plate-formes progradantes de la Méditerranée occidentale. *Strata* **1**, 59–61.

Aloïsi, J.C., Monaco, A., Planchais, N., Thommeret, J. & Thommeret, Y. (1978) The Holocene transgression in the Golfe du Lion, southwestern France, Paleogeographic and paleobotanical evolution. *Géogr. Phys. Quat.* **XXXII** (2), 145–162.

Berryhill, H.L., Jr (Ed.) (1986) Sea level lowstand features of the shelf margin off southwest Louisiana. In: *Late Quaternary Facies and Structure, Northern Gulf of Mexico.* Am. Assoc. Petrol. Geol. Stud. Geol. Tulsa. 23, 225–240.

Bertrand, J.P. & L'Homer, A. (1975) Les deltas de la Méditerranée du Nord — Le delta du Rhône. *IX ème Congrès International de Sédimentologie, Nice* **16**, 65 pp.

Boyd, R., Suter, J. & Penland, S. (1989) Relation of sequence stratigraphy to modern sedimentary environments. *Geology* **17**, 926–929.

Canals, M. & Got, H. (1986) La morphologie de la pente continentale du Golfe du Lion: une résultante structuro-sédimentaire. *Vie et Milieu* **36**, 153–163.

Chappell, J. & Shackleton, N.J. (1986) Oxygen isotopes and sea level. *Nature* **324**, 137–140.

Coutellier, V. (1985) Mise en évidence et rôle des mouvements gravitaires dans l'évolution de la marge continentale. Exemple des marges du Golfe du Lion et de la Provence Occidentale. Thèse troisième cycle University of P. and M. Curie, Paris, 189 pp.

Cravatte, J., Dufaure, P.H., Prim, M. & Rouaix, S. (1974) Les sondages du Golfe du Lion: stratigraphie, sédimentologie. *Notes et Mém. C.F.P.* **11**, 209–274.

Droz, L. (1983) L'éventail sous-marin profond du Rhône (Golfe du Lion): grands traits morphologiques et structure semi profonde. Thèse troisième cycle University of P. and M. Curie, Paris, 195 pp.

Ellenberger, F. & Gottis, M. (1967) Sur les jeux des failles pliocènes et quaternaires dans l'arrière pays narbonnais. *Rev. Géogr. Phys. Géol. Dyn.* **IX** (f 2), 153–159.

Fairbanks, R.G. & Mathews, R.K. (1978) The marine oxygen isotope record in Pleistocene coral, Barbados, West Indies, *Quat. Res.* **10**, 181–196.

Farran, M. & Maldonado, A. (1990) The Ebro continental shelf: quaternary seismic stratigraphy and growth patterns. In: *The Ebro Margin* (Eds Nelson, C.H. & Maldonado, A.) *Mar. Geol.* **95**: 289–312.

Gensous, B., El Hmaidi, A., Williamson, D. & Taieb, M. (1989) Caractérisation chronologique et sédimentologique des dépôts récents de la marge rhôdanienne. *2 éme Congr. Fra. Sédimentol., Paris*, pp. 113–114.

Got, H., Aloïsi, J.C., Leenhardt, O., Monaco, A., Serra-Raventos, J. & Theilen, F. (1979) Structures sédimentaires sur les marges du Golfe du Lion et de Catalogne. *Rev. Géog. Phys. Géol. Dyn.* **21**(4), 267–280.

Got, H., Aloïsi, J.C. & Monaco, A. (1985) Sedimentary processes on deltas and shelves of the Mediterranean Sea. In: *Geological Evolution of the Mediterranean Basins* (Eds Wezel, F.C. & Stanley, D.J.), Pergamon Press, Oxford, pp. 355–376.

Haq B.U., Hardenbol, J. & Vail, P.R. (1987) The chronology of fluctuating sea level since the Triassic. *Science* **194**, 1121–1132.

Kindinger, J.L. (1988) Seismic stratigraphy of the Mississippi–Alabama shelf and upper continental slope. *Mar. Geol.* **83**, 79–94.

Kindinger, J.L. (1989) Depositional history of the Lagniappe Delta, Northern Gulf of Mexico. *Geo-Mar. Lett.*, **9**, 59–66.

Kruit, C. (1955) Sediments of the Rhône delta — grain size and microfauna. *Verhand. Konoink. Neder. Geol. Mijnbouw* **15** (2), 357–514.

Labeyrie, J., Lalou, C., Monaco, A. & Thommeret, J. (1976) Chronologie des niveaux eustatiques sur la côte du Roussillon de -33000 ans B.P. à nos jours. *C.R. Acad. Sci. Paris* **282**, 349–352.

Leenhardt, O., Pierrot, S., Rebuffatti, A. & Sabatier, R. (1969) Etude sismique de la zône de Planier (Bouches du Rhône). *Rev. IFP* **XXIV**, 1261–1287.

Lefebvre, D. (1980) Evolution morphologique et structurale du Golfe du Lion. Essai de traitement statistique des données. Thèse d'Etat, University of Pierre et Marie Curie, Paris, 163 pp.

L'Homer, A., Bazile, F., Thommeret, J. & Thommeret, Y. (1981) Principales étapes de l'édification du delta du Rhône de 7000 ans BP à nos jours; variations du niveau marin. *Océanis* **7** (f 4), 389–408.

Monaco, A. (1971) Contribution à l'étude géologique et sédimentologique du plateau continental du Roussillon (Golfe du Lion). Thèse d'Etat, University of Perpignan, 295 pp.

Moore, W.S. (1982) Late Pleistocene sea level history. In: *Uranium Series Disequilibrium: Application to Environmental Problems* (Eds Ivanovich, M. & Harmon, R.S.) Clarendon Press, Oxford, pp. 481–496.

Oomkens, E. (1967) Depositional sequences and sand distribution in a deltaic complex. *Geol. Mijnbouw*, **46**, 265–278.

Oomkens, E. (1970) Depositional sequences and sand distribution in the postglacial Rhône delta complex. In: *Deltaic Sedimentation* (Ed. Morgan, J.P.) Spec. Publ. Soc. Econ. Paleontol. Mineral. 15, 198–212.

Plint, A.G. (1988) Sharp-based shoreface sequences and 'offshore bars' in the Cardium formation of Alberta:

their relationship to relative changes in sea level. In: *Sea-level Changes – An Integrated Approach* (Eds Wilgus, C.K., Hastings, B.S., Kendall, C.G.St.C., Posamentier, H.W., Ross, C.A. & Van Wagoner, J.C.), Spec. Publ. Soc. Econ. Paleontol. Mineral, Tulsa, 42, 357–370.

Posamentier, H.W. & Vail, P.R. (1988) Eustatic controls on clastic deposition II. Sequences and system tract models. In: *Sea-level Changes – An Integrated Approach* (Eds Wilgus, C.K., Hastings, B.S., Kendall, C.G.St.C., Posamentier, H.W., Ross, C.A. & Van Wagoner, J.C.), Spec. Publ. Soc. Econ. Paleontol. Mineral, Tulsa, 42, 125–154.

Posamentier, H.W., Allen, G.P., James, D.P. & Tesson, M. (1992) Forced regressions in a sequence stratigraphic framework. *Am. Assoc. Petrol. Geol.* **76**, 1687–1709.

Posamentier, H.W., Jervey, M.T. & Vail, P.R. (1988) Eustatic controls on clastic deposition I. Conceptual framework. In: *Sea-level Changes – An Integrated Approach* (Eds Wilgus, C.K., Hastings, B.S., Kendall, C.G.St.C., Posamentier, H.W., Ross, C.A. & Van Wagoner, J.C.), Spec. Publ. Soc. Econ. Paleontol. Mineral, Tulsa, 42, 109–124.

Savelli, D., Tramontana, M. & Wezel, F.C. (1990) Cyclic sedimentation and erosion of Quaternary sedimentary wedges off the Gargano promontary (Southern Adriatic Sea). *Bulletin di Oceanologia Teorica* ed Applicata, Vol. VIII, **3**, 163–174.

Suter, J.R. & Berryhill, H.L. Jr. (1985) Late Quaternary shelf-margin deltas, Northwest Gulf of Mexico. *Bull. Am. Assoc. Petrol. Geol.* **69** (1): 77–91.

Suter, J.R., Berryhill, H.L. & Penland, S. (1987) Late Quaternary sea-level fluctuations and depositional sequences, Southwest Louisiana Continental Shelf. In: *Sea Level Fluctuations and Coastal Evolution* (Eds Nummedal, D., Pilkey, O.H. & Howard, J.P.) Spec. Publ. Soc. Econ. Paleontol. Mineral. 41, 199–219.

Swift, D.J.P. (1975) Barrier island genesis: evidence from the central Atlantic shelf, eastern U.S.A. *Sediment. Geol.* **14**, 1–43.

Tesson, M., Ravenne, C. & Allen, G.P. (1989) Application des concepts de l'analyse en stratigraphie séquentielle aux prismes de bas-niveau marin sur la plate-forme rhôdanienne. *Strata* **1**, 55–57.

Tesson, M., Ravenne, C. & Allen, G.P. (1990a) Application des concepts de stratigraphie sequentielle à un profil sismique haute résolution transverse à la plate-forme rhôdanienne. *C. R. Acad. Sci. Paris* **310** (II), 565–570.

Tesson, M., Gensous, B., Allen, G.P. & Ravenne, C. (1990b) Late Quaternary deltaic lowstand wedges on the Rhône continental shelf, France. *Mar. Geol.* **9**, 325–332.

Vail, P.R. (1987) Seismic stratigraphy interpretation using sequence stratigraphy, Part 1: seismic stratigraphy interpretation procedure. In: *Atlas of Seismic Stratigraphy* (Ed. Bally, A.W.), Am. Assoc. Petrol. Geol. Stud. Geol., 27, 1–10.

Vail, P.R., Mitchum, R.M., Jr., Todd, R.G. *et al.* (1977) Seismic stratigraphy and global changes of sea level. In: *Seismic Stratigraphy – Applications to Hydrocarbon Exploration* (Ed. Payton, C.E.), Mem. Am. Assoc. Petrol. Geol. 26, 49–212.

Van Straaten, L.M.J.U. (1959) Littoral and submarine morphology of the Rhône delta. *Sec. Coastal Geogr. Conf.* Baton Rouge, Proc., Nat. Acad. Sci. Nat. Res. Council, 233–264.

Van Wagoner, J.C., Mitchum, R.M., Campion, K.M. & Rahmanian, V.D. (1990) Siliciclastic sequence stratigraphy in well logs, cores, and outcrops: techniques for high-resolution correlation of time and facies, *Am. Assoc. Petrol. Geol. Meth. Expl. Ser.* **7**, 55 pp.

Van Wagoner, J.C., Posamentier, H.W., Mitchum, R.M. Jr., *et al.* (1988) An overview of the fundamentals of sequence stratigraphy and key definitions. In: *Sea-level Changes – An Integrated Approach* (Eds Wilgus, C.K., Hastings, B.S., Kendall, C.G.St.C., Posamentier, H.W., Ross, C.A. & Van Wagoner, J.C.), Spec. Publ. Soc. Econ. Paleontol. Mineral, Tulsa, 42, 39–46.

Spec. Publs Int. Ass. Sediment. (1993) **18**, 197–211

Late-Quaternary transgressive and highstand deposits of a deltaic shelf (Rhône delta, France)

B. GENSOUS*, D. WILLIAMSON† *and* M. TESSON*

**Laboratoire de Sédimentologie et Géochimie Marines, Université de Perpignan, Perpignan 66000, France; and*
†Laboratoire de Géologie du Quaternaire, Université d'Aix-Marseille II, Marseille, 13288, France

ABSTRACT

Sequence analysis of high-resolution seismic profiles, supplemented by piston core sampling, illustrates the organization and characteristics of late-Quaternary deposits of the Rhône delta plain and adjacent continental shelf. Chronological data, provided by radiocarbon dating and the record of geomagnetic secular variations, allow a tentative correlation with post-glacial relative sea-level fluctuations.

Deposits of the outer continental shelf consist of stacked progradational wedge-shaped units interpreted as 'shelf-perched' lowstand wedges (Tesson *et al.*, 1990a). The uppermost wedge is overlain by backstepping parasequences of the post-glacial transgressive systems tract. The highstand systems tract is restricted to the inner shelf and subaerial Rhône delta plain.

Parasequences of the transgressive systems tract that develop on the outer shelf form complex, elongated prograded bodies derived from shoreface erosion of lowstand deposits and on the inner–middle shelf where sand shoals mark the position of ancient barrier shorelines associated with early-Holocene shelf-phase deltas. On the mid–outer shelf, transgressive and lowstand deposits are covered by a thin, discontinuous veneer of hemipelagic muds and relict shelly sands forming a condensed section associated with the post-glacial relative sea-level rise.

Highstand regressive facies have developed since the period of reduced rate of relative sea-level rise at the end of the Holocene. During this period, the Rhône river has built three major delta complexes and prograded about 15 km over the shelf.

INTRODUCTION

The development of sequence stratigraphy has provided a new perspective for basin-fill evolution. It was observed that the rock record comprised a succession of sequences bounded by unconformities. The cause of this punctuated succession was suggested to be the interaction between relative sea-level change and sedimentation (Vail *et al.*, 1977; Haq *et al.*, 1987; Posamentier & Vail, 1988; Posamentier *et al.*, 1988).

Modern depositional environments represent good natural laboratories to test these new concepts and provide models for comparison with ancient sequences, but at the present time few studies have been published (Suter *et al.*, 1987; Boyd *et al.*, 1989). The Rhône continental margin is well suited for the study of the sedimentary response to high-frequency cyclic variations of sea level. This area is characterized by active progradation with major fluvial input from the Rhône River and occurs in a microtidal setting.

Recent papers (Tesson *et al.*, 1990a,b) have described the organization of Quaternary deposits emphasizing the sediments accumulated during periods of eustatic lowstand. The objective of the present study is to provide a description of deposits accumulated during the late transgressive and highstand intervals. This study is based on a synthesis of previous studies carried out on the deltaic plain (Oomkens, 1967, 1970; Rapport 'CAMARGUE', 1970) and original data recently acquired on the continen-

(a)

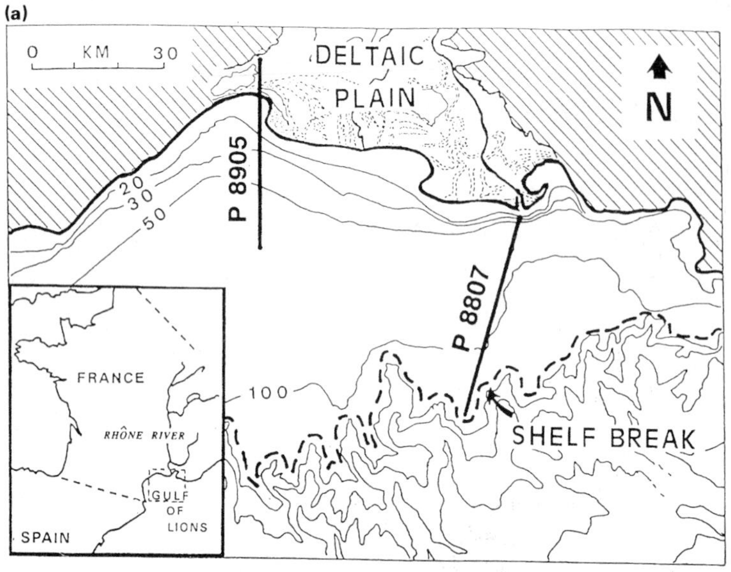

(b)

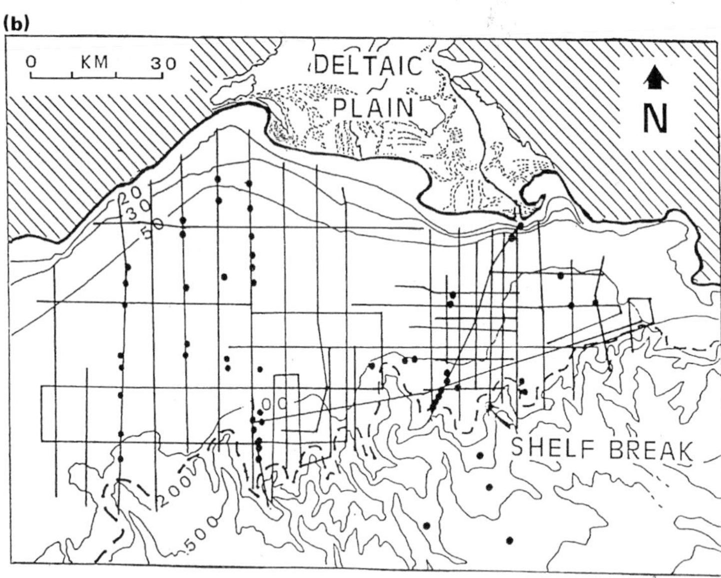

Fig. 1. (a) Location map and bathymetry of studied area (water depths are in metres). (b) High-resolution seismic track lines (1987–90) and location of coring sites.

tal shelf (Fig. 1). Determination of stratal geometry was based on analyses of seismic data; lithologic descriptions were based on analyses of core data collected by Kullenberg core samplers.

REGIONAL SETTING

On the French Mediterranean coast adjacent to the Rhône delta plain (Fig. 1), the continental shelf forms a relatively flat platform (gradient of 1 to 3% in the inner zone and 0.1 to 0.5% in the mid–outer zone) about 60–70 km wide. It is a microtidal environment affected by relatively low wave energy. Under fair-weather conditions, significant sediment transport is restricted to the upper shoreface (5 to 7 m depth); in winter, waves generated by easterly storms can remove deposits as deep as 30 m (Bertrand & L'Homer, 1975; Blanc, 1977).

Previous studies (Kruit, 1955; Van Straaten,

1959; Monaco, 1971; Got *et al.*, 1985; Aloïsi, 1986) provide valuable information on morphology and surficial sedimentary cover of the subaerial delta complex and adjacent shelf (Fig. 2). On the upper delta plain, fluvial deposits (crevasse, point bars, levee) dominate the morphology. On the lower delta plain, brackish to highly saline lakes and ponds are enclosed in an extensive system of coastal ridges; these ridges are dissected by narrow strips of active and abandoned distributaries. Deposits of the delta front and barrier beaches are well-sorted, horizontally bedded sands. They grade seaward from interbedded silts and sands to finely laminated silts and clays of the shoreface area. Prodeltaic deposits are found in front of the main distributaries.

The mid-continental shelf is blanketed by an extensive mud belt between the 50 and 90 m isobaths. In the western part of the shelf, the mud belt is interrupted by sandy shoals between the 30 and 60 m isobaths. The outer shelf is covered by relict

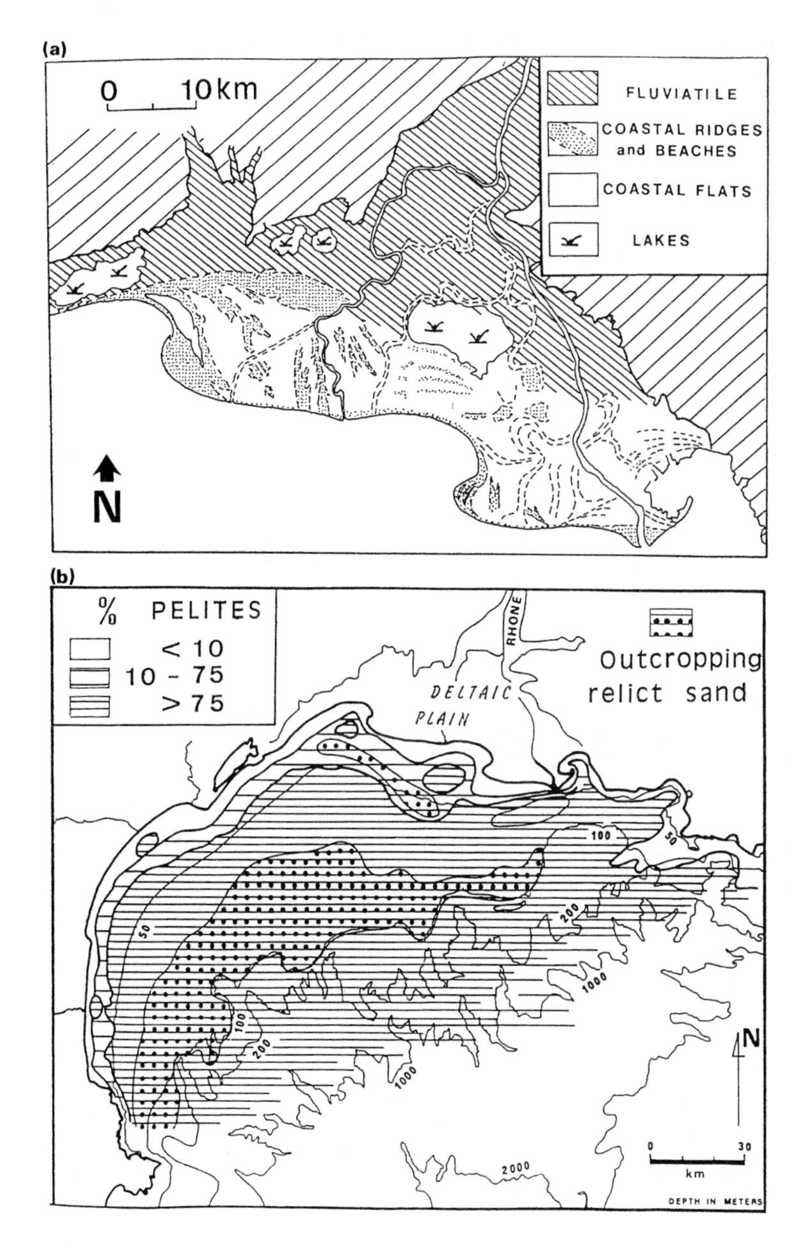

Fig. 2. Distribution of sediments on the Rhône river delta plain (a) and adjacent continental shelf (b). (Data from Kruit, 1955; Van Straaten, 1959; Monaco, 1971; Got *et al.*, 1985; Aloïsi, 1986.)

sands locally overlain by a thin veneer of hemipelagic muds. The shelf edge occurs at 120 m depth delimiting a steep continental slope affected by large-scale slumping and sediment failure (Canals & Got, 1986).

RESULTS

Seismic stratigraphy

Tesson *et al.* (1990a,b) presented the initial results of a regional study of the Rhône deltaic shelf and described the geometry and stratal patterns of Quaternary wedges that accumulated on the outer shelf. The grid of high-resolution seismic profiles was completed in August 1989 and July 1990 over the entire shelf. Two types of seismic profiling systems were used: 50-J minisparker and 3.5 kHz sub-bottom profiler.

Mid- and outer-shelf deposits consist of stacked progradational wedge-shaped units (Fig. 3) thickening seaward up to 50–60 m at the shelf edge and pinching out landward by onlap onto the mid-shelf at a constant water depth of 80 m. To the west of the present Rhône River mouth the continuity of the wedges is interrupted by a large, incised valley system, cutting across the entire shelf, connecting seaward with the Rhône submarine canyon, the conduit for the Rhône deep-sea fan. The wedges contain seaward-prograding clinoforms bounded by erosional toplap and downlap surfaces. The clinoforms are episodically truncated by small-scale erosion surfaces associated with a downward shift of the toplap surface. The upper surface of the wedge is locally incised by erosional channels 0.5 km wide and 10 to 15 m deep filled with irregular to horizontal bedded sediments representing distributary channels. Both internal downward shift and erosional channels are erosional features affecting the prograding wedges during small-scale falls in sea level (Tesson *et al.*, this volume, pp. 183–196). They have been interpreted as prograding wedges accumulated during late-Quaternary glacio-eustatic lowstand. These have been termed 'shelf-perched lowstand wedges' as defined by Posamentier *et al.* (1988).

The uppermost wedge is cut by an erosional surface representing the post-glacial transgressive (i.e. ravinement) surface. Overlying this surface are progradational units successively backstepping across the shelf; they represent parasequences com-

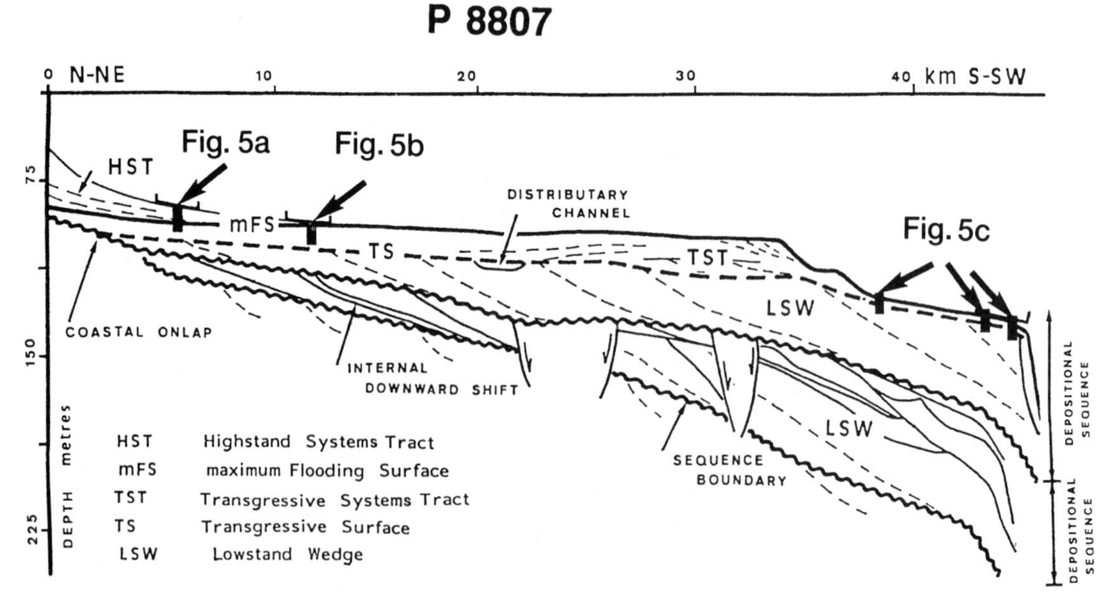

Fig. 3. Regional transect (see Fig. 1 for location) illustrating the sequence-stratigraphic interpretation of late-Quaternary deposits of the Rhône continental shelf. (From Tesson *et al.*, 1990.) Arrows indicate position of cores and sub-bottom profiles in Fig. 5.

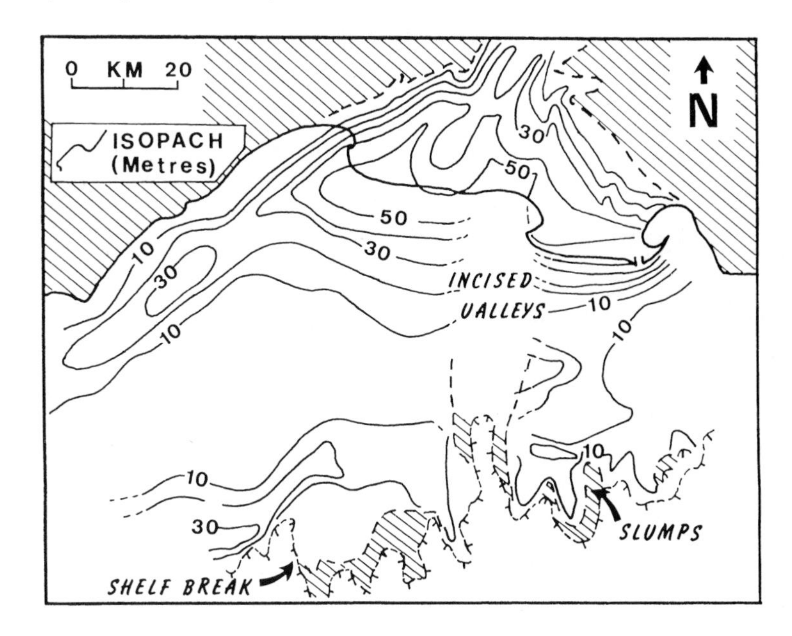

Fig. 4. Isopach map of transgressive and highstand post-glacial deposits. Database from high-resolution seismic profiles on the continental shelf and core drilling on the Rhône delta plain.

posing a transgressive systems tract as illustrated by Van Wagoner *et al.* (1988). A similar pattern was observed by Boyd *et al.* (1989) for the Mississippi delta.

On the inner shelf and near the present coastline, modern Rhône delta-front deposits downlap onto transgressive deposits; they represent the distal extremity of the highstand systems tract making up the modern delta plain.

An isopach map of deposits overlying the upper lowstand wedge (Fig. 4) shows that transgressive highstand deposits develop on the outer shelf at 100–110 m depth, forming complex bodies parallel to isobaths; they are very thin on the middle shelf (< 5 m) but thicken on the western part of the inner shelf and below the Rhône delta plain where their maximum thickness is about 50 m.

Description of deposits

On the continental shelf

Lithological and sedimentological characteristics of transgressive and highstand deposits were obtained from Kullenberg cores located on seismic lines. A complementary sub-bottom profiler survey (3.5 kHz) was carried out before coring. Cores collected on a representative seismic section, in front of the present Rhône delta mouth, contain the following facies:

Delta-front deposits are well-sorted horizontal bedded sands grading to interbedded silt and sand layers. They form a coarsening-upward sequence about 10 to 15 m thick (Bertrand & L'Homer, 1975).

Modern prodeltaic deposits on the inner shelf develop seaward of the main distributaries. Sediments cored are rich organic silty clays (Fig. 5(a), core K-8802). The upper part (0–5 m) exhibits varved structures with alternation of centimetre to millimetre scale grey-beige mottled layers and black layers. The lower part (5–10 m) comprises fluidized dark-grey silty clays. The relatively high carbonate content (between 35 to 50%) is the result of clastic supply from carbonate Mesozoic rocks outcropping in the lower part of the drainage basin of the Rhône River.

Because of the lack of shell materials for radiocarbon dating, the method of geomagnetic field secular variations recorded in deposits was used (Williamson, 1991; Williamson *et al.*, in press). The peak inclination directions were compared with European Holocene-type curves (Turner & Thompson, 1981; Creer, 1985), providing a detailed chronostratigraphy (Fig. 6). The upper part of the core (varved silty clays) was deposited during the last 6000 years and based on seismic data is correlative with the highstand Rhône delta complex. In the lower deposits, the high water content precludes palaeomagnetic sampling; none the less, these

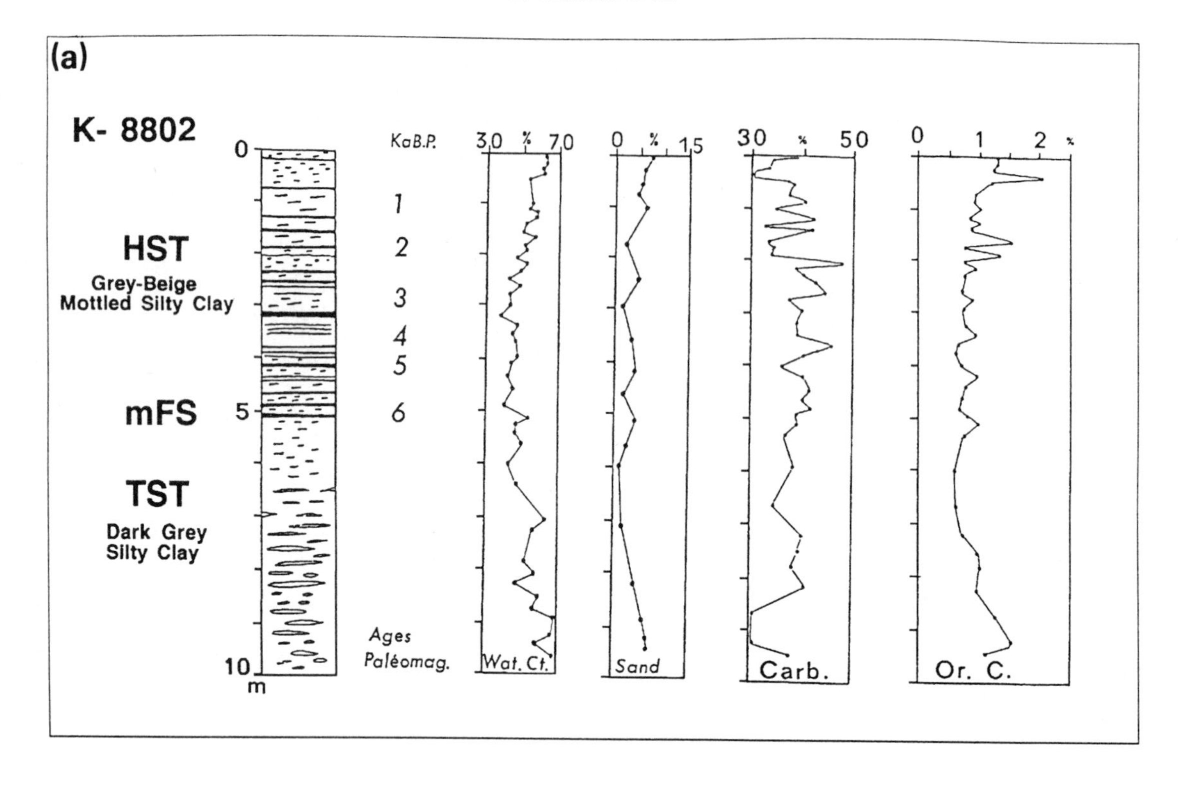

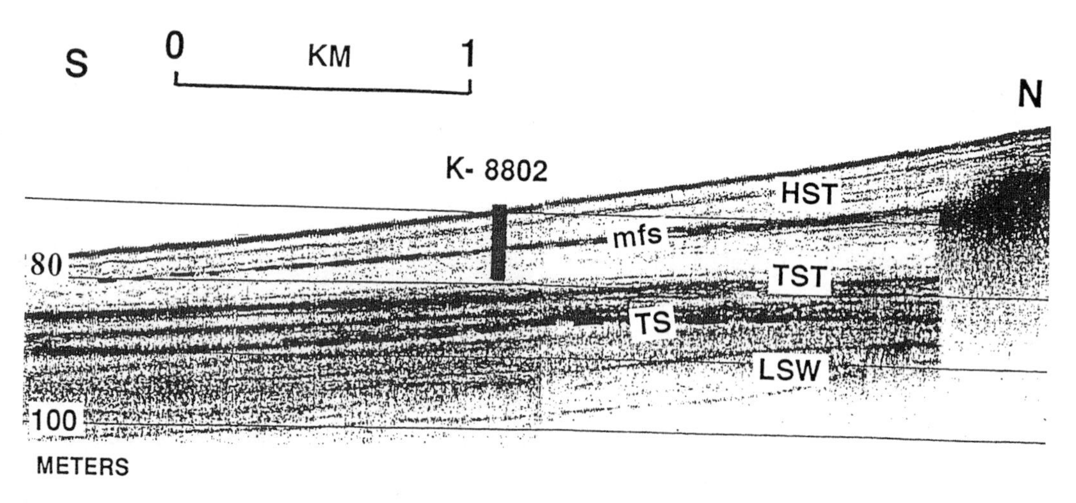

Fig. 5. Lithology and composition of cores collected on the continental shelf (see Fig. 3 for location) and correlation with 3.5 kHz sub-bottom profiles. (a) Inner shelf. (b) Middle shelf.

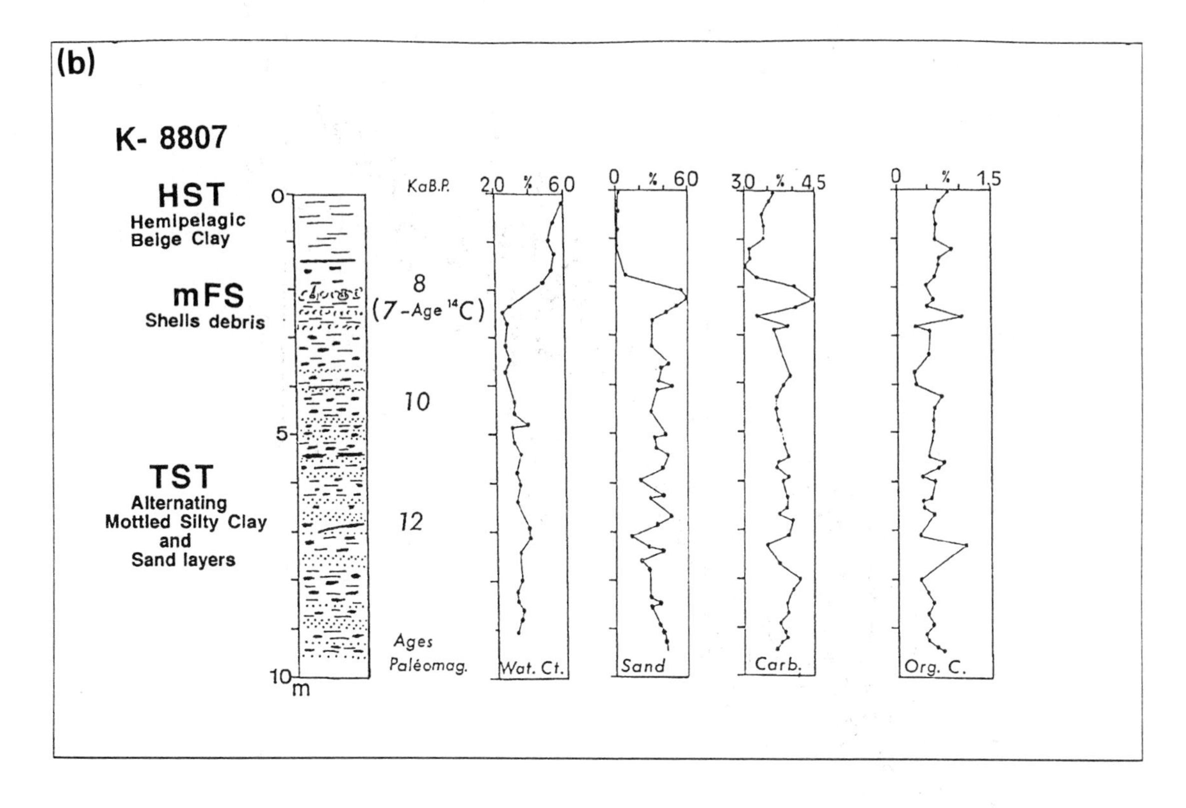

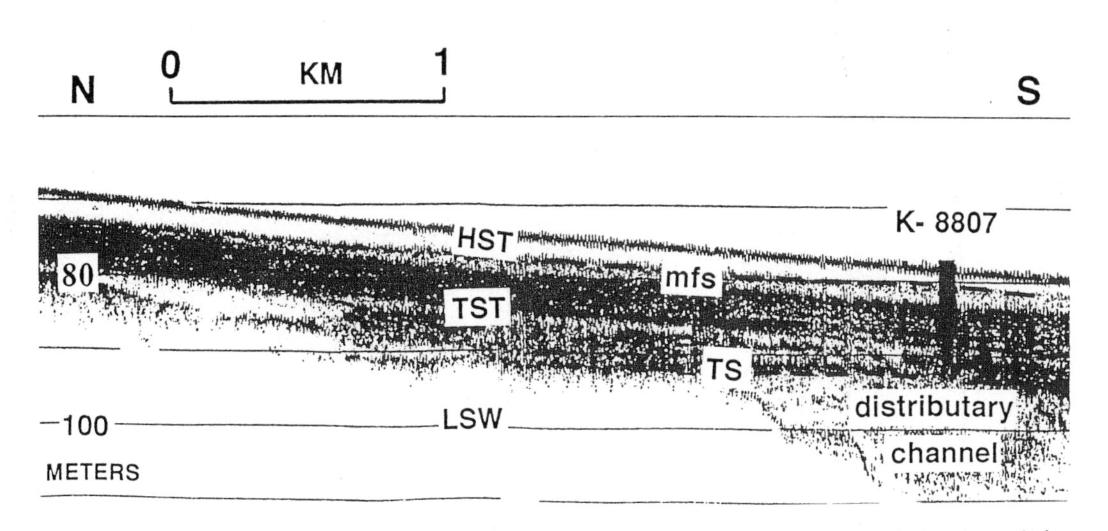

Fig. 5. Lithology and composition of cores collected on the continental shelf (see Fig. 3 for location) and correlation with 3.5 kHz sub-bottom profiles. (b) Middle shelf.

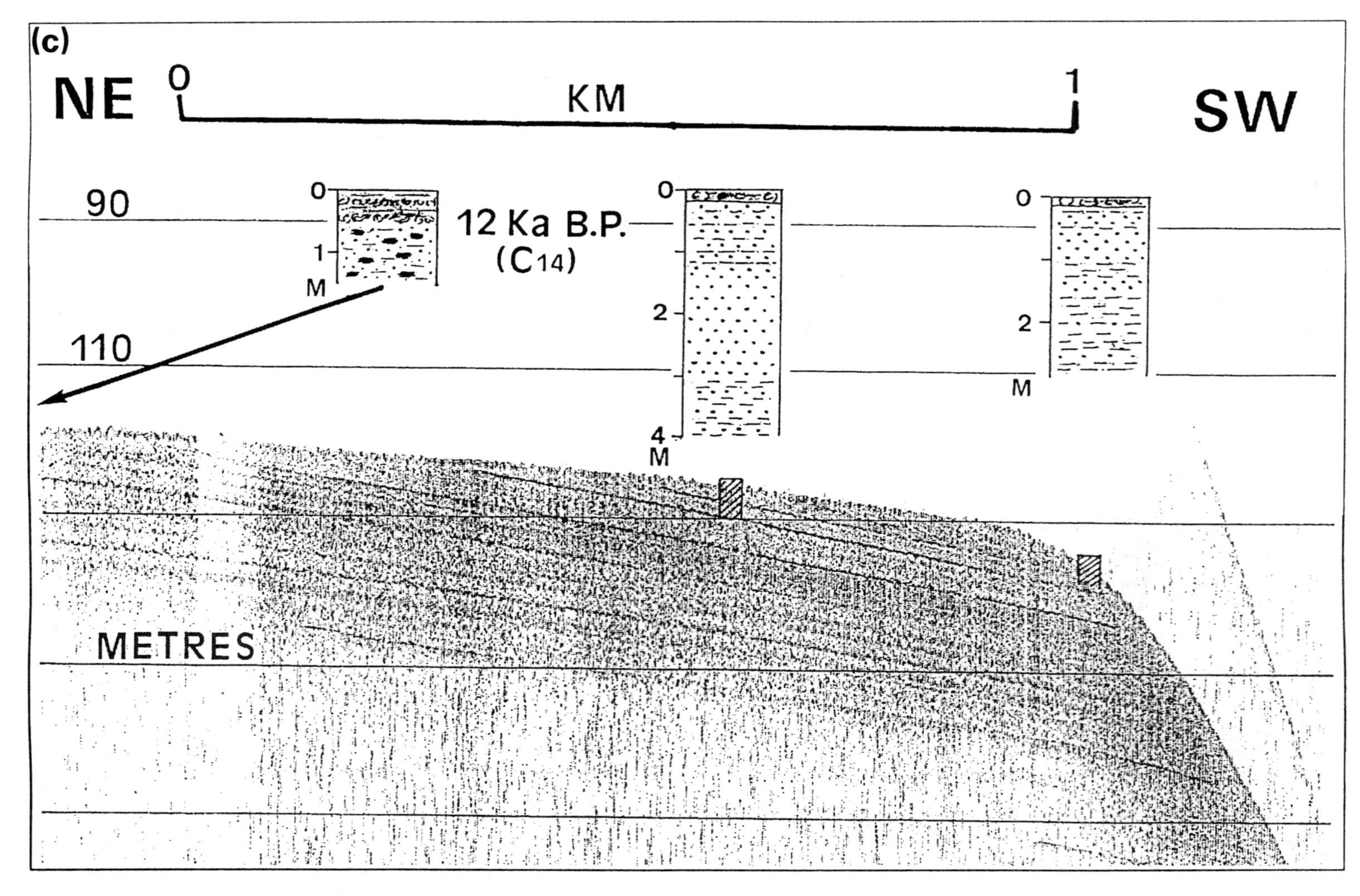

Fig. 5. Lithology and composition of cores collected on the continental shelf (see Fig. 3 for location) and correlation with 3.5 kHz sub-bottom profiles. (c) Outer shelf.

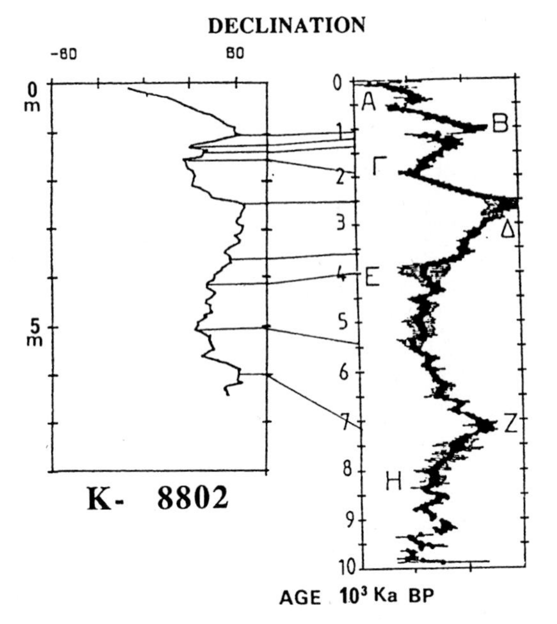

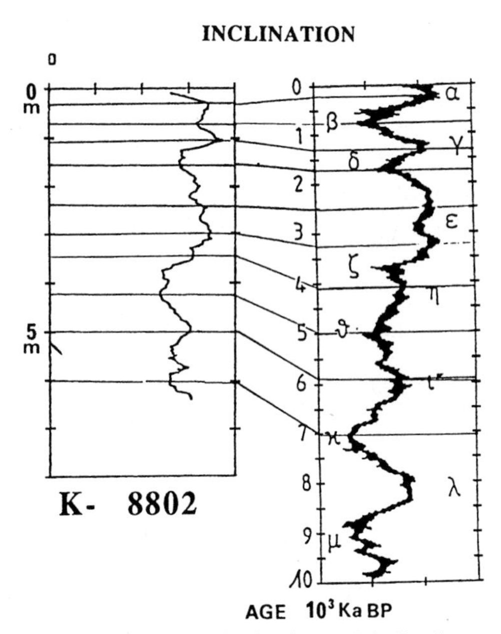

Fig. 6. Palaeomagnetic record of the core K-8802 (see Fig. 5(a)). Comparison with the UK inclination and declination type curves from Turner and Thompson (1981). Detailed chronological markers are provided by the identification of characteristic inclination and declination features. (From Williamson, 1991.)

deposits are correlative with gas-rich acoustically transparent prodeltaic muds of the late transgressive systems tract.

On the middle shelf

Two different sediment facies separated by shell layers are recognized regionally. In core K-8807 (Fig. 5(b)), the lower part (2–10 m) consists of sand layers grading upward to laminated silty clays and mottled clayey silts; their composition is similar to modern prodeltaic deposits and suggests an overall upward-deepening trend. The upper deposits (0–2 m) are homogeneous beige clays with pelagic microfauna, progressively thinning seaward. Examination of seismic data (Fig. 5(b)) indicates that sediments of the lower part of the core correspond to transgressive systems tract deposits. Radiocarbon dating of shell layers in different cores indicates ages varying from 7 to 10 Ky BP (Aloïsi *et al.*, 1976; Ausseil-Badie, 1978; Gensous *et al.*, 1989) representing a period of non-deposition during a time of rapidly rising relative sea level. The upper hemipelagic muds represent the distal part of highstand deposits forming a condensed section produced since the time of maximum transgression of the shoreline.

On the outer shelf

At about the 100 m isobath, the lowstand wedges are overlain by isobath-parallel elongate prograded units exhibiting prominent terrace-like features. Seismic sections display complex organization with several stacked structures. Each unit is characterized by high-angle clinoforms, a sharp basal contact with underlying deposits and poor penetration on sub-bottom profiler records. Kullenberg cores sampled medium to coarse shelly sands.

Seaward of the shoals, the lowstand surface is covered by a thin veneer of shelly sands and beige clay with glauconite and abundant pelagic microfauna (Fig. 5(c)). This represents a condensed section associated with post-glacial sea-level rise. Radiocarbon dating of shell debris provides ages between 10 to 15 Ky BP.

Core data from the Rhône delta plain were acquired by the Shell Company and the Compagnie Nationale d'Aménagement du Bas-Rhône et du Languedoc (Rapport 'CAMARGUE', 1970). Detailed lithological and faunal analyses have been performed on core samples by Lagaaj and Kopstein (1964) and Oomkens (1967, 1970).

A schematic cross-section oriented north–south across the central part of the delta plain is shown in

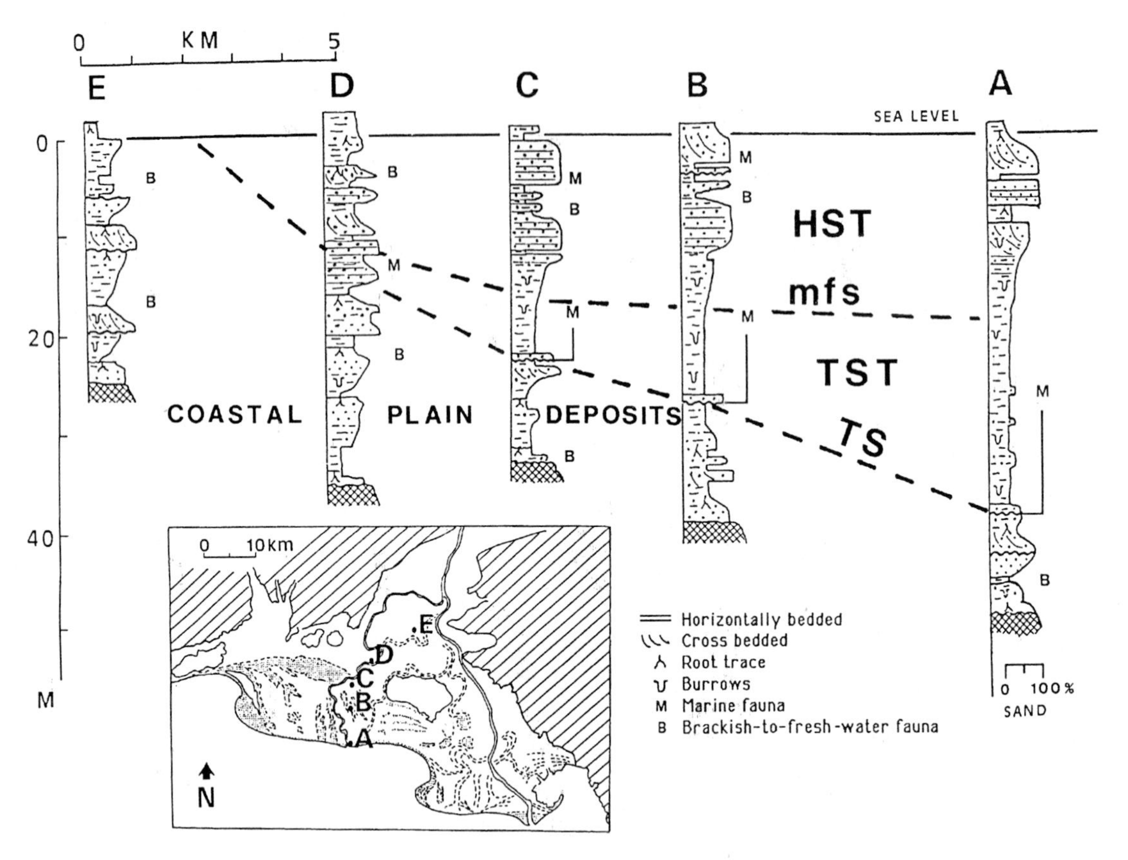

Fig. 7. Cross-section from the central Rhône River delta plain showing lithology distribution and stratigraphic interpretation of post-glacial deposits. (Data from Rapport 'CAMARGUE', 1970 and Oomkens, 1970.)

Fig. 7. The lower deposits overlying the alluvial fill consist of clays with brackish to fresh-water fauna and alternating silty clays and sand beds with shell debris, root traces and soil horizons. These coastal-plain deposits are truncated by an erosional surface overlain by a coarse-grained lag deposit less than 1 m thick. A seaward-thickening wedge of marine deposits lies above the erosional surface. The lower part consists of burrowed clays with marine fauna; the upper part coarsens upward (increase in number and thickness of silt and sand beds) and grades in the uppermost part to well-sorted horizontally bedded sands interpreted as a coastal-barrier environment. This topmost member is commonly eroded by channel scours filled with cross-bedded sands.

From these lithofacies relationships it can be deduced (Fig. 7) that the erosional surface overlain by coarse lag deposits represents the late transgressive (ravinement) surface. Above are marine deposits of the late transgressive systems tract overlain by the regressive facies of the highstand systems tract.

On the delta plain, the shoreline of maximum transgression is about 15 km landward of the present coastline. It consists of barrier beach deposits with shell layers dated between 5–6000 BP (L'Homer *et al.*, 1981). Morphological studies of the well-preserved coastal ridge patterns on the delta plain, supplemented by radiocarbon and archaeological data (Kruit, 1955; L'Homer *et al.*, 1981), indicate that the Rhône river has built three major highstand delta complexes over the last 6000 years (Fig. 8). Major shifts were followed by reworking and longshore transport of the distributary mouth deposits forming extensive beach ridges and coastal barriers. The active mouth, located in the eastern part of the delta plain, dates from the end of the 19th century.

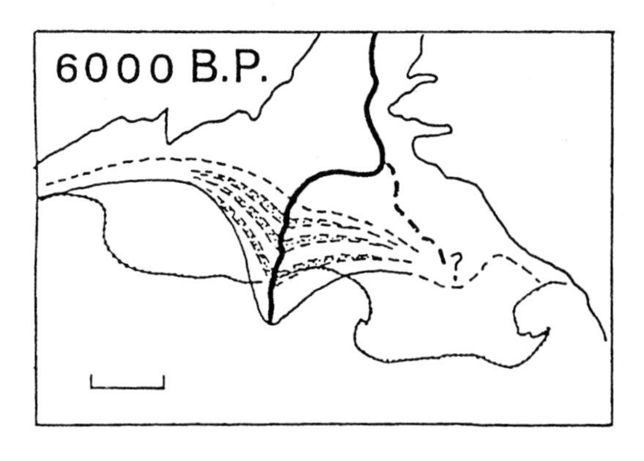

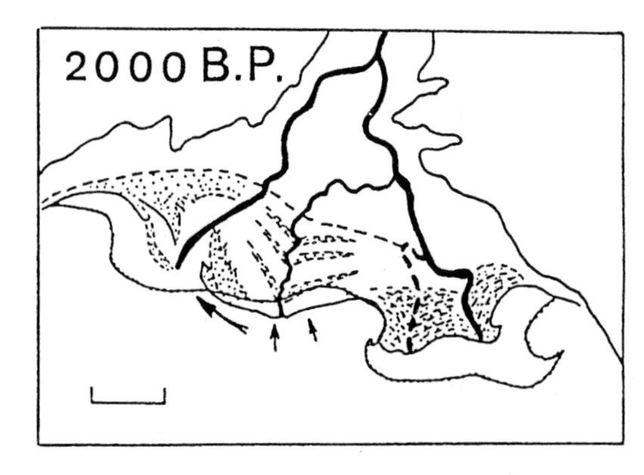

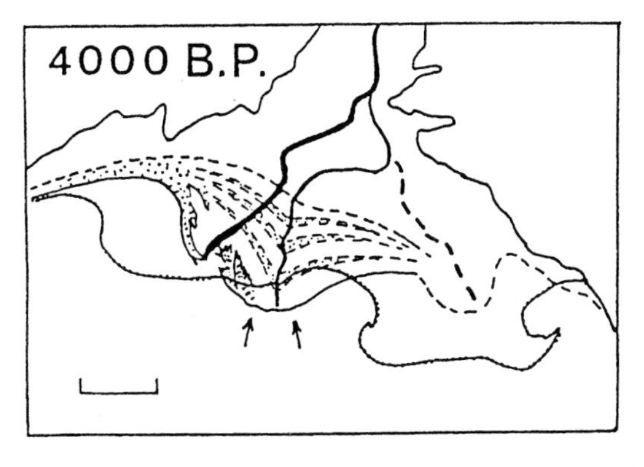

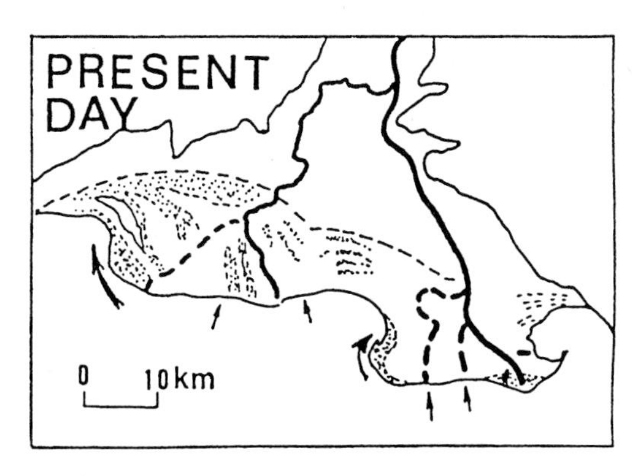

Fig. 8. Evolution of the Rhône delta plain since 6000 BP and major highstand delta complexes. (From Bertrand & L'Homer, 1975.)

On the inner delta shelf

West of the area of incised valleys, a series of shore-parallel sand shoals occur in water depths shallowing westward from 60 to 30 m (Fig. 2(b)). Seismic data show that these sand shoals correspond to the upper part of prograding units about 15 to 20 m thick (Fig. 9). Each unit is composed of a lower part with parallel reflectors onlapping landward onto the sequence boundary and an upper part exhibiting well-defined clinoform reflectors downlapping seaward onto the lower reflectors. Cores collected across the shoals show a seaward progression at the top of the cores from clean sands to organic-rich silty clays and clays similar to upward-coarsening sequences of modern barrier beaches and shorefaces. These units are overlain landward by the highstand Rhône delta deposits and seaward by shelly layers and thin, hemipelagic clays. Radiocarbon dating of the shell layers indicates ages between 8 to 10 ky BP, becoming younger westward.

A strike section (Fig. 10), across the delta plain near the present coastline, shows distinct coastal-plain and channel-fill deposits grading laterally and vertically into coastal-barrier deposits. They are characterized by upward and westward migration and can be correlated on dip sections with the prograding units of the inner shelf.

DISCUSSION AND CONCLUSIONS

The post-glacial deposits of the Rhône continental shelf are organized into a succession of backstepping shore-parallel progradational deposits. In sequence-stratigraphic terms this stacking pattern is

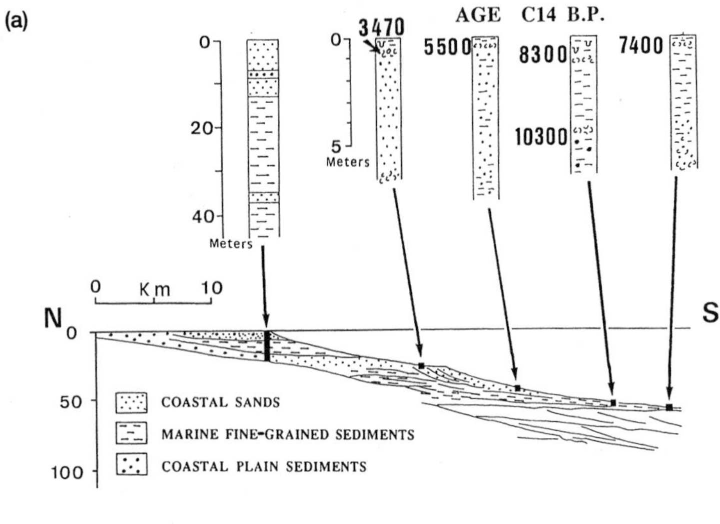

Fig. 9. Line drawing from a mini-sparker profile located in the western part of the delta shelf (see Fig. 1 for location) showing the prograding parasequences associated with sand shoals on the inner–middle shelf. (a) Lithology deduced from Kullenberg cores (Ausseil–Badie, 1978) and core drilling. (b) Sequence-stratigraphic interpretation.

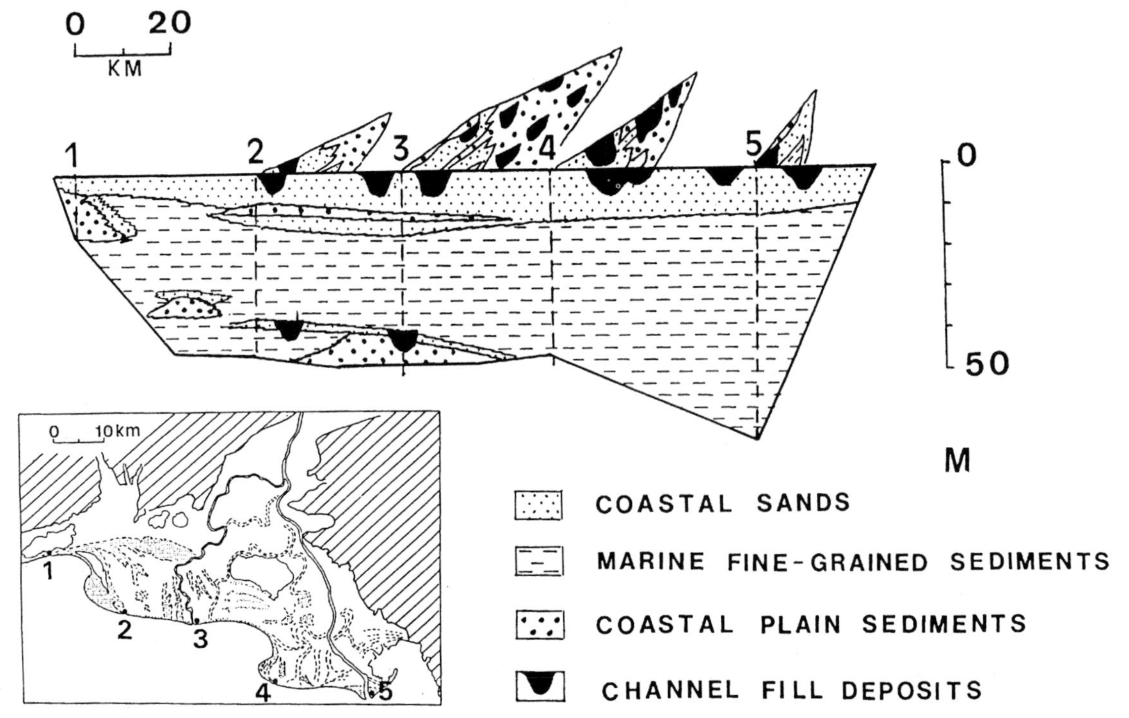

Fig. 10. Strike section across the Rhône River delta plain at the present shoreline. (Data from Rapport 'CAMARGUE', 1970 and Oomkens, 1970.)

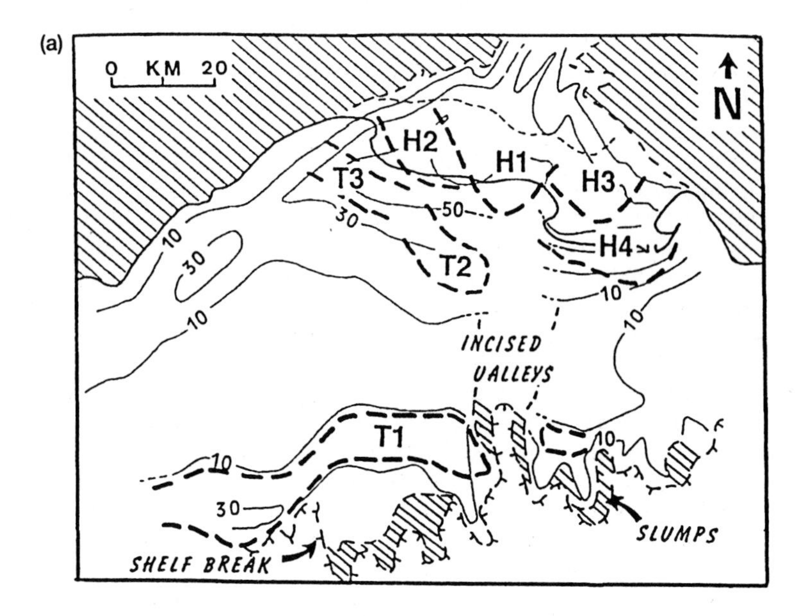

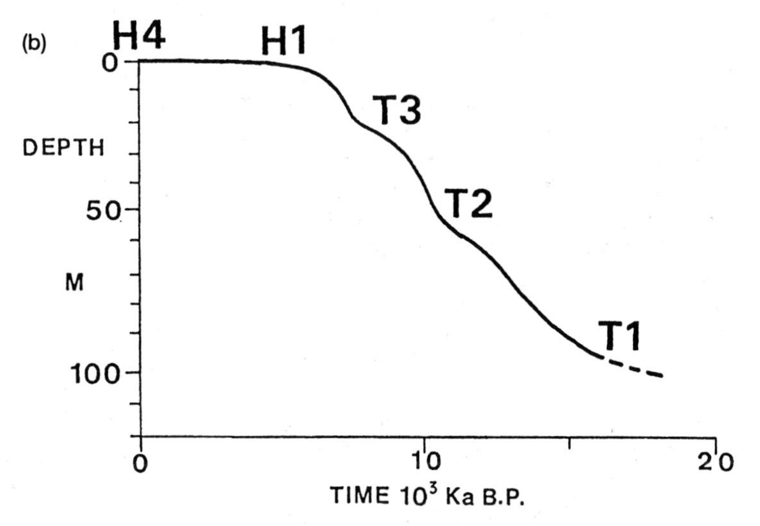

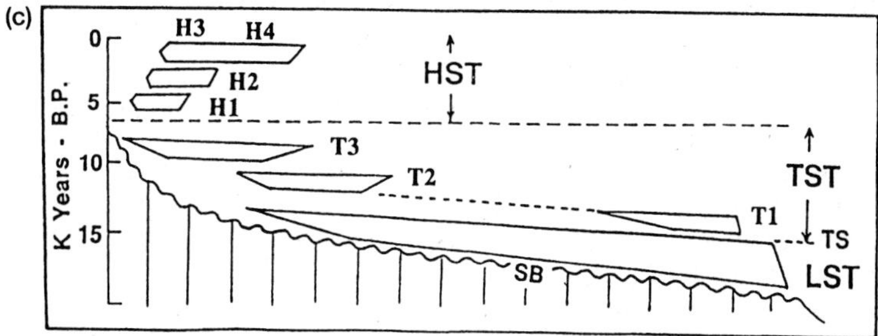

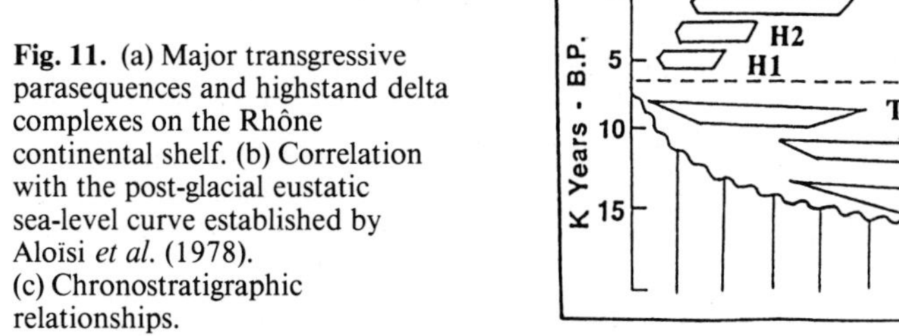
Fig. 11. (a) Major transgressive parasequences and highstand delta complexes on the Rhône continental shelf. (b) Correlation with the post-glacial eustatic sea-level curve established by Aloïsi *et al.* (1978). (c) Chronostratigraphic relationships.

characteristic of the transgressive systems tract. The deposits are correlated with periods of reduced rate of relative sea-level rise or stillstand positions during the post-glacial transgression.

Oxygen-isotope analysis of deep-sea cored deposits indicates that the last deglaciation occurred in discrete steps corresponding to stillstands during the overall sea-level rise; two or three major steps are recognized (Mix & Ruddiman, 1985). In the Golfe du Lion, Aloïsi *et al.* (1978) have established a post-glacial sea-level curve, based on compilation of palynological and radiocarbon data. It reveals that relative sea-level rise took place from 14 to 6 ky BP with two periods of accelerated rate (14–12 and 8–6 ky BP) and two marked inflections between 12 to 8 ky BP. Although the exact age of the backstepping parasequences of the Rhône shelf cannot yet be definitively established, we can tentatively correlate the stillstand interval and associated parasequences with the pattern of sea-level change that occurred during deglaciation (Fig. 11).

On the outer shelf, the parasequences with terrace-like features and sharp basal contacts are interpreted as nearshore sand bodies resulting from transgressive erosion and longshore transport of the lowstand-wedge deposits. They probably formed during an early stillstand of the post-glacial transgression when sedimentation was restricted to infilling of the incised valley. More detailed seismic data are needed to determine whether this complex parasequence stacking pattern results from short-term sea-level fluctuations during the last glacial period or interstadial sea-level fluctuations. Furthermore, relative-sea-level variations could have been affected by local events such as shelf edge slumping or subsidence (Tesson *et al.*, this volume, pp. 183–196).

On the inner–middle shelf the prograding parasequences are mainly located in the western area and differ from those of the outer shelf in several respects. They are geometrically and stratigraphically similar to modern coastal-barrier environments. They are coupled landward with coastal plain and channel deposits, suggesting that they are linked to the development on the shelf of individual delta complexes during late stillstand periods of the Holocene transgression (between 12 and 8 ky BP). Alternating periods of sea-level rise led to the submergence and reworking of deltaic headlands and formation of inner-shelf shoals (Penland *et al.*, 1988). Preservation was favoured by relatively low-energy conditions and/or rapid relative sea-level rise. Because of their present bathymetric location below storm-wave base they are not subjected to reworking and are consequently preserved.

Reduced rate of sea-level rise at the end of the Holocene caused the formation of prograding delta complexes making up a highstand systems tract downlapping onto the late transgressive deposits. Major shifts of the delta complexes may be attributed to autocyclic events. Nevertheless, archaeological data (L'Homer *et al.*, 1981) indicate that during the last 2000 years subsidence has been more important in the western part of the delta plain and may have favoured the migration of distributaries to their present position.

ACKNOWLEDGEMENTS

This work was in part supported by the French INSU (DBT Programme) of the CNRS. The Institut Français du Pétrole and Total Exploration Laboratory also provided support. We thank the crews of the N.O. *G. Petit* and *C. Laurence*. Thanks are also due to the reviewers and B. Deniaux for their helpful comments and suggestions. Contribution no. 375 to the DBT Programme.

REFERENCES

Aloïsi, J.C. (1986) Sur un modèle de sédimentation deltaïque. Contribution à l'étude des marges passives. Thèse d'ètat, University of Perpignan, 162 pp.

Aloïsi, J.C., Monaco, A., Planchais, N., Thommeret, J. & Thommeret, Y. (1978) The Holocene transgression in the Golfe du Lion, Southwestern France, paleogeographic and paleobotanical evolution. *Geogr. Phys. Quat.* **XXXII**(2), 145–162.

Aloïsi, J.C., Monaco, A., Thommeret, J. & Thommeret Y. (1976) Interprétation paléogéographique du plateau continental languedocien dans le cadre du golfe du Lion. Analyse comparée des données sismiques, sédimentologiques et radiométriques concernant le Quaternaire récent. *Rev. Géogr. Phys. Géol. Dyn.* **16**(f5), 13–22.

Ausseil-Badie, J. (1978) Contribution à l'étude paléoécologique des foraminifères du Quaternaire terminal sur le plateau continental languedocien. Thèse troisième cycle University of Toulouse, 164 pp.

Bertrand, J.P. & L'Homer, A. (1975) Les deltas de la Méditerrannée du Nord — Le delta du Rhône. *Ninth Int. Congr. Sédimentol, Nice*, 16, 65 pp.

Blanc, J.J. (1977) Recherches de sédimentologie appliquée au littoral du delta du Rhône, de Fos au Grau du Roi. *Contrat C.N.E.X.O* **75**, 1193.

Boyd, R., Sutter, J. & Penland, S. (1989) Relation of sequence stratigraphy to modern sedimentary environments. *Geology* **17**, 926–929.

Canals, M. & Got, H. (1986) La morphologie de la pente continentale du Golfe du Lion: une résultante structuro-sédimentaire. *Vie et Milieu* **36**(3), 153–163.

Creer, K.M. (1985) Review of lake sediment paleomagnetic data. *Geophys. Surv.* **7**, 125–160.

Gensous, B., El Hmaidi, A., Williamson, D. & Taieb, M. (1989) Caractérisation chronologique et sédimentologique des dépôts récents de la marge rhodanienne. *2ème Congr. Fra. Sédimentol. Paris*, Nov. 1989, pp. 113–114.

Got, H., Aloïsi, J.C. & Monaco, A. (1985) Sedimentary processes in Mediterranean deltas and shelves. In: *Geological Evolution of the Mediterranean Basin* (Eds Stanley, D.J. & Wezel, F.C.) Springer Verlag Stroudsburg, Pennsylvania, pp. 355–376.

Haq, B.U., Hardenbol, J. & Vail, P.R. (1987) The chronology of fluctuating sea level since the Triassic. *Science* **194**, 1121–1132.

Kruit, C. (1955) Sediments of the Rhône delta — grain size and microfauna. *Verhand. Konoink. Neder. Geol. Mijnbouw.* **15**(2), 357–514.

Lagaaj, R. & Kopstein, F.P.M.W. (1964) Typical features of a fluviomarine offlap sequence, in deltaic and shallow marine deposits. In: *Sixth International Sedimentology Congress 1963* (Ed. Van Straaten, L.M.J.V.), Elsevier, Amsterdam, pp. 216–226.

L'Homer, A., Bazile, F., Thommeret, J. & Thommeret, Y. (1981) Principales étapes de l'édification du delta du Rhône de 7000 ans BP à nos jours; variations du niveau marin. *Océanis* **7**(f4), 389–408.

Mix, A.C. & Ruddiman, W.F. (1985) Structure and timing of the last deglaciation: oxygen-isotope evidence. *Quat. Sci. Rev.* **42**, 59–108.

Monaco, A. (1970) Contribution à l'étude géologique et sédimentologique du plateau continental du Roussillon (Golfe du Lion). Thèse d'état, University of Perpignan, 295 pp.

Oomkens, E. (1967) Depositional sequences and sand distribution in a deltaic complex. *Geol. Mijnbouw,* **46**, 265–278.

Oomkens, E. (1970) Depositional sequences and sand distribution in the post glacial Rhône delta complex. In: *Deltaic Sedimentation.* Spec. Publ. Soc. Econ. Paleontol. Mineral. 15, 198–212.

Penland, S., Boyd, R. & Suter, J.R. (1988) Transgressive depositional systems of the Mississippi delta plain: a model for barrier shoreline and shelf sand development. *J. Sediment. Petrol.* **58** (6), 932–949.

Posamentier, H.W. & Vail, P.R. (1988) Eustatic controls on clastic deposition II — sequence and systems tract models. In: *Sea-Level Changes — An Integrated Approach* (Eds Wilgus, C.K., Hastings, B.S., Kendall, C.G.St.C., Posamentier, H.W., Ross, C.A. & Van Wagoner, J.C.) Spec. Publ. Soc. Econ. Paleontol. Mineral. 42, 125–154.

Posamentier, H.W., Jervey, M.T. & Vail, P.R. (1988) Eustatic controls on clastic deposition I — conceptual framework. In: *Sea-Level Changes — An Integrated Approach* (Eds Wilgus, C.K., Hastings, B.S., Kendall, C.G.St.C., Posamentier, H.W., Ross, C.A. & Van Wagoner, J.C.) Spec. Publ. Soc. Econ. Paleontol. Mineral. 42, 109–124.

Rapport 'Camargue' (1970) Etude hydrogéologique pédologique et de salinité. Dir. Dep. Agri. & Cie Nat. Aménag. Reg. Bas-Rhône et Languedoc, Nîmes, France (unpublished results).

Suter, J., Berryhill, H.L. & Penland, S. (1987) Late Quaternary sea level fluctuations and depositional sequences, Southwest Louisiana Continental Shelf. In: *Sea Level Fluctuations and Coastal Evolution* (Eds Nummedal, D., Pilkey, O.H. & Howard, J.P.) Spec. Publ. Soc. Econ. Paleontol. Mineral. 41, 199–219.

Tesson, M., Gensous, B., Allen, G.P. & Ravenne, C. (1990a) Late Quaternary deltaic lowstand wedges on the Rhône Continental Shelf, France. *Mar. Geol.* **91**, 325–332.

Tesson, M., Ravenne, C. & Allen, G.P. (1990b) Application des concepts de stratigraphie séquentielle à un profil sismique haute résolution transverse et à la plateforme rhodanienne. *C.R. Acad. Sci. Paris*, **310**, Série II, 565–570.

Turner, G.M. & Thompson, R. (1981) Lake sediment record of geomagnetic secular variation in Britain during Holocene times. *Geophys. J. Roy. Abstr. Soc.* **65**, 703–725.

Vail, P.R., Mitchum, R.M., Todd, R.G., *et al.* (1977) Seismic stratigraphy and global changes of sea level. In: *Seismic Stratigraphy — Applications to Hydrocarbon Exploration* (Ed. Payton, C.E.) Mem. Am. Assoc. Petrol. Geol. 26, 49–212.

Van Straaten, L.M. (1959) Littoral and submarine morphology of the Rhône delta. *Sec. Coastal Geogr. Conf., Bâton Rouge* Proc., Nat. Ac. Sc. Nat. Res. Council, 233–264.

Van Wagoner, J.C., Posamentier, H.W., Mitchum, R.M., Jr., *et al.* (1988) An overview of the fundamentals of sequence stratigraphy and key definitions. In: *Sea-Level Changes — An Integrated Approach* (Eds Wilgus, C.K., Hastings, B.S., Kendall, C.G.St.C., Posamentier, H.W., Ross, C.A. & Van Wagoner, J.C.) Spec. Publ. Soc. Econ. Paleontol. Mineral. 42, 39–46.

Williamson, D. (1991) Propriétés magnétiques de séquences sédimentaires de Méditerranée et d'Afrique intertropicale. Implications environnementales et géomagnétiques pour la pèriode 30-0 Ka B.P. Thèse, University of Aix-Marseille II, 230 pp.

Williamson, D., Gensous, B., El Hmaidi, A., Taieb, M. & Thouveny, N. (in press) Deltaic platform deposits as recorders of geomagnetic oscillations. Submitted to *Geoph. J. Int.*

Spec. Publs Int. Ass. Sediment. (1993) **18**, 213–232

Seismic stratigraphy of Quaternary stacked progradational sequences in the southwest Japan forearc: an example of fourth-order sequences in an active margin

Y. OKAMURA* *and* P. BLUM†

**Marine Geology Department, Geological Survey of Japan, Tsukuba, Ibaraki 305, Japan; and*
†Joides Office, The University of Texas at Austin, TX 78759-8345, USA

ABSTRACT

Single-channel seismic profiles reveal the presence of stacked progradational sequences beneath three forearc basins off southwest Japan. Stacking patterns of the sequences are retrogradational or progradational depending on tectonic movements and sediment dispersal in the basins. The sequences are generally composed of aggradational and progradational units without a retrogradational unit. The lack of a retrogradational unit is interpreted to be due to rapid sea-level rise during the Quaternary. Major sequences are defined by pronounced oblique clinoforms, suggesting faster and longer sea-level falls. No encroachments on the oblique clinoforms imply no or very short duration of sea-level lowstand. The stacks in the three basins show similar successions of seven progradational sequences. The bases of the stacks are inferred to be late Pliocene in age, based on regional geology. The sequences therefore are of fourth or higher order, and their resolution is the result of rapid subsidence of forearc basins. A tentative correlation between the sequences and Quaternary oxygen-isotope curves suggests that the major sequences were formed during oxygen-isotope stages 6, 12, 16 and 22. Comparison of this Quaternary stack of high-order sequences with the sea-level curve of Haq *et al.* (1987) from third-order and lower order sequences shows that the apparently most important Quaternary unconformity at ~0.8 Ma of Haq *et al.* (1987) merely presents the first of several similarly extreme events, and its detection from lower resolution profiles may be due to the general change to high-amplitude sea-level changes of higher order.

INTRODUCTION

The Pacific coast of southwest Japan is the location of the most active tectonic movement in the Japanese island arc system. The topography of the coastal areas shows rapid uplift and subsidence at the rate of 10 times faster than that of passive margins. In the offshore of these tectonically active coasts, multiple progradational sequences are observed on single-channel seismic profiles (Okamura, 1989). Sequences of fourth or higher order can be differentiated due to combination of high subsidence rates in the basins, high uplift rates of structural highs bounding individual basins and sediment supply patterns controlled by this complex structural setting. They provide a good opportunity to study the impact of high-frequency/high-amplitude Quaternary sea-level changes on the construction of forearc basin stratigraphy. In this paper, detailed structures and stacking patterns of these progradational sequences are described, and their origin and possible timing are discussed. The comparison between these higher-order sequences and a general sequence stratigraphy model implies significance of fourth-order or higher order cycles.

†Present address: Ocean Drilling Program, Texas A&M University, College Station, TX 77845–9547, USA.

GEOLOGICAL BACKGROUND

The area surveyed is off the Pacific coast of the Shikoku and Kyushu islands in the southwest Japan

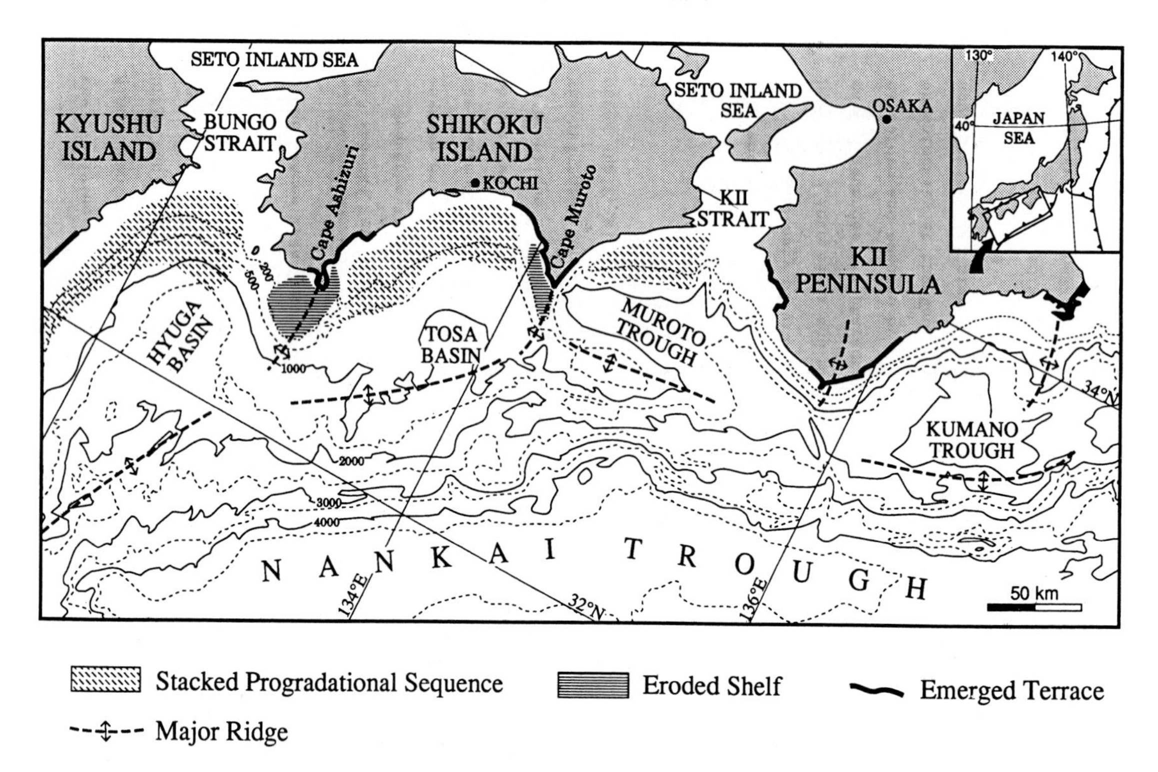

Fig. 1. Bathymetric map of the southwest Japan forearc. Forearc basins are divided by N–S trending structural ridges which include eroded shelves and emerged terraces. Stacked progradational sequences develop under the landward slope of the forearc basins. Distribution of stacked progradational sequences and uplifting shelves to the east of the Kii Strait is not clear.

forearc (Fig. 1). The islands are composed mainly of Mesozoic and Tertiary accretionary complexes. On Kyushu, Neogene igneous rocks and Quaternary volcanoes are widely distributed. To the south of the islands, an accretionary complex has been growing along the Nankai Trough, presumably since the late Neogene and throughout the Quaternary (Fig. 1; Taira, 1985). On top of the landward part of the offshore accretionary complex, a chain of forearc basins 50 km wide has developed. Four major basins are separated by N–S-trending ridges that have formed in the early Quaternary (Okamura, 1990) — the Hyuga Basin, the Tosa Basin, the Muroto Trough and the Kumano Trough (from west to east; Fig. 1). The two ridges separating the Hyuga, Tosa and Muroto Basins, respectively, are composed mainly of Pliocene and early Pleistocene deformed sedimentary rocks (Okamura & Joshima, 1986; Okamura *et al.*, 1987). To the north, the two ridges extend into the mountains of Shikoku island through Cape Muroto and Cape Ashizuri, respectively (Fig. 1). The coasts around the capes are characterized by emerged marine terraces up to a few hundred metres high. The uplift rate of the terraces was estimated to be up to 2 m/ky during the late Pleistocene (Yoshikawa *et al.*, 1964). In contrast, the basin areas extend northward as the Bungo Strait, the Tosa Bay and the Kii Strait, respectively. The coasts around the straits and the bay are characterized by submerged mountains and valleys (Fig. 1), indicating subsidence of the basins. Fluvial systems are draining the islands towards the subsiding areas, and clastic sediment supply is maximized toward the axis of the basins. Stacked prograding wedges have been formed in the landward margins of the subsiding basins. Beneath the stacked progradational sequences, the acoustic basement presumably composed of Tertiary accretionary complexes and igneous rocks is widely recognized. The basement is characterized by a relatively smooth or irregular surface, suggesting subaerial erosion before deposition of the progradational sequences.

DATA

The principal data set is a grid (5 to 6 km spacing, Fig. 2) of single-channel seismic profiles. Over 15,000 km, profiles were collected off the Shikoku and Kyushu islands, including shelves, slopes and basin floors of the forearc basins, by the R/V *Hakureimaru*. Two 1.97 l (120 in^3) air guns were used as sound source. The shot pulse was shortened by a wave-shaped kit installed in the air guns. The profiles were not processed but show clear data typically to 1.0–2.0 s sub-bottom, except where masked by multiples. Because sound velocities in the sediments are unknown, depth and thickness are presented as two-way time travel in seconds.

Sediment samples were obtained from more than 500 localities by grab, core and dredge samplers (e.g. Ikehara, 1988). Beneath the subsiding basin

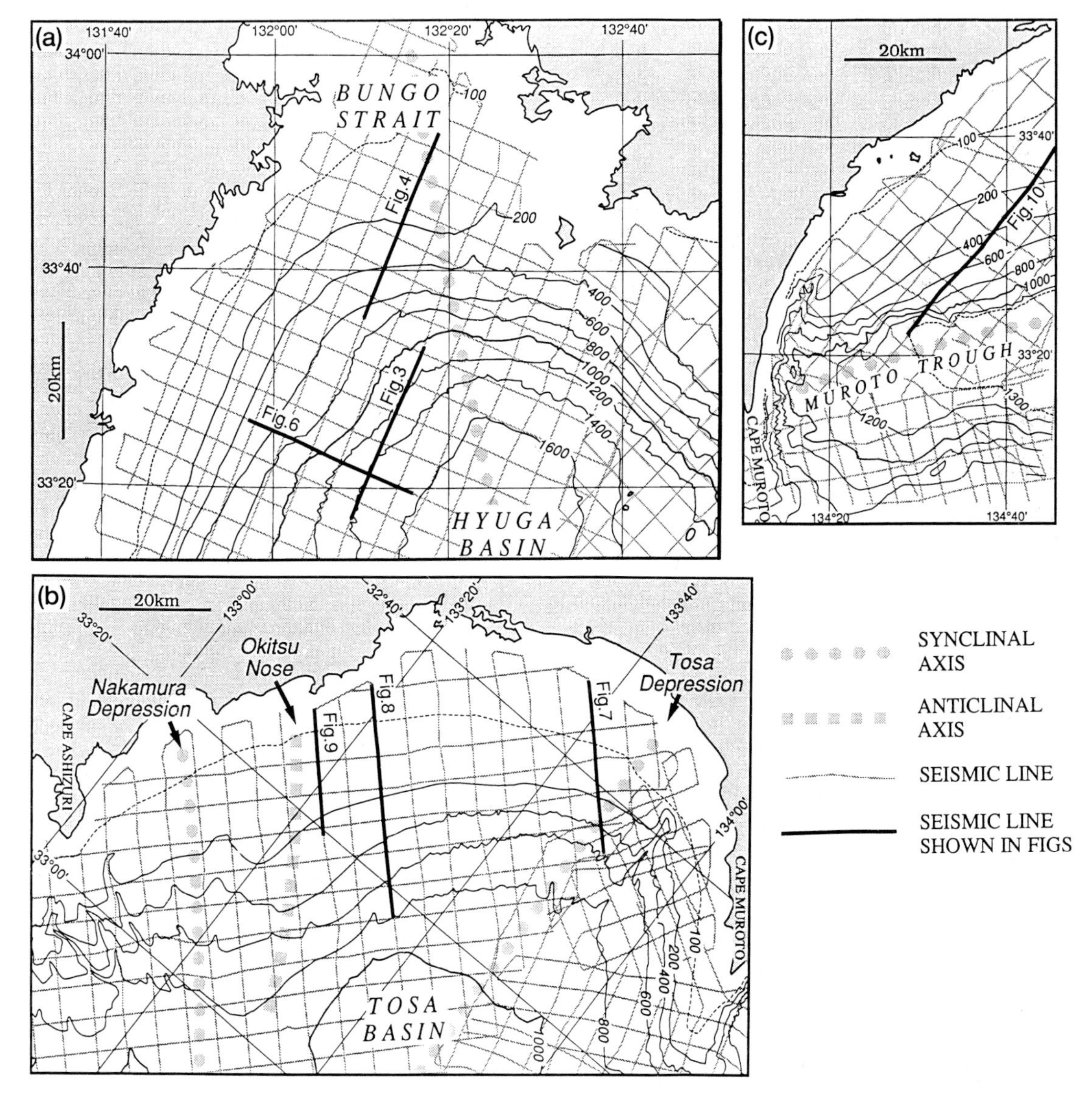

Fig. 2. Detailed bathymetry and seismic survey lines of the northern margin of the forearc basins. (a) Hyuga Basin; (b) Tosa Basin; and (c) Muroto Trough.

floors, Recent soft sediments were recovered. Tectonic ridges (arc-parallel outer ridges and N–S-directed ridges dividing the forearc basins) expose Pliocene and Miocene sedimentary rocks, dated by nannofossil and planktonic foraminifera analysis (Okamura & Joshima, 1986; Okamura *et al.*, 1987).

SEISMIC STRATIGRAPHY OF STACKED PROGRADATIONAL SEQUENCES

Bungo strait — Hyuga basin

The Bungo Strait, 30 km wide, 50 km long and 80–100 m deep, connects the Pacific Ocean to the Seto Inland Sea (Fig. 1). The strait is thought to have emerged and to have supplied a large quantity of clastics from the Seto Inland Sea area to the Hyuga Basin during the ice ages. The strait extends southward to the Hyuga Basin (1300–1800 m deep) along a large-scale syncline (Fig. 2(a)).

Seven progradational sequences, bounded by a toplap and/or erosional unconformities, are identified (Table 1; Figs 3 & 4). Three of the unconformities are distinct as compared with the others, and the four formations defined by these unconformities are Bn1, Bn2, Bn3 and Bn4 Formations in ascending order (Okamura, 1989). These sequences show a retrogradational stacking pattern, i.e. younger sequences are successively displaced landward (Figs 3, 4 & 5). Generally, each sequence has a maximum thickness in the axial zone of the syncline and extends westward for tens of kilometres, while its dimensions rapidly decrease eastward from the axis (Figs 5(a) & (b)).

Bn1 Formation

The Bn1 Formation is composed of two sequences (Fig. 3). The lower sequence Bn1a consists of smooth, nearly horizontal, sub-parallel reflections onlapping against the basement and includes no prograding wedges. To the east, this sequence con-

Table 1. Summary of progradational sequences in the three basins. Maximum thickness of sequences above underlying shelves in seconds. Maximum thickness of sequences prograded seaward of an underlying shelf break in parentheses

Oxygen isotopic stage	Hyuga Basin	Tosa Basin (Line 10)	Muroto Trough
2 ←			
	Bn4b 0.12 s (0.2 s)	Tw2e 0.06 s	Tk4 0.09 s
6 ←			
	Bn4a 0.2 s	Tw2d 0.08 s	Tk3b 0.15 s
		Tw2c 0.05 s	Tk3a 0.06 s
12 ←			
	Bn3b 0.23 s (0.27 s)	Tw2b 0.11 s	Tk2b 0.08 s
	Bn3a 0.17 s		Tk2a 0.15 s
16 ←		?	
	Bn2b 0.12 s (0.4 s)	Tw2a 0.05 s ?	Tk1 0.4 s
22 ←			
	Bn2a 0.55 s	Tw1b 0.2 s	
Late Plio.? ←			
	Bn1b 0.15 s	Tw1a 0.05 s	Tosabae F.
	Bn1a		

(——) Major unconformity, (------) minor unconformity, (——) downlap surface

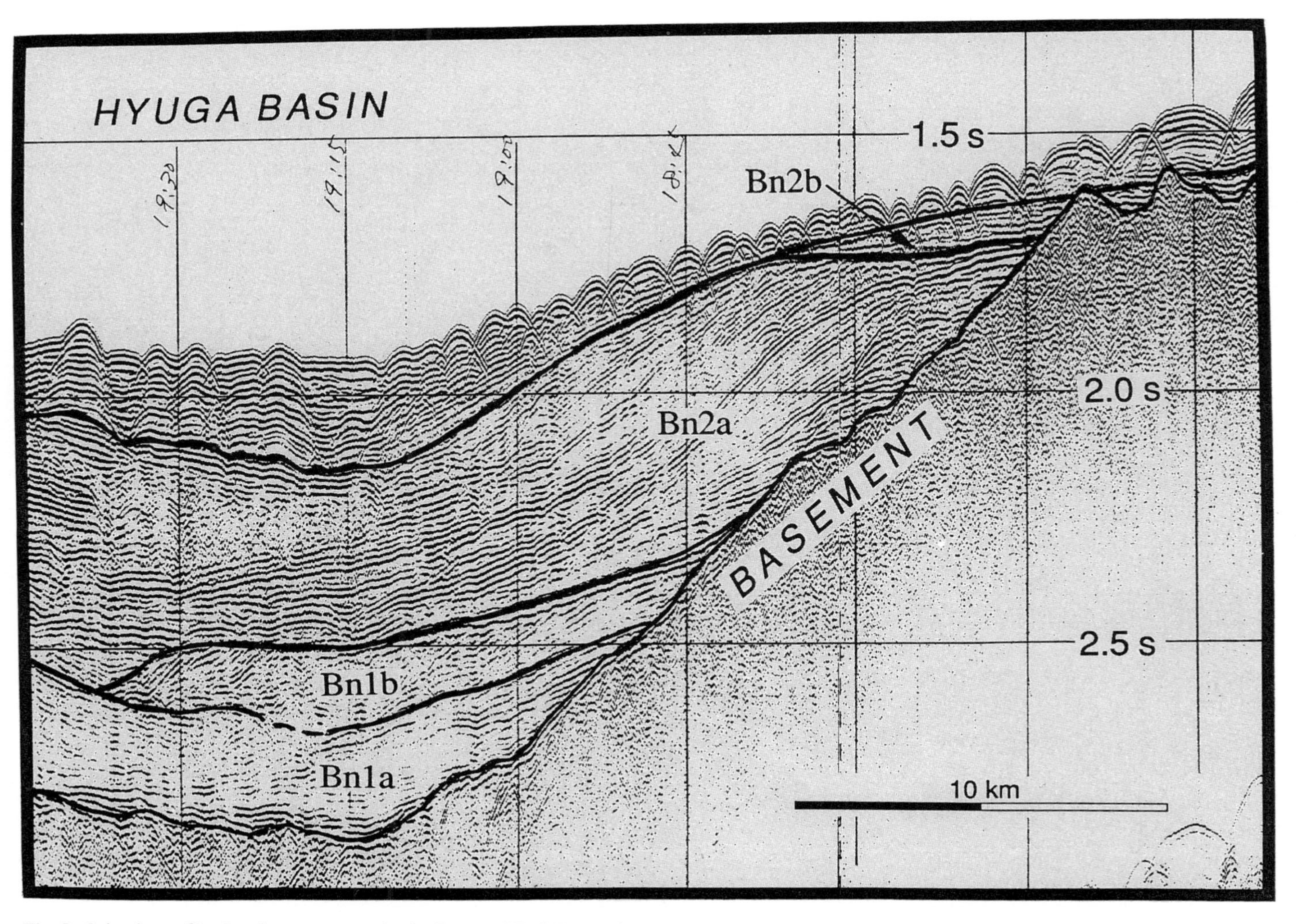

Fig. 3. Seismic profile showing sequences in the Bn1 and Bn2 Formations. The base of sequence Bn1b is defined by a downlap surface. The reflection pattern in sequence Bn2a shows an upward change from retrogradational to an aggradational and oblique clinoform. The oblique clinoform is continuous basinward with a chaotic mass.

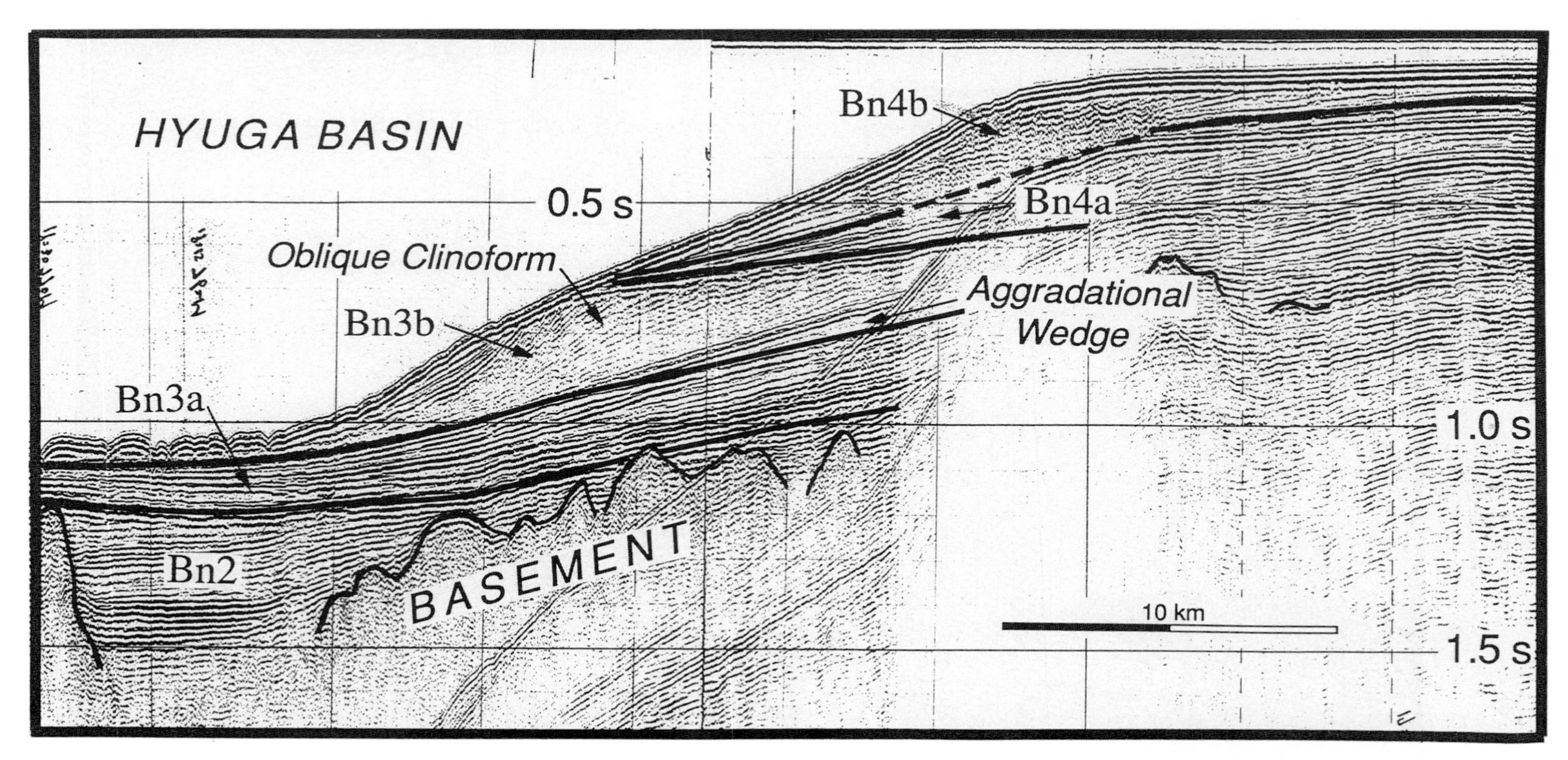

Fig. 4. Sequences in the Bn3 and Bn4 Formations. Much of sequence Bn4a and the aggradational wedge in sequence Bn3b are masked by multiples. Their internal structures can be partly observed on the western margin of the syncline where the dimensions of the sequences are smaller. Bn2 includes landward extension of sequences Bn2a and Bn2b onlapping against the basement.

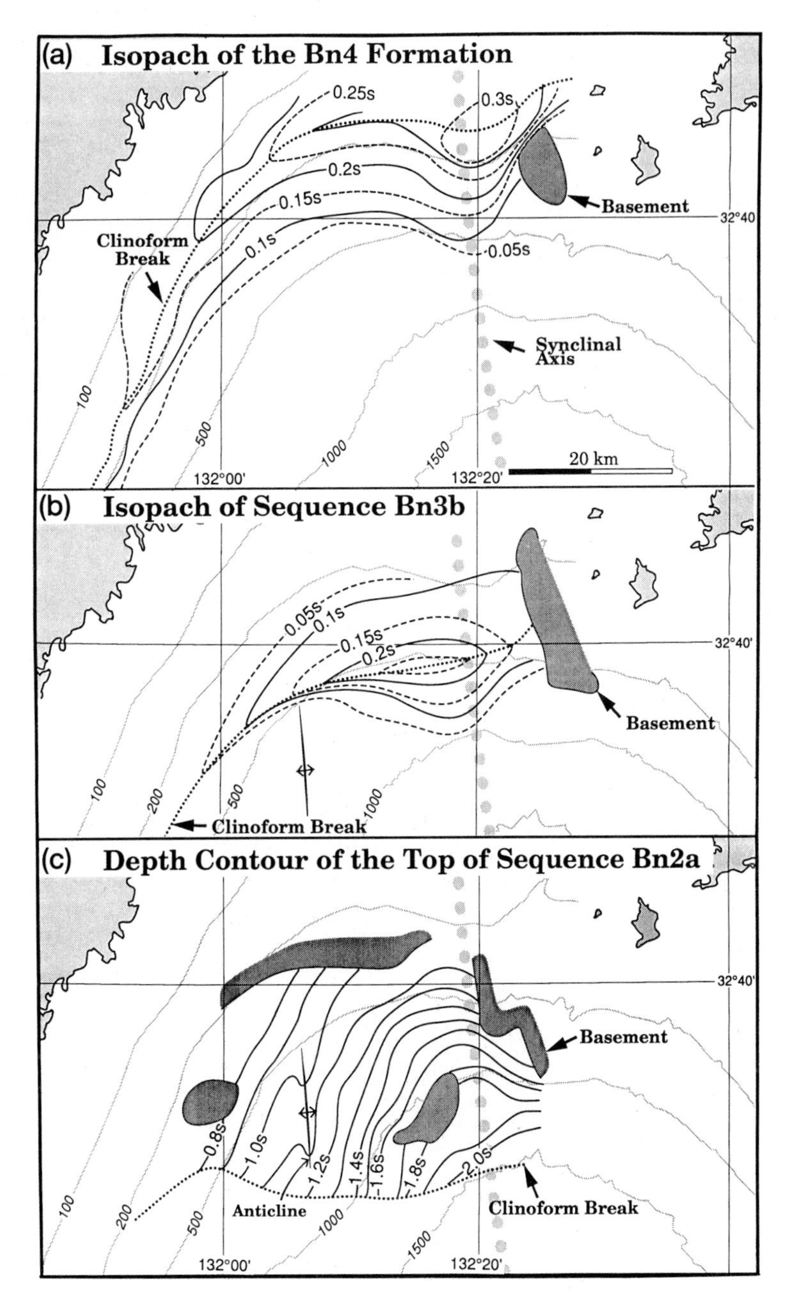

Fig. 5. (a) Isopach of the Bn4 Formation. The sequence has maximum thickness in the synclinal axis and extends westward for over 50 km, while the sequence is limited by a basement high just east of the synclinal axis. (b) Isopach of sequence Bn3b. (c) Depth contour of the toplap and/or truncational unconformity of sequence Bn2a. Note the synclinal deformation of the unconformity.

tinues to the upper Pliocene Ashizurioki Formation exposed on the shelf off Cape Ashizuri and outer ridges of the Tosa Basin (Okamura *et al.*, 1987), although reflections are partly obscured due to structural disruption. The upper sequence (Bn1b) consists mainly of progradational reflection patterns (Fig. 3). An oblique clinoform can clearly be recognized in the axial part of the syncline, while it becomes unclear both eastward (due to a decrease in its dimensions) and westward (due to sound attenuation by increased thickness of overlying sediments). A truncational unconformity, correlative to the toplap unconformity at the top of sequence Bn1b, is recognized for 35 km northwest-

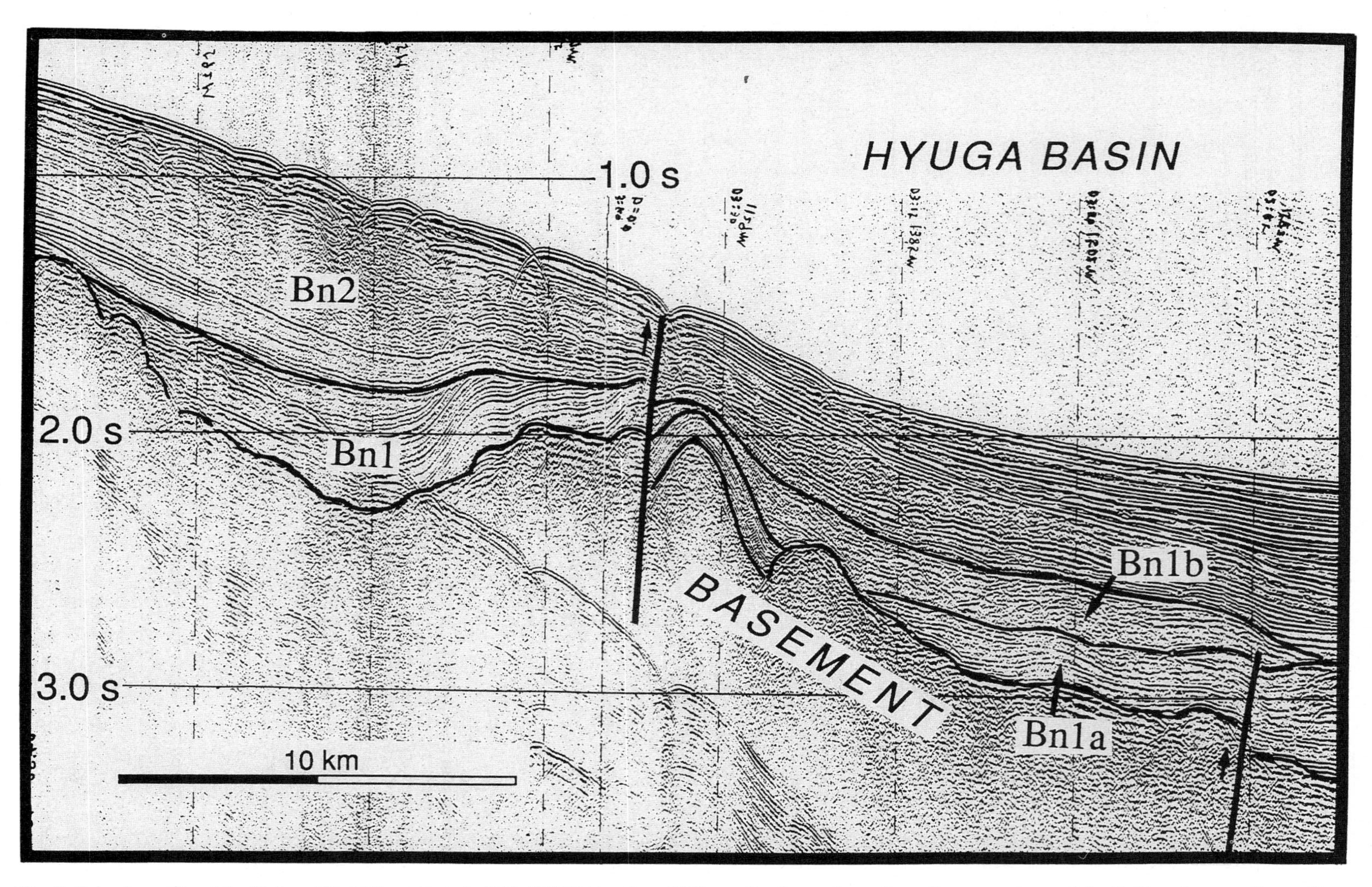

Fig. 6. Seismic profile of the Bn1 and Bn2 Formations in ESE–WNW direction. Oblique clinoforms show discontinuous, sub-parallel reflections because the profile is sub-parallel to the depositional strike. The Bn1 Formation and its upper truncational unconformity are folded and faulted.

ward from the synclinal axis (Fig. 6). Reflections under the unconformity are continuous, smooth and parallel (Fig. 6) and probably include the landward extension of sequences Bn1a and Bn1b. The widespread unconformity strongly suggests a large sea-level fall, presumably during late Pliocene. The unconformity is faulted, folded and tilted (Fig. 6) and attains 2.6 s in maximum depth, indicating active deformation and rapid subsidence (more than 1 m/ky) of the syncline during the Quaternary.

Bn2 Formation

The Bn2 Formation can be divided into two sequences designated as sequences Bn2a and Bn2b in ascending order. Sequence Bn2a is the thickest prograding wedge (about 0.55 s in maximum) in this area (Fig. 3). The sequence rapidly thins eastward from the synclinal axis, while it extends westward for about 45 km (Fig. 5(c)), gradually decreasing in thickness. It generally shows upward and seaward change of reflection patterns from retrogradational to aggradational and oblique progradational (Fig. 3). In the axial zone of the syncline, an oblique clinoform progrades southward for up to 6 km in maximum, while the dimensions of the clinoform decrease westward, suggesting that the sediments were mainly supplied to the axial zone of the syncline. The lower retrogradational and aggradational reflections extend northward as continuous, smooth and sub-parallel reflections decreasing in thickness and onlapping against the basement (Figs 3 & 4). Along the synclinal axis, reflections can be traced northward for over 40 km from the oblique clinoform and then become masked by multiples. Incised valleys are not clear in the landward extension.

Sequence Bn2b is characterized by a complex oblique clinoform that prograded seaward of the previous, underlying shelf edges of sequence Bn2a. Maximum progradation of about 5 km occurred on the western limb of the syncline, extending southwestward for over 30 km along strike while decreasing in thickness. Eastern extension of the clinoform is limited by a N–S-trending anticline along 132°07′ E (Fig. 5(c)). In the axial zone of the syncline, sequence Bn2b is composed of thin, sub-parallel reflections (Fig. 3).

The clinoform pattern of sequences Bn2a and Bn2b is clear on the NNE–SSW-directed profiles (Fig. 3), while WNW–ESE-directed profiles show sub-parallel reflections (Fig. 6). In addition, the shelf edge of sequence Bn2a trends E–W (Fig. 5(c)), indicating that the clinoforms prograded southward, although the upper unconformity of the sequences dips to the southeast at present (Fig. 5(c)). The clinoforms show the active growth of the syncline after the formation of the unconformities.

Bn3 Formation

The Bn3 Formation is composed of sequences Bn3a and Bn3b in ascending order (Fig. 4). Sequence Bn3a is wedge-shaped and composed mainly of sub-parallel or aggradational reflections except for an upper, thin, low-angle oblique clinoform (Fig. 4). Reflections are generally continuous, clear and smooth (Fig. 4). The sequence attains a maximum thickness of about 0.17 s in the synclinal axis, while the thickness decreases and the oblique clinoform is obscured westward. Sequence Bn3b is composed of a lower, thin aggradational wedge and an upper oblique clinoform (Fig. 4). The aggradational wedge consists of smooth, continuous and gently inclined reflections. It extends 55 km westward from the axis of the syncline. The oblique clinoform is characterized by steeply inclined, weak reflections (Fig. 4). The maximum inclination of the internal reflections is 4–5°. In the axial zone of the basin, the clinoform is up to 0.27 s thick and prograded ~15 km southward. The width of the oblique clinoform wedge along strike is ~35 km. The oblique clinoform is smaller than the aggradational wedge along the strike but wider across the strike, and the reflection pattern abruptly changes from the aggradational wedge to the oblique clinoform. These clinoforms probably indicate an abrupt change of sediment supply and/or sea-level fall.

Bn4 Formation

The Bn4 Formation is divided into the lower Bn4a and upper Bn4b sequences (Fig. 4). Both sequences are composed of an oblique clinoform in their seaward section and a basal, landward part which is unclear due to masking by multiples. Each sequence has a maximum thickness of ~0.2 s on the synclinal axis and extends westward for over 70 km, while no prograding wedge is observed to the east of a basement high just east of the synclinal axis (Fig. 5(a)). The detailed internal structure of sequence Bn4a is not clear, because the overlying sequence Bn4b prograded over the clinoform front of sequence Bn4a and much of the sequence is masked by multiple reflections, especially in the axial zone (Fig. 4). Sequence Bn4b forms the present shelf at 100–150 m water depth. In the axial zone of the syncline, the sequence shows an oblique clinoform pattern ~14 km across the strike. Reflections in the clinoform are highly discontinuous and rough, obscuring the boundary between Bn4a and Bn4b (Fig. 4). The reflections become more continuous and smooth towards the west. Sequence Bn4b extends northward for over 35 km from the shelf edge into the Bungo Strait. This landward extension, about 0.1 s thick, is composed of parallel, continuous and horizontal reflections in which V-shaped incisions are recognized. The incisions are presumably a seaward extension of fluvial systems from the Seto Inland Sea during the last ice age (Japan Association for Quaternary Research, 1987). The channels become unclear towards the shelf edge, where the prograding pattern becomes distinct.

Tosa Bay — Tosa Basin

Tosa Bay includes two tectonic depressions — the Tosa and Nakamura depressions (Fig. 2(b); Okamura *et al.*, 1987). The Tosa depression is larger and located in the east of the bay, and extends southward to the Tosa Basin (Fig. 2(b)). The Nakamura depression, elongated in a NW–SE direction, lies in the west of the bay. The two depressions are divided by a basement high known as the Okitsu Nose, a broad anticline sloping southwards (Fig. 2(b); Okamura *et al.*, 1987). Seven Quaternary progradational sequences are recognized beneath the shelf and slope across the depressions and the

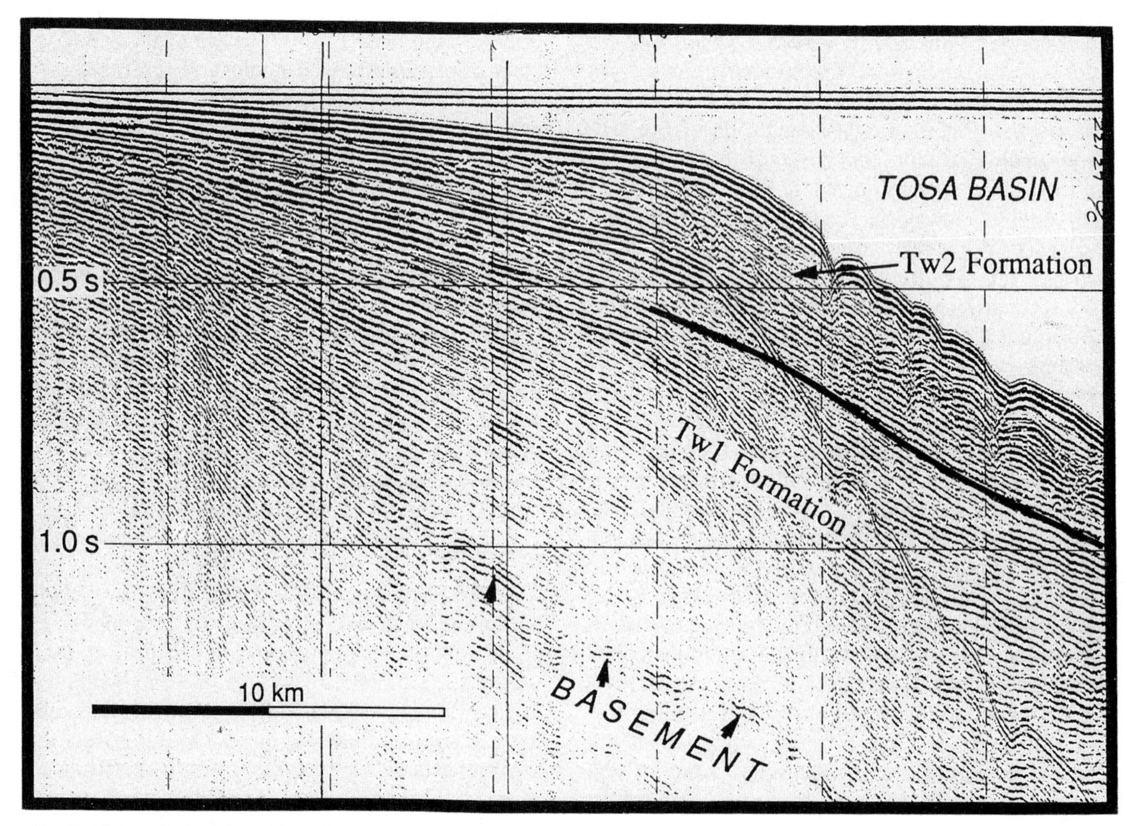

Fig. 7. Progradational stacking of prograding sequences in the Tosa depression. Although the sequences are thickest in the Tosa depression, older sequences were completely buried by younger sequences under the inner shelf. The boundary between the Tw1 and Tw2 Formations can be traced in slope-apron sediments, while its landward extension beneath the shelf is obscured by multiples.

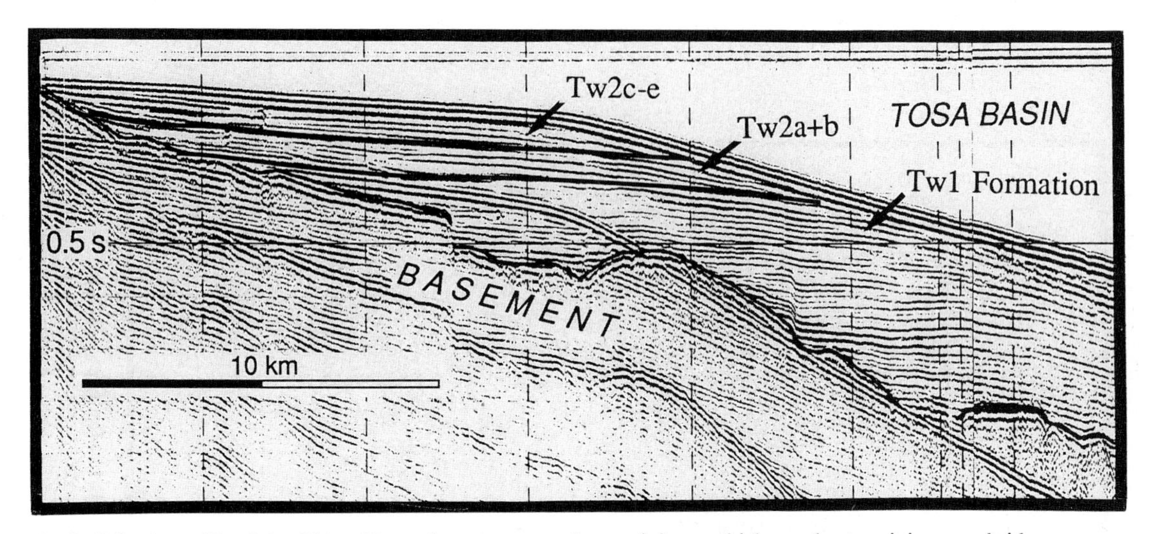

Fig. 8. Seismic profile of the Okitsu Nose where sequences have minimum thickness due to minimum subsidence rate in the Tosa Bay. Because the thickness of the sequences is below the resolution of profiles, only two major unconformities are recognized and others are completely obscured. The Tw1 Formation probably includes the Pliocene Ashizurioki Formation.

Okitsu Nose (Table 1), but their stacking pattern varies across the slope (Figs 7, 8 & 9) depending on subsidence rate and sediment supply (Okamura, 1989).

In the Tosa depression, the sequences are thickest in Tosa Bay because of the high rate of sediment supply and subsidence. Because any sequences prograded seaward of the previous shelf edges form a prograding stacking pattern, the older sequences are completely buried by younger ones and masked by multiples (Fig. 7). Therefore, the detailed configuration and internal structure of the sequences are unclear except that the youngest sequence is composed mainly of an oblique clinoform. At the shelf edge, the oblique clinoform continues onwards to basin fills in the Tosa Basin covering the entire slope as a slope-apron (Fig. 7; Blum & Okamura, 1992). In the eastern part of the Tosa depression, the subsidence rate decreases and the size of the sequences decreases eastward. At the eastern margin of the depression near Cape Muroto, the Pliocene to early Pleistocene Tosabae Formation is exposed on a shelf and no Quaternary progradational sequences are stacked (Okamura & Joshima, 1986), indicating the uplift of the shelf (Okamura, 1989). To the west of the Tosa depression, the thickness of the sequence and its width along the slope gradually decrease. Therefore, the stacking pattern changes from progradational to retrogradational westward (Figs 8 & 9) on the eastern limb of the Okitsu Nose. On the axis of the Okitsu Nose, most of the sequences are thinner than the resolution of the seismic profiles, and only two major unconformities can be recognized (Fig. 8). The Nakamura depression is also underlain by progradational sequences showing a progradational stacking pattern, and sequences beneath the shelf are not observed on the seismic profiles.

The entire stack of sequences can be observed as prograding wedges on a few profiles on the eastern limb of the Okitsu Nose due to retrogradational stacking. The following descriptions of the stacked sequences are therefore based on the profile along line 10, which has been selected as the best one along which to observe the stacked prograding wedges (Fig. 9).

Tw1 and Tw2 Formations

The sedimentary succession under the shelf and shelf slope was divided into the lower Tw1 and upper Tw2 Formations by clear toplap or truncational unconformities (Fig. 9; Okamura *et al.*, 1987). The Tw1 Formation is underlain by a

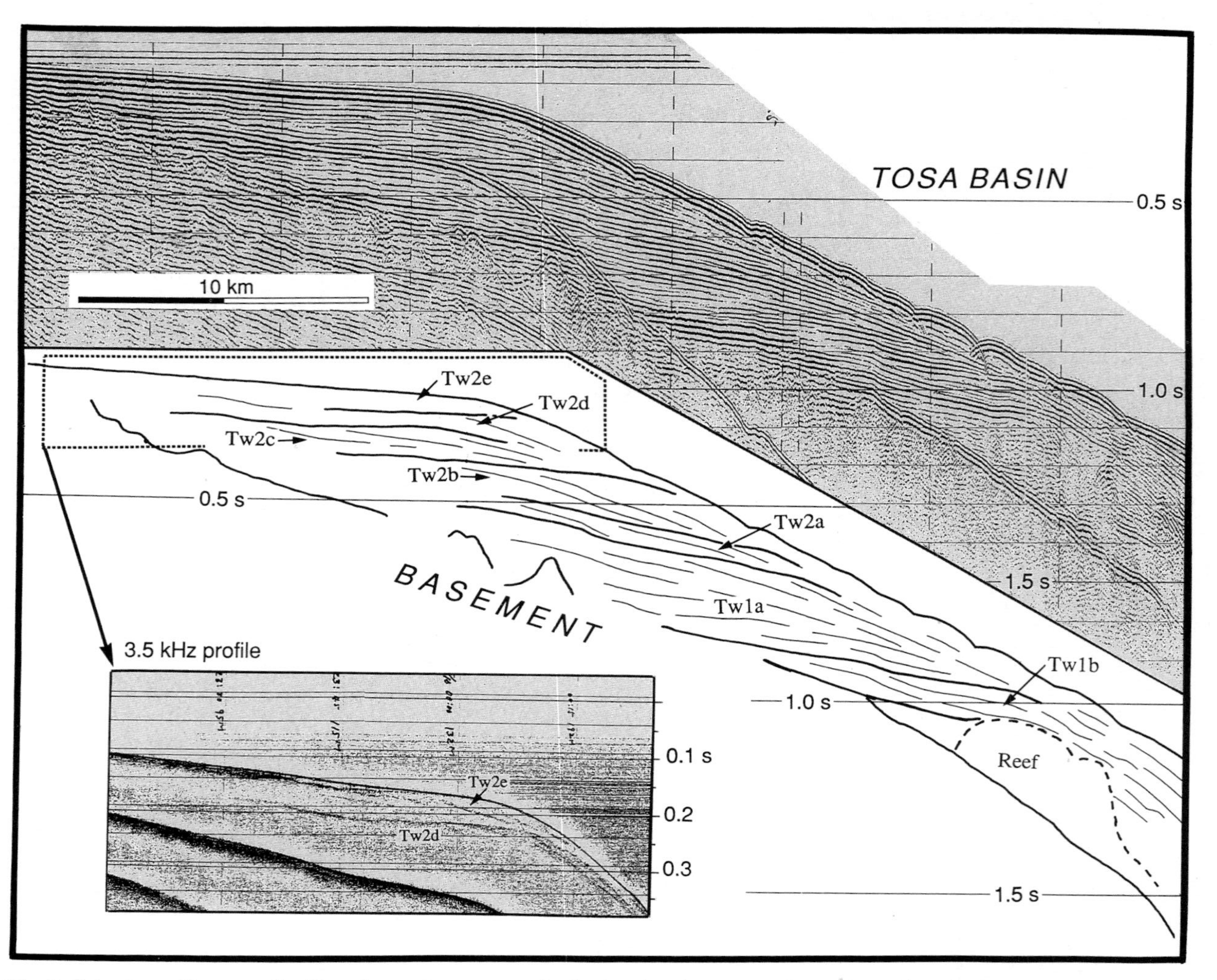

Fig. 9. Seismic profile across the Tosa depression, western flank, showing the retrogradational stacking of prograding sequences. This stacking pattern, which is favourable for differentiating individual sequences, is the result of an insufficient supply of sediment to fill the space created by subsidence.

reflection-free mound interpreted as a reef complex by Okamura *et al.* (1987) (Fig. 9). The presumed reef and basal part of the Tw1 Formation is estimated to be Pliocene in age based on tracing reflections from the Pliocene to early Pleistocene Tosabae Formation exposed on the shelf around Cape Muroto. The Tw1 Formation is composed of a ~0.05 s thick lower sequence Tw1a and a ~0.2 s-thick upper sequence Tw1b (Table 1, Fig. 9). Sequence Tw1b is the thickest one along line 10. The reflection pattern in sequence Tw1b changes from retrogradational at the base to aggradational and oblique progradational upward and seaward (Fig. 9). The retrogradational reflections onlap against the basement landward.

In the Tw2 Formation, five prograding sequences were identified: Tw2a, Tw2b, Tw2c, Tw2d and Tw2e sequences in ascending order (Table 1, Fig. 9). No sequence exceeds 0.1 s and the total thickness of the formation is ~0.3 s. An oblique clinoform pattern can be observed in the seaward part of the sequences but the detailed structure of their landward part is unknown because their thickness is close to the resolution of the seismic profiles.

Muroto Trough

The Muroto Trough is an E–W elongated basin about 150 km long and 50 km wide, maximum (Fig. 1). The shelf slope descends steeply to the Muroto Trough and is associated with slump scarps in the western-most basin (Okamura & Joshima, 1986). The Kii Strait, another gateway from the Pacific Ocean to the Seto Inland Sea is located north of the central part of the shelf. The strait is 50 to 90 m deep, and probably emerged and supplied large quantities of sediment to the Muroto Trough during the ice ages. Three major canyons are incised into the shelf slope south of the Kii Strait. The area surveyed is limited to the western half of the Muroto Trough (Fig. 2(c)).

The sedimentary succession under the shelf and shelf slope was divided into the Pliocene to early Pleistocene Tosabae Formation and the Quaternary Toki Group (Okamura & Joshima, 1986). The Toki Group was divided into four formations by unconformities: the Tk1, Tk2, Tk3 and Tk4 Formations in ascending order (Fig. 10). The Tk1 and Tk4 Formations are single progradational sequences. The Tk2 and Tk3 Formations consist of two sequences, Tk2a and Tk2b and Tk3a and Tk3b, respectively (Table 1, Fig. 10). These sequences show a similar retrogradational stacking pattern across the shelf gradually decreasing in size westward.

Tosabae Formation

A thin wedge-shaped sequence (less than 0.1 s thick) occurs in the upper part of the Tosabae Formation (Fig. 10). The clinoform pattern of the sequence is unclear because its thickness is very close to the resolution of seismic profiles. Above the wedge-shaped sequences, the Tosabae Formation consists of sub-parallel reflections gradually decreasing their thickness landward.

Tk1 Formation

Sequence Tk1 at the base of the Toki Group is the thickest sequence (~0.4 s) in the Muroto Trough. The reflection pattern of sequence Tk1 changes from an aggradational to an oblique progradational one upward (Fig. 10). The upper boundary is defined by clear toplap or truncational unconformity. The lower aggradational reflections continue landward for over 20 km as sub-parallel reflections decreasing in thickness and onlapping against the basement.

Tk2 Formation

Sequence Tk2a is composed mainly of aggradational reflections and includes a thin, low-angle oblique reflection pattern in the eastern part of the area surveyed (Fig. 10). Its upper boundary is defined by a low-angle toplap unconformity that becomes unclear to the west as the thickness of the sequence decreases. Sequence Tk2b is an oblique clinoform (Fig. 10). Both sequences continue landward for about 20 km decreasing in thickness and onlapping against the basement. The maximum thicknesses of sequences Tk2a and Tk2b are 0.15 and 0.08, respectively.

Tk3 and Tk4 Formations

Sequence Tk3a is a low-angle oblique clinoform (Fig. 10) in the east of the surveyed area where the sequence attains the maximum thickness of 0.06 s, but the sequence boundary becomes obscure further west due to decreasing thickness. Sequence Tk3b is an oblique clinoform (Fig. 10). Its maximum thickness is ~0.15 s. The Tk4 Formation consists of an

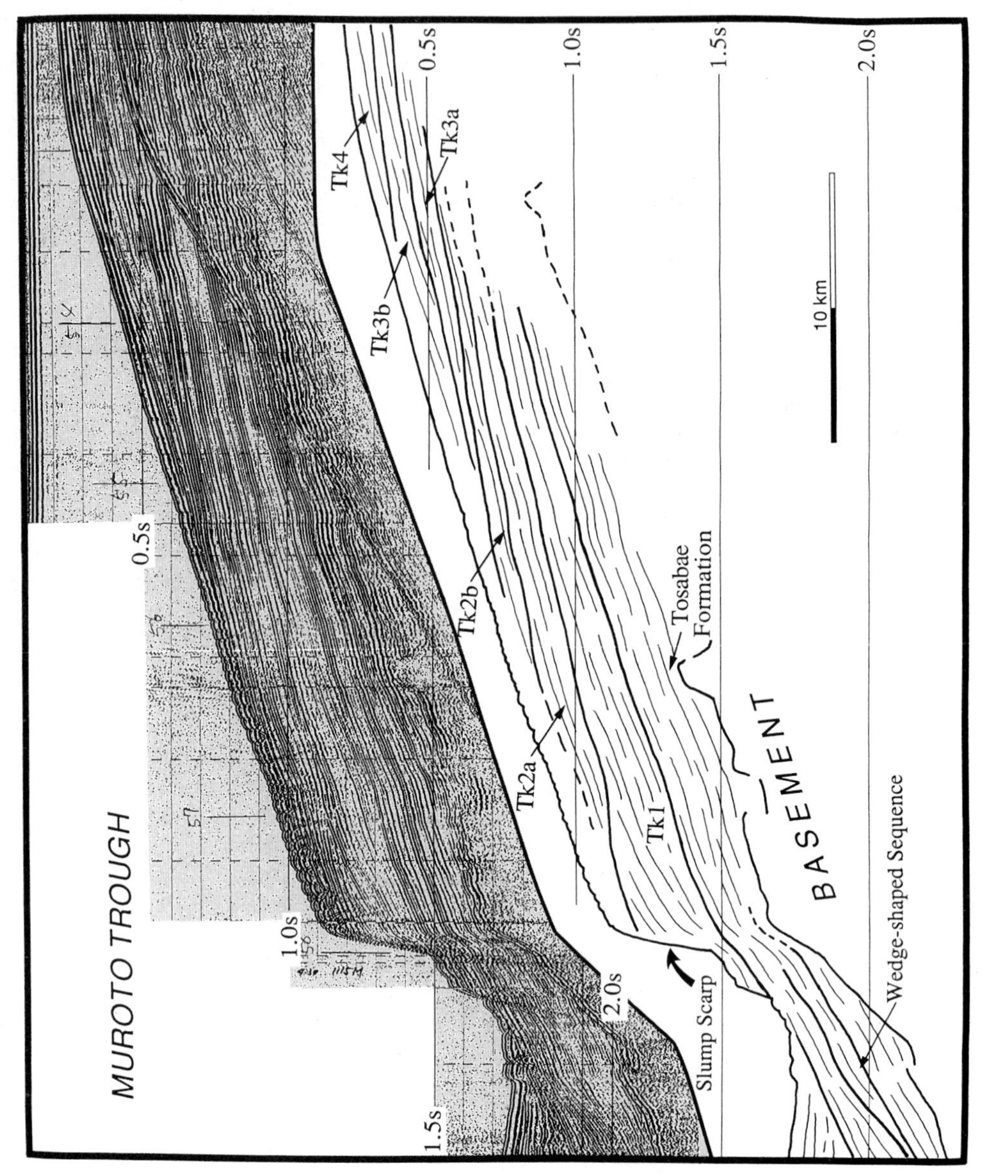

Fig. 10. Seismic profile showing the stacked progradational sequences in the Muroto Trough. The lowermost sequence in the Tosabae Formation defined by external shape, but its prograding pattern is unclear. Although the seaward part of an oblique clinoform in sequence Tk1 was lost by mass wasting on this profile, the oblique clinoform is widely recognized in other profiles.

oblique clinoform beneath the outer shelf, with a maximum thickness of ~0.09 s.

DISCUSSION

Stacking pattern of the sequences

A retrogradational stacking pattern is predominant in the Hyuga Basin (Figs 3 & 4) and the Muroto Trough (Fig. 10), while a progradational stacking pattern is present in the depressions in the Tosa Bay (Fig. 7). The stacking pattern variations observed in seismic profiles are a function of the ratio of rate of accommodation to rate of deposition (Van Wagoner *et al.*, 1988). In the area surveyed, accommodation and sediment supply are largely controlled by the style of deformation of the basins.

The Hyuga Basin is a wide and rapidly subsiding syncline, providing plenty of accommodation. The retrogradational stacking pattern indicates that the increase of accommodation exceeds the sediment supply. Sediment supply is inferred to be large, because the progradational sequences are apparently thickest and most extensive when compared with the other two basins. The large amount of sediment supply is probably caused by extensive Quaternary volcanism on the Kyushu island. The retrogradational stacking pattern must result from the rapid subsidence. The sequences in the Hyuga Basin have their maximum thickness in the region of the synclinal axis and extend mainly westward (Figs 5(a) & (b)), suggesting that westward currents along the coast played an important role during deposition of the progradational sequences. Lateral dispersal of sediments decreased seaward growth of the sequence, which might have contributed to the formation of retrogradational stacking.

The sequences in the Tosa and Nakamura depressions show a progradational stacking pattern (Fig. 7). The growth rate of the accommodation in the depressions is presumed to be lower than that of the Hyuga Basin because the depressions are smaller and shallower. In addition, the sediment supply by the fluvial systems has been restricted to the depressions, thus promoting a progradational stacking pattern. Towards the Okitsu Nose, the decreasing sediment supply has resulted in a retrogradational stacking pattern and suggests that longshore currents played a minor role in redistributing sediments away from the synclinal axis.

In the Muroto Trough, a retrogradational stacking pattern is predominant. Three large canyons are located to the south of the Kii Strait and presumably have been the major sediment source of the Muroto Trough. Most sediments probably bypassed the shelf and shelf slope through these canyons.

Rapid subsidence in active margins is an important factor for promoting retrogradational stacking of sequences. Retrogradational stacking provides a good opportunity to study Quaternary progradational sequences at high resolution. In passive margins, lower Pleistocene sequences are generally deeply buried and masked by multiples (Suter *et al.*, 1987; Farrain & Maldonado, 1990; Tesson *et al.*, 1990). Quaternary sequences in active margins are therefore important in the study of Quaternary sea-level changes and sequence stratigraphy of higher order cycles.

Type and origin of sequences

Quaternary progradational sequences in the Hyuga, Tosa and Muroto Basins can be divided into major and minor sequences (Fig. 11). Major sequences are characterized by oblique clinoforms and are bounded by a toplap or truncational unconformity at the top. Minor sequences consist mainly of aggradational or sigmoid clinoform patterns with less-extensive oblique clinoforms and unconformities.

An oblique or sigmoid style of clinoform is mainly controlled by the position of the equilibrium

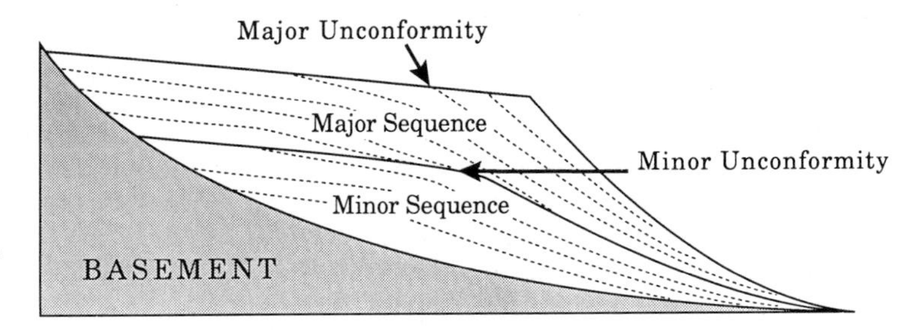

Fig. 11. Schematic drawing of major and minor sequences. Major sequences are characterized by a pronounced oblique clinoform and a toplap or truncational unconformity. The minor sequences have smaller scale, low-angle oblique clinoforms on aggradational units and are generally buried by major sequences.

point, i.e. the point at which the subsidence rate and rate of sea-level change are equal (Raynolds *et al.*, 1991). If the equilibrium point (Posamentier *et al.*, 1988) is located seaward of the depositional shelf edge (i.e. rate of sea-level fall > subsidence rate on the shelf), no accommodation is created on the shelf, and oblique clinoforms are promoted (Raynolds *et al.*, 1991). Sigmoid clinoforms associated with encroachment of onlap are formed if the equilibrium point is located landward of the depositional shelf edge (Raynolds *et al.*, 1991). Therefore, development of pronounced oblique clinoforms of the major sequences in the SW Japan forearc probably shows that eustatic sea level falls faster than the subsidence rate of the basins continued for a relatively longer period in the cycles, and the unconformities at the top of the sequences can be regarded as type 1 unconformities (Posamentier *et al.*, 1988; Raynolds *et al.*, 1991). In contrast, formation of minor sequences suggests that the rate of fall in eustatic sea level was comparable to the subsidence rate of the basins and the unconformities at the top of the sequences would be compared to type 2 unconformities (Posamentier *et al.*, 1988; Raynolds *et al.*, 1991). No encroachment of onlap on the oblique clinoforms suggests either no or a very short period of eustatic sea-level lowstand compared with the duration of preceding falls in sea level, because eustatic sea-level lowstands presumably result in the growth of sigmoid clinoforms associated with encroachments of onlap due to relative rise of sea level in the rapidly subsiding basins (Raynolds *et al.*, 1991). The oblique clinoforms may be called regressive systems tracts rather than lowstand systems tracts.

It is not clear whether landward onlap of the aggradational units is coastal onlap (Vail *et al.*, 1977) or not. If it is coastal onlap, the formation of aggradational units can be attributed to be the result of relative rise of sea level. Another possibility is that the onlap is apparent onlap formed below sea level due to sideways growth of progradational sequences from the synclinal axes. In this case, the aggradational units can be regarded as seaward extensions of progradational wedges, and the units do not necessarily indicate relative sea-level rise.

Most of the sequences apparently have either no or a very thin transgressive systems tract represented by a retrogradational pattern (Figs 3, 4, 9 & 10). In Quaternary sequences reported from other areas, a Holocene transgressive systems tract is generally recognized in the inner shelf but is unclear on the outer shelf (Suter *et al.*, 1987; Boyd *et al.*, 1989; Farrain & Maldonado, 1990; Tesson *et al.*, 1990; Lykousis, 1991). Present outer shelves are generally covered by a Holocene thin, transgressive sand sheet (Suter *et al.*, 1987; Ikehara, 1988; Saito, 1991; Blum & Okamura, 1992) that is too thin to be recognized on seismic profiles. The absence of a transgressive systems tract on the outer shelves can be explained by the rapid rises in sea level in the Quaternary era (Broecker & van Donk, 1970). The apparent lack of a transgressive systems tract suggests that sequence boundaries in the outer shelf include transgressive surfaces and maximum flooding surfaces. The thickest sequences in the Hyuga and Tosa Basins (Bn2a and Tw1b) include a retrogradational unit at the base, suggesting a slower and/or more extreme rise of sea level during the early stage of the sequence formation.

Correlation of the sequences in the three basins and oxygen isotope change curve

Tectonic movement of Japan during the Quaternary is considered to have been constant or gradually accelerated under the E–W or ESE–WNW compressional stress field (Sugimura, 1967; Huzita *et al.*, 1973). Late Pleistocene emerged terraces along the coast near Cape Muroto and Cape Ashizuri were interpreted to have formed by combination of a constant or slightly accelerated rate of uplift and eustatic change (Yoshikawa *et al.*, 1964; Ota, 1975). The uplifting pattern of the late Pleistocene terraces is similar to the growth pattern of the anticlines dividing the forearc basins, estimated to have been constant throughout the Quaternary (Okamura, 1990). Therefore, higher order cycles of Quaternary stacked progradational sequences must be caused by eustatic sea-level change approximately represented by oxygen-isotope curves. Okamura (1989) attempted to correlate the oblique clinoforms across the three basins and glacial stages of the Quaternary oxygen-isotope curve. A revised correlation is presented here (Table 1).

Each of the stacks in the three basins begins with a thin progradational sequence at the base (Table 1, Figs. 3, 9 & 10), followed by the thickest sequence. The thickest sequences in the Hyuga (Bn2a) and Tosa (Tw1b) Basins show the common change of reflection pattern from retrogradational at the base to aggradational and oblique clinoform upward (Figs 3 & 9). These sequences are more than twice as thick as the other sequences (Table 1). The toplap or truncational unconformity at the top of the thickest sequences (Bn2a, Tw1b and Tk1) is one of the most pronounced unconformities in each basin (Figs 3, 9 & 10). These similarities between the thickest sequences in each basin suggest that they are correlative. Above the thickest sequences five sequences are stacked (Table 1). They generally lack retrogradational units and include two or three major unconformities. The most pronounced unconformities are always located near the middle of the stacks, at the top of sequences Bn3b, Tw2b and Tk2b, respectively (Figs 4, 9 & 10). Based on these common successions of sequences and unconformities across the three basins, sequences are correlated with each other as shown in Table 1, although a few minor unconformities cannot be correlated.

The amplitude and frequency of the oxygen-isotope curve (Shackleton & Opdyke, 1973; Feeley *et al.*, 1990) abruptly change at 0.8 Ma (stage 22). Before stage 22, the curve is characterized by smaller and high-frequency fluctuations except for a few larger peaks in the late Pliocene (Feeley *et al.*, 1990). Larger and lower frequency changes predominate after stage 22 (Williams *et al.*, 1988; Feeley *et al.*, 1990). The pre-stage 22 fluctuations may have caused thin sequences below the resolution of the seismic profiles, appearing as a conformable succession of reflections (Okamura, 1989). This succession could be represented by the thickest sequence in the stacks (Bn2a, Tw1b, Tk1). The pronounced clinoforms and unconformities at the top of the thickest sequences are correlated to the extreme falls in sea level at stage 22 (Table 1, Fig. 12). Haq *et al.* (1987) showed a type 1 unconformity at 0.8 Ma probably corresponding to stage 22. Feeley *et al.* (1990) demonstrated a dramatic increase of turbidites in the Mississippi fan since 0.8 Ma. If this correlation is correct, the small sequences at the base of the stacks can be correlated to one of the falls in sea level in the late Pliocene (Fig. 12).

The stacks of the progradational sequences above the thickest sequences can be interpreted by larger amplitude cycles of about 100 ky period after stage 22. Shackleton (1987) estimated that sea-level falls at isotopic stages 6, 12 and 16 were larger than those of the other glacial ages after stage 19. Piper and Perissoratis (1991) inferred that major delta sequences were formed at isotopic stages 6, 12, 16 and 22 in the North Aegean continental margin. These stages are believed to correlate to the major unconformities in the stack above the thickest sequences as shown in Table 1 and Fig. 12.

Recognition of unconformities on seismic profiles

The number of unconformities recognized on seismic profiles decreases with decreasing sequence thickness and/or decreasing resolution of seismic-profiling systems. It is generally assumed that sea-level curves based on third-order and lower order sequences, such as those of Haq *et al.* (1987), represent the major sea-level changes, and that higher order sequences would consist of less extreme and higher order frequency fluctuations (Mitchum & Van Wagoner, 1991). However, Quaternary high-resolution sequences caused by high-frequency/high-amplitude sea-level changes illustrate that the curve from low-resolution sequence stratigraphy may not be an accurate representation of significant sea-level changes. Haq *et al.* (1987) showed two unconformities in the Quaternary, at the age of 0.8 Ma and 1.6 Ma, respectively. The unconformity at 0.8 Ma presumably corresponds to stage 22, but no clear fall is recognized on the oxygen-isotope curve at 1.6 Ma. In this study, one of the major unconformities correlates with stage 22, but additional major unconformities not shown in the curve of Haq *et al.* (1987) are also recognized. The dimensions of the oblique clinoforms topped by these additional unconformities are comparable with one another, suggesting that the sea-level fall at stage 22 was not significantly larger than those at stages 6, 12 and 16 (Fig. 12). The thicknesses of the sequences Bn2a, Tw1b and Tk1 beneath the presumed stage 22 unconformity are significantly larger than the other major sequences and this may be the reason why the unconformity correlative to stage 22 is shown in Haq *et al.*'s oxygen-isotope curve. As discussed above, the thick sequences are inferred to be composed of thin sequences formed by lower amplitude and higher frequency cycles. Therefore, stage 22 does not represent the largest sea-level fall during the Quaternary, but the first high-amplitude fall in the middle and late Pleistocene. Subsequent fluctuations of similarly high amplitude are not detected on lower resolution profiles because of their high frequency, which makes these sequences appear as one composite. Major sea-level falls of third-order cycles therefore may partly represent changes in frequency and amplitude of fourth-order or higher order cycles, such as the Milankovitch cycles, rather than individual major events.

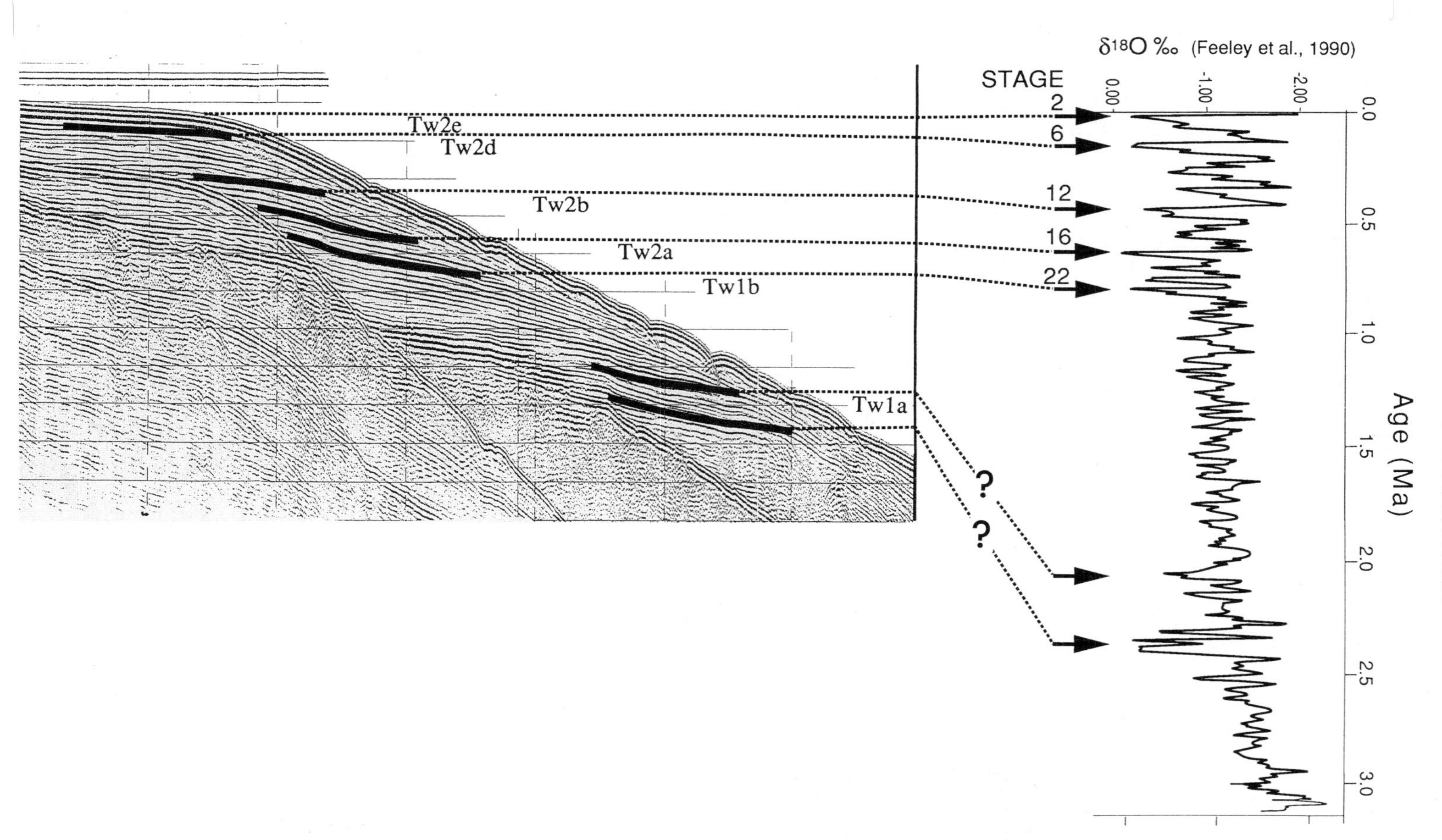

Fig. 12. Correlation of oxygen-isotope curve with the sequences in the Tosa Bay shown in Fig. 9. The major unconformities are correlated to stages 6, 12, 16 and 22, respectively. Correlations between the minor sequences including the lowest sequence and isotopic peaks are not clear. Sub-parallel correlation lines between the major unconformities and isotopic peaks above stage 22 imply nearly constant subsidence rate after stage 22 and comparable sea-level falls at stages 6, 12, 16 and 22, respectively.

CONCLUSIONS

1 High-frequency/high-amplitude changes during the Quaternary and rapid subsidence resulted in stacked, fourth-order or higher order sequences in three forearc basins of southwest Japan. Seven progradational sequences are recognized in single-channel seismic profiles due to retrogradational stacking.

2 The stacking pattern of the sequences varies from aggradational to progradational across the basins. The rapid increase of accommodation due to wide and rapid subsidence of the basins causes a retrogradational stacking pattern. Progradational stacking is prompted in more confined basins, where sediment supply from fluvial systems is generally directed towards the basin axes. Longshore currents and submarine canyons transport sediments laterally and basinward, respectively, and may contribute to the formation of retrogradational stacking.

3 The internal structures of the sequences represent the style of sea-level change curves. Each sequence consists of lower aggradational and upper progradational reflection patterns and generally lacks a retrogradational pattern. The sequence boundaries formed by a sea-level fall apparently include a transgressive surface and maximum flooding surface. The lack of the retrogradational unit is interpreted by a rapid rise of sea level in the Quaternary sea-level fluctuation (Broecker and van Donk, 1970). Oblique clinoforms of major sequences suggest faster and longer sea-level falls. Lack of encroachments of onlap on the top of the clinoforms implies no or very short duration of sea-level lowstands.

4 The sequence stacks begin with a late Pliocene thinner sequence at the base, followed by the thickest sequence interpreted to be composed of very thin sequences below resolution formed by high-frequency/low-amplitude sea-level fluctuation before oxygen-isotope stage 22. Above the thickest sequences, five sequences are identified. Four of them include clear oblique clinoforms presumably formed by larger falls of sea level and correlated with oxygen-isotope stages 2, 6, 12 and 16, respectively.

5 The number of unconformities recognized on seismic profiles generally decreases with decreasing sequence thickness (period of sea-level fluctuation) and/or decreasing resolution of the profiling system. Sequences resulting from high-frequency fluctuations may not be resolved although the fluctuations may have similar high amplitude as other lower frequency fluctuations which produced thicker sequences. Major sea-level falls of third order, therefore, may partly represent changes in frequency and amplitude of fourth-order or higher cycles, such as Milankovitch cycles, rather than individual major events.

ACKNOWLEDGEMENTS

All data were obtained during the research cruise for the national programme 'Marine Geological Study of Continental Shelves around Southwest Japan' supported by the Geological Survey of Japan. We thank Captain Okumura and the crew of the research vessel *Hakureimaru* for their help during the cruises of GH82-1, 83-1 and 83-2. We would also like to thank Dr M. Arita and colleagues in the Marine Geology Department for their help and encouragement. We thank M. Tesson and J.R. Suter for their constructive criticism of an earlier version of the manuscript. P. Jervis kindly reviewed the manuscript. The nannofossils and foraminifera were identified by Dr S. Nishida of Nara Educational University and Dr S. Maiya of the Japan Petroleum Exploration Co, Ltd, respectively.

REFERENCES

Blum, P. & Okamura, Y. (1992) Pre-Holocene sediment dispersal systems and effects of structural controls and Holocene sea-level rise from acoustic facies analysis: SW Japan forearc. *Mar. Geol.* **108**, 297–324.

Boyd, R., Suter, J. & Penland, S. (1989) Relation of sequence stratigraphy to modern sedimentary environments. *Geology* **17**, 926–929.

Broecker, W.S. & Van Donk, J. (1970) Insolation changes, ice volumes and the ^{18}O record in deep-sea cores. *Rev. Geophys. Space Phys.* **8**, 169–198.

Farran, M. & Maldonado, A. (1990) The Ebro continental shelf: Quaternary seismic stratigraphy and growth patterns. *Mar. Geol.* **95**, 289–312.

Feeley, M.H., Moore, T.C. Jr., Loutit, T.S. & Bryant, W.R. (1990) Sequence stratigraphy of Mississippi Fan related to oxygen isotope sea level index. *Am. Assoc. Petrol. Geol.* **74**, 407–424.

Haq, B.U., Hardenbol, J. & Vail, P.R. (1987) Chronology of fluctuating sea levels since the Triassic. *Science* **235**, 1156–1167.

Huzita, K., Kishimoto, Y. & Shiono, K. (1973) Neotectonics and seismicity in the Kinki area, southwest Japan *J. Geosci., Osaka City Univ.* **16**, 93–119.

Ikehara, K. (1988) Sedimentological map of Tosa Wan, (Scale 1:200,000), *Mar. Geol. Map Ser.* **34**, Geological

Survey of Japan, 29 pp. (in Japanese with English abstract).

Japan Association for Quaternary Research (1987) *Quaternary Maps of Japan*, University of Tokyo Press, 119 pp. (in Japanese).

Lykousis, V. (1991) Sea-level changes and sedimentary evolution during the Quaternary in the northwest Aegean continental margin, Greece. In: *Sedimentation, Tectonics and Eustacy: Sea-level Changes at Active Margins* (Ed. Macdonald, D.) Spec. Publ. Int. Assoc. Sediment. 12, 123–131.

Mitchum, R.M. Jr. & Van Wagoner, J.C. (1991) High-frequency sequences and their stacking patterns: sequence-stratigraphic evidence of high-frequency eustatic cycles. *Sediment. Geol.* **70**, 131–160.

Okamura, Y. (1989) Multi-layered progradational sequences in the shelf and shelf-slope of the southwest Japan forearc. In: *Sedimentary Facies in the Active Plate Margin* (Eds Taira, A. & Masuda, F.) Terra Publ., Tokyo, pp. 295–317.

Okamura, Y. (1990) Geologic structure of the upper continental slope off Shikoku and Quaternary tectonic movement of the outer zone of southwest Japan. *J. Geol. Soc. Japan* **96**, 223–237 (in Japanese with English abstract).

Okamura, Y. & Joshima, M. (1986) Geological map of Muroto Zaki (Scale 1:200,000), *Mar. Geol. Map Ser.* **28**, Geological Survey of Japan, 31 pp. (in Japanese with English abstract).

Okamura, Y., Kishimoto, K., Murakami, F. & Joshima, M. (1987) Geological map of Tosa Wan (Scale 1:200,000), *Mar. Geol. Map Ser.* **29**, Geological Survey of Japan, 32 pp. (in Japanese with English abstract).

Ota, Y. (1975) Late Quaternary vertical movement in Japan, estimated from deformed shorelines. In: *Quaternary Studies* (Eds Suggate, R.P. & Cresswell, M.M.) Bull. Roy. Soc. New Zealand, 13, 231–239.

Piper, D.J. & Perissoratis, C. (1991) Late Quaternary sedimentation on the north Aegean continental margin, Greece. *Am. Assoc. Petrol. Geol.* **75**, 46–61.

Posamentier, H.W., Jervey, M.T. & Vail, P.R. (1988) Eustatic controls on clastic deposition I — conceptual framework. In: *Sea-level Changes: An Integrated Approach* (Eds Wilgus, C.K., Hastings, B.S., Kendall, C.G.St.C., Posamentier, H.W., Ross, C.A. & Van Wagoner, J.C.) Soc. Econ. Paleontol. Mineral. Spec. Publ. 42, 109–124.

Raynolds, D.J., Steckler, M.S. & Coakley, B. (1991) The role of the sediment load in sequence stratigraphy: the influence of flexural isostasy and compaction. *J. Geophys. Res.* **96**, 6931–6949.

Saito, Y. (1991) Sequence stratigraphy on the shelf and upper slope in response to the latest Pleistocene–Holocene sea-level changes off Sendai, northeast Japan. In: *Sedimentation, Tectonics and Eustacy: Sea-level changes at Active Margins* (Ed. Macdonald, D.), Spec. Publ. Int. Assoc. Sediment. 12, 133–150.

Shackleton, N.J. (1987) Oxygen isotopes, ice volume and sea level. *Quat. Sci. Rev.* **6**, 183–190.

Shackleton, N.J. & Opdyke, N.D. (1973) Oxygen isotope and paleomagnetic stratigraphy of equatorial Pacific core V28-238: oxygen isotope temperatures and ice volumes on a 10^5 year and 10^6 year scale. *Quat. Res.* **3**, 39–55.

Sugimura, A. (1967) Uniform rates and duration period of Quaternary earth movements in Japan. *J. Geosci. Osaka City Univ.* **10**, 25–35.

Suter, J.R., Berryhill, H.L. Jr. & Penland, S (1987) Late Quaternary sea-level fluctuations and depositional sequences, southwest Louisiana continental shelf. In: *Sea-level Fluctuation and Coastal Evolution* (Eds Nummedal, D., Pilkey, O.H. & Howard, J.D.) Soc. Econ. Paleontol. Mineral. Spec. Publ. 41, 199–219.

Taira, A. (1985) Sedimentary evolution of Shikoku subduction zone: the Shimanto belt and Nankai Trough. In: *Formation of Active Plate Margin* (Eds Nasu, N., Kobayashi, K., Uyeda, S., Kushiro, I. & Kagami, H.), Terra Publ. Tokyo, pp. 835–851.

Tesson, M., Gensous, B., Allen, G.P. & Ravenne, C. (1990) Late Quaternary deltaic lowstand wedges on the Rhône continental shelf, France. *Mar. Geol.* **91**, 325–332.

Vail, P.R., Mitchum, R.M. & Thompson, S., III. (1977) Seismic stratigraphy and global changes of sea level, part 3: relative changes of sea level from coastal onlap. In: *Seismic Stratigraphy — Applications to Hydrocarbon Exploration* (Ed. Payton, C.E.) Mem. Am. Assoc. Petrol. Geol. 26, 63–81.

Van Wagoner, J.C., Posamentier, H.W., Mitchum, R.M. *et al.* (1988) An overview of the fundamentals of sequence stratigraphy and key definition. In: *Sea-level Changes: An Integrated Approach* (Eds Wilgus, C.K., Hastings, B.S., Kendall, C.G.St.C., Posamentier, H.W., Ross, C.A. & Van Wagoner, J.C.) Soc. Econ. Paleontol. Mineral. Spec. Publ. 42, 39–45.

Williams, D.F., Thunell, R.C., Tappa, E., Rio, D. & Raffi, I. (1988) Chronology of the Pleistocene oxygen isotope record: 0–1.88 m.y.b.p. *Palaeogeogr. Palaeoclimatol., Palaeoecol.* **64**, 221–240.

Yoshikawa, T., Kaizuka S. & Ota, Y. (1964) Mode of crustal movement in the late Quaternary on the southeast coast of Shikoku, Southwestern Japan. *Geogr. Rev. Japan* **37**, 627–648 (in Japanese with English abstract).

Pre-Quaternary Applications of Sequence-stratigraphic Concepts

EUROPE

Spec. Publs Int. Ass. Sediment. (1993) **18**, 237–246

The Devonian–Carboniferous boundary in southern Belgium: biostratigraphic identification criteria of sequence boundaries

M. VAN STEENWINKEL

Petroleum Development Oman, PO Box 81, Muscat, Sultanate of Oman.

ABSTRACT

A biostratigraphic hiatus in the latest Devonian of the Dinant synclinorium, southern Belgium, is characterized by the extinction of several species, followed by a period in which index fauna are absent or scarce, and by a later, rapid diversification of evolved Carboniferous species.

This biostratigraphic hiatus is shown to be associated with a basinward facies shift. A relative sea-level fall in the latest Devonian is held responsible for both the biostratigraphic hiatus and the facies shift, causing the late-Devonian highstand sedimentation with its great faunal diversity to cease. This resulted in a period of non-deposition, expressed in the rock record as a hiatal sequence boundary. Sedimentation eventually resumed, but in a much shallower, lowstand depositional setting. These lowstand sediments do not contain the deep-water index species necessary to define the Devonian–Carboniferous boundary biostratigraphically. Deeper water environments were re-established only later, during a transgressive rise in relative sea level. These deeper water deposits contain newly evolved and diversified Carboniferous index fossils suitable for biostratigraphic subdivision.

Comprehending such geologic events by integration of bio- and lithostratigraphic data in a sequence-stratigraphic framework allows the correlation of small-scale flooding events (parasequence-bounding surfaces) within well-dated systems tract boundaries, and hence, lateral correlations in hiatal intervals can be significantly refined.

INTRODUCTION

Worldwide temporal correlations near the Devonian–Carboniferous (D/C) boundary have been thoroughly studied through biostratigraphy over the last century. Biostratigraphic analyses of numerous sections have allowed the earliest appearance of fossil species to be identified with reasonable accuracy and confidence (Paproth & Sevastopulo, 1988). However, non-sequences and marked changes in facies have sometimes resulted in incomplete or non-conclusive biostratigraphic control around the time of the boundary between the Devonian and Carboniferous systems, hampering precise lateral correlations and resulting in ambiguous stratigraphic conclusions.

Global sea-level changes (Johnson *et al.*, 1985, 1986; Ross & Ross, 1987), tectonically enhanced, are held responsible for the biostratigraphic complexity near this boundary. In the Belgian Dinant synclinorium (Fig. 1), missing conodont zones and the sudden faunal change from a Devonian to a Carboniferous 'style' of fossils have led to a sequence-stratigraphic analysis of the D/C transition (Van Steenwinkel, 1988, 1990). This study was based on sedimentary facies analysis and combined with all the biostratigraphic control known from previous studies. The current paper presents the effect of relative sea-level changes near the D/C boundary on the sedimentation and the biostratigraphic signal in the Dinant synclinorium. It also shows how the cyclicity from a larger to smaller scale was used as a tool to refine lateral correlations in intervals beyond the available biostratigraphic resolution. The lithostratigraphic formations involved are the Etrœungt Limestone, the Hastière Limestone, the Pont d'Arcole Shale and the Landelies Limestone.

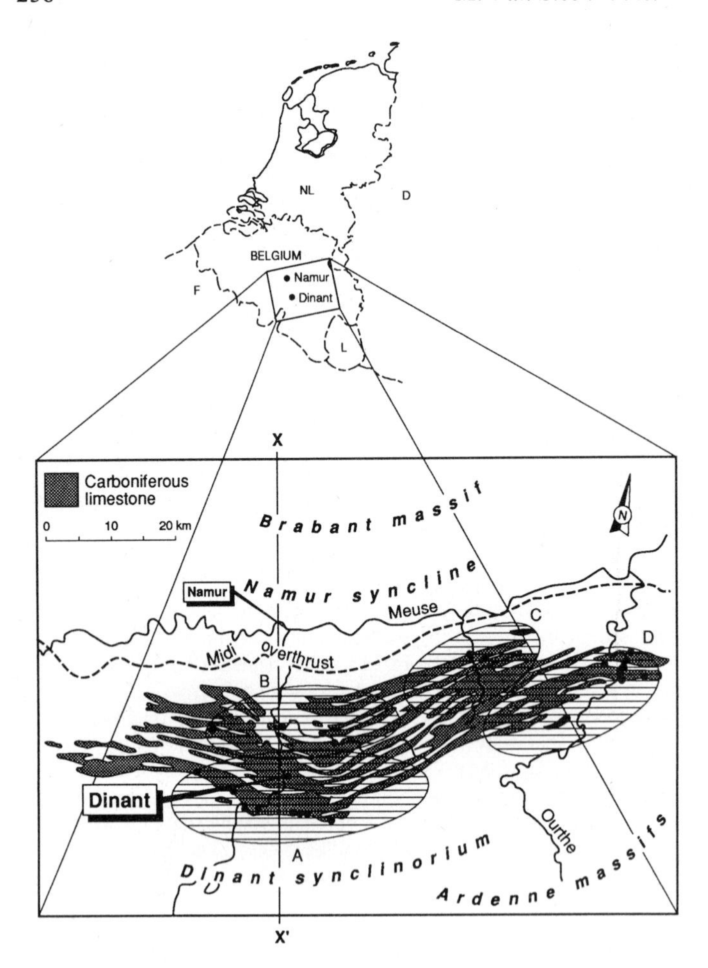

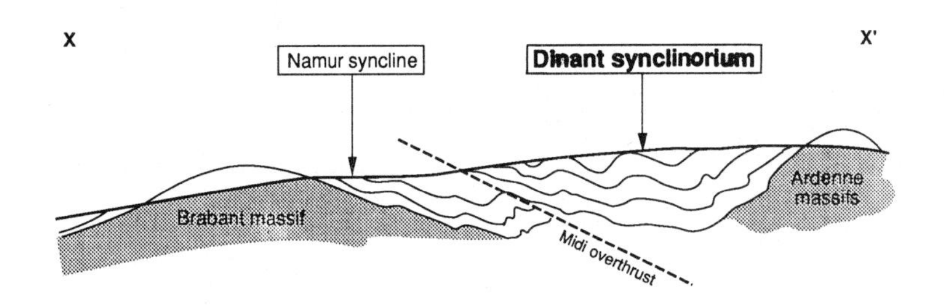

Fig. 1. Study area: the Dinant synclinorium in southern Belgium. Four depositional areas (A, B, C and D). Depositional dip is to the south. Dots indicate the studied outcrop sections. Bottom: simplified cross-section of the main Belgian geological units indicating the position of the allochthonous Dinant synclinorium, overthrust over the autochthonous Namur syncline.

PALAEOGEOGRAPHIC AND STRUCTURAL CONTEXT

The Dinant synclinorium is an allochthonous belt of Devonian and Carboniferous rocks in southern Belgium (Fig. 1) that was displaced northwards by the Midi overthrust during the Variscan orogeny. Palaeogeographically, this area is a ramp setting at the northern edge of the epeiric Cornwall–Rhenish Basin. During the late Devonian, sedimentation in the Dinant synclinorium was predominantly siliciclastic, with material supplied by erosion of the old massifs in the north. During early Carboniferous times the siliciclastic supply diminished and carbonates developed. From Tournaisian time onwards, repeated and gradually greater inundations of the northern Cornwall–Rhenish Basin margin took place. This led to the development of a wide carbonate platform during the Visean (the 'Kohlenkalk Platform'), which sloped into a euxinic, siliceous black shale basin towards the south and east (Paproth & Zimmerle, 1980). In the Dinant synclinorium the development of the platform was still in an early phase in the uppermost Devonian (Strunian) and lower Tournaisian and was characterized by a ramp-like (Ahr, 1973) configuration, sloping to the south and east.

Structurally, the Cornwall–Rhenish Basin was part of the Variscan geosynclinal system that was characterized by back-arc extension and rapid subsidence during the late Devonian and earliest Carboniferous (Leeder, 1987). This basin, which later developed into the Variscan Foreland Basin, was gradually incorporated in the northward-migrating Variscan deformation front during the late Carboniferous.

BIOSTRATIGRAPHIC BACKGROUND

The D/C boundary was defined (Lane *et al.*, 1980) 'at the first appearance of the conodont *Siphonodella sulcata* within the evolutionary lineage from *Siphonodella praesulcata* to *Siphonodella sulcata*'. The *praesulcata–sulcata* transition is considered to be the best marker for the D/C boundary on a worldwide scale. These conodonts, however, are absent in the shallow-water facies of D/C basin margins. In deeper water basinal facies (Paproth & Streel, 1971, 1984) the *praesulcata–sulcata* transition is often obliterated: either a condensed section occurs or the upper part of the *praesulcata* range is truncated by shallow water strata carrying a different conodont biofacies.

Walliser (1984) summarized these worldwide faunal changes around the D/C transition (Fig. 2). He believed that some kind of 'D/C' event was associated with the extinction of many fossil

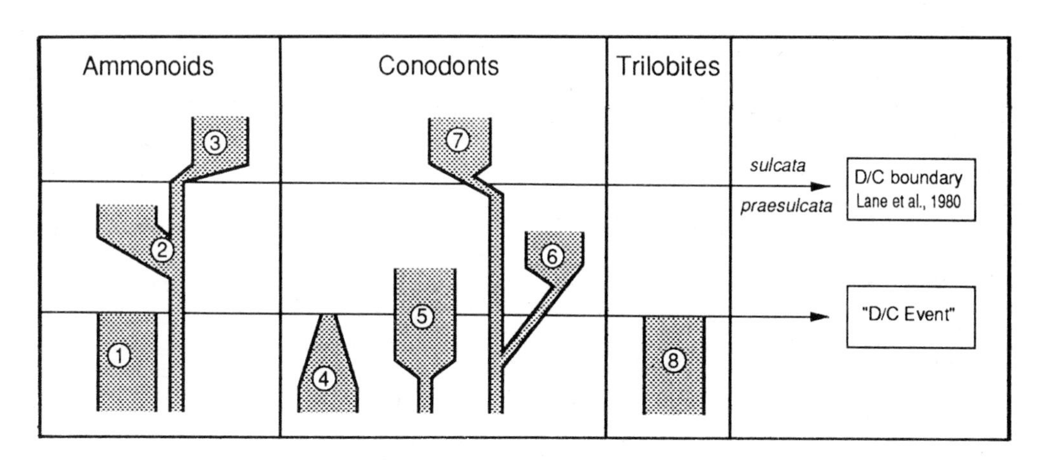

① *Wocklumeria sphaeroides* : Sudden extinction
② *Imitoceras* : Survival and subsequent radiation
③ *Gattendorfia subinvoluta* : First appearance
④ *Palmatolepis* : Extinction after gradual reduction
⑤ *Bispathodus* : Survival without disturbance
⑥ *Protognathodus* : Radiation soon after the crisis
⑦ *Siphonodella* : Radiation
⑧ *Phacopida* : Extinction

Fig. 2. Outline of faunal changes near the Devonian–Carboniferous (D/C) boundary showing 'the D/C event' associated with the extinction of many fossil groups and with the rapid diversification of the ones that survived the crisis (based on Walliser, 1984). Note that the position of the *praesulcata–sulcata* boundary, chosen as the D/C boundary (Lane *et al.*, 1980), is at a younger level than this event, but still prior to the zone of faunal re-establishment.

groups and the subsequent rapid diversification ('radiation') of the ones that survived the 'event'. Besides conodonts, ammonoids, trilobites, ostracods, echinoderms, foraminifera and corals were also strongly influenced by the D/C event. The position of the *praesulcata–sulcata* boundary, however, is at a younger level than this D/C event but still prior to the zone of faunal re-establishment (Fig. 2). This implies that the faunal crisis may be within the latest Devonian.

APPROACH

The approach used for lateral correlation in the D/C transition interval consists of five steps, discussed below. It was applied to 26 outcrop sections in different palaeogeographic positions (areas A, B, C and D; Fig. 1).

1 The first step is a biostratigraphic age determination at the smallest possible scale, and the subsequent deduction of the interval of problematic age.

2 Relative palaeodepths were deduced from sedimentary facies, inferred from field data, rock slabs, acetate peels and *c.* 800 thin sections in and around the interval of problematic age.

3 Systems tracts were delineated on the basis of aggradational, progradational and retrogradational facies trends and abrupt changes in the palaeowater depth.

4 Parasequences, represented by small-scale shallowing-upward units, were distinguished between systems tract bounding surfaces.

5 Finally, the parasequences were correlated between the different areas, using the maximum flooding surface and the transgressive surface as references.

RESULTS

1 The biostratigraphic age determination and the delineation of the interval of problematic age in the Dinant synclinorium were mainly based on conodonts (Bouckaert & Groessens, 1974; Sandberg *et al.*, 1978; Sandberg & Ziegler, 1979; Van Steenwinkel, 1984), foraminifera (Conil *et al.*, 1964; Conil, 1968), corals (Poty, 1984, 1986) and spores (Streel, 1971, 1983).

A reference section (Anseremme, near Dinant) with the conodont distribution near the D/C transition is shown in Fig. 3. Typical Devonian species (e.g. *Pelekysgnathus* sp., *Bispathodus costatus*, *Pseudopolygnathus graulichi* and *Protognathodus kockeli*) occur no later than during the deposition of bed 159, which is the basal bed of the Hastière Limestone in the Anseremme section. Carboniferous species (e.g. *Siphonodella duplicata*, *S. quadruplicata* and *S. cooperi*) are recorded from bed 174 upwards and show an upwardly increasing diversification. Part of the Devonian *praesulcata* zone, the Carboniferous *sulcata* zone and possibly part of the *duplicata* zone are missing.

Generally, the biostratigraphic complexity is characterized in the Dinant synclinorium by the extinction of many Devonian fossils at the top of the Etrœungt Limestone (middle-*praesulcata* zone), followed by an interval lacking index fauna (lower part of Hastière Limestone) in which lateral correlation is impossible. Some rare Devonian species, however, do extend into the basal bed of the Hastière Limestone. This is followed (in the upper part of the Hastière Limestone and in the Pont d'Arcole Shale) by rapid diversification of the surviving Carboniferous taxa. The *praesulcata–sulcata* boundary cannot be established precisely because of the absence of these index species, but it corresponds to an unidentified level in the lower ('problematic') part of the Hastière Limestone, between 'known Devonian' and 'known Carboniferous' (Figs 3–6).

2 The sedimentary facies encountered range from foreshore and shoreface crinoidal grainstones to bioclastic and peloidal storm deposits (from proximal to distal) and lime–wackestone/marl couplets deposited below storm-wave base. Sometimes local oolitic shoals in the north separated quiet water conditions from the open sea. Hence, the model envisaged represents a shallow ramp with local shoals, dropping off into a deeper basin to the south and east (Van Steenwinkel, 1988). Figure 4 outlines the occurrence of the lithofacies in the profiles representing the four palaeogeographic positions.

3 Relative depth profiles were deduced (Fig. 4) for the different palaeogeographical positions (areas A, B, C and D in Fig. 1) from the succession of sedimentary facies. Trends in relative depth show that a gradual progradation in the offshore facies of the Etrœungt Limestone changes suddenly into much shallower facies at the base of the Hastière Limestone. The last consists of generally shallow-water crinoidal facies. It is composed of shallowing-upward units (Van Steenwinkel, 1988), which are stacked in an aggradational to progradational way, resulting in a prograding wedge thickening offshore.

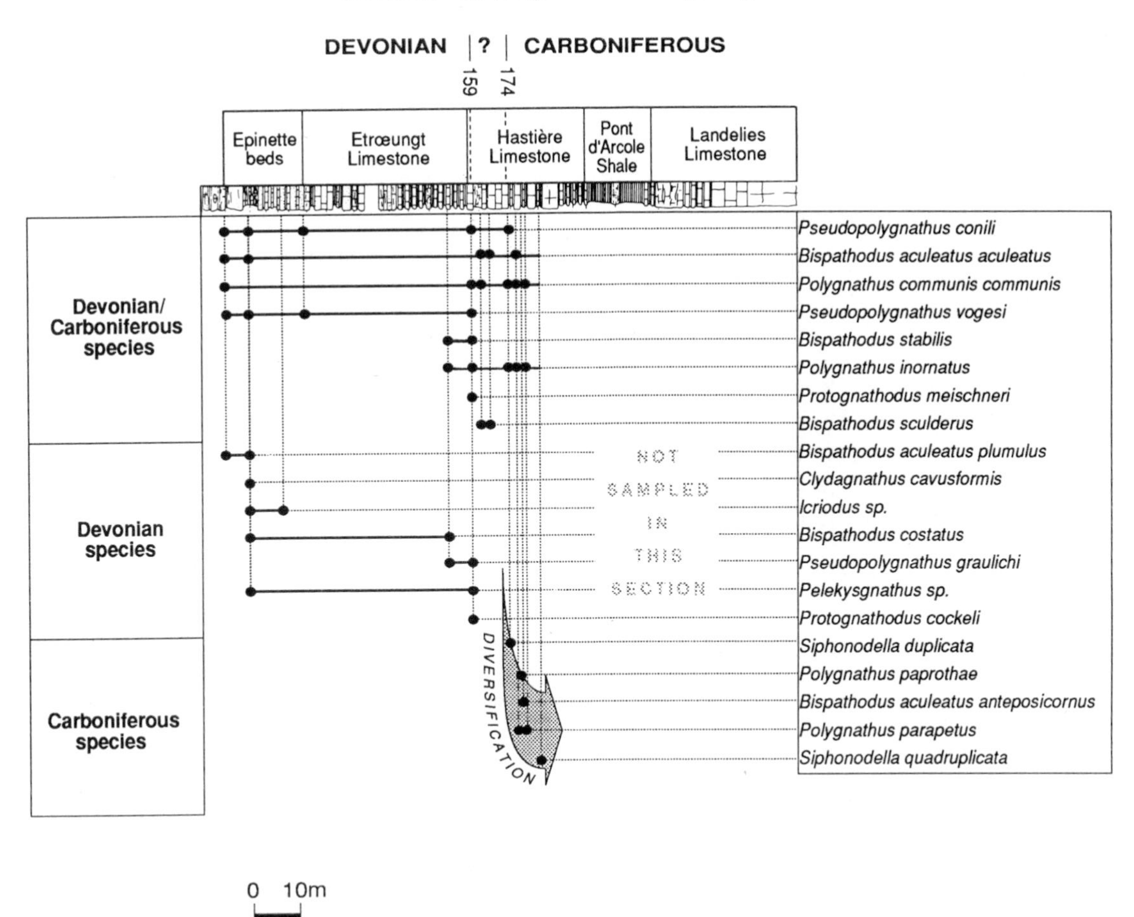

Fig. 3. Conodont distribution near the D/C transition in the reference section (Anseremme) of area A (see Fig. 1). (Based on Bouckaert & Groessens, 1976; Sandberg & Ziegler, 1979; and Van Steenwinkel, 1984.) Typical Devonian species occur no later than bed 159 (basal bed of Hastière Limestone); Carboniferous species are recorded from bed 174 upwards and, although still very scarce in the Hastière Limestone, show an upwardly increasing diversification towards the overlying Pont d'Arcole Shale (shaded arrow). The latter formation (not sampled in detail in this section) contains abundant and diverse Carboniferous fauna.

The upper part of the Hastière Limestone is characterized by a sudden deepening of the depositional environment, followed by increasingly deeper water facies into the Pont d'Arcole Shale, indicating retrogradation. Overlying the deepest interval in the Pont d'Arcole Shale, shallowing-upward facies occur, culminating in the Landelies Limestone. These trends and the major changes in relative depth have led to the distinction of systems tracts (Figs 4–6).

The maximum drowning is characterized by open-marine, bioturbated shales in the Pont d'Arcole Formation. Somewhere within these shales there is a maximum flooding surface (mfs), above which progradation occurred. The unit below these shales is characterized by progressive deepening upwards due to retrogradation: foreshore crinoidal grainstones are drowned above a distinct transgressive surface and overlain by proximal and then distal storm deposits, offshore, fairweather limestone–marl couplets, marls and finally the bioturbated shales. It is the first Carboniferous transgressive systems tract, which correlates in the Rhenish Slate Mountains with the condensed 'Liegende Alaunschiefer'. The base of this unit is datable and can be correlated over the entire Dinant synclinorium. It is the transgressive surface (TS),

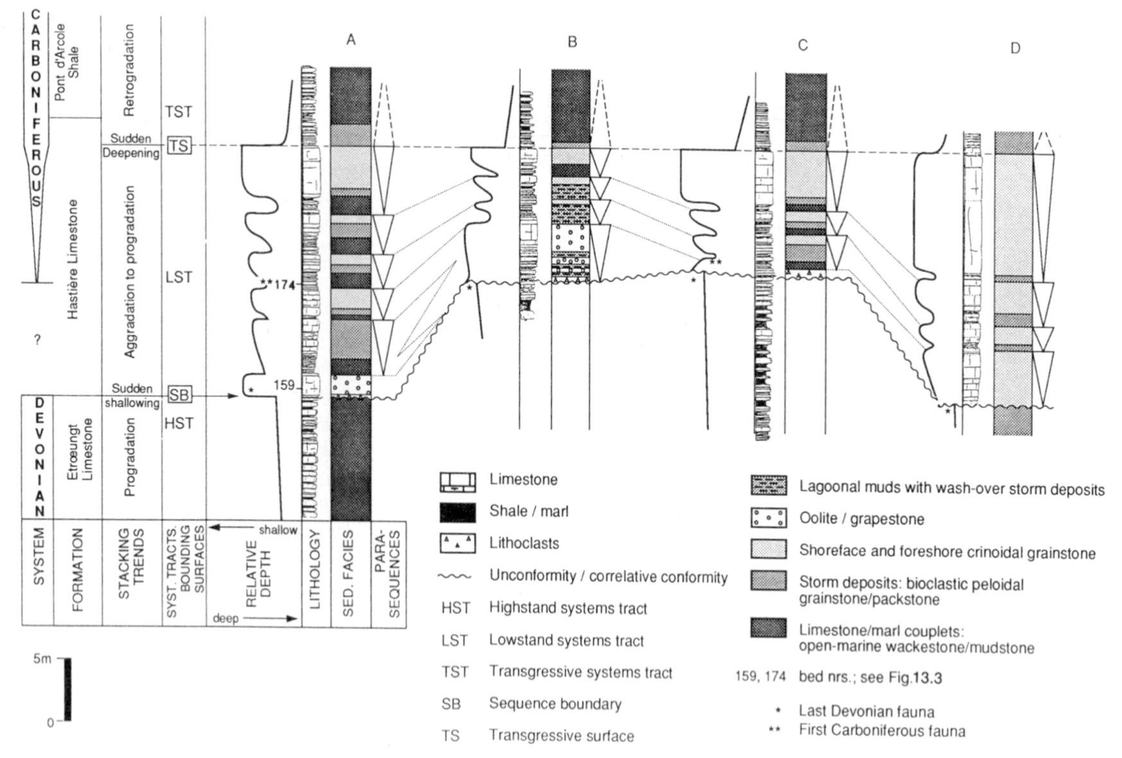

Fig. 4. Type sections of the four depositional areas (A, B, C and D in Fig. 1) and their lateral correlation, based on five steps: (1) biostratigraphic determination of known Devonian and known Carboniferous; (2) deduction of relative palaeodepth from sedimentary facies; (3) delineation of systems tracts and their bounding surfaces on the basis of aggradational, progradational and retrogradational facies trends; (4) parasequence distinction between transgressive surface (TS) and sequence boundary (SB); (5) parasequence correlation from the transgressive surface (TS) downwards. Note locally absent parasequences in lower part of Hastière Limestone, interpreted as differential onlap, and increasingly better correlation in upper part.

coinciding with the top of the uppermost and most prominent shallowing-upward unit (foreshore crinoidal grainstones) of the Hastière Limestone. Below this level, only sporadic Carboniferous conodonts were found, too long ranging for reliable lateral correlation.

The most abrupt shallowing is represented by a sudden basinward shift of facies between the Etrœungt and the Hastière Limestone (Figs 4–6). It is characterized in the north (landward) by subaerial exposure and in the south by an abrupt contact between deep- and shallow-water facies. These facies, and therefore the contact between deep and shallow deposits, vary between proximal and distal palaeogeographic positions. In the southernmost outcrops limestone–marl couplets, interpreted as sediments below storm-wave base, are overlain by foreshore oolitic grainstones with bird's-eye structures. This level of abrupt shallowing and subsequent accommodation increase is interpreted as a sequence boundary (SB in Figs 4–6). It corresponds in time (middle-*praesulcata* zone) with an incision and channel-fill sediments 95 m thick (Seiler channel : Paproth, 1986) in the Rhenish Slate Mountains of Germany (Steenwinkel, 1993).

4 Stacked shallowing-upward facies units ('parasequences': Van Wagoner *et al.*, 1989, 1990) some tens of centimetres up to 5 m thick are identified between the transgressive surface and the sequence boundary (main part of the Hastière Limestone; Figs 4 & 5). They are thought to be the sedimentary response to higher order sea-level fluctuations. The effect of sea level varies with distance from the shoreline, as reflected by variations in the facies, thickness and configuration of the parasequences. These parasequences are bounded by marine flooding surfaces, which are the result of small-scale deepening pulses, and reflected by small upward

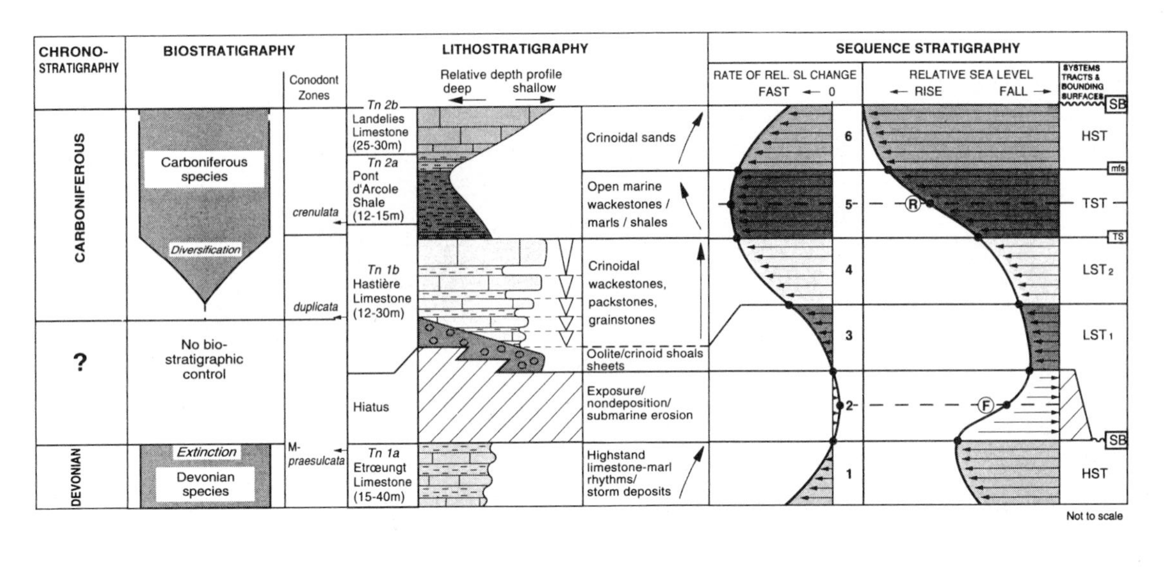

Systems tracts:

HST : Highstand Systems Tract
TST : Transgressive Systems Tract
LST2 : Lowstand Systems Tract (upper part)
LST1 : Lowstand Systems Tract (lower part)

Bounding surfaces:

SB : Sequence Boundary
mfs : maximum flooding surface
TS : Transgressive Surface

Ⓡ Rise inflection point
Ⓕ Fall inflection point
1-6 Time Units (see text)
▽ Parasequence
↗ Stacking pattern trend

Fig. 5. Summary of chrono-, bio- and lithostratigraphy in a sequence-stratigraphic framework for the Belgian Devonian–Carboniferous boundary. The biostratigraphic hiatus between the Devonian and the Carboniferous is interpreted as a consequence of a relative sea-level fall at the end of the Devonian.

shifts of facies. These deepening pulses become increasingly more prominent upward in the Hastière Limestone. The few conodonts recorded in the Hastière Limestone are associated with the deeper water facies of parasequence bases; conodonts are absent in the shallower, upper parts. The crinoidal grainstone bed directly below the transgressive surface is distinctly thicker than the others in all the outcrops studied.

5 Parasequence correlation between the four depositional areas (Fig. 4) is possible because marine flooding surfaces can still be recognized, regardless of local facies changes or differing subsidence rates. They are used as isochronous correlation markers. The dated transgressive surface is used as a datum horizon, from which the parasequence-bounding flooding surfaces can be correlated downward. The upper three parasequences were easy to correlate; even individual beds could be traced between the outcrops. However, the lower part of the Hastière Limestone is much more differentiated; the lower parasequences of the Meuse area cannot be traced laterally. The abrupt shallowing (sequence boundary) at the base of this formation is recorded in all the areas. Locally absent parasequences are interpreted as the result of differential onlap (Fig. 4).

RELATIVE SEA-LEVEL CHANGES

The fact that marine flooding surfaces and stacking-pattern trends can be recognized regardless of the palaeogeographic setting (i.e. with facies variations down the depositional slope) points to relative sea-level changes as the common factor controlling the stratigraphic pattern; this is independent of local differences in the sedimentation or subsidence rates.

The overall section can be subdivided into six time units (Fig. 5). Each unit is the sedimentary response to various rates of relative sea-level change (Vail *et al.* 1977) and corresponds to particular systems tracts (Vail, 1987; Posamentier & Vail, 1989; Posamentier *et al.*, 1989). The amplitude and rate curves of relative sea-level change (Fig. 5) deduced from the above facies analysis are representative for the D/C transition in the Dinant syn-

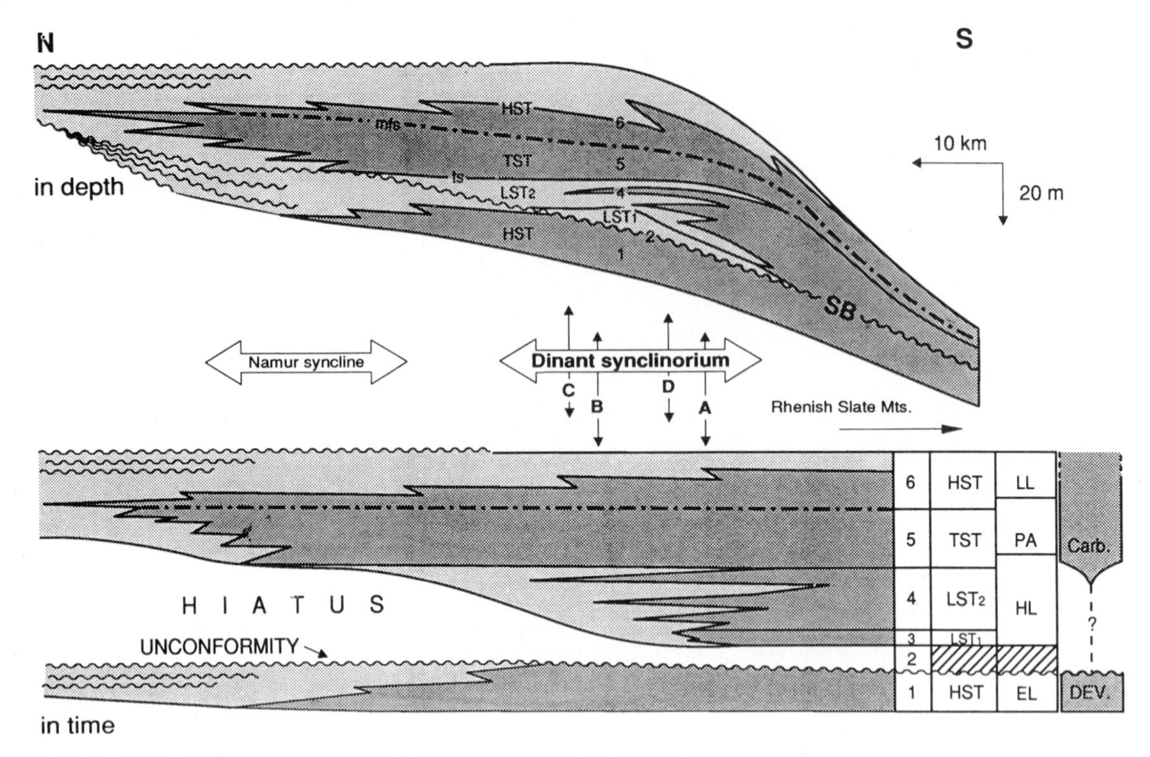

Fig. 6. Depositional sequence of the Belgian Devonian–Carboniferous boundary. Light raster shows above wave base facies; dense raster shows below wave base facies; palaeogeographic positions A, B, C, D of Dinant synclinorium (see Fig. 1); other symbols, see Fig. 5. (Modified after Van Steenwinkel, 1990.)

clinorium. Whether they are similar in other D/C basins can only be determined by studying these basins using a similar approach (Steenwinkel, 1993).

Figure 5 shows that the two major events — the overall drowning (mfs) and the most abrupt shallowing (SB) — correspond, respectively, to the intervals near the rise and fall inflection points, where the rising and falling limbs of the amplitude curve are the steepest and where the rate curve attains a maximum in rise and fall, respectively. The units in between are related to different relative sea-level phases between these major events. The detailed stratigraphic response to these relative sea-level changes in the four palaeogeographic positions is discussed by Van Steenwinkel (1988, 1990). The general trend is summarized below (Figs 5 & 6):

1 Time unit 1 (Etrœungt Limestone, uppermost Devonian) represents a highstand of sea level, when most of the area was subjected to open-marine, offshore conditions. The sedimentation is characterized by below wave base limestone–marl couplets and distal to proximal storm deposits.

2 After the late-Devonian highstand, accommodation space was eliminated by a relative sea-level fall (time unit 2). The resulting sequence boundary is represented in the northern (landward) areas by an unconformity, characterized by subaerial exposure, and in the southern areas by its correlative conformity, which is characterized by an abrupt juxtaposition of deep- and shallow-water facies. In these areas intertidal oolite and foreshore sands immediately overlie near and below storm wave base limestone–marl couplets. A local lithoclastic layer occurs on top of this surface. It is not clear whether this reworking occurred at the time of the fall or during the initial rise (transgressive ravinement).

3 Sedimentation resumed during the following slow relative sea-level rise. Shallow-water oolitic and crinoidal grainstones occur immediately above the deeper water sediments of the previous highstand (time unit 3).

4 Later (time unit 4: main part of Hastière Limestone), the rate of relative sea-level rise increased, although the environment of deposition was still generally shallow. Smaller-scale, relative sea-level fluctuations superimposed upon this rising trend were responsible for the parasequence stacking on which the lateral correlation was based. The small-scale rises increased upward in importance (increasing rate) but could also be increasingly compensated by sedimentation (establishment of 'carbonate factory'). The first parasequences are only present in the south. Hence, time unit 4 represents both an onlapping and prograding wedge complex reflecting the sedimentary fill that levelled the topographic relief before the subsequent rapid rise in sea level. The first Carboniferous fauna were scarce, and are only found in the deeper water parts of the parasequences.

5 During the subsequent phase of rapid relative sea-level rise (time unit 5) sedimentation could not keep pace with rising sea level. The resulting transgressive systems tract is characterized by backstepping depositional patterns and by the establishment of deeper marine Carboniferous fauna with biostratigraphically significant markers. The base of this unit (transgressive surface) is well dated and used as a reference level for lateral correlation of the underlying parasequences.

6 After the maximum transgression (maximum flooding surface), progradation occurred again (time unit 6). A shallowing-upward crinoidal sand sequence developed, culminating in the Landelies Limestone (highstand systems tract).

CONCLUSIONS

1 The biostratigraphic gap and the associated unconformity near the D/C boundary in the Dinant synclinorium can be explained by a fall in relative sea level at the end of the Devonian. The late-Devonian highstand sedimentation, with its large, open-marine faunal diversity, was brought to a halt by this fall. After a period of non-deposition, sedimentation resumed but in a much shallower, lowstand setting, devoid of the deep-water conodonts that currently define the D/C boundary. Only after increased deepening (transgressive systems tract) was a deeper water fauna with more evolved Carboniferous assemblages re-established and able to provide good biostratigraphic control.

2 The recognition of marine flooding surfaces and stacking-pattern trends, regardless of the palaeogeographic setting and of locally differing subsidence rates (i.e. with facies variations down the depositional slope), provides the link between bathymetric changes and relative sea-level changes. Understanding cyclicity in the rock record from a larger to smaller scale promotes its use as a tool for refining lateral correlation beyond biostratigraphic resolution.

3 The importance of eustasy and subsidence in the relative sea-level curve of the Dinant synclinorium can only be deduced after comparing the relative sea-level curves of different tectonic settings. However, since the D/C boundary is associated with similar biostratigraphic complexity in all the Devonian–Carboniferous basins of the world, the fall in relative sea level at the end of the Devonian may be assumed to be due to a eustatic fall. Consequently, the sequence boundary associated with this fall should be found worldwide and could be used as a reference marker where the biostratigraphic *praesulcata–sulcata* marker is not recorded.

ACKNOWLEDGEMENTS

This study is part of a PhD thesis under the supervision of Prof. Dr J. Bouckaert, carried out at the University of Leuven (Belgium), and was supported by a grant from IWONL. I am indebted to the Devonian–Carboniferous Working Group, especially to E. Paproth and M. Streel, for the generous interest and support, and to the late Prof. Dr R. Conil, whose pioneering studies on the Lower Carboniferous and the Devonian–Carboniferous boundary constitute the indispensable basis for this work. I thank H.J. Droste, B.U. Haq, B.D. Macurda, B. Pratt, J. Senior and A. Strasser for reviewing the manuscript.

REFERENCES

Ahr, W.M. (1973) The carbonate ramp: an alternative to the shelf model. *Trans. Gulf Coast Assoc. Geol. Soc.* **13**, 221–225.

Bouckaert, J. & Groessens, E. (1976) *Polygnathus paprothae, Pseudopolygnathus conili, Pseudopolygnathus graulichi:* espèces nouvelles à la limite Dévonien-Carbonifère. *Ann. Soc. Géol. Belgique* **99** (2), 587–599.

Conil, R. (1968) Le calcaire carbonifère depuis le Tn1a jusqu'au V2a. *Ann. Soc. Géol. Belgique*, **90**, B687–726.

Conil, R. avec la coll. de Lys, M. & Paproth, E. (1964) Localités et coupes types pour l'étude du Tournaisien inférieur (Révision des limites sous l'aspect micropaléontologique). *Acad. Roy. Belg., Cl. Sci.* **15**(4), 1–87.

Johnson, J.G., Klapper, G. & Sandberg, C.A. (1985) Devonian eustatic fluctuations in Euramerica. *Bull. Geol. Soc. Am.* **96**, 567–587.

Johnson, J.G., Klapper, G. & Sandberg, C.A. (1986) Late Devonian eustatic cycles around margin of Old Red Continent. *Ann. Soc. Géol. Belgique* **109**, 141–148.

Lane, H.R., Sandberg, C.A. & Ziegler, W. (1980) Taxonomy and phylogeny of some Lower-Carboniferous conodonts and preliminary standards post-*Siphonodella* zonation. *Geol. Palaeontol.* **14**, 117–164.

Leeder, M.R. (1987) Tectonic and palaeogeographic models for Lower Carboniferous Europe. In: *European Dinantian Environments* (Eds Miller J., Adams, A.E. & Wright V.P.) Spec. Issue Geol. J., 12, 1–20.

Paproth, E. (1986) An introduction to a field trip to the late Devonian outcrops in the northern Rheinisches Schiefergebirge (Federal Republic of Germany). *Ann. Soc. Géol. Belgique* **109**, 275–284.

Paproth, E. & Sevastopulo (1988) The search for a stratotype for the base of the Carboniferous. In: *Devonian–Carboniferous Boundary–Results of Recent Studies* (Eds Flajs, G., Faist R. & Ziegler W.). Cour. Forsch.-Inst. Senckenberg, 100. Frankfurt a. M.

Paproth, E. & Streel, M. (1971) Corrélations biostratigraphiques près de la limite Dévonien–Carbonifère entre les faciès littoraux ardennais et les faciès bathyaux rhénans. In: *Colloque sur la Stratigraphie du Carbonifère.* Congr. et coll. Univ. Liège, 55, 365–398.

Paproth, E. & Streel, M. (Eds) (1984) The Devonian–Carboniferous Boundary. *Cour. Forsch.- Inst. Senckenberg,* **67**, 258 pp.

Paproth, E. & Zimmerle, W. (1980) Stratigraphic position, petrography and depositional environment of phosphorites from the Federal Republic of Germany. In: *Pre-Permian around the Brabant Massif in Belgium, the Netherlands and Germany* (Eds Bless, M.J.M., Bouckaert, J. & Paproth, E.) Meded. Rijks Geol. Dienst, 32 (11), 81–95.

Posamentier, H.W. & Vail, P.R. (1989) Eustatic controls on clastic deposition II—sequence and systems tract models. In: *Sea-level Changes—An Integrated Approach* (Eds Wilgus, C.K., Hastings, B.S., Kendall, C.G.St.C., Posamentier, H.W., Ross, C.A. & Van Wagoner, J.C.) Spec. Publ. Soc. Econ. Paleontol. Mineral. 42, 125–154.

Posamentier, H.W., Jervey, M.T. & Vail, P.R. (1989) Eustatic controls on clastic deposition I—conceptual framework. In: *Sea-level Changes—An Integrated Approach* (Eds Wilgus, C.K., Hastings, B.S., Kendall, C.G.St.C., Posamentier, H.W., Ross, C.A. & Van Wagoner., J.C.) Spec. Publ. Soc. Econ. Paleontol. Mineral. 42, 109–124.

Poty, E. (1984) Rugose corals at the Devonian–Carboniferous boundary. In: *The Devonian–Carboniferous Boundary* (Eds Paproth E. & Streel, M.). Cour. Forsch.-Inst. Senckenberg, 67, 29–36.

Poty, E. (1986) Late Devonian to early Tournaisian rugose corals. *Ann. Soc. Géol. Belgique,* **109**, 65–74.

Ross, C.A. & Ross, J.R.P. (1987) *Late Paleozoic Sea Levels and Depositional Sequences.* Cushman Foundation for Foraminiferal Res. Spec. Publ., 24, 137–149.

Sandberg, C.A. & Ziegler, W. (1979) Taxonomy and biofacies of important conodonts of late Devonian *styriacus*-zone, United States and Germany. *Geol. Palaeontol.* **13**, 173–212.

Sandberg, C.A., Ziegler, W., Leuteritz, K. & Brill, S. (1978) Phylogeny, speciation and zonation of *Siphonodella* (*Conodonta*), Upper Devonian and Lower Carboniferous. *Newsl. Stratigr.* **7**(2), 102–120.

Steenwinkel, M. Van (1984) *The Devonian–Carboniferous Boundary in the Vicinity of Dinant, Belgium.* Cour. Forsch.-Inst. Senckenberg, 67, 57–70.

Steenwinkel, M. Van (1988) The sedimentary history of the Dinant Platform during the Devonian-Carboniferous transition. PhD thesis, Katholieke Universiteit Leuven, 173 pp.

Steenwinkel, M. Van (1990) Sequence stratigraphy from 'spot' outcrops: example from a carbonate-dominated setting: Devonian-Carboniferous transition, Dinant synclinorium (Belgium). *Sediment. Geol.* **69**, 259–280.

Steenwinkel, M. Van (1993) The Devonian–Carboniferous boundary: Comparison between the Dinant synclinorium and the northern border of the Rhenish Slate Mountains — A sequence-stratigraphic view. *Ann. Soc. Géol. Belgique* **115**(2), in press.

Streel, M. (1971) Biostratigraphie des couches de transition Dévono-Carbonifère et limite entre les deux systèmes. *C.R. 7e Congrès Intern. Strat. et Géol. du Carbonifère, Krefeld.*

Streel, M. (1983) Bio- and lithostratigraphic subdivisions of the Dinantian in Belgium, a review. *Ann. Soc. Géol. Belgique,* **106**, 192–193.

Vail, P.R. (1987) Seismic stratigraphy interpretation using sequence stratigraphy. In: *Atlas of Seismic Stratigraphy* (Ed. Bally, A.W.). Am. Assoc. Petrol. Geol., Stud. Geol., 27, 1–10.

Vail, P.R., Mitchum, R.M. Jr., Todd, R.G. *et al.* (1977) Seismic stratigraphy and global changes of sea level. In: *Seismic Stratigraphy — Applications to Hydrocarbon Exploration* (Ed. Payton, C.E.) Mem. Am. Assoc. Petrol. Geol., Tulsa, 26, 49–212.

Van Wagoner, J.C., Mitchum, R.M., Campion, K.M. & Rahmanian, V.D. (1990) Siliciclastic sequence stratigraphy in well logs, core, and outcrops: concepts for high-resolution correlation in time and facies. *Am. Assoc. Petrol. Geol. Meth. Expl. Ser.* **7**, 1–55.

Van Wagoner, J.C., Mitchum, R.M. Jr., Posamentier, H.W. & Vail, P.R. (1987) The key definitions of sequence stratigraphy. In: *Atlas of Seismic Stratigraphy* (Ed. Bally, A.W.) Am. Assoc. Petrol. Geol., Stud. Geol. 27.

Van Wagoner, J.C., Posamentier, H.W., Mitchum, R.M. Jr., *et al.* (1989) An overview of the fundamentals of sequence stratigraphy and key definitions. In: *Sea-level Changes—An Integrated Approach* (Eds Wilgus, C.K., Hastings, B.S., Kendall, C.G.St.C., Posamentier, H.W., Ross, C.A. & Van Wagoner, J.C.) Spec. Publ. Soc. Econ. Paleontol. Mineral. 42, 39–46.

Walliser, O.H. (1984) Pleading for a natural D/C-boundary. *Cour. Forsch.-Inst. Senckenberg* **67**, 241–246.

Spec. Publs Int. Ass. Sediment. (1993) **18**, 247–281

Namurian (late Carboniferous) depositional systems of the Craven–Askrigg area, northern England: implications for sequence-stratigraphic models

O.J. MARTINSEN*

Geologisk Institutt Avd. A, Universitetet i Bergen, Allegt. 41, 5007 Bergen, Norway

ABSTRACT

Sequence-stratigraphic models of Exxon Production Research (EPR) and Galloway (1989a) have been applied to Namurian E1c–H2c (Upper Carboniferous) sedimentary successions of the Craven–Askrigg area, northern England. These sediments were deposited in an extensional setting characterized by a transition from a slowly subsiding basin margin fault block into a more rapidly subsiding initially deep basin. The basin margin fault block experienced reactivations which had impacts on sequence-stratigraphic development of the area. The basin fill is punctuated at several levels by goniatite-bearing marine bands which carry their own distinctive species, occur basin-wide and form excellent correlation markers. The goniatite bands were probably controlled by eustatic fluctuations in sea level.

While sedimentation on the basin margin fault block for the entire time period was dominated by shallow-marine and fluviodeltaic deposition, the basin was initially filled by turbidites which were followed by slope deposits and several subsequent deltaic episodes. In both areas the sedimentation was episodic with numerous progradations taking place punctuated by transgressions represented by the basin-wide marine bands and their correlatives. All of these progradational episodes are termed minor cycles in the traditional approach. The minor cycles form stacking patterns, which may be described as major cycles, separated by multi-minor cycle mudstone units.

Only in part does the EPR sequence-stratigraphic approach seem to be a reasonable predictor for the basin-fill stratigraphy, and it cannot account fully for lateral variability caused by tectonic influence on sedimentation patterns by reactivations of the basin margin fault block. In addition, evidence suggests that EPR type 1 sequences developed at more than one time scale and this severely constrains the potential of interregional correlations. Moreover, defining boundaries at unconformities within the basin fill is problematical since most 'unconformable' surfaces occur at the bases of channel sandstones which are localized and sometimes apparently linked genetically to the sediments below the erosion surfaces.

Galloway's (1989a) model provides a reasonable approach to the basin fill, at the scale of minor cycles. The goniatite bands form widely correlative genetic sequence boundaries, and appear to separate distinct phases of progradation and retreat of depositional systems. However, it does not predict deposits formed during lowstands of base level, which are present in several minor cycles. Moreover, at the larger scale of major cycles, the Galloway (1989a) model does not provide a satisfactory prediction of the basin fill.

INTRODUCTION

The last few years have seen the publication of several different models of sequence stratigraphy, a new method which may provide a better understanding of how sedimentary basins fill and their major controls (e.g. Embry, 1988; Embry & Podruski, 1988; Posamentier & Vail, 1988; Posamentier *et al.*, 1988; Van Wagoner *et al.*, 1988; Galloway, 1989a). These models have had different origins, being based on seismic data (the Exxon model; cf. Wilgus *et al.*, 1988 and references therein), well-log and core data (Galloway, 1989a) and outcrop data (Embry, 1988).

*Present address: Norsk Hydro Research Centre, Sandsliveien 90, 5049 Sandsli, Norway.

More well-documented case studies are needed so that the sequence-stratigraphic concepts can be rigorously tested down to the level of the sedimentary process (cf. however, Nummedal & Swift, 1987; James & Leckie, 1988 and references therein). In particular, there is a definite need to apply the sequence-stratigraphic models to outcrops and core data. In addition, many of the hitherto published studies tend to be model-driven rather than data-driven (e.g. Coterill *et al.*, 1990; Dockery, 1990), and are rather uncritical of the models, although these still need to be better founded and modified.

The aim of the present paper is to compare two existing models of sequence stratigraphy—the Exxon Production Research approach (cf. Posamentier *et al.*, 1988; Posamentier & Vail, 1988; Van Wagoner *et al.*, 1988) and the Galloway (1989a) approach. These sequence stratigraphy models were applied to the uppermost Pendleian to Alportian (Namurian E1c–H2c; Upper Carboniferous) sedimentary successions of the Craven–Askrigg area in northern England and their conceptual framework tested against detailed sedimentological studies (Martinsen, 1990a). A third model of sequence stratigraphy, of Embry (1988), seemed unsuitable to apply because of the general lack of transgressive deposits in the area (see below).

The Craven–Askrigg area (Fig. 1) seems generally suited for such a purpose. First, the area was part of a broad extensional province during the early and early late Carboniferous (see below) where clastic sediment was fed from the north across an active fault block (the Askrigg Block) into a southerly initially deep basin (the Craven Basin). Second, the biostratigraphic framework with discrete, often basin-wide bands of goniatites and their correlatives dissecting the basin fill at numerous levels (Figs 2 & 3) provides an excellent basis for correla-

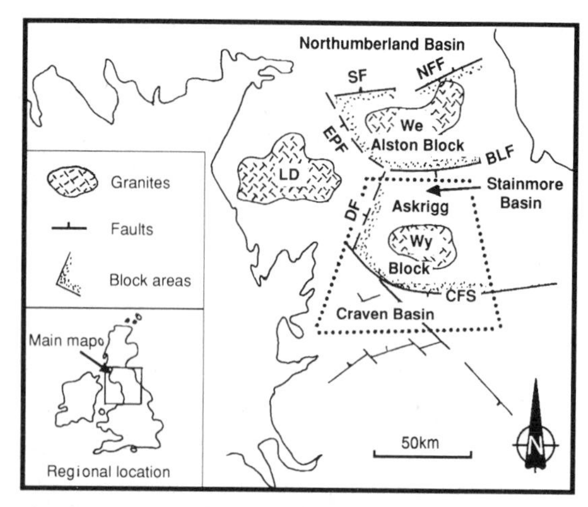

Fig. 1. Location and simplified structural map of northern England. The stippled area shows the region studied in this paper. Note the granite bodies at the cores of several fault blocks. LD, Lake District Granite; Wy, Wensleydale Granite; We, Weardale Granite; DF, Dent Fault; CFS, Craven Fault System; EPF, Eden–Pennine Fault; SF, Stublick Fault; NFF, Ninety Fathom Fault; BLF, Butterkowle–Lunedale Fault. (Modified from Gawthorpe *et al.*, 1989.)

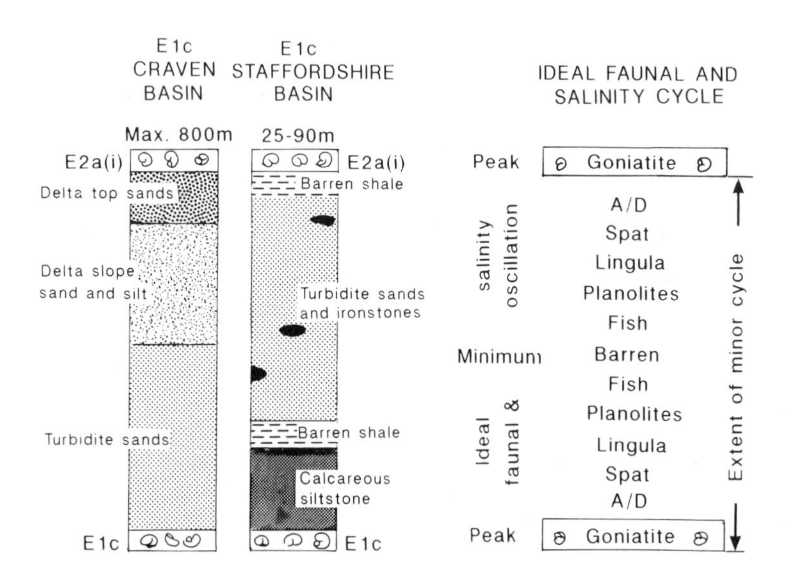

Fig. 2. Right, Ideal faunal salinity cycle of goniatite bands. Left, Two examples of the E1c minor cycle from Staffordshire and from the Craven Basin to illustrate variability in thickness and facies. (Modified from Holdsworth & Collinson, 1988.)

Part of Namurian (Upper Carboniferous)		MINOR CYCLE	GONIATITE SPECIES
	Alportian	H2c(ii)	*Ht. prereticulatus*
		H2c(i)	*V. eostriolatus*
		H2b	*H. undulatum*
		H2a(ii)	*H. smithi-undulatum*
		H2a(i)	*H. smithi/Hd. proteus*
	Chokierian	H1b(iv)	*H. aff. beyrichianum*
		H1b(iii)	*H. aff. beyrichianum*
		H1b(ii)	*H. beyrichianum*
		H1b(i)	*H. beyrichianum*
		H1a(iv)	*Homoceras sp. nov.*
		H1a(iii)	*I. subglobosum*
		H1a(ii)	*I. subglobosum*
		H1a(i)	*I. subglobosum*
	Arnsbergian	E2c2(iii)	*N. nuculum*
		E2c2(ii)	*N. nuculum*
		E2c2(i)	*N. nuculum*
		E2c1	*N. stellarum*
		E2b3	*Ct. nititoides*
		E2b2(ii)	*F. holmesi*
		E2b2(i)	*Ct. nitidus*
		E2b1(iii)	*C. subplicatum Grp.*
		E2b1(ii)	*Ct. edalensis*
		E2b1(i)	*C. subplicatum Grp.*
		E2a(iii)	*E. yatesae*
		E2a(ii)	*E. bis. ferrimontanum*
		E2a(i)	*C. cowlingense*
	Pendleian (part)	E1c	*C. malhamense*

Explanation: C=*Cravenoceras*; E=*Eumorphoceras*; Ct= *Cravenoceratoides*; F=*Fayettevillea*; N=*Nuculoceras*; I=*Isohomoceras*; H=*Homoceras*; V=*Vallites*; Ht=*Homoceratoides*

Fig. 3. Stratigraphic column showing investigated stages, minor cycles, and goniatite species. (Based on Holdsworth & Collinson, 1988 and Riley *et al.*, 1987.)

tion. Around 60 such bands were deposited during the 11 Ma-long Namurian epoch in northern England, yielding an average age of around 180,000 years for each intervening interval.

The E1c–H2c interval (containing 27 goniatite bands in the standard succession; see below) of the Namurian was selected because this part of the stratigraphy showed the best and most extensive outcrops. Sixty-five sections were measured at selected localities to construct detailed depositional models of the cycles (Martinsen, 1990a). All localities did not cover the entire E1c–H2c interval (see Appendix, p. 277). Most of the outcrops were extensive crags or continuous, relatively well-exposed vertical sections in streams.

The following is one attempt at applying sequence stratigraphy to a complex area subject to several large-scale controls on sedimentation. One main intention is to stimulate a discussion of sequence stratigraphy in the Upper Carboniferous, as well as to point out some significant problems about sequence-model application in the study area which may have general importance. Application on a basinal scale is emphasized to provide a framework and basis for discussion for the Craven–Askrigg area. Work presented elsewhere will deal with the high-resolution aspects of the sequence stratigraphy of the area.

STRUCTURAL FRAMEWORK

There is a general consensus that Carboniferous basin formation in northern England was controlled by a N–S extension north of the northwards-migrating Variscan front (Leeder & McMahon, 1988; Gawthorpe *et al.*, 1989). This extensional deformation gave rise to a highly differentiated block and basin bathymetry (Fig. 1) which appears to have been most pronounced at the start of the Namurian or possibly the latest Dinantian (Collinson, 1988). Northern England at this time was dominated by tilt blocks and half-grabens (Leeder, 1982, 1988; Gawthorpe *et al.*, 1989).

The Askrigg Block is one such tilt block (Fig. 1) which in the Dinantian and earliest Namurian experienced northwards tilting. This gave rise to the Stainmore half-graben at the northern end of the block (Gawthorpe *et al.*, 1989), which was mainly filled during the earliest Namurian so that no differential bathymetry existed in this area at later times and fluviodeltaic systems could freely prograde southwards across the Askrigg Block (Martinsen, 1990a).

The Askrigg Block is bounded to the west by the Dent Line (Fig. 1). This is an early Palaeozoic lineament which experienced several later reactivations (Underhill *et al.*, 1988). In the early Carboniferous, it probably represented a major transfer fault which linked the extension in the Stainmore half-graben to the north with movement on the Craven Faults, which bound the Askrigg Block to the south (cf. Underhill *et al.*, 1988; Gawthorpe *et al.*, 1989).

The Craven Fault system at the southern margin of the Askrigg Block forms the boundary between the Askrigg Block and the Craven Basin (Fig. 1). This fault system has had a very complex history with several different senses of movement (e.g. Dunham & Stubblefield, 1945; Gawthorpe *et al.*,

1989), but in early Carboniferous and early late Carboniferous times there was most likely a general downthrow to the south maintaining the Askrigg Block as a structural high and the Craven Basin area as a more rapidly subsiding area to the south.

Several of the northern England tilt blocks are underlain by granite batholiths (e.g. Bott, 1961; Holland & Lambert, 1970) probably of late Caledonian origin, their age being estimated at 400–410 Ma (cf. Dunham, 1974). The Askrigg Block is underlain by the Wensleydale Granite which has been proven in boreholes and by gravity studies (Dunham, 1974). These granites probably had a significant influence in buoying up the fault blocks and maintaining them as structural highs throughout Carboniferous times.

Active extension is believed to have occurred in Dinantian and possibly early Namurian times and the remainder of the Namurian is thought to have been characterized by post-extensional thermal sag (e.g. Leeder, 1988; Leeder & McMahon, 1988; Gawthorpe *et al.*, 1989). Recently, it was postulated that regional variability in depositional styles and sediment thicknesses across both the Askrigg Block and the Craven Basin for parts of the lower and mid-Namurian successions indicates tectonic readjustments of the Askrigg Block with possible reactivations of the block-bounding faults (Martinsen, 1990a). These readjustments of the block can possibly be related to a final phase of the main extension, and had a profound influence on the localization of sediment input across the block and therefore into the Craven Basin. Consequently, these tectonic movements had also important impacts on the development of stratigraphic sequences and their components.

BIOSTRATIGRAPHIC FRAMEWORK

Throughout the Namurian of the British Isles and several other places in Europe, goniatite-bearing shales dissect the sedimentary successions at numerous levels. These fossiliferous shales are usually discrete bands, most frequently from a few centimetres up to tens of centimetres thick. In their most ideal form, the fossil bands are zoned with their base being characterized by a transition from underlying barren sediment into fauna representing low salinity which is overlain by progressively more saline fauna and at the centre of the band by a thick-shelled goniatite species (Fig. 2). The goniatites are then overlain by progressively less saline fauna and finally by barren sediment (Ramsbottom *et al.*, 1962; Holdsworth & Collinson, 1988). It is thus likely that the goniatite bands represent the most marine parts of the successions, while the interbedded sediments were deposited in periods of lower salinity. The thickness of the interbedded sediments between successive goniatite bands may vary from less than 1 m of barren mudstones to several hundred metres of coarsening-upward deltaic successions (Fig. 2; cf. Collinson, 1988; Holdsworth & Collinson, 1988).

Over 60 such goniatite bands occur in the Namurian succession in northern England (e.g. Ramsbottom *et al.*, 1978), yielding an average duration of approximately 180,000 years for the deposition of the goniatite band and the overlying sediments below the next band in the standard succession. Each of these bands has its own goniatite species making the bands fairly reliable chronostratigraphic markers and powerful tools for local as well as regional and interregional correlation. For instance, the *C. cowlingense* band of E2a(i) (Arnsbergian; cf. Fig. 3) can be identified both in the Craven Basin of northern England as well as in the Donetz Basin of the former Soviet Union (Aisenverg *et al.*, 1979). In addition, 18 goniatite bands can be directly correlated from County Clare, western Ireland to the Craven Basin (Ramsbottom *et al.*, 1978). Thus, many of the goniatite bands are not merely local, intrabasinal features. The marine bands represent significant condensed sections related to important transgressive periods. As suggested by Loutit *et al.* (1988), they represent the first important step in stratigraphic breakdown and correlation.

Many Namurian workers agree that the cyclicity between goniatite band and barren, interbedded sediment most likely was caused by eustatic fluctuations in sea level (e.g. Ramsbottom, 1977, 1979; Holdsworth & Collinson, 1988; Leeder, 1988), probably induced by Gondwanan glaciations (Veevers & Powell, 1987). Thus, the good correlation potential, the high time resolution and the eustatic influence favour testing sequence-stratigraphic models in the Upper Carboniferous of northern England.

Nevertheless, problems do exist. Towards the basin margins, the goniatite bands often are replaced by bands of brachiopods, limestone beds and less saline fauna, such as *Lingula*. This causes difficulty with some correlations. However, key goniatite bands do occur at some levels in these settings and together with reconstructions of depositional environments and mapping for each of the cycles investigated, a reliable correlation scheme can be established (Martinsen, 1990a).

LITHOSTRATIGRAPHIC AND SEDIMENTARY FRAMEWORK

Stratigraphic interval investigated

This study concentrates on early to mid-Namurian (early late Carboniferous) cyclicity displayed in the 'Millstone Grit' of Yorkshire and Cumbria, northern England (cf. Holdsworth & Collinson, 1988). Following on from the detailed goniatite biostratigraphic work of Bisat (1924) and Ramsbottom *et al.* (1962), Holdsworth and Collinson (1988) described the Namurian of northern England in terms of *minor cycles* (Figs 2 & 3), and created a nomenclature for the cycles. Thus, in Holdsworth and Collinson's (1988) terminology, the present paper describes minor cycles E1c through H2c(ii), which represent the upper part of the Pendleian through the Alportian stages of the Namurian (Fig. 3).

Throughout the area, the lithological naming varies a great deal (cf. Figs 13 & 14), making correlation exercises confusing for the unfamiliar reader. In addition, only the sandstones have been formally named, although finer lithologies make up the bulk of the stratigraphic column in several areas. These problems are largely overcome by the detailed biostratigraphy and by reconstructions of the palaeogeographies for each of the succeeding cycles.

The depositional pattern can be divided into two parts: (i) an early phase characterized by a pronounced bathymetric contrast between the Askrigg Block and the Craven Basin ('block-margin' phase); and (ii) a late phase when this bathymetric contrast had been largely eliminated and apparently only a very lowly inclined southerly dipping 'ramp' existed. This significant change probably owed its origin to a final cessation of the extension with time and quiescence on the block and basin-bounding faults, causing infill of the basin-floor topography.

Early depositional patterns on the block and in the basin ('block-margin' phase)

Introduction

This period encompasses the E1c to E2a(iii) minor cycles of Holdsworth and Collinson (1988). In the earliest Namurian, northwards tilting of the Askrigg Block caused a prominent unconformity to be developed across its southern portions (Fig. 4(a); Dunham & Wilson, 1985). The tectonic origin and angular character of this unconformity are well established since progressively older strata are truncated towards the south of the block and appear successively directly below the unconformity (Rowell & Scanlon, 1957a; cf. Fig. 13 in Dunham & Wilson, 1985). On the northern end of the Askrigg Block, this unconformity is not present, and an apparently continuous succession occurs. This variable relationship was probably a direct result of the northwards tilting of the Askrigg Block and a base level which remained more or less at the level of the tilt axis, causing only the southerly portions of the block to be subaerially exposed (Fig. 4(a)). The northern portions seem to have dropped below base level, thus being generally protected from erosion and at the same time functioning as a small basin within which organic-rich shales were deposited (Martinsen, 1990a).

In the Craven Basin, no unconformable relationship has been confidently proven at this stratigraphic level, although some authors have argued for an angular unconformity at a slightly higher level (Mundy & Arthurton, 1980; Arthurton *et al.*, 1988). The evidence for this unconformity is at best arguable (cf. Martinsen, 1990a), and will not be discussed further here.

E1c cycle (Fig. 4(a))

The E1c minor cycle represents the earliest depositional phase on the Askrigg Block following generation of the unconformity and deposition of the organic-rich shales on the northern portions of the block. Fluvial-channel and overbank deposits dominate (Grassington Grit and lateral equivalents; cf. Martinsen, 1990a), but important lateral differences exist. In the northwest of the block, the E1c cycle comprises only a few metres of mudstone and occasionally thin, rooted crevasse splay-type sandstones. Within this thin interval, a *relatively* mature palaeosol is developed, the Mirk Fell Palaeosol. Towards the southeast of the block, the cycle increases progressively in thickness, and the number of major fluvial–distributary channel sandstones and palaeosols increases, while the maturity of the palaeosols generally decreases, implying a much more frequent input of clastics from the active channel belts (cf. Bown & Kraus, 1987). The chan-

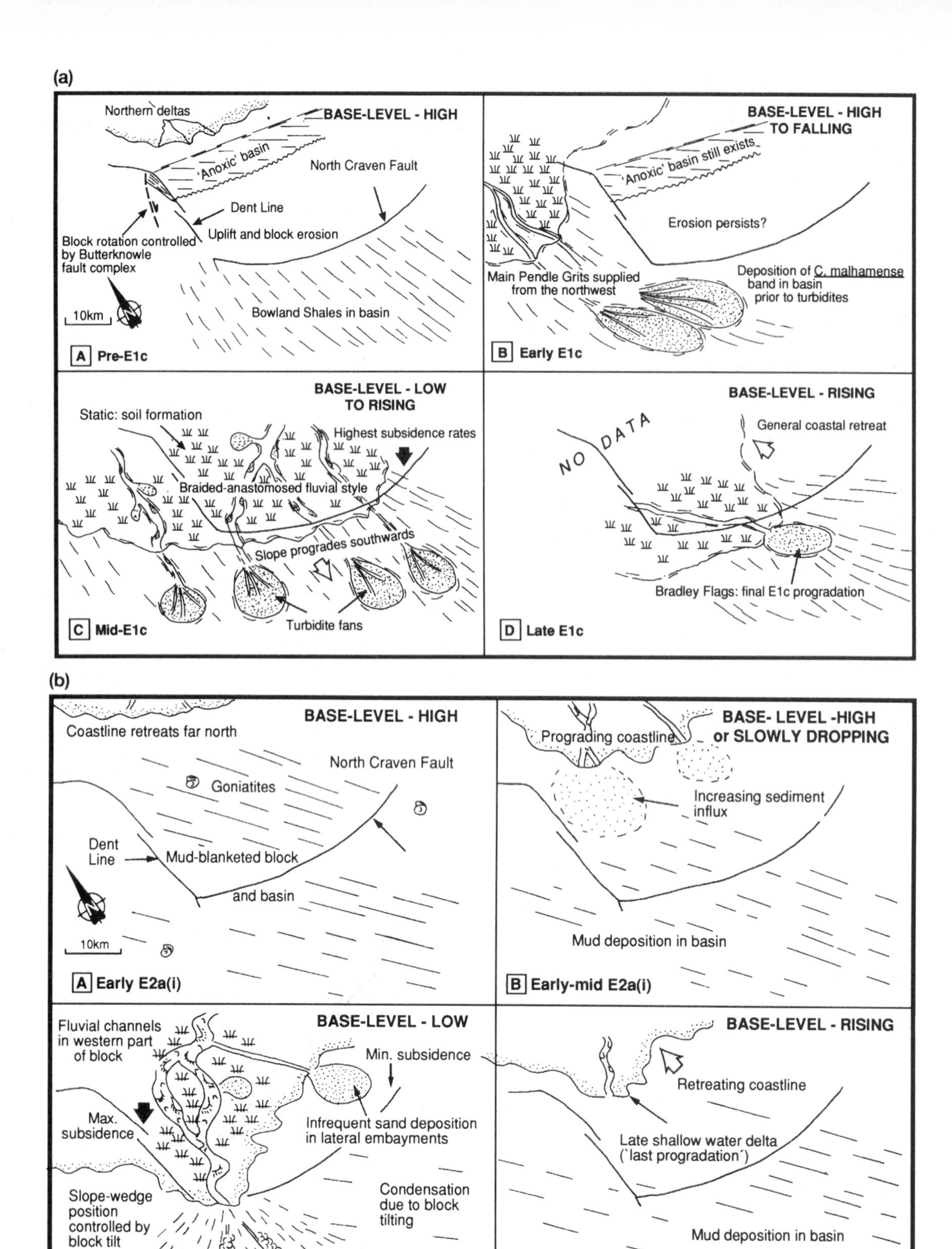

Fig. 4. Palaeogeographic sketches of (a) the E1c minor cycle; and (b) the E2a(i) minor cycle. Inferred base-level conditions are given in relation to stages in the evolution of the depositional systems.

nels appear to have been braided and low sinuous, although the preservation of relatively thick flood plain suites suggests major interchannel areas existed between active channels. Thus, the overall fluvial pattern may have compared with a 'braided-anastomosed' pattern, as e.g. seen in the Brahmaputra (Bristow, 1987).

Both the thickness and facies changes across the Askrigg Block imply a differential subsidence control on the deposition, with the highest rates occurring in the southeast of the block (Fig. 4(a)), while the northwest of the block remained relatively static. This contrasts markedly with the earlier northwards tilt which caused the pre-E1c unconformity to develop.

Although the distribution of the fluvial deposits on the block is generally related to the differential subsidence, local and important variations occur which cannot be explained by this. In the central, northern parts, where the thickness of the fluvials is generally very thin, lenticular sandbodies occur at the base of the cycle and appear to be incised into the underlying, pre-unconformity deposits (cf. Fig. 13 in Dunham & Wilson, 1985). These fluvial sandbodies may reflect the filling of incised small 'valleys' caused by a drop of base level. Such a drop of base level is also documented elsewhere, for instance in the northwest of the block, where the major E1c Mirk Fell Palaeosol (see above) is developed on a substratum of basinal mudstones.

The final phase of deposition in the E1c cycle on the Askrigg Block was the deposition of a locally developed, very shallow water delta in the southeast of the block on top of the fluvial succession (Fig. 4(a)).

In the Craven Basin, a different E1c succession occurs both in terms of facies and thickness, reflecting the pronounced bathymetric change from the perched Askrigg Block into the deep Craven Basin. A turbidite succession, the Pendle Grits, overlies the basal goniatite band. The turbidites have been interpreted as a stacked submarine fan channel complex that initially was supplied from the west *around* the Askrigg Block (Fig. 4(a); Sims, 1988). At a later stage, supply of sediment started coming across the block, thus prograding beyond the southern edge of the block and causing a major slope succession, the Pendle Shales, to develop. The slope succession coarsens further upwards into a major fluviodeltaic unit, the Warley Wise Grit and lateral equivalents (cf. Collinson, 1988; Sims, 1988). This major upward-shallowing succession, which attains a thickness up to 800 m, contrasts markedly with the maximum 65 m thick and laterally variable fluvial succession on the Askrigg Block.

The final phase of deposition in the basin was marked by the southeasterly progradation of a delta, the Bradley Flags. This deltaic succession is different in lithology and general grain size to the sediments below (Baines, 1977), and may represent an entirely different source area. Nevertheless, it occurs on top of the fluvial succession and probably represents general retreat of the E1c depositional system (Fig. 4(a), part A).

E2a(i) cycle (Fig. 4(b))

The goniatite band at the base of this cycle is present over much of the Askrigg Block, and directly overlies the E1c palaeosol in the northwest of the block and the thin deltaic succession on top of the E1c fluvials in the southeast. In the Craven Basin, the band is found in the extreme east and west (Fig. 4(b)), and here directly overlies fluvial-channel deposits.

On the Askrigg Block, a poorly developed coarsening-upward succession of basinal to prodeltaic (?) mudstones and siltstones overlies the band in the northwest. This succession is directly overlain by single-storey, lenticular and localized fluvial-channel sandstones with clearly erosive bases (Rowell & Scanlon, 1957b). A regionally extensive fluvial sheet sand complex overlies the lenticular sandstone, and this complex represents deposition in braided-anastomosed, low-sinuosity channels. The fluvial channels are localized to the western part of the Askrigg Block. In addition, palaeocurrent data show a consistent palaeoflow towards the south–southwest, i.e. along the trace of the Dent Fault (Fig. 4(b); see also Fig. 11).

In the southeastern portions of the block, no major fluvial-channel sandstones occur, and only a thin upward-coarsening succession occurs which culminates in a rooted and burrowed flood plain succession.

These lateral facies contrasts across the Askrigg Block imply increased subsidence rates took place in the west of the block along the Dent Fault. It is worth noting that this differential subsidence pattern is opposite to that inferred for the underlying E1c cycle.

In the Craven Basin, the E2a(i) sediments seem to reflect the block patterns quite well. In the west of the basin, a large delta system, represented by the lower parts of the Roeburndale Formation prograded southwards during E2a(i) time. This delta's slope in the early phase was dominated by the deposition of sand from channelized turbidity currents (Fig. 4(b)). This phase may reflect the incision of the early, lenticular fluvial channels on the Askrigg Block (see above) with a general bypass of sediment across the block. The channelized turbidites coarsen up into a distributary channel succession, which is further overlain by a palaeosol. Thus, the Craven Basin in this area filled to base level, as was the case for the preceding E1c cycle. The palaeosol is overlain by offshore mudstones, which form the base of the next cycle (see below).

In the east of the basin, only a condensed mudstone interval is found at the E2a(i) level, reflecting the character of the E2a(i) sediments on the corresponding southeastern part of the Askrigg Block where mudstones dominate (see above).

E2a(ii) cycle (Fig. 5(a))

A base-level rise at the start of this cycle caused flooding and introduction of goniatites in the basin and deposition of a condensed fossil band, while on the Askrigg Block, a shelly limestone with only sporadic goniatites was laid down (Fig. 5(a); Burgess & Ramsbottom, 1970; Dunham & Wilson, 1985). This facies change of the fossil band probably reflects a shallower water depth and more nearshore conditions on the block.

Overlying the E2a(ii) fossil band in the northwest of the Askrigg Block is a localized shallow-marine sandwave complex, the Fossil Sandstone, which is one of the very few documented shallow marine sands of substantial volume in the Namurian in the British Isles (cf. Brenner & Martinsen, 1990). The sandwave complex has been interpreted to result from transgressive reworking of a former westerly delta lobe (Brenner & Martinsen, 1990), possibly of E2a(i) age. Occasional storms caused migration of the sandwaves eastward where the complex became moribund in the northwest corner of the Askrigg Block (Fig. 5(a)). Probable storm deposits are also observed in the northeastern portion of the Craven Basin (Martinsen, 1990a), thus making the early part of the E2a(ii) cycle rather distinctive in British Namurian terms.

A poorly developed upward-coarsening succession occurs above the Fossil Sandstone in the northwest of the block, which at the top is truncated by a major, coarse-grained multi-storey fluvial-channel complex (Pickersett Edge Grit) which represents the deposits of southwesterly flowing, braided, low sinuosity channels (Fig. 5(a)). This channel unit is in places seen to cut into thin, shallow marine sands, perhaps suggesting a base-level drop. In the southeast, mudstones predominate at this level. Only near the top of the cycle are occasional, rooted, crevasse splay-type sandstones observed which are further overlain by one single-storey distributary channel unit and associated mouth-bar deposits (top of Nidderdale Shales). This is then overlain by a much coarser-grained, multi-storey fluvial unit (Red Scar Grit), which probably correlates timewise with the coarse-grained fluvial unit in the west of the block, although it is substantially thinner and less extensive laterally (Wilson, 1960).

This relationship suggests that subsidence during E2a(ii) time was consistent with that of the E2a(i), with increased subsidence rates in the west of the Askrigg Block causing the major rivers to flow along the trace of the Dent Fault towards the south–southwest (Fig. 5(a)). Nevertheless, the differential subsidence pattern may have been less pronounced than for the E2a(i) cycle since fluvial deposits also occur in the east of the block, in contrast to the E2a(i).

In the Craven Basin, E2a(ii) deposition largely mimics that of the preceding E2a(i) cycle (Figs 4(b) & 5(a)). In the west, channelized turbidites which were fed from the Askrigg Block occur at the base of the cycle in the middle parts of the Roeburndale Formation. The turbidites are overlain and in part form the base of a coarsening-upward succession which culminates in a distributary channel sandstone and an overlying palaeosol. In the east of the basin, no turbidites occur at the base of the cycle, and a coarsening-upward succession is observed which at the top has a major fluvial-channel sandstone, the Marchup Grit. This contrast between the east and the west may reflect, as within the E2a(i) cycle, an increased sediment input during the early parts of the cycle into the west of the basin due to the localization of fluvial channels on the west of the block. This was probably a direct result of a westward tilt of the Askrigg Block.

Sand input in the form of turbidity currents may have been enhanced by a base-level drop (Fig. 5(a)), which is also suggested by the apparent incision by the major fluvial complex in the west of the block into shallow-marine sands (see above).

(a)

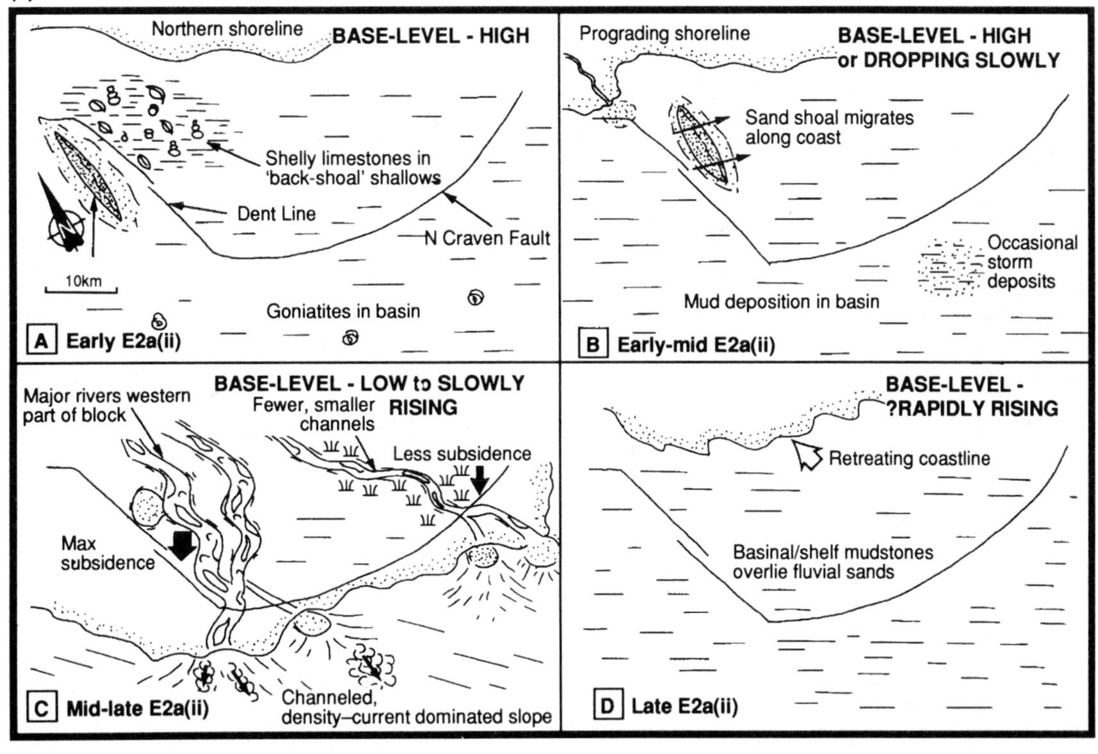

(b)

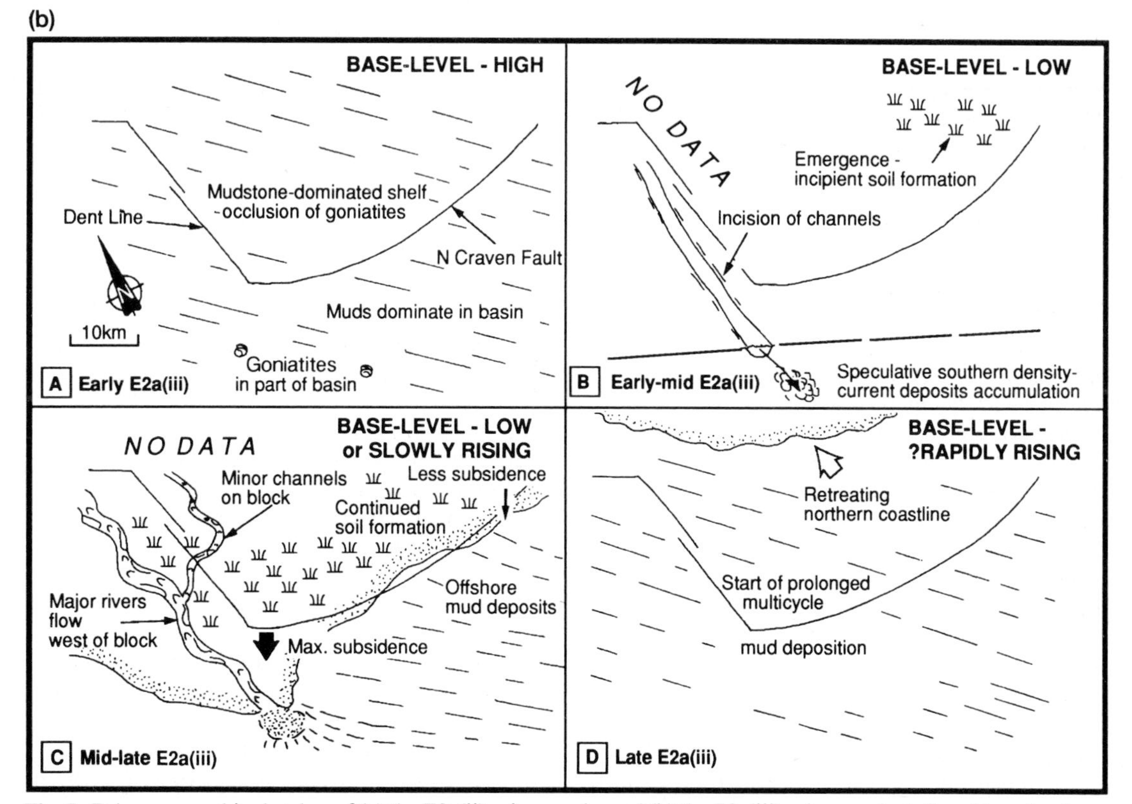

Fig. 5. Palaeogeographic sketches of (a) the E2a(ii) minor cycle; and (b) the E2a(iii) minor cycle. Inferred base-level conditions are given in relation to stages in the evolution of the depositional systems.

E2a(iii) cycle (Fig. 5(b))

Only thin mudstones overlain by a palaeosol occur on the eastern part of the Askrigg Block in this cycle (Fig. 5(b)). In the west of the block, a relatively thin (up to 5 m thick) coarse-grained fluvial-channel succession occurs directly and erosively above a thin (around 5 m thick), black and homogeneous, probably offshore mudstone unit. Thus, a significant portion of the facies in between seem to have been truncated by the channel.

In the Craven Basin, a 40 m-thick fluvial succession occurs in the extreme west represented by the Ward's Stone Sandstone. These fluvial sands represent deposition from braided, low-sinuosity rivers which were flowing from the northwest towards the southeast, i.e. they originated west of the Askrigg Block and only marginally flowed onto the block (Fig. 5(b); Martinsen, 1990b). It is important to note that the Ward's Stone Sandstone channels in places entirely truncate underlying E2a(iii) deposits and erode into the E2a(ii) cycle (cf. Wilson *et al.*, 1989). Eastwards, the E2a(iii) cycle thins markedly, and in the east of the Craven Basin, the entire cycle is apparently represented by a thin mudstone interval. This pattern again accords well with the pattern on the Askrigg Block where only offshore mudstones and subsequently a palaeosol were developed during the cycle, and consequently little or no sediment was transported across the block.

Late depositional patterns on the block and in the basin ('ramp' phase)

Introduction

This period encompasses the E2b1(i)–H2c(ii) cycles and represents a time when deposition in the Craven Basin and on the Askrigg Block tended to be much more uniform than hitherto. Variations existed, both in terms of facies and thickness, but the extreme differences seen in earlier cycles were absent. Above, this was ascribed to a generally decreasing subsidence rate, which probably, sometime during the E2b1(i)–E2b2(ii) cycles, attained a fairly uniform level throughout the region, both on the block and in the basin.

E2b1(i)–E2b2(ii) cycles (Fig. 6(a))

These cycles represent one of three multi-cycle periods when the entire Craven–Askrigg area was entirely dominated by fine-grained sedimentation. This is manifested in multi-cycle condensed mudstone units which are found over the entire study area (cf. Martinsen, 1990a). These mudstone units are extremely complex, and despite being very fine-grained often display an intricate pattern of several stacked and diverse faunal bands (some of which contain goniatites) interbedded with barren mudstones, organic-rich shales or even impure coal (Wilson & Thompson, 1959). Thus, despite the fact that coarser lithologies are not present, the basic cyclicity pattern is retained, probably reflecting changes in base level (Fig. 6(a)).

During relative lowstand periods, non-deposition or soil formation occurred in emergent areas while barren mudstones were laid down in submerged areas. During relative highstands, fauna was introduced (cf. also Collinson, 1988; Holdsworth & Collinson, 1988) and faunal-rich bands of mudstone deposited (Fig. 6(a)).

The reason for the prolonged absence of clastic sediment from the Craven–Askrigg is disputable. Two likely mechanisms could separately or collectively have generated such a development: (i) a major diversion of sediment supply away from the region; or (ii) an unusually high base level could have caused the coastline to sit far to the north, thus inhibiting the transport of coarser sediment to the Craven–Askrigg region. This argument is also relevant to the two overlying multi-cycle mudstone units and is discussed further below.

E2b3 cycle (Fig. 6(b))

This cycle commenced with a regionally important phase of fine-grained offshore deposition. The basal goniatite band is only found in the Craven Basin, but probable correlatives occur on parts of the Askrigg Block (Fig. 6(b); Martinsen, 1990a).

In the southeast of the block, a fluviodeltaic succession, the Scar House Beds, overlies the basal fine-grained unit (Martinsen, 1990b). This delta appears to have been very localized and prograded from the northeast (Fig. 6(b); cf. Wilson, 1960; Martinsen, 1990b). It only reached the southeastern portions of the block since the equivalent stratigraphic interval in the Craven Basin is mudstone-dominated and thinner. However, sharp-based, probably turbiditic sands (although thin and relatively few) occur immediately south of the block in the northeast of the Craven Basin (Wilson, 1977). These sands were probably distal representatives of

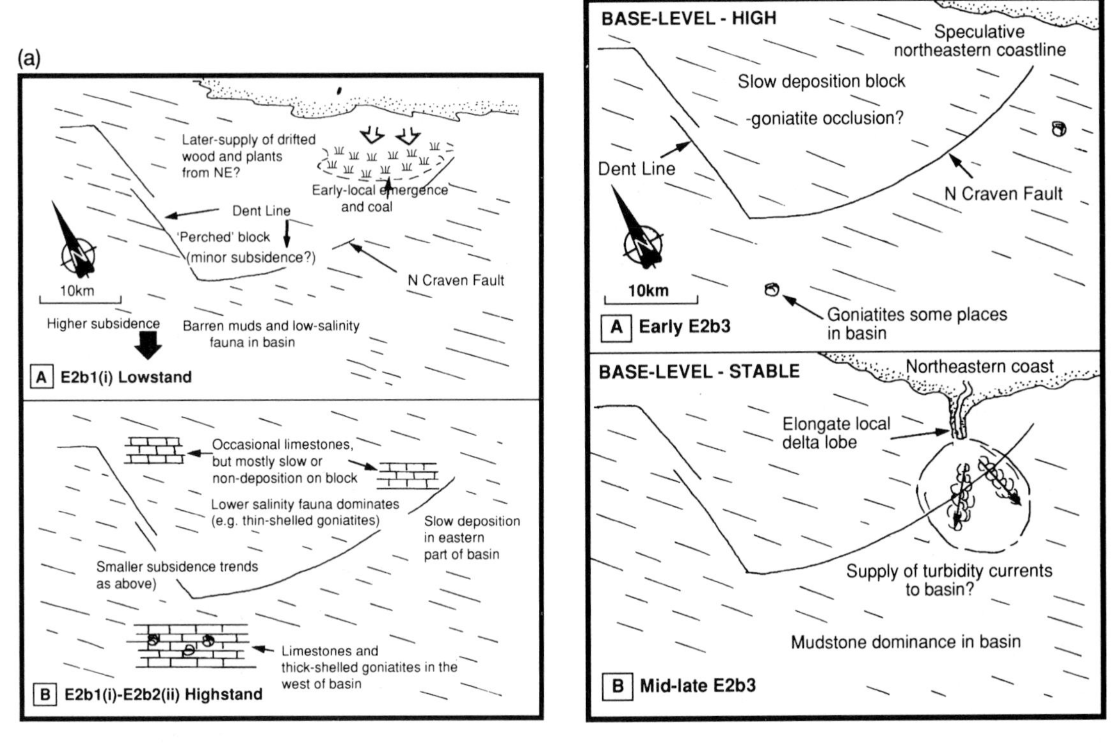

Fig. 6. Palaeogeographic sketches of (a) the E2b1(i)–E2b2(ii) minor cycles; and (b) the E2b3 minor cycle. Inferred base-level conditions are given in relation to stages in the evolution of the depositional systems. In (a), the sketch could represent any of the five cycles.

density current deposits which appear to have dominated this delta's front (Fig. 6(b); cf. Martinsen, 1990b).

Nowhere else within the Craven–Askrigg area are sediments coarser than mudstone observed, also suggesting the E2b3 progradation was localized. This may have been caused by the shoreline being positioned north of the studied area with only the most southerly delta lobe progradation being recorded.

E2c1 cycle (Fig. 7(a))

Flooding during the early part of the E2c1 cycle transgressed the E2b3 delta in the southeast of the block and mudstones were deposited over the entire Craven–Askrigg area (Fig. 7(a)). On the Askrigg Block, E2c1 deposits are only found in the northwest, where a thin, deltaic succession occurs. In the southeast, there appears to be no representative of the E2c1 cycle, suggesting that the sediments were either eroded by the overlying cycle's depositional system (see below), or that non-deposition occurred. The latter seems unlikely, because in the northeast of the Craven Basin, i.e. in a basinward direction, a 30 m-thick, wave-influenced deltaic succession, the Nesfield Sandstone, occurs within the E2c1 cycle (Fig. 7(a)). This delta was fed from the Askrigg Block (Martinsen, 1990a), probably during a period of relative lowstand or stability of base level.

A major fluviodeltaic succession occurs in the E2c1 cycle in the west of the basin, the Heversham House Sandstone (cf. Wilson *et al.*, 1989), and this was probably fed from the northwest. Thus, prograding deltas reached further south and were more widely developed in E2c1 times than in E2b3. However, in the central area of the Craven Basin, only fine-grained muds with occasional sharp-based sands were deposited during E2c1 (Fig. 7(a)). This

(a)

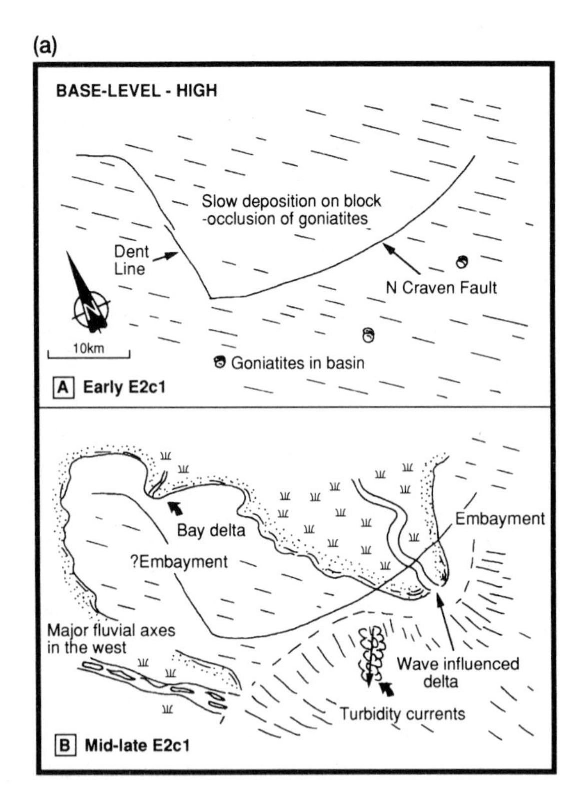

Fig. 7. Palaeogeographic sketches of (a) the E2c1 minor cycle; and (b) the E2c2(i) minor cycle. Inferred base-level conditions are given in relation to stages in the evolution of the depositional systems.

(b)

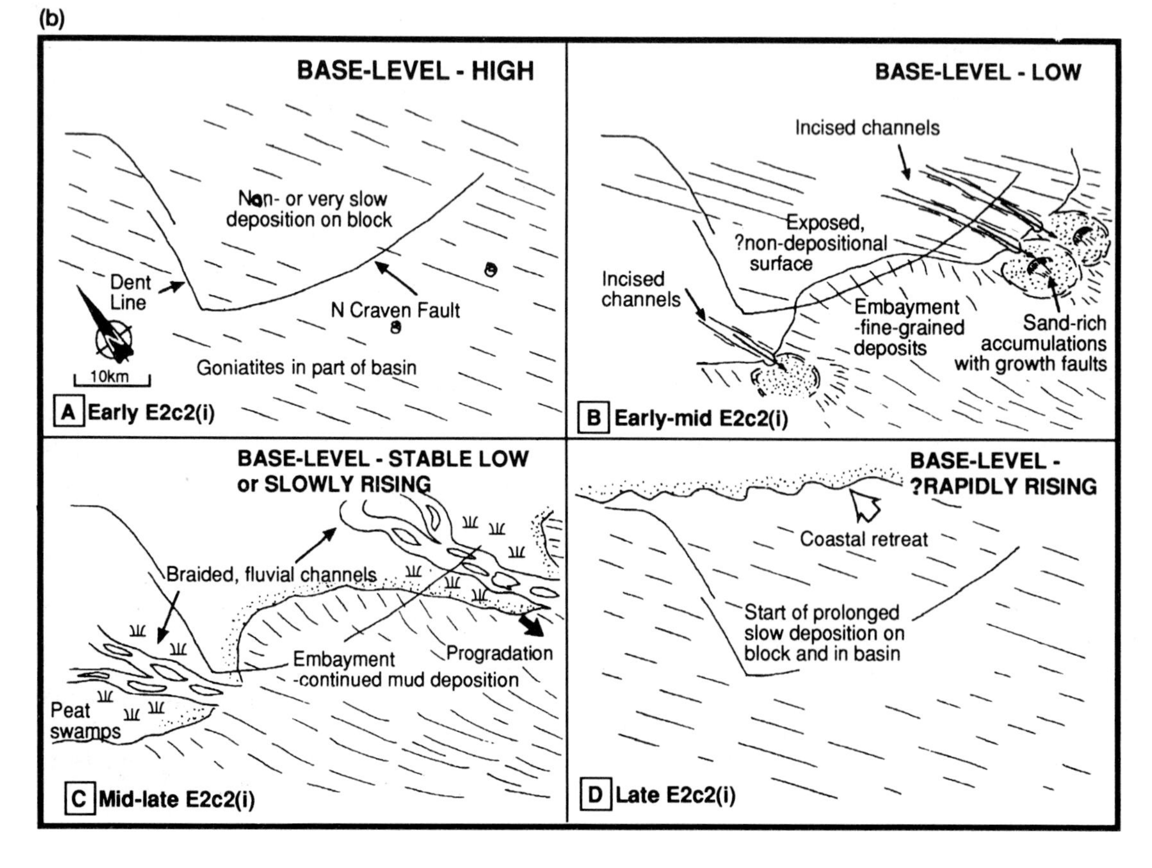

lack of progradation into the north-central parts of the basin persisted until the end of the studied time interval, the H2c(ii), although deltaics are seen both to the east and the west in several cycles.

E2c2(i) cycle (Fig. 7(b))

A base-level rise at the start of the cycle caused flooding of the E2c1 deltas and retreat of active, clastic deposition north of the Craven–Askrigg area (Fig. 7(b)). This led to a phase of fine-grained and slow deposition across the region.

In the southeast of the Askrigg Block, the base of the cycle is marked by a prominent erosion surface below the fluvial Lower Follifoot Grit. Thus, no offshore or prodeltaic deposits are seen at the base of the cycle there, and the erosion surface sits directly on top of the E2b3 Scar House Beds (see above). No representatives of the E2c1 cycle are seen in the area. The Lower Follifoot Grit has an uneven distribution, because locally, lenticular fluvial sands up to 15 m thick occur below a main sheet sandstone body. The sheet sand can be correlated throughout the area (Wilson, 1960), and represents deposition from braided, low-sinuosity channels (Martinsen, 1990a). Consequently, it appears that the Lower Follifoot Grit incised into and truncated the E2c1 deposits on the Askrigg Block (Fig. 7(b)), probably because of a prominent base-level drop. Since the deposits of a whole cycle were eroded away, it seems that this drop of base level must have been more substantial than those inferred for other cycles.

In the east of the Craven Basin, a deltaic succession, the Middleton Grit, is the E2c2(i) representative (Fig. 7(b)). This is only slightly thicker than the Lower Follifoot Grit on the Askrigg Block, suggesting there was little difference in subsidence and thus probably bathymetry between block and basin. Near the top of the Middleton Grit, coarse-grained fluvial-channel sands occur, but these do not seem to be erosive at the scale of their counterparts on the Askrigg Block. Growth faulting influenced the deltaic part of the Middleton Grit (Martinsen, 1990a), a feature which is not seen elsewhere within the studied region.

Towards the west, in the north-central parts of the Craven Basin, the E2c2(i) cycle is only represented by a thin barren mudstone above the basal marine band, and no representative of the prominent base-level drop is seen. Further west, however, a fluviodeltaic succession occurs, the Silver Hills Sandstone. At the top of this, a coarse-grained fluvial-channel unit sits with prominent erosional contact on the deltaics below.

Thus, it appears that E2c2(i) clastic deposition reached even further south than the previous E2c1 system. Consequently, the E2b3–E2c2(i) depositional systems reached successively further south with time (further discussion below) with fluvial facies only being present in the uppermost cycle (cf. Figs 6(a), 6(b) & 7(a)).

E2c2(ii)–H1b(iii) cycles (Fig. 8(a))

These nine cycles represent a prolonged period of predominantly fine-grained sedimentation within which coarser clastics were generally absent in the studied region. This gave rise to deposition of condensed, multi-cycle mudstone units in which several of the goniatite bands are missing, particularly on the Askrigg Block.

The basic cyclicity is apparent in several areas, with faunal bands being interbedded with barren sediments, suggesting alternating periods of high and low base level (Fig. 8(a)). However, the very thin nature of the multi-cycle mudstone unit on the Askrigg Block is either explained by condensation or non-deposition. It is suggested that non-deposition occurred at times of relatively high base-level (Fig. 8(a)), when active clastic systems were very far removed from the region, while periods of relatively lower base level were dominated by slow deposition. Another explanation is that the non-deposition took place during periods of relatively low base level, but nowhere within the Craven–Askrigg region is there evidence of pedogenic alteration or erosion during this period.

An unusually high base level may have existed during the H1b(i) cycle (Fig. 8(a)) because the cycle's goniatite band is found also on structural 'highs' such as the Askrigg Block and the Bowland Block (cf. Arthurton *et al.*, 1988; Martinsen, 1990a) in the north-central parts of the Craven Basin.

It is likely that a major shift in sediment supply directions was the direct cause of the prolonged absence of coarse clastic influx into the area. This is supported by the fact that the multi-cycle mudstone unit thickens progressively and quite dramatically in the extreme west of the Craven Basin. There, it also includes occasional sharp-based sandstone beds, which may be turbiditic in origin, within an otherwise entirely basinal mudstone succession (Moseley, 1954; Wilson *et al.*, 1989). These sedi-

(a)

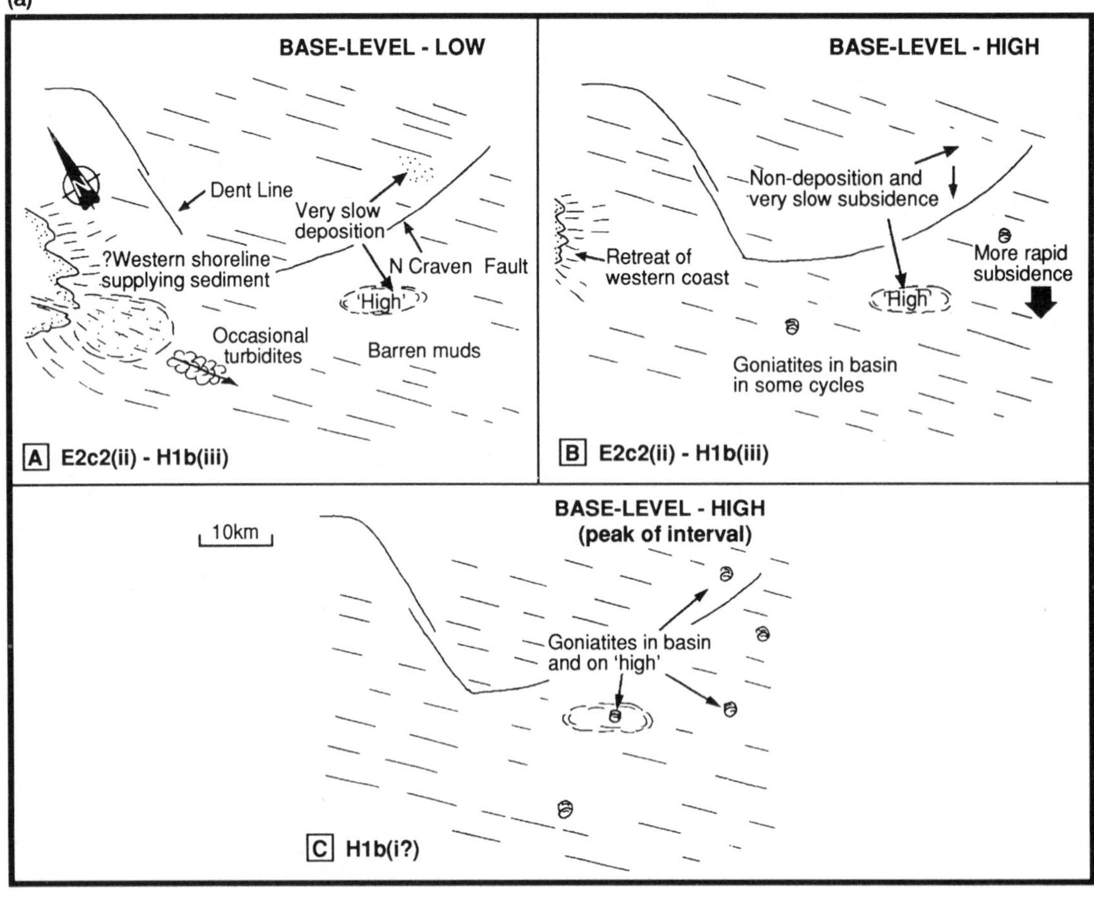

BASE-LEVEL - LOW
Dent Line
Very slow deposition
?Western shoreline supplying sediment
N Craven Fault
'High'
Occasional turbidites
Barren muds
A E2c2(ii) - H1b(iii)
BASE-LEVEL - HIGH
Non-deposition and very slow subsidence
Retreat of western coast
More rapid subsidence
'High'
Goniatites in basin in some cycles
B E2c2(ii) - H1b(iii)
BASE-LEVEL - HIGH (peak of interval)
10km
Goniatites in basin and on 'high'
C H1b(i?)

(b)

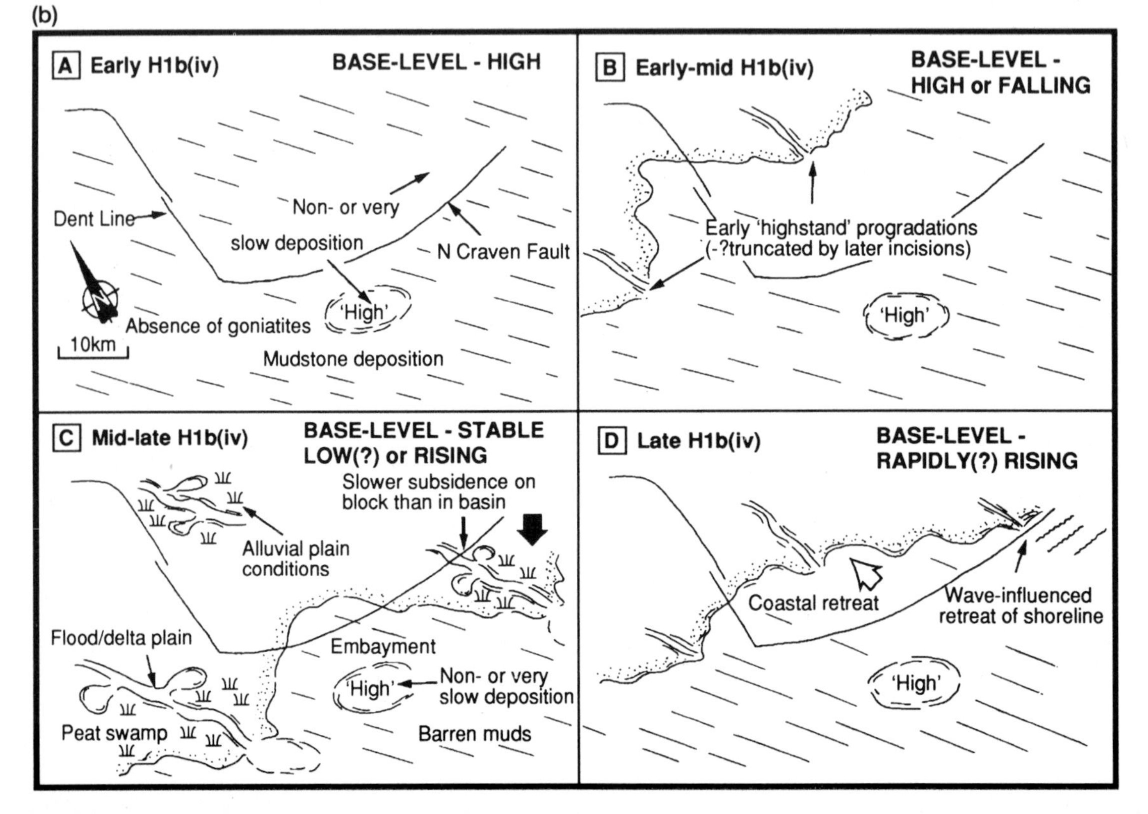

A Early H1b(iv)
BASE-LEVEL - HIGH
Dent Line
Non- or very slow deposition
N Craven Fault
'High'
Absence of goniatites
10km
Mudstone deposition
B Early-mid H1b(iv)
BASE-LEVEL - HIGH or FALLING
Early 'highstand' progradations (-?truncated by later incisions)
'High'
C Mid-late H1b(iv)
BASE-LEVEL - STABLE LOW(?) or RISING
Slower subsidence on block than in basin
Alluvial plain conditions
Flood/delta plain
Embayment
'High'
Non- or very slow deposition
Peat swamp
Barren muds
D Late H1b(iv)
BASE-LEVEL - RAPIDLY(?) RISING
Coastal retreat
Wave-influenced retreat of shoreline
'High'

ments were probably fed from the west–northwest, suggesting that major clastic deposition took place in that area (Fig. 8(a)). An unusually high base level may have contributed to the absence of coarse clastics, but the effect of this is uncertain, particularly since active clastic deposition took place in the west.

H1b(iv) cycle (Fig. 8(b))

This cycle was the last during the E1c–H2c(ii) period when active clastic deposition took place in the Craven–Askrigg region. The basal goniatite band does not occur in the region, and the reason for assigning the interval to the H1b(iv) cycle is that this marks the only pronounced desalination 'event' (and hence probably lowstand) within a truly basinal and faunally complete succession further south at Ashover, Derbyshire (cf. Ramsbottom *et al.*, 1962; Holdsworth & Collinson, 1988). In the interval between E2c2(ii) and H2c(ii), there is only one imprecisely dated fluviodeltaic unit in the Craven–Askrigg area. This unit probably corresponds in age to the desalination event (cf. Collinson, 1988 for a broader discussion on cyclic desalination in the British Namurian basins).

In the southeastern portions of the Askrigg Block, the H1b(iv) sediments are erosively based, lenticular fluvial channels and associated, pedogenically modified overbank deposits of the Upper Follifoot Grit. These directly overlie the offshore muds of the underlying multi-cycle mudstone unit. Thus, any sediments representing intervening depositional environments are missing, suggesting truncation by the fluvial channels (Fig. 8(b)), possibly because of a base-level drop.

In the northeast of the basin, the Brocka Bank Grit is a more complete fluviodeltaic succession, and apparently the Clintsfield Grit and the Wellington Crag Sandstone in the far west were also deposited in a deltaic to fluvial environment based on the occurrences of palaeosols (Fig. 8(b); cf. Moseley, 1954). The area in between, in the north-central parts of the basin, was dominated by mudstone deposition and probably represented an embayment. This trend continued that observed in underlying cycles, that clastics 'avoided' the north-central parts of the basin and the area was dominated by mudstone deposition for an extremely prolonged time period.

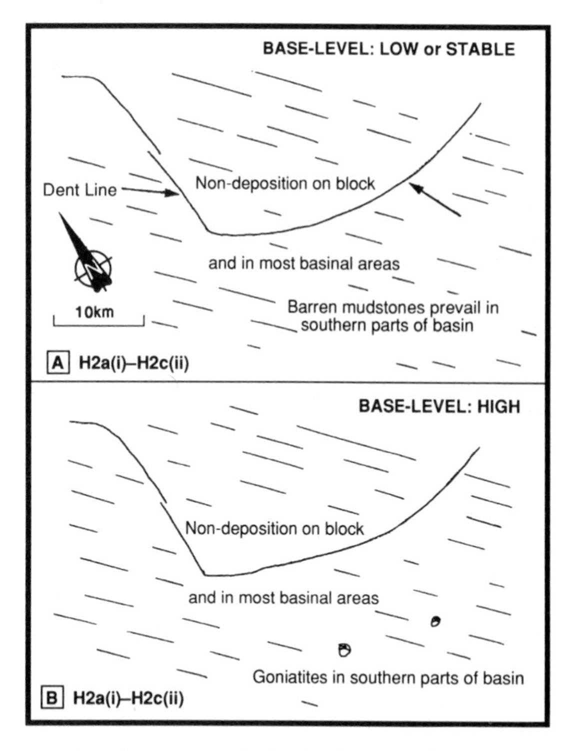

Fig. 9. Palaeogeographic sketch of the H2a(i)–H2c(ii) minor cycles. The relation between base-level and depositional conditions is indicated. The sketch could represent any of the five cycles.

H2a(i)–H2c(ii) cycles (Fig. 9)

These five cycles form the upper part of the studied interval. They form the third time period when only fine-grained deposition and non-deposition were important in the entire Craven–Askrigg region (Fig. 9; cf. Figs 6(a) & 8(a)). However, the basic minor cyclicity may still be recognized for parts of the interval in some areas, primarily in the Craven Basin. On the Askrigg Block, no deposits of H2a(i)–

Fig. 8. (*opposite*) Palaeogeographic sketches of (a) the E2c2(ii)–H1b(iii) minor cycles; and (b) the H1b(iv) minor cycle. Inferred base-level conditions are given in relation to stages in the evolution of the depositional systems. In (a), the sketch could represent any of the cycles, although as shown, the H1b(i) cycle probably represented a particularly high base-level.

H2c(ii) age can be recognized with confidence, due to the lack of goniatite bands. However, no major erosion surfaces or pedogenic modification of underlying sediments have been identified anywhere, except for perhaps in the northwest of the block (cf. Martinsen, 1990a). Thus, the H2a(i)–H2c(ii) cycles remain enigmatic, the only certain factor being the absence of coarse clastics.

Like the two underlying multi-cycle condensed mudstone units, the absence of coarse clastics may have been a major diversion of supply to adjacent regions. In addition, a very high base level may have contributed to keeping active deposition further north. However, without sufficient data, this remains speculative.

CONTROLS ON DEPOSITIONAL PATTERNS

Tectonics

Three factors made tectonism important in the development of the Namurian Craven–Askrigg succession: (i) inherited bathymetric contrast from pre-Namurian extension; (ii) transition of the southern end of the Askrigg Block from a pronounced 'edge' to a subdued 'ramp'; and (iii) readjustments of the Askrigg Block.

Most tectonic activity had ceased by the start of the Namurian (cf. Leeder & McMahon, 1988; Gawthorpe *et al.*, 1989), leaving a pronounced bathymetric contrast between the Askrigg Block and the Craven Basin (Collinson, 1988). Consequently, progradation of substantial, clastic sedimentary systems into the Craven Basin led to the generation of a turbidite succession at the base of the slope, subsequent progradation of the slope and ultimately filling of the basin to base level by fluvial systems as supply persisted. This trend continued at least in three cycles until E2a(ii) times, probably because the differential subsidence between block and basin was substantial enough to create sufficient accommodation space for turbidite-fronted systems following abandonment or retreat of the previous system.

With time, the pronounced bathymetric contrast between block and basin became less important, probably as a result of decaying thermal, post-extensional subsidence in the Craven Basin. Thus, the southern end of the Askrigg Block transformed from a clearly defined edge ('shelf edge') to a

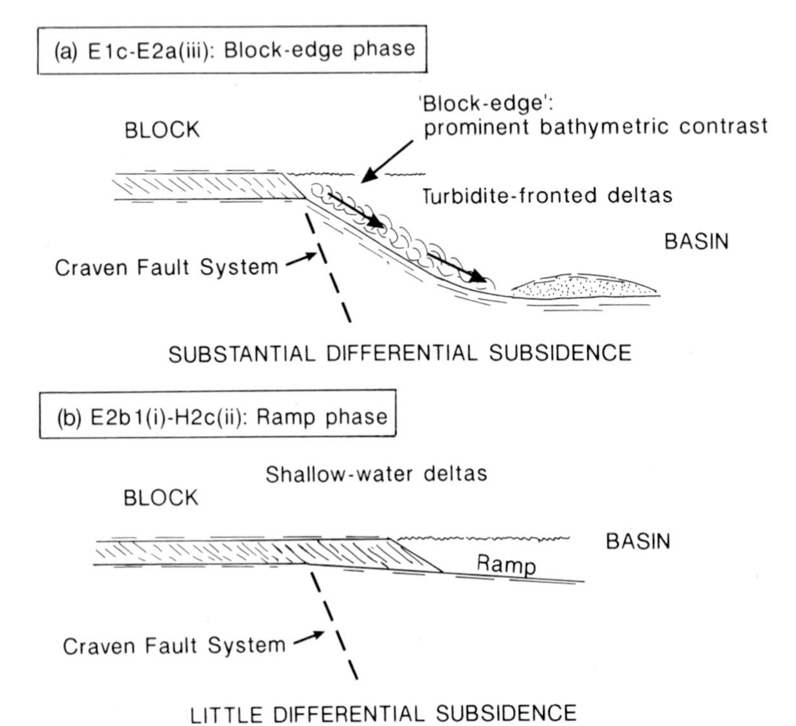

Fig. 10. Schematic drawing showing the transition of the southern margin of the Askrigg Block from a pronounced 'edge' to a 'ramp'. This was accompanied by a transition from turbidite-fronted, deep-water deltas to shallow-water deltas.

shallowly inclined ramp, dipping southwards (Fig. 10). This is evidenced by the much more uniform nature of the deposits on the Askrigg Block and in the Craven Basin both in terms of depositional environments and in terms of thickness of individual cycles from the E2b1(i) to the H2c(ii) cycle (cf. Martinsen, 1990a for details).

There seem to have been minor readjustments of the Askrigg Block during the four earliest cycles, E1c to E2a(iii), also implying reactivations of the block-bounding faults. The apparent movement of the block was heterogeneous and appears to have been different for each successive cycle, with the relationship to the preceding N–S extensional regime being unclear. It is possible that the buoyancy of the Wensleydale Granite coring the Askrigg Block (cf. Fig. 1) played an important role, but this needs clarification.

Sedimentation

The nature and type of depositional systems in the Craven–Askrigg area were a response to the tectonic framework. Only early (E1c–?E2a(iii)) did turbidite-fronted deltas develop because of sufficient accommodation space formed by inherited bathymetry and differential subsidence. Later, shallow-water deltas were important (Fig. 10), probably mainly because the bathymetric contrast between block and basin was insignificant.

Nevertheless, in the early cycles, there seems to have been a two-stage development of deltas within each cycle. During relative highstand at the earliest stages in a cycle, it is probable that deltas developed on the Askrigg Block. This is suggested by the truncated coarsening-upward successions below the main fluvial sandstones there (Fig. 11). These

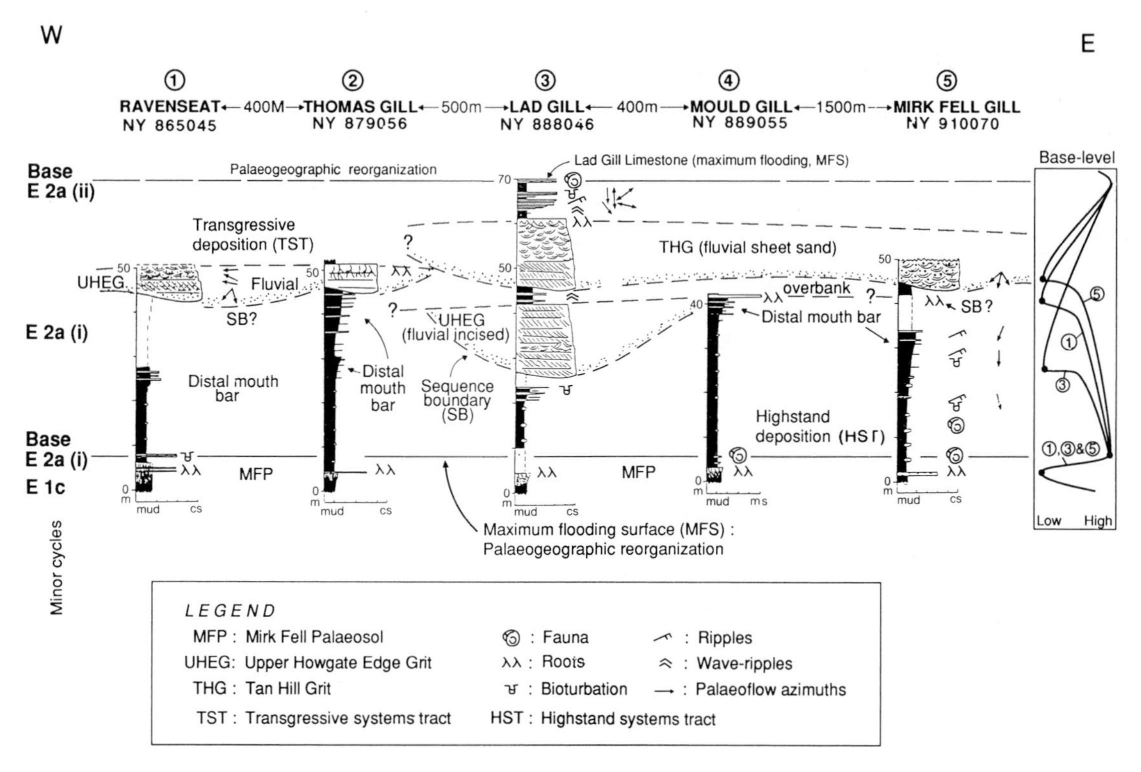

Fig. 11. Selected vertical sections of the E2a(i) cycle deposits from the northwest corner of the Askrigg Block. The British National Grid references of the localities are given for location. See also Martinsen (1990a). Note the laterally restricted development of the inferred sequence boundary below the Upper Howgate Edge Grit in Section 3 and that it cannot be traced into Sections 1 and 2, and only with difficulty into Section 5. Possible base-level curves for three of the sections (1, 3 & 5) are given based on what stratigraphic levels can be used as indicators of base-level lows and highs. Note that the base-level high points are fixed while several possibilities exist for location of the base-level lows. (See text for discussion.)

deltas, which loosely may be termed 'block-deltas', were probably truncated at the base-level fall in each cycle, and deltaic sedimentation shifted to the edge of the block (see, for example, Fig. 4(b)). There, turbidite-fronted 'block-edge' deltas developed. Such a scenario is essentially similar to the Quaternary development of shelf and shelf-edge deltas on the Gulf of Mexico shelf (Suter & Berryhill, 1985), and will be dealt with in more detail elsewhere.

Fluvial and deltaic depositional environments dominate the sandy E1c–H2c successions in the Craven–Askrigg area. The shallow-marine environments are very fine-grained, suggesting that mud-dominated shelves were important in several time periods (cf. Martinsen, 1990a). Only in one cycle are shallow-marine sands found (the E2a(ii); Fig. 5(a)). Namurian basins in the British Isles were generally not subject to a significant influence from storms, ocean currents and tides (cf. Besly & Kelling, 1988 and references therein). There were therefore only minor modifications of fluvial and deltaic deposits by marine reworking upon abandonment.

The multi-cycle mudstone units which occur at three separate stratigraphic levels within the Craven–Askrigg successions imply that major switches of sediment supply took place. Where active deposition moved to during these periods is uncertain, but the thickening of beds in the E2c2(ii)–H1b(iii) unit towards the west suggests relocation in that direction for that period. Eustasy may have contributed in causing this apparent 'major' cyclicity. However, the fact that the E2c2(ii)–H1b(iii) unit thickens westward and that the separate units include very different numbers of minor cycles suggest major switches of sediment supply were important.

It is likely that overall sediment supply decreased with time during the E1c–H2c period. This is evidenced by the fact that although differential subsidence and subsidence overall probably decreased with time, yielding progressively less accommodation space, the depositional systems did not increase their extent toward the south (cf. Figs 14 & 15). This would have been expected if sediment supply remained constant. It is possible that a general wearing down of the source area could have caused this. Another and possibly more likely explanation is that since the Craven–Askrigg area was at the brink of being filled completely by E2a(iii) time (only relatively thin successions were deposited after this compared to previously; cf. Martinsen, 1990a), general sediment supply may have been diverted by a higher gradient to less filled areas further to the west.

Eustasy

Eustasy is probably the only mechanism which sensibly can explain the minor cyclicity (*sensu* Holdsworth & Collinson, 1988). Both the cyclicity frequency (*c.* 180,000 years) and the wide distribution of goniatites are most easily explained by a glacial-eustatic mechanism (cf. Holdsworth & Collinson, 1988; Leeder, 1988; Martinsen, 1990a for a detailed discussion).

The goniatite bands and their correlatives probably represent the eustatic highs, and these are the least present on the slowly subsiding Askrigg Block. Since subsidence apparently was consistently taking place throughout the studied period, the eustatic falls may be marked by the erosion surfaces below the lowermost fluvial sandstones in any minor cycle (see Fig. 11). In contrast to the marine bands, the erosion surfaces are best developed on the Askrigg Block. Particularly where no major changes of palaeogeography across the erosion surfaces can be documented, and there is a lack of palaeosols in interfluve areas, the interpretation of base-level falls is ambiguous (Fig. 11). However, on the Askrigg Block, which appears to have remained a low-subsidence area throughout the studied interval, it is unlikely that a sea-level drop would have been sufficiently balanced by subsidence. Since there seems to be overwhelming evidence for base-level rises, most likely related to eustatic rises, there ought to be some evidence for base-level falls in such a low-subsidence area. The channel-base erosion surfaces are the most likely candidates, yet some are more convincing than others (e.g. Section 3 (Lad Gill) in Fig. 11).

The intervals between the goniatite bands and the fluvial erosion surfaces probably represent a relatively high or slowly falling sea level (highstand systems tract in Exxon terminology), while those above the erosion surface but below the overlying marine band represent a gradually rising sea level (Fig. 11; transgressive systems tract in Exxon terms). It is interesting to note that while the position of the relative high of sea level is fixed vertically in Fig. 11, the position of the apparent low varies between sections because of lateral variability of the depth of the incised channels and their

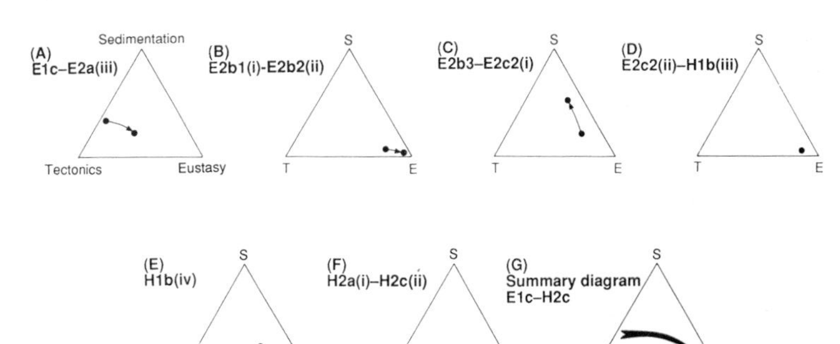

Fig. 12. Succession of ternary diagrams showing the interpreted qualitative relative change of the major controls eustasy, sediment supply and tectonics on the depositional conditions in the Craven–Askrigg area. (Based on an idea of Galloway, 1989a.)

correlatives. This seems to have implications for the correlation potential of the important surfaces. The marine flooding surface can easily be correlated across the five sections, while it is much more uncertain how to trace the base of channel erosion surfaces and possibly correlative palaeosols across (see Fig. 11). Particularly in areas where the incised channels are single storey and separated by interfluves, and the incision was polyphasal, this may be a significant problem.

While the eustatic influence at the scale of minor cycles is fairly convincing, such an influence is less readily documented at larger scales. However, the occurrence of the multi-cycle mudstone units at three separate stratigraphic levels points to an external control mechanism for their generation. Major switches in sediment supply are the most plausible control mechanism.

Variability with time

Considering qualitatively the temporal relative influence of the major controls — eustasy, tectonism and sedimentation/sediment input — with time (Fig. 12), it seems as though eustasy increased in relative importance. This was probably not because eustatic change *increased* in magnitude. Rather, the decaying influence from tectonics in the post-extensional phase and the apparently decreasing sediment supply with time combined to cause a *relative* increase in eustatic influence. It is interesting from a sequence-stratigraphic point of view that this relative increase in eustatic influence was accompanied by a reduced rate of clastic deposition (Fig. 12; see also above), showing rather clearly the importance of sediment supply and tectonism.

APPLICATION OF SEQUENCE-STRATIGRAPHIC MODELS

Introduction

Based on the detailed sedimentological interpretations and palaeoenvironmental reconstructions discussed in the previous sections, sequence-stratigraphic models of Galloway (1989a) and Exxon Production Research (EPR; e.g. Posamentier *et al.*, 1988; Van Wagoner *et al.*, 1990) are applied to the Craven–Askrigg successions. The main difference between these models is where the sequence boundary is placed. The EPR *depositional sequence model* emphasizes 'unconformities and their correlative conformities bounding relative conformable packages of genetically related strata', because these tend to mark important hiatuses in successions in marginal areas of basins, and tend to be well seen on seismic sections. This is firmly rooted in the work by Sloss (1963) and in seismic stratigraphy (e.g. Mitchum *et al.*, 1977). Galloway's (1989a) *genetic stratigraphic-sequence* approach is 180° phase shifted, emphasizing marine flooding surfaces as the important boundaries because these generally tend to mark major reorganizations of the sedimentary systems and are claimed to be easier to correlate both in the sub-crop and outcrop. Galloway's view is based on Frazier's (1974) work on the Gulf Coast Quaternary depositional systems.

Recently, Walker (1990) has discussed the relevance of the sequence models in relation to traditional facies schemes. In particular, he considers the position of important stratigraphic boundaries related to packages of genetically related sediments to be important. These views are highly relevant also for the following discussion, because the interpretation of apparent key surfaces as sequence boundaries and important stratigraphic breaks has importance for the overall interpretation of the basin fill. This is further discussed at the end of the section.

The EPR model and the Craven–Askrigg area

Sequence boundaries

In this model, the only candidates for sequence boundaries within the E1c–H2c succession are the erosion surfaces at the bases of the lowest fluvial-channel sandstones in each cycle (see Fig. 11). At least some of these erosion surfaces may represent incision related to drops of base level (cf., for example, the E2a(iii) cycle; Fig 5(b)). On the Askrigg Block, in particular, the channel sandstones may be sedimentologically unrelated to the sediments below (thus breaking Walther's Law), and represent significant basinward shift of facies.

However, as shown above, the erosion surfaces are generally localized and correlative horizons are difficult to recognize in the interchannel facies. Consequently, they are difficult to trace beyond the extent of individual outcrops (Fig. 11). The general absence of reliable palaeosols with which to correlate the erosion surfaces in interfluve areas means that channel-base erosion surfaces are inappropriate as major boundaries, although base-level falls can be inferred across them. For example, within the E2a(i) cycle, the erosion surface at the base of the lowest channel unit in the Upper Howgate Edge Grit on the northwestern part of the block cannot be traced with confidence between closely spaced outcrops (Fig. 11). The position of the sequence boundary is ambiguous in the adjacent sections. If this variability occurs in such a local area, extreme care is needed when correlating sequence boundaries over larger distances and in thick successions.

The unconformity at the base of the E1c cycle on the Askrigg Block is another candidate for a sequence boundary (Fig. 4(a)). This was caused by tectonic tilting of the Askrigg Block northward, although possibly modified locally by a later base-level drop causing channel incision and a marked basinward shift of facies (see above). This unconformity could compare with a more regional type 1 sequence boundary since it separates distinctly different sediments and marks a prominent boundary (cf. Dunham & Wilson, 1985).

It is difficult to trace the bases of the fluvial-channel sandstones into the basinal areas, or rather, find their correlative conformities. For the lowest cycles, the generation of the erosion surface may have corresponded with initiation of turbidite deposition in the basin. However, this is not the case in the E1c cycle, as the first turbidites in the basin were not fed across the Askrigg Block (Sims, 1988), and the unconformity on the block was mainly generated by tectonic tilting northward and only locally modified by later river incision (Fig. 4(a); see above).

Sequence components

Application of the EPR model will be attempted at the scale of the basin fill first, using apparent stacking patterns of single minor cycles. Taking a composite cross-section from the Askrigg Block into the Craven Basin (Fig. 13; this makes no consideration of 3-D geographic variability of individual cycles; see below), it seems possible to find apparent representatives for most components of the depositional sequence model. Taking the type 1 sequence boundary at the sub-E1c unconformity on the Askrigg Block (Figs 13 & 14), a turbidite complex, the Pendle Grits, overlies this, and possibly corresponds to a *lowstand fan* in the model (Fig. 13). A progradational slope and delta system, the E1c Pendle Shales and Grassington Grit and equivalents, further overlies the turbidites, and may represent a *lowstand wedge* and *lowstand delta.* The slope deposits are probably the first equivalents of fluvial deposition on the Askrigg Block (cf. Fig. 4(a)), perhaps in relation to a rising base level, while the delta represents the first filling of the Craven Basin to base level (cf. Collinson, 1988).

Three minor cycles (E2a(i) cycle to E2a(iii) cycle), separated by single goniatite bands, overlie the E1c deposits (Fig. 13). Judging from the most southerly known extent of fluvial deposition, these cycles show an irregular but relatively aggradational to slightly progradational stacking pattern, in relation to the E1c cycle below. Following the EPR scheme, a *transgressive systems tract* should occur at this level, although the overall stacking pattern of the minor cycles is not particularly retrogradational (Fig. 13). This inconsistency is important, and it may be related to a relatively high sediment supply.

An alternative interpretation is to view only the lower part of the E1c delta as part of the lowstand. The E1c delta seems to develop a backstepping

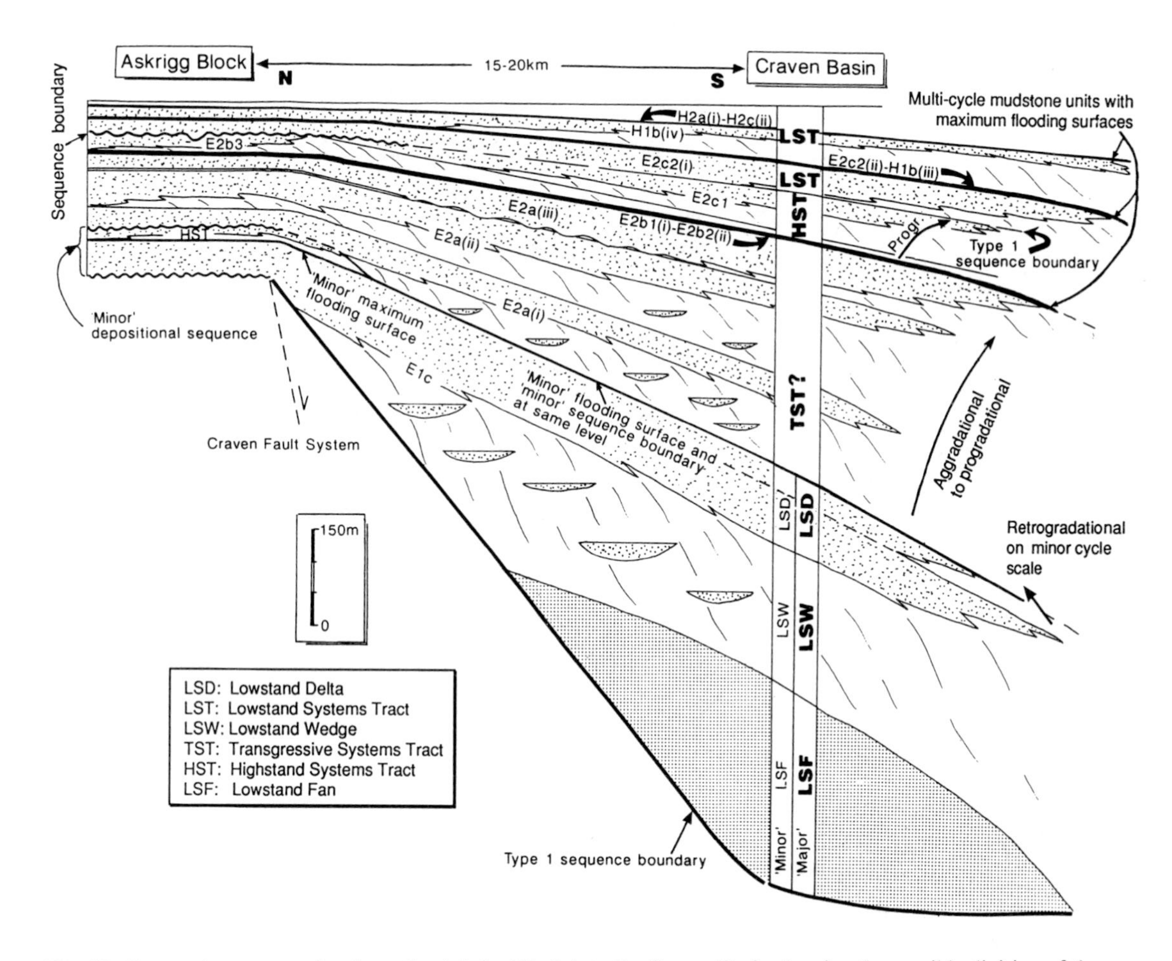

Fig. 13. Composite cross-section from the Askrigg Block into the Craven Basin showing the possible division of the stratigraphy into EPR depositional sequences. Note the two scales of sequences.

pattern internally, perhaps suggesting an initial transgressive trend. Maybe more likely, this transgressive trend is related to base-level rise at the scale of minor cycles, not the entire basin fill (see also below).

Overlying the E2a(iii) cycle is the multi-cycle mudstone unit extending from minor cycle E2b1(i) to minor cycle E2b2(ii). The position of this unit corresponds to the level of *maximum flooding* in the EPR model (Fig. 13). The unit is not a single maximum flooding surface, since it is very complex (see above). It is not unrealistic to assume that a time of maximum flooding at the scale of the basin fill occurred at some point during the deposition of this unit. Nevertheless, a major switch of sediment supply can also fully explain the position of the unit, although its position is intriguing.

The expected response to the sea-level curve following maximum flooding is a gradually decreasing accommodation space and deposition of progressively progradational minor cycles in the *highstand systems tract*. In fact, the E2b3 and the E2c1 minor cycles show exactly this tendency where the E2c1 cycle reaches further south than the E2b3 cycle (Fig. 13). These two cycles then mark the top of the type 1 E1c–E2c1 sequence.

On the Askrigg Block, the overlying E2c2(i) Lower Follifoot Grit seems to incise markedly more than other fluvial sandstones. In addition, there appears to be no representative of the E2c1 cycle on the block (Fig. 13), probably as a result of truncation by the E2c2(i) fluvials (see above). Thus, it appears that the base of the E2c2(i) sandstones may represent another type 1 sequence boundary (Fig. 13).

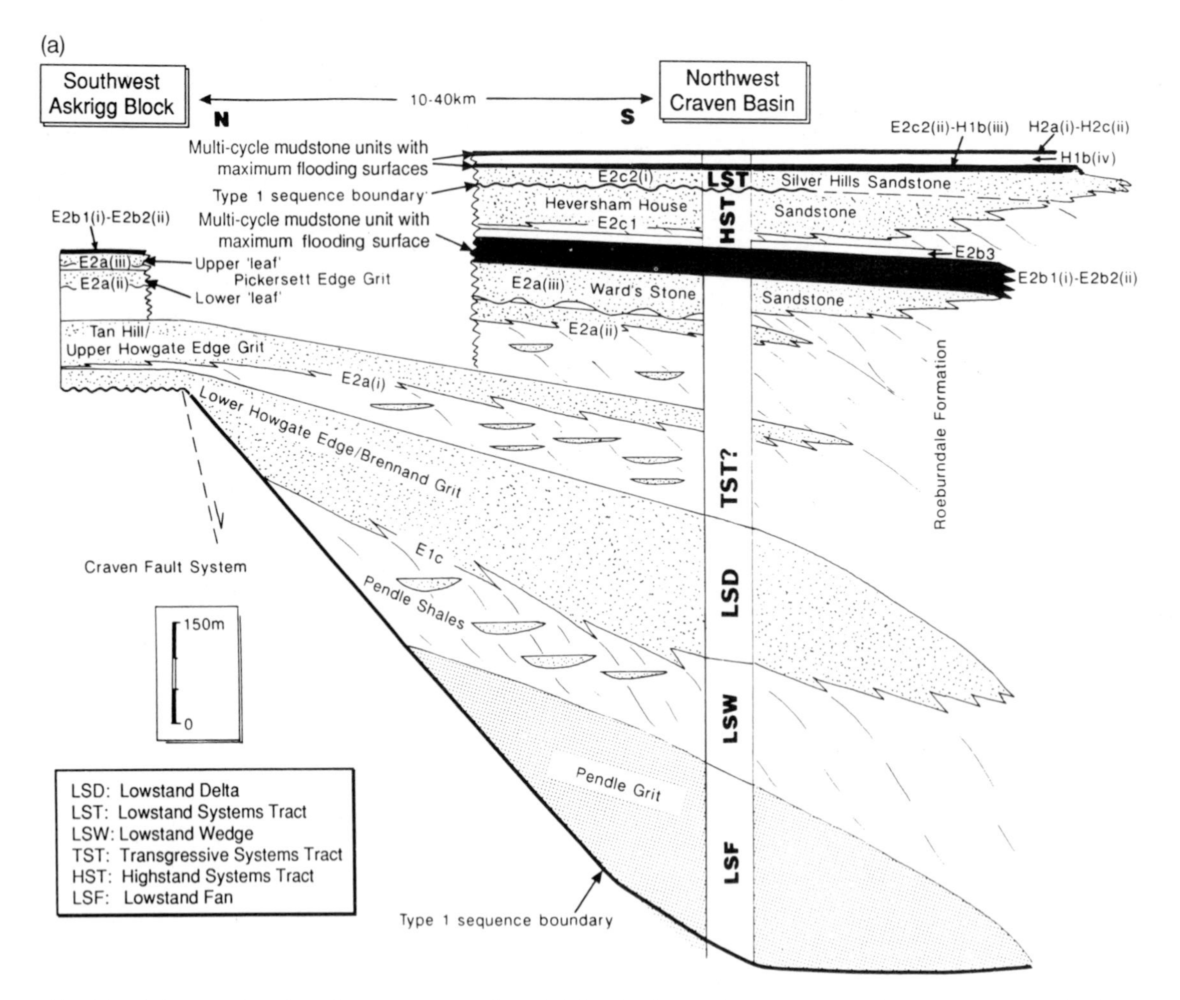

Fig. 14. (a) Cross-section from the western end of the Askrigg Block into the western part of the Craven Basin showing possible division into EPR depositional sequences. Contrast the interpreted transgressive and highstand systems tracts with Fig. 13 and with (b).

No turbidite fan can be found as a counterpart of the sequence boundary. Instead, a deltaic succession occurs in the basin. It is important to recall that at this time, the southern margin of the Askrigg Block functioned as a ramp rather than as a prominent 'edge'. Thus, it is possible that a *lowstand deltaic wedge* developed, rather than a fan.

To this point, the Craven–Askrigg composite data seem to correspond to a fairly satisfactory degree to the EPR depositional sequence model. The remainder of the studied interval, however, is less clear.

A multi-cycle mudstone unit overlies the E2c2(i) deposits. Following the interpretation above, this ought to correspond to a period of maximum flooding. If so, no transgressive systems tract occurs in the area. The multi-cycle mudstone unit is further overlain by the H1b(iv) fluviodeltaic system, above considered to reflect a particular lowstand period and hence, in EPR terms, probably another lowstand deltaic wedge. This means that no highstand systems tract is present in the Craven–Askrigg area for this particular sequence (Fig. 13).

The explanation for these inconsistencies may be found in modern sedimentary environments. In the Mississippi delta area, for example, large areal variations occur in the locations of components of sequences due to substantial shifts in the river

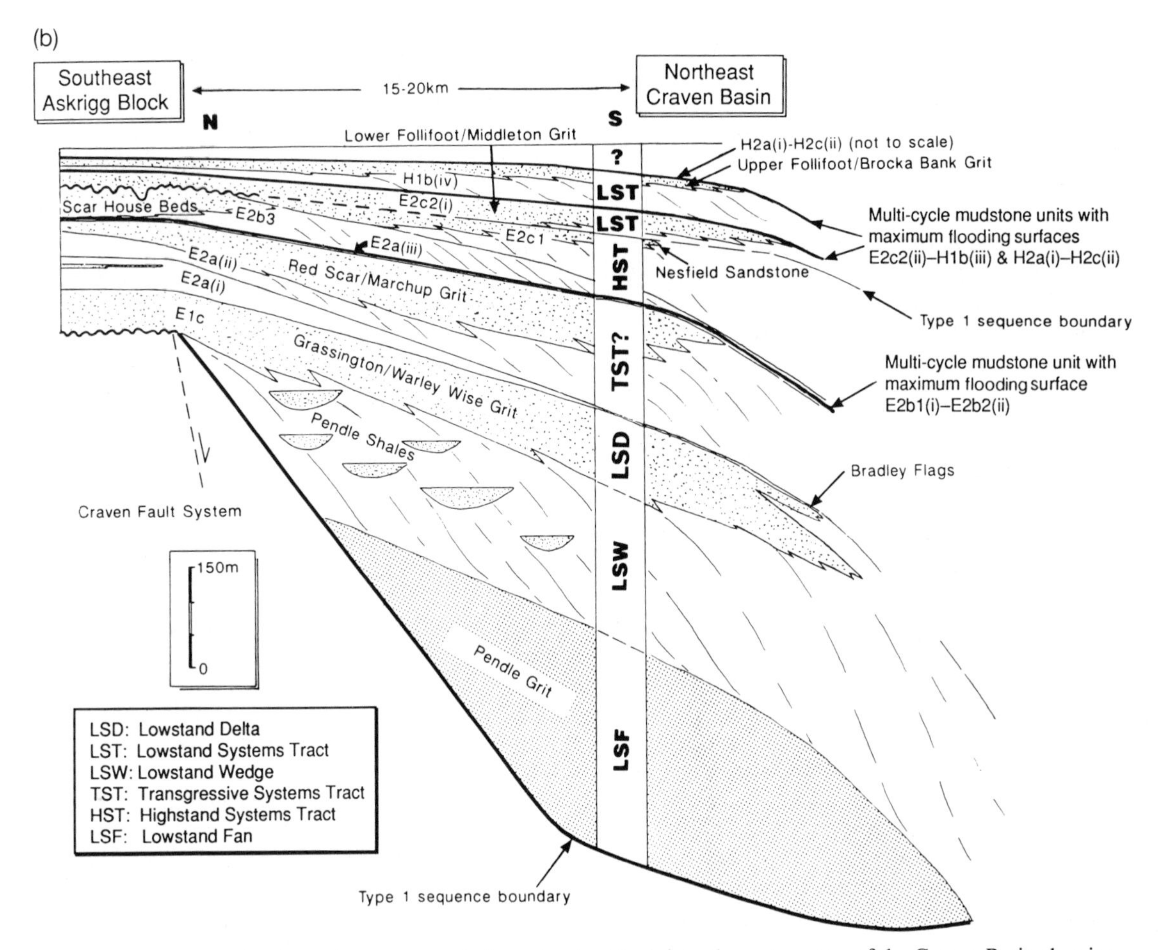

Fig. 14. (b) Cross-section from the eastern end of the Askrigg Block into the eastern part of the Craven Basin showing possible division into EPR depositional sequences. Contrast the interpreted transgressive and highstand systems tracts with Fig. 13 and with (a).

course (Boyd *et al.*, 1989). For instance, the highstand systems tract, represented by the St Bernard, Lafourche, Modern and Atchafalaya delta lobes, is offset more than 250 km eastwards in relation to the transgressive systems tract represented by the Outer Shoal, Maringouin and Teche delta lobes (cf. Fig. 3 in Boyd *et al.*, 1989). This means that in no single dip-oriented two-dimensional cross-section from basin margin towards basin centre will all parts of the sequence be developed.

Similarly, large shifts in the sediment supply routes could have caused substantial parts of sequences overlying the E1c–E2c1 sequence to be deposited elsewhere, beyond the Craven–Askrigg region. Following the same line of reasoning, important parts of the E1c–E2c1 sequence (i.e. any clastic systems active during E2b1(i)–E2b2(ii) times, which was interpreted as the time of maximum flooding) could have been deposited elsewhere. This could significantly distort the sequence architecture of the E1c–E2c1 sequence.

Finally, the possibility also exists that the missing parts of the sequence never were deposited. Given that the Craven–Askrigg region probably is only one part of a greater basinal area, this cannot be controlled with presently available information. Nevertheless, it remains as likely as the explanation given above. If verified, the sequence architecture of EPR is merely one of several ways of viewing basin fills in extensional settings.

Minor cycle scale sequence stratigraphy

It also seems possible to apply the sequence concepts at the scale of individual minor cycles, shown for instance by the E1c and E2a(i) minor cycles (Fig. 13). The basal sequence boundary is the same as for the 'major' sequence, i.e. the sub-E1c unconformity on the block. As in the 'major' sequence, the Pendle Grits, Pendle Shales and Grassington Grit (and equivalents) would probably correspond to the *lowstand systems tract* representing the lowstand fan, lowstand prograding wedge and a lowstand delta.

The *transgressive systems tract* could be represented by the 'final' progradational pulse of the E1c, the Bradley Flags, because this extends less far to the south than the underlying delta (see above). Furthermore, the E2a(i) goniatite band probably reflects maximum flooding and transgression (Fig. 13; as suggested by the minor cycle model; cf. Holdsworth & Collinson, 1988; see also Fig. 11). The overlying, poorly developed coarsening-upward succession on the Askrigg Block in the E2a(i) cycle may represent the *highstand systems tract* (Figs 11 & 13). The reason for its poor development may be related to low subsidence. The low subsidence rate of the Askrigg Block probably made relative highstand deposits there much more susceptible to truncation by following lowstand systems than equivalent deposits in the basin.

The E2a(i) lowermost lenticular channels on the western parts of the block may represent lowstand, incised channels of the overlying 'minor' sequence, making the lowermost erosion surface a sequence boundary. Across this surface, there is a marked basinward shift in facies from prodeltaic muds to fluvial sands (cf. Figs 4(b) &11).

This means that 'minor' sequences are 180° phase shifted in comparison with minor cycles. It will be discussed below which seems to be the most practical stratigraphic unit to use.

Since no smaller scale stratigraphic unit of significance than the 'minor sequences' and minor cycles has been recorded, and particularly since at least some of them show evidence of base-level falls, usage of the term 'parasequences' seems inappropriate in the present study. Parasequences are simple shoaling-upward successions of sediment. Since minor cycles appear not to be, these cannot be equated. Thus, the minor cycles are the building blocks of the EPR model at basinal scale in the present case.

Three-dimensional variability of major sequences

Rather prominent three-dimensional variability of the sedimentary systems is observed in the Craven–Askrigg area (see above; Figs 4(a)–9). To this point in the sequence-stratigraphic analysis, this has not been considered, and it seems to impose complications.

A cross-section from the southwest of the Askrigg Block into the northwest of the Craven Basin (Fig. 14(a)) is contrasted with a cross-section from the southeast of the block into the northeast of the basin (Fig. 14(b)). In the western cross-section, a 'complete' succession of the interpreted lowstand and transgressive systems tracts is observed with deposits from all cycles (E1c–E2a(iii)) being present. In the east, the lowstand systems tract is complete, but the transgressive systems tract contains only one minor cycle, the E2a(ii) (Fig. 14(b)).

The same problem arises with the highstand systems tract. In the east, two minor cycles are present, the E2b3 and the E2c1 (Fig. 14(b)). In the west, only one occurs, the E2c1 cycle (Fig. 14(a)). Thus, exact correlation at minor cycle scale is not straightforward purely based on sequence-stratigraphic signatures. Furthermore, in the north-central parts of the basin, no highstand systems tract occurs (cf. palaeogeographic sketches; Figs 4(a) to 9). For this entire time period, no clastic systems entered the north-central parts of the basin, maintaining this area as a large embayment. This may suggest a tectonic control by the Askrigg Block, localizing sediment input to the east and the west of the basin.

Correlation potential

The fact that sequences may develop on two temporal and spatial scales, as shown by the Craven–Askrigg data (Fig. 13), is essential for the regional and interregional correlation potential of EPR sequences. In particular, the construction of sea-level curves or coastal onlap curves on an interregional scale (e.g. Haq *et al.*, 1988) is speculative when major variability in sequence development occurs within a relatively small area as the Craven–Askrigg region. Such reservation was also expressed by Boyd *et al.* (1989) and Fulthorpe and Carter (1989).

Since it is difficult to trace sequence-bounding erosion surfaces beyond the areas of channel influence (cf. Fig. 11), the correlation potential of these is limited. This also influences the correlation potential of entire sequences, and is a major disadvantage of application of the EPR model in the Craven–Askrigg area.

Galloway's model and the Craven–Askrigg data

Sequence boundaries

The goniatite bands form boundaries which reflect transgressions and flooding and palaeogeographic reorganizations. These bands correspond reasonably well to the condensed sections of Galloway's model and probably also contain the maximum flooding surfaces. Most of the bands are regionally extensive, and many can be correlated on a basinal and interbasinal scale (see above).

On a larger scale, the multi-cycle mudstone units may be regarded as more important boundaries, separating multi-cycle periods of coastal outbuilding and withdrawal. These can be correlated on a basinal scale, but their interregional context is speculative.

Both with the single goniatite bands and the multi-cycle mudstone units there are obvious problems when traced in the direction of the basin margin (i.e. northwards). On the Askrigg Block, the goniatite fauna may be replaced by lower salinity, non-diagnostic fauna, and also, some bands may disappear entirely. Detailed sedimentological analyses, which form the basis of this study, can resolve this problem to a certain degree. The goniatite bands will probably correlate with the horizons where palaeogeographic reorganizations take place, and these levels can be identified with confidence for most cycles (cf. Martinsen, 1990a). Therefore, a detailed analysis of the sediments above a particular goniatite band can aid the correlation where the fauna disappears, although circular reasoning must be avoided. Consequently, the bands where goniatites are present are the only unambiguous data points for correlation on the Askrigg Block, except where classic mapping and walking out of beds is possible.

Genetic stratigraphic-sequence components

1 *Minor sequences.* Minor cycles (*sensu* Holdsworth & Collinson, 1988) in the Craven–Askrigg Namurian show several similarities to genetic stratigraphic sequences. The E2a(i) cycle is used as an example (Fig. 15).

Two cross-sections of the cycle are shown, from the southwest of the Askrigg Block into the northwest of the Craven Basin, and from the southeast of the block into the northeast of the basin. The genetic sequence is bounded by goniatite bands, respectively the *C. cowlingense* band at the base (E2a(i)) and the *E. bisulcatum ferrimontanum* band (E2a(ii)) on top. The latter is replaced by a bioclastic limestone bed, the Lad Gill Limestone, on the northwestern part of the block.

The sequence shows marked areal variability. The western representative is the better developed (Fig. 15; see above). Progradational (fluvial-incision phase and slope progradation), aggradational (main fluvial phase) and retrogradational (localized deltaic phase on Askrigg Block and biogenic reworking) elements may be recognized. In most cycles, transgressive elements are not recognized. For the E2a(i) sequence, however, the top flooding surface is overlain by the shallow-marine Fossil Sandstone (Fig. 15). This sandwave complex may represent reworking of E2a(i) deposits elsewhere and migration into the Craven–Askrigg during the early phase of the E2a(ii) cycle (cf. Brenner & Martinsen, 1990 for a detailed discussion).

Probably, the reason for the general absence of the transgressive component and minimal reworking in most cycles was a general lack of sufficient basinal energy as tides and waves (cf. Collinson, 1988 and references therein). Thus, the transgressive deposits at the base of the E2a(ii) cycle represent an anomaly.

In the east of the area, a much attenuated genetic sequence is observed (Fig. 15). Nevertheless, it still seems possible to recognize progradational, aggradational and retrogradational sequence components. No transgressive component can be identified.

Problems exist with the application of the Galloway scheme. It does not properly account for deposits formed during lowstands of base level. It was shown above that base-level lowstands are quite likely to have taken place during deposition of the minor cycles. Within the E2a(i) cycle, a turbidite apron occurs in the western part of the Craven Basin, possibly formed during base-level lowstand and bypass of the Askrigg Block (cf. Martinsen, 1990a). Admittedly, this could equate with a resedimented apron at the foot of the prograding slope, in Galloway's model, formed during transgression and reworking of the previous abandoning depositional system. However, the fact that, seemingly everywhere, the E2a(i) goniatite band separates the E1c deposits from the turbidites at the base of the

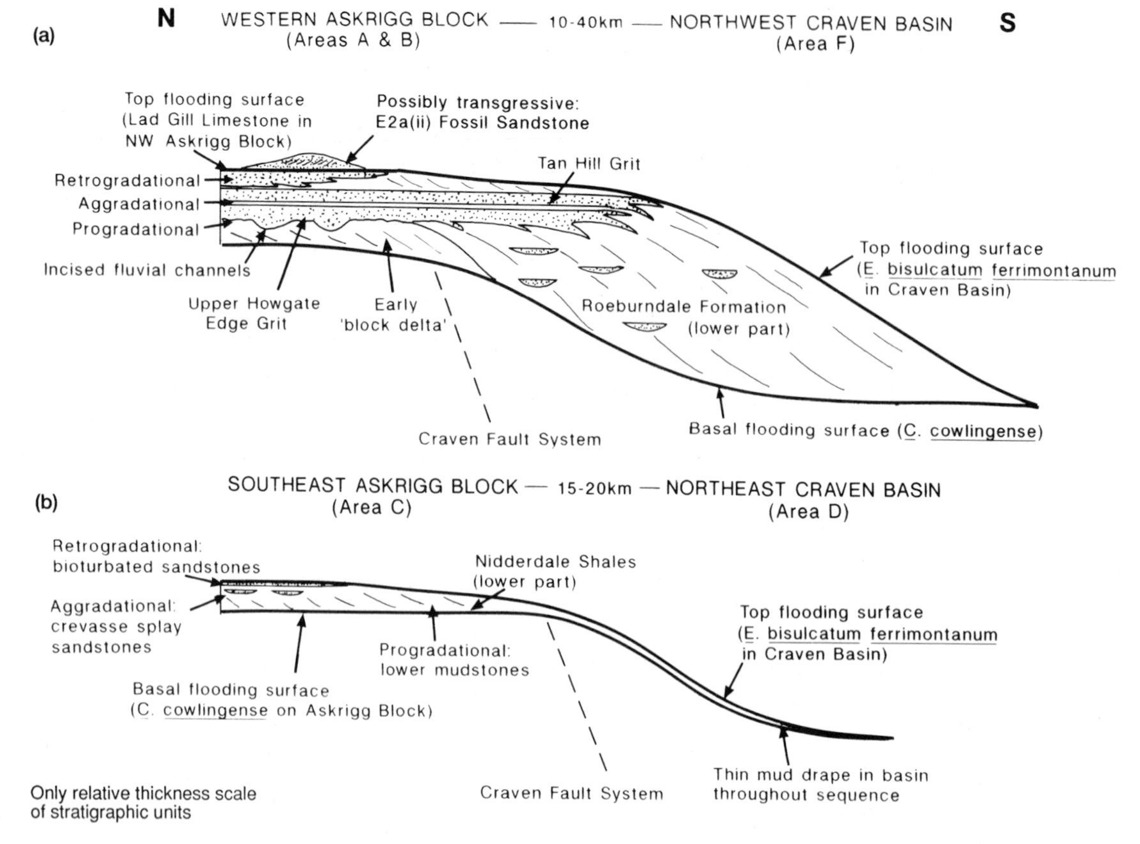

Fig. 15. Example of the E2a(i) 'minor' genetic stratigraphic sequence and its three-dimensional variability illustrated by the E2a(i) sequence (see text for discussion). (a) cross-section from western part of Askrigg Block to northwestern part of Craven Basin, (b) Cross-section from southeastern part of Askrigg Block to northeastern part of Craven Basin.

E2a(i), and that the lack of evidence of significant removal of sediment from the top of the E1c deposits (the removal seems rather to have been during incision of fluvial channels), suggest that the turbidites formed during a relative lowstand in E2a(i).

2 *Major genetic sequences.* Each 'major' cycle, i.e. each stratigraphic interval bounded below and/or above by multi-cycle mudstone units, can perhaps be viewed as a major genetic stratigraphic sequence based on the nature of the boundaries (Fig. 16). However, there is no clear architecture of these 'major' sequences in that progradational, aggradational and retrogradational as well as transgressive patterns are not recognized. The fact that they are separated by extensive mudstone units suggests that each cycle represents a separate, major polyphasal progradation. Exactly why these sequences lack an ordered architectural arrangement is uncertain, but it presents a problem for strict application of the Galloway model at this scale. The 'major' genetic sequences seem to reflect different stages of the basin fill in relation to a generally decreasing sediment input and decaying differential subsidence. Their boundaries most likely represent major shifts in sediment supply.

Three-dimensional variability of genetic sequences

The goniatite bands and their correlatives bound separate phases of coastal progradation and retreat (cf. Figs 4(a)–9) which varied laterally. Hence, the development of the components of the genetic sequences varied accordingly. Provided that the genetic sequence boundaries are picked correctly, this will yield valuable information on how the entire sedimentary system varied in three-dimensions.

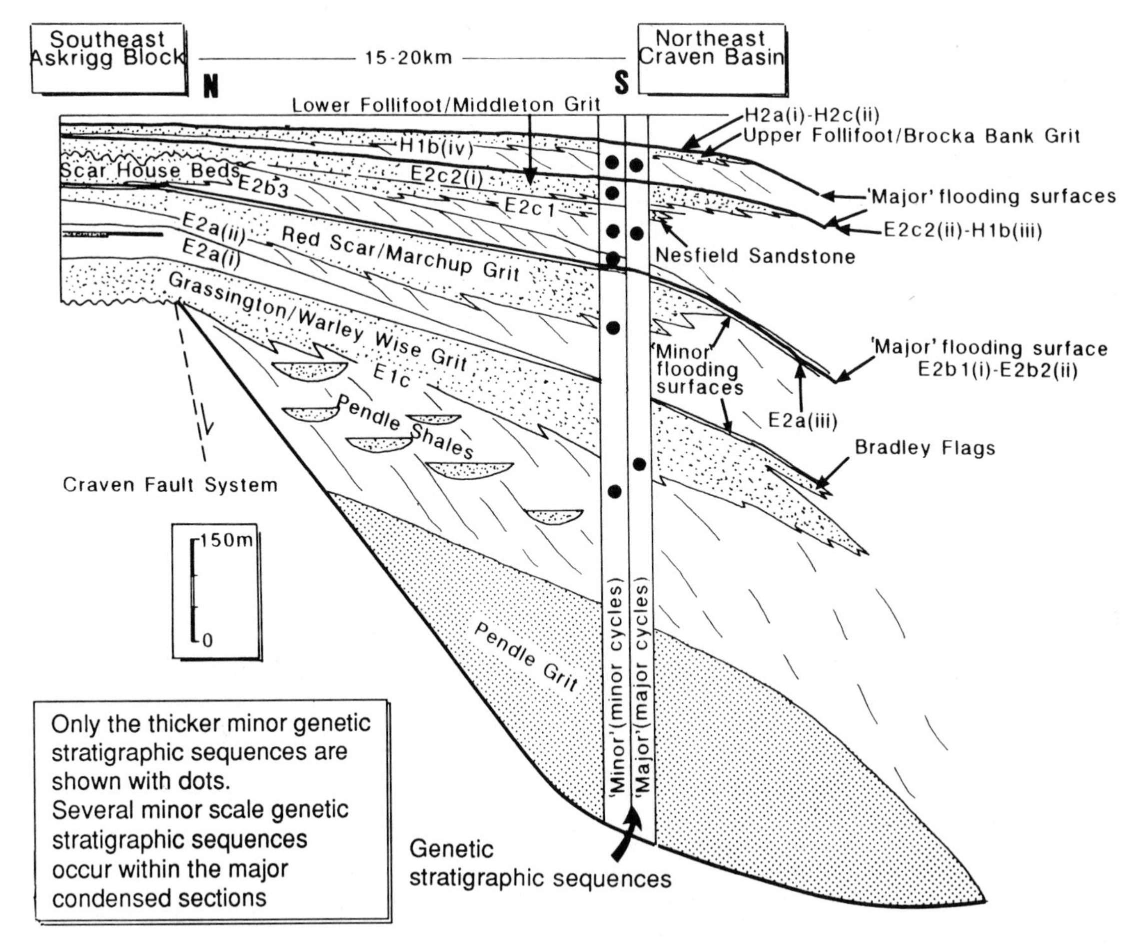

Fig. 16. Cross-section from the eastern end of the Askrigg Block into the eastern part of the Craven Basin showing the division of the stratigraphy into 'minor' and 'major' genetic stratigraphic sequences. Only the thicker 'minor' genetic sequences are shown by black dots. In reality, however, every minor cycle is a genetic sequence.

Major differences occurred between as well as within depositional systems in the Craven–Askrigg region (see above). In some cycles, e.g. the E2a(i) (Fig. 15), the entire genetic sequence may be represented by a condensed mudstone unit in some areas. In many areas where some or all of the sequence components are absent, the goniatite bands (i.e. the sequence boundaries) tend to occur. Their presence helps correlation and gives information on the variability of the system.

Correlation potential

The correlation potential of the genetic sequences in the Craven–Askrigg area relies mostly on the correlation potential of the goniatite bands (see above). On the Askrigg Block, many bands are missing, inhibiting the correlation. Detailed depositional models and understanding of how the individual depositional systems of each genetic sequence operated aid correlation and reduce the number of correlation possibilities. This has been fairly successful in the present study where also classical mapping is possible. Nevertheless, correlation of genetic sequences at any scale is fully dependent on a solid biostratigraphic framework and closely spaced sections to construct detailed depositional models.

Contrasting models of sequence stratigraphy: an evaluation

Introduction

The application of different sequence-stratigraphic models to the Craven–Askrigg succession generates several important questions about the validity of these models. In the following, these questions are addressed thus providing an evaluation of the models based on the data from the Craven–Askrigg region.

What type of stratigraphic surface forms the most practical and easily recognizable sequence boundary?

The goniatite bands appear to be the most easily correlative surfaces in the Craven–Askrigg area. Many of these occur basin-wide, have particular goniatite species associated with them and are interpreted to separate different periods of coastal outbuilding. Their correlation potential is often not entirely satisfactory toward the basin margins, however.

Erosion surfaces at the bases of fluvial channel sandstones may also be important surfaces if it can be documented that these surfaces represent periods of relative sea-level drop and basinwards shift of facies (e.g. Fig. 4(b)). In addition, they are difficult to correlate laterally (Fig. 11). Furthermore, palaeosols are often poorly developed in interfluve areas. The last argument also questions the interpretation that erosion surfaces represent base-level falls and thus sequence boundaries. In basinal areas, correlative conformities of unconformable surfaces are difficult to identify with ease (cf. Embry, 1988).

Walker (1990) suggested that packages of genetically related sediments do not extend across either maximum flooding surfaces or unconformities. Implicitly, it is unlikely that packages of genetically related strata should extend from e.g. a maximum flooding surface to the next maximum flooding surface provided a likely candidate for an unconformity is recognized in between. This is ambiguous for the Craven–Askrigg data, because some erosion surfaces at the bases of channel sandstones seem to represent base-level falls, whilst others are more unclear. In any matter, the goniatite bands seem to represent more significant breaks than do the erosion surfaces at the bases of the channel sandstones. Also, significant palaeogeographic breaks are difficult to prove across the erosion surfaces (perhaps apart from the E2c2(i) basal surface on the Askrigg Block). Therefore, in most cases, it seems sensible to view the sediments between two goniatite bands as genetically related.

Where do the most important changes in palaeogeography occur within the stratigraphic successions?

Apparently, the most important changes occur at the basin-wide goniatite bands. Careful sedimentological analysis and palaeogeographic reconstructions have suggested that significant palaeogeographic changes took place across the marine bands, not just in terms of spatial reorganizations, but also in terms of changes in e.g. delta types (cf. Martinsen, 1990a).

Evidently, changes also take place across at least some of the channel-base erosion surfaces since these reflect basinward shifts of facies and truncation of parts of the underlying stratigraphy (e.g. the E2c2(i) base on the Askrigg Block). However, evidence for major changes in types of depositional environments and sediment supply routes across these surfaces has not been found.

What is the predictive potential of the sequence models considering the Craven–Askrigg data?

An important advance of sequence stratigraphy is the potential to predict basin-fill patterns. The EPR model forms only in part a reasonably good predictor for sedimentation patterns in the Craven–Askrigg area from E1c–E2c2(i) times (Fig. 13). No clearly developed transgressive systems tract occurs, suggesting that the controls on the sedimentation at least for this period differed from those of the EPR model.

From E2c2(ii)–H2c(ii) times, no typical EPR-type sequences developed on major cycle scale. Only parts of the sequences may be recognized. Thus, the major controls on sedimentation probably differed in some significant way from those in the EPR model. A decreasing and variable sediment supply may have been important, perhaps combined with tectonism. Consequently, it seems that complete EPR sequences can be recognized when the controls on sedimentation are close to those preconditions set out in the model (cf. Posamentier & Vail, 1988; Posamentier *et al.*, 1988). Perhaps the two factors of sustained subsidence (at any one location, although increasing basinwards) and constant sediment supply are most difficult to accept. In any basin, as in the Craven–Askrigg area, these factors are likely to change significantly with time.

The predictive potential of the Galloway model seems highest at the scale of minor genetic stratigraphic sequences. Progradational, aggradational and retrogradational elements may be recognized where sediment supply is sufficient. In areas between major progradations, only fine-grained sediments accumulated, mainly because of the lack of a significant basinal wave regime to create interdeltaic strandplain systems as in the case of the Gulf of Mexico Cenozoic (Galloway, 1989b). The lack of a basinal wave climate which could rework coastlines was also the primary cause for the general lack of transgressive deposits within minor genetic sequences. The Galloway model fails to predict deposits formed during base-level lowstands which occur in several of the minor cycles. Although these deposits may be considered part of the progradational part of a genetic sequence, their genetic link to a particular state of base level is important and cannot be ignored.

At the scale of major genetic stratigraphic sequences, the Galloway model seems less reliable as a predictor when considering the internal architecture of the sequences. No clear elements reflecting separate stages of the progradation and retreat of these sequences are seen. Instead, the major genetic sequences seem to reflect progressive stages of the infill of an initially highly differential bathymetry, punctuated by prolonged periods of absence of sediment supply.

Causal mechanisms: what is the relationship between the Craven–Askrigg data and the models' preconditions?

The sedimentary successions of the Craven–Askrigg region were controlled by an interaction of variability in sediment supply, tectonism and eustasy (see Fig. 12). Any model that is expected to yield a 'best approximate' of the data should also be based on the same controls as those of the real data.

The proponents of the EPR model lately have claimed that this originally eustatically tied scheme can cope with tectonic influences and that variations in sediment supply are not important, provided the model is adjusted to the local data (Posamentier *et al.*, 1988). However, conceptual studies need to be published where the areal variability of the EPR sequence model is shown for various tectonic settings and where sediment supply is variable (W. Helland-Hansen, pers. comm.).

The EPR model remains tied to eustasy as the main causal factor for sequence development since sediment supply is regarded as a constant and subsidence is pictured as being constant although increased basinwards along the depositional profile (e.g. Posamentier & Vail, 1988; Van Wagoner & Mitchum, 1989; Mitchum & Van Wagoner, 1990). At the scale of minor cycles, it is likely that eustasy was the main controlling factor for the signature of barren sediment–goniatite band cyclicity, although their final expression was modified by sediment supply and subsidence.

The eustatic influence at the larger scale of the basin fill is questionable, however. Variations in sediment supply and subsidence due to source area and basin tectonics are more likely explanations. Nevertheless, the overall coherence of the lower part of the Craven–Askrigg succession to the EPR model (not considering three-dimensional variability) suggests that at least subsidence and sediment supply were similar to that assumed in the EPR model.

The Galloway model favours no particular causal mechanism for the development of genetic stratigraphic sequences (Galloway, 1989a). Any combination of three main variables (tectonics, sediment supply and eustasy) can, according to the model, cause the development of sequences. However, the model may have to be modified in basins where there is insufficient wave energy to create transgressive deposits, as for instance in the Craven–Askrigg area. For example, basin wave climate may be added as a fourth variable to tectonics, eustasy and sediment supply.

Furthermore, it does not easily incorporate sedimentary products of e.g. relatively low sediment supply and base-level lowstand. Galloway (1989a) assumes that, at most basin margins, subsidence is sufficiently rapid to overprint any effect of sea-level drop. This is questionable, particularly in settings like the Askrigg Block where subsidence was relatively slow for a prolonged period of time.

CONCLUSIONS

1 The Exxon Production Research and Galloway sequence-stratigraphic models were applied to Namurian E1c–H2c (Upper Carboniferous) sedimentary successions of the Craven–Askrigg area, northern England. The area and the stratigraphy seemed appropriate for this purpose, particularly because of the extensional setting, refined biostratigraphy and cyclic framework with an average cycle duration of 180,000 years.

2 The Galloway genetic stratigraphic-sequence model seems to best incorporate the important variability which took place in Namurian E1c–H2c times in the Craven–Askrigg area. This is largely because it emphasizes putting sequence boundaries at marine flooding surfaces representing transgressions, which in the present data are recorded by widely extensive goniatite bands, and because major palaeogeographic reorganizations seem to have occurred across these goniatite bands. In addition, three-dimensional variability in sequence development is predicted by the Galloway model, and it favours no particular causal mechanism for sequence development, both of which are significant for the Craven–Askrigg data. However, the model fails to predict deposits formed during lowstands of base level, and does not account for basins where basinal wave climate is insufficient to create transgressive deposits.

3 The EPR model forms only in part a reasonable predictor for the basin-fill patterns at the scale of major cycles, and it does not incorporate significant three-dimensional variability of the depositional systems. This is particularly significant for the interpretation of two-dimensional seismic sections. At the scale of minor cycles, the model is a better predictor.

Furthermore, it is difficult to correlate laterally candidates for sequence boundaries (erosion surfaces at the bases of fluvial sandstones) because these occur at the bases of laterally restricted fluvial channel sandstones. Their interpretation as sequence boundaries is thus questionable in many cases. It is also difficult to document palaeogeographic changes across the erosion surfaces.

4 Application of sequence models to basins where the controls on sedimentation differ from the passive, extensional settings for which the models were developed may be risky and lead to wrong conclusions. The Craven–Askrigg region is one area that, at least apparently, seemed suited for model application because of the overall extensional setting and highly refined biostratigraphy. Nevertheless, inconsistencies are observed which suggest that the controls on sedimentation in the Craven–Askrigg area differed from the preconditions in the models. This inevitably leads to questioning of model universality. Therefore, more detailed studies are needed both in extensional and other tectonic settings to test the models to broaden the understanding of their applicability and ultimately strengthen them.

ACKNOWLEDGEMENTS

This work forms part of a Dr Sc thesis completed at Geologisk Institutt, University of Bergen, Norway. The work was entirely sponsored by TOTAL Norway. John Collinson and Brian Holdsworth are thanked for guidance throughout the work. Ron Steel, William Helland-Hansen and Nick Riley are thanked for reading through drafts of the manuscript. The referees Andy Pulham, Colin Jones and Janok Bhattacharya are thanked for critically constructive reviews. This paper was written during a post-doctoral research fellowship sponsored by Norsk Hydro A/S at the University of Bergen.

APPENDIX

Table A1 Localities investigated in this work, and the stratigraphic intervals exposed at each locality

Locality name	Grid reference	Exposed stratigraphic units *(and minor cycles)*
Northwest Askrigg Block		
Mousegill Beck, South Stainmore	NY 830126–NY 837125	Stricegill Grit, Upper Felltop Limestone, High Wood Grit, High Wood Marine Beds *(E1c–H2c(ii))*
Mallerstang Edge, Upper Vale of Eden	NY 802038–SD 796993	Upper Howgate Edge Grit, Tan Hill Grit, Fossil Sandstone, Pickersett Edge Grit *(E2a(i)–E2c2(ii))*
Wild Boar Fell, Upper Vale of Eden	SD 764993–SD 761978	Upper Howgate Edge Grit, Tan Hill Grit, Fossil Sandstone, Pickersett Edge Grit *(E2a(i)–E2a(ii))*
Washer Gill, nr Garsdale Head	SD 798957–SD 802959	Mirk Fell Palaeosol, Upper Howgate Edge Grit, Fossil Sandstone, Pickersett Edge Grit *(E1c–E2a(iii))*
Uldale Beck, Birkdale	NY 813034–NY 808029	Mirk Fell Palaeosol, Fossil Sandstone *(E1c–E2a(ii))*
Crook Seal, Birkdale Common	NY 835024	Mirk Fell Palaeosol *(E1c)*
Lops Wath, Birkdale Common	NY 842019–NY 844017	Upper Howgate Edge Grit *(E2a(i))*
Birkdale Tarn Quarry	NY 857018	Tan Hill Grit *(E2a(i))*
Hoods Bottom Beck, Ravenseat	NY 865045	Mirk Fell Palaeosol, Upper Howgate Edge Grit *(E1c–E2a(ii))*
Thomas Gill, West Stonesdale	NY 879056–NY 873054	Mirk Fell Palaeosol, Upper Howgate Edge Grit *(E1c–E2a(i))*
Lad Gill, West Stonesdale	NY 888046–NY 900050	Mirk Fell Palaeosol, Upper Howgate Edge Grit, Tan Hill Grit, Lad Gill Limestone, Fossil Sandstone *(E1c–E2a(ii))*
Mould Gill, West Stonesdale	NY 889055	Mirk Fell Palaeosol and strata immediately above *(E1c–E2a(i))*
Mirk Fell Gill and Tan Hill area	NY 910070, and area around Tan Hill Inn	Mirk Fell Palaeosol, Mirk Fell Ironstones, Tan Hill Grit *(E1c–E2a(i))*
Southwest Askrigg Block		
Long Scar, Buttertubbs Pass	SD 875958	Lower Howgate Edge Grit *(E1c)*
Humesett, Great Shunner Fell	SD 843941	Lower Howgate Edge Grit *(E1c)*
Whernside	SD 738817–SD 722817	Lower Howgate Edge Grit, Upper Howgate Edge Grit *(E1c–E2a(i))*
Pen-y-ghent	SD 837739–SD 837729	Grassington Grit *(E1c)*
Southeast Askrigg Block		
How Stean Beck, Nidderdale	SE 070740–SE 050722	Grassington Grit *(E1c)*
Hag Dyke, west Great Whernside	SD 990736, SD 992743	Grassington Grit *(E1c)*

Locality name	Grid reference	Exposed stratigraphic units *(and minor cycles)*
West Gill Dike, Upper Nidderdale	SE 033748–SE 028738	Grassington Grit, Nidderdale Shales, Red Scar Grit *(E1c–E2a(ii))*
Backstean Gill, Nidderdale	SE 058723–SE 054712	Cockhill Marine Band, Nidderdale Shales, Red Scar Grit *(E2a(i)–E2a(ii))*
Great Blowing Gill Beck, Nidderdale	SE 049727–SE 035723	Nidderdale Shales *(E2a(i)–E2a(ii))*
Birk Gill, Colsterdale	SE 125822–SE 134818	Nidderdale Shales, Red Scar Grit *(E2a(i)–E2a(ii))*
Penhill, Wensleydale	SE 050867–SE 044867	Nidderdale Shales, Red Scar Grit *(E2a(i)–E2a(ii))*
Woodale Scar, Upper Nidderdale	SE 080766	Nidderdale Shales, Red Scar Grit *(E2a(i)–E2a(ii))*
Ulfers Gill, Coverdale	SE 095827, SE 098832	Nidderdale Shales, Red Scar Grit *(E2a(i)–E2a(ii))*
Carle Side Quarry, Upper Nidderdale	SE 063776	Scar House Beds *(E2b3)*
Rain Stang Quarry, Upper Nidderdale	SE 074764	Scar House Beds *(E2b3)*
Backstone Gill, Colsterdale	SE 095806	Scar House Beds *(E2b3)*
Beldin Gill, Colsterdale	SE 095799	Scar House Beds, Lower Follifoot Grit, Upper Follifoot Grit *(E2b3–H1b(iv))*
Great Whernside	SE 0074 area	Lower Follifoot Grit *(E2c2(i))*
Shooting House Crag, Coverdale	SE 082824	Lower Follifoot Grit *(E2c2(i))*
Slipstone Crags, Colsterdale	SE 138821	Lower Follifoot Grit *(E2c2(i))*
Thrope Edge, Nidderdale	SE 106752	Lower Follifoot Grit *(E2c2(i))*
Northeast Craven Basin		
Birk Crag, nr Harrogate	SE 280548	Almscliff Grit *(E1c)*
Cononley Beck, Airedale	SD 986468	Top Bradley Flags, Edge Marine Band *(E1c–E2a(ii))*
Kildwick, Airedale	SE 014463	Beds between Edge Marine Band and Marchup Grit *(E2a(ii))*
Scargill Reservoir, nr Harrogate	SE 233533	Scargill Grit *(E2a(ii))*
Thruscross Reservoir, Washburn Valley	SE 155575	Red Scar Grit *(E2a(ii))*
Crag Farm, Washburn Valley	SE 204386	Marchup Grit *(E2a(ii))*
Bank End, Rombald's Moor	SE 044503	Marchup Grit *(E2a(ii))*
Silsden Reservoir	SE 044474	Marchup Grit *(E2a(ii))*
Nesfield, Wharfedale	SE 093493	Nesfield Sandstone *(E2c1)*

Locality name	Grid reference	Exposed stratigraphic units *(and minor cycles)*
Nesfield to Upper Austby, Wharfedale	SE 097496–SE 098501	Strata above Nesfield Sandstone *(E2c1)*
Ling Park, Wharfedale	SE 104503	Middleton Grit *(E2c2(i))*
Brooks Crag Quarry, Cringles	SE 052484	Middleton Grit *(E2c2(i))*
Horn Crag Quarry, Silsden	SE 053480	Middleton Grit *(E2c2(i))*
Swartha, Airedale	SE 052468	Middleton Grit *(E2c2(i))*
Tivoli, Wharfedale	SE 105494	Marine beds between Middleton Grit and Brocka Bank Grit *(E2c2(ii)–H1b(iii))*
Crag House, south, Silsden	SE 056474	Brocka Bank Grit *(H1b(iv))*
Crag House, north, Silsden	SE 056477	Brocka Bank Grit *(H1b(iv))*
Brocka Bank, Wharfedale	SE 073485	Brocka Bank Grit *(H1b(iv))*
North-central Craven Basin		
Warley Wise and Stone Head	SD 944436, SD 946433	Warley Wise Grit, Sabden Shales *(E1c–H2c(ii))*
Roughlee, Pendle Waters	SD 846404	Sabden Shales *(H2a(i)–H2c(ii))*
Northwest Craven Basin		
Buck Haw Brow, Giggleswick	SD 796657	Roeburndale Formation, Caton Shales *(E2a(i)–E2b2(ii))*
West Rome Farm, nr Giggleswick	SD 782626	Roeburndale Formation *(E2a(i)–E2a(ii))*
Kettles Beck	SD 743626–SD 745633	Roeburndale Formation, Caton Shales, Keasden Flags, Silver Hills Sandstone *(E2a(i)–E2c2(i))*
Cowsen Gill, nr Keasden Beck	SD 725625	Roeburndale Formation, middle parts *(E2a(i)–E2a(ii))*
Keasden Beck	SD 721627–SD 723653	Roeburndale Formation, Caton Shales *(E2a(i)–E2b2(ii))*
Whitray Beck, Lythe Fell	SD 675620	Botton Head Grit (middle of Roeburndale Formation, *E2a(i))*
Roeburndale	SD 600638–SD 602639	Roeburndale Formation, lower part *(E2a(i))*
Baines Crag, Littledale	SD 543618	Ward's Stone Sandstone (top of Roeburndale Formation, *E2a(iii))*

REFERENCES

Aisenverg, D.E., Brazhnikova, N.E., Vassilyuk, N.P., Reitlinger, E.A., Fomina, E.V. & Einor, O.L. (1979) The Serpukhovian Stage of the Lower Carboniferous of the USSR. In: *The Carboniferous of the USSR* (Ed. Wagner, R.H., Higgins, A.C. & Meyen, S.V.) *Occ. Publ. Yorks. Geol. Soc.* **4**, 43–59.

Arthurton, R.S., Johnson, E.W. & Mundy, D.J.C. (1988) Geology of the country around Settle. *Mem. Brit. Geol. Surv.* **60**, 147.

Baines, J.G. (1977) The stratigraphy and sedimentology of the Skipton Moor Grits (Namurian E1c) and their lateral equivalents. Unpublished PhD thesis, Keele University, UK.

Besly, B.M. & Kelling, G. (Eds) (1988) *Sedimentation in a Synorogenic Basin Complex.* Blackie, Glasgow, 276pp.

Bisat, W.S. (1924) The Carboniferous goniatites of the north of England and their zones. *Proc. Yorks. Geol. Soc.* **20**, 40–124.

Bott, M.H.P. (1961) Geological interpretation of magnetic anomalies over the Askrigg Block. *Quart. J. Geol. Soc. Lond.* **117**, 481–495.

Bown, T.M. & Kraus, M.J. (1987) Integration of channel and floodplain suites I. Developmental sequence and lateral relations of alluvial paleosols. *J. Sediment. Petrol.* **57**, 587–601.

Boyd, R., Suter, J. & Penland, S. (1989) Relation of sequence stratigraphy to modern sedimentary environments. *Geology* **17**, 926–929.

Brenner, R.L. & Martinsen, O.J. (1990) The Fossil Sandstone: a shallow-marine sandwave complex in the Namurian of Cumbria and North Yorkshire, England. *Proc. Yorks. Geol. Soc.* **48**, 149–162.

Bristow, C.S. (1987) Brahmaputra River: channel migration and deposition. In: *Recent Developments in Fluvial Sedimentology* (Eds Ethridge, F.G., Flores, R.M. & Harvey, M.D.) Spec. Publ. Soc. Econ. Paleontol. Mineral. **39**, 63–74.

Burgess, I.C. & Ramsbottom, W.H.C. (1970) A new goniatite horizon in the Hearne Beck Limestone (Namurian E2) near Lovely Seat, upper Wensleydale. *J. Earth Sci.* **8**, 143–147.

Collinson, J.D. (1988) Controls on Namurian sedimentation in the Central Province basins of northern England. In: *Sedimentation in a Synorogenic Basin Complex: The Upper Carboniferous of Northwest Europe* (Eds Besly, B.M. & Kelling, G.) Blackie, Glasgow, pp. 85–101.

Coterill, K., Allen, S., De Tagle, F., *et al.* (1990) Well-log/seismic sequence stratigraphy of Miocene-Pleistocene depositional sequences, High Island area, offshore Texas: AAPG transect-phase II. In: *Sequence Stratigraphy as an Exploration Tool: Concepts and Practices in the Gulf Coast.* Programs and Abstracts, 11th Annual Research Conference, Gulf Coast Sect. Soc. Econ. Paleontol. Mineral., Houston, pp. 135–138.

Dockery, D.T., III (1990) The Eocene–Oligocene boundary in the northern Gulf—a sequence boundary. In: *Sequence Stratigraphy as an Exploration Tool: Concepts and Practices in the Gulf Coast.* Programs and Abstracts, 11th Annual Research Conference, Gulf Coast Sect. Soc. Econ. Paleontol. Mineral., Houston, pp. 135–138.

Dunham, K.C. (1974) Granite beneath the Pennines in North Yorkshire. *Proc. Yorks. Geol. Soc.* **40**, 191–194.

Dunham, K.C. & Stubblefield, C.J. (1945) The stratigraphy, structure and mineralisation of the Greenhow mining area, Yorkshire. *Ouart. J. Geol. Soc. Lond.* **100**, 209–268.

Dunham, K.C. & Wilson, A.A. (1985) Geology of the Northern Pennine Orefield, Vol. 2, Stainmore to Craven. *Econ. Mem. Brit. Geol. Surv.* 247.

Embry, A.F. (1988) Triassic sea-level changes: evidence from the Canadian Arctic Archipelago. In: *Sea-Level Changes: An Integrated Approach* (Eds Wilgus, C.K., Hastings, B.S., Kendall, C.G.St.C., Posamentier, H.W., Ross, C.A. & Van Wagoner, J.C.) Spec. Publ Soc. Econ. Paleontol. Mineral. **42**, 249–259.

Embry, A.F. & Podruski, J.A. (1988) Third-order depositional sequences of the Mesozoic succession of Sverdrup Basin. In: *Sequences, Stratigraphy, Sedimentology: Surface and Subsurface* (Eds James, D.P. & Leckie, D.A.) Mem. Can. Soc. Petrol. Geol. **15**, 73–84.

Frazier, D.E. (1974) Depositional episodes: their relationship to the Quaternary stratigraphic framework in the northwestern portion of the Gulf basin. *Geol. Circ. University of Texas, Bureau of Economic Geology,* **74–1**, 28pp.

Fulthorpe, C.S. & Carter, R.M. (1989) Test of seismic sequence methodology on a Southern Hemisphere passive margin: the Canterbury Basin, New Zealand. *Mar. Petrol. Geol.* **6**, 348–359.

Galloway, W.E. (1989a) Genetic stratigraphic sequences in basin analysis I: architecture and genesis of flooding-surface bounded depositional units. *Bull. Am. Assoc. Petrol. Geol.* **73**, 125–142.

Galloway, W.E. (1989b) Genetic stratigraphic sequences in basin analysis II: application to northwest Gulf of Mexico Cenozoic basin. *Bull. Am. Assoc. Petrol. Geol.* **73**, 143–154.

Gawthorpe, R.L., Gutteridge, P. & Leeder, M.R. (1989) Late Devonian and Dinantian basin evolution in northern England and North Wales. In: *The Role of Tectonics in Devonian and Carboniferous Sedimentation in the British Isles* (Eds Arthurton, R.S., Gutteridge, P. & Nolan, S.C.) *Occ. Publ. Yorks. Geol. Soc.* **6**, 1–23.

Haq, B.U., Hardenbol, J. & Vail, P.R. (1988) Mesozoic and Cenozoic chronostratigraphy and cycles of sea-level change. In: *Sea-level Changes: An Integrated Approach* (Eds Wilgus, C.K., Hastings, B.S., Kendall, C.G.St.C., Posamentier, H.W., Ross, C.A. & Van Wagoner, J.C.) Spec. Publ. Soc. Econ. Paleontol. Mineral. **42**, 71–108.

Holdsworth, B.K. & Collinson, J.D. (1988) Millstone Grit cyclicity revisited. In: *Sedimentation in a Synorogenic Basin Complex: The Upper Carboniferous of Northwest Europe* (Eds Besly, B.M. & Kelling, G.). Blackie, Glasgow, pp. 132–152.

Holland, J.G. & Lambert, R.St.J. (1970) Weardale Granite. *Trans. Nat. Hist. Soc., Northumberland* **41**, 103–123.

James, D.P. & Leckie, D.A. (eds) (1988) *Sequences, Stratigraphy, Sedimentology: Surface and Subsurface.* Mem. Can. Soc. Petrol. Geol. **15**, 586 pp.

Leeder, M.R. (1982) Upper Palaeozoic basins of the British Isles: Caledonide inheritance versus Hercynian plate margin processes. *J. Geol. Soc., London* **139**, 479–491.

Leeder, M.R. (1988) Recent developments in Carboniferous geology: a critical review with implications for the British Isles and N.W. Europe. *Proc. Geol. Assoc.* **99**, 73–100.

Leeder, M.R. & McMahon, A.H. (1988) Upper Carboniferous (Silesian) basin subsidence in northern Britain. In: *Sedimentation in a Synorogenic Basin Complex: The Upper Carboniferous of Northwest Europe* (Eds Besly, B.M. & Kelling, G.) Blackie, Glasgow, pp. 43–52.

Loutit, T.S., Hardenbol, J., Vail, P.R. & Baum, G.R. (1988) Condensed sections, the key to age determination and correlation of continental margin sequences. In: *Sea-Level Changes: An Integrated Approach* (Eds Wilgus, C.K., Hastings, B.S., Kendall, C.G.St.C., Posamentier, H.W., Ross, C.A. & Van Wagoner, J.C.) Spec. Publ Soc. Econ. Paleontol. Mineral. **42**, 183–213.

Martinsen, O.J. (1990a) Interaction between eustasy, tectonics and sedimentation with particular reference to the Namurian E1c-H2c of the Craven–Askrigg area, northern England. Unpublished Dr. Sc thesis, University of Bergen, Norway, 495pp.

Martinsen, O.J. (1990b) Fluvial, inertia-dominated deltaic deposition in the Namurian of northern England. *Sedimentology* **37**, 1099–1113.

Mitchum, R.M. Jr. & Van Wagoner, J.C. (1990) High-frequency sequences and eustatic cycles in the Gulf of Mexico Basin. In: *Sequence Stratigraphy as an Exploration Tool: Concepts and Practices in the Gulf Coast.* Programs and Abstracts, 11th Annual Research Conference, Gulf Coast Sect. Soc. Econ. Paleontol. Mineral., Houston, 257–267.

Mitchum, R.M. Jr., Vail, P.R. & Thompson, S. III (1977) Seismic stratigraphy and global changes of sea level, Part 2: the depositional sequence as a basic unit for stratigraphic analysis. In: *Seismic Stratigraphy Applications to Hydrocarbon Exploration* (Ed. Payton, C.E.) Mem. Am. Assoc. Petrol. Geol. **26**, 53–62.

Moseley, F. (1954) The Namurian of the Lancaster Fells. *Quart. J. Geol. Soc. London* **109**, 423–454.

Mundy, D.J.C. & Arthurton, R.S. (1980) Field report of field meeting Settle and Flasby, 1st and 2nd July, 1978. *Proc. Yorks. Geol. Soc.* **43**, 32–36.

Nummedal, D. & Swift, D.J.P. (1987) Transgressive stratigraphy at sequence-bounding unconformities: some principles derived from Holocene and Cretaceous examples. In: *Sea-Level Fluctuation and Coastal Evolution* (Eds Nummedal, D., Pilkey, O.H. & Howard. J.D.) Spec. Publ. Soc. Econ. Paleontol. Mineral. **41**, 241–260.

Posamentier, H.W. & Vail, P.R. (1988) Eustatic controls on clastic deposition II—sequence and systems tract models. In: *Sea-level Changes: An Integrated Approach* (Eds Wilgus, C.K., Hastings, B.S., Kendall, C.G.St.C., Posamentier, H.W., Ross, C.A. & Van Wagoner, J.C.) Spec. Publ. Soc. Econ. Paleontol. Mineral. **42**, 125–154.

Posamentier, H.W., Jervey, M.T. & Vail, P.R. (1988) Eustatic controls on clastic deposition I — conceptual framework. In: *Sea-level Changes: An Integrated Approach* (Eds Wilgus, C.K., Hastings, B.S., Kendall, C.G.St.C., Posamentier, H.W., Ross, C.A. & Van Wagoner J.C.) Spec. Publ Soc. Econ. Paleontol Mineral. **42**, 109–124.

Ramsbottom, W.H.C. (1977) Major cycles of transgression and regression (mesothems) in the Namurian. *Proc. Yorks. Geol. Soc.* **41**, 261–291.

Ramsbottom, W.H.C. (1979) Rates of transgression and regression in the Carboniferous of NW Europe. *J. Geol. Soc., London* **136**, 147–154.

Ramsbottom, W.H.C., Calver, M.A., Eagar, *et al.* (1978) A correlation of Silesian rocks in the British Isles. *Spec. Rep. Geol. Soc. London* **10**, 81.

Ramsbottom, W.H.C., Rhys, G.H. & Smith, E.G. (1962) Boreholes in the Carboniferous of the Ashover district, Derbyshire. *Bull. Geol. Surv. Great Britain* **19**, 75–168.

Riley, N.J., Varker, W.J., Owens, B., Higgins, A.C. & Ramsbottom, W.H.C. (1987) Stonehead Beck, Cowling, North Yorkshire, England: a British proposal for the Mid-Carboniferous boundary stratotype. *Cour. Forsch. Inst. Senckenberg* **98**,159–177.

Rowell, A.J. & Scanlon, J.E. (1957a) The relation between the Yoredale Series and the Millstone Grit on the Askrigg Block. *Proc. Yorks. Geol. Soc.* **31**, 79–90.

Rowell, A.J. & Scanlon, J.E. (1957b) The Namurian of the north-west quarter of the Askrigg Block. *Proc. Yorks. Geol. Soc.* **31**, 1–38.

Sims, A.P. (1988) The evolution of a sand-rich basin-fill sequence in the Pendleian (Namurian, E1c) of north-west England. Unpublished PhD thesis, University of Leeds.

Sloss, L.L. (1963) Sequences in the cratonic interior of North America. *Bull. Geol. Soc. Am.* **74**, 93–114.

Suter, J.R. & Berryhill, H.L. (1985) Late Quaternary shelf-margin deltas, northwest Gulf of Mexico. *Bull. Am. Assoc. Petrol. Geol.* **69**, 77–91.

Underhill, J.R., Gayer, R.A., Woodcock, N.H., Donnelly, R., Jolley, E.J. & Stimpson, I.G. (1988) The Dent Fault System, northern England—reinterpreted as a major oblique-slip fault zone. *J. Geol. Soc. London* **145**, 303–316.

Van Wagoner, J.C. & Mitchum, R.M., Jr (1989) High-frequency sequences and their stacking patterns. *Abstr. 28th Int. Geol. Congress,* Washington, D.C., 3–284.

Van Wagoner, J.C., Mitchum, R.M., Jr., Campion, K.M. & Rahmanian, V.D. (1990) Siliciclastic sequence stratigraphy in well-logs, cores and outcrops. *Am. Assoc. Petrol. Geol. Meth. Expl. Ser.* **7**, 55pp.

Van Wagoner, J.C., Posamentier, H.W., Mitchum, R.M., Jr *et al.* (1988) An overview of the fundamentals of sequence stratigraphy and key definitions. In: *Sea-level Changes: An Integrated Approach* (Eds Wilgus, C.K., Hastings, B.S., Kendall, C.G.St.C., Posamentier, H.W., Ross C.A. & Van Wagoner, J.C.) Spec. Publ. Soc. Econ. Paleontol. Mineral. **42**, 39–45.

Veevers, J.J. & Powell, C. McA (1987) Late Paleozoic glacial episodes in Gondwanaland reflected in transgressive-regressive depositional sequences in Euramеriea. *Bull. Geol. Soc. Am.* **98**, 475–487.

Walker, R.G. (1990) Facies modeling and sequence stratigraphy. *J. Sediment. Petrol.* **60**, 777–786.

Wilgus, C.K., Hastings, B.S., Kendall, C.G.St.C., Posamentier, H.W., Ross, C.A. & Van Wagoner, J.C. (Eds) (1988) *Sea-level Changes: An Integrated Approach.* Spec. Publ. Soc. Econ. Paleontol. Mineral. **42**.

Wilson, A.A. (1960) The Millstone Grit Series of Colsterdale and neighbourhood, Yorkshire. *Proc. Yorks. Geol. Soc.* **32**, 429–452.

Wilson, A.A. (1977) The Namurian rocks of the Fewston area. *Trans Leeds Geol. Assoc.* **9**, 1–42.

Wilson, A.A. & Thompson, A.T. (1959) Marine bands of Arnsbergian age (Namurian) in the south-eastern portion of the Askrigg Block, Yorkshire. *Proc. Yorks. Geol. Soc.* **32**, 45–67.

Wilson, A.A., Brandon, A. & Johnson, E.W. (1989) Geological context of the Wyresdale Tunnel methane explosion. *Brit. Geol. Surv. Tech. Rep.* WA/89/30.

Spec. Publs Int. Ass. Sediment. (1993) **18**, 283–294

Sequence stratigraphy of the Clansayesian (uppermost Aptian) formations in the western Pyrenees (France)

J.L. LENOBLE *and* J. CANÉROT

URA (CNRS) no. 1405 (Stratigraphie séquentielle et Micropaléontologie) and Laboratoire de Géologie sédimentaire et Paléontologie, Université Paul Sabatier, 39, Allées Jules Guesde, 31062 Toulouse, France

ABSTRACT

A diversified siliciclastic and calcareous Clansayesian (uppermost Aptian) facies association is observed in the Basque and Bearn areas of the western Pyrenees. Through the application of sequence-stratigraphic concepts, three principal systems tracts can be identified:

1 Lowstand systems tracts: calcareous mudmounds interfingering with black spicule marls (0–100 m);
2 Trangressive systems tracts: conglomerates, sandstones, siltstones and shales (0–20 m);
3 Highstand systems tracts: marls, carbonate bars, reef complexes, overlain by shelf limestones (up to 150 m).

The lowstand systems tracts deposits are located on the lowest part of crustal blocks tilted towards the southwest or southeast that occur on the Iberian margin of the Pyrenees. The transgressive systems tracts deposits backstep landward and aggrade over the preceding systems tracts. Finally, the highstand systems tracts deposits prograde regularly basinward and are separated from the underlying transgressive systems tracts by a maximum flooding surface (condensed section). Subaerial erosional unconformities separate the sequence from the sedimentary units above and below.

This sequence can be correlated with the third-order ZB 4.2 sequence of the global sea-level cycle chart of Haq *et al.* (1987). In the study area, this interval corresponds to a period of strong extensional (transtensional) tectonics leading to the development of Albian flysch basins on the underlying Aptian carbonate platform. Nevertheless, in spite of this tectonic activity, the facies associations and distribution observed here support an interpretation suggesting the strong influence of eustatic changes on the sedimentary succession.

INTRODUCTION

In the Pyrenees, the Clansayesian (uppermost Aptian) interval corresponds to the break-up of the wide Urgonian (Aptian) platforms and the establishment of several 'pull-apart' Albian flysch basins on the plate boundaries. This break-up is associated with the sinistral east–west motion of Iberia in the south, with respect to Europe in the north (Choukroune & Mattauer, 1978; Souquet, 1988). The precise location of the depocentres within these basins is a function of the regional tectonics, but the general facies organization and stratal architecture have been influenced strongly by contemporaneous relative sea-level changes (Canérot, 1991).

The purpose of this paper is to examine, using sequence-stratigraphy concepts (Haq *et al.*, 1987; Haq *et al.*, 1988; Posamentier & Vail, 1988; Sarg, 1988; Van Wagoner *et al.*, 1988), the local tectonic and eustatic effects on the Clansayesian deposits. The Lichançumendy and Arudy examples (Iberian margin of the western Pyrenees) provide critical insights into the aforementioned issues.

THE LICHANÇUMENDY CLANSAYESIAN SEQUENCE

The Lichançumendy (also known as Lichans Mountain in Basque) example corresponds to the Jurassic and Cretaceous section overlying a piercement diapir consisting of Triassic evaporites

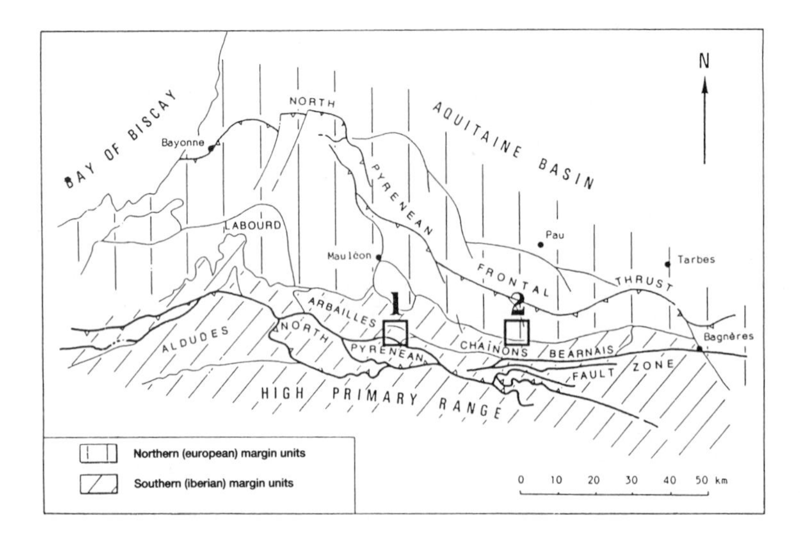

Fig. 1. Location map of the study area. (1) Lichançumendy Mountain section; and (2) Arudy section.

(Canérot & Lenoble, 1989). It is located at the northern border of the North Pyrenean Fault Zone adjoining the 'Chainons Bearnais' area to the Arbailles Basque Massif (location 1, Fig. 1).

The Clansayesian series can be observed under favourable conditions in the 'Chapeau de Gendarme' section (location A, Fig. 2). Subsequent observations to the west and east yield a comprehensive sequence-stratigraphic interpretation.

The Chapeau de Gendarme section

On the south side of Chapeau de Gendarme Mountain, the Lower Cretaceous series begins with the Etchebar breccia, a unit related to diapiric activity during the Neocomian and Aptian period (Canerot *et al.*, 1990). Overlying these deposits, five lithological units can be observed (Fig. 3):

1 *The Arhansus conglomerates* (15 m thick) comprise reworked Jurassic limestones and dolostones (70% of the clasts), Triassic sandstones and ophite (30%) and finally rare Palaeozoic quartzites. Towards the upper part of the formation, these rocks are associated with dark sandstones and siltstones and include shell (oyster-type) fragments. This unit is characterized by poor cementation, well-rounded reworked clasts, marine fauna and a progressive fining-upward grain size, indicating pebble beach deposits deposited under the influence of wave energy (Postma & Nemec, 1990) followed by inner-shelf sedimentation under transgressive conditions.

The boundary with the underlying Etchebar breccia correlates with the 'd1' unconformity, which has been interpreted as a subaerial erosional surface related to karstification and collapse during the Lower Cretaceous diapiric period (Canérot & Lenoble, 1989).

2 *The Lichançumendy black marls* (40 m thick). Located within the lowermost 5 m of the series the faunal assemblage is a very thin layer (consisting exclusively of oyster shells). Overlying this unit is a grey-coloured calcareous condensed layer (10–20 cm thick) containing centimetre scale ferruginous nodules, followed by a layer of marls that gradually become richer in (often fragmented) wide and flat corals (microsolenidids) associated with rare, isolated calyx forms. Towards the top, the calcareous fraction increases and the microsolenidids become thicker, less numerous and progressively more silicified. The composition of these marly deposits suggests an outer-shelf environment under regressive conditions.

3 *Bioclastic limestones* (50 m thick) gradationally overlie the Lichançumendy marls. They are organized into obliquely inclined decametric-scale bars prograding westward. The silicified microsolenidids (occurring rarely, but up to 50 cm wide) are associated with red algal debris including *Agardhiellopsis cretacea, Kimalithon belgicum, Pseudolithothamnium album and Paraphyllum primaevum* that are commonly observed within the 'Vinport facies' (N'da Loukou, 1984). These calcareous bars are interpreted as being outer-shelf regressive deposits close to reef areas.

4 *The Chapeau de Gendarme reef complex* (50 m thick) consists of three principal coral formations

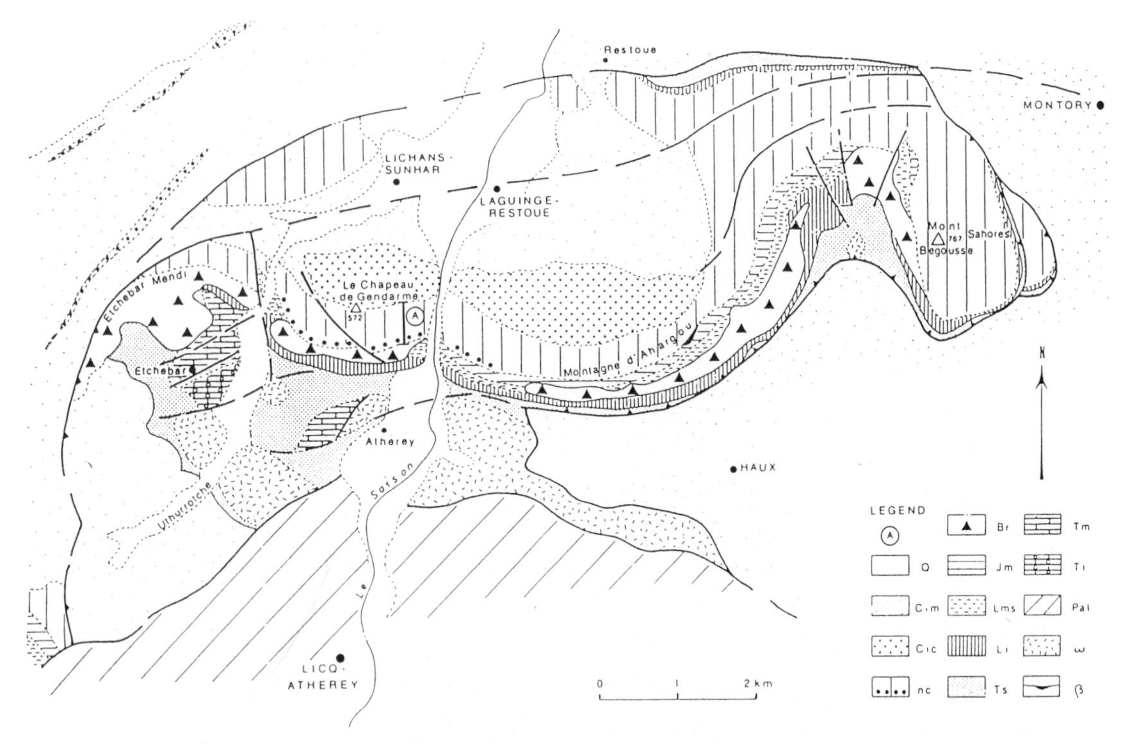

Fig. 2. Geological map of the Lichançumendy area (modified from Canerot & Lenoble, 1989). B, bauxites; W, ophite; Pal., Palaeozoic basement and Permo-Triassic 'revêtement'; Ti, Lower Trias; Tm, Middle Trias; Ts, Upper Trias; Li, Lower Lias; Lms, Middle and Upper Lias; Jm, Middle Jurassic; Br, Etchebar breccia (Lower Cretaceous); nc, Lichançumendy Formation (Clansayesian), with Arhansus conglomerates; Clc, carbonate mudmound (lowermost Albian); Clm, black spicule marls (Lower Albian); Q, Quaternary; A, Chapeau de Gendarme section.

that prograde westward across the underlying bars and interfinger eastward with inner-shelf, lagoon-type, massive grey mudstones containing rudistids (*Toucasia*) and benthic foraminifera (*Mesorbitolina*).

5 *Carbonate 'mudmound-type' features* (100 m thick) comprise large-sized microsolenidids and red algal debris in a micritic beige-coloured matrix laterally equivalent to black spicule marls (Lichans–Sunhar Formation). In the adjoining Bearn area, cementation of similar formations is thought to have been brought about by bacterial activity (Van der Plaetsen *et al.*, 1988), possibly as a result of warmer conditions generated by distensive faulting. The basal contact between the mudmound and the underlying Chapeau de Gendarme reef limestones corresponds to the d3 unconformity (karstic surface) well expressed along the eastern side of the Saison Valley.

These units have been considered to be Clansayesian (uppermost Aptian) in age as they contain *Mesorbitolina texana* associated with the 'Vinport' flora and occur below the Albian spicule marls (N'da Loukou, 1984).

Lateral variations

The Clansayesian series of the Chapeau de Gendarme section thins rapidly towards the west. Near Uthurrotche Valley (Fig. 2), the Arhansus Formation is represented by a thin conglomerate layer (4 m thick) although here the Palaeozoic quartzite elements are significantly more common than at the Chapeau de Gendarme section (Fig. 4). The grain-supported matrix as well as the absence of marine shell fragments suggest a debris-flow type of sedimentation in a non-marine environment. Further to the west, the conglomerates and the overlying Lichançumendy marls pinch out over the d1 unconformity. Consequently, on the western slopes of

Fig. 3. The Chapeau de Gendarme Clansayesian section. The diapiric collapse Etchebar breccia (E) overlies thin-bedded Jurassic (lower Lias to Dogger) deposits (J). Above unconformity d1, the Clansayesian succession comprises two systems tracts: (a) a transgressive systems tract including the Arhansus thinning-upward unit conglomerates/sandstones/siltstones (1) and the lower part of the Lichançumendy black marls (2); (b) a highstand systems tract including the upper part of the Lichançumendy black marls (2), the bioclastic limestones topped by oblique outer-shelf bars (3) and the Chapeau de Gendarme reef complex interfingering with inner-shelf limestones to the right (4). To the left, the westward direction of sedimentation is distinctly shown by bars and reef progradation.

the Etchebarmendy, the diapiric Etchebar breccia is directly overlain by outer-shelf thin-bedded limestones, less than 20 m thick. A ferruginous crust can be observed along the bounding surface between these two units (Fig. 4).

Eastward, a gradual thickening as well as an important facies change can be observed. The Arhansus marine conglomerate first thickens (up to 20 m thick on the western side of Ahargou Mountain) then pinches out. The Lichançumendy marls do not occur in the Mont Begousse area; in this area inner-shelf bioclastic limestones with rudistids and benthic foraminifera overlie directly a Liassic calcareous series (Fig. 5). Moreover, in this area also, the Chapeau de Gendarme reef complex is absent. Consequently, at Ahargou Mountain outer-shelf limestones interfinger with lagoon-type ones. At the same time, overlying Albian mudmounds interfinger eastward with black spicule marls (Laguinge Restoue area on Fig. 2).

Sequence-stratigraphic analysis

Through the application of sequence-stratigraphy concepts (Haq *et al.*, 1987; Haq *et al.*, 1988; Posamentier & Vail, 1988; Sarg, 1988), the various units of the Clansayesian series described here can be interpreted as occurring within a single eustatic (third-order) cycle (Figs 6 & 7). The succession is bounded at its base by unconformity d1 (Fig. 6), which is a subaerial erosional surface interpreted as a sequence boundary. The unconformity overlies undated karst and collapse materials (Etchebar breccia) corresponding to several superimposed or merged unconformities. The deposits corresponding to the hiatal breaks associated with these uncon-

Fig. 4. The Etchebarmendy Clansayesian section. Above the Triassic materials of the diapiric core (T), the Etchebar breccia (E) is overlain by a thinned Clansayesian section (CL) consisting of a highstand systems tract only (marls associated to the right with outer-shelf limestones).

formities can be observed westward in the Lower Cretaceous Arbailles Formation (Canerot, 1989).

The part of the succession comprising the Arhansus conglomerates, the dark-coloured sandstones, the shelly siltstones and the poorly fossiliferous marls (lower part of the Lichançumendy Formation on Fig. 3) is interpreted as being a transgressive systems tract. This unit is characterized by thinning-upward bedding style, backstepping (landward) aggradation and a low rate of accumulation (based on an inferred catch-up depositional style).

The conglomeratic bed, rich in Palaeozoic clasts, observed in the Uthurrotche Valley, could indicate the beginning of the sedimentation process during late lowstand. At this time erosional and sedimentary bypass conditions could have characterized the emerged Jurassic platform with sediment being supplied directly from distant Palaeozoic and Triassic highs.

Occurring above the conglomerates, the calcareous level (at d2, Figs 6 & 8), which contains ferruginous nodules in the Chapeau de Gendarme area, can be interpreted as being a maximum flooding surface (i.e. the core of the condensed section) suggesting the turnaround from transgression to regression during an interval of relatively rapid sea-level rise (Posamentier & Vail, 1988).

Subsequently, the succession comprising the Lichançumendy marls, the bioclastic limestones (outer-shelf regressive bars) and finally the reef complex and the equivalent landward-occurring lagoonal limestones could represent the highstand systems tract. Its relatively high rate of accumulation suggests a keep-up carbonate system and its upward shoaling character is typical of such a systems tract (Sarg, 1988).

The d3 erosional unconformity characterized by karstic development is interpreted as a sequence boundary. The overlying mudmound consequently belongs to the lowstand systems tract of the following sequence. This mudmound carbonate build-up is characterized by a high rate of sedimentation (i.e.

Fig. 5. The Mont Begousse Clansayesian section. In the eastern part of the Lichançumendy area, inner-shelf limestones corresponding to the highstand systems tract (HST) directly overlie the Etchebar diapiric breccia to the left (E) and eroded Liassic limestones to the right (landward) (L).

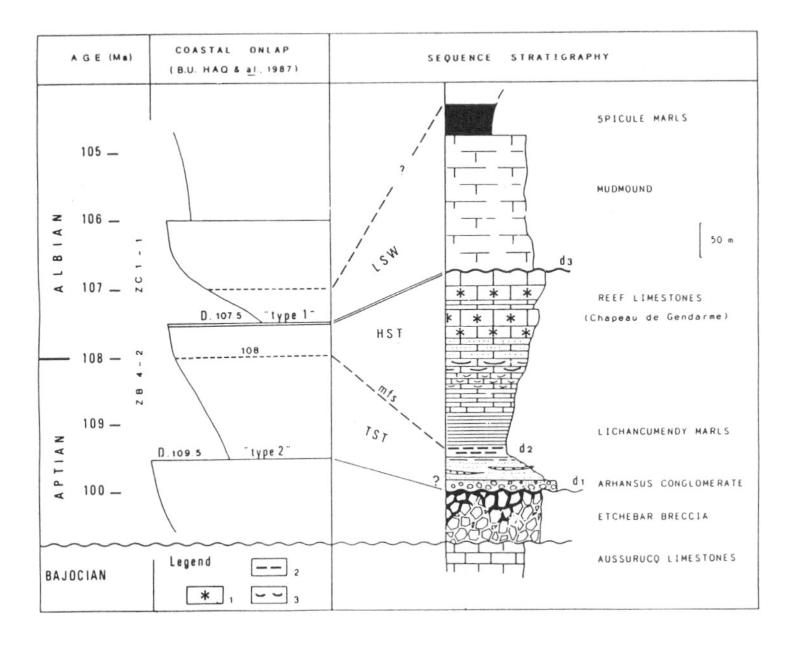

Fig. 6. Sequence stratigraphy of the Aptian–Albian section in the Lichançumendy area (Chapeau de Gendarme section): (1) coral cores; (2) condensed section; and (3) silicified microsolenidids. TST, transgressive systems tract; mfs, maximum flooding surface; LSW, lowstand (wedge) systems tract.

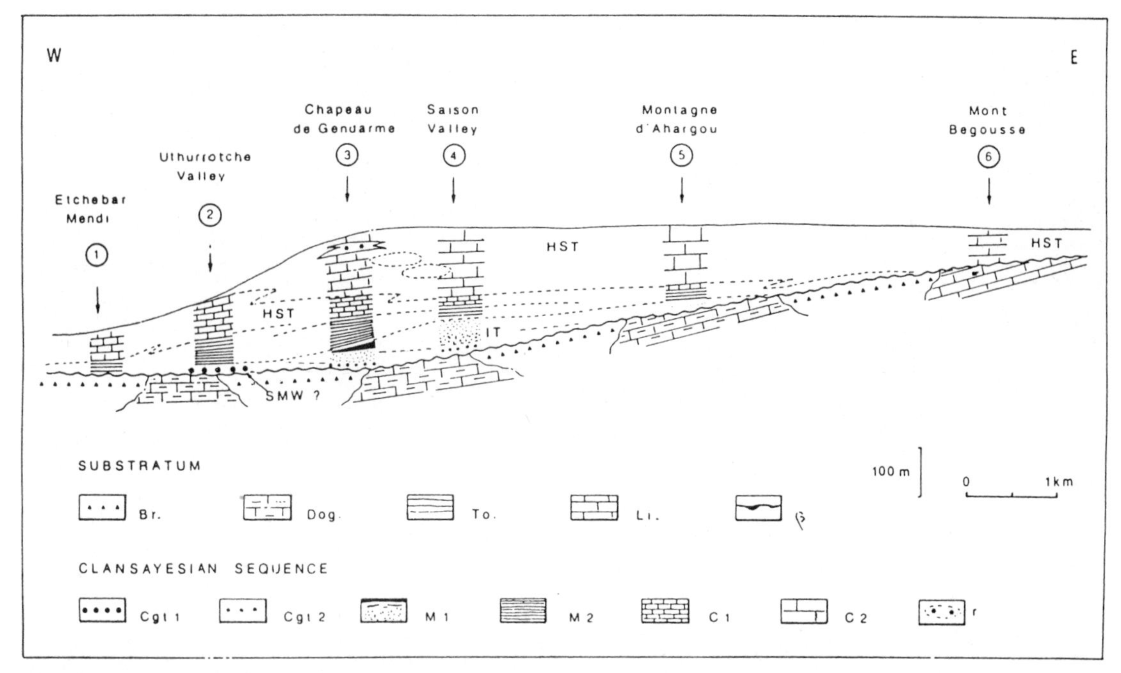

Fig. 7. Correlated sections showing the Clansayesian succession in the Lichançumendy area. Cgt 1, Uthurrotche conglomerate (shelf-margin wedge?); Cgt 2, Arhansus conglomerate; M1, sandstones, siltstones and black marls (transgressive systems tract); M2, Lichançumendy marls; C1, outer-shelf bars; C2, inner-shelf limestones; r, reef complexes; Br, Etchebar breccia (Aptian); Dog, Dogger; To, Toarcian; Li, Lias; B, bauxite.

a keep-up system) and interfingers with black spicule, basin-type marls. This build-up has no landward equivalent on the platform (Canérot *et al.*, 1990) as is typical of transgressive or highstand mounds. Similar carbonate build-ups recently have been described in the lowstand systems tracts of Albian formations near Soba, northern Spain (Garcia-Mondejar & Fernandez-Mendiola, 1990).

Because the sediments are Clansayesian (uppermost Aptian) in age (N'da Loukou, 1984), we suggest that the Lichançumendy sequence that underlies the mudmound could correspond to the ZB 4.2 cycle of Haq *et al.* (1987). Consequently, the d1, d2 and d3 surfaces could be correlated respectively with the 109.5 Ma unconformity (type 2) occurring at the base of the sequence, the 108 Ma maximum flooding surface occurring within the sequence and the 107.5 Ma unconformity (type 1) occurring at the top of the sequence.

DISCUSSION

The facies organization within the ZB 4.2 cycle seems to be closely linked to relative sea-level changes. The contribution of tectonics to relative sea-level rise, and hence accommodation, is associated with crustal extension occurring at the beginning of the Albian period. The backstepping stratal geometry (eastward) within the transgressive section, the prograding stratal geometry (westward) of the highstand wedge (Fig. 7) and the location of the depocentre also are a function of the westward tilt of the Lichançumendy crustal block during the formation of the Mauleon pull-apart basin (Canérot, 1991) due to east–west-oriented, sinistral wrench faulting. Consequently, the Lichançumendy area provides a good example of the interrelationship between eustasy and tectonics as it affects the sedimentary record at the Aptian–Albian boundary in the western Pyrenees.

THE ARUDY CLANSAYESIAN SEQUENCE

In the Ossau Valley, 30 km east of Lichançumendy Mountain, the Clansayesian series is again well exposed in the vicinity of Arudy (location 2 in Fig. 1; Fig. 9), where Clansayesian limestones are actively

Fig. 8. Maximum flooding surface (MFS) in the Chapeau de Gendarme Clansayesian succession. This surface comprises carbonate sediments with ferruginous nodules within the Lichançumendy black marls. Transgressive and highstand systems tracts, respectively occur under (to the left) and over (to the right) the MFS.

quarried. In this area, the Urgonian Formation, Upper Aptian in age (N'da Loukou, 1984), culminates in a ferruginous surface correlated with the 109.5 Ma unconformity of Haq *et al.* (1987) (Lenoble & Canérot, 1988). On the western side of the valley, up river from the town of Izeste, this surface is overlain by Urgonian limestones whose coarsening upwards organization indicates lowstand condi-

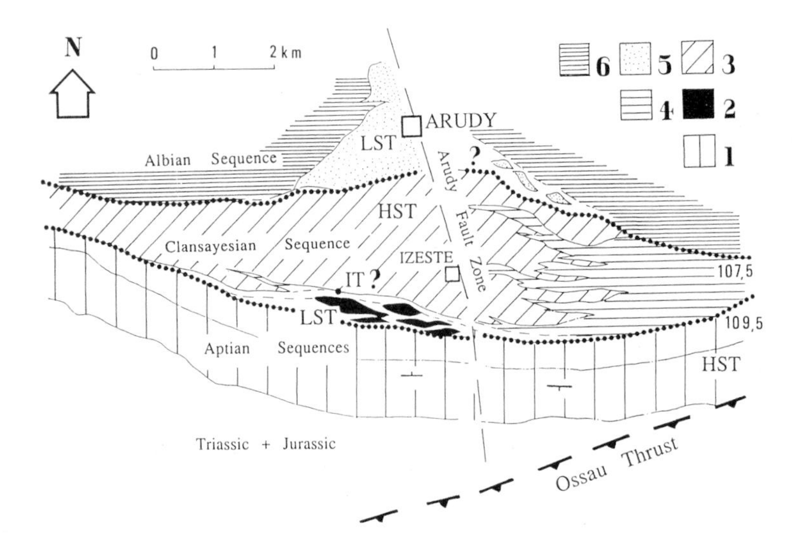

Fig. 9. Geological sketch map of the Arudy area. (1) Urgonian platform; (2) Clansayesian Izeste mudmounds; (3) Clansayesian carbonate platform; (4) Clansayesian black marls; (5) Albian Arudy mudmounds; (6) Albian black spicule marls. LST, lowstand systems tract; IT, transgressive systems tract; HST, highstand systems tract.

Fig. 10. The Izeste Clansayesian mudmounds. Mound-shaped build-ups are shown that represent the transgressive systems tract of the Clansayesian sequence as they interfinger with marls (left and right sides of the quarry) and overlie the Urgonian limestones of the lowstand systems tract.

tions. These limestones are followed by carbonate build-ups comprising *Toucasia* shell-bearing mudstones (Fig. 10) that are associated with black marls (Canérot *et al.*, 1990; Canérot, 1991). These limited (less than 100 m thick) build-ups can be interpreted as the transgressive systems tract of the Clansayesian sequence.

Subsequent to deposition of the black marls and mudmounds, a thick (200 m) carbonate succession develops. Westward, it overruns the Izeste build-ups and overlies directly the older Urgonian (Aptian) limestones. Eastward, it interfingers with black spicule marls. The detailed west to east evolution of the carbonate facies is characterized by: limestones with *Orbitolinid, Toucasia* and algal associations (Vinport lagoon-type facies); reef complexes (including microsolenidids); crinoidid limestones (outer-shelf environments); and marly limestones (transition from shelf to basin). This carbonate succession and the associated black marls characterize the eastward-prograding highstand systems tract of the Clansayesian sequence (Fig. 11).

Several microsolenidid build-ups overlie the aforementioned highstand Clansayesian section at Arudy. These deposits are Albian in age (Bouroullec *et al.*, 1979) and are very similar to the mudmound of the Lichançumendy section (Fig. 12). These build-ups, whose thickness can exceed 100 m, are correlated westward and eastward with black marls (Fig. 13). They consistently occur within the lower part of a tilted fault block and are believed to represent the lowstand wedge of the Albian sequence. Other synchronous build-ups have been observed recently in the same position on different tilted blocks of the Iberian margin (Lenoble, pers. comm.).

Consequently, the Arudy Clansayesian succession is very similar to that observed at Lichançumendy. In contrast with the Lichançumendy section however, the lowstand systems tract is represented by carbonate sediments and the transgressive systems tract (comprising carbonate build-ups) is much more developed.

This geodynamic interpretation suggests a very

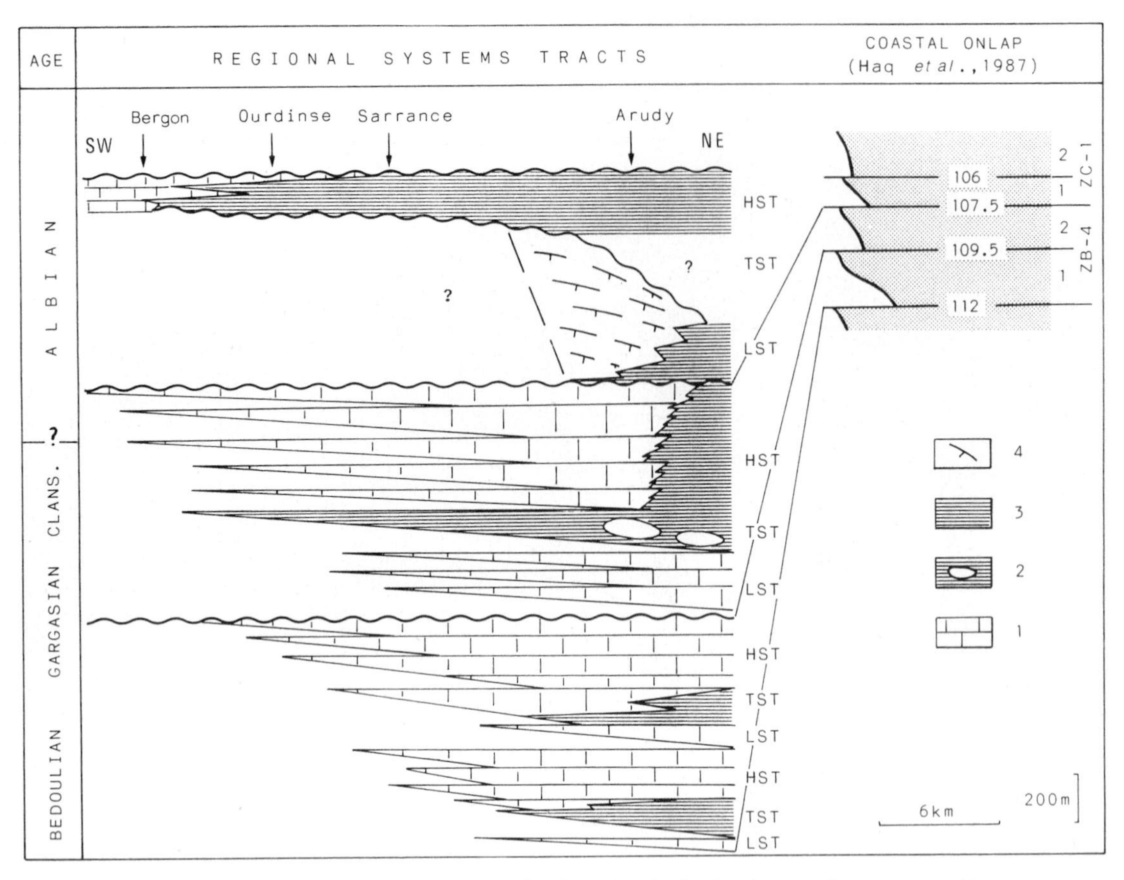

Fig. 11. Sequence-stratigraphic units at the Aptian–Albian boundary in the Arudy area: (1) arenaceous limestones (Aptian); (2) mudmounds (Clansayesian); (3) black marls (Aptian to Albian); (4) mudmounds (lowermost Albian). LST: Lowstand systems tract; TST: Transgressive systems tract; HST: Highstand systems tract.

high accommodation rate characterizing the interval around the Aptian–Albian boundary and is thought to be associated with the opening of the adjoining Mauleon flysch basin.

CONCLUSIONS

The Lichançumendy and Arudy sections are characterized by a Clansayesian sequence corresponding probably to the ZB 4.2 cycle of Haq *et al.* (1987). Within this sedimentary unit, the facies organization can be explained by analysing the interaction between sedimentation style and relative sea-level variations: carbonate mudmounds, terrigenous thinning- and deepening-upward deposits and shelf limestones interfingering with marls being respectively related to lowstand, transgressive, and highstand systems tracts.

In this area, accommodation rate increases, as well as the location of depocentres and progradation direction, seem to be strongly influenced by local extensional tectonics. Consequently, the sequence discussed here may be part of the initial deposition within the mid-Cretaceous Pyrenean pull-apart basins induced by a gradual wrench faulting and tilting down of crustal blocks on the Iberian margin of the Pyrenees, previously covered by the wide Urgonian (Aptian) platform. Nonetheless, the eustatic signal seems to be expressed in the sections studied, even if it has been overprinted by local tectonics.

Fig. 12. The Arudy lowermost Albian mudmound. Flat corals (S) (microsolenidids) occur within a mudstone matrix wherein cementing has been generated by bacterial activity (horizontal scale bar is 40 cm).

ACKNOWLEDGEMENTS

I wish to thank Elf Aquitaine Oil Company in Boussens (France) for the facilities and support provided during this research on the western Pyrenees. Special thanks are extended to G.P. Allen for very useful comments that have improved the manuscript and to the reviewer, B. Granier. Thanks are also extended to H.W. Posamentier for constructive advice and to Mrs P. Eychene for technical assistance.

REFERENCES

Bouroullec, J., Delfaud, J. & Deloffre, R. (1979) Organisations sédimentaire et paléoécologique de l'Aptien supérieur a faciès urgonien dans les Pyrénées occidentales et l'Aquitaine méridionale. *Mem. Geobios* **3**, 25–43.

Canérot, J. (1989) Rifting éocrétacé et halocinèse sur la marge Ibérique des Pyrénées occidentales (France). Consequences structurales. *Bull. CREP Elf Aquitaine*, **13**, 87–99.

Canérot, J. (1991) Comparative study of the Mesozoic Basins from the Eastern Iberides (Spain) and the Western Pyrenees (France). In: *Tethyan Paleogeography and Paleoceanography* (Eds Channell, J., Winterer, E.L. and Jansa, L.F.) Palaeo 3, 87, n. 1–4, 1–28.

Canérot, J. & Lenoble, J.L. (1989) Le diapir du Lichançumendy (Pyrenees Atlantiques), nouvel élément de la marge ibérique des Pyrénées occidentales. *C.R. Acad. Sci. Paris* **308** (II), 1467–1472.

Canérot, J., Villien, A. & Lenoble, J.L. (1990) Sequence stratigraphy and sedimentary unconformities: Western Pyrenean (field data) and Southern Aquitanian Basin (subsurface data) examples. *Bull. Soc. Geol. Fr., Paris* (8), VI, pp. 995–1000.

Choukroune, P. & Mattauer, M. (1978) Tectonique des plaques et Pyrenees: sur le fonctionnement de la faille transformante nord-pyreneenne; comparaisons avec les modeles actuels. *Bull. Soc. Geol. Fr., Paris* **7** (XX), 689–700.

Garcia-Mondejar, J. & Fernandez-Mendiola, P.A. (1990) Comparative study of carbonate mounds in the context of systems tracts analysis: Albian of Soba, Northern Spain. *13th Int. Sediment. Congress, Nottingham* (Abstr.), 181.

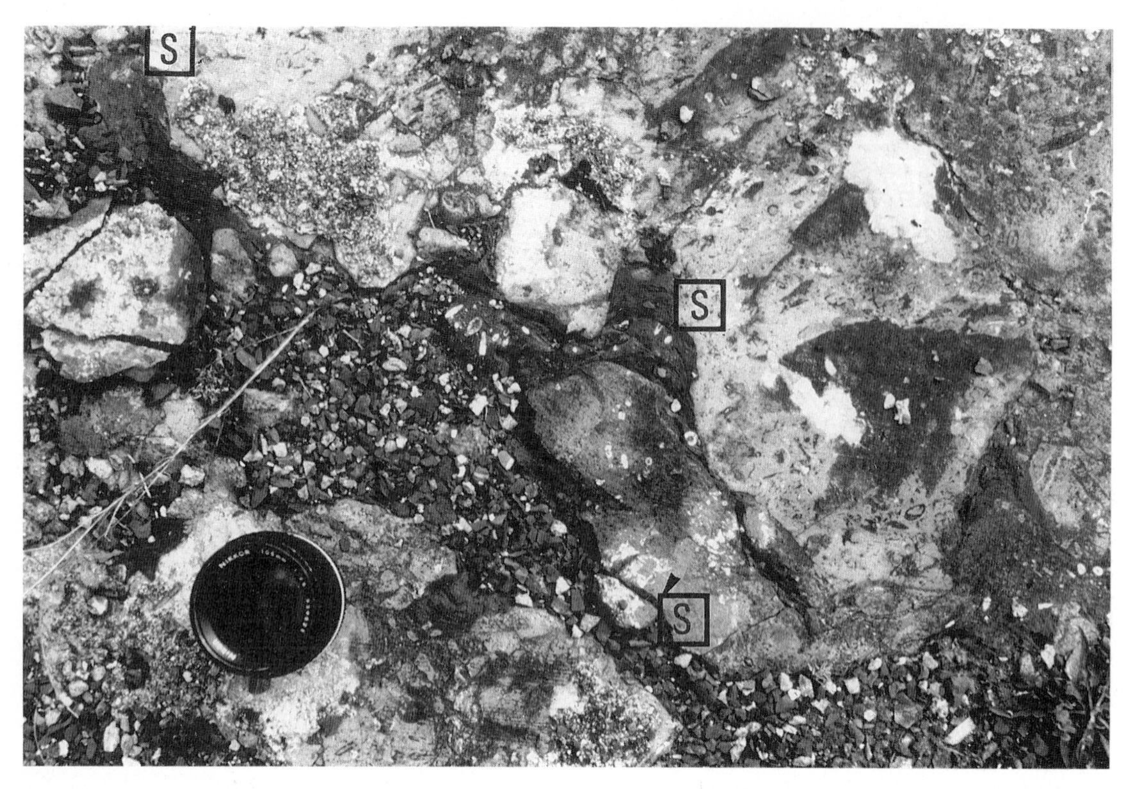

Fig. 13. The Arudy lowermost Albian mudmound (detail). The brecciated carbonate build-up (upper right) is separated from black spicule marls (centre of photograph) by an irregular erosion surface (S) indicating unstable conditions during the constructional stage.

Haq, B.U., Hardenbol, J., & Vail, P.R. (1988) Mesozoic and Cenozoic chronostratigraphy and cycles of sea-level change. In: *Sea-level Changes: An Integrated Approach* (Eds Wilgus, C.K., Hastings, B.S., Kendall, C.G.St.C., Posamentier, H.W., Ross, C.A. & Van Wagoner, J.C.) Spec. Publ. Soc. Econ. Paleontol. Mineral. 42, 71–108.

Haq, B.U., Hardenbol, J., Vail, P.R. *et al.* (1987) Mesozoic Cenozoic cycle chart. In: *Atlas of Seismic Stratigraphy* (Ed. Bally, A.W.) Am. Assoc. Petrol. Geol. Stud. Geol. 1 (27), 124.

Lenoble, J.L. & Canerot, J. (1988) Stratigraphie séquentielle au passage Aptien — Albien dans les Pyrénées occidentales. Exemples du Lichançumendy et d'Arudy *Strata* **I** (4), 131–135.

N'Da Loukou, V. (1984) Urgonien des Pyrénées occidentales. Synthèse paléoécologique, micropaléontologique et paléogéographique. Thesis, University of Pau, 225 pp.

Posamentier, H.W. & Vail, P.R. (1988) Eustatic controls on clastic deposition II—sequence and systems tract models. In: *Sea-level Changes: An Integrated Approach* (Eds. Wilgus, C.K., Hastings, B.S., Kendall, C.G.St.C., Posamentier, H.W., Ross, C.A. & Van Wagoner, J.C.) Spec. Publ. Soc. Econ. Paleontol. Mineral. 42, 125–154.

Postma, G. & Nemec, W. (1990) Regressive and transgressive sequences in a raised Holocene gravelly beach, southwestern Crete. *Sedimentology* **37**, 907–920.

Sarg, J.F. (1988) Carbonate sequence stratigraphy. In: *Sea-level Changes: An Integrated Approach* (Eds. Wilgus, C.K., Hastings, B.S., Kendall, C.G.St.C., Posamentier, H.W., Ross, C.A. & Van Wagoner, J.C.) Spec. Publ. Soc. Econ. Paleontol. Mineral. 42, 155–181.

Souquet, P. (1988) Evolucion del margen noriberico en los Pirineos durante el Mesozoico. *Rev. Soc. Geol. Esp.* **I** (3–4), 349–356.

Van Der Plaetsen, L., Arnaud-Vanneau, A. & Delfaud, J. (1988) Mudmound en période transgressive: la série clansayesienne d'Arudy (Aquitaine méridionale). 12e Réunion Sc. de la Terre, Lille (Ed. Soc. Geol. Fr.) 129.

Van Wagoner, J.C., Posamentier, H.W., Mitchum, R.M., Jr, *et al.* (1988) An overview of the fundamentals of sequence stratigraphy and key definitions. In: *Sea-level Changes: An Integrated Approach* (Eds Wilgus, C.K., Hastings, B.S., Kendall, C.G.St.C., Posamentier, H.W., Ross, C.A. & Van Wagoner, J.C.), Spec. Publ. Soc. Econ. Paleontol. Mineral. 42, 39–45.

Spec. Publs Int. Ass. Sediment. (1993) **18**, 295–305

Allochthonous deep-water carbonates and relative sea-level changes: the Upper Jurassic–Lowermost Cretaceous of southeast France

G. DROMART, S. FERRY *and* F. ATROPS

Université de Lyon, Centre des Sciences de la Terre, URA CNRS no. 11, 43 bd du 11 novembre, 69622 Villeurbanne Cedex, France

ABSTRACT

The subpelagic, limestone-dominant Kimmeridgian–Berriasian series of the Subalpine Basin is examined from two sets of sections, each representative of a slope/margin and a basin depositional setting. Massive, polymictic limestone conglomerates (olistostromes), together with turbidites and slide-slumps, show an extensive development at the base of the lowstand systems tracts. However, such depositional features are not unique to intervals of relative sea-level fall and do occur within both highstand and lowstand units. In contrast, monomictic limestone conglomerates (lag-type deposits) seem to uniquely characterize the transgressive systems tracts. Finally, it is suggested that depositional sequence development may be controlled by tectono-eustasy (change in basin volume and physiography) in this basin area.

INTRODUCTION

In siliciclastic environments, the occurrence of sediment gravity flows is enhanced during lowstands of sea level (Mitchum, 1985; Mutti, 1985; Kolla & Macurda, 1988; Posamentier & Vail, 1988; Shanmugan & Moiola, 1988, Vail *et al.*, 1991). In contrast, transportation downslope of carbonate shelves has been much more diversely assigned with respect to relative changes of sea level.

On the basis of studies of modern and recent carbonate series located offshore of the Bahamas, downslope transfers have been related primarily to highstand shedding (Kier & Pilkey, 1971; Lynts *et al.*, 1973; Schlager & Ginsburg, 1981; Boardman & Neuman, 1984; Droxler & Schlager, 1985; Reymer *et al.*, 1988) even though some large debris sheets have been specifically shown to be deposited when relative sea level was falling (Crevello & Schlager, 1980; Droxler & Schlager, 1985).

In contrast, for ancient carbonate slopes, major erosion and gravity-related transfers often have been interpreted to occur during falls and lowstands of sea level (Biddle, 1984; Bosellini, 1984; Shanmugan & Moiola, 1984; Sarg, 1988; Cook & Taylor, 1991; Jacquin *et al.*, 1991; Vail *et al.*, 1991). However, it has been stated that an actively growing transgressive and highstand bank margin also is capable of shedding debris as it progrades, especially if it builds a steep margin where slope instability becomes a factor (Sarg, 1988). In addition, major slumps and olistolites have been documented to be features of transgressive systems tracts, reflecting tectonically induced margin collapse associated with relative sea-level rise (Ferry & Rubino, 1989b).

Our purpose here is to present an example of deep-water carbonate sedimentation during a sea-level cycle. It is based on the Upper Jurassic–Lowermost Cretaceous (Kimmeridgian–Berriasian) series of the Subalpine Basin in southeastern France. The study specifically aims at recognizing the occurrence and importance (volume, frequency) of distinct depositional lithofacies (limestone conglomerates, slumps, turbidites) through well-assigned systems tracts and surfaces within a sequence-stratigraphic framework. This is accomplished first through a set of sections representative of an outer-shelf/upper-slope setting wherein depositional sequences are fully developed. The depositional model is then extended to basinal sections for which systems tracts are identified on the basis of biostratigraphic correlations with the basin margin area. Subsequently, discussion is focused on the potential genetic connections between the distinct features associated with carbonate excavation and relative sea-level change. Finally, third-order sequences recognized in this study are tentatively compared with the Exxon global cycle chart (Haq *et al.*, 1988).

The succession of autochthonous deposits that frames the depositional sequences is described and then interpreted within a systems-tract framework. Deposits related to submarine erosion are further described, identified and partitioned within the depositional tracts.

BASIN MARGIN SEQUENCES

The area of investigation is the western margin of the Subalpine Basin (Ardèche Area). Our study addresses sections located in the northern part of the Ardèche Area (Fig. 1). The stratigraphic interval of interest is the Upper Kimmeridgian–Lowermost Berriasian. The sections are fully developed and accurately age-dated based on ammonite and calcareous nannofossil biostratigraphy (Le Hegarat, 1973; Atrops, 1984; Cecca *et al.*, 1989a,b).

Autochthonous deposits and systems tracts

The total section shown in Fig. 2 has been compiled from several exposures located in the close vicinity of Le Pouzin village. The section is presented below in a stratigraphically ascending order.

Upper Kimmeridgian (Acanthicum–Lower Eudoxus) deposits are interbedded calcareous shales and fine-grained, grey-coloured limestones. The limestones are skeletal wacke to mudstones with bioclasts referred to as *filaments* (Dromart & Atrops, 1988), along with calcite-replaced tests of radiolaria. Overlying Upper Eudoxus–Beckeri massive limestones are occasionally and variably enriched in shelf-derived bioclasts, such as shell fragments of benthic molluscs and echinids. Some of these bio-

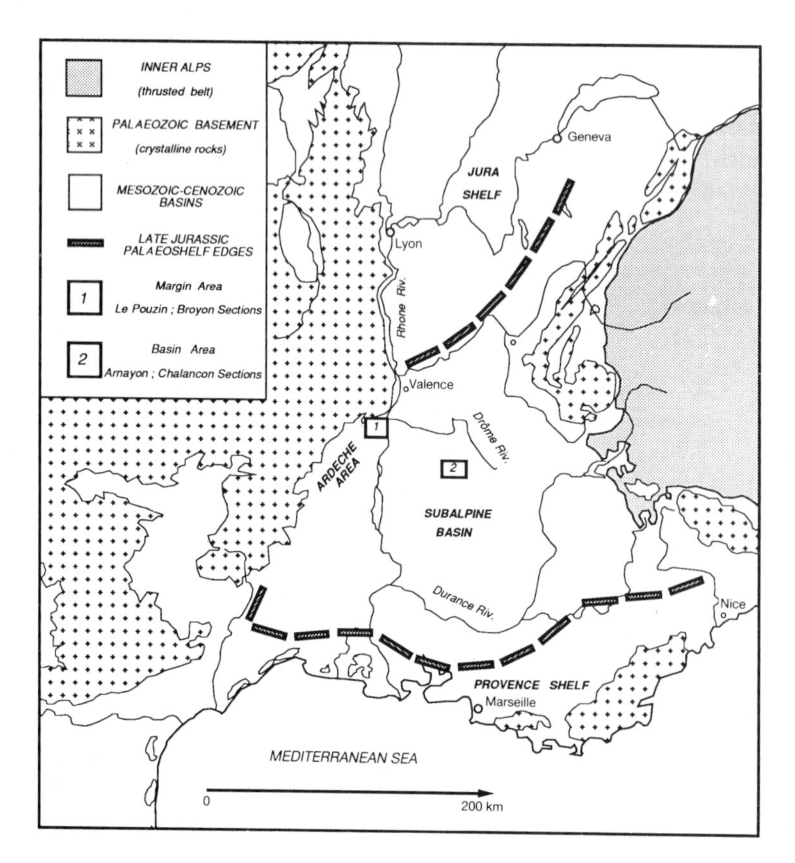

Fig. 1. Simplified geological map of southeastern France with location of sections (bold squares) examined in the Subalpine Basin.

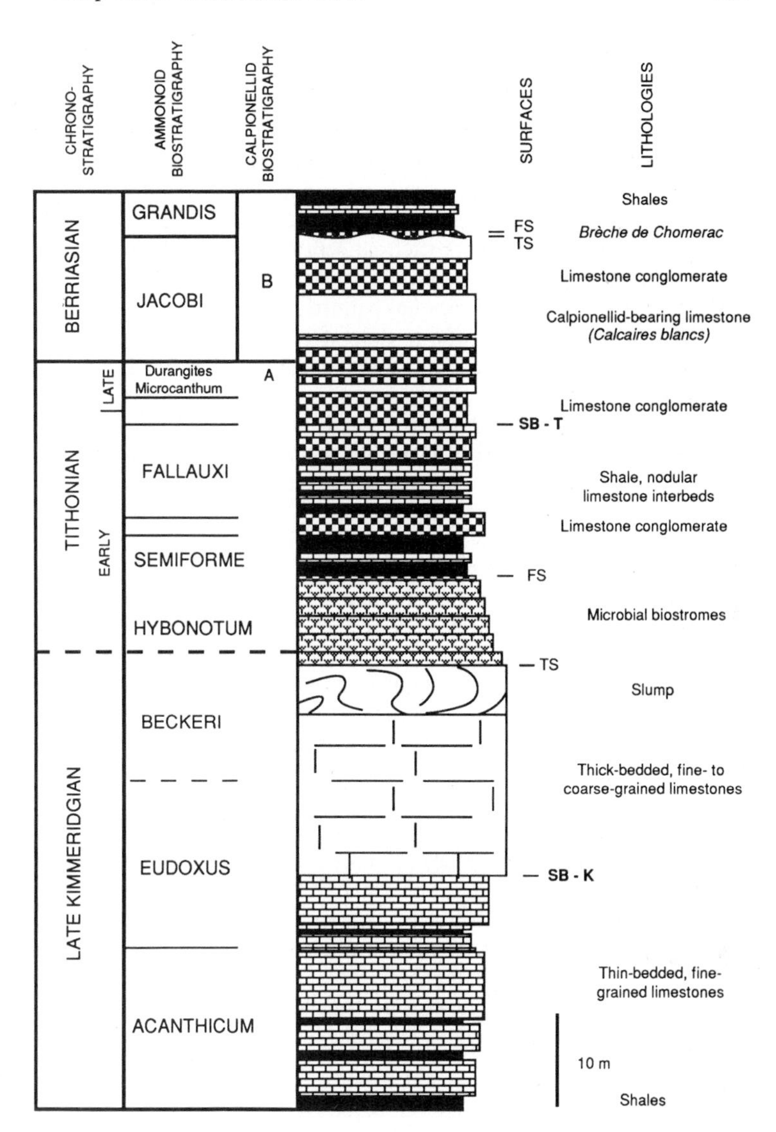

Fig. 2. Stratigraphic column for the Kimmeridgian–Berriasian of Le Pouzin area (western margin of the Subalpine Basin) along with the sedimentary surfaces. SB, sequence boundary; TS, transgressive surface; FS, flooding surface.

clasts are coated whereas others are not. Depositional textures are wacke to packstones.

It has been suggested that Oxfordian–Kimmeridgian fine-grained carbonates of this area are composed in part of off-bank deposits (Dromart, 1989a). The progressive upward enrichment of fine-grained carbonates throughout the Acanthicum–Lower Eudoxus interval is thus assumed to reflect progradation of a highstand systems tract. Higher up, the stacking pattern, which consists of a conformable but relatively abrupt transition from thin- to thick-bedded, massive limestones of the Upper Eudoxus — Beckeri interval, likely reflects a basinward shift of the shelf-derived deposits, with a higher rate of sedimentation. This change is interpreted to mark a sequence boundary. This assignment is consistent with the observation of Ferry and Rubino (1989a,b) that sequence boundaries of depositional sequences cut across Klüpfel-type sequences (i.e. the Upper Kimmeridgian lithocline as a whole). Whether this sharp change represents a type I or a type II unconformity such as defined by Posamentier and Vail (1988) and Sarg (1988) cannot be resolved here because precise correlations with the platform/bank margin (Jura Area, Fig. 1) have not yet been established.

The massive lowstand limestones are overlain by Lowermost Tithonian pseudonodular limestones (thin, wavy bedding) comprising pelagic products that include saccocomid debris, radiolarians, *Globochaete* and calcispherulids. Besides their structure, these limestones are characterized by a mottled appearance due to the presence of intraclasts and microbial carbonate elements such as centimetre-sized oncolites plus low stromatolitic columns (Dromart, 1989a). Such a compositional pattern reflects low rates of sedimentation and is characteristic of transgressive to condensed intervals (Dromart, 1989b, 1992).

The overlying 'Middle' Tithonian highstand deposits consist of interbedded dark-grey shales and argillaceous limestones. The limestones are characterized by a pseudonodular structure and a mudstone depositional texture. This shaly lithologic unit is sharply overlain by calpionellid-bearing, light-coloured limestones of a late Tithonian to earliest Berriasian age, regionally referred to as *Calcaires blancs*. The surface occurring at the abrupt lithologic change from shale dominant to limestone dominant is interpreted as a sequence boundary and the *Calcaires blancs* unit as a part of either a shelf-margin wedge or a lowstand complex. Finally, the uppermost part of the section (Grandis subzone *pro parte*) comprises brown-coloured shales and argillaceous limestones and again marks a highstand interval.

The Kimmeridgian–Tithonian sequence presented here is several tens of metres thick (about 70 m) and its duration is approximately 5–6 Ma assuming that each ammonoid-zone time span is 1 Ma. Consequently, it is believed that the sequence is a third-order sequence even though the time span appears long relative to age ranges proposed by Vail and Eisner (1989) (0.8–4.0 Ma) and Vail *et al.* (1991) (0.5–5 Ma).

Reworked deposits and systems tracts

Three lithofacies characterize submarine excavation: slumps, turbidite beds and limestone conglomerates. The limestone conglomerates are further split into two distinct types: common 'resedimentation breccias' and singular deposits regionally referred to as *Brèche de Chomerac* (Le Hegarat, 1973). The occurrences of these lithofacies within systems tracts are summarized in Fig. 3.

Slumps

Massive slide and slumps are observed as local occurrences a few metres thick, localized in the uppermost part of the lowstand massive limestones. Only one slump has been encountered within the Lowermost Tithonian transgressive tract as observed all across the margin.

Turbidites

Coarse-grained turbidite beds are 0.2 to 0.6 m thick and are characterized by sharp basal contacts with unidirectional flute-casts. Grading often is well developed and cross-laminations are common. Components consist of lime–mudstone pebbles, peloids, coated grains and rare shoal-sourced bioclasts, including foraminifers and fragments of skeletal algae.

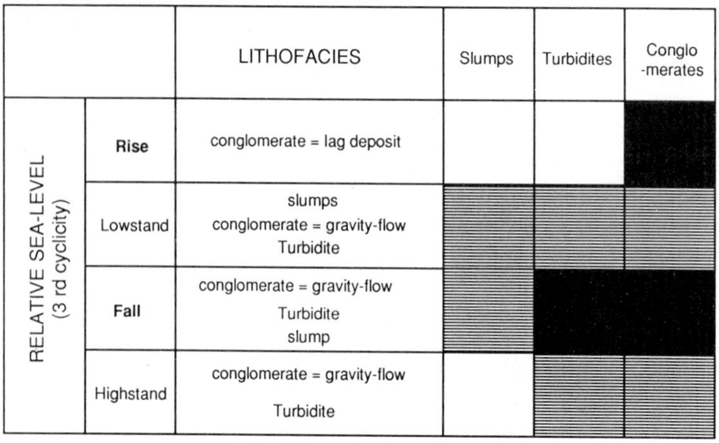

RELATIVE SEA-LEVEL (3 rd cyclicity)		LITHOFACIES	Slumps	Turbidites	Conglo-merates
	Rise	conglomerate = lag deposit			■
	Lowstand	slumps conglomerate = gravity-flow Turbidite	▤	▤	▤
	Fall	conglomerate = gravity-flow Turbidite slump	▤	■	■
	Highstand	conglomerate = gravity-flow Turbidite		▤	▤

Fig. 3. Type and importance of distinct carbonate lithofacies related to deep-water removal along with a relative sea-level transit cycle (third order). Constructed from the Tithonian–Lower Berriasian sections of the basin margin area, northern Ardèche Area, France. The lithofacies frequency is indicated by: solid squares, abundant — stacked and widespread occurrences; banded squares, rare — scattered and local occurrences; blank squares, non-occurrence.

Turbidite beds recur throughout the depositional sequences with the exception of within transgressive tracts. All over the area, such beds are regularly encountered at the boundary between the highstand shaly deposits and the lowstand limestones as well as within the lowstand limestones themselves. On the other hand, turbidite beds are rare within the highstand shaly tracts. In summary, when present, turbidite beds preferentially occur near the base of the depositional sequence, within the lowstand systems tract (Fig. 4).

Conglomerates

Based on sedimentologic evidence, the Tithonian–Berriasian limestone conglomerates can clearly be subdivided into two types of deposits: allochthonous olistostromes (gravity flows) and para-autochthonous lag deposits (Table 1). Olistostromes are distributed within systems tracts as follows:

1 Climax units (close stacking and widespread development) at the boundary between highstand shaly deposits and lowstand limestones (Mid–Upper Tithonian only);

2 Vertically scattered and restricted in areal distribution within highstand shales and lowstand limestones;

3 Absent in transgressive systems tracts.

In contrast, lag deposits (*Brèche de Chomérac*, Fig. 2) comprise the transgressive systems tract occurring between lowstand *Calcaires blancs* and highstand shales of the Grandis subzone. These lag deposits are highly variable in thickness (0.1–1 m) and are sporadically observed across the area (i.e. patchy distribution). They directly overlie a highly irregular surface that cuts down into the *Calcaires blancs*. Pebbles are similar in composition to the underlying unit and have a para-autochthonous origin (i.e displaced a short distance). The ammonite assemblage within the lag deposits is homogeneous and representative of the underlying Jacobi subzone (Le Hegarat, 1973). The transgressive erosional surface below can be described as a ravinement surface (Weimer, 1984).

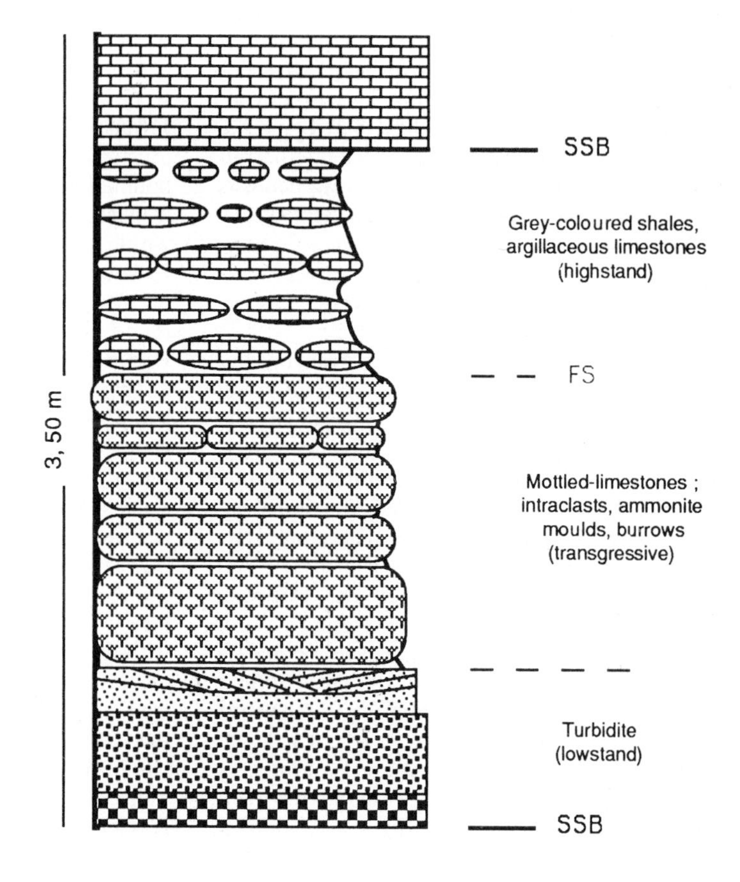

Fig. 4. High-frequency (i.e. suborder) sequences within the third-order, 'Middle' Tithonian highstand tract. Broyon quarry, Ardèche Area, southeast France. Note the basal development of the turbidite bed and the general sequence self-similarity (similar stacking pattern of suborder and depositional sequences). SSB, suborder sequence boundary; FS, flooding surface.

Table 1. Tithonian–Berriasian deep-water limestone conglomerates of the marginal area of the Subalpine basin; comparative sedimentologic attributes of *resedimentation breccias* and *Brèche de Chomérac*.

	Limestone conglomerates (resedimientation breccias)	'Brèche de Chomérac'
Texture	Matrix to clast supported	Clast supported
Grading	Common no grading, rare normal grading	No grading
Clast size	Sand to boulder	Pebble to cobble
Clast shape	Subangular to rounded	Rounded
Clast nature	Polymictic (allochthonous)	Monomictic (para-autochthonous)
Stratal pattern	Massive sheet, channel fill (0.5–3 m thick)	Lens-shaped (depression fill) (0–1 m thick)
Occurrence	Extensive to local; any ST but transgressive ST	Extensive (patchy); restricted to transgressive ST
Nature	Olistostromes (gravity flows)	Lag deposits

Types and occurrences of carbonate deposits related to deep-water excavation versus relative sea-level changes are grouped in Fig. 3. Two main points are as follows: (1) downslope gravity-related transfers are associated with low- and highstand systems tracts concentrated directly overlying sequence boundaries; (2) limestone conglomerate lithofacies can be associated with both falling and rising sea level. Nonetheless, sedimentologic attributes permit clear distinction of olistostromes and lag deposits that identify sea-level fall and rise, respectively.

BASIN SUCCESSIONS

Two basinal sections, referred to as Arnayon and Chalancon, illustrated in Fig. 5, are exposed in the vicinity of La-Motte-Chalancon village in the Drôme Area (see Fig. 1). Upper Jurassic–Lowermost Cretaceous sections here are composed mainly of displaced carbonate material. Associated depositional facies include turbidites, conglomerates (breccias plus pebbly mudstones; Remane, 1970) and slumps, similar in composition and texture to their marginal counterparts (see Table 1). Conversely, autochthonous subpelagic sediments are poorly developed here, presumably because of their original low rate of deposition along with repetitive erosion by gravity-related flows.

The recognition of depositional sequences through delineation of sequence boundaries (SB-Kimmeridgian; SB-Tithonian; Fig. 5) is indirect and based on biostratigraphic time correlations with the marginal sections of the Ardèche Area. Age assignments for the basinal sections were established on the basis of both microfacies (Dromart & Atrops, 1988) and ammonite biostratigraphy.

A number of conclusions can be drawn from the sections:

1 Massive, polymictic limestone conglomerates (olistostromes) seem to be associated with sequence boundaries (BR1, SB-K; BR3, SB-T);

2 Olistostromes together with turbidites occur within highstand systems tracts (i.e. BR2, Lower Fallauxi to Darwini);

3 Olistostromes, turbidites and slumps can be found within lowstand systems tracts (i.e. BR4, Upper Tithonian–? Lowermost Berriasian).

The Berriasian consists of a predominantly limestone lithology. However, these strata are not clearly exposed here and age dating is quite poor, precluding any subdivision into systems tracts. The Berriasian shows an abrupt decrease in frequency of occurrence of olistostromes, turbidites and slumps that appears to be concurrent with the deposition of a singular pebbly mudstone that resembles a transgressive lag-type deposit and that may correlate with the *Brèche de Chomérac* of the Ardèche Area.

Development of carbonate sediments transported from upslope and development of systems tracts in this basinal setting are fully consistent with what has been described throughout the marginal sections (Fig. 3). It should be further noted that gravity-related deposits commonly are vertically

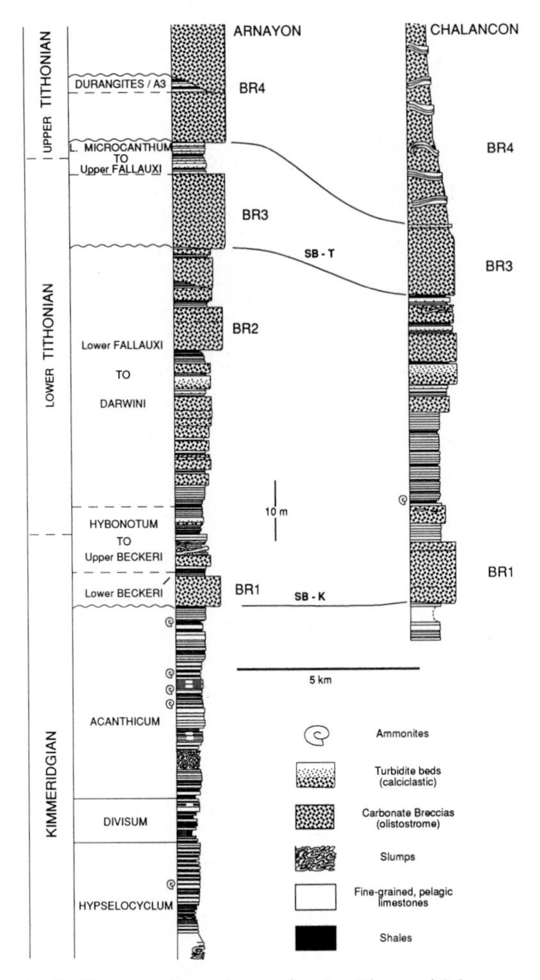

Fig. 5. Stratigraphic columns for the Kimmeridgian–Tithonian of La-Motte-Chalancon area (Drôme Area; Subalpine Basin) along with sequence boundaries. The sequence boundaries have been defined on the basis of biostratigraphic correlation with the marginal section (see Fig. 2). SB-K, Kimmeridgian sequence boundary; SB-T, Tithonian sequence boundary; BR, 'resedimentation breccia'.

amalgamated in the basinal setting. As a result of this amalgamation any attempt to identify directly depositional sequences here is not possible.

RELATIVE SEA LEVEL AND TRIGGERING OF GRAVITY-RELATED TRANSFERS

Within Upper Jurassic–Lowermost Cretaceous depositional sequences, deep-water carbonate olistostromes are interpreted to climax with a relative fall of the sea level. Our objective in this discussion is to address the role of relative sea-level change in the downslope transfer of sediments.

The Broyon quarry in the Ardèche area provides clear evidence of syndepositional tectonics at the passage from the highstand shaly deposits to the lowstand limestones. The development and geometry of the carbonate olistostromes (i.e. localization of the submarine erosional channels) are here controlled by normal faults (Fig. 6). The carbonate gravity flows may have been triggered merely by related earthquakes. A variation of this scenario is possible; activity of normal faults in this marginal area may be evidence of accentuated differential subsidence between the shelf and the basin with subsequent increase in the basin accommodation space, withdrawal of water from shelves and increase of declivity and consequently instability of the slope. It should be stressed here that carbonate gravity flows that mark a sequence boundary may be consistently related to tectono-eustatic events. The overall interpretation is still speculative and would imply that development of third-order depositional sequences is controlled by tectono-eustasy (i.e. change in basin volume and physiography), in contrast to glacio-eustasy (change in water volume) as argued by Vail and Eisner (1989).

Gravity-displaced deposits are interspersed within lowstand systems tracts. Their non-periodic recurrence there appears to be in response to occasional partial collapse of the shelf margin wedge because of overbuilding due to the high rate of the lime–mudstone sedimentation.

Finally, gravity-related transfers of sediment have been documented within highstand systems tracts. It is possible that they occupy a position at the base of higher frequency sequences (Fig. 4) that comprise the systems tract. Such depositional events may therefore reflect minor and repetitive relative sea-level falls during the overall stillstand of the late highstand.

COMPARISON WITH THE EXXON GLOBAL CYCLE CHART

A number of differences occur between the age model derived from the stratigraphy in this study area and the Exxon global cycle chart (Haq *et al.*, 1988). During the interval covering the Kimmeridgian to early Berriasian (Fig. 7):

Fig. 6. Field view in the Broyon quarry (Ardèche area) showing a normal, extensive fault affecting the highstand shaly deposits and in turn sealed by limestone conglomerates (resedimentation breccias) and turbidite beds associated with an interval of sea-level fall. CB, lowstand *Calcaires blancs*; Tr, turbidite beds; BR, resedimentation breccias — limestone conglomerates; SB-T, Tithonian sequence boundary; SS, (highstand) suborder sequence (shown in Fig. 4).

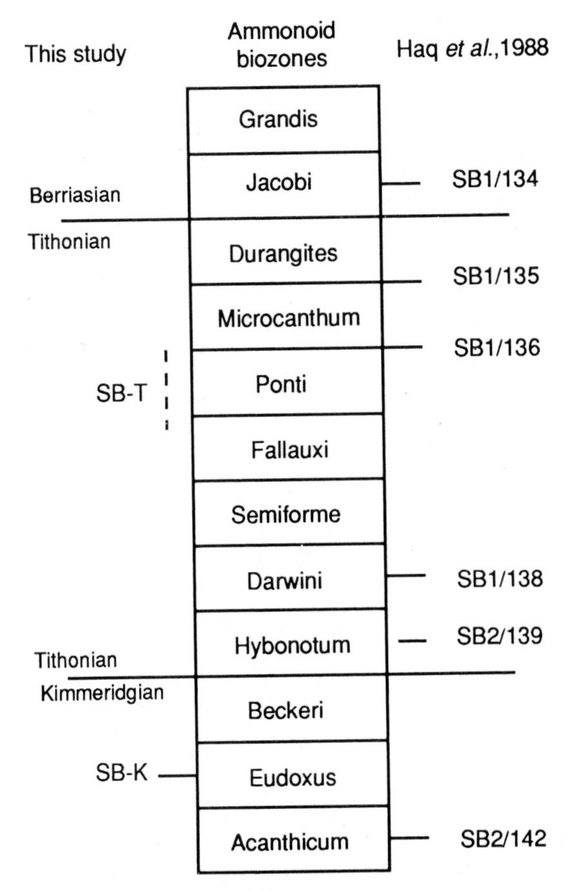

Fig. 7. Comparative occurrences of sequence boundaries throughout the latest Kimmeridgian—earliest Berriasian: Ardèche area (this study) versus Exxon chart (Haq *et al.*, 1988). The ammonoid biochronozones presented here are from Enay and Geysant (1975); sequence boundaries of the Exxon chart have been placed on the basis of zonal equivalences available in Cecca *et al.* (1989a).

1 The number of sequence boundaries observed by Haq *et al.* (1988) is six, but two in this study area;
2 The Upper Kimmeridgian sequence boundary is offset (one ammonoid zone) relative to that shown on the chart of Haq *et al.* (1988);
3 Two sequence boundaries occur within the Lower Tithonian of the global cycle chart but there are none in this study area;
4 Three sequence boundaries are reported in the Exxon global cycle chart in the Upper Tithonian–Lowermost Berriasian but one only in this study area; SB1/136 of the Exxon chart possibly correlates with the SB-T (Tithonian sequence boundary) in this study.

These differences could reflect an amalgamation of unrelated causes:
1 Poor direct age assignments;
2 Erroneous correlation of the biochronologic subdivisions;
3 Failure to identify depositional sequence boundaries, either through confusion in distinguishing orders of sequences or through misinterpretation of surfaces and lithofacies;
4 As suggested earlier, all or a number of the depositional sequences defined here are controlled by tectono-eustasy which may have only a regional, basin-scale expression.

It is noteworthy that Hallam (1988) similarly, and independently, has already pointed to the unusually 'noisy' Kimmeridgian–Tithonian part of the Exxon curve and suggested that the feature may be related to regional tectonics.

SUMMARY AND CONCLUSIONS

Coupled stratigraphic/sedimentologic examination of both marginal and basinal sites of the Upper Jurassic–Lowermost Cretaceous of southeast France has resulted in the following conclusions:
1 Deep-water limestone conglomerates as a whole can be associated with both sea-level rise and fall; however, closer sedimentologic inspection permits recognition of olistostromes and lag deposits that tie to sea-level fall and rise respectively.
2 Even though olistostromes and turbidites are primarily associated with intervals of sea-level fall, they can occur also within high- and lowstand systems tracts.
3 Delineation of sequence boundaries must be done through comprehensive studies in areas where autochthonous deposits are fully developed, and not through studies of gravity-related deposits as these deposits are not unique to intervals of relative sea-level fall and they may occur as amalgamated deposits in the basin.
4 Carbonate gravity flows, concurrent with third-order sequence boundaries, can be tied to tectono-eustatic events and suggest that depositional sequence cycles can be caused by tectono-eustasy (i.e. changes in basin volume and physiography).

ACKNOWLEDGEMENTS

The authors gratefully acknowledge the support of

the CNRS-URA n°11 in this study. Claude Bacchiana and François Guillocheau provided initial suggestions and comments on the manuscript that further benefited from a constructive review by J. Frederick Sarg. Thanks are due to Noël Podevigne for assistance with photography.

REFERENCES

Atrops, F. (1984) Jurassique supérieur, bordure ardéchoise. In: *Synthèse Géologique du Sud-Est de la France* (Eds Debrand-Passard, S., Courbouleix, S. & Lienhard, M.-J.). Mém. Bur. Rech. Géol. Min., Orléans 125, 247–249.

Biddle, K. (1984) Triassic sea-level change and the Ladinian–Carnian stage boundary. *Nature*, **308** (5960), 631–633.

Boardman, M.R. & Neumann, A.C. (1984) Sources of periplatform carbonates: Northwest Providence Channel, Bahamas. *J. Sediment. Petrol.* **54**, 1110–1123.

Bosellini, A. (1984) Progradation geometries of carbonate platforms: examples from the Triassic of the Dolomites, northern Italy. *Sedimentology* **31**, 1–24.

Cecca, F., Enay, R. & Le Hegarat, G. (1989a) L'Ardescien (Tithonique supérieur) de la région stratotypique: séries de référence et faune (ammonites, calpionelles) de la bordure ardéchoise. *Doc. Lab. Géol. Lyon* **107**, 115 pp.

Cecca, F., Enay, R. & Le Hegarat, G. (1989b) The Tithonian of Ardèche (S–E France). *Newsl. Stratigr.* **20**(3), 115–119.

Cook, H.E. & Taylor, M.E. (1991) Carbonate-slope failures as indicators of sea-level lowerings. *Bull. Am. Assoc. Petrol. Geol.* (Convention issue) **75**(3), 556.

Crevello, P.O. & Schlager, W. (1980) Carbonate debris sheets and turbidites, Exuma Sound, Bahamas. *J. Sediment. Petrol.* **50**, 1121–1148.

Dromart, G. (1989a) Deposition of Upper Jurassic fine-grained limestones in the Western Subalpine Basin. *Palaeogeogr., Palaeoclim., Palaeoecol.* **69**, 23–43.

Dromart, G. (1989b) Deep-water microbial biostromes, depositional sequences, and sea-level fluctuations: the Upper Jurassic of the Western Subalpine margin. In: *Book of Abstracts of 2ème Congrès Français de Sédimentologie, Mesozoic Eustacy Record on Western Tethyan Margins.* Spec. Publ. Assoc. Sediment. Franç. 11, 25–26.

Dromart, G. (1992) Jurassic deep-water microbical biostromes as Flooding markers in carbonate sequence stratigraphy. *Palaeogeogr. Palaeoclim., Palaeoecol. 91,* 219–228.

Dromart, G. & Atrops, F. (1988) Valeur stratigraphique des biomicrofaciès du Jurassique supérieur de la Téthys occidentale. *Comptes Rendus Acad. Sci. Paris* **2**(4), 303. 311–316.

Droxler, A.W. & Schlager, W. (1985) Glacial versus interglacial sedimentation rates and turbidite frequency in the Bahamas. *Geology* **13**, 799–802.

Enay, R. & Geysant, J.R. (1975) Les faunes tithoniques des chaînes bétiques (Espagne méridionale), Colloque limite Jurassique-Crétacé, Lyon-Neuchâtel, 1973. *Mém. Bur. Rech. Géol. Min., Orléans* **86**, 39–55.

Ferry, S. & Rubino, J.-L. (1989a) Correspondance entre séquence de Klüpfel et séquence de Vail en système de dépôt carbonaté. *Strata, Toulouse,* **1**(5), 9–11.

Ferry, S. & Rubino, J.-L. (1989b) Climatic–eustatic mixed control on carbonate deposition, (Mesozoic, S-E France). In: *Book of Abstracts of 2ème Congrès Français de Sédimentologie, Mesozoic Eustacy Record on Western Tethyan Margins.* Spec. Publ. Assoc. Sediment. Franç. 11, 30–32.

Hallam, A. (1988) A reevaluation of Jurassic eustasy in the light of new data and the revised Exxon curve. In: *Sea-level Changes: An Integrated Approach* (Eds Wilgus, C.K., Hastings, B.S., Kendall, C.G.St.C., Posamentier, H.W., Ross, C.A. & Van Wagoner, J.C.) Spec. Publ. Soc. Econ. Paleontol. Mineral. 42, 261–274.

Haq, B.U., Hardenbol, J. & Vail, P.R. (1988) Mesozoic and Cenozoic chronostratigraphy and cycles of sea-level change. In: *Sea-level Changes: An Integrated Approach* (Eds Wilgus, C.K., Hastings, B.S., Kendall, C.G.St.C., Posamentier, H.W., Ross, C.A. & Van Wagoner, J.C.) Spec. Publ. Soc. Econ. Paleontol. Mineral. 42, 71–108.

Jacquin, T., Arnaud-Vanneau, A., Arnaud, H., Ravenne, C. & Vail, P.R. (1991) Systems tracts and depositional sequences in a carbonate setting: a study of continuous outcrops from platform to basin at the scale of seismic lines. *Mar. Petrol. Geol.* **8**, 122–137.

Kier, J.S. & Pilkey, O.H. (1971) The influence of sea-level changes on sediment carbonate mineralogy, Tongue of the Ocean, Bahamas. *Mar. Geol.* **11**, 189–200.

Kolla, V. & Macurda, Jr, D.B. (1988) Sea-level changes and timing of turbidity-current events in deep-sea fans systems. In: *Sea-level Changes: An Integrated Approach* (Eds Wilgus, C.K., Hastings, B.S., Kendall, C.G.St.C., Posamentier, H.W., Ross, C.A. & Van Wagoner, J.C.) Spec. Publ. Soc. Econ. Paleontol. Mineral. 42, 381–392.

Le Hegarat, G. (1973) Le Berriasian du Sud-Est de la France. *Doc. Lab. Géol. Lyon* **43**, 576 pp.

Lynts, G.W., Judd, J.B. & Stehman, C.F. (1973) Late Pleistocene history of the Tongue of the Ocean, Bahamas. *Bull. Geol. Soc. Am.* **84**, 2665–2684.

Mitchum, R.M., Jr (1985) Seismic stratigraphic expression of submarine fan. In: *Seismic Stratigraphy II: An Integrated Approach to Hydrocarbon Exploration* (Eds Berg, O.R. & Woolverton, D.G.) Mem. Am. Assoc. Petrol. Geol. 39, 117–138.

Mutti, E. (1985) Turbidite systems and their relations to depositional sequences. In: *Provenance of Arenites* (Ed. Zuffa, G.G.) NATO ASI Series, Reidel Publishing, Dordrecht, Netherlands, 65–93.

Posamentier, H.W. & Vail, P.R. (1988) Eustatic controls on clastic deposition II — sequence and systems tract model. In: *Sea-level Changes: An Integrated Approach* (Eds Wilgus, C.K., Hastings, B.S., Kendall, C.G.St.C., Posamentier, H.W., Ross, C.A. & Van Wagoner, J.C.) Spec. Publ. Soc. Econ. Paleontol. Mineral. 42, 125–154.

Remane, J. (1970) Die Entstehung der resedimentären Breccien im Obertithon des subalpinen Ketten Frankreichs. *Eclog. Geol. Helv.* **63**(3), 685–739.

Reymer, J.C., Schlager, W. & Droxler, A.W. (1988) Site 632: Pliocene–Pleistocene sedimentation cycles in a Bahamian basin. In: *Proc. ODP, Sci. Results, 101* (Eds

Austin, J.A. & Schlager, W.). College Station Texas (Ocean Drilling Program), pp. 213–220.

SARG, J.F. (1988) Carbonate sequence stratigraphy. In: *Sea-level Changes: An Integrated Approach* (Eds Wilgus, C.K., Hastings, B.S., Kendall, C.G.St.C., Posamentier, H.W., Ross, C.A. & Van Wagoner, J.C.) Spec. Publ. Soc. Econ. Paleontol. Mineral. 42, 155–181.

SCHLAGER, W. & GINSBURG, R.N. (1981) Bahama carbonate platform — the deep and the past. *Mar. Geol.* **44**, 1–24.

SHANMUGAN, G. & MOIOLA, R.J. (1984) Eustatic control of calciclastic turbidites. *Mar. Geol.* **56**, 1–24.

SHANMUGAN, G. & MOIOLA, R.J. (1988) Submarine fans: characteristics, models, classification and reservoir potential. *Earth Sci. Rev.* **24**, 383–408.

VAIL, P.R. & EISNER, P.N. (1989) Stratigraphic signatures separating tectonic, eustatic and sedimentologic effects on sedimentary sections. In: *Book of Abstracts of 2ème Congrès Français de Sédimentologie, Mesozoic Eustacy Record on Western Tethyan Margins.* Spec. Publ. Assoc. Sediment. Franç 11, 62–64.

VAIL, P.R. AUDEMARD, F., BOWMAN, S.A., EISNER, P.N. & PEREZ-CRUZ, C. (1991) The stratigraphic signatures of Tectonics, Eustasy and Sedimentology - an Overview. In: *Cycles and Events in Stratigraphy* (Eds Einsele, G., Ricken, W. & Seilacher, A.) Springer-Verlag, pp. 617–659.

WEIMER, R.J. (1984) Relation of unconformities, tectonics, and sea-level changes in Cretaceous of Western Interior. In: *Unconformities and Hydrocarbon Accumulation.* Mem. Am. Assoc. Petrol. Geol. 36, 7–35.

Spec. Publs Int. Ass. Sediment. (1993) **18**, 307–341

Sequence stratigraphy of carbonate shelves with an example from the mid-Cretaceous (Urgonian) of southeast France

D. HUNT* *and* M.E. TUCKER

Department of Geological Sciences, University of Durham, Durham DH1 3LE, UK

ABSTRACT

Carbonate depositional systems differ fundamentally from their siliciclastic equivalents and these differences have been considered and incorporated to develop new concepts and models for carbonate shelves in open ocean settings. Stratal patterns developed by carbonate platforms can be similar to those reported from siliciclastic settings, but in the majority of cases they are different.

Two systems tracts are associated with falling and lowstands of relative sea level, respectively, and these are: the *forced regressive wedge systems tract* formed during falling relative sea level, bounded below by the *'basal surface of forced regression'* and above by the *sequence boundary* representing the lowest point of sea-level fall, and the *lowstand prograding wedge systems tract*, developed as relative sea level begins to rise after sequence-boundary formation. This systems tract downlaps the basin-floor forced regression deposits in a basinward direction and onlaps forced regressive wedge sediments on the slope. For many carbonate platforms, lowstand sedimentation is greatly reduced and the position, size and geometry of deposits reflect the inherited foreslope architecture. Two end members of slope exist: low-angle mud-dominated slopes and high-angle grain-supported slopes. Characteristic deposits and stratal patterns are developed upon each type during the lowstand systems tract.

Carbonate transgressive systems tracts can also develop a number of different stratal patterns which reflect the complex interplay of relative sea-level rise, sedimentation rate and environmental change. Two different types of geometric stacking pattern are distinguished: *type 1 geometries*, developed when the rate of relative sea-level rise is greater than sedimentation, and *type 2 geometries*, formed when sedimentation rates of shelf margin facies are equal to or greater than rates of relative sea-level rise. The highstand systems tract is the time of maximum productivity of carbonate platforms and is normally associated with rapid basinward progradation. Two different highstand foreslope depositional patterns are distinguished: *slope aprons* leading to the formation of accretionary margins and *toe-of-slope aprons* developing along bypass margins.

The differences between carbonate and siliciclastic depositional systems suggest that simple application of previously published sequence-stratigraphic models for carbonate depositional systems to subsurface and surface data can lead to incorrect interpretation of systems tracts, sequences and ultimately relative sea-level curves. The models developed in this paper are used in a sequence-stratigraphic interpretation of the mid-Cretaceous Urgonian carbonate platform of the French Alps where many of the models and concepts developed can be illustrated.

INTRODUCTION

In recent years application of sequence-stratigraphic models to sedimentary basin fills has increased drastically with the general acceptance of the sequence-stratigraphic approach (Van Wagoner *et al.*, 1990). The driving controls upon individual sequences, notably the role of eustatic, tectonic, depositional and/or environmental factors, however, remain controversial (Hubbard, 1988; Posamentier & Vail, 1988; Galloway, 1989; Schlager, 1991). The application of sequence stratigraphy to carbonate depositional systems has, to date, been

* Present address: Department of Geology, The University, Manchester, M13 9PL.

based upon the models derived from studies of passive-margin siliciclastic depositional systems (Sarg, 1988), although much literature documents the differences between siliciclastic and carbonate depositional systems. Such studies have been incorporated here to produce new sequence-stratigraphic models for carbonate shelves. Ramps, isolated platforms, intracratonic platforms and carbonate–evaporite basins are discussed in Calvet *et al.* (1990), Tucker (1991), Tucker *et al.* (this volume, pp. 397–415). Diagenetic patterns associated with the sequence-stratigraphic evolution of carbonate platforms are explored by Tucker (1992).

SEISMIC EXPRESSION OF CARBONATE PLATFORMS

Carbonate platforms in seismic sections are frequently more difficult to interpret with the same resolution as siliciclastic depositional systems. This is often the result of the high seismic velocities of carbonates which give fewer reflections per unit thickness. Diagenetic alteration, especially dolomitization, can also homogenize large parts of a platform. High-resolution studies of Quaternary carbonate platforms (Ravenne *et al.*, 1987; Mullins *et al.*, 1988; Brooks & Holmes, 1989) can distinguish individual sedimentological events, whereas low-resolution studies, as part of regional exploration programmes (Gamboa *et al.*, 1985; May & Eyles, 1985), can generally only recognize platform evolution down to third-order level (10^6 years, see Table 1 in Tucker *et al.*, this volume, p. 399).

Typical reflection characteristics for ramp-type platforms are low-angle shingled oblique clinoforms, with low basinward dips. Upon ramps mounded structures can develop in deep or shallow water as nearshore barrier sand and/or reef complexes or down-ramp mud mounds/pinnacle reefs, respectively. Shelves tend to be more complex, and the type of reflection observed depends upon position on the shelf. Inner-shelf reflectors are typically laterally continuous and parallel. However, if the shelf top has been exposed, dolines may have developed, which can be difficult to recognize and interpret on seismic sections. Upon platform tops, patch reefs tend to have mounded reflection characteristics that will be onlapped. Similar reflection patterns can also develop around deeper water mud mounds formed on the foreslope to rimmed shelves. Laterally, inner-platform parallel–parallel reflections pass to the shelf margin region where reflection patterns can indicate the type of rim to the shelf. Sand-dominated shelf margins tend to have coherent reflections that dip up to 25° basinwards, and may show either sigmoidal or oblique geometries (Eberli & Ginsburg, 1989). Reef-dominated shelf margins are different in that the shelf margin reflections tend to be chaotic, passing basinwards into dipping reflectors that are more coherent. Seismic modelling experiments have rarely been undertaken at outcrop where the actual geometry is observable. However, experiments on the Triassic of the Alps indicate that complex reflection geometries can develop from apparently simple depositional relationships (Rudolph *et al.*, 1989; Biddle *et al.*, 1992).

Recent seismic-stratigraphic studies of subsurface carbonate platforms have almost ubiquitously interpreted cutoff, onlap, downlap and toplap relationships in terms of sequence stratigraphy, defining sequences as one would in siliciclastic sections (Rudolph & Lehmann, 1989). It is hoped here to show that some of these geometries and termination patterns do not necessarily correspond to the interpretations of Vail *et al.* (1977) and Haq *et al.* (1987), etc.

CONTROLS UPON CARBONATE SEDIMENTATION

One of the most fundamental differences between carbonate and siliciclastic depositional systems is the sensitivity of carbonates to environmental change. Carbonates unlike siliciclastics are largely produced in their depositional environment whereas siliciclastic sediments are introduced from external sources.

Production of sediment upon a carbonate platform tends to reflect several controlling factors, the most important of which are: clastic contamination, temperature, light, nutrient supply and water energy (see reviews in Tucker & Wright, 1990; Schlager, 1991). The maximum potential for sediment production is typically within 10 m of sea level, and it decreases rapidly below these depths. Thus, production potential for a platform (other factors being equal) can be directly related to the area with water depths less than 10 m. The potential for sediment production is also a function of water energy. High-energy areas of platforms, e.g. inner ramps and shelf margins, have the greatest current activity, and this promotes biogenic and abiogenic grain production. Away from these regions potential sedimentation rates tend to decrease, both basinwards and landwards.

EXTERNAL CONTROLS UPON SEDIMENT-BODY GEOMETRY

There are factors, other than subsidence and eustasy, that can modify the stacking patterns of depositional units (parasequences, systems tracts and sequences) upon a platform. Normally these are related to currents, the most important of which are driven by winds and/or tides and normally interact with the inherited topography to control both the sites of maximum sediment production and the redeposition of sediments from these areas (Mullins *et al.*, 1988; Brooks & Holmes, 1989). Also, certain sedimentary systems such as reefs have a marked polarity, with the maximum potential for growth facing the open ocean (for example, the Arabian Gulf (Purser, 1973); the Bahamas (Gebelein, 1974); the Great Barrier Reef (Davies *et al.*, 1989)).

Currents affecting carbonate platforms are in most cases not uniformly bimodal but have a preferred direction; their effects and relative importance can vary along the strike of a platform. Changes in the orientation of a platform rim will result in changes of current intensity and, correspondingly, the sites of maximum growth potential. Tidal currents may be strongest in either an onshore or offshore direction. Wind-driven currents are of considerable importance as windward platform margins tend to be steep and reef dominated, promoting sedimentary bypass, whereas leeward margins tend to be low-angle, sand-dominated, accretionary slopes as can be seen both in the Bahamas today (Hine *et al.*, 1981) and the subsurface (Eberli & Ginsburg, 1989). The influence of external controls means that almost all platforms but particularly the shelf margin have a polarity and as a consequence so do the sedimentary units. This factor appears to be much more important in carbonate than siliciclastic systems.

SEQUENCE STRATIGRAPHY OF CARBONATE SHELVES

Carbonate shelves are shallow carbonate platforms with an abrupt change in gradient that marks their outer edge. The shelf margin is subject to high wave energy and current activity and is typically rimmed by a continuous–semicontinuous barrier of reefs and/or carbonate sand shoals. Landwards of the margin is the inner shelf; protected from the vigorous current activity it may be one to hundreds of kilometres wide. Faunal and textural changes shorewards across the shelf lagoon reflect the increasingly restricted circulation (see Fig. 1(a)).

Three categories of shelf can be distinguished: *rimmed shelves*, where the shelf margin is marked by high-energy facies and backed by a protected lagoon (e.g. Fig. 1(a)); *aggraded shelves*, where the whole expanse of the inner shelf behind the margin is at or within a few metres of sea level (Fig. 1(b)); and *'drowned' shelves*, an intermittent stage in platform evolution where the shelf has become 'drowned' after a particularly rapid sea-level rise and/or some environmental change (Schlager, 1981). Rimmed shelves, with their more pronounced on-shelf topography associated with the shelf lagoon, where water depths may be several tens of metres, are characterized by a greater differentiation of facies across the shelf, compared with aggraded shelves. Upon rimmed shelves, facies belts generally prograde across all or part of their width, whereas on an aggraded shelf, facies generally accrete vertically, and so are broadly time specific (Fig. 1(b)).

Individual platforms evolve through time, such as from aggraded to rimmed to drowned types, in response to relative sea level and/or environmental changes. Aggraded shelves would appear to be preferentially developed during times of low-amplitude sea-level changes (as occur during greenhouse times), whereas rimmed shelves should be more typical of greater amplitude sea-level changes (as during icehouse times) (see Tucker, 1992).

The shelf margin passes to the basin floor via a slope which can be several kilometres in height (more typically a few hundred metres), and varies in angle between 2 and 90°, most commonly between 4 and 40° (Kenter, 1990). Slope sedimentation is usually dominated by gravity-driven processes (grainflows, turbidity currents, debris flows, slumps, etc.) which transport shelf-margin and upper-slope sediments downslope. Gravity-driven processes are the main mechanism by which a platform progrades. Slopes may be divided into three types depending upon their sediment budget: accretionary, bypass and erosional types (Fig. 2) (Schlager &

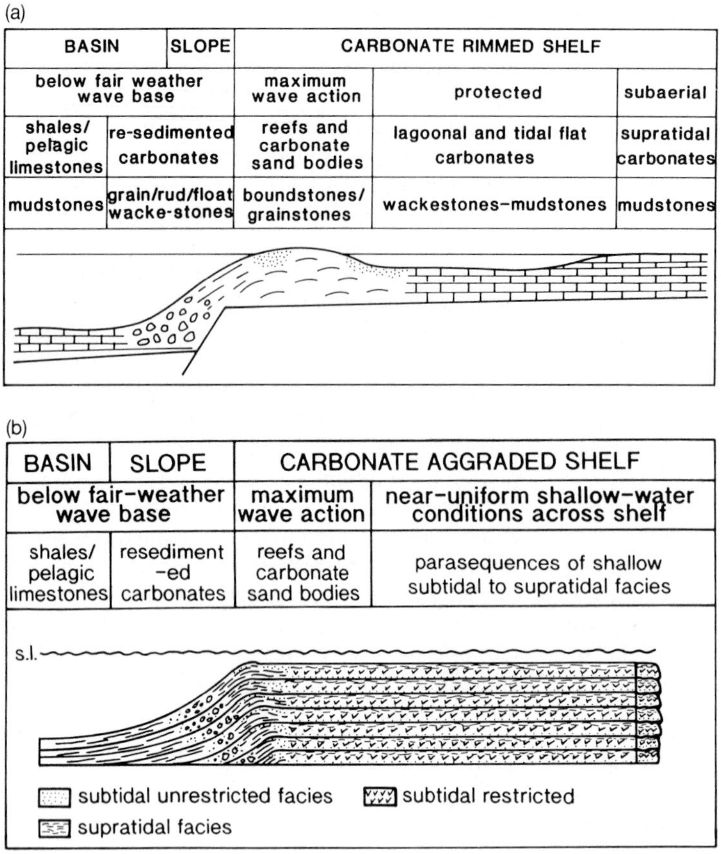

(a)

BASIN	SLOPE	CARBONATE RIMMED SHELF		
below fair weather wave base		maximum wave action	protected	subaerial
shales/ pelagic limestones	re-sedimented carbonates	reefs and carbonate sand bodies	lagoonal and tidal flat carbonates	supratidal carbonates
mudstones	grain/rud/float wacke-stones	boundstones/ grainstones	wackestones–mudstones	mudstones

(b)

BASIN	SLOPE	CARBONATE AGGRADED SHELF	
below fair-weather wave base		maximum wave action	near-uniform shallow-water conditions across shelf
shales/ pelagic limestones	resediment -ed carbonates	reefs and carbonate sand bodies	parasequences of shallow subtidal to supratidal facies

s.l.

subtidal unrestricted facies
subtidal restricted
supratidal facies

Fig. 1. Idealized cross-sections across a rimmed shelf (a) (from Tucker & Wright, 1990) and an aggraded platform (b).

Ginsburg, 1981). Each type has a characteristic profile that may result from a number of variables, the most important of which are tectonics, sea level, shelf to basin relief, slope angle, sediment type and fabric, early diagenesis and the presence or absence of frame-building organisms (Schlager & Camber, 1986; Kenter, 1990).

Facies models for carbonate shelves are well-documented and understood from both the Recent and ancient (Fig. 1; and see review in Tucker & Wright, 1990) and, like other carbonate platform deposits, shelf limestones are commonly organized into metre-scale packages. Ideally they pass from subwave base lime mudstones in their lower part into intertidal fenestral mudstones or keystone grainstones in their upper part, which may be capped by an exposure surface. Such shallowing-upward cycles (parasequences) are thought to result from fourth- or fifth-order relative sea-level changes (10^4 to 10^5 years). The composition and development of a parasequence depends upon position on the shelf, rate of sea-level change, water depth relative to the amplitude of sea-level variation and sediment production rate. Upon siliciclastic shelves, parasequences mostly form by facies belt migration. However, with aggraded shelves, parasequences may result purely from aggradation. The case for parasequences upon rimmed shelves where there is significant original topography is more ambiguous.

The nature and geometry of parasequence stacking patterns upon carbonate shelves are controlled by second–third-order (10^6 to 10^7 yrs) relative sea-level changes, which control the development of individual sequences. Application of sequence-stratigraphic models to carbonate shelves is considerably more complex than for ramps (Tucker *et al.*, this volume, pp. 397–415) and siliciclastic depositional systems, mainly as a result of (i) the high

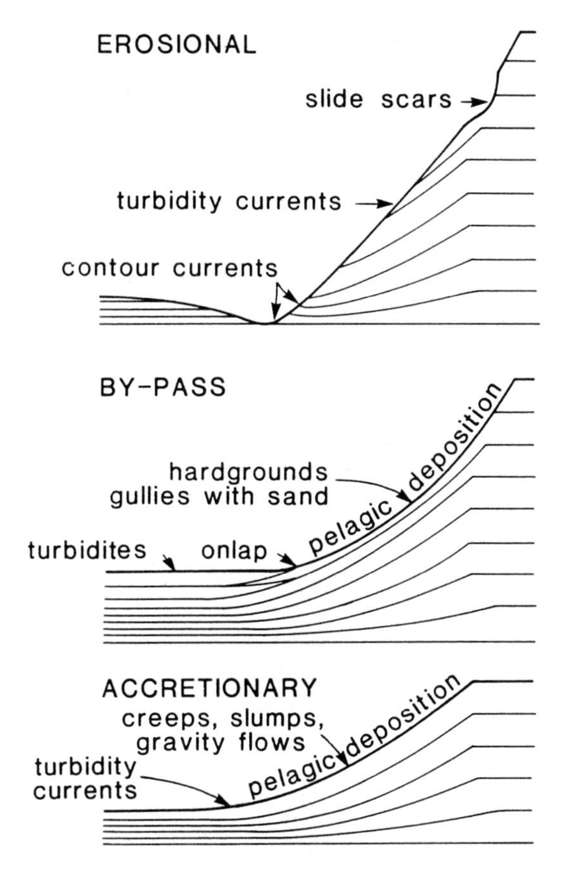

Fig. 2. Accretionary, bypass and erosional types of shelf margin. Note the increasing shelf to basin-floor topography from accretionary to erosional margins and the onlap of basin-floor turbidites in bypass margins. (From Tucker & Wright, 1990 after Schlager & Ginsburg, 1981.)

growth potential of shelf rims; (ii) the large difference in area of sediment production for when sea level is above the shelf compared with below; and (iii) the presence of the break-of-slope at the shelf margin, causing much shallow-water sediment to be redeposited on the foreslope and in the basin.

At its most elementary, a sequence boundary results from a downward or basinward shift of relative sea level, resulting in exposure of the shelf. In siliciclastic depositional systems different rates of relative sea-level fall can be determined. Lower rates and/or lower magnitude relative sea-level falls give rise to type 2 sequence boundaries when, gradually, sea level falls, increasingly exposing the inner shelf and shifting the shoreline to the vicinity of, but not below, the offlap break. (In siliciclastic systems, the offlap break occurs close to fairweather wave base, water depths of 5 to 20 m, Fig. 3). A shelf margin wedge systems tract may develop upon this sequence boundary. Higher rates and/or larger amplitude falls drop sea level below the offlap break to develop a type 1 sequence boundary when the whole shelf is exposed. The shelf may undergo river incision, and contemporaneously sediment may be supplied to the new depocentre below the offlap break (lowstand fan systems tract). In complete contrast to siliciclastic shelves, carbonate rimmed-shelf margins commonly build up to, or very close to, sea level (Fig. 3), so that almost any fall regardless of the rate will lower sea level to below the shelf rim. Thus, aggraded shelves and the margins of rimmed shelves are less sensitive to the rate and magnitude of relative sea-level fall, and will in most cases experience complete exposure with low-amplitude falls. Sequence boundaries on carbonate shelves will generally be type 1 boundaries rather than type 2 sequence boundaries. It would follow from this that the shelf margin wedge (SMW) systems tract should be rare in carbonate shelf systems. Type 2 sequence boundaries should be more common upon distally steepened ramps, where there is a break-of-slope (ramp 'shoulder') in deeper water, 30 to 100 m (Tucker *et al.*, this volume, pp. 397–415). For rimmed shelves, type 1 sequence boundaries could be defined on the basis of complete exposure of the shelf, whereas type 2 sequence boundaries could be for the situation of exposure of the shelf rim, but not of the deeper water shelf lagoon. Not all sea-level falls exposing a carbonate platform will be associated with a sequence boundary; areally less extensive unconformities produced by short-lived sea-level falls are associated with parasequence boundaries.

Some rimmed shelves have a break-of-slope in front of the shallowest point of the platform ('armoured' shelves). The Guadalupian reefs of west Texas, for example, occur at interpreted depths of 25 to 40 m in front of shelf margin sand complexes (Ward *et al.*, 1986). The development of 'armoured' shelves is usually a reflection of the preferred habitats of the organisms forming the reefs at that particular time. Upon these 'armoured' shelves, type 2 sequence boundaries could be developed with relatively low-amplitude sea-level falls.

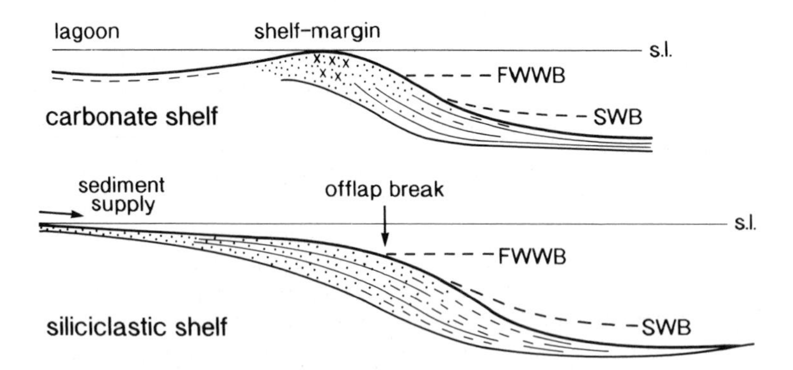

Fig. 3. Sketch illustrating the difference in geometry of a rimmed carbonate shelf (with shelf margin reefs and/or oolite shoals) and a siliciclastic shelf (with delta progradation), both drawn for deposition during a highstand systems tract. The difference in the depth profile is a reflection of the dominantly *in situ* source of sediment and highest production rates at depths of less than 10 m but mostly less than 5 m at a carbonate shelf margin versus the more distant source of sediment and purely physical control on sediment deposition in the case of a siliciclastic shelf. In both instances water depth at fairweather wave base (FWWB) would be in the range of 5 to 20 m, depending on the fetch and shelf orientation relative to prevailing winds. Storm wave base (SWB) would be at depths of several to many tens of metres.

DEPOSITIONAL PROCESSES UPON CARBONATE SHELVES ASSOCIATED WITH FALLING AND LOWSTAND OF RELATIVE SEA LEVEL

Exposure of carbonate shelf tops rarely results in mechanical reworking of the shelf, but more typically a chemical 'reworking' (cementation/dissolution) in the form of early meteoric diagenesis that will tend to be climatically controlled (e.g. humid karstification, arid dolomitization). Exposure of the shelf thus does not result in an increased sediment supply to the adjacent slope/basin, but the reverse, as negligible sediment is supplied off the shelf top (Crevello & Schlager, 1980; Goldhammer & Harris, 1989; Schlager, 1991). During the lowstand, basinal pelagic sedimentation rates may decrease drastically as the shelf top produces no carbonate mud (Crevello & Schlager, 1980; Droxler & Schlager, 1985; Boardman *et al.*, 1986; Wilber *et al.*, 1990). A condensed section can thus be deposited during the lowstand, especially if there is no increased supply of siliciclastics.

Carbonate systems are characterized by two genetically different types of sediment generated during falling and lowstands of relative sea level: *allochthonous material*, calciclastic sediment derived mechanically from the preceding highstand, and *autochthonous material*, formed *in situ* on the modified/unmodified foreslope to the preceding highstand (Sarg, 1988) (Fig. 4). Two types of allochthonous debris can be differentiated: the *turbidite fan* and/or *apron*, which can be analogous to the

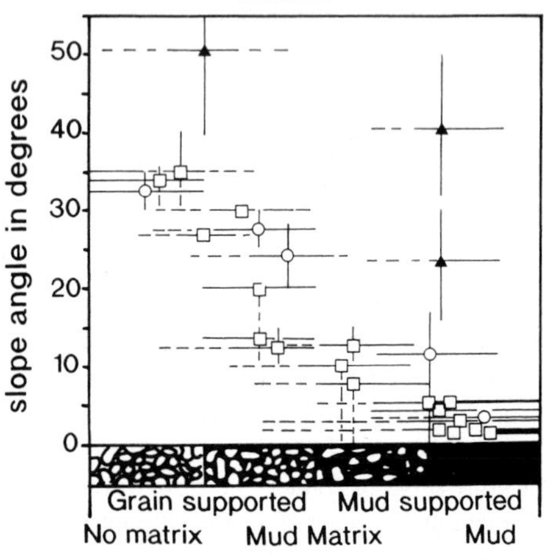

Fig. 4. Relationship of slope angle to the dominant type of sedimentary fabric. Grain-supported slopes tend towards higher slope angles than mud-supported slopes in the absence of early cementation and/or frame-building organisms. (After Kenter, 1990.)

lowstand fan of siliciclastic systems (Vail, 1987; Van Wagoner *et al.*, 1990), and *megabreccia* or *slump sheets*, formed by catastrophic collapse of the preceding highstand.

Upon siliciclastic shelves, type 1 sequence boundaries result in exposure and incision of the preceding highstand strata by fluvial processes, and sediments are supplied via a canyon to the basin-floor fan. Turbidite-fan depositional systems are relatively rare in carbonate systems but they have been documented (Wright & Wilson, 1984; Watts, 1987). Examples of incision and development of basin-floor carbonate fans associated with falling sea level occur in the lower Barremian of the French Subalpine Chains (Ferry & Flandrin, 1979; Ferry & Rubino, 1989; Jacquin *et al.*, 1991) (as discussed in the example later) and the Triassic of Arabia (Watts, 1987). Both are interpreted to be formed as sea level fell very near to or below the break-of-slope upon distally steepened ramps where incised canyons fed basin-floor fans. Mixed siliciclastic–carbonate depositional systems respond in a similar manner to purely siliciclastic depositional systems, with lowstand siliciclastic sands incising the preceding carbonate highstand to supply a toe-of-slope fan/apron. This is well seen at the margin of the Guadalupian platform of the Delaware basin, Texas, in the Cherry and Brushy Canyon Sandstone Formations (Sarg, 1988; Saller *et al.*, 1989).

Mechanical reworking of the preceding highstand (and 'stranded' parasequences, if developed — see later discussion) upon carbonate shelves to generate allochthonous debris more commonly takes the form of megabreccia sheets as the shelf margin and/or upper foreslope undergoes catastrophic failure. Megabreccias tend to form upon steep slopes (>25°), and as such are more likely to form upon mud-free, grain-supported slopes, or those subject to early cementation/frame-building (see Fig. 4) (Kenter, 1990). Examples of such lowstand deposits include the lower Carnian (Triassic) of the Dolomites in northern Italy (Bosellini, 1984; Doglioni *et al.*, 1990), the late Cretaceous–Eocene platforms of southern Italy (Bosellini, 1989) and the 80–120,000 years BP debrite of Exuma Sound described by Crevello and Schlager (1980). Such allochthonous debris appears to have been formed during the sea-level fall and at the lowest point of the relative sea-level curve (Fig. 5). A possible example of a scar at the shelf margin is seen in the 'classic' face of

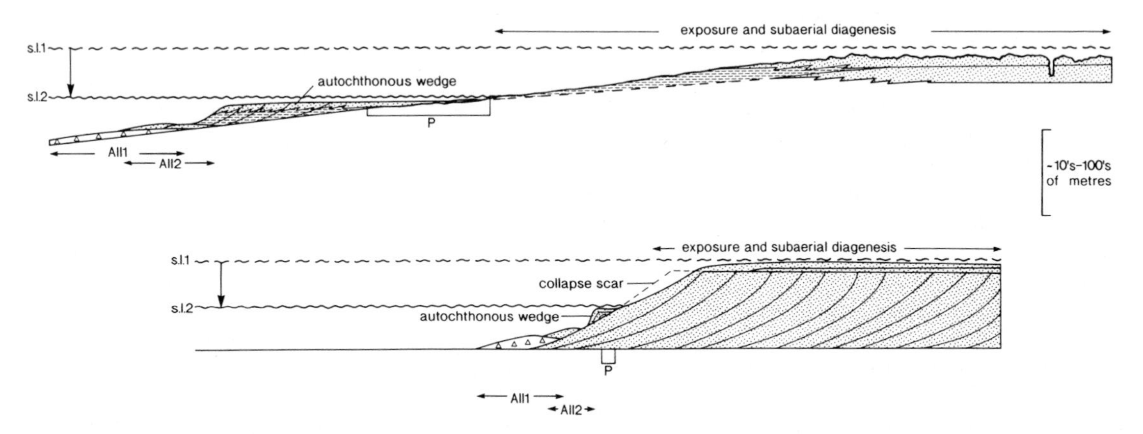

Fig. 5. Rimmed shelf LST models showing the control of inherited slope morphology. Two end-member models for carbonate regressive and lowstand systems tracts as modelled for an arid and a humid platform with grain-supported and mud-supported foreslopes, respectively. As sea-level falls slopes undergo mass wasting as slumps/debris flows (All1). Upon exposure, the shelf undergoes chemical reworking. As sea level reaches its lowest point the sequence boundary is formed and an autochthonous wedge is developed and builds out from the slope. The potential for sediment production is directly related to the available source area (P), and increases as slope angles decrease. If, as modelled here, carbonate production rates remain constant then as the wedge builds out, topography increases and a bypass margin to the autochthonous wedge can develop (All2). Such secondary basin-floor allochthonous debris formed after the lowest point of relative sea level sits above the sequence boundary. The two slope end-members, high and low angle, develop contrasting styles of lowstand wedge. Low-angle slopes develop wide, volumetrically significant lowstand wedges whereas high-angle slopes develop narrow autochthonous wedges. Note that autochthonous wedges onlap the slope and downlap basinwards. (From Hunt, 1992.)

Windjana Gorge, Australia, where a palaeokarstic surface developed on subhorizontal limestones passes basinwards into a subvertical erosion surface (see Fig. 5 of Playford, 1980).

Caution should be used if attempting to use megabreccias as lowstand 'predictors' as they are not specific to times of falling or lowstands of relative sea level. Aggradation during the transgressive systems tract (TST discussed later) can lead to oversteepening and collapse of the shelf margin (McIlreath, 1977; Saller *et al.*, 1989); faulting can also generate megabreccias, especially in rift basins (Colacicchi *et al.*, 1975; Eberli, 1987; see also sequence AP2 of the Urgonian platform of the French Alps, described later). On lower-angle mud-dominated slopes, allochthonous debris will tend to take the form of disorganized slumps and debrites (Hilbrecht, 1989), but such redeposited units upon low-angle slopes are likely to be volumetrically less important than the autochthonous lowstand wedge which may also develop there (Fig. 5).

The occurrence and volume of sediments deposited during a lowstand as an autochthonous wedge (a lowstand, prograding wedge systems tract, see next section) owe much to the morphology of the preceding highstand foreslope and to any subsequent modification during sea-level fall (as discussed above). Lowering of sea level below the shelf top drastically reduces the area available for the production of shallow-water carbonates (Mullins, 1983; Droxler & Schlager, 1985; Goldhammer & Harris, 1989), as illustrated in Fig. 5. Critical factors in determining the volume of an autochthonous wedge are the angle and profile of the slope. As mud-dominated foreslopes (in the absence of framebuilders) tend to have lower gradients than grain/clast-supported foreslopes (Kenter, 1990), the potential sediment production area available during a lowstand will be significantly greater upon mud-dominated than grain-supported slopes. Thus, two end-member scenarios for lowstand sedimentation can be proposed, as depicted in Fig. 5.

A steep accretionary and bypass/erosional slope of the preceding highstand systems tract (HST) (see later section) is more likely to produce allochthonous megabreccias (slope wedges) during times of falling and lowstand of relative sea level, with a volumetrically small (e.g. the Triassic of Italy (Bosellini, 1989; Doglioni *et al.*, 1990)) or even non-existent (e.g. Zechstein of northeast England (Tucker, 1991)) autochthonous wedge, whereas a mud-dominated accretionary/low-angle bypass slope of the preceding HST will tend to be the site of more extensive carbonate production during a lowstand, leading to a significant autochthonous wedge without megabreccia (e.g. the Urgonian of the French Alps, see later example). A lowstand wedge should show a pattern of progradation, with downlap on to the basin floor, but this will become more aggradational as the rate of relative sea-level rise increases and the TST is approached. Secondary basin-floor sediments may be deposited in association with an autochthonous lowstand wedge if the wedge builds up steep angles (All2 in Fig. 5), and itself becomes a bypass system. Lowstand autochthonous wedges may be grain-dominated (e.g. the Urgonian of SE France (see later section)), peritidal mudstone-dominated (e.g. the Durrenstein Formation of the Dolomites, Italy (Sarg, 1988)) and/or have reefal developments (e.g. the Natuna Platform, Miocene of the Sarawak Basin (Rudolph & Lehmann, 1989)).

SYSTEMS TRACTS AND DYNAMICS OF SEDIMENTATION DURING FORCED REGRESSION AND LOWSTAND OF RELATIVE SEA LEVEL

During third-order relative sea-level falls, higher order cycles (e.g. fourth- to fifth-order) may be superimposed upon the general fall resulting in acceleration and deceleration of the third-order signature (e.g. Fig. 2, Tucker *et al.*, this volume, pp. 399–400). During times of decelerated fall, parasequence-scale bodies may be deposited and then subsequently abandoned, exposed and incised as the third-order fall continues. Such deposits are termed 'stranded' parasequences, and the relative sea-level fall is a forced regression. Since the stranded parasequences are deposited during the sea-level fall but prior to the lowest point of sea level, they are placed *below* the sequence boundary (Van Wagoner *et al.*, 1990). In carbonate systems two types of 'stranded' deposit can be recognized: *toe-of-slope/basin-floor allochthonous debris* and *autochthonous slope wedges* (Fig. 6). These have a high preservation potential due to their position in the basin (e.g. Marmolada conglomerate on the basin floor, Triassic, Dolomites, Italy (Doglioni *et al.*, 1990)) or early cementation (e.g. Miocene Níjar Reefs of Spain (Dabrio *et al.*, 1981; Fransen & Mankiewicz, 1991)).

The stratal patterns and chronostratigraphy of both 'stranded' parasequences and the lowstand systems tract in relation to the 'Exxon' sequence boundary are summarized in Fig. 6(a). On the slope, 'stranded' parasequences are placed below the sequence boundary which is developed at the lowest point of sea level. However, in the scheme of Van Wagoner *et al.* (1990, p. 36) basin-floor time-equivalent deposits (allochthonous debris) to the 'stranded' slope parasequences are placed *above* the sequence boundary so that, by way of contrast, the formation of the sequence boundary on the basin floor occurs prior to the lowest point of sea level (S12 in Fig. 6(a)). Thus, in this model, the position of the sequence boundary in relation to geological time is somewhat contradictory and ambiguous, requiring a revision of the model. This is provided here in Fig. 6(b) (see also Hunt & Tucker, 1992).

The revised scheme is based on the models of a type 1 sequence of Van Wagoner *et al.* (1990) and earlier authors. The most significant revisions are the subdivision of the current lowstand systems tract into two newly named systems tracts and the alteration of the sequence boundary position on the basin floor to *above* deposits formed as sea level fell so that it is now everywhere coincident with the lowest point of relative sea level. The new systems tract boundaries are chosen to coincide with changes of both the rate and direction of relative-sea-level change. In reality this new scheme is academic and no easier to use than its predecessors for real geological situations, but it does eliminate the contradictions and ambiguities associated with falling and lowstands of relative sea level in the previous models.

Sediments deposited during *forced regression* (i.e. falling relative sea level), but prior to the lowest point of relative sea level, are placed within the *forced regressive wedge systems tract* (FRWST) (Fig. 6(b)). The base of this systems tract is the *basal surface of forced regression* (BSFR) (Fig. 6(b)), a chronostratigraphic surface separating older sediments of the preceding highstand systems tract deposited during slowing rates of relative sea-level rise and stillstand from younger sediments deposited during the relative sea-level fall (Fig. 6(b)). The systems tract has a slope component, termed the *forced regressive slope wedge* (after which the systems tract is named), and a basin-floor component, the *forced regressive basin-floor fan/apron*. Both components are schematically illustrated in Fig. 6(b).

The slope wedge component of the FRW systems tract consists of one or more 'stranded' parasequences bound below by the basal surface of forced regression and above by the sequence boundary (Fig. 6(b)). Thus, the upper and lower surfaces are common to both components of the systems tract (Fig. 6(b)). Slope and basin-floor elements of the FRW systems tract (as depicted in Fig. 6(b)) are not necessarily developed together during an individual forced regression; the systems tract may be represented by just the slope wedge or just the basin-floor apron. Alternatively, the systems tract may be totally absent due to no sedimentation during the forced regression (e.g. sea-level fall too rapid) or post-depositional erosion (e.g. upon the slope).

The upper surface of the FRW systems tract is the *sequence boundary*; this represents the lowest point of relative sea level, and is thus the most extensive unconformity (Van Wagoner *et al.*, 1990) (Figs 6(a) & (b)). The position of the sequence boundary on the shelf top and slope is unchanged from previous models, but on the basin floor it should be placed above sediments (if any) deposited during the forced regression so that it is now truly a chronostratigraphic surface in that all sediments below it are older and those above younger (e.g. compare positions of the sequence boundary in Figs 6(a) and (b)). Any sediments deposited at or after sea level has reached its lowest position (e.g. All2 of Fig. 5) are above the sequence boundary and are thus part of the *lowstand prograding wedge* (LPW) *systems tract*, developed from the time relative sea level is at its lowest point and is beginning to rise (bottom left, Fig. 6(b)), but prior to the transgressive systems tract. The LPW systems tract downlaps the sequence boundary in a basinwards direction and onlaps it landwards (Fig. 6(b)). The size of the LPW systems tract reflects slope angles (as previously discussed) and the ratio of the rate of relative sea-level rise to that of sedimentation. Its upper surface marks the beginning of the transgressive systems tract.

Since the FRW systems tract lies *below* the sequence boundary, it becomes the fourth and final systems tract of a sequence. The first three systems tracts (lowstand prograding wedge, transgressive and highstand systems tracts) of a sequence are now all formed during times of rising relative sea level (Fig. 6(b)) *after* the lowest point of relative sea level (represented by the sequence boundary in Fig. 6(b)). The development and distinction of each of these systems tracts (LPW, TST and HST) will depend upon the ratio of sedimentation rate to the rate of

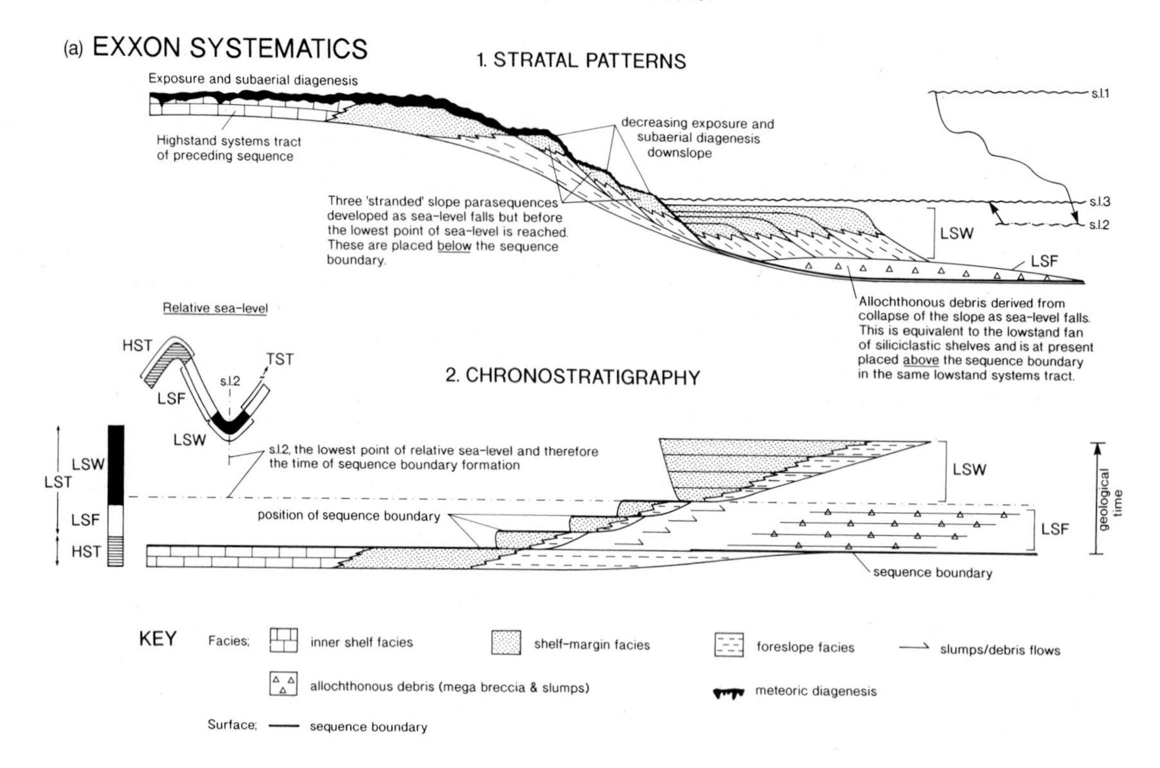

(a) EXXON SYSTEMATICS
1. STRATAL PATTERNS
Exposure and subaerial diagenesis
Highstand systems tract of preceding sequence
decreasing exposure and subaerial diagenesis downslope
Three 'stranded' slope parasequences developed as sea-level falls but before the lowest point of sea-level is reached. These are placed below the sequence boundary.
s.l.1
s.l.3
s.l.2
LSW
LSF
Allochthonous debris derived from collapse of the slope as sea-level falls. This is equivalent to the lowstand fan of siliciclastic shelves and is at present placed above the sequence boundary in the same lowstand systems tract.
Relative sea-level
HST
TST
s.l.2
LSF
LSW
2. CHRONOSTRATIGRAPHY
LST
s.l.2, the lowest point of relative sea-level and therefore the time of sequence boundary formation
position of sequence boundary
sequence boundary
geological time
KEY Facies; inner shelf facies
shelf-margin facies
foreslope facies
slumps/debris flows
allochthonous debris (mega breccia & slumps)
meteoric diagenesis
Surface; sequence boundary

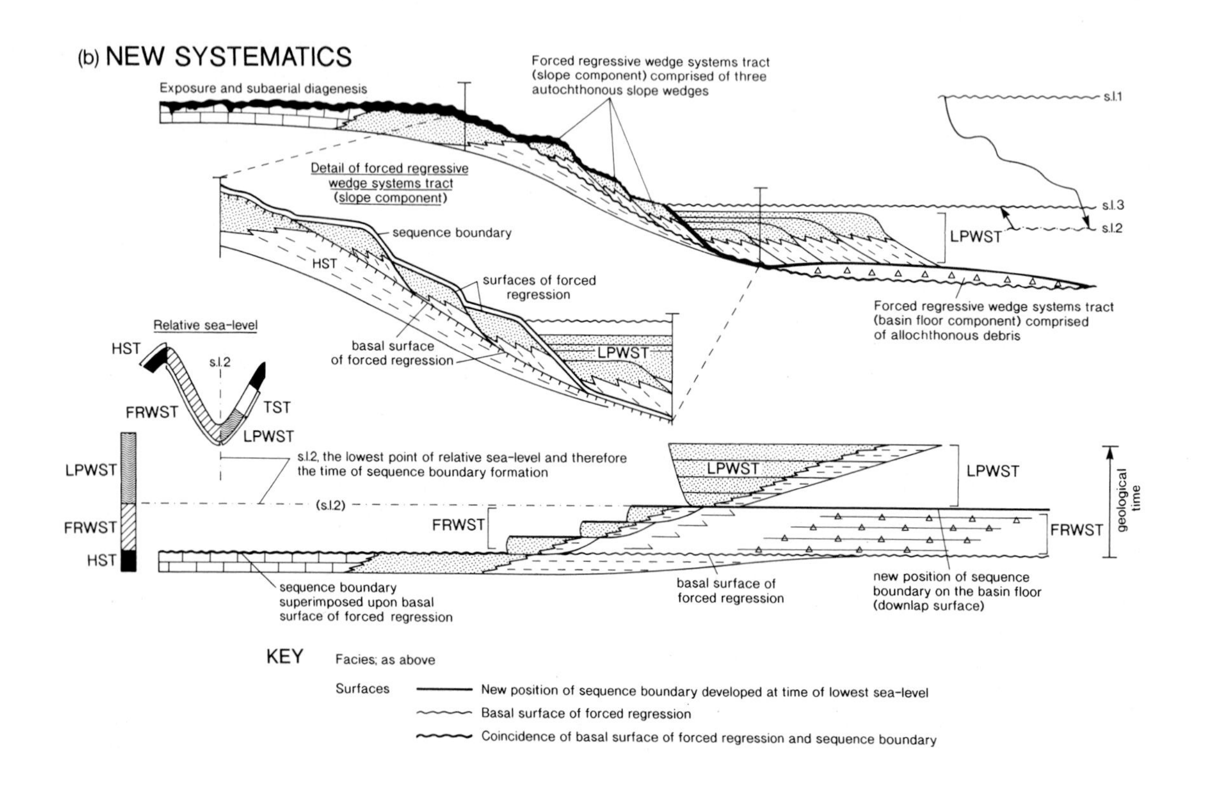

(b) NEW SYSTEMATICS
Exposure and subaerial diagenesis
Forced regressive wedge systems tract (slope component) comprised of three autochthonous slope wedges
s.l.1
s.l.3
s.l.2
LPWST
Detail of forced regressive wedge systems tract (slope component)
sequence boundary
HST
surfaces of forced regression
basal surface of forced regression
LPWST
Forced regressive wedge systems tract (basin floor component) comprised of allochthonous debris
Relative sea-level
HST
s.l.2
FRWST
TST
LPWST
s.l.2, the lowest point of relative sea-level and therefore the time of sequence boundary formation
(s.l.2)
FRWST
geological time
sequence boundary superimposed upon basal surface of forced regression
basal surface of forced regression
new position of sequence boundary on the basin floor (downlap surface)
KEY Facies; as above
Surfaces
New position of sequence boundary developed at time of lowest sea-level
Basal surface of forced regression
Coincidence of basal surface of forced regression and sequence boundary

relative sea-level rise, so their boundaries can form at different stages of the relative sea-level rise. The fourth systems tract of a sequence (FRW) forms during times of falling relative sea level (forced regression) and is terminated by the sequence boundary representing the lowest point of relative sea level (Fig. 6(b)). Thus, the upper and lower bounding surface to a sequence remains the sequence boundary which is now more precisely defined *everywhere* to form at the lowest point of relative sea level.

It could be argued that the megabreccias which commonly develop during the lowstand of a carbonate rimmed shelf should be attributed to the FRW systems tract. In most cases, these megabreccias are derived entirely from the highstand shelf margin and they probably form *during* the relative sea-level fall. They also represent the final stage in a cycle of deposition, before any new sediment is generated. They should therefore be regarded as the end of one sequence rather than the beginning of the next. However, these distinctive sediments will doubtless continue to be referred to as lowstand megabreccias.

In siliciclastic depositional systems the times of falling and lowstand of relative sea level are synonymous with increased sedimentation rates and the deposition of extensive basin-floor sediments. In carbonate depositional systems the response is typically more complex, and in terms of sedimentation rates is often 180° out of phase, with times of falling and lowstand of relative sea level normally associated with a decrease or even cessation of sedimentation.

THE TRANSGRESSIVE SYSTEMS TRACT OF CARBONATE SHELVES

The transgressive systems tract (TST) is defined at its base by the first backstepping parasequence, two or more of which form a retrogradational parasequence set as the rate of accommodation increase/relative sea-level rise is greater than that of deposition/sediment supply (Vail *et al.*, 1977; Posamentier & Vail, 1988; Van Wagoner *et al.*, 1990). In carbonate systems, and in particular carbonate shelves, such a definition is often inadequate as the rates of sediment production for shelf margin reefs and/or sands are commonly sufficiently high to outpace rates of relative sea-level rise (Schlager, 1981). Hunt and Tucker (1991) and Hunt (1992) distinguished between two types of geometric stacking pattern that can develop during a TST — types 1 and 2. The different geometries reflect the ratio of *rate of relative sea-level rise to sedimentation rate.* When rates of relative sea-level rise are greater than sedimentation, type 1 geometries are developed. Type 2 geometries are formed when the rate of relative sea-level rise is less than or equal to sedimentation rates at the shelf margin, although the rate of sea-level rise may be greater than sedimentation rates for inner-shelf facies. The different type 1 and type 2 geometries described below can be developed at a variety of scales representing a parasequence, systems tract, sequence or even a sequence set (e.g. Miocene Natuna build-up (Rudolph & Lehmann, 1989)), although the emphasis here is at the systems tract and parasequence scale.

Before full discussion of type 1 and 2 geometries, it is worth noting that in all the published sequence-stratigraphic models (for example Vail *et al.*, 1987; Haq *et al.*, 1987; Van Wagoner *et al.*, 1990, etc.) lowstand sedimentation (LPW systems tract) of a type 1 sequence is *always* depicted to 'fill' up to the shelf break of the preceding sequence prior to the commencement of the TST. This arrangement is a product of the assumptions of the model and, in reality, the LPW will not always reach this point before the beginning of the TST. Indeed, the lowstand prograding wedge can 'underfill' (e.g. autochthonous Triassic San Cassian lowstand, as interpreted from the sequence stratigraphy of Doglioni *et al.* (1990) superimposed upon geometry of Sarg (1988)) or 'overfill' as in the case of the

Fig. 6. (*opposite*) Cross-section of a carbonate sand-shoal rimmed shelf showing facies, stratal patterns, chronostratigraphy and relationship to relative sea level of sediments deposited during falling and lowstand of relative sea level. As relative sea level falls, slope collapse supplies sediment to the basin floor (allochthonous debris) and autochthonous wedges site progressively lower upon the slope. Allochthonous debris can form at the sequence boundary or during the sea-level lowstand (Fig. 5), but is here omitted for clarity. Because carbonate shelves normally aggrade to within a few metres of sea level shallow-water 'stranded' parasequences are normally only developed upon the upper slope. Facies and stratal patterns of (a) and (b) are identical, but the timing and the positioning of the sequence boundary and systems tracts are significantly different. See text for further discussion.

Cretaceous Urgonian 'general' lowstand wedge (Fig. 11, discussed later). In such cases the geometry of the TST can be significantly modified. For simplicity, the following description of geometries begins when the TST follows filling of the lowstand prograding wedge to the preceding shelf break or when a TST follows an SMW/HST (e.g. Fig. 8).

Both type 1 and 2 geometries are subdivided on the basis of stacking pattern (Fig. 7). Type 1a geometries are characterized by reef/shoal drowning with little or no aggradation so that no backstepping units are developed at the shelf margin and the whole TST is represented by a single surface or thin bed (e.g. Pleistocene of Barbados (Humphrey & Kimbell, 1990)). Type 1b geometric patterns are similar to those depicted for siliciclastic shelves in that the shelf margin facies aggrade and backstep prior to drowning (Fig. 7) (e.g. upper part of sequence BA2 (i.e. stratigraphic unit Bs1) of the Urgonian platform, SE France, discussed later). When type 1 geometries characterize the whole TST, there may be an associated 'jump' of outer-shelf-type facies (reefs/sands) to the shoreline and condensed sections will be developed on the slope and outer shelf. Because of the capacity of high-energy shelf margin sediments to keep up with most rates of relative sea-level rise (Schlager, 1981) type 1 geometries suggest glacio-eustatic rates of relative sea-level change (e.g. Humphrey & Kimbell, 1990) and/or environmental deterioration suppressing carbonate production rates (e.g. Hallock & Schlager, 1986).

Geometries of the type 2a category develop when there is a balance of sedimentation rates at the shelf margin and relative sea-level rise. As the shelf margin aggrades to keep pace with rates of relative sea-level rise inner-shelf facies may well drown as a result of the lower carbonate production rate compared with the shelf margin. A condensed section can then be deposited on the shelf (Fig. 7). An example of this geometry, developed throughout the whole TST and part of the HST, is seen in the Devonian of Alberta, Canada illustrated in Fig. 8. During the development of type 2b geometries, rates of sedimentation at the shelf margin are greater than those of relative sea-level rise so that an aggrading–prograding margin is backed by a drowned inner shelf in which the development of condensed sections can be complex (Fig. 7). Several different type 2b geometries can develop, and these reflect both the type of shelf margin facies and the interaction of the shelf margin with predominant currents developing polarity (Fig. 7). Along windward margins for example, there may be extensive backshedding on to the shelf, whereas along leeward margins, excess sediments may generate clinoforms.

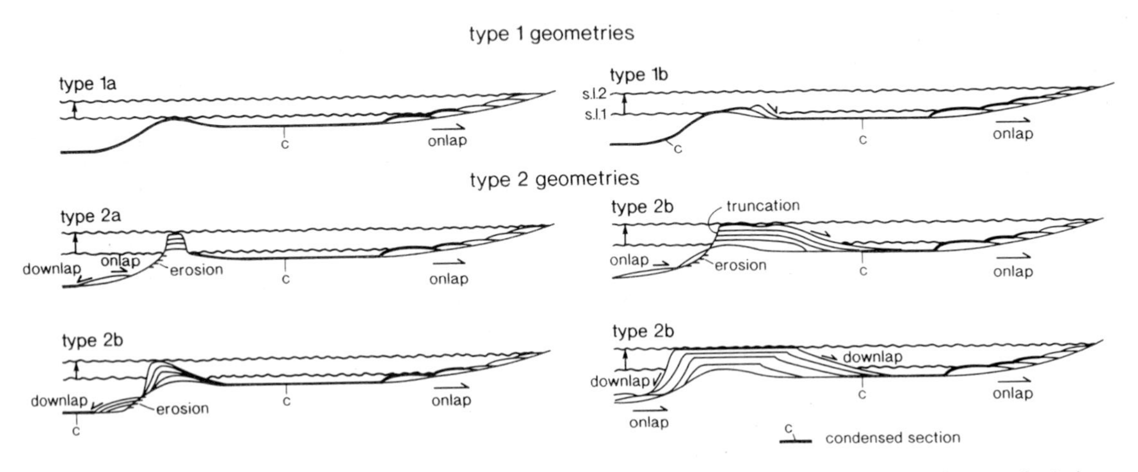

Fig. 7. Theoretical geometries that can be developed upon carbonate shelves during the TST. Controls: rate of relative sea-level rise, shelf margin sedimentation rates, (environmentally sensitive), type of shelf margin sedimentation, current polarity. The different geometries reflect the interplay of rates of sedimentation and relative sea-level rise. Individual geometries are depicted to have developed throughout the whole TST. Note that the starting point for each of these models is at the point when either lowstand sedimentation has 'filled' to the shelf slope break of the preceding sequence or there has been no lowstand. Different geometries can develop if the starting point for the TST is below this interface. See text for further discussion.

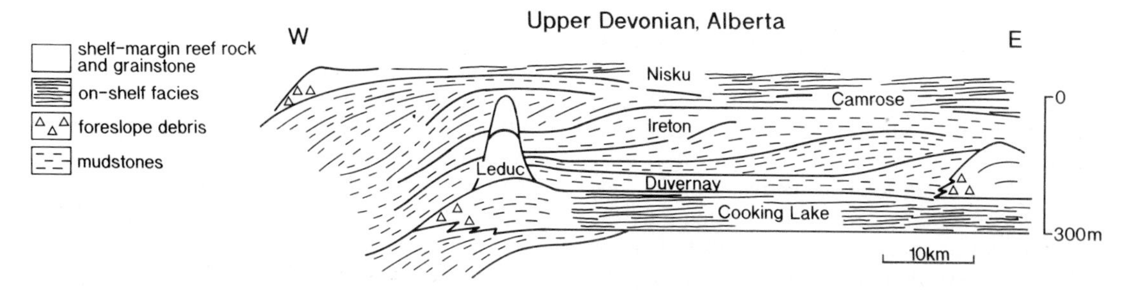

Fig. 8. Cross-section of the Upper Devonian of Alberta, Canada (adapted from Stoakes, 1980). The TST represented by the Leduc Formation follows the Cooking Lake Formation apparently without a lowstand. The Leduc is an example of a type 2a geometry developed throughout the whole TST, and is associated with vertical aggradation at the shelf margin and a facies jump with development of a new secondary shelf margin with an aggrading backstepping geometry. The slope sediments to the new, backstepped shelf margin (Duvernay) downlap on the preceding HST Cooking Lake shelf, and onlap the aggraded shelf margin. The HST is represented by the Ireton Formation, Camrose Member and Nisku Formations and consists of five parasequences, forming a prograding parasequence set. Each parasequence developed after a 20 m relative sea-level rise (Stoakes, 1980). Each parasequence of the Ireton Formation onlaps the preceding parasequence and the aggraded shelf margin. Such onlap is a drowning unconformity. Note continued growth of the aggraded Leduc reef into the highstand. Termination of reef growth is at least in part the result of contamination by muds of the Camrose Member.

Examples of these different types of geometry include the Cretaceous Las Pilas Member of northern Cohuila, Mexico (Bay, 1977) and the Miocene build-ups of the Pearl River Mouth Basin, offshore China (Erlich *et al.*, 1990).

Type 2 TST geometries accentuate shelf margin relief with respect to both the shelf and basin floor. When type 2 geometries characterize the whole or most of the TST, building of relief steepens slopes and this can lead to their evolution from accretionary to bypass/erosional types with associated foreslope erosion and deposition of toe-of-slope and/or basin-floor allochthonous debris (Figs 2 & 7). In such a situation, sand-dominated margins may redeposit sands on the basin floor via a gullied-bypass slope, possibly even capped by a megabreccia sheet if the topography becomes excessive (Saller *et al.*, 1989). Reef-dominated margins will tend to produce base-of-slope talus aprons and megabreccia sheets from catastrophic collapse.

The combination of continued sedimentation and possible erosion at the shelf margin can lead to stratal patterns on the foreslope/toe-of-slope similar to a lowstand systems tract (LST), indicating the presence of a type 1 sequence boundary. However, the presence of a condensed section on the shelf and continuing shoreline sedimentation would show that the systems tract was transgressive. Depending on the fetch of the platform, both type 2 geometry TST's shoreline facies may be unchanged or a 'jump' of outer-shelf facies to this position may occur. When a facies 'jump' occurs the shoreline may simply backstep, aggrade vertically (Fig. 8) or aggrade and prograde (e.g. Cretaceous Las Pilas Member of northern Cohuila, Mexico described by Bay, 1977) depending on the rates of shoreline sedimentation compared to relative sea-level rise. Downlap of the drowned-shelf condensed section can be developed by both vertically aggrading (Fig. 8) and prograding shorelines, and also by 'backshedding' of sediments on to the shelf from the aggrading shelf margin (Fig. 7).

Continuity of sedimentation at the shelf margin during development of type 2 geometries would suggest that the role of environmental factors upon sedimentation was secondary and that the rates of sea-level rise were not those of glacio-eustasy which would have drowned the shelf margin (Schlager, 1981). During an individual TST the ratio of sedimentation to relative sea-level rise may change and such changes will be recorded by the superimposition of different geometries developed at the parasequence scale during a TST. Such changes of geometry do not, however, necessarily simply reflect changes in the rate of sea-level rise as sedimentation rates cannot be assumed to be constant (Vail, 1987; Sarg, 1988) due to the strong environmental control upon carbonate sedimentation (Schlager, 1981, 1991). This sensitivity to environmental change demands caution if attempting to relate transposition of geometric stacking patterns to acceleration(s) and/or deceleration(s) of relative sea-level rise. However, it does appear that the drowning of type 2 TST geometries is often associated with environmental deterioration (e.g. Fig. 8).

If the TST began after a relative sea-level lowstand, at some point during the systems tract the shelf top will be transgressed. If the rate of rise is equal to or less than the rate of 'start up', the platform will become productive, and platform-top sedimentation will resume (e.g. the West Florida shelf in Recent times (Brooks & Holmes, 1989)). However, if the shelf does not have the ability to aggrade sufficiently fast to 'keep up' with the rate of rise, or does not 'seed' at all, both normally the result of deleterious environmental factors such as high suspended sediment load or anomalous nutrient levels, it will become drowned, and hardgrounds or condensed horizons will form there (Fig. 7). Often, during TSTs characterized by type 2 geometries the shelf drowns and the shelf lagoon becomes overdeepened. In such situations organic-rich muds are commonly deposited on the shelf, and they may form an intraplatform source rock to which the shelf margin reefs or sand bodies could be potential reservoirs (e.g. Duvernay and Ireton Formations between the Leduc Formation reefs (Fig. 8)). Carbonate bodies with type 2 geometries developed preferentially during second-order relative sea-level rises and account for approximately 70% of the world's known carbonate build-up hydrocarbon reserves (Greenlee & Lehmann, 1990). Pinnacle reefs and mud mounds would appear to initiate upon foreslopes to rimmed shelves during the TST. They commonly form upon a skeletal sand bank deposited during the lowstand.

Another stratal geometry peculiar to the TST of carbonate platforms is the 'drowning' unconformity (Schlager & Camber, 1986; Schlager, 1989). These develop when basinal lime mudstones or siliciclastic sediments with lower angles of repose (due to lower internal shear strength (Kenter, 1990)) are deposited directly upon a drowned, higher-angle carbonate slope (Schlager & Camber, 1986; Schlager, 1989). Stratal patterns similar to those at a type 1 sequence boundary are produced, i.e. strata onlapping on to the shelf foreslope. Such unconformities have previously been misinterpreted as type 1 sequence boundaries from the subsurface, e.g. the Berriasian–Valanginian boundary off Morocco (Schlager, 1989). Other examples can be seen in Fig. 8, and are also developed against the Urgonian platform of south-east France in sequences BAI (Figs 16 & 17) and AP2 (Fig. 23), as discussed later.

Stratal patterns developed during the TST upon carbonate shelves can be similar to those reported from siliciclastic shelves, and these are often associated with environmental deterioration and/or glacio-eustatic rates of relative sea-level rise. However, as also shown here stratal patterns developed can also be very different, and in some cases the facies associations and/or stratal patterns on the slope/basin floor are very similar to those normally ascribed to falling and lowstands of relative sea level.

THE HIGHSTAND SYSTEMS TRACT OF CARBONATE SHELVES

The HST is the third systems tract developed in a sequence, and overlies the maximum flooding surface. It is the last systems tract to be developed as relative sea-level rises. Highstand sedimentation begins when sedimentation rates exceed those of relative-sea-level rise for *both* outer-shelf and inner-shelf facies. Classically, the highstand marks the return to the normal 'bankfull' stage of the platform (e.g. Boardman *et al.*, 1986; Wilber *et al.*, 1990), although after particularly large rises of sea level it may take some time for normal inner-shelf sedimentation to resume across the shelf top (e.g. Camrose Member to Nisku Formations (Fig. 8)). The topography inherited by this systems tract can reflect both that developed during the lowstand, as through karstification (Purdy, 1974), and/or any developed during the TST (type 2 geometries) as previously discussed, a reflection of climate and the amplitude and rates of relative sea-level change.

In general, highstands tend to be periods of maximum sediment production as the area of shallow water suitable for carbonate sediment production tends to be the largest at this time (Mullins, 1983). Correspondingly, platforms tend to expand most rapidly during the highstand by clinoform progradation at the shelf margin, which can be spectacular (Fig. 10) (see also Bosellini, 1984; and Doglioni *et al.*, 1990). Two different styles of progradation can be distinguished as portrayed in Fig. 9, and these are: *slope aprons* and *toe-of-slope aprons*. The latter type is particularly well developed in the modern Bahamas where banktop-derived sediments bypass the upper slope to form the major component of basin-floor sedimentation. These observations have led to the concepts of 'highstand shedding' (Mullins, 1983; Droxler & Schlager, 1985; Boardman *et al.*, 1986; Wilber *et al.*, 1990), where basin-floor redeposition is 180° out of phase to that of siliciclastic depositional systems and has become the subject of intense controversy (Mullins, 1983; Jacquin *et al.*, 1991; Schlager, 1991). Highstand shedding appears to be a function of both foreslope morphology and high rates of banktop production. The toe-of-slope apron progradational pattern results in basin-floor shallowing and an ascending geometry developed by clinoform packages similar to that described from the Carnian Sella platform, Italy (Bosellini, 1984).

The second distinctive pattern of highstand progradation is the slope apron, where shallow banktop-derived sediment is mainly deposited on the upper-mid foreslope. Toe-of-slope sedimentation is dominated by periplatform muds (Fig. 9), and the overall pattern is to develop subhorizontal to descending lower boundaries to clinoform packages (Fig. 10). The upper surface of the HST is the

(a) Foreslope/slope apron: major progradation

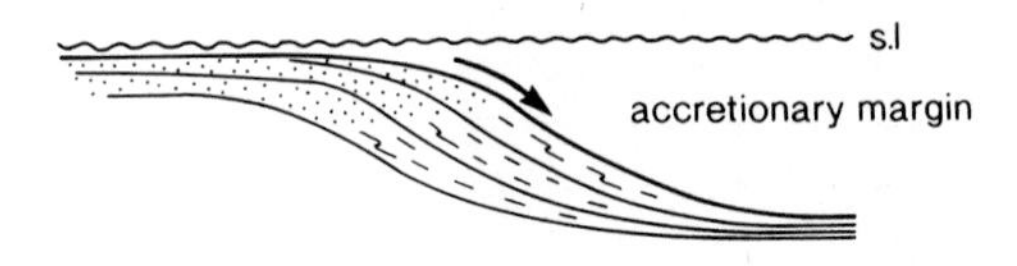

(b) Base-of-slope apron : e.g. reef-rim

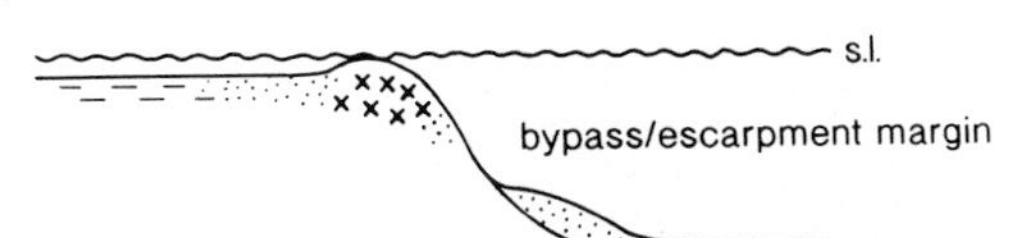

Fig. 9. Two different patterns of highstand slope progradation. Controls: water depth, slope angle, cementation, sediment grain size, reef/shoal type, leeward versus windward. (a) Slope-apron pattern is characterized by the trapping of shallow-water grains on the upper to mid slope. Sedimentation rates decrease downslope with periplatform muds deposited at the toe-of-slope. This pattern is characterized by horizontal to descending clinoform packages. (b) Toe-of-slope patterns are characterized by shallow bank-derived sediment bypassing the upper slope to the basin floor which is the locus of highstand deposition resulting in basin-floor shallowing and is typified by ascending clinoform packages.

basal surface of forced regression that represents the turn-around point of relative sea level, from times of rising sea level (LPWST, TST & HST) to times of falling relative sea level and formation of the forced regressive wedge systems tract (FRWST) as already discussed (Fig. 6(b)). On the shelf itself, the basal surface of forced regression (BSFR) generally coincides with the sequence boundary.

SEQUENCE STRATIGRAPHY OF THE URGONIAN CARBONATE PLATFORM, SE FRANCE

Introduction

The Urgonian platform of the French Subalpine Chains is one of the best exposed carbonate platforms in the world and provides an ideal testing ground for sequence-stratigraphic models. Stratal patterns are similar to those described from siliciclastic settings and these in part reflect the dominance of bioclastic carbonates deposited at and below the shelf margin. The interpretation of these patterns though is often different. The Urgonian limestones were deposited on a shelf-type platform which developed between the uppermost Hauterivian and mid-Aptian upon the early Jurassic–mid-Cretaceous continental margin to Ligurian Tethys (Lemoine *et al.*, 1986). The Urgonian platform is divisible into two parts: (i) lower Barremian Borne (<120 m thick) and lower–upper Barremian Glandasse Bioclastic Limestone Formations and their lateral equivalents (<2000 m thick); and (ii) the upper Barremian–mid-Aptian Urgonian Limestone Formation, consisting of inner-platform rudistid facies (typically 300 m thick) and correlative shelf margin, slope and basinal facies (<1500 m thick) (Fig. 11).

The Glandasse Formation constitutes a 'general' lowstand prograding wedge which is itself divisible into two major sequences: a lower type 1 sequence (BA1) and an upper type 2 sequence (BA2) of which the SMW and TST form the upper part of the Glandasse Limestone Formation (Fig. 12). The upper surface of the Glandasse Formation corresponds to the maximum flooding surface (mfs) of the TST to sequence BA2 and is also the upper surface of the 'general' lowstand prograding wedge (Figs 12 & 13). The Urgonian platform *sensu stricto* (platform with rudistid facies) has, at its base, the lateral correlatives of the flooding surface at the top

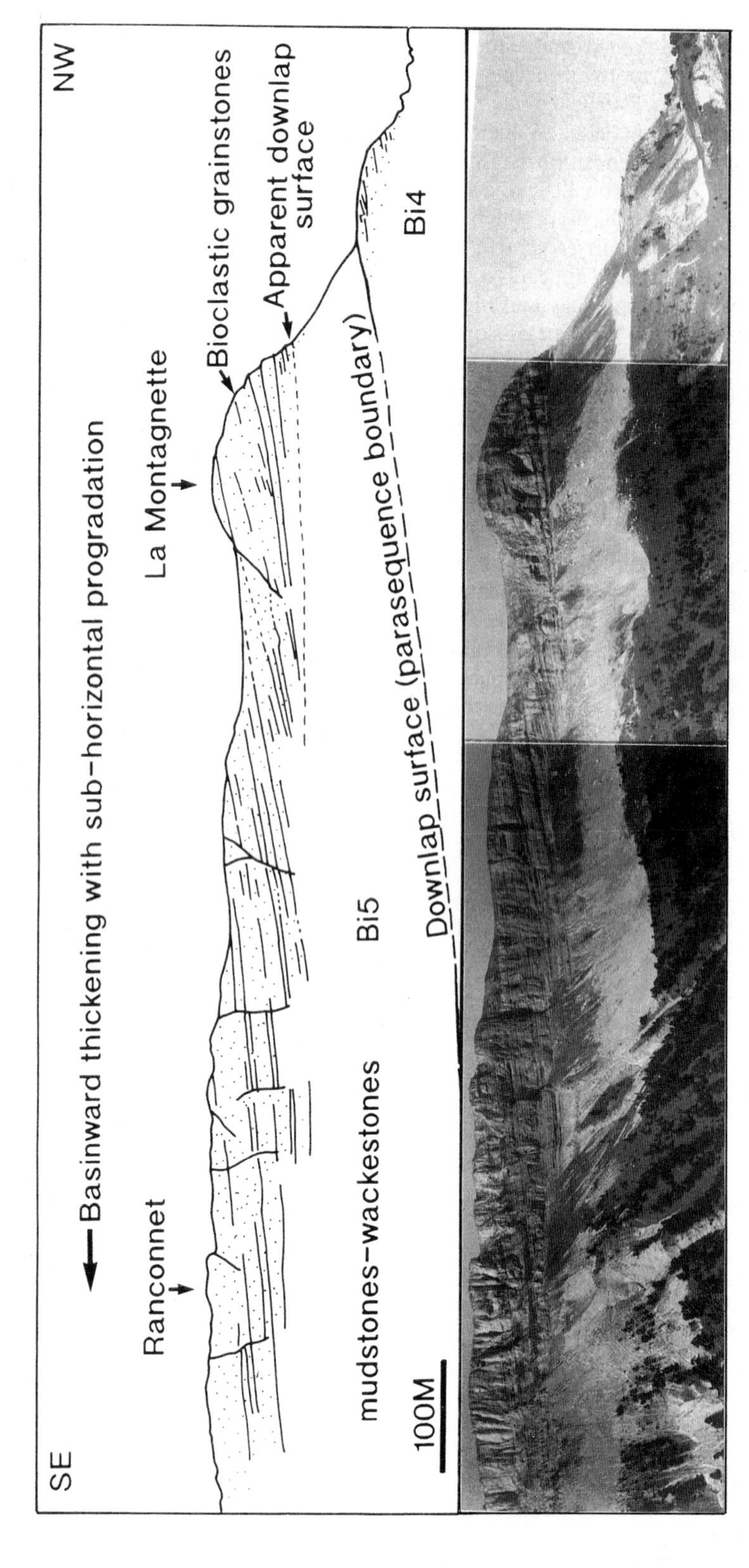

Fig. 10. Spectacular HST prograding slope-apron clinoforms of sequence BA1 (stratigraphic unit Bi5), developed during a relative stillstand of sea level. Lower Cretaceous Urgonian carbonate platform, southeast France. See text for further discussion.

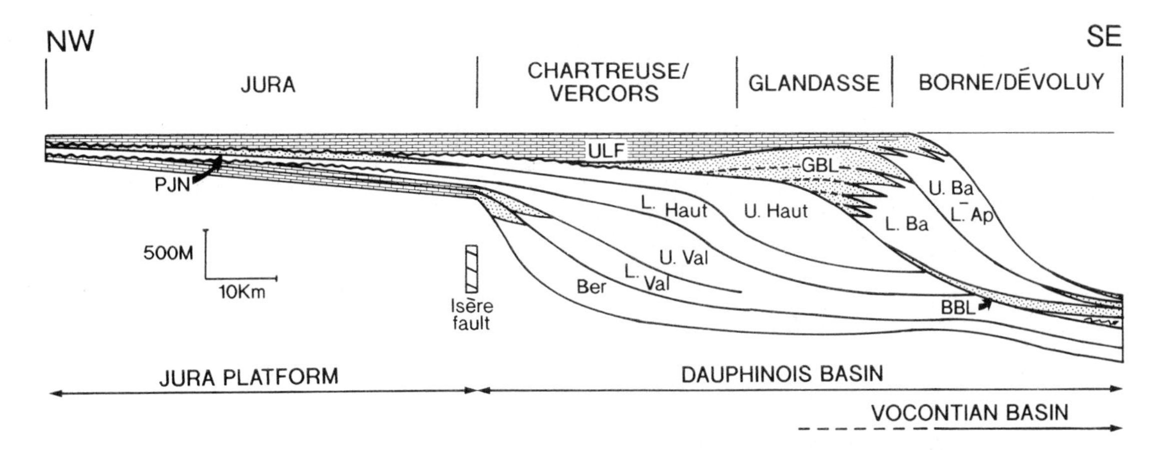

Fig. 11. Synoptic cross-section of the Lower Cretaceous platform of southeast France. Note the 'whaleback' form of the lower Barremian 'general' lowstand prograding wedge. The Urgonian platform is the uppermost unit and is shown to rest on progressively older units to the north. Blocked ornament represents inner-platform facies, stippled bioclastics. Hemipelagic and pelagic sediments have no ornamentation. PJN, Pierre Jaune de Neuchâtel Limestone Formation; BBL, Borne Bioclastic Limestone Formation; GBL, Glandasse Bioclastic Limestone Formation; and ULF, Urgonian Limestone Formation. (Modified after Arnaud-Vanneau & Arnaud, 1990.)

Stage and biostratigraphy			Depositional units	Basin-floor sands	Major slumps /debrites	Formations and Members	Previous sequences	Sequences and sequence boundaries	
APTIAN	LOWER	Bowerbanki	Ai3			Upper Orbitolina Beds	AP2	AP2	
		Deshayesi							SbAP2
		Forbesi	Ai2			Urgonian Lst. Fm. – U. Member	AP1	AP1	
		Fissicostatus	Ai1			L. Orb Beds			SbAP1
BARREMIAN	UPPER	Colchidites sp	BsAi			Lower Member	BA2	BA5	SbBA5
		Astieri	Bs3					BA4	SbBA4
								BA3	SbBA3
		Feraudi	Bs2					BA2	
		Barremense	Bs1			Glandasse Bioclastic Limestone Formation	?		
	LOWER	Moutoniceras sp	Bi6						SbBA2
		Compressissima	Bi5, Bi4, Bi3				BA1	BA1	
		Hugii	Bi2, Bi1			Borne Bio. Lst. Fm.			SbBA1
HAUTERIVIAN	UPPER	Angulicostata	HsBi					HB	BSFR
		Balearis				Pierre Jaune De Neuchatel			
		Sayni							
	LOWER	Cruasense							
		Nodosoplicatum							
		Loyri							
		Radiatus							

Fig. 12. Chronostratigraphic correlation of the Urgonian platform *sensu lato*, and its component formations, members, depositional units and sequence boundaries. Note the timing of basin-floor sedimentation in relation to interpreted sequence boundaries. Depositional units according to Arnaud-Vanneau (1980) and Arnaud (1981). Previous sequences are those of Arnaud-Vanneau and Arnaud (1990).

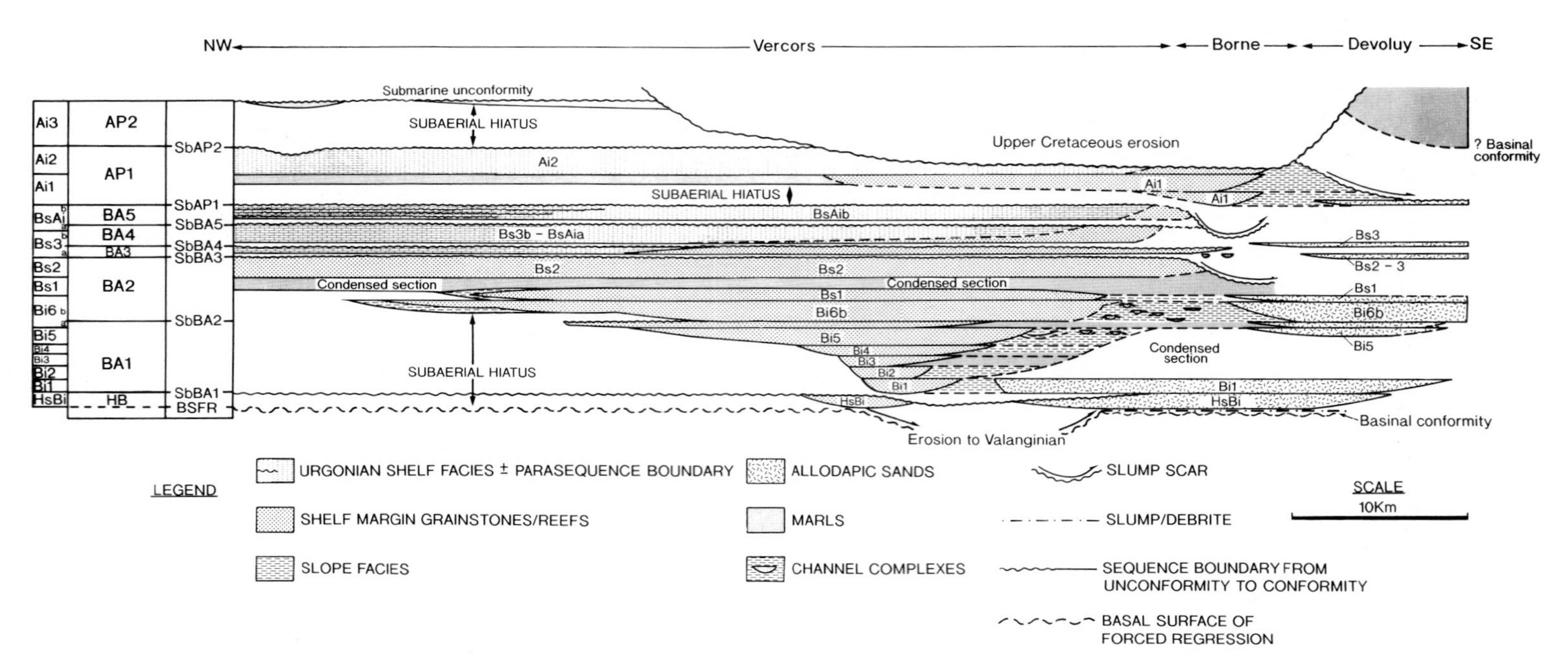

Fig. 13. Chronostratigraphic cross-section of the Urgonian platform in the southern Vercors and Dévoluy, located in Fig. 14. It should be noted that some sequences are better developed off this line of section and that sequences BA2–5 have been projected to some extent from other sections due to faulting and folding of these units along this section at the shelf margin.

of the Glandasse Formation, and is composed of all or a part of six major sequences: BA2, BA3, BA4, BA5, AP1 and AP2 (Fig. 12), each composed of one (e.g. BA3) or more parasequences (e.g. BA1). The regressive and lowstand deposits of these sequences are represented by basin-floor allochthonous debris and the variable development of autochthonous bioclastic wedges on the slope. The TST and HSTs are interpreted to be the main component of sequences on the shelf. The Urgonian Limestone Formation evolved from an open, locally rimmed shelf (BA2), to an aggraded shelf (BA3–5) and, finally, a rimmed shelf (AP2), which was exposed and then drowned.

Considerable debate still surrounds both the palaeogeographic and micropalaeontological development of the Urgonian platform (Clavel *et al.*, 1987; Schroeder *et al.*, 1989; A. Everets pers. com. 1991). Correspondingly different interpretations of palaeogeography and sequence stratigraphy are more fully explored by Hunt (1992). The view presented here reflects the present consensus upon palaeogeography and stratigraphy (Jacquin *et al.*, 1991) based upon the work of Arnaud-Vanneau (1980) and Arnaud (1981) and builds upon but differs from theirs (Arnaud-Vanneau & Arnaud, 1990) and other sequence stratigraphies (Jacquin *et al.*, 1991).

Geological setting

The Urgonian platform is the uppermost platform developed upon the Jurassic–Cretaceous continental margin to Ligurian Tethys. Passive margin sedimentation was dominated by carbonate platforms which prograded, aggraded or retreated in response to relative sea level and/or major oceanic/environmental changes. The platforms can be divided into either synrift or postrift megasequences from Liassic to Upper Jurassic and Lower to Mid-Cretaceous, respectively, separated by a 'breakup' unconformity interpreted to be within the uppermost Jurassic/lowermost Cretaceous strata (Lemoine *et al.*, 1986).

Along the passive margin, extension was accommodated by large displacements along several major lineaments trending NE–SW or NW–SE. One such lineament, the Isère fault, trends NE–SW and separates two distinct palaeogeographic domains: the Jura platform to the northwest and the highly subsident Dauphinois basin to the southeast (Figs 11 & 14). The Jura platform is dominated by shallow-water platform facies and the Dauphinois basin by their laterally equivalent periplatform and basinal facies. Locally, along its northwest edge, the Dauphinois basin received interpreted lowstand deposits from the Jura platform. The Dauphinois basin is itself cut by the more highly subsident E–W trending Vocontian basin, to the south of the Vercors (Fig. 14). This basin, interpreted to be a failed rift branching off the Ligurian Tethys, was up to 2 km deep and is filled by rhythmically bedded pelagic marls and limestones cut by major canyon systems which supplied the basin with shallow-water shelf and slope sediments (e.g. Ferry & Rubino, 1989; Joseph *et al.*, 1989).

Hauterivian architecture and sedimentation

Study of the preserved upper Hauterivian sediments and geometry of basal units to the Urgonian platform suggests that at this time platform sedimentation was restricted to the Jura platform (Figs 11 & 14), where it is represented by 5–10 m thick parasequences passing from subwave-base hemipelagic lime mudstones–wackestones to tidal cross-bedded oo-bioclastic grainstones (Arnaud-Vanneau & Arnaud, 1990). Across the Isère structure platform facies pass into periplatform interbedded marls and nodular limestones (Fig. 11), interpreted to have been deposited on a hemipelagic ramp, dipping at less than 1° basinwards (i.e. towards the southeast, Arnaud-Vanneau & Arnaud, 1990). These distal ramp sediments thicken towards the southeast on the margin of the Vocontian basin, where they reach a maximum thickness of 900 m, before thinning rapidly into the basin as pelagic facies. The thickest part of this ramp is believed to coincide with the Hauterivian slope break, basinwards of which the slope dipped at up to 5°. Thus, the Hauterivian platform can be interpreted to have had the overall geometry of a distally steepened ramp (Arnaud-Vanneau & Arnaud, 1990). This antecedent topography became modified during the uppermost Hauterivian and lower Barremian in the Jura, NW Vercors and NW Chartreuse by subaerial exposure, and in the southeast, upon the Vocontian slope, by mass wasting processes and incision. Thus, the architecture inherited by the Urgonian platform reflected both sea-level changes and structural elements inherited from both Jurassic rifting and minor early Cretaceous extension (Arnaud, 1981).

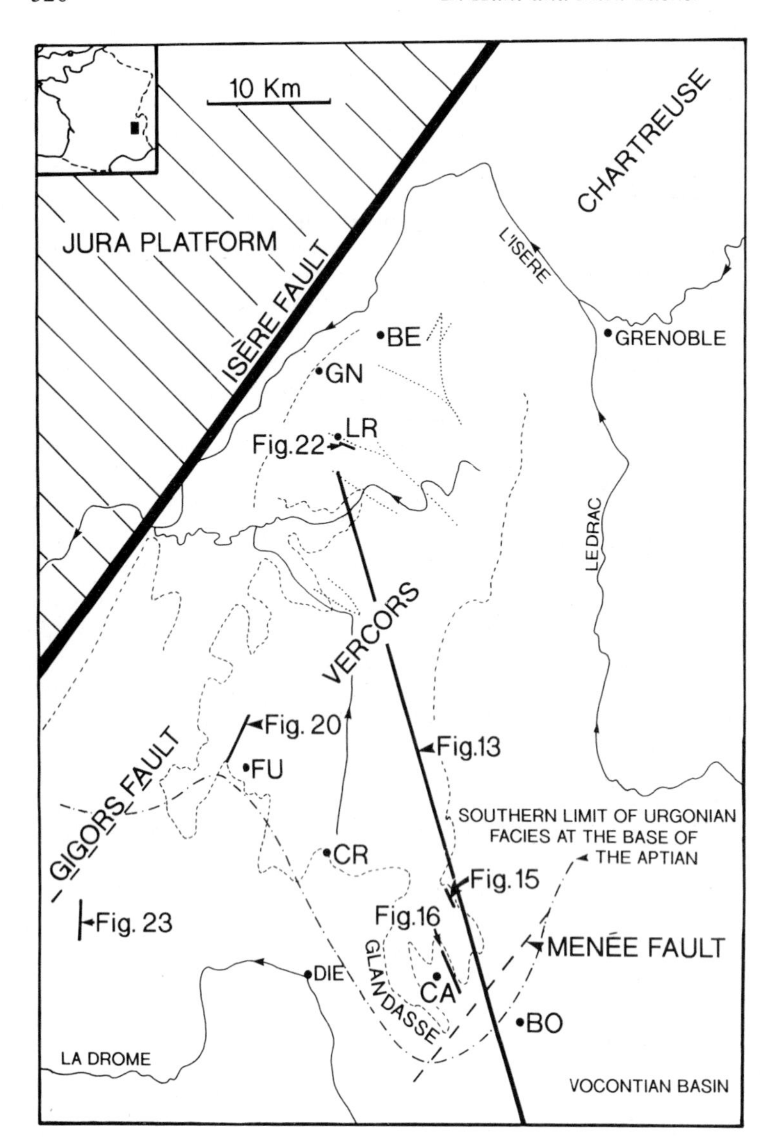

Fig. 14. Location of the Urgonian platform, southeast France, with cross-sections, localities and major faults mentioned in text. The northwest–southeast trending dotted lines, one of which passes through Les Rimets in the northern Vercors, are the mapped courses of palaeovalleys associated with SbAP2 (Arnaud-Vanneau & Arnaud, 1990). BE, Balcon des Ecouges; BO, Borne; CA, Cirque d'Archiane; CR, Col du Rousset; FU, Font d'Urle; GN, Gorge du Nant, LR, Les Rimets.

The 'general' lowstand wedge (lower Barremian)

The 'general' lowstand prograding wedge (Glandasse Limestone Formation (Figs 11 & 12)) is up to 2 km thick and is restricted to the margins of the Vocontian basin, covering only a small area of the shelf. It consists of stratigraphic units HsBi to Bs1 which offlap basinwards and onlap the slope and shelf as schematically shown in Fig. 11. The wedge is itself divisible into three sequences (Figs 12 & 13): HB (unit HsBi), BA1 consisting of units Bi1 to Bi6a and the lowstand prograding wedge and transgressive systems tracts of sequence BA2 (units Bi6b–Bs1) (Arnaud-Vanneau & Arnaud, 1990). As the 'general' lowstand wedge is divided by sequence boundaries SbBA1 and SbBA2 its constituent sequences can be described in terms of systems tracts.

Stratigraphic unit HsBi, a forced regressive wedge, and sequence BA1

Shallow-water carbonate sedimentation continued upon the Jura platform until the uppermost Hauterivian/lowermost Barremian (Pierre Jaune de

Neuchâtel Limestone (Figs 11 & 12)). Sediments of lower Barremian age are not known from the Jura platform (Arnaud-Vanneau & Arnaud, 1990), although good sedimentological evidence for exposure is absent. During latest Hauterivian times an interpreted stepped relative sea-level fall shifted outer-platform bioclastic facies 60–70 km basinwards (southeast) to sit upon the upper Hauterivian hemipelagic–pelagic slope of the Vocontian basin in the Glandasse region (Figs 11 & 13).

Deposits of uppermost Hauterivian/lowermost Barremian are found on the slope (southern Vercors) and the basin floor (Borne and Dévoluy) and are here included into unit HsBi (modified after Arnaud, 1981) (Figs 12 & 13). On the upper to mid slope HsBi is superbly exposed at Tête Chevalière where its base is abrupt (Fig. 15), as it is at Col du Rousset where it marks the change from pelagic to fine bioclastic sediments deposited as distal lobes on a turbidite apron/fan system (Hunt, 1992). On the basin floor in Dévoluy, slumps and debrites are the lowermost deposits of this age and are overlain by up to 120 m of coarse bioclastic sands of which the lower 60 m are ascribed to HsBi (Arnaud, 1981). Slumps and debrites are interpreted to have been derived from the slope just prior to deposition of HsBi bioclastics (Arnaud, 1981; Jacquin *et al.*, 1991; Hunt, 1992).

Jacquin *et al.* (1991) placed two sequence boundaries within the exposure at Tête Chevalière (Fig. 15), one coincident with the abrupt change in lithology (base of unit HsBi (Fig. 15)) and another at the upper erosional surface of unit HsBi against which unit Bi1 onlaps. On the basin floor, sequence boundaries are correspondingly placed below basin-floor deposits of HsBi age, and at the base of unit Bi1 allochthonous sands. The second sequence boundary of Jacquin *et al.* (1991) at the base of Bi1 is interpreted to represent the lowest point of relative sea level.

An alternative interpretation based upon the new model discussed earlier (e.g. Fig. 6(b)) is that unit HsBi is not a true sequence but the fourth systems tract (FRW) of the preceding sequence (HB) developed during forced regression, prior to the lowest point of relative sea level (base of unit Bi1). Thus, the base of unit HsBi is interpreted to be the BSFR which passes basinwards from Tête Chevalière (Fig. 15) along the top of slump scars on the slope and beneath HsBi slumps/debrites and sands on the basin floor (Fig. 13). The sequence boundary *sensu* Hunt (1992) (Fig. 6(b)) is developed at the lowest point of relative sea level interpreted to be coincident with the base of unit Bi1. SbBA1 is thus an areally more extensive unconformity than the BSFR (see Hunt, 1992 for further discussion). The sequence boundary passes above all sediments interpreted to have been deposited during the forced regression (i.e. unit HsBi) and in this way differs from earlier interpretations; it also appears to be supported by the short, parasequence scale duration of this unit of 0.3 to 0.5 Ma (Hunt, 1992).

Sequence boundary SbBA1 is associated with erosional truncation and the formation of submarine canyons incised up to 100 m into the slope which facilitated bypass of LST bioclastic sands to the basin floor. The erosional truncation of unit HsBi at Tête Chevalière is not, however, associated with an increase of grain size and/or incoming of shallow bank derived sediments but the converse (Fig. 15): the erosional topography is draped by hemipelagic sediments, suggesting that erosional truncation of HsBi was associated with gravity collapse during a pause in bioclastic slope sedimentation (Hunt, 1992). Bioclastic slope sediments of the LST unit Bi1 were initially contained within a channel complex at least 1.5 km wide and 150 m deep which contains several internal erosive, onlap surfaces (Fig. 15). The uppermost part of Bi1 on the slope south and west of Tête Chevalière is an areally extensive sand sheet with feeder channels containing hemipelagic lithoclasts (Hunt, 1992).

On the basin floor, unit Bi1 is a lowstand fan complex (the Borne Bioclastic Limestone Formation) and its upper surface is a 'flooding' surface, marking a return to hemipelagic/pelagic sedimentation on the basin floor. On the slope it is associated with a reduction in the areal extent of bioclastic sands. Succeeding units thus belong to the TST. The stacking pattern of slope sand bodies reflects the complex interplay of rates of sedimentation and relative sea-level rise as sand bodies initially backstepped (units Bi1–2) and subsequently aggraded (units Bi2–3). During this time the Fontaine Graillère marls were deposited on the lower slope, and they represent a condensed section and maximum flooding (Fig. 13). The TST of sequence BA1 thus has a lower type 1 and upper type 2a geometry.

The HST of sequence BA1 is composed of stratigraphic units Bi4, Bi5 and part of Bi6a and is associated with a change from an accretionary slope apron (Fig. 10) to a toe-of-slope apron pattern of progradation. In basinal sections, parasequences

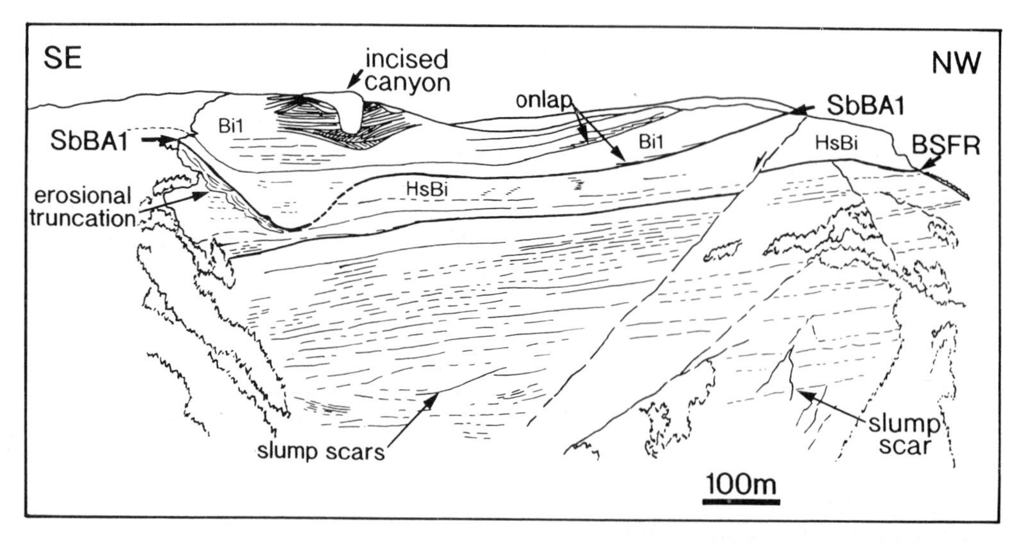

Fig. 15. Spectacular outcrop of stratigraphic units HsBi and Bi1 at Tête Chevalière in the southern Vercors. An abrupt lithological change marks the base of HsBi at this locality, which contrasts with the draped erosional truncation and onlap of HsBi at the base of Bi1. Bi1 coarsens upwards and contains several internal, onlapped erosion surfaces. Note that obliquity of the photograph to the southern end of the cliff exaggerates dips. (See text for further discussion.)

Bi2–4 are represented by hemipelagic or pelagic facies up to 15 m thick (condensed sedimentation). Parasequence Bi5 is an exception in that during the later part of its deposition a second, but relatively thin (10 m) basin-floor sand unit was deposited (Arnaud, 1981) (Fig. 13). Classically, this renewed bypassing of sands to the basin floor would be interpreted to have resulted from a relative sea-level fall and a sequence boundary would be placed at the base of the basin-floor sands. In the Cirque d'Archiane, parasequence Bi5 is aggradational to progradational in its lower half, but strongly progradational in its upper part where cliff-forming shallow-water bioclastic grainstones prograde spectacularly basinwards (Figs 16 & 17). These bioclastic sands change their progradational geometry from initial horizontal to a descending progradational geometry. The upper part of the Bi5 cliff is essentially aggradational and draped by the Lower Fontaine Colombette marls and equivalent hemipelagic limestones (Figs 16, 17 & 18(a)). Overlying foreslope sediments of Bi6a dip quite steeply basinwards ($<15°$) but onlap the slope in Cirque d'Archiane. This stratal pattern tends to support the hypothesis of a sequence boundary at the top of Bi5, as suggested by deposition of Bi5 basin-floor sands

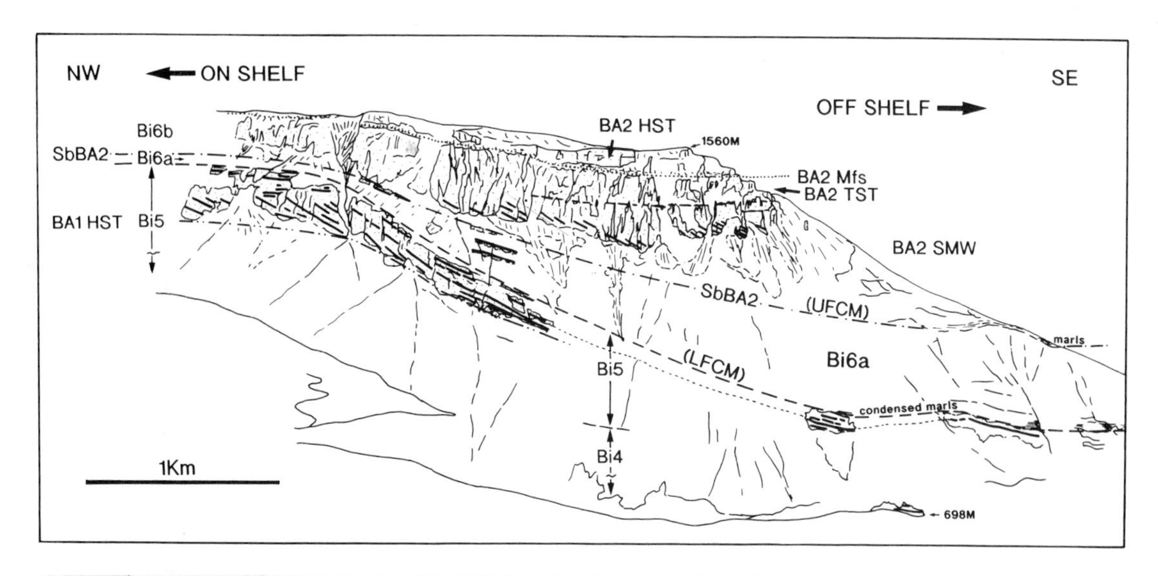

Fig. 16. Photopanorama and line drawing with interpretation of the eastern side of the Cirque d'Archiane in the southern Vercors. This section shows the stratal relationships within sequence BA1 and those at its upper boundary SbBA2. Note the basinward change of progradational geometry in unit Bi5 which exerts a strong control upon the stratal relationships and geometry of younger units.

(Jacquin *et al.*, 1991) although sedimentologically the opposite relationship is observed: a flooding surface where an exposure surface would be predicted (Fig. 18(a)). Such a relationship suggests that the placing of a sequence boundary at this level is incorrect. An alternative interpretation is that during the HST stillstand Bi5 sands prograded into a topographic low (possibly a collapse scar), and this geometric change rather than relative sea-level variations facilitated the change from an accretionary slope apron to a toe-of-slope apron associated with slope bypass. After a relative stillstand, relative sea level rose at a rate equal to sedimentation to develop the aggradational package forming the upper part of the Bi5 cliff in Cirque d'Archiane (Figs 16 & 17). Bi5 is terminated by either an increase in the rate of relative sea-level rise and/or a decrease in sedimentation rates causing a backstepping of facies associated with development of Lower Fontaine Colombette marls on the slope

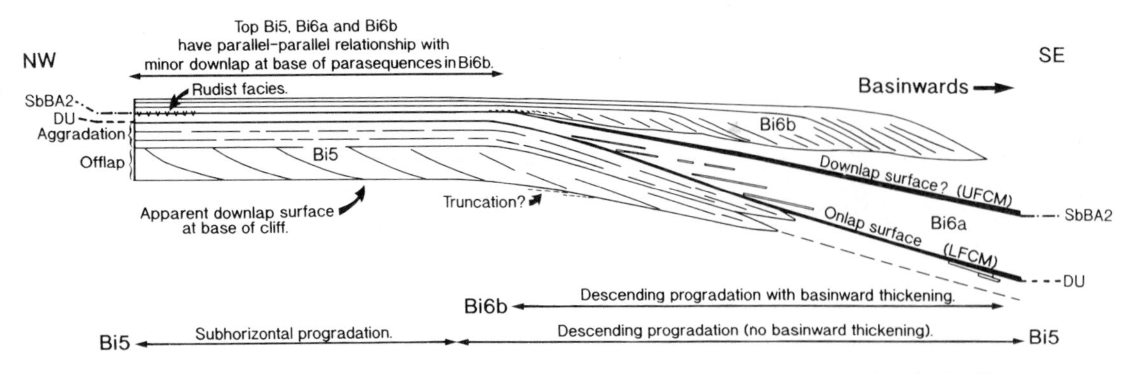

Fig. 17. Simplified cross-section showing stratal relationships and interpreted sequence boundary in the Cirque d'Archiane. LFCM, Lower Fontaine Colombette marls; UFCM, Upper Fontaine Colombette marls; DU, drowning unconformity.

in the southern Archiane valley and hemipelagic mudstones above bioclastic sands in the north (Fig. 18(a)). Thus, the observed onlap of Bi6a bioclastics in the southern Archiane valley (Figs 16 & 17) against the draping Lower Fontaine Colombette marls is analogous to the drowning unconformity of Schlager and Camber (1986) and Schlager (1989).

The final unit of sequence BA1 is Bi6a, with a thickness of 25 to 30 m of bioclastic grainstones in the north of the Archiane valley. It thickens dramatically southwards into relatively steeply dipping wackestones containing prominently weathering packstones–wackestones in lobes and channels, best developed above the descending sands of Bi5 where they onlap the Lower Fontaine Colombette marls (Figs 16 & 17). In the north of the Archiane valley parasequence Bi6a consists of four to five shallowing-up units and, as a whole, shallows upwards very rapidly from hemipelagic mudstones at its base (Fig. 18(a)) to a conspicuous rudistid facies marking the sequence boundary SbBA2 at its top (Fig. 18(b)). The rapid shallowing of cycles below the rudistid facies strongly suggests a part of Bi6a developed during forced regression, before the lowest point of relative sea level was reached and is thus part of the FRW systems tract.

Sequence BA2

This sequence is composed of units Bi6b, Bs1 and Bs2 (Figs 12 & 13) and like the previous sequence BA2 is dominated by outer-platform bioclastic facies and periplatform equivalents. Development of true Urgonian facies is limited to the base and top of the sequence. The sequence reaches a maximum thickness of 600 m on the slope, as can be seen in the southern Archiane valley (Fig. 16) and its surroundings. On-shelf in the northern Vercors/Chartreuse only TST and HST of the sequence are represented and a thickness of 50 to 80 m is typical. The bioclastics of the SMW (unit Bi6b) and the TST (unit Bs1) are a part of the Glandasse Limestone Formation and as such form the uppermost part of the 'general' lowstand wedge (Figs 12 & 13).

The exact classification of BA2, as to whether it is a type 1 or type 2 sequence, is somewhat problematic. It is here considered to be a type 2 sequence on the basis of its relationship to the underlying sequence. However, if viewed on the scale of the platform the boundary could be argued to be of type 1 affinity as it is contained within the 'general' lowstand wedge when relative sea level was considered to be below the shelf break of the Urgonian platform *sensu lato*. Sequence boundary SbBA2 is marked by rudistid facies in the northern Archiane valley (Fig. 18(b)) and appears to pass basinwards to the Upper Fontaine Colombette marls above unit Bi6a (Figs 16 & 17). Thus, relatively condensed sedimentation is associated with the sequence boundary. As relative sea level began to rise bioclastic sedimentation resumed and the SMW systems tract developed with a descending geometry of progradation into the inherent topographic low in the southern Cirque d'Archiane where it is interpreted to downlap onto the Upper Fontaine Colombette marls, contrasting with a parallel–parallel relationship in the north of the valley (Fig. 17). The rapid basinward progradation of unit Bi6b reflects the large area of sediment production, but relatively

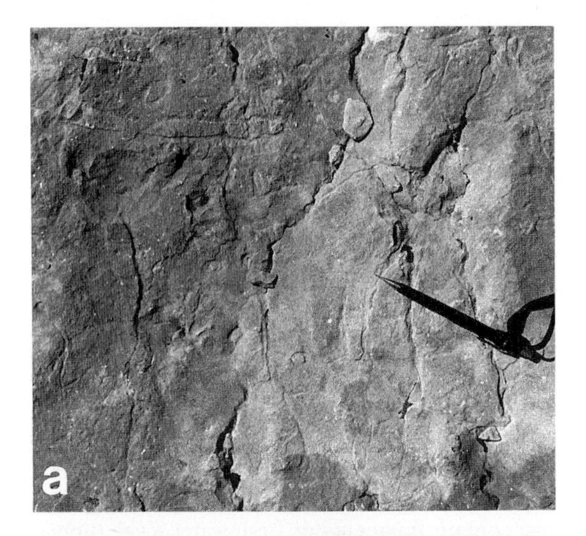

Fig. 18. Two key surfaces from the northern Cirque d'Archiane. (a) Bioturbated dark-grey hemipelagic mudstones at the top of unit Bi5/base of unit Bi6a. Pencil for scale. (b) Rudistid wackestone to packstone towards the top of unit Bi6a. Bottle top for scale.

small space to accommodate it northward of the slope break visible in Cirque d'Archiane (Fig. 17).

The sequence boundary is associated with a basin-floor sand body although continued basin-floor sand deposition throughout units Bi6b and Bs1 does suggest that the deposition of basin-floor sands is not directly related to lowstand of relative sea level but to inherited slope morphology. Both the continued bypassing of sands to the basin floor and the locus of unit Bi6b descending sands in southern Archiane (Fig. 17) suggest that the slope geometry developed by unit Bi5 is an extremely important inherited architectural element which controlled position of the depocentre and patterns of slope sedimentation during BA2.

The whole geometry of the shelf margin wedge cannot be directly observed as exposure is incomplete in the southern Cirque d'Archiane and vicinity. Analysis of both thicknesses and facies of Bi6b shoaling-up cycles in the northern Archiane show that they evolve from backstepping (20 m) to aggrading (75 m) to prograding (40 m). This may reflect the geometry at the shelf margin although the relationship does not appear to correspond directly where geometries are observed (Hunt, 1992). The thickness of 135 m for unit Bi6b in the north of Cirque d'Archiane suggests a relative sea-level rise of this amount during this time. The strong component of aggradation and relative stasis of facies within unit Bi6b suggest a balance between rates of sedimentation and relative sea-level rise. Secondly, thinning of BA2 highstand and sequence BA3 over the Glandasse area also suggests the building of topography in this region as schematically shown in Fig. 11. Such an association is characteristic of type 2 geometries developed during a TST (Fig. 7). The facies, shallowing-up cycle patterns and stratal patterns for units Bi6a to Bs1 suggest development from a type 2a (aggrading) to 1b (aggrading and backstepping) geometry for the TST. The backstepping–aggrading geometry of unit Bs1 is well developed at Rocher du Combau, but not in the Cirque d'Archiane and this is interpreted to indicate higher sedimentation rates in this latter area.

The upper surface of the Glandasse Formation is the mfs above Bs1, represented across the Glandasse area by hemipelagic mudstones. Equivalent slope sedimentation is pelagic and locally glauconitic as a result of reduced sedimentation rates. Upon the shelf top, the TST is represented by 5 to 10 m of reworked Hauterivian mudstone facies, typically unevenly bedded and commonly containing a faunal mélange overlain by <0.5 m of hemipelagic mudstones, the mfs. There is no evidence of exposure but this may have been removed by reworking of mudstone facies on the shelf during the TST.

The HST is the first shallow-water carbonate unit to cover the whole shelf top, but unlike carbonate highstands in general it did not prograde significantly further than the underlying lowstand over which it thins. This pattern is also common to sequences BA3, 4 and 5 which overall are essentially aggradational sequences (Fig. 13). The HST of BA2 is itself unusual as almost no facies differentiation is discernible over a very large area of the shelf top. Across the shelf above the mfs, facies shallow from subwave-base mudstones–packstones to oobioclastic grainstones and their widespread similarity and facies suggest open high-energy conditions with little or no restriction (Hunt, 1992). Only in the upper few metres do restricted inner-platform facies develop in the inner parts of the platform, such as in the northwest Vercors and northwest Chartreuse (see Fig. 19). Due to the apparent low topography inherited by the shelf and aggradation rather than progradation of facies over the shelf top, this unit does not appear to have a downlapping contact at its base across the shelf, but a subparallel to parallel–parallel contact. This is typical of sequence boundaries upon carbonate shelves developed in tectonically quiescent settings.

Sequence BA3

This is a fourth-order sequence (10^5 years) with a thickness (20 to 40 m) (Fig. 19) and duration (200–400,000 years) more normally associated with parasequences. On the shelf, it is the stratigraphic unit Bs3a of Arnaud (1981). The sequence boundary SbBA3 is represented by a widespread exposure surface on the shelf associated with local dissolution, dolomitization or appearance of fresh-water limestones which are common across the Glandasse plateau (Arnaud, 1981). No autochthonous wedge is discernible for this sequence which may reflect a short duration of exposure. Basin-floor marls of this age contain fragments of fresh-water algae (uppermost Bs2 of Arnaud, 1981). Differentiation of

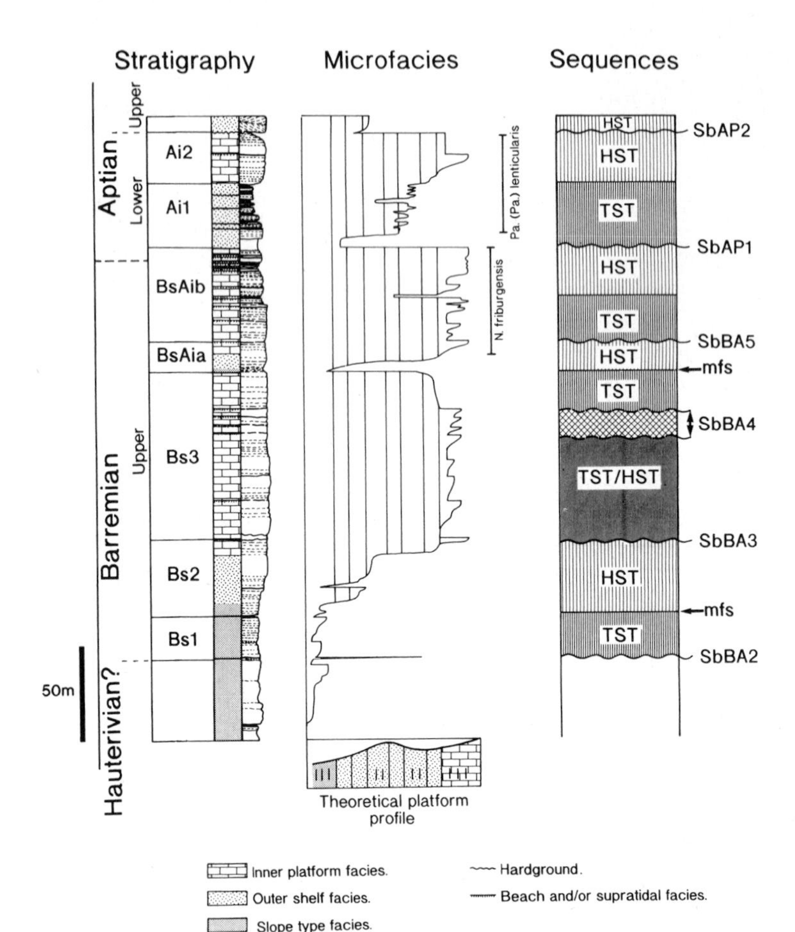

Fig. 19. Gorge du Nant section in the northern Vercors. (Reinterpreted and adapted from Arnaud-Vanneau, 1980.)

transgressive and highstand sediments is difficult, if not impossible. On-shelf facies are essentially aggradational and dominated by restricted inner-shelf rudist facies (Fig. 19). The indistinctive development of the TST and HSTs suggests low rates of relative sea-level rise compared to sedimentation rates.

The amplitudes of the relative sea-level change bounding the sequence above and below were probably less than 10 to 20 m, similar to those typically associated with parasequences (Van Wagoner *et al.*, 1990). The overlap of time, thicknesses and amplitude would appear characteristic of aggraded shelves where fourth- or even fifth-order 'sequences' can develop on the basis of the widespread nature of exposure and fall of sea level below the shelf break. Such 'aggraded' sequences are best developed when the rates and amplitudes of relative sea-level rise are low in comparison to those of sedimentation. In such situations, a platform can aggrade to within a few metres of sea level across much, if not all of its width, with the result that small-scale, relative sea-level changes normally associated with parasequences form 'amalgamated' sequence boundaries which are commonly regional unconformities.

Sequence BA4

Sequence BA4 is composed of stratigraphic units Bs3b and BsAia (Figs 12 & 13). The sequence boundary (SbBA4) is of type 2 affinity on the shelf associated with a stacking of exposure surfaces in the middle part of unit Bs3. The exact position of the sequence boundary is problematic (Fig. 19). In the southern Vercors, northeast of Col du Rousset (Glandasse plateau region) the surface is well defined and often marked by a sharp, low-topography surface (<0.1 m relief) overlain by thin beds (<0.1 m) of greenish fresh-water limestones bearing *Chara*. Characean fragments are also redeposited on to the basin floor, east of Borne (Dèvoluy) at this time where they are closely associated with allochthonous debris in the form of debrite and slump complexes (Arnaud, 1979) which may represent the forced regressive wedge of the preceding sequence. The best preserved shelf margin of this age is to the west of Font d'Urle where the sequence boundary is associated with submarine erosion of 10 to 50 m (Fig. 20). This incision is significantly greater than would be predicted from the development of the sequence boundary elsewhere and suggests that other factors, such as high sedimentation rates, may have played an important role in developing the magnitude of incision at Font d'Urle.

The TST is relatively thick and associated with aggradation rather than retrogradation at the shelf margin as sedimentation rates approximately matched those of relative sea-level rise (type 2a geometry). On the shelf, sedimentation rates could not match those of relative sea-level rise and the TST is recorded by a gradual opening up of conditions, with the development of bioclastic dasycladacean sands representing the mfs, which is coincident with the base of unit BsAia (Fig. 19). Differences in stacking pattern between the shelf

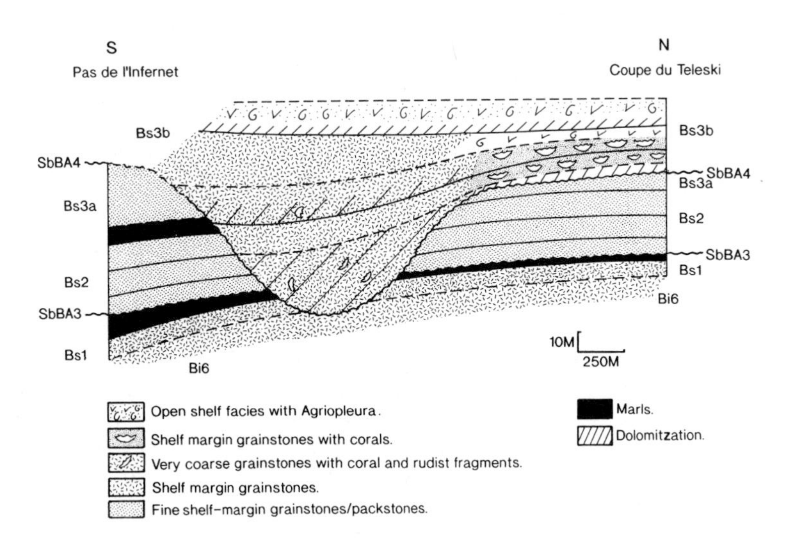

Fig. 20. Interpreted shelf margin of sequences BA2 to BA4 at Font d'Urle. Incision of up to 50 m is associated with sequence boundary SbBA4. This section is oblique to the shelf margin. (Modified after Arnaud, 1981.)

margin and inner shelf appear to reflect differential sedimentation rates across the platform, which are often highlighted during relative sea-level rises developing type 2 geometries (see earlier discussion).

The HST (unit BsAia) is a single parasequence between 10 and 40 m thick, which thins towards the northwestern Vercors (e.g. Balcon des Ecouges, Gorge du Nant region) and shallows from bioclastic and oolitic grainstones to typical inner-platform rudist facies (Fig. 19). Conditions upon the shelf top generally became increasingly restricted, culminating in a widely developed exposure surface, SbBA5.

Sequence BA5

Sequence BA5 is composed of unit BsAib. The exact nature of the sequence boundary is difficult to determine due to the poor exposure of sediments of this age at the platform margin. The sequence boundary is once again most evident upon the shelf where it is represented by fresh-water limestones commonly with mudrocks preserved in small depressions (1 to 2 m wide by 0.1 m deep) and dissolution breccias (Hunt, 1992). These sediments are equivalent to incision of 40 to 50 m at the shelf margin and the superposition of inner-platform rudistid facies above outer-shelf grainstones in the southwestern Vercors, northwest of Font d'Urle (Arnaud, 1981). The basinwards facies jump and incision in the western Vercors suggest a significant fall of relative sea level (20 to 40 m) and hence a type 1 sequence boundary.

On the shelf the TST is represented by a 20 to 25 m shallowing-upward unit representing one or two amalgamated parasequences approximately 20 to 25 m thick capped by an emergence horizon. It is overlain by *Agriopleura* rudistid facies which are the least restricted facies and represent the mfs (Fig. 19). At the shelf margin (west of Font d'Urle) the TST is reef-dominated and essentially aggradational with only minor backstepping at the base of the TST.

The HST is also strongly aggradational at the shelf margin and is associated with only limited facies progradation in the western Vercors where it is reef-dominated. On the shelf, the HST consists of four 5–10 m thick fourth-order parasequences that form a 20 to 50 m unit. These distinctive parasequences can be recognized over a large part of the Vercors (Hunt, 1992) and are most readily recognized at the Balcon des Ecouges where their component fifth-order shallowing-upward cycles are also easily distinguished in a shallow subtidal to supratidal sequence (Fig. 21). Pentacycles can be recognized, but the often low rates of relative sea-level change compared to sedimentation resulted in complex stacking patterns with the base of a parasequence not necessarily coincident with the least restricted conditions (Fig. 21). Elsewhere, in predominantly subtidal facies, distinction of fourth-order parasequences is more difficult, although their upper surface is normally marked by supratidal facies.

Sequence AP1

This sequence consists of stratigraphic units Ai1–Ai2 and the upper part of unit BsAi at the shelf margin (Arnaud-Vanneau, 1980; Arnaud, 1981), here included into Ai1. The sequence boundary is one of the clearest horizons within the Urgonian platform *sensu stricto* and is associated with an autochthonous lowstand wedge (Fig. 13, the lower Ai1 unit). It is characterized by the overlap of species *Neotrocholina friburgensis* and *Palorbitolina (Palorbitolina) lenticularis*, two species which are never seen together upon the shelf (Arnaud-Vanneau, 1980; Arnaud, 1981) (Fig. 19) where they are separated by SbAP2. On the shelf the regressive and lowstand systems tracts are commonly represented by fresh-water limestones and, locally in the northern Vercors, with approximately 30 m of fresh-water-restricted lagoonal organic-rich facies (5 km northeast of Gorge du Nant (Arnaud-Vanneau & Medus, 1977)). More generally, the boundary on the shelf is a compound surface reworked as a hardground surface during the TST (Hunt, 1992). At the shelf margin a volumetrically significant autochthonous lowstand prograding wedge in the Borne region developed contemporaneously to the fresh-water limestones on the shelf (Fig. 13). The exact stratal relationships of the autochthonous wedge to the slope cannot be directly observed due to subsequent movement and disruption along the Menée fault. Prior to development of the slope wedge, slope failure transported debris to the basin floor and this is closely associated with basin-floor fans in Dévoluy. These deposits may be the basin-floor component of the FRW systems tract strictly of the preceding sequence.

The TST of sequence AP1 is very well developed on the shelf in the western Vercors (e.g. Gorge du Nant and Balcon des Ecouges) and Chartreuse-

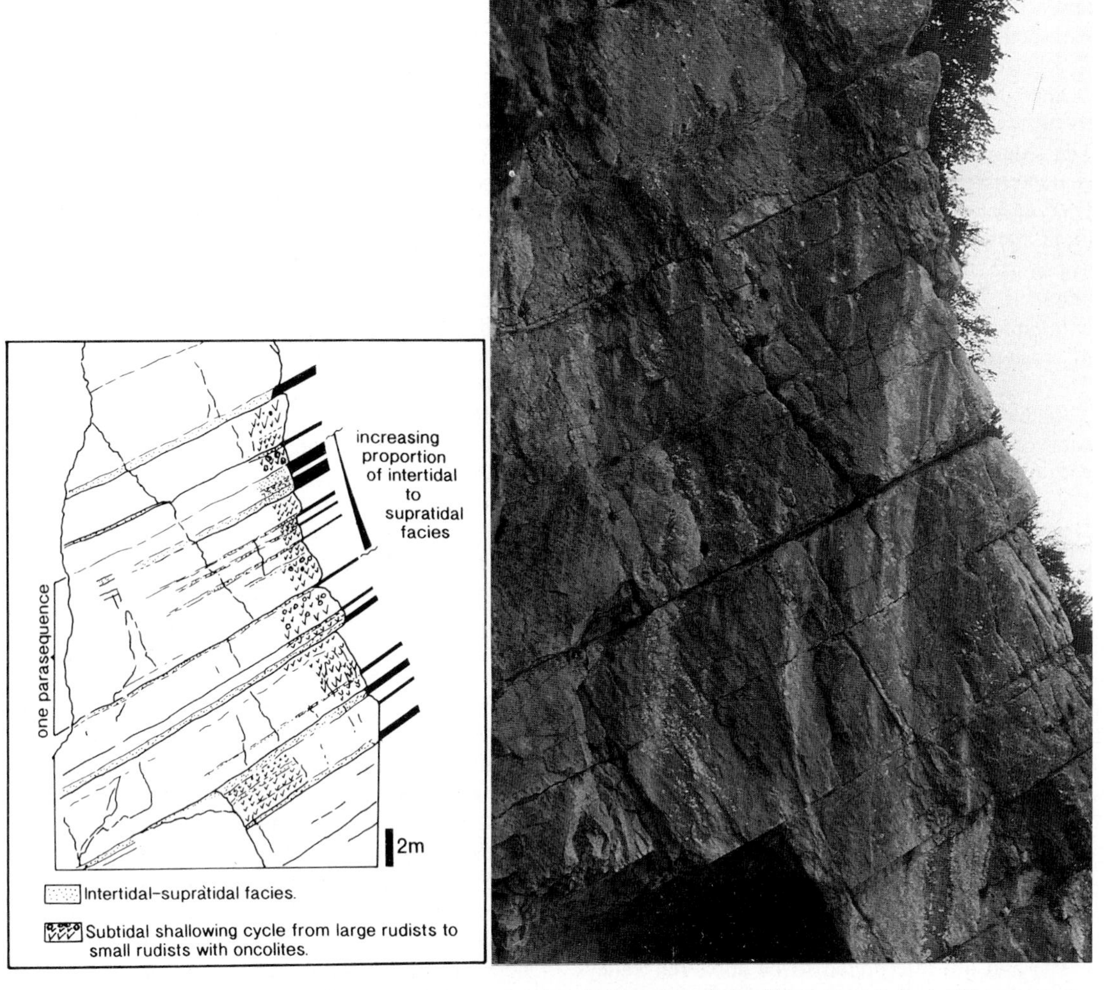

Fig. 21. Complex stacking of fifth-order shallowing-up cycles within BA5 (unit BsAib) at the Balcon des Ecouges in the northern Vercors. Fourth-order parasequence boundaries are marked by siliciclastic clays. The complexity of stacking patterns at this locality probably reflects high rates of sedimentation compared to those of relative sea-level change. One fourth-order parasequence (labelled) has a classical stacking pattern, containing five higher order cycles that contain a higher proportion of inter- to supratidal facies upwards. In the next parasequence, however, the pattern is not as asymmetric as the most open conditions do not occur at the base of the parasequence but some way into the parasequence. This pattern is interpreted to develop because of a close match between rates of relative sea-level rise and sedimentation.

where it is represented by normally 30 m (but up to 50 m) of carbonates with a variable proportion of siliciclastic facies, the Lower *Orbitolina* Beds of the Urgonian Limestone Formation (Fig. 12). The TST consists of four to six, 4–5 m thick shallowing-upward units from shales to limestones typically capped by a mineralized hardground surface. Facies of this TST are atypical and contain anomalous facies such as *Orbitolinid* packstones and micro-ooidal grainstones (Arnaud-Vanneau *et al.*, 1987). This is a type 2 style TST as sediment production rates approximately equalled those of sea-level rise, allowing development of a thick TST package on the shelf. Well-developed asymmetric shallowing-upward cycles also distinguish this TST, interpreted to result from sedimentation rates being slightly lower than initial rates of sea-level rise for each parasequence (start-up to catch-up cycles of Kendall & Schlager, 1981).

The HST is up to 40 m thick and marks a return to normal carbonate sedimentation on the shelf. The shut-down of siliciclastic sedimentation that characterizes the TST is abrupt at the base of the HST, and may be related to changes of the sediment input pattern and/or to relative sea-level change. These are the last true Urgonian facies of the platform. The HST appears to consist of at least two parasequences, variably preserved on the shelf due to erosion at the base of the overlying sequence. The HST typically begins with the establishment of high-energy grainstones, commonly containing corals, that shallow rapidly into inner-shelf facies with frequent emergence horizons.

Sequence AP2

The sequence is represented by units Ai2–3 and has the most spectacular sequence boundary on the shelf top interpreted to result from siliciclastic influx during subaerial exposure. At the shelf margin an autochthonous wedge developed in the western Vercors but it is not preserved (if it were developed) in the eastern Vercors due to Upper Cretaceous erosion (Fig. 13). On the shelf, the sequence boundary is defined by an exposure surface and associated NW–SE-trending incised valleys in the northern Vercors and Chartreuse (Figs 14 & 22). These are up to 200 m wide and 50 m deep incised into the preceding highstand (AP1) and are interpreted to be subaerial valleys (Arnaud-Vanneau & Arnaud, 1990). The amplitude of the relative sea-level fall associated with the development of the sequence boundary is suggested by the amount of downcutting in the palaeovalleys (approximately 50 m). At the shelf margin, bioclastic grainstones sit directly on hemipelagic mudstones and fine bioclastics in the southwestern Vercors (Fig. 23). Such an amplitude of sea-level fall dropped sea level beyond the shelf break of the platform to produce a type 1 sequence boundary.

The TST of this sequence is represented by the second appearance of the *Orbitolina* facies, particularly notable in the northern Vercors where these facies fill the depressions cut into the shelf and have been locally eroded to exhume the valleys (e.g. Fig. 22). The TST is associated with hardground formation across the shelf and frequently reworking of the sequence boundary, forming an encrusted and mineralized compound surface. Shallowing-upward cycles like those that characterize the Lower *Orbitolina* Beds are not developed. At the shelf margin the TST in several localities is a time of collapse and the deposition of basin-floor megabreccias such as in Dévoluy and in the western Vercors Gigors region (Fig. 23) (Arnaud, 1979; Ferry & Flandrin, 1979). The widespread slope collapse at this time appears to be related to reactivation of basement lineaments as collapse is notably localized to the proximity of Gigors and Menée faults (Arnaud, 1979; Ferry & Flandrin, 1979) (Fig. 14). The occurrence of these megabreccias on the basin floor highlights their limited use as lowstand 'predictors' as discussed earlier. The scalloped slope of the platform in the Gigors region forms a prominent onlap surface to organic-rich pelagic facies during the Upper Aptian and Albian (Fig. 23).

The HST of sequence AP2 is complex and terminates the Urgonian platform in the Subalpine Chains. In the basin it is represented by stratigraphically complete dark grey to black interbedded pelagic limestones and shales, that onlap the slope (Fig. 23). Contrastingly, on the shelf there is a significant unconformity at the base of the Lumachelle, a 10 to 12 m package of sandy glauconitic grainstones with a low faunal diversity (mainly bryozoans) which represents the upper Aptian. This gap is interpreted to be a submarine unconformity developed by current winnowing often associated with deposition of glauconitic sediments upon current-dominated shelves, where short periods of deposition alternate with longer time spans of non-deposition and/or erosion (Föllmi, 1990). The Lumachelle is itself overlain by condensed phosphatic hardground surfaces of Albian age. The occurrence of phosphatic sediments and black shales above glauconitic sediments suggests raising of the oxygen minimum zone on to the shelf. The termination of the Urgonian platform appears to be related to major environmental changes associated with rearrangement of oceanic currents (Delamette, 1988).

CONCLUSIONS

Carbonate shelves can differ significantly from their siliciclastic equivalents, so that the direct application of existing sequence-stratigraphic models to

Fig. 22. Exhumed, incised subaerial valleyside at Les Rimets in the northwestern Vercors. This cliff-like feature is interpreted to have developed during falling and lowstand of relative sea level during unit Ai3 times, associated with formation of sequence boundary AP2. The prominent lineation dipping from top left to bottom right is the intersection of bedding with the valley wall. There is a structural dip of approximately 20° to the southeast. Scale bar is approximately 10 m.

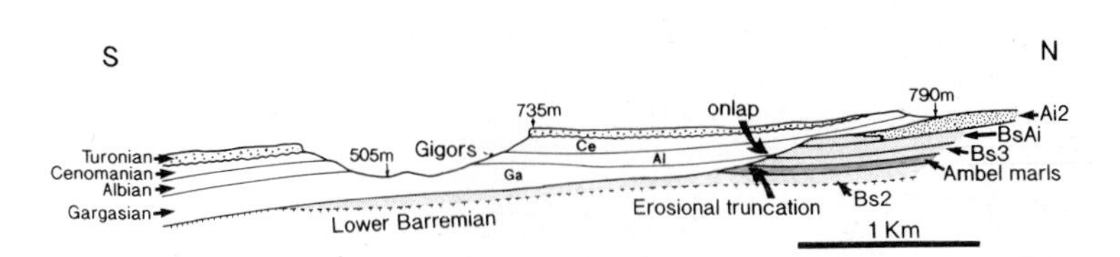

Fig. 23. Cross-section from the Gigors region where a slump scar truncates unit Ai2 lowstand sediments down to the lower Barremian. The collapse scar forms a prominent onlap surface to pelagic basinal shales. (From Arnaud, 1981.)

carbonate shelves is often an oversimplification which can lead to serious errors in assignment of geometric relationships to specific stands of sea level. Examples have been given in this paper from the Urgonian platform of southeast France.

The development of systems tracts on carbonate shelves is the result of a complex interplay of slope angle, rates of sedimentation compared to relative sea-level rise, environmental changes and the relative mobility of the depositional system itself. Development of the LST tends to reflect the slope angles of the preceding highstand. Two end-member lowstand models are envisaged. Low-angle systems of autochthonous grainstone wedges developed upon mud-dominated slopes tend to be deposited over wide areas and are volumetrically significant. Grain-supported slopes tend to attain higher angles and, upon these, autochthonous wedges tend to be narrow and volumetrically small or even absent. Both types of lowstand deposit may be preceded by collapse of the slope, forming allochthonous sheets and fans. Secondary allochthonous debris can be deposited if sedimentation rates remain constant, leading to the development of a bypass slope to the lowstand wedge.

Potential sedimentation rates vary across an individual platform and are highest at the shelf margin, decreasing back into the lagoon. Such factors are particularly important during the TST when the inner platform can become drowned. Geometries developed during the TST reflect the complex interplay of changing rates of relative sea-level rise and sedimentation. The nature of the shelf-rimming complex and polarity of this system also play an important role. Unlike siliciclastic systems carbonate sedimentation rates cannot be assumed to be constant due to the strong environmental control upon sedimentation. Thus, changes in geometric stacking patterns do not necessarily directly reflect changes in the rates of relative sea-level rise as environmental changes are not always intimately linked to changes of sea level. The HST is the time when the maximum growth potential of the platform is realized and the shelf margin typically progrades rapidly basinwards.

ACKNOWLEDGEMENTS

D. Hunt would firstly like to thank his parents for their considerable encouragement and continuing assistance. The considerable time and knowledge of Annie and Hubert Arnaud is also greatly appreciated. Jenny Robb was also of great assistance during fieldwork. We would like to extend our appreciation to Karen Atkinson for drafting, Gladys K. Lawson for text reading, Alan Carr and Gerry Dresser for photographic work and Lynne Gilchrist for typing the references. The constructive criticism of J.F. Sarg and an anonymous reviewer improved the manuscript. D. Hunt gratefully acknowledges the receipt of NERC grant GT4/88/GS/30.

REFERENCES

Arnaud, H. (1979) Surfaces d'ablation sous-marines et sédiments barrémo-bédouliens remaniés par gravité du Barrémian au Cénomanien entre le Vercors et le Dévoluy (SE de la France). *Géol. Alpine* **55**, 5–21.

Arnaud, H. (1981) De la plate-forme urgonienne au bassin vocontien: le Barrémo-Bédoulien des Alpes occidentales entre Isère et Buëch (Vercors méridional, Diois oriental et Dévoluy). *Mém. Géol. Alpine.* **11**, 804 pp.

Arnaud-Vanneau, A. (1980) Micropaléontologie, paléoécologie et sédimentologie d'une plate-forme carbonatée de la marge passive de la Téthys: l'Urgonien du Vercors septentrional et de la Chartreuse (Alpes occidentales). *Mém. Géol. Alpine* **10**, 874 pp.

Arnaud-Vanneau, A. & Arnaud, H. (1990) Hauterivian to Lower Aptian carbonate shelf sedimentation and sequence stratigraphy in the Jura and northern Subalpine chains (southeastern France and Swiss Jura). In: *Carbonate Platforms* (Eds Tucker, M.E., Wilson, J.L., Crevello, P.D., Sarg, J.F. & Read, J.F.) Int. Assoc. Sediment. Spec. Publ. 9, 203–233.

Arnaud-Vanneau, A. & Medus, J. (1977) Palynoflores barrémo-aptiennes de la plate-forme urgonienne du Vercors Palynostratigraphie de quelques formes de

Classopollis et de quelques pollens angiosperms. *Géol. Alpine* **53**, 35–55.

ARNAUD-VANNEAU, A., ARNAUD, H., MEUNIER, A.- R. & SEGUIN, J.-C. (1987) Caractères des transgressions du Crétacé inférieur sur les marges de l'océan ligure (Sud-Est de la France et Italie centrale). *Mém. Géol. Université de Dijon* **11**, 167–182.

BAY, T.A. (1977) Lower Cretaceous stratigraphic models from Texas and Mexico. In: *Cretaceous Carbonates of Texas and Mexico* (Eds Bebout, D.G. & Loucks, R.G.) Bureau of Economic Geology, Rept of Investigations 89, Austin, Texas, pp. 12–30.

BIDDLE, K.T., SCHLAGER, W., RUDOLPH, K.W. & BUSH, T.L. (1992) Seismic model of a progradational carbonate platform, Picco di Vallandro, the Dolomites, Northern Italy. *Bull. Am. Assoc. Petrol. Geol.* **76**, 14–30.

BOARDMAN, M.R., NEUMANN, A.C., BAKER, P.A., *et al.* (1986) Banktop response to Quaternary fluctuations in sea-level recorded in periplatform sediments. *Geology* **14**, 28–31.

BOSELLINI, A. (1984) Progradation geometries of carbonate platforms: examples from the Triassic of the Dolomites, northern Italy. *Sedimentology* **31**, 1–24.

BOSELLINI, A. (1989) Dynamics of Tethyan carbonate platforms. In: *Controls on Carbonate Platform and Basin Development* (Eds Crevello, P.D., Wilson, J.L., Sarg, J.F. & Read, J.F.) Spec. Publ. Soc. Econ. Paleontol. Mineral. 44, 3–13.

BROOKS, G.R. & HOLMES, C.W. (1989) Recent carbonate slope sediments and sedimentary processes bordering a non-rimmed platform: southwest Florida continental margin. In: *Controls on Carbonate Platform and Basin Development* (Eds Crevello, P.D., Wilson, J.L., Sarg, J.F. & Read, J.F.) Spec. Publ. Soc. Econ. Paleontol. Mineral. 44, 259–272.

CALVET, F., TUCKER, M.E. & HENTON, J.M. (1990) Middle Triassic carbonate ramp systems in the Catalan Basin, northeast Spain: facies, systems tracts, sequences and controls. In: *Carbonate Platforms* (Eds Tucker, M.E., Wilson, J.L., Crevello, P.D., Sarg, J.F. & Read, J.F.) Int. Assoc. Sediment. Spec. Publ. 9, 79–108.

CALVET, F., TUCKER, M.E. & HUNT, D. (1993) Sequence stratigraphy of Carbonate Ramps: systems tracts, models and application to the Muschelkalk carbonate platforms of eastern Spain. In: *Sequence Stratigraphy and Facies Associations* (Eds Posamentier, H.W., Summerhayes, C.P., Haq, B.U. & Allen, G.P. Int. Assoc. Sediment. Spec. Publ. **18**, 397–415.

CLAVEL, B., CHAROLLAIS, J. & BUSNARDO, R. (1987) Donnèes biostratigraphiques nouvelles sur l'apparation des facies urgoniens du Jura au Vercors. *Eclogae Gèol. Helv.* **80**, 59–68.

COLACICCHI, R., PIALLI, G. & PRATURLON, A. (1975) Megabreccias as a product of tectonic activity along a carbonate platform margin. *11th Int. Sediment. Congress, Nice,* pp. 61–70.

CREVELLO, P.D. & SCHLAGER, W. (1980) Carbonate debris sheets and turbidites, Exuma Sound, Bahamas. *J. Sediment. Petrol.* **50**, 1121–1148.

DABRIO, C.J., ESTEBAN, M. & MARTIN, J.M. (1981) The coral reef of Níjar, Messinian (uppermost Miocene), Almería Province, SE Spain. *J. Sediment. Petrol.* **51**, 521–539.

DAVIES, P.J., SYMONDS, P.A., FEARY, D.A. & PIGRAM, C.J. (1989) The evolution of the carbonate platforms of northeast Australia. In: *Controls on Carbonate Platform and Basin Development* (Eds Crevello, P.D., Wilson, J.L., Sarg, J.F. & Read, J.F.) Spec. Publ. Soc. Econ. Paleontol. Mineral. 44, 233–258.

DELAMETTE, M. (1988) Relation between the condensed Albian deposits of the Helvetic domain and the oceanic-current influenced continental margin of the northern Tethys. *Bull. Soc. Géol. France* **5**, 739–745.

DOGLIONI, C., BOSELLINI, A. & VAIL, P.R. (1990) Stratal patterns: a proposal of classification and examples from the Dolomites. *Basin Res.* **2**, 83–95.

DROXLER, A. & SCHLAGER, W. (1985) Glacial versus interglacial sedimentation rates and turbidite frequency in the Bahamas. *Geology* **13**, 799–802.

EBERLI, G.P. (1987) Carbonate turbidite sequences deposited in rift-basins of the Jurassic Tethys Ocean (eastern Alps, Switzerland). *Sedimentology* **34**, 363–388.

EBERLI, G.P. & GINSBURG, R.N. (1989) Cenozoic progradation of northwestern Great Bahama Bank, a record of lateral platform growth and sea-level fluctuations. In: *Controls on Carbonate Platform and Basin Development* (Eds Crevello, P.D., Wilson, J.L., Sarg, J.F. & Read, J.F.) Spec. Publ. Soc. Econ. Paleontol. Mineral. 44, 339–351.

ERLICH, R.N., BARRETT, S.F. & GUO BAI JU (1990) Seismic and geological characteristics of drowning events on carbonate platforms. *Bull. Am. Assoc. Petrol. Geol.* **74**, 1523–1537.

FERRY, S. & FLANDRIN, J. (1979) Mégabrèches de resédimentation, lacunes mécaniques et pseudo 'hardgrounds' sur la marge vocontienne au Barrémien et à l'Aptien inferieur (Sud-Est de la France). *Géol. Alpine* **55**, 75–92.

FERRY, S. & RUBINO, J.L. (1989) *Mesozoic Eustasy Record on Western Tethyan Margins. Post-meeting Field Trip in the Vocontian Trough.* Spec. Publ. Assoc. Sediment. Franc. pp. 141.

FÖLLMI, K.B. (1990) Phosphatic sediments, indicators of non-steady-state depositional and diagenetic environments (Abstr.). *10th IAS regional meeting, Nottingham, England,* pp. 172–173.

FRANSEN, E.V. & MANKIEWICZ, C. (1991) Depositional sequences and correlation of middle to late Miocene carbonate complexes, Las Negras and Níjar areas, southeastern Spain. *Sedimentology* **38**, 871–898.

GALLOWAY, W.E. (1989) Genetic stratigraphic sequences in basin analysis. 1: architecture and genesis of flooding-surface bounded depositional units. *Bull. Am. Assoc. Petrol. Geol.* **73**, 125–142.

GAMBOA, L.A., TRUCHAN, M. & STOFFA, P.L. (1985) Middle and Upper Jurassic depositional environments at outer shelf and slope of Baltimore Canyon Trough. *Bull. Am. Assoc. Petrol. Geol.* **69**, 610–621.

GEBELEIN, C.D. (1974) *Guidebook for Modern Bahamian Platform Environments.* Geol. Soc. Am. Annual Meeting Fieldtrip Guide, pp. 93.

GOLDHAMMER, R.K. & HARRIS, M.T. (1989) Eustatic controls on the stratigraphy and geometry of the Latemar buildup (Middle Triassic), the Dolomites of northern Italy. In: *Controls on Carbonate Platform and Basin Development* (Eds Crevello, P.D., Wilson, J.L., Sarg, J.F. & Read, J.F.) Spec. Publ. Soc. Econ. Paleontol. Mineral. 44, 323–338.

GREENLEE, S.M. & LEHMANN, P.J. (1990) Stratigraphic framework of productive carbonate buildups (Abstr.). *Bull. Am. Assoc. Petrol. Geol.* **74**, 618.

HALLOCK, P. & SCHLAGER, W. (1986) Nutrient excess and the demise of coral reefs and carbonate platforms. *Palaios* **1**, 389–398.

HAQ, B.U., HARDENBOL, J. & VAIL, P.R. (1987) Chronology of fluctuating sea levels since the Triassic. *Science* **235**, 1156–1167.

HILBRECHT, H. (1989) Redeposition of Late Cretaceous pelagic sediments controlled by sea-level fluctuations. *Geology* **17**, 1072–1075.

HINE, A.C., WILBERT, R.J. & NEUMANN, A.C. (1981) Carbonate sand-bodies along contrasting shallow-bank margins facing open seaways; northern Bahamas. *Bull. Am. Assoc. Petrol. Geol.* **65**, 261–290.

HUBBARD, R.J. (1988) Age and significance of sequence boundaries on Jurassic and early Cretaceous rifted continental margins. *Bull. Am. Assoc. Petrol. Geol.* **72**, 49–72.

HUMPHREY, J.S. & KIMBELL, T.N. (1990) Sedimentology and sequence stratigraphy of Upper Pleistocene carbonates of S.E. Barbados, West Indies. *Bull. Am. Assoc. Petrol. Geol.* **74**, 1671–1684.

HUNT, D. (1992) Application of sequence stratigraphic concepts to the Urgonian carbonate platform, SE France. Unpublished PhD thesis, University of Durham.

HUNT, D. & TUCKER, M.E. (1991) Responses of rimmed shelves to relative sea-level rises; a proposed sequence stratigraphic classification (Abstr.). *Proceedings Dolomieu Conference, the Dolomites, Italy,* p. 114–115.

HUNT, D. & TUCKER, M.E. (1992) Stranded parasequences and the forced regressive wedge systems tract; deposition during base-level fall. *Sediment. Geol.* **81**, 1–9.

JACQUIN, T., VAIL, P.R. & RAVENNE, C. (1991) Systems tracts and depositional sequences in a carbonate setting: study of continuous outcrops from platform to basin at the scale of seismic sections. *Mar. Petrol. Geol.* **8**, 122–139.

JOSEPH, P., BEAUDOIN, B., FRIES, G. & PARIZE, O. (1989) Les vallées sous-marines enregistrent au Crétacé inférieur le fonctionnement en blocs basculés de domaine vocontien. *C. R. Acad. Sci. Paris* **309**, 1031–1038.

KENDALL, C.G. & SCHLAGER, W. (1981) Carbonates and relative changes in sea-level. *Mar. Geol.* **44**, 181–212.

KENTER, J.A.M. (1990) Carbonate platform flanks: slope angle and sediment fabric. *Sedimentology* **37**, 777–794.

LEMOINE, M., BAS, T., ARNAUD-VANNEAU, A., *et al.* (1986) The continental margin of the Mesozoic Tethys in the Western Alps. *Mar. Petrol. Geol.* **3**, 179–199.

MCILREATH, I.A. (1977) Accumulation of a Middle Cambrian, deep-water limestone debris apron adjacent to a vertical submarine carbonate escapement, southern Rocky Mountains, Canada. In: *Deep-water Carbonate Environments* (Eds Cook, H.E. & Enos, P.) Spec. Publ. Soc. Econ. Paleontol. Mineral. 25, 113–124.

MAY, J.A. & EYLES, D.R. (1985) Well log and seismic character of Tertiary Terumbu carbonate, South China Sea, Indonesia. *Bull. Am. Assoc. Petrol. Geol.* **69**, 1339–1358.

MULLINS, H.T. (1983) Comment on 'Eustatic control of turbidites and winnowed turbidites'. *Geology* **11**, 57–58.

MULLINS, H.T., GARDULSKI, A.F., HINE, A.C., *et al.* (1988) Three-dimensional sedimentary framework of the carbonate ramp slope of central west Florida: a sequential seismic stratigraphic perspective. *Bull. Geol. Soc. Am.* **100**, 514–533.

PLAYFORD, P.E. (1980) Devonian 'Great Barrier Reef' of Canning Basin, Western Australia. *Bull. Am. Assoc. Petrol. Geol.* **64**, 814–840.

POSAMENTIER, H.W. & VAIL, P.R. (1988) Eustatic controls on clastic deposition. II: sequence and systems tract models. In: *Sea-level Changes, An Integrated Approach* (Eds Wilgus, C.K., Hastings, B.S., Kendall, C.G.St.C., Posamentier, H.W., Ross, C.A. & Van Wagoner, J.C.) Spec. Publ. Soc. Econ. Paleontol. Mineral. 42, 125–154.

PURDY, E.G. (1974) Reef configurations: cause and effect. In: *Reefs in Space and Time* (Ed. Laporte, L.F.) Spec. Publ. Soc. Econ. Paleontol. Mineral. 18, 9–76.

PURSER, B.H. (1973) Sedimentation around bathymetric highs in the southern Persian Gulf. In: *The Persian Gulf* (Ed. Purser, B.H.), Springer-Verlag, Berlin, pp. 157–177.

RAVENNE, C., LEQUETTFE, P., VALLERY, P. & VIALLY, R. (1987) Deep clastic carbonate deposits of the Bahamas — comparison with Mesozoic outcrop of the Vercors and Vocontian trough. In: *Atlas of Seismic Stratigraphy* (Ed. Bally, A.V.) Am. Assoc. Petrol. Geol. Stud. Geol. 27, 104–140.

RUDOLPH, K.W. & LEHMANN, P.J. (1989) Platform evolution and sequence stratigraphy of the Natuna Platform, South China. In: *Controls on Carbonate Platform and Basin Development* (Eds Crevello, P.D., Wilson, J.L., Sarg, J.F. & Read, J.F.) Spec. Publ. Soc. Econ. Paleontol. Mineral. 44, 353–361.

RUDOLPH, K.W., SCHLAGER, W. & BIDDLE, K.T. (1989) Seismic models of a carbonate foreslope-to-basin transition, Picco di Vallandro, Dolomite Alps, northern Italy. *Geology* **17**, 453–456.

SALLER, A.H., BARTON, J.W. & BARTON, R.E. (1989) Slope sedimentation associated with a vertically-building shelf. Bone Spring Formation, Mescalero Escarpe Field, southeastern New Mexico. In: *Controls on Carbonate Platform and Basin Development* (Eds Crevello, P.D., Wilson, J.L., Sarg, J.F. & Read, J.F.) Spec. Publ. Soc. Econ. Paleontol. Mineral. 44, 275–288.

SARG, J.F. (1988) Carbonate sequence stratigraphy. In: *Sea-level Changes: An Integrated Approach* (Eds Wilgus, C.K., Hastings, B.S., Kendall, C.G.St.C., Posamentier, H.W., Ross, C.A. & Van Wagoner, J.C.) Spec. Publ. Soc. Econ. Paleontol. Mineral. 42, 155–181.

SCHLAGER, W. (1981) The paradox of drowned reefs and carbonate platforms. *Bull. Geol. Soc. Am.* **92**, 197–211.

SCHLAGER, W. (1989) Drowning unconformities on carbonate platforms. In: *Controls on Carbonate Platform and Basin Development* (Eds Crevello, P.D., Wilson, J.L., Sarg, J.F. & Read, J.F.) Spec. Publ. Soc. Econ. Paleontol. Mineral. 44, 15–25.

SCHLAGER, W. (1991) Depositional bias and environmental change—important factors in sequence stratigraphy. *Sediment. Geol.* **70**, 109–130.

SCHLAGER, W. & CAMBER, O. (1986) Submarine slope angles, drowning unconformities and self-erosion of limestone escarpments. *Geology* **14**, 762–765.

Schlager, W. & Ginsburg, R.N. (1981) Bahama carbonate platforms — the deep and the past. *Mar. Geol.* **44**, 1–24.

Schroeder, R., Busnardo, R., Clavel, B. & Charollais, J. (1989) Position des couches à *Valserina brönnimanni* Schroeder et Conrad (Orbitolinidés) dans la biozonation du Barrémien. *C. R. Acad. Sci. Paris* **309**, 2093–2100.

Stoakes, F.A. (1980) Nature and control of shale basin fill and its effect on reef growth and termination: Upper Devonian Duvernay and Ireton Formations of Alberta, Canada. *Can. Petrol. Geol. Bull.* **28**, 345–410.

Tucker, M.E. (1991) Sequence stratigraphy of carbonate-evaporite basins: the Upper Permian (Zechstein) of northeast England and adjoining North Sea. *J. Geol. Soc. London* **148**, 1019–1036.

Tucker, M.E. (1992) Carbonate diagenesis and sequence stratigraphy. In: *Sediment. Rev.* (Ed. Wright, V.P.). **1**, 51–72.

Tucker, M.E. & Wright, V.P. (1990) *Carbonate Sedimentology.* Blackwell Scientific Publications, Oxford, 482 pp.

Vail, P.R. (1987) Seismic stratigraphy interpretation using sequence stratigraphy. Part 1: seismic stratigraphy interpretation procedure. In: *Atlas of seismic stratigraphy* **1** (Ed. Bally, A.W.) AAPG Studies in Geology 27, 1–10.

Vail, P.R., Mitchum, R.M., Todd., R.G., Widmeir, J.M., Thompson, S. & Hatfield, W.G. (1977) Seismic-stratigraphy and global changes of sea-level. In: *Seismic Stratigraphy — Application to Hydrocarbon Exploration* (Ed. Payton, C.E.) Am. Assoc. Petrol. Geol. Mem. **26**, 49–212.

Van Wagoner, J.C., Mitchum, R.M., Campion, K.M. & Rahmanian, V.D. (1990) Siliciclastic sequence stratigraphy in well logs, cores and outcrops. *Am. Assoc. Petrol. Geol. Meth. Expl. Ser.* **7**, pp. 45.

Ward, R.F., Kendall, C.G.St.C. & Harris, P.M. (1986) Upper Permian (Guadalupian) facies and their association with hydrocarbons — Permian Basin, west Texas and New Mexico. *Bull. Am. Assoc. Petrol. Geol.* **70**, 239–262.

Watts, K.F. (1987) Triassic carbonate submarine fans along the Arabian platform margin, Sumeini Group, Oman. *Sedimentology* **34**, 43–71.

Wilber, R.J., Milliman, J.D. & Halley, R.B. (1990) Accumulation of bank-top sediment on the western slope of Great Bahama Bank: rapid progradation of a carbonate megabank. *Geology* **18**, 970–974.

Wright, V.P. & Wilson, R.C.L. (1984) A carbonate submarine fan sequence from the Jurassic of Portugal. *J. Sediment. Petrol.* **54**, 394–412.

Spec. Publs Int. Ass. Sediment. (1993) **18**, 343–368

Sedimentary evolution and sequence stratigraphy of the Upper Jurassic in the Central Iberian Chain, northeast Spain

M. AURELL *and* A. MELÉNDEZ

Departamento Geología, Universidad de Zaragoza, 50009-Zaragoza, Spain

ABSTRACT

Successive low-angle slope carbonate ramps developed during the Upper Jurassic in the Iberian Basin (northeast Spain). The transition between these ramps was punctuated by lower-order tectonic events, that involved the uplift of the edges of the basin and a correlative basinwards coastal shift, generating unconformities that are detected at least in the proximal areas of the ramps. These unconformities allow the Upper Jurassic to be subdivided into three unconformity-bounded units or depositional sequences. An extensive facies analysis across the basin and the lateral tracing of sedimentary units and bounding surfaces (favoured by an accurate ammonite biozonation) resulted in the identification of the lowstand, transgressive and highstand systems tracts of the lower two sequences (i.e. Oxfordian and Kimmeridgian sequences). These systems tracts define six successive episodes in the sedimentary evolution of the basin. On the basis of an intermediate unconformity, two further episodes are distinguished in the upper Upper Jurassic sequence or Tithonian–Berriasian sequence. The successive facies associations and the interpreted depositional settings are illustrated through these eight episodes in the evolution of the basin.

The observed distribution of facies into systems tracts is a function of the interplay between sedimentation and accommodation rates. Information about the amount and temporal variation of these two elements was obtained by taking into account the duration of each episode, the sediment thickness and the estimated depositional depth. The accommodation rate is a function of eustasy and subsidence. The roles of these two factors were evaluated on the basis of the geographical extension of the recognized unconformities and marine flooding events. Eustasy is thought to be the main factor responsible for the evolution of the accommodation during the Lower and Middle Oxfordian, when the subsidence is interpreted as slow and regularly distributed across the basin, whereas the Kimmeridgian–Middle Berriasian sediments would mainly record the local subsidence history of the basin.

INTRODUCTION

The application of sequence-stratigraphy concepts has been proved useful in the documentation of the evolution of the sedimentary basins, which are considered to be filled by successive unconformity-bounded units or depositional sequences. The history of the infill of the sedimentary basin must be understood as the interplay of several related factors such as subsidence, local synsedimentary tectonics, possible eustatic fluctuations of several orders and the sedimentary input and/or carbonate production. Depositional sequences can be further subdivided into systems tracts and parasequences. Predictive models of facies and systems tracts distribution in these sequences have been proposed for both carbonate rimmed-type platforms and carbonate ramps (Haq *et al.*, 1987; Vail *et al.*, 1987; Van Wagoner *et al.*, 1988).

The Iberian Chain, a mountainous system which stretches in a northwest–southeast direction in the northeastern part of the Iberian Peninsula (Fig. 1), contains well-preserved and continuous outcrops of Mesozoic sedimentary rocks. Field-based sequence-

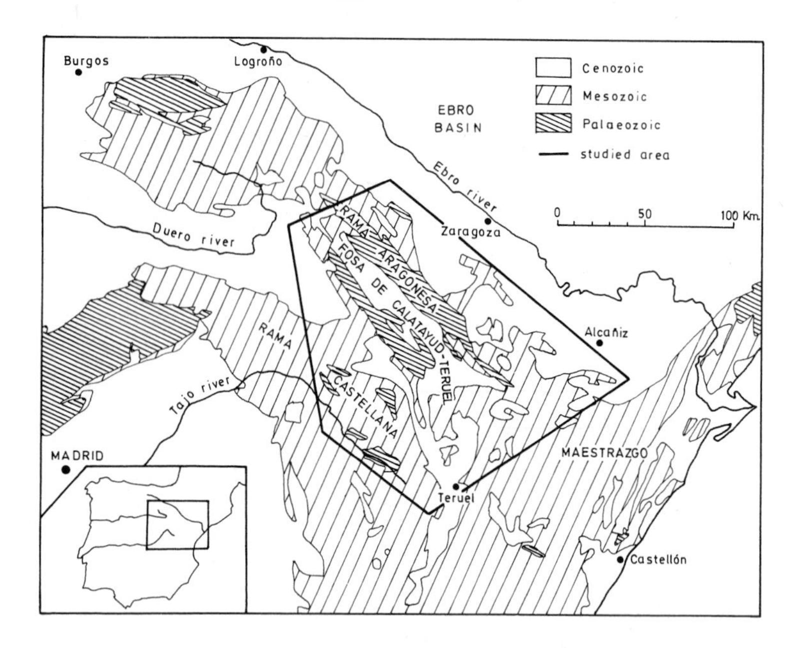

Fig. 1. Geological map of the Iberian Chain.

stratigraphic studies in the Iberian Chain have yielded information on the distribution of depositional sequences and systems tracts in low-angle carbonate ramp settings, e.g. Muschelkalk facies (Middle Triassic) (Calvet *et al.*, 1990) and the successive Upper Jurassic ramps (Salas, 1989; Aurell, 1991). The determination of the thickness and facies distribution in these ramps is favoured by the presence of laterally continuous outcrops, which allowed us to examine several cross-sections to the basin of some hundred kilometres in length, linking the more proximal zones (towards the west) with the more distal ones (towards the east). The Upper Jurassic sequence stratigraphy was placed in a precise biostratigraphic framework, using the data of several previous analyses of the ammonite biozones of the Callovian–Lower Tithonian interval (Bulard, 1972; Goy *et al.*, 1979; Meléndez & Brochwicz-Lewinski, 1983; Meléndez, 1984; Atrops & Meléndez, 1985).

The aims of this paper are: (i) to present a revised stratigraphic description of the Upper Jurassic in the Central Iberian Chain in which the previously defined lithostratigraphic units are incorporated into the new sequence-stratigraphic framework; (ii) to illustrate the eight episodes of sedimentary evolution, which consist of successive genetically related units (i.e. depositional sequences and systems tracts), documenting the facies associations and distribution in a carbonate-dominated and low-angle ramp setting; (iii) to quantify the sedimentation rates and the amount and evolution of the accommodation in the basin, illustrating how the interplay of these two factors results in the observed distribution of facies in the successive systems tracts; (iv) to discuss the participation of eustasy and subsidence creating the accommodation.

METHODS

The results of this study are based on an extensive outcrop analysis, consisting of the measurement of 65 stratigraphic sections. The geographical distribution of these sections is shown in Fig. 2. Field analysis includes the lateral tracing of sedimentary units and bounding surfaces. Major facies types were assessed by interpretation of depositional settings in outcrop and from petrographic studies. Age determinations and correlations for the Callovian–Lower Tithonian interval were made on the basis of standard ammonite biozonations (see also Fig. 6). Biostratigraphic information for the Tithonian–Berriasian interval was obtained using lituolids and charophytes. The time scale used is that of Haq *et al.* (1987).

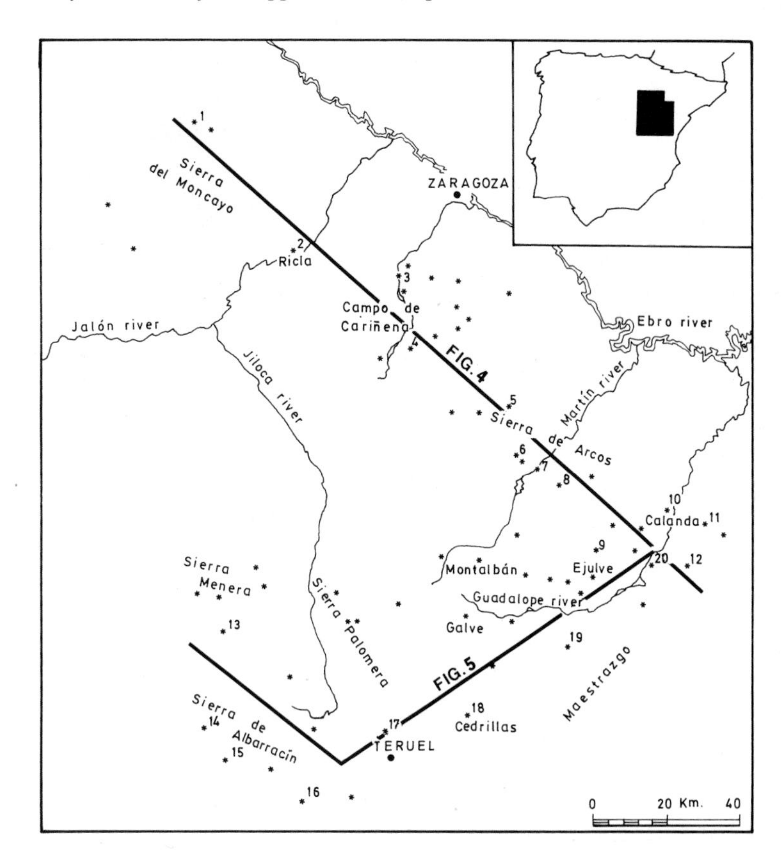

Fig. 2. Location map indicating positions of the studied sections (asterisks) and the geographical terms referred to in the text. Continuous lines show situation of the lower cross-sections in Figs 4 and 5.

PALAEOGEOGRAPHICAL DESCRIPTION

During the Mesozoic, sedimentation in the area east of the so-called Iberian Massif took place in intracontinental basins, which were eventually flooded by shallow epicontinental seas (Fig. 3). The genesis and evolution of these basins were controlled by the tilting of blocks along normal faults (Salas, 1987). The evolution of these Mesozoic basins was characterized by the alternation of periods in which the activity of these basement faults was predominant and more tectonically stable periods, which favoured the presence of more homogeneous and laterally consistent depositional settings.

Our study concentrates on one of the more tectonically stable periods. Sedimentation in the northeast of the Iberian Peninsula occurred in shallow and extensive ramp settings (several hundreds of kilometres) during Upper Jurassic times. The ramps were very homogeneous and possessed gentle slope angles. Phases of local activity of some basement faults occurred only at the Oxfordian–Kimmeridgian and Jurassic–Cretaceous boundaries. These minor order tectonic phases involved the uplift of the marginal areas, a correlative basinwards coastal shift and major changes in the sedimentation on these ramps, namely, a transition from the sponge-dominated Oxfordian ramp to the mud-dominated Kimmeridgian ramp and, later, to the oncoid-dominated Tithonian ramp.

At the onset of the Upper Jurassic a series of palaeogeographic highs are detected, such as the so-called Ejulve high (Bulard, 1972; Gómez, 1979; see Fig. 3). The Ejulve high, which is located to the east of the study area, had a major effect on sedimentation during the Upper Jurassic. This palaeogeographic high, which was developed during the Middle Jurassic, was a passive palaeogeographic element during the Upper Jurassic, when it was progressively covered by marine units deposited during the successive floodings of the basin (especially those sediments linked with the major trans-

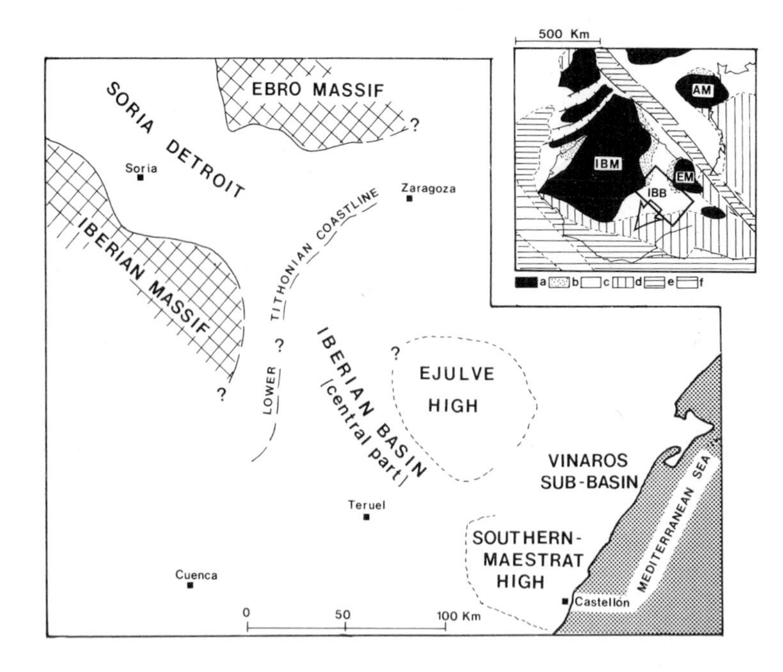

Fig. 3. Middle Upper Jurassic palaeogeographic map of the Iberian Basin. (Compiled from Bulard, 1972 and Gómez, 1979.) Map in the inset displays this palaeogeography for the Iberian Peninsula. (Modified from Ziegler, 1988.) (a) Positive areas (IBM, Iberian Massif; EM, Ebro Massif; AM, Armorican Massif). (b) Deltaic shallow marine, mainly sands. (c) Shallow marine, carbonates and clastics (IBB, Iberian Basin). (d) Shallow marine, mainly carbonates. (e) Deeper marine, clastics or carbonates. (f) Basin floored by oceanic crust.

gressions of the Middle Oxfordian and Lower Kimmeridgian).

STRATIGRAPHIC UNITS

The Upper Jurassic sediments that crop out in the central part of the Iberian Chain have been subdivided previously into several litho- and biostratigraphical units (Bulard, 1972; Gómez & Goy, 1979; Meléndez, 1984). The recent identification and definition of unconformity-bounded units provides a new stratigraphic framework. The subdivision of the Upper Jurassic and Lower Cretaceous of the Iberian Basin into depositional sequences was proposed by Salas (1987, 1989), as a result of the analysis of the strata in the eastern part of the chain (i.e. Vinaros sub-basin and surrounding areas, (Fig. 3)). In this study, the Upper Jurassic was further subdivided into three depositional sequences. These are the so-called J3.1 or Oxfordian sequence, the J3.2 or Kimmeridgian sequence and the J3.3 or Tithonian–Berriasian sequence. These sequences and their bounding unconformities were also recognized in the central part of the Iberian Basin and in the northwestern sector of the basin or Soria Detroit (Aurell, 1990 and Alonso *et al.*, 1989, respectively; see Fig. 3 for location).

The main stratigraphic features of the Upper Jurassic of the Central Iberian Chain are shown through two cross-sections of the basin, from their western or more proximal zones to their eastern or more distal ones. Figure 4 shows the section for the north part of the basin (from the Sierra del Moncayo to El Maestrazgo), whereas Fig. 5 corresponds to the south part (from Sierra Menera to the proximity of Ejulve).

SEDIMENTARY EVOLUTION AND PALAEOENVIRONMENTAL RECONSTRUCTIONS

The evolution of the Upper Jurassic of the central part of the Iberian Basin consisted of eight successive depositional episodes. The six lower episodes correspond to the systems tracts of the Oxfordian and Kimmeridgian sequences. Extensive facies analysis across the basin and precise biostratigraphic dating allowed us to determine the position of the systems tracts within these two basal sequences (Aurell, 1991). The interpreted stratigraphical extent of these systems tracts is illustrated in Fig. 6. Two additional episodes are interpreted to be part of the uppermost Upper Jurassic sequence, or Tithonian–Berriasian sequence, on the basis of a lower unconformity recorded only in the marginal areas of the basin. This middle unconformity must be of uppermost Tithonian age based on the presence of several associations of lituolids.

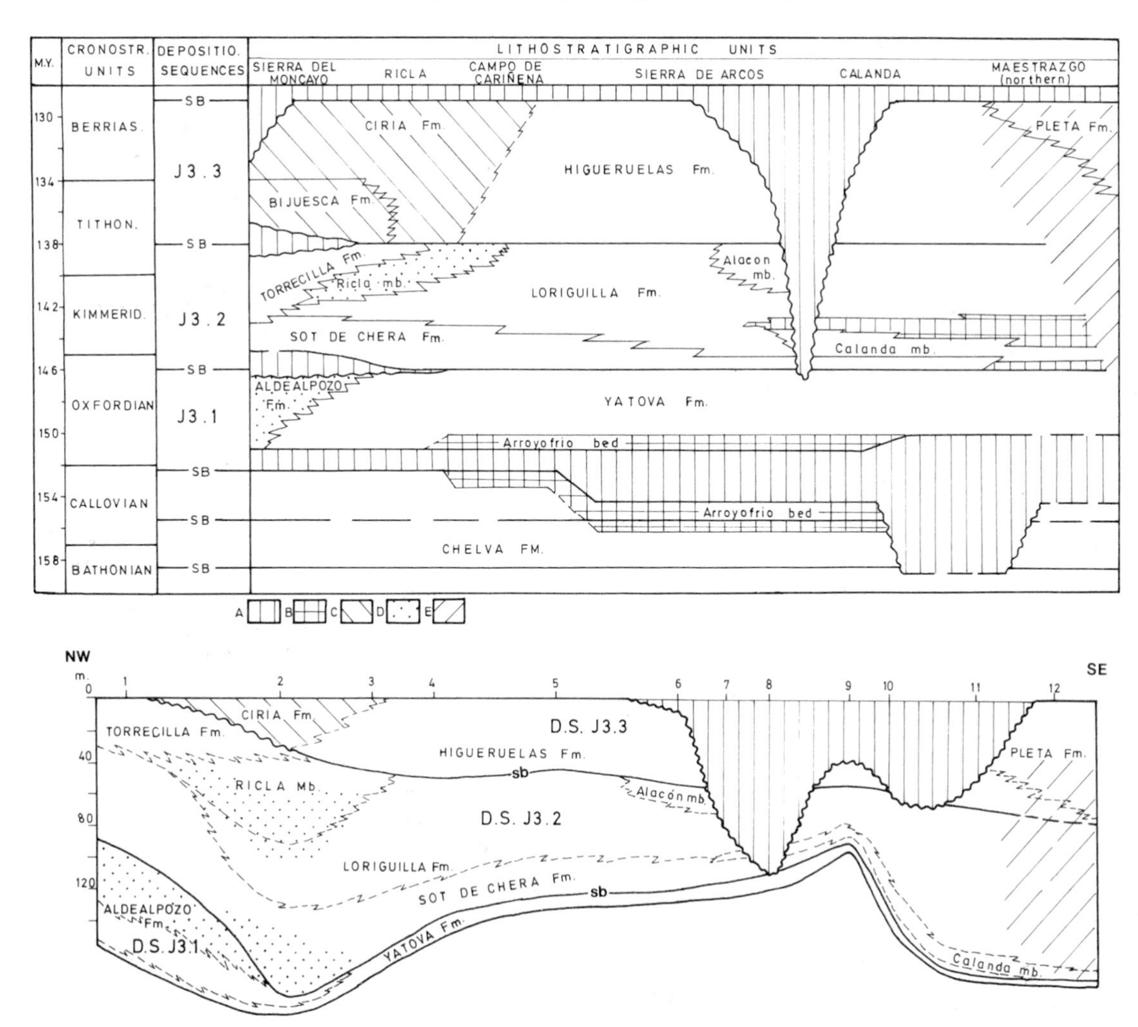

Fig. 4. Stratigraphic cross-sections of the northern part of the basin, between Sierra del Moncayo and El Maestrazgo. Geological time is plotted on the vertical axis in the upper section. This panel shows the extent of the stratigraphic lacunae associated with the unconformities, which are the boundaries of the successive depositional sequences. The lower panel takes into account the thicknesses observed in different sections across the basin (reference numbers correspond to measured sections; see Fig. 2 for location). A, hiatus. B, thickness very reduced. C, continental deposits. D, predominance of siliciclastic facies. E, dolomitized unit.

Below we summarize the observed facies associations and depositional setting for each of these eight episodes, after a brief review of the boundaries of these episodes or systems tracts.

Lower Oxfordian (Lamberti p.p.–Plicatilis biozones)

Boundaries

This basal episode corresponds to the lowstand systems tract of the Oxfordian sequence, which is interpreted to have been deposited from the end of the Callovian (upper part of the Lamberti biozone) to the lower Middle Oxfordian (Plicatilis biozone) (Fig. 6). The base of this episode is the unconformity which marks the lower boundary of the Oxfordian sequence. This unconformity is widespread and has an associated stratigraphic lacuna which comprises at least the last Callovian biozone (Lamberti biozone) and the first Oxfordian biozone (Mariae biozone) (see Figs 4 & 5). The maximum duration

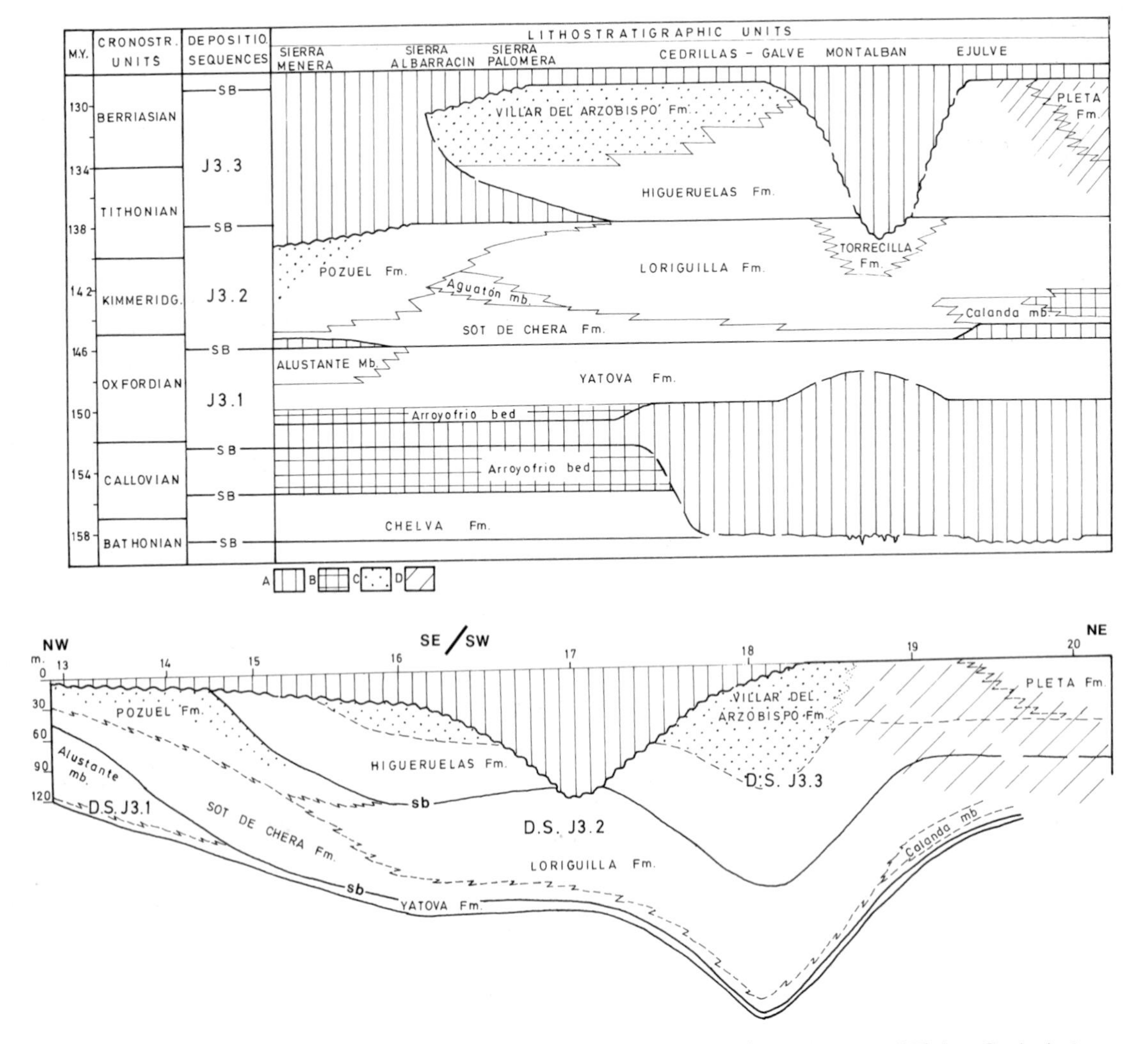

Fig. 5. Stratigraphic cross-sections of the southern part of the basin, between Sierra Menera and Ejulve. Geological time is plotted on the vertical axis in the upper section. This panel shows the extent of the stratigraphic lacunae associated with the unconformities, which are the boundaries of the successive depositional sequences. The lower panel takes into account the thicknesses observed in different sections across the basin (reference numbers correspond to measured sections; see Fig. 2 for location). (A–D, see Fig. 4.)

of this hiatus occurred to the east, where the Middle Oxfordian bioclastic limestones rest directly on Bathonian ooid grainstones through a subaerial karstification surface (Cedrillas). There are some local indications of erosional truncation associated with this surface (Galve, Ejulve) (Fig. 7(a)). An uneven surface found in the Sierra del Moncayo area at the bottom of the Oxfordian sequence is also regarded as evidence for subaerial exposure (Fig. 7(b)).

This basal unconformity is interpreted to have involved the subaerial exposure of the entire basin area studied. This subaerial exposure was not a synchronous event; it occurred in the most easterly zones (under the influence of the Ejulve high) from the Bathonian onwards, whereas it commenced in the west during the Upper Callovian time.

The top of this systems tract is a transgressive surface. Ammonite associations found in these deposits provide evidence that the flooding occurred

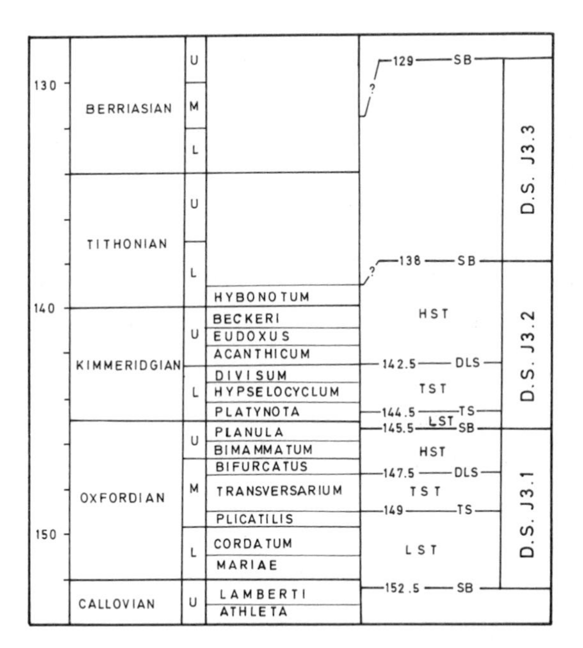

Fig. 6. Stratigraphic units, ammonite biozones and interpreted systems tract distribution in the Oxfordian and Kimmeridgian sequences (i.e. J3.1 and J3.2, respectively). LST, lowstand systems tract. TST, transgressive systems tract. HST, highstand systems tract.

as three distinct episodes separated by a biostratigraphic hiatus (Meléndez *et al.*, 1990). The first flooding occurred in the middle part of the Lower Oxfordian (Cordatum biozone, Claromontanus subzone); the second one occurred at the beginning of the Middle Oxfordian (Plicatilis biozone, Vertebrate subzone); the third one occurred in the middle of the Middle Oxfordian (base of the Transversarium biozone). The first two floodings represent a transition to shallow-marine conditions in the west. The eastern Ejulve high was reached only by the third transgressive pulse, during which the Bathonian karst surface was covered by marine sediments. In addition to flooding this palaeogeographic high, the third transgressive pulse reached other proximal areas in the northwest. This flooding surface was interpreted as the transgressive surface of the Oxfordian sequence because of its widespread occurrence and is therefore regarded as the top of this first episode.

Facies distribution and depositional setting

The sediments representing this episode are very thin (from 0.1 to 0.3 m) or even absent, and correspond to the so-called Arroyofrío bed (see Figs 4 & 5).

Figure 8 summarizes the distribution of the main known facies. Facies A consists of fossiliferous mudstones and wackestones. The fossils are benthic (brachiopods, molluscs, echinoids) and nectoplanktonic (ammonites, belemnites). Facies B and C contain iron ooids. The iron ooids are well sorted and have diameters ranging between 1 to 2 mm in facies B. The ooids float in a micritic matrix (wackestone to packstone textures), or fill the internal fossil shells. The ooids consist of regular and concentric layers, 10 to 35 μm thick, of goethite, with traces of kaolinite and carbonate hydroxyapatite.

The iron particles are poorly sorted in facies C. In addition to the above-described ooids, facies C also contains iron pisoids with diameters of up to 1 to 2 cm and irregular coatings. Marine fossils are relatively scarce in this facies with only occasional brachiopods and ammonites occurring. The cores of these iron ooids and pisoids in both facies are fragments of broken ooids or quartz grains; no skeletal fragments are present, despite their relative abundance in the micritic matrix. These data suggest an allochthonous origin for the iron ooids, particularly for those which are well sorted in facies B. To the southeast (Fig. 8, facies D), the sedimentary record is scarce, and only centimetre-thick iron crusts with a considerable amount of kaolinite are found. The areas of non-deposition are interpreted as having been emergent during this episode, as a consequence of their location on palaeogeographic highs (i.e. Ejulve high).

At the end of the Callovian and during the Lower Oxfordian, the sedimentary setting is believed to have been an extensive and homogeneous ramp, that dipped towards the northeast, with emerged areas to the southeast. Iron ooids and pisoids are interpreted as having formed mainly in subaerial exposed environments (pedogenetic processes: Siehl & Thein, 1978; Nahon *et al.*, 1980) located adjacent to the Ejulve high or perhaps on the high itself. The iron ooids and pisoids were transported towards the centre of the basin, mainly during the above-mentioned flooding episodes. The homogeneity and flat topography of this depositional setting enabled floating ammonite shells to enter during these punctuated highstands.

In addition to the sedimentological evidence discussed above, there is taphonomic evidence for

Fig. 7. (a) The Oxfordian sequence in the Ejulve section. The lower unconformity of the Oxfordian sequence is an even erosional surface (arrow in centre of photograph) resting on Bathonian ooid grainstones (1). The Oxfordian sequence consists of intraclastic and bioclastic limestones (2), and its upper boundary is also an unconformity (arrow at left side). On this surface rests the well-bedded ammonite wackestones (3) that correspond to the bottom of the transgressive systems tract of the Kimmeridgian sequence.

Fig. 7. (b) The lower part of the Oxfordian sequence in the Ricla section. The Upper Callovian sandy limestones (1) are unconformably overlain by the Oxfordian sequence (lower arrow). Above this unconformity are the lowstand systems tract deposits (2), that are in turn overlain by a transgressive surface (flat surface pointed to by the upper arrow). On this surface rests the sponge wackestones, corresponding to the bottom of the transgressive systems tract (3).

this ooid iron level having been deposited during a sea-level lowstand (Meléndez *et al.*, 1990; Aurell *et al.*, 1991). The ammonites, which are generally preserved as inner calcareous moulds, have iron crusts and truncational or erosional surfaces which are interpreted as having formed under extremely shallow conditions (Fernandez López, 1984). Furthermore, evidence of the allochthony of the ammonites is found in this level, namely a mixture of elements of diverse biogeographic origins and the predominance of adult forms, with a larger proportion of macroshells than microshells.

Middle Oxfordian (Transversarium biozone)

Boundaries

The second episode corresponds to the transgressive systems tract of the Oxfordian sequence, and belongs in the Middle Oxfordian Transversarium biozone (Fig. 6). The lower boundary of this episode is the transgressive surface. The transgressive surface was largely colonized by sponges and other benthic groups, giving rise to the lithofacies which are more characteristic of the so-called Yátova Formation (Figs 4 & 5). A hardground with a high

Fig. 7. (c) The Oxfordian sequence in the Veruela section (Sierra del Moncayo). The transgressive systems tract is very reduced in thickness and consists of 2.5 m of sponge and ammonite wackestones (1). The highstand systems tract is much thicker and shows two main lithologies: sandy marls at the bottom (2) and well-cemented sandstones and micro-conglomerates at the top (3). The arrow at the left side points to the boundary between these two systems tracts or maximum flooding surface.

Fig. 7. (d) The Oxfordian and Kimmeridgian sequences in the Pozuel del Campo section (Sierra Menera). In the Oxfordian sequence, the transgressive systems tract consists of 2 m of sponge and ammonite wackestone facies (1), whereas the highstand systems tract consists of 40 m of marl-dominated succession (2), crowned by a sandy, bioclastic bank. On top of this bank is the unconformable boundary (arrow) of the Kimmeridgian sequence. The Kimmeridgian sequence consists of marls at the bottom and massive ooid grainstones at the top (3).

concentration of iron oxides developed within this formation in the middle and distal areas of the basin or at its top in proximal areas of the basin. This hardground rests on a highly condensed interval affecting the top of the Transversarium biozone (Schilli and Rotoides subzones). The reduction in thickness of these subzones increases basinwards (Fontana & Meléndez, 1990). This hardground surface is interpreted as being the maximum flooding surface of the Oxfordian sequence and therefore marks the top of this second episode.

Facies distribution and depositional setting

The horizontal distribution of the main recognized facies is shown in Fig. 9. Facies C, which is the most widespread lithofacies, consists of burrowed wackestones with an abundant and diverse open-marine fauna, including ammonites, sponges, brachiopods, echinoids, belemnites, forams (planktonics and benthics), serpulids, ahermatypic corals and algae. The wackestones are massive to decimetres thick bedded and have local intercalations of fossiliferous

Fig. 7. (e) The boundary between the Kimmeridgian (1) and Tithonian–Berriasian sequences (2) in the Aguilón section. This boundary is a low-relief unconformity surface located between the rhythmic series and the massive oncolitic grainstones.

Fig. 7. (f) The highstand systems tract of the Kimmeridgian sequence in the Ricla section; Decimetric-thick sandy mudstone banks (top of the rhythmic series (1)) are conformably overlain by cross-bedded ooid grainstones and sandstones (2).

marls with fossils. Facies C is a condensed interval with its thickness across the basin ranging from 3 to 6 m. Some decimetric beds that consist of accumulations of sponge debris and have a sharp and planar lower boundary are interpreted as being tempestites. Facies C is interpreted as having been deposited in an open, relatively deep subtidal environment, that was occasionally influenced by storm-induced waves.

Areas to the southeast, which were under the influence of the Ejulve high, consist predominantly of well-sorted intraclastic and bioclastic facies (generally packstones, with debris of sponges, tuberois, echinoids, planktonic and benthic forams, gastropods, serpulids, bryozoans and algae), which are indicative of shallower and more waved conditions (facies D).

Facies A, which was deposited in the northwest, consists of mudstones with gastropods, benthic forams, charophytes and bivalves as well as algal-laminated mudstones and skeletal grainstone facies. Both of the subfacies of facies A are arranged in shallowing-upward metre-thick sequences, interpreted as having been deposited in a lagoon tidal

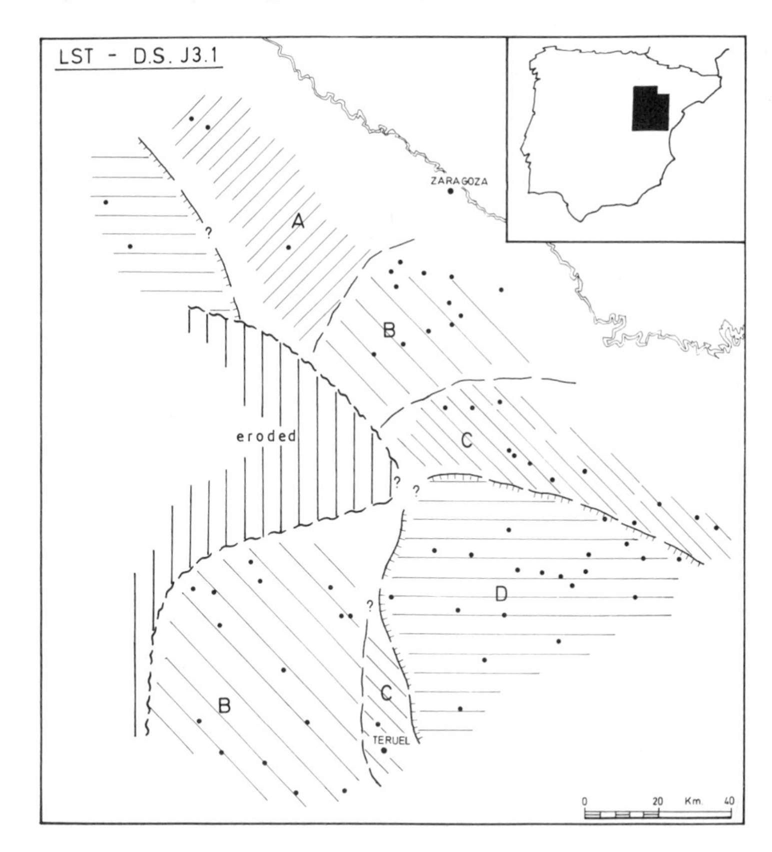

Fig. 8. Lower Oxfordian facies map (i.e. lowstand systems tract, Oxfordian sequence). A, fossil mudstones to wackestones. B, packstone with iron oolites (well sorted). C, wackestones with iron ooids to pisoids (poorly sorted). D, areas with no sediments (interpreted as emerged).

flat environment. Facies B consists of skeletal wackestone facies, with abundant ahermatypic coral, crinoids and sponges. Facies B was deposited in a subtidal setting, with an intermediate depth between that of facies A and C.

In summary, the depositional setting for this second episode was a low-angle carbonate ramp, with shallower areas towards the southeast. Open subtidal conditions were predominant and most of the ramp was colonized by sponges and other open-marine benthic groups. Carbonate tidal flats are present to the northwest, adjoining the emerged areas of the Iberian Meseta.

Upper Oxfordian (Bifurcatus to Planula p.p. biozones)

Boundaries

The third episode developed at the end of the Oxfordian sequence corresponds to its highstand systems tract. This systems tract is interpreted as extending from the upper part of the Middle Oxfordian (Bifurcatus biozone) to the Upper Oxfordian (Bimammatum and Planula biozone) (Fig. 6). The lower boundary or maximum flooding surface is the hardground at the top of the sponge and ammonite facies described above. The upper boundary is the unconformity developed at the end of the Oxfordian sequence. In the Sierra de Albarracín there is a minor unconformity in the upper part of this systems tract, affecting the boundary between the Hypselum and Bimammatum subzones (Corbalán & Meléndez, 1986). This minor unconformity, which is only present in the proximal areas of the basin, occurs below the widespread unconformity at the boundary between the Oxfordian and Kimmeridgian sequences.

Facies distribution and depositional setting

The third episode is the thickest sedimentary interval of the Oxfordian sequence, ranging from up to 60 m thick in the proximal areas to the west (Figs

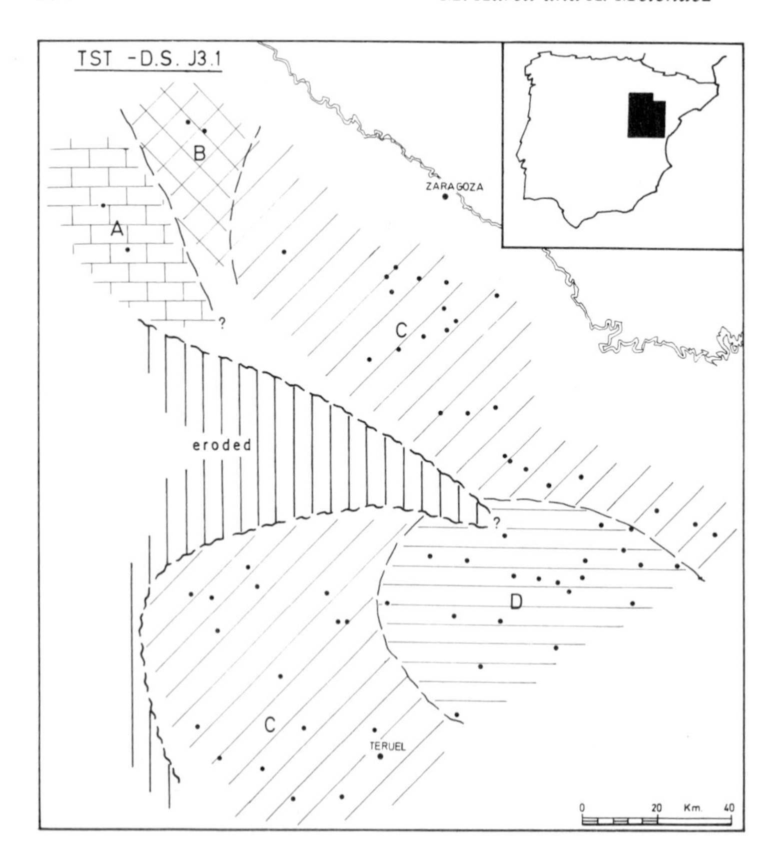

Fig. 9. Middle Oxfordian facies map (i.e. transgressive systems tract, Oxfordian sequence). A, laminated mudstones and ooid grainstones. B, sponge and ahermatypic coral wackestones. C, sponge and ammonite wackestones. D, bioclastic and intraclastic packstones.

7(c) & (d)) to generally less than 5 m thick in the distal areas to the east.

Figure 10 summarizes the distribution of facies in the upper part of this episode. Facies A is similar to that described in the former episode (i.e. mudstones with charophytes, laminated mudstones and skeletal grainstones, deposited in a lagoon to tidal flat environment). These carbonate facies pass laterally into siliciclastic facies (facies B and C). Facies B consists of alternating lithologies of well-cemented sandstone with planar cross-bedding and poorly sorted microconglomerates with channel morphologies and trough cross-lamination. The source for these clastic sediments was the uplift of the northern marginal areas (i.e. Ebro Massif, Fig. 2). Facies B corresponds to a shallow-water high-energy environment, interpreted as forming part of a delta plain complex.

Facies C consists predominantly of sandy marls with plant remains and fossils, including ahermatypic corals, crinoids, ammonites, brachiopods, benthic forams and gastropods. The total thickness of facies C ranges from 20 to 60 m. The marls of facies C are located basinwards of the more coarse siliciclastic facies B and are interpreted to be the distal part of the delta complex or prodelta marls.

A comparatively thin interval of rhythmically alternating marls and limestones (facies E) was deposited in the centre of the basin. The limestones, which are arranged in decimetric and continuous beds, consist of skeletal wackestones (sponges, ammonites, echinoids, crinoids, algae, benthic forams) and peloidal packstone facies. The marls may contain glauconite (the dashed line in Fig. 10 shows the extent of these glauconitic facies). The skeletal packstone facies described in the former episode (facies F in Fig. 10) was deposited in the eastern areas of the basin which were still under the influence of the Ejulve high.

This episode shows a progradational arrangement of facies. The deltaic facies B and C pass laterally into carbonate tidal flats (facies A) in proximal (western) areas that were located out of the influence of the clastic input. The delta system progrades

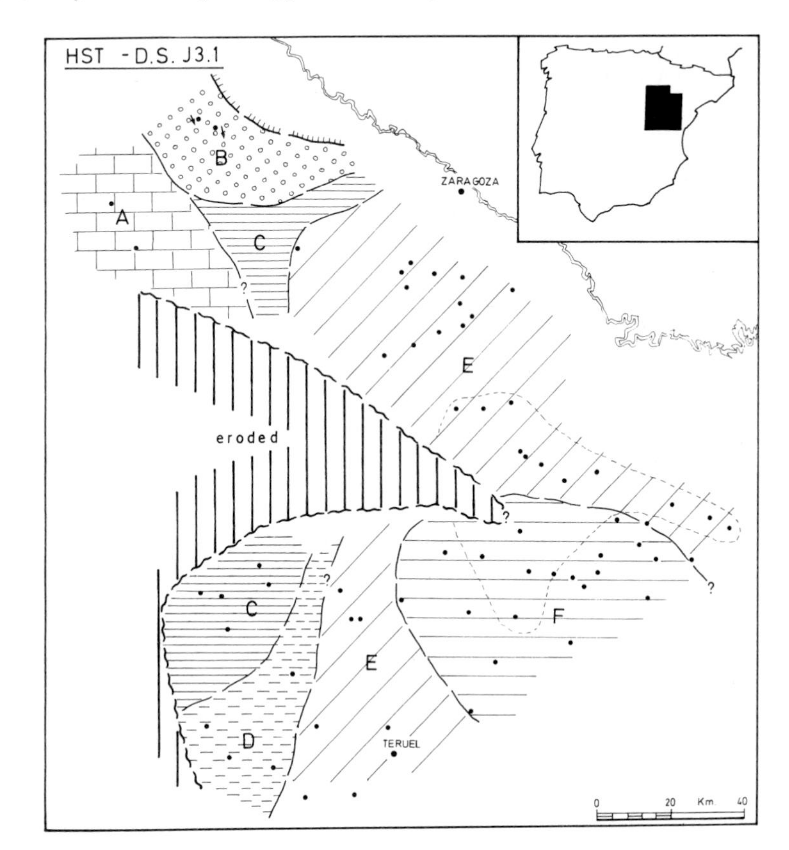

Fig. 10. Upper Oxfordian facies map (highstand systems tract, Oxfordian sequence). A, laminated mudstones and ooid grainstones. B, cross-bedded sandstones and microconglomerates. C, D, sandy marls with benthic fossils. E, alternation of marls and sponge and ammonite wackestones. F, skeletal packstones. The broken line (at east) shows the areas with glauconite-bearing facies.

onto the deeper water and distal condensed facies, which contain an open-marine fauna (e.g. sponges and ammonites) and locally glauconite (facies E). The maximum thickness reduction is towards the southeast, where the influence of the Ejulve high persists as indicated by the geographical distribution of the intraclastic and bioclastic facies (facies F).

Uppermost Oxfordian to lowermost Kimmeridgian (Planula p.p. to Platynota p.p. biozones)

Boundaries

This episode is a lowstand systems tract at the base of the Kimmeridgian sequence. This episode includes strata ranging in biostratigraphic age from the top of the Planula biozone (Galar subzone) to the base of the Platynota biozone (Orthosphinctes subzone) (Fig. 6).

The base of this episode is the unconformity at the boundary between the Oxfordian and Kimmeridgian sequences. A stratigraphic lacuna is associated with this unconformity which affects at least the boundary between the Galar and Planula subzones (Planula biozone, Upper Oxfordian). This lacuna increases landward where the entire Planula subzone may be absent (between Ricla and the Sierra de Arcos (Fig. 4)). Karstic surfaces occur at the top of the Oxfordian sequence in the most marginal areas of the basin at the uppermost Oxfordian indicating subaerial exposure (Sierra de la Demanda, Alonso & Mas, 1990).

The uppermost Oxfordian unconformity changes basinwards into an iron-rich planar surface. An ammonite wackestone facies of basal Kimmeridgian (Platynota biozone, Desmoides subzone) rests on this unconformity in the more distal areas of the basin (Fig. 7(a)). The ammonite wackestone facies marks the recovery of the open-marine conditions in the ramp. The surface below this ammonite wackestone facies is interpreted to be the transgressive surface of the Kimmeridgian sequence, and therefore marks the top of this lowstand of sea level episode. The sequence boundary and transgressive surface may merge at the edges of the basin, where lowstand sediments may be absent.

Facies distribution and depositional setting

Figure 11 shows the geographical extent of the main recognized facies in this episode. As mentioned above, these deposits are absent towards the edges of the basin (A in Fig. 11). The strata corresponding to this episode consist mainly of marly lithofacies. The main differences between the distinguished facies B, C and D are their total thicknesses and their coarse siliciclastic material concentrations. The marls are thickest (40 to 50 m) towards the west, where marls with a high concentration of mica and plant remains and scattered fossils (ostracods, molluscs) are the main lithology (facies B). Some intercalated decimetric-thick sandstones occur in this western area. Sandstone beds are normal-graded and have low-angle planar cross-lamination at their base and even lamination and current ripples towards their top. The siliciclastic beds are absent to the east (i.e. facies C and D). In these eastern areas, the facies consist of 10 to 20 m marls with mica and plant remains in the intermediate areas of the basin (facies C) and less than 5 m marls in the east (facies D).

The depositional setting of this lowstand episode is interpreted to be a shallow water low-angle ramp of great lateral extent (more than 100 km). The lithologies and fossil assemblages deposited on this episode suggest a restricted and shallow water environment. The low-energy and partially anoxic conditions are interpreted to have been produced by the damping of the waves and currents across this shallow ramp. Decimetric-thick graded beds on the margin of the ramp are interpreted to be flood deposits that originated in the emerged areas and spilled over the adjoining marine areas (e.g. inundites, see Seilacher, 1985). Only the silty and clay-sized detritic material and some larger size components, such as plant remains and micaceous

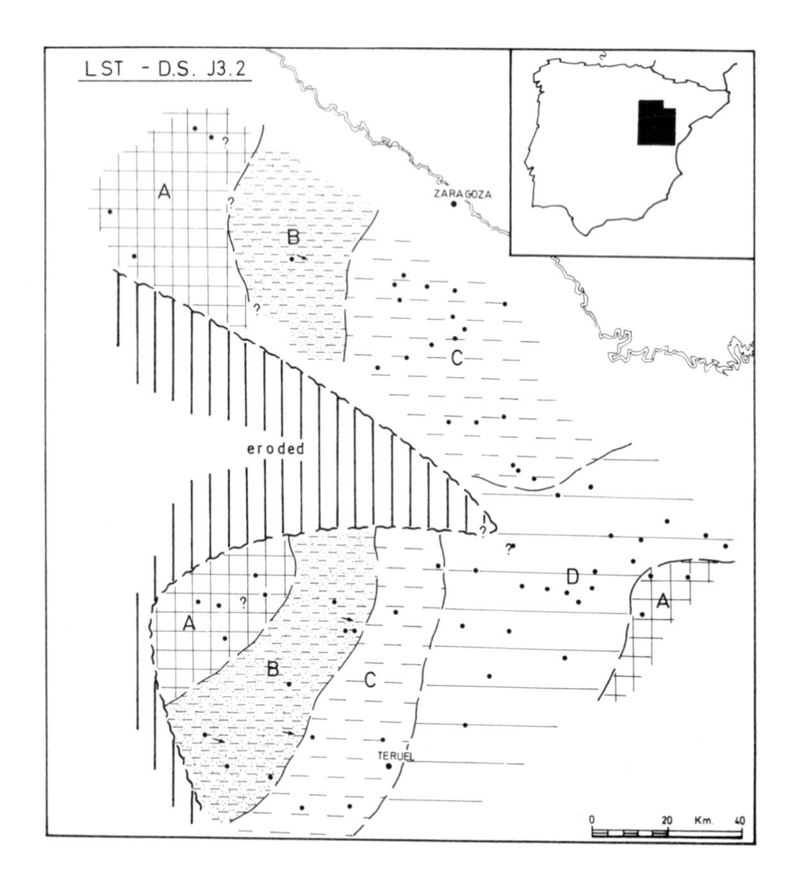

Fig. 11. Uppermost Oxfordian to lowermost Kimmeridgian facies map (lowstand systems tract, Kimmeridgian sequence). A, unit absent. B, marls with interbedded sandstones. C, marls with plant remains and scattered benthic fossils (15 to 20 m thick). D, as facies C, but reduced in thickness (2 to 5 m thick).

minerals, were deposited via suspension in the low-energy subtidal plain of the more distal areas.

Lower Kimmeridgian

Boundaries

This episode corresponds to the transgressive systems tract of the Kimmeridgian sequence and encompasses the Lower Kimmeridgian, from the middle Platynota biozone (i.e. Desmoides subzone) to the top of the Divisum biozone (Fig. 6).

The lower boundary of this episode is marked by the first significant marine flooding across the basin or transgressive surface. The initial transgressive deposits belong to the middle part of the Platynota biozone. The top of this episode is the maximum flooding surface, which in the distal parts of the ramp is a hardground developed on top of a sedimentary condensed facies (i.e. ammonite wackestones of the Calanda Mbr., see Figs 4 & 5).

Facies distribution and depositional setting

The facies distribution during the Lower Kimmeridgian episode is illustrated in Fig. 12. In contrast with the previous lowstand episode, marine sediments were deposited on the more proximal parts of the basin, as well as the more shallow remains of the Ejulve high, which became submerged during the Lower Kimmeridgian.

Coraline boundstone, arranged in stacked coral biostromes (facies A), was found in the more proximal areas, to the northwest. Facies A is interpreted to have been deposited in littoral and sublittoral settings (see also Alonso *et al.*, 1986). Oolitic grainstones deposited in similar environments (facies B) are the main deposits towards the south. Facies B consists of well-sorted ooids, approximately 2 mm in diameter, and marine fossils including benthic forams, coral debris, echinoids, red and green algae, bryozoans and gastropods.

A zone of siliciclastic influence (facies E, F) is

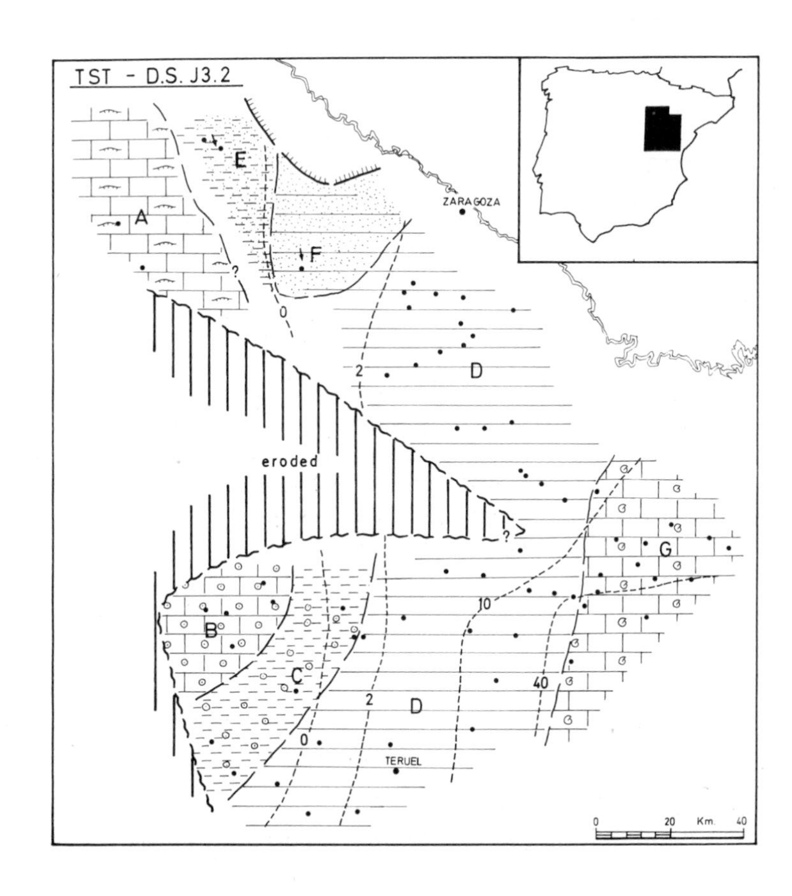

Fig. 12. Lower Kimmeridgian facies map (transgressive systems tract, Kimmeridgian sequence). A, coraline boundstones. B, ooid grainstones. C, marls with calcareous ooids. D, mudstone and marls (rhythmic series). E, sandy marls. F, sandy mudstones and marls. G, ammonite wackestones. The broken lines show ratio between thickness of the rhythmic series (i.e. facies D and F) and marly facies (i.e. facies E and C).

present towards the north. These facies consist mainly of sandy marls with benthic fossils (molluscs, crinoids, benthic forams), and intercalated beds of sandstone and sandy mudstone that contain scattered benthic fossils and display current structures, such as asymmetric ripples and low-angle cross-lamination. These facies were deposited in a shallow subtidal environment, with the clastic input interpreted as having been derived from the northern emerged zones corresponding to the Ebro Massif.

A rhythmic alternation of marls and decimetre-thick, even-bedded micritic facies (mudstones with silty grains and very scarce fossils such as brachiopods, molluscs, ostracods and crinoids (facies D)) is the predominant deposit in the middle part of the basin. The mudstones display a very consistent *Chondrites* bioturbation. Graded bioclastic or fine-grained siliciclastic layers in the mudstones, interpreted to be tempestites, are locally present. This rhythmic series passes landward into sandy marls with plant remains and scattered benthic fossils. The overlapping of the rhythmic series present in the middle of the ramp (i.e. Loriguilla Formation) over the more proximal, marly facies (i.e. Sot de Chera Formation) during the Lower Kimmeridgian marks a retrogradational facies arrangement (see Figs 4 & 5). This point is also illustrated in Fig. 12, which shows a series of broken lines joining points at which the ratio of the thicknesses of the marly facies (to the bottom) and rhythmic series (to the top) is constant. These values increase progressively basinwards, where the marly facies have almost disappeared.

The Lower Kimmeridgian sediments are relatively thin (10 to 15 m) in the more distal zones of the ramp to the southeast (facies G). Facies G consists of marls and wackestones with ammonites as well as benthic fossils (brachiopods, echinoids, molluscs, forams and serpulids). Several hardground horizons are present on these limestone beds. The facies is interpreted as having been deposited under distal open-marine conditions.

To summarize, the depositional setting during the Lower Kimmeridgian was a low-slope angle ramp, with the more open, deeper water and distal areas located to the east, where the ammonite facies occur and there is a maximum reduction in thickness. Reef and oolitic banks prevailed in the littoral and proximal areas. Most of the ramp was dominated by a muddy and marly facies, deposited in a shallow and partly restricted environment. As in the previous episode, the restricted conditions of the middle part of the ramp are related to the damping of waves and currents, the consequence of the geometry of this shallow, low-angle and extensive ramp.

Upper Kimmeridgian–lowermost Tithonian (Hibonotum biozone?)

Boundaries

This sixth episode considered in the evolution of the Upper Jurassic is the highstand systems tract at the top of the Kimmeridgian sequence. This systems tract is interpreted to be of Upper Kimmeridgian to lowermost Tithonian age (Fig. 6). The lower boundary of this episode is the maximum flooding surface of the Kimmeridgian sequence.

Its upper boundary is the upper sequence boundary. The upper sequence boundary is an unconformity in the western proximal areas, where continental deposits (i.e. palustrine or lacustrine facies of the Ciria and Bijuesca Formations) locally overlie the shallow-marine facies of the Upper Kimmeridgian sequence (see Ricla–Sierra del Moncayo region, Fig. 4). The boundary between the two sequences in the middle areas of the basin is a planar to low relief unconformity surface, with carbonate crusts and concentrations of iron oxides, developed on top of the well-bedded, middle-ramp micritic facies (e.g. Loriguilla Formation). This unconformity is overlain by the massive oncolitic facies of the Higueruelas Formation (Fig. 7(e)). This unconformity grades basinwards into a correlative conformity; in distal areas there is a gradational change between the rhythmic series and the massive oncoid limestones, consisting of an intermediate set of shallowing-upward parasequences, with micritic limestone towards the bottom and limestone with oncoids towards the top.

Facies distribution and depositional setting

The horizontal distribution of facies at the onset of this episode is illustrated in Fig. 13. The facies located in the proximal or western areas of the basin are, like in the former episode, coraline boundstones and ooid grainstones (facies A and B), deposited in littoral to sublittoral settings. Facies C also consists of ooid grainstones, but also contains intercalated planar cross-bedded sandstones and trough cross-bedded microconglomerates. In con-

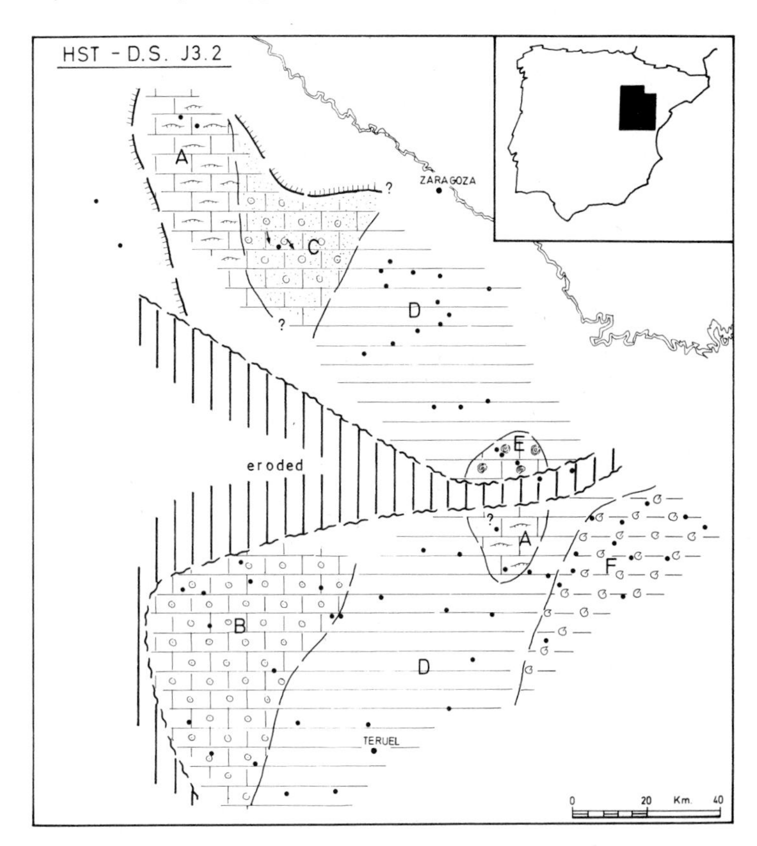

Fig. 13. Upper Kimmeridgian facies map (highstand systems tract, Kimmeridgian sequence). A, coraline boundstones. B, ooid grainstones. C, ooid grainstones and cross-bedded sandstones. D, mudstone and marls (rhythmic series). E, peloidal and oncolitic packstones. F, marls with ammonites.

trast with the former episode, the basinwards extension of these marginal facies is greater.

As in the Lower Kimmeridgian, a rhythmic alternation of mudstones and marls was deposited in most of the middle part of the Kimmeridgian ramp during the Upper Kimmeridgian (Fig. 13, facies D). The mudstones contain scattered benthic fossils, including molluscs, ostracods, benthic forams, ahermatypic corals, sponges, echinoids, red and green algae, serpulids and crinoids. These mudstones in the marginal areas of the ramp contain sand-sized siliciclastic grains, and are overlain by the cross-bedded ooid grainstones and sandstones of facies C (Fig. 7(f)). Shallower water and wave-agitated facies (i.e. coral boundstones (facies A) and peloid and oncoid packstones and wackestones (facies E)) occurred locally in the centre of the ramp. The presence of higher energy deposits may be related to a sedimentary high.

Marls containing ammonites and benthic fossils (molluscs, forams and brachiopods) and interbedded ammonite wackestones were deposited to the eastern part of the basin at the onset of the Upper Kimmeridgian (i.e. Acanthicum biozone) (facies F). These marly facies are covered by the benthic fossil-poor restricted micritic facies (D).

The depositional setting during the Upper Kimmeridgian was a low-slope angle carbonate ramp similar to that of the former episode. Unlike the Lower Kimmeridgian, however, the facies distribution reveals a prograding carbonate system. The marginal oolitic and coraline facies belt prograded over the mudstones that covered the middle part of the ramp. These facies prograded in turn over the marls with ammonite facies, that are in the most distal areas.

Tithonian p.p.

Boundaries

This episode corresponds with the so-called lower episode of the Tithonian–Berriasian sequence. In contrast to the two previous sequences, it is difficult to determine precisely the distribution of systems tracts within this uppermost sequence. This is due mainly to three reasons: (i) unlike the preceding sequences, the study area does not reach the most open-marine facies belt, which might enable us to be precise about the stratigraphic position and definition of the intermediate marine flooding surfaces as either transgressive or maximum flooding surfaces; (ii) the vertical facies distribution generally reveals a single, general progradational pattern (e.g. Aurell & Meléndez, 1987), so that facies evolution criteria cannot be used to characterize the systems tracts (only the highstand systems tract is represented in this proximal part of the basin); evidence of transgressive deposits at the onset of the sequence is found only locally, such as in the Sierra de Albarracín (Giner & Barnolas, 1979; Aurell, 1990); (iii) the deposition of this unit is linked to tectonic activity (i.e. uplift of the western marginal areas (Alonso *et al.*, 1986; Aurell, 1990)), which makes this sequence offlap and is probably the main cause for the observed progradational pattern.

Nevertheless two episodes, bounded by a minor unconformity (detected only in the proximal areas of the basin), are considered in the Tithonian–Berriasian sequence. The lower episode, which in the main corresponds to the Higueruelas Formation, is mostly Tithonian in age. The upper episode extends from the Upper Tithonian (?) to the Middle Berriasian and consists of predominantly siliciclastic facies to the west (Villar del Arzobispo Formation) and carbonate facies to the east (La Pleta Formation).

Facies distribution and depositional setting

The onset of the Tithonian–Berriasian sequence is characterized by a large basinwards coastal shift. In contrast with the former sequence, the marine deposits of this sequence are not found in the western proximal zones of the basin.

Three generic facies occur in the basal Tithonian episode, the distribution of which at the onset of this episode is illustrated in Fig. 14. The more widespread facies A consists of massive limestones

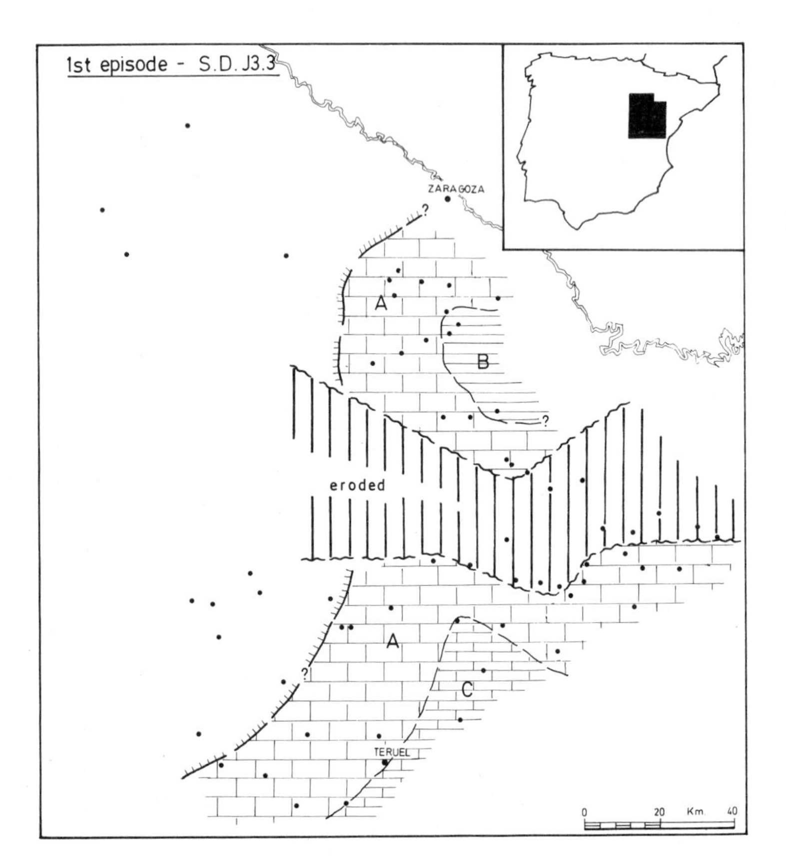

Fig. 14. Tithonian p.p. facies map (i.e. episode 1 of Tithonian–Berriasian sequence). A, oncolitic and peloidal grainstones. B, mudstones and wackestones with fossils. C, wackestones with oncoids and fossils.

with a variety of grain-supported lithologies, such as oncoid and skeletal rudstones (molluscs, hermatypic corals, benthic forams, echinoids), peloidal grainstones and packstones, well-sorted ooid grainstones and packstones with ostracods and lituolids and coraline boundstones (i.e. scattered metre-thick patch reef). Facies B and C are dominated, respectively, by skeletal wackestones and mudstones (molluscs, hermatypic corals, benthic forams and echinoids) and oncoid floatstones. These two facies form the base of this episode in the eastern areas, where they are overlain by the grain-supported lithologies of facies A.

The depositional setting is interpreted to be a shallow carbonate ramp with a very slight slope, open to the east. The vertical succession reveals a shallowing-upward arrangement of facies. The low-energy micritic facies to the east are directly overlain by oncolitic and peloidal shoals, with scattered coral patch reef, which were deposited in a shallow water wave-agitated environment. The oncolitic and peloidal facies are, in turn, overlain by well-sorted ooid grainstones, which were deposited in littoral and sublittoral environments. Packstones with ostracods and lituolids facies occur at the top of this shallowing-upward succession and are interpreted as deposited in restricted lagoons.

Lower to Middle Berriasian

Boundaries

The last episode in the evolution of the Upper Jurassic of the Iberian Basin is the so-called second episode of the Tithonian–Berriasian sequence, which is uppermost Tithonian to Lower–Middle Berriasian in age. The upper boundary of this episode is an important unconformity which marks the end of the so-called 'Jurassic supersequence' (Salas, 1989). This upper boundary is usually an angular unconformity with a stratigraphic lacuna which involves at least all the Valanginian and is also marked by the overlapping of continental units onto the shallow-marine facies of the upper Upper Jurassic sequence.

Facies distribution and depositional setting

The interpreted position of the coastline at the onset of the Lower to Middle Berriasian episode is illustrated in Fig. 15. Continental facies consisting of marls with molluscs and charophytes (facies A), deposited in shallow lacustrine and palustrine environments, occur in the emerged zones (Martín i Closas, 1990).

The geographical limits of this study include only the most shallow water and marginal marine facies belt of this episode. Mixed clastic and carbonate lithologies were deposited in the west (facies D). The sandstones are often well bedded, and display such structures as wave ripples, herring-bone and low-angle cross-lamination and planar lamination. The carbonate facies consists of wackestones and packstones with molluscs (ostreids, gastropods), benthic forams (lituolids, miliolids), charophytes and ostracods. These carbonate and siliciclastic facies were deposited in tidal flats and restricted to lagoonal depositional environments. The sandstones also may form lenticular-shaped bodies of up to 10 m thick with channels and trough cross-bedding. Siliciclastic lenticular bodies fill tidal channels and form part of the bars of a delta complex (see Mas *et al.*, 1984; Diaz *et al.*, 1985).

The most broad elemental parasequence in the east is metric in thickness (facies B) and contains the transition from bioturbated mudstones–wackestones of low-energy subtidal environments to laminated mudstones deposited in carbonate tidal flats. This upper facies is occasionally capped by subaerial exposure surfaces. The peritidal facies progrades onto the lagoon biomicritic facies which predominates the central sector, beyond the siliciclastic influence (facies C).

THE AMOUNT OF ACCOMMODATION

The sedimentary evolution of the Iberian Basin during the Upper Jurassic has been described through eight successive episodes, which correspond to genetically related sedimentary units. The measured sections provide a control of the facies and thickness distribution of each of these units across the basin, from the proximal to the distal areas. This control is not possible, however, in the last episode, because of the limited geographical extent of our study. The other seven episodes are composed of a set of facies belts, each facies belt having a similar lithology and thickness distribution and spreading across the basin in a roughly north–south direction. This homogeneity allows us to select some representative data in order to estimate the amount of accommodation created during the Upper Jurassic in the Iberian Basin.

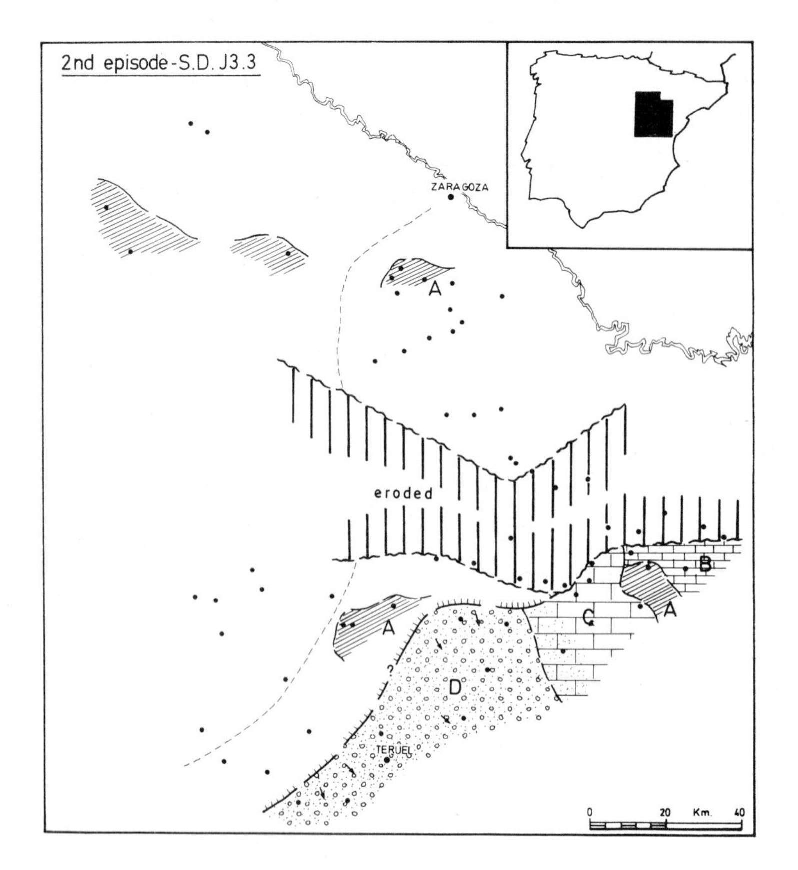

Fig. 15. Lower to Middle Berriasian facies map (i.e. episode 2 of Tithonian–Berriasian sequence). A, marls with charophytes. B, laminated mudstones and fossil wackestones. C, wackestones with fossils and peloidal grainstones. D, wackestones with fossils, marls and cross-bedded sandstones.

A synthetic cross-section of the basin is illustrated in Fig. 16, which is based on the average thickness values for each episode. The table included in Fig. 16 gives for each episode a set of figures across the proximal (A), middle (B) and distal areas (C) of the basin.

In order to obtain a more accurate idea of the amount of sediment (and therefore a more realistic idea of the amount of accommodation created in the basin) it is necessary to consider the effects of compaction. The observed thicknesses were decompacted by considering the percentage of water per volume in the original sediment, according to the predominant lithology (data from Bond & Kominz, 1984).

Average sedimentation rates were calculated by dividing the decompacted sediment thickness by the duration of each episode. Our calculated sedimentation rates correspond with the rates in modern and some ancient sedimentary deposits (e.g. Galloway, 1989). Carbonate sedimentation rates on the Iberian Basin ramps ranged from 0.3 to 4 cm/1000 years (from the marine condensed sections of the Middle Oxfordian and the Lower Kimmeridgian to the rhythmic series deposited in the middle ramp during the Kimmeridgian). The predominantly siliciclastic sedimentary systems, which developed in the middle part of the ramp during lowstands or which prograded into the marginal areas to form delta systems during highstands, have maximum rates of 2 to 6 cm/1000 years.

The accommodation is the space made available for potential sediment accumulation and is defined as the sum of the thickness of the sediments plus the water depth in which they were deposited. The average water depths shown in Fig. 16 were based on the available sedimentological data. In order to calculate the new accommodation created in each episode it is necessary to consider the possible accommodation inherited from former episodes (i.e. water depth at the onset of the episode). The accommodation rates for each episode were determined by dividing the amount of new accommodation created by the duration of the episode.

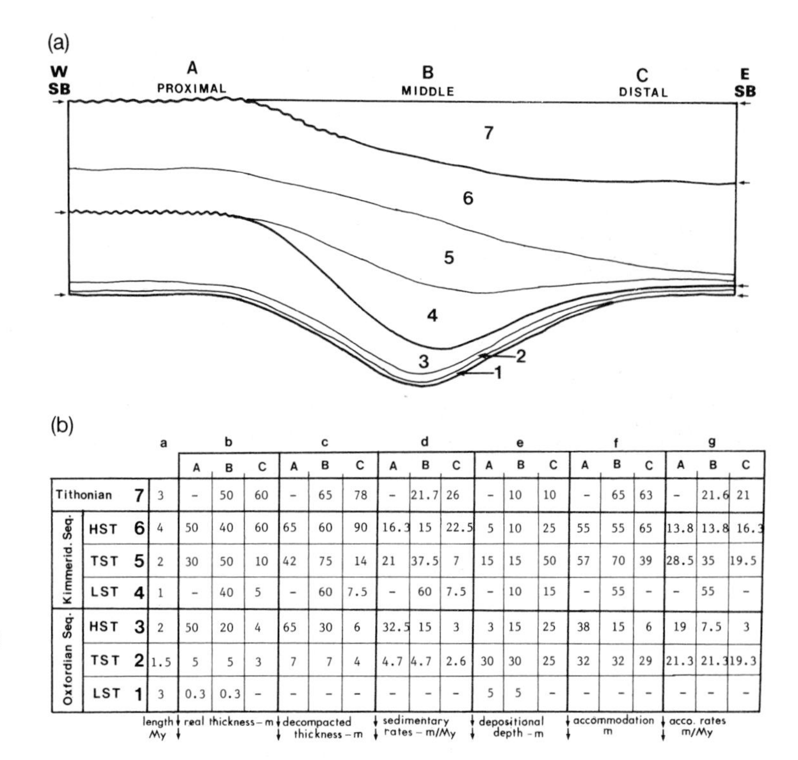

		a	b			c			d			e			f			g		
			A	B	C	A	B	C	A	B	C	A	B	C	A	B	C	A	B	C
Tithonian	7	3	–	50	60	–	65	78	–	21.7	26	–	10	10	–	65	63	–	21.6	21
Kimmerid. Seq.	HST 6	4	50	40	60	65	60	90	16.3	15	22.5	5	10	25	55	55	65	13.8	13.8	16.3
	TST 5	2	30	50	10	42	75	14	21	37.5	7	15	15	50	57	70	39	28.5	35	19.5
	LST 4	1	–	40	5	–	60	7.5	–	60	7.5	–	10	15	–	55	–	–	55	–
Oxfordian Seq.	HST 3	2	50	20	4	65	30	6	32.5	15	3	3	15	25	38	15	6	19	7.5	3
	TST 2	1.5	5	5	3	7	7	4	4.7	4.7	2.6	30	30	25	32	32	29	21.3	21.3	19.3
	LST 1	3	0.3	0.3	–	–	–	–	–	–	–	5	5	–	–	–	–	–	–	–
		length My	real thickness – m			decompacted thickness – m			sedimentary rates – m/My			depositional depth – m			accommodation m			acco. rates m/My		

Fig. 16. Synthetic cross-section showing the Upper Jurassic systems tracts in the Iberian Basin (not to scale). The table gives some data from these systems tracts through the proximal (A), middle (B) and distal (C) areas of the basin.

DISCUSSION

The observed facies distribution in systems tracts

Systems tracts are considered here as descriptive entities and were defined on the character of their bounding surfaces, their position within a sequence and their facies distribution patterns (Aurell, 1991). Systems tracts are not necessarily interpreted in terms of eustatic fluctuations.

The facies distribution in systems tracts is determined by the accommodation rates and the rates of sedimentation and/or carbonate production. Since excess sediment is transported basinwards, progradational systems would occur when sedimentation rates exceed the accommodation rates, as occurred in the Iberian Basin during the phases interpreted as stillstand of sea level (highstand or lowstand deposits: compare the rates in sections (d) and (g) in the table (Fig. 16)). Retrogradational or transgressive systems would be a consequence of fast rates of accommodation, which could produce moderate sedimentary starvation, at least in the distal areas of the ramp (in the Iberian Basin, sedimentation rates in these condensed sections range between 0.2 and 0.7 cm/1000 years), as well as a correlative landwards shift of the open marine facies belt. It should be noted that there is a sharp increase in the rates of accommodation across the entire basin during the episodes interpreted to be transgressive systems tracts (Fig. 17). Subsequent episodes, interpreted to be highstand systems tracts, show a general decrease in their accommodation rates.

The irregular distribution of the accommodation in episode 4 or lowstand systems tract of the Kimmeridgian sequence (Fig. 17), where large and local accommodation produced a sedimentary furrow in the middle areas of the basin, is related to the tectonic reactivation of some basement faults by the end of the Oxfordian (Salas, 1987; Aurell & Meléndez, 1989). The basinwards coastal shift detected at the onset of the Tithonian–Berriasian sequence may be also related with another phase of tectonic activity that involved the uplift of the marginal areas of the basin.

Eustasy versus subsidence creating the accommodation

The accommodation depends on the rates of

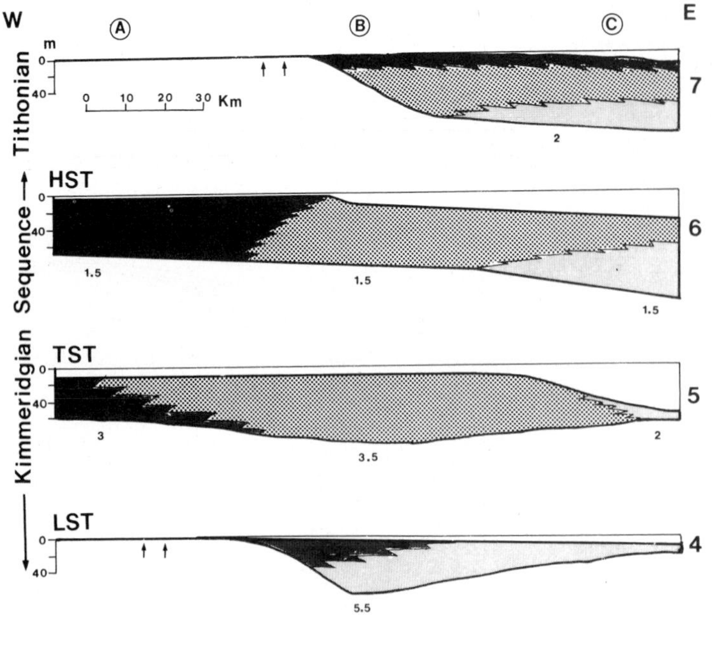

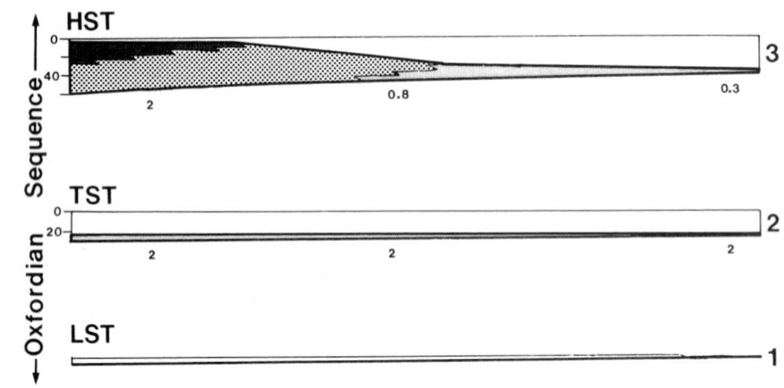

Fig. 17. Cross-sections showing the evolution of the Upper Jurassic in the Iberian Basin through seven successive episodes (constructed from the data of sediment thickness and water depth in Fig. 16). Data in the lower part of the sections indicate the average accommodation rates in cm/1000 years (simplified from section (g) in table, Fig. 16). Vertical exaggeration × 200.

eustasy and subsidence. In practice it is impossible to disentangle the effect of each process. It is possible, however, to obtain reasonable geological models to explain basin history (Kendall *et al.*, 1990). The models used in sequence stratigraphy are based on the study of passive margins, where subsidence is assumed to have a simple and continuous history across the basin. Variations in the creation of the accommodation would, therefore, be a consequence of eustatic fluctuations, which would result in cycles of varying magnitude or order (Vail *et al.*, 1987; Van Wagoner *et al.*, 1988). The subsidence history in intracratonic basins, such as the Iberian Basin, can be complex, with the subsidence rates varying over both time and space. We cannot, therefore, assume in our study a direct relationship between eustasy and depositional sequences and systems tracts as is usually proposed in the sequence-stratigraphy concepts.

Some data are available on the evolution of the subsidence during the Upper Jurassic in the Iberian Basin. The calculated figures by Salas (1987) for the subsidence during the Oxfordian, Kimmeridgian and Tithonian–Berriasian sequences in the eastern Iberian Chain (i.e. Vinaros sub-basin (Fig. 3)), where the sedimentary record is thicker and more continuous, are, respectively, 0.2, 4.3 and 2.2 cm/1000 years (the author used the backstripping techniques described in Bond & Kominz, 1984). Using this data, we may expect a major role of eustasy creating the accommodation in the Oxfordian sequence, when the subsidence was much lower.

The stratigraphic position of the unconformities and maximum flooding surfaces recognized in the Central Iberian Chain during the Upper Jurassic is illustrated in Fig. 18(A). Besides the basin-wide unconformities that define the sequence boundaries, we also consider three minor unconformities within the Oxfordian sequence. The two lower ones correspond to the boundaries of two punctuated highstands that occurred during the lowstand systems tract (i.e. lower Cordatum and lower Plicatilis biozones). The upper one is a minor unconformity at the top of the highstand systems tract (intra-Bimammatum biozone). The relative change of coastal onlap between the successive depositional sequences is also shown in Fig. 18(A). This onlap curve was deduced from the relative expansion towards the edges of the basin of the marine facies during the depositional sequences and systems tracts (see Figs 8–15).

The geographical extent of the unconformities recognized in the Iberian Chain may be tested by comparing them with unconformities recognized in other basins or proposed to be global in origin. Our results are also compared with the global eustatic curves proposed by Haq *et al.* (1987) and Hallam (1988) in Fig. 18. We also considered data from local and regional studies of other European basins (e.g. Cariou *et al.*, 1985; Gabilli *et al.*, 1985; Gygi, 1986; Atrops & Ferry, 1987; Floquet *et al.*, 1989; Oloriz & Marques, 1990). Comparison of the above data set with our observations in the Iberian Basin reveals a reasonable correlation between unconformities and flooding events recorded in the Oxfordian strata in several European basins (see Fig. 18 and discussion in Aurell, 1991). The above correlation does not occur in the Kimmeridgian–Berriasian interval, in which no minor order unconformities were detected in the Iberian Chain. These observations can yield some information on the contributions of subsidence and eustasy towards the creation of accommodation during the successive depositional sequences. Below, we consider two episodes in the evolutionary history of the Iberian Basin:

1 The Oxfordian sequence appears to have developed during a period of very slow subsidence in which the accommodation in the basin would have been essentially a function of the eustatic fluctuations. The contribution of eustasy should be more evident in the two punctuated highstands, developed during the lowstand systems tract as well as during the successive marine floodings related with the transgressive systems tract, that leave a condensed section on the entire basin (i.e. the Transversarium biozone). Indeed, deposition during the Lower and Middle Oxfordian was very homogeneous, with no evidence of formation of local subsident furrows, which would indicate the activity of some basement faults (see the regular

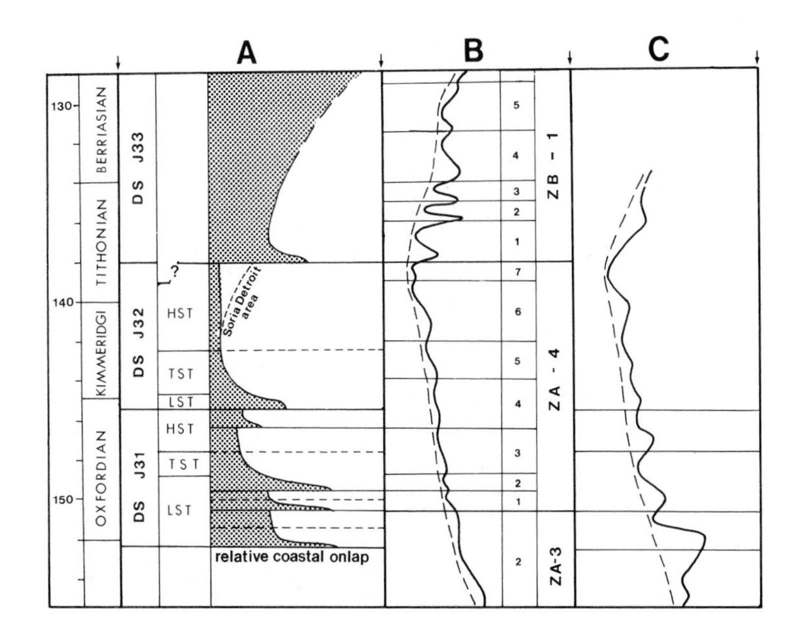

Fig. 18. Relative change of coastal onlap and stratigraphical position of the unconformities (continuous lines) and maximum flooding surfaces (dashed lines) during the Callovian–Berriasian in the Central Iberian Chain (A), compared with the eustatic curves of Haq *et al.* (1987) and Hallam (1988) ((B) and (C), respectively).

distribution of the accommodation in episodes 1 and 2, Fig. 17).

2 From the Kimmeridgian onwards, there is a significant increase in the accommodation rates. Furthermore, a synsedimentary tectonic activity has been detected at the onset of the Kimmeridgian sequence (Fig. 17). We assume a model for the Kimmeridgian and Tithonian–Berriasian sequences in which the creation of the space in the basin is mainly a function of the subsidence. Higher order eustatic signals would be diluted and would not appear in the sedimentary record. It should be noted, however, that the relative sea-level curve in the Iberian Chain (i.e. the largest coastal onlap during the Upper Jurassic is during the Kimmeridgian, and is followed by a progressive offlap) is in general agreement with the long-term eustatic curve proposed by the Exxon group and the Hallam curve for the Upper Jurassic (see dashed lines in Fig. 18).

SUMMARY AND CONCLUSIONS

The study of the Upper Jurassic ramps by means of two cross-sections to the Iberian Basin allowed the documentation of the evolution of the rates of sedimentation and accommodation. The relationship between these two factors defines the observed facies distribution. The evolution of the accommodation is governed by the interplay of eustasy and subsidence. Some conclusions about their relative role in the successive sequences and systems tracts of the Upper Jurassic were made by comparing our results with those obtained in other sedimentary basins. In synthesis, we proposed the following basin evolution history (Fig. 17):

1 *Uppermost Callovian (Lamberti biozone p.p.) to lower Middle Oxfordian (Plicatilis biozone) (i.e. lowstand systems tract, Oxfordian sequence).* The depositional setting is interpreted to be a homogeneous plain that was eventually covered by a thin layer of water. The deepest water was towards the west. A palaeogeographic high (i.e. the Ejulve high) was present to the southeast, which remained emerged throughout this episode. The iron ooids and pisoids, which are generally found within marine skeletal micritic facies, are interpreted to have formed on or adjacent to the emerged areas. Two punctuated eustatic highstands (interpreted as possible eustatic signals) occur within this episode and involved the partial marine inundation of the basin during the lower Cordatum and lower Plicatilis chrons.

2 *Middle Oxfordian (Transversarium biozone) (i.e. transgressive systems tract, Oxfordian sequence).* At the onset of this episode there was a fast rise in the sea level, which involved the recovery of the open-marine conditions across the ramp and its colonization by benthic groups (sponges, echinoids, brachiopods, algae). The Ejulve high was submerged and became the site of deposition of shallow-water bioclastic and intraclastic facies. The palaeogeography was an extensive and homogeneous low-angle slope ramp. Carbonate production was not rapid enough to compensate for the created accommodation (which was caused mainly by a eustatic rise), resulting in condensed sections.

3 *Middle Oxfordian (Bifurcatus biozone) to Upper Oxfordian (Planula biozone p.p.) (i.e. highstand systems tract, Oxfordian sequence).* Sedimentary rates in the proximal areas of the basin were greater than accommodation rates during this episode, which is reflected in the progressive shallowing-upwards vertical facies distribution, as well as the lateral progradation of the siliciclastic and marly proximal deltaic facies over the open ammonite and spongiolitic facies of the middle ramp. Condensed sequences bearing open-marine fossils and glauconite are still found basinwards.

4 *Uppermost Oxfordian (Planula biozone p.p.)–lowermost Kimmeridgian (Platynota biozone p.p.) (i.e. lowstand systems tract, Kimmeridgian sequence).* The reactivation of some basement faults at the end of the Oxfordian resulted in the uplift of the edges of the basin, as well as the creation of a subsident furrow in the central areas of the basin. Large clastic supply allowed the correlative infilling of this tectonic furrow. Sedimentation took place in a very shallow-water and partly restricted low-angle ramp environment. Subsidence was lower in the distal zones, where the influence of the Ejulve high persisted. The palaeogeography at the end of this episode was a very homogeneous ramp with a slight dip to the east.

5 *Lower Kimmeridgian (i.e. transgressive systems tract, Kimmeridgian sequence).* A relative rise of sea level during this episode resulted in the creation of a considerable amount of accommodation across the basin, with the recovery of the marine conditions on the edges. This accommodation is correlated with an increase of the subsidence rates. The facies distribution reveals a retrogradational pattern, with the progressive overlaying of the middle-ramp rhythmic series on the more proximal and restricted marls. High rates of carbonate production nearly compensated for the accommodation in the marginal and middle zones of the ramp. Sedimentation rates were much lower in the distal zones, as reflected by the reduced thickness of the open-marine facies.

6 *Upper Kimmeridgian–Lower Tithonian p.p. (i.e. highstand systems tract, Kimmeridgian sequence).* At the beginning of this episode, sedimentation rates increased relative to the accommodation rate, resulting in the progradation of the marginal ooid and coraline banks over the rhythmic carbonate series of the middle ramp. The accommodation rate was similar throughout the ramp but lower than in the previous episode and is interpreted to have been produced mainly by homogeneous subsidence.

7 *Tithonian p.p. (episode 1, Tithonian–Berriasian sequence).* There was a relative sea-level fall during the Lower Tithonian which involved a considerable basinwards coastal shift at the onset of this episode. This coastal shift is interpreted to have been caused by a moderate tectonic uplift of the edges of the basin. This tectonic uplift, combined with what may be a long-term sea-level fall and sedimentation rates large enough to compensate the accommodation, explains the progressive offlap and the observed progradational facies distribution. Shallow carbonate coraline and oncoid facies are dominant during the Tithonian. These facies prograded over lower energy and deeper water mudstone and fossiliferous wackestone facies.

8 *Lower to Middle Berriasian (episode 2, Tithonian–Berriasian sequence).* The sedimentation during the Berriasian was influenced by a new phase of reactivation of some basement faults at the Jurassic–Cretaceous boundary, which involved an important basinwards coastal shift and the creation of a very subsident furrow in the western part of the basin that was infilled by a siliciclastic sediment supply. Carbonate sedimentation dominated the more tectonically stable tidal flat lagoon settings to the east. The end of this episode involved a general withdrawal of the sea basinwards (to the east of the studied area).

ACKNOWLEDGEMENTS

We are indebted to Guillermo Meléndez and Francois Atrops for the palaeontological determinations and for many helpful discussions and suggestions. Some of the ideas presented here have also benefited from discussion with Angel Gonzalez, Ramón Mas and Ramón Salas. The manuscript was significantly improved by the reviews of Serge Ferry, Marc Floquet and Robert G. Maliva. Financial support was provided by MEC (Ministerio de Educación y Ciencia, Spain, PB89–0230 Project) and DGA (Diputación General de Aragón, PCB-G/89 Project). The final version of this paper was completed during a postdoctoral fellowship of the MEC at the Comparative Sedimentology Laboratory (University of Miami) of Marc Aurell: special gratitude is expressed to Robert N. Ginsburg, Jeroen Kenter, Flavio Anselmetti and María E. Lara for continuous help.

REFERENCES

Alonso, A. & Mas, J.R. (1990) El Jurásico superior en el sector Demanda-Cameros (La Rioja-Soria). *Cuad. Geol. Ibérica* **14**, 173–198.

Alonso, A., Aurell, M., Mas, J.R., Melendez, A. & Nieva, S. (1989) Estructuración de las plataformas del Jurásico superior en la zona de enlace entre la cuenca Ibérica y el estrecho de Soria. *X Congreso Español de Sedimentología, Bilbao* **1**, 175–178.

Alonso, A., Mas, J.R. & Melendez, N. (1986) Los arrecifes coralinos del Malm en la Sierra de Cameros (La Rioja, España). *Acta Geol. Hispánica* **21–22**, 293–306.

Atrops, F. & Ferry, S. (1987) Les glissements sous-marins kimmeridgiens du bassin subalpin: témoins possibles de variations eustatiques négatives. *Mém. Géol Alpine* **13**, 179–185.

Atrops, F. & Melendez, G. (1985) Kimmeridgian and lower Tithonian from the Calanda-Berge area (Iberian Chain, Spain): some biostratigraphic remarks. *Proc. Ist. Int. Symp. on Jurassic Stratigraphy, Earlangen, 1984*, 377–392.

Aurell, M. (1990) El Jurásico superior en la Cordillera Ibérica Central (provincias de Zaragoza y Teruel). Análisis de cuenca. PhD thesis, University of Zaragoza, 389 pp.

Aurell, M. (1991) Identification of systems tracts in low angle carbonate ramps: examples from the Upper Jurassic of Iberian Chain (Spain). *Sediment. Geol.* **73**, 101–115.

Aurell, M. & Melendez, A. (1987) Las bioconstrucciones de corales y sus facies asociadas durante el Malm en la Cordillera Ibérica Central (prov. de Zaragoza). *Est. Geol.* **43**, 261–269.

Aurell, M. & Melendez, A. (1989) Influencia de la falla del Jiloca durante la sedimentación del Malm en la Cordillera Ibérica Central (prov. de Teruel): relación tectónica-sedimentación. *Rev. Soc. Geol. de Esp.* **2**, 65–75.

Aurell, M., Fernandez Lopez, S. & Melendez, G. (1991) The Middle–Upper Jurassic oolitic ironstone facies: eustatic implications. *3rd Int. Symp. on Jurassic Stratigraphy, Poitiers, Sept 1991, Geobios. esp. vol.* (in press).

Bond, G.C. & Kominz, M.A. (1984) Construction of tectonic subsidence curves for the early Paleozoic Miogeocline, southern Canadian Rocky Mountains: implications for subsidence mechanisms, age of breakup and crustal thinning. *Bull. Geol. Soc. Am.* **95**, 155–173.

Bulard, P.F. (1972) Le Jurassique moyen et supérieur de la Chaine Ibérique sur la bordure du bassin de l'Ebre (Espagne). PhD thesis, University of Nice, 702 pp.

Calvet, F., Tucker, M.E. & Henton, J.M. (1990) Middle Triassic ramp systems in the Catalan Basin, northeast Spain: facies, systems tracts, sequences and controls. *Spec. Publ. Int. Assoc. Sediment.* **9**, 79–108.

Cariou, E., Contini, J., Dommergues, J. *et al.* (1985) Biogéographie des Ammonites et évolution structurale de la Téthys au cours de Jurassique. *Bull. Geol. Soc. France* **8** (5), 679–697.

Corbalan, F. & Melendez, G. (1986) Nuevos datos bioestratigráficos sobre el Jurásico superior del sector central de la Cordillera Ibérica. *Acta Geol. Hispánica* **21–22**, 555–560.

Diaz, M., Yebenes, A., Goy, A. & Sanz, J.L. (1984) Landscapes inhabited by Upper Jurassic–Lower Cretaceous archosaurs (Galve, Teruel, Spain). *3rd. Symp on Mesozoic Terrestrial Ecosystems, Tübingen,* 67–72.

Fernandez Lopez, S. (1984) Criterios elementales de reelaboración tafonómica en ammonites de la Cordillera Ibérica. *Acta Geol. Hispánica* **19**, 105–116.

Floquet, M., Laurin, B., Laville, P., *et al.* (1989) Les systémes sédimentaires bourguignons d'âge Bathonien terminal-Callovien. *Bull. C.R.E.P. Elf-Aquitaine,* **13**(1), 133–165.

Fontana, B. & Melendez, G. (1990) Caracterización bioestratigráfica de la Biozona Transversarium (Oxfordiense medio) en el sector oriental de la Cordillera Ibérica. *Geogaceta* **8**, 76–78.

Gabilly, J., Cariou, E. & Hantzpergue, P. (1985) Les grandes discontinuités stratigraphiques au Jurassique: témoins d'événements eustatiques, biologiques et sédimentaires. *Bull. Soc. Géol. France* **1,3**(8), 391–401.

Galloway, W.E. (1989) Genetic stratigraphic sequences in basin analysis I: architecture and genesis of flooding-surface bounded depositional units. *Bull. Am. Assoc. Petrol. Geol.* **73**, 125–142.

Giner, J. & Barnolas, A. (1979) Las construcciones arrecifales del Jurásico superior en la Sierra de Albarracín. *Cuadernos de Geol.* **10**, 73–82.

Gomez, J.J. (1979) El Jurásico superior en facies carbonatadas del sector levantino de la Cordillera Ibérica. PhD thesis, University Complutense de Madrid, Seminarios de Estratigrafía (serie monografías), **4**, 683 pp.

Gomez, J.J. & Goy, A. (1979) Las unidades litoestratigráficas del Jurásico medio y superior, en facies carbonatadas del sector levantino de la Cordillera Ibérica. *Est. Geol.* **35**, 569–598.

Goy, A., Melendez, G., Sequeiros, L. & Villena, J. (1979) El Jurásico superior del sector comprendido entre Molina de Aragón y Monreal del Campo (Cordillera Ibérica). *Cuadernos de Geol.* **10**, 95–106.

Gygi, R.A. (1986) Eustatic sea level changes of the Oxfordian (late Jurassic) and their effect documented in sediments and fossil assemblages of an epicontinental sea. *Eclog. Geol. Helv.* **79**(2), 455–491.

Hallam, A. (1988) A reevaluation of Jurassic eustacy in the light of new data and the revised Exxon curve. In: *Sea Level Changes: An Integrated Approach* (Eds Wilgus, C.K., Hastings, B.S., Kendall, C.G.St.C., Posamentier, H.W., Ross, C.A. & Van Wagoner, J.C.) Spec. Publ. Soc. Econ. Paleontol. Mineral. 42, 261–274.

Haq, B.H., Hardenbol, J. & Vail, P.R. (1987) Chronology of fluctuating sea levels since the Triassic. *Science* **235**, 1156–1167.

Kendall, C.G.St.C., Moore, P. & Cannon, R. (1990) A challenge: is it possible to determine eustacy? *Geol. Soc. Am. 1990 Ann. Meet., Dallas (Abstr. with Progr.)* **A 28**.

Martin i Closes, C. (1990) Els caròfits del Cretaci inferior de las conques periferiques del bloc de l'Ebre. PhD thesis, University of Barcelona, 581 pp.

Mas, J.R., Alonso, A. & Melendez, N. (1984) La Formación Villar del Arzobispo: un ejemplo de llanuras de marea siliciclásticas asociadas a plataformas carbonatadas. Jurásico terminal (NW de Valencia y E de Cuenca). *Publ. Geol. (Univ. Autónoma Barcelona)* **20**, 175–188.

Melendez, G. (1984) El Oxfordiense en el sector central de la Cordillera Ibérica. I. Bioestratigrafía. II. Paleontología (Perisphinctidae, Ammonoidea). PhD thesis, University Complutense of Madrid, 825 pp.

Melendez, G. & Brochwicz-Lewinski, W. (1983) El Oxfordiense Inferior en el sector central de la Cordillera Ibérica (S Zaragoza–N Teruel). *Teruel* **69**, 211–226.

Melendez, G., Aurell, M., Fontana, B. & Lardies, M.D. (1990) El Tránsito Dogger-Malm en el sector nororiental de la Cordillera Ibérica: análisis tafonómico y reconstrucción paleogeográfica. *Com. Reunión de Tafonomía y Evolución, Univ. Comp. Madrid, 1990,* pp. 221–229.

Nahon, D., Carozzi, A.V. & Parron, C. (1980) Lateritic weathering as a mechanism for the generation of ferruginous ooids. *J. Sediment. Petrol.* **50**(4), 1287–1298.

Oloriz, F. & Marques, B. (1990) La marge Sud-Ouest d'Iberia pendant le Jurassique supérieur (Oxfordien–Kimméridgien). Essai de reconstruction géobiologique. *Cuad. Geol. Ibérica* **13**, 237–250.

Salas, R. (1987) El Malm i el Cretaci inferior entre el Massís de Garraf i la Serra d'Espadá. Analisi de conca. PhD thesis, University of Barcelona, 345pp.

Salas, R. (1989) Evolución estratigráfica secuencial y tipos de plataformas de carbonatos del intervalo Oxfordiense–Berriasiense en las Cordilleras Ibérica Oriental y Costero Catalana Meridional. *Cuad. Geol. Ibérica* **13**, 121–157.

Seilacher, A. (1985) Storm beds: their significance in event stratigraphy. In: *Stratigraphy: Quo vadis?* (Eds Seibold, E. & Meulenkamp, J.D.) Am. Assoc. Petrol. Geol. Stud. Geol. 16, 49–54.

Siehl, A. & Thein, J. (1978) Geochemische trends in der Minette (Jura, Luxenburg/Lothringen). *Geol. Rdsch.* **67** (3), 1052–1077.

Vail, P.R., Colin, J.P., Chene, J., Kuchly, J., Mediavilla, F. & Trifilief, F. (1987) La stratigraphie séquentielle et son application aux corrélations chronostratigraphiques dans le Jurassique du bassin de Paris. *Bull. Soc. Geol. France* **III,7** (8), 301–321.

Van Wagoner, J.C., Posamentier, H.W., Mitchum, R.M., Jr, *et al.* (1988) An overview of the fundamentals of sequence stratigraphy and key definitions. In: *Sea Level Changes: An Integrated Approach* (Eds Wilgus, C.K., Hastings, B.S., Kendall, C.G.St.C., Posamentier, H.W., Ross, C.A. & Van Wagoner, J.C.) Spec. Publ. Soc. Econ. Paleontol. Mineral. 42, 39–45.

Ziegler, P.A. (1988) Evolution of the Arctic-North Atlantic and the western Tethys. *Mem. Am. Assoc. Petrol. Geol.* **43**, 189 pp.

Spec. Publs Int. Ass. Sediment. (1993) **18**, 369–395

Shelf-to-basin Palaeocene palaeogeography and depositional sequences, western Pyrenees, north Spain

V. PUJALTE*, S. ROBLES*, A. ROBADOR†,
J.I. BACETA* *and* X. ORUE-ETXEBARRIA*

**Departmento de Estratigrafía, Geodinámica y Paleontología, Universidad del País Vasco, Ap. 644, 48080 Bilbao, Spain; and*
†División de Geología, Instituto Tecnológico Geominero de España, Rios Rosas 23, 28003 Madrid, Spain

ABSTRACT

Palaeocene depositional sequences and their constituent *systems tracts* (ST) have been recognized and studied in the deep-water interplate Basque basin and in the adjacent north Iberian shelf (western Pyrenees, north Spain). In addition, re-examination of the Palaeocene of the north Pyrenean basin strongly suggests that a similar set of sequences also exists there.

The age of shelfal sequences is established with benthic foraminifera, whereas in basinal settings they are accurately dated with planktonic foraminifera. In the north Iberian shelf sequences are dominated by the transgressive ST (mostly sandy grainstones) and the highstand ST (reefal limestones), which grades landwards to lagoonal and continental red shales and dolomites. In addition, the shallow (proximal) part of the lowstand prograding complexes of two sequences is preserved in the outer platform margin.

The bulk of basinal deposits occur within the lowstand ST; the transgressive ST and highstand ST are either thin or even absent in the deep-water setting. Basinal sequences from southern (Iberian) sources are entirely made up of carbonates and were accumulated in a base-of-slope carbonate apron. They are usually composed of a lower slope fan complex (resedimented breccias, calcidebrites and thick-bedded carbonate turbidites) and an upper lowstand prograding complex (thick- and thin-bedded carbonate turbidites and pelagic limestones). Sequences from western (north Pyrenean) sources have a mixed carbonate–siliciclastic nature and include: basin-floor fan complexes (thick-bedded, massive siliciclastic turbidites; not present in all sequences); slope-fan complexes (resedimented breccias, thick-bedded carbonate turbidites and marls); and lowstand prograding complexes (hemipelagic limestones or alternations of hemipelagic limestone and marls). The coarse-grained deposits of these basinal complexes infill a deep-sea channel system.

The succession was deposited during an early phase of the Pyrenean convergence, the change from extension to compression being recorded in the area by a strong reduction of siliciclastic influx and sedimentation rates, denoting sediment-starved conditions. In spite of the changing tectonics and low sediment supply, predictions of the Exxon Group's models are reasonably met, most notably: (i) the number and age of depositional sequences can be matched with third-order sea-level cycles of the 1988 version of the sea-level chart; (ii) relative magnitudes of global sea-level changes are also recorded, since the major sequence boundary of the succession can be chronostratigraphically tied to the 58.5 Ma eustatic fall, one of the more significant Palaeocene events; (iii) good agreement with current sequence-stratigraphic models is observed that relate stratal geometries and eustasy, regardless of whether carbonate or siliciclastic sediments are involved.

INTRODUCTION

The Palaeocene system outcrops extensively in the western Pyrenees and surrounding areas (Fig. 1(a)), and it is also well known in numerous boreholes in the Aquitaine Basin (Faber, 1961; Kieken, 1973; BRGM *et al.*, 1974). Palaeocene deposits are represented by a wide variety of facies, from continental siliciclastic deposits to hemipelagic limestones; they record an almost continuous environmental spectrum, from terrestrial to deep-marine settings. This area therefore affords an excellent opportunity to test sequence-stratigraphy concepts.

The general stratigraphy and the palaeogeographic features of these sediments have been previously worked out by several authors, the contributions of Mangin (1959), Leon (1972) and, especially, Plaziat (1975a, 1981) being perhaps the most important. There have also been extensive biostratigraphic studies, involving planktonic foraminifera and calcareous nanoplankton in basinal successions (Von Hillebrandt, 1965; Kapellos, 1974; Van Vliet, 1982; Orue-etxebarría, 1983) and benthic species within shallow-water deposits (see Robador, 1991, for a compilation). More recently, the study of the latest Cretaceous–Palaeocene–early Eocene succession has been resumed using a sequence-stratigraphic approach (Pujalte *et al.*, 1988, 1989; Baceta *et al.*, 1991; Robador *et al.*, 1991).

This paper presents an updated synthesis of our present knowledge. It is based primarily on extensive field studies of the southwestern Pyrenean and Basque areas (Fig. 1), integrated with published and new biostratigraphic data. Also, it incorporates literature information from the north Pyrenean and Aquitaine areas. Along with palaeogeographical and sedimentological improvements, it will be demonstrated that the Palaeocene succession is composed of five depositional sequences that can be matched, in age and relative magnitude, with third-order cycles of the sea-level chart of Haq *et al.* (1988). This is particularly well expressed in the basin, where the record is more complete and the sequences can be dated with better accuracy.

SETTING

Both in shelfal and basinal settings (Figs 1 & 2), the Palaeocene of the western Pyrenees consists of a comparatively thin package of sediments (up to

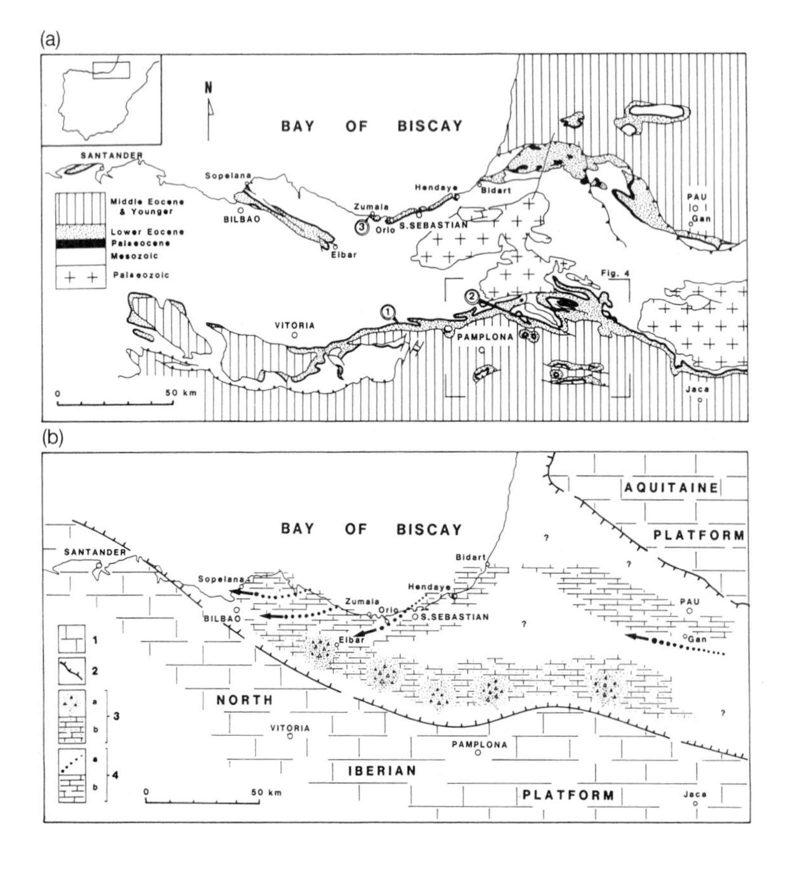

Fig. 1. (a) Location and geological map of the western Pyrenees–Basque Cantabrian area (partly based on Plaziat, 1975a) Encircled 1, 2 and 3 are locations of sections in Fig. 2, (b) Palaeogeographic reconstruction of the same area for the Palaeocene (note that the actual basin must have been wider than shown here, since no palinspastic restoration has been attempted that would eliminate the important post-Palaeocene north–south tectonic shortening). 1, shelf area, mostly carbonate platforms; 2, shelf-basin transition; 3, north Iberian carbonate slope apron basinal system, with zones respectively dominated by breccias (a) and turbidite/ hemipelagites (b); 4, north Pyrenean basinal systems, with zones respectively dominated by coarse-grained resedimented facies (a, channel-fill facies) and fine-grained autochthonous facies (b, interchannel facies).

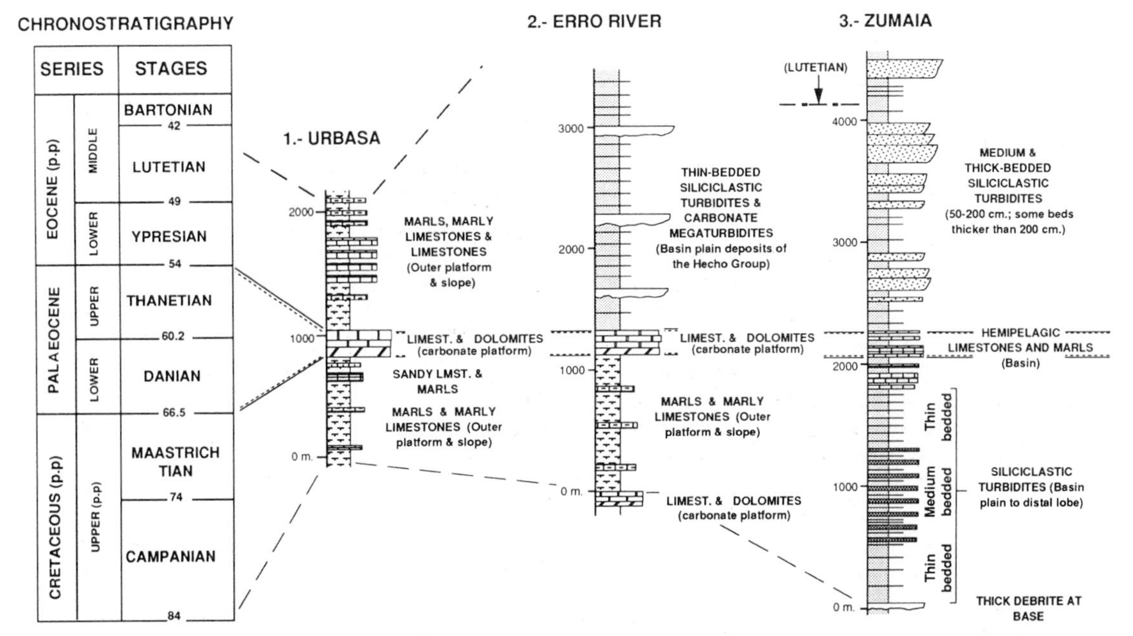

Fig. 2. Representative columnar sections of Upper Cretaceous–Lower Eocene successions of different sectors of the study area. See Fig. 1(a) for location.

300 m, but usually less) sandwiched between the much thicker Senonian and earliest Eocene successions (1000–3000 m each). The Palaeocene is composed of many types of lithologies but carbonate deposits predominate by far (Fig. 2 and below). This fact implies a strong reduction in siliciclastic input which, coupled with regionally low sedimentation rates, clearly suggests a period of relatively starved conditions. Puigdefabregas and Souquet (1986) include these Palaeocene sediments in their tectono-sedimentary cycle no. 8 ('transition to foreland basins'), which marks the onset of plate-convergence conditions on the Bay of Biscay. According to them, the first foreland basin geometry became established in the eastern Pyrenees during the Palaeocene, at which time the western part still maintained much of the previous Mesozoic features.

Palaeogeographically, the study area during the Palaeocene was a narrow interplate gulf opening westwards to the proto-Atlantic Ocean. Within this basin, the different domains were arranged in a broadly symmetrical manner (Fig. 1(b)): a central deep basin was rimmed on the north, east and south sides by shelf areas, mostly occupied by carbonate platforms. These, in turn, graded further landward into continental environments.

Basinal deposits outcrop in the Basque country, in the north of the Navarra province and in the Bearn region (France) (Fig. 1(a)). Platform sediments of the north Iberian shelf lie along an almost continuous band of outcrops stretching for more than 300 km, from the central part of the Pyrenees to Santander (Fig. 1). Continental deposits (i.e. Garumnian facies, Tremp Formation) are widely exposed on the eastern Pyrenees. As the latter will not be considered here, interested readers are directed to the works of Freytet and Plaziat (1982) and Eichenseer (1988).

SUMMARY OF SEDIMENTARY SYSTEMS

The major Palaeocene sedimentary systems of the western Pyrenees are shown in the palaeogeographic reconstruction of Fig. 1(b) which, being based on the present-day position of the outcrops, does not take into account the important post-Palaeocene north–south shortening of the area (cf. Razin, 1989). This reconstruction is very similar to that of Plaziat (1975a), although the latter's reconstruction is based on independent evidence. Data from the Aquitaine and the north Pyrenees areas (see below) are from literature sources and will be referred to only briefly. Distinctive features of these reconstructions are discussed below.

North Iberian systems

Shelf

The Palaeocene of the north Iberian shelf is represented almost exclusively by shallow-water limestones and dolomites, deposits of successive carbonate platforms (i.e. depositional sequences) which evolved there from Danian to earliest Eocene times. These deposits form a composite sedimentary body of relatively uniform thickness (100–150 m) tens of kilometres wide and hundreds of kilometres long (Fig. 1(b)); geomorphic features that occur here are those typical of pericratonic settings. Figure 3 depicts the overall reconstructed geometry of two of these sequences (i.e. 1.4 and 2.1), along with the different subenvironments that can be recognized in them. Such geometry bears some likeness to that of the 'distally steepened ramp' model of Read (1985). It is doubtful, however, whether the three remaining sequences had the same character and therefore the more general term 'carbonate platform' will be used subsequently.

Each of the different platform subenvironments is typified by its own litho- and biofacies, the more representative being, according to Serra-Kiel *et al.* (1991), the following: (i) *restricted inner platform,* primary dolomites with small miliolids and ammodiscacids, mudstones with ostracods, small miliolids and dasycladacean algae and siliciclastic beds with oysters and other bivalves; (ii) *protected inner platform*, mainly represented by wackestones and packstones, including the 'alveolina limestones'; (iii) *inner 'reefal' platform margin*, characterized by the colonization of the sea floor by rhodophytes, bryozoans, corals and attached foraminifera; and (iv) *outer platform margin* identified by limestones with microbenthic and planktonic foraminifera in the Danian and Lower Thanetian and, later on, by nummulitid limestones. In addition to these, there are variable amounts of secondary dolomites. They are usually observed in the landward portion of the carbonate platforms and are indicative of lagoonal conditions and/or of emergences of the platform during lowstand periods.

Shelf basin transition

In the Navarra province, where they are preserved in neighbouring outcrops, north Iberia shelfal and basinal successions are seen to be neatly separated by a comparatively narrow (2–4 km) transitional zone, here interpreted as a bypass upper slope (Fig. 4). This is a zone of reduced sedimentation and widespread hiatuses, which sometimes comprised the whole Palaeocene. Around the village of Erro, for example (Fig. 4), the Palaeocene is represented exclusively by a thin (14 m) succession of micritic limestones and marls, rich in planktonic foraminifera of the *M. velascoensis* biozone (latest Palaeocene), that sits directly on late Cretaceous marlstones. More diagnostic still is the local appear-

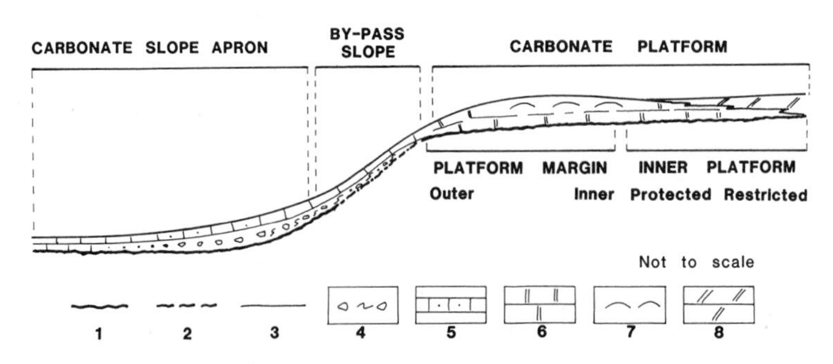

Fig. 3. Simplified sketch illustrating the geometry and facies of an idealized depositional sequence of the north Iberian margin, and the terminology used to describe them in the text. 1, lower sequence boundary in the basin and lower slope; 2, lower erosional boundary in the upper slope (i.e. base-of-slope gullies); 3, upper boundary, prior to erosion by overlying sequence; 4, carbonate breccias and slumps; 5, foraminiferal wackestone/mudstone, with or without intercalated carbonate turbidites; 6, shallow-water limestones; 7, reefal limestones; 8, dolomites.

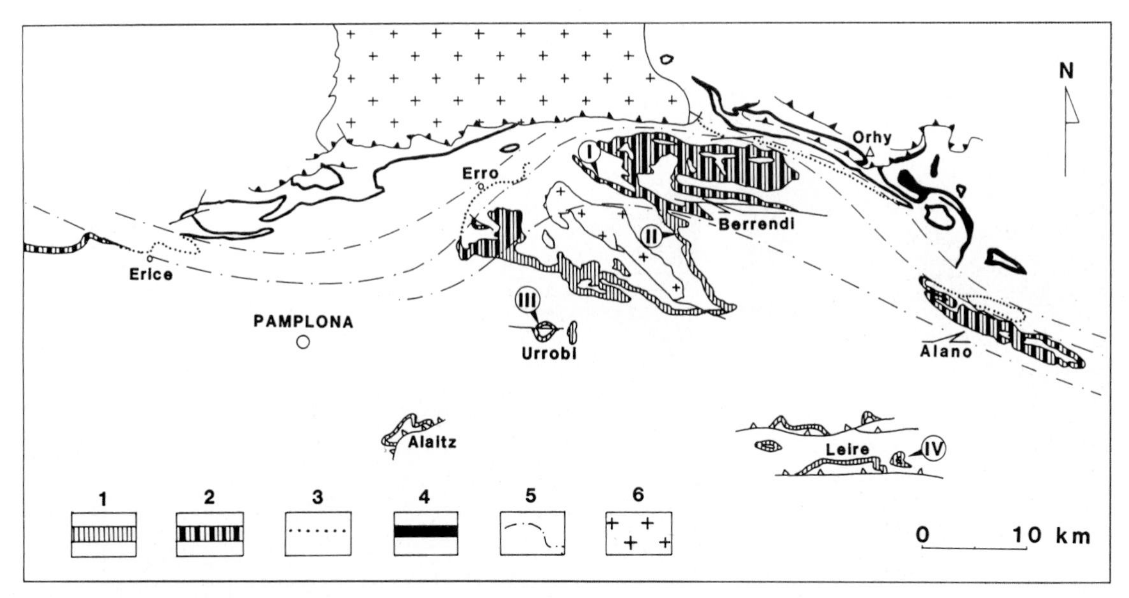

Fig. 4. Palaeocene outcrop map, north of Navarra and Huesca provinces (for location, see Fig. 1(a)). 1, Inner-platform facies; 2, platform margin facies; 3, shelf–basin transition facies; 4, basinal facies; 5, limits of sedimentary domains; 6, Palaeozoic and Triassic basement. Encircled I, II, III and IV, location of sections of Fig. 6.

ance of coarse-grained carbonate breccias and coarse-grained turbidites infilling erosional gullies excavated into late Cretaceous marlstones. One of these gullies, near the village of Erice, is 800 m wide and 30 m deep (Fig. 4), and in some examples erosion of 300 m of Cretaceous marlstones can be demonstrated. As described by Mullins and Cook (1986), such relatively small but deep submarine gullies abound in recent and ancient bypass slopes, being the main conduits for resedimentation into the basin.

Basin

Iberian Palaeocene basinal deposits are made up largely of carbonates, being composed of two broad groups of facies associations which, for descriptive purposes, will be described as 'autochthonous' and 'resedimented'. The former is made up mainly of pelagic limestones, usually intercalated with variable amounts of marls and/or very fine-grained carbonate turbidites. They are considered to be the result of the slow settling of biochemical materials and also of dilute turbidite currents. The resedimented association is characterized by carbonate breccias, calcidebrites and coarse-grained carbonate turbidites. A good proportion of its constituent clasts are fragments of contemporaneous and older shallow-water deposits, clear evidence of mechanical transfer of sediments from the shelf into the basin.

These sediments usually occur in the field as two carbonate subunits separated by a marly interval (Fig. 5(a)), laterally continuous but of variable thickness (40 to 210 m). The thickest segments of the successions contain the highest proportion of coarse-grained deposits but otherwise there is not a well-marked lateral segregation between the resedimented and autochthonous associations. The Eibar outcrop, the best-studied example of these systems (see below), is situated in the southeastern tip of the Biscay synclinorium (Fig. 1(a)). Similar deposits are much more abundant, however, in the southern Pyrenean area, in the north of the Navarra province and along the Spanish–French border, such as in the Orhy peak (Figs 1(a) & 4). In that area, they fringe the north Iberian carbonate platform, from which they were clearly shed, since palaeocurrents indicate consistently northwards and northwestwards sediment transport. Because of the above features and also of the nature of the shelf–basin transition described above (i.e a bypass slope), the system is interpreted here as a carbonate slope apron, akin to the 'base-of-slope' model of Mullins and Cook (1986).

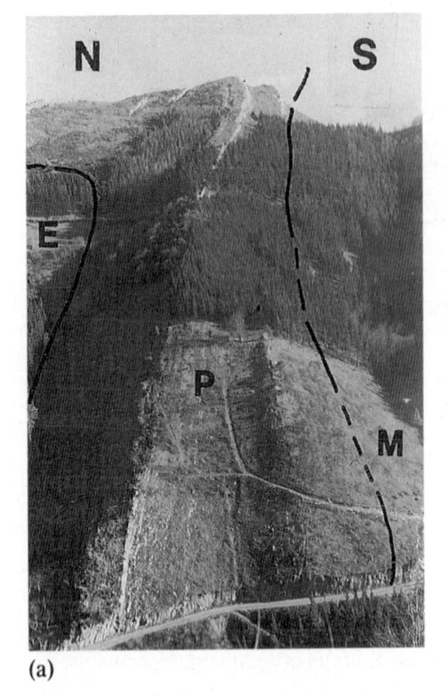

(a) (b)

Fig. 5. Field expression of the two contrasting sedimentary systems recognized in the basin. (a) North–south view of an outcrop of the north Iberian system, Eibar, Guipuzcoa province (M, Maastrichtian; P, Palaeocene, two laterally continuous packages of carbonate turbidites and debrites separated by a more marly interval; E, Eocene). (b) Deep-sea channel-fill deposits of the north Pyrenean system situated near Markina, Biscay province (S, siliciclastic turbidites; C, carbonate turbidites and debrites. Note wedging of these deposits towards observer). M, Maastrichtian; P, Palaeocene; E, Eocene.

South Aquitaine–north Pyrenean systems

Shelf

To the north of Pau (Fig. 1(a)), local outcrops and numerous boreholes demonstrate a Palaeocene succession that mirrors that of north Iberia. It is formed by a sedimentary body of exclusively shallow-water dolomites and limestones, containing algae, corals and benthic foraminifera and is up to 200 m thick and laterally extensive (30 to 50 km wide, 80 km long, according to BRGM *et al.*, 1974). However, during Thanetian time, sandy sediments were deposited also in the easternmost part of the Aquitaine platform (BRGM *et al.*, 1974; Plaziat, 1981).

Shelf–basin transition

All the north–south cross-sections produced by BRGM *et al.* (1974) show a rather abrupt transition between shallow- and deep-water Palaeocene deposits. This evidence is compatible with a distally steepened ramp margin or might well indicate a by-passed upper slope similar to that of north Iberia. The latter interpretation is also supported by the east–west reconstruction of Fig. 2 in Kieken (1973).

Basin

Unlike their Iberian counterparts, deposits of the south Aquitaine or north Pyrenean basinal system exhibit a neat spatial separation between resedimented and autochthonous facies associations. The former is further typified by a hybrid carbonate–siliciclastic nature, since it may include up to 40% of siliciclastic turbidites, although still mainly composed of resedimented carbonates.

In the north Pyrenean area itself, the coarse-grained, resedimented association is observed in outcrop only to the southeast of Pau (Bearn region, Fig. 1; see, for example, Flicoteaux, 1972 and Plaziat *et al.*, 1975), but has also been encountered in several boreholes along the east–west-trending 'Flysch Trough' (BRGM *et al.*, 1974). The autochthonous association is well exposed (cf. Seyue, 1984, 1990; Razin, 1989), with the best examples occurring in the coastal cliffs near Hendaye and Bidart (Fig. 1(a)).

Deposits of the same basinal system are extensively exposed in the Basque country, both in the Guipuzcoa monocline and in the Biscay synclinorium (Fig. 1(a)). Evidence suggesting their former connection with those of the north Pyrenean zone are: their geographical position (Fig. 1(a)), their similar stratigraphic succession (see below) and their mean transport directions towards the west–southwest, averaging 240° in the Guipuzcoa monocline.

As a rule, resedimented successions are thicker (150 to 300 m) than autochthonous ones (10 to 180 m). They are also more resistant to present-day erosion and stand out in the landscape (Fig. 5(b)), being therefore easily mapped. Consequently, it can be demonstrated that the resedimented association always infills morphostructural or erosional depressions of the lower slope and/or basin floor, interpreted as long-lived deep-sea channels, in the sense of Carter (1988). The width of such channels ranges between 1 to 5 km and their maximum observable length is 12 km. Based on palaeogeography a total length of 30 to 50 km seems a reasonable guess (Fig. 1(b)). The autochthonous association, more widespread, must therefore represent deposition onto relative highs or interchannel areas (Pujalte *et al.*, 1989, and below).

SEQUENCE ANALYSIS

Depositional sequences were established for three of the four sedimentary systems accessible in the study area, namely the north Iberian platform and both the north Iberian and north Pyrenean basinal systems. The fourth one, the north Iberian bypass slope, was left out because of its discontinuous record. As explained below, different procedures had to be used for each of the three systems, but in general they included: (i) the identification of sequence boundaries, which were picked at discontinuity surfaces, either physical (i.e. important erosional boundaries, subaerial exposure surfaces, meaningful facies changes, etc.), biostratigraphical, or both; (ii) the assessment of the internal and external geometrical relationships of stratal packages, in some cases by direct mapping, in others through correlation of neighbouring sections; and (iii) delineation of systems tracts by their facies association, vertical patterns and relative position within the sequences.

Age calibration was attempted biostratigraphically, with uneven results. Most of the basinal sequences were accurately dated with planktonic foraminifera (see Appendix 1, p. 394–395), in some cases integrated with published nanoplankton data. They form, therefore, the main basis for the correlation presented here. It was found that each of the dated basinal sequences matched with a third-order sea-level cycle of the Exxon sea-level chart (Haq *et al.*, 1988) and, accordingly, they are given the same code (i.e. 1.3, 1.4, etc., see below).

Danian–early Thanetian shelfal sequences are less well dated, partly due to poor palaeontological record and/or dolomitization, but mainly for want of a sound, benthic foraminifera biostratigraphic scale for this interval. The age of the rest of the platform sequences has been reasonably constrained and, again, they are found to be correlatable with third-order eustatic cycles. Therefore, although in some cases tentatively, shelfal sequences are named in the same way as that used for basinal ones.

The main features of the sequences and systems tracts of selected sections of each sedimentary system are described below. It must be stressed, however, that the same or similar features also have been observed at many other outcrops not mentioned here.

North Iberian shelf (carbonate-platform system)

The sequential analysis of this system is based on the detailed sedimentological and biostratigraphic study of four sections located in the north of Navarra (Figs 4 & 6(a)), integrated with mapping of several large-scale exposures such as those of the Berrendi and Alano ridges (Figs 7 & 8). This zone was chosen because it contains the widest facies spectrum, from inner to outer settings, being in fact the only known sector of the study area where the sensitive platform margin is well preserved and exposed. It should be noted, however, that there is a lack of continuity between outcrops (Fig. 4) precluding direct correlation of the different sections. This has partly been overcome by biostratigraphic correlation, although some uncertainty in the resulting reconstruction, shown in Fig. 6(b), is inevitably introduced. A brief description of the five successive platform sequences is offered below.

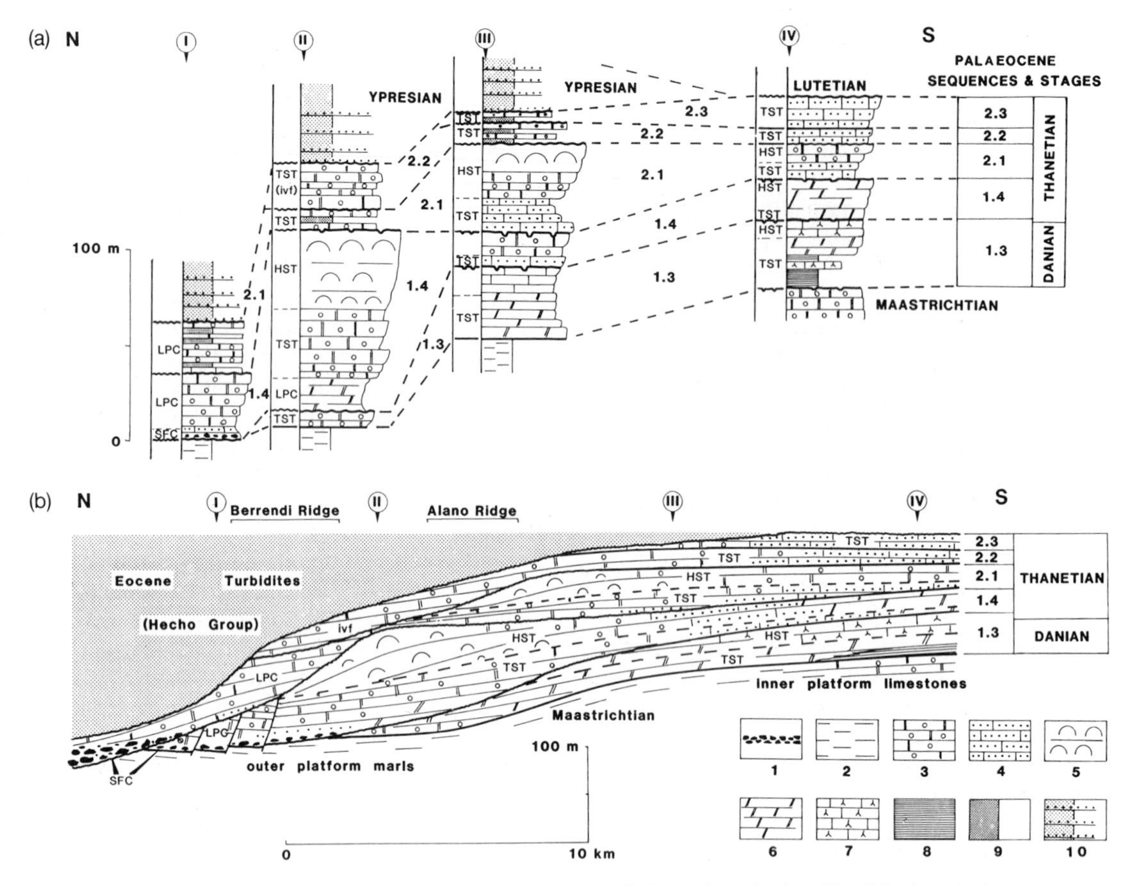

Fig. 6. (a) Simplified columnar sections of Palaeocene carbonate platform deposits (see Fig. 4 for location). (b) Tentative cross-section and sequential interpretation. SFC, slope fan complex; ivf, incised valley fill; LPC, lowstand prograding complex; TST, transgressive systems tract; HST, highstand systems tract. 1, carbonate breccias; 2, marls; 3, bioclastic limestones; 4, sandy limestones; 5, reefal limestones; 6, dolomites; 7, lacustrine limestones, usually intercalated with dolomites; 8, continental red marls; 9, marls, shales, or sandy marls; 10, siliciclastic turbidites and marls. Encircled I, II, III and IV indicate studied sections (see Fig. 4 for location).

Sequence 1.3

This sequence is well represented only in the inner-platform setting (sections III and IV), being absent in section I, and poorly developed in section II (Fig. 6). Its lower boundary is always an abrupt transition between outer-platform Maastrichtian marls and shallow-water Palaeocene carbonates in sections II and III, and between Maastrichtian shallow-water grainstones and continental 'Garumnian' red siltstones in section IV. These features clearly demonstrate an important basinward shift of facies, indicative of a type 1 sequence boundary.

In section II, sequence 1.3 begins with a glauconite-rich bioclastic limestone bed with abraded fragments of echinoderms, bryozoans, oysters and even planktonic foraminifera, and then continuing with dark, fining-upward lime mudstones with sparse bioclasts. The basal bed is thought to be a ravinement lag deposit, while the bulk of the succession, about 7 m thick, is interpreted as a *transgressive systems tract* (TST) because of its retrogradational or backstepping character. In section III, sequence 1.3 consists of two parts, the lower one formed by thinning-upward, dolomitized algal/coral boundstones, the upper one by an alternation of micritic limestones (mudstones and wackestones), with ostracods and/or charophytes, a typical lagoonal association according to Serra-Kiel *et al.* (1991). The lower secondary dolomites are

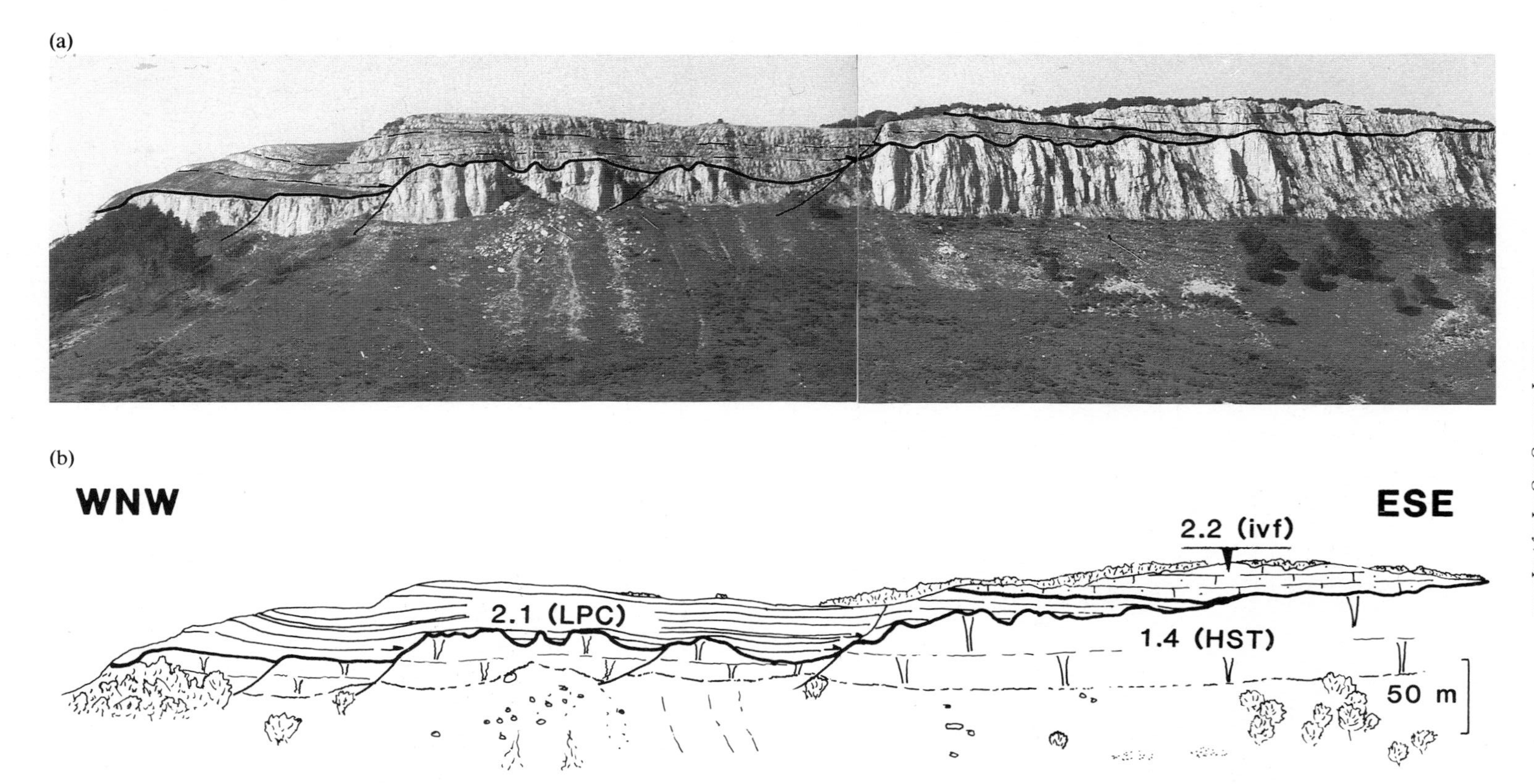

Fig. 7. (a) Outcrop photograph of the Berrendi ridge (see Fig. 4 for location) showing the stratigraphic relationships of some of the Palaeocene sequences at the outer-platform margin. (b) Schematic diagram of the same outcrop with indication of sequences and systems tracts. Note major boundary at the base of sequence 2.1 and fossilized down-to-the basin faults.

Fig. 8. Outcrop photograph of the Alano ridge (see Fig. 4 for location), showing stacking of three Palaeocene carbonate-platform sequences. Note progradational clinoforms in highstand deposits of sequence 2.1.

tentatively ascribed to a TST because of their sequential position and vertical trend. The upper limestones are, very likely, *highstand systems tract* (HST) deposits, as they record the time of maximum regression and progradation of the sequence. In section IV, the succession is similar to that just described except for the occurrence of a basal interval of 'Garumnian' red siltstones with intercalations of lacustrine/palustrine limestones, and is considered a part of the TST (Fig. 6).

Sequence 1.3 is ascribed to the Danian mainly because of its stratigraphic position, above well-dated Maastrichtian marls or limestones, and below Thanetian limestones. In any event, it should be noted again that no shallow-water fossils of stratigraphic value are known to have existed during the Danian.

Sequence 1.4

This sequence sits on the outer-platform margin overlying an important erosional surface, which in section I is associated with the absence of sequence 1.3 (Fig. 6). In section II, the lower sequence boundary is characterized by an abrupt shallowing of facies (i.e. from marine limestones to dolomites (Fig. 6)), a change suggesting another downward shift of facies. Both features suggest a type 1 boundary.

In section I, sequence 1.4 begins with a 2 m thick interval of breccias, directly overlain by 0.5 m of graded bioclastic rudstones and grainstones. The breccias are composed mainly of fragments of Palaeocene platform limestones of sequence 1.3 up to 1 m in diameter, the rest being chips of the underlying Maastrichtian marls. Because of this composition and also its stratigraphic position in the uppermost part of the by-pass slope (Fig. 6), the breccias are considered the record of the final stages of collapse of the previous Danian carbonate-platform margin. The rudstones and grainstones, which include a large proportion of abraded shallow-water bioclasts such as miliolids, red algae and echinoderms, are carbonate turbidites. These basal deposits are interpreted therefore as a low-stand systems tract deposit, probably a part of the *slope fan complex* (SFC). (See also discussion and interpretation of the carbonate breccias of sequence 2.1 that occur in similar sequential position.) The remainder of the sequence in this section consists of stacked parasequences. Individual parasequences are a few metres thick and have a shoaling-up character, beginning with outer-platform mudstones and/or wackestones and grading up to bioclastic grainstones with miliolids and red algae. The vertical trend of the parasequence set is also clearly progradational and it is interpreted therefore as a *lowstand prograding complex* (LPC).

Sequence 1.4 reaches its thickest development at section II, where it consists of three parts. The lower part is formed exclusively of dolomites, and probably represents the more proximal (landward) segment of the LPC. The middle part is made up of shoaling-up parasequences, characterized by a retrogradational stacking pattern, and it is interpreted as a TST. Individual TST parasequences are thicker and have a more 'proximal' character than those of section I, usually beginning with wackestones containing red algae, bryozoans and foraminifera, grading upward into grainstones containing corals, oysters and calcareous algae. Finally, the upper part is formed largely by coral–algae–bryozoan boundstones, including individual reef bodies up to 35 m in diameter. It is attributed to an HST.

To the south of this point, sequence 1.4 is thinner and is represented by a much more restricted facies. At section III, it is made up of miliolid and bryozoan mudstones and wackestones, with occasional dolomite intercalations. At section IV, however, the succession consists almost exclusively of dolomites and sandy dolomites. While systems tracts are difficult to differentiate in sections III and IV, the bulk of their sediments is tentatively assumed to represent HST deposits because of their reconstructed geometries (Fig. 6).

The only biostratigraphically useful foraminifera found in sequence 1.4 so far is the rotalid *Miscellanea juliettae*. It has been observed in several samples of limestones of sections I and II, and it suggests an early Thanetian age for this sequence (cf. Serra-Kiel *et al.*, 1991).

Sequence 2.1

The lower bounding surface of this sequence is, by far, the most prominent sequence boundary of the Palaeocene platform succession. Its more dramatic expression occurs on the platform margin, where it is represented by an erosional unconformity, which in the outermost part of the margin is associated with erosion of the underlying sequence 1.4 (Fig. 6). In the inner margin, the unconformity is eroded deeply although irregularly into the underlying limestones, producing a rapid basinward reduction in the thickness of the underlying sequence 1.4, from near 120 m at section II to less than 50 m at section I. In the large-scale Berrendi ridge outcrop, which is situated between the two sections, it can further be observed that the unconformity fossilized a group of small, down to the basin normal faults, clearly recording the destabilization and partial failure of the platform margin. As a result, the unconformity surface has preserved a rugged residual relief, including some palaeoscarps, now buried under the onlapping deposits of the overlying sequence 2.1 (Fig. 7). Finally, in sections III and IV, and in many other inner-platform successions, the boundary is characterized by a subaerial exposure surface with karstification features and extensive microcodium biocorrosion. The development of this important boundary, clearly a type 1 unconformity, can be constrained to having been formed during the early Thanetian (see below). It is therefore at least approximately contemporaneous with the 58.5 Ma sea-level fall, the most prominent of the Palaeocene according to Haq *et al.* (1988).

Sequence 2.1 begins in many parts of the outermost platform margin with a disorganized accumulation of carbonate megabreccias that rest on an erosional surface underlain by Maastrichtian marls (Fig. 6). The megabreccias comprise mainly angular fragments of Palaeocene limestones derived from sequence 1.4, but include also a smaller proportion of Maastrichtian marly clasts. The distribution and development of the megabreccias are irregular and they are absent locally. In the Navarra pass (at the border between the Navarra and Huesca provinces), the megabreccias reach almost 50 m in thickness; their individual clasts commonly exceed

5 m in length, with a few slabs larger than 25 m. In contrast, in the Mintxate river section (northwest of the village of Isaba) the megabreccias consist of only a discontinuous string of isolated blocks, each a few metres across. In most places, however, the megabreccias range in thickness between 5 and 10 m, with clast sizes ranging from 1 to 4 m.

The megabreccias are always deposited in the immediate proximity of the remnant edge of the former platform margin, and consequently are deduced to have been transported a short distance only (several tens to a few hundred metres). Further supporting this interpretation is the lack of internal organization and the great angularity of constituent clasts within the megabreccias. Because of these attributes, these units are thought to record an almost *in situ* collapse of parts of the outer margin of the previous sequence, probably through retrogressive slump and sliding. The small synsedimentary faults observed further landward (Fig. 7) would represent early phases of the same process. The collapse was probably triggered by contemporaneous and/or earlier erosion of the by-pass slope, and certainly was ended just after the reflooding of the platform margin, since overlying sediments are not disturbed. Some of the megabreccias may actually represent the final stage of infilling of the head of slope gullies. Because of this, the megabreccias are considered as the final (uppermost) part of the SFC.

Sequence 2.1 continues with a well-defined package of sandy marls and sandy or silty limestones. These sediments sit on the megabreccias in the outer-platform margin, and on the eroded top of sequence 1.4 in sections I, II and in the intervening Berrendi area (Figs 6 & 7). The package reaches a maximum thickness of about 100 m around the Berrendi geodesic survey point, where it comprises a shallowing-up, progradational set of stacked parasequences. From this point, its thickness diminishes both towards the north (i.e. seaward) and towards the south (i.e. landward). In the former case, the reduction is gradational and is coupled with a gradual change to more open platform facies (foraminiferal wackestones). In the latter, thinning is rapid, mainly by onlap onto the basal sequence boundary as well as by subsequent erosion so that this unit measures just a few tens of metres in section II (Fig. 7). This sedimentary package is interpreted as an LPC because of its overall geometry, basal relationships and vertical and lateral trends.

In section III, sequence 2.1 consists of two parts, the lower one formed by fossil-rich sandy limestones and the upper one characterized by the progressive encroachment of coralgal reefal limestones (Fig. 6). The same two parts can be recognized in the Alano ridge outcrop, where low-angle, seaward-dipping clinoforms are clearly visible in the upper part (Fig. 8), a diagnostic feature of HST deposits. The lower sandy limestones consequently are ascribed to the TST. In section IV, sequence 2.1 also has two parts, the lower one made up of sandy limestones and the upper one of bioclastic–oolithic limestones. They are interpreted as TST and HST, respectively, mainly by comparison with the succession of section III

Sequence 2.1 is characterized by a rich fossil assemblage. The more diagnostic associations have been observed in TST sandy limestones of section III (*Orbitoclypeus seunesi* and *Operculina heberti*), and in HST deposits of several other inner-platform sections (*Alveolina (Glomoalveolina) primaeva, Fallotela alavensis* and *Coskinon rajkae*). Both associations are characteristic of the *A. (G.) primaeva* biochronozone (Middle Thanetian).

Sequence 2.2

The lower boundary of this sequence coincides with an erosional surface at the platform margin (sections II and III, and Berrendi outcrop), and with an emersion surface characterized by microcodium biocorrosion, in section IV. It is thus a type 1 boundary. Within the study area, sequence 2.2 is never thick (reaching a maximum of 15 m) and is composed largely of bioclastic sandy limestones. In the Berrendi outcrop they infill a concave-up erosional depression (Fig. 7), possibly an incised valley or the headmost part of a submarine slope gully. In section III, this sequence is made up of burrowed sandy limestones with open platform foraminifera, and arranged in a retrogradational (deepening-up) succession. In section IV it is represented by protected inner-platform facies: wackestone–packstones with alveolinids and miliolids. In both cases, these sediments are provisionally interpreted as TST. Highstand deposits were either not fully developed or eroded by the overlying sequence.

The sandy limestones of section III contain a rich fossil assemblage with *Operculina azilensis, Assilina yvettae* and *Orbitoclypeus seunesi.* This is a typical association of the *A. (G.) levis* biozone, probably correlatable with the lower part of the *M. velascoensis* biozone (cf. Serra-Kiel *et al.*, 1991).

Sequence 2.3

This sequence corresponds with part of the 'Alveolina limestone', an informal lithostratigraphic unit well developed in the central and eastern Pyrenees (Ferrer *et al.*, 1973; Luterbacher, 1973; Eichenseer, 1988), in the southern and western part of the Basque area (Plaziat & Manguin, 1969) and even in the north Pyrenean zone (Plaziat, 1975b). Sequence 2.3 is poorly represented, however, in the sector of the platform considered here, being absent in sections I and II. In section III a residual thickness of 2 m is preserved and it is composed only of nummulite-rich marls, probably comprising a reworked transgressive lag. Finally, in section IV, sequence 2.3 consists of about 20 m of low-angle cross-bedded grainstones intercalated with wackestone/packstone couplets, the former interpreted as beach deposits, the latter interpreted as cyclic lagoonal sediments. This succession, interpreted as a TST, rests conformably on the underlying sequence, from which it is separated by a thin interval of glauconite-rich grainstones. These grainstones are interpreted as a possible transgressive (ravinement) lag deposit. The age of sequence 2.3 as latest Thanetian is well established with many alveolinid species of the *A. ellipsoidalis* and *A. cucumiformis* biozones.

The platform succession is bounded at its top by a surface of erosion and/or of non-deposition abruptly separating shallow-water Palaeocene carbonates from the overlying Eocene turbidites of the Hecho Group (Fig. 6). This dramatic change records the final drowning of the Palaeocene carbonate platform system and its progressive burying under southwards-advancing clastic turbiditic wedges (Labaume *et al.*, 1983; Seguret *et al.*, 1984). Such profound palaeogeographic inversion was initiated by the development of the foreland basin in this area (Puigdefabregas & Souquet, 1986) and was due therefore to tectonic causes (see below).

North Iberian basinal system (carbonate slope apron)

The sequence analysis of this sedimentary system involves the southern limb of the tight syncline characterizing the eastern tip of the Tertiary Biscay synclinorium, near the village of Eibar (Fig. 9(a)). This limb strikes at almost 90° with respect to palaeocurrents, affording therefore a transverse section of the system. There exist two main reference sections, both with 100% exposure, situated on the Egoarbitza peak and to the north of the Aixola reservoir (Fig. 9). The former is a scenic escarpment (Fig. 10(a)) that has been studied previously (Rat, 1959; Plaziat, 1975a; Plaziat *et al.*, 1975); the latter occurs in an abandoned quarry. The study of these sections, as well as of seven other less complete intervening sections, has led to the reconstruction of the internal facies architecture of the segment of the outcrop occurring between the reference sections. Reference sections were sampled for planktonic foraminifera, the results being summarized in Fig. 9(b) and in Appendix 1 (pp. 394–395).

The reconstructed Palaeocene succession comprises five stacked unconformity-bounded units, considered to be depositional sequences because of their internal configuration (see below). These depositional sequences range in thickness between 20 and 50 m, each generally consisting of just two parts ('lower' and 'middle'), except for the youngest one that is capped by an additional marly interval ('upper part').

The lower parts of the sequences are typified by a variable proportion (50–95%) of carbonate debris flow deposits (debrites) (Fig. 10(c)). These deposits are polygenetic; the following groups of clast lithologies are recognized in decreasing order of abundance: (i) Palaeocene intraformational mudstone and grainstones; (ii) Maastrichtian slope marls, usually deformed and contorted between limestone clasts; (iii) Palaeocene shelf-derived clasts, including coral–algal boundstones; (iv) a varied group of slope- and shelf-derived Cretaceous limestones, sandstones and volcanics. The maximum observed clast size is 9 m, although clasts larger than 2 m are comparatively rare. In most cases, they show a completely disorganized clast fabric (Fig. 10(c)), though some are normally graded and capped by thin grainstone beds. The debrites of sequence 1.3 occur as numerous isolated channel fills vertically and laterally intercalated with overbank turbidites and pelagic mudstones. The debrites of the other four sequences form amalgamated accumulations, convex-up at their tops, that fill in the irregular erosional topography characterizing the underlying sequence boundary (Fig. 9). They represent more proximal deposits than those of sequence 1.3. These carbonate debrites clearly are the record of impor-

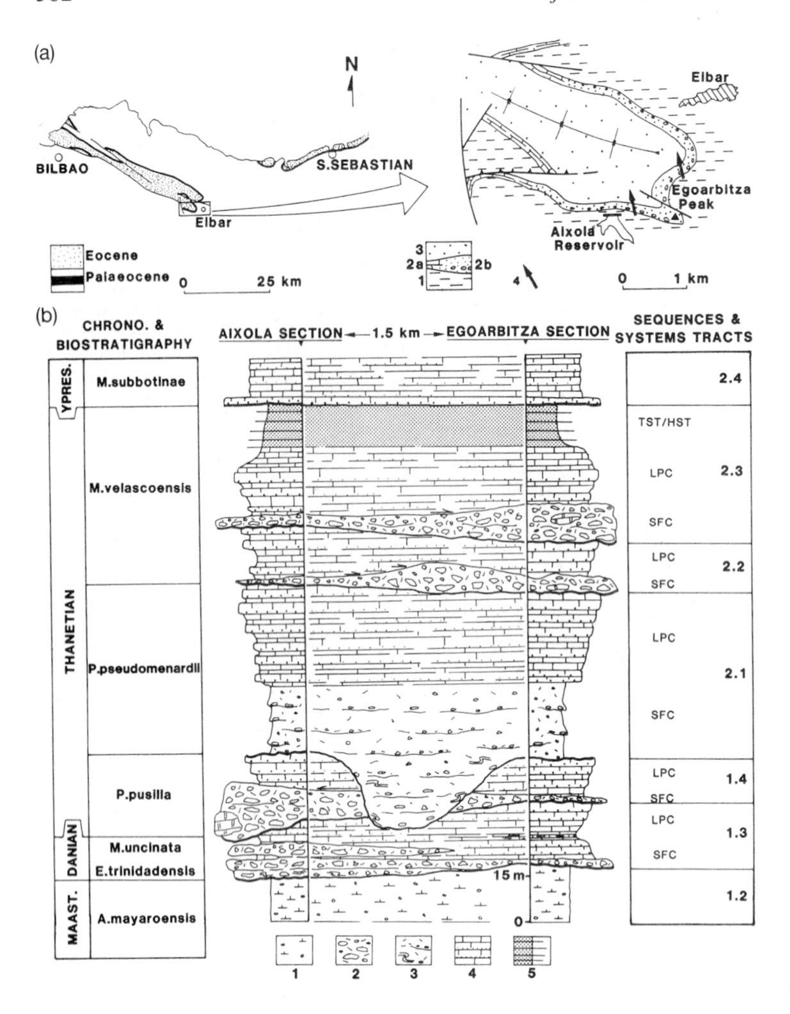

Fig. 9. (a) Location and geological map of the north Iberian carbonate slope apron deposits of the Eibar outcrop. 1, Maastrichtian; 2a, Palaeocene, predominance of coarse-grained deposits; 2b, Palaeocene, predominance of hemipelagic limestones; 3, Eocene; 4, palaeocurrents. (b) Cross-section of the southeast part of the outcrop and representative columnar sections. Note that the orientation of the cross-section is perpendicular to palaeocurrents. 1, Maastrichtian marls, with intercalation of debrites; 2, carbonate breccias; 3, muddy debrites; 4, thin- and thick-bedded carbonate turbidites and hemipelagic limestones; 5, marls and shales with intercalated thin-bedded turbidites (SFC = slope fan complex; LPC, TST and HST, as in Fig. 6).

tant resedimentation processes that remoulded and mixed a collection of coeval and older sediments, probably through the action of large-scale retrogressive slumps. Because of this and of their stratigraphical position, the lower parts of the sequences are interpreted as lowstand SFC.

Variations in the lithological composition of the SFC debrites within the different sequences do exist and can be related to the relative magnitude of the erosion and resedimentation processes. Thus, the debrites of sequence 1.3 are formed predominantly by Palaeocene intraformational clasts and suggest minor resedimentation. Those of sequences 1.4, 2.2 and 2.3 record somewhat more vigorous erosion, as they include a significant proportion (10–15%) of Palaeocene shallow-water clasts that attest to the collapse of earlier platform margins. Finally, the composition of the SFC debrites of sequence 2.1 reveals important submarine downcutting, since they are mostly formed by soft marly clasts and have a very large proportion of marly matrix (up to 60% of the bulk composition), derived both from Maastrichtian outer-platform as well as slope deposits. The great extent of submarine erosion linked to sequence 2.1 is further supported by the prominent erosional character of its basal boundary in the Eibar succession that in places is associated with the entire removal of underlying sequence 1.4 (Fig. 9(b)). This surface, which has an irregular scalloped geometry in cross-section, is completely crisscrossed by striations, grooves and flute-like depressions of different sizes oriented parallel to palaeocurrents (Fig. 10(b)). This boundary has been observed with similar features in several

Fig. 10. Outcrop photographs of the north Iberian carbonate slope apron deposits of the Eibar outcrop. (a) Egoarbitza peak outcrop, showing the stacking of three Palaeocene sequences. Note that the lower boundary of sequence 1.4 is a flat erosional surface (discontinuously strewn with Maastrichtian clasts), while that of sequence 2.1 has an irregular topography and is overlain by a relatively thick accumulation of muddy debrites (grass-covered). SFC, slope fan complex (note retrogradational stack in sequence 1.3); LPC, lowstand prograding complex (note progradational stack). (b) Close-up of the grooved, irregular surface of the lower boundary of sequence 2.1 and of the SFC muddy debrites that overlie it. (c) Disorganized SFC carbonate breccias of sequence 2.3 (scale = 1 m).

other outcrops as well. Its age has been narrowly constrained to the very latest portion of the *P. pusilla* biozone (Robador *et al.*, 1991) and must correspond therefore to the 58.5 Ma sea-level fall, the most pronounced during the Palaeocene (cf. Haq *et al.*, 1988).

The middle parts of the depositional sequences comprise an irregular alternation of globigerinid mudstones/wackestones with thin- and thick-bedded turbidites. The globigerinid mudstones/wackestones may occur as individual beds or amalgamated stacks, but more commonly appear as gradational units overlying the thin-bedded turbidites. They are interpreted, therefore, as hemipelagic limestones and/or very dilute turbiditic deposits. Thin-bedded turbidites range in thickness between 1 and 20 cm, and are characterized by base-missing Bouma sequences. They are fine-grained intrabasinal bioclastic grainstones, largely or exclusively composed of reworked globigerinids tests. Occasional small-scale slumps and slides involving both the mudstones and the thin-bedded turbidites attest to deposition on gentle slopes. Thick-bedded turbidites are non-channelized, laterally continuous tabular beds 0.30 to 5 m thick, that usually display good grading and sometimes complete Bouma sequences. They are bioclastic rudstones and grainstones, mostly shelf-derived (coraline algae, benthic foraminifera, bryozoan, crinoid and coral fragments). Such predominance of bioclasts further indicates resedimentation of unconsolidated sediments. The attributes of these thick-bedded turbidites are therefore analogous with the so-called 'shingled turbidites' of Vail *et al.* (1991).

The middle parts of the sequences are interpreted as LPC because of their facies and stratigraphic position. This conclusion is also supported by the following observations: (i) the turbidites of the middle parts of the depositional sequences generally bury the underlying carbonate debrites by progressive downlap; (ii) the lower and middle parts of sequence 1.3 are separated in the Egoarbitza section by a 2 m-thick interval of highly bioturbated marls and mudstones, probably the expression of a condensed section (Fig. 10(a)); (iii) with the only exception of sequence 2.3, in which no clear vertical trend is apparent, the thickness and frequency of occurrence of the thick-bedded (shingled) turbidites increase upwards (Figs 9 & 10(a)), suggesting progradation.

The upper part of sequence 2.3 is composed of dark-grey marls with common intercalations of thin-bedded bioclastic turbidites. Although of relatively modest thickness (15 m), it has great lateral extent, serving as a key horizon for the separation of Palaeocene and Eocene sediments in the Biscay synclinorium. This marly part is thought to be the basinal equivalent of the TST and/or HST, because of its sequential position and widespread distribution.

Palaeocene sequences are covered in Eibar, as in the whole Biscay synclinorium, by a thick (1500–2000 m) early–middle Eocene succession (Orue-etxebarria, 1983). It consists of a stack of turbiditic sediments (mostly SFC deposits) of several sequences, each a few hundred metres thick and laterally extensive. The lowest of them (sequence 2.4) comprises several beds of thick-bedded bioclastic turbidites overlying a *basal erosional boundary* (BFC?). The remainder of the section is represented by an alternation of thin-bedded carbonate turbidites and mudstones (distal SFC). This sequence is dated as earliest Ypresian (Fig. 9(b) and Appendix 1, pp. 394–395)), and records a sudden, important increase of sedimentation rates, the result being the development of turbiditic systems of much greater thickness and areal extent than those of the Palaeocene.

North Pyrenean basinal system (deep-sea channel system)

The sequence stratigraphy of this Palaeocene system, and of the sediments immediately underlying and overlying it, was inspected at two separate outcrops, 'Zumaia' and 'Orio', both situated in the western part of the Guipuzcoa monocline (Fig. 11(a)). The Zumaia outcrop is a very well known sea-cliff section that has been the subject of numerous studies mainly, but not exclusively, from a biostratigraphic perspective (Herm, 1965; von Hillebrandt, 1965; Crimes, 1973; Kapellos, 1974; Kruit *et al.*, 1975; Van Vliet, 1982; Mount & Ward, 1986; Lamolda *et al.*, 1988; Ward, 1988; Pujalte *et al.*, 1989). It is, in fact, one of the global reference sections for calcareous nanoplankton and planktonic foraminifera biostratigraphy. The Orio outcrop is a 12 km-long inland exposure that extends from this locality to the east (Baceta *et al.*, 1991). It has received much less attention than the Zumaia section, although some references to it were made by Gómez de Llarena (1954), Hanisch and Pflug (1974) and Van Vliet (1982).

Our own data and interpretations are depicted in

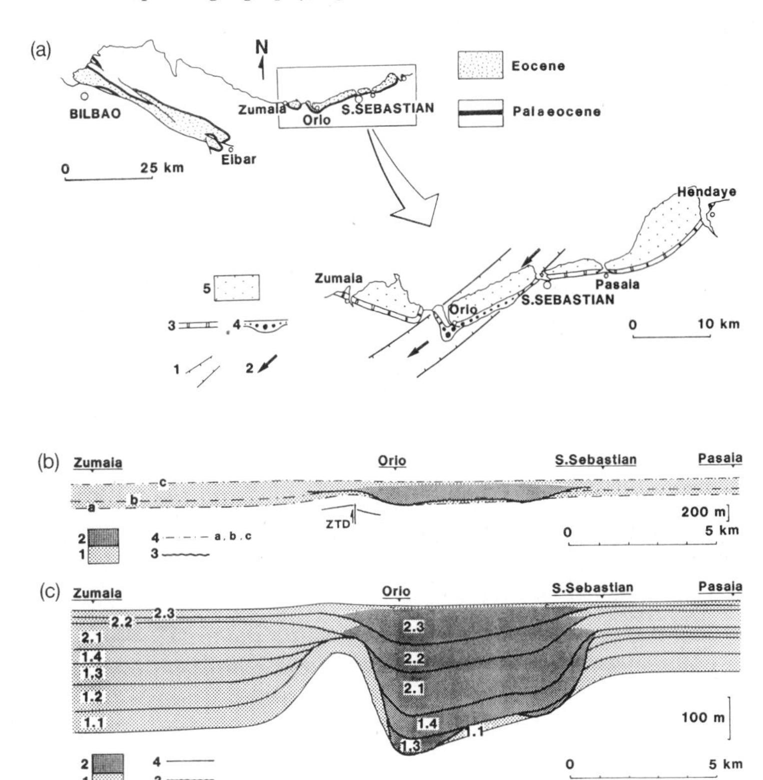

Fig. 11. (a) Location and outcrop map of the north Pyrenean basinal system in the Guipuzcoa monocline. 1, reconstructed position of the deep-sea channel margins; 2, palaeocurrents; 3, Palaeocene autochthonous interchannel deposits; 4, Palaeocene resedimented channel-fill deposits; 5, Eocene. (b) Reconstructed cross-section of the late Maastrichtian–Palaeocene succession. Note the relative (×2) vertical exaggeration: 1, pre-channel and interchannel deposits; 2, channel-fill deposits; 3, erosional boundaries; 4, isochronous lines (a, *gansseri/mayaroensis* biochronozones boundary; b, Cretaceous–Tertiary boundary; c, Palaeocene–Eocene boundary). (c) Reconstructed cross-section of the same succession, with vertical exaggeration (×10), showing the geometry of depositional sequences. 1 and 2, as above; 1.1 to 2.3, depositional sequences; 3, erosional boundaries; 4, conformable boundaries; ZTD = Zarautz tectonic disturbance.

Figs 11 and 12 and discussed below. The biostratigraphy of the Zumaia section is primarily based on published information, although critical boundaries were additionally re-examined for this study; the biostratigraphy of the Orio section is based entirely on our own data (Fig. 12 and Appendix 1, pp. 394–395). Latest Cretaceous and earliest Eocene sequences are included in the description below because they show significant differences with those of the Palaeocene. It should also be noted that the occurrence of a tectonic disturbance, probably a deep-seated fault (Pujalte *et al.*, 1989), is expressed as a diapiric piercement of Keuper marls just to the east of Orio ('Zarautz diapir' of Hanisch & Pflug, 1974). This structural event yielded a local hiatus in the latest Cretaceous–Palaeocene succession and determined locally the position of the northwest margin of the deep-sea channel (Fig. 11(b)).

Latest Maastrichtian–earliest Danian sequences

The succession of this interval in the Zumaia section is composed of two depositional sequences correlatable with global cycles 1.1 (latest Maastrichtian) and 1.2 (latest Maastrichtian–earliest Danian) (cf. Haq *et al.*, 1988). Each depositional sequence is composed of three parts (Figs 12 & 13(a)).

The lower part consists primarily of purple and greenish limey marls, with frequent intercalations of thin-bedded turbidites. The marls are structureless and, according to Mount and Ward (1986), have insoluble residue values that range from 46–72%, averaging 55% (including illite clays, quartz silt, iron oxides and pyrite). The turbidites are base-missing Bouma sequences (Tbc or Tc) and have great lateral continuity. They are very frequent (up to 25 turbidites per metre) but thin, in general just several millimetres to a few centimetres thick. There are, however, occasional thicker beds (10–25 cm) that occur preferentially in the lower and middle parts of the marly intervals, diminishing in thickness and frequency upwards. These thicker turbidites are usually bioclastic grainstones, sometimes with a small percentage of detrital quartz. The fossil content of the interval is dominated by benthic and planktonic foraminifera, with some

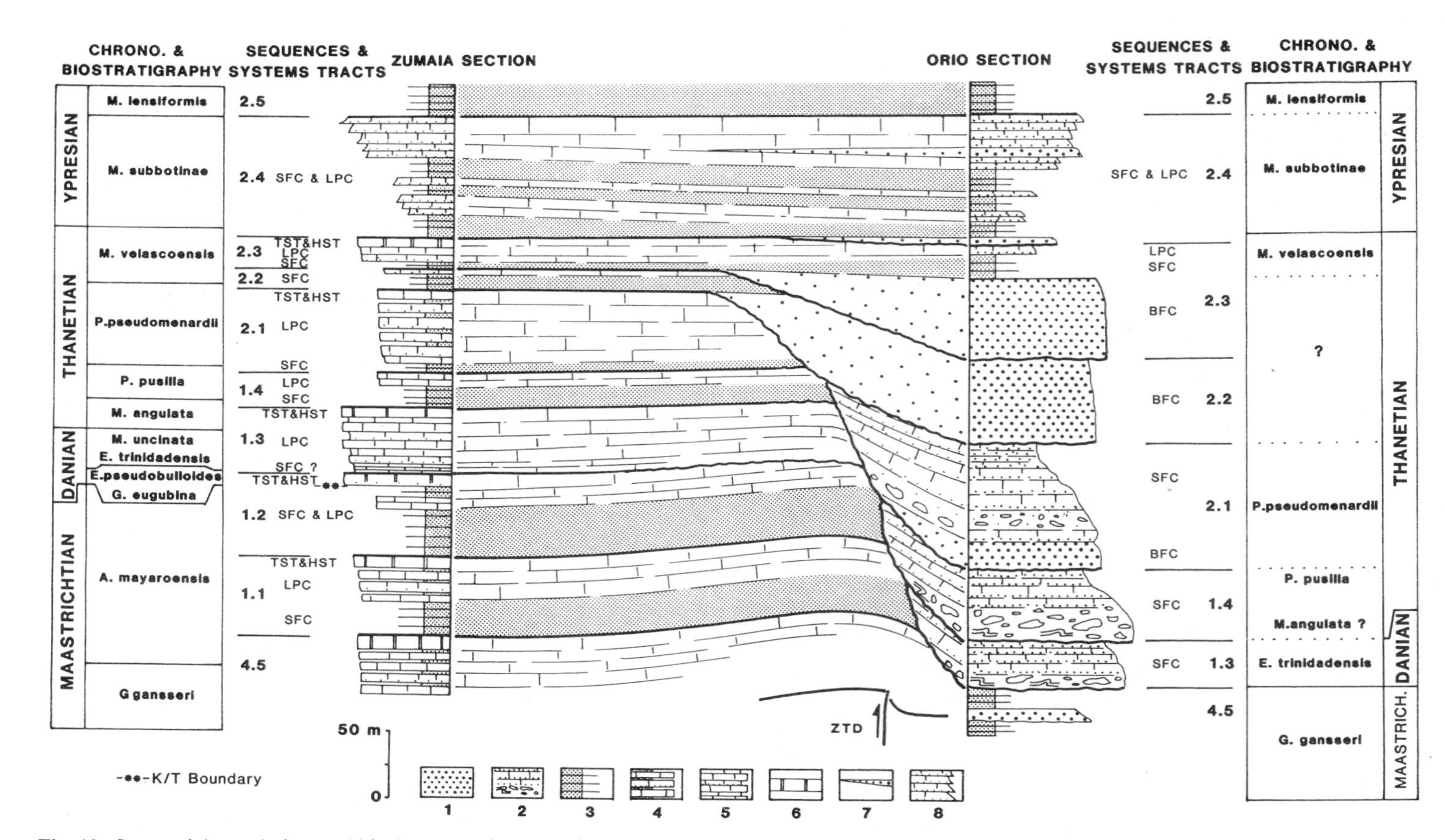

Fig. 12. Sequential correlation and biochronostratigraphy of the two sections that typify the north Pyrenean basinal system in the Guipuzcoa monocline, offering a reconstructed transverse cross-section of the system (see Fig. 11 for location). BFC, basin-floor fan complex; SFC, LPC, TST and HST as in Figs 6 and 9. 1, Amalgamated stack of coarse-grained siliciclastic turbidites; 2, slump and debrites with large blocks at base, grading up into coarse-grained carbonate turbidites; 3, marls with intercalations of thin-bedded turbidites; 4, pelagic limestone/marl alternations, with variable proportion of thin-bedded turbidites; 5, stack of pelagic limestone with intercalated thin marls and occasional thin-bedded turbidites; 6, pelagic limestones, usually condensed ('bunched'); 7, isolated thick-bedded siliciclastic turbidites; 8, stack of thin-bedded turbidites with intercalated pelagic limestones, marls and/or shales. ZTD, Zarautz tectonic disturbance.

Fig. 13. Outcrop photographs of the Zumaia section. Continuous lines indicate sequence boundaries, dotted lines, systems tract boundaries. (a) Late Maastrichtian–earliest Palaeocene sequences at San Telmo point (note K/T boundary). (b) Palaeocene sequences, eastern part of Zumaia beach. (c) Latest Palaeocene–earliest Eocene sequences, central part of Zumaia beach (sequences 2.2 and 2.3 are tectonically duplicated). T–H, transgressive and highstand deposits; SFC and LPC, as in Fig. 9.

ammonites, echinoderms and bivalves. These lower marly parts of the sequences are interpreted as distal SFC, because of their largely terrigenous composition, slightly retrogradational character and stratigraphic position. The basal boundaries of both SFC units are abrupt but essentially non-erosional, neither of them suggesting a hiatus, at least one that is biostratigraphically recognizable. Therefore, these well-defined surfaces probably represent the distal, conformable surfaces of the respective sequence boundaries.

The middle parts of the depositional sequences are mostly made up of light-grey limestones and purple marls, which in sequence 1.1 occur rhythmically interbedded and in similar proportion. In contrast, in sequence 1.2 the marls dominate the section (Figs 12 & 13). Thin-bedded bioclastic turbidites occur occasionally within each sequence. The limestones are pelagic/hemipelagic foraminiferal mudstones and wackestones, with insoluble residues that range from 20–29%, averaging 23% (data from Mount & Ward, 1986). Individual beds range in thickness from 10 to 40 cm and do not show any clear vertical trend. The marls are lithologically similar to those of the lower parts of the sequences. Because of their lithology, aggradational character and stratigraphic position, the middle parts of the sequences are interpreted as LPC deposits. Similar alternations of pelagic limestones and marls do indeed typify the LPC in other deep pelagic realms (i.e. Magniez-Jannin & Jacquin, 1990).

The upper parts are formed in both sequences by thin caps (1 to 2 m) of pelagic limestone, either amalgamated or separated by very thin shale partings. In sequence 1.2, at least, this cap is demonstrably the result of very low resedimentation rates, for this unit comprises two planktonic foraminiferal zones spanning 3.5 Ma (Fig. 12). Similar extremely starved conditions are more difficult to prove in sequence 1.1, but it is interesting to note that these particular beds used to be relatively enriched in ammonites, though now largely removed by collectors. The upper parts of both sequences are therefore interpreted as very condensed basinal equivalents of transgressive and/or highstand systems tracts.

Sequences 1.1 and 1.2 are largely missing in the Orio section, due to earliest Palaeocene erosion associated with formation of the Palaeocene deep-sea channel (see below). However, the sequences occur with similar features and thickness in widely separated coastal sections of the Bay of Biscay, from Sopelana in the western tip of the Biscay synclinorium to Bidart on the French coast (cf. Ward, 1988; Razin, 1989).

Danian–Thanetian sequences

Five successive depositional sequences have been identified within the Danian–Thanetian interval corresponding to cycles 1.3, 1.4, 2.1, 2.2 and 2.3 of the global chart (Fig. 12 and Appendix 1, pp. 394–395; Haq *et al.*, 1988). These, however, lack the lateral continuity and homogeneity of the underlying sequences 1.1 and 1.2, having a very different character in Orio (channel-axis succession) and in Zumaia (interchannel succession) (Figs 11, 12 & 13).

The sequences are commonly characterized in the Orio section by significant basal erosional surfaces, deeply cut into the underlying deposits. This erosion, which is thought to have been produced by highly competent gravity flows bypassing the area, repeatedly deepened the floor of the submarine channel. The principal evidence for this interpretation is the observation of erosional truncation of sequences 1.1 and 1.2 in the Orio section. Sequence 1.3 (Danian) directly overlies middle Maastrichtian turbidites of the *G. gansseri* biochronozone (see Appendix 1, pp. 394–395); the apparent hiatus suggests downcutting of at least 150 m (i.e. the thickness in the Zumaia section of the missing interval). Regional mapping and geometrical reconstruction suggest similar magnitudes of erosion occurring at the other four sections as well (Figs 11 & 12).

The sediments directly overlying the lower erosional boundaries of sequences 2.1, 2.2 and 2.3 are the only substantial accumulations of coarse-grained siliciclastic deposits (middle- and coarse-sized sandstones and pebbly sandstones) of the studied succession. These deposits comprise siliciclastic turbidites stratified in metre-scale beds, either amalgamated or separated by thin shale partings and stacked aggradationally. They have sharp, lower and upper boundaries, and are confined to the channel-axis area. These deposits are, in short, sand-prone discrete sedimentary bodies with clear-cut limits, their features being diagnostic of basin-floor fan complexes (BFC) (cf. Posamentier & Vail, 1988).

Sequence 1.3 begins in the Orio section with a discontinuous accumulation of cobbles and boulders of up to 6 m diameter. Most clasts are irregular and their lithologies can be correlated readily with a part of the underlying missing interval (i.e. late Maastrichtian turbidites, purple marls and pelagic mudstones, or earliest Danian pelagic limestones). The lower part of sequence 1.4 consists of an accumulation of muddy debris-flow deposits, mainly composed of large contorted clasts of late Maastrichtian marls, but also with occasional inclusions of intraformational and shallow-water clasts. These two disorganized accumulations clearly are the result of collapse, slide and slump processes on the eroded slope and/or on the channel-side walls. They are the record of the transition from the erosional to the depositional phase in the channel, and can be considered therefore as lower or proximal SFC deposits.

The remainder of sequences 1.3, 1.4 and 2.1 in the Orio section consists of thick- and thin-bedded carbonate turbidites, with lesser intercalation of (hemi-) pelagic limestones and marls. Thick-bedded turbidites are normally graded rudstones and/or coarse-grained grainstones. The grainstones are largely bioclastic, being formed mainly by calcareous algae (50%) as well as fragments of echinoderms, bryozoa, oyster, corals and benthic foraminifera. Rudstones have a similar composition but also include intraformational and shallow-water lithoclasts. They range in thickness between 0.5 and 10 m, and are stacked with a somewhat irregular fining-upwards trend. Hemipelagic limestones and thin-bedded turbidites are lithologically identical with those of Eibar (north Iberian basinal system), being respectively the result of autochthonous sedimentation and minor intrabasinal reworking. They are more abundant in sequence 1.3 and in the upper part of sequence 2.1. Together, these carbonate sediments account for the bulk of the preserved portions of the three sequences (Fig. 12), and must record therefore the main channel-fill stages of sedimentation. They are thus analogous to the type II turbidites that fill the Eocene channel-like erosional features near Ainsa in the south-central Pyrenees (Mutti *et al.*, 1985), and interpreted therefore as SFC.

The character of the Danian–Thanetian sequences is very different in the Zumaia section. Firstly, they are there separated by abrupt but non-erosional boundaries, at which no biostratigraphically recognizable hiatus can be observed. Secondly, they primarily consist of hemipelagic limestones and marls intercalated with thin-bedded turbidites. Three parts generally can be recognized in these sequences: the lower part is dominated by marls with thin-bedded turbidites; the middle and upper parts are composed either of hemipelagic limestones (foraminiferal mudstones and wackestones) or of cyclically arranged limestone/marl alternations (Figs 12 & 13). These three parts are lithologically similar to those of the underlying sequences 1.1 and 1.2, and occur in identical sequential position. They are therefore interpreted in the same way, namely as SFC, LPC and TST/HST deposits, respectively. It should be noted, however, that the relative development of SFC deposits varies greatly, being comparatively thick only in sequences 1.4 and 2.2 (Figs 12 & 13). Since SFC marls and thin-bedded turbidites are probably the result of overspilling from the Orio channel (levees?), the extent of their development is thought to be related to the depth of channel incision.

Early Ypresian sequences

Early Eocene sequences are similar in both the Zumaia and Orio sections, and indeed in the whole of the Guipuzcoa monocline. The oldest of these sequences, for example, is a well-defined cartographic unit that can be traced with few changes from the Spanish–French border in the east to Zumaia in the west (i.e. 'pre-fan deposits' of Kruit *et al.*, 1975; 'pre-megacycle 1' deposits of Van Vliet, 1982; 'Hondarribia sequence' of Rossel *et al.*, 1985). It consists of an irregular alternation of thin-bedded turbidites, marls and hemipelagic limestones (Figs 12 & 13) unanimously interpreted as basin-plain deposits by the authors referred to above. We concur as well, tentatively ascribing these sediments to the distal part of a large-scale slope fan complex. They are well dated in both the Orio and Zumaia sections, occurring within sea-level cycle 2.4 of Haq *et al.* (1988).

This basal Ypresian sequence is somewhat thicker in Orio than in Zumaia, and containing several sections of interbedded thick-bedded siliciclastic turbidites in the former locality that are not present in the latter (Fig. 12). Both features suggest that contemporaneous residual relief and/or slightly different subsidence rates continued to persist between the Orio and Zumaia areas during the earliest Eocene. Such differences are not recorded in the rest of the preserved Eocene succession (cf. Van Vliet, 1982).

MAIN CONTROLS OF SEDIMENTATION

Eustasy

The Palaeocene of the western Pyrenees has been observed to be composed of five depositional sequences, both in shelfal and basinal settings. This is precisely the same number of third-order sea-level cycles observed by Haq *et al.* (1988) for this interval. Where dated, these sequences can be matched further with their corresponding third-order cycles. Finally, the major sequence boundary of the succession can be tied to the most pronounced sea-level drop of the Palaeocene and, in the basin, the thickest sequence (i.e. 2.1) corresponds to the longest lasting cycle. We believe, therefore, that there is ample evidence in the studied succession to suggest strongly a cause–effect relationship between eustasy and sequence development.

Tectonics

Owing to the oblique collision between the Iberian and European plates, a diachronous transition from extensional to compressional tectonics took place in the Pyrenees from the late Cretaceous to the early Tertiary. Tectonic unrest began in the eastern Pyrenees in the late Cretaceous, and was clearly recorded in the Palaeocene by the development of the first thrust sheet, and by the subsequent emergence of inner parts of the chain with concomitant deposition of a thick red-bed continental succession (Plaziat, 1981; Muñoz *et al.*, 1986; Puigdefabregas & Souquet, 1986).

In contrast, the study area experienced a phase of relative tectonic quiescence during the Palaeocene. Thus, on the north Iberian shelf, the stack of comparatively thin carbonate platforms clearly suggests low subsidence rates. A closer inspection of these units reveals, however, subtle influences of changing tectonics, including the relative displacement of successive carbonate platforms. In effect, the overall (first-order) fall of sea level had determined the development of each of the latest Cretaceous platforms in a more basinward position than the preceding one. Although the general sea-level lowering continued well into the early Eocene, this trend was arrested and even slightly reversed during the Palaeocene. Successive platform sequences were stacked with a minor backstepping pattern (Fig. 6). The enhancement of sequence boundaries, all of which exhibit features typical of type 1 boundaries, also can be ascribed to tectonics.

The Pyrenean chain was fully transformed into a thrust and foldbelt foreland basin during the earliest Eocene (Seguret, 1972; Puigdefabregas & Souquet, 1986). This tectonic event is recorded in the study area by a major change in the subsidence patterns and/or sedimentation style. Thus, the former Palaeocene carbonate platform system was drowned and buried by thick wedges of turbidites (i.e. Hecho Group), while subsequent Eocene carbonate platforms experienced a significant landward retreat (Labaume *et al.*, 1983; Seguret *et al.*, 1984; Puigdefabregas & Souquet, 1986). A contemporaneous rapid increase in subsidence occurred in the basinal setting. This was coupled with an expansion and development of large-scale turbiditic systems, coincident with the initial subdivision of the basin into different domains by large-scale folds (Pujalte *et al.*, 1989).

Sedimentary supply

Probably as a direct outcome of the latest Cretaceous to Palaeocene period of tectonic quiescence, the basinal setting became progressively starved. In the Zumaia section, for instance, sedimentation rates dropped from 200 m/Ma (compacted) for the Lower Maastrichtian section to less than 80 m/Ma for the Upper Maastrichtian one (Ward, 1988). As noted before, however, the late Maastrichtian sequences 1.1 and 1.2 have great lateral persistence of thickness and facies, clear evidence that the sedimentary supply was still sufficient to mantle most of the basin floor.

Sedimentary supply was further reduced, however, during the Palaeocene, the lowest rates occurring during the Danian. This may, in part, explain the stability and very low siliciclastic content of the north Iberian carbonate platform–slope apron systems. But the more important effect was felt in the central parts of the basin, where the whole of the Palaeocene is locally represented only by a few tens of metres of pelagic limestones (Agirre *et al.*, 1987).

The initial development of the north Pyrenean deep-sea channel system coincides with the time of maximum starvation of the basin (i.e. Danian, sequence 1.3), and it lasted while starvation persisted (i.e. until the end of the Palaeocene). Consequently, a genetic link between these processes is strongly suspected. In effect, according to Carter (1988), the fundamental control on the development of a deep-sea channel is the submarine base level, which operates in much the same way as the sea level does for drainage systems on land. Therefore, a sedimentary supply unable to compensate for differential subsidence would favour the entrenchment of turbidite flows (faster subsidence rates are assumed to the west of the study area because of the persistence of westwards directed palaeocurrents from Campanian to middle Eocene times, even during intervals of large sedimentary input; see, for instance, Van Vliet, 1982 and Mathey, 1986).

The process occurred in a cyclic way. Submarine base level was lowered during high sea-level positions, when sedimentary influx was most drastically reduced, while at lowstands turbidite currents first excavated the channels and, later on, partially backfilled them. The oldest turbidite flows were probably funnelled through the basin via structural and/or morphological depressions. This possibility is suggested, for instance, by the presence of a contemporaneous tectonic event in the northwest margin of the Orio channel. In any case, the stacking of channel-fill deposits of successive sequences (Figs 11 & 12) demonstrates that the courses of the younger channels were dictated by the location of the oldest one.

As would be expected, the massive arrival of sediments at the beginning of Eocene times ended the development of the deep-sea channel system.

CONCLUSIONS

The Palaeocene palaeogeography of the western Pyrenees was largely inherited from Mesozoic times, with a central deep-water basin flanked north, east and south by shallower shelf areas. Most of the shelf areas were covered with carbonate platforms, while in the deep basin at least two different types of sedimentary systems were developed: (i) base-of-slope carbonate aprons, linked to the north Iberian carbonate platform, and (ii) long-lived, multi-phase submarine channels, that were nourished predominantly by axial flows sourced from the eastern and perhaps also from the northern margin.

Eustatically controlled depositional sequences have been differentiated in all of these different settings. The facies association within systems tracts and their vertical arrangement within depositional sequences can also be explained within the context of sea-level changes, reinforcing the applicability of the sequence-stratigraphy models of Sarg (1988), Haq (1991) or Vail *et al.* (1991) to this essentially carbonate setting.

Thus, the carbonate platform consists mainly of retrogradational TST and progradational HST, the former being composed primarily of grainstones or sandy grainstones, the latter by reefal limestones. The shallow portion of the LPC is a major component of the outer-platform margin in two of the sequences. Lowstand systems tracts, on the other hand, account for the bulk of basinal sediments, TST and HST being there always thin and condensed. No carbonate BFC has been positively recognized. However, siliciclastic BFCs, largely composed of coarse-grained turbidites, occur at the base of three of the sequences in the deep-sea channel system. SFCs and LPCs appear in all of the sequences. In the north Iberian carbonate slope apron, SFCs are made up primarily of resedimented breccias, while the LPCs are formed by stacks of thick-bedded (shingled) turbidites, thin-bedded turbidites and hemipelagic limestones. In the deep sea channel system, SFCs occur both as a channel-fill facies (thick- and thin-bedded carbonate turbidites) and as a channel-margin facies (marls with thin-bedded turbidites). LPCs are represented by cyclic alternation of hemipelagic limestones and marls, with widespread distribution in channel-margin and interchannel areas.

A transition from extensional to compressional conditions occurred diachronously in the Pyrenees from the latest Cretaceous to the early Tertiary, influencing sedimentation in several ways. For instance, patterns of subsidence were modified determining the backstepping of Palaeocene carbonate platforms as well as final drowning and enhancement of sequence boundaries. Also, tectonics controlled the sedimentary supply to the basinal area, a most important factor in the development and evolution of the north Pyrenean basinal system. It is important to note, in any event, that the eustatic signature of the succession remained in spite of this tectonic overprint. Variations in the configuration of the sequences are, in fact, a clue to understanding tectonically induced palaeogeographic modifications.

ACKNOWLEDGEMENTS

We are grateful to B. Haq, J. Rosell, J. Smit, P.R. Vail and J. Van Hinte for stimulating discussions in the field. G.P. Allen, P. Crumeyrolle and P. Razin are warmly thanked for their helpful reviews, and H.W. Posamentier for making significant improvements to the text. Finally, we are indebted to J. Serra-Kiel and co-workers for the identification of benthic faunas and to B. Bernedo for the typing of the manuscript. This research was supported by the Basque Country University projects no. 121.310-0158/89 and 121.310-EO55/90, and by the IGTE project 'Geological Study of the South-Pyrenean Marine Palaeocene'.

REFERENCES

Agirre, A., Orue-Etxebarria, X. & Arriola, A. (1987) Contribución a un mejor conocimiento del tránsito Cretácico-Terciaro y del Paleoceno en el flanco norte del Sinclinorio de Bizkaia mediante los foraminíferos planctónicos. *Kobie* **XVI**, 185–214.

Baceta, J.I., Pujalte, V., Robles, S. & Orue-Etxebarria, X. (1991) Influencia del Diapiro de Zarautz sobre los procesos de resedimentación paleocenos de Orio (Guipuzcoa, Cuenca Vasca). *Geogaceta* **9**, 57–60.

BRGM, ELF-Re, ESSO-REP & SNPA (1974) Géologie du bassin d'Aquitaine. Atlas. BRGM. ed.

Carter, R.M. (1988) The nature and evolution of deep-sea channel systems. *Basin Res.* **1**, 41–54.

Crimes, T.P. (1973) From limestones to distal turbidites: a facies and trace fossil analysis in the Zumaya flysch (Paleocene–Eocene), North Spain. *Sedimentology* **20**, 105–131.

Eichenseer, H. (1988) Facies geology of late Maastrichtian to early Eocene coastal and shallow marine sediments, Tremp-Graus Basin, northeastern Spain. PhD. thesis. In: *Arb. Inst. u. Mus. Geol. Pal. Universität Tübingen* **1**, 273 pp.

Faber, J. (1961) Paléogéographie et sédimentologie du Danien et du Paléocène de la région de Pau. *Rev. de l'Instit. Français du Pétrole* **XVI**, 907–913.

Ferrer, J., Calvez, Y., Luterbacher, H.P. & Premoli Silva, I. (1973) Contribution à l'étude des foraminifères de la région de Tremp (Catalogne). *Mém. Muséum Nat. Hist. Nat.*, **XXIX**, 1–107.

Flicoteaux, R. (1972) L'analyse stratonomique, méthode d'étude des différents faciès du flysch paléocène de Pau et de leur milieu de dépôt. Application aux problèmes posés par les corrélations entre affleurements et sondages. In: *Coll. Méthodes et Tendances de la Stratigraphie. Mém. B.R.G.M.* **77**, 615–626.

Freytet, P. & Plaziat, J.C. (1982) Continental carbonate sedimentation and pedogenesis. Late Cretaceous and Early Tertiary of Southern France. *Contr. Sediment.* **12**, 213 pp.

Gomez De Llarena, J. (1954) Observaciones geológicas en el Flysch Cretaceo-Nummulitico de Guipuzcoa. *Monogr. Inst. Lucas Mallada de Invest. Geol.* **15**, 98pp.

Hanisch, J. & Pflug, R. (1974) The interstratified breccias and conglomerates in the Cretaceous Flysch of the Northern Basque Pyrenees: submarine outflow of diapiric mass. *Sediment. Geol.* **12**, 287–296.

Haq, B.U. (1991) Sequence stratigraphy, sea-level change, and significance for the deep sea. In: *Sedimentation, Tectonics and Eustasy: Sea-level Changes at Active Margins* (Ed. Mcdonald, D.) Spec. Publ. Int. Assoc. Sediment. 12, 3–39.

Haq, B.U., Hardenbol, J. & Vail, P.R. (1988) Mesozoic and Cenozoic chronostratigraphy and cycles of sea-level change. In: *Sea-Level Change: An Integrated Approach* (Eds Wilgus, C.K., Hastings, B.S., Kendall, C.G.St.C., Posamentier, H.W., Ross, C.A. & Van Wagoner, J.C.) Spec. Publ. Soc. Econ. Paleontol. Mineral., Tulsa, 42, 71–108.

Herm, D. (1965) Mikropaläontologisch-stratigraphische untersuchungen in Kreideflysch zwischen Deva und Zumaya (prov. Guipuzcoa, Nordspanien). *Z. dt. Geol. Ges.* **115**, 227–348.

Kapellos, C. (1974) Uber das Nannoplankton im Alttertiär des Profils von Zumaya-Guetaria (provinz Guipuzcoa, Nordspanien). *Eclog. Geol. Helv.* **67**, 435–444.

Kieken, M. (1973) Evolution de l'Aquitaine au cours du Tertiaire. *Bull. Soc. Géol. France* **XV**, 40–50.

Kruit, C., Brouwer, J., Knox, G., Schollinberger, W. & Van Vliet, A. (1975) An excursion to the Tertiary deep-water fan deposits near San Sebastian (province of Guipuzcoa, Spain). *9th International Congress of Sedimentology*, Nice, 80 pp.

Labaume, P., Mutti, E., Seguret, M. & Rosell, J. (1983) Mégaturbidites carbonatées du bassin turbiditique de l'Eocène inférieur et moyen sud-pyrénéen. *Bull. Soc. Géol. France* **XXV**, 927–941.

Lamolda, M.A., Mathey, B. & Wiedmann, J. (1988) Field-guide excursion to the Cretaceous–Tertiary boundary section at Zumaya (Northern Spain). *Rev. Esp. de Paleontologia* No. Extraord., 141–155.

Leon, L. (1972) Síntesis Paleogeográfica y Estratigráfica del Paleoceno del Norte de Navarra. Paso al Eoceno. *Bol. Geol. Min.* **83**, 234–241.

Luterbacher, H.P. (1973) La sección tipo del Piso Ilerdiense. In: *XIII Col. Europ. de Micropaleontología, España* (Ed. Perconig, E.), Enadimsa, 113–140.

Magniez-Jannin, F. & Jacquin, T. (1990) Validité du découpage séquentiel sensu Vail en domaine de bassin: arguments fournis par les foraminifères dans le Crétacé inférieur de Sud-Est de la France. *C.R. Acad. Sci., Paris,* **310**, 263–269.

Mangin, J.P. (1959) Le Nummulitique sud-pyrénéen à l'Ouest de l'Aragon. *Pirineos,* **45**, 631 pp.

Mathey, B. (1986) Les flyschs Crétacé supérieur des Pyrénées Basques. Age, anatomie, origine du matérial, milieu de dépôt et rélations avec l'ouverture du golfe de Gascogne. Phd thesis, University of Bourgogne, Dijon, 403 pp.

Mount, J.F. & Ward, P. (1986) Origin of limestone/marl alternations in the Upper Maastrichtian of Zumaya, Spain. *J. Sediment. Petrol.* **56**, 228–236.

Mullins, H.T. & Cook, H.E. (1986) Carbonate apron models: alternatives to the submarine fan model for paleoenvironmental analysis and hydrocarbon exploration. *Sediment. Geol.* **48**, 37–79.

Muñoz, J.A., Martinez, A. & Verges, J. (1986) Thrust sequences in the eastern Spanish Pyrenees. *J. Struct. Geol.* **8**, 399–405.

Mutti, E., Remacha, E., Sgavetti, M., Rosell, J., Valloni, R. & Zamorano, M. (1985) Stratigraphy and facies characteristics of the Eocene Hecho Group Turbidite systems, south-central Pyrenees. In: *Excursion Guidebook of the 6th IAS Europ. Meet.* (Eds Mila, M.D. & Rosell, J.) Institut d'Estudis Ilerdences, Lleida, 519–576.

Orue-Etxebarria, X. (1983) Los foraminíferos planctónicos del Paleógeno del sinclinorio de Bizkaia (corte de Sopelana-Punta Galea). Parte 1. *Kobie* **XIII**, 175–249.

Plaziat, J.C. (1975a) Signification paléógéographique des 'calcaires conglomérés', des brèches et des niveaux à Rhodophycées dans la sédimentation carbonatée du bassin Basco-Béarnais a la base du Tertiaire (Espagne, France). *Rév. Géog. Phys. Géol. Dynam.* **XVIII,** 239–258.

Plaziat, J.C. (1975b) L'Ilerdien à l'intérieur du Paléogène languedocien; ses relations avec le Sparnacien, l'Ilerdien sud-pyrénéen, l'Yprésien et le Paléocène. *Bull. Soc. Géol. France* **XVII**, 168–181.

Plaziat, J.C. (1981) Late Cretaceous to Late Eocene palaeogeographic evolution of southwest Europe. *Palaeogeogr., Palaeoclim., Palaeoecol.* **36**, 263–320.

Plaziat, J.C. & Manguin, J.P. (1969) Données nouvelle sur l'Eocène inférieur du Bassin de Villarcayo et de ses annexes (Prov. de Burgos, Espagne). *Bull. Soc. Géol. France* **7**, 367–372.

Plaziat, J.C., Toumarkine, M. & Villatte, J. (1975) L'age des calcaires pélagiques et néritiques de la base du Tertiaire (Danien, Paléocene). Bassin basco-cantabrique et béarnais (Espagne, France). Mise au point sur leurs faunes d'échinides. *Eclog. Geol. Helv.* **68**, 613–647.

Posamentier, H.W. & Vail, P.R. (1988) Eustatic controls on clastic deposition II — sequence and systems tract models. In: *Sea Level Changes: An Integrated Approach* (Eds Wilgus, C.K., Hastings, B.S., Kendall, C.G.St.C., Posamentier, H.W., Ross, C.A. & Van Wagoner, J.C.). Spec. Publ. Soc. Econ. Paleontol. Mineral. 42, 125–154.

Puigdefabregas, C. & Souquet, P. (1986) Tectosedimentary cycles and depositional sequences of the Mesozoic and Tertiary from the Pyrenees. *Tectonophysics* **129**, 173–203.

Pujalte, V., Robles, S., Orue-Etxebarria, X. & Zapata, M. (1988) Secuencias deposicionales del tránsito Cretácico–Terciario del surco flysch de la cuenca Vasco-Cantábrica: relaciones con la tectónica y los cambios del nivel marino. *Proc. II Congr. Geol. España. Vol. Simposios*, 251–259.

Pujalte, V., Robles, S., Zapata, M., Orue-Etxebarria, X. & Garcia-Portero, J. (1989) Sistemas sedimentarios, secuencias deposicionales y fenómenos tectoestratigráficos del Maastrichtiense superior-Eoceno inferior de la cuenca vasca (Guipuzcoa y Vizcaya). In: *Libro Guía Excursiones Geol. XII Congr. Español de Sediment.* (Ed. Robles, S.), pp. 47–88.

Rat, P. (1959) Les Pays Crétacés Basco-Cantabriques (Espagne). PhD thesis, University of Dijon, XVIII, 525 pp.

Razin, P. (1989) Evolution tecto-sédimentaire alpine des Pyrénées Basque à l'Ouest de la transformante de Pamplona (province de Labourd). PhD thesis, University of Bordeaux, 464 pp.

Read, J.F. (1985) Carbonate platforms facies models. *Bull. Am. Assoc. Petrol. Geol.* **69**, 1–21.

Robador, A. (1991) Early Paleogene stratigraphy. In: *Introduction to the Early Paleogene of the South Pyrenean Basin. Field-Trip Guidebook (Early Paleogene Benthos)* (Ed. Inst. Tec. Geominero España, Madrid). IGCP Project N.286P, pp. 31–68.

Robador, A., Pujalte, V., Orue-Etxebarria, X., Baceta, J.I. & Robles, S. (1991) Una importante discontinuidad estratigráfica en el Paleoceno de Navarra y del Pais Vasco: caracterización y significado. *Geogaceta* **9**, 62–65.

Rosell, J., Remacha, E., Zamorano, M. & Gabaldon, V. (1985) Estratigrafía de la cuenca turbidítica terciaria de Guipúzcoa. Comparación con la cuenca turbidítica prepirenaica central. *Bol. Geol. Min.* **XCVI,**, 471–482.

Sarg, J.F. (1988) Carbonate sequence stratigraphy. In: *Sea-level Changes: An Integrated Approach* (Eds Wilgus, C.K., Hastings, B.S., Kendall, C.G.St.C., Posamentier, H.W., Ross, C.A. & Van Wagoner, J.C.). Spec. Publ. Soc. Econ. Paleontol. Mineral., 42, 155–182.

Seguret, M. (1972) Etude tectonique des nappes et séries décollées du versant sud des Pyrénées. PhD thesis, University of Science and Technology, Languedoc Publ., Montpellie, n2, 115 pp.

Seguret, M., Labaume, P. & Madariaga, R. (1984) Eocene seismicity in the Pyrenees from megaturbidites of the south Pyrenean basin (Spain). *Mar. Geol.* **55**, 117–131.

Serra-Kiel, J., Robador, A., Samso, J.M. & Tosquella, J. (1991) Biostratigraphic aspects to the early Cenozoic south-Pyrenean Basin. In: *Introduction to the Early Paleogene of the South Pyrenean Basin. Field-trip Guidebook (Early Paleogene Benthos)* IGCP Project N. 286, (Ed. Inst. Tec. Geominero España, Madrid). pp. 69–77.

Seyve, C. (1984) Le passage Crétacé–Tertiaire à Pont-Labau (Pyrénées Atlantiques, France). *Bull. C.R.E.P. Elf-Aquitaine* **8**, 385–423.

Seyve, C. (1990) Nannofossil biostratigraphy of the Cretaceous–Tertiary boundary in the French Basque Country. *Bull. C.R.E.P. Elf-Aquitaine* **14**, 553–572.

Vail, P.R., Audemard, F., Bowman, S.A., Eisner, P.N. & Perez-Cruz, C. (1991) The stratigraphic signatures of tectonics, eustasy and sedimentation. In: *Cycles and Events in Stratigraphy* (Eds Einsele, G. Ricken, W. & Seilacher, A.) Springer-Verlag, Berlin–Heidelberg, 617–659.

Van Vliet, A. (1982) Submarine fans and associated deposits in the Lower Tertiary of Guipuzcoa (Northern Spain). PhD thesis, University of Utrecht, 45pp.

Von Hillebrandt, A. (1965) Foraminiferen-Stratigraphie in alttertiar von Zumaya (proviz Guipuzcoa, NW Spanien) und ein Vergleich mi anderen Tethis-Gebieten. *Bayr. Akad. Wiss. Math.-Natur. Kl. Abh. N F* **123**, 66pp.

Ward, P.D. (1988) Maastrichtian Ammonite and Inoceramid ranges from Bay of Biscay Cretaceous–Tertiary Boundary sections. *Rev. Esp. Paleontol.* No. Extraord., 119–126.

APPENDIX

Appendix 1

Table A1. Planktonic foraminifera data (Pujalte *et al.*, 1989)

Sequence	Diagnostic species*		Biochronozone
	Orio section	Eibar section	
4.5	1, 2, 4, 5, 6, 7, 13, 15, 16, 17, 18, 19, 20, 21, 22, 24, 25, 27, 29, 30, 32, 33, 35		*G. gansseri*
1.1 and **1.2**		2, 3, 6, 8, 9, 10, 11, 12, 13, 14, 17, 18, 19, 20, 21, 22, 23, 24, 25, 26, 27, 28, 29, 30, 31, 32, 33, 34, 35, 36	*A. mayaroensis*
1.3	37, 39, 41, 43, 44, 45, 49, 79	37, 39, 43, 49	*E. trinidadensis*
1.3		39, 43, 44, 45, 49, 67, 68	*M. uncinata*
1.4	38, 40, 43, 45, 46, 50, 52, 56, 59, 60, 65	40, 43, 45, 46, 50, 52, 56, 59, 60	*P. pusilla*
2.1	40, 46, 48, 50, 52, 54, 56, 57, 58, 65, 68, 70, 71, 79	40, 46, 47, 48, 50, 52, 54, 56, 57, 58, 65, 68, 70, 71	*P. pseudomenardii*
2.2 and **2.3**	40, 46, 48, 53, 57, 58, 64, 65, 68, 70, 73, 75, 79, 81	40, 46, 48, 52, 57, 58, 65, 66, 68, 70, 71, 73	*M. velascoensis*
2.4	40, 42, 46, 53, 58, 64, 67, 71, 72, 75, 76, 78, 79, 80, 81	40, 46, 51, 53, 55, 58, 61, 64, 66, 69, 70, 71, 72, 73, 75, 76, 81	*M. subbotinae* (Subzone *M. subbotinae*)
2.5	40, 42, 46, 53, 55, 58, 62, 64, 66, 69, 71, 73, 74, 75, 77, 79, 81	40, 46, 51, 53, 58, 62, 63, 66, 69, 71, 73, 74, 75, 76, 79	*M. subbotinae* (Subzone *M. lensiformis*)

* Key to species:
1 *Globotruncana aegyptiaca* Nakkady, 1950
2 *Globotruncana arca* (Cushman, 1926)
3 *Globotruncana dupeublei* Caron *et al.*, 1983–84
4 *Globotruncana falsostuarti* Sigal, 1952
5 *Globotruncana linneiana* (d'Orbigny, 1839)
6 *Globotruncana mariei* Banner & Blow, 1960
7 *Globotruncana obliqua* Herm, 1965
8 *Globotruncana orientalis* El Naggar, 1966
9 *Globotruncana rosetta* (Carsey, 1926)
10 *Abathomphalus mayaroensis* (Bolli, 1951)
11 *Globotruncanita conica* (White, 1928)
12 *Globotruncanita stuarti* (de Lapparent, 1918)
13 *Globotruncanita stuartiformis* (Dalbiez, 1955)
14 *Rosita contusa* (Cushman, 1926)
15 *Rosita fornicata* (Plummer, 1931)
16 *Rosita patelliformis* (Gandolfi, 1955)
17 *Rosita walfischensis* (Todd, 1970)
18 *Globotruncanella havanensis* (Voorwijk, 1937)
19 *Globotruncanella petaloidea* (Gandolfi, 1955)
20 *Rugoglobigerina hexacamerata* (Brönnimann, 1952)
21 *Rugoglobigerina milamensis* Smith & Pess, 1973
22 *Rugoglobigerina pennyi* Brönnimann, 1952
23 *Rugoglobigerina scotti* (Brönnimann, 1952)
24 *Pseudotextularia deformis* (Kikoine, 1948)
25 *Pseudotextularia elegans* (Rzehak, 1891)
26 *Planoglobulina acervulinoides* (Egger, 1899)
27 *Planoglobulina multicamerata* (De Klasz, 1953)
28 *Racemiguembelina fructicosa* (Egger, 1899)
29 *Heterohelix glabrans* (Cushman, 1938)
30 *Heterohelix globulosa* (Ehrenberg, 1840)
31 *Heterohelix navarroensis* Loeblich, 1951
32 *Heterohelix planata* (Cushman, 1938)
33 *Pseudoguembelina costulata* (Cushman, 1938)
34 *Gublerina cuvillieri* Kikoine, 1948
35 *Globigerinelloides subcarinatus* (Brönnimann, 1952)
36 *Globigerinelloides yaucoensis* (Pessagno, 1960)
37 *Eoglobigerina edita* (Subbotina, 1953)
38 *Eoglobigerina ferreri* Orue-etxebarria & Apellaniz, 1991
39 *Eoglobigerina inconstans* (Subbotina, 1953)
40 *Eoglobigerina linaperta* (Finlay, 1939)
41 *Eoglobigerina pseudobulloides* (Plummer, 1926)
42 *Eoglobigerina pseudoeocaena* Subbotina, 1953
43 *Eoglobigerina triloculinoides* (Plummer, 1926)
44 *Eoglobigerina trinidadensis* (Bolli, 1957)
45 *Eoglobigerina varianta* (Subbotina, 1953)
46 *Eoglobigerina velascoensis* (Cushman *sensu* Bolli, 1957)
47 *Planorotalites albeari* (Cushman & Bermudez, 1949)
48 *Planorotalites chapmani* (Parr, 1938)
49 *Planorotalites compressa* (Plummer, 1926)
50 *Planorotalites ehrenbergi* (Bolli, 1957)
51 *Planorotalites indiscriminata* (Mallory, 1959)
52 *Planorotalites laevigata* (Bolli, 1957)
53 *Planorotalites planoconica* (Subbotina, 1953)
54 *Planorotalites pseudomenardii* (Bolli, 1957)
55 *Planorotalites pseudoscitula* (Glaessner, 1937)
56 *Planorotalites pusilla* (Bolli, 1957)
57 *Morozovella acuta* (Toulmin, 1941)

58 *Morozovella aequa* (Cushman & Renz, 1942)
59 *Morozovella angulata* (White, 1928)
60 *Morozovella conicotruncata* (Subbotina, 1947)
61 *Morozovella edgari* (Premoli Silva & Bolli, 1973)
62 *Morozovella gracilis* (Bolli, 1957)
63 *Morozovella lensiformis* (Subbotina, 1953)
64 *Morozovella marginodentata* (Subbotina, 1953)
65 *Morozovella occlusa* (Loeblich & Tappan, 1957)
66 *Morozovella subbotinae* (Morozova, 1939)
67 *Morozovella uncinata* (Bolli, 1957)
68 *Morozovella velascoensis* (Cushman, 1925)
69 *Acarinina broedermanni* (Cushman & Bermudez, 1949)
70 *Acarinina mckannai* (White, 1928)
71 *Acarinina nitida* (Martin, 1943)
72 *Acarinina primitiva* (Finlay, 1947)
73 *Acarinina pseudotopilensis* (Subbotina, 1953)
74 *Acarinina quetra* (Bolli, 1957)
75 *Acarinina soldadoensis* (Brönnimann, 1952)
76 *Acarinina wilcoxensis* (Cushman & Ponton, 1932)
77 *Pseudohastigerina wilcoxensis* (Cushman & Ponton, 1932)
78 *Chiloguembelina crinita* (Glaessner, 1937)
79 *Chiloguembelina midwayensis* (Cushman, 1940)
80 *Chiloguembelina paralela* (Beckmann, 1957)
81 *Chiloguembelina wilcoxensis* (Cushman & Ponton, 1932)

Spec. Publs Int. Ass. Sediment. (1993) **18**, 397–415

Sequence stratigraphy of carbonate ramps: systems tracts, models and application to the Muschelkalk carbonate platforms of eastern Spain

M.E. TUCKER*, F. CALVET† *and* D. HUNT‡

**Department of Geological Sciences, University of Durham, Durham, DH1 3LE, UK;*
†Departamento GPPG, Facultat de Geologia, Universitat de Barcelona,
Zona Universitaria de Pedralbes, 08028 Barcelona, Spain; and
‡Department of Geology, University of Manchester, Manchester, M13 9PL, UK

ABSTRACT

Carbonate ramps are a type of carbonate platform where there is no major break of slope from the shoreline into deeper water. They are characterized by inner-ramp sands and outer-ramp muddy sands and muds. Storm processes are generally important on ramps, but resedimentation of material to slope fan and aprons by turbidity currents and debris flows is only important on distally steepened ramps. Large reefal structures are generally not well developed on ramps, but pinnacle reefs and mud mounds do develop in deeper water on ramps.

In terms of sequence stratigraphy, the *transgressive* and *highstand systems tracts* (TST and HST) are the most important depositional periods for carbonate ramps. During the TST, different geometries are produced, depending on the relative rates of sea-level change and carbonate sedimentation. The latter is controlled by other factors than just sea-level change. When the rate of relative sea-level rise is greater than the rate of carbonate sedimentation, backstepping of the shoreline and drowning of the earlier inner-ramp sands take place, or the shoreline retrogrades to produce an onlapping, retrogradational set of parasequences or a transgressive sheet sand. When the rate of sea-level rise is equal to or less than the rate of carbonate sedimentation, aggradation and even progradation of the inner-ramp sands take place to produce a thick sand body. During the highstand, ramp carbonates aggrade and prograde, downlapping on to the earlier TST sediments. On a homoclinal ramp, there is a downward shift of facies belts during the lowstand systems tract (LST), and exposure of the former inner ramp produces a sequence boundary. A significant sand body may develop at the new lowstand shoreline. On a distally steepened ramp, there may be much resedimentation of outer- and inner-ramp material to a slope apron if the sea-level fall is great.

The concepts and models developed in this paper for carbonate-ramp sequence stratigraphy are applied to the mid-Triassic Muschelkalk carbonate platforms of eastern Spain. Both are dominated by TST and HST sediments of third-order sequences and contain fourth-/fifth-order, shallowing-upward parasequences. Beneath each platform, in the uppermost Buntsandstein and in the Middle Muschelkalk, respectively, there is an LST of marls and evaporites, deposited at a time of tectonic extension within the Catalan Basin. The carbonate platforms formed during the succeeding phase of regional subsidence.

INTRODUCTION

Carbonate ramps are a type of carbonate platform where there is no major break of slope in shallow water, water depth gradually increases away from the shoreline and carbonate sands are the main shallow-water ('updip') facies (Fig. 1) (Ahr, 1973). Two main categories of ramp are generally distinguished: *homoclinal ramps*, where there is a gentle gradient into the basin, and *distally steepened ramps*, where there is an increase in gradient in the outer-ramp region (Read, 1982, 1985). Facies

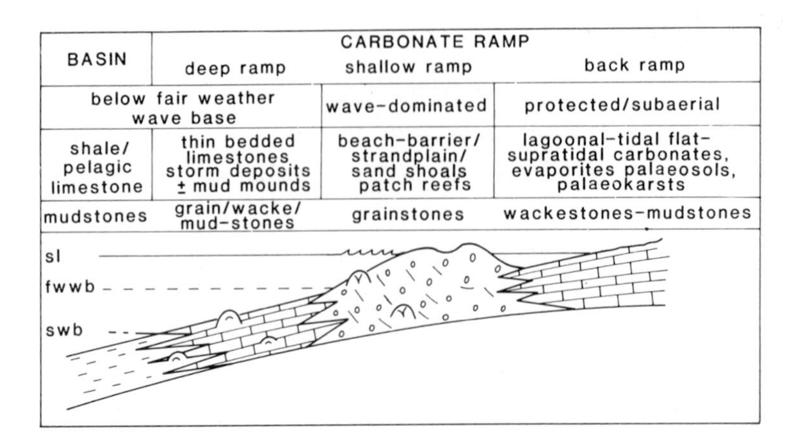

Fig. 1. General facies model for a homoclinal ramp. (From Tucker, 1991a.) Abbreviations: sl, sea level; fwwb, fairweather wave base; swb, storm wave base.

models for these ramp types are well documented and many ancient examples have been described (see review in Tucker & Wright, 1990). However, the sequence stratigraphy of carbonate ramps and how the sequence patterns differ from carbonate shelves, siliciclastic shelves and siliciclastic ramps have not been explored. The sequence stratigraphy of carbonate ramps is discussed in this paper and the models and points raised are applied to the Muschelkalk carbonate platforms of eastern Spain (Calvet *et al.*, 1990).

Carbonate ramps can be divided into inner (shallow-water) and outer (deeper-water) regions. Variations in facies patterns do occur on ramps, particularly in the inner ramp. Two common inner-ramp types are those dominated by a strandplain complex and those with a barrier–lagoon shoreline. These two are high to moderate energy ramps, but a low-energy type of ramp does occur, where the shoreline is dominated more by tidal flats and lagoons, with some sand shoals. The deeper water outer ramp generally consists of muddy sands and muds, but storm processes are important in transporting sand out from the shoreline, to produce sand bodies, lenses and beds. Again, the role of storms does vary between ramps, depending on ramp orientation with respect to storm winds. Reefs are generally poorly developed on ramps. Patch reefs may occur in shallow water and pinnacle reefs and mud mounds on the outer ramp. Where ramps are distally steepened, resedimentation of outer-ramp sediments to a slope apron or fan can take place.

Modern ramp-type carbonate platforms occur in the Arabian Gulf, especially along the Trucial Coast (papers in Purser, 1973), where a barrier with lagoon and tidal flats behind is well displayed. The eastern Yucatan coast of Mexico is a good example of a carbonate ramp with a strandplain developed in the inner-ramp region (Ward & Brady, 1979; Ward *et al.*, 1985). Well-documented, ancient ramp carbonates include the Jurassic Smackover of the subsurface of the Gulf Rim, southern USA (Ahr, 1973; Moore, 1984), the Cambro-Ordovician of the Appalachians, where ramps developed around an intracratonic basin (Markello & Read, 1981), the Lower Carboniferous of South Wales (Wright, 1986; Burchette *et al.*, 1990) and the Triassic Muschelkalk of Germany (Aigner, 1984). Distally steepened ramps developed along the eastern USA continental margin in the Lower Palaeozoic (Read, 1985) and occur within the British Lower Carboniferous (e.g. Gutteridge, 1989).

As with other carbonate-platform limestones, some thick ramp successions are organized into 1–10 m-scale cycles, which generally are shallowing-upward units of lime mudstone, with storm beds, passing up into shallow-water grainstone deposited in the shoreface–foreshore. If the cycle becomes exposed, there may be an emergence horizon capping the unit, with minor palaeokarst, palaeosol, dolomite or, less commonly, an evaporite bed, depending on the climate. Such ramp parasequences are well developed within the Triassic Muschelkalk of Germany (Aigner, 1984).

SEQUENCE STRATIGRAPHY: GENERAL POINTS

Sequences are defined as a conformable succession of genetically related strata, bound at the top and

Table 1. The orders and mechanisms of relative sea-level change. (From Tucker, 1991a.)

Order	Time scale	Mechanism			
First order	10^8 years	Tectono-eustatic (first–second order)			
Second order	10^7 years		Rifting and thermal subsidence (second–third order)		Global eustatic (second–third order)
Third order	10^6 years			In-plate stress (third–fourth order)	
Fourth order	10^5 years	Glacio-eustatic, tectonic, sedimentary (fourth–fifth order)			
Fifth order	10^4 years				

bottom by unconformities and their correlative conformities (Van Wagoner *et al.*, 1988, 1990). The unconformities are defined as surfaces of erosion or non-deposition and represent a significant time gap. The major control on deposition is relative sea-level change, determined by rates of eustatic sea-level variation and tectonic subsidence. Particular depositional systems tracts are developed during specific phases of the sea-level curve: lowstand (LST), transgressive (TST), highstand (HST) and shelf-margin wedge (SMW) systems tracts. Different orders of sea-level change are recognized (Table 1), although there is still much debate on the mechanisms behind the various changes. Sequences, as defined above, are generated by high-amplitude, sea-level changes, with the bounding unconformities produced during the relative sea-level falls. In carbonate rimmed shelf systems, the unconformities are mainly the sites of exposure and karstification, but on carbonate ramps, with their higher gradient relative to carbonate shelves, fluvial incision and siliciclastic influx are more likely to occur during sequence-boundary formation.

High-amplitude sea-level changes producing sequence boundaries occur on several time scales, depending basically on the presence or absence of polar ice caps (see Fig. 2). When there are no polar ice caps (i.e. greenhouse times (Fischer, 1980; Veevers, 1990)), high-amplitude sea-level changes on the scale of tens of metres are produced by second-/third-order processes (tectono-eustatic or tectonic) while fourth-/fifth-order processes (e.g. simple temperature-controlled ocean volume changes driven by Milankovitch rhythms and auto-

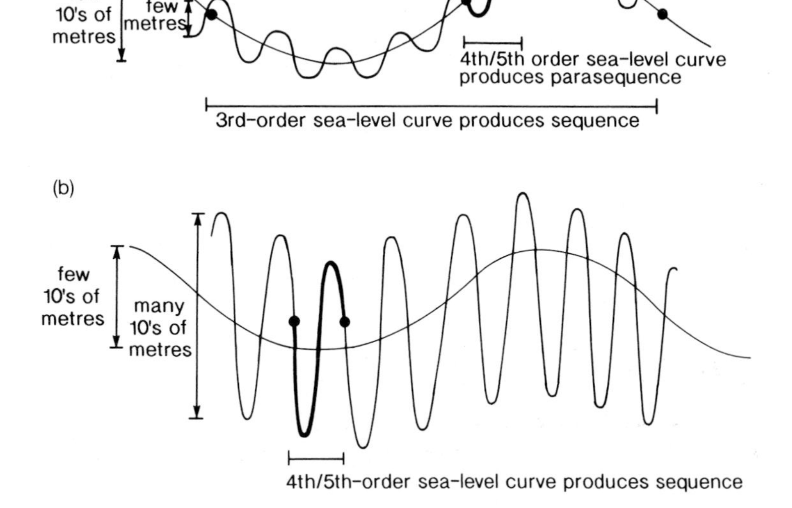

Fig. 2. Schematic third-order and fourth/fifth-order relative sea-level curves for (a) greenhouse and (b) icehouse periods. During greenhouse times, the third-order (1–10 Ma) relative-sea-level curve has a greater amplitude than the fourth/fifth-order (10,000 year–1 Ma) relative sea-level curve, whilst during icehouse times, the fourth/fifth-order relative sea level curve has a greater amplitude than the third-order curve. Thus, third-order sequences with fourth/fifth-order parasequences are typical of greenhouse periods, whilst fourth-order sequences are produced during icehouse periods.

cyclic mechanisms) produce only metre-scale relative sea-level changes. However, when there are polar ice caps (icehouse times), the high-amplitude sea-level changes, up to 100 m or more, are brought about by fourth-/fifth-order processes, namely changes in ocean volume arising from changes in ice volume, driven again by Milankovitch rhythms. During icehouse times, the longer term second-/third-order sea-level changes are on a smaller scale (lower amplitude) than the short-term fourth-/fifth-order sea-level changes (see Fig. 2). Thus, during icehouse times, fourth-order sequences are produced, which may be arranged into sequence sets, while during greenhouse times third-order sequences consist of fourth-/fifth-order parasequences. The Quaternary sedimentary record, for example, consists of fourth-order sequences (i.e. unconformity-bound stratal packages) and these are relatively thin (5 to 30 m) (see, for example, Humphrey & Kimbell, 1990). By way of contrast, the Triassic to Cretaceous and early Palaeozoic, times of no or only small ice caps (i.e. greenhouse episodes), generally consist of much thicker sequences (third-order, 50 to hundreds of metres thick) made up of fourth-/fifth-order parasequences (1 to 10 m thick) (e.g. Koerschner & Read, 1989; Goldhammer *et al.*, 1990).

SEQUENCE STRATIGRAPHY OF CARBONATE RAMPS

Application of the concepts of sequence stratigraphy to carbonate ramps is relatively straightforward compared to carbonate shelves, where the second-/third-order relative sea-level changes produce some distinctive facies types and geometries, especially during lowstands (see Hunt & Tucker, this volume, pp. 312–317). The relatively simple geometry of a homoclinal ramp basically means that facies belts simply move up and down the ramp in response to long-term relative sea-level changes. The situation is a little more complicated for a distally steepened ramp during the lowstand as discussed later.

A sequence-stratigraphic model for a homoclinal ramp is shown in Fig. 3, where two third-order sequences are represented, with constituent parasequences. In the lower sequence, the inner ramp consists of a strandplain grainstone package, with no finer grained facies to landward, except for some siliciclastics of restricted, coastal-plain origin deposited during the late TST–HST. In the upper sequence, the inner ramp is a barrier shoreline with a back-ramp lagoon where lime muds and possibly evaporites are precipitated. In both cases the inner-

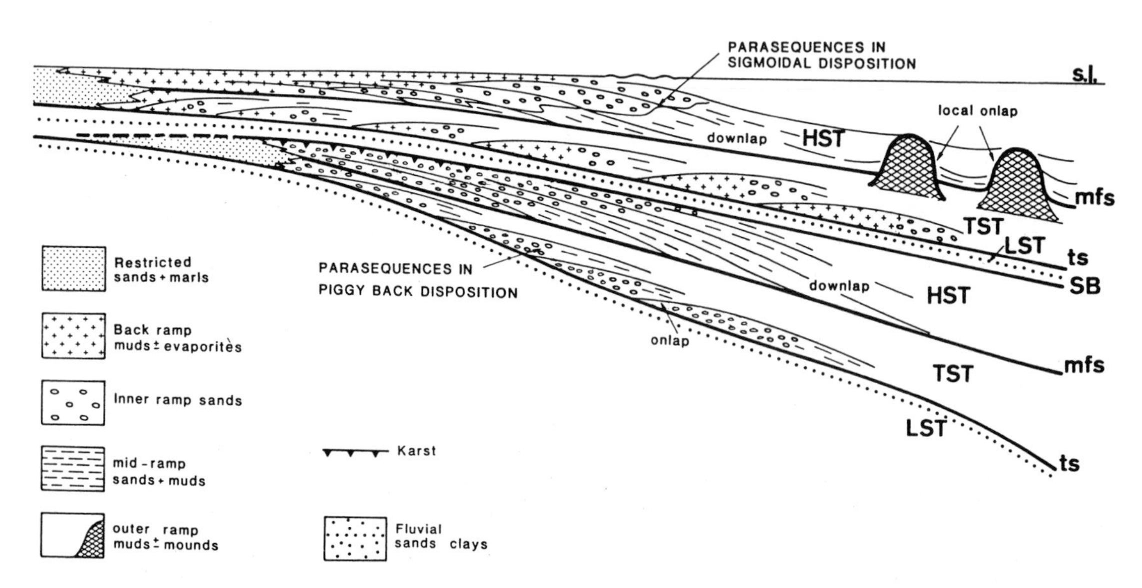

Fig. 3. Sequence-stratigraphic model for a carbonate ramp and its systems tracts. Abbreviations: LST, lowstand systems tract; TST, transgressive systems tract; HST, highstand systems tract; mfs, maximum flooding surface; ts, transgressive surface; SB, sequence boundary; s.l., sea level.

ramp grainstones pass downramp into muddy sands and muds. Two mud mounds are depicted in the upper sequence, developing during the TST, perhaps upon drowned sand shoals, and being buried by the downlapping HST muds. As with all carbonate sequences, climate exerts a strong control on the nature of the supratidal emergence horizons and sequence boundaries, determining whether evaporites and dolomites form on the one hand, when the climate is arid, or palaeokarsts and calcretes on the other, when the climate is more humid. Sequence boundaries will vary across a ramp, as from a major palaeokarst, perhaps with palaeosols, in a proximal, landward setting to a less deeply cut palaeokarst in a more distal situation, close to the lowest position of sea level. Under a humid climate, parasequence boundaries would each be a minor palaeokarst passing basinwards into a discontinuity surface. Deposition on a carbonate ramp mainly takes place during the transgressive and highstand parts of the relative sea-level change curve. Lowstand deposition is more important on distally steepened ramps where major resedimentation is possible.

During the TST, there are several possible responses of a carbonate ramp, depending on the rate of sea-level rise relative to the rate of carbonate sedimentation (Fig. 4). For carbonate shelves, Hunt and Tucker (1992) distinguished two types of TST geometry: type 1 when the rate of relative sea-level rise is greater than the carbonate sedimentation rate, and type 2 when the rate of relative sea-level rise is equal to or less than the rate of carbonate sedimentation. Type 1a and 1b geometries are distinguished for when the rate of sea-level rise is much greater (1a) and greater (1b) than the sedimentation rate. The same scheme can be applied to ramps.

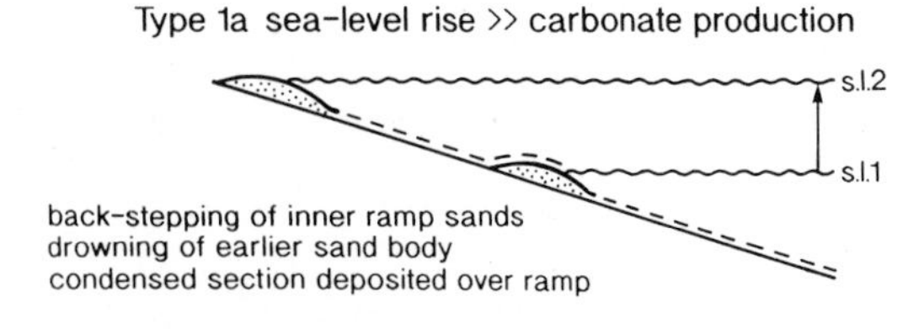

Type 1b sea-level rise > carbonate production

s.l.2

s.l.1

retrogradation of inner ramp sands
generation of onlapping parasequences
or transgressive sheet sand
progressive drowning of earlier sand bodies

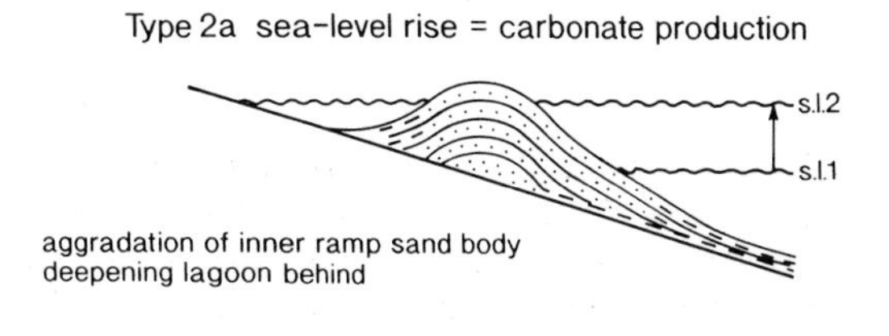

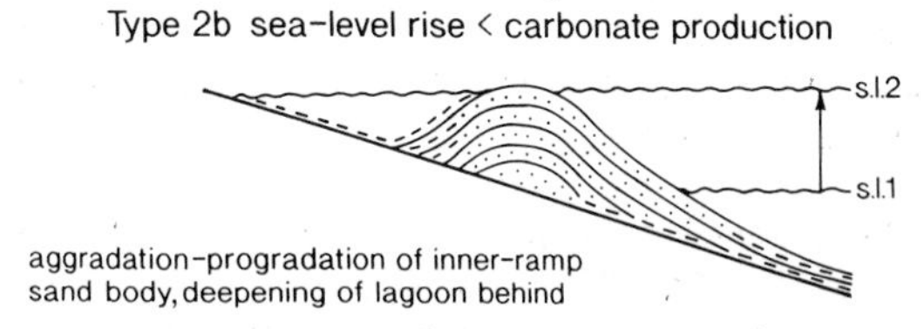

Fig. 4. Schematic sketches of the common types of transgressive systems tract (TST) geometry for a carbonate ramp.

Type 1a. When sea-level rise is much greater than sedimentation rate, backstepping of the inner ramp takes place and a new shoreline of barrier–lagoon or beach ridges is established. This facies jump leaves the former inner-ramp sands drowned on the new outer ramp, where they will be buried by pelagic or hemipelagic muds. Much of the TST is represented by condensed mid–outer ramp facies. At the new shoreline aggradation may take place, before the main phase of sand body growth and progradation during the HST.

Type 1b. When sea-level rise is greater than carbonate sedimentation, retrogradation of the inner-ramp sands takes place and the stratal package onlaps the ramp. This situation is shown for both sequences in the ramp model of Fig. 3, drawn for a third-order sequence with parasequences. In this case, the parasequences of the TST will form a parasequence set with a systematic upward change in character of the parasequences at any one locality, with slightly deeper water facies occurring in successive parasequences. An alternative, especially in a fourth-order sequence, is for a transgressive sheet sand to be deposited. In both situations, there is a drowning of sand bodies formed earlier in the TST.

Type 2. When sea-level rise is equal to carbonate sedimentation rate, aggradation of the inner, ramp sand bodies can take place (type 2a geometry (Fig. 4)). A back-barrier lagoon is likely to develop and deepen in view of differential carbonate production rates. This situation of aggradation is not uncommon with carbonates because of potential

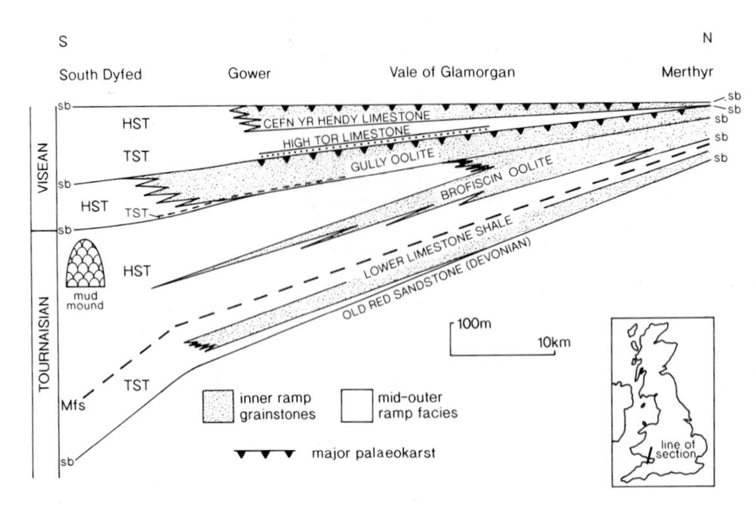

Fig. 5. Sequence-stratigraphic framework of the Lower Carboniferous ramp carbonates of South Wales. (Based on Burchette *et al.*, 1990.) The lowest sequence has a well-developed transgressive systems tract (TST) with several major carbonate sand bodies. The third sequence has an onlapping bioclastic sand body (the High Tor Limestone) which formed during the TST as a retrogradational onlapping unit. All three sequences have prominent oolites developed in the highstands which prograded downramp. The lowstands are mostly represented by palaeokarst and palaeosol horizons, best developed in the inner-ramp area to the north.

high skeletal and abiogenic (ooid) grain production rates. In a siliciclastic system, sedimentation rates are governed by sediment supply, normally from outside the basin, whereas most carbonate sediments are produced *in situ*. Where carbonate sedimentation rates exceed relative-sea-level rise, an aggrading–prograding sand body may form (type 2b geometry (Fig. 4)).

The Lower Carboniferous ramp sediments of South Wales (Wright, 1986; Burchette *et al.*, 1990) show type 1a and type 1b TST geometries (Fig. 5). In the first sequence, the Lower Limestone Shale shows a type 1b TST geometry with grainstones well developed in several parasequences. The third sequence has a type 1b TST with a transgressive sheet grainstone (High Tor Limestone) overlain by outer-ramp mudrocks deposited when the ramp shoreline had migrated landwards in response to a relative sea-level rise. The second sequence has a type 1a TST; the ramp was rapidly flooded and then the Gully Oolite (HST) prograded across the ramp. A similar situation of thin, muddy, type 1a TSTs succeeded by thick HST prograding oolites occurs in the two Great Oolite sequences of the Middle Jurassic of the Wessex Basin (Sellwood *et al.*, 1985). The Lincolnshire Limestone, Middle Jurassic of eastern England (Ashton in Tucker, 1985) is a thick retrogradational TST package of barrier–lagoonal facies.

During the HST, the ramp shoreline undergoes aggradation and then progradation as the stillstand is reached and sea level begins to fall (Fig. 6). With a lengthy phase of highstand deposition, ramp carbonates can offlap considerable distances (e.g. some 20 km in the Upper Jurassic of the Baltimore Canyon region, subsurface of the eastern US shelf (Gamboa *et al.*, 1985)). The stratal package, often best seen on seismic-reflection profiles, may show a change from oblique to sigmoidal stratal geometry, through the higher sedimentation rates of the shallower water inner ramp compared to the outer ramp, and so the preferential progradation of the inner-ramp facies over the outer-ramp facies. This may lead to an increase in topography on the sea floor and, during a subsequent sea-level rise, it could herald the evolution of a rimmed shelf or a distally steepened ramp.

If the sequence is third order (as in Fig. 3), HST parasequences are likely to be present. They should show a systematic upward trend of more shallow-water facies in each parasequence through the package.

There are many examples of ramp carbonates with well-developed HST grainstones. The Great

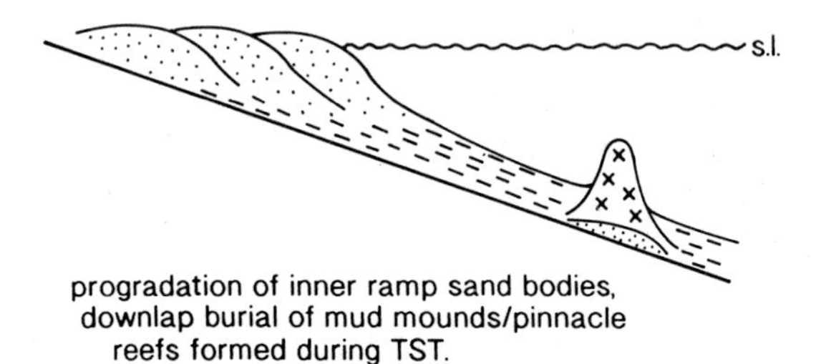

s.l.
progradation of inner ramp sand bodies, downlap burial of mud mounds/pinnacle reefs formed during TST.

Fig. 6. Schematic sketch of the highstand systems tract (HST) for a homoclinal ramp.

Oolite of the Wessex Basin has already been mentioned, where two HST oolite bodies migrated some 40 km downramp, and the HST oolites of the Lower Carboniferous in South Wales are shown in Fig. 5. The oolites of the Smackover Formation of the Gulf Coast Rim also largely formed during the highstand.

For the *lowstand systems tract on homoclinal carbonate ramps*, no specific depositional environments related to the relative sea-level fall are created (Fig. 7). On this type of ramp, there is insufficient gradient to generate slides, slumps, rockfalls or megabreccias, which are generally taken as a feature of LSTs especially in siliciclastic shelf systems (Van Wagoner *et al.*, 1990) and on some carbonate rimmed shelves (see Hunt & Tucker, 1992). On a homoclinal ramp, there will simply be a basinward shift in facies belts, and similar inner-ramp facies can develop at the new shoreline, on the previous mid–outer ramp. The extent of the new area of shallow-water (inner-ramp) deposition need not change; this contrasts with the situation for a rimmed shelf, where lowstand carbonate deposition is generally minor, unless autochthonous wedges are

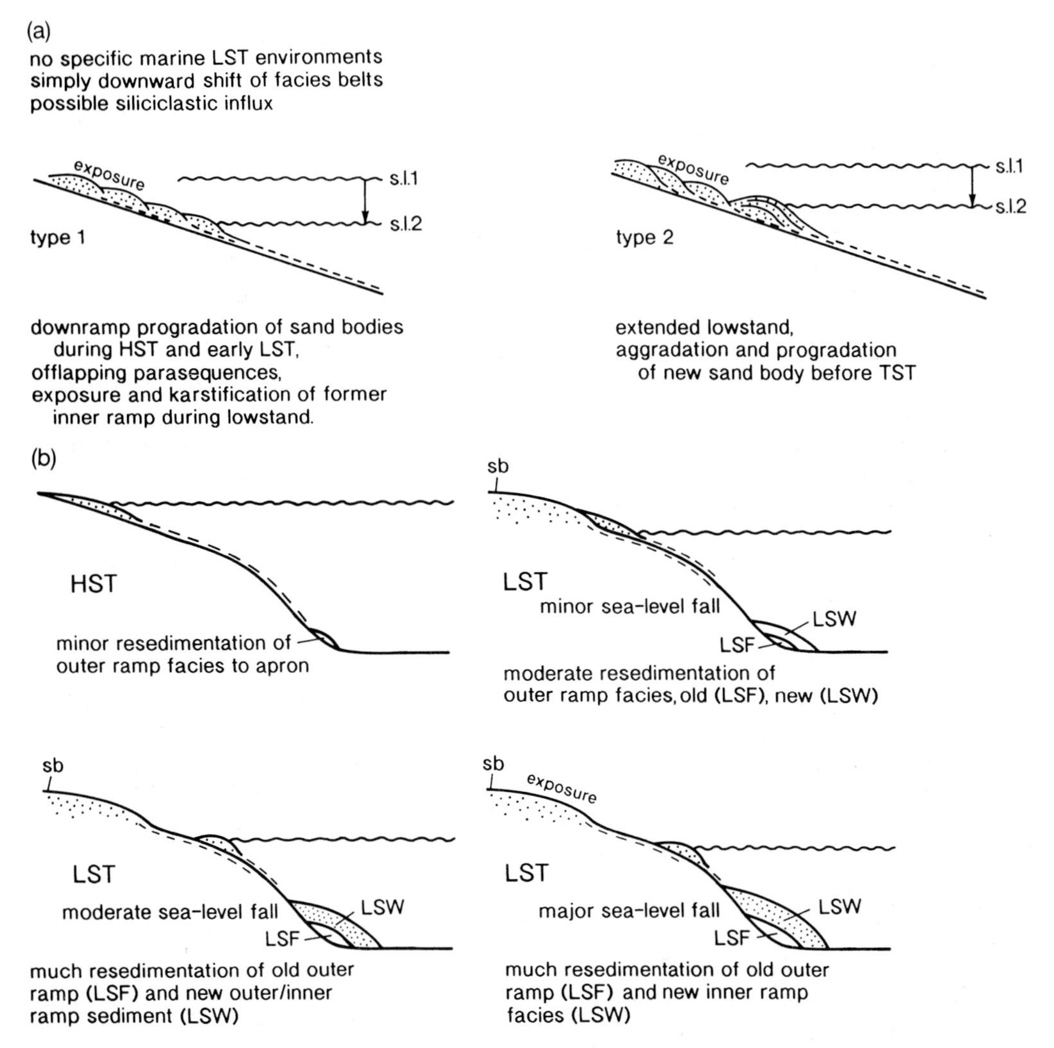

Fig. 7. Schematic sketches of the lowstand systems tract (LST) for (a) homoclinal and (b) distally steepened carbonate ramps.

developed below the shelf break (Hunt & Tucker, 1993). If there is a lengthy stillstand during the lowstand, inner-ramp carbonate sands may aggrade and even prograde to form a significant sand body (Fig. 7), before the subsequent sea-level rise of the TST leads to drowning or retrogradation, as discussed earlier. Such a sand body is analogous to a shelf-margin wedge (SMW), developed on siliciclastic shelves during a type 2 sea-level fall, when the position of sea level does not go below the shelf break (Van Wagoner *et al.*, 1988, 1990).

In most cases, the former inner-ramp area will be exposed during the lowstand and the sediments subjected to erosion, karstic dissolution, calcretization, dolomitization or evaporite precipitation and replacement, depending largely on the climate. If there is some uplift in the hinterland which coincides with the relative sea-level fall and lowstand (it could even be the cause of it), an influx of siliciclastic material may occur and the lowstand deposits may consist of fluvial sandstones, which may be incised into the former inner-ramp carbonates.

For the *lowstand systems tract on a distally steepened carbonate ramp*, the situation may be rather different. In this case, the important point is the depth at which the break of slope occurs on the ramp relative to the sea-level fall (Fig. 7). During the highstand, there may be some resedimentation off the shoulder of the distally steepened ramp into the basin to form a wedge or apron. The sediment involved in the slides, slumps and sediment gravity flows will generally be outer-ramp facies, since the ramp-slope break of a distally steepened ramp generally occurs in deep water on the outer ramp. During the sea-level fall, the frequency of the resedimentation of the highstand outer-ramp deposits may increase to give a lowstand fan (or more likely apron). If the relative sea-level fall is minor (Fig. 7), so that outer-ramp facies continue to be deposited in the region of the ramp-slope break, sediment gravity flows may transport the new outer-ramp deposits into the basin to form a lowstand wedge. The frequency of such flows is likely to be higher, through the effects of higher sedimentation rates and storm wave pounding on the substrate, at this time of lower sea level. If the magnitude of the relative sea-level fall is moderate (Fig. 7), and the inner-ramp environments shift basinward close to the ramp-slope break, inner- and outer-ramp material may be redeposited into the basin to form the lowstand wedge. With a major sea-level fall (Fig. 7), the shoreline may drop below the ramp-slope break. In this case, the area of shallow water available for carbonate sediment production will decrease, as a result of the increased gradient on the slope. Nevertheless, shallow-water debris may now contribute directly to the construction of a thick lowstand wedge. The ramp itself, now exposed, may be severely incised by fluvial processes or dissected by karstic dissolution in the new, low base-level situation. This was the case on the West Florida ramp during the Pleistocene lowstand (Mullins *et al.*, 1988). A good example of collapse of a distally steepened ramp during a relative sea-level fall is seen in the Raisby Formation of the Upper Permian (Zechstein) strata of northern England (Tucker, 1991b). The megabreccia consists largely of clasts of outer-ramp facies.

Reefs and mud mounds on ramps

Large reef complexes are not a feature of carbonate ramps, but patch reefs, pinnacle reefs and mud mounds do occur, with the last two in particular being a feature of Palaeozoic carbonate ramps (e.g. the Silurian pinnacle reefs of the Michigan Basin (Shaver, 1977; Petta, 1980) and the Lower Carboniferous mud mounds of western Europe (Lees & Miller, 1985)). If patch reefs develop during a lowstand, they may grow into pinnacle reefs in the outer, deeper-water ramp during the transgressive systems tract. In these situations, growth of the pinnacles is able to keep pace with rising sea level, so that shallow-water reef organisms continue to exist on the tops of the pinnacles. Mud mounds are not uncommon in outer-ramp settings, and although there is much argument about many aspects of mud mounds, some at least appear to have developed upon muddy sand banks, which formed during a lowstand–early transgressive systems tract. The mud mounds themselves then formed as the rate of sea-level rise increased. Many mud mounds appear to have stayed within deeper waters for most of their existence. However, towards the top of some, shallow-water facies do occur, and some even have exposure horizons at the top (e.g. the Niagaran pinnacle reefs with mud mound cores, Michigan Basin (Petta, 1980)). These features will probably have formed when the relative sea-level rise was slowing or reaching a stillstand, and highstand deposition was taking over, or during higher frequency sea-level changes. The exposure reflects the subsequent lowstand. Both pinnacle reefs and mud mounds may be partially to completely buried by downlapping HST muds as the inner ramp progrades downramp (as shown in Figs 3 & 6).

Type 1 and type 2 sequence boundaries

For carbonate and siliciclastic sediments deposited on shelves, the distinction has been made between type 1 and type 2 sequence boundaries (Van Wagoner *et al.*, 1988), reflecting the magnitude of the relative sea-level fall (below the shelf break in the case of type 1, but only towards the shelf margin in the case of type 2). The distinction between these two types of boundary for third-order sequences deposited on a carbonate ramp may not be easy, unless the ramp is distally steepened. For both types on a homoclinal ramp, there will be a basinward shift of inner-ramp facies, but for a type 1 sequence boundary, with the greater relative sea-level drop, the sequence boundary itself should be more conspicuous compared with a type 2. There may be a more prominent palaeokarst or greater depth of dolomitization for a type 1. There should also be a greater contrast in the nature of the facies above and below a type 1 boundary, compared to a type 2, with deeper water facies of the outer ramp below the boundary and shallower water facies of the inner ramp above, with several microfacies belts missing. Sabkha, lacustrine, fluvial and other supratidal/non-marine environments are more likely to be developed during a type 1 relative sea-level fall to produce lowstand deposits, than during a type 2 relative sea-level fall, when an exposure horizon, without subaerial deposits, may be the only feature. In both types of sequence boundary, there will be lateral changes in the nature of the boundary itself, as a result of the decreasing time available for subaerial processes to operate in a basinward direction. A high-relief, upramp palaeokarst may lose its topography downramp and pass into a minor discontinuity surface. Some sequence boundaries within the Lower Carboniferous ramp deposits of South Wales (Fig. 5) are defined just by palaeokarsts and may be the equivalent of type 2 sequence boundaries of rimmed shelves; however, in at least one instance, fluvial sediments and palaeosols are associated, probably reflecting a more significant relative sea-level fall and the equivalent of a type 1 sequence boundary.

CARBONATE-RAMP SEQUENCE STRATIGRAPHY: AN EXAMPLE FROM THE TRIASSIC OF EASTERN SPAIN

The first half of this paper has explored various aspects of carbonate-ramp sequence stratigraphy. A case history is now presented where many of the situations discussed can be seen.

The Middle Triassic Muschelkalk carbonates of the Catalan Basin in eastern Spain (Fig. 8) were deposited in ramp-type carbonate platforms. Two platforms developed, the Lower Muschelkalk in the Anisian and the Upper Muschelkalk in the Ladinian, separated by marls and gypsum (50 to 130 m thick) of the Middle Muschelkalk (Figs 8, 9 & 10). Both carbonate platforms are around 120 m thick

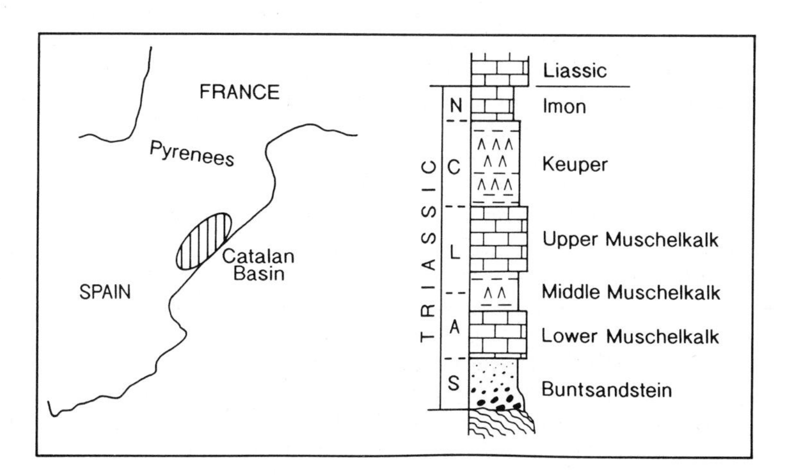

Fig. 8. Sketch location map and lithostratigraphy for the Triassic of the Catalan Basin, eastern Spain.

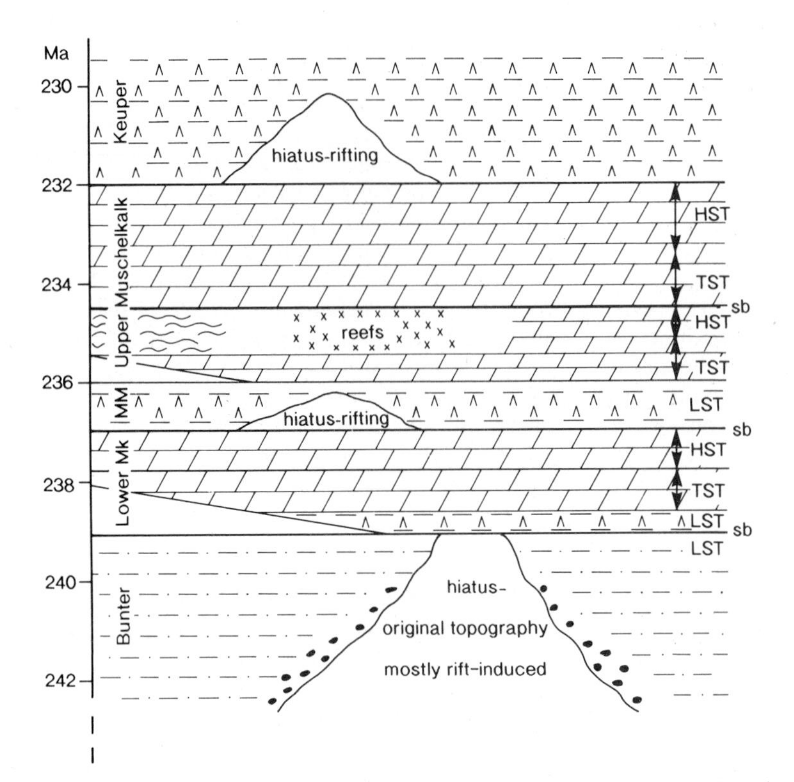

Fig. 9. Chronostratigraphy for the Triassic of the Catalan Basin, Spain.

and apart from the central part of the Upper Muschelkalk platform there is much lateral continuity of facies along strike in the basin. Both carbonate platforms were interpreted by Calvet *et al.* (1990) as consisting of the transgressive and highstand deposits of third-order sequences. In each case the lower part of the sequence, the LST, is represented by marls and evaporites, the uppermost Buntsandstein occurring beneath the Lower Muschelkalk platform and the Middle Muschelkalk below the Upper Muschelkalk platform. The Lower Muschelkalk carbonate platform is readily divisible into two parts: (i) a lower unit of onlapping, retrogradational strata of broadly transgressive facies (the TST); and (ii) an upper unit of aggradational to progradational, offlapping strata (the HST). However, further consideration of the Ladinian Upper Muschelkalk platform suggests that it should be divided into two sequences. There is a major palaeokarst within the Upper Muschelkalk which is interpreted here as a sequence boundary. The Lower Muschelkalk sequence lasted for around 3 million years, the Middle Muschelkalk–lower Upper Muschelkalk sequence for around 3 Ma and the upper Upper Muschelkalk sequence for about 1 million years. Each sequence contains parasequences produced by fourth-/fifth-order sea-level changes.

The Lower Muschelkalk

In the Lower Muschelkalk (Fig. 11), the TST consists of three stratigraphic units, the El Brull, Olesa and Vilella Baixa units. The El Brull Unit (6 to 14 m thick) consists of laminated and stromatolitic dolomites, bioturbated lime mudstones and local rippled oolites and intraclast breccias. They are interpreted as the deposits of a generally low-energy tidal-flat environment. The Olesa Unit (4 to 12 m thick) mostly consists of lime mudstones to packstones, with a limited marine fauna, interpreted as lagoonal to quiet water inner-ramp deposits. The succeeding Vilella Baixa Unit, up to 90 m thick, is made up of bioturbated and massive lime mudstones in the lower part, and skeletal and oolitic wackestones through grainstones, interbedded with mudstones, in the upper part. This unit shows a passage from lagoonal mudbanks through to sand-shoal deposits at the top. Thus, on a broad scale (i.e. of the whole carbonate platform), these three units

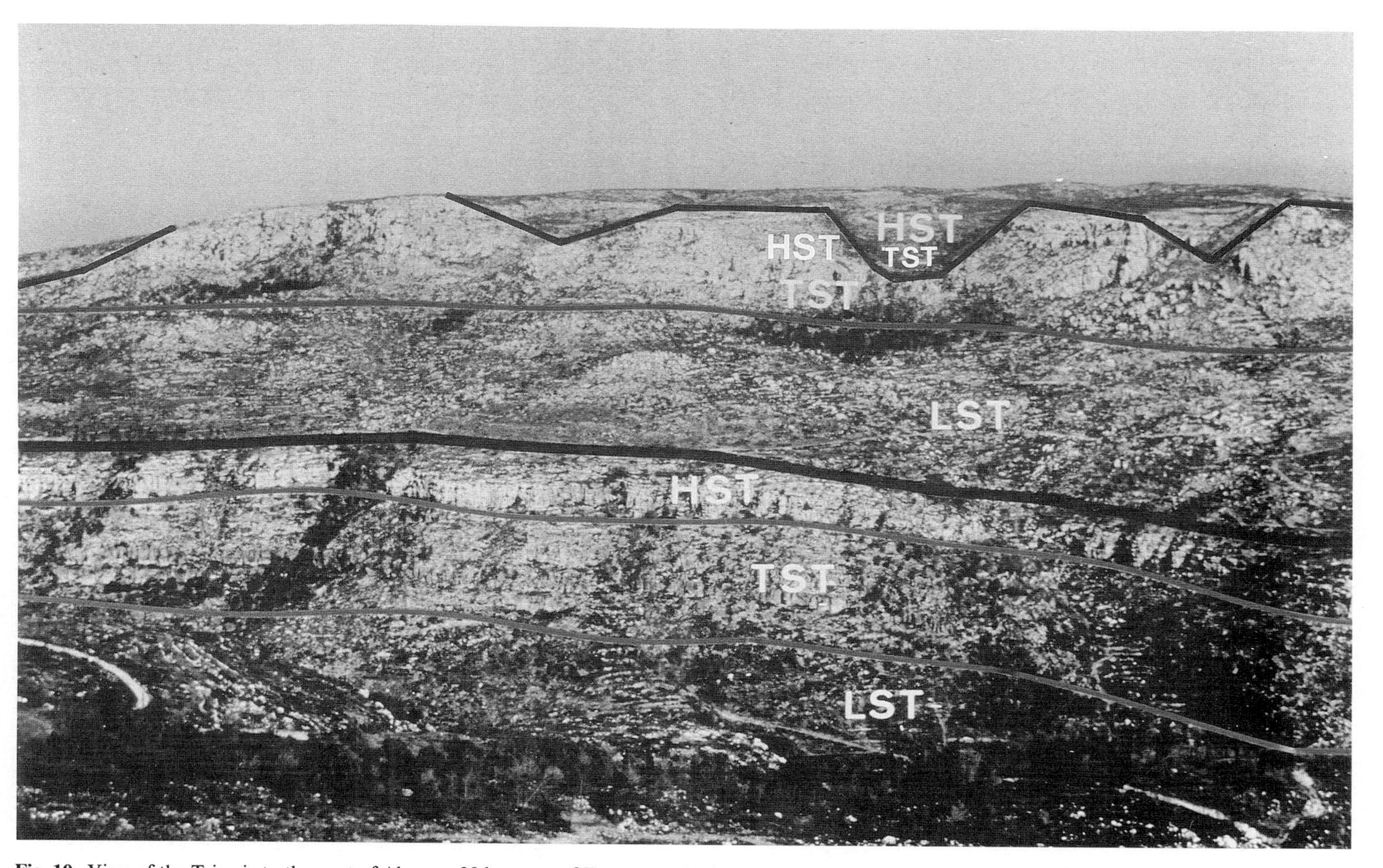

Fig. 10. View of the Triassic to the west of Alcover, 30 km west of Tarragona, Spain, showing the two Muschelkalk carbonate platforms. The Lower Muschelkalk is the lower TST + HST; the Upper Muschelkalk consists of two sequences — L1 and L2. The Middle Muschelkalk is the LST to sequence L1. The Bunter Sandstone occurs in the lower part of the photograph (lower LST). Note the topography between sequences L1 and L2 due to the La Riba buildups.

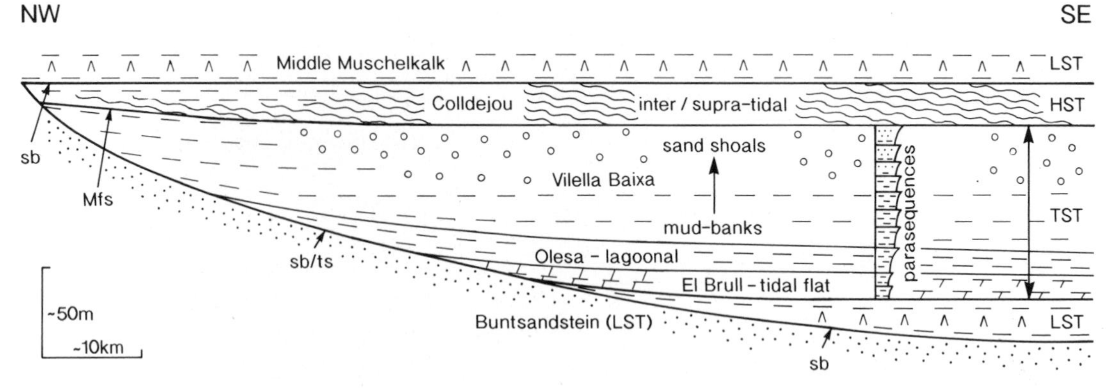

Fig. 11. Lower Muschelkalk lithostratigraphy and systems tracts. Abbreviations as in Fig. 3.

constitute a transgressive package, from tidal-flat to lagoonal to sand-shoal facies. The three units in addition show a gradual onlap towards the west/northwest, the direction of the hinterland. This supports the transgressive systems tract interpretation. The TST geometry is the type 1b discussed earlier, where relative sea level was rising faster than sedimentation rate, but not so fast so as to drown the ramp. Retrogradation of facies belts is the response in this case.

The fourth unit of the Lower Muschelkalk platform, the Colldejou (Fig. 11), is a laminated, stromatolitic dolomicrite with replaced evaporites, up to 40 m thick. It was deposited in an intertidal–supratidal environment, under an arid climate. The Colldejou Unit represents a highstand systems tract. On a basin scale, the lateral equivalent of the Colldejou dolomites towards the hinterland (to the west/northwest) is a red evaporitic marl, indicating a broad, offlapping progradational stratal pattern.

On a smaller scale, parasequences occur within the El Brull and Vilella Baixa units. Up to fifteen, 3–6 m-thick, shallowing-upward cycles can be recognized, although in detail there are differences in the facies types present (see Calvet *et al.*, 1990 for details). However, looked at on a broad scale, there is a general change in character of the parasequences up through the Lower Muschelkalk. Those in the El Brull Unit are dominated by intertidal and inter/supratidal facies (some 80–90% of each cycle), occurring above subtidal mudstones; those of the lower part of the Vilella Baixa Unit consist of 50–70% intertidal facies above subtidal mudstones, whereas those in the upper part of the Vilella Baixa Unit consist of 80% subtidal facies overlain by oolitic–bioclastic packstones/grainstones. This parasequence stacking pattern, showing a systematic upward decrease in the amount of inter/supratidal facies in each parasequence and an increase in subtidal facies, is consistent with the whole stratal package, El Brull–Olesa–Vilella Baixa units, being a transgressive systems tract of a third-order relative-sea-level change. Parasequences are not obviously developed in the Colldejou Unit.

The Lower Muschelkalk carbonate platform consists of the TST and HST of one depositional sequence. The lowstand systems tract of this sequence is represented by a unit of red marls and evaporites, some 10 m thick, which was deposited in continental playas and sabkhas, fed by marine waters. This unit can be traced all over western Europe. It is equivalent to the Röt of the German and North Sea basins, and was deposited during the earliest stages of relative sea-level rise, before the major transgression which initiated the Lower Muschelkalk carbonate platform. The transgressive surface (ts) of this sequence is readily identified in the field as the base of the Lower Muschelkalk. The maximum flooding surface (mfs) is the distinctive surface occurring between the Vilella Baixa and Colldejou Units and represents the change from retrogradation to aggradation/progradation, that is TST to HST.

The Upper Muschelkalk

The Upper Muschelkalk carbonate platform (Fig. 12) also neatly falls into two parts, but each is interpreted here as a separate sequence (L1 and L2, L = Ladinian). This platform is notable for the

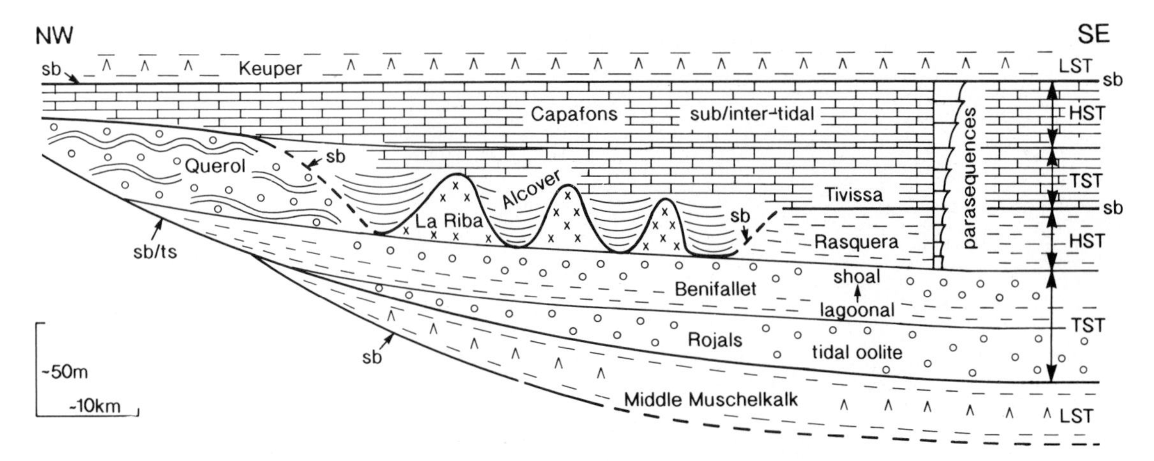

Fig. 12. Upper Muschelkalk lithostratigraphy and systems tracts. Abbreviations as in Fig. 3.

occurrence of mud mounds, which developed into more reefal structures (seen in Fig. 10 (Calvet & Tucker, 1993)). The sequence boundary, a prominent palaeokarst, occurs on the top of the mud mound–reefal complexes. In addition, there was some tectonic control on sedimentation during the later part (HST) of the lower sequence, when the mud mounds were developing, so that there are strong facies variations across the basin (Fig. 12).

The Upper Muschelkalk carbonate platform begins with an oolitic–bioclastic grainstone (Rojals Unit, 6–17 m), which was deposited in sand-shoal and sand-flat environments with associated tidal channels and tidal flats. The succeeding Benifallet Unit (20 to 50 m thick) is dominated by bioturbated lime mudstones and wackestones, deposited in an extensive lagoon, and these grade up into oolitic grainstones, deposited in an inner-ramp sand belt. These first two units, the Rojals and Benifallet, form a broadly transgressive package of high-energy shoreline sands, lagoonal muds and then barrier sands, and belong to the transgressive systems tract of sequence L1. The transgressive surface (ts) is again at the base of the carbonate platform and to the west coincides with the lower sequence boundary.

Facies differentiation now occurred across the platform with a series of shallowing-upward parasequences (the Rasquera Unit) succeeded by marlstones and dolomites (the Tivissa Unit) in the southern part (Baix Ebre-Priorat domain), the mud mounds developing into reefal complexes (La Riba Unit) succeeded by laminated dolomicrites (Alcover Unit) in the central part (Prades domain) and interbedded metre-scale domal stromatolites and grainstones (Querol Unit) in the more northern Gaià-Montseny domain. Across the whole platform, there was then deposited the Capafons Unit of dolomites in a variety of peritidal facies, which grade up into the Keuper red marls and gypsum.

The Rasquera Unit of the southern domain consists of five outer-ramp parasequences, 1.5 to 12 m thick (Fig. 13; Calvet & Tucker, 1988). In most of the parasequences, marlstone passes up into marlstone with thin limestones containing nektoplanktonic bivalves, which in turn passes up into a thin-bedded, bioturbated limestone with rare graded beds (tempestites). The parasequences are capped by wackestones and packstones containing a range of planktonic and benthic fossils, notably '*Tubiphytes*'. The parasequences are shallowing-upward cycles which were deposited on the outer part of the ramp in moderate to deep water. They are not capped by very shallow-water grainstones or emergent facies; thus each represents progradation of the mid-ramp area into deeper water.

There are lateral variations in the internal facies patterns of the parasequences. Towards the south, the direction of the basin centre, there is an increase in the thickness of the marlstone units (interpreted as a basinal facies) and marlstone interbedded-limestone facies (deep distal ramp). Towards the north/northwest, the direction of the inner ramp, the facies become more calcareous and the number and thickness of graded beds (tempestites) increase. On a broad scale, there is a vertical change up

Fig. 13. Two shallowing-upward parasequences from the Rasquera Unit of the Upper Muschelkalk, 5 km east of the village of Rasquera, 55 km southwest of Tarragona.

through the parasequence set at any one locality, with more shallower water facies, wackestones and packstones, in the higher parasequences. This indicates that the parasequence set has a progradational geometry and so represents the highstand systems tract of Ladinian sequence L1. The upper surface of the top parasequence is a prominent fossil lag horizon, with ammonoids, other molluscs and some iron mineralization. It is overlain by marls of the succeeding Tivissa Unit. This horizon is interpreted as the sequence boundary, equivalent to the palaeokarst of the adjacent Prades domain, and can be recognized throughout the southern part of the basin.

The succeeding Tivissa Unit consists of outer-ramp facies with some graded beds passing up into inner-ramp bioclastic packstones. This unit is the lower part of the succeeding sequence and reflects a deepening of the depositional environment. It would represent the lowstand, but particularly the transgressive systems tract of the next Ladinian sequence (L2). In the deeper water outer-ramp/basin environment, the distinction between these systems tracts is not easy, and also in a ramp environment, with no break of slope, lowstand deposits à la carbonate rimmed shelf will not form. The overlying peritidal–sabkha Capafons Unit is the progradational highstand systems tract of sequence L2.

One interesting general point about the Rasquera–Tivissa units is that they show a poor development of storm beds. These are a feature of many ramp formations, notably the Muschelkalk of the German Basin (Aigner, 1984), but many others including the Lower Carboniferous ramp of South Wales (Faulkner, 1988). The southern part of the Catalan Basin thus appears to have been a low-energy ramp, with few major onshore storms to generate offshore surges which could transport material out to the deeper ramp. There are two possible reasons for this: the Catalan Basin, at 10–20° north, was not in the hurricane belt where storms are most frequent and/or the ramp was located in a leeward situation so that storm winds were mostly blowing offshore, and so would be unable to set up the strong sea floor waves and currents.

The central part of the Catalan Basin is noteworthy for the occurrence of mud mounds which developed into reefal complexes (Figs 10 & 14) (Calvet & Tucker, 1993). The La Riba Unit build-ups are discrete structures, circular to elongate in plan view, and up to 60 m high. The mud mounds are massive bodies of peloidal and homogeneous dolomicrite, which have grown upon low banks

Fig. 14. La Riba mud mound (R) showing massive nature, original topography and onlap and overlap by the Alcover Unit (A). The lagoonal Benifallet Unit (B) occurs below the mud mound and below the Alcover Unit where the reef is absent and is represented by only a 10 cm-thick bed. The dashed line represents the boundary between the Benifallet and La Riba Alcover units. 4 km west of the town of La Riba, 25 km north of Tarragona.

of peloidal–bioclastic wackestone–grainstone. The areas between the buildups were starved of sediment, with only some 0.1 m of sediment deposited on the Benifallet Unit. The upper parts of the structures consist of a coral–algal–sponge framework with marine cements, and a quite varied biota of molluscs, foraminifera, crinoids and others. Some buildups show a lateral progradation of several hundred metres, with clinoform-type structures dipping up to 20°. They appear to have developed preferentially on the northeastern sides of the structures. The buildups are capped by bedded dasyclad grainstones and peritidal dolomites which contain tepees and black pebbles. The buildups are bounded by a prominent surface, which truncates the structures on the reef flanks in some cases. This is interpreted as a palaeokarst horizon. The overall succession of facies within the buildups is one of growth into shallower water and finally exposure. This suggests that the buildups represent the highstand systems tract, with the palaeokarst being the sequence boundary to the succeeding L2 sequence. The maximum flooding surface would occur within the lower part of the buildups.

The overlying Alcover Unit is an organic-rich laminated dolomicrite which fills the topography between the reefs, onlapping and overlapping them (Fig. 15). The Alcover unit passes up into shallow-water facies, with oolites and tidal flat sediments, and small-scale parasequences. Of note is the occurrence of exquisitely preserved fossils in the lower part of the Alcover unit (Fig. 16), and some horizons of replaced evaporites. Deposition of the lower part of the Alcover unit was in a stratified, anoxic and hypersaline basin. The Alcover unit represents the lowstand and transgressive systems tract of sequence L2. Again it is difficult to separate these systems tracts in a ramp depositional setting. The whole of the Alcover could be the TST, with the LST represented just by the palaeokarst on the top of the buildups. Alternatively, the lower part of the Alcover could be the late LST, before the more rapid relative sea-level rise that characterizes the TST. The overlying Capafons Unit, peritidal with evaporites, is aggradational and probably progradational too, and represents the highstand systems tract of sequence L2.

In the northern, more marginal part of the Cata-

Fig. 15. Near-horizontal Alcover Unit in far distance resting on Benifallet Unit (B), with just 10 cm-thick bed equivalent to the La Riba reef between, rising towards observer to onlap and overlap a reef mound (R). Dashed line shows top of Benifallet Unit. 10 km west of La Riba.

lan Basin (Gaià-Montseny domain), where the exposure is not good, the thin Querol Unit (2 to 12 m thick) is apparently equivalent to the La Riba unit. The Querol grainstones are interpreted as tidal in origin, with bipolar cross-bedding, reactivation surfaces and channels. They probably represent the HST of sequence L1. The succeeding Capafons unit in this domain represents the TST and HST of sequence L2.

The lowstand systems tract of sequence L1 is represented by the Middle Muschelkalk, consisting of local sandstones and siltstones, marls, laminated gypsum and nodular gypsum, deposited on a coastal plain of sabkhas, evaporitic lagoons and ponds with distal marine influence.

The Keuper evaporitic sediments succeeding the Upper Muschelkalk carbonates are all lowstand deposits of the next depositional sequence, resulting from another third-order relative sea-level change. They are followed by another carbonate platform, the Norian Imon Formation. This platform was more of an aggraded shelf, deposited during transgressive and highstand systems tracts.

The Lower and Upper Muschelkalk carbonate-ramp sequences illustrate many of the sequence-stratigraphic features of carbonate ramps discussed earlier. Both carbonate ramps are preceded by lowstand siliciclastics and evaporites (the uppermost Buntsandstein and Middle Muschelkalk), and these show great variations in thickness as a result of synsedimentary extensional tectonics (noted in Fig. 9). In this respect, with well-developed LSTs, these ramp sequences are different from many others (such as the Lower Carboniferous of South Wales, the Smackover and Great Oolite discussed earlier), but this is a reflection of the tectonic regime of eastern Spain, and western Europe generally, in this Triassic time. As discussed in Calvet *et al.* (1990), the carbonate platforms developed during the periods of gentle regional subsidence following the extension and lowstand deposition. The relative third-order sea-level changes (TST and HST) recorded in the platforms could thus be attributed more to regional tectonic causes, rather than to a global eustatic mechanism, although there is a good correlation with the Haq *et al.* (1987) global sea-level curve (Calvet *et al.*, 1990).

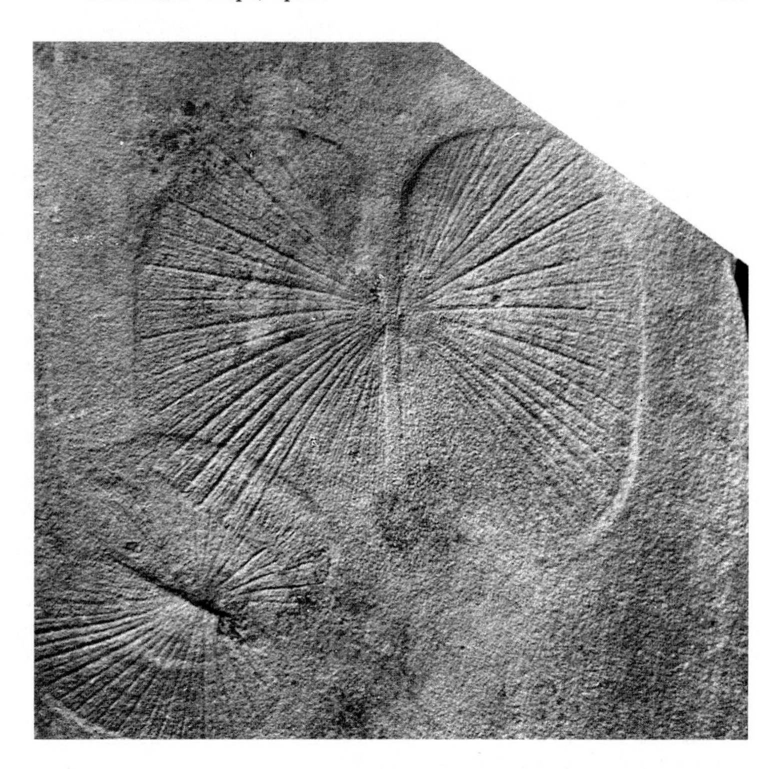

Fig. 16. Well-preserved thin-shelled bivalves, *Daonella*, from the laminated dolomicrites of the Alcover Unit.

DISCUSSION

There are several important differences between the sequence stratigraphy of carbonate rimmed shelves and ramps, which are outlined here. The major difference of course will be in the development of the lowstand systems tract, which is generally a major part of the sequence in the case of rimmed shelves during type 1 sea-level falls. The autochthonous and allochthonous lowstand wedges are distinctive sedimentary bodies, in some cases on a large scale, as in the Urgonian of the Vercors, French Alps (Arnaud-Vanneau & Arnaud, 1990; Hunt & Tucker, this volume, pp. 326–330). However, it is worth noting here that lowstand deposits are generally much more important in siliciclastic systems than carbonate systems, where much siliciclastic sediment is derived from erosion of the exposed shelf. On carbonate ramps, there is normally little resedimentation during a sea-level fall and at a lowstand unless the ramp is distally steepened and the sea-level fall is substantial. In most cases, there will simply be a downramp migration of facies belts, with palaeokarsts and palaeosols developing on the exposed inner ramp.

The transgressive systems tract may be a more important part of a ramp sequence in view of the higher gradient compared with a rimmed shelf. The flat top of a shelf leads to rapid flooding and there may be a delay (the concept of lag time, e.g. Hardie, 1986) before carbonate sedimentation gets back to normal production rates. Landward facies jumps and backsteps can be produced in this way. Drowning unconformities (Schlager, 1989) will be more prominent against rimmed shelves than on ramps. Carbonate sedimentation is likely to be more continuous on a ramp during a third-order sea-level rise, as a result of the higher gradient compared to shelves, and onlapping parasequences may well develop. Complete drowning of a rimmed shelf can occur more easily than that of a ramp during a more rapid sea-level rise. Although retrogradation is a common response of a carbonate platform, both ramp and shelf, to a third-order relative sea-level rise, both may aggrade and even prograde during the TST, if the rate of sea-level rise is less than the carbonate production rate. This is another major difference from siliciclastic systems, which are unable to prograde during the TST unless sediment supply is exceedingly high at this time. Other environmental factors, such as nutrient supply, anoxia and turbulence, are also major factors in carbonate production which can affect the geometries of facies.

The highstand deposits of both ramps and rimmed shelves are commonly the most conspicuous, with major clinoform progradation taking place at this time. With rimmed shelves, however, much resedimentation of shallow-water material takes place into the basin through turbidity currents and debris flows. Such HST resedimentation does not take place on ramps, unless they are distally steepened, and then it is relatively deep-water (outer-ramp) sediments which are supplied to the adjoining basin.

CONCLUSIONS

Carbonate ramps are relatively simple depositional systems compared with carbonate rimmed shelves and siliciclastic shelves in terms of sequence stratigraphy. Deposition mainly takes place during the transgressive and highstand systems tracts and several types of TST geometry can be distinguished depending on the rate of sea-level rise relative to carbonate production. Environmental factors other than sea-level change are also important in carbonate sedimentation. Lowstand systems tracts are generally poorly developed on carbonate ramps unless the relative lowstand is for a prolonged time and then an aggrading sand body can form. If the ramp is distally steepened, then slope aprons and wedges may evolve from resedimentation off the ramp shoulder. The concepts of carbonate-ramp sequence stratigraphy presented here have been applied to the Triassic Muschelkalk carbonate platforms of eastern Spain, where the sequences are third order consisting of fourth-/fifth-order parasequences.

ACKNOWLEDGEMENTS

Support for the fieldwork in Spain was kindly supplied through the Natural Environment Research Council, the Acciones Integradas Hispano-Britanica/British Council and the CICYT project number PB 0322. We are grateful to Karen Atkinson for drafting the figures and Gerry Dresser and Alan Carr for photography.

REFERENCES

AHR, W.M. (1973) The carbonate ramp: an alternative to the shelf model. *Trans. Gulf-Coast Assoc. Geol. Socs* **23**, 221–225.

AIGNER, T. (1984) Dynamic stratigraphy of epicontinental carbonates, Upper Muschelkalk (M. Triassic), South German Basin. *Neues Jb. Geol. Palaont. Abh.* **169**, 127–159.

ARNAUD-VANNEAU, A. & ARNAUD, H (1990) Hauterivian to lower Aptian carbonate shelf sedimentation and sequence stratigraphy in the Jura and northern Subalpine chains (southeastern France and Swiss Jura). In: *Carbonate Platforms* (Eds Tucker, M.E. *et al.*). Int. Assoc. Sediment. Spec. Publ. 9, 203–233.

BURCHETTE, T.P., WRIGHT, V.P. & FAULKNER, T.J. (1990) Oolitic sandbody depositional models and geometries, Mississippian of southwest Britain: implications for petroleum exploration in carbonate ramp settings. *Sediment. Geol.* **68**, 87–115.

CALVET, F. & TUCKER, M.E. (1988) Outer ramp carbonate cycles in the Upper Muschelkalk, Catalan Basin, N.E. Spain. *Sediment. Geol.* **57**, 185–198.

CALVET, F. & TUCKER, M.E. (1993) Mud-mound and reef complexes, Upper Muschelkalk, Triassic, Spain. In: *Mud Mounds* (Eds Monty, C.L.V., Bosence, D., Bridges, P. & Pratt, B.), Int. Assoc. Sediment. Spec. Publ. (in press).

CALVET, F., TUCKER, M.E. & HENTON, J.M. (1990) Middle Triassic carbonate ramp systems in the Catalan Basin, northeast Spain: facies, systems tracts, sequences and controls. In: *Carbonate Platforms* (Eds Tucker, M.E. *et al.*) Int. Assoc. Sediment. Spec. Publ. 9, 79–108.

FAULKNER, T.J. (1988) The Shipway Limestone of Gower: sedimentation on a storm-dominated early Carboniferous ramp. *J. Geol.* **23**, 85–100.

FISCHER, A.G. (1980) Long-term climatic oscillations recorded in stratigraphy. In: *Climate in Earth History* (Eds Berger, W.H. & Crowell, J.C.) National Academy Press, Washington DC, pp. 97–104.

GAMBOA, L.A., TRUCHAN, M. & STOFFA, P.L. (1985) Middle and Upper Jurassic depositional environments at outer shelf and slope of Baltimore Canyon Trough. *Bull. Am. Assoc. Petrol. Geol.* **69**, 610–621.

GOLDHAMMER, R.K., DUNN, P.A. & HARDIE, L.A. (1990) Depositional cycles, composite sea-level changes, cycle stacking patterns, and the hierarchy of stratigraphic forcing: examples from Alpine Triassic platform carbonates. *Bull. Geol. Soc. Am.* **102**, 535–562.

GUTTERIDGE, P. (1989) Controls on carbonate sedimentation in a Brigantian intrashelf basin, Derbyshire. In: *The Role of Tectonics in Devonian and Carboniferous Sedimentation in the British Isles* (Eds Arthurton, R.S., *et al.*), Yorks. Geol. Soc. Occ. Publ. 6, 171–187.

HAQ, B.U., HARDENBOL, J. & VAIL, P.R. (1987) Chronology of fluctuating sea-levels since the Triassic. *Science* **235**, 1156–1167.

HARDIE, L.A. (1986) Stratigraphic models for carbonate tidal-flat deposition. *Quart. J. Colorado Sch. Mines* **81**, 59–74.

HUMPHREY, J.S. & KIMBELL, T.N. (1990) Sedimentology and sequence stratigraphy of Upper Pleistocene carbonates of S.E. Barbados, West Indies. *Bull. Am. Petrol. Geol.* **74**, 1671–1684.

Hunt, D. & Tucker, M.E. (1993) Sequence stratigraphy of carbonate shelves with an example from the mid-Cretaceous (urgonian) of Southeast France. In: *Sequence Stratigraphy and Facies Associations* (Eds Posamentier, H.W., Summerhayes, C.P., Haq, B.U. & Allen, G.P.) *Int. Assoc. Sediment. Spec. Publ.* **18**, 307-341.

Koerschner, W.F. & Read, J.F. (1989) Field and modelling studies of Cambrian carbonate cycles, Virginia Appalachians. *J. Sediment. Petrol.* **59**, 654–687.

Lees, A. & Miller, J. (1985) Facies variation in Waulsortian buildups, Part 2; Mid-Dinantian buildups from Europe and North America. *J. Geol.* **20**, 159–180.

Markello, J.R. & Read, J.F. (1989) Carbonate ramp to deeper shale shelf transitions of an upper Cambrian intrashelf basin, Nolichucky Formation, Southwest Virginia Appalachians. *Sedimentology* **28**, 573–597.

Moore, C.H. (1984) The Upper Smackover of the Gulf Rim: depositional systems, diagenesis, porosity evolution and hydrocarbon production. In: *Jurassic of the Gulf Rim* (Eds Ventress, W.P.S., Bebout, D.G., Perkins, R.F. & Moore, C.H.) Soc. Econ. Paleontol. Mineral. Gulf. Coast Sections Publ. 283–308.

Mullins, H.T. *et al.* (1988). Three dimensional sedimentary framework of the carbonate ramp slope of central west Florida: a sequential seismic stratigraphic perspective. *Bull. Geol. Soc. Am.* **100**, 514–533.

Petta, T.J. (1980) Silurian pinnacle reef diagenesis — Northern Michigan; effects of evaporites on pore space distribution. In: *Carbonate Reservoir Rocks* (Eds Halley, R.B. & Loucks, R.G.) Soc. Econ. Paleontol. Mineral. Core Workshop 1, 32–42.

Purser, B.H. (Ed.) (1973) *The Persian Gulf: Holocene Carbonate Sedimentation and Diagenesis in a Shallow Epicontinental Sea.* Springer-Verlag, Heidelberg, pp. 471.

Read, J.F. (1982) Carbonate platforms of passive (extensional) continental margins: types, characteristics and evolution. *Tectonophysics* **81**, 195–212.

Read, J.F. (1985) Carbonate platform facies models. *Bull. Am. Assoc. Petrol. Geol.* **66**, 860–878.

Schlager, W. (1989) Drowning unconformities on carbonate platforms. In: *Controls on Carbonate Platform and Basin Development* (Eds Crevello, P.D. *et al.*). Spec. Publ. Soc. Econ. Paleontol. Mineral. 44, 15–25.

Sellwood, B.W., Scott, J., Mikkelsen, P. & Akroyd, P. (1985) Stratigraphy and sedimentology of the Great Oolite Group in the Humbly Grove Oilfield, Hampshire, S. England. *Mar. Petrol. Geol.* **2**, 44–55.

Shaver, R.H. (1977) Silurian reef geometry — new dimensions to explore. *J. Sediment. Petrol.* **47**, 1409–1424.

Tucker, M.E. (1985) Shallow marine carbonate facies and facies models. In: *Sedimentology Recent Developments and Applied Aspects* (Eds Brenchley, P.J. & Williams, B.P.J.). Spec. Publ. Geol. Soc. London 139–161.

Tucker, M.E. (1991a) *Sedimentary Petrology: an Introduction to the Origin of Sedimentary Rocks.* Blackwell Scientific Publications, Oxford.

Tucker, M.E. (1991b) Sequence stratigraphy of carbonate-evaporite basins: models and application to the Upper Permian (Zechstein) of northeast England and adjoining North Sea. *J. Geol. Soc. London* **148**, 1019–1036.

Tucker, M.E. & Wright, V.P. (1990) *Carbonate Sedimentology.* Blackwell Scientific Publications, Oxford.

Van Wagoner, J.C., Mitchum, R.M., Campion, K.M. & Rahmanina, V.D., (1990) Siliciclastic sequence stratigraphy in well logs, cores and outcrops. *Am. Assoc. Petrol. Geol. Meth. Expl. Ser.* **7**, 45.

Van Wagoner, J.C., Posamentier, H.W., Mitchum, R.M., Jr, *et al.* (1988) An overview of the fundamentals of sequence stratigraphy and key definitions. In: *Sea-level Changes — an Integrated Approach* (Eds Wilgus, C.K., Hastings, B.S., Kendall, C.G.St.C., Posamentier, H.W., Ross, C.A. & Van Wagoner, J.C.). Spec. Publ. Soc. Econ. Paleontol. Mineral. 42, 39–45.

Veevers, J.J. (1990) Tectonic climatic supercycles in the billion-year plate-tectonic eon: Permian Pangean icehouse alternates with Cretaceous dispersed continents greenhouse. *Sediment. Geol.* **68**, 1–16.

Ward, W.C. & Brady, M.J. (1979) Strandline sedimentation of carbonate grainstones, Upper Pleistocene, Yucatan Peninsula, Mexico. *Bull. Am. Assoc. Petrol. Geol.* **63**, 362–369.

Ward, W.C., Weidie, A.E. & Back, W. (1985) *Geology and Hydrogeology of the Yucatan.* New Orleans Geological Society, 160 pp.

Wright, V.P. (1986) Facies sequences on a carbonate ramp: the Lower Carboniferous of South Wales. *Sedimentology* **33**, 221–241.

GREENLAND

Spec. Publs Int. Ass. Sediment. (1993) **18**, 419–448

Cyclic sedimentation in a large wave- and storm-dominated anoxic lake; Kap Stewart Formation (Rhaetian–Sinemurian), Jameson Land, East Greenland

G. DAM* *and* F. SURLYK†

**Geological Survey of Greenland, Øster Voldgade 10, DK-1350 Copenhagen K, Denmark; and*
†Geological Institute, University of Copenhagen, Øster Voldgade 10, DK-1350 Copenhagen K, Denmark

ABSTRACT

During Rhaetian–Sinemurian times a large wave- and storm-dominated lake was situated in the Jameson Land Basin, East Greenland. Sandy and pebbly alluvial material was transported into the lake from source areas to the east, west and north. Lake deposits consist of alternating black unfossiliferous mudstones and sheet sandstones. Anoxic conditions dominated at the lake bottom during deposition of the muds and the relatively deep-water column probably was stratified. The sandstones were deposited during progradation of wave- and storm-dominated deltas in a water depth less than 15 m. A high-resolution sequence-stratigraphic interpretation of the succession suggests that the mudstones were deposited in periods of rising and large highstands of lake level, whereas progradation of the deltaic sheet sandstones took place during forced regressions, caused by significant falls. The lake thus experienced abundant high-frequency and fairly high-amplitude changes in level. These high-frequency lake-level changes may reflect Milankovitch-type climatic cycles. The high-frequency cycles can be grouped into a number of low-frequency cycles. These cycles show the same number of large-scale high- and lowstand points as published eustatic sea-level curves for the Rhaetian–Sinemurian, and it is suggested that the long-term trends in lake level mirrored and probably were controlled by eustasy. The lake-level curve is accordingly interpreted to show a high-periodicity climatically induced signal superimposed on a long-periodicity eustatic signal. A sequence-stratigraphic analysis of the Kap Stewart Formation shows that it consists of two third-order sequences with a maximum duration of 15 to 20 Ma.

INTRODUCTION

A major stimulation to the study of lakes and their deposits in recent years has been undoubtedly their sensitivity to climate. Ancient lake deposits probably contain our best indicators of palaeoclimate. Moreover, lacustrine petroleum source rocks have come more into focus as exploration has shifted to new areas and as more detailed analysis of known petroleum provinces has become an exploration necessity (Katz, 1990a). The large variability in lacustrine facies requires a multidisciplinary approach, including clastic sedimentology, organic and inorganic geochemistry, palynostratigraphy, well-log analyses and seismic reflection interpretation.

The sediments of the Rhaetian–Sinemurian Kap Stewart Formation of East Greenland were deposited in a large wave- and storm-dominated anoxic lake. Most studies of lacustrine clastic sediments have focused on 'Gilbert-type' and fluvial-dominated deltas. Descriptions of modern and ancient examples of Gilbert-type deltas are numerous in the literature (Aario, 1971; Boothroyd & Ashley, 1975; Gustavson, 1975; Gustavson *et al.*, 1975; Stanley & Surdam, 1978; Sneh, 1979;

Surdam & Stanley, 1979; Vessell & Davies, 1981; Clemmensen & Houmark-Nielsen, 1981; Bogen, 1983; Syvitski & Farrow, 1983; Weirich, 1985, 1986; Ori & Roveri, 1987; Massari & Colella, 1988; Postma & Cruickshank, 1988; Postma *et al.*, 1988; Wood & Ethridge, 1988). Such deltas occur mainly in fjords, ice-dammed lakes and tectonically active basins characterized by relatively steep slopes, whereas fluvial-dominated deltas are characteristic of low-relief lakes in rift valleys, intermontane basins and delta plains (Axelsson, 1967; Hyne *et al.*, 1979; Fielding, 1984; Haszeldine, 1984; Tye & Coleman, 1989a, b). Contemporaneous but spatially separated formation of both delta types is known from the late Mesozoic lacustrine Camp Hill Beds of Antarctica and from Lake Hazar of Turkey (Farquharson, 1982; Dunne & Hempton, 1984).

Studies of modern shoreline processes and ancient shoreface successions show that wave- and storm-induced processes may be of great significance in major lakes with large fetches (e.g. Van Dijk *et al.*, 1978; Allen, 1981a, b; Duke, 1984; Eyles & Clark, 1986, 1988; Greenwood & Sherman, 1986, 1989; Talbot, 1988; Katz, 1990b; Martel & Gibling, 1991). However, only a few examples of lacustrine deltaic successions dominated by these processes have been described. Thus Eyles and Clark (1988) identified a late Pleistocene storm-influenced glacio-lacustrine delta on the margin of the enlarged, ice-dammed ancestral Lake Ontario.

The basinal parts of the Upper Triassic–Lower Jurassic succession of the Jameson Land Basin, East Greenland consist of alternating anoxic lacustrine mudstones and wave- and storm-dominated delta-front sandstones (Dam, 1991). In this study special emphasis is placed on the internal geometry and origin of wave- and storm-dominated deltaic sediment bodies and their relation to cyclic lake-level fluctuations. Similar contemporaneous cyclic lake-level fluctuations have been described from the Newark and Hartford Basins of eastern North America, and have been related to climatic cycles (Olsen, 1986, 1988, 1990). Correlative successions in other North Atlantic basins (e.g. the Åre Formation of Halten Banken, offshore Norway) should be investigated for the presence of similar lacustrine systems. The importance of the sedimentary effect of wave and storm processes in lakes with large fetches is another subject worthy of further investigations. A wide array of methods have been used in the study, including sedimentology, organic geochemistry, phytopalaeontology, seismic reflection and sequence stratigraphy.

REGIONAL SETTING

An extensive hydrologically closed lake was situated in the Jameson Land Basin of East Greenland during late Triassic–early Jurassic times (Dam, 1991). The basin is located in the southern end of the exposed part of the East Greenland rift system (Surlyk, 1990). The basin was bounded to the west by a major N–S fault and to the east by a NNE–SSW elongated landmass covering the present-day Liverpool Land area. To the north, the Jameson Land Basin was bounded by a NW–SE cross-fault zone in Kong Oscar Fjord (Surlyk, 1977, 1978, 1990). The southern boundary is unknown but the basin may have extended south of Scoresby Sund (Fig. 1). Mesozoic subsidence was due primarily to thermal contraction after a long period of late Palaeozoic rifting (Surlyk *et al.*, 1986; Larsen & Marcussen, 1992). However, minor phases of fault-controlled subsidence took place throughout the Mesozoic (Surlyk, 1978, 1990; Clemmensen, 1980; Surlyk *et al.*, 1981).

Establishment of the Kap Stewart lake took place during a marked change from semi-arid to humid conditions. Palaeomagnetic data suggest that this climatic change was caused by a slow northwards drift of the Laurasian continent (Smith *et al.*, 1981). Early Triassic continental red beds with subordinate carbonates and evaporites were thus gradually succeeded by middle–late Triassic playa mudstones, lacustrine carbonates and flood-plain deposits indicative of a more humid climate (Bromley & Asgaard, 1979; Clemmensen, 1980) (Fig. 2). By latest Triassic, Rhaetian time, the climate became warm, temperate and humid and the perennial lacustrine basin of the Kap Stewart Formation developed. In the early Pliensbachian the lacustrine basin was transgressed and a shallow-marine embayment was formed, probably reflecting the important eustatic sea-level rise during the late Triassic–early Jurassic (cf. Hallam, 1988).

At outcrop, the Kap Stewart Formation is 155 to 300 m thick increasing from north to south and from basin margins to central parts. Seismic data show that the formation may exceed 700 m in thickness in the most central parts of the basin (including *c.* 100 to 200 m of Tertiary sills) (Dam, 1991; C. Marcussen, pers. comm., 1991).

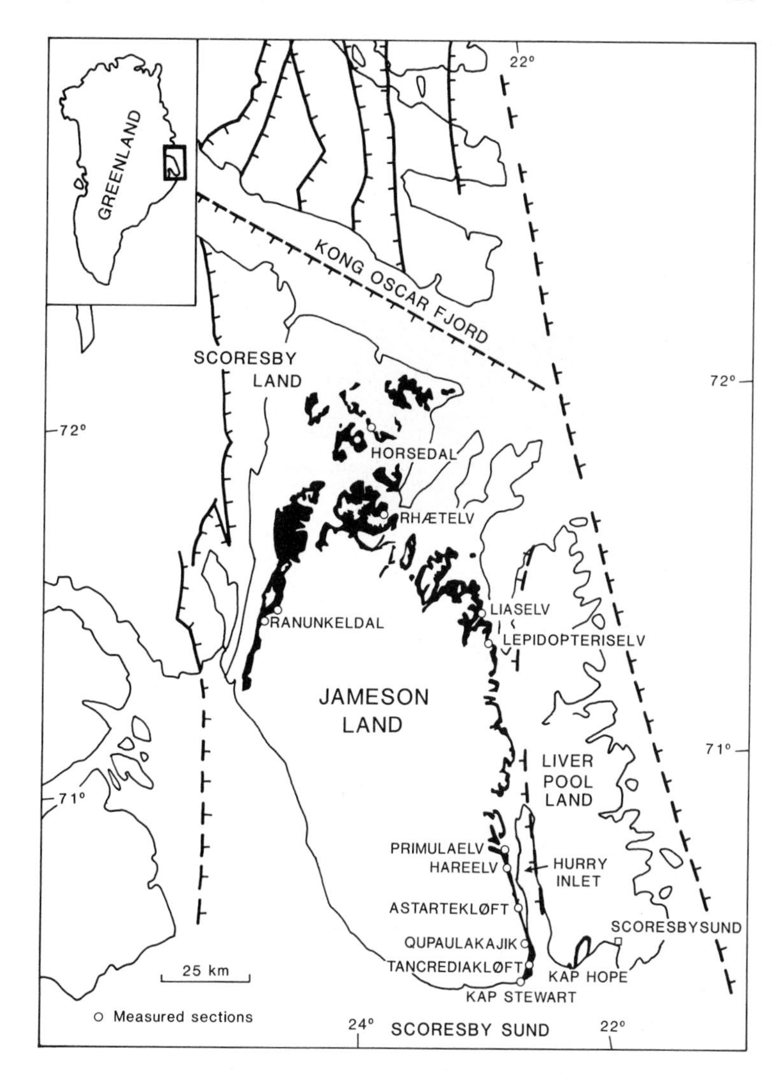

Fig. 1. Map of Jameson Land showing outcrops of the Kap Stewart Formation and location of studied sections. (Based on maps from the Geological Survey of Greenland.)

STRATIGRAPHY

In the southeastern part of the basin the base of the Kap Stewart Formation is taken by the lithological change from red, marly shales, grey sandstones and carbonates of the fluvial and lacustrine Ørsted Dal Member (Flemming Fjord Formation), to the grey arkosic sandstones, conglomerates and dark plant-bearing shales of the alluvial-plain, deltaic and lacustrine Kap Stewart Formation. Along Hurry Inlet the lower boundary is covered by scree, but at Kap Hope the Kap Stewart Formation rests unconformably upon the Ørsted Dal Member or the older Klitdal Member and is characterized by basin-margin onlap (Birkenmajer, 1976). Along Hurry Inlet the Kap Stewart Formation can be subdivided into the lower 'barren sandstone' (85 m) composed mainly of coarse alluvial-plain deposits and the 'plant-bearing series' composed of delta-plain deposits (90 m thick) (Harris, 1937) (Fig. 3). The flora indicates that the barren sandstone and the lowest 30 m of the plant-bearing series belong to the *Lepidopteris* zone (lower part of the Rhaetian). After a 5 m 'transitional region' the remaining 55 m of the plant-bearing series belong to the Hettangian *Thaumatopteris* zone (Harris, 1937). The two floras have only very few species in common and Upper Rhaetian macro- or microfloras are completely missing. This suggests the presence of a hiatus at the base of the *Thaumatopteris* zone (Harris, 1937;

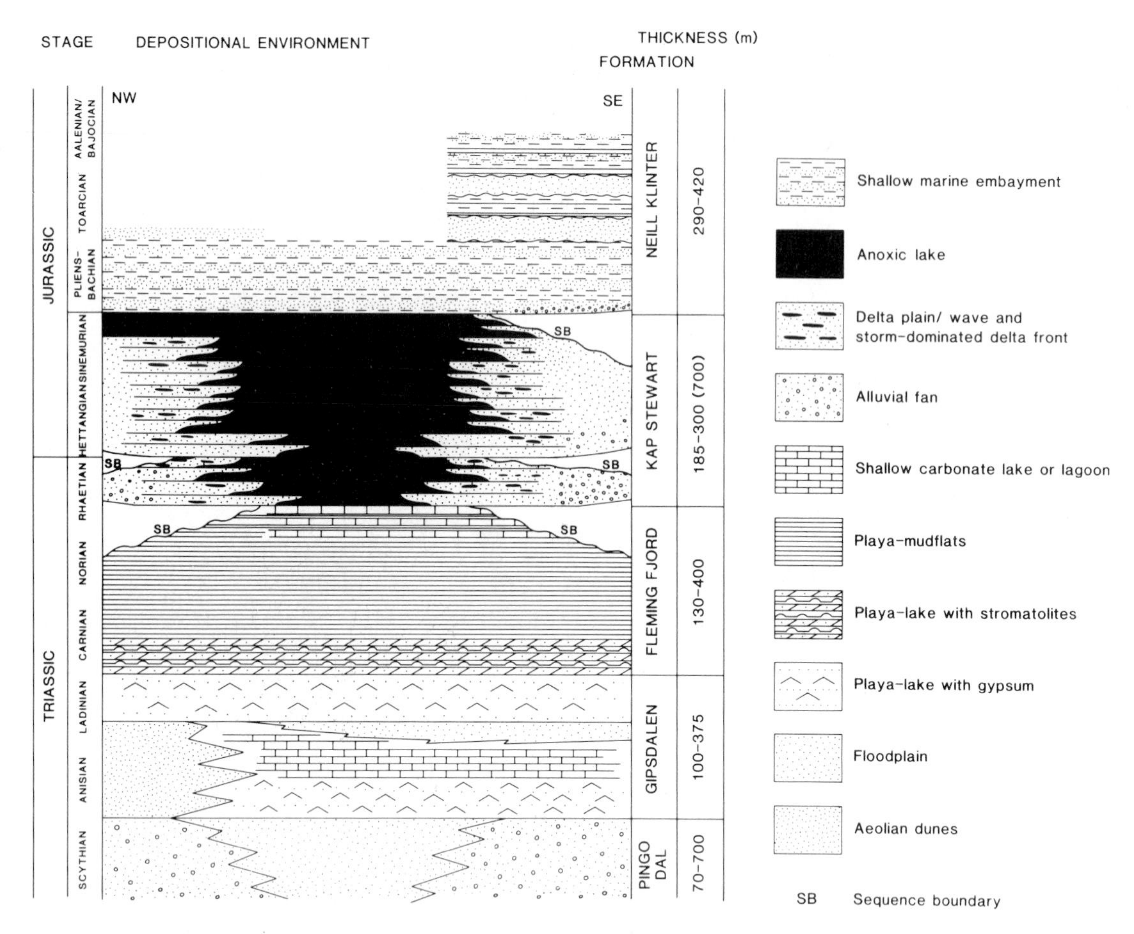

Fig. 2. Stratigraphy and depositional environments of the Middle Triassic–Lower Jurassic succession in the Jameson Land Basin. Note the Kap Stewart lacustrine complex between the Triassic red beds and the marine Jurassic sandstones. (Based on Surlyk *et al.*, 1973; Clemmensen, 1980; and Dam, 1991.)

Pedersen & Lund, 1980). Two distinct microfloral zones are recognized. Zone 1 corresponds to the macrofloral *Lepidopteris* zone and the transitional region, whereas zone 2 is equivalent to the *Thaumatopteris* zone. Zone 1 is correlated with the *Rhaetipollis–Limbosporites* zone from the lower part of the Rhaetian (Middle Rhaetian *sensu germanico*) in NW Europe, and zone 2 with the Hettangian *Pinuspollenites–Trachysporites* zone (Pedersen & Lund, 1980). The barren sandstone and the plant-bearing series have not been recognized as separate members (Surlyk *et al.*, 1973; Dam, 1991). However, they do reflect a lithological change from dominantly conglomeratic alluvial-plain deposits below to delta-plain deposits with well-developed fining-upward cycles above. Along Hurry Inlet the formation is unconformably overlain by the basal Pliensbachian Rævekløft Member (Rosenkrantz, 1934; Surlyk *et al.*, 1973).

In the western part of the basin the Kap Stewart Formation is 300 m thick. In Ranunkeldal the formation was subdivided into a lower '*Sandsteinserie*' of coarse-grained sandstones and an upper '*pflanzenführenden und mehr schieferigen Serie*' of more shaly plant-bearing deposits by Stauber (1940). The lower *Sandsteinserie* is 140 m thick and consists of alluvial-plain deposits. The upper *pflanzenführenden und mehr schieferigen Serie* is 150 m thick and consists of intercalated open-lacustrine mudstones and wave- and storm-dominated delta-front sandstones (Dam, 1991) (Fig. 3). The plant fossils of the plant-bearing deposits include both Rhaetian and early Jurassic species (Harris, 1946). The lower boundary of the formation is not well exposed in this area, but upwards the formation shows a gradual transition into the marine deposits of the Neill Klinter Formation (Surlyk *et al.*, 1973; Dam, 1991).

In the northern (Horsedal), central (Rhætelv) and northeastern (Lepidopteriselv) parts of the basin the thickness of the formation is approximately 300 m (Fig. 3). In these areas the formation is made up of intercalated open-lacustrine mudstones and wave- and storm-dominated delta-front sandstones (Fig. 4), with a general tendency for the mudstone intervals to thicken towards the basin centre (Dam, 1991). Only Hettangian plant fossils have been found in these parts of the basin (Harris, 1946). However, ostracods from the bottom of the formation in the northern and northeastern parts of the basin are suggestive of an early Rhaetian age (Defretin-Lefranc *et al.*, 1969; O. Michelsen, pers. com., 1990).

At these localities there is no apparent unconformity at the contact between the early Rhaetian Tait Bjerg Beds (top Flemming Fjord Formation) and the Kap Stewart Formation, but a hiatus in the early Rhaetian has been suggested by Perch-Nielsen *et al.* (1974). Upwards the formation shows a gradual transition into the marine Neill Klinter Formation (Surlyk *et al.*, 1973; Dam, 1991).

Fossils indicative of a Sinemurian age have not been observed in the Kap Stewart Formation, but the gradual upwards transition to the Pliensbachian–Toarcian/(?)Aalenian Neill Klinter Formation and the great thickness of the Kap Stewart Formation in the basinal areas suggest that most or all of the Sinemurian Stage is represented here.

The Kap Stewart Formation thus probably spans the early Rhaetian–Sinemurian time interval where it is thickest developed. The base of the formation is unconformable in the Kap Hope area where it also is characterized by basin-margin onlap (Birkenmajer, 1976). The hiatus probably corresponds to a minor time interval in the early Rhaetian below present biostratigraphic resolution. Basinwards the unconformity passes into a correlative conformity. An important facies change occurs in the middle part of the formation along the basin margins, where alluvial-plain deposits pass into delta-plain and intercalated open-lacustrine mudstones and wave- and storm-dominated delta-front sandstones. This facies change is not associated with any unconformable relations. A major hiatus occurs in the upper part of the formation corresponding to the late Rhaetian time interval. This hiatus is marked by the change from the *Lepidopteris* flora to the *Thaumatopteris* flora, but is not associated with an unconformity.

The upper boundary is developed as an important basin-margin unconformity and the associated hiatus corresponds to all of the Sinemurian Stage. Basinwards it passes into a correlative conformity and most or all of the Sinemurian is thought to be represented in the basinal areas (Surlyk, 1990; Dam, 1991).

FACIES ASSOCIATIONS

The sediments of the Kap Stewart Formation are subdivided into the open-lacustrine, delta-front, delta-plain, alluvial-plain and delta-abandonment facies associations.

Open-lacustrine facies association

Lithology and structures

This association is dominated by dark-grey to black, parallel-laminated mudstones ('paper shales'), in places with siderite concretions, and beds of coarse-grained sandstones with symmetrical ripples. The mudstone intervals, which are up to 35 m thick, have been followed continuously for up to 125 km along the exposed parts of the basin margins. They thicken towards the basin centre, suggesting that they are also continuous in the subsurface, covering more than 12,000 km^2.

The mineralogical composition of the mudstones is uniform and the dominant constituent is kaolinite with minor amounts of quartz, mica, feldspar, calcite, siderite and vermiculite/chlorite. Laterally persistent bands of early diagenetic siderite concretions occur in several mudstone intervals (Fig. 3). Concretions in the uppermost, thick mudstone interval commonly contain bivalves, viviparid gastropods and rare insects, fish scales and spines. Bivalve and gastropod densities are high, whereas diversities are low. Smooth-shelled ostracods and abundant bivalves also occur in the mudstones at the bottom of the formation in Liaselv and Rhætelv (Fig. 3). Outside these intervals the mudstones are very sparse in both body and plant fossils.

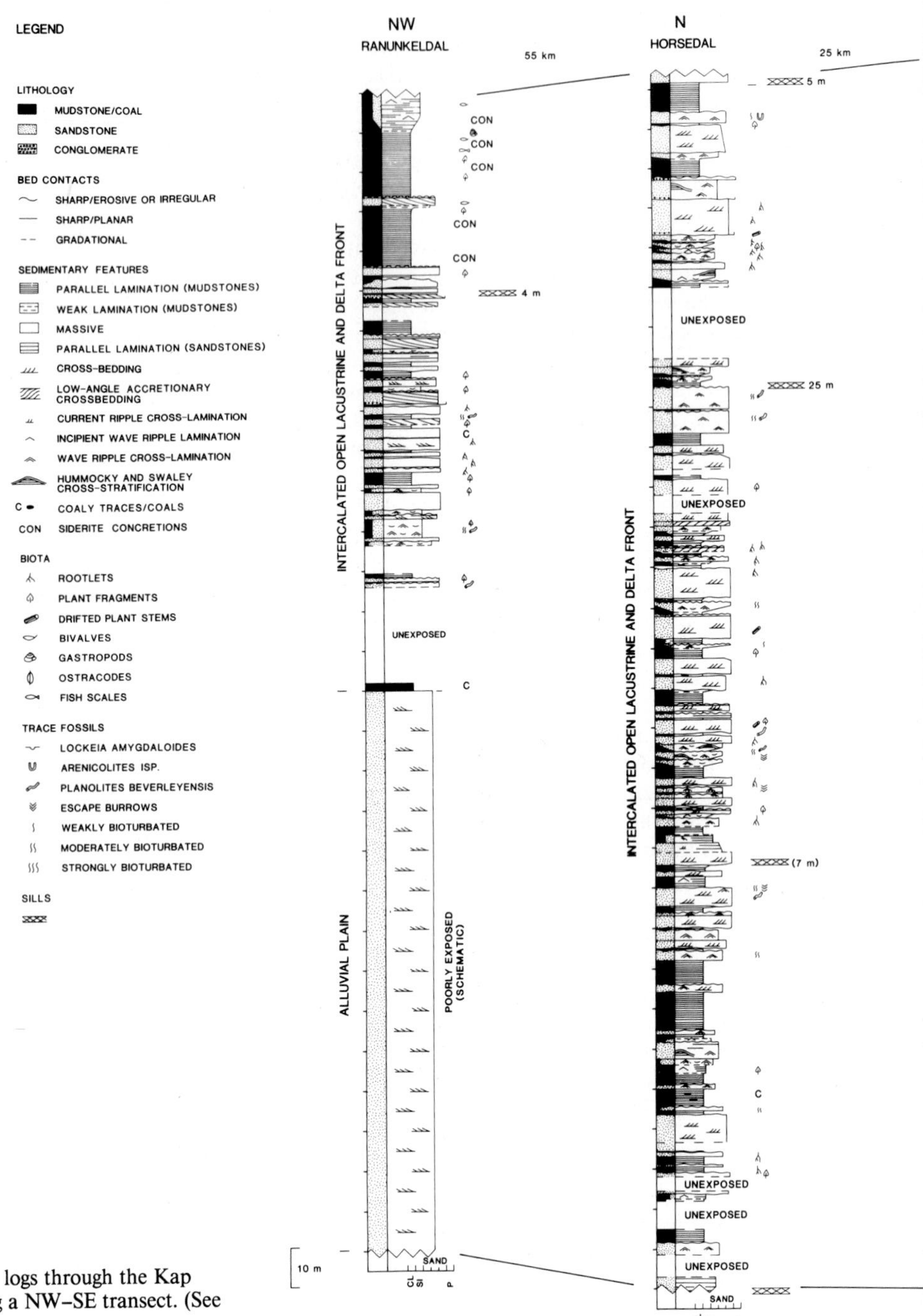

Fig. 3. Sedimentological logs through the Kap Stewart Formation along a NW–SE transect. (See Fig. 1 for locations.)

Coarse to very coarse-grained sandstones with symmetrical ripples (CGR) occur in sharply based, laterally persistent beds, averaging 25 cm in thickness (Fig. 5). Ripple height is up to 20 cm and wave lengths may reach 1 m. Crest lines are straight to slightly sinuous, commonly with bifurcations. Some forms are apparently sediment-starved and CGR appear as trains of regularly spaced, but

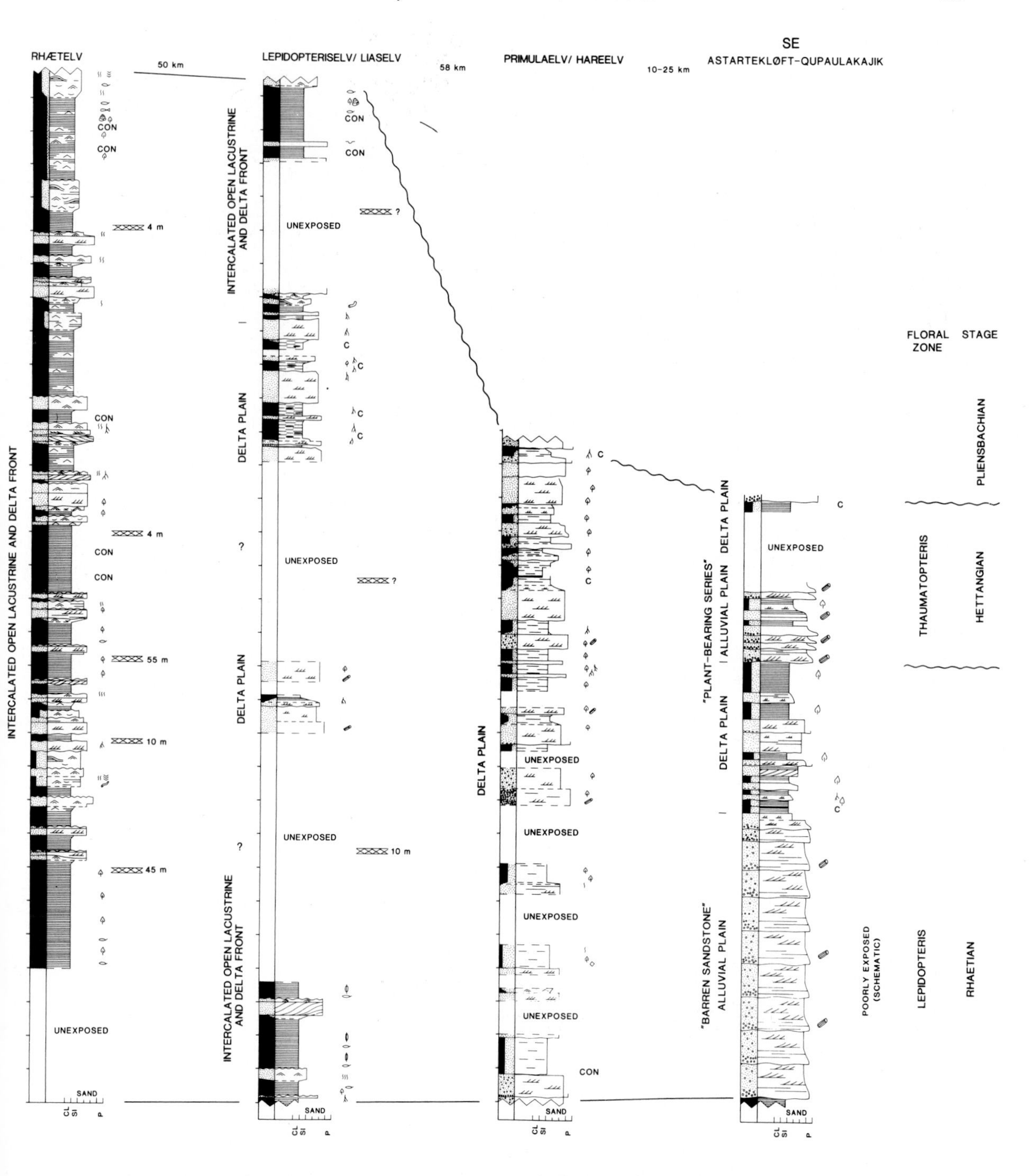

isolated, convex-upward bedforms overlying planar basal surfaces. Where stratification is visible there is a dominance of cross-strata dipping in one direction. The CGR may occur together with hummocky cross-stratification (HCS) and other structures or bedforms attributable to wave and storm processes. CGR are commonly moderately to strongly burrowed with *Planolites beverleyensis*.

Fig. 4. Alternating open-lacustrine mudstones and delta-front sheet sandstones from the central part of the basin. The sheet sandstones can be followed laterally for several kilometres with only very small changes in thickness. The middle black mudstone is 30 m thick. From the middle part of the Ranunkeldal section (see Figs 1 & 3 for location).

Organic matter

The mudstones typically contain 2 to 10% total organic carbon (TOC). The hydrogen index (HI) ranges from approximately 100 to more than 700. The organic matter is a mixture of fresh-water algal (*Botryococcus*) and higher plant remains (type I and III kerogen). The organic matter of mudstones with low HI values is completely dominated by terrestrial-derived detritus, whereas increasing HI

Fig. 5. Coarse-grained symmetrical ripples interbedded with fine-grained storm sandstones and open-lacustrine mudstones. Lens cap 55 mm across. From upper part of the Horsedal section (see Figs 1 & 3 for location).

values are noted in association with increasing amounts of liptinic amorphous material, occasionally with remains of *Botryococcus*-like algal material (Dam & Christiansen, 1990).

Sulphur

All of the samples studied display very low total sulphur (TS) contents (below 0.31%). However, a good linear correlation is observed between TOC and TS (Dam & Christiansen, 1990). The low content of TS and the good linear correlation between TS and TOC is typical of fresh-water sedimentary rocks (Berner & Raiswell, 1984).

Gas chromatography

A remarkable predominance of odd-numbered *n*-alkanes in the high molecular weight region (nC_{21} to nC_{31}) is observed in many samples, and nC_{25} and nC_{27} are particularly abundant compared to even-numbered *n*-alkanes in the same range (Dam & Christiansen, 1990). This pattern is commonly reported from lacustrine source rocks and supports the interpretation of the organic matter being of non-marine origin (Connan & Casou, 1980; Tissot & Welte, 1984; Moldowan *et al.*, 1985; Powell, 1986).

The pristane to phytane ratios are high (2.5 to 8, averaging 3.9) (Dam & Christiansen, 1990). This is typical for oil and extracts from lacustrine, fluvial or deltaic source rocks, with significant amounts of terrestrial organic matter, which have undergone some oxidation prior to preservation (e.g. Powell, 1986).

The well-preserved lamination of the mudstones suggests deposition from suspension in a static water column below fair-weather wave base in the absence of bottom-living burrowing invertebrates. Water was fresh, as indicated by the total absence of marine palynomorphs, the probably non-marine invertebrate fauna, the occasional presence of the fresh-water algae *Botryococcus*, gas-chromatography data, the low content of TS and the good linear correlation between TS and TOC. The significant quantities of preserved algal and herbaceous remains and the high TOC values indicate deposition in anoxic bottom waters in a permanently stratified lake. This is supported by the general absence of bioturbation and of a bottom-living fauna. Periodic oxygenation of bottom waters is, however, indicated by the presence of bivalves and gastropods in certain narrow stratigraphic intervals. Waters above the bottom did at least at times support fish and pelagic ostracods. The kaolinitic composition of the mudstones suggests intense weathering and leaching conditions under temperate, humid climatic conditions in the sediment source area (cf. Potter *et al.*, 1980; Chamley, 1989).

Phosphate nodules form in modern sediments in the transition zone between oxygenated and anoxic waters, or where the water column is oxygenated but conditions below the sediment–water interface are anoxic and strongly reducing (Cook, 1976). The calcium phosphate nodules may therefore have been formed near the redox potential discontinuity surface in the mudstones.

The symmetry and bifurcations of the ripple forms and the association with hummocky and swaley cross-stratified fine-grained sandstones indicate that the CGR are wave ripples (Clemmensen, 1976). However, the unidirectional foreset dips of the cross-strata at each locality suggest some unidirectional flow component. The presence of isolated CGR within the lacustrine mudstones and their unimodal crest line orientation suggest that CGR formed at depths below fair-weather wave base produced by large-amplitude, long-period storm waves. Coarse-grained sandstones covered with large symmetrical ripples occur today mainly on broad open shelves at depths up to 100 m (e.g. Newton & Werner, 1972; Hunter *et al.*, 1982; Cacchione *et al.*, 1984; Leckie, 1988). However, they have been recorded also from bays with a fetch of only a few tens of kilometres, at water depths of 2 to 20 m (e.g. Masuda & Makino, 1987; Hunter *et al.*, 1988). The coarse sandstones probably were derived from storm erosion of beach sands at the distributary mouths, and may have been transported offshore by near-bottom seaward flows.

The high density of *Planolites beverleyensis* in the CGR suggests that more aerated conditions at times were introduced to the lake bottom during storms. This monospecific trace fossil assemblage may represent an opportunistic fauna of worm-like deposit feeders that colonized the lake bottom immediately after the storms.

Delta-front facies association

This association is composed of heterolithic sediments and sandstones forming coarsening-upward successions, up to 9 m thick. Three types of facies are distinguished, representing wave- and storm-dominated shoreface, terminal-lobe and distributary channel environments.

Wave- and storm-dominated shoreface

The shoreface deposits are arranged in coarsening-upward successions, the lower part of which is composed of micaceous mudstones with silty and fine-grained incipient sandy ripples (cf. Facies M_1 of De Raaf *et al.*, 1977). The sandstone streaks and incipient ripples are less than 1 cm thick, have sharp bases and are normally graded (Fig. 6(a)). The upper part of the shoreface deposits is characterized by lenticular beds which comprise sandstone layers with continuous and discontinuous mudstone drapes (cf. Facies M_2 of De Raaf *et al.*, 1977). The lenticular beds grade into parallel- and cross-laminated sandstones (cf. Facies S_1 of De Raaf *et al.*, 1977) (Fig. 6(b)). Symmetrical ripple-form sets and structures typical of wave generation are common in climbing ripple cross-sets. Ripple wave lengths and amplitudes are 8.9 to 12.4 cm and 0.7 to 1.5 cm, respectively. The ripple symmetry index ranges from 1.1 to 1.7.

The shoreface deposits may contain medium- to coarse-grained cross-bedded sandstones that form co-sets, up to 2 m thick. The co-sets show features characteristic of wave action such as opposed unidirectional cross-bedded lenses, offshooting and draping foreset laminations and symmetrical ripple-form sets (cf. De Raaf *et al.*, 1977). The cross-bedded sandstones occur usually below or above distributary-channel and terminal-lobe sandstones.

Coarse-grained ripples (CGR) and parallel-laminated, hummocky (HCS) and swaley cross-stratified (SCS) very fine to fine-grained sandstones occur interbedded with the shoreface facies. HCS and SCS occur in single or amalgamated beds, up to 5 m thick. HCS beds contain low-angle discordant

Fig. 6. (a) Mica-rich lacustrine mudstones with silty and fine-grained sandy streaks and incipient ripples from the offshore lower shoreface transition.
(b) Parallel- and wave-ripple cross-laminated upper-shoreface sandstones. Hammer 32 cm long. From the lower part of the Horsedal section (see Figs 1 & 3 for location).

undulating and truncated laminae, that pinch and swell over distances up to 1 m. Amplitudes are less than 20 cm (Fig. 7(a)). SCS occur intercalated with HCS and consist of amalgamated beds in which truncated laminae dip less than 10° (Fig. 7(b)). Amplitudes are less than 15 cm. The laminae of parallel-laminated HCS and SCS beds are draped by comminuted plant fragments and mica grains.

Fig. 7. (a) Amalgamated hummocky cross-stratification in a storm-dominated delta-front succession. Lens cap 55 mm across. (b) Amalgamated swaley cross-stratification in a storm-dominated delta-front succession. From the upper part of the Horsedal section (see Figs 1 & 3 for location).

Plant remains, unioid bivalves and trace fossils commonly occur in the shoreface facies. Trace fossils include abundant *Lockeia amygdaloides*, *Planolites beverleyensis*, escape burrows and rare *Palaeophycus heberti*, *Arenicolites* isp. and *Diplocraterion parallelum*. The former two ichnospecies are confined mainly to beds that are completely bioturbated.

All internal structures of the shoreface facies show features diagnostic of wave and storm action (cf. De Raaf *et al.*, 1977). The general upwards increase in sand content and change in wave-generated structures reflect a shallowing-upward tendency and an increase in wave energy. The cross-bedded medium-to coarse-grained sandstones are attributed to shoreward migration of swash bars. The close association with terminal-lobe and distributary sandstones suggests that bars formed at wave-influenced river mouths by strong wave reworking.

Hummocky and swaley cross-stratification are formed by aggradation and/or translation in a combined or oscillatory flow regime (e.g. Dott & Bourgeois, 1982; Swift *et al.*, 1983; Allen, 1985; Surlyk & Noe-Nygaard, 1986; Nøttvedt & Kreisa, 1987). Amalgamation of hummocky beds is caused by scouring and erosion above storm wave base with the removal of fine-grained caps (Dott & Bourgeois, 1982). Swaley cross-stratification is probably formed by storm waves in waters shallower than that for HCS, and shallower than fair-weather wave base (Leckie & Walker, 1982). Ancient examples of lacustrine HCS and SCS are described by Van Dijk *et al.* (1978), Duke (1984), Eyles and Clark (1986, 1988) and Martel and Gibling (1991). Modern examples, mainly formed at very shallow depths, are described by Greenwood and Sherman (1986, 1989).

The depth and fair-weather wave climate during deposition are estimated on the basis of wave-ripple data, according to the method of Diem (1985). The calculated depths are 1 to 15 m and wave lengths 2 to 24 m. Calculations based on wave-ripple features associated with HCS and SCS indicate that these structures formed mainly at water depths less than 2 m. Wave periods ranged from 2 to 7 s and wave heights ranged between 0.5 to 4.5 m.

Terminal lobe

Terminal-lobe facies consist of single foreset beds composed of medium to very coarse-grained micaceous sandstones. The foreset beds drape the underlying shoreface deposits and are succeeded erosively by distributary channel or shoreface deposits (Fig. 8(b)). Less commonly the beds have a basal erosive bounding surface (Fig. 9). The beds are 0.60 to 4.95 m thick and foresets show basinward dips of 3 to 29°. Foresets are tangential, angular or convex-upward and may show reactivation surfaces and backflow ripples. Foresets locally contain unioid bivalves and logs.

The presence of single basinwards dipping foreset beds, with a conformable contact with the underlying coarsening-upward shoreface deposits, and an erosive contact with the overlying distributary channel deposits, suggests that the beds are formed by progradation of small terminal lobes in front of distributary channels. The sets are similar to Gilbert-type foreset beds described from lacustrine deltas in a Mesozoic alluvial succession from Camp Hill, Antarctica (Farquharson, 1982).

Distributary channel

The distributary channel facies is composed of single fining-upward sandstone sheets, 2.5 to 11 m thick. The base is an erosive surface overlain by a thin lag conglomerate of logs, mud and coal intraclasts and pebble extraclasts.

The grain size decreases upwards from very coarse pebbly to medium-grained sandstones. The sandstones are parallel-laminated, cross-bedded or cross-laminated. Both planar and trough cross-bedded sets, 0.1 to 2 m thick, occur in co-sets up to 9 m thick. Foreset dips are 17 to 27°. Foresets are concave-upward and commonly contain plant and coal fragments. Generally, the set thickness decreases upwards and cross-bedding gives way to cross-lamination. Palaeocurrent directions determined from cross-bed foreset orientation show a very narrow range within each sand body.

The unidirectional palaeocurrent directions within individual channel successions, the lack of lateral accretion deposits and the coarseness of the sediments suggest deposition in low-sinuosity gravelly and sandy distributary rivers. Lateral migration of channels caused superimposition of successively higher terrace levels and generation of the fining-upward successions (cf. Miall, 1985). Parallel-laminated sandstones are attributed to upper-stage plane-bed transportation and cross-bedded and cross-laminated sandstones to lower flow regime migration of dunes, transverse bars and ripples (Allen, 1984).

(a)

(b)

Fig. 8. (a) Gradational-based delta-front sheet sandstones. (b) Sharp-based delta-front sheet sandstones. Fine-grained lower- and coarse-grained upper-shoreface sandstones are separated by an erosion surface (arrow). OL, open lacustrine; WLS, wave-worked lower shoreface; WUS, wave-worked upper shoreface; TL, terminal lobe; DC, distributary channel. From the middle part of the Horsedal section (see Figs 1 & 3 for location).

Delta-plain facies association

This association consists of interbedded distributary channel and interdistributary deposits (Fig. 10). The distributary channel facies consist of single sheet-like fining-upward pebbly conglomerates and sandstones very similar to those of the delta-front association. The channel sandstones have a basal, erosive bounding surface overlain by a thin lag conglomerate, and commonly followed by clast-supported massive, crudely horizontally stratified or cross-bedded conglomerates, up to 2.3 m thick. The clasts are well-rounded quartzite and metamorphic pebbles, 0.5 to 8 cm in diameter. The cross-bedded sets are generally more fine-grained than the massive conglomerates and occur in single sets, 0.5 to 2 m thick. Foreset dips are 18 to 26°. Upward the grain size decreases to very coarse- and medium-grained parallel-laminated, cross-bedded and cross-laminated sandstones. Palaeocurrent directions show a very narrow range within each sandstone body.

Lateral accretion (epsilon) cross-bedding has been observed only in Astartekløft in the lower part of the plant-bearing series of Harris (1937). It is represented by a single set forming isolated

Fig. 9. Sharp-based delta-front sheet sandstones. Open-lacustrine mudstones and coarse-grained terminal-lobe sandstones are separated by an erosion surface (arrow). Person for scale. From the upper part of the Ranunkeldal section (see Figs 1 & 3 for location).

medium-grained sandstone body. Cross-strata are sigmoidal and dip less than 15° at right angles to the ripple cross-lamination.

The interdistributary deposits consist of black, non-laminated to laminated kaolinitic mudstones interbedded with thin sheet sandstones. The organic material of the mudstones is completely dominated by terrestrially derived detritus of coaly and woody matter, commonly with large numbers of sporomorphs. Abundant plant fossils, drifted logs, allochthonous coal seams and rootlet beds are interbedded with the mudstones. Harris (1926, 1931a, b, 1932a, b, 1935, 1937) and Pedersen and Lund (1980) recorded a very diverse floral assemblage from the interdistributary deposits, dominated by cycadophytes, ginkgophytes, conifers, pteridosperms and ferns. Sheets of silty fine-grained to medium-grained sandstones, 0.25 to 1.75 m thick, are arranged in small coarsening–fining-upward or fining-upward successions. Fining-upward successions always possess a sharp erosive base succeeded by a lag conglomerate. The sandstones are either structureless or show cross-bedding, passing upwards into parallel-laminated and cross-laminated sandstones. Mudstone drapes and rootlet beds may occur at the top of the sandstones.

The channel sandstones were deposited in low-sinuosity gravelly and sandy rivers, very similar to those of the delta front. Massive conglomerates were deposited from shallow high-velocity flow over longitudinal bars and shallow intervening channels (cf. Rust, 1984). The finer grained cross-bedded conglomerates are attributed to deposition during the falling stage, when the longitudinal bars emerged, causing flow to diverge from bar axes into adjacent channels (cf. Rust, 1975). The mudstones were deposited from suspension in protected lacustrine or lagoonal environments on the delta plain. The thin coal seams and rootlet beds represent compressed plant remains deposited above seat earths in the backswamp of the delta plain. The sheet geometry and the internal facies arrangement of the interdistributary delta-plain sandstones suggest deposition in erosively or non-erosively based unidirectional sheet splays into shallow, unconfined areas. The presence of structureless sandstones indicates rapid deposition from ephemeral hyperconcentrated flows during periods of intense overbank flooding.

Alluvial-plain facies association

Conglomeratic sheet sandstone bodies arranged in multistorey fining-upward successions constitute the alluvial-plain association. The sheets can be followed laterally for more than 1 km, both parallel and perpendicular to the palaeoslope, without showing any lateral termination. Sedimentary facies are the same as in the low-sinuosity distributary channel deposits of the delta plain and delta front, except being generally more conglomeratic.

The multistorey conglomeratic sandstone bodies

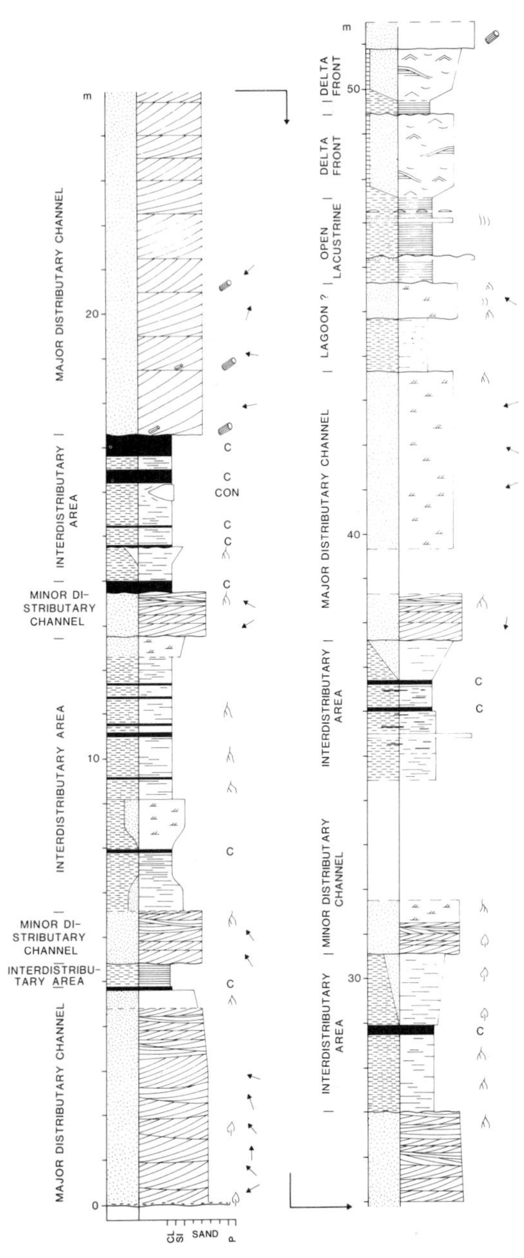

Fig. 10. Vertical section showing the internal facies organization of the delta-plain deposits and the vertical transition into delta-front and open-lacustrine deposits. From the upper part of the Lepidopteriselv section (see Figs 1 & 3 for location).

were deposited in low-sinuosity to braided rivers sweeping an alluvial plain, representing the more proximal, fluvially dominated part of the deltaic distributaries.

Delta-abandonment facies association

The delta-abandonment facies association consists of sheets of wave-worked shoreface deposits (as described in the delta-front facies association) and coal seams overlying rootlet beds. Shoreface deposits range from a single wave-ripple train, less than 5 cm thick, to beds more than 4.5 m thick, and are commonly strongly bioturbated.

The thin coal seams and rootlet beds represent compressed autochthonous plant remains deposited above seat earths when the backswamp spread over the delta front during abandonment.

GEOMETRY AND INTERNAL ORGANIZATION OF DELTA-FRONT SANDSTONES

The boundary between the open-lacustrine mudstones and the delta-front sandstones is gradational or erosional. The gradational-based units start with open-lacustrine laminated mudstones, locally interbedded with coarse-grained sandstones showing wave-ripple form sets (Fig. 8(a)). The mudstones pass gradually upward into wave- and storm-dominated shoreface sandstones. The units are topped by shoreface, terminal-lobe, distributary channel or delta-abandonment deposits. The sharp-based type has a sharp erosional base, and the open-lacustrine mudstones or fine-grained shoreface sandstones are abruptly overlain by upper-shoreface, terminal-lobe or distributary channel sandstones (Figs 8(b) & 9).

Both the gradational and sharp-based sandstones form sheet-like bodies, up to 15 m thick, that can be followed laterally for several kilometres both along dip and strike. The sheet sandstones only show very small changes in thickness (Fig. 4). There is a clear tendency for the sharp-based sandstones to dominate in basin-margin areas, whereas the gradational-based sandstones are more common in the more central parts of the basin. Some of the units can be traced throughout the basin on the basis of geochemical and fossil characteristics.

Delta-front sandstones have very uniform thicknesses and facies development over distances of many hundreds of metres as seen in extensive strike sections (sections parallel to the inferred palaeo-coastline). In the example in Fig. 11, open-lacustrine mudstones are overlain erosionally by upper-shoreface sandstones. Local thickening of

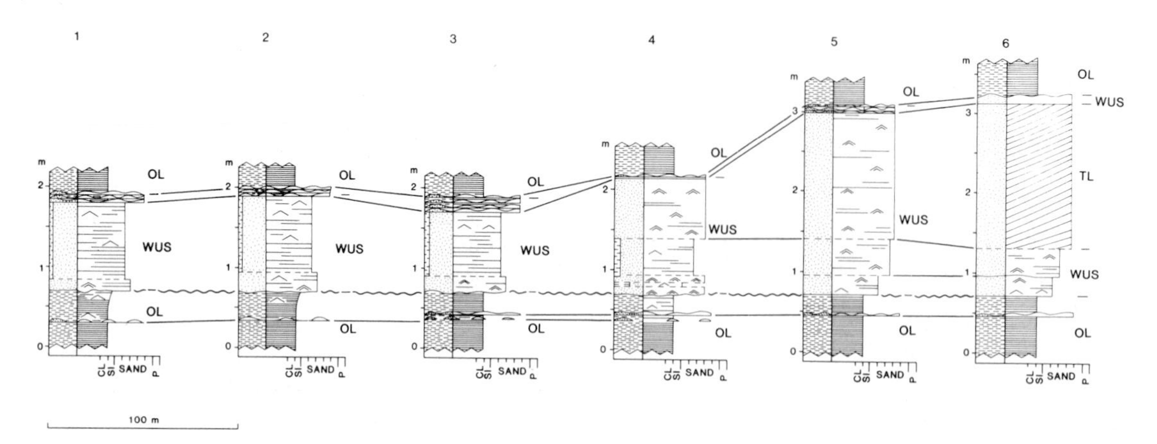

Fig. 11. Strike section of a sharp-based delta-front sheet sandstone showing uniform lateral thickness with an increase in the terminal-lobe area. A laterally continuous erosion surface with small scours separates the open-lacustrine mudstones from overlying shoreface sandstones. Wavy line indicates high-resolution sequence boundary. Key to letter symbols, see Fig. 8. From the upper part of the Ranunkeldal section (see Figs 1 & 3 for location). (Modified from Dam, 1991.)

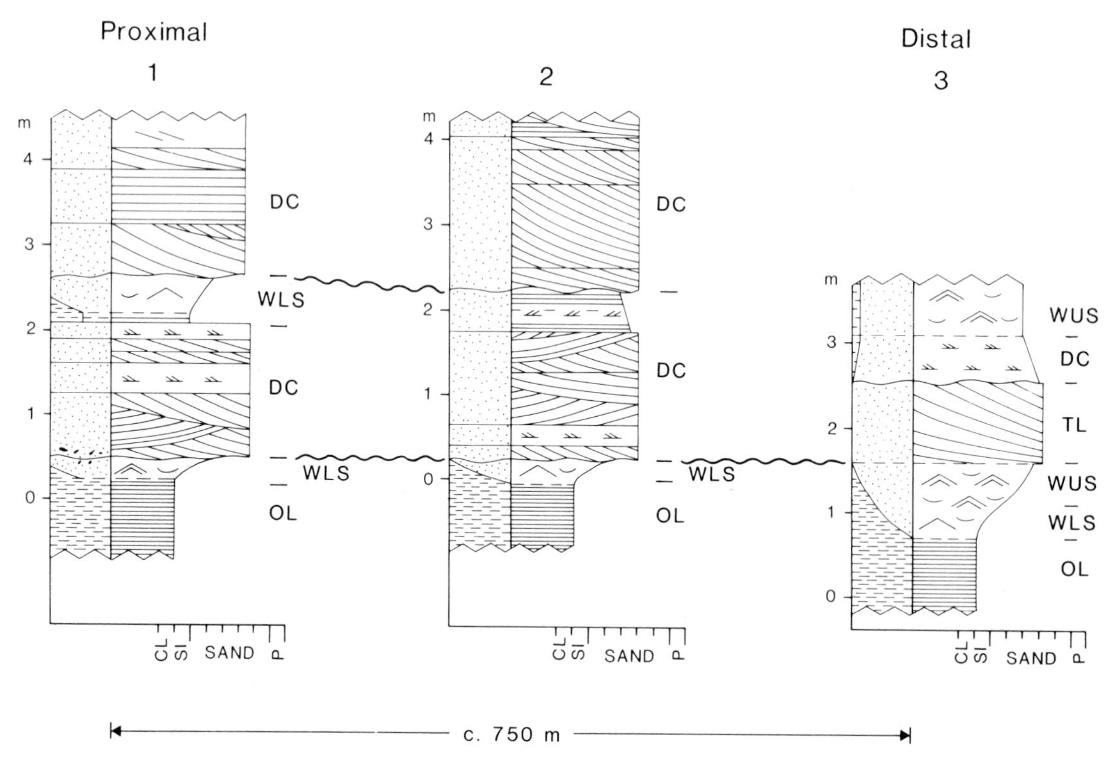

Fig. 12. Dip sections of a sharp-based delta-front sheet sandstone. In the proximal part (left) the shoreface sandstones of the lower deltaic sheet are erosionally overlain by distributary-channel sandstones and in the distal part (right) by terminal-lobe sandstones. This sheet is followed by a second sharp-based sheet consisting of a thin shoreface unit overlain by distributary channel sandstone. High-resolution sequence boundaries are laterally continuous and marked by a wavy line. Key to letter symbols, see Fig. 8. From the lower part of the Horsedal section (see Figs 1 & 3 for location). (Modified from Dam, 1991.)

sandstone sheets only occurs around terminal-lobe deposits and is probably due to differential compaction. Sandstone sheets are topped by a few coarse- to very coarse-grained wave-ripple trains representing a transgressive lag deposit formed by reworking during the delta-abandonment phase.

Dip sections (sections perpendicular to the inferred palaeocoastline) show a more complex picture. The sand sheet pictured in Fig. 12 is composite in nature. Lacustrine mudstones coarsen upwards into wave-influenced lower-shoreface sandstones. Proximally, the shoreface sandstones are erosionally overlain by distributary channel sandstones and, more distally, by terminal-lobe sandstones. It is overlain by a second, sharp-based sheet consisting of shoreface and distributary channel sandstones. Open-lacustrine mudstones are virtually missing between the two sandy delta-front sheet sandstones either because they were never deposited or because they were removed by erosion preceding deposition of the upper sheet.

The amalgamated nature of the stacked delta-front sheet sandstones makes it difficult to identify and interpret correctly the intervening erosion surfaces. The dip section in Fig. 13 highlights the composite nature in the middle section of what at first sight appears as a simple coarsening-upward deltaic unit. Close examination reveals, however, the presence of a lower relatively fine-grained sharp-based but non-erosional shoreface sandstone overlain by a thin mudstone or directly by a second sharp-based channel and lobe-dominated deltaic unit. Taken alone, the middle section seems only to show a simple coarsening-upward succession. Careful lateral tracing of the units and their bounding surfaces shows, however, that the section is an amalgamated package consisting of a lower sharp, but erosive, shoreface sandstone and an upper erosively based channel and lobe sandstone. The whole amalgamated unit is topped by a single wave-ripple train representing the delta-abandonment phase, and is overlain by open-

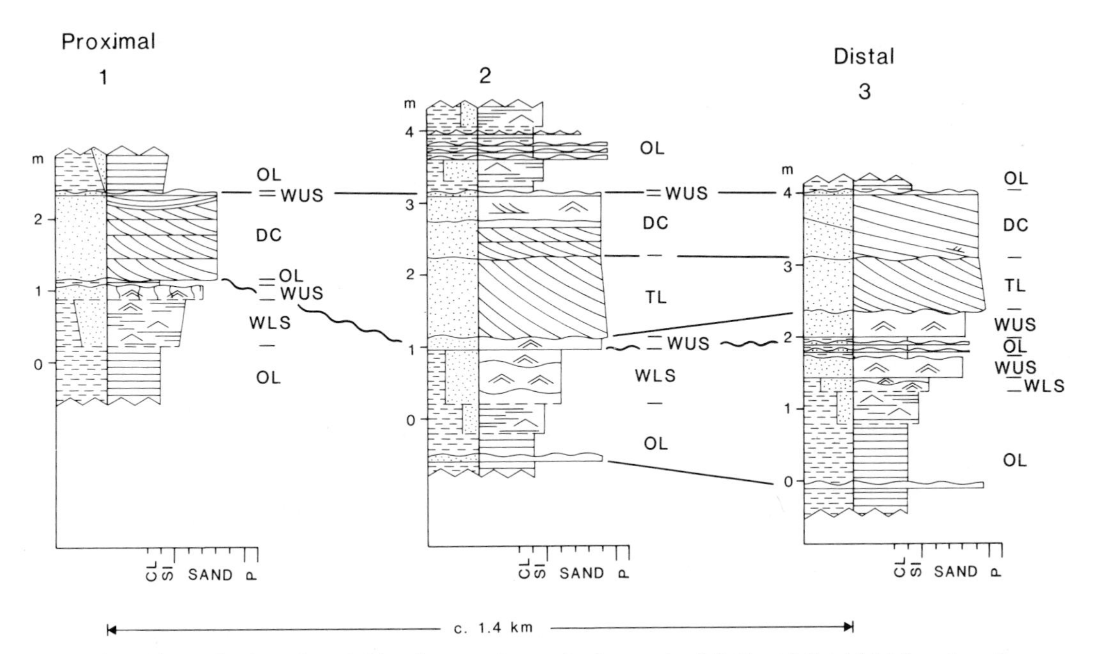

Fig. 13. Dip sections of a sharp-based delta sheet sandstone. In the proximal (left) and distal (right) sections the amalgamated package consists of a lower sharp, but non-erosive shoreface sandstone and an upper erosively based channel and upper-shoreface sandstone. The middle section seems only to include a simple sharp-based coarsening-upward section. The open-lacustrine mudstones separating the two sandstone sheets are missing in this section because they were removed by erosion preceding deposition of the upper sandstone sheet. Wavy line indicates high-resolution sequence boundary. Key to letter symbols, see Fig. 8. From the lower part of the Horsedal section (see Figs 1 & 3 for location). (Modified from Dam, 1991.)

lacustrine mudstones reflecting drowning of the delta.

The open-lacustrine mudstones and shoreface sandstones show the most extensive lateral distribution. Terminal-lobe and distributary channel deposits have not been followed for more than 2 km in strike sections. Swash-bar sandstones only occur associated with terminal-lobe and distributary channel deposits. Distributary channel sandstones show a basinward transition into either terminal-lobe or shoreface deposits. Distributary channel deposits only occur very locally in the most distal parts of the sand sheets, where they are incised into the more extensive shoreface sheet sandstones.

The delta-front sandstones are always succeeded by thin sheets of either wave-worked shoreface deposits and/or coal seams and rootlet beds of the delta-abandonment facies association. Dip sections show that delta-abandonment shoreface facies usually increase in thickness in a basinward direction (Fig. 12). In the more distal parts it may be impossible to distinguish between shoreface facies deposited during the progradational phase and those formed during abandonment.

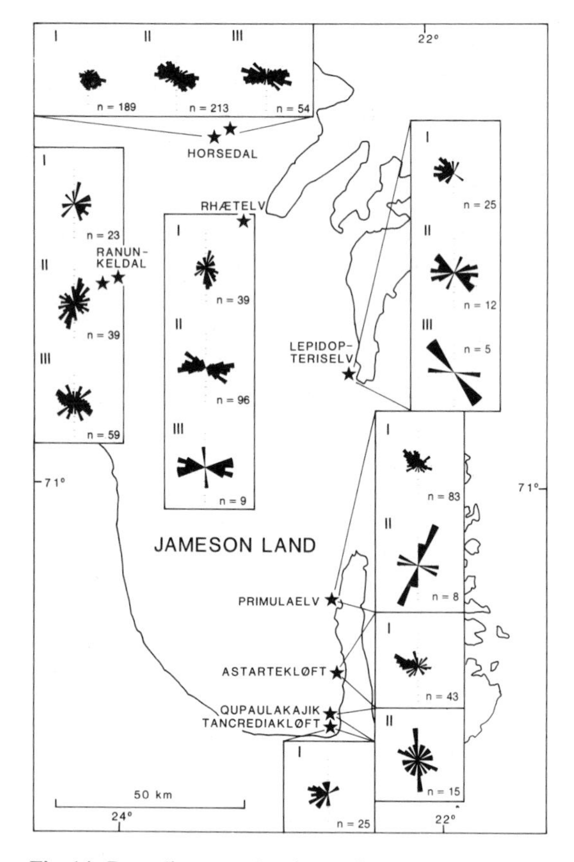

Fig. 14. Rose diagrams showing: I, fluvial palaeocurrent directions; II, crestline orientations of wave ripples; III, crestline orientation of symmetrical coarse-grained ripples. Rose diagrams shown as true area plots. Marks at 5, 10, 20 and 30% frequency. n, number of measurements.

PROVENANCE, CURRENT AND WAVE-RIPPLE PATTERNS

Palaeocurrent directions are interpreted from foreset dip directions of fluvial cross-bedding and cross-lamination. Directions of wave oscillations are interpreted from wave-ripple crestline orientation. All palaeocurrent and wave-ripple data are plotted in area-true rose diagrams (Fig. 14). Petrographical studies were made on 34 thin sections. The average composition of the sandstones was estimated by 300 point counts of each thin section. A triangular plot of sandstone composition within the Kap Stewart Formation is shown in Fig. 15.

Along the eastern basin margin, distributary-channel deposits show palaeocurrents towards WNW, roughly perpendicular to the Liverpool Land High, which formed the eastern margin of the Kap Stewart depositional basin. The palaeocurrent data show no indications of northwards flowing rivers as suggested by Sykes (1974), although the sections were measured at the same localities. Wave-ripple crestline orientations indicate a NNE–SSW-trending shoreline, parallel to the inferred basin margin and the present orientation of the Liverpool Land High.

In the northwestern part of the basin palaeocurrent directions indicate a general fluviatile transport towards ESE. Wave-ripple crestlines suggest a NNE–SSW-trending shoreline, parallel to the main basin-margin fault.

In the northern part of the basin fluvial palaeocurrent measurements generally are towards SW–SE. Wave-ripple crestlines suggest a NW–SE-trending shoreline, parallel to the basin-margin cross-fault in Kong Oscar Fjord. Palaeocurrent measurements and crestline orientation of wave ripples are generally in agreement with those figured

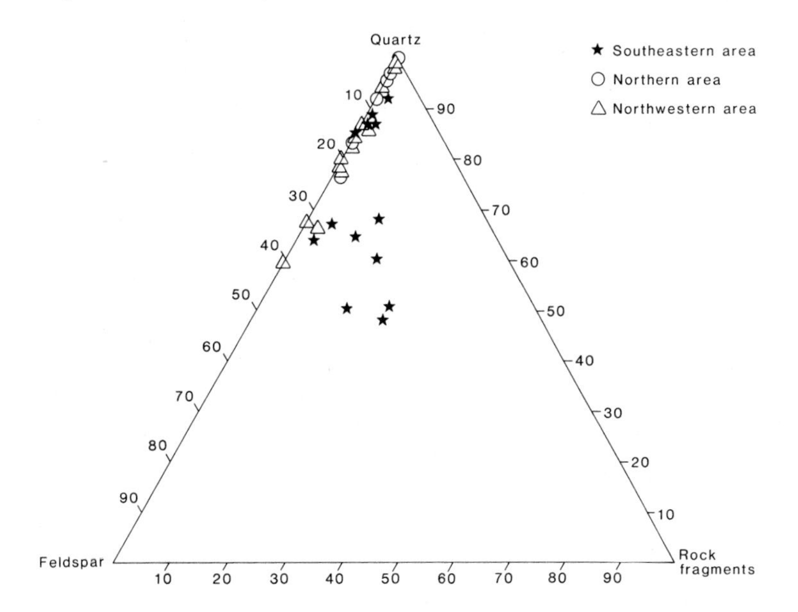

Fig. 15. Triangular plot of sandstone compositions within the Kap Stewart Formation.

from the same area by Clemmensen (1976), who interpreted the sediments as tidally influenced deltaic deposits. We have found no evidence for marine nor tidal influence in the succession.

Palaeocurrent measurements closer to the basin centre show a more polydirectional distribution than in the northern part of the basin, but a dominant direction towards the south is recognized. The total palaeocurrent distribution in the basinal areas suggests that deltas prograded from all sides of the basin, still, however, with the main source areas situated towards the north. Crestline orientations of wave ripples also suggest a dominant WNW–ESE shoreline orientation, conforming with the northern coastline.

Crestline orientations of the storm-produced coarse-grained symmetrical ripples (CGR) strike between E–W and SE–NW throughout the basin. This unidirectional crestline orientation possibly reflects storm tracks towards the south or north.

The sandstones in the southeastern part of the basin have a higher content of lithic grains and are more immature than the sandstones in the northern and northwestern areas, suggesting a shorter transport distance from a source area in the plutonic and migmatitic rocks of the present-day Liverpool Land High (Fig. 15). The sandstones of the northern and northwestern areas are very uniform in composition (Fig. 15). They lie on the quartz–feldspar line with polycrystalline quartz grains making up a major proportion. In the crystalline rocks of Stauning Alper, west of Jameson Land, polycrystalline quartz grains dominate over monocrystalline varieties. These rocks form the source areas for the sandstones of the northwestern part of the basin. Monocrystalline grain varieties constitute the bulk of the quartz grains in the northern part of the basin. The sandstones in these areas are better sorted, more mature and finer grained than those of the northwestern and eastern parts, suggesting a third source area. This is in accordance with the heavy mineral assemblages that indicate two different source areas for the sandstones of the eastern and northern parts of the basin (Bromley *et al.*, 1970). The heavy mineral assemblages are similar to the assemblages in older, Triassic rocks suggesting that the sandstones from the northern part of the basin were derived mainly by reworking of middle and late Triassic sediments north of Kong Oscar Fjord or had the same source area as these sediments.

PHYTOPALAEONTOLOGY

Almost 200 species of fossil plants have been described from the delta-plain deposits of the Kap Stewart Formation along Hurry Inlet (Harris, 1926,

1931, 1932a,b, 1935, 1937). Interdistributary mudstones contain large and well-preserved leaves commonly representing only a few autochthonous species. Coarser sediments deposited from ephemeral hyperconcentrated overbank flows and in distributary channels contain leaves and wood of plants representing many species that were fragmented and probably brought together from a large area (allochthonous). The *Lepidopteris* and *Thaumatopteris* floral zones contain about 100 species each, and are dominated by gymnosperms represented by cycadophytes, ginkgophytes, pteridosperms, conifers and common ferns. Pedersen and Lund (1980) identified about 80 species of spores and pollen from the plant-bearing series along Hurry Inlet.

The micro- and macrofloras from the delta-plain deposits of the Kap Stewart Formation indicate good conditions for growth and a humid, rather warm climate. The horsetails reach a considerable size and the ferns have large, commonly rather thin-walled, leaves. The leaves of the gymnosperms range from types with a thick cuticle to forms with a thin cuticle, but mainly the latter. Annual growth rings (1 to 3 mm across) are present in gymnosperm trees showing climatic seasonality with a favourable growing season (Harris, 1937). The non-marine depositional environment is clearly reflected by the total absence of marine palynomorphs. A markedly northern geographical position of the Scoresby Sund area in relation to contemporaneous floras from Europe is indicated by the low numbers of *Corollina* (*Classopollis*) (Pedersen & Lund, 1980).

The plant beds in the western and northern parts of the basin are mainly allochthonous and occur in open-lacustrine mudstones and wave- and storm-dominated delta-front shoreface sandstones. Autochthonous plant beds occur associated with delta-abandonment deposits. Only 16 species have been recorded from these areas and most localities provided less than three species. The very low diversities stand in marked contrast to the diversity of the plant beds along Hurry Inlet, where almost every section provides at least one bed with a dozen or more species. It is, however, possible that these differences reflect collection biases.

Rhaetian species have only been found at one locality (Ranunkeldal) represented only by three species (Harris, 1946). It is difficult to compare the floras of the western and northern localities with those along Hurry Inlet, because each locality has its own assemblage of species. However, they resemble those of Hurry Inlet forms more than any other flora and the differences in species composition and diversity throughout the basin may reflect differences in the environment (Harris, 1946). Marine palynomorphs are also absent in the northern and western localities (K. R. Pedersen, pers. comm., 1991).

Rootlet beds and coal seams of varying thickness are common in the Kap Stewart Formation. Most coal seams are only a few centimetres thick, but beds are locally up to about 50 cm thick. Impressions of stems more than half a metre in diameter are present.

PALAEOGEOGRAPHIC RECONSTRUCTION AND DEPOSITIONAL MODEL

During Rhaetian–Sinemurian times a large wave- and storm-dominated lake was situated in the Jameson Land Basin, East Greenland. Coarse alluvial material was transported into the lake from land areas covering present-day Liverpool Land and Stauning Alper, whereas a northern source area supplied better sorted, mature, fine-grained sand (Fig. 16). Evidence for a fresh-water lake includes a non-marine fauna, the total absence of marine palynomorphs, the presence of the fresh-water algae *Botryococcus*, the low total sulphur content of the mudstones and the good linear correlation between total sulphur and total organic carbon (Dam & Christiansen, 1990).

Anoxic conditions dominated in the lower water masses and at the lake bottom during deposition of the muds, and the water column was probably stratified. These conditions resulted in accumulation and preservation of large quantities of algal and herbaceous remains.

In modern lakes within the East African rift system one of the principal controls on stratification is wind regime. The Kap Stewart lake was comparable in size to modern Lake Turkana, a water body that is also much affected by waves. Despite a maximum depth of 130 m, the Turkana water column is holomictic, bottom waters are oxygenated and the sediments consequently are organic matter-poor (M. R. Talbot, pers. comm., 1991). Lake Albert is of similar size and slightly less windy.

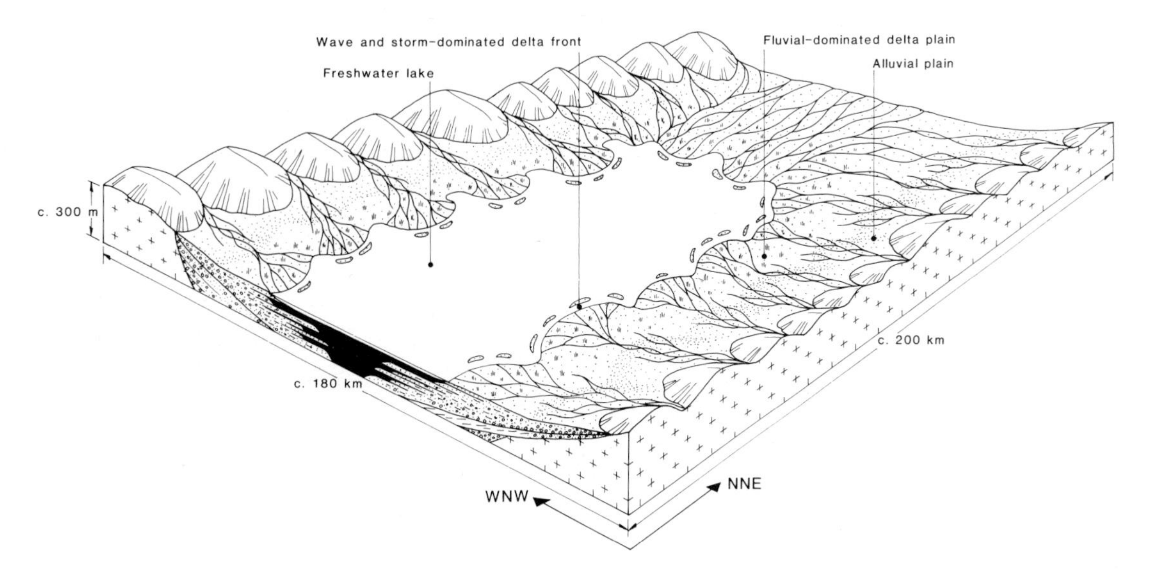

Fig. 16. Palaeogeographical reconstruction of the Jameson Land Basin during Rhaetian times.

More or less permanent mixing of this lake is due to the orientation of the basin parallel to the dominant wind direction, and the lake is generally mixed to the bottom (*c.* 60 m), and consequently the sediments are dominated by hydrogen-poor, woody organic matter (Talbot, 1988; Katz, 1990b). Deposition of the open-lacustrine mudstones of the Kap Stewart Formation must by comparison be expected to have occurred at depths of several tens of metres and perhaps as much as one hundred metres.

Major storms initiated near-bottom basinward-directed flows that were capable of transporting coarse-grained sands from the shoreface far into the Kap Stewart lake and storm-related waves reworked them into symmetrical coarse-grained ripples.

The presence of widespread uniformly thick delta-front sand sheets indicates large progradation distances (probably several tens of kilometres) across the marginal lake bottom which was characterized by low slope gradient and low relief. Delta-front sands were effectively redistributed by waves and the shoreline advanced over a broad front along the basin margin. The delta-front sands were generally deposited in water depths less than 15 m as shown by wave-ripple data. Larger storm waves were clearly capable of eroding deeply into the delta-front sediments, producing amalgamated hummocky and swaley cross-stratified sheet sand bodies. Wave- and storm-dominated shoreface facies may be succeeded by Gilbert-type foreset beds, representing progradation of terminal lobes across the shoreface in front of the distributary river mouths during fair-weather periods (Fig. 17). The lobes formed in response to rapid deceleration and loss of sediment transporting capability within short distances from the river outlets (cf. Wright, 1977). Shoaling waves created swash bars in front of the river outlets and on top of the terminal lobes.

The presence of widespread sheet-like shoreface sandstones in lateral continuity with river mouth deposits indicates that wave and storm processes were capable of redistributing most of the sediment supplied to the delta front. The sediment was transported along the coast and progradation involved the entire shoreline. Crestline orientations of wave ripples indicate that shorelines were linear, with only slight deflections at river outlets where terminal lobes developed.

The nature of the abandonment facies suggests that wave reworking after delta abandonment was appreciable (with much of the delta-front facies being reworked), producing a laterally continuous sand sheet across the entire delta front. This resulted in a predominance of incomplete, vertically attenuated delta-front successions. Fluvial distributary channel sandstones were only locally preserved in the distal parts of the basin. A relative

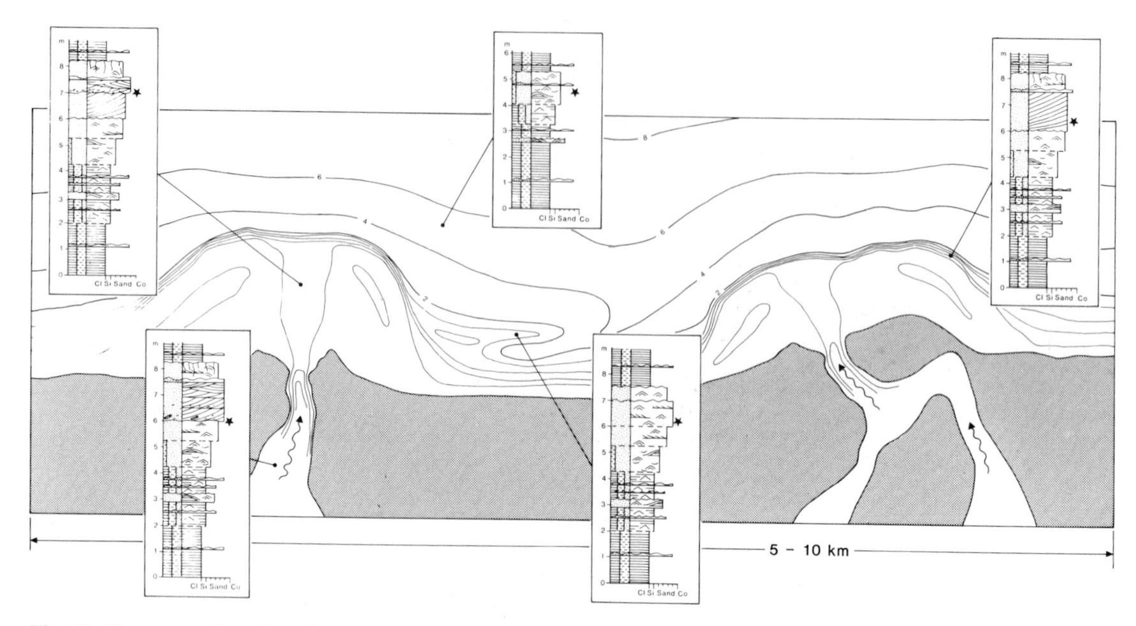

Fig. 17. Reconstruction of vertical sections produced by delta progradation and abandonment. Asterisks mark the position of the sediment surface at the time represented by the 'map'.

rise in lake level resulted in delta abandonment and rapid spreading of the delta-plain backswamp across the drowning delta. *In situ* organic matter accumulated above the rootlet beds and thin autochthonous coal seams developed.

The apparently contradicting water depth estimates derived from the anoxic mudstones and the interbedded wave- and storm-dominated deltaic sandstone sheets can be interpreted as reflecting major fluctuations in lake level. Dam and Surlyk (1992) interpreted the sheet sandstones as representing forced regressions occurring in response to relative lake-level fall. The base of the delta-front sandstones is sharp and erosional in the proximal part and sharp or gradual, erosional or non-erosional in the distal part. This is exactly what has been observed in the Upper Albian Viking Formation and is one of the recognition criteria for forced regressions (Posamentier *et al.*, 1992). The proximal basal erosive surface of the delta-front sandstones of the Kap Stewart Formation was formed by lowering of storm wave base resulting in erosion of the sea floor in front of the prograding delta (cf. Plint, 1988). The extensive, sheet-like geometry of the Kap Stewart delta sandstones reflects very rapid basin-wide lakeward shift of shoreline during continued fall (Dam & Surlyk, 1992) (Fig. 18). During subsequent lake-level rise the delta front probably was inundated very rapidly due to the very low slope gradients and relief of the lake bottom. The deltaic lowstand deposits are topped by a lacustrine ravinement surface formed by shoreface erosion during transgression (Fig. 19). In this setting a minor rise in lake level would cause a shoreface retreat of several kilometres. This interpretation is supported by the presence of widespread delta-abandonment facies and basin-wide lacustrine mudstone intervals.

LAKE-LEVEL FLUCTUATIONS

A detailed lake-level curve for the Kap Stewart lake can be constructed on the basis of the alternation between deep-water anoxic mudstones and shallow-water deltaic sandstones (Dam & Surlyk, 1992) (Fig. 19). The curve is based on detailed vertical logs representing the complete Kap Stewart Formation in the central part of the basin, where it is thickest developed (Fig. 19, inset log example). Two orders of lake-level fluctuation are recognized. The dating of the Kap Stewart Formation is based mainly on palynology and macro-plants, and the precision is not great. The high-frequency fluctua-

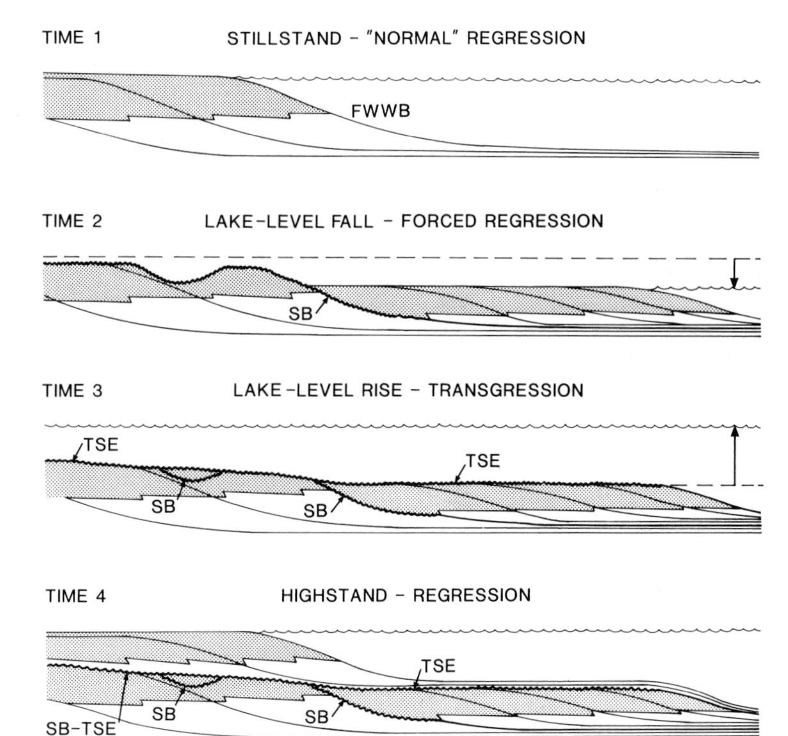

Fig. 18. Schematic cross-section showing the evolution of key surfaces in the delta-front sheet sandstones. Time 1, regression of shoreline during lake-level stillstand. Time 2, a fall in lake level produces a high-resolution sequence boundary developed as a proximal fluvial incision surface and a distal unconformity at the base of the shoreface sands and an associated lowstand shoreline (forced regression). Time 3, a rise in lake level results in the formation of a transgressive surface of erosion succeeded by delta-abandonment facies. Time 4, regression of shoreline during lake-level highstand. FWWB, fair-weather wave base; SB, sequence boundary; TSE, transgressive surface of erosion.

tions are reminiscent of those documented for the contemporaneous lacustrine deposits of the Newark Supergroup of eastern North America (Olsen, 1986, 1988, 1990) and may be related to lake-level changes controlled by Milankovitch-type climatic cycles. In the Lockatong, Passaic and East Berlin Formations of the Newark and Hartford Basins, changes in precipitation resulted in dramatic changes in lake level from water depths of perhaps 200 m or more to complete exposure, producing repetitive sedimentary cycles 1.5 to 35 m thick, showing extreme lateral continuity (Olsen, 1986). The East African lakes are probably the closest modern analogue to the Kap Stewart lake in terms of dramatic lake-level fluctuations. Seismic data show that water level in Lake Malawi was 250 to 500 m lower than today, before about 25,000 years ago. Water levels in Lake Tanganyika at that time were more than 600 m below the current lake level (Scholz & Rosendahl, 1988). In Lake Turkana the water level was about 60 m lower than today, at a time probably just prior to 10,000 years before present (Johnson *et al.*, 1987). A drier climate appears to have been the main cause for these lowstands, although tectonics also may have been a contributing factor.

The high-frequency cycles of the Kap Stewart Formation can be grouped into a number of low-frequency cycles (Fig. 19). Despite the poor dating of the Kap Stewart Formation, it is remarkable that the sea-level curves of Hallam (1988) and Haq *et al.* (1988) show the same trend and number of fluctuations as the Kap Stewart low-frequency lake-level curve during the Rhaetian–Sinemurian (Fig. 19). This suggests that eustasy may have had a direct influence on the long-term trend in the Kap Stewart lake level (Dam & Surlyk, 1992). The possibility of eustatic influence on the Kap Stewart lake is not easy to explain. It could be due to direct contact between the lake and the open sea so that the lake, like the Black Sea, turned into a land-locked marine basin at times of high sea level. There is, however, no evidence for marine or brackish interludes in the Kap Stewart Formation so this hypothesis is considered unlikely. Periods of high sea level may, as in the Quaternary, also be periods of enhanced humidity. This in turn may be reflected by overall highstand of lake level. The lake was situated in the wide N–S-oriented extensional rift-basin complex between Greenland and Norway. Part of this basin was flooded by a shallow sea during Hettangian–Sinemurian times (e.g. Ziegler, 1988; Pedersen

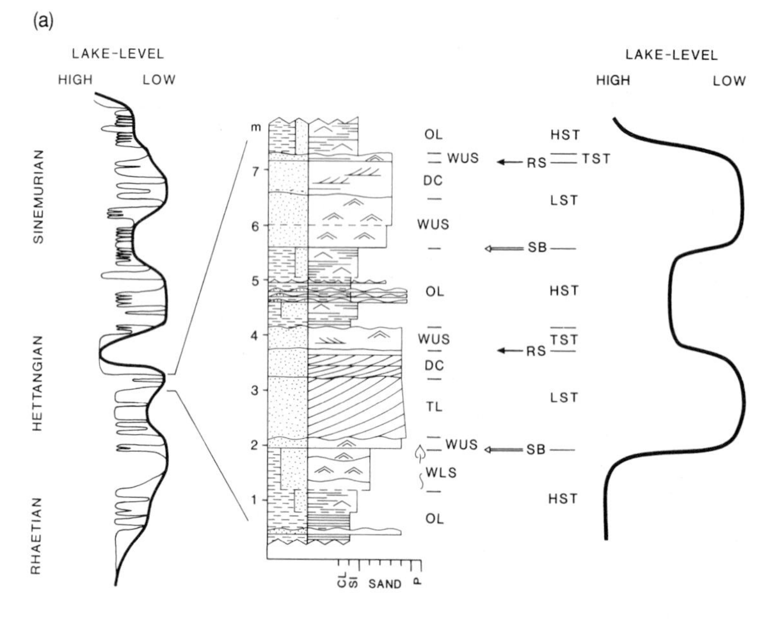

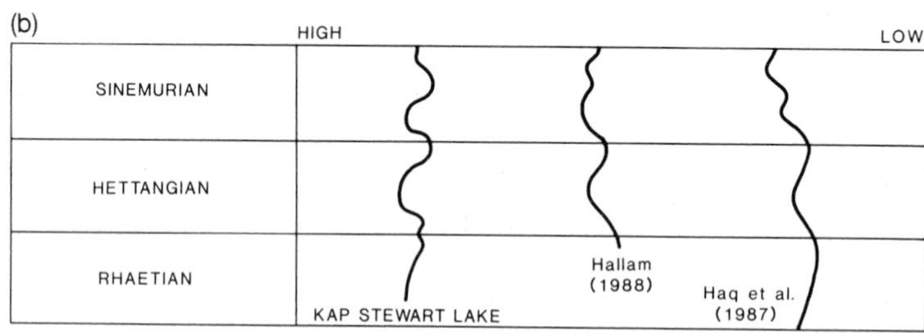

Fig. 19. (a) Kap Stewart lake-level curve (fine line) constructed on the basis of high-resolution sequence-stratigraphic interpretation of detailed sedimentological logs covering the complete thickness of the formation (example inset). OL, open lacustrine; WLS, wave-worked lower shoreface; WUS, wave-worked upper shoreface; TL, terminal lobe; DC, distributary channel; SB, sequence boundary; RS, ravinement surface; LST, lowstand systems tract; TST, transgressive systems tract; HST, highstand systems tract. It is assumed that subsidence rates were uniform during the early Rhaetian–latest Sinemurian time span of the formation. The curve shows numerous closely spaced, probably climatically controlled fluctuations. The bold line curve is a smoothed version of the first curve. It shows long-term trends in Kap Stewart lake level. (b) The trend of this curve is remarkably similar to the eustatic sea-level curves of Hallam (1988) and Haq *et al.* (1988) suggestive of a causal link between long-period eustasy and lake level. (Modified from Dam & Surlyk, 1992.)

et al., 1989; Richards, 1990). The Jameson Land Basin was transgressed during mid-Triassic time and again in the Pliensbachian. The basin thus appears to have been low lying and close to the open sea also during deposition of the Kap Stewart Formation. It may be speculated that a change in relative sea level would lead to a change in water table and in base level. This is clearly an oversimplification, but base-level effects are important in the lower reaches of many river systems and certainly in coastal lakes. Eustatic sea-level changes thus may have had a direct influence on the long-term trends in the Kap Stewart lake level.

SEQUENCE STRATIGRAPHY

Application of sequence-stratigraphic concepts to non-marine successions is notoriously difficult. Fluvial deposits characteristically show numerous erosive surfaces, mainly representing relatively localized channel scours. These may, however, be difficult to distinguish from regionally extensive erosional unconformities. Sequence boundaries can be located by recognition of major incised valleys, basin-margin unconformities and onlap and to some extent, also, significant changes in channel geometry and sedimentary fill. Laterally extensive, multistorey channel sandstones reflect a low rate of base-level rise and may overlie sequence-bounding unconformities (e.g. Shanley & McCabe, 1991). Changes in lithology may, however, also reflect climatic changes controlling changes in fluvial discharge or tectonic uplift of the source area.

Parasequences are mainly recognized in shallow-marine deposits, and analogous units occur in lacustrine deposits, whereas it is doubtful if they can be recognized in fluvial successions. It is thus not likely that parasequence stacking patterns can be used to identify systems tracts and implicitly sequence boundaries in non-marine successions.

The Kap Stewart Formation is underlain by the Ørsted Dal Member (Flemming Fjord Formation) throughout the basin except in the Kap Hope area where the latter wedges out and the Kap Stewart Formation rests unconformably on the Klitdal Member (Pingo Dal Formation). Seismic reflection data from the eastern part of the basin clearly show this unconformable relationship between the two formations and the onlapping nature of the base of the Kap Stewart Formation interpreted as a type 1 sequence boundary (cf. Van Wagoner *et al.*, 1988) following an early Rhaetian hiatus (Fig. 20). The sequence boundary formed during a period where climatic conditions changed from arid to humid. The boundary between the coarse-grained 'barren sandstones' and the overlying 'plant-bearing series' separates coarse-grained alluvial-plain deposits from delta-plain and lacustrine deposits which are overall finer grained and contain a higher proportion of shale beds. The facies change is well developed along the eastern and especially the western basin margin. It may reflect a climatically controlled decrease in runoff, resulting in a rise in base level, but also may be the expression in the alluvial–lacustrine succession of a sea-level rise. We prefer the latter explanation and the barren sandstones are thus interpreted as a lowstand systems tract topped by a 'transgressive' surface overlain by a transgressive systems tract represented by the lower part of the plant-bearing series. In the basinal areas the maximum flooding surface occurs in a condensed section where siderite concretions are developed.

The dramatic change in macroflora and palynology at the boundary of the *Lepidopteris* and *Thaumatopteris* zones, and of palynozones 1 and 2, was interpreted as caused by eustatic fall and non-

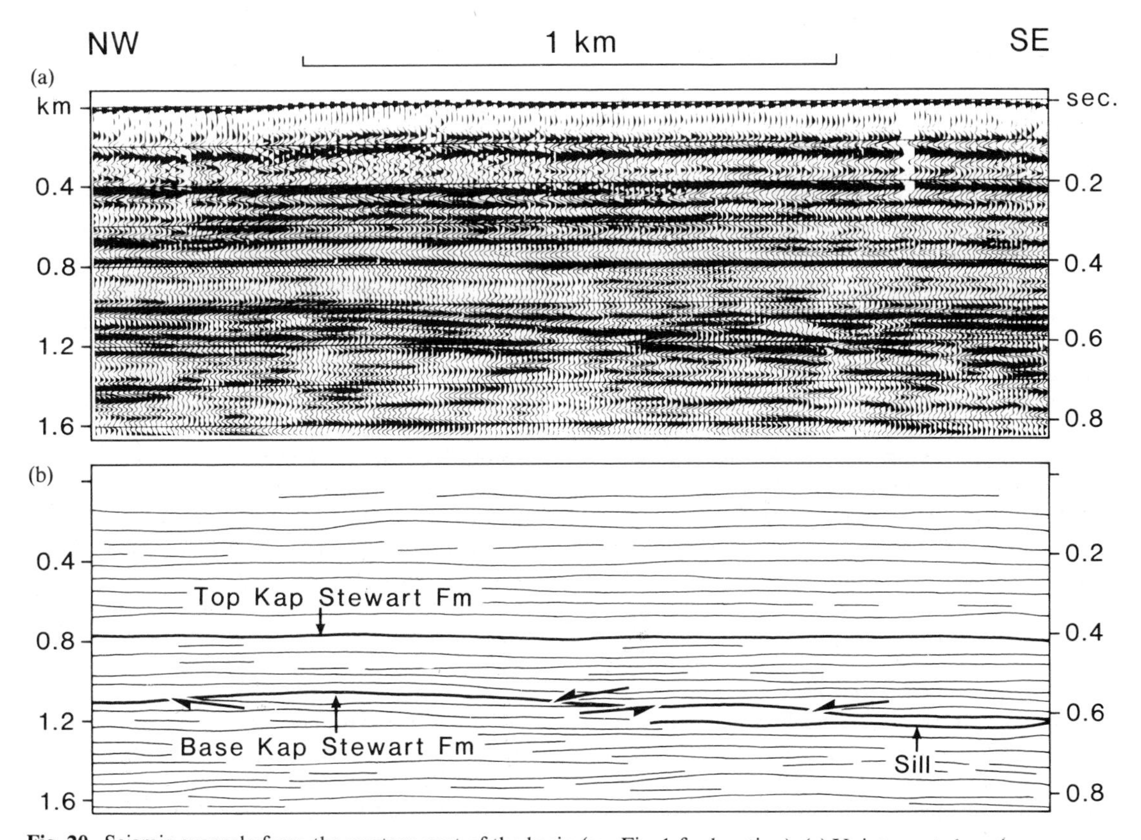

Fig. 20. Seismic example from the western part of the basin (see Fig. 1 for location). (a) Uninterpreted section. (b) Interpreted section showing the erosive truncation of the Ørsted Dal Member and the onlapping nature of the basal Kap Stewart Formation. The seismic interpretation is linked to outcrops in the eastern part of the basin.

deposition as early as in 1937 by Tom Harris. Biostratigraphic data suggest that all of the Upper Rhaetian is missing along Hurry Inlet. The boundary is marked by a change in facies at Astartekløft where muddy interdistributary and meandering river deposits are overlain by coarser grained, conglomeratic low-sinuosity river deposits. The basinal sections have a relatively high content of sandstone at this level.

The presence of an important hiatus and the absence of major erosional unconformity suggest that the zonal boundary represents a type 2 sequence boundary.

The overlying relatively coarse part of the plant-bearing series may thus represent a second lowstand systems tract (Van Wagoner *et al.*, 1988), but the differentiation between lowstand and shelf margin systems tracts is not considered meaningful in a fully continental succession where shelf margins are absent.

The top part of the plant-bearing series represents the most widespread distribution of shales in the formation and is interpreted as a transgressive systems tract.

This sequence-stratigraphic scheme for the Kap Stewart Formation thus includes two sequences with a maximum duration of 15 to 20 Ma. The Sinemurian is missing along the eastern basin margin. It is probably represented in the basinal areas, but biostratigraphic data are lacking here preventing a satisfactory sequence-stratigraphic interpretation for this interval.

A highly tentative comparison with the sequence chart of Haq *et al.* (1988) suggests the following correlations: the basal type 1 sequence boundary (SB) corresponds to the 215 Ma SB; the maximum flooding surface (MFS) above the 'barren sandstone' corresponds to the 211.5 Ma MFS. The type 2 SB between the *Lepidopteris* and *Thaumatopteris* zones corresponds to the 211 Ma SB, whereas the upper type 1 SB separating the Kap Stewart and Neill Klinter Formations corresponds to the 195 Ma SB.

Changes in lake level are responsible for the numerous forced regressions recognized in the Kap Stewart lake. It is thus possible to undertake a high-resolution sequence-stratigraphic interpretation within the framework of the large-scale sequences described above. Most of the deltaic sheet sandstones are interpreted as representing lowstand systems tracts. The sequence boundary occurs at the base of the sandstones (Fig. 19). The lowstand deposits are topped by a lacustrine ravinement surface formed by erosion during renewed transgression. The delta-abandonment deposits represent the basal part of the transgressive systems tract. The occasional occurrence of a thin coarsening-upwards sandy part in the top of the lacustrine mudstones is interpreted as a poorly developed highstand systems tract.

CONCLUSIONS

The sediments of the Kap Stewart Formation were deposited in a wave- and storm-dominated lacustrine-deltaic environment flanked by fluvially dominated delta plains and alluvial plains. The fresh-water lake 'was extensive and periodically covered areas exceeding 12,000 km^2. Facies analysis, palaeocurrent and wave-ripple data and petrographical analyses indicate source areas to the east, west and north. Wave-ripple data show that the sandstones were deposited in shallow waters, at depths not exceeding 15 m. The mudstones were deposited under anoxic conditions in a permanently stratified lake. This is incompatible with very strong wave and storm activity and shallow water indicated by the delta-front sheet sandstones. Conventional sequence-stratigraphic data, such as recognition of major unconformities, hiatuses and basin-margin onlap, show that the formation includes two third-order sequences with a maximum duration of 15 to 20 Ma. High-resolution sequence-stratigraphic analysis suggests that the deltaic sandstones represent forced regressions caused by lake-level falls, whereas the mudstones were deposited in relatively deep water during periods of high lake level. The Kap Stewart Formation thus experienced numerous high-frequency fluctuations in lake level similar to those documented from contemporaneous lacustrine deposits of the Newark Supergroup of eastern North America (Olsen, 1986, 1988, 1990). The cycles are probably related to lake-level changes controlled by Milankovitch-type climatic variations. The long-term trend in lake level shows a marked similarity to the eustatic sea-level curves of Hallam (1988) and Haq *et al.* (1988). Although the Kap Stewart Formation is too poorly dated to allow any detailed comparison, the parallel trend of the curves is suggestive of a causal link between eustasy and lake level. The lake-level curve is accordingly interpreted to show a high-frequency climatically induced signal superimposed on a low-frequency eustatic signal. The occurrence of numerous sandstone bodies with sheet geometry lying isolated in an important source rock is of obvious relevance for petroleum exploration in the Jameson Land Basin and in comparable settings on the Norwegian shelf. The Kap Stewart lake is one of the few ancient examples of a large wave- and storm-dominated anoxic lake, and may also serve as an exploration model for other similar lacustrine basins.

ACKNOWLEDGEMENTS

Work by Dam was supported by BP Exploration Operating Company Limited London, the Carlsberg Foundation and the Geological Survey of Greenland. Field work by Surlyk was supported by the Carlsberg Foundation and Norsk Hydro. We thank H. Olsen, H.W. Posamentier, T.C.R. Pulvertaft, R.K. Suchecki and M.R. Talbot for suggestions and constructive criticism of early manuscript versions, B. Sikker Hansen and J. Halskov for drafting and J. Lautrup and J.C. Nymose for dark-room work. Thanks also to C. Marcussen for discussion and interpretation of seismic data on the Kap Stewart Formation. Dam publishes with the permission of the Geological Survey of Greenland.

REFERENCES

Aario, R. (1971) Associations of bed forms and palaeocurrent patterns in an esker delta, Haaparvi, Finland. *Ann. Acad. Sci. Fennicae, Ser. A., pt. III*, **111**, 55 pp.

Allen, J.R.L. (1984) Sedimentary structures; their character and physical basis. *Dev. Sediment.* **30**, 1254 pp.

Allen, P.A. (1981a) Wave-generated structures in the Devonian lacustrine sediments of south-east Shetland and ancient wave conditions. *Sedimentology* **28**, 369–379.

Allen, P.A. (1981b) Devonian lake margin environments and processes, SE Shetland, Scotland. *J. Geol. Soc. London* **138**, 1–14.

Allen, P.A. (1985) Hummocky cross-stratification is not produced purely under progressive gravity waves. *Nature* **313**, 562–564.

Axelsson, V. (1967) The Laiture delta — a study of deltaic morphology and processes. *Geogr. Ann.* **49A**, 127 pp.

Berner, R.A. & Raiswell, R. (1984) C/S method for distinguishing freshwater from marine sedimentary rocks. *Geology* **12**, 365–368.

Birkenmajer, K. (1976) Middle Jurassic nearshore sediments at Kap Hope. East Greenland. *Bull. Geol. Soc. Denmark* **25**, 107–116.

Bogen, J. (1983) Morphology and sedimentology of deltas in fjord and fjord valley lakes. *Sediment. Geol.* **36**, 245–267.

Boothroyd, J. & Ashley, G.M. (1975) Processes, bar morphology, and sedimentary structures on braided outwash fans, northeastern Gulf of Alaska. In: *Glaciofluvial and Glaciolacustrine Sedimentation* (Eds Jopling, A.V. & McDonald, B.C.) Spec. Publ. Soc. Econ. Paleontol. Mineral. 23, 193–222.

Bromley, R.G. & Asgaard, U. (1979) Triassic freshwater ichnocoenoses from Carlsberg Fjord, East Greenland. *Palaeogeogr., Palaeoclim., Palaeoecol.* **28**, 39–80.

Bromley, R.G., Bruun-Petersen, J. & Perch-Nielsen, K. (1970) Preliminary results of mapping in the Palaeozoic and Mesozoic sediments of Scoresby Land and Jameson Land. *Rapp. Grønlands Geol. Unders.* **30**, 17–30.

Cacchione, D.A., Drake, D.E., Grant, W.D. & Tate, G.B. (1984) Rippled scour depressions on the inner continental shelf of central California. *J. Sediment. Petrol.* **54**, 1280–1291.

Chamley, H. (1989) *Clay Sedimentology*. Springer-Verlag, Berlin, 623 pp.

Clemmensen, L.B. (1976) Tidally influenced deltaic sequences from the Kap Stewart Formation (Rhaetic-Liassic), Scoresby Land, East Greenland. *Bull. Geol. Soc. Denmark* **25**, 1–13.

Clemmensen, L.B. (1980) Triassic rift sedimentation and palaeogeography. *Bull. Grønlands Geol. Unders.* **136**, 1–72.

Clemmensen, L.B. & Houmark-Nielsen, M. (1981) Sedimentary features of a Weichselian fluviolacustrine delta. *Boreas* **10**, 229–245.

Connan, J. & Cassou, A.M. (1980) Properties of gases and petroleum liquids derived from terrestrial kerogen at various maturation levels. *Geochim. Cosmochim. Acta* **44**, 1–23.

Cook, P.J. (1976) Sedimentary phosphate deposits. In: *Handbook of Strata-bound and Stratiform Ore Deposits* (Ed. Wolfe, K.H.) Elsevier, Amsterdam/New York, pp. 505–535.

Dam, G. (1991) A sedimentological analysis of the continental and shallow marine Upper Triassic to Lower Jurassic succession in Jameson Land, East Greenland. Unpublished PhD thesis, 6 parts, University of Copenhagen, 243 pp.

Dam, G. & Christiansen, F.G. (1990) Organic geochemistry and source potential of the lacustrine shales of the Upper Triassic—Lower Jurassic Kap Stewart Formation. *Mar. Petrol. Geol.* **7**, 428–443.

Dam, G. & Surlyk, F. (1992) Forced regressions in a large wave- and storm-dominated anoxic lake, Rhaetian–Sinemurian Kap Stewart Formation, East Greenland. *Geology* **20**, 749–752.

Defretin-Lefranc, S., Grasmück, K. & Trümphy, R. (1969) Notes on Triassic stratigraphy and paleontology of north-eastern Jameson Land (East Greenland). *Meddr Grønland* **168**(2), 137 pp.

De Raaf, J.F.M., Boersma, J.R. & Van Gelder, A. (1977) Wave generated structures and sequences from a shallow marine succession, Lower Carboniferous, County Cork, Ireland. *Sedimentology* **4**, 1–52.

Diem, B. (1985) Analytic method for estimating paleowave climate and water depth from wave ripple marks. *Sedimentology* **32**, 705–720.

DOTT, R.H., JR. & BOURGEOIS J. (1982) Hummocky stratification: significance of its variable bedding sequences. *Bull. Geol. Soc. Am.* **93**, 663–680.

DUKE, W.L. (1984) Paleohydraulic analysis of hummocky cross-stratified sands indicates equivalence with wave-formed flat bed: Pleistocene Lake Bonneville deposits, Northern Utah. *Bull. Am. Assoc. Petrol. Geol.* **68**, 472.

DUNNE, L.A. & HEMPTON, M.R. (1984) Deltaic sedimentation in the Lake Hazar pull-apart basin, south-eastern Turkey. *Sedimentology* **31**, 401–412.

EYLES, N. & CLARK, B.M. (1986) Significance of hummocky and swaley cross-stratification in late Pleistocene lacustrine sediments of the Ontario basin, Canada. *Geology* **14**, 679–682.

EYLES, N. & CLARK, B.M. (1988) Storm-influenced deltas and ice scouring in a late Pleistocene glacial lake. *Bull. Geol. Soc. Am.* **100**, 793–809.

FARQUHARSON, G.W. (1982) Lacustrine deltas in a Mesozoic alluvial sequence from Camp Hill, Antarctica. *Sedimentology* **29**, 717–725.

FIELDING, C.R. (1984) Upper delta plain lacustrine and fluviolacustrine facies from the Westphalian of the Durham Coalfield, NE England. *Sedimentology* **33**, 119–140.

GREENWOOD, B. & SHERMAN, D. (1986) Hummocky cross-stratification in the surf zone: flow parameters and bedding genesis. *Sedimentology* **33**, 33–46.

GREENWOOD, B. & SHERMAN, D. (1989) Hummocky cross-stratification and post-vortex ripples: length scales and hydraulic analysis. *Sedimentology* **36**, 981–986.

GUSTAVSON, T.C. (1975) Sedimentation and physical limnology in proglacial Malaspina Lake, southeastern Alaska. In: *Glaciofluvial and Glaciolacustrine Sedimentation* (Eds Jopling, A.V. & McDonald, B.C.) Spec. Publ. Soc. Econ. Paleontol. Mineral. 23, 249–263.

GUSTAVSON, T.C., ASHLEY, G.M. & BOOTHROYD, J.C. (1975) Depositional sequences in glaciolacustrine deltas. In: *Glaciofluvial and Glaciolacustrine Sedimentation* (Eds Jopling, A.V. & McDonald, B.C.) Spec. Publ. Soc. Econ. Paleontol. Mineral., 23, 264–280.

HALLAM, A. (1988) A reevaluation of Jurassic eustasy in the light of new data and the revised Exxon curve. In: *Sea-Level Changes — An Integrated Approach* (Eds Wilgus, C.K., Hastings, B.S., Kendall, C.G.St.C., Posamentier, H.W., Ross, C.A. & Van Wagoner, J.C.) Spec. Publ. Soc. Econ. Paleontol. Mineral., 42, 261–273.

HAQ, B.U., HARDENBOL, J. & VAIL, P.R. (1988) Mesozoic and Cenozoic chronostratigraphy and cycles of sea-level changes. In: *Sea-level Changes — An Integrated Approach* (Eds Wilgus, C.K., Hastings, B.S., Kendall, C.G.St.C., Posamentier, H.W., Ross, C.A. & Van Wagoner, J.C.) Spec. Publ. Soc. Econ. Paleontol. Mineral, 42, 71–108.

HARRIS, T.M. (1926) The Rhaetic flora of Scoresby Sound East Greenland. *Meddr Grønland* **68**(2), 45–147.

HARRIS, T.M. (1931a) Rhaetic floras. *Biol. Rev.* **6**, 133–162.

HARRIS, T.M. (1931b) The fossil flora of Scoresby Sound East Greenland. 1. Cryptogams. *Meddr Grønland* **85**(2), 1–102.

HARRIS, T.M. (1932a) The fossil floras of Scoresby Sound East Greenland. 2. Descriptions of seed plants *incertae sedis* together with a discussion of certain cycadophyte cuticles. *Meddr Grønland* **85**(3), 1–112.

HARRIS, T.M. (1932b) The fossil flora of Scoresby Sound East Greenland. 3. Caytoniales and Bennittitales. *Meddr Grønland* **85**(5), 1–133.

HARRIS, T.M. (1935) The fossil flora of Scoresby Sound East Greenland. 4. Ginkgoales, Coniferales, Lycopodiales and isolated fructifications. *Meddr Grønland* **112**(1), 1–176.

HARRIS, T.M. (1937) The fossil flora of Scoresby Sound East Greenland. 5. Stratigraphic relations of the plant beds. *Meddr Grønland* **112**(2), 1–114.

HARRIS, T.M. (1946) Liassic and Rhaetic plants collected in 1936–38 from East Greenland. *Meddr Grønland* **114**(9), 41 pp.

HASZELDINE, R.S. (1984) Muddy deltas in freshwater lakes, and tectonism in the Upper Carboniferous Coalfield of NE England. *Sedimentology* **31**, 811–822.

HUNTER, R.E., DINGLER, J.R., ANIMA, J. & RICHMOND, B.M. (1988) Coarse-sediment bands on the inner shelf of southern Monterey Bay, California. *Mar. Geol.* **80**, 81–98.

HUNTER, R.E., THOR, D.R. & SWISHER, M.L. (1982) Depositional and erosional features of the inner shelf, northeastern Bering Sea. *Geol. Mijnbouw* **16**, 49–62.

HYNE, N.J., COOPER, W.A. & DCIKEY, P.A. (1979) Stratigraphy of intermontane, lacustrine delta, Catatumbo River, Lake Maracaibo, Venezuela. *Bull. Am. Assoc. Petrol. Geol.* **63**, 2042–2057.

JOHNSON, T.C., HALFMAN, J.D., ROSENDAHL, B.R. & LISTER, G.S. (1987) Climatic and tectonic effects on sedimentation in a rift-valley lake. Evidence from high-resolution seismic profiles, Lake Turkana, Kenya. *Bull. Geol. Soc. Am.* **98**, 439–447.

KATZ, B.J. (1990a) Introduction. In: *Lacustrine Basin Exploration — Case Studies and Modern Analogs* (Ed. Katz, B.J.) Mem. Am. Assoc. Petrol. Geol., 50, vii–ix.

KATZ, B.J. (1990b) Controls and distribution of lacustrine source rocks through time and space. In: *Lacustrine Basin Exploration — Case Studies and Modern Analogs* (Ed. Katz, B.J.) Mem. Am. Assoc. Petrol. Geol., 50, 61–76.

LARSEN, H.C. & MARCUSSEN, C. (1992) Sill-intrusion, flood basalt emplacement and deep crustal structure of the Scoresby Sund region, East Greenland. In: *Magmatism and the Causes of Continental Break-up*. Geol. Soc. Spec. Publ. 68, 365–386.

LECKIE, D.A. (1988) Wave-formed, coarse-grained ripples and their relationship to hummocky cross-stratification. *J. Sediment. Petrol.* **58**, 607–622.

LECKIE, D.A. & WALKER, R.G. (1982) Storm- and tide-dominated shorelines in Cretaceous Moosebar-Lower Gates Interval — outcrop equivalents of deep basin gas trap in western Canada. *Bull. Am. Assoc. Petrol. Geol.* **66**, 138–157.

MARTEL, A.T. & GIBLING, M.R. (1990) Wave-dominated lacustrine facies and tectonically controlled cyclicity in the Lower Carboniferous Horton Bluff Formation, Nova Scotia, Canada. In: *Lacustrine Facies Analysis* (Eds Anádon, P., Cabrera, L.I. & Kelts, K.) Spec. Publ. Int. Assoc. Sediment. **13**, 223–243.

MASSARI, F. & COLELLA, A. (1988) Evolution and types of fan-delta systems in some major tectonic settings. In: *Fan Deltas: Sedimentology and Tectonic Settings* (Eds Nemec, W. & Steel, R.J.) Blackie, Glasgow, pp. 103–122.

Masuda, F. & Makino, Y. (1987) Palaeo-wave conditions reconstructed from ripples in the Pleistocene paleo-Tokyo Bay deposits. *J. Geogr.* **96**, 23–45.

Miall, A.D. (1985) Architectural-element analysis: a new method of facies analysis applied to fluvial deposits. *Earth Sci. Rev* **22**, 261–308.

Moldowan, J.M., Seifert, W.K. & Gallogos, E.J. (1985) Relationship between petroleum composition and depositional environment of petroleum source rocks. *Bull. Am. Assoc. Petrol. Geol.* **69**, 1255–1268.

Newton, R.S. & Werner, F. (1972) Transitional-size ripple marks in Kiel Bay (Baltic Sea). *Meyniana* **22**, 89–96.

Nøttvedt, A. & Kreisa, R.D. (1987) Model for the combined-flow origin of hummocky cross-stratification. *Geology* **15**, 357–361.

Olsen, P. (1986) A 40-million year lake record of early Mesozoic orbital climatic forcing. *Science* **234**, 842–848.

Olsen, P. (1988) Continuity of strata in the Newark and Hartford basins. *Bull. U.S. Geol. Surv.* **1776**, 6–18.

Olsen, P. (1990) Tectonic, climatic, and biotic modulation of lacustrine ecosystems — examples from Newark Supergroup of eastern North America. *Mem. Am. Assoc. Petrol. Geol.* **50**, 209–224.

Ori, G.G. & Roveri, M. (1987) Geometries of Gilbert-type deltas and large channels in the Meteora Conglomerate, Meso-Hellenic basin (Oligo-Miocene), central Greece. *Sedimentology* **34**, 845–859.

Pedersen, K.R. & Lund, J.J. (1980) Palynology of the plant-bearing Rhaetian to Hettangian Kap Stewart Formation, Scoresby Sund, East Greenland. *Rev. Palaeobot. Palynol.* **31**, 1–69.

Pedersen, T., Harms, J.C., Harris, N.B., Mitchell, R.W. & Toby, K.M. (1989) The role of correlation in generating the Heidrun Field geological model. In: *Correlation in Hydrocarbon Exploration* (Ed. Collinson, J.D.) Norwegian Petroleum Society. Graham & Trotman, London, pp. 231–241.

Perch-Nielsen, K., Birkenmajer, K., Birkelund, T. & Aellen, M. (1974) Revision of Triassic stratigraphy of the Scoresby Land and Jameson Land, East Greenland. *Bull. Grønland Geol. Unders.* **48**, 39–59.

Plint, A.G. (1988) Sharp-based shoreface sequences and 'offshore bars' in the Cardium Formation of Alberta: their relationship to relative changes in sea-level. In: *Sea-level Changes — An Integrated Approach* (Eds Wilgus, C.K., Hastings, B.S., Kendall, C.G.St.C., Posamentier, H.W., Ross, C.A. & Van Wagoner, J.C.) Spec. Publ. Soc. Econ. Paleontol. Mineral. 42, 357–370.

Posamentier, H.W., Allen, G.P. & James, D.P. (1992) Forced regressions in a sequence stratigraphic framework: concepts, examples, and exploration significance. *Bull. Am. Assoc. Petrol. Geol.* **76**, 1687–1709.

Postma, G. & Cruickshank, C. (1988) Sedimentology of a late Weichselian to Holocene terraced fan delta, Varangerfjord, northern Norway. In: *Fan Deltas: Sedimentology and Tectonic Settings* (Eds Nemec, W. & Steel, R.J.) Blackie & Son, Glasgow, pp. 144–157.

Postma, G., Babic, L., Zupanic, J. & Røe, S.-L. (1988) Delta-front failure and associated bottomset deformation in a marine, gravelly Gilbert-type fan delta. In: *Fan Deltas: Sedimentology and Tectonic Settings* (Eds Nemec, W. & Steel, R.J.) Blackie & Son, Glasgow, pp. 91–102.

Potter, P.E., Maynard, J.B. & Pryor, W.A. (1980) *Sedimentology of Shale*. Springer Verlag, New York, 306 pp.

Powell, T.G. (1986) Petroleum geochemistry and depositional setting of lacustrine source rocks. *Mar. Petrol. Geol.* **3**, 200–219.

Richards, P.C. (1990) The early to mid-Jurassic evolution of the northern North Sea. In: *Tectonic Events Responsible for Britain's Oil and Gas Reserves* (Eds Hardman, R.F.P. & Brooks, J.) Geol. Soc. London Spec. Publ. 55, 191–205.

Rosenkrantz, A. (1934) The Lower Jurassic rocks of East Greenland. *Meddr Grønland* **110**(1), 146 pp.

Rust, B.R. (1975) Fabric and structure of glaciofluvial gravels. In: *Glaciofluvial and Glaciolacustrine Sedimentation* (Eds Jopling, A.V. & McDonald, B.C.) Spec. Publ. Soc. Econ. Paleontol. Mineral. 23, 238–248.

Rust, B.R. (1984) Proximal braidplain deposits in the Middle Devonian Malbai Formation of Eastern Gaspé, Quebec, Canada. *Sedimentology* **31**, 675–695.

Scholz, C.A. & Rosendahl, B.R. (1988) Low lake stands in Lakes Malawi and Tanganyika, East Africa, delineated with multifold seismic data. *Science* **240**, 1645–1648.

Shanley, K.W. & McCabe, P.J. (1991) Predicting facies architecture through sequence stratigraphy — an example from the Kaiparowits Plateau, Utah. *Geology* **19**, 742–745.

Smith, A.G., Hurley, A.M. & Briden, J.C. (1981) *Phanerozoic Paleocontinental World Maps*. Cambridge University Press, Cambridge, 102 pp.

Sneh, A. (1979) Late Pleistocene fan deltas along the Dead Sea Rift. *J. Sediment. Petrol.* **49**, 541–552.

Stanley, K.O. & Surdam, R.C. (1978) Sedimentation on the front of Eocene Gilbert-type deltas, Washakie. *J. Sediment. Petrol.* **48**, 557–573.

Stauber, H. (1940) Stratigraphisch-geologische untersuchungen in der ostgrönländischen senkungszone des nördlichen Jamesonlandes. *Meddr Grønland* **114**(7).

Surdam, R.C. & Stanley, K.O. (1979) Lacustrine sedimentation during the culmination phase of Eocene Lake Gosiute, Wyoming (Green River Formation). *Bull. Geol. Soc. Am.* **90**, 93–110.

Surlyk, F. (1977) Mesozoic faulting in East Greenland. In: *Fault Tectonics in NW Europe* (Eds Frost, R.T.C. & Dikkers, A.J.). Geologie en Mijnbouw, 56, 311–327.

Surlyk, F. (1978) Jurassic basin evolution of East Greenland. *Nature* **274** (5667), 130–133.

Surlyk, F. (1990) Timing, style and sedimentary evolution of Late Palaeozoic–Mesozoic extensional basins of East Greenland. In: *Tectonic Events Responsible for Britain's Oil and Gas Reserves* (Eds Hardman, R.F.P. & Brooks, J.) Geol. Soc. London Spec. Publ. 55, 107–125.

Surlyk, F. & Noe-Nygaard, N. (1986) Hummocky cross-stratification from the Lower Jurassic Hasle Formation of Bornholm, Denmark. *Sediment. Geol.* **46**, 259–273.

Surlyk, F., Callomon, J.H., Bromley, R.G. & Birkelund, T. (1973) Stratigraphy of the Jurassic–Lower Cretaceous sediments of Jameson Land and Scoresby Land, East Greenland. *Bull. Grønlands Geol. Unders.* **105**, 76 pp.

Surlyk, F., Clemmensen, L.B. & Larsen, H.C. (1981) Post-Paleozoic evolution of the East Greenland continental margin. In: *Geology of the North Atlantic Borderlands* (Eds Kerr, J.W. & Ferguson, A.J.) Mem. Can. Soc. Petrol. Geol. 7, 611–645.

Surlyk, F., Hurst, J.M., Piasecki, S., *et al.* (1986) The Permian of the western margin of the Greenland Sea — a future exploration target. In: *Future Petroleum Provinces of the World* (Ed. Halbouty, M.T.) Mem. Am. Assoc. Petrol. Geol. 40, 629–659.

Swift, D.J.P., Hudelson, P.M., Brenner, P.M. & Thompson, P. (1983) Hummocky cross-stratification and megaripples: a geological double standard? *J. Sediment. Petrol.* **53**, 1295–1317.

Sykes, R.M. (1974) Sedimentological studies in southern Jameson Land, East Greenland. I. Fluviatile sequences in the Kap Stewart Formation (Rhaetic–Hettangian). *Bull. Geol. Soc. Denmark,* **23**, 203–212.

Syvitski, J.P.M. & Farrow, G.E. (1983) Structures and processes in bayhead deltas, Knight and Bute Inlet, British Columbia. *Sediment. Geol.* **36**, 217–244.

Talbot, M.R. (1988) The origins of lacustrine oil source rocks: evidence from the lakes of tropical Africa. In: *Lacustrine Petroleum Source Rocks* (Eds Fleet, A.J., Kelts, K. & Talbot, M.R.) Geol. Soc. London Spec. Publ. 40, 29–43.

Tissot, B. & Welte, D.H. (1984) *Petroleum Formation and Occurrence.* Springer Verlag, Berlin, 699 pp.

Tye, R.S. & Coleman, J.M. (1989a) Evolution of Atchafalaya lacustrine deltas, south-central Louisiana. *Sediment. Geol.* **65**, 95–112.

Tye, R.S. & Coleman, J.M. (1989b) Depositional processes and stratigraphy of fluvially dominated lacustrine deltas: Mississippi delta plain. *J. Sediment. Petrol.* **59**, 973–996.

Van Dijk, D.E., Hobday, D.K. & Tankard, A.J. (1978) Permo-Triassic lacustrine deposits in the Eastern Karoo Basin, Natal, South Africa. In: *Modern and Ancient Lake Sediments* (Eds Matter, A. & Tucker, M.E.) Spec. Publ. Int. Assoc. Sediment. 2, 225–239.

Van Wagoner, J.C., Posamentier, H.W., Mitchum, R.M., Jr, *et al.* (1988) An overview of the fundamentals of sequence stratigraphy and key definitions. In: *Sea-level Changes — An Integrated Approach* (Eds Wilgus, C.K., Hastings, B.S., Kendall, C.G.St.C., Posamentier, H.W., Ross, C.A. & Van Wagoner, J.C.) Spec. Publ. Soc. Econ. Paleontol. Mineral. 42, 39–45.

Vessell, R.K. & Davies, D.K. (1981) Nonmarine sedimentation in an active orearc basin. *Spec. Publ. Soc. Econ. Paleontol. Mineral.* **31**, 31–45.

Weirich, F.H. (1985) Sediment budget for a high energy glacial lake. *Geogr. Ann.* **67A**, 83–99.

Weirich, F.H. (1986) The record of density-induced underflow in a glacial lake. *Sedimentology* **33**, 261–277.

Wood, M.L. & Ethridge, F.G. (1988) Sedimentology and architecture of Gilbert- and mouth bar-type fan deltas, Paradox Basin, Colorado. In: *Fan Deltas: Sedimentology and Tectonic Settings* (Eds Nemec, W. & Steel, R.J.). Blackie & Son, Glasgow, pp. 251–263.

Wright, L.D. (1977) Sediment transport and deposition at river mouths: a synthesis. *Bull. Geol. Soc. Am.* **88**, 857–868.

Ziegler, P.A. (1988) Evolution of the Arctic-North Atlantic and the Western Tethys. *Mem. Am. Assoc. Petrol. Geol.* **43**, 1–198.

NORTH AMERICA

Spec. Publs Int. Ass. Sediment. (1993) **18**, 451–467

Origin of an erosion surface in shoreface sandstones of the Kakwa Member (Upper Cretaceous Cardium Formation, Canada): importance for reconstruction of stratal geometry and depositional history

B.S. HART* *and* A.G. PLINT

Department of Geology, University of Western Ontario, London, Ontario, Canada N6A 5B7

ABSTRACT

The Upper Cretaceous Cardium Formation of western Canada is a shallow marine clastic unit deposited along the western margin of a foreland basin occupied by the Western Interior Seaway. The Kakwa Member comprises the main shoreface sandbody of the Cardium Formation and is traceable for at least 150 km perpendicular to the basin margin, and for over 500 km along the depositional strike. The Kakwa Member is divisible throughout its extent into two main parts: a lower, very fine to fine-grained sandstone portion dominated by swaley cross-stratification and an upper, coarser-grained portion (locally conglomeratic) dominated by medium-scale cross-bedding. Both units are, at least locally, capped by beach-laminated sandstone with root traces. The lower and upper portions of the Kakwa are everywhere separated by an erosion surface, locally mantled with mudstone or pebbles. Petrographic evidence indicates siderite precipitation in the lower Kakwa *prior* to deposition of the upper sandbody.

Previously, the erosional break within the Kakwa shoreface was interpreted to reflect scour in rip channels on a barred shoreline. Another explanation suggested the shoreface prograded as a series of 'shingled' lenticular sandbodies, separated by seaward-dipping erosion surfaces generated during minor relative sea-level fluctuations. The work described here indicates that the Kakwa Member consists essentially of two superimposed shoreface sandstones which are separated by an originally near-planar ravinement surface over which at least 150 km of transgression can be demonstrated. Only the *lower* fine-grained part of the Kakwa is shingled; the *upper*, coarser grained portion is an *entirely separate*, sheet-like sandbody which prograded into water too shallow to allow preservation of offshore muds.

INTRODUCTION

Proper delineation of the physical surfaces (sequence boundaries, condensed sections, ravinement surfaces, etc.) which bound depositional packages in any stratigraphic succession is critical to the reconstruction of that succession's depositional history. Whereas superimposed coarsening-upward successions in heterolithic shelf deposits are readily recognizable as representing successive progradational events (e.g. the parasequences of Van Wagoner *et al.*, 1988), distinction of stratal geometry and hence reconstruction of relative sea-level changes from thick sandstone successions can be more ambiguous.

Figure 1(a) shows four measured sections from a hypothetical clastic shelf to shoreface succession. In each measured section, fine-grained shoreface deposits *gradationally* overlie sandier upward shelf deposits, but are separated by an *erosion surface* from an overlying, coarser grained package representing shoreface to beach environments. At least three plausible stratal geometries (with correspondingly different depositional histories) could be inferred from the evidence provided by the four

* Present address: Pacific Geoscience Centre, PO Box 6000, Sidney, BC V8L 4B2.

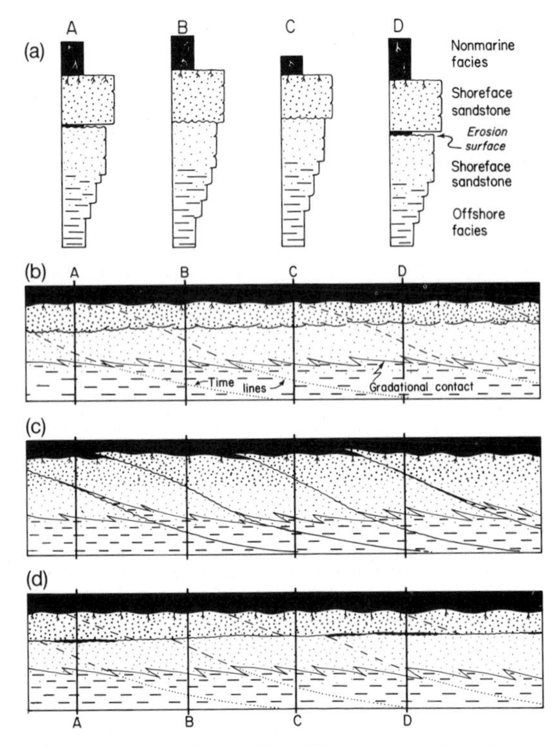

Fig. 1. Schematic illustration of four measured sections through a hypothetical shoreface sandbody (a); (b), (c) and (d) illustrate three different depositional scenarios that depend upon the interpretation placed on the erosion surface that occurs within each succession. (See text for detailed discussion.)

measured sections: (i) a single prograding shoreface (Fig. 1(b)); (ii) a series of 'shingled'* sandstone lenses separated by erosion surfaces (Fig. 1(c)); or (iii) two 'stacked' near-tabular shoreface sandstone units of differing grain size (Fig. 1(d)).

In the first case, the erosion surface is interpreted as a diachronous, purely sedimentological feature, generated by scour in rip-current channels of a barred shoreface (cf. Hunter *et al.*, 1979; Plint & Walker, 1987; Rahmani & Smith, 1988) and has no chronostratigraphic significance. In the second case, the erosion surfaces could be interpreted as seaward-dipping ravinement surfaces demarcating discrete phases of marine transgression and regression. In this case each erosion surface has *local* chronostratigraphic significance. In the third case, the erosion surface separating finer from coarser grained sandstones is interpreted as the result of a single episode of transgressive erosion and has *regional* chronostratigraphic significance.

The Upper Cretaceous Cardium Formation of the Alberta Basin has been the subject of intense stratigraphic and sedimentological study in recent years (Duke, 1985; Plint *et al.*, 1986, 1988; Plint & Walker, 1987; Plint, 1988; Hart, 1990a). Plint *et al.* (1986, 1988) proposed a primarily allostratigraphic approach, based on the recognition of disconformity-bounded packages (Fig. 2(a)) interpreted as representing shoreline progradation subject to a succession of relative sea-level fluctuations each probably of about 10^5 years duration (Plint *et al.*, 1988; Hart, 1990a).

The Kakwa Member forms the main shoreface sandbody of the Cardium Formation (Fig. 2). The member can be traced for at least 150 km perpendicular to the basin margin, and for over 500 km along the depositional strike. Previously, this part of the formation has been interpreted (Duke, 1985; Plint & Walker, 1987; Plint *et al.*, 1988) as being comprised of a series of shingled sandbodies (Fig. 2(a)) which represent shoreface progradation punctuated by relatively minor marine regressions and transgressions which generated seaward-dipping, sigmoidal erosion surfaces. In this paper we re-examine this interpretation, based on the results of new work (Hart, 1990a) into the sedimentology and stratigraphy of the Cardium Formation (Fig. 2(b)). This work demonstrates the importance of distinguishing between erosion surfaces generated by facies shifts and erosion surfaces related to relative sea-level changes (e.g. flooding surfaces, sequence boundaries), as well as the need to define accurately the geometry of the latter. Our results emphasize that caution is needed when trying to interpret the geometry of thick shoreface sandbodies in areas of sparse data.

DATA BASE

The results presented here are based on both outcrop and subsurface observations. The data include (Fig. 3) 57 borehole cores, 51 outcrop sections and over 1100 geophysical well logs (generally consisting of both γ-ray and resistivity logs) from an area of about 44,000 km^2. Stratigraphic relationships were derived using a regional grid of cross-sections that incorporated both subsurface and outcrop data.

Petrographic analyses of over 50 sandstone samples from the Kakwa Member were completed in an

* *Webster's Third New International Dictionary* defines the verb 'to shingle' to mean 'to lay or dispose so as to overlap'.

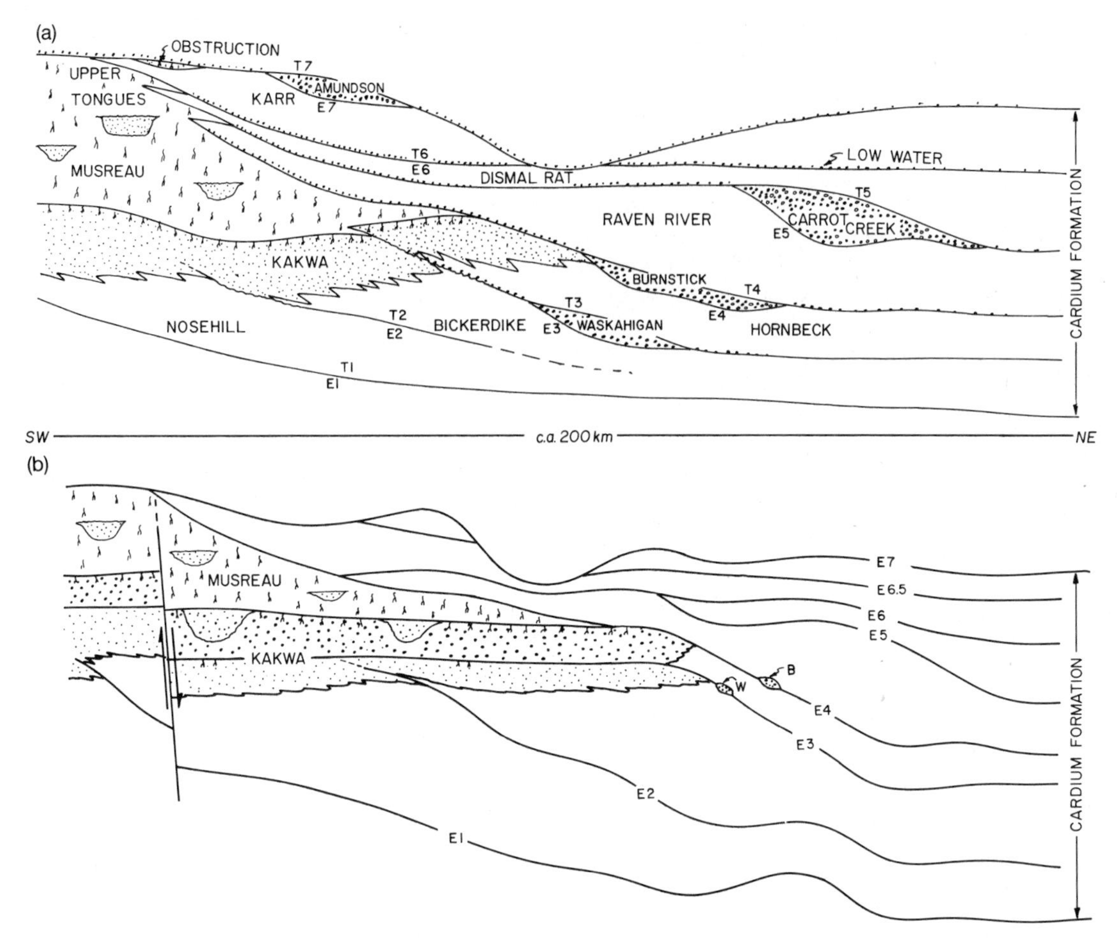

Fig. 2. Stratigraphy of the Cardium Formation. (a) From Plint *et al.* (1988). (b) From Hart (1990a). Note particularly the differences in correlation of surfaces E1–E4 with respect to the Kakwa Member (stippled), and interpretation of faulting in the west (b). W, Waskahigan Member; B, Burnstick Member in (b). Other differences are explained in Hart (1990a).

attempt to recognize provenance, diagenetic effects and indicators of depositional environment. Conclusions pertinent to the depositional history of the Kakwa Member will be presented here, with further petrographic information presented in Hart (1990a) and Hart *et al.* (1992).

SEDIMENTOLOGY OF THE KAKWA MEMBER

Facies

A comprehensive facies scheme for the Cardium Formation, incorporating both marine and non-marine facies, was developed by Walker (1983) and Plint and Walker (1987) with additional description and interpretation by Hart (1990a). Only the facies of immediate interest will be reviewed here.

Heterolithic shelf facies

Facies 7 (Walker, 1983) consists of dm-scale beds of fine to very fine-grained hummocky cross-stratified sandstones interbedded with bioturbated mudstone. Facies 7 is commonly interbedded with facies 15 (Plint & Walker, 1987) which comprises cm-scale interbeds of dark mudstone and lenses of fine- to very fine-grained sandstone displaying ripple cross-lamination, wave-ripple form sets and

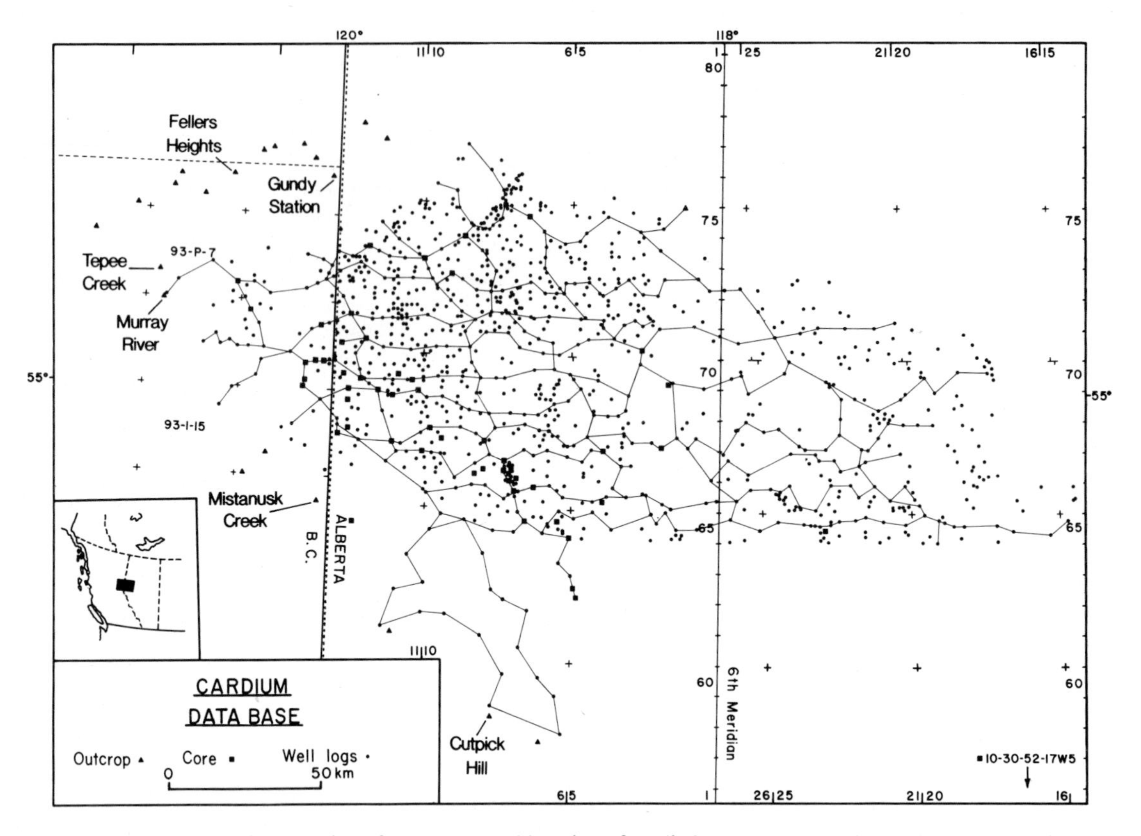

Fig. 3. Map showing regional setting of study area and location of studied cores, outcrops, borehole logs and regional cross-section grid.

parallel to undulose lamination. These facies are interpreted to represent a storm-influenced, (?)middle- to inner-shelf environment.

Shoreface sandstones

Facies 16 comprises very fine to fine-grained, sparsely bioturbated, swaley cross-stratified sandstone lacking mud interbeds. Broad, shallow pebble-lined scours are present locally. Facies 17 consists of fine- to coarse-grained sandstone characterized by medium-scale cross-bedding and parallel lamination. Swaley cross-stratified sandstones can be readily distinguished from medium-scale cross-bedded sandstones in core (Hart, 1990b). Both the swaley cross-stratified and medium-scale cross-bedded sandstones are the product of deposition on the shoreface; the differences in sedimentary structures are attributed here to differences in grain size. Nottvedt and Kreisa (1987) and Duke (1990) have reported a direct relationship between grain size and the type of bedform produced under the combined influence of oscillatory wave motion and unidirectional currents (conditions typical of high-energy shoreface environments). Angle of repose cross-bedding will develop in fine upper and coarser grained sands, whereas low-angle to hummocky cross-stratification typifies finer sands and coarse silts. Gently inclined planar-laminated sandstones, locally with root traces, are clearly identifiable as beach deposits.

Non-marine sediments

Facies 19 through 22 represent a variety of non-marine environments (Plint & Walker, 1987; Hart, 1990a). Facies 19 comprises black, laminated lagoonal and lacustrine mudstones. Facies 20 comprises very fine-grained sandstone and black mudstone and is interpreted to represent a variety of lagoon/lake subenvironments including distributary mouth bars, crevasse splays, etc. Facies 21 comprises cross-bedded and ripple cross-laminated fine- to medium-grained sandstones interpreted as fluvial-channel deposits. Mudstones of facies 22 have a distinctive waxy and slickensided appearance and often contain carbonaceous root traces; this facies represents palaeosol horizons.

Conglomerates

Facies 8 consists of various types of chert pebble conglomerates (Walker, 1983; Plint & Walker, 1987; Hart, 1990a). Conglomerates are present locally in the Kakwa Member and represent shoreface, beach and fluvial settings (Hart & Plint, 1989, 1991; Hart, 1990a).

Facies successions

Outcrop and core sections through the Kakwa Member are remarkably consistent. Two distinct shoreface sandstone units can be distinguished, with the lower dominated by facies 16 and the upper dominated by facies 17 sandstones. These two units are everywhere separated by a sharp contact.

In the lower part of the Kakwa, swaley cross-stratified shoreface sandstones of facies 16 (Fig. 4(a)) gradationally overlie inner-shelf deposits of facies 7 and 15. This lower unit is characterized by an upward coarsening from very fine to fine-grained sandstone. In a few localities, facies 16 sandstones pass up into parallel-laminated beachface sandstone (Fig. 4(b)) and in two sections the beachface deposits are capped by a rooted horizon (Fig. 4(c)). The thickness of the lower part of the Kakwa can range from over 20 m to less than 2 m depending upon the position of the section along the basin margin to centre transect (Fig. 2(b)).

The surface separating the upper and lower portions of the Kakwa Member is commonly defined by an abrupt change from fine- to medium- or coarse-grained, locally pebbly sandstone (Figs 4(d) & (e)). This surface can have a relief of several centimetres to decimetres. Where well exposed, the surface may display erosional furrows which may have overhanging walls (Figs 4(d) & (e)), suggesting at least partial lithification of the underlying sandstone. The contact may be overlain by undulose-laminated sandstones with abundant comminuted plant debris, discontinuous black carbonaceous mudstones up to a few centimetres thick (Fig. 4(f)), a thin bed (up to a few clasts thick) of extraformational (chert and quartz) pebbles or intraformational (sandstone, siltstone, mudstone) clasts or by a combination of these lithologies. In some outcrops, coarse-grained wave ripples are present above the contact.

Above the basal heterolithic portion, a predictable succession of sedimentary structures occurs in the upper part of the Kakwa Member. This includes medium-scale cross-bedding, overlain by a horizon of intense *Macaronichnus* bioturbation (Fig. 4(g)) capped by parallel-laminated sandstone with root traces (Fig. 4(h)). The thickness (usually 7 to 8 m, but locally at least 12 m) of this part of the member tends to be less variable than that of the lower part. The top of the Kakwa is conformably overlain by non-marine strata of the Musreau Member (Fig. 2).

Selected sections through the Kakwa Member, plus corresponding well logs, are illustrated in Fig. 5 (based on core) and Fig. 6 (based on outcrop). It can be seen that the contact between the upper and lower portions of the Kakwa can be correlated with characteristic deflections in the geophysical logs. These core–log correlations provide a basis for the identification, in wells lacking core, of the contact between the lower and upper portions of the Kakwa Member (e.g. Fig. 7).

Both the lower and upper portions of the Kakwa Member are *locally* conglomeratic, and shoreface to beach successions are identifiable in the conglomerates (e.g. Hart & Plint, 1989, 1991). In addition, some core sections show that the upper portion of the member is replaced by heterolithic strata attributable to facies 20 and 21; these are interpreted as the deposits of fluvial channels incised into the shoreface.

Petrography of the Kakwa Member

Petrographic analyses indicate quartz and sedimentary rock fragments to be the principal constituents of all samples (Table 1, p. 467). In general, feldspars and carbonate grains each constitute less than 2% of the detrital fraction and thus the sandstones can be approximated as a two-component system (quartz–rock fragments). Statistical analyses (t-test, 90% confidence level) of sandstone composition (% quartz) from the upper and lower portions of the Kakwa indicate that there is *no* significant difference in their mean compositions.

(b)
15 cm

(a)
15 cm

(e)

(d)

(c)

Fig. 4. Core and outcrop photographs of Kakwa Member. (a) Facies 16, swaley cross-stratified sandstones at Gundy Station. (b) Beach-laminated sandstones, conformably overlying facies 16 at Gundy Station. (c) Root traces from beach-laminated sandstones conformably overlying facies 16 sandstones at Fellers Heights. A rooted horizon has also been observed at this level in core from 14-5-63-5W6, 1880 m. (d) Sharp contact (E3 surface) between lower and upper portions of Kakwa Member, with distinct grain-size break, erosional truncation of underlying laminae and steep-sided scour walls (arrows). Core from 10-7-70-11W6, 823 m. (e) Sharp contact (E3 surface) between upper and lower portions of Kakwa Member with overhanging wall (arrow) suggestive of lithification of underlying sandstone prior to erosion. Core from 7-10-68-10W6, 1087 m. (f) Heavily sideritized facies 16 sandstone (base) overlain by black shale (behind scale bar), sharply overlain by facies 17 sandstone. Core from 6-8-70-13W6, 891 m. (g) Intense *Macaronichnus* bioturbation, facies 17 sandstone at Mistanusk Creek. (h) Root traces from top of beach-laminated facies 17 sandstone at Mistanusk Creek. See Fig. 3 for location of outcrops.

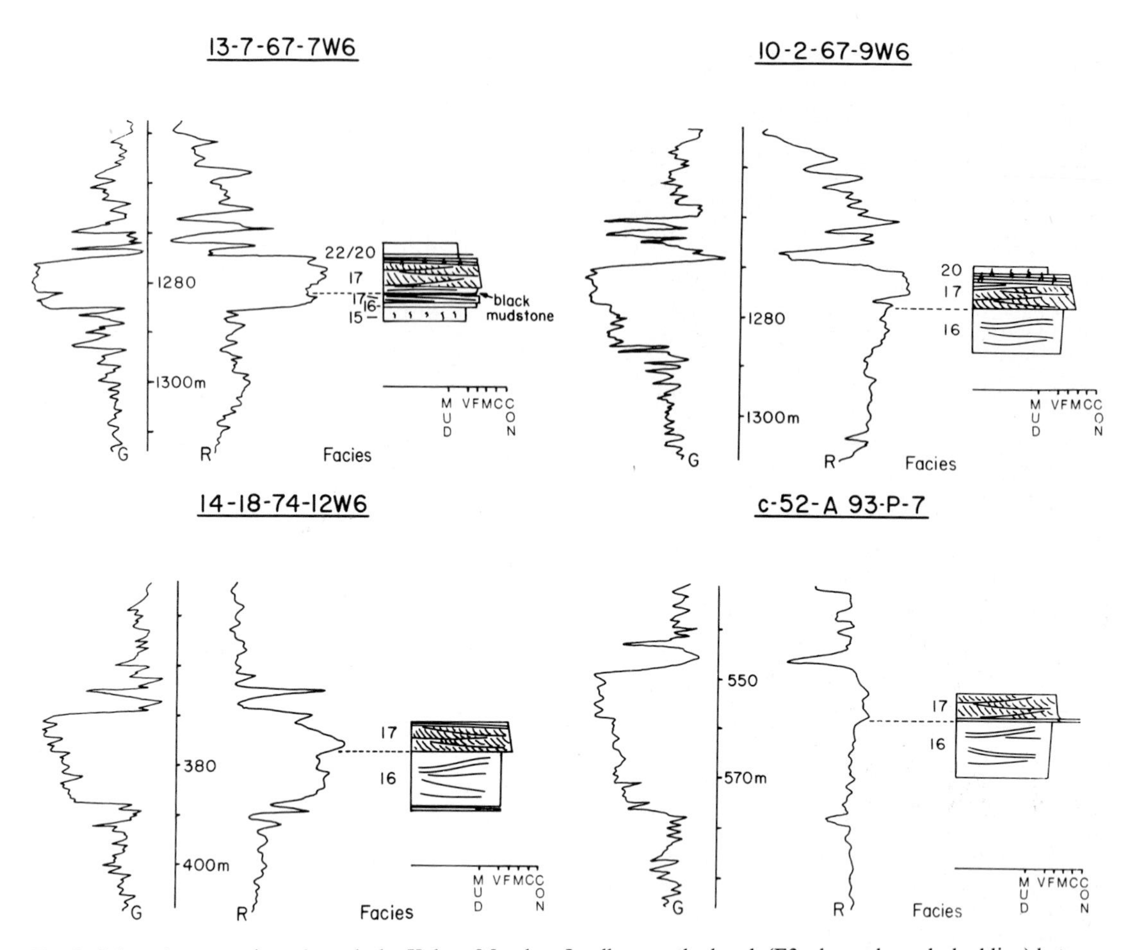

Fig. 5. Selected core sections through the Kakwa Member. In all cases, the break (E3, shown by a dashed line) between the lower part of the member (dominated by facies 16) and the upper part (dominated by facies 17) can be associated with a distinct γ-ray (G) or resistivity (R) deflection.

Despite the similarity in detrital composition between the lower and upper sandstones, there are important diagenetic differences (Hart, 1990a; Hart *et al.*, 1992). Sandstones forming the lower portion of the Kakwa are characterized by silt-sized siderite rhombs (representing the earliest diagenetic cement) and well-developed quartz overgrowths (Fig. 8(a)). Hart *et al.* (1992) provide isotopic and textural evidence which indicates that siderite precipitation probably occurred very shortly after deposition of the host sandstone. In the upper sandstone, siderite cement is *absent*, quartz overgrowths are not abundant and compaction (pressure solution and deformation of labile grains) tends to be the most conspicuous diagenetic effect (Fig. 8(b)). Given these observations, it seems likely that the petrographic differences between the upper and lower portions of the Kakwa Member reflect different early diagenetic histories rather than control by detrital mineralogy.

Sedimentological interpretation of the Kakwa Member

Previously, Plint and Walker (1987) interpreted the contact between the lower (facies 16-dominated) and upper (facies 17-dominated) portions of the Kakwa to represent an erosion surface produced by the lateral migration of rip-current channels on the shoreface (Hunter *et al.*, 1979). This model is simplistic, as more recent studies indicate that shoreface morphology (presence/absence and num-

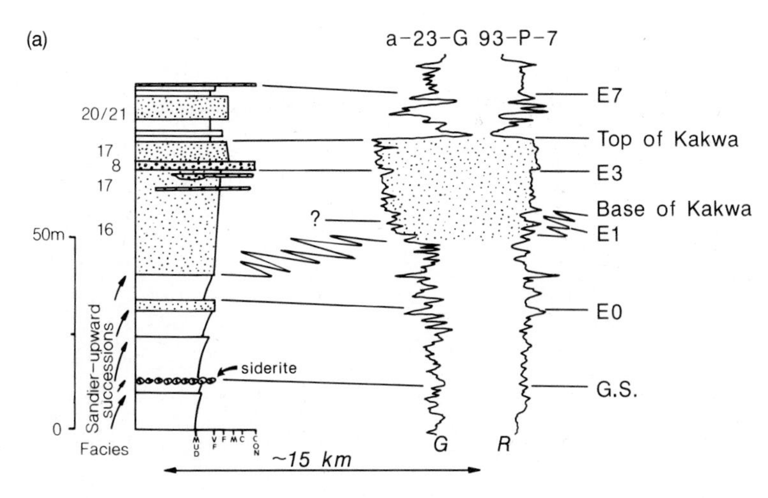

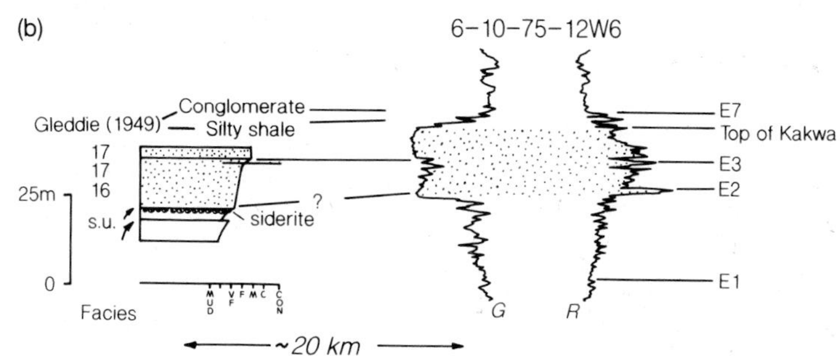

Fig. 6. Outcrop to well-log correlations from: (a) Murray River/Tepee Creek; (b) Gundy Station. In both cases, note γ-ray (G) or resistivity (R) well-log signature associated with erosion surface (E3) separating upper and lower portions of the Kakwa Member. Facies 17 beach-laminated sandstones are found below the E3 surface at both locations. Location of outcrops shown in Fig. 3.

ber of bars and rip-current channels, etc.) is a complex function of wave and tidal conditions and sediment grain size (Wright & Short, 1984). Variations in these parameters will lead to differences in the character of the preserved stratigraphic successions (Short, 1984).

The observations described here indicate that the 'barred-shoreface' interpretation of Plint and Walker (1987) can be rejected. The consistent differences in diagenetic history across the contact (including evidence of early lithification of the lower part of the member), the local presence of a mudstone between the upper and lower parts of the member and the thickness and textural variability (e.g. thick shoreface conglomerates in places) of the two portions of the member are all considered evidence *against* the 'single bar' shoreface model. Most important, however, is the preservation of

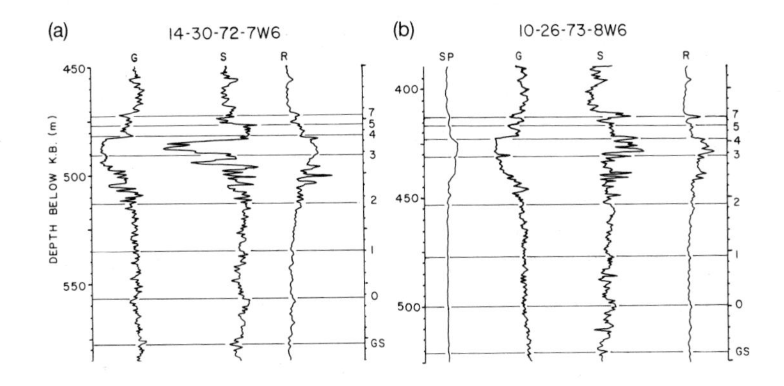

Fig. 7. Examples of how well logs alone can be used to identify contact (E3) between lower and upper portions of the Kakwa Member. Well logs are spontaneous potential (SP), γ-ray (G), sonic (S) and resistivity (R). GS and 0–7 refer to erosion surfaces in the Cardium and underlying Kaskapau Formations (Hart, 1990a).

(a)

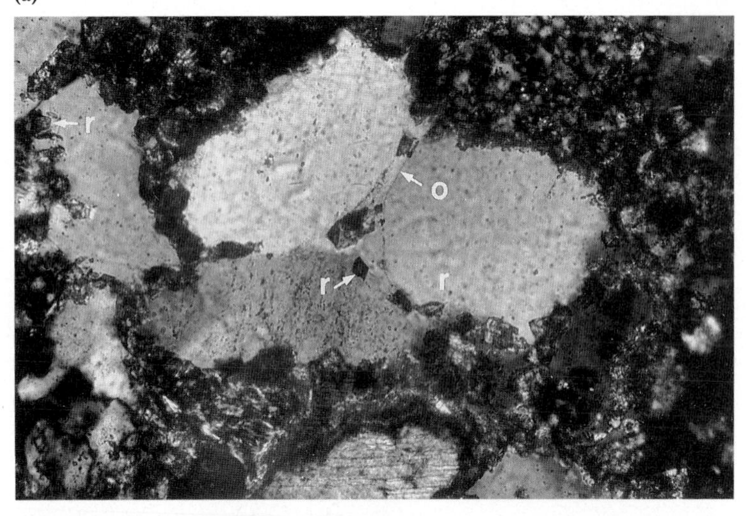

(b)

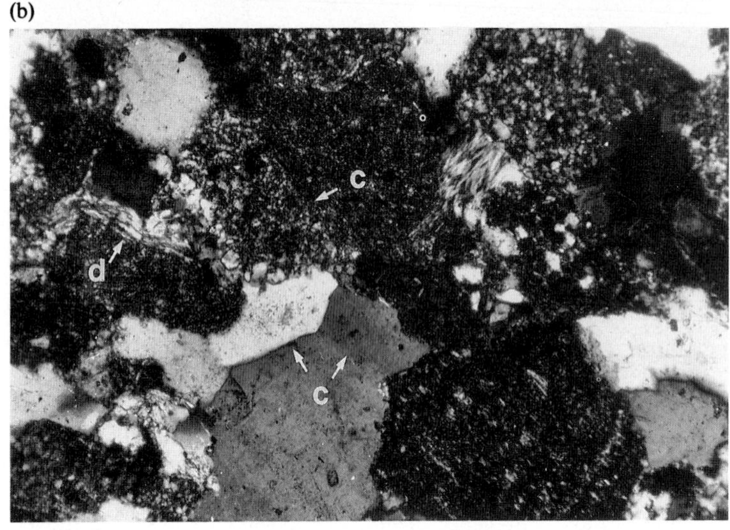

Fig. 8. Representative thin sections from lower and upper portions of the Kakwa Member showing different diagenetic constituents. (a) Sandstone in the lower part of the member is characterized by early diagenetic siderite rhombs (r) and well-developed quartz overgrowths (o). (b) Sandstone in the upper part of the member lacks siderite and well-developed quartz overgrowths, and is dominated by evidence of compaction such as line contacts (c) and deformed labile grains (d).

beach-laminated sandstones and rooted horizons *below* the contact (e.g. Figs 4(b), (c) & 6) in several of the measured sections. This last evidence testifies to the preservation of superimposed shoreface to beach successions at these localities.

INTERNAL 'GEOMETRY' OF THE KAKWA MEMBER

Based upon the evidence provided by the core and outcrop sections, the Kakwa Member could represent either of the possibilities illustrated in Figs 1(c) or 1(d). However, if sufficient outcrop or subsurface control is available, it should be possible to distinguish between 'shingled' and 'stacked' geometries (illustrated in Figs 1(c) and 1(d), respectively) by varying the distance between control points, as shown in Fig. 9.

The hypothetical shingled sandbody shown in Fig. 9(a) is the same as that illustrated in Fig. 1(c). In Figs 9(b)–(d), the effect of changing the spacing between control points is illustrated. If closely spaced sections are used (A, A′, A″ in Fig. 9(b)), it should be possible to trace out the seaward-dipping contact between two disconformity-bounded shin-

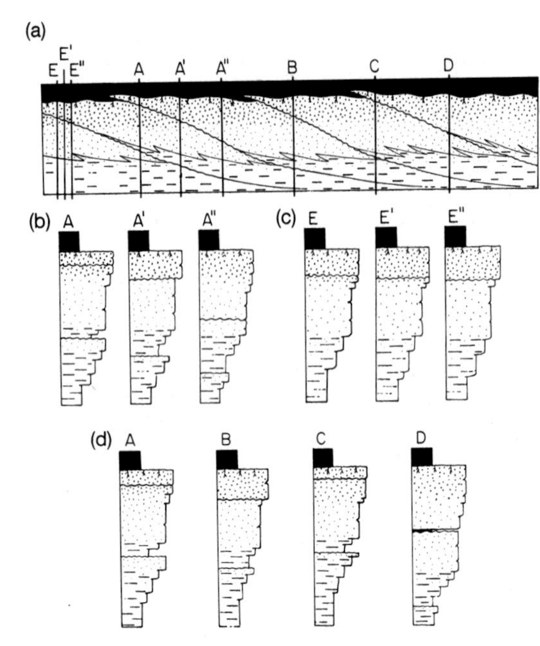

Fig. 9. Cartoon illustrating the influence of well spacing on the interpretation of the internal architecture of a shoreface sandbody. Closely spaced wells (A, A′, A″) are more likely to provide a true picture of the geometry of internal erosion surfaces than are more widely spaced wells (A, B, C, D). See text for further discussion.

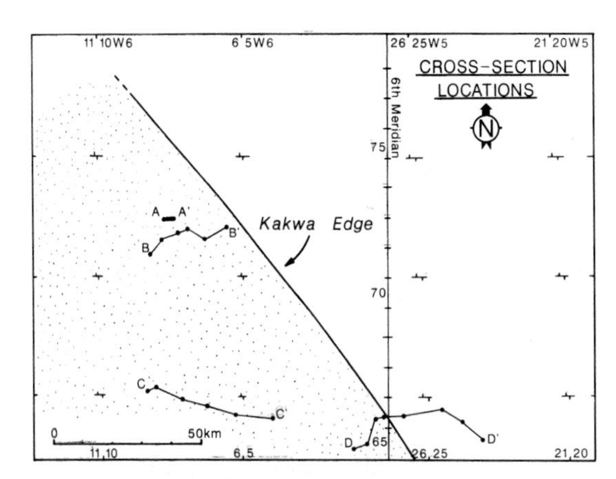

Fig. 10. Location map for cross-sections illustrated in Figs 11, 12 and 13. Area underlain by Kakwa Member is stippled. The abrupt, progradational limit of the sandbody is labelled 'Kakwa Edge'.

gles. However, if only a few, very closely spaced sections are selected, and the disconformity dips very gently, it is possible that the surface might be interpreted as horizontal, as shown in Fig. 9(c). A more widely spaced series of sections (e.g. A, B, C, D in Fig. 9(d)) is unlikely to permit detailed mapping of a single bounding surface, and instead may intersect one or more boundaries, most probably at different depths.

Figure 10 shows the location of four well-log cross-sections, illustrated in Figs 11–13. The spacing of wells has been varied in order to determine whether this has any effect on the apparent geometry of the sandbodies that comprise the Kakwa Member. The datum used in the cross-sections is a marker ('M-1') in the overlying Muskiki Formation which was inferred by Plint (1990) to represent an essentially planar surface in the present study area.

Figure 11 is constructed using the most closely spaced well control available. The distance between wells is 0.8 km (0.5 miles), and the whole section is 3.2 km long. The contact between the upper and lower shoreface units is clearly seen in all the log responses and is traceable as a planar surface essentially parallel to the datum. Note that this result does not rule out the existence of shingled shoreface lenses, provided that either the width of

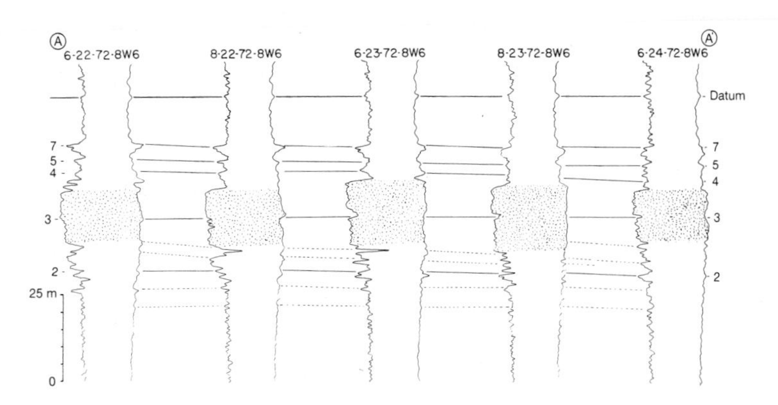

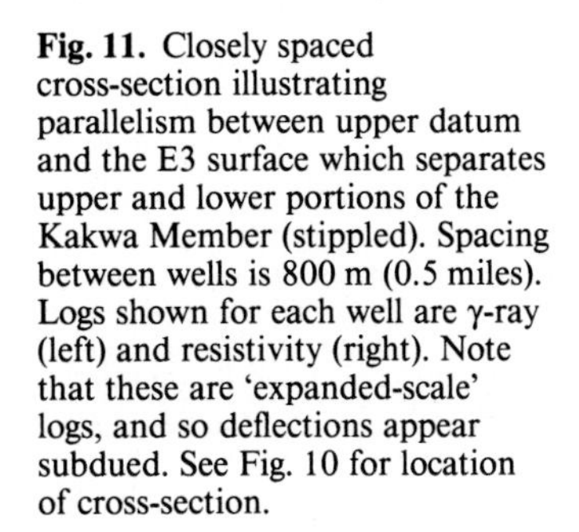

Fig. 11. Closely spaced cross-section illustrating parallelism between upper datum and the E3 surface which separates upper and lower portions of the Kakwa Member (stippled). Spacing between wells is 800 m (0.5 miles). Logs shown for each well are γ-ray (left) and resistivity (right). Note that these are 'expanded-scale' logs, and so deflections appear subdued. See Fig. 10 for location of cross-section.

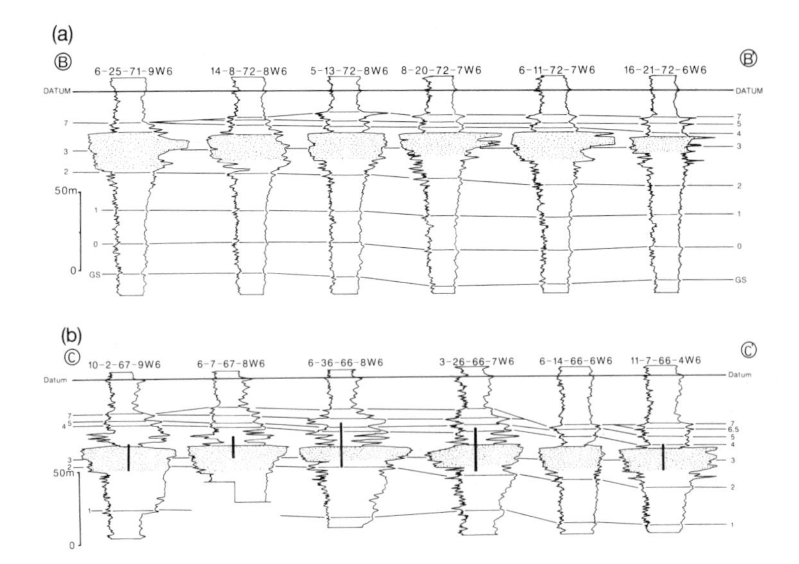

Fig. 12. Cross-sections with several km spacing between wells. Note parallelism between datum and E3 erosion surface within the Kakwa Member (stippled). Logs shown for each well are γ-ray (left) and resistivity (right). Measured core intervals shown by black bars in C–C′. See Fig. 10 for location of cross-sections.

each lens is less than 0.8 km (e.g. Fig. 1(c)) or much greater than that distance (e.g. Fig. 9(d)).

The two sections illustrated in Fig. 12 are constructed with wells spaced several km apart. Logs, supplemented by good core control in section C–C′, again show that the contact between the upper and lower sandstone units is an essentially planar surface, parallel to the datum. Indeed, cross-sections *throughout* the Kakwa Member consistently show that the contact between its upper and lower portions forms a nearly planar surface with respect to the upper datum. Only in the western part of the study area are there abrupt changes in the spacing between the M-1 datum and the contact between upper and lower sandstone units. This deviation was attributed by Hart and Plint (1990) to late Turonian to early Coniacian block faulting associated with basement-fault movement on the southern flank of the Peace River Arch. We therefore conclude that the contact between the upper and lower portions of the Kakwa *was* an originally planar surface that separated two vertically stacked shoreface sandbodies.

Correlation with the E3 surface

The origin of the erosion surface separating the upper and lower portions of the Kakwa Member becomes clear when it is traced northeast to the progradational limit of the Kakwa Member (Fig. 10). There, the contact drops down and merges with the E3 surface of Plint *et al.* (1986) (Figs 2(b) & 13). The upper and lower shoreface sandstones of the Kakwa Member are therefore separated by the E3 surface which is inferred to have been generated by a significant relative sea-level fall and subsequent rise (Plint *et al.*, 1986).

Figure 14 shows the topography on the E3 surface with respect to the upper datum. A distinct NW–SE trending break in slope (termed the 'Kakwa Edge') occurs on this surface, the location of which appears to have been controlled by underlying structural

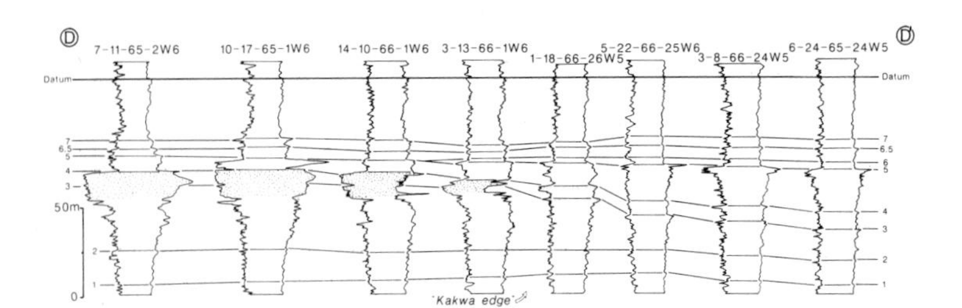

Fig. 13. Correlation of E3 surface from basinward portions of Cardium Formation (right) up and into sandstones of Kakwa Member (stippled, left). Note position of Kakwa Edge at depositional limit of shoreface sandstones. See Fig. 10 for location of cross-section.

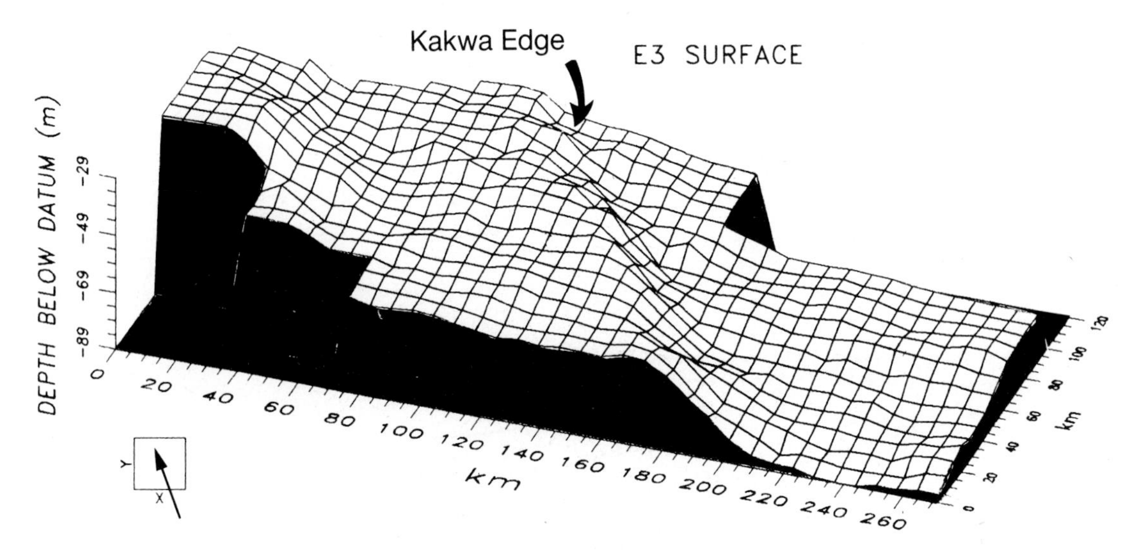

Fig. 14. Three-dimensional mesh diagram illustrating topography on E3 surface with respect to the upper datum. View from the southeast. Note NW–SE-trending break in slope (Kakwa Edge) at progradational limit of Kakwa shoreface sandstones. The E3 surface is essentially flat throughout the Kakwa Member (SW of Kakwa Edge), except where offset by faulting in the extreme west.

elements (Hart, 1990a; cf. Cant, 1988). This break corresponds to the NE depositional limit of the Kakwa Member. West of the Kakwa Edge the E3 surface is nearly planar throughout the Kakwa Member except, as noted, in the far west. Lowstand shoreface deposits of the Waskahigan Member (where present) are located along the more steeply dipping portion of the E3 surface just seaward of the Kakwa Edge (Fig. 2(b)).

DEPOSITIONAL HISTORY OF THE KAKWA MEMBER

The depositional history of the Kakwa Member is summarized in Fig. 15. Although the base of the Cardium Formation was defined by Plint *et al.* (1986) at the E1 erosion surface, it is evident that similar surfaces exist in the upper part of the underlying Kaskapau Formation. Surfaces designated E0 and GS (Figs 6(a) & 15(a)) approximately parallel E1 and, like the latter, probably converged with the base of the Kakwa Member further to the SW. Progradation of the lower part of the Kakwa Member thus appears to have been punctuated by relatively short-lived (fourth-order) relative sea-level fluctuations (Figs 15(a) & (b)). The amount of shoreline progradation cannot be evaluated for the intervals below E1, but over 50 km of shoreface progradation can be demonstrated for both the E1–E2 and E2–E3 intervals (Figs 2(b) & 15(b)).

Subsequent to progradation of the E2–E3 interval, relative sea-level fall caused the shoreline to drop below the Kakwa Edge and stabilize at a position about 10 m below the top of the adjacent Kakwa strandplain (Fig. 15(b)). Pebbly sandstones of the Waskahigan Member accumulated at this lowstand shoreline position (Plint, 1988). During subsequent relative sea-level rise, the shoreline rose up the depositional slope until it reached the top of the Kakwa Member, after which it migrated rapidly landwards at least 150 km (the distance from the Kakwa Edge to the furthest outcrop in the west) over a nearly horizontal surface. In almost all areas, wave-induced erosion planed off the upper few metres of the shoreface sandbody, removing any primary depositional topography together with most direct evidence of subaerial exposure (Fig. 15(c)). The depth to which the Kakwa was eroded may have been limited by the development of the early siderite cement. The resulting ravinement surface was essentially planar, and was veneered with a transgressive lag of wave-rippled conglomerates and carbonaceous sandstones. Although locally a few centimetres of mudstone overlie the ravinement surface, it appears that as a

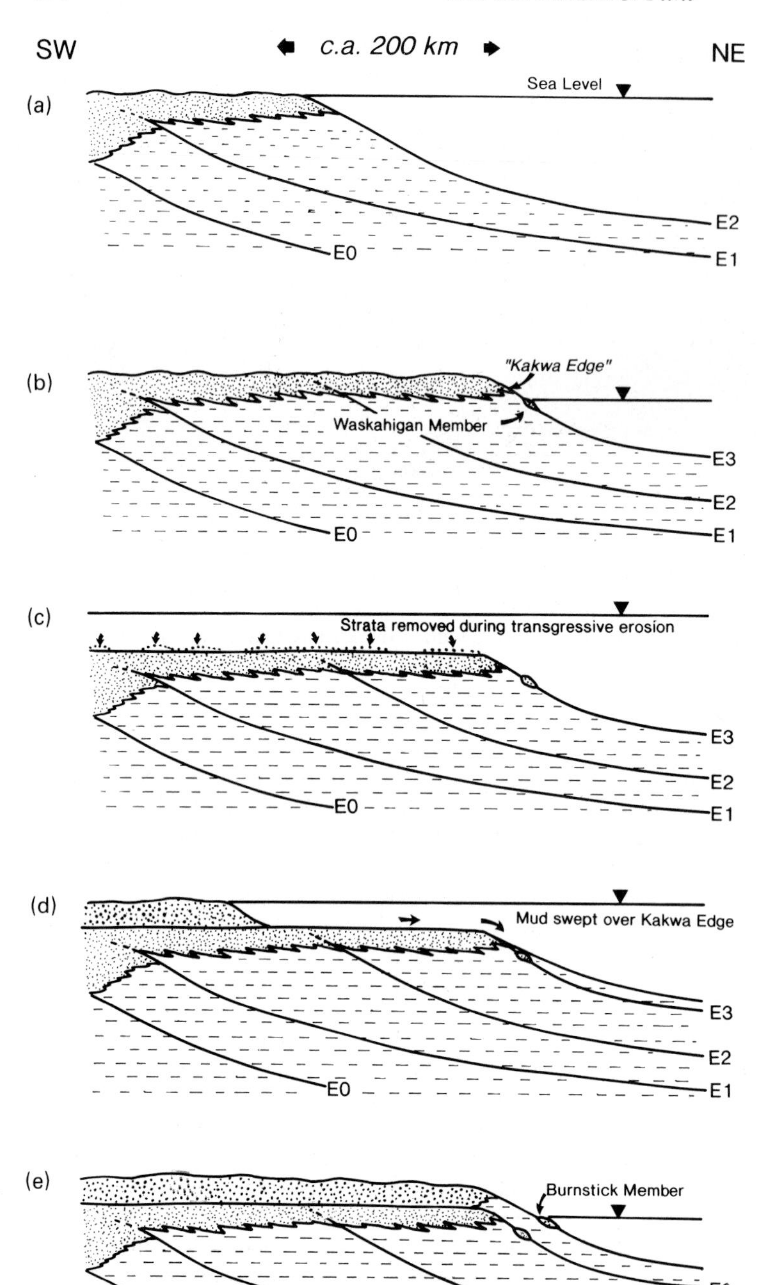

Fig. 15. Schematic reconstruction of the depositional history of the Kakwa Member, as a function of fourth-order relative sea-level fluctuations. (a) Progradation of lower portion of member is punctuated by short-lived regressions and transgressions, generating E0, E1 and E2 horizons. Shoreface sandstones interfinger with shelf muds at base of Kakwa Member. (b) Shoreline position drops below Kakwa Edge, generating linear shoreface deposits of Waskahigan Member. (c) During ensuing relative sea-level rise, shoreline position moves upslope to top of Kakwa Edge, then rapidly landward across the near-horizontal upper surface of lower Kakwa strandplain, generating a ravinement surface. (d) Ensuing phase of progradation occurs across a platform in water depths too shallow to permit preservation of shelf muds seaward of the prograding shoreface. (e) Shoreline position again drops below Kakwa Edge. Coarse-grained deposits of Burnstick Member were deposited in a lowstand shoreface.

whole, the water was too shallow to permit deposition (or preservation) of a significant thickness of offshore muds seaward of the next prograding shoreface. In consequence, shoreface sands of the upper part of the Kakwa Member prograded directly over the underlying shoreface sandstones (Fig. 15(d)).

Progradation of the upper portion of the Kakwa was terminated by a relative sea-level fall that again caused the shoreline to drop down below the Kakwa Edge, where coarse-grained lowstand shoreface deposits of the Burnstick Member accumulated (Fig. 15(e); Pattison, 1988; Plint, 1988). At this time, the upper surface of the Kakwa was locally incised by rivers. Lowstand deposition was terminated by relative sea-level rise that initiated aggradation of the non-marine Musreau Member to the landward of a transgressive (probably barrier) shoreline (Hart, 1990a).

DISCUSSION

Demarest and Kraft (1987) suggested that a ravinement surface, cut during marine transgression, may be recognized by truncation of stratification below the surface of erosion, the presence of intra- or extraformational clasts or a shell lag above and by the lack of interfingering of lithosomes across the surface. They also suggested that 'a dramatic change in lithofacies, palaeo water-depth and palaeoenvironment' should be seen across a ravinement surface. Although the E3 surface that divides the Kakwa Member clearly satisfies the first set of criteria, there is no obvious change in palaeoenvironment or water depth: shoreface sandstones are directly juxtaposed. The lack of obvious palaeoenvironmental break is due to the removal of (most) beach deposits capping the lower unit during the E3 transgression, and because water depths *following* transgression were too shallow to allow accumulation of offshore mudstone in advance of the prograding upper Kakwa shoreline.

The general subtlety of the facies changes across the E3 surface helps explain previous misinterpretation of its origin as representing rip-current channel scour. Use of the top of the Kakwa Member as a datum for correlation (Duke, 1985; Plint *et al.*, 1988) has obscured the internal geometry of the member. The work presented here (see also Hart, 1990a) indicates that thickness variations of the upper part of the member result from depositional topography at the top of the sandbody. Furthermore, examination of modern prograding shorelines (Dominguez *et al.*, 1987) indicates that the top of coastal sandbodies typically has a few metres relief representing beach ridges produced by high-frequency fluctuations in relative sea level, sediment supply or wave energy ('storminess').

Many stratigraphic models for shelf–shoreline systems predict (either implicitly or explicitly) lenticular geometries for shoreface sandbodies and interfingering relationships with inner-shelf deposits (McCubbin, 1982; Demarest & Kraft, 1987). Although such stratal geometries are probably the general rule, it is important to recognize that vertically stacked, near-tabular units may also be developed under certain circumstances.

CONCLUSIONS

Outcrop and subsurface analyses of shoreface sandstones of the Kakwa Member of the Cardium Formation have shown that this member comprises two genetically distinct units, separated by a near-horizontal erosion surface. The lower unit is dominated by very fine and fine-grained swaley cross-stratified sandstones which gradationally overlie heterolithic shelf deposits. In contrast, the upper sandstone unit consists mainly of medium-grained to conglomeratic sandstone, dominated by medium-scale cross-bedding. In most places, this upper sandbody rests directly on the underlying, fine-grained sandbody and the interface may show an irregular erosional relief with evidence of lithification prior to erosion. Thin, muddy carbonaceous or conglomeratic transgressive deposits are locally present above the erosion surface. The upper and lower sandbodies represent two distinct phases of progradation, and are separated by a near-horizontal ravinement surface which is contiguous with the E3 surface defined in offshore areas. The lower sandbody underwent very early cementation by siderite and quartz overgrowth; in contrast, the upper unit experienced very little cementation. About 150 km of transgressive shoreline displacement can be demonstrated across the E3 surface.

Criteria which might be useful in recognizing vertically stacked shoreface sandbodies in areas without good well control include: (i) the widespread presence of two (or more) shoreface sandstone units (each of which may be characterized by distinct granulometric properties and association of sedimentary structures), separated by sharp contacts; (ii) the presence of beach lamination or root traces below a contact separating two such shoreface sandbodies; (iii) consistent differences in early diagenetic history between successive sandstone units.

ACKNOWLEDGEMENTS

This paper is based on the senior author's doctoral dissertation, and was prepared for publication during tenure of a Visiting Fellowship at the Pacific Geoscience Centre. Our work was supported by grants to AGP from the National Sciences and Engineering Research Council, the Department of Energy, Mines and Resources and the University of Western Ontario. We thank Canadian Hunter Exploration Ltd, Esso Resources Canada Ltd, Home Oil Co. Ltd and Unocal Canada Ltd for generous technical assistance. Discussions with Bill Arnott, Bill Duke, Henry Posamentier and Rod Tillman and reviews by Doug Cant and David James have helped clarify our ideas and arguments although responsibility for facts and interpretations remains with us.

REFERENCES

Cant, D.J. (1988) Regional structure and development of the Peace River Arch, Alberta: a Paleozoic failed-rift system? *Bull. Can. Petrol. Geol.* **36**, 284–295.

Demarest, J.M. & Kraft, J.C. (1987) Stratigraphic record of Quaternary sea levels: implications for more ancient strata. In: *Sea-level Fluctuation and Coastal Evolution* (Eds Nummedal, D., Pilkey, O.H. & Howard, J.D.) Spec. Publ. Soc. Econ. Paleontol. Mineral., 41, 223–239.

Dominguez, J.M.L., Martin, L. & Bittencourt, A.C.S.P. (1987) Sea-level history and Quaternary evolution of river mouth-associated beach-ridge plains along the east-southeast Brazilian coast: a summary. In: *Sea-level Fluctuation and Coastal Evolution* (Eds Nummedal, D., Pilkey, O.H. & Howard, J.D.) Spec. Publ. Soc. Econ. Paleontol. Mineral., 41, 115–127.

Duke, W.L. (1985) Sedimentology of the Upper Cretaceous Cardium Formation in southern Alberta, Canada. Unpublished PhD thesis, McMaster University, Hamilton, Canada, 724pp.

Duke, W.L. (1990) Geostrophic circulation or shallow marine turbidity currents? The dilemma of paleoflow patterns in storm-influenced prograding shoreline systems. *J. Sediment. Petrol.* **60**, 870–883.

Gleddie, J. (1949) Upper Cretaceous in the Western Peace River Plains, Alberta. *Bull. Am. Assoc. Petrol. Geol.* **33**, 511–532.

Hart, B.S. (1990a) The sedimentology and stratigraphy of the Upper Cretaceous Cardium Formation in northwestern Alberta and adjacent British Columbia. Unpublished PhD thesis, University of Western Ontario, London, Canada, 505pp.

Hart, B.S. (1990b) Discussion on: swaley cross-stratification produced by unidirectional flows, Bencliff Grit (Upper Jurassic), Dorset, UK. *J. Geol. Soc. London* **147**, 396–400.

Hart, B.S. & Plint, A.G. (1989) Gravelly shoreface deposits: a comparison of modern and ancient facies sequences. *Sedimentology* **36**, 551–557.

Hart, B.S. & Plint, A.G. (1990) Upper Cretaceous warping and fault movement on the southern flank of the Peace River Arch, Alberta. *Bull. Can. Petrol. Geol.* **39A**, 190–195.

Hart, B.S. & Plint, A.G. (1991) Conglomeratic shoreface deposits from the Cretaceous Cardium Formation, Alberta, Canada. *Coastal Sediments '91*, Am. Soc. Civ. Eng., 949–959.

Hart, B.S., Longstaffe, F.J. & Plint, A.G. (1992) Evidence for relative sea level change from isotopic and elemental composition of siderite in the Cardium Formation, Rocky Mountain Foothills. *Bull. Can. Petrol. Geol.* **40**, 52–59.

Hunter, R.E., Clifton, H.E. & Phillips, R.L. (1979) Depositional processes, sedimentary structures and predicted vertical sequences in barred nearshore systems, southern Oregon coast. *J. Sediment. Petrol.* **49**, 711–726.

McCubbin, D.G. (1982) Barrier-island and strand plain facies. In: *Sandstone Depositional Environments* (Eds Scholle, P.A. & Spearing, D.). Mem. Am. Assoc. Petrol. Geol. 31, 247–279.

Nottvedt, A. & Kreisa, R.D. (1987) Model for the combined-flow origin of hummocky cross-stratification. *Geology* **15**, 357–361.

Pattison, S.A.J. (1988) Transgressive, incised shoreface deposits of the Burnstick Member (Cardium 'B' sandstone) at Caroline, Crossfield, Garrington, and Lochend; Cretaceous Cardium Formation, Western Interior Seaway, Alberta, Canada. In: *Sequences, Stratigraphy, Sedimentology: Surface and Subsurface* (Eds James, D.P. & Leckie, D.A.) Mem. Can. Soc. Petrol. Geol. 15, 155–166.

Plint, A.G. (1988) Sharp-based shoreface sequences and 'offshore bars' in the Cardium Formation: their relationship to relative sea-level changes. In: *Sea Level Changes: An Integrated Approach* (Eds. Wilgus, C.K., Hastings, B.S., Kendall, C.G.St.C., Posamentier, H.W., Ross, C.A. & Van Wagoner, J.C.). Spec. Publ. Soc. Econ. Paleontol. Mineral., 42, 357–370.

Plint, A.G. (1990) An allostratigraphic correlation of the Muskiki and Marshybank formations (Coniacian–Santonian) in the foothills and subsurface of the Alberta Basin. *Bull. Can. Petrol. Geol.* **38**, 288–306.

Plint, A.G. & Walker, R.G. (1987) Cardium Formation 8. Facies and environments of the Cardium shoreline and coastal plain in the Kakwa Field and adjacent areas, northwestern Alberta. *Bull. Can. Petrol. Geol.* **35**, 48–64.

Plint, A.G., Walker, R.G. & Bergman, K.M. (1986) Cardium Formation 6. Stratigraphic framework of the Cardium in subsurface. *Bull. Can. Petrol. Geol.* **34**, 213–225.

Plint, A.G., Walker, R.G. & Duke, W.L. (1988) An outcrop to subsurface correlation of the Cardium Formation in Alberta. In: *Sequences, Stratigraphy, Sedimentology: Surface and Subsurface* (Eds James, D.P. & Leckie, D.A.) Mem. Can. Soc. Petrol. Geol. 15, 167–184.

Rahmani, R.A. & Smith, D.G. (1988) The Cadotte Member of northwestern Alberta: a high-energy barred shoreline.

In: *Sequences, Stratigraphy, Sedimentology: Surface and Subsurface* (Eds James, D.P. & Leckie, D.A.) Mem. Can. Soc. Petrol. Geol. 15, 431–438.

SHORT, A.D. (1984) Beach and nearshore facies: southeast Australia. *Mar. Geol.* **60**, 261–282.

VAN WAGONER, J.C., POSAMENTIER, H.W., MITCHUM, R.M., JR., *et al.* (1988) An overview of the fundamentals of sequence stratigraphy and key definitions. In: *Sea Level Changes: An Integrated Approach* (Eds Wilgus, C.K., Hastings, B.S., Kendall, C.G.St.C., Posamentier, H.W., Ross, C.A. & Van Wagoner, J.C.) Spec. Publ. Soc. Econ. Paleontol. Mineral., 42, 39–45.

WALKER, R.G. (1983) Cardium Formation 3. Sedimentology and stratigraphy in the Caroline–Garrington area. *Bull. Can. Petrol. Geol.* **31**, 213–230.

WRIGHT, L.D. & SHORT, A.D. (1984) Morphodynamic variability of surf zones and beaches: a synthesis. *Mar. Geol.* **56**, 93–118.

Table 1. Detrital composition of Kakwa sandstones[1]

		Quartz	Rock fragments[2]	Feldspar	Carbonate
Lower Kakwa	mean	48.4	48.5	1.4	1.3
$n = 31$	st. dev.	19.5	19.7	1.4	2.5
Upper Kakwa	mean	43.8	55.2	0.8	0.2
$n = 20$	st. dev.	18.8	19.1	0.7	0.6

[1] All values expressed in per cent.
[2] Dominant components: chert, low-grade argillites and shales.

Spec. Publs Int. Ass. Sediment. (1993) **18**, 469–485

Sequence-stratigraphic analysis of Viking Formation lowstand beach deposits at Joarcam Field, Alberta, Canada

H.W. POSAMENTIER* *and* C.J. CHAMBERLAIN†

**ARCO Oil and Gas Company, 2300 West Plano Parkway, Plano, TX 75075, USA; and*
†Esso Resources Canada Ltd, 237 4th Avenue SW, Calgary, Alberta T2P 0H6, Canada

ABSTRACT

Sequence-stratigraphic principles applied to core and wireline-log information have aided in the development of a geologic model for the Viking Formation at Joarcam Field, Alberta, Canada. Joarcam is a linear northwest/southeast trending oil field *c.* 4 km wide and 40 km long. Three units within the Viking Formation contain hydrocarbons: the 'upper', 'main' and 'third' sands. Of these, the main sand is the most prolific producer. Previously, the main sand had been described as a prograding beach shaling out to the northeast. This study suggests that the main sand can be split into two units: a lowstand systems tract and a transgressive systems tract, *each with unique reservoir attributes.*

Deposition of the areally restricted regressive lowstand beach occurred during a relative sea-level stillstand following an interval of relative-sea-level fall. This unit consists of medium to coarse-grained sandstone deposited in a lower- to upper-shoreface environment and contains minor but potentially significant local, vertical permeability barriers. These permeability barriers are associated with shingling/progradation of the lowstand shoreline and are oriented parallel to the slope of the shoreface. Beach progradation ended when a relative sea-level rise caused shoreline transgression. During transgression the upper part of the lowstand beach was sheared off resulting in a ravinement surface capping the shoreface deposits. Sediments cannibalized from the upper part of the lowstand systems tract were redeposited as storm deposits seaward of as well as above the lowstand shoreface. These deposits, interpreted as comprising the transgressive systems tract, consist of alternating clean sands and muds deposited in an offshore environment. In this setting, little sediment mixing by bioturbation occurs, preserving both the reservoir properties of the sandstone beds (up to 1.5 m thick) as well as the continuity of the impermeable shale interbeds.

INTRODUCTION

The stratigraphy and sedimentology of the Upper Albian Viking Formation, Alberta, Canada (Fig. 1) have been the subject of numerous papers (e.g., Jardine, 1953; Hunt, 1954; Evans, 1970; Amajor & Lerbekmo, 1980; Beaumont, 1984; Amajor, 1986; Hein *et al.*, 1986; Leckie, 1986; Downing & Walker, 1988; Power, 1988; Reinson *et al.*, 1988). A variety of depositional environments have been interpreted to occur within this formation in different areas including tidal (Evans, 1970; Amajor, 1986; Leckie, 1986), offshore storm-dominated shelf (Koldijk, 1976), estuarine and fluvial (Reinson *et al.*, 1988) and shoreface (Downing & Walker, 1988; Posamentier & Chamberlain, 1989). Recently the recognition of the effects of sea-level changes on stratigraphic successions (Plint *et al.*, 1987; Posamentier & Vail, 1988; Posamentier *et al.*, 1988) has led to a reinterpretation of Viking stratigraphy (Beaumont, 1984; Posamentier & Chamberlain, 1989). This has resulted in the interpretation of lowstand shorelines in lieu of offshore bars and incised valleys (Allen & Posamentier, 1991) in lieu of tidal inlets.

The Viking Formation is a prolific producer of oil and gas in the western Canadian Sedimentary Basin (Fig. 2). To date, this formation has produced over 39.9×10^6 m^3 (251 mmbbls) of oil and over 139×10^9 m^3 (4.9 tcf) of gas in Alberta (Energy Resources Conservation Board, 1988). The Joarcam Field

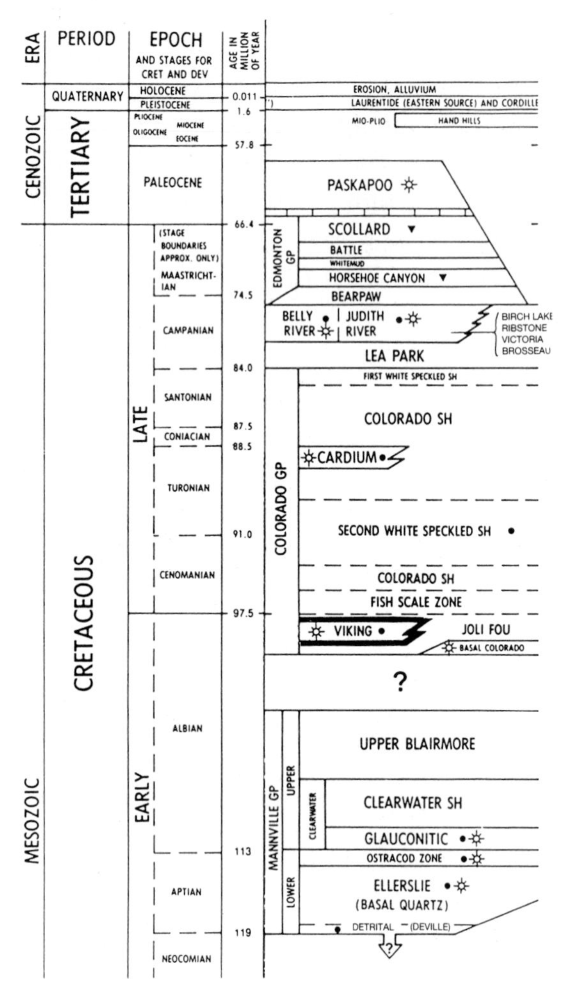

Fig. 1. Stratigraphic chart for Cretaceous of western Canada. (Compiled by AGAT Laboratories, Calgary, Canada.)

(Fig. 3), located just southeast of Edmonton, Alberta, Canada, produces both oil and gas from the Viking Formation. The play is a stratigraphic/structural trap within a northeast prograding shoreline (Posamentier & Chamberlain, 1989). The field was discovered in 1949 while drilling for the deeper Nisku and Leduc Formations at 11-22-50-22W4 and is estimated to contain 18.6×10^6 m^3 (116.7×10^6 bbls) recoverable reserves of which 16.2×10^6 m^3) (102.1×10^6 bbls) or 90% has been produced through 1987 (Energy Resources Conservation Board, 1988). Over 850 wells at quarter-section spacing have been drilled at this field.

Typical reservoir permeabilities range from 800 to 1500 millidarcies (mD) with porosities ranging from 16 to 24%. The thickness of the producing unit ranges from 7 to 10 m (Fig. 4). The reservoir lithology is a medium- to very coarse-grained, clean quartzose sandstone. Typical initial production for many of the wells has ranged from 7 to 14 m^3 per day of 37 to 38° API oil. The recovery factor approaches 47% where influenced by waterflood.

This study incorporates analyses of core and wireline well-log data. Over 75 cores were examined both within and outside Joarcam Field. Wireline logs from all the wells in and immediately surrounding the field were examined and both dip (oriented southwest to northeast) and strike (oriented northwest to southeast) stratigraphic cross-sections were constructed at 5-mile spacing.

FACIES ASSOCIATIONS

Three units within the Viking Formation, informally referred to as the 'upper', 'main' and 'third' sands, contain hydrocarbons. Figure 4 illustrates the wireline log response of the Viking Formation at 12-8-49-21W4. The principal producing interval is the main sand, which is the focus of this paper. Descriptions of the facies associations within the entire Viking Formation are presented in Power (1988). Members 'B' and 'C' of Power (1988) comprise the main sand in the Joarcam Field described here.

Member 'B'

This member comprises the bulk of the 'main sand' and is composed of a coarsening- and cleaning-upward quartzose sandstone (Power, 1988). It is heavily bioturbated towards the base and only sparsely bioturbated towards the top. The lowest part of this member consists of fine-grained marine mudstone that is heavily bioturbated so that no primary sedimentary structures are recognizable. Typically, two bentonitic beds approximately 10 cm thick occur within the marine-mudstone facies primarily in the northeast part of the study area. The fine-grained marine mudstone grades upward to a bioturbated muddy very fine-grained sandstone, which, in turn, is overlain by intercalated bioturbated muddy very fine-grained sandstone and medium- to coarse-grained sparsely bioturbated sandstone. The sandstone is character-

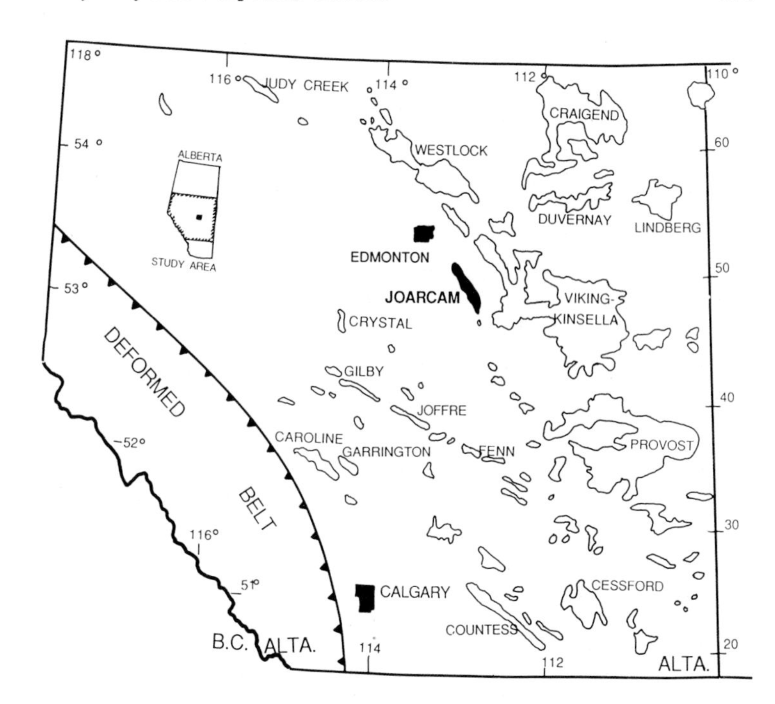

Fig. 2. Location of Joarcam Field as well as other oil and gas producing fields in south-central Alberta, Canada.

ized by high-angle trough (?) cross-bedding. Towards the top of Member 'B' this succession grades into massively bedded and sometimes trough cross-bedded coarse-grained sandstone with only rare evidence of burrowing. Porosities in the upper parts of Member 'B' range from 16–24% and permeabilities range from 800 to 1500 mD. The top of this succession is marked by the abrupt occurrence of marine mudstones of Member 'C'.

This succession is interpreted as the deposits of a high-energy shoreface which prograded northeastward (Power, 1988). This succession is similar to those described elsewhere in both modern as well as ancient settings (e.g. Clifton *et al.*, 1971; McCubbin, 1982).

Member 'C'

Member 'C' sharply overlies Member 'B' and is composed of alternating fine-grained marine siltstones and mudstones and medium- to coarse-grained quartzose sandstones. Porosities in the sandstone beds are as high as 24% and permeabilities up to 1500 millidarcies are common. This member is characterized by a low-diversity and low-abundance trace fossil assemblage that includes *Planolites* and *Zoophycus*. The siltstones are characterized by occasional wave-ripple laminations and the sandstones up to 1.5 m thick are characterized by high-angle cross-beds. The overall trend seems to fine upward in the vicinity of the principal part of the field (e.g. 12-8-49-21W4 in Fig. 4) and coarsen upward northeastward of the field (e.g. 14-11-48-20W4 in Fig. 5).

Member 'C' is interpreted as having been deposited offshore predominantly by fair-weather and storm processes. The offshore setting is suggested by the abundance of low-energy mudstone and the occurrence of *Zoophycus*, typical of a low-energy shelf environment, that characterizes this member. The medium- to coarse-grained sandstone beds that are interbedded with the mudstones are inferred to have been transported and deposited a short distance away from the shoreline (<10 km) by storm processes. The sediment is interpreted to have been derived from the erosion of the immediately preceding shoreline (i.e. the upper part of Member 'B') during transgression. These underlying Member 'B' sandstones are of similar grain size and lithology to those that occur as discrete sandstone beds in Member 'C'.

Member 'C' is characterized by a fining-upward trend in certain areas and a coarsening-upward in others. The fining-upward trend occurs only where Member 'C' directly overlies the preceding shoreline and suggests an environment characterized by a

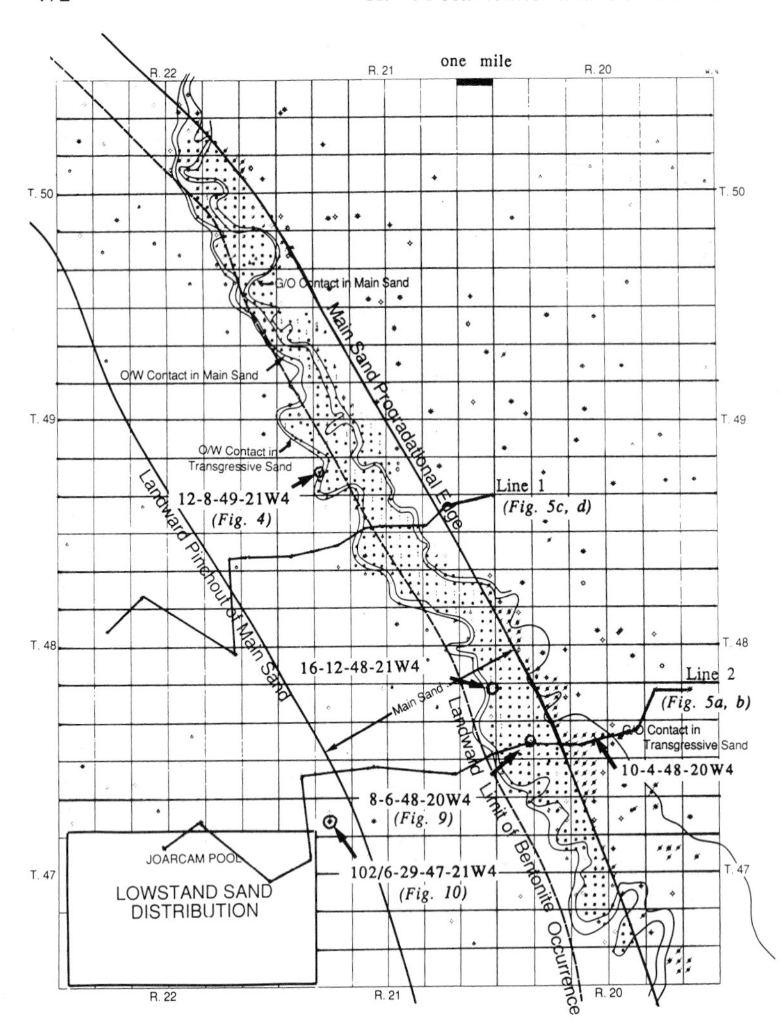

Fig. 3. Joarcam Field, Alberta, Canada. The updip and downdip limits of the sandy facies of the lowstand systems tract are shown — depositional strike is north–northwest and depositional dip is east–northeast. Structural dip is to the southwest. The landward limit of two marker bentonite beds is shown.

gradual increase of water depth and a progressively more distal location through time as the shoreline continued to migrate landward.

The coarsening-upward trend occurs only seaward of the preceding shoreline. At that location, the coarsening-upward trend overlies the distal offshore marine equivalent of the underlying shoreline. Coarse-grained storm-associated sediments occur only near the top of Member 'C'. A narrow transition zone between coarsening- and fining-upward trends occurs within 2 km of the seaward limit of the prograding shoreface. Seaward of this transition zone, the coarse-grained sandstones are interpreted not to have been deposited until aggradation with finer grained sediments first filled the 'hole' or low area just seaward of the prograding shoreface (Figs 5 & 6(b)). Gradually, as this topographically low area filled, storm deposits were able to be carried successively further out across this now-featureless surface. This resulted in an apparent coarsening-upward trend towards the top of this member.

GEOLOGIC MODEL

In the past, the producing interval (i.e. the 'main sand') at Joarcam has been described as an offshore tidal bar or ridge (Amajor, 1986; Amajor and Lerbekmo, 1980) or as a regressive prograding beach complex that gradually shales out in a north-easterly direction (Power, 1988) (Fig. 6(a)). The

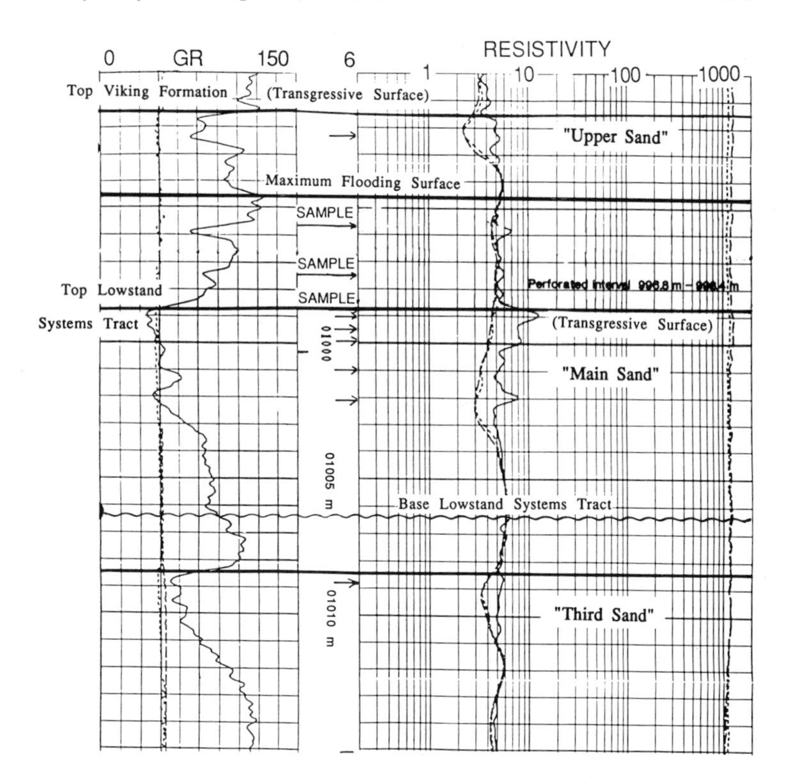

Fig. 4. Well 12-8-49-21W4 (location shown in Fig. 3). *Lithostratigraphic* classification of sandstone units into third sand, main sand and upper sand as well as *chronostratigraphic* classification within the main sand into lowstand systems tract and transgressive systems tract are shown. Member 'B' of Power (1988) occurs between the base and top lowstand systems tract; member 'C' of Power (1988) occurs between the top lowstand systems tract and the maximum flooding surface.

results of this study concur with those of Power (1988) in so far as the depositional environment is concerned but suggest that the stratigraphy is somewhat more complex. *Two sub-units* of the 'main sand' have been identified (Fig. 6(b)) and mapped. They are referred to as the lowstand systems tract (LST) and transgressive systems tract (TST) in the sense of Posamentier *et al.* (1988) and correspond to some extent with Members 'B' and 'C', respectively of Power (1988). The sands with the best porosity and permeability occur within the LST although some good sands of limited thickness do occur within the TST. The LST occurs as a northwest/southeast trend approximately 10 km wide (Fig. 3) and at least 60 km long. The overlying TST is thin (< 2 m) where it overlies the shoreface depositional system and thickens abruptly to the northeast just beyond the seaward-most position of the last shoreline of the LST (Fig. 6(b)).

The evolution of the LST and TST sands at Joarcam is interpreted to be associated with a relative sea-level fall followed by a stillstand and subsequently a relative sea-level rise. A seaward shift of depocentre responsible for the basin-isolated position of the Joarcam shoreline occurs initially in response to a relative sea-level fall. An apparent basinward jump in the position of the shoreline has occurred between the underlying progradational succession and the shoreline at Joarcam. This apparent basinward jump seems to be the result of rapid shoreline regression with little or no preserved record of deposition between the time of highstand progradation and the time of lowstand (or stillstand) progradation. This process has been referred to as a *forced regression* (Posamentier *et al.*, 1990, 1992b)*.

The LST shoreline is located just seaward of a subjacent highstand progradational unit. The seaward limit of the shoreline of this highstand progradational unit occurs between wells 5-19-47-21W4 and 10-31-47-21W4 on Line 2 (Figs 5(a) & (b)) and wells 6-28-48-22W4 and 16-14-48-22W4 on Line 1

*Plint (1988), Posamentier and Vail (1988) and Posamentier *et al.* (1990, 1992a,b) have described this process as a seaward shift of the shoreline occurring in response to relative sea-level fall and independent of sediment flux variations that may be associated with 'normal' regressions. Plint (1988) and Posamentier and Vail (1988) suggested that the stratigraphic expression of the forced regression may be a sharp-based shoreface succession.

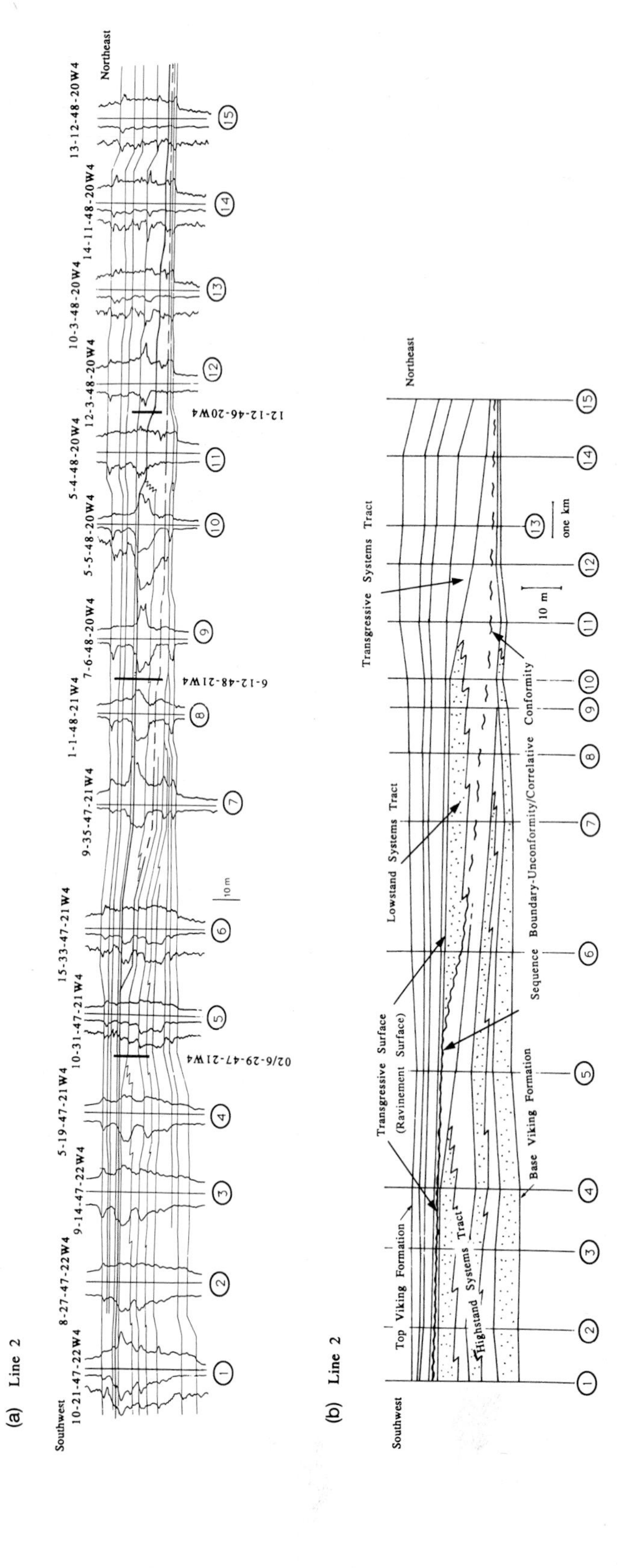
(a) Line 2
Southwest
Northeast
10-21-47-22W4
8-27-47-22W4
9-14-47-22W4
5-19-47-21W4
10-31-47-21W4
15-33-47-21W4
9-35-47-21W4
1-1-48-21W4
7-6-48-20W4
5-5-48-20W4
5-4-48-20W4
12-3-48-20W4
10-3-48-20W4
14-11-48-20W4
13-12-48-20W4
02/6-29-47-21W4
6-12-48-21W4
12-12-46-20W4
10 m
(b) Line 2
Southwest
Northeast
Top Viking Formation
Base Viking Formation
Transgressive Surface
(Ravinement Surface)
Highstand Systems Tract
Lowstand Systems Tract
Transgressive Systems Tract
Sequence Boundary-Unconformity/Correlative Conformity
10 m
one km

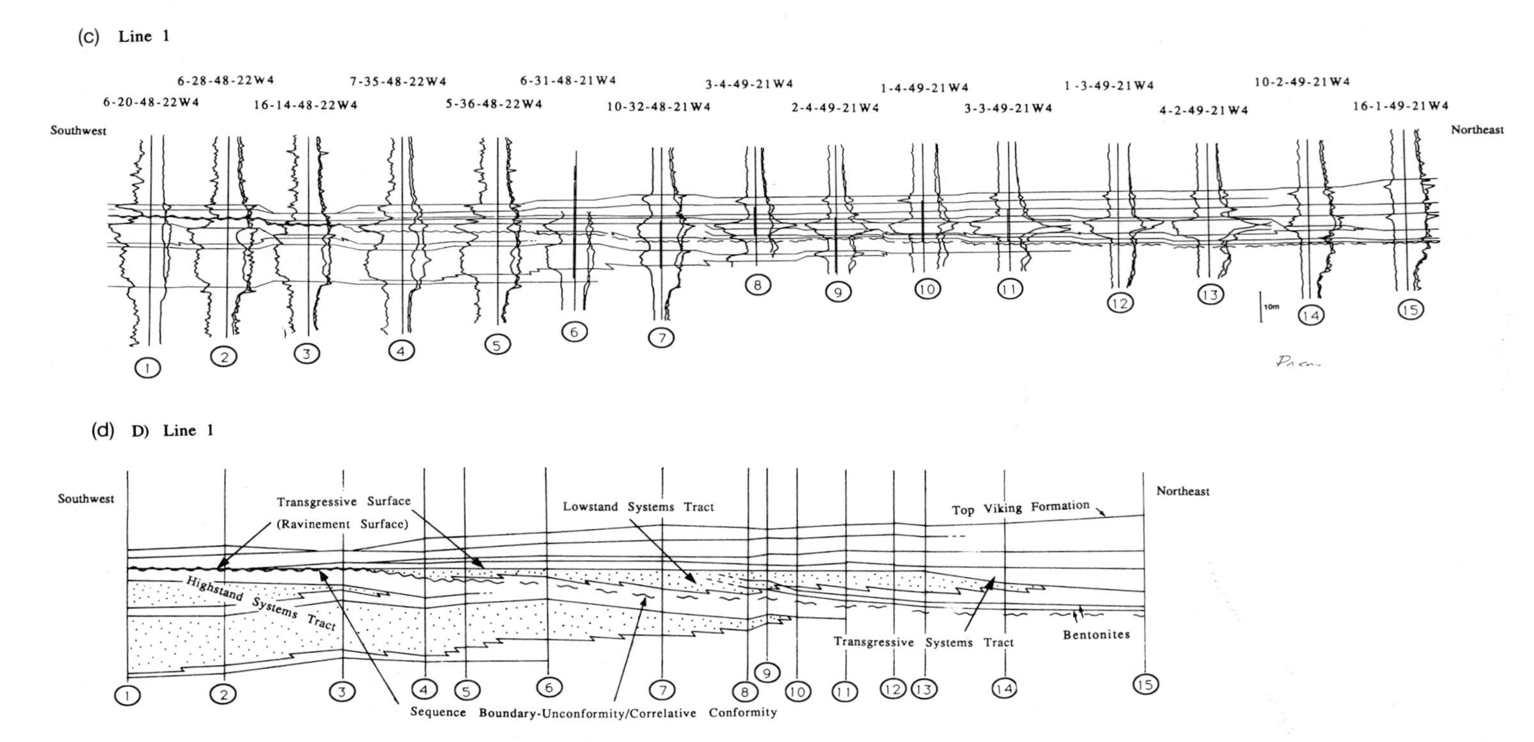

Fig. 5. Dip-oriented well-log cross-sections across Joarcam Field. Locations shown in Fig. 3. In the proximal part of the LST, the sequence boundary at the base of the shoreface succession is characterized by sandy beach deposits sharply overlying offshore muds. In the central and distal parts of the LST, the sequence boundary becomes a correlative conformity and is difficult to identify. (a) Cross-section A–A′ (see Fig. 3 for location). (b) Stickplot of section A–A′ showing true dip spacing (i.e. projected onto straight dip line). (c) Cross-section B–B′ and (d) stickplot of section B–B′ showing true dip spacing (i.e. projected onto straight dip line). Note the bentonitic beds climbing abruptly up into the upper shoreface on section B–B′.

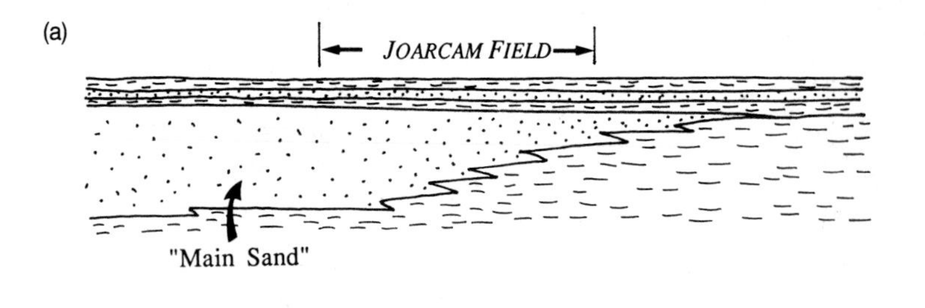

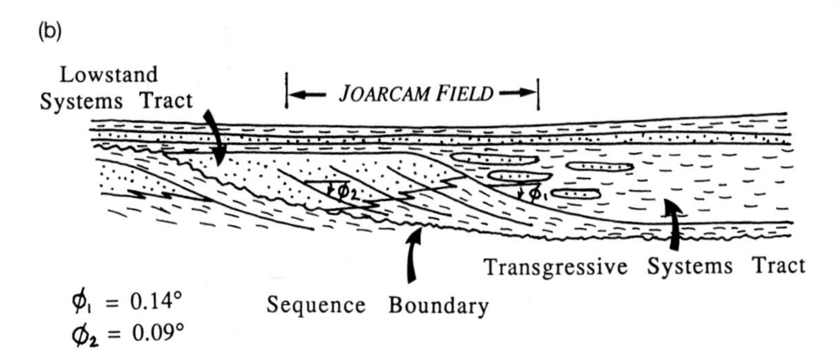

Fig. 6. Schematic dip sections across Joarcam Field.
(a) Interpretation of the principally producing 'main sand' as a northeastwardly prograding shoreline resulting in deposition of a sheet sand across the area.
(b) Interpretation of the 'main sand' as two discrete depositional units: the lowstand systems tract and transgressive systems tract. The lowstand systems tract comprises coarsening-upward shoreface deposits isolated from and seaward of the underlying highstand shoreline and overlying the sequence boundary. The internal physical stratigraphy of this unit is characterized by a shingled geometry. The overlying transgressive systems tract comprises alternating offshore shales and storm-deposited sandstones. The physical stratigraphy of this unit is characterized by essentially horizontal bedding.

(Figs 5(c) & (d)). The pinchout or onlap of the LST shoreline occurs between 10-31-47-21W4 and 15-33-47-21W4 (Figs 5(a) & (b)). This pinchout geometry was not recognized by Power (1988), who instead correlated the sands at Joarcam southwestward with the sands of the underlying highstand systems tract (HST). At the LST pinchout landward, the underlying progradational unit, which maintains a relatively uniform thickness of 10 to 14 m east–southeastward of Joarcam, thins dramatically from 9 m to 1 m over a distance of 5.2 km, yielding a depositional slope of approximately 0.09° (Figs 5(a) & (b)). It is suggested that it is the steep slope of this underlying unit that sets up or localizes the location of the LST lowstand beach. The mapped landward pinchout of the LST beach has a linear pattern (Fig. 3).

During the ensuing relative sea-level stillstand, progradation of the shoreline occurred developing a 'lowstand strandplain' within the LST (Fig. 7(a)). A sequence boundary is interpreted at the base of this lowstand beach. Based on examination of core and well logs, this surface is characterized by a sharp-based shoreface succession or unconformity (Plint, 1988) only in the most proximal areas of the LST (e.g. at 15-33-47-21W4 in Fig. 5), and appears to be an indistinct or gradational contact where the sequence boundary grades from an unconformity to a correlative conformity. The facies association within this unit is interpreted as a prograding shoreface succession.

Within the LST, thin (<10 cm) bentonitic beds occur within the lowest-energy lower shoreface and offshore facies. These beds were deposited nearly horizontally in the offshore environment but climb in the section within the shoreface setting (this climbing geometry occurs between wells 1-4-49-21W4 and 3-4-49-21W4 on Figs 5(c) & (d)). Due to poor preservation potential, they generally do not occur within the relatively high-energy upper-shoreface environment. This location represents the point at which sufficiently high energy precludes preservation of these deposits. The landward-most occurrence (i.e. west–southwestward) of these bentonites has been mapped and is observed to be linear (Fig. 3). This pattern of disappearance suggests a *linear* shoreline pattern at the time of deposition, further supporting an interpretation of a wave-dominated shelf.

Progradation of the LST shoreface ends (i.e. shales out) abruptly (e.g. from 4-2-49-21W4 to 10-2-49-21W4 (Figs 5(c) & (d))), over a distance of approximately 1.3 km, in response to a subsequent relative sea-level rise. Basinward of this shale-out,

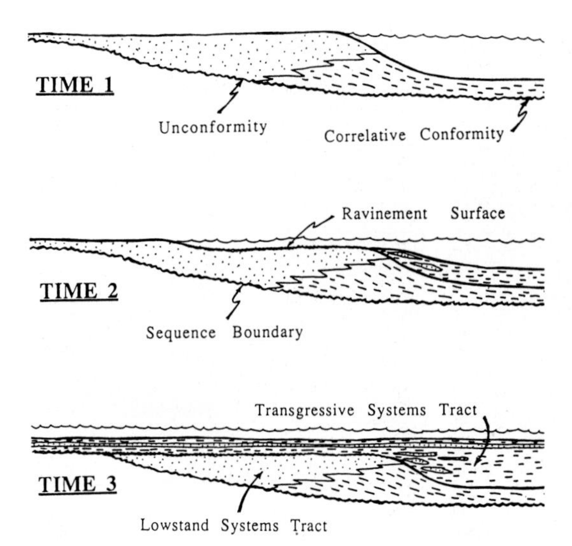

Fig. 7. Schematic evolution of the lowstand and transgressive systems tracts. (a) Time 1, stillstand of relative sea level at a lowstand position. Note the occurrence of bentonites within the offshore shale facies. The bentonites end where they extend into the higher energy of the mid to upper shoreface. These sloping marker beds are evidence supporting the interpretation of a shingling internal strata/geometry. (b) Time 2, a snapshot in time during the transgression across the lowstand beach. The sediments derived from cannibalization of the lowstand beach are redeposited as storm deposits interbedded with offshore marine shales. (c) Time 3, transgression is complete and the transgressive systems tract composed of horizontally bedded storm deposits interbedded with offshore marine shales overlies the lowstand systems tract composed of shingled coarsening- and cleaning-upward lower to upper shoreface deposits.

time-equivalent facies to the LST shoreface consist of bioturbated silty shale with few primary sedimentary structures preserved (Member 'B') (Fig. 8). A sharp contact with the overlying relatively unbioturbated TST offshore facies characterizes the upper bounding surface of the LST in this distal setting (Fig. 8).

LST beach progradation ended when relative sea level rose and caused shoreline transgression (Figs 7(b) & (c)). The processes associated with shoreface erosion during transgression are inferred to have resulted in shearing of the upper 5 to 10 m of the original LST and produced a ravinement surface at the top of the lowstand shoreface*. Sediments cannibalized from the upper part of the LST were redeposited as storm deposits by combined wave action and storm-induced currents seaward (northeastward) of, as well as overlying, the lowstand shoreface. The grain size of the storm-deposited sediments decreases markedly in a basinward direction; coarse-grained storm-deposited beds persist little more than 4–6 km offshore. These deposits are interpreted to comprise the TST.

The transgressive deposits consist of alternating sandstones and mudstones (facies association Member 'C') deposited as sea level is rising. Because of the relative absence of marine fauna, little sediment mixing occurs at the sea floor, leaving sands and muds well segregated. Thus, the good reservoir properties of the sandstone beds (up to 1.5 m thick) are preserved, though the vertical permeability barriers created by mudstone deposits also remain intact. Preserved primary sedimentary structures dominated by wave-oscillation ripples are common in the sandstone beds.

As shoreline transgression shifted the sediment source progressively further landward, storm-deposited sands became less common towards the upper part of this systems tract where it overlies the beach depositional system of the LST. Figure 4 illustrates the apparent *fining-upward* trend of the TST due to the thinning-upward nature of the storm-deposited beds. This trend characterizes the TST only in the area where it directly overlies the beach facies of the LST.

Seaward of the depositional limit of the LST beach, an apparent *coarsening-upward* log response within the TST is observed (see logs for wells 5-4, 10-3, 14-11, and 13-12-48-20W4 on Figs 5(a) & (b)). This is inferred to be associated with the filling of a bathymetric low situated just seaward of the LST beach as discussed above. This low is apparent when stratigraphic cross-sections are flattened on either the downlap surface at the base of the LST or the flooding surface at the top of the TST.

Initially, at the onset of transgression, storm beds are restricted to a relatively narrow zone just seaward of the LST shoreline and are not transported into the deeper waters just distal of this beach. In this quieter deeper water setting (*c.* 5–7 m deeper), lower wave energy at the sea floor would not have been capable of transporting storm deposits far from the LST beach. As this deeper setting gradually filled in up to the level of the LST beach with mud/silt-prone sediments, coarser grained storm-deposited beds interbedded with offshore muds

*Demarest and Kraft (1988) document *c.* 10 m of erosion associated with Holocene transgression across the Atlantic shelf, offshore US.

Fig. 8. Core from well 12-12-46-20W4 (located along strike southeast of 12-3-48-20W4, Fig. 5(a)). This well is situated northeast or seaward of the field. The upper part of the lowstand systems tract and the lower part of the transgressive systems tract are shown. The lowstand systems tract is characterized by heavily bioturbated silty shales and is abruptly overlain by comparatively poorly bioturbated silty shales. Detail of the lowstand/transgressive systems tract contact is shown in (b).

would have been increasingly more common, and consequently, an apparent coarsening-upward wireline-log response would result.

The contact between the LST and TST is well expressed in core. Basinward of the lowstand beach, this contact is characterized by an abrupt change from bioturbated offshore silty shale with few preserved primary sedimentary structures to relatively unburrowed interbedded sands and muds with well-preserved primary sedimentary structures (Fig. 8). This surface represents the final depositional surface associated with the prograding LST beach. It is this surface that is subsequently onlapped by the horizontally bedded alternating sands and muds of the TST. Directly overlying the LST beach facies, the contact between the two systems tracts is a ravinement surface characterized by a change from massive and large-scale cross-bedded sandstone, below, to relatively unburrowed, alternating storm-deposited sands and muds above (Fig. 9). This same surface can be correlated landward where it is expressed as an E/T surface (erosion/transgression of Plint *et al.*, 1987) or a flooding surface merged with a sequence boundary or unconformity

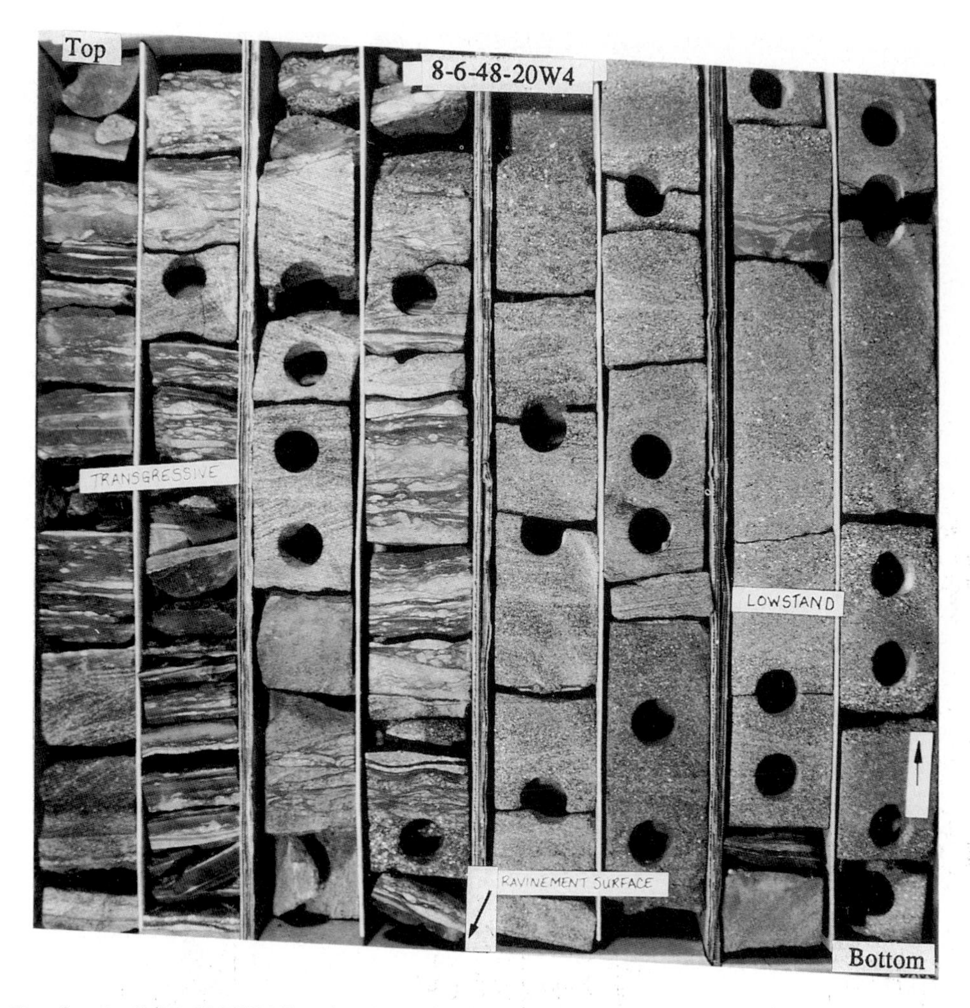

Fig. 9. Core from well 8-6-48-20W4 (location shown in Figs 3 & 5(a)). This well is situated in the central part of the field. The top of the lowstand systems tract and the base of the transgressive systems tract are shown. The two units are separated by the transgressive surface (i.e. a ravinement surface). Note the high-angle cross-bedded sand-rich storm deposits interbedded with mud-rich offshore shale beds within the transgressive systems tract.

(Fig. 7(c)). This contact separates the underlying pre-Joarcam HST from the overlying TST. Consequently, the Joarcam LST is absent here. Any evidence for subaerial exposure associated with deposition of the LST has been removed by erosion associated with transgression across the LST beach. The ravinement surface is overlain by a thin (<4 cm) grit layer (transgressive lag) (Fig. 10). It is this surface that eventually gives rise to the LST beach at Joarcam, approximately 1.5 km to the northeast. The overlying TST in this landward location is, again, characterized by relatively unburrowed alternating sandstones and mudstones. The underlying pre-Joarcam HST is characterized by bioturbated silty mudstones with poor preservation of primary sedimentary structures.

STRATAL ARCHITECTURE OF PRODUCING SANDS AT JOARCAM

The stratal architecture and subsequent hydrodynamics are significantly different in the two units (i.e. the LST and TST) that comprise the principal producing sand at Joarcam Field. The LST consists of a prograding shoreface succession with an inter-

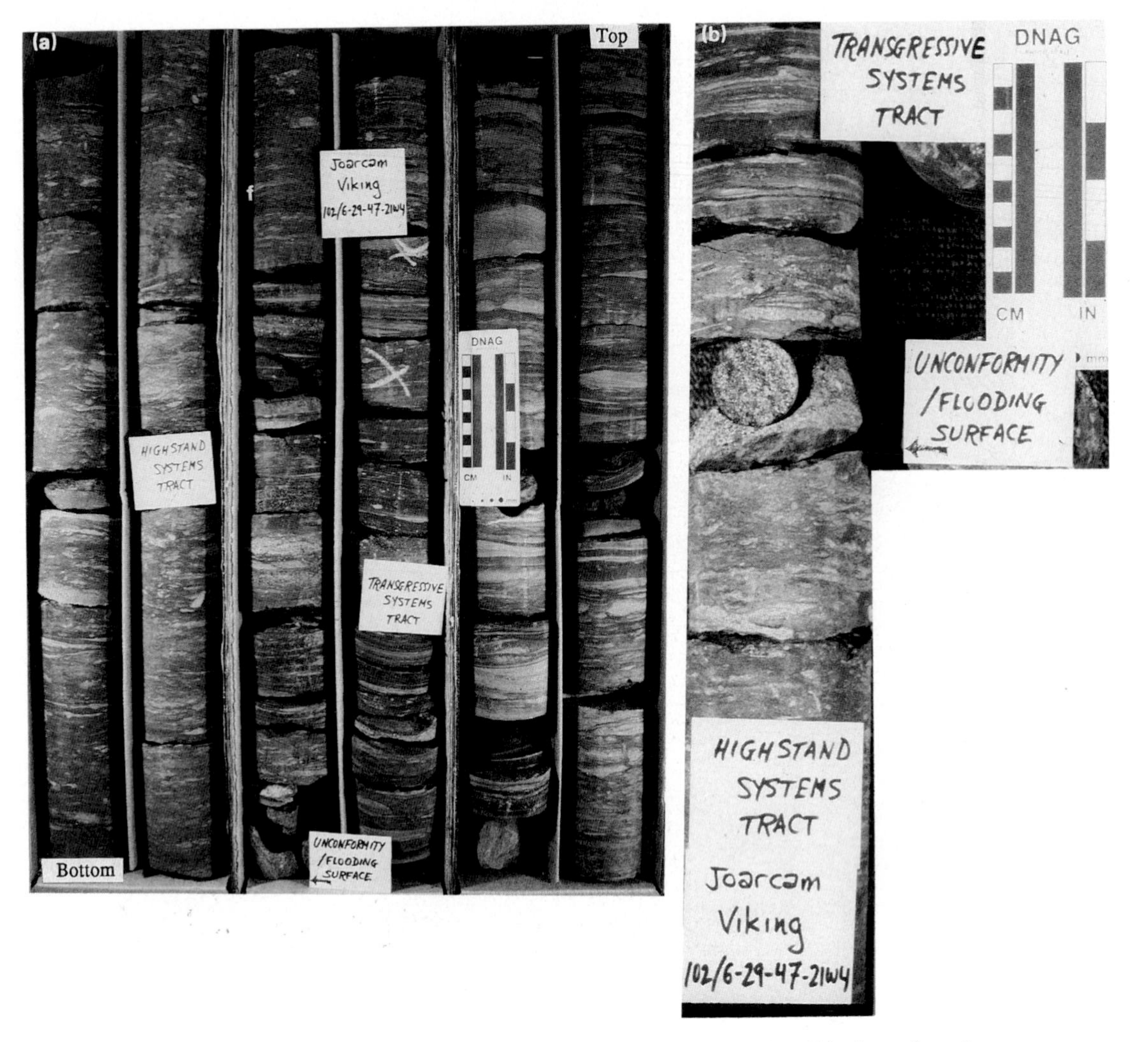

Fig. 10. Core from well 102/6-29-47-21W4 (location shown in Figs 3 & 5(a)). This well is situated southwest or landward of the field. The lowstand systems tract has pinched out northeast of this location and is therefore absent here. The transgressive systems tract directly overlies the preceding highstand systems tract. The contact between these two units is a ravinement or E/T surface (in the sense of Plint *et al.*, 1987) overlain by a thin grit layer. Detail of this contact is shown in (b).

nal *shingled* geometry made up of beds dipping at approximately 0.09° to 0.14° to the northeast. Vertical permeability barriers exist and are most common in the lower parts of this succession. In contrast, extensive vertical permeability barriers exist throughout the TST. This systems tract consists of predominantly *horizontally bedded* sediments deposited in an offshore setting. The occurrence of minimal bioturbation within this unit results in preservation of discrete shale as well as sandstone beds. The shale beds act as local permeability barriers and the sandstone beds as conduits or migration pathways. The reason why these shale beds are only partially effective as barriers to fluid migration lies in the nature of the depositional process associated with the intercalated sandstone beds. These deposits occurred as a result of storm events that transported beach-derived sediments onto the inner shelf. Associated with these storm events, scour of the muddy substrate (i.e. the shale beds) may have resulted in patchy preservation of the shale beds between successive storm deposits.

Consequently, storm-deposited sand beds may be interconnected to a limited degree via tortuous migration pathways. Recognition of these differences in stratal architecture and consequent differences in fluid migration potential within the two units enhances efficient exploitation of each unit.

Pressure and production data from Joarcam Field suggest that permeability barriers between the LST and TST have been ineffective over the long term (i.e. *over geologic time*) but effective over the short term (i.e. *over the life of the field* or *real time*). The LST and TST share the same original oil/water contact indicating that hydrocarbons have migrated between the two systems tracts over geologic time. However, current pressure and production data suggest that, over the life of the field, permeability barriers between the two units have provided partially effective hydrodynamic isolation. A recently drilled well (12-8-49-21W4) (Fig. 4) confirms the isolation of the TST over real time. In this well RFT pressures were taken throughout the Viking interval. From these samples a regional Viking pressure gradient was established (Fig. 11). Samples taken specifically in the LST and TST were plotted against the regional gradient. The pressures from the LST were considerably lower than the regional gradient indicating significant oil depletion of this zone. Pressures in the TST were only slightly lower than the regional gradient suggesting that these sands are only partially depleted. Fluid samples were taken from the TST and the LST. The lower sandstones of the TST sample contained 100% oil. The lowstand sample, taken from a position only 1.5 m lower in the well bore, contained 25% oil, 75% water. This indicates that the TST is isolated hydronamically from the in the LST.

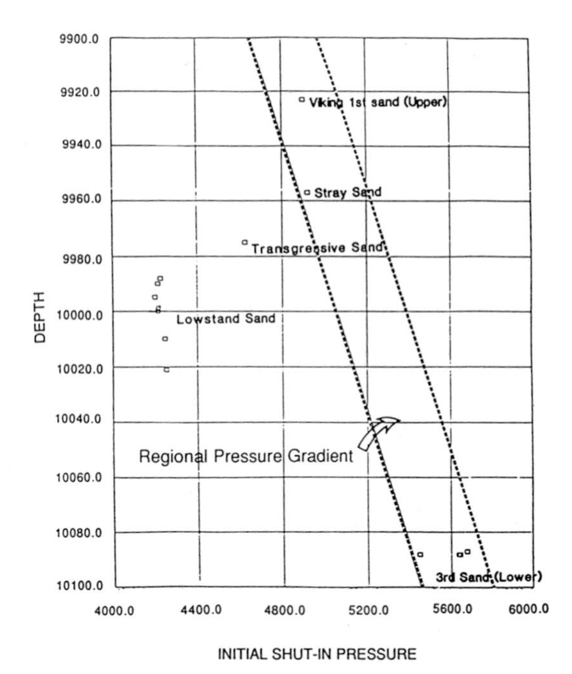

Fig. 11. Pressure data from RFT samples from 12-8-49-21W4. The regional pressure gradient is determined by pressure information from the upper and third sands. Note that the lowstand systems tract sandstone seems depleted, whereas the transgressive sand seems only partially depleted. The uppermost sandstone sampled within the transgressive systems tract contains 100% water suggesting that the uppermost part of the transgressive systems tract is not in communication with either the lower part of this systems tract or the underlying lowstand systems tract.

GEOCHEMICAL ANALYSIS

Samples were taken for pyrolysis analyses from shaly zones within two cores — one from a well drilled through the TST and into the lowstand beach (well 16-12-48-21W4) and the other from a well drilled through the TST and into the offshore equivalent of the lowstand beach (well 10-4-48-20W4) — with the objective of geochemically characterizing each systems tract and determining whether this type of analysis would be useful in identifying the contact between the two units. Tables 1 and 2 summarize the results of the pyrolysis analyses. A sharp contact with regard to hydrogen index (HI, measured in milligrams of hydrocarbon per gram of organic carbon) was observed at the LST/TST boundary at each location. In contrast, total organic content (TOC), while distinctively lower beneath the sequence boundary in the 10-4 well does not vary as markedly across the same surface in the 16-12 well. These observations, and in particular the observations regarding HI, confirm the identification of the LST/TST contact recognized in core initially on the basis of sedimentologic and ichnologic criteria.

Figure 12 illustrates the chronostratigraphic relationship between the samples from the two wells. Comparison between the HI values within the LST shows that HI values are significantly higher proximally (average of 164.3 mg/g) than distally (average

Table 1. Summary of results of pyrolysis analysis on core samples from well 16-12-48-21W4

Sample depth (m)	S1	S2	T_{MAX}	Hydrogen index (OMT S2/TOC*100)	TOC (%)
973.0	0.57	8.12	431	240.9	3.37
975.9	0.77	8.92	435	288.7	3.09
976.9	0.98	8.58	431	303.2	2.83
979.0	2.71	13.98	431	287.7	4.86
982.7	1.55	7.13	431	146.1	4.88
984.8	0.71	3.67	431	138.0	2.66
987.4	0.27	1.93	429	154.4	1.25
987.7	0.21	2.80	431	218.7	1.28

of 70.5 mg/g). This suggests that the sediments are progressively more oxidized distally within the LST (Creaney, 1988, pers. comm.). Comparison of the HI values within the TST reveals some significant relationships when viewed within a chronostratigraphic framework. The basal TST at 10-4 is interpreted, based on geologic parameters, to be onlapping the LST beach. Consequently, the basal TST samples at the two wells would not be co-eval. The observation that the basal samples within the TST at 10-4 have significantly lower HI values than the basal samples at 16-12 (average of 149.0 mg/g versus 280.1 mg/g) and that the basal TST HI values are more akin to the LST HI values at 16-12 suggests that the basal TST sediments were derived from cannibalization during transgression of the nearby LST beach/coastal plain and were not significantly oxidized.

The distinct decrease in oxidation of sediments overlying the LST/TST contact is consistent with the distinct decrease of trace fossil diversity and abundance across this contact. One possible explanation is that anoxic waters existed basinward of this area throughout the interval analysed and merely expanded, moving landward as water depth increased subsequent to initiation of the TST (Creaney, 1988, pers. comm.). Plint and Norris (1991) observe a similar relationship of dark shales immediately overlying ravinement surfaces within the Upper Cretaceous Marsheybank and Muskiki Formations. They also suggest that the region of anoxic waters may have spread landward in conjunction with rising sea level and shoreline transgression resulting in relatively unbioturbated shales overlying ravinement surfaces. Davis and Byers (1989) also observe an unbioturbated, siltstone and sandstone lithofacies immediately overlying a sandstone within the Mowry Shale in Wyoming. Another possible explanation is that, during transgression, erosion of the upper approximately 5 to 10 m of section (presumably comparatively organic-rich coastal plain and upper shoreface sediments)

Table 2. Summary of results of pyrolysis analysis on core samples from well 10-4-48-20W4

Sample depth (m)	S1	S2	T_{MAX}	Hydrogen index (OMT S2/TOC*100)	TOC (%)
967.0	0.94	7.55	424	225.4	3.35
970.0	0.52	5.44	428	180.1	3.02
971.0	0.97	8.21	426	187.9	4.37
972.6	0.37	2.20	429	105.3	2.09
973.0	0.55	7.08	428	200.6	3.53
974.0	0.29	3.63	431	134.4	2.70
974.4	0.33	2.91	431	116.9	2.49
974.8	0.16	0.44	425	58.7	0.75
975.1	0.19	0.69	421	63.3	1.09
975.4	0.20	0.91	421	83.5	1.09
975.8	0.65	1.34	422	76.6	1.75

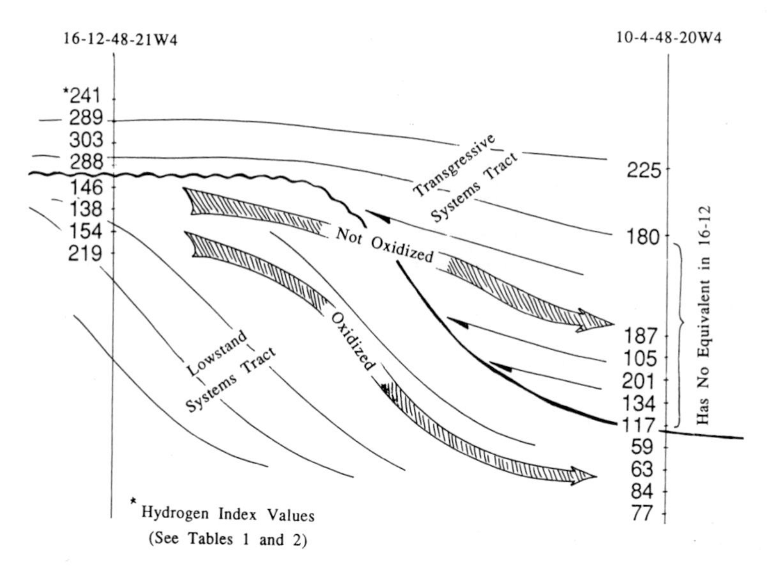

Fig. 12. Schematic cross-section relating hydrogen index (HI) values in 16-12-48-21W4 and 10-4-48-20W4 within a chronostratigraphic framework (see Fig. 3 for locations). The lowstand systems tract sediments in 10-4 are thought to be oxidized time-equivalent sediments to those encountered within the lowstand systems tract at 16-12. The basal transgressive systems tract sediments at 10-4 are thought to have been derived from the upper section of the lowstand systems tract but not oxidized. Consequently HI values of the basal transgressive systems tract at 10-4 are equivalent to those in the upper part of the lowstand systems tract at 16-12. It is suggested that the basal transgressive systems tract at 10-4 onlaps the shoreface of the lowstand systems tract. The upper part of the transgressive systems tract at 10-4 is thought to be continuous with the transgressive systems tract at 16-12.

may have resulted in increased bacterial activity in the water column, thus producing oxygen-deficient waters. These oxygen-deficient waters would have resulted in decreased total faunal population as well as diversity.

SIGNIFICANCE OF VERTICAL PERMEABILITIES WITHIN THE LOWSTAND SYSTEMS TRACT

Depositional dip of beds within the LST shoreface has been determined based on the dip of the bentonitic beds, which are readily identified on wireline logs, and the depositional slope of the last shoreface deposits at the top of the LST. The dips range from 0.09° (as measured from the bentonites) to 0.14° (as measured from the top of the LST) (see Fig. 6) to the northeast in an offlapping geometry.

The dipping geometry of both the bentonitic beds and the last shoreface deposits at the top of the LST is compelling evidence supporting an interpretation of clinoforming or shingling physical stratigraphy within the lowstand unit (Fig. 5). The oil production implication of this shingling pattern is that water injected during waterflooding may tend to travel most efficiently parallel to bedding planes along depositional dip rather than directly across the top of the lowstand shoreline across depositional dip, and yield anomalous unanticipated results. Consequently, even though impermeable beds in the LST may have limited areal distribution the efficiency of the waterflood nonetheless may be affected. This opens up the possibility of recompletions for some wells where waterflood efficiency may not have been optimal. Figure 13 illustrates the zone of maximum sweep efficiency set up by the dipping bedding geometry. Note that perforation of the 5 m thick section dipping at 0.09° to 0.14° results in a sweep radius of 2.0 to 3.2 km.

CONCLUSIONS

A geological analysis utilizing sequence-stratigraphic principles has led to a revised strategy for exploitation of a mature oilfield — the Joarcam Field — in central Alberta, Canada. Recognition

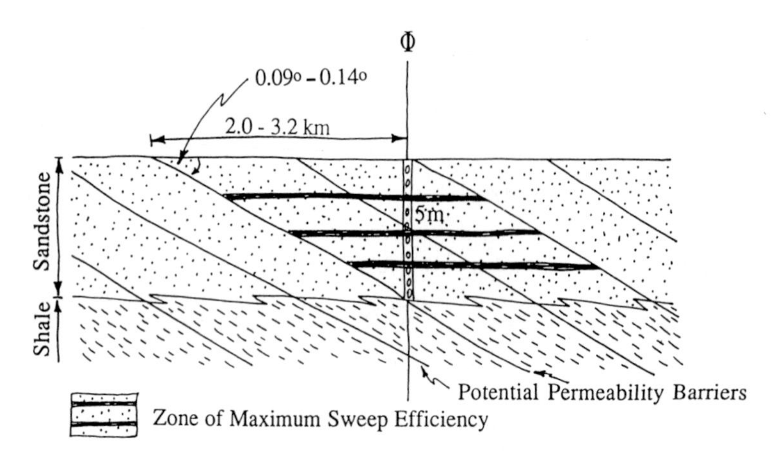

Fig. 13. Sweep radius as a function of depositional dip. Injection of a 5 m zone of permeable sand dipping at 0.09° to 0.14° results in a sweep radius of 2.0 to 3.2 km. The dip angle of 0.09° is based on the measured dip of the bentonitic beds and the dip of 0.14° is based on the measured dip of the final shoreface slope of the lowstand systems tract.

that the principal producing sand comprised two discrete units — an LST and a TST, each with its own hydrodynamic attributes, has led to identification of a number of recompletion and workover candidates and a subsequent increase in daily production.

The LST is a narrow lowstand beach *c.* 10 m thick, characterized by an upward-coarsening and cleaning lithology. Internally, the physical stratigraphy is characterized as offlapping or shingling. Vertical permeability barriers are restricted to occasional shale or bentonitic beds with local distribution. The depositional dip of 0.09–0.14° in this unit can be incorporated into an analysis and enhanced understanding of waterflood efficiency.

The TST is composed of alternating muds and sands overlying the LST. This unit is relatively thin (<3–4 m) where it overlies the lowstand beach and thicker (<10 m) where it overlies the offshore equivalent of the lowstand beach. The reservoir sands within the transgressive unit are separated from the lowstand unit by relatively impermeable and widespread shale beds. These sands are in fluid communication with sands of the LST *over geological time* but are isolated *over the life of the field.* Consequently, significant bypassed pay may exist here.

ACKNOWLEDGEMENTS

The authors gratefully acknowledge Esso Resources Canada Ltd for permission to publish this study. We thank Steve Creaney for his assistance with the geochemical analyses and for sharing his insights regarding interpretation of the geochemical data. We also thank David James for many enlightening discussions about Viking Formation stratigraphy. Dale Leckie and David James reviewed the manuscript and provided very useful feedback that resulted in significant improvement.

REFERENCES

Allen, G.P. & Posamentier, H.W. (1991) Facies and stratal patterns in incised valley complexes: examples from the recent Gironde Estuary (France) and the Cretaceous Viking Formation (Canada). *Annual Meeting, Dallas, Texas, April 7–10, 1991, Abstracts with Programs*, p. 70.

Amajor, L.C. (1986) Patterns of hydrocarbon occurrences in the Viking Formation, Alberta, Canada. *J. Petrol. Geol.* **9**, 53–70.

Amajor, L.C. & Lerbekmo, J.F. (1980) Subsurface correlation of bentonite beds in the Lower Cretaceous Viking Formation of south-central Alberta. *Bull. Can. Soc. Petrol. Geol.* **28**, 149–172.

Beaumont, E.A. (1984) Retrogradational shelf sedimentation; Lower Cretaceous Viking Formation, central Alberta. In: *Siliciclastic Shelf Sediments* (Eds Tillman, R.W. & Siemers, C.T.) Spec. Publ. Soc. Econ. Paleontol. Mineral. Tulsa 34, 163–177.

Clifton, H.E., Hunter, R.E. & Phillips, R.L. (1971) Depositional structures and processes in the non-barred, high energy nearshore. *J. Sediment. Petrol.* **41**, 651–670.

Davis, H.R. & Byers, C.W. (1989) Shelf sandstones in the Mowry Shale: evidence for deposition during Cretaceous sea level falls. *J. Sediment. Petrol.* **59**, 548–560.

Demarest, J.M. & Kraft, J.C. (1987) Stratigraphic record of Quaternary sea levels: implications for more ancient strata. In: *Sea-level Fluctuation and Coastal Evolution* (Eds. Nummedal, D., Pilkey, O.H. & Howard, J.D.) Spec. Publ. Soc. Econ. Paleontol. Mineral. Tulsa 41, 223–239.

Downing, K.P. & Walker, R.G. (1988) Viking Formation, Joffre Field, Alberta: shoreface origin of long, narrow sand body encased in marine mudstones. *Bull. Am. Assoc. Petrol. Geol.* **72**, 1212–1228.

Energy Resources Conservation Board (1988) *Alberta's Reserves of Crude Oil, Oil Sands, Gas, Natural Gas Liquids, and Sulphur.* Energy Resources Conservation Board, Reserve Report Series ERCB-18, Calgary, Alberta, pp. 2:1–2:191, 4:1–4:167.

Evans, W.E. (1970) Imbricate linear sandstone bodies of Viking Formation in Dodsland–Hoosier area of southwestern Saskatchewan, Canada. *Bull. Am. Assoc. Petrol. Geol.* **54**, 469–486.

Hein, F.J., Dean, M.E., Deiure, A.M., Grant, S.K., Robb, G.A. & Longstaffe, F.J. (1986) The Viking Formation in the Caroline, Garrington, Harmatton East Fields, western south-central Alberta: sedimentology and paleogeography. *Bull. Can. Petrol. Geol.* **34**, 91–110.

Hunt, W.C. (1954) The Joseph Lake–Armena–Camrose producing trend, Alberta. In: *Western Canada Sedimentary Basin* (Ed. Clark, L.M.) pp. 452–463.

Jardine, D. (1953) The Joseph Lake–Armena–Camrose Viking Oil Fields. *Can. Min. Metal. Bull.* **57**, 184–186.

Koldijk, W.S. (1976) Gilby Viking 'B': a storm deposit. In: *The Sedimentology of Selected Oil and Gas Reservoirs in Alberta* (Ed. Lerand, M.M.), Can. Soc. Petrol. Geol. Core Conf. Proc. 62–77.

Leckie, D.A. (1986) Tidally influenced, transgressive shelf sediments in the Viking Formation, Caroline, Alberta. *Bull. Can. Petrol. Geol.* **34**, 111–125.

Plint, A.G. (1988) Sharp-based shoreface sequences and 'offshore bars' in the Cardium Formation of Alberta; their relationship to relative changes in sea level. In: *Sea Level Change — An Integrated Approach* (Eds Wilgus, C.K., Hastings, B.S., Kendall, C.G.St.C., Posamentier, H.W., Ross, C.A. & Van Wagoner, J.C.) Spec. Publ. Soc. Econ. Paleontol. Mineral. Tulsa 42, 357–370.

Plint, A.G. & Norris, B. (1991) Anatomy of a ramp margin: facies successions, paleogeography, and sediment dispersal patterns in the Muskiki and Marshybank Formations, Alberta Foreland Basin. *Bull. Can. Petrol. Geol.* **39**, 18–42.

Plint, A.G., Walker, R.G. & Bergman, K.M. (1987) Cardium Formation 6. Stratigraphic framework of the Cardium in subsurface. *Bull. Can. Petrol. Geol.* **34**, 213–225.

Posamentier, H.W. & Chamberlain, C.J. (1989) Viking lowstand beach deposition at Joarcam Field, Alberta. *Can. Soc. Petrol. Geol./Can. Soc. Explor. Geophys.* Exploration Update, June 11–15, 1989, Programs and Abstracts, pp. 96–97.

Posamentier, H.W. & Vail, P.R. (1988) Eustatic controls on clastic deposition II — sequence and systems tract models. In: *Sea Level Change — An Integrated Approach* (Eds Wilgus, C.K., Hastings, B.S., Kendall, C.G.St.C., Posamentier, H.W., Ross, C.A. & Van Wagoner, J.C.) Spec. Publ. Soc. Econ. Paleontol. Mineral. Tulsa 42, 125–154.

Posamentier, H.W., Allen, G.P. & James, D.P. (1992a) High resolution sequence stratigraphy — the East Coulee Delta. *J. Sediment. Petrol.* **62**, 310–317.

Posamentier, H.W., Allen, G.P., James, D.P. & Tessen, M. (1992b) Forced regressions in a sequence stratigraphic framework: concepts, examples, and exploration significance. *Am. Assoc. Petrol. Geol.* **76**, 1687–1709.

Posamentier, H.W., James, D.P. & Allen, G.P. (1990) Aspects of sequence stratigraphy: recent and ancient examples of forced regressions. *Bull. Am. Assoc. Petrol. Geol.* **74**, 742.

Posamentier, H.W., Jervey, M.T. & Vail, P.R. (1988) Eustatic controls on clastic deposition I — conceptual framework. In: *Sea Level Change — An Integrated Approach* (Eds Wilgus, C.K., Hastings, B.S., Kendall, C.G.St.C., Posamentier, H.W., Ross, C.A. & Van Wagoner, J.C) Spec. Publ. Soc. Econ. Paleontol. Mineral. Tulsa 42, 110–124.

Power, B.A. (1988) Coarsening-upwards shoreface and shelf sequences: examples from the lower Cretaceous Viking Formation at Joarcam, Alberta, Canada. In: *Sequences, Stratigraphy, Sedimentology: Surface and Subsurface* (Eds James, D.P. & Leckie, D.A.) Mem. Can. Soc. Petrol. Geol. 15, 185–194.

Reinson, G.E., Clark J.E. & Foscolos, A.E. (1988) Reservoir geology of Crystal Viking Field, Lower Cretaceous estuarine tidal channel-bay complex, south-central Alberta. *Bull. Am. Assoc. Petrol. Geol.* **72**, 1270–1294.

Spec. Publs Int. Ass. Sediment. (1993) **18**, 487–499

Limits on late Ordovician eustatic sea-level change from carbonate shelf sequences: an example from Anticosti Island, Quebec

D.G.F. LONG

Department of Geology, Laurentian University, Sudbury, Ontario, Canada P3E 2C6

ABSTRACT

Short-term fluctuations in global sea level during late Ordovician and early Silurian times are thought to be directly related to volume changes within a continental ice sheet, centred on the Gondwana super-continent. The extent and frequency of eustatic sea-level change are best recorded in shallow-marine strata deposited far from the loading effects of the ice sheet, where the marine record is sufficiently long to permit calculation of tectonic subsidence rates.

The St Lawrence Platform of eastern Canada contains a thick, continuous sequence of early Ordovician to early Silurian marine carbonates, with minor mudstones and quartz-rich sandstones which record the depositional history of a carbonate ramp to platform sequence on the northwestern margin of Iapetus ocean. Five transgressive–regressive cycles can be recognized in the late Rawtheyan and Hirnantian strata which form the uppermost Vauréal and Ellis Bay Formations on Anticosti Island. Mean shelf subsidence rates for this part of the St Lawrence Platform fell from a mean of 26 cm/1000 yr in the Ashgill to 4 cm/1000 yr in the Wenlock. Comparison of estimated subsidence rates with the distribution of depth-dependent facies in Hirnantian strata of Anticosti Island indicates a maximum synglacial sea-level change of 32 to 36 m in the latest Ordovician.

INTRODUCTION

During the Pleistocene, major changes in global sea level of 20 to 180 m, with interglacial highstands at periods of close to 100,000 yrs (Peltier, 1987; Allen & Allen, 1990), along with the $\delta^{18}O$ record from deep-sea cores (Shackleton & Opdyke, 1977), reflect changes in ice volume in both the northern and southern hemispheres. The presence of a major ice cap in the southern hemisphere in the late Ordovician (Caputo & Crowell, 1985; Hambrey, 1985) must have had a similar impact on global sea level, even if the scale of such changes may have been less pronounced.

The major problem in estimating the magnitude of past sea-level changes is to isolate the global sea-level component (eustasy) from base-level changes due to local tectonic and load-induced subsidence (isostasy). To separate these effects it is necessary to examine parasequences associated with cycles of sea-level rise and fall which could be associated with expansion and contraction of the ice sheet in areas well away from the effects of glacial isostatic loading, in settings where the rate of tectonic subsidence is negligible. Possible target areas include the interior parts of continents (Cisne *et al.*, 1984; Cisne & Gildner, 1988) or areas with a well-defined, prolonged subsidence history, as is the case for the St Lawrence Platform.

Brenchley (1988) and Brenchley *et al.* (1991) note that major depressions of global sea level, which can be related to expansion of the late Ordovician continental ice sheet, are confined to the Hirnantian Stage of the Ashgill. This stage may have had a duration of only 0.5 Ma (Harland *et al.*, 1989). On Anticosti Island, this interval is represented by strata at the top of the Vauréal Formation and in the Ellis Bay Formation. In this paper the bathymetric signature of these formations is examined in conjunction with the subsidence history of the St Lawrence Platform in order to provide a more accurate estimate of sea-level change.

ANTICOSTI ISLAND

Tectonic framework and subsidence history

The thick succession of limestones, mudstones and minor sandstones now exposed on Anticosti Island, Quebec (Fig. 1) were deposited as part of a carbonate ramp to platform sequence on the western side of Iapetus ocean, at about 12°S (McKerrow & Cocks, 1977; Seguin & Petryk, 1986; Kent & Van der Voo, 1990). While this sector of the St Lawrence Platform was well away from any direct influence of isostatic loading by the Gondwana ice sheet, it was not completely isolated from the effects of tectonic activity which continued after the partial closure of Iapetus ocean during the Caradocian (Pickering, 1987). Late Ordovician and Silurian strata on the St Lawrence Platform accumulated in a residual foreland basin initiated during the Taconic orogeny. Differential subsidence of the platform can be related to asymmetric obductive loading of the Laurentian plate margin by rising highlands in the Taconic mountains and Miramichi areas to the southwest and the Humber terrain to the east (Fyffe, 1982; James & Stephens, 1982), combined with sediment-induced loading. These tectonic highlands controlled the palaeogeography of the St Lawrence Platform, producing, at least during lowstands, a number of marine embayments along the coast of Laurentia (Long & Copper, 1987b). At highstands, a greater portion of the craton was flooded and an open seaway may have developed between the Anticosti basin and the continental interior (Long & Copper, 1987b).

Net tectonic plus load-induced subsidence of the St Lawrence Platform in the vicinity of Anticosti Island averaged 26 cm/1000 yrs in the Ashgill (represented by 1019 m of strata in the Vauréal Formation and 52 to 72 m of strata in the Ellis Bay Formation), falling to 4 cm/1000 yrs in the Wenlock (366 m of strata). If deposition of the Ellis Bay Formation took 500,000 yrs (the duration of the Hirnantian Stage according to Harland *et al.*, 1989), subsidence (accommodation) rates at the west end of the basin may have been 30% higher (15 cm/1000 yrs) than at the east end (10 cm/1000 yr).

Sedimentology

Strata in the uppermost part of the Vauréal Formation and in the Ellis Bay Formation in the eastern part of Anticosti Island show marked facies changes which can be used to delimit parasequences associated with cycles of sea-level rise and fall (Long, 1988). Long and Copper (1987a,b) provided a detailed description of this late Ordovician sequence, and introduced the member names used in this paper. They recognized four main lithofacies associations within this sequence: (i) a muddy facies of terrigenous or calcareous mudrocks with minor sandstones; (ii) a sandy facies of bioclastic or siliciclastic sandstones; (iii) a conglomeratic facies of bioclasts with minor siliciclastics and biohermal material; and (iv) a fine-grained, laminated carbon-

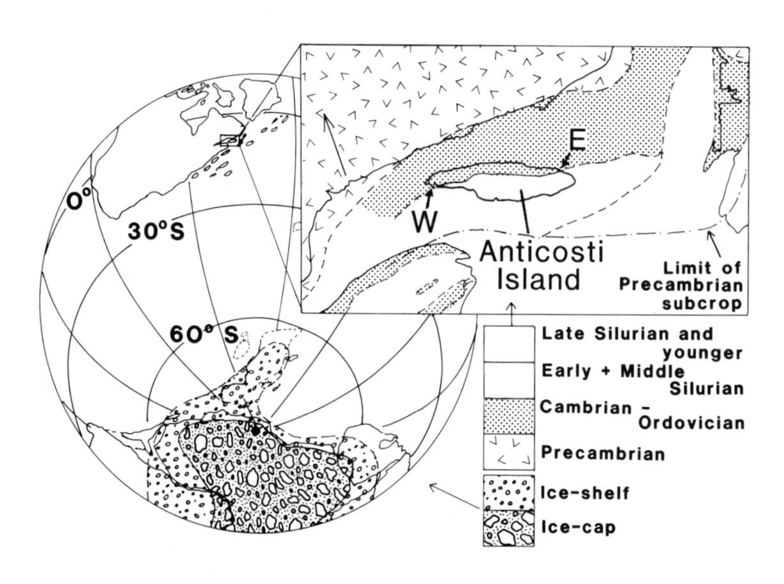

Fig. 1. Location of the St Lawrence Platform in late Ordovician time, based on map by Kent and Van der Voo (1990). Extent of continental ice sheet and ice shelf from Brenchley *et al.* (1991). Insert shows geology and location of eastern (E) and western (W) sections.

Fig. 2. Isolated sharp-based hummocky cross-stratified sandstones in muddy facies of Velleda Member (scale divisions = 10 cm).

ate facies. These lithofacies assemblages (labelled M, S, C and L, respectively in Fig. 8) are described below as a basis for palaeobathymetric analysis.

Muddy facies

The muddy lithofacies assemblage is dominated by laminated to thinly bedded calcareous mudstones and muddy limestones with minor laterally discontinuous beds of sandstone and intraformational mud-chip conglomerates. Sandstone beds may be plane or wavy-bedded, ripple cross-laminated or hummocky cross-stratified (Fig. 2). The latter typically occurs in units 5 to 25 cm thick, with a sharp erosional base, with hummocks 1.2 to 2 m apart. Most units lack a rippled top and pass abruptly into plane-bedded or massive lime mudstones (HFM, HFbM, and HFbMb sequences of Dott & Bourgeois, 1982). Pseudonodules of silt to fine sand grade associated with laminated to thin-bedded mudrocks are common and are interpreted as floundered starved ripples.

The muddy facies is characterized by the absence of abundant calcareous algae. This combined with the presence of trace fossils belonging to the *Cruziana* ichnofacies, and combined with the presence of trilobites and gastropods, aulacerids, small subspherical tabulate and cup corals and nests of small brachiopods in life position suggests deposition in the lower levels of the photic zone. The abundance of plane bedding combined with the presence of stands of upright, cylindrical aulacerid stromatoporoids (Fig. 3) indicates that the muddy facies accumulated on a relatively undisturbed open shelf or ramp well below fair-weather wave base. The presence of discontinuous sandstone beds, starved ripple horizons and intraformational mud-chip conglomerates indicates that the shelf was at times affected by storms with sufficient energy to rework the sea floor and transport material from shallower regimes.

Swift *et al.* (1983) suggested that hummocky cross-stratification might be generated on the inner shelf as irregular megaripples, in a setting below normal fair-weather wave base, where wave-generated orbital currents are superimposed on, but are subordinate to, unidirectional currents generated by a wind set-up of coastal waters during storms. Duke *et al.* (1991) downplay the role of unidirectional currents and suggest that isotropic hummocky cross-stratification, as seen in the muddy facies, is a product of long-period oscillatory flow or very strongly oscillatory-dominant combined flow. Hummocky cross-stratification can form in water depths as great as 30 (Harms, 1979) to 80 m (Campbell in Dott & Bourgeois, 1982). While they can be formed in water depths as shallow as 2 m (Hunter & Clifton, 1982), the preservation potential of these structures is low in shallow-water settings due to the potential for reworking by wind- and tide-driven currents.

Sandy facies

Strata of the sandy lithofacies assemblage are dom-

Fig. 3. Base of vertical aulacerid stromatoporoid in growth position. Muddy facies, Prinsta Member.

inated by trough cross-bedded fine- and very fine-grained sandstone, with lesser amounts of planar cross-bedded, plane-laminated, ripple-laminated and hummocky cross-stratified sandstone (Fig. 4). In places these have been remobilized after deposition as channel fill slump units (Fig. 5). Medium- and coarse-grained sandstone, intraformational mud-chip conglomerate, biorudite and siltstone laminae are common (Fig. 6). Hummocky units are typically amalgamated (HF type of Dott & Bourgeois, 1982) and near isotropic (Duke *et al.*, 1991). While ripple-laminated units are locally associated with hummocky strata, interbedding with trough cross-stratified units is more common.

Strata of the sandy facies typically overlie, or are interbedded with, rocks of the muddy facies. They appear to have accumulated between fair-weather and storm wave base, at depths shallower than the muddy facies. This interpretation is supported by the abundance of traction current features, scour surfaces, bioclastic and intraformational lags, combined with the presence of minor hummocky cross-stratification. Deposition above fair-weather wave base would have reworked most of the hummocky

Fig. 4. Hummocky cross-stratification in sandy facies, Velleda Member.

Fig. 5. Retrogressive flow slide containing abundant pillows, sandy facies, Grindstone Member.

Fig. 6. Plane, trough and hummocky cross-stratification in the sandy facies, Grindstone Member. Note intraformational conglomerate above the mudstone interbeds.

cross-stratified sandstones and the associated siltstone laminae. Most of the cross-bedding was produced by periodic migration of sinuous crested dunes, giving rise to unidirectional palaeocurrent distributions. Bimodal opposed cross-bed distributions in the lower Grindstone Member suggest local influence of tidal currents. The absence of bioturbation in all but the flat to wavy laminated beds, combined with the fragmental character of fossils in the bioclastic lags, indicates a very mobile substrate. The presence of bioclastic and intraformational lags and low-relief reactivation surfaces (Long & Copper, 1987b) points to erosion during bedform migration. The presence of mudstone drapes indicates that quiescent conditions must have prevailed at intervals below fair-weather wave base.

An early, pre-burial origin is indicated for pillow structures in both channel-fill and unconfined sandstone units as the tops of pillows commonly protrude from the top of these beds. The pillowed units probably reflect instantaneous liquefaction of the sand unit by wave-induced shock (Allen, 1982), while channel-fill units reflect liquefaction associated with development of retrogressive flow slides on slopes as low as 1 or 2° (Andresen & Berrum, 1967; Long & Copper, 1987b).

Conglomeratic facies

Strata of the conglomeratic facies assemblage are present in the Grindstone and Laframboise Members. They include biointrarudites and mixed

Fig. 7. Channel-fill conglomerate, with large overturned *Palaeofavosites* and boulders of reworked sandy facies; conglomeratic facies at top of Grindstone Member.

bioclastic–siliciclastic conglomerates and conglomeratic sandstones.

In the Grindstone Member, the conglomerates have both sheet and channel geometries. Early lithification of the underlying sandy facies is indicated by the presence of channel margins with slopes as high as 80° and abundant cobble and boulder grade clasts of sandstone in the conglomerates (Fig. 7). This, in conjunction with the presence of overturned colonies of *Palaeofavosites* (up to 60 cm thick), may indicate that the conglomeratic facies may have developed in shallower water than most of the underlying sandy facies. Alternatively they may represent infrequent storm reworking of a partly lithified surface in slightly deeper water. The absence of silt or mud drapes indicates sustained current flow.

Conglomeratic facies in the Laframboise Member include both small pebble intraclast conglomerates and very coarse calcarenites with abundant algal oncolites (containing *Wetheredella* and *Girvanella*) along with abundant broken coral and shell debris. These beds are associated with bioherms throughout much of Anticosti Island, and are interpreted as inter-reef or peri-reefal strata that accumulated on a shallow carbonate platform above fair-weather wave base.

Laminated carbonate facies

Thinly bedded, plane to wavy laminated micrites with minor intraclast-rich horizons dominate much of the upper half of the Ellis Bay Formation. They commonly occur in sets 3–10 cm thick interbedded with laminated argillaceous micrite and calcareous mudstone. This facies probably accumulated near to or within fair-weather wave base, in shallower water than the muddy facies and under less agitated conditions than the sandy and conglomeratic facies.

Palaeobathymetry

Coarsening-upward cycles in the sequence examined are of three types (Fig. 8). The first involves up-section transitions from muddy to sandy facies (M > S); the second, transition from muddy to sandy then conglomeratic facies (M > S > C); and the third, transitions from muddy facies to laminated carbonate then conglomeratic peri-reefal facies (M > L > C). The first two types, involving mixed carbonate–siliciclastic facies, are interpreted as shoaling sequences related to migration of subtidal sandbanks or sand-wave complexes which developed as detached offshore bars in an inner- to mid-shelf setting, while the third type reflects shoaling by progradation of an open carbonate platform (Long & Copper, 1987b).

In modern seas, sand-ridge complexes are generated by tidal activity, by storms, or by a combination of these processes (Stride, 1982). Most tidal sand bodies are dominated by cross-stratification, while storm-generated bodies are dominated by smaller-scale features. Sedimentary structures

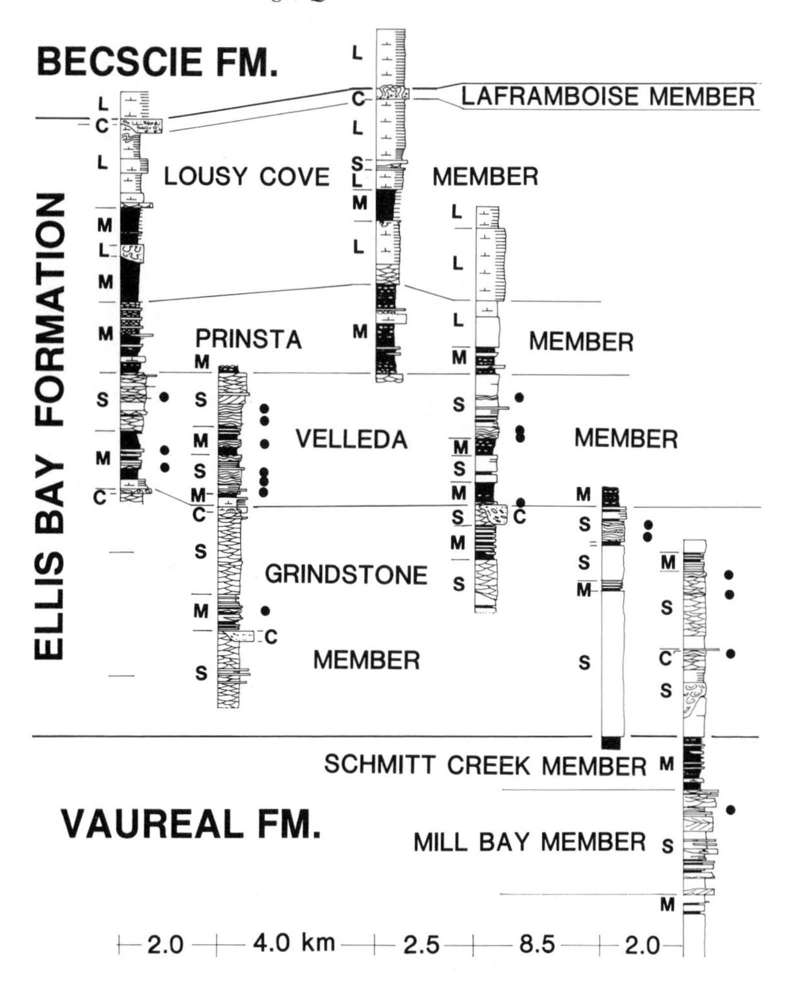

Fig. 8. Stratigraphic sections in the uppermost Vauréal Formation and Ellis Bay Formation, eastern Anticosti Island. (See Long & Copper, 1987a,b for further details.) Black dots indicate hummocky cross-stratification. M, muddy facies; S, sandy facies; C, conglomeratic facies; L, laminated carbonate facies.

within the mixed carbonate—siliciclastic sequences in the Vauréal and Ellis Bay Formations indicate the influence of both tide- and storm-generated processes in ridge generation. A major storm influence is indicated by the abundance of hummocky cross-stratification in both the sandy and muddy facies and the presence of sharp-based sandstone units, including starved ripples, in the muddy facies. A minor tidal influence is indicated by the presence of local bipolar opposed cross-bedding and low-relief reactivation surfaces in the sandy facies.

Comparison of the mixed carbonate–siliciclastic sequences in the Vauréal and Ellis Bay Formations with modern siliciclastic detached shoreface-bar complexes (Swift & Field, 1981; Belderson, 1986) allows limits to be placed on palaeodepths. Thicker sequences of the muddy facies reflect deposition at the downcurrent end of the sediment dispersal path at depths possibly in excess of 30 m. In this setting deposition of muds from suspension would have alternated with periodic deposition of sands (derived from nearby sand ridges) during storms. The presence of hummocky cross-stratification in some of the sharp-based sandstone units indicates deposition in water depths of no more than 80 m and probably less than 30 m (Harms, 1979; Dott & Bourgeois, 1982). Thinner muddy sequences may have accumulated in swales between sand ridges at depths of 20 to 27 m (Swift & Field, 1981).

The lower parts of the sandy facies accumulated on the flanks and distal slopes of advancing sand-wave complexes, with flat to wavy lamination developing in response to storm and to a lesser extent tidal processes. The presence of hummocky cross-stratification within the lower parts of sandy units suggests water depths of less than 30 m. Local depositional slopes in excess of 1° are indicated by the presence of channel fill slump units. The presence of planar and trough cross-stratification in the upper parts of sandy facies indicates deposition from straight and sinuous crested dunes on the upper flanks and crest of the sand-wave complex. Local mudstone interbeds and intraformational conglomerates in the sandy facies indicate that sand-wave migration was periodic, with deposition from suspension occurring below wave base in fair-weather conditions. Comparison with modern sand-wave complexes off the coast of North Carolina (Swift & Field, 1981) indicates that the crests of the Ellis Bay sand waves may have been as shallow as 10 to 12 m.

Conglomerates at the top of some cycles cannot be directly related to progressive accumulation of lag deposits by migration of sand waves on the crest or backslope of an active sand-wave complex (Brenner *et al.*, 1985; Cotter, 1985). The presence of locally derived sandstone intraclasts of pebble and cobble grade, in conjunction with the steep walls of the channels, indicates that the crests of the sand-wave complexes must have been partially lithified at the time the conglomerates formed. They may reflect the effects of storm winnowing during a moribund stage of ridge evolution as this would allow sufficient time for partial lithification to occur. Drowning of the ridge crests may also explain the presence of large, hemispherical *Palaeofavosites* colonies which would not have developed on the ridge crests if these were active as growth would have been inhibited by detrital sediment. Further deepening to below fair-weather wave base provided a focal point for growth of aulacerid stromatoporoids. By comparison with modern, inactive tidal sand-wave complexes in the North Sea (Stride, 1982), the sand waves may have become moribund when water depths above the crests exceeded 15 m, and may have become completely inactive in water depths greater than 35 m.

The presence of abundant corals and algal oncolites in the Laframboise Member indicates accumulation in water depths of less than 10 m. Local intertidal exposure of the top of the formation may be indicated by blackening of inter-reef areas and reef tops (Fig. 9). Water depths for the laminated carbonates are more difficult to establish. Strata in the Ellis Bay Formation above the Velleda Member are 21 to 24 m thick. Post-depositional compaction of the limestones appears to have been minimal, due to development of early cements. Compaction of more argillaceous units around pseudonodules indicates a volume loss of about 55% in the associated mudrocks. If strata in the Prinsta, Lousy Cove and Laframboise Members (Fig. 8) represent a simple shoaling sequence, the maximum water depth above the crests of the sand waves in the Velleda Member was probably about 40 m.

Using the criteria outlined above, it is possible to develop a palaeodepth model for strata at the east end of Anticosti Island, which involves maximum changes of local water depths of 40 m or less (Fig. 10).

Fig. 9. Blackened top of inter-reef strata. Laframboise Member at the western end of Anticosti Island.

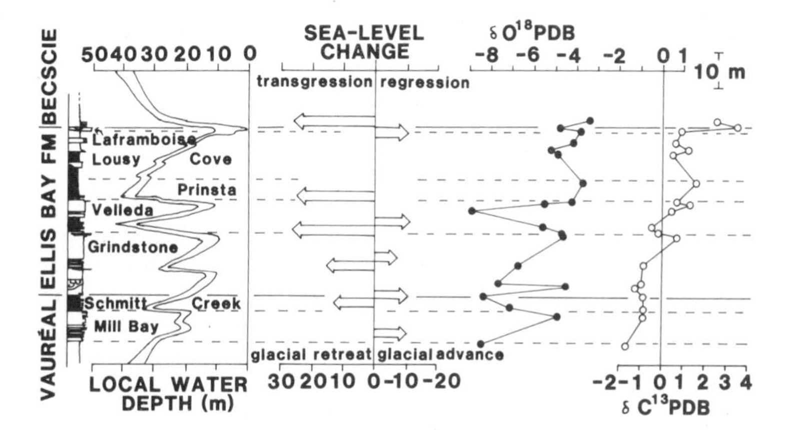

Fig. 10. Relation between stratigraphy, local sea level, sea-level change and isotope abundance in micrites, eastern Anticosti Island.

Isotopic record

The isotopic record of late Ordovician glaciation is as yet poorly defined. Orth *et al.* (1986) examined $^{13}C/^{12}C$ and $^{18}O/^{16}O$ ratios in strata at the top of the Lousy Cove Member and in the Laframboise Member of the Ellis Bay Formation at the west end of Anticosti Island. They noted a sudden decrease in $\delta\,^{18}O$ and $\delta\,^{13}C$ in a clay-rich mudstone parting corresponding to the top of McCracken and Barnes (1981) conodont Fauna 13 (within the lower part of the Laframboise Member), which they interpreted as a product of fresh-water dilution of the carbonate platform by rivers.

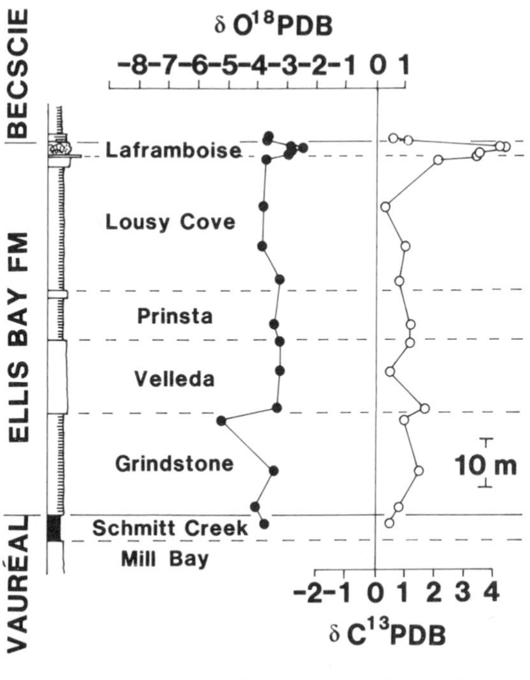

Fig. 11. Isotope stratigraphy, western Anticosti Island.

The results of new $\delta\,^{18}O$ and $\delta\,^{13}C$ analysis of micritic carbonate from the upper part of the Ordovician sequence on Anticosti Island are given in Figs 10 and 11. Several isotopic peaks can be seen in the data from the west end of the island (Fig. 11), the most noticeable being in the Laframboise Member, which shows a positive excursion of 1.4‰ of the $\delta\,^{18}O$ ratios and a positive shift of about 2‰ in the $\delta\,^{13}C$ curve. The greater range of values seen in samples from the east end of the island may reflect diagenetic overprinting as both the carbon- and oxygen-isotope ratios show similar positive increases upsection (Fig. 10). Despite the diagenetic overprinting a number of excursions from the general trends can be seen. These excursions may indicate from three to five periods of oxygen-isotope depletion which may be directly related to glacial advances, although the spacing of samples is as yet too broad to provide a comprehensive picture.

CONCLUSIONS, GLOBAL COMPARISONS

Late Ordovician glacial deposits are widely distributed in Africa (Beuf *et al.*, 1971; Biju-Duval *et al.*, 1981; Destombes, 1981; Deynoux & Trompette, 1981; Hambrey, 1981, 1985; Rust, 1981; Tucker & Reid, 1981: Deynoux, 1985) and once adjacent parts of Saudi Arabia (McClure, 1978, 1988) and Europe (Dangeard & Doré, 1971; Doré, 1981; Doré *et al.*, 1985; Robardet & Doré, 1988; Young, 1988; Brenchley & Storch, 1989).

Many of these glacial sequences show evidence of regressive–transgressive cycles which can be related to growth and decay of a continental ice cap. For example, at least two glacial advances are required to explain the stratigraphy of early Palaeozoic glacial deposits in the Tindouf Basin of Morocco (Destombes, 1981), the Pakhuis Tillite in South Africa (Rust, 1981) and the Prague Basin of Czech Republic (Brenchley & Storch, 1989). The Tamadjert Formation of Central Sahara contains three parasequences, each beginning with an unconformity showing signs of glacial erosion (Biju-Duval *et al.*, 1981). Three glacial advances are also apparent from the stratigraphy of glacial deposits in Normandy (Doré, 1981). Up to four parasequences can be identified in glacial strata in the Taoudeni Basin of West Africa (Deynoux & Trompette, 1981).

As isostatic response to glacial loading has a major influence on periglacial sea levels (Peltier, 1987), the magnitude of global sea-level fluctuations is best sought in tropical areas well away from the isostatic effects of the ice sheet. Brenchley and Newall (1980) used facies distributions in the Oslo District to indicate maximum (local) sea-level changes of 40 to 170 m, with limits of 50 to 100 m being most likely. Conversely both Geng (1982) and Chen (1984) suggest that sea-level changes must have been substantially less than 100 m to explain the continued existence of the Yangtze Basin, of China, in late Ashgill times and may have been less than 20 m. Gildner and Cisne (1987) note that sea-level changes of as little as 20 m may be all that is required to explain changes in shelf area in the late Ordovician. Both Bjørlykke (1985) and Gildner and Cisne (1987) considered that Brenchley's and Newall's (1980) depth estimates are significantly higher than global sea-level changes due to the effects of local tectonic subsidence on the Oslo sequence.

Brenchley and Newall (1980), Brenchley (1984, 1988) and Brenchley *et al.* (1991) indicate that two main regressive phases can be recognized in the Hirnantian. This is supported by Barnes's (1986) interpretation of oxygenic and anoxygenic facies in the Dob's Lin section, in Scotland. Petryk (1981), using lithological criteria from the Anticosti sequence, suggested that up to five regressive–transgressive cycles could be recognized, a conclusion supported by the present study.

The sedimentology of the Upper Ordovician sequence at the east end of Anticosti Island (Fig. 8) provides evidence for five cycles of regression and transgression, resulting in local sea-level changes of 10 to 40 m (Fig. 10). Most cycles begin with evidence of rapid deepening, indicated by sharp conformable contacts between the mudstone facies and underlying sandstone or conglomerate (Fig. 8). Transgressive events of this type appear more pronounced than regressive events. This reflects the problem of distinguishing the effects of shoaling from the effects of regression. As with the Oslo sequence such local palaeobathymetric curves are affected by tectonic subsidence, sediment loading and geoid deformation as well as a host of other variables (Peltier, 1987; Allen & Allen, 1990).

The extent to which long-term subsidence rates affect local sea levels is a function of the time involved. If sea-level changes were as rapid as the 8 m/1000 yrs suggested for the late Pleistocene and Holocene (Matthews, 1984), a sea-level change of 40 m could be achieved in 5000 . The effects of long-term subsidence (at 10 cm/ 1000) would be minimal on the local sea level (Fig. 12). Assuming Airy isostasy an absolute change in global sea levels of 1 m would result in local subsidence of 1.43 m (Matthews, 1984). If this load was fully compensated, a 40 m change in local sea level would be produced by a global sea-level rise of only

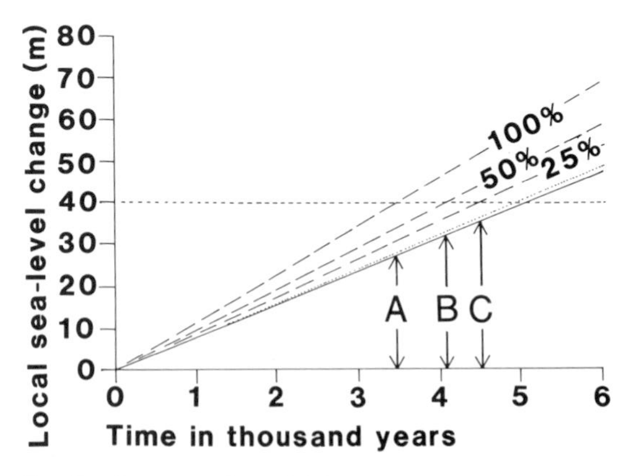

Fig. 12. Interpretation of global sea-level change (solid line) from local palaeodepths, using corrections for tectonic subsidence (dotted line) and isostatic adjustment for added water load (dashed lines, with degree of relaxation). Global sea-level change required to explain a local sea-level change of 40 m is given by A for 100% relaxation, B for 50% relaxation and C for 25% relaxation.

28 m (Fig. 12). As the crust behaves elastically in response to short-period loads and relaxation times (the half-life for subsidence predicted using Airy theory) for crust of normal thickness are of the order of 4400 yrs (Allen & Allen, 1990), it is unlikely that isostatic equilibrium would be achieved within the first few thousand years after the load was applied. From this it appears likely that the maximum global sea-level change required to produce a local sea-level change of 40 m would fall between 33 m (50% relaxation) and 36 m (25% relaxation, Fig. 12). These figures would be further reduced if significant sediment load was applied at times of sea-level change.

Brenchley and Newall (1980) and Brenchley (1988) noted that karst features in the late Ordovician Boda Limestone penetrate to depths in excess of 30 m. Brenchley (1988) cited this, and the presence of 45 m relief at the base of the Silurian sequence in Iowa (Johnson, 1975), as evidence for minimum sea-level changes. Assuming isostatic equilibrium (100% relaxation), and no tectonic overprint, this suggests a minimum sea-level fall of 21 and 32 m in the central Scandinavian Shield and Iowa, respectively. The maximum sea-level rise probably coincides with strata immediately above the Ellis Bay Formation, with the blackened zone on the top of the Laframboise Member representing a major transgressive flooding surface (cf. Nummedal & Swift, 1987).

REFERENCES

Allen, J.R.L. (1982) *Sedimentary Structures, Their Character and Physical Basis*, vol. 2. Developments in Sedimentology, 30B, Elsevier, Amsterdam, 663 pp.

Allen, P.A. & Allen, J.R. (1990) *Basin Analysis, Principles and Applications*. Blackwell Scientific Publications, Oxford, 451 pp.

Andresen, A. & Berrum, L. (1967) Slides in subaqueous slopes in loose sand and silt. In: *Marine Geotechnique* (Ed. Richards, A.F.) University of Illinois Press, Urbana, IL, pp. 221–229.

Barnes, C.R. (1986) The faunal extinction event near the Ordovician–Silurian boundary: a climatically induced crisis. In: *Global Bio-events, Facts — Hypothesis — Tasks* 5th Alfred Wegener — Conference and 1st International Workshop of the IGCP Project 216. *Summaries*, pp. 3–8.

Belderson, R.H. (1986) Offshore tidal and non-tidal sand ridges and sheets: differences in morphology and hydrodynamic setting. In: *Shelf Sands and Sandstones* (Eds Knight, R.K. & McLean, J.R.) Mem. Can. Soc. Petrol. Geol. 11, 293–301.

Beuf, S., Biju-Duval, B., de Charpal, O., Rognon, P., Gariel, O. & Bennacef, A. (1971) Les grès du Paléozoïque inférieur au Sahara, sedimentation et discontinuités évolution structurale d'un craton. *Publications de l'Institut Français du Pétrole, collection 'Science et Technique du Pétrole'* **18**, Editions Technip. Paris, 464 pp.

Biju-Duval, B., Deynoux, M. & Rognon, P. (1981) Late Ordovician tillites of the Central Sahara. In: *Earth's Pre-Pleistocene Glacial Record* (Eds Hambrey, M.J. & Harland, W.B.) Cambridge University Press, Cambridge, UK, pp. 99–107, 1004 pp.

Bjørlykke, K. (1985) Glaciations, preservation of their record and sea-level changes — a discussion based on the Late Precambrian and Lower Paleozoic sequence in Norway. *Palaeogeogr. Palaeoclim., Palaeoecol.* **51**, 197–207.

Brenchley, P.J. (Ed.) (1984) Late Ordovician extinctions and their relationship to the Gondwana Glaciation. In: *Fossils and Climate*, Wiley, Chichester, UK, pp. 291–315, 352 pp.

Brenchley, P.J. (1988) Environmental changes close to the Ordovician–Silurian boundary. *Bull. Brit. Mus. (Nat. Hist.) Geol.* **43**, 377–385.

Brenchley, P.J. & Newall, G. (1980) A facies analysis of Upper Ordovician regressive sequences in the Oslo region, Norway — a record of glacio-eustatic changes. *Palaeogeogr. Palaeoclim. Palaeoecol.* **31**, 1–38.

Brenchley, P.J. & Storch, P. (1989) Environmental changes in the Hirnantian (Upper Ordovician) of the Prague basin, Czechoslovakia. *J. Geol.* **24**, 165–181.

Brenchley, P.J., Romano, M., Young, T.P. & Storch, P. (1991) Hirnantian glaciomarine diamictites — evidence for the spread of glaciation and its effect on Upper Ordovician faunas. In: *Advances in Ordovician Geology* (Eds Barnes, C.R. & Williams, S.H.) Geol. Surv. Can. Paper 90-9, 325–336.

Brenner, R.L., Swift, D.J.P. & Gaynor, G.C. (1985) Reevaluation of coquinoid sandstone depositional model, Upper Jurassic of Wyoming and Montana. *Bull. Geol. Soc. Am.* **84**, 1685–1697.

Caputo, M.V. & Crowell, J.C. (1985) Migration of glacial centres across Gondwana during Paleozoic Era. *Bull. Geol. Soc. Am.* **96**, 1021–1036.

Chen Xu (1984) Influence of the late Ordovician glaciation on basin configuration of the Yangtze Platform in China. *Lethaia* **17**, 51–59.

Cisne, J.L. & Gildner, R.F. (1988) Measurement of sea-level change in epeiric seas: the Middle Ordovician transgression in the North American Midcontinent. In: *Sea-level Changes: An Integrated Approach* (Eds Wilgus, C.K., Hastings, B.S., Kendall, C.G.St.C., Posamentier, H.W., Ross, C.A. & Van Wagoner, J.C.) Spec. Publ. Soc. Econ. Paleontol. Mineral. 42, 217–225.

Cisne, J.L., Gildner, R.F. & Rabe, B.D. (1984) Epeiric sedimentation and sea level: synthetic ecostratigraphy. *Lethaia* **17**, 267–288.

Cotter, E. (1985) Gravel-topped bar sequences in the Lower Carboniferous of Southern Ireland. *Sedimentology* **32**, 195–213.

Dangeard, L. & Doré, F. (1971) Faciès glaciaires de l'Ordovicien Supérieur en Normandie. *Mem. Bureau de Recherches Géologiques et Minières* **73**, 119–127.

Destombes, J. (1981) Hirnantian (Upper Ordovician) tillites on the north flank of the Tindouf Basin, Anti-Atlas, Morocco. In: *Earth's Pre-Pleistocene Glacial Record* (Eds. Hambrey, M.J. & Harland, W.B.) Cambridge University Press, Cambridge, UK, pp. 84–88, 1004 pp.

Deynoux, M. (1985) Terrestrial or waterlain glacial diamictites? Three case studies from the Late Precambrian and Late Ordovician glacial drifts in West Africa. *Palaeogeogr., Palaeoclim., Palaeoecol.* **51**, 97–142.

Deynoux, M. & Trompette, R. (1981) Late Ordovician tillites of the Taoudeni Basin, West Africa. In: *Earth's Pre-Pleistocene Glacial Record* (Eds Hambrey, M.J. & Harland, W.B.) Cambridge University Press, Cambridge, UK, pp. 89–96, 1004 pp.

Doré, F. (1981) The late Ordovician tillite in Normandy (Armoricain Massif). In: *Earth's Pre-Pleistocene Glacial Record* (Eds Hambrey, M.J. & Harland, W.B.) Cambridge University Press, Cambridge, UK, pp. 582–585, 1004 pp.

Doré, F., Dupret, L. & LeGall, J. (1985) Tillites et tilloïdes du Massif Armoricain. *Palaeogeogr., Palaeoclim., Palaeoecol.* **51**, 85–96.

Dott, R.H. & Bourgeois, J. (1982) Hummocky cross-stratification: significance of its variable bedding sequence. *Bull. Geol. Soc. Am.* **93**, 663–680.

Duke, W.L., Arnott, R.W.C. & Cheel, R.C. (1991) Shelf sandstones and hummocky cross-stratification: new insights on a stormy debate. *Geology* **19**, 625–628.

Fyffe, L.R. (1982) Taconian and Acadian structural trends in central and northern New Brunswick. In: *Major Structural Zones and Faults of the Northern Appalachians* (Eds St-Julien, P. & Bélande, J.), Geol. Assoc. Can. Spec. Pap. **24**, 117–130.

Geng Liang-Yu (1982) Late Ashgillian glaciation — effects of eustatic fluctuations on the Upper Yangtze Sea. In: *Stratigraphy and Palaeontology of Systemic Boundaries in China. Ordovician - Silurian Boundary* (Eds. Nanjing Institute of Geology and Paleontology, Academia Sinica) vol. 1. Anhui Science Technical Publishing House, China, pp. 269–286.

Gildner, R.F. & Cisne, J.L. (1987) Comparison of albedo changes due to shelf exposure and ice during the Late Ordovician glacioeustatic regression. *Paleoceanography* **2**, 117–183.

Hambrey, M.J. (1981) Paleozoic tillites in northern Ethiopia. In: *Earth's Pre-Pleistocene Glacial Record* (Eds. Hambrey, M.J. & Harland, W.B.) Cambridge University Press, Cambridge, UK, pp. 38–40, 1004 pp.

Hambrey, M.J. (1985) The late Ordovician–early Silurian glacial period. *Palaeogeogr., Palaeoclim., Palaeoecol.* **51**, 273–289.

Harland, W.B., Armstrong, R.L., Cox, A.V., Craig, L.E., Smith, A.G. & Smith, D.G. (1989) A geological time scale 1989. Card published by British Petroleum Company plc, by arrangement with Cambridge University Press, July 1989.

Harms, J.C. (1979) Primary sedimentary structures. *Ann. Rev. Earth Planet. Sci.* **7**, 227–284.

Hunter, R.E. & Clifton, H.E. (1982) Cyclic deposits and hummocky cross-stratification of probable storm origin in Upper Cretaceous rocks of the Cape Sebastian area, southwestern Oregon. *J. Sediment. Petrol.* **52**, 127–143.

James, N.P. & Stephens, R.K. (1982) Anatomy and evolution of a lower Paleozoic continental margin, western Newfoundland. 11th International Congress of the International Association of Sedimentologists, Hamilton, Ontario, *Field Excursion Guidebook* **2B**, 75 pp.

Johnson, M.E. (1975) Recurrent community patterns in epeiric seas: the lowest Silurian of eastern Iowa. *Proc. Iowa Acad. Sci.* **82**, 130–139.

Kent, D.V. & Van der Voo, R. (1990) Palaeozoic palaeogeography from palaeomagnetism of the Atlantic-bordering continents. In: *Palaeozoic Palaeogeography and Biogeography* (Eds McKerrow, W.S. & Scotese, C.R.) Mem. Geol. Soc. 12, 49–56.

Long, D.G.F. (1988) Amplitude and frequency of Late Ordovician glacio-eustatic sea level fluctuations. In: *Program and Abstracts* (Eds Williams, S.H. & Barnes, C.R.) Fifth International Symposium on the Ordovician System, August 9–12, 1988. Memorial University of Newfoundland, St. John's, Newfoundland, Canada, p. 53.

Long, D.G.F. & Copper, P. (1987a) Stratigraphy of the Upper Ordovician, upper Vauréal and Ellis Bay Formations, eastern Anticosti Island, Quebec. *Can. J. Earth Sci.* **24**, 1807–1820.

Long, D.G.F. & Copper, P. (1987b) Late Ordovician sand-wave complexes on Anticosti Island, Quebec: a marine tidal embayment. *Can. J. Earth Sci.* **24**, 1821–1832.

McClure, H.A. (1978) Early Paleozoic glaciation in Arabia. *Palaeogeogr., Palaeoclim., Palaeoecol.* **25**, 315–336.

McClure, H.A. (1988) The Ordovician–Silurian boundary in Saudi Arabia. *Bull. Brit. Mus. (Nat. Hist.), Geol.* **43**, 155–163.

McCracken, A.D. & Barnes, C.R. (1981) Conodont biostratigraphy and paleoecology of the Ellis Bay Formation, Anticosti Island, Quebec, with special reference to late Ordovician–early Silurian chronostratigraphy and the systematic boundary. *Bull. Geol. Surv. Can.* **329**, Part 2, pp. 51–134, 137–145.

McKerrow, W.S. & Cocks, L.R.M. (1977) The location of the Iapetus Ocean suture in Newfoundland. *Can. J. Earth Sci.* **14**, 448–495.

Matthews, R.K. (1984) *Dynamic Stratigraphy, An Introduction to Sedimentation and Stratigraphy*, 2nd edn. Prentice Hall, New Jersey, 489 pp.

Nummedal, D. & Swift, D.J.P. (1987) Transgressive stratigraphy at sequence-bounding unconformities: some principles derived from Holocene and Cretaceous examples. In: *Sea-level Fluctuation and Coastal Evolution* (Eds Nummedal, D., Pilkey, O.H. & Howard, J.D.) Spec. Publ. Soc. Econ. Paleontol. Mineral. 41, 241–260.

Orth, C.J., Gilmore, J.S., Quintana, L.R. & Sheehan, P.M. (1986) Terminal Ordovician extinction: geochemical analysis of the Ordovician/Silurian boundary, Anticosti Island, Quebec. *Geology* **14**, 433–436.

Peltier, W.R. (1987) Glacial isostasy, mantle viscosity and Pleistocene climatic change. In: *North America and Adjacent Oceans during the Last Deglaciation* (Eds Ruddiman, W.F. & Wright, H.E) Geological Society of America, The Geology of North America, K3, pp. 155–182, 619 pp.

Petryk, A.A. (1981) Upper Ordovician glaciation: effects of eustatic fluctuations on the Anticosti Platform succession, Quebec. In: *Field Meeting, Anticosti-Gaspé, Québec, 1981. Volume II Stratigraphy and Paleontology*

(Ed. Lespérance, P.J.) Departement de Géologie, Université de Montréal, pp. 81–85.

Pickering, K.T. (1987) Deep-marine foreland basin and forearc sedimentation: a comparative study from the Lower Palaeozoic Northern Appalachians, Quebec and Newfoundland. In: *Marine Clastic Sedimentology* (Eds Leggett, J.K. & Zuffa, G.G.) Graham & Trotman, London, pp. 190–211, 211 pp.

Robardet, M. & Doré, F. (1988) The Late Ordovician diamictic formations from southwestern Europe: north Gondwana glaciomarine deposits. *Palaeogeogr., Palaeoclim., Palaeoecol.* **66**, 19–31.

Rust, I.C. (1981) Early Paleozoic Pakhuis Tillite, South Africa. In: *Earth's Pre-Pleistocene Glacial Record* (Eds. Hambrey, M.J. & Harland, W.B.) Cambridge University Press, Cambridge, UK, pp. 113–117, 1004 pp.

Seguin, M.K. & Petryk, A.A. (1986) Paleomagnetic study of the late Ordovician–early Silurian platformal sequence of Anticosti Island, Quebec. *Can. J. Earth Sci.* **23**, 1880–1890.

Shackleton, N.J. & Opdyke, N.D. (1977) Oxygen isotope and palaeomagnetic evidence for an early northern hemisphere glaciation. *Nature* **270**, 216–219.

Stride, A.H. (1982) *Offshore Tidal Sands, Processes and Deposits.* Chapman & Hall, London, 222 pp.

Swift, D.J.P. & Field, M.E. (1981) Evolution of a clastic sand ridge field, Maryland sector, North American inner shelf. *Sedimentology* **28**, 461–482.

Swift, D.J.P., Figueiredo, A.G., Freeland, G.L. & Oertel, G.F. (1983) Hummocky cross-stratification and megaripples; a geological double standard? *J. Sediment. Petrol.* **53**, 1295–1317.

Tucker, M.E. & Reid, P.C. (1981) Late Ordovician glaciomarine sediments, Sierra Leone. In: *Earth's Pre-Pleistocene Glacial Record* (Eds Hambrey, M.J. & Harland, W.B.) Cambridge University Press, Cambridge, UK, pp. 97–98, 1004 pp.

Young, T.P. (1988) The lithostratigraphy of the upper Ordovician of central Portugal. *J. Geol. Soc. London* **145**, 337–392.

Spec. Publs Int. Ass. Sediment. (1993) **18**, 501–520

The effect of tectonic and eustatic cycles on accommodation and sequence-stratigraphic framework in the Upper Cretaceous foreland basin of southwestern Wyoming

W.J. DEVLIN, K.W. RUDOLPH, C.A. SHAW *and* K.D. EHMAN

Exxon Production Research Co., PO Box 2189, Houston, TX, 77252-2189, USA

ABSTRACT

Both tectonics and eustasy contribute to a basin's accommodation, and hence, to observed cyclic stratal patterns, but distinguishing between the relative contribution of these components is not always straightforward. This paper addresses this problem with an example from the Upper Cretaceous foreland basin of southwestern Wyoming. Three large-scale sedimentation cycles, 500 m to 1500 m in thickness, are present in the Upper Cretaceous section, and these cycles consist of two parts. The lower part is characterized by a relatively deep-water shale succession that exhibits an abruptly deepening or retrogradational stacking pattern. The upper part consists of shallow-marine to non-marine sandstone units which exhibit an overall progradational stacking pattern. Geohistory analysis from a stratigraphic section at Rock Springs, Wyoming indicates that the onset of deposition of the lower, shale-prone part of each cycle corresponds with an acceleration in subsidence rate. The increases in subsidence rate were caused by thrust movements in the Sevier Fold-Thrust Belt. Deposition of the upper, progradational part of each cycle occurred during times of decreased subsidence rate and can be tied to periods of relative quiescence in the thrust belt. Independent evidence for the timing of thrust movements is provided by palynologically dated, syntectonic deposits in the thrust belt.

Basement-involved, Laramide-style uplifts also affected the foreland basin in the Late Cretaceous, and were responsible for dramatic changes in depositional style and palaeogeography. The relative timing between thrusting in the fold-thrust belt and basement-involved tectonism in the foreland basin determined stratigraphic patterns within the sedimentary fill. When the uplift of basement structures occurs during times of relative quiescence in the thrust belt, the tectonic movements are said to be 'out-of-phase'. In basins where thrusting and basement-involved movements occur simultaneously, the events are said to occur 'in-phase'. Thick, aggradational braided stream sediments were deposited in the vicinity of Rock Springs in association with out-of-phase movements. A period of in-phase tectonism was associated with the progradation of deltaic complexes that were supplied with abundant sediment from uplifted areas and deposited in adjacent, differentially subsiding sub-basins. The sub-basins were also the site of basinally restricted lowstand deposition during higher order sea-level fluctuations.

Superimposed on the larger scale, subsidence-related stratal packages are third- and fourth-order depositional sequences. The expression of the higher order sequences and their component systems tracts was controlled by the larger scale, subsidence-related accommodation cycles. However when both the regional and local tectonic controls in the basin are accounted for, a shorter duration, higher order control on relative sea-level fluctuations is still needed to account for the deposition of these sequences. Our working model is that the controlling mechanism, at least for the third-order sequences, is eustasy. However, other mechanisms may account for some of the observed depositional cyclicity, and these are discussed. Regardless of the mechanism, the observed stratigraphic relationships indicate that higher order fluctuations in relative sea level occurred in the Late Cretaceous that cannot be directly attributed to thrust-belt tectonics or intra-foreland basement-involved uplifts.

INTRODUCTION

Sequence stratigraphy relates sedimentary strata to variations in accommodation through the interpretation of stratal geometry and stacking patterns within a basin. Accommodation, defined as the space available for sediment to be deposited, is a function of basin tectonics (subsidence and uplift), eustasy and sediment supply (Jervey, 1988; Posamentier *et al.*, 1988)*. The effects of subsidence on accommodation have been considered for thermally subsiding, passive margins by Jervey (1988), Posamentier *et al.* (1988) and Posamentier and Vail (1988). However, a similar analysis for basins in compressional regimes, specifically in foreland basins, has not been previously addressed, nor have the effects of tectonic controls on observed, cyclic sedimentation patterns in these basins been fully considered within a sequence-stratigraphic framework.

The major tectonic controls on accommodation in a foreland basin include the flexural response of the basin to periods of thrusting and quiescence in the thrust belt and intra-foreland tectonic movements. In the Upper Cretaceous foreland of southwestern Wyoming, intra-foreland tectonic movements include basement-involved, Laramide-style uplifts. Another potential tectonic control on accommodation in compressional basins includes variations in in-plane stress (Cloetingh *et al.*, 1986). In this paper we isolate and document the tectonic components of basin accommodation within the Upper Cretaceous foreland basin of southwestern Wyoming, and examine the stratal response to basin tectonics in the stratigraphic section exposed in the southeastern Rock Springs Uplift. We then briefly examine the stratigraphic expression of smaller scale, third- and fourth-order changes in relative sea level that are superimposed on the large-scale tectonically influenced stratal patterns. Finally, with the working hypothesis that eustasy is the driving mechanism for at least the third-order cycles, we evaluate the contribution of the tectonic and eustatic components of accommodation to the Upper Cretaceous stratigraphy in the Rock Springs Uplift.

* Fundamental sequence-stratigraphic concepts and terminology applied in this paper can be found in Vail *et al.* (1977), Jervey (1988), Posamentier and Vail (1988), Posamentier *et al.* (1988) and Van Wagoner *et al.* (1988, 1990).

REGIONAL GEOLOGY AND DATA BASE

The Upper Cretaceous foreland basin of southwestern Wyoming was formed as a result of flexural loading of the lithosphere by the emplacement of several major thrust sheets (Jordan, 1981). In the Idaho–Wyoming Thrust Belt, the timing for the emplacement of these thrust sheets is relatively well known (Oriel & Armstrong, 1966; Royse *et al.*, 1975; Jacobson & Nichols, 1982; Wiltschko & Dorr, 1983). The Wyoming foreland also has an additional level of complexity in that basement-involved, Laramide-style, structural movements have tectonically partitioned the basin since mid-Campanian time (Fig. 1). Stratigraphic sections in the basin consequently reflect this tectonic activity in both their stratal architecture and distribution of sedimentary facies. Throughout this paper the effects of tectonic and eustatic changes in accommodation are related to the stratigraphic section exposed in the southeastern Rock Springs Uplift (Fig. 1).

The total area included in this study consists of the portions of the Greater Green River Basin that include the Rock Springs and Rawlins Uplifts, the Washakie and Red Desert Basins and the southern Green River Basin (Fig. 1). An integrated data base of seismic, well-log and outcrop data was inter-

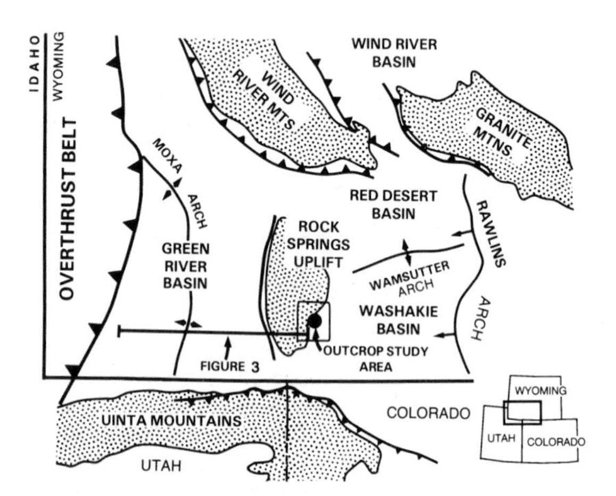

Fig. 1. Location map of southwestern Wyoming. Stippled areas represent Laramide basement-involved uplifts. The location of the cross-section of Fig. 3 is indicated. The location of the stratigraphic column and geohistory analysis from the southeastern Rock Springs Uplift (Figs 2 & 4, respectively) indicated by the dot.

preted throughout this area. The study utilized a detailed sequence-stratigraphic framework established in exposures of the Upper Cretaceous section in the southeastern Rock Springs Uplift by Rahmanian *et al.* (1990) (Fig. 2), a framework that was extended into the subsurface by correlation into nearby wells and seismic data. The subsurface data were also tied into biostratigraphically dated strata on the eastern side of the Washakie Basin (Gill *et al.*, 1970). Ammonite biostratigraphy and palynology were used for age control (Hale, 1950; Smith, 1961; Cobban, 1969; Gill *et al.*, 1970; Miller, 1977; Roehler, 1978; Chen *et al.*, 1991). Regional stratigraphic relationships across the southwestern part of Wyoming are summarized in the cross-section of Fig. 3. This regional cross-section was constructed, in part, from seismic and well-log data, and extends approximately 500 km (300 miles) from the overthrust belt, across the Moxa Arch and the Rock Springs Uplift, to the western Washakie Basin (Fig. 1). The interpretation of the integrated subsurface data base was essential to understanding the regional stratigraphy. These data: (i) allowed for the correlation of chronostratigraphic packages and their attendant facies; (ii) provided a means of correlating non-fossiliferous, non-marine strata into age-constrained strata; and (iii) led to a detailed understanding of structural and depositional timing, especially related to intra-foreland movements.

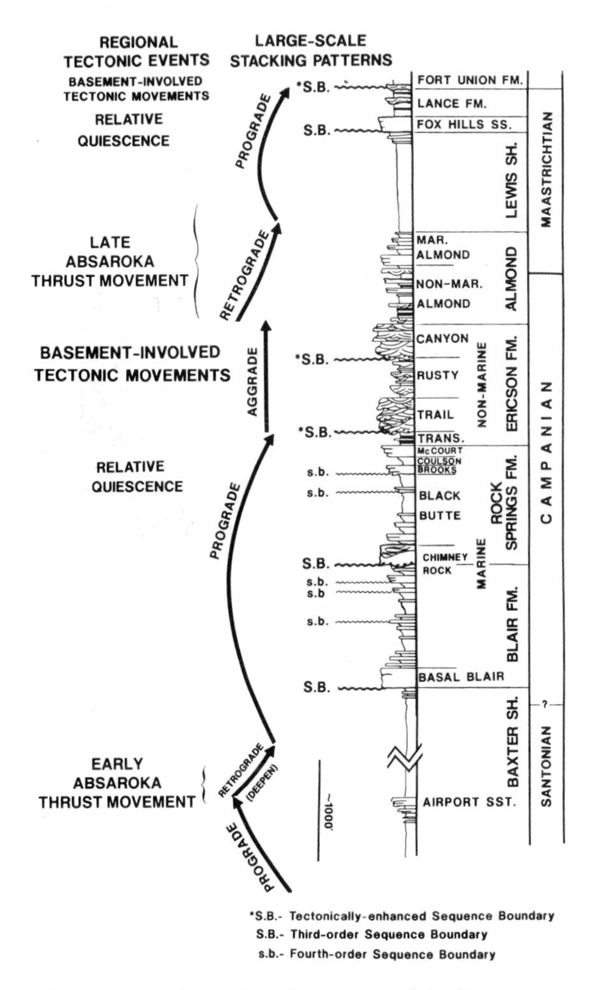

Fig. 2. Schematic stratigraphic column of the Upper Cretaceous section of the southeastern Rock Springs Uplift (schematic column from Rahmanian *et al.*, 1990). The large-scale stratal stacking patterns and their relationship to regional tectonic events are shown. Times of thrusting shown indicate *onset* of thrust-sheet emplacement. Smaller-scale, third- and fourth-order sequence boundaries are also denoted.

Structural movements on basement-involved Laramide structures such as the Rock Springs Uplift and the Moxa Arch can be recognized on seismic data by truncation, onlap and thinning of reflections over the structures. The timing of the movements was determined from the geometric relationships of the biostratigraphically dated seismic reflections. Three periods of significant late Cretaceous, basement-involved tectonism are recognized on the seismic data. Two of these movements occurred in middle to late Campanian time, and the third occurred in the latest Maastrichtian to early Palaeocene. In the exposed strata of the Rock Springs Uplift, unconformities associated with these movements are found at the base of the Trail and Canyon Members of the Ericson Formation and at the base of the Fort Union Formation, respectively (Figs 2 & 3). Miller (1977) and Hendricks (1990) have recognized truncation beneath the Trail Member of the Ericson Formation, and widespread truncation of strata has been mapped below the base of the Fort Union Formation (Love & Christiansen, 1985). Although significant uplift has not been previously associated with the unconformity at the base of the Canyon Member of the Ericson Formation, marked erosion and stratigraphic discordance have been associated with the equivalent surface east of the Washakie Basin (Gill & Cobban, 1966a; Reynolds, 1967, 1976; Roehler, 1989).

West of the Rock Springs Uplift, an angular unconformity is present across the Moxa Arch as a result of basement-involved uplift along an east-

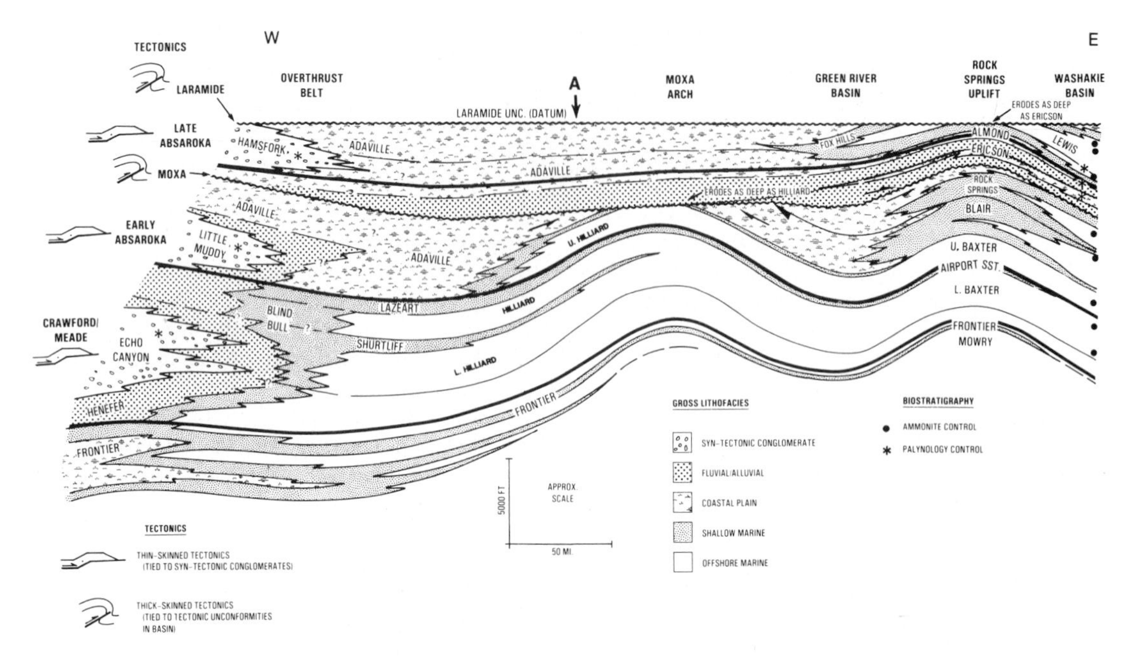

Fig. 3. Schematic east–west cross-section depicting stratigraphic relationships of Upper Cretaceous strata in southwestern Wyoming. Location of the cross-section is indicated in Fig. 1. The cross-section is constructed from interpreted regional seismic and well data east of point A. West of point A the section is schematic and based on published biostratigraphic and lithostratigraphic data (Royse *et al.*, 1975; Jacobson & Nichols, 1982; Wiltschko & Dorr, 1983; Lawrence, 1984). Tectonic events are noted at left. Palynological age control is denoted by asterisks, ammonite age control is indicated by dots. Heavy lines mark the bases of the three large-scale, subsidence-related sedimentation cycles (discussion in text). Deposition of the Ericson Formation is not considered part of the middle cycle, but instead is related to sedimentation associated with intra-foreland basement-involved movements.

dipping thrust (Kraig *et al.*, 1987). Stratigraphic relationships on the crest of the arch indicate that the middle Campanian Ericson Formation overlies Santonian Hilliard Shale across the unconformity (Fig. 3). Furthermore, there is evidence that other basement-involved uplifts in the region were active during the late Cretaceous, including the Lost Soldier Anticline (Reynolds, 1967), the ancestral Wind River and Gros Ventre Ranges (Wiltschko & Dorr, 1983; Cerveny, 1990), the Wamsutter Arch and the Blacktail–Snowcrest Uplift of southwestern Montana (Nichols *et al.*, 1985).

BASIN SUBSIDENCE, THRUST BELT TECTONICS AND THE TECTONIC COMPONENT OF ACCOMMODATION

Tectonic subsidence analysis, also referred to as geohistory (Van Hinte, 1978), was used to isolate and document the tectonic component of basin accommodation in a composite outcrop/subsurface section from the southeastern Rock Springs Uplift (Figs 1 & 4). In the tectonic subsidence curve of Fig. 4, three deflections in the curve indicate an acceleration in the rate of tectonic subsidence and, therefore, an increase in the tectonic component of accommodation. The *onset* of each of these subsidence events occurs at the late Coniacian, late Santonian and earliest Maastrichtian* (Fig. 4). Each of these deflections is followed by a flattening of the subsidence curve, indicating a deceleration in the rate of tectonic subsidence and a decrease in subsidence-related accommodation. There are also three periods of uplift indicated on the curve that are associated with the three late Cretaceous, basement-involved tectonic movements of the Rock Springs Uplift discussed above.

* The data used to construct the geohistory curve are presented in an Appendix (p. 516–518) at the end of the paper.

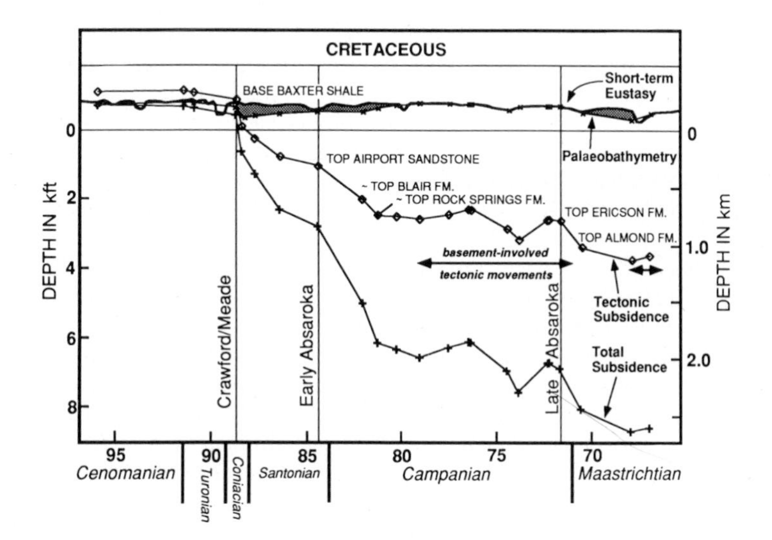

Fig. 4. Geohistory plot for the Upper Cretaceous section of the Rock Springs Uplift showing curves for total and tectonic subsidence, short-term eustasy and palaeobathymetry. Annotations indicate the times for the *onset* of thrusting events in the Idaho–Wyoming Thrust Belt, movement of the Rock Springs Uplift and tops of selected lithostratigraphic units (discussion of age constraints in Appendix (p. 516–518)). Sensitivity analysis of decompaction and the removal of palaeobathymetry and eustatic estimates do not affect the form of the subsidence curves. Furthermore, the assumption of Airy isostasy in the analysis (in a flexurally subsiding basin) does not affect the timing of the subsidence events.

A foreland basin is the product of regional isostatic adjustments that result from the emplacement of a tectonic load in the geodynamically coupled mountain belt (Beaumont, 1981). Although there are several rheological models that attempt to describe the lithosphere's flexural response to loading, the important feature shared by all of the models is that the lithosphere responds initially to the emplacement of a load with an instantaneous elastic deformation (Quinlan & Beaumont, 1984). The elastic deformation consists of the flexural downbending of the lithosphere beneath and surrounding the load produced by the thrust sheet. The flexural downbending caused by the emplacement of the load is manifest in the foreland basin by an abrupt acceleration in subsidence. Since subsidence is a major contributing factor to a basin's accommodation, there will be an attendant increase in the accommodation rate as well. The three subsidence events mentioned above are interpreted to have resulted from thrust loading in the orogenic belt (Devlin *et al.*, 1990; Devlin & Shaw, 1990). The deceleration in subsidence is interpreted in terms of the cessation of loading during a period of relative tectonic quiescence. The subsidence acceleration/deceleration couplets are collectively referred to as a subsidence cycle.

Independent evidence for the timing of thrust loading in the orogenic belt is provided by sediments in the proximal portion of the foreland that have been interpreted as syntectonic deposits. These units include the Echo Canyon Conglomerate, the Little Muddy Creek Conglomerate and the Hams Fork Conglomerate (Fig. 3), and have been linked with the creation of relief and sediment source areas resulting from the emplacement of the Crawford/Meade, early Absaroka and late Absaroka Thrust Sheets, respectively (Oriel & Armstrong, 1966; Royse *et al.*, 1975; Wiltschko & Dorr, 1983). From oldest to youngest, these have been dated palynologically as Coniacian/Santonian, Santonian and late Campanian/early Maastrichtian (Jacobson & Nichols, 1982; Nichols & Bryant, 1990; see Appendix, pp. 516–518). These ages are in excellent agreement with the timing of the *onset* of the three subsidence cycles in the Rock Springs area (Fig. 4). Note that the ages of thrusting interpreted from the palynologically dated syntectonic deposits are totally independent from the timing of thrusting interpreted from the subsidence analysis at Rock Springs, which is based on stratigraphic thicknesses and age control from that locality.

STRATIGRAPHIC EXPRESSION OF BASIN TECTONICS

Three large-scale sedimentation cycles*, ranging from 500 to 1500 m in thickness, are recognized in the Upper Cretaceous section of the Green River Basin. Each of these cycles consists of a lower deeper-water shale-prone succession and an upper shallow-marine to non-marine sandstone-prone succession. The lower shale-prone part of each cycle exhibits an abruptly deepening or retrogradational stacking of depositional facies and stratal units. The upper sandstone-prone part represents a shift in the locus of coarse clastic deposition into the basin, and exhibits a progradational stacking of depositional facies and stratal units. In the Rock Springs area, the three sedimentation cycles are represented by (i) the lower Baxter Shale and overlying Airport Sandstone; (ii) the upper Baxter Shale and overlying Blair and Rock Springs Formations; and (iii) the upper Almond Formation/Lewis Shale and overlying Fox Hills and Lance Formations (Figs 2 & 3). The non-marine strata of the Ericson Formation were deposited in association with basement-involved, Laramide-style tectonic movements (Figs 2 & 3), and are discussed in a following section.

The onset of rapid subsidence in each of the subsidence cycles in the geohistory analysis (Fig. 4) corresponds with the onset of deposition of the lower part of each sedimentation cycle. Furthermore, biostratigraphic correlation indicates that these shale-prone strata are equivalent to the coarse clastic syntectonic strata in the proximal foreland (Fig. 3). This tectonically active phase of deposition in the two lower cycles is characterized by the abrupt introduction of deeper water shales within the Baxter Shale. However, the syntectonic portion of the uppermost cycle is marked by a well-developed retrogradational succession consisting of non-marine alluvial/coastal plain strata, through backstepping nearshore marine sandstones, to offshore marine shale (the Almond and Lewis Formations) (Figs 2 & 3). The difference in the sedimentological expression between the two lower cycles and the upper cycle is interpreted to have been related to greater sediment supply during deposition of the upper cycle. The enhanced sediment supply was a result of uplift of basement-involved structures which occurred immediately prior to and during the emplacement of the late Absaroka Thrust and, hence, the deposition of the upper sedimentation cycle. In addition, the thrust belt and its supply of sediment had been advancing further eastward with each successive thrust emplacement.

As indicated on the geohistory analyses, the upper progradational and coarser grained part of each cycle was deposited during times of decreased subsidence rate. No syntectonic deposits were deposited updip of these strata, and these times are interpreted as periods of relative tectonic quiescence in the thrust belt when the effect of thrust loading and subsidence-related accommodation in the basin decreased. In the Rock Springs area, the Airport Sandstone overlies the lower Baxter Shale, and was deposited following the active emplacement of the Crawford/Meade Thrust (Figs 2 & 3). The relative thinness of the Airport Sandstone is attributed to renewed subsidence associated with early movement on the Absaroka Thrust and, consequently, a shift in the locus of coarse clastic sedimentation back to the proximal foreland basin.

Following early emplacement of the Absaroka Thrust (and deposition of the upper Baxter Shale), the Upper Cretaceous section entered an overall progradational phase with deposition of the Blair and Rock Springs Formations (Figs 2 & 3). The uppermost subsidence cycle begins with the retrogradational succession in the upper Almond through Lewis Formations, and is related to late emplacement of the Absaroka Thrust (Figs 2 & 3). The change to a progradational stacking pattern occurs within the upper Lewis Shale and continues with marine to non-marine deposition of the Fox Hills and Lance Formations, respectively (Fig. 2). The upper part of this cycle is truncated by the Laramide unconformity at the base of the Fort Union Formation (Fig. 3).

* In this paper the term 'large-scale' sequence or sedimentation cycle is often used to indicate a scale of sedimentary cyclicity that is larger than identified third-order sequences in terms of thickness (thicknesses between 500–1500 m). In Wyoming, these larger scale sequences are not second-order sequences in the sense of Vail *et al.* (1977) because they are less than 10 Ma in duration.

THRUST BELT–FORELAND RESPONSE MODEL

The basic conceptual model to describe thrust belt–foreland interaction and deposition of the larger-scale stratigraphic packages has been previously presented by several authors (Covey, 1986; Beck *et al.*, 1988; Blair & Bilodeau, 1988; Heller *et al.*, 1988; Flemings & Jordan, 1990). In southwestern Wyoming, we can document the thrust belt–foreland basin interaction over three separate subsidence and sedimentation cycles. During periods of thrusting (Fig. 5(a)), rapid flexural subsidence occurs in the foreland as a result of the emplacement of the thrust load. Consequently, there is a rapid increase in accommodation related to this subsidence. Sediment supply is typically high because of erosion in the rising mountain belt, but most of the sediment is deposited as thick accumulations proximal to the mountain front because of the large amount of space being generated there via subsidence. Stratal stacking in proximal areas tends to be aggradational to retrogradational, with little progradation of sediment into the basin (Fig. 5(a)).

In a medial position of the foreland basin (e.g. at the position of Rock Springs), the stratigraphic section displays an abrupt deepening or overall retrogradational stacking pattern because the rate of subsidence exceeds the rate of sediment supply. Moreover, since most of the coarse-grained sediments are being deposited in the most proximal portion of the foreland, any sediment deposited in medial to distal positions during tectonically active times is commonly fine-grained (shale-prone).

During times of relative tectonic quiescence, there is a decrease in the rate of subsidence related to the cessation of thrust loading. The possibility also exists for relative uplift associated with isostatic rebound. As a result, the rate of change of subsidence-related accommodation decreases and incoming sediment quickly fills any accommodation that is left over from the active thrusting phase. A progradational sedimentation pattern results because very little new space is being generated in the proximal portion of the foreland and incoming coarse clastic sediment is bypassed out into more distal parts of the basin.

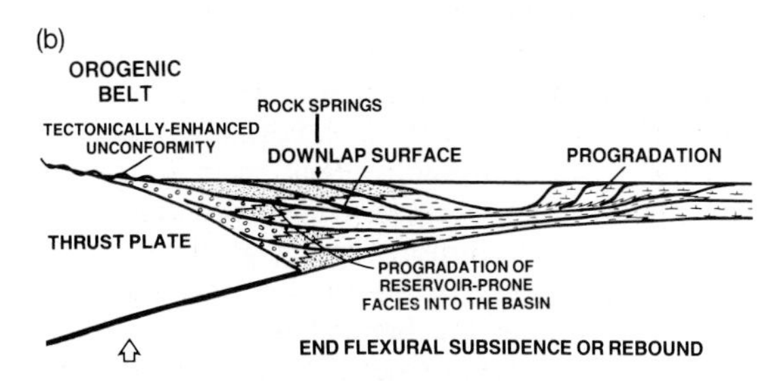

Fig. 5. Schematic cross-sections depicting basin-scale stratal patterns resulting from thrust belt foreland basin interaction. (a) Foreland stratal patterns during active thrusting in the thrust belt. (b) Foreland stratal patterns during relative tectonic quiescence in the thrust belt. The approximate position of the stratigraphic section deposited at Rock Springs is indicated (Figs 2 & 4).

THE INTERACTION OF THRUST BELT AND INTRA-FORELAND MOVEMENTS AND THEIR EFFECT ON SEDIMENTATION

An important variation on the model outlined above occurs in basins where basement-involved, Laramide-style tectonism produces significant uplift within portions of the foreland. Structural highs tectonically partition the basin into discrete subbasins and can significantly affect palaeogeography, sediment dispersal patterns, sequence stacking and lithofacies distribution. The relative timing between thrusting in the fold thrust belt and basement-involved tectonism in the foreland basin plays a major role in determining stratigraphic patterns in the sedimentary fill (Devlin & Rudolph, 1991). When the uplift of basement structures occurs during times of relative quiescence in the thrust belt, the tectonic movements are said to be 'out of phase' (Fig. 6(b)). In basins where thrusting and basement-involved movements occur simultaneously, the events are said to occur 'in phase' (Fig. 6(a)).

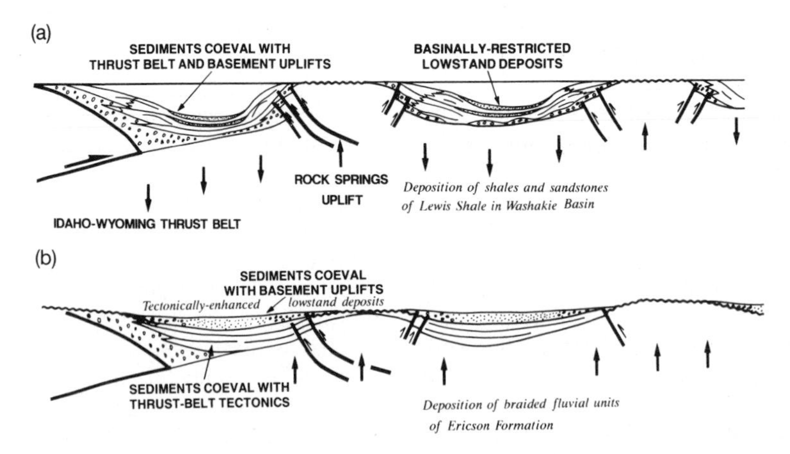

Fig. 6. Schematic cross-sections depicting the interaction of tectonic events in the thrust belt with basement-involved tectonic movements in the foreland basin, and the effect on foreland sedimentation patterns. (a) Stratigraphic response of foreland basins that experience thrusting (and its attendant increase in accommodation) coincident with basement-involved uplifts within the foreland basin. In this case, the two tectonic movements are said to be 'in phase' with each other. (b) Stratigraphic response of foreland basins that experience intra-foreland basement-involved uplift during times of relative quiescence in the orogenic belt. In this case, basement-involved movements are 'out of phase' with thrusting in the orogenic belt. In this figure, if the uplift in the centre of the diagram represents the Rock Springs Uplift and the basin to the right is the Washakie Basin, then deposition of the Lewis Shale and associated sandstones would be represented by the situation in (a), and deposition of the Ericson Formation would be represented by (b). See text for further discussion.

Out-of-phase tectonism

During out-of-phase tectonic movements there tends to be a decrease in regional accommodation. This results because relative quiescence in the thrust belt is marked by a lack of thrust loading, and hence, flexural subsidence in the basin. A further reduction in accommodation occurs on local to semi-regional scales within the basin because of uplift associated with the basement-involved structures although some flexural subsidence can be expected adjacent to the basement uplifts, especially along those with low-angle bounding faults (Hagen *et al.*, 1985; Jie *et al.*, 1990). Sediment supply tends to be high during out-of-phase tectonism, being derived from both the eroding thrust belt and the basement uplifts. The combined effect of reduced accommodation and high sediment supply can result in the deposition of what we term tectonically enhanced lowstand deposits. These lowstand strata are said to be tectonically enhanced because of the increased sediment supply and enhanced drop in relative sea level associated with the tectonic uplift. Marked angular unconformities can also underlie these strata. Aggradational, non-marine strata are commonly associated with these deposits in the depositionally updip positions of a basin. In addition, during out-of-phase tectonism the limited accommodation available quickly fills with incoming sediment, and continued sediment input leads to marked progradation into the basin.

Sedimentation associated with out-of-phase tectonism in the southern Wyoming foreland basin

During the Late Cretaceous, aggradational braided stream sedimentation occurred over two distinct periods of time, as represented by the Trail and Canyon Members of the Ericson Formation (Figs 2 & 3). Deposition of the Ericson Formation occurred during a period of low regional basin subsidence associated with a phase of relative tectonic quiescence in the fold thrust belt (Fig. 4). Uplift of the Rock Springs structure occurred prior to the deposition of each of the braided fluvial deposits as indicated by truncation beneath the Trail and Canyon Members on seismic data. Well logs, tied to the seismic data with synthetic seismograms, indicate that approximately 500 and 800 ft of strata were removed beneath the Trail and Canyon Members, respectively, on the eastern flank of the Rock Springs Uplift. The combined effect of low regional subsidence and uplift associated with the Rock Springs structure, as well as other Laramide structures in the region (Wiltschko & Dorr, 1983; Cerveny, 1990), led to a marked, tectonically enhanced downward shift of facies into the basin. Continued input of sediment led to pronounced progradation east of the Washakie Basin (Fig. 1).

Significant basinward progradation during deposition of the Ericson Formation is indicated by regional stratigraphic relationships. Seismic and well data in the southern Washakie Basin indicate that correlative deposits to the braided fluvial strata of the Ericson Formation were dominated by nonmarine alluvial-plain to coastal-plain strata (see also Roehler, 1989). Correlative nearshore marine strata at this time, as indicated by ammonite biostratigraphy, were being deposited east of the Rawlins Arch (Fig. 1) in the central and eastern Hanna Basin. These nearshore marine strata correspond to strandlines mapped by Zapp and Cobban (1962) and Gill and Cobban (1973) of the *Baculites perplexus* through (at most) the *Baculites scotti* ammonite Zones for the Trail Member and the *Baculites cuneatus* and perhaps part of the *Baculites reesidei* Zones for the Canyon Member (see Appendix, p. 516). These facies trends indicate a marked shift of sediment into the basin and a corresponding shift of the shoreline approximately 140 to 180 km to the east of the Rock Springs Uplift. Although the initial basinward shift is related to a tectonically enhanced drop in sea level, the subsequent basinward shift in shoreline position is interpreted as the result of tectonically enhanced lowstand deposition in a prograding lowstand wedge (or lowstand sequence set). Measured sections in the central Hanna Basin demonstrate gross progradational stacking patterns during at least deposition of the Trail Member (sections D, E and F of Gill *et al.*, 1970). The depositional model of Posamentier and Vail (1989) may explain the deposition of the aggradational braided stream deposits of the Ericson Formation. In their model progradation results in the basinward migration of the point to which river systems are graded. This, in turn, leads to widespread alluvial aggradation in depositionally updip areas because of necessary adjustments in the equilibrium stream profiles.

In-phase tectonism

During times of in-phase tectonism there tends to be high subsidence-related accommodation in the basin associated with thrust emplacement and flexural loading. However, in the vicinity of the basement-involved structures, there is a tendency towards decreased accommodation because of tectonic uplift. The difference in accommodation between the uplifts and the newly formed sub-basins can create dramatic changes in bathymetry and facies over short distances. If the uplifts remain submerged, then sediment deposited over such a feature exhibits onlap and thinning over the structure. If the basement uplifts are emergent, they can be cut by one or more unconformities that record their structural growth and/or the effects of superimposed eustatic fluctuations. Sediments eroded from the uplifts can be redeposited as restricted lowstand deposits in the adjacent sub-basin (Fig. 6(a)). Strata within the centre of the sub-basins, however, record an overall deepening during the early stages of in-phase basin tectonism because of the regional increase in subsidence rate related to flexural loading (Fig. 6(a)). With continued erosion of both the uplifts and the active fold thrust belt, large volumes of sediment can be supplied to the basin by progradational, deltaic complexes. Consequently, relatively steep depositional shelf edges may also form along the margins of the adjacent sub-basins. Any higher order sea-level fluctuations superimposed on this depositional regime could result in the deposition of basinally restricted lowstand deposits within the sub-basins (Fig. 6(a)).

Sedimentation associated with in-phase tectonism in the southern Wyoming foreland basin

After the onset of braided fluvial deposition of the Canyon Member, late emplacement of the Absaroka Thrust occurred in the fold thrust belt, initiating another subsidence cycle in the basin (Figs 3 & 4). Higher subsidence rates in the proximal portion of the basin relative to those in the vicinity of Rock Springs caused the locus of coarse-grained sedimentation to shift towards a more proximal position in the basin (Figs 3 & 5(a)). Even though subsidence rates at Rock Springs eventually overtook the high rate of sediment supply to the basin, sedimentation rates remained high enough in the initial portion of the subsidence cycle to deposit the thick, retrogradational, transgressive succession in the Almond through lower Lewis Formations (Figs 2 & 3).

However, the southern Wyoming foreland by the time of Lewis deposition had become tectonically partitioned as a result of uplift of the intra-foreland structures. The Rock Springs Uplift, although submerged, was still a relative bathymetric high as indicated on seismic data by thinning of the unit via onlap. The thickness of the Lewis Shale exposed today on the southeastern flank of the uplift measures approximately 700 ft. In the adjacent Washakie Basin, which at this time was a differentially subsiding sub-basin, approximately 2000 ft of strata were deposited. Little, if any, sediment was probably being eroded from the Rock Springs Uplift at this time.

However, abundant sediment was being supplied to the Red Desert and Washakie sub-basins from the north, and possibly, during the later stages of deposition, from the south (Asquith, 1970; Weimer, 1970; Winn *et al.*, 1987; McMillen & Winn, 1988; Perman, 1990). A major source for the sediment was probably the uplifted Wind River Range and Granite Mountains (Fig. 1). Detailed well-log correlations and seismic geometry indicate significant progradation of sediment in huge deltaic depositional systems (Asquith, 1970). Depositional shelf edges associated with these progradational systems were of the order of a few hundred metres in height (Winn *et al.*, 1987). In front of the deltas, in the Red Desert and Washakie sub-basins, sand deposition occurred as basinally restricted lowstand deposits associated with higher order fluctuations in sea level. Hence, previously existing palaeogeography and sediment dispersal patterns changed dramatically during in-phase tectonism. In addition, the shelfal parts of the basin changed to a shelf edge from a ramp depositional geometry as indicated by the stratal geometry observed in the Lewis Shale and underlying formations, respectively.

THIRD- AND FOURTH-ORDER SEQUENCE-STRATIGRAPHIC FRAMEWORK OF THE ROCK SPRINGS UPLIFT

The sequence-stratigraphic framework of Upper Cretaceous strata of the southeastern Rock Springs Uplift has been documented by means of an integrated data base that includes outcrop, well-log, seismic and biofacies data (Rahmanian *et al.*, 1990; Chen *et al.*, 1991). Moreover, within the thick Upper Cretaceous section there is a high degree of stratal resolution that facilitates the interpretation of third- and fourth-order depositional cyclicity from the observed stratal stacking patterns and physical surfaces.

The identification of third- and fourth-order sequences depends, in part, upon the scale of observations made. For example, the large-scale outcrop architecture of the Rock Springs section observed in the field, on aerial photos or on outcrop panoramas displays a repetitive stratal pattern in which recessive-weathering, shale-dominated valleys alternate with resistant, sandstone-dominated hogbacks (Fig. 7). At the scale of observation shown in Fig. 7, over approximately 3300 ft of section, stratal stacking patterns are broadly retrogradational into the shale valleys and progradational into the stratigraphically overlying hogbacks. Similarly, gross parasequence stacking patterns of the same stratigraphic succession in well logs (Fig. 8) exhibit an overall retrogradational parasequence stacking into shale-dominated intervals, and overall progradational stacking patterns above the shale-dominated intervals into sandstone-dominated intervals (Fig. 8). Age control provided by ammonite biostratigraphy (Smith, 1961; Roehler, 1978) indicates that the cyclic alternation observed at this large outcrop scale occurs over intervals of ~1.0 to 2.0 Ma, and, in the sense of Vail *et al.* (1977), is an expression of third-order depositional cyclicity. The shale-dominated intervals in the valleys are interpreted as the outcrop expression of third-order condensed sections and downlap surfaces. In other words, these stratigraphic intervals represent the times of maximum landward encroachment of the shoreline (Fig. 3) (Loutit *et al.*, 1988). Third-order sequence boundaries occur within the sandstone-dominated ridges (Figs 2 & 7), and are identified by basinward shifts in facies and reversals in parasequence stacking patterns. Third-order sequence boundaries and downlap surfaces are expressed on nearby seismic data, and biofacies data also emphasize the third-order cyclicity in the Upper Cretaceous section (Chen *et al.*, 1991).

When more detailed observations are made on outcrops and in well logs, a higher degree of stratal resolution can be discerned. These observations lead to the recognition of more detailed stratal stacking patterns that are indicative of a higher order sedimentary cyclicity than the third-order sequences. In other words, within the third-order sequences, a higher frequency repetition of stratal stacking patterns exists (Figs 2, 7 & 8). This higher frequency stratal repetition is interpreted as the expression of fourth-order sequences. The physical

Fig. 7. Aerial photograph of part of the Upper Cretaceous section in the southeastern Rock Springs Uplift. The Ericson Formation is approximately 1000 ft thick. Compare with Fig. 2.

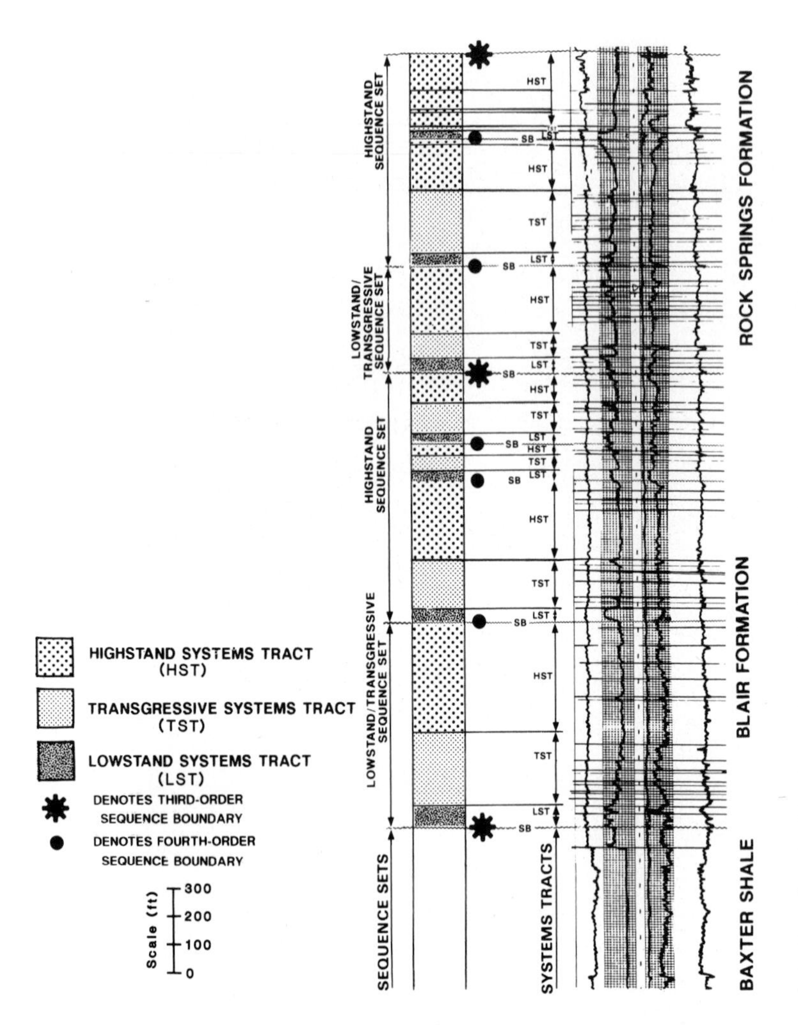

Fig. 8. Well log from the southeastern Rock Springs Uplift showing third- and fourth-order stratal stacking patterns and sequence-stratigraphic interpretation for the Blair and Rock Springs Formations (Fig. 2). Fourth-order systems tracts are keyed to different patterns to indicate easily the abundance and thickness of each. This section was deposited during a period of relatively rapid subsidence associated with the early emplacement of the Absaroka Thrust (Fig. 4). Consequently, sequence set and systems tract development is skewed towards transgressive and early highstand deposition at the expense of lowstand deposition (see Fig. 9). The transgressive systems tracts and sequence sets are characterized by retrogradational parasequence and sequence stacking patterns, respectively. Early highstand systems tracts and sequence sets are characterized by aggradational parasequence and sequence stacking patterns, respectively. Well-log interpretation from Rahmanian *et al.* (1990).

expression and criteria used to identify the fourth-order sequences are identical to those of third-order sequences (Van Wagoner & Mitchum, 1989; Van Wagoner *et al.*, 1990). Just as in third-order sequences, the fourth-order sequences and their component systems tracts are identified by the stacking patterns of parasequences and parasequence sets and the recognition of their attendant bounding surfaces (Van Wagoner *et al.*, 1990) (Figs 7 & 8).

Significantly, the third- and fourth-order sequences observed in outcrop and subsurface data are contained *within* the larger scale sedimentation cycles that are tied to periods of tectonism and quiescence in the fold thrust belt (Figs 2 & 3). Moreover, the higher order sequences observed in the Rock Springs Uplift were deposited at a time when basement-involved intra-foreland uplifts were not active in the basin.

HIGHER ORDER SEQUENCES—POSSIBLE MECHANISMS

Third- and fourth-order depositional sequences associated with the Blair and Rock Springs Formations are presented in the well log of Fig. 8. These sequences are part of the large-scale sedimentation cycle associated with the early emplacement of the Absaroka Thrust (Figs 2, 3 & 4). The presence of these higher order sequences indicates that even when thrust-belt and intra-foreland tectonics have been accounted for in the basin, additional fluctuations in relative sea level of ~10 to 50+ m (as indicated by benthonic foraminifera) (Chen *et al.*, 1991) occurred over time periods ≤ 1.0 to 2.0 Ma. Possible mechanisms responsible for these relative sea-level fluctuations and deposition of the higher order sequences include variations in sediment supply, eustasy or some form of short-period tectonic movement such as those proposed to be related to variations in in-plane or whole-plate stress within a basin (Cloetingh, 1986; Cathles & Hallam, 1990).

Variation in sediment supply is not considered a viable mechanism for all of the higher order sequences for several reasons. Basinward shifts in depositional facies across sequence boundaries tend to be abrupt. In other words, relatively proximal depositional facies overlie more basinal facies across an unconformity with no strata representative of intermediate depositional environments. In addition, the downward shifts in lithofacies are sometimes accompanied by a distinctive change in depositional environment, such as from wave-dominated facies to tidally dominated facies across the sequence boundary. Furthermore, the downward shifts occur numerous times over two orders of depositional cyclicity (Fig. 8) (Rahmanian *et al.*, 1990), and would require abrupt changes in sediment supply. Most importantly, the observed stratal stacking patterns in both the third- and fourth-order sequences consist of regular systematic changes in parasequence stacking patterns and attendant development of systems tracts that are interpreted to reflect directly systematic changes in accommodation (Fig. 8). These systematic changes in accommodation can be readily tied to a sinusoidally varying change in relative base level (Posamentier *et al.*, 1988; Posamentier & Vail, 1988) but not to changes in sediment supply.

For eustatic and short-period tectonic mechanisms, available data do not allow an unequivocal distinction between these two possibilities. A eustatic origin for the higher order sequences could be tested through the analysis of additional Upper Cretaceous sections in different areas of the world. If time-equivalent higher order sequences are present that cannot be directly attributed to either local or regional tectonics, then a eustatic mechanism can be more rigorously argued. The plausibility of an in-plane stress mechanism could be evaluated by forward modelling (Cloetingh, 1986), although modelling alone would not prove that such a mechanism is in fact controlling the relative sea-level changes. Like the in-plane stress mechanism, there is the possibility that some other tectonic mechanism could have been operating in the Cretaceous foreland below the resolution of the biostratigraphic data used to date the tectonic movements and the foreland stratigraphy. However, such mechanisms are not readily apparent and would be difficult to document given our current understanding of the thrust belt and foreland basin.

For the remaining discussions in this paper we will use a eustatic mechanism as our working hypothesis for the deposition of at least the third-order sequences. Although we acknowledge the difficulty in documenting a eustatic mechanism, especially in pre-Tertiary sections, this limitation does not detract from the fact that observed stratigraphic relationships indicate that significant, higher order fluctuations in relative sea level occurred in the Late Cretaceous that cannot be directly attributed to thrust-belt tectonics or intra-foreland basement-involved uplifts.

CONTROL ON HIGHER ORDER CYCLES BY LARGER SCALE CYCLES

Because accommodation represents the interaction of eustasy, tectonics and sediment supply, the expression and distribution of third- and fourth-order eustatically related sequences and systems tracts are controlled, in part, by subsidence and uplift patterns. In the Upper Cretaceous section at Rock Springs, the expression of higher order sequences and systems tracts is controlled by the larger scale subsidence-related accommodation cycles. Furthermore, the sequence and parasequence set development is affected by the particular phase of the larger scale subsidence cycle in which they develop. Figure 9 illustrates this concept. At the top of this figure, a higher order eustatic signal is given which represents intervals of relatively high and low accommodation potential, associated with eustatic rises and falls. The centre curve represents total subsidence and reflects the longer term accommodation change related to tectonically active and quiescent subsidence phases. When the shorter term eustatic signal is combined with the longer term tectonic signal the bottom curve results (Fig. 9). The composite curve represents the new accommodation added by the combined effects of subsidence and eustasy. Note that the composite curve does not include any potential effects related to isostasy (Christie-Blick, 1991; Reynolds *et al.*, 1991).

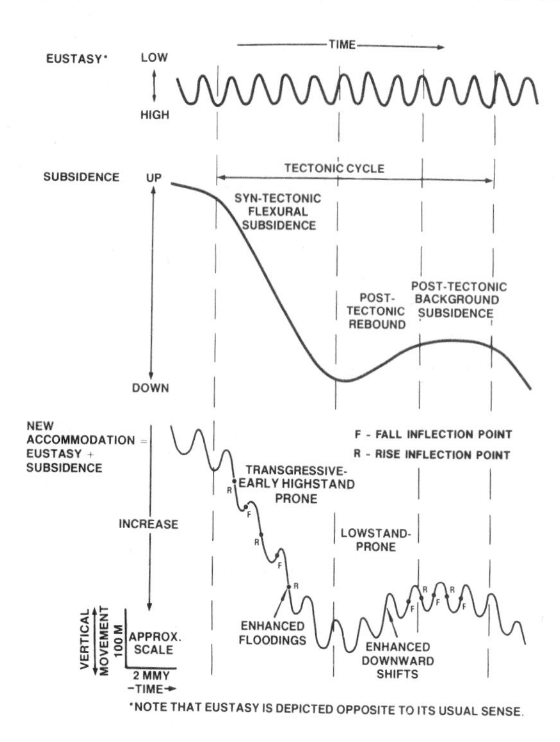

Fig. 9. Effect of long-term total subsidence patterns on shorter term sequence set and systems tract development. Flooding events tend to be enhanced during periods of rapid subsidence (i.e. the syntectonic phase) and downward shifts tend to be enhanced during periods of relative uplift (post-tectonic phase). The eustatic curve (top) is depicted with eustatic lowstands at the peaks and eustatic highstands at the troughs. Note that these same patterns could be produced by superimposing higher order (i.e. third- and fourth-order) eustatic changes on lower order (i.e. second-order) eustatic changes. Although this can be the case, in the southwestern Wyoming foreland basin the longer term signal can be demonstrated to be dominated by regional subsidence patterns. R and F refer to the rise and fall inflection points (Posamentier & Vail, 1988). The figure does not account for potential isostatic effects (Reynolds *et al.*, 1991).

In the higher order sequences, periods of rapid subsidence associated with the high-subsidence syntectonic phase lead to deposition of enhanced transgressive and early highstand systems tracts and sequence sets. This results because of the additive effect of long-term, subsidence-related accommodation and the shorter term eustatic increase in accommodation. Note the asymmetry of the composite curve of Fig. 9, with the portion representing rising sea level being greater than that part of the curve representing falling relative sea level. If sedimentation rates are high, thick transgressive sequence sets and systems tracts can be deposited, such as in the Almond Formation of the Rock Springs area (Figs 2 & 3). If sedimentation rates are low at a given position in the basin, a condensed section should result because of the creation of sediment-starved conditions. Note that both of these situations could exist laterally within the same basin where thick transgressive and early highstand strata are deposited in proximal shelfal areas and thin condensed strata are deposited in more distal areas.

Eustatically related lowstands are typically not as well developed during times of rapid subsidence as they are during times of slower subsidence. High rates of basinal subsidence mean that higher rates of eustatic fall must occur in order for relative sea level to drop below the depositional shoreline break and form a type I sequence boundary (Posamentier *et al.*, 1988; Van Wagoner *et al.*, 1990). Valley incision and potential lowstand fan deposition will tend to be inhibited during these times of rapid subsidence. For more moderate rates of eustatic fall, downward shifts of facies across sequence boundaries may be subtle, and sequence boundaries will be more typically recognized by reversals in parasequence stacking patterns. If the rate of eustatic fall does not exceed the rate of total subsidence at the depositional shoreline break, a type II sequence boundary forms (Posamentier *et al.*, 1988; Van Wagoner *et al.*, 1990).

The effect of periods of high subsidence on sequence and systems tract development is well expressed in the Blair and Rock Springs Formations (Fig. 8). At the time of deposition of these strata, rates of subsidence were high because of flexural loading associated with the early emplacement of the Absaroka Thrust (Figs 2 & 4). Hence, any superimposed eustatic fluctuations that occurred at this time would result in a relative sea-level curve that looks like the rising limb of the composite curve in Fig. 9. Significantly, during deposition of the Blair and Rock Springs Formations there is enhanced development of transgressive systems tracts and the early, aggradational part of highstand systems tracts. There is also relatively poor development of lowstand systems tracts. Sequence boundaries in this part of the section do not tend to be associated with extensive incision, but rather, are marked by basinward shifts in facies and the regular reversal of parasequence stacking patterns across the sequence boundary (Figs 7 & 8).

During times of relative quiescence in the thrust

belt there are reduced subsidence rates because of the lack of flexural loading. A higher order eustatic signal superimposed on the overall low basinal subsidence rate will potentially enhance eustatic lowstands and inhibit the development of transgressive systems tracts. Because very little space is being created by subsidence at these times, even a eustatic fall of moderate rate could result in an enhanced downward shift in facies. The lack of subsidence-related space on the shelf means that sediment will tend to bypass this area and be deposited further in the basin. Hence, the potential for progradation will be enhanced and, consequently, will result in well-developed lowstand wedge deposits and the late progradational portion of highstand systems tracts. The lack of space in relatively proximal shelfal areas will also be responsible for the poor development of the transgressive systems tract.

DISCUSSION AND CONCLUSIONS

Large-scale stratal patterns in the Upper Cretaceous foreland basin of southwestern Wyoming can be tied directly to periods of tectonic activity and quiescence in the Idaho–Wyoming Thrust Belt. Geohistory analysis shows three separate subsidence cycles (Fig. 4), each of which consists of a rapid increase, then decrease, in subsidence rates. The three subsidence cycles on the geohistory plot correspond precisely with the three large-scale sedimentation cycles in the Green River Basin (Figs 2 & 3). The sedimentation cycles consist of a lower, shale-prone, abruptly deepened or retrogradational succession and an upper sandstone-prone, progradational succession. The stratal characteristics of the lower and upper parts of the cycles can be tied directly to the subsidence-related changes in accommodation. Basement-involved tectonic movements complicate these sedimentation patterns and are responsible for dramatic changes in depositional style and palaeogeography. The style of sedimentation that occurs at these times will be dependent upon whether intra-foreland and thrust movements occur synchronously (in-phase), or whether the intra-foreland uplifts occur during relative quiescence in the thrust belt (out-of-phase).

In the southwestern Wyoming foreland, the subsidence-related accommodation changes range from 300 m to 500 m over periods of ~3 to 4 Ma (Fig. 4). Second-order eustatic changes in the Late Cretaceous are reported to reach magnitudes of 70 to 80 m over periods of 10 to 12 Ma (Haq *et al.*, 1987). Third-order eustatic changes are estimated to range from 50 to 70 m over 1 to 4 Ma (Haq *et al.*, 1987), although these estimates are controversial (Christie-Blick *et al.*, 1990). Basement-involved movements of the Rock Springs Uplift are estimated to have produced local changes in relative sea level of ~150 m and possibly up to 240 m over periods of 1.5 to 3.0 Ma (Fig. 4). Note that for the southern Wyoming foreland basin, all of the tectonically related accommodation changes occur over time periods ranging from 1.5 to 4.0 Ma, or time scales typical of third-order eustatic fluctuations. The geohistory analyses, in conjunction with timing data for movements of intra-foreland structures obtained from seismic geometric relationships, provide an effective means of identifying the tectonic controls on basin accommodation. The large changes in subsidence-related accommodation in the foreland are easily identified on the geohistory plots. Once identified, their effect on basin stratigraphy can be readily evaluated. Furthermore, in the Idaho–Wyoming Thrust Belt independent evidence for structural timing allows for the correlation of each of the large accommodation events with specific structural movements. Within the resolution of the biostratigraphic dating, this correlation establishes a direct cause and effect relationship between large-scale basin tectonics and the time-equivalent stratigraphic architecture of the basin.

In the Upper Cretaceous section at Rock Springs, the presence of shorter duration higher order depositional sequences reflects the effect of high-frequency fluctuations in relative sea level. These higher frequency fluctuations, however, cannot be directly tied to any of the documented tectonic events in either the thrust belt or within the foreland basin. In the case of the Blair and Rock Springs Formations (Figs 2 & 8), third-order relative sea-level changes of ~10 to 50+ m have been documented from biofacies data (Chen *et al.*, 1991), and occur over time periods of approximately 1.0 to 2.0 Ma. Fourth-order fluctuations in sea level can have similar magnitudes, but occur over time intervals less than 1 Ma. Eustasy is a viable mechanism for these higher order sea-level fluctuations, and is our favoured interpretation, at least for the third-order sequences. However, all that can be definitively said with the available data is that some other mechanism is working to produce the higher order sea-level changes in the basin besides the documented thrust and intra-foreland movements. In order to understand more fully the depositional mechanism behind the higher order sea-level fluctuations we will have to understand better such additional factors as the effect of potential short-period tectonism in the foreland, transverse and longitudinal sediment supply to the basin and the effect of variations in the Earth's orbital parameters and climate on clastic sedimentation.

ACKNOWLEDGEMENTS

The authors acknowledge the contribution of V.D. Rahmanian to the understanding of the sequence-stratigraphic framework of the Rock Springs Uplift, and we thank him for the use of the stratigraphic column shown in Fig. 2 and the well-log interpretation in Fig. 8. Reviews by N. Christie-Blick, T.R. Garfield and D. Leckie improved the manuscript. We also thank G.R. Bergan, R.M. Lindholm, F.L. Wehr and E. White for reviews of an early version of the manuscript. The authors thank Exxon Production Research Co. for permission to release this paper.

APPENDIX

Cretaceous time scale

The Cretaceous time scale used in this paper is that of Devlin (unpublished data). Duration of fossil range zones from Fouch *et al.* (1983) after Cobban and Reeside (1952a) and Merewether and Cobban (1981), but modified for new isotopic age data. Campanian/Maastrichtian boundary after Jeletsky (1951) and Hancock and Kaufman (1989). Cretaceous age boundaries and fossil zone ages are modified after data in Obradovich and Cobban (1975) corrected for new ^{40}K decay constants (W.A. Cobban, pers. com. to F.B. Zelt (EPR), 1987). Absolute time scale also modified for new ^{40}Ar/^{39}Ar data from bentonites in basal Sharon Springs Member of the Redbird section (Gill & Cobban, 1966b) of eastern Wyoming (F.B. Zelt, pers. com., 1990).

Dating of tectonic movements in the fold thrust belt

Cross-cutting stratigraphic relationships in the fold thrust belt provide only broad age constraints on Late Cretaceous thrust movements (Oriel & Armstrong, 1966; Royse *et al.*, 1975). The age constraints provided by palynological dating of syntectonic deposits (see e.g. Royse *et al.*, 1975) are more tightly constrained. These ages, as reported in Jacobson and Nichols (1982) and Nichols and Bryant (1990), are as follows: Echo Canyon Conglomerate–Santonian–Coniacian; Little Muddy Conglomerate–Santonian; Hams Fork Conglomerate–late Campanian–early Maastrichtian. Based on the palynological ages, stratigraphic superposition and biostratigraphic and geologic correlation with strata further in the basin the interpreted ages for the onset of deposition of these units are as follows: Echo Canyon–middle Coniacian; Little Muddy–late Santonian; and Hams Fork–earliest Maastrichtian.

Age constraints on stratigraphic surfaces, Upper Cretaceous, southeastern Rock Springs Uplift

Acknowledging the time-transgressive nature of formation boundaries, the following age constraints have been established for the Upper Cretaceous section in the southeastern Rock Springs Uplift.

Top Fox Hills

Upper age constraint: up to 2000 ft of upper Maastrichtian Lance Formation is present between top Fox Hills and Tertiary Fort Union in Washakie Basin (Hettinger & Kirschbaum, 1991). Lower age constraint: *Baculites clinolobatus* in underlying Lewis Shale (Gill *et al.*, 1970). Hettinger and Kirschbaum (1991) report *B. clinolobatus* in Fox Hills Sandstone. Interpreted age: top *B. clinolobatus* zone.

Top Lewis Shale

Age constraint: *B clinolobatus* in upper Lewis Shale (Gill *et al.*, 1970) and in overlying Fox Hills Sandstone (Hettinger & Kirschbaum, 1991). Interpreted age: lower part of *B. clinolobatus* zone.

Top Almond Formation

Upper age constraint: *Baculites grandis* in overlying Lewis Shale (Gill *et al.*, 1970). *Baculites baculus* in uppermost Almond Formation on east side of Rock Springs Uplift (Van Horn, 1979; Hendricks, 1983). Lower age constraint: palynological Campanian/Maastrichtian boundary 800 ft below top of Almond Formation (Y.Y. Chen, pers. comm., 1989). Interpreted age: middle *B. baculus* zone.

Top Ericson Formation (top Canyon Member)

Upper age constraint: palynological Campanian/Maastrichtian boundary in overlying Almond Formation. Regional correlation indicates top Ericson equals top Pine Ridge Sandstone on east side of Washakie Basin (Gill *et al.*, 1970; Roehler, 1990; Devlin, unpublished data). *Baculites reesidei* is present just above Pine Ridge. Lower age constraint: based on regional correlations, strata beneath the Pine Ridge Sandstone or its stratigraphic equivalents are within *Didymoceras stevensoni* zone (eastern Washakie Basin; Gill *et al.*, 1970) or within *Exiteloceras jennyi* zone (Powder River Basin; Gill & Cobban, 1966a), but could be younger because of erosion beneath Canyon Member and its stratigraphic equivalents. Interpreted age: middle *B. reesidei* zone.

Base Canyon Member

Upper age constraint: *B. reesidei* present above Ericson equivalent. Lower age constraint: regional correlations indicate youngest strata beneath stratigraphic equivalents of Ericson Formation are within *D. stevensoni* zone (Gill *et al.*, 1970) or *E. jennyi* (Gill & Cobban, 1966a), but could be younger because of erosion. Interpreted age: middle *B. cuneatus* zone.

Top of pre-Canyon deposition

Seismic and well-log data in Washakie Basin indicate approximately 800 ft of strata were deposited, then eroded, between top Rusty surface and base Canyon surface on the southeastern Rock Springs Uplift (Devlin *et al.*, unpublished data). Upper age constraint: regional correlations indicate youngest strata beneath stratigraphic equivalents of Canyon Member are within *D. stevensoni* zone (Gill *et al.*, 1970) or *E. jennyi* (Gill & Cobban, 1966a), but could be younger because of erosion. Lower age constraint: base *D. stevensoni* zone because 370 ft of strata are present between occurrence of fossil and base of Pine Ridge Sandstone (Gill *et al.*, 1970). Interpreted age: top *D. stevensoni* zone.

Top Rusty Member

Upper age constraint: *D. stevensoni* zone. Lower age constraint: regional correlations indicate *Baculites perplexus* (form?) is found beneath equivalent of Trail Member in Powder River Basin (Gill & Cobban, 1966a) and *B. asperiformis* in eastern Washakie Basin (Gill *et al.*, 1970). Interpreted age: lower part of *Didymoceras nebrascence* zone based on possible correlation of Rusty Member with unnamed marine member of the Allen Ridge Formation containing *D. nebrascence.*

Base Ericson Formation (base Trail Member)

Constraints based on regional correlation of Trail Member with Parkman Sandstone (Gill & Cobban, 1966a; Gill *et al.*, 1970). Upper age constraint: *Baculites scotti* is present above Parkman Sandstone (Gill & Cobban, 1966a). Lower age constraint: *B. perplexus* (form?) is present beneath Trail equivalent in eastern Wyoming (Gill & Cobban, 1966a); *B. asperiformis* is found beneath Trail equivalent in eastern Washakie Basin (Gill *et al.*, 1970). Interpreted age: middle of *B. perplexus* (late form) zone.

Top of pre-Trail deposition

Seismic and well-log data in Washakie Basin indicate approximately 500 ft of strata were deposited, then eroded, between top Transition Member and base Trail surface on the southeastern Rock Springs Uplift (Devlin *et al.*, unpublished data). Upper age constraint: *B. perplexus* (form?). Lower age constraint: *B. asperiformis.* Interpreted age: top *B. asperiformis* zone.

Top Transition Member

Seismic/well-log correlations tie base Trail Member of Ericson Formation and top McCourt Member of Rock Springs Formation from Rock Springs Uplift into eastern Washakie Basin (Devlin *et al.*, unpublished data). *Baculites obtusus* and *B. asperiformis* are identified within this interval (Gill *et al.*, 1970). Upper age constraint: top *B. asperiformis* zone. Lower age constraint: base *B. obtusus* zone. Interpreted age: middle *Baculites mclearni* zone.

Approximate top Black Butte Member

Upper age constraint: *Baculites* sp. (weak flanks) found in basal Brooks Member on the west side of Rock Springs Uplift (Roehler, 1978). Lower age constraint: *Scaphites hippocrepis* III found in upper

Blair Formation (Roehler, 1978). Interpreted age: middle of *Baculites* sp. (weak) zone.

Approximate top Blair Formation

Upper age constraint: *Baculites* sp. (weak) in basal Brooks Member (Roehler, 1978). Lower age constraint: *S. hippocrepis* III found 60 to 100 m below Chimney Rock Tongue (Cobban, 1969; Roehler, 1978). *S. hippocrepis* III also reported from the upper Blair and Chimney Rock Tongue by Smith (1961), but localities not reported. Interpreted age: top *S. hippocrepis* III zone.

Top Airport Sandstone Member of Baxter Shale

Desmoscaphites bassleri was reported by Hale (1950) and Smith (1961) in the Airport Sandstone. Interpreted age: middle of *D. bassleri* zone.

Upper part of lower Baxter Shale

Clioscaphites vermiformis is reported by Smith (1961) to be present in the lower Baxter Shale 300 to 350 ft below the base of the Airport Sandstone in the Rock Springs Uplift. To the south, in Spring Creek Gap, the fossil is reported 2140 ft above the Frontier Formation (Smith, 1961). Interpreted age: middle *C. vermiformis* zone.

Lower part of lower Baxter Shale

Scaphites depressus is reported in the lower Hilliard Shale in the Kemmerer area of western Wyoming (Smith, 1961). Interpreted age indicates tentative correlation of these strata into the Baxter Shale of the Rock Springs area. Interpreted age: middle *S. depressus* zone.

Top Frontier Formation

Upper age constraint: *S. depressus* in overlying Hilliard Shale in Kemmerer area (Smith, 1961). Lower age constraint: *Inoceramus erectus* in uppermost Frontier Formation (Cobban & Reeside, 1952b). Interpreted age: top *I. erectus* zone.

REFERENCES

Asquith, D.O. (1970) Depositional topography and major marine environments, Late Cretaceous, Wyoming. *Bull. Am. Assoc. Petrol. Geol.* **54**, 1184–1224.

Beaumont, C. (1981) Foreland basins. *Roy. Astron. Soc. Geophys. J.* **65**, 291–329.

Beck, R.A., Vondra, C.F., Filkins, J.E. & Olander, J.D. (1988) Syntectonic sedimentation and Laramide basement thrusting, Cordilleran foreland: timing of deformation. *Mem. Geol. Soc. Am.* **171**, 465–487.

Blair, T.C. & Bilodeau, W.L. (1988) Development of tectonic cyclothems in rift, pull-apart, and foreland basins: sedimentary response to episodic tectonism. *Geology*, **16**, 517–520.

Cathles, L.M. & Hallam, A. (1990) Stress changes and rapid, non-glacial sea-level fluctuations. *Geol. Soc. Am. (Abstr. with Prog.)* **22** (7), A237.

Cerveny, P. (1990) Fission track thermo-chronology of the Wind River Range and other basement cored uplifts in the Rocky Mountain Foreland. Unpublished PhD thesis, University of Wyoming.

Chen, Y.Y., Pflum, C.E. & Wright, R.C. (1991) Biofacies expression of Upper Cretaceous sequences in the Rock Springs Uplift. *Bull. Am. Assoc. Petrol. Geol.* **75/3**, 552.

Christie-Blick, N. (1991) Onlap, offlap, and the origin of unconformity-bounded depositional sequences. *Mar. Geol.* **97**, 35–56.

Christie-Blick, N., Mountain, G.S. & Miller, K.G. (1990) Seismic stratigraphic record of sea-level change. In: *Studies in Geophysics: Sea-Level Change,* National Research Council, National Academy Press, Washington, DC, pp. 116–140.

Cloetingh, S. (1986) Intraplate stresses: a new tectonic mechanism for fluctuations of relative sea level. *Geology* **14**, 617–620.

Cobban, W.A. (1969) The Late Cretaceous ammonites *Scaphites leei* Reeside and *Scaphites hippocrepis* (DeKay) in the Western Interior of the United States. *U. S. Geol. Surv. Prof. Pap.* **619**, 29 pp.

Cobban, W.A. & Reeside, J.B., Jr (1952a) Correlation of the Cretaceous formations of the Western Interior of the United States. *Bull. Geol. Soc. Am.* **63**, 1011–1044.

Cobban, W.A. & Reeside, J.B., Jr (1952b) Frontier Formation, Wyoming and adjacent areas *Bull. Am. Assoc. Petrol. Geol.* **36**, 1913–1961.

Covey, M. (1986) The evolution of foreland basins to steady state: evidence from the western Taiwan foreland basin. In: *Foreland Basins* (Eds Allen, P.A. & Homewood, P.) Int. Assoc. Sediment. Spec. Publ. 8, 77–90.

Devlin, W.J. & Rudolph, K.W. (1991) The interaction of thrust loading, flexural subsidence and basement-involved Laramide movements, and their effect upon sequence stratigraphic framework, Upper Cretaceous of southwestern Wyoming. *Geol. Soc. Am. Cordilleran Sect. (Abstr. with Prog.)* **23** (2), 18.

Devlin, W.J. & Shaw, C.A. (1990) Controls on accommodation in the Upper Cretaceous foreland basin of southwestern Wyoming. *Geol. Soc. Am. (Abstr. with Prog.)* **22** (7), A238.

Devlin, W.J., Rudolph, K.W., Ehman, K.D. & Shaw, C.A. (1990) The effect of tectonic and eustatic cycles on accommodation and sequence stratigraphic framework in the southwestern Wyoming foreland basin. (Abstr.) 13th International Sedimentological Congress, Nottingham, England, p. 131.

Flemings, P.B. & Jordan, T.E. (1990) Stratigraphic modeling of foreland basins: interpreting thrust deformation and lithosphere rheology. *Geology* **18**, 430–434.

Fouch, T.D., Lawton, T.F., Nichols, D.J., Cashion, W.B. & Cobban, W.A. (1983) Patterns and timing of synorogenic sedimentation in the Upper Cretaceous of central and northeast Utah. In: *Mesozoic Paleogeography of West-central United States, Rocky Mountain Section.* (Eds Reynolds, M.W. & Dolly, E.D.) Soc. Econ. Paleontol. Mineral. pp. 305–336.

Gill, J.R. & Cobban, W.A. (1966a) Regional unconformity in Late Cretaceous, Wyoming. *U. S. Geol. Surv. Prof. Pap.* **550-B**, B20–B27.

Gill, J.R. & Cobban, W.A. (1966b) The Redbird section of the Upper Pierre Shale in Wyoming, with a section on a new echinoid from the Cretaceous Pierre Shale of eastern Wyoming. *U. S. Geol. Surv. Prof. Pap.* **393-A**, 73 pp.

Gill, J.R. & Cobban, W.A. (1973) Stratigraphy and geologic history of the Montana Group and equivalent rocks, Montana, Wyoming and North and South Dakota. *U. S. Geol. Surv. Prof. Pap.* **776**, 37 pp.

Gill, J.R., Merewether, E.A. & Cobban, W.A. (1970) Stratigraphy and nomenclature of some Upper Cretaceous and lower Tertiary rocks in south-central Wyoming. *U. S. Geol. Surv. Prof. Pap.* **667**, 53 pp.

Hagen, E.S., Shuster, M.W. & Furlong, K.P. (1985) Tectonic loading and subsidence of intermontane basins: Wyoming foreland province. *Geology* **13**, 585–588.

Hale, L.A. (1950) Stratigraphy of the Upper Cretaceous Montana Group in the Rock Springs Uplift, Sweetwater County, Wyoming. *Wyoming Geol. Assoc. Fifth Annual Field Conference Guidebook* pp. 49–59.

Hancock, J.M. & Kaufman, E.G. (1989) Use of eustatic sea-level changes to fix the Campanian/Maastrichtian boundary in the Western Interior of the U.S.A. *28th Int. Geol. Congress (Abstr. with Prog.)* pp. 2–23.

Haq, B.U., Hardenbol, J. & Vail, P.R. (1987) Chronology of fluctuating sea levels since the Triassic. *Science* **235**, 1156–1167.

Heller, P.L., Angevine, C.L., Winslow, N.S. & Paola, C. (1988) Two-phase stratigraphic model of foreland basin sequences. *Geology* **16**, 501–504.

Hendricks, M.L. (1983) Stratigraphy and tectonic history of the Mesaverde Group (Upper Cretaceous), east flank of the Rock Springs Uplift, Sweetwater County, Wyoming. Unpublished PhD thesis, Colorado School of Mines, 213 pp.

Hendricks, M.L. (1990) Evidence for an unconformity at the base of the Ericson Formation along the east flank of the Rock Springs Uplift. *Bull. Am. Assoc. Petrol. Geol.* **74**, 1327.

Hettinger, R.D. & Kirschbaum, M.A. (1991) Chart showing correlations of some Upper Cretaceous and lower Tertiary rocks, from the east flank of the Washakie Basin to the east flank of the Rock Springs Uplift, Wyoming. *U. S. Geol. Surv. Misc. Investi. Ser. Map* I-2152.

Jacobson, S.R. & Nichols, D.J. (1982) Palynological dating of syntectonic units in the Utah–Wyoming thrust belt: the Evanston Formation, Echo Canyon Conglomerate and Little Muddy Creek Conglomerate. In: *Geologic Studies of the Cordilleran Thrust Belt* (Ed. Powers, R.B.) Rocky Mountain Assoc. Geol., 1982 Symp., pp. 735–750.

Jeletsky, J.A. (1951) The place of the Trimmingham and Norwhich Chalk in the Campanian/Maastrichtian succession. *Geol. Magazine* **88**, 197–208.

Jervey, M.T. (1988) Quantitative geological modeling of siliciclastic rock sequences and their seismic expression. In: *Sea Level Change–An Integrated Approach* (Eds Wilgus, C.K., Hastings, B.S., Kendall, C.G.St.C., Posamentier, H.W., Ross, C.A. & Van Wagoner, J.C.) Spec. Publ. Soc. Econ. Paleontol. Mineral. 42, 47–69.

Jie, T., Lerche, I. & Cogan, J. (1990) Elastic flexure with compressive thrusting of the Green River Basin, Wyoming, U.S.A. *Geol. Magazine* **127**, 349–359.

Jordan, T.E. (1981) Thrust loads and foreland basin evolution, Cretaceous, western United States. *Bull. Am. Assoc. Petrol. Geol.* **65**, 291–329.

Kraig, D.H., Wiltschko, D.V. & Spang, J.H. (1987) Interaction of basement uplift and thin-skinned thrusting, Moxa Arch and the Western Overthrust Belt, Wyoming: a hypothesis. *Bull. Geol. Soc. Am.* **99**, 654–722.

Lawrence, D.T. (1984) Patterns and dynamics of late Cretaceous marginal marine sedimentation; overthrust belt, southwestern Wyoming. Unpublished PhD dissertation, Yale University, 488 pp.

Loutit, T.R., Hardenbol, J., Vail, P.R. & Baum, G.R. (1988) Condensed sections: the key to age dating and correlation of continental margin sequences. In: *Sea-level Change–An Integrated Approach* (Eds Wilgus, C.K., Hastings, B.S., Kendall, C.G.St.C., Posamentier, H.W., Ross, C.A. & Van Wagoner, J.C.) Spec. Publ. Soc. Econ. Paleontol. Mineral. 42, 183–213.

Love, J.D. & Christiansen, A.C. (1985) *Geological Map of Wyoming (1:500,000).* U.S. Geol. Surv.

McMillen, K.J. & Winn, R.D., Jr. (1988) Seismic stratigraphy of Lewis Shale deltaic and deep-water clastics, Red Desert–Washakie Basins, Wyoming. In: *Atlas of Seismic Stratigraphy,* Vol. 3 (Ed. Bally, A.W.) Am. Assoc. Petrol. Geol. Stud. Geol. **27**, 134–139.

Merewether, E.A. & Cobban, W.A. (1981) Mid-Cretaceous formations in eastern South Dakota and adjoining areas—stratigraphic, paleontologic, and structural interpretation. In: *Cretaceous Stratigraphy and Sedimentation in Northwest Iowa, Northeast Nebraska, and Southeast South Dakota.* Iowa Geol. Surv. Guidebook Ser. 4, 43–56.

Miller, F.X. (1977) Biostratigraphic correlation of the Mesaverde Group in southwestern Wyoming and northwestern Colorado. *Rocky Mountain Assoc. Geol. 1977 Symp.* 117–137.

Nichols, D.J. & Bryant, B. (1990) Palynologic data from Cretaceous and lower Tertiary rocks in the Salt Lake City 30' × 60' quadrangle. *U.S. Geol. Surv. Misc. Invest. Ser. Map* I–1944.

Nichols, D.J., Perry, W.J., Jr & Haley, J.C. (1985) Reinterpretation of the palynology and age of Laramide syntectonic deposits, southwestern Montana, and revision of the Beaverhead Group. *Geology* **13**, 149–153.

Obradovich, J.D. & Cobban, W.A. (1975) A time scale for the Late Cretaceous of the Western Interior of North America. *Geol. Assoc. Can. Spec. Pap.* **13**, 13–54.

Oriel, S.S. & Armstrong, F.C. (1966) Times of thrusting in Idaho–Wyoming thrust belt: reply. *Bull. Am. Assoc. Petrol. Geol.* **50**, 2614–2621.

Perman, R.C. (1990) Depositional history of the Maastrichtian Lewis Shale in south-central Wyoming: deltaic and interdeltaic, marginal marine through deep-water marine environments. *Bull. Am. Assoc. Petrol. Geol.* **74** (11), 1695–1717.

Posamentier, H.W. & Vail, P.R. (1988) Eustatic controls on clastic deposition II. In: *Sea Level Change–An Integrated Approach* (Eds Wilgus, C.K., Hastings, B.S., Kendall, C.G.St.C., Posamentier, H.W., Ross, C.A. & Van Wagoner, J.C.) Spec. Publ. Soc. Econ. Paleontol. Mineral. **42**, 125–154.

Posamentier, H.W., Jervey, M.T. & Vail, P.R. (1988) Eustatic controls on clastic deposition I. In: *Sea-Level Change–An Integrated Approach* (Eds Wilgus, C.K., Hastings, B.S., Kendall, C.G.St.C., Posamentier, H.W., Ross, C.A. & Van Wagoner, J.C.) Spec. Publ. Soc. Econ. Paleontol. Mineral. **42**, 109–124.

Quinlan, G.M. & Beaumont, C. (1984) Appalachian thrusting, lithospheric flexure, and the Paleozoic stratigraphy of the Eastern Interior of North America. *Can. J. Earth Sci.* **21**, 973–996.

Rahmanian, V.D., Devlin, W.J. & Rudolph, K.W. (1990) *Sequence Stratigraphic Framework of the Upper Cretaceous Section of the Rock Springs Uplift, Wyoming, U.S.A.* (Abstr.) 13th International Sedimentological Congress, Nottingham, England, p. 445.

Reynolds, D.J., Steckler, M.S. & Coakley, B.J. (1991) The role of the sediment load in sequence stratigraphy: the influence of flexural isostasy and compaction. *J. Geophys. Res.* **96**, 6931–6949.

Reynolds, M.W. (1967) Physical evidence for Late Cretaceous unconformity, south-central Wyoming. *U.S. Geol. Surv. Prof. Pap.* **575-D**, D24–D28.

Reynolds, M.W. (1976) Influence of recurrent Laramide structural growth on sedimentation and petroleum accumulation, Lost Soldier area, Wyoming. *Bull. Am. Assoc. Petrol. Geol.* **60**, 12–33.

Roehler, H.W. (1978) Correlation of coal beds in the Fort Union, Almond, and Rock Springs Formations, Sweetwater County, Wyoming. *U.S. Geol. Surv. Open-File Rep.* OF-78-395.

Roehler, H.W. (1989) Surface subsurface correlations of the Mesaverde Group and associated Upper Cretaceous Formations, Rock Springs to Atlantic Rim, southwest Wyoming. *U.S. Geol. Surv. Misc. Field Stud. Map* MF-2078.

Roehler, H.W. (1990) Stratigraphy of the Mesaverde Group in the central and eastern Greater Green River Basin, Wyoming, Colorado, and Utah. *U.S. Geol. Surv. Prof. Pap.* **1508**, 52 pp.

Royse, F., Jr., Warner, M.A. & Reese, D.L. (1975) Thrust belt structural geometry and related stratigraphic problems Wyoming–Idaho–northern Utah. In: *Deep Drilling Frontiers of the Central Rocky Mountains* (Ed. Bolyard, D.W.) Rocky Mountain Assoc. Geol. 1975 Symp. 41–54.

Smith, J.H. (1961) A summary of stratigraphy and paleontology of the Upper Colorado and Montanan Groups, southcentral Wyoming, northeastern Utah, and northwestern Colorado. *Wyoming Geol. Assoc. 19th Annual Field Conference Guidebook* pp. 13–26.

Vail, P.R., Mitchum, R.M., Todd, R.G. *et al.* (1977) Seismic stratigraphy and global changes in sea level. In: *Seismic Stratigraphy–Applications to Hydrocarbon Exploration* (Ed. Payton, C.E.) Mem. Am. Assoc. Petrol. Geol. 26, 49–212.

Van Hinte (1978) Geohistory analysis—application of micropaleontology in exploration geology. *Bull. Am. Assoc. Petrol. Geol.* **62**, 201–222.

Van Horn, M.D. (1979) Stratigraphy of the Almond Formation, east flank of the Rock Springs Uplift, Sweetwater County Wyoming: a mesotidal–shoreline model for the late Cretaceous. Unpublished MSc thesis, Colorado School of Mines, 150 pp.

Van Wagoner, J.C. & Mitchum, R.M. (1989) High-frequency sequences and their stacking patterns. (Abstr. with Programs), *28th International Geological Congress* **3**, 3–284.

Van Wagoner, J.C., Mitchum, R.M., Campion, K.M. & Rahmanian, V.D. (1990) Siliciclastic sequence stratigraphy in well-logs, cores and outcrops. *Am. Assoc. Petrol. Geol. Meth. Expl. Ser.* **7**, 55 pp.

Van Wagoner, J.C., Posamentier, H.W., Mitchum, R.M., Jr., *et al.* (1988) An overview of the fundamentals of sequence stratigraphy and key definitions. In: *Sea-level Change–An Integrated Approach* (Eds Wilgus, C.K., Hastings, B.S., Kendall, C.G.St.C., Posamentier, H.W., Ross, C.A. & Van Wagoner, J.C.) Spec. Publ. Soc. Econ. Paleontol. Mineral. **42**, 39–46.

Weimer, R.J. (1970) Rates of deltaic sedimentation and intrabasin deformation, Upper Cretaceous of Rocky Mountain region. In: *Deltaic Sedimentation–Modern and Ancient.* (Eds Morgan, J.P. & Shaver, R.H.) Spec. Publ. Soc. Econ. Paleontol. Mineral. 15, 270–292.

Wiltschko, D.V. & Dorr, J.A., Jr. (1983) Timing of deformation in overthrust belt and foreland basin of Idaho, Wyoming, and Utah. *Bull. Am. Assoc. Petrol. Geol.* **67**, 1304–1332.

Winn, R.D., Bishop, M.G. & Gardner, P.S. (1987) Shallow water and sub-storm-base deposition of Lewis Shale in Cretaceous Western Interior Seaway, south-central Wyoming. *Bull. Am. Assoc. Petrol. Geol.* **71**, 859–881.

Zapp, A.D. & Cobban, W.A. (1962) Some Late Cretaceous strand lines in southern Wyoming. *U.S. Geol. Surv. Prof. Pap.* **450-D**, D52–D55.

Spec. Publs Int. Ass. Sediment. (1993) **18**, 521–535

The sequence-stratigraphic significance of erosive-based shoreface sequences in the Cretaceous Mesaverde Group of northwestern Colorado

D.F. HADLEY* *and* T. ELLIOTT

Department of Earth Sciences, The University of Liverpool, Brownlow St., PO Box 147, Liverpool L69 3BX, UK

ABSTRACT

The Upper Cretaceous (Campanian age) Mesaverde Group of northwestern Colorado forms an eastward-thinning clastic wedge that partially infills the Western Interior Foreland Basin. The wedge comprises progradational and aggradational parasequence sets that are separated by discrete flooding surfaces. Parasequences are composed of delta-front and delta-plain facies.

Two types of parasequence are recognized. Type A parasequences include an entirely gradational delta-front coarsening-upward facies sequence. Type B parasequences include a distinctive, storm-generated, heterolithic facies in the mid- to lower parts of the delta-front facies sequence which contains deeply erosional but highly localized gutter-casts. This gutter cast facies is truncated by a prominent erosion surface which is overlain by amalgamated, swaley cross-stratified sandstones of the mid- to upper shoreface. Erosion surfaces can be traced for 20 to 50 km down depositional dip, indicating that they are of regional scale.

The erosion surface separates facies which accumulated below fair-weather wave base from facies which accumulated above fair-weather wave base. This contact represents an abrupt shallowing and a concomitant basinward shift of facies belts within the parasequence, and is interpreted to be the result of a fall of relative sea level on the scale of a single parasequence. The gutter-cast facies is interpreted to be the result of a change in sediment supply which was the initial response of the system to the fall of relative sea level. The overlying erosion surface formed by submarine storm erosion during the period of maximum rate of relative sea-level fall and was associated with a period of sediment bypass. Type B parasequences prograded further into the basin than type A parasequences and may be associated with more sandstone-prone intervals in the shelf part of the facies tract.

INTRODUCTION

The application of sequence-stratigraphic concepts to surface exposures of siliciclastic sequences is essential if the concepts are to be rigorously tested and further developed. These concepts were developed initially using mainly subsurface data sets from passive margins (Vail *et al.*, 1977), and have since been refined considerably (Wilgus *et al.*, 1988). Sequence-stratigraphic concepts, however, are still largely based on subsurface data, with only a few examples of their application to surface exposures (Baum & Vail, 1988; Plint, 1988a; Van Wagoner *et al.*, 1990).

Deltas are excellent depositional systems for testing sequence stratigraphy due to their sensitivity to controls on sedimentation such as variations in eustatic sea level, subsidence rate and sediment supply. Numerous interpretations of delta systems have emphasized the role of autocyclic controls, in particular, the avulsion of distributary channels or more upstream fluvial channels, in the production of the observed stacking patterns of depositional sequences (Frazier, 1967; Elliott, 1974). In contrast, Boyd *et al.* (1989) have argued that lobes of the

*Present address: LASMO North Sea Plc., Broadwalk House, 5 Appold St., London EC2A 2NS, UK.

Mississippi Delta can be grouped into transgressive and highstand systems tracts. Their work suggests that allocyclic controls may be at least as important as autocyclic controls in generating deltaic stratigraphy. The importance of delta systems in sequence stratigraphy is further enhanced by the fact that their presence or absence and their location, character and moderate- to long-term behaviour are a prime control on more basinward depositional systems.

The Cretaceous Mesaverde Group of the Western Interior, USA, comprises a series of large-scale clastic wedges which include fluvio-deltaic and inter-deltaic, strandplain depositional systems and their shelf equivalents (Kiteley & Field, 1984; Siepman, 1986). This paper is concerned with the Mesaverde Group of northwestern Colorado which is considered to be deltaic in view of the high ratio of sand to shale, the abundance of erosive-based channel sandstone bodies and the large distances over which shoreline/delta-front environments prograde. This study focuses on the sequence-stratigraphic significance of erosive-based shoreface sandstones. These facies sequences differ from the more conventional, entirely gradational delta-front coarsening-upwards facies sequences in several important respects, and are considered to reflect a fall of relative sea level and a basinward shift of facies at the parasequence scale. The erosive-based shoreface sandstones resemble examples described in subsurface studies of the Cardium Formation, Alberta (Plint, 1988b) and the Viking Formation, Alberta (Posamentier & Chamberlain, this volume, pp. 469–485). Examination of surface-exposed examples confirms the importance and regional extent of erosive-based shoreface sandstones and permits further observations to be made which contribute to an understanding of their genesis.

This paper represents one aspect of a continuing research project concerned with testing the concepts of sequence stratigraphy in regional scale, surface exposures of the Cretaceous Mesaverde Group in the Western Interior, USA.

THE MESAVERDE GROUP, NORTHWESTERN COLORADO

During the late Cretaceous, the Western Interior of the USA was a foreland basin related to overall mountain building and, more specifically, to eastward directed thrusting during the Sevier deformation phase of the Rocky Mountain orogeny. The seaway occupying the basin had oceanic connections both to the north and south, and was supplied with copious amounts of clastic sediment from the evolving mountain belt to the west. The late Cretaceous was a period of relatively high eustatic sea level. During the period of deposition of the Mesaverde Group three eustatic sea-level falls are interpreted to have occurred at 77.5, 75 and 71 Ma (Haq *et al.*, 1988). Other authors, more specifically concerned with the Western Interior, have one (Weimer, 1988) or three (Kiteley, 1983) sea-level fall(s) within this time period.

The basin filled asymmetrically, with the thickest deposits flanking the thrust front, in response to higher subsidence rates (Jordan, 1981). The basin fill comprises a series of clastic wedges composed of fluvial, coastal plain deltaic and shelf deposits. Each wedge extends several tens of kilometres down depositional dip and thins basinwards to the east (Fig. 1). Although Asquith (1970) interpreted shelf, shelf-break, slope and basin-floor topography within the Western Interior Basin, Van Wagoner *et al.* (1990) have interpreted depositional surfaces as shingled and sigmoidal clinoforms with dips of 0.5° or less which are indicative of a ramp-type basin margin.

In northwestern Colorado the Mesaverde Group was deposited in a 7.5 Ma time period of the Campanian, from 78 to 70.5 Ma (*Baculites* sp. (smooth) to *Baculites baculus*) (Cobban & Reeside, 1952; Gill & Cobban, 1973).

In this area, the Mesaverde Group is extensively exposed south of Craig (Fig. 2) and has been previously studied by Boyles and Scott (1982) and Boyles (1983). Continuous exposures extend for 60 km along depositional dip and in addition to this dip section, a series of canyons normal to the direction of progradation provide serial strike sections 1 to 2 km wide, spaced 2 km apart on average. The Mesaverde Group clastic wedge has an average thickness of 600 m, is underlain by the Mancos Shale and overlain by the Lewis Shale (Fig. 1). It is composed of parasequences that are grouped into progradational and aggradational sets. Correlations of parasequences are based on transgressive flooding surfaces of various orders.

The Mesaverde delta system is considered to be sand-rich and has previously been interpreted to be storm- and wave-dominated (Balsey, 1983; Siepman, 1986). Whilst accepting the significance of storm and wave processes in the delta front, we consider that tidal current processes were also significant, both in the delta front and the delta plain.

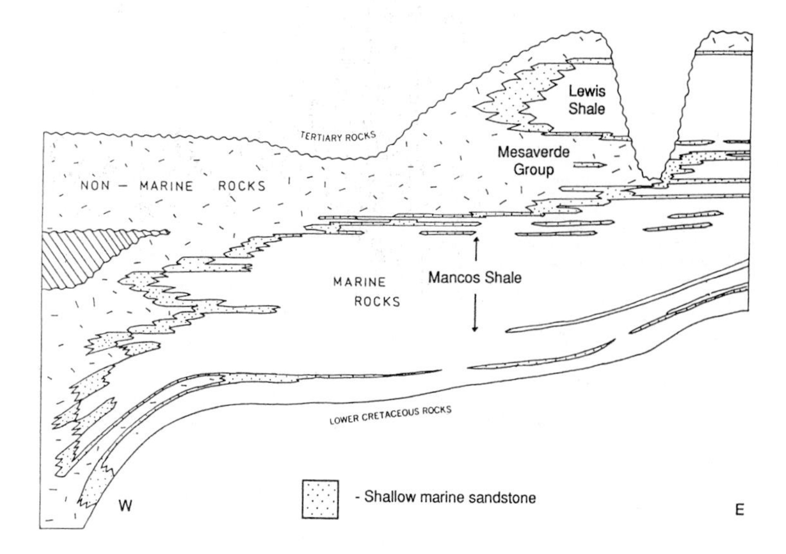

Fig. 1. Restored diagrammatic east–west cross-section through the Upper Cretaceous (Coniacian through Maastrichtian) of northern Utah and Colorado. Clastic wedges thin basinwards to the east, and are thickest adjacent to the thrust front, in the west. Length of section approximately 650 km. (Modified from McGookey, 1972.)

This paper focuses on the lower Iles Formation (Fig. 3), which comprises two progradationally stacked sets of small- to medium-scale parasequences, each of which is overlain by a relatively thick delta-plain interval. Progradational parasequence sets are considered to reflect highstand systems tract deposition, whilst the delta-plain intervals are interpreted to be incised valley fills, de-

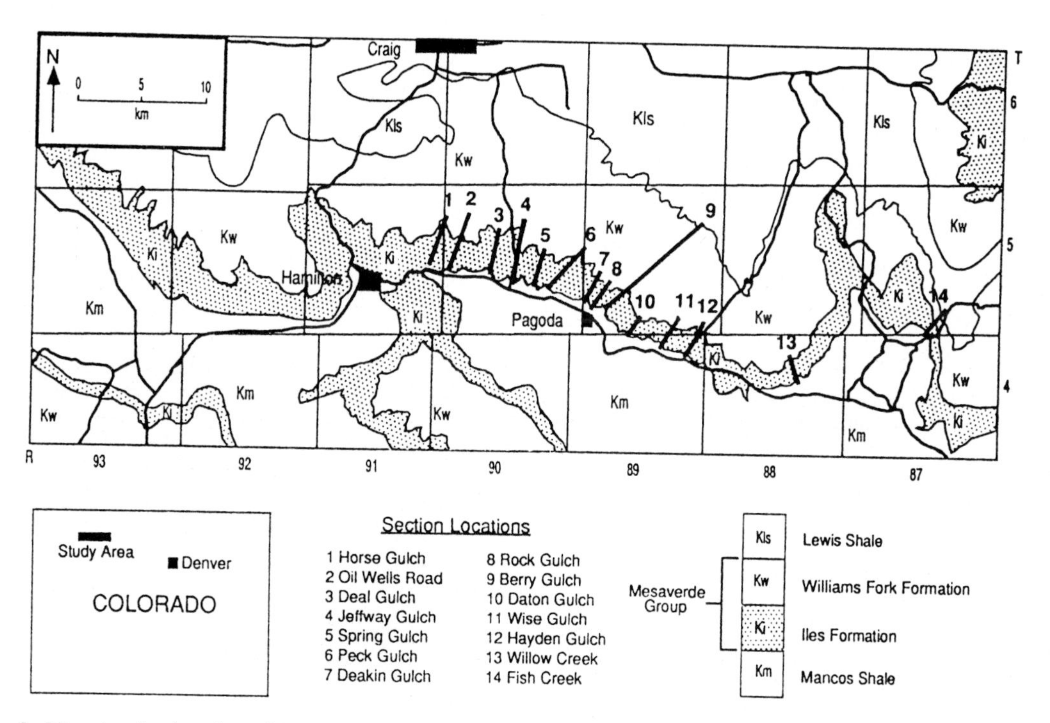

Fig. 2. Map showing location of the study area and the locations of the measured sections through the Iles Formation (stippled) referred to in Fig. 3. The measured sections lie to the north of county road 317. (Modified from Tweto, 1976.)

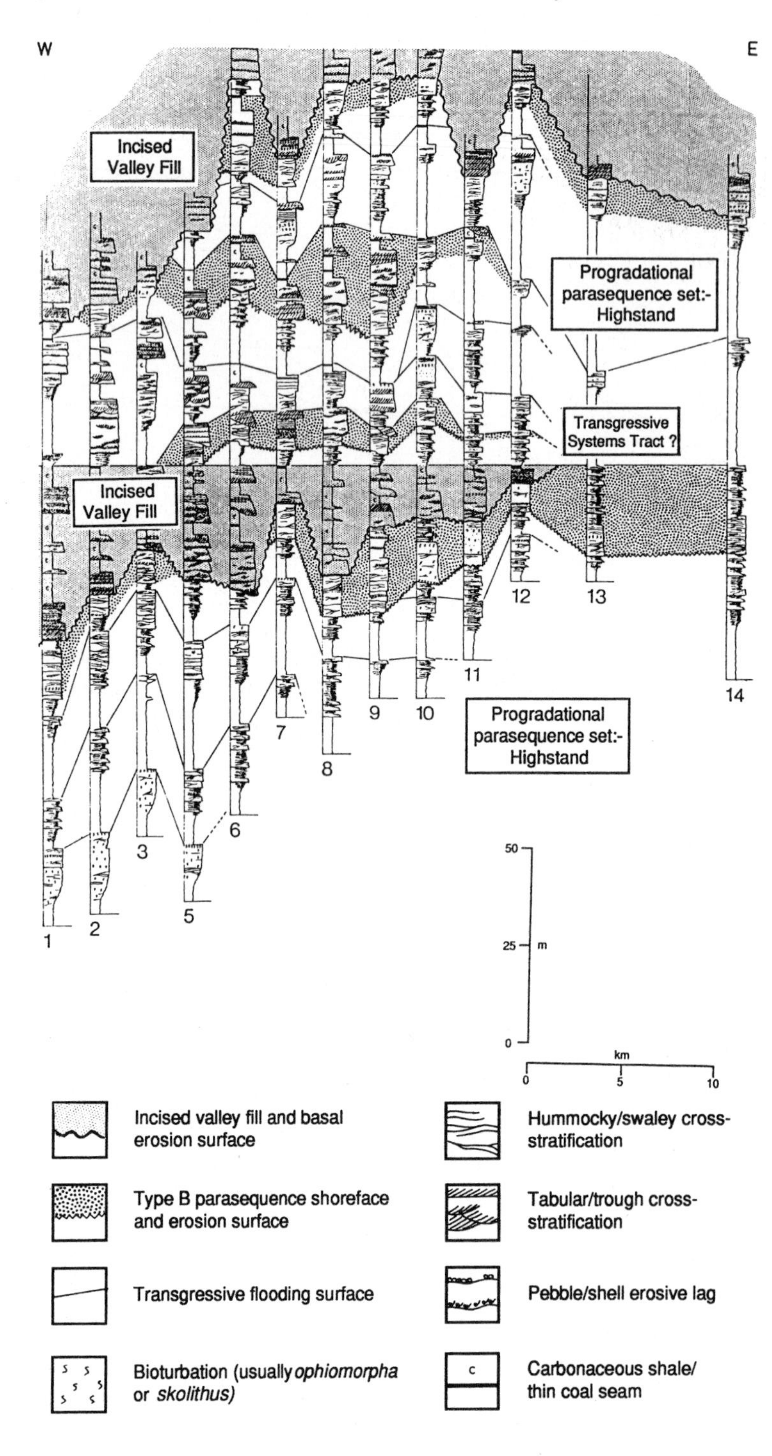

Fig. 3. Correlation of measured sections through the lower Iles Formation, Mesaverde Group, northwestern Colorado. Recognition of depositional systems tracts is based on work outside the scope of this study. Location of sections given in Fig. 2, section #4 omitted from this figure. Length of section approximately 40 km.

posited as part of the lowstand wedge systems tract. The transgressive systems tract is possibly represented by two thin, retrogradationally stacked parasequences, overlying the lower incised valley fill (Fig. 3). Parasequences are bounded by marine flooding surfaces and their correlative surfaces in the coastal plain and the basin. Flooding surfaces are represented by sharp contacts seaward of the upper shoreface of the underlying parasequence. Landward of here, they are represented by a thin (average 20 cm) shell or mud–pebble lag, indicating limited transgressive reworking of the underlying deposits. No significant transgressive deposits are found in association with the flooding surfaces. Two types of parasequence are recognized. Type A parasequences are dominated by a gradational, coarsening-upwards delta front facies sequence, which may be capped by lower delta plain deposits. Type B parasequences contain a regionally extensive erosion surface that separates lower delta front from upper delta front deposits. Both type A and B parasequences are described and interpreted below.

TYPE A PARASEQUENCES

These parasequences comprise delta-front and delta-plain facies and are bounded by discrete transgressive flooding surfaces in the depositional dip section under consideration (Fig. 4). The lower delta front facies are widespread, but the overlying delta plain facies tend to be absent downdip. The occurrence of delta-plain facies over delta-front facies is interpreted to reflect the continuation of the underlying progradational trend, rather than a basinward shift of facies caused by a fall of relative sea level.

Delta front

Description

Delta-front environments are represented by 5–30 m-thick gradational coarsening-upward facies sequences (Figs 4 & 5) that commence with grey-black shales characterized by millimetre-scale horizontal lamination defined by variations in siltstone content. This facies grades upwards into a heterolithic facies comprising thinly interbedded, fine-grained sandstones and bioturbated silty shales. Individual sandstone beds are 1 to 2 cm thick, erosively based and exhibit well-defined, gently undulating laminations with an average wavelength of 10 cm, interpreted to be hummocky cross-stratification (HCS). Mid-parts of the facies sequences are composed of hummocky cross-stratified sandstone in 30–100 cm-thick beds, separated by heterolithic facies. Sandstone beds in the upper parts of the facies sequences are often amalgamated into composite units up to 5 m thick with minimal preservation of the finer grained

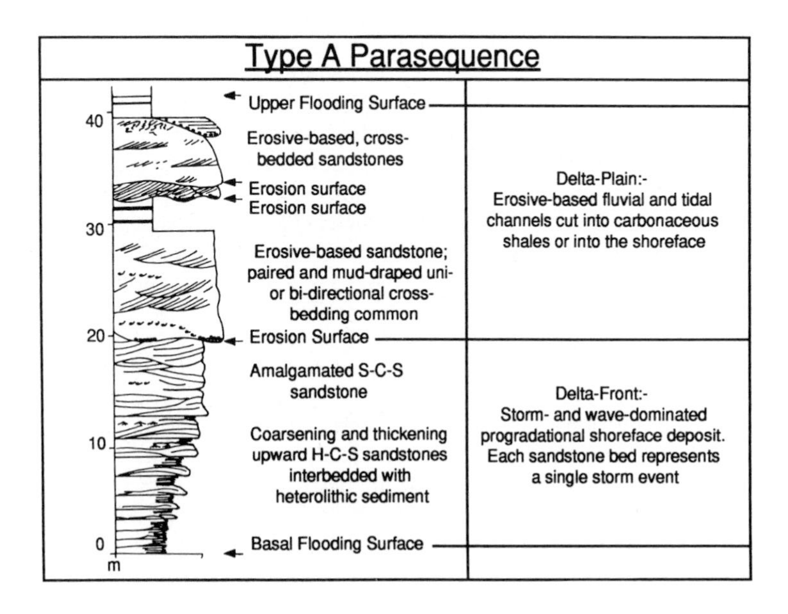

Fig. 4. Log through a type A parasequence from the mid-part of the study section. The parasequence represents the progradation of a storm- and wave-dominated shoreface with tidal influence in the upper delta front and lower delta plain.

Fig. 5. Gradational, storm-dominated delta front coarsening-upward facies sequence. Base of exposure is heterolithic sandstones and shales deposited just above storm wave base, top is amalgamated swaley cross-stratified sandstone deposited above fair-weather wave base. Twentymile Sandstone, US highway 40, near Mount Harris T6N R87W.

heterolithic facies. In the absence of shale interbeds, amalgamation surfaces can be identified by concentrations of *Inoceramus* shells or by structureless sandstones immediately overlying well-laminated sandstones. These amalgamated sandstone units display cross-cutting sets of gently dipping, concave-up laminations that are interpreted to be swaley cross-stratification (SCS) (defined by Leckie & Walker, 1982). The upper few centimetres of the beds may contain cross-lamination sets, interpreted to be wave-produced.

Where preserved, the tops of these gradational facies sequences often display 0.5 to 1.0 m sets of unidirectional tabular cross-bedding with mud drapes on the lower parts of the foresets. Alternatively, unidirectional tabular cross-bed sets lacking mud drapes may alternate with horizontally laminated sandstone. Where the top of the delta front facies sequence is not the parasequence boundary, it is either cut by an erosive-based channelized sandstone, or is directly overlain by carbonaceous shale.

Interpretation

Hummocky cross-stratification and swaley cross-stratification reflect storm-dominated deposition above storm and fair-weather wave bases, respectively (Harms *et al.*, 1975; Leckie & Walker, 1982). The occurrence of these structures in a coarsening-upward facies sequence is indicative of a shoaling trend, interpreted as being due to progradation of the delta front.

Variations in the upper parts of the facies sequences reflect the positions of sections with respect to tidally influenced distributary mouths. Cross-stratification with mud drapes is indicative of tidal processes (Visser, 1980) and is interpreted to occur in the vicinity of distributary mouths in ebb-tidal deltas or estuarine shoals. Alternation of this tidal facies with storm- and wave-influenced facies suggests that evidence of tidal influence is preserved only during fair-weather periods. Sandstone units in which sets of cross-bedding without mud drapes alternate with parallel lamination are interpreted as foreshore deposits laterally adjacent to distributary mouth settings.

Delta plain

Description

The delta plain comprises carbonaceous shales, thin coals and erosive-based channel sandstones (Fig. 4). Channel sandstones are of two types: (i) wholly tidally dominated; and (ii) mixed tidal/fluvial influenced.

Tidally dominated channel fills are represented by erosive-based lenticular to sheet-like units of cross-bedded medium- to fine-grained sandstone. The erosion surface may have a basal lag of shale pebbles, usually altered to ironstone, which is, on average, 20 cm thick (range 0 to 1 m), and usually found only in the troughs of the erosion surface. Single-storey channel fills are, on average, around 7 m thick, whereas multi-storey sandstone bodies can attain thicknesses of 25 m. In the latter case internal erosion surfaces are defined by *Inoceramus* or mud–pebble lags. Their lateral extent varies from 40 m to 2 km +, with the smaller examples commonly displaying a channel geometry. Trough cross-bedding is pervasive throughout the sandstone, with individual sets being 0.2 to 1 m thick. Individual cross-laminae are draped by mudstone, and are often paired. In rare cases the sandstone between the two mudstone laminae of a pair, near to the base of the foresets, may show ripple lamination in an opposed direction to the main flow sense. Bimodal cross-bedding in which alternate (herringbone) sets have opposite flow directions is rare, but palaeocurrent directions at different sites are frequently opposed. In several cases, sandstone units contain large-scale gently dipping surfaces extending through their entire thickness. These lateral accretion surfaces are defined by finer grained layers and dip at right angles to the main flow direction. These tidally dominated channel sandstones tend to cut into the shoreface or foreshore deposits, though they may occur in isolation within carbonaceous shale, higher up in the delta plain.

Channel sandstones occurring in isolation within carbonaceous shale represent the upstream continuations of those which cut the shoreface. Multi-storey sandstones are of a similar scale, though individual channels tend to be smaller scale (2 to 5 m thick, 30 to 40 m wide). They are erosive-based, often into a subjacent coal, and have a locally thick basal lag of mud pebbles, altered to ironstone. Tidally cross-bedded intervals, described above, may alternate with intervals displaying unidirectional trough and tabular cross-bedding without mud drapes, and ripple lamination. The thickness of cross-bed sets tends to decrease upwards through the channel fill, and ripple lamination becomes more prevalent. Large-scale, gently dipping lateral accretion surfaces are still common.

The carbonaceous shale occasionally contains nodular horizons in which rootlets may be preserved. With increasing amounts of carbonaceous material, the shale may develop into a thin (less than 30 cm), impure coal. Other dominantly fine-grained facies present include a 'pinstripe' facies in which grey mudstone alternates with siltstone/fine sandstone. Each 'stripe' is around 2 mm thick, bedding tends to be undulatory, though no cross-cutting relationships are seen.

Interpretation

The erosive base and generally lenticular nature of the sandstone units suggest that they are channel fills. Cross-bed sets in which the foresets are paired and draped by mud are indicative of tidally influenced deposition (Visser, 1980). Mud drapes are deposited on foresets from suspension during relatively slack water periods in the tidal cycle when sand transport is not possible. The paired nature of the foresets is due to the dominance of either flood or ebb flows at a given location, the thicker sand interval between pairs being deposited during the dominant flow phase. Small-scale ripple lamination in an opposed direction to the main flow sense found between the two laminae of a pair is the result of limited sand transport up the leeside of the bedform during the subordinate flow phase (Visser, 1980). This style of cross-bedding dominates some of the channel fills totally, especially those which cut into shoreface deposits. In other channel sandstones, this tidally produced cross-bedding alternates with unidirectional cross-bedding that does not display these tidal features, and which is interpreted to be the product of fluvial processes.

The channels are interpreted to be tidal/estuarine channels, becoming increasingly fluvially dominated upstream. Large-scale gently dipping surfaces interpreted to be lateral accretion surfaces produced by point-bar migration indicate that the channels are sinuous. Alternating fluvial and tidal dominance within the channel fills is thought to be related to short-term, possibly seasonal, discharge variations of the fluvial system. Sandstone bodies may be the isolated deposits of individual channels, or may be multi-storey bodies resulting from channel-belt deposition. The erosion of channels on the delta plain is often limited by thin coal seams, due to the high resistance of peat to erosion. Heterolithic 'pinstripe' facies deposits represent deposition in an intertidal environment.

In view of the close relationship of these channels with upper delta front (shoreface) and lower delta plain (e.g. tidal flat) facies, and the absence of a regionally extensive channel complex, they are interpreted as a continuation of the underlying progradational trend, and not as the result of a basinward shift of facies belts caused by a fall in relative sea level. The argument against a relative sea-level fall at the base of the delta-plain facies is further strengthened by the degree of tidal influence present, indicating deposition on the *lower* delta plain, close to the shoreline, and not in more landward, fluvial systems.

TYPE B PARASEQUENCES

Four parasequences in the lower Iles Formation (Fig. 3) differ from those described above in two important respects. Firstly, the mid-parts of the delta front coarsening-upward facies sequences include a very distinctive, heterolithic facies, characterized by deeply erosional and highly localized gutter casts. Secondly, these type B parasequences include a regionally significant erosion surface that truncates the heterolithic gutter-cast facies and is overlain by amalgamated, swaley cross-stratified sandstones (Fig. 6). Similar to the type A parasequences, shoreface deposits of type B parasequences may be overlain by lower delta plain facies.

These observations are derived from the mid-part of the facies tract in the lower Iles Formation (Fig. 2), which is dominated by stacked delta front facies sequences. Correlation of the parasequences throughout the exposed length of the facies tract suggests that, in some cases, type B parasequences prograde further into the basin than type A parasequences (Fig. 3).

Description

Heterolithic gutter-cast facies

This facies abruptly overlies storm-dominated heterolithic facies in the lower parts of the delta front facies sequence and is immediately recognizable by the fact that it is more fine-dominated, and includes sandstone-filled gutter casts (Fig. 6). The gutter casts occur in bioturbated silty shales, with occasional thin, erosive-based sandstones (Figs 7 & 8). This facies represents a pronounced change in character of storm-dominated sedimentation.

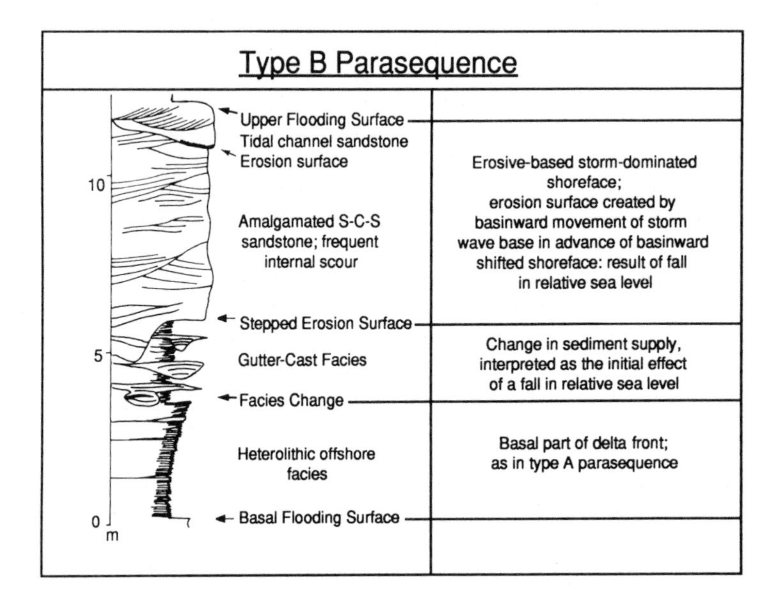

Fig. 6. Log through a type B parasequence illustrating the gutter-cast facies and overlying erosion surface at the base of the shoreface sandstone. The gutter-cast facies represents a sudden change in sediment supply, from relatively sand-dominated to relatively mud-dominated, and is interpreted to be the initial response of the depositional system to a minor fall of relative sea level. Subsequently, a submarine erosion surface is created by the lowering and basinward translation of storm wave base, in advance of deposition of the basinward shifted shoreface. Second type B parasequence, section #6 (Peck Gulch).

Fig. 7. Sandstone-filled gutter cast with hummocky cross-stratification (tape measure is 1 m long). Uppermost type B parasequence, section #6 (Peck Gulch).

The gutter casts trend to the southeast, are typically 1.0 m wide and 0.8 m deep (range 0.5 to 5 m wide, 0.3 to 1 m deep) and have highly irregular and very steep margins, with one example displaying overhanging sections (Fig. 8). The sandstone fill of the gutter casts exhibits well-defined parallel laminations with a shallow convex-upwards shape. Laminae terminate abruptly against the margins of the gutter cast and are often concordant with a dome-shaped form preserved on the upper surface of the feature.

In the absence of gutter casts, the facies remains distinctive by virtue of the occurrence of structureless sandstone beds, which are appreciably thicker (20 to 30 cm) than the underlying storm-dominated sandstones (2 to 4 cm). The facies is also more disorderly than the underlying facies.

The erosion surface

The erosion surfaces are generally planar (Fig. 9), but can include local undulatory relief of up to 20 cm and occasional steep, metre-scale downward-cutting steps. Where the erosion surfaces are planar, they may be difficult to recognize, though they remain distinctive by virtue of the nature of the overlying sandstone. No basal lags have been observed, but the surfaces are commonly ornamented by erosional tool marks, which indicate southeasterly transport directions. The erosion surfaces can be traced for 20 to 50 km down depositional dip and for the entire width of exposures. The erosion surfaces are confined to individual parasequences, i.e. they do not cut into underlying parasequences.

Swaley cross-stratified sandstone

The sandstones overlying the erosion surfaces are 5–10 m-thick units composed of fine-grained, amalgamated swaley cross-stratified beds (Fig. 9). The swales are concave-up features, 0.5 m deep and 1 to 2 m wide with many cross-cutting contacts between sets. Laminae are poorly to well defined, usually by slight variations in grain size or by mineralogical differences. Laminae defining the swales are gently curved, following the basal erosive

Fig. 8. Sandstone-filled gutter cast with overhanging margins set in thinly bedded, heterolithic sediment (trowel is 28 cm long). Uppermost type B parasequence, section #6 (Peck Gulch).

surface of the swale, and dip at 2 to 10°. The swales have the same geometry regardless of the orientation of the exposure (similar to Leckie & Walker, 1982). Individual depositional events are more difficult to identify in this facies than in the equivalent facies of type A parasequences, due to the absence of fine-grained interbeds and clear erosive contacts between beds.

The erosive-based shoreface described above is complemented farther basinwards by more subtle facies changes, where shoreface sandstones become thinner and the erosion surface less pronounced (Fig. 10). Still farther basinwards, mudstones or siltstones deposited from suspension may be erosively overlain by storm wave-influenced heterolithic facies, which would indicate that storm-generated facies accumulated farther basinward than was previously the case. Up depositional dip, the gutter-cast facies is absent, presumably due to erosion at the base of the swaley cross-stratified facies.

Interpretation

Swaley cross-stratification is indicative of storm-dominated deposition above fair-weather wave base in the mid- to upper shoreface (Leckie & Walker, 1982; Plint, 1988b). The swaley cross-stratified facies found in type B parasequences differs markedly from that in type A parasequences, being largely devoid of fine-grained interbeds and clear erosive surfaces.

The erosion surface separates storm-dominated facies which accumulated some distance below fair-weather wave base (normal heterolithic and gutter-cast hummocky cross-stratified facies) from facies which accumulated above fair-weather wave base (amalgamated, swaley cross-stratified facies). This contact therefore represents an abrupt shallowing and, by implication, a basinward shift of facies. This is interpreted to be the result of a lowering of relative sea level on the scale of an individual parasequence. The erosion surface was formed by a

Fig. 9. Type B parasequence showing erosive-based (arrowed) shoreface sandstone and underlying gutter-cast facies. Shoreface sandstone approximately 11 m thick. Third type B parasequence, section #6 (Peck Gulch).

succession of storm events as the nearshore zone of high-energy erosive processes was pulled basinwards during the period of relative sea-level fall. The regionally extensive nature of the erosion surface is probably favoured by the gently dipping ramp-like morphology (Van Wagoner *et al.*, 1990) of the basin margin, as a minor fall of relative sea level would cause a large basinward shift of facies. The final erosion surface is the composite result of net erosion during this period. There is no evidence of emergence associated with these surfaces.

In the context of this interpretation of the erosion surface, the heterolithic gutter-cast facies is interpreted as the *initial response* of the system to a fall of relative sea level (Fig. 11(b)). The heterolithic gutter-cast facies is interpreted to reflect an abrupt increase of both fine-grained suspended sediment and storm-generated sand supply, coupled with a change in the nature of storm sedimentation. The gutter casts are erosional features which must have been cut and filled during a single storm event in order to preserve their steep to overhanging margins. The prevalence of gutter casts in this interval may result from the increased supply of muddy sediment, which forms a more cohesive substrate in which the casts can be sculpted. This facies is interpreted to have accumulated between storm and fair-weather wave bases on the delta front. The increase in sediment supply implied by this facies may reflect an increase in river gradient with enhanced deposition of mud-grade sediment from river-generated suspended sediment plumes creating the muddy substrate in which storm flows produce the gutter casts. At this time, there is limited shoreface erosion in the updip part of the

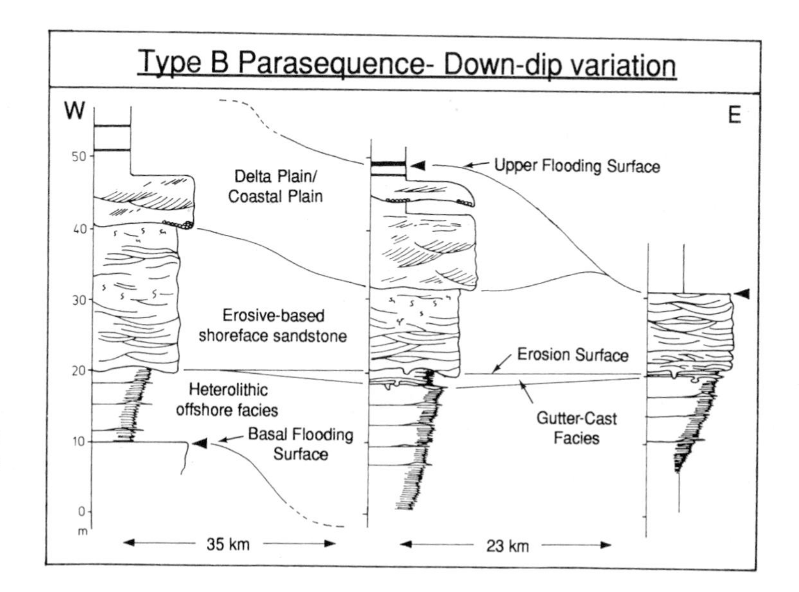

Fig. 10. Logs to illustrate down depositional dip variation in a type B parasequence, using the erosion surface at the base of the shoreface as the datum, for reasons of clarity. The delta/coastal plain is seen to thin to zero basinwards (eastwards). The shoreface also thins, but the basal erosion surface is still discernible.

(a)

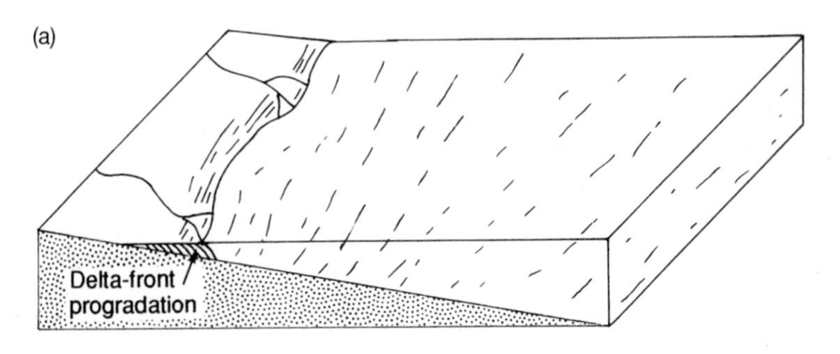

(b)

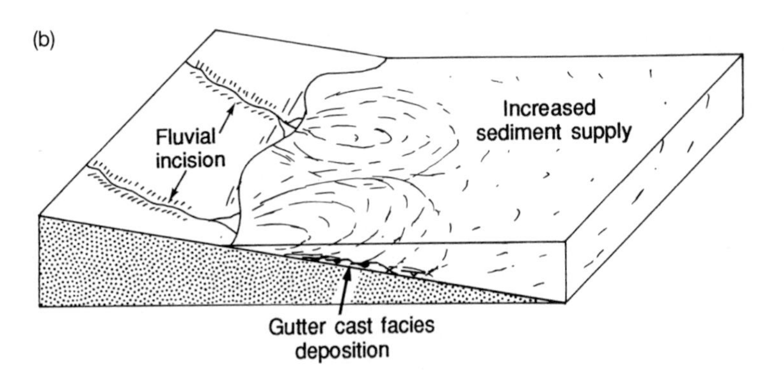

(c)

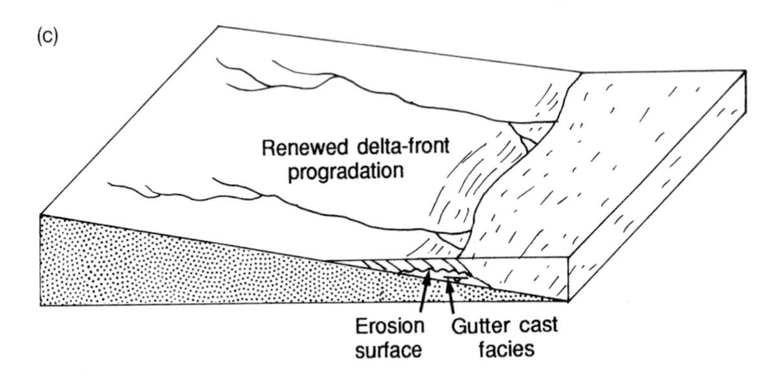

Fig. 11. Block diagrams to illustrate the development of a type B parasequence as a response of the depositional system to a fall of relative sea level on a parasequence scale. (a) Under conditions of stable relative sea level (slowly rising), progradation of the storm- and wave-dominated shoreline occurs. (b) As the initial response of the system to a fall of relative sea level, fluvial systems incise, thus increasing the sediment supply, resulting in deposition from suspended sediment plumes. The gutter-cast facies results from this increase of sediment supply, coupled with the lowering of storm wave base. (c) At the time of maximum rate of relative sea-level fall, the nearshore zone of high-energy erosive processes is pulled basinwards, producing an erosion surface which truncates the gutter-cast facies. During the latter parts of the relative sea-level fall, and the early parts of the relative sea-level rise, shoreline progradation is resumed, in its basinward shifted position. Shoreline progradation continues until a rapid rise of relative sea level creates a flooding surface, and terminates deposition of the parasequence.

facies tract. *Subsequently*, widespread wave and storm wave erosion occur as sea level continues to fall and accommodation space decreases, forming the prominent erosion surface (Fig. 11(c)). During this time of maximum rate of relative sea-level fall, it is thought that there is little or no gutter cast facies deposition; instead, parts of the facies tract basinward of the shoreface are characterized by sediment bypass. Deposition of the upper-shoreface swaley cross-stratified facies above the erosion surface is thought to occur during the lowstand and the initial period of relative sea-level rise (Fig. 11(c)). The sandstone-dominated nature of the swaley cross-stratified facies reflects increased sand supply during the lowstand period. Progradation continues, as reflected by the deposition of delta-plain facies above the shoreface. The overlying flooding surface is formed at the time of the maximum rate of relative sea-level rise, terminating deposition of the parasequence.

DISCUSSION

The interpretation of type B parasequences as including evidence for a fall of relative sea level on a parasequence scale has important implications for the position of sandstone-prone depositional systems on the clastic ramp. Firstly, due to the basinward shift of facies belts, shoreface deposition above the erosion surface occurred in a more basinward position than would have been the case without a fall of sea level. Type B parasequences are therefore seen to extend further into the basin than type A parasequences. Secondly, during the period of falling sea level, widespread erosion of the shelf occurred, at least as far as the final downshifted limit of storm wave base. During this erosional phase, significant sediment bypass must have occurred across the inner shelf, by implication increasing sand supply to the mid- to outer shelf. The southeasterly orientations of the gutter casts and transport direction indicators from the bases of erosive-based shoreface sandstones contrast with the easterly directions of type A parasequences, and may signify increasing alignment in the flow direction of the shelf systems which are directed southwards.

The above evidence of increased sand supply and the more southerly transport directions during the period of falling sea level may provide a better explanation of the 'pro-delta shelf sandstone' bodies described by Swift *et al.* (1987), from the Mesaverde Group in the Book Cliffs, Utah. In depositional dip sections, these bodies are lenticular and indicate southerly, along-shelf transport directions (e.g. Hatch Mesa Lentil) (Swift *et al.*, 1987). They are tentatively interpreted as plume-shaped bodies, generated by storm-driven flows, stripping sand from the shoreface of cuspate wave-dominated deltas, and transporting it along the depositional strike of the shelf. Alternatively, these shelf sandstone bodies may be regarded as a depositional product of erosional bypass associated with periods of sea-level fall outlined above.

Enhanced sand supply to the shelf during periods of falling sea level may also assist in explaining the larger scale sandstone bodies which occur in this basin at this period of geological time. These facies sequences occur in the mid- to outer parts of the shelf as 25–30 m coarsening-upward sequences which reflect shallowing of the shelf. Individual facies sequences comprise shelf mudstones–siltstones, thinly bedded storm-generated sandstones and a cross-bedded facies which has been interpreted as storm-generated (Swift & Rice, 1984) or tidal current-generated (Elliott, 1987). These may be interpreted to reflect shelf-sand deposition as a result of erosional bypassing during falling sea level, indicating that they are significant in sequence-stratigraphic terms, being the basinward time-equivalent deposits to shorefaces of type B parasequences. Others may be interpreted as the deposits of basinward displaced shoreline systems, later isolated by shoreface retreat in a manner similar to the 'forced regressions' of Posamentier and Chamberlain (this volume, pp. 469–485).

CONCLUSIONS

1 Parasequences in the lower part of the Mesaverde Group in northwestern Colorado can be divided into: (i) type A parasequences, characterized by gradational storm wave-dominated delta front facies sequences; and (ii) type B parasequences, which include a regional-scale erosion surface separating fine-grained heterolithic offshore-transition deposits from mid- to upper-shoreface sandstones. The erosion surface is immediately preceded by a distinctive storm-generated, gutter-cast facies.
2 Type B parasequences contain evidence for a fall of relative sea level, the effects of which are confined to the individual parasequence.
3 The facies tract of type B parasequences differs from that associated with type A parasequences. Type B parasequences include a basinward displaced shoreline system, a higher sediment load due to increased gradient of the fluvial system and, possibly, a more sand-prone shelf system which results from sediment bypass during falling relative sea level.
4 The genesis of the 'pro-delta shelf sandstone bodies' of Swift *et al.* (1987) and other shelf sandstone bodies in the Cretaceous Western Interior Basin may be associated with periods of relative sea-level fall and generation of type B parasequences.

ACKNOWLEDGEMENTS

Research for this paper was carried out during tenure of a British Petroleum PhD Studentship (D.F. Hadley), which is gratefully acknowledged. We thank Tony Tollitt for preparing the photographs. We also thank K.W. Shanley and R.W.W. Lovell for their critical comments which improved the final manuscript.

REFERENCES

Asquith, D.O. (1970) Depositional topography and major marine environments, Late Cretaceous, Wyoming. *Bull. Am. Assoc. Petrol. Geol.* **54**, 1184–1224.

Balsey, J.K. (1980) *Cretaceous Wave-dominated Delta Systems: Book Cliffs, East Central Utah.* Am. Assoc. Petrol. Geol. Cont. Ed. Field Guide, 86 pp.

Baum, G.R. & Vail, P.R. (1988) Sequence stratigraphic concepts applied to Paleogene outcrops, Gulf and Atlantic Basins. In: *Sea-Level Changes: An Integrated Approach* (Eds Wilgus, C.K., Hastings, B.S., Kendall, C.G.St.C., Posamentier, H.W., Ross, C.A. & Van Wagoner, J.C.) Spec. Publ. Soc. Econ. Paleontol. Mineral., Tulsa, 42, 309–327.

Boyd, R., Suter, J. & Penland, S. (1989) Relation of sequence stratigraphy to modern sedimentary environments. *Geology* **17**, 926–929.

Boyles, J.M. (1983) Depositional history and sedimentology of Upper Cretaceous Mancos Shale and Lower Mesaverde Group, northwestern Colorado: migrating shelf-bar and wave-dominated shoreline deposits. PhD thesis, University of Texas, Austin, 270 pp.

Boyles, J.M. & Scott, A.J. (1982) A model for migrating shelf-bar sandstones in Upper Mancos Shale (Campanian), northwestern Colorado. *Bull. Am. Assoc. Petrol. Geol.* **66**, 491–508.

Cobban, W.A. & Reeside, J.B. (1952) Correlations of the Cretaceous formations of the Western Interior of the United States. *Bull. Geol. Soc. Am.* **63**, 1011–1044.

Elliott, T. (1974) Abandonment facies of high-constructive lobate deltas, with an example from the Yoredale Series. *Proc. Geol. Assoc.* **85**, 359–365.

Elliott, T. (1987) Tidally-produced sigmoidal cross-bedding in Cretaceous shelf sandstone bodies, Western Interior Basin, U.S.A. (Abstr.) *Soc. Econ. Paleontol. Mineral. Ann. Midyear Meet. 20–23 August, 1987, Austin, Texas.*

Frazier, D.E. (1967) Recent deltaic deposits of the Mississippi Delta: their development and chronology. *Trans. Gulf Coast Assoc. Geol. Soc.* **17**, 287–315.

Gill, J.R. & Cobban, W.A. (1973) Stratigraphy and geologic history of the Montana Group and equivalent rocks, Montana, Wyoming, and North and South Dakota. *U.S. Geol. Surv. Prof. Pap.* **776**, 37 pp.

Haq, B.U., Hardenbol, J. & Vail, P.R. (1988) Mesozoic and Cenozoic chronostratigraphy and cycles of sea level change. In: *Sea-Level Changes: An Integrated Approach* (Eds Wilgus, C.K., Hastings, B.S., Kendall, C.G.St.C., Posamentier, H.W., Ross, C.A. & Van Wagoner, J.C.) Spec. Publ. Soc. Econ. Paleontol Mineral., Tulsa 42, 71–108.

Harms, J.C., Southard, J.B., Spearing, D.R. & Walker, R.G. (1975) *Depositional Environments as Interpreted from Primary Sedimentary Structures and Stratification Sequences.* Soc. Econ. Paleontol. Mineral., Short Course 2, 161 pp.

Jordan, T.E. (1981) Thrust loads and foreland basin evolution, Cretaceous, Western United States. *Bull. Am. Assoc. Petrol. Geol.* **65**, 2506–2520.

Kiteley, L.W. (1983) Paleogeography and eustatic-tectonic model of late Campanian Cretaceous sedimentation, southwestern Wyoming and northwestern Colorado. In: *Mesozoic Paleogeography of West-Central United States* (Eds Reynolds, M.W. & Dolly, E.D.) Rocky Mountain Section, Soc. Econ. Paleontol. Mineral., Denver, pp. 273–303.

Kiteley, L.W. & Field, M.E. (1984) Shallow marine depositional environments in the Upper Cretaceous of northern Colorado. In: *Siliciclastic Shelf Sediments* (Eds Tillman, R.W. & Siemers, C.T.) Spec. Publ. Soc. Econ. Paleontol. Mineral., Tulsa 34, 179–204.

Leckie, D.A. & Walker, R.G. (1982) Storm- and tide-dominated shorelines in Cretaceous Moosebar–Lower Gates interval — outcrop equivalents of Deep Basin gas trap in western Canada. *Bull. Am. Assoc. Petrol. Geol.* **66**, 138–157.

McGookey, D.P. (1972) Cretaceous System. In: *Geologic Atlas of the Rocky Mountain Region* (Ed. Mallory, W.W.) Rocky Mountain Assoc. Geol., Denver, pp. 190–228.

Plint, A.G. (1988a) Global eustacy and the Eocene sequence in the Hampshire Basin, England. *Basin Res.* **1**, 11–22.

Plint, A.G. (1988b) Sharp-based shoreface sequences and 'offshore bars' in the Cardium Formation of Alberta: their relation to relative changes in sea level. In: *Sea-level Changes: An Integrated Approach* (Eds Wilgus, C.K., Hastings, B.S., Kendall, C.G.St.C., Posamentier, H.W., Ross, C.A. & Van Wagoner, J.C.) Spec. Publ. Soc. Econ. Paleontol. Mineral., Tulsa, 42, 357–370.

Siepman, B.R. (1986) Facies relationships in Campanian wave-dominated coastal deposits in Sand Wash Basin. In: *New Interpretations of Northwest Colorado Geology* (Ed. Stone, D.S.) Rocky Mountain Assoc. Geol., Denver, pp. 157–164.

Swift, D.J.P. & Rice, D.D. (1984) Sand bodies on muddy shelves: a model for sedimentation in the Western Interior Cretaceous Seaway, North America. In: *Siliciclastic Shelf Sediments* (Eds Tillman, R.W. & Siemers, C.T.) Spec. Publ. Soc. Econ. Paleontol. Mineral., Tulsa 34, 43–62.

Swift, D.J.P., Hudelson, P.M., Brenner, R.L. & Thompson, P. (1987) Shelf construction in a foreland basin: storm beds, shelf sandbodies, and shelf-slope depositional sequences in the Upper Cretaceous Mesaverde Group, Book Cliffs, Utah. *Sedimentology* **34**, 423–457.

Tweto, O. (1976) Geologic Map of the Craig 1° × 2° Quadrangle, Northwestern Colorado. *U.S. Geol. Surv., Denver, Colorado.*

Vail, P.R., Mitchum, R.M. Jr., Todd, R.G. *et al.* (1977) Seismic stratigraphy and global changes of sea level. In: *Seismic Stratigraphy — Applications to Hydrocarbon Exploration* (Ed. Payton, C.E.) Mem. Am. Assoc. Petrol. Geol. 26, 49–212.

Van Wagoner, J.C., Mitchum, R.M., Campion, K.M. & Rahmanian, V.D. (1990) Siliciclastic sequence stratigraphy in well logs, cores, and outcrops: concepts for high resolution correlation of time and facies. *Am. Assoc. Petrol. Geol. Meth. Expl. Ser.* **7**, 55 pp.

VISSER, M.J. (1980) Neap-spring cycles reflected in Holocene subtidal large-scale bedform deposits: a preliminary note. *Geology* **8**, 543–546.

WEIMER, R.J. (1988) Record of relative sea level changes, Cretaceous of western Interior, USA. In: *Sea-level Changes: An Integrated Approach* (Eds Wilgus, C.K., Hastings, B.S., Kendall, C.G.St.C., Posamentier, H.W., Ross, C.A. & Van Wagoner, J.C.) Spec. Publ. Soc. Econ. Paleontol. Mineral., Tulsa, 42, 285–288.

WILGUS, C.K., HASTINGS, B.S., KENDALL, C.G.ST.C., POSAMENTIER, H.W., ROSS, C.A. & VAN WAGONER, J.C. (Eds) (1988) *Sea-level Changes: An Integrated Approach.* Spec. Publ. Soc. Econ. Paleontol. Mineral., Tulsa, 42, 407 pp.

Spec. Publs Int. Ass. Sediment. (1993) **18**, 537–561

Sequence stratigraphy of the onshore Palaeogene, southeastern Atlantic Coastal Plain, USA

W.B. HARRIS, V.A. ZULLO *and* R.A. LAWS

Department of Earth Sciences, University of North Carolina at Wilmington, Wilmington, North Carolina 28403, USA

ABSTRACT

Application of sequence-stratigraphic concepts to basin margin sediments in the Albemarle embayment and Cape Fear arch areas of North Carolina, and the Santee and Savannah River areas of the Southeast Georgia embayment of South Carolina allows recognition of Palaeogene depositional sequences. Through their contained micro- and megafossils the sequences are correlated with standard biostratigraphic zonations and global cycles of coastal onlap. The Albemarle and Southeast Georgia embayments contain two Palaeocene siliciclastic depositional sequences (Danian and Thanetian) separated by the global 58.5 Ma type 1 unconformity. The Danian sequence correlates to the TA1.3 cycle in the Albemarle embayment and the Savannah River area and the TA1.2 cycle in the Santee River area. The Thanetian sequence correlates to the TA2.1 cycle in the Albemarle and Southeast Georgia embayments. The Albemarle embayment and the Santee and Savannah River areas preserve several early, middle and late Eocene depositional sequences. Siliciclastic and carbonate Ypresian sequence(s) are separated from the Lutetian–Priabonian sequences by the global 49.5 Ma unconformity. They are correlated to one cycle between the TA2.4 and TA2.9 cycles in the Albemarle embayment, the TA2.5 cycle in the Santee River area and the TA2.4 and TA2.5 cycles in the Savannah River area. The Lutetian–Priabonian sequences are carbonate in the Albemarle embayment, siliciclastic and carbonate in the Santee River area and predominantly siliciclastic in the Savannah River area. They are correlated to the TA3.3 through TA4.3 cycles in the Albemarle embayment and the Santee River area and the TA3.2 through TA4.3 cycles in the Savannah River area.

One Rupelian and four Chattian siliciclastic and carbonate depositional sequences are recognized in the Albemarle embayment and two Chattian sequences in the Santee River area. Erosion during the major Oligocene sea-level fall at 30 Ma is probably responsible for the absence of early Oligocene sequences in the onshore Southeast Georgia embayment. In the Albemarle embayment they are correlated to the TA4.4 through TB1.4 cycles, and in the Santee River area to the TB1.1 cycle and either the TB1.2 or TB1.3 cycle. Although Oligocene sediments are reported in the Savannah River area, we have not recognized them.

INTRODUCTION

The southeastern Atlantic Coastal Plain province of the United States extends from the Norfolk arch in Virginia for about 800 km to the South Georgia–Florida Peninsula arch in Georgia. It is traditionally divided from north to south into the Albemarle embayment, the Cape Fear arch (or Carolina Platform) and the Southeast Georgia embayment. The Albemarle embayment centred in North Carolina is bounded by the Norfolk and the Cape Fear arches; the Southeast Georgia embayment is bounded by the Cape Fear and South Georgia–Florida Peninsula arches. The Coastal Plain province is bounded on the northwest by crystalline rocks of the Piedmont province and is generally a northeast to southwest trending monocline of marine sediments that thicken to the southeast. Throughout this paper the area between the Piedmont and the Orangeburg scarp is referred to as the

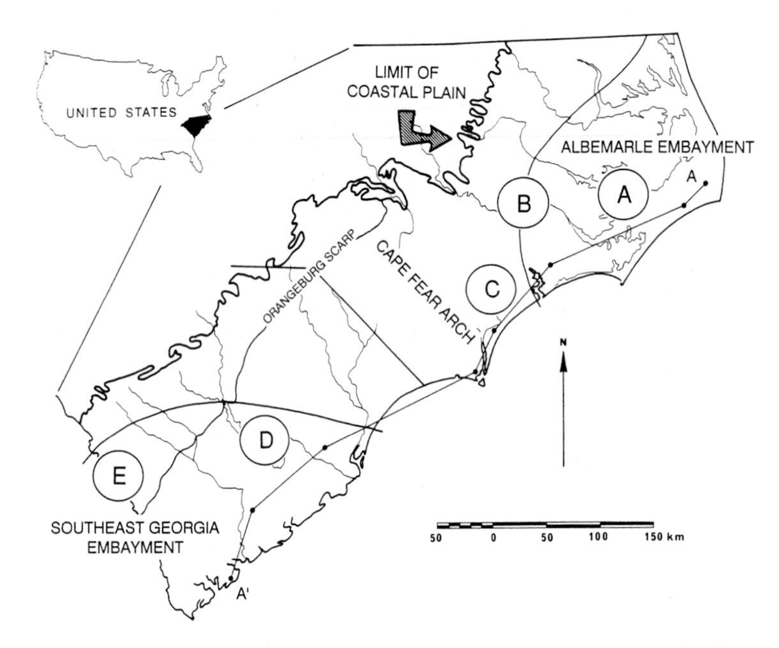

Fig. 1. Major structural features of the southeastern Atlantic Coastal Plain and locations of control used in this study. Letters refer to areas of control; specific locations are given in the Appendix (p. 558).

inner Coastal Plain, and the area seaward of the Orangeburg scarp as the middle and outer Coastal Plain (Fig. 1).

This paper synthesizes and identifies Palaeogene depositional sequences in the onshore part of the Albemarle embayment of North Carolina and the northern (Santee River area) and western (Savannah River area) parts of the Southeast Georgia embayment in South Carolina (Fig. 1). Eleven depositional sequences ranging in age from Danian to Chattian are recognized in the Albemarle embayment and the Santee River area. Eleven depositional sequences ranging in age from Danian to Priabonian are recognized in the Savannah River area. The contained micro- and megafossils permit correlation of the depositional sequences to standard zonations and to the coastal-onlap cycles of Haq *et al.* (1987). Nannofossil zonations follow Martini (1971) and Okada and Bukry (1980); planktonic foraminiferal zones are after Blow (1969) and Berggren (1971). Although complete successions of systems tracts are seldom preserved in any sequence at one exposure, we have identified transgressive and highstand sediments in most sequences. Lowstand sediments have only been identified in one middle/upper Eocene depositional sequence. Locations of major outcrops and subsurface data used in this study are given in the Appendix, pp. 558–559. Figure 2 summarizes the lithostratigraphic terminology for the areas. Figure 3 illustrates the structural relations of some of the sequences from the Albemarle embayment to the Savannah River area of the Southeast Georgia embayment.

DEPOSITIONAL SEQUENCES, ALBEMARLE EMBAYMENT

Palaeocene

Lithostratigraphy

Palaeocene strata in the Albemarle embayment are referred to the Beaufort Formation, which consists of the lower Jericho Run Member and an upper member formally referred to in this paper as the Moseley Creek Member (Fig. 2) with the section described by Harris and Baum (1977) the holostratotype. These two members are recognized in outliers in the upper part of the middle Coastal Plain and are separated from the main body of Palaeocene strata that occurs in the subsurface in the outer part of the Coastal Plain. In outcrop, the Beaufort Formation reaches a maximum thickness of about 26 m (Brown *et al.*, 1978), whereas in the subsurface it has a maximum thickness of almost 100 m (Zarra, 1989).

The Jericho Run Member is siliceous indurated claystone and shale up to about 8.5 m in thickness

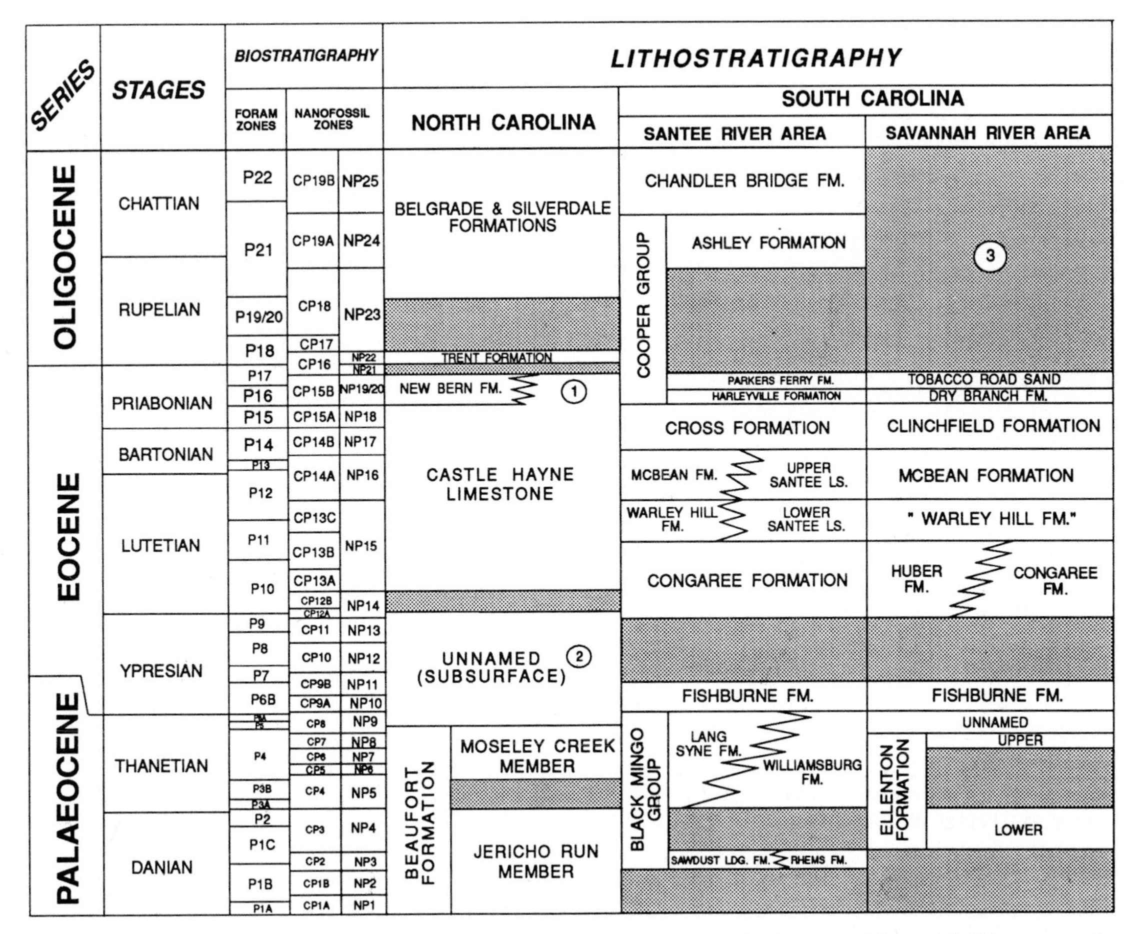

Fig. 2. Palaeogene lithostratigraphic units for the Albemarle embayment and the Santee and Savannah River areas of the Southeast Georgia embayment. Notes: (i) Zarra (1989) recognized an unnamed upper Eocene unit in the Albemarle embayment; (ii) Zarra (1989) recognized an unnamed latest Palaeocene to early Eocene unit in the Albemarle embayment; (iii) Edwards (1992) recognized Oligocene sediments in the Savannah River area.

(Brown *et al.*, 1978). Opaline-cemented claystone and siliceous mudstone are not widespread and appear to be restricted to surface outliers. The Jericho Run Member is correlated downdip in the outer Coastal Plain to a lower Palaeocene sequence that has a thickness between 40 and 60 m (Zarra, 1989). This sequence, which is predominantly clastic, consists of glaucony-rich, very fine-grained sandstone to sandy siltstone with calcareous cement; in some places silty shale is common.

The Moseley Creek Member consists of 3 to 4 m of alternating, unconsolidated, sandy, foraminiferal–glauconitic sediments and thinner, slightly glauconitic, foraminiferal biosparite and sandy biosparite (Harris & Baum, 1977). In the subsurface of the outer Coastal Plain, an upper Palaeocene sequence which is correlated to the Moseley Creek Member consists of lower thin, medium- to coarse-grained sand that is overlain by silty clay and very fine-grained sand (Zarra, 1989). This sequence obtains a thickness of about 40 m.

Sequence stratigraphy and correlation

Beaufort Formation sediments represent two depositional sequences both in outcrop and in the subsurface of the Albemarle embayment (Fig. 4). Brown *et al.* (1977) assigned the Jericho Run Member to planktonic foraminiferal zone P1, based on the foraminiferal assemblage. Zarra (1989), on the presence of a sparse but well-preserved planktonic assemblage, assigned the subsurface sequence to the

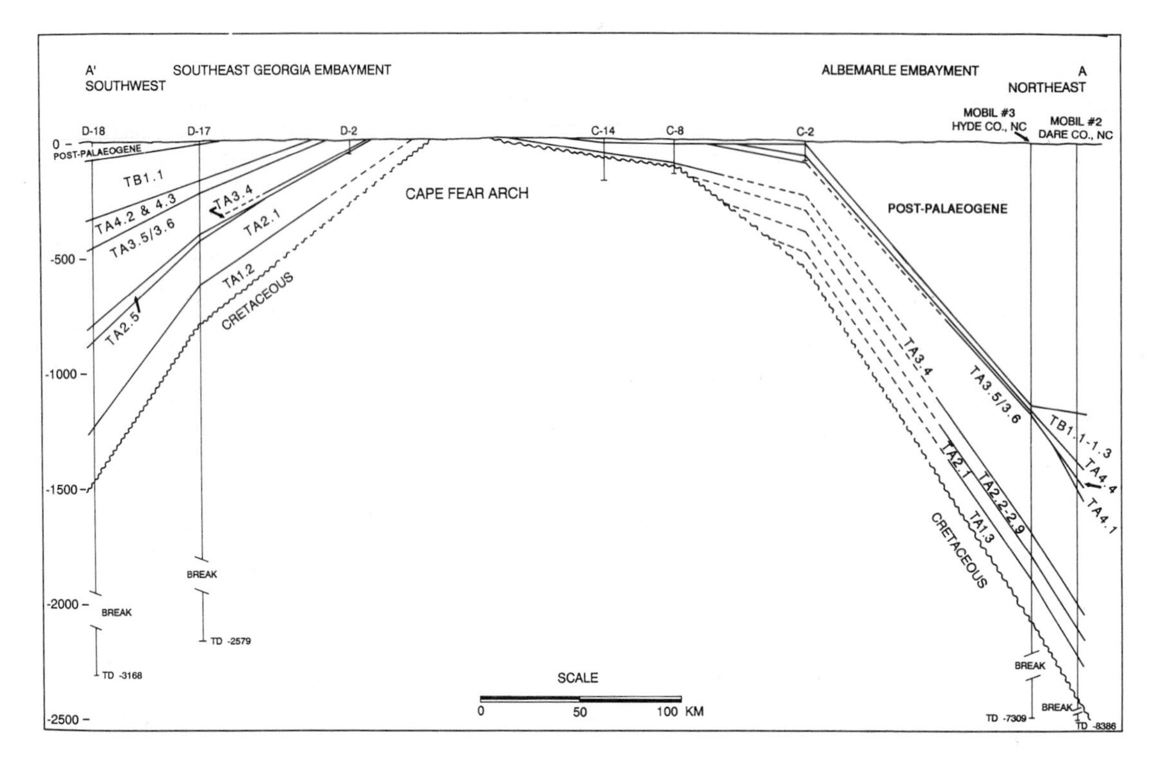

Fig. 3. Cross-section from the Albemarle embayment to the Savannah River area of the Southeast Georgia embayment; the line of section is located in Fig. 1. Section locations are designated by letter and numbers that refer to control points given in the Appendix (p. 558). Depths are in feet.

P2 planktonic foraminiferal zone. Based on these data, the Jericho Run Member and the downdip lower Palaeocene sequence are assigned to the TA1.3 cycle. The lower phosphate pebble conglomerate and glaucony-rich fine-grained sandstone to silty sandstone described by Brown *et al.* (1978) and Zarra (1989) represent a transgressive systems tract and initial transgression of the Danian sea onto the late Cretaceous shelf (Fig. 4). Reworked Upper Cretaceous micro- and megafossils and phosphatized Cretaceous lithoclasts in the conglomerate and the abundance of glaucony support this interpretation. Upper siliceous claystone and shale updip and sandy siltstone, silty shale and claystone downdip are interpreted to represent highstand deposits (Fig. 4). Carlson (1976) and Zarra (1989) suggested on the basis of foraminifers that the Jericho Run Member was deposited in the innerneritic environment. Carlson (1976) also inferred a slight shoaling-upward of the Jericho Run Member which supports assignment of the unit to a highstand systems tract.

Harris and Baum (1977) identified an abundant and diverse benthonic and planktonic foraminiferal assemblage in the Moseley Creek Member indicative of the P4 planktonic foraminiferal zone. Worsley and Turco (1979) identified a diverse nannoflora indicative of zones NP5 or NP6. Later Turco *et al.* (1979) restricted the unit to calcareous nannofossil zone, NP6. Based on these data the Moseley Creek Member is assigned to the TA2.1 cycle. The lower phosphate pebble conglomerate and the overlying interbedded unconsolidated sandy foraminiferal sediments and consolidated biosparites described by Harris and Baum (1977) represent transgressive and highstand systems tracts, respectively (Fig. 4). Zarra (1989) identified planktonic species in his upper Palaeocene sequence in the subsurface of the Albemarle embayment that he assigned to planktonic foraminiferal zone P4. He delineated the base of this sequence on reflection seismic lines by truncated underlying reflectors and onlap of overlying reflectors. Zarra (1989) also recognized a persistent, coarse basal sand in the upper Palaeocene

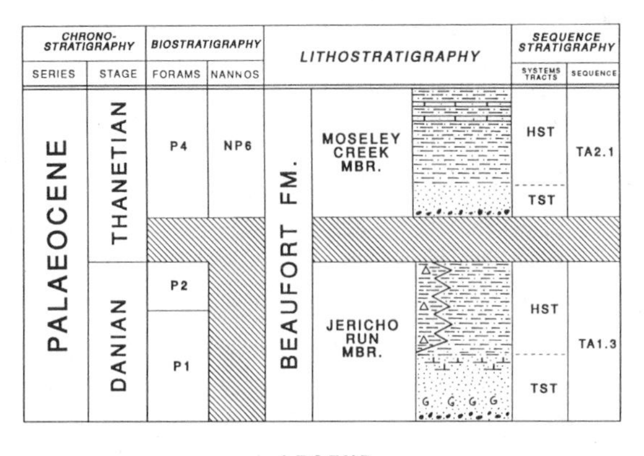

Fig. 4. Generalized biostratigraphy, lithostratigraphy and sequence stratigraphy of Palaeocene sediments in the Albemarle embayment of North Carolina. HST, the highstand systems tract; TST, the transgressive systems tract; and LST, the lowstand systems tract; dashed line between LST and TST is the transgressive surface and between TST and HST is the condensed section.

sequence that was overlain by an interval of silty clay to very fine-grained sand. The basal sand is interpreted to represent a transgressive systems tract and the silty clay and very fine-grained sand a highstand systems tract. Based on foraminifera, Zarra (1989) interpreted the depositional setting of this sequence as inner- to outer-neritic.

Palaeocene/Eocene

Lithostratigraphy

Zarra (1989) recognized a thin (6 to 30 m thick), wedge-shaped depositional sequence in the subsurface of the onshore Albemarle embayment (Fig. 2). He indicated that this depositional sequence consists of a basal thin, ubiquitous medium-grained sand overlain by a thicker silty clay that, based on resistivity log signatures, fines and then coarsens upwards. No sequence has been recognized in outcrop between the late Palaeocene and the middle Eocene.

Sequence stratigraphy and correlation

Although foraminifera are rare in this sequence, Zarra (1989) suggested a latest Palaeocene to early Eocene age because of the occurrence of *Subbotina soldadoensis soldadoensis*, which ranges from planktonic zones P5 to P9. This sequence, therefore, is assigned to one cycle between the TA2.4 and TA2.9 cycles. The lower sand and the lower part of the silty clay are interpreted to represent a transgressive systems tract and the remaining silty clay a highstand systems tract (Fig. 5). Benthonic foraminifers in this sequence indicate an inner-neritic to shallow middle-neritic environment of deposition (Zarra, 1989).

Eocene

Lithostratigraphy

Outcropping Eocene sediments in the Albemarle embayment and on the north flank of the Cape Fear arch are referred to the Castle Hayne Limestone and the New Bern Formation (Fig. 2). Zullo and Harris (1986, 1987) identified five depositional sequences (0–4) in the Castle Hayne Limestone from outcrop studies. Three sequences (0–2) were interpreted as exclusively middle Eocene in age, one (3) middle to late Eocene in age and one (4) late Eocene in age. They placed the middle–upper Eocene boundary within sequence 3 at a zone of phosphatized and glauconitized sponges interpreted as a marine hardground within the condensed section. The maximum thickness of about 15 m for the three (0–2) middle Eocene sequences occurs in an outlier in Duplin County. Sequence 1 in Duplin County contains a 60 cm-thick bed characterized by an irregular dendritic growth of smectite in crystal aggregates with minor euhedral biotite and apatite interpreted as an altered ash fall (Thayer *et al.*, 1986). Harris and Fullagar (1989) determined K–Ar and Rb–Sr dates of 46.2 ± 1.8 and 45.7 ± 0.7 Ma, respectively, on biotite from the ash fall.

The upper parts of sequences 3 and 4 of the Castle Hayne Limestone consist of lower sponge-bearing biomicrite containing abundant molluscs, brachiopods and pectinids that grade upward into sponge-bearing biomicrite and sponge-bearing bryozoan biomicrudite. They achieve a maximum thickness of about 9 m in New Hanover County. North of the White Oak River, the carbonate-dominated lithofacies of the upper Eocene sequence (4) becomes siliciclastic with sandy pelecypod mould biosparite and biosparrudite being the dominant rock types (Baum, 1981). This lithofacies

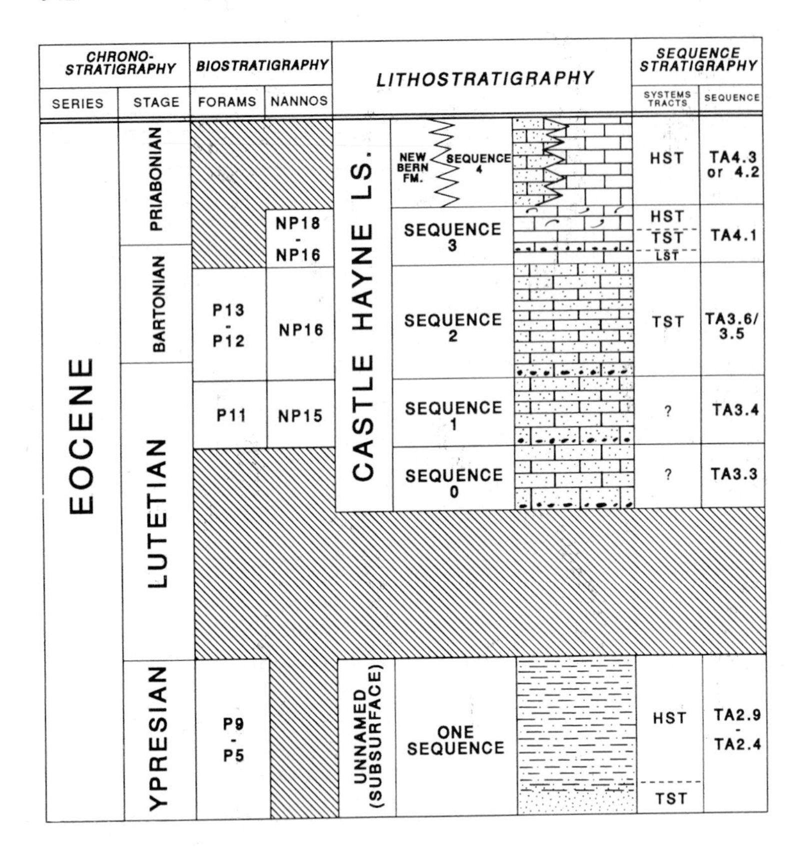

Fig. 5 Generalized biostratigraphy, and sequence stratigraphy of Eocene sediments in the Albemarle embayment and on the Cape Fear arch of North Carolina. Lithologic symbols as in Fig. 4.

is the New Bern Formation and has a maximum thickness of about 10 m in Craven County. In the subsurface of the Albemarle embayment Zarra (1989) recognized two Eocene depositional sequences; a middle one and an upper one.

Sequence stratigraphy and correlation

Sequence 0 of the Castle Hayne Limestone has not yielded age-diagnostic fossils and is included in the middle Eocene on the basis of lithologic similarity to the overlying sequences. It is tentatively assigned to the TA3.3 cycle (Fig. 5). Sequence 1 contains echinoids and a small oyster which is conspecific with a form from the lower part of the Santee Limestone in South Carolina. Worsley and Laws (1986), on the basis of calcareous nannofossils, correlated this sequence to nannofossil zones CP13b and NP15 and planktonic foraminiferal zone P11. This sequence is assigned to the TA3.4 cycle (Fig. 5).

Powell and Baum (1982) and Zullo and Harris (1986, 1987) used echinoids, oysters and pectinids to establish correlation of sequence 2 to the upper Santee Limestone of South Carolina and the upper Lisbon Formation of Alabama. Worsley and Laws (1986) assigned this sequence to calcareous nannofossil zone NP16. Jones (1983), from the study of surface and subsurface samples, assigned all of the Castle Hayne Limestone to the *Morozovella lehneri* (P12) and *Orbulinoides beckmanni* (P13) zones which is consistent with the biostratigraphy proposed by Zullo and Harris (1987) for the middle Eocene part (1, 2) of the Castle Hayne Limestone. This sequence is assigned to the TA3.5/3.6 cycle (Fig. 5).

The middle to upper Eocene depositional sequence (3) contains characteristic mega-invertebrates in the lower part which include the echinoid *Protoscutella plana* and the earliest known occurrence of the barnacle *Euscapellum carolinensis.* Worsley and Laws (1986) identified a calcareous nannofossil assemblage from the lower part of the sequence that they assigned to calcareous nannofossil zone CP14a or NP16. The upper part of this sequence is characterized by diagnostic echinoids and pectinids, the barnacle *Arcoscalpellum jacksonense*, and the oyster *Pycnodonte trigonalis.* Worsley and Laws (1986) identified a calcareous nannofossil assemblage from the upper part of this sequence which included *Chiasmolithus oamaruensis* without *Ismolithus recurvus*, and assigned it to nannofossil zone NP18. This sequence is assigned to the TA4.1 cycle (Fig. 5). Although the TA4.1 cycle is considered as only Priabonian in age, faunal evidence from North Carolina indicates that Bartonian and Priabonian calcareous nannofossils occur in the same cycle above the 39.5 Ma sequence boundary. Therefore, in Figs 5 and 12, we place the Bartonian–Priabonian boundary within the TA4.1 cycle.

The upper Eocene sequence (4) has not yielded age-diagnostic microfossils and contains the same megafauna identified in the upper part of sequence 3. The New Bern Formation contains large campanulate specimens of the echinoid, *Periarchus lyelli*, and the pectin, *Chlamys (Aequipecten)* n. sp.; these forms also occur in sequence 4 of the Castle Hayne Limestone. This sequence is assigned to the TA4.2 and 4.3 cycles (Fig. 5).

In sequences 0 and 1 of the Castle Hayne Limestone, systems tracts could not be identified by Zullo and Harris (1987). Although sequence 2 is thicker, better developed and more widespread, it is also difficult to identify specific systems tracts. The overall nature of this sequence, which shows an upward decrease in grain size and lime mud and an abundance of glaucony, suggests that it may represent a transgressive systems tract. Sequence 3 of the Castle Hayne Limestone is the most complete depositional sequence exposed in the Coastal Plain of North Carolina. Sediments in this sequence generally represent transgressive and highstand systems tracts; however, Zullo and Harris (1987) identified thin (0.5 m) lowstand deposits at one locality (East Coast Limestone Quarry, Pender County). Lowstand deposits at this location, which represent the most downdip exposure of sequence 3, are interpreted as a proximal lowstand wedge within an incised valley. Well-washed cross-bedded bryozoan biosparrudite interpreted as a transgressive systems tract abruptly overlies the lowstand systems tract and is widespread. The upper part of this sequence overlies a thin persistent zone of phosphate- and glauconite-coated sponges which displays characteristics of a marine hardground, is interpreted as a marine flooding event associated with maximum sea-level rise and is considered to represent the condensed section (Harris *et al.*, 1986; Zullo & Harris, 1987). Above this hardground, sediments consist of deeper water biomicrite that contain a diverse microfossil assemblage that gradually coarsens and shoals upward; it is interpreted to represent the highstand systems tract and is the most widespread part of sequence 3. Sequence 4 and the New Bern Formation are tentatively assigned to the highstand systems tract; however, both shelf margin and highstand systems tracts may be represented (Fig. 5).

In wells from the onshore Albemarle embayment Zarra (1989) identified one middle Eocene depositional sequence with a thickness between 112 and 158 m. The lower part of the sequence consists of highly calcareous, fine-grained sandstone that grades upward into sandy bioclastic limestone containing abundant bryozoan fragments. Although foraminiferal assemblages were generally sparse and poorly preserved, Zarra (1989) assigned an age of P9 to possible P12, suggesting correlation to one cycle between the TA3.1 and TA3.5/3.6 cycles.

Zarra (1989) identified one upper Eocene sequence in the Albemarle embayment. This thin, 15 m-thick sequence occurred only in the most downdip well examined and consists of lower sandstone and upper, silty glauconitic clay. Based on abundant and diverse benthonic and planktonic foraminifers, he correlated the sequence to zones P15 to lowermost P16. The lower sandstone and upper, silty glauconitic clay recognized by Zarra (1989) are interpreted to represent transgressive and highstand systems tracts, respectively, and were deposited in an outer-neritic environment. This sequence is assigned either to the TA4.1 or TA4.2 cycle.

Oligocene

Lithostratigraphy

Stratigraphic revisions of outcropping North Carolina Coastal Plain units by Baum *et al.* (1978) and Ward *et al.* (1978) established different lithostratigraphic frameworks for Oligocene sediments. In this paper, we follow the framework established by Baum *et al.* (1978) and modified by Zullo and Harris (1987); a complete discussion of the differences in the lithostratigraphy and age assignments is presented in Zullo and Harris (1987).

The Trent Formation which is confined to the outer Coastal Plain between the Neuse and New Rivers consists of three ascending lithofacies: sandy echinoid biosparite; sandy pelecypod mould biomicrudite; and barnacle pelecypod mould biosparrudite (Baum *et al.*, 1978; Baum, 1981). The unit obtains a maximum thickness of about 8 m. Zarra (1989) recognized a thin wedge-shaped silty clay in the subsurface of the Albemarle embayment north of the Neuse River, and assigned it to the lower Oligocene. The unit has a maximum thickness of about 36 m and thins to the north onto the Norfolk arch and to the south onto the Cape Fear arch (Zarra, 1989).

The Belgrade Formation is known only from quarries along the Onslow and Jones County Line where it consists of 8 m of sandy pelecypod-mould biomicrudite with minor interbeds of quartz sand. In the subsurface, at least another 20 m of biomicrudite and quartz sand are assigned to the Belgrade Formation. The top of the Belgrade Formation is picked at the phosphatized surface within the condensed section that separates the unit from the overlying *Crassostrea* beds. The Silverdale Formation is known from the surface only in the area east of the Belgrade Quarries. At the stratotype the unit consists of about 3 m of mollusc-rich quartz sand which is occasionally partially lithified and mouldic; however, several additional metres of this sequence occur in the subsurface. In the subsurface of the Albemarle embayment, Zarra (1989) recognized a thin wedge-shaped depositional sequence which he placed in the upper Oligocene. The sequence, which thins to the north and to the south onto the Norfolk and Cape Fear arches, respectively, has a maximum thickness of about 67 m and consists of lower coarse- to medium-grained quartz or phosphatic sand overlain by silty clay, sandy mouldic limestone or barnacle shell hash (Zarra, 1989). Riggs *et al.* (1985) also recognized Oligocene sediments in Onslow Bay which they correlated to the Trent, Belgrade and Silverdale Formations.

Sequence stratigraphy and correlation

The Trent Formation contains the archaeobalanid barnacle *Lophobalanus kellumi*, the pectinid *Chlamys trentensis* and another resembling *Pecten poulsoni* (Baum *et al.*, 1978; Zullo & Harris, 1987). Rossbach and Carter (1991) indicate that the molluscan fauna of the Trent Formation has considerable affinities to early Vicksburgian (Rupelian) strata in Mississippi. Zarra (pers. comm.) assigned the Trent Formation to *Globigerina ampliapertura* zone (P19/20), except for one locality where he found fauna indicative of the *Globergerina ciperoensis* zone. Worsley and Turco (1979) suggested that the Trent includes calcareous nannofossil zones NP21 and NP22. In the Albemarle embayment, Zarra (1989) assigned the lower Oligocene sequence to the *Globigerina ampliapertura* zone on the presence of *Globigerina ampliapertura* and *Globorotalia increbescens* and the absence of *Globorotalia opima opima* and *Pseudohastigerina micra.* The sequence represented by the Trent Formation is assigned to the TA4.4 cycle (Fig. 6).

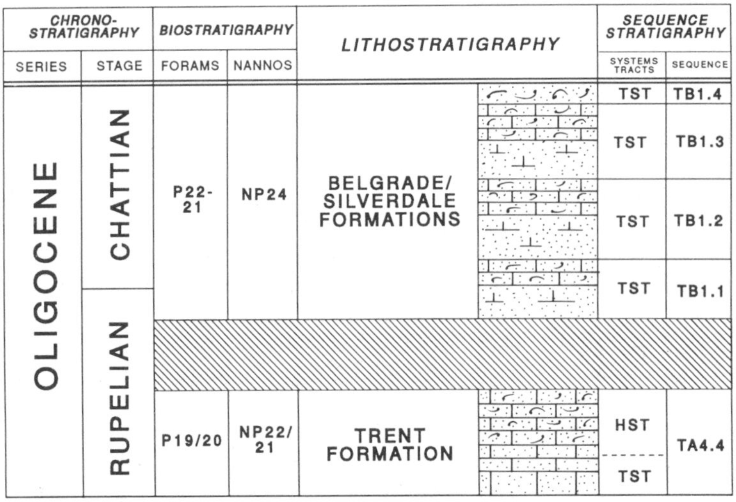

CHRONO-STRATIGRAPHY		BIOSTRATIGRAPHY		LITHOSTRATIGRAPHY	SEQUENCE STRATIGRAPHY	
SERIES	STAGE	FORAMS	NANNOS		SYSTEMS TRACTS	SEQUENCE
OLIGOCENE	CHATTIAN	P22-21	NP24	BELGRADE/ SILVERDALE FORMATIONS	TST	TB1.4
					TST	TB1.3
					TST	TB1.2
					TST	TB1.1
	RUPELIAN					
		P19/20	NP22/ 21	TRENT FORMATION	HST / TST	TA4.4

Fig. 6. Generalized biostratigraphy, lithostratigraphy and sequence stratigraphy of Oligocene sediments in the Albemarle embayment and on the Cape Fear arch of North Carolina. Lithologic symbols as in Fig. 4.

The Trent Formation, although thin, illustrates a complete Oligocene depositional sequence. The basal sandy biosparite is interpreted to represent a transgressive systems tract and the overlying pelecypod mould biomicrudite and barnacle pelecypod mould biosparrudite, early and late highstand systems tracts, respectively (Fig. 6).

The Belgrade and Silverdale Formations have close megafaunal affinities. The balanid barnacle *Concavus belgradensis* occurs in both units with the oyster *Crassostrea blanpiedi.* Zullo and Harris (1987) indicated that the Belgrade and Silverdale Formations span planktonic foraminiferal zones, P21 and P22. Laws and Worsley (1986), Laws (1990) and Parker and Laws (1991) suggested that the Silverdale Formation correlates to calcareous nannofossil zone NP24. Zarra (pers. comm.) identified *Globorotalia ampliapertura* and *G. ciperoensis* zones in the Belgrade Formation and the *Globorotalia kugleri* zone in the Silverdale Formation. Zarra (1989) indicated that the distribution and preservation of foraminifers from the upper Oligocene sequence were variable in the subsurface of the Albemarle embayment, but assigned the sequence to the *Globigerina ciperoensis* zone (P22). Parker and Laws (1991) correlated the Belgrade Formation to zones NP24–25 on the basis of *Helicosphaera recta.*

Zullo and Harris (1987) suggested that the Belgrade and Silverdale Formations and the *Crassostrea* beds represented four depositional sequences that range in age from Chattian to Aquitanian. They assigned these sequences to the TB1.1, TB1.2, TB1.3 and TB1.4 cycles, and indicated that the contact between the Belgrade Formation and *Crassostrea* beds is within the condensed section of the TB1.4 cycle (Fig. 6). Although Zullo and Harris (1987) identified several phosphate-coated surfaces in the Belgrade Formation and suggested that they may represent marine hardgrounds associated with the condensed section or sequence boundaries, they suggested that these were parasequence boundaries and that the unit only represented transgressive systems tracts (Fig. 6). The distribution and thickness of the units indicate that they are restricted to an incised valley that formed during the 30 Ma sea-level fall. Highstand systems tracts for these sequences probably occur in the submerged Coastal Plain in Onslow Bay.

Zarra (1989) interpreted the upper Oligocene sequence in the Albemarle embayment as consisting of a lower transgressive systems tract of coarse-grained sand with abundant phosphate pellets. He interpreted the upper part of the sequence which consists of silty clay, limestone or barnacle shell hash as a highstand systems tract. Zarra (1989) suggested from a study of the planktonic foraminifers that the depositional setting of this sequence was inner to outer shelf. This sequence is assigned to either the TB1.1, the TB1.2 or the TB1.3 cycles.

DEPOSITIONAL SEQUENCES – SANTEE RIVER AREA

Palaeocene

Lithostratigraphy

Palaeocene sediments in the Santee River area are referred to the Black Mingo Group (Fig. 2). In the outer Coastal Plain of South Carolina, Van Nieuwenhuise and Colquhoun (1982) recognized two Palaeocene units in the Black Mingo Group — the Danian Rhems and Thanetian Williamsburg Formations, and an unnamed Eocene (Ypresian) unit. Subsequently, the unnamed Eocene unit was designated the Fishburne Formation by Gohn *et al.* (1983) and removed from the Black Mingo Group by Nystrom *et al.* (1989). Muthig and Colquhoun (1988) extended the Rhems Formation into the inner Coastal Plain and recognized Sloan's (1908) Sawdust Landing Beds and Lang Syne Beds as updip members. Nystrom *et al.* (1989) also recognized two Palaeocene units in the inner Coastal Plain — the Danian Sawdust Landing Formation and the Thanetian Lang Syne Formation of the Black Mingo Group. The principal difference between the interpretations deals with recognition of Thanetian age sediments in the inner Coastal Plain. Sediments that Nystrom *et al.* (1989) refer to as Thanetian in age (Lang Syne Formation), Muthig and Colquhoun (1988) place as a member of the Danian Rhems Formation. In this paper, the terminology of Nystrom *et al.* (1989) is used for Palaeocene sediments in the inner Coastal Plain of the Santee River area and the terminology of Van Nieuwenhuise and Colquhoun (1982) is used for the middle and outer Coastal Plain.

Two depositional sequences are recognized in the Black Mingo Group. In the inner Coastal Plain they are the Danian Sawdust Landing Formation and the Thanetian Lang Syne Formation; in the outer Coastal Plain they are the Danian Rhems Formation and the Thanetian Williamsburg Formation. The Sawdust Landing Formation disconformably overlies Cretaceous sediments and disconformably underlies the Lang Syne Formation. It consists primarily of feldspathic, micaceous, clayey quartzose sand with quartz granules and pebbles interbedded with thin silts and sandy grey clay. The unit reaches a thickness of about 15 m at the stratotype (Muthig & Colquhoun, 1988) and is interpreted to represent fluvial deposition in an upper delta plain (alluvial) environment.

To the east in the middle and outer South Carolina Coastal Plain, the Sawdust Landing Formation is correlated to the Rhems Formation. The Rhems Formation consists of lower arenaceous shale and argillaceous sand (Browns Ferry Member) and upper pelecypod-rich clayey sand (Perkins Bluff Member). The two members interfinger, with their contact either sharp or gradational (Van Nieuwenhuise & Colquhoun, 1982). In outcrop, the Rhems Formation obtains a thickness of almost 15 m; in the Clubhouse Crossroads core #1 in Dorchester County, the Rhems is about 50 m thick (Van Nieuwenhuise & Colquhoun, 1982).

The Lang Syne Formation consists of laminated, black to grey fuller's earth interlayered with green glauconitic quartzose sand, micaceous quartzose sand and clayey sand and sandy clay (Nystrom *et al.*, 1989). The Williamsburg Formation consists of lower arenaceous shale, fuller's earth and fossiliferous argillaceous sand (Lower Bridge Member) and fossiliferous, argillaceous sand and molluscan-mould biomicrudite (Chicora Member). Van Nieuwenhuise and Colquhoun (1982) described the Lower Bridge and Chicora Members as grading from fine siliciclastics below to coarser siliciclastics and carbonate above. At the stratotype of the Chicora Member on the Santee River, repetitive sedimentary packages consisting of lower argillaceous sand and upper pelecypod-mould biomicrudite occur. In outcrop, the Williamsburg Formation reaches a thickness of about 6 m, whereas in the Clubhouse Crossroads core, it is about 150 m thick.

Sequence stratigraphy and correlation

Although no fossils have been recovered from the Sawdust Landing Formation, Colquhoun *et al.* (1983) and Nystrom *et al.* (1989) consider the unit to be Danian in age based on regional correlation and superposition. Van Nieuwenhuise and Colquhoun (1982) indicated that the lower part of the Rhems Formation contains ostracod and pollen assemblages of late Danian age. The argillaceous sand and arenaceous clay of the lower part of the Rhems Formation are interpreted to represent an inner-neritic environment, whereas the pelecypod-rich sand of the upper part of the Rhems Formation is considered to represent a littoral environment (Van Nieuwenhuise & Colquhoun, 1982). This depositional sequence is assigned to the TA1.2 cycle (Fig. 7).

In the Clubhouse Crossroads core #1, Van Nieuwenhuise and Colquhoun (1982) identified about 50 m of sediment that they assigned to the Rhems Formation. Hazel *et al.* (1977) indicated that on the

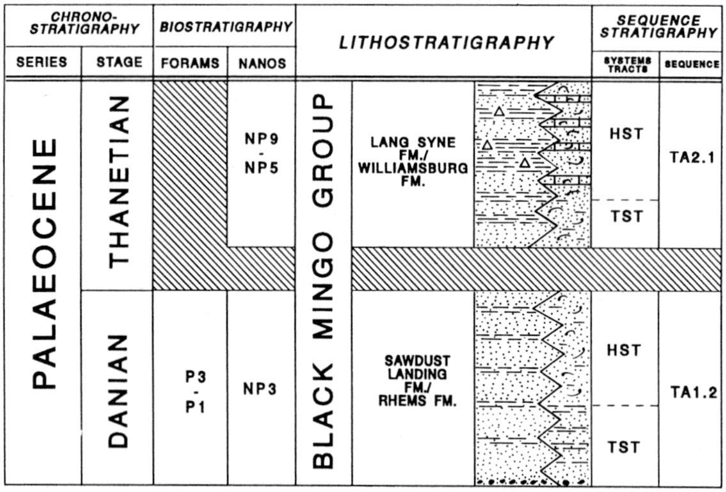

CHRONO-STRATIGRAPHY		BIOSTRATIGRAPHY		LITHOSTRATIGRAPHY		SEQUENCE STRATIGRAPHY	
SERIES	STAGE	FORAMS	NANOS			SYSTEMS TRACTS	SEQUENCE
PALAEOCENE	THANETIAN		NP9 - NP5	BLACK MINGO GROUP	LANG SYNE FM./ WILLIAMSBURG FM.	HST	TA2.1
PALAEOCENE	THANETIAN		NP9 - NP5	BLACK MINGO GROUP	LANG SYNE FM./ WILLIAMSBURG FM.	TST	TA2.1
PALAEOCENE	DANIAN	P3 - P1	NP3	BLACK MINGO GROUP	SAWDUST LANDING FM./ RHEMS FM.	HST	TA1.2
PALAEOCENE	DANIAN	P3 - P1	NP3	BLACK MINGO GROUP	SAWDUST LANDING FM./ RHEMS FM.	TST	TA1.2

Fig. 7. Generalized biostratigraphy, lithostratigraphy and sequence stratigraphy of Palaeocene sediments in the Santee River area of South Carolina. Lithologic symbols as in Fig. 4.

basis of the calcareous nannofossils this entire interval is Danian in age and assigned it to zone NP3. Based on planktonic foraminifers, Hazel *et al.* (1977) assigned this same interval to zones P1, P2 and P3 which span the Danian and the lower part of the Thanetian. This inconsistency in age between the calcareous nannofossils and the planktonic foraminifers has not been resolved.

The Sawdust Landing and Rhems Formations represent a single depositional sequence which preserves transgressive and highstand systems tracts. Updip, fluvially deposited feldspathic quartzose sand and interbedded sandy kaolinitic clay of the Sawdust Landing Formation were deposited during the cycle. Downdip, inner-neritic arenaceous shale and argillaceous sand of the lower part of the Rhems Formation (Browns Ferry Member) and littoral pelecypod-rich sand of the upper part of the Rhems Formation (Perkins Bluff Member) represent transgressive and highstand systems tracts, respectively (Fig. 7). The sharp contact between the Browns Ferry and the Perkins Bluff Members is interpreted to be within the condensed section. In the Clubhouse Crossroads core #1, the lower 11 m of the sediments correlated to the Browns Ferry Member consist of glauconitic muddy sand which is interpreted to represent a transgressive systems tract. However, based on electric-log characteristics the condensed section may occur within the Browns Ferry Member.

Although, Muthig and Colquhoun (1988) assigned the Lang Syne Beds to the Rhems Formation and considered them Danian in age, Nystrom *et al.* (1989) consider the unit Thanetian in age on the basis of dinoflagellates and calcareous nannofossils. Muthig and Colquhoun (1988) interpreted the depositional environment of the Lang Syne Formation as consisting of barrier islands and backbarrier lagoons and estuaries, tidal creeks and deltas. Nystrom *et al.* (1989) recognized evidence for more open-marine offshore environments, and consider it to be the updip equivalent of the downdip Williamsburg Formation.

The Lower Bridge Member of the Williamsburg Formation is about 24 m thick in the Clubhouse Crossroads core and contains planktonic foraminifers which are referable to the P4 zone (Hazel *et al.*, 1977). Calcareous nannofossils that range from zone NP5 through zone NP9 also occur throughout the interval assigned to the Lower Bridge and Chicora Members (Hazel *et al.*, 1977). The Chicora Member in outcrop contains numerous molluscan fragments, foraminifers and ostracods and is assigned a Thanetian age by Van Nieuwenhuise and Colquhoun (1982).

The updip Lang Syne Formation contains transgressive and highstand systems tracts which consist of lower quartz sand and upper siliceous claystone and shale. The Lower Bridge and Chicora Members of the Williamsburg Formation represent inner-neritic and littoral environments, respectively (Van Nieuwenhuise & Colquhoun, 1982). Lithologic and electric-log analysis of the Williamsburg Formation stratotype and correlative sediments that occur in the Clubhouse Crossroads core #1 indicate both members represent the highstand systems tract; however, the lowermost part of the Lower Bridge Member is tentatively assigned to the transgressive systems tract with little supporting evidence. The Lang Syne and Williamsburg Formations are assigned to the TA2.1 cycle (Fig. 7).

Eocene

Lithostratigraphy

Lower Eocene sediments of the Fishburne Formation disconformably overlie the Thanetian Williamsburg Formation in the subsurface from just north of the Santee River south to at least Fripp and Parris Islands in Beaufort County. This thin, widespread subsurface unit consists of glauconitic, clayey, microfossil–mollusc biomicrite with a maximum thickness of almost 23 m in a well at Fripp Island (Gohn *et al.*, 1983). Gohn *et al.* (1983) suggested that the Fishburne Formation was deposited in an inner sublittoral environment in a warm temperate climate.

Middle Eocene sediments in the inner Coastal Plain are referred to the Orangeburg Group and include the Congaree and Huber Formations, the Warley Hill Formation and the McBean Formation (Nystrom *et al.*, 1989). In the outer Coastal Plain the group includes the Congaree Formation, the lower and upper Santee Limestone and the lower part of the Cross Formation (Fig. 2).

In the inner Coastal Plain (Fig. 1), the Congaree Formation unconformably overlies the Thanetian part of the Black Mingo Group (Nystrom *et al.*, 1989) and unconformably underlies glauconitic sand of the Warley Hill Formation. In the middle and outer Coastal Plain the Congaree Formation underlies sandy molluscan carbonates of the Santee Limestone. In the inner Coastal Plain the Congaree

Formation is composed mainly of pale-yellow, loose quartz sand with distinctive indurated light-green thin-bedded silicified claystone and siltstone layers and reaches a thickness of about 15 m; it thins to the southeast. In the middle and outer Coastal Plain the Congaree Formation consists of muscovite-rich quartz sand, sometimes glauconitic, that is interbedded with dark-grey to light-grey clay, fuller's earth and silicified zones. It reaches a maximum thickness of about 10 m and rapidly thins to the southeast until it becomes absent. The Congaree Formation accumulated on a shallow-marine shelf dominated by land-derived coarse siliciclastic sediments.

In the inner Coastal Plain, the middle middle Eocene Warley Hill Formation disconformably overlies the lower middle Eocene Congaree Formation and underlies the upper middle Eocene McBean Formation. The Warley Hill Formation consists of green, quartzose and glauconitic sand and glauconitic silt with minor interstitial clay or green clay beds (Nystrom *et al.*, 1989). The unit is restricted to the areas of Calhoun and northwestern Orangeburg Counties and obtains a thickness of about 6 m. The Warley Hill Formation is correlated to fossiliferous, green glauconitic calcareous sediments and fossiliferous biosparrudite of the lower Santee Limestone in the middle and outer Coastal Plain.

The lower Santee Limestone is separated from the overlying upper Santee Limestone and the underlying Black Mingo Group by regional disconformities (Powell & Baum, 1982). It consists of basal phosphate-pebble biomicrudite (0.6 m) and upper bryozoan biomicrudite which reach a combined thickness of about 5 m in Georgetown County. In the Clubhouse Crossroads core in Dorchester County the Santee Limestone is not separated into lower and upper parts.

In the inner Coastal Plain the upper middle Eocene McBean Formation, as used in the sense of Nystrom *et al.* (1989), includes sediments equivalent to the upper Lisbon and Cook Mountain Formations of the Gulf Coastal Plain. The McBean Formation consists of non-calcareous to calcareous sandy clay or clayey sand with thin zones of quartz sand and interlayers of light-grey mouldic limestone. In the middle and outer Coastal Plain the upper part of the Santee Limestone is correlated to the McBean Formation (Nystrom *et al.*, 1989). The unit reaches a maximum thickness of about 15 m in Orangeburg County, but throughout its area of outcrop, it is generally 6 to 9 m thick (Nystrom *et al.*, 1989).

In the middle and outer Coastal Plain the upper part of the Santee Limestone consists of laterally equivalent bryozoan biosparrudite, bryozoan biomicrudite and foraminiferal biomicrite (Chapel Branch Member of Powell, 1984) and is separated from the lower part of the Santee Limestone by a regional disconformity. Lithofacies of the upper Santee Limestone probably have a thickness from about 5 to 10 m in updip areas, and reach a thickness of between 20 and 25 m in the Clubhouse Crossroads core in Dorchester County (Hazel *et al.*, 1977).

The middle Eocene part of the Cross Formation is thin, less than 3 m thick, and occurs at several localities just south of the Santee River in Berkeley, Orangeburg and Dorchester Counties. It consists of molluscan mould foraminiferal biomicrudite and disconformably overlies the Santee Limestone. In the Clubhouse Crossroads core, the Cross Formation consists of pale-yellowish grey sandy biomicrite (Gohn *et al.*, 1977) and reaches a thickness of over 30 m, most of which is assigned to the late Eocene.

Late Eocene sediments are absent in the inner Coastal Plain in the Santee River area. In the middle and outer Coastal Plain they are assigned to the upper part of the Cross Formation and the Harleyville and Parkers Ferry Formations of the Cooper Group. Each lithostratigraphic unit represents a separate depositional sequence.

The upper part of the Cross Formation of Baum *et al.* (1980) occurs in quarries in Orangeburg and northern Dorchester Counties and pinches out on the south flank of the Cape Fear arch in the vicinity of Lake Marion. The unit has a maximum thickness of almost 11 m in outcrop and 41 m in the Clubhouse Crossroads core (Ward *et al.*, 1979).

The Harleyville Formation of the Cooper Group consists of compact, phosphatic, calcareous clay and clayey calcarenite that disconformably overlies the Cross Formation (Ward *et al.*, 1979). The unit is 12 m thick in the Clubhouse Crossroads core (Hazel *et al.*, 1977), thickens to the south to about 30 m and thins to the north. The Parkers Ferry Formation is only known from the subsurface along the east side of the Edisto River and disconformably overlies the Harleyville Formation. The unit consists of glauconitic, clayey, fine-grained limestone, and contains abundant microfauna as well as mollusc and bryozoan fragments (Ward *et al.*, 1979). The unit is only a few metres thick.

Sequence stratigraphy and correlation

The Fishburne Formation contains a calcareous nannofossil assemblage in the Clubhouse Crossroads core that indicates placement in the lower Eocene (Gohn *et al.*, 1983). Based on planktonic foraminifers, they assigned the unit to the *Morozovella subbotinae* zone (P6b). Lithologic and electric-log characteristics of the unit indicate that in the northern part of its extent (Dorchester County), the unit principally represents a highstand systems tract with the transgressive systems tract thin to absent, whereas to the south (Parris Island) both transgressive and highstand systems tracts are present. Transgressive deposits consist of glauconitic fine-grained biomicrite and highstand deposits consist of biomicrite with less clay and glauconite. Updip unnamed strata were assigned by Van Nieuwenhuise and Colquhoun (1982) to the *Haplocytheridea wallacei* ostracode assemblage zone which they correlated to the lower Eocene Fishburne Formation. The Fishburne Formation is assigned to the TA2.5 cycle (Fig. 8).

The Congaree Formation contains the bivalve guide fossil *Anodontia augustana*, which is restricted to the lower middle Eocene Tallahatta Formation of Alabama (Toulmin, 1977), and is assigned to the TA3.3 cycle. The Congaree Formation is tentatively assigned to a highstand systems tract on the basis of its geographic distribution and alternating lithology of coarse siliciclastics, silicified claystone and siltstone. Although the type Warley Hill Formation contains no age-diagnostic fossils, correlative units in the middle and outer Coastal Plain contain the middle middle Eocene guide fossil *Cubitostrea lisbonensis*. The abundance of authigenic glauconite and burrow structures suggests that the unit was deposited as a transgressive systems tract under moderate energy conditions periodically washed by strong currents (Fig. 8). The presence of *Cubitostrea lisbonensis* in the lower Santee Limestone establishes correlation of the unit to the middle middle Eocene lower Lisbon Formation of Alabama and beds in North Carolina (sequence 1) that contain a calcareous nannofossil assemblage indicative of zone NP15. In outcrop, the unit is

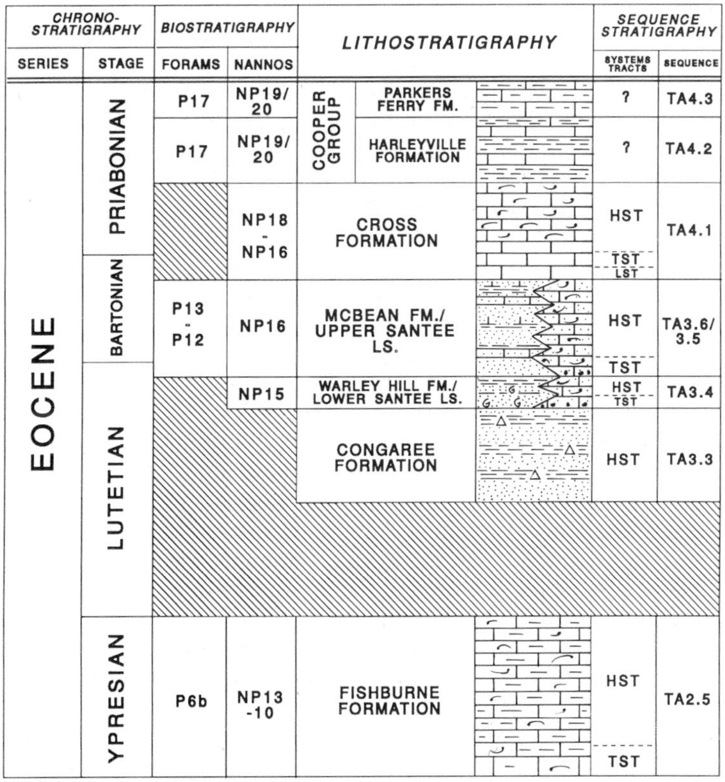

CHRONO-STRATIGRAPHY		BIOSTRATIGRAPHY		LITHOSTRATIGRAPHY	SEQUENCE STRATIGRAPHY	
SERIES	STAGE	FORAMS	NANNOS		SYSTEMS TRACTS	SEQUENCE
EOCENE	PRIABONIAN	P17	NP19/20	COOPER GROUP: PARKERS FERRY FM.	?	TA4.3
EOCENE	PRIABONIAN	P17	NP19/20	COOPER GROUP: HARLEYVILLE FORMATION	?	TA4.2
EOCENE	PRIABONIAN / BARTONIAN		NP18 - NP16	CROSS FORMATION	HST / TST / LST	TA4.1
EOCENE	BARTONIAN	P13 - P12	NP16	MCBEAN FM./ UPPER SANTEE LS.	HST / TST	TA3.6/ 3.5
EOCENE	LUTETIAN		NP15	WARLEY HILL FM./ LOWER SANTEE LS.	HST / TST	TA3.4
EOCENE	LUTETIAN			CONGAREE FORMATION	HST	TA3.3
EOCENE	YPRESIAN	P6b	NP13 -10	FISHBURNE FORMATION	HST / TST	TA2.5

Fig. 8. Generalized biostratigraphy, lithostratigraphy and sequence stratigraphy of Eocene sediments in the Santee River area of South Carolina. Lithologic symbols as in Fig. 4.

strea lisbonensis. Lower phosphate-pebble biomicrudite of the lower Santee Limestone is assigned to a transgressive systems tract whereas upper bryozoan biomicrudite is assigned to a highstand systems tract (Fig. 8). The decrease in micrite and increase in calcite spar in stratigraphically younger parts of the lower Santee Limestone support this interpretation. The lower Santee Limestone was deposited on a shallow inner-neritic carbonate shelf. The depositional sequence represented by the Warley Hill Formation and the lower Santee Limestone is assigned to the TA3.4 cycle (Fig. 8).

The McBean Formation and the upper Santee Limestone contain the late-middle Eocene guide fossils *Cubitostrea sellaeformis* and *Pteropsella lapidosa* and pectinids, barnacles and nannofossils which establish their correlation to the upper middle Eocene Lisbon Formation of Alabama (Powell & Baum, 1982) and to sequence 2 of the Castle Hayne Limestone in North Carolina (Fig. 2). The McBean Formation and the upper part of the Santee Limestone represent a shallow-water siliciclastic to carbonate transition. Although it is difficult to recognize specific systems tracts within the McBean Formation because of the paucity of exposure, it is tentatively assigned to transgressive and highstand systems tracts. Lower bryozoan biosparrudite of the upper Santee Limestone represents a transgressive systems tract and bryozoan biomicrudite and foraminiferal biomicrite of the upper Santee Limestone represent a highstand systems tract. The McBean Formation and the upper Santee Limestone are assigned to the TA3.5/3.6 cycle (Fig. 8). Zullo and Harris (1987), who considered the Cross Formation as both middle and upper Eocene, place the lower part in the middle Eocene on the basis of echinoids and barnacles and assigned it to lowstand and transgressive systems tracts. Zullo and Harris (1987) considered the upper part of the Cross Formation to be late Eocene in age on the basis of megafauna and assigned it to the highstand systems tract because of the overall increase in carbonate content and decrease in siliciclastics. The Cross Formation is assigned to the TA4.1 cycle (Fig. 8). The Harleyville Formation contains the pecten *Chlamys cocoana* which is principally found in upper Eocene deposits in Alabama and Georgia (Ward *et al.*, 1979). Planktonic foraminifers and calcareous nannofossils are indicative of zones P17 and NP19/20, respectively (Ward *et al.*, 1979; Laws, 1988). The Harleyville Formation is assigned to the TA4.2 cycle (Fig. 8). The Parkers Ferry Formation contains planktonic foraminifers which indicate a late Eocene age for the unit and assignment to zones P17 and NP19/20 (Ward *et al.*, 1979). It is assigned to the TA4.3 cycle (Fig. 8). Because of the thin nature and limited distribution of the Harleyville and Parkers Ferry Formations, no systems tracts are assigned.

Oligocene

Lithostratigraphy

No lower Oligocene rocks are recognized in the Santee River area of the inner, middle and outer Coastal Plain of South Carolina. Two upper Oligocene depositional sequences are recognized in the outer Coastal Plain in South Carolina – the Ashley Formation of the Cooper Group and the Chandler Bridge Formation.

The Ashley Formation fills channels that are cut into the underlying Eocene Parkers Ferry and Harleyville Formations and is disconformably overlain by the Chandler Bridge Formation. The Ashley Formation consists of 'phosphatic, muddy, calcareous, very fine-grained sand' (Ward *et al.*, 1979). In the Clubhouse Crossroads core the Ashley is 50 m thick (Hazel *et al.*, 1977), but may reach 60 m thick in the vicinity of Charleston (Ward *et al.*, 1979). The Chandler Bridge Formation is a thin arenaceous unit that has a maximum thickness of about 5 m, and consists of soft, very fine- to fine-grained quartz-phosphate sand with minor clay (Sanders *et al.*, 1982). It crops out only in the area just north of Charleston and occurs mainly in low areas developed on top of the Ashley Formation.

Sequence stratigraphy and correlation

Calcareous nannofossils from the Ashley Formation in the Clubhouse Crossroads core indicate that the unit is late Oligocene in age and assignable to calcareous nannofossil zone NP24 (Hazel *et al.*, 1977). It is correlated to the TB1.1 cycle (Fig. 9) and is assigned to a highstand systems tract because of its overall lithologic nature. However, sediments representing a transgressive systems tract probably occur in the immediate subsurface. The Chandler Bridge Formation contains an odontocete whale fauna that indicates an age for the unit of late Oligocene (Sanders *et al.*, 1982). This sequence is assigned to either the TB1.2 or TB1.3

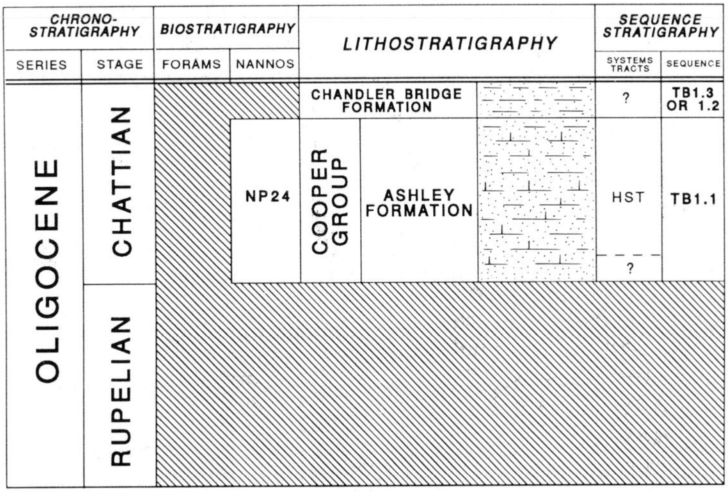

CHRONO-STRATIGRAPHY		BIOSTRATIGRAPHY		LITHOSTRATIGRAPHY		SEQUENCE STRATIGRAPHY	
SERIES	STAGE	FORAMS	NANNOS			SYSTEMS TRACTS	SEQUENCE
OLIGOCENE	CHATTIAN			CHANDLER BRIDGE FORMATION		?	TB1.3 OR 1.2
			NP24	COOPER GROUP	ASHLEY FORMATION	HST ?	TB1.1
	RUPELIAN						

Fig. 9. Generalized biostratigraphy, lithostratigraphy and sequence stratigraphy of Oligocene sediments in the Santee River area of South Carolina. Lithologic symbols as in Fig. 4.

cycles (Fig. 9); because of its limited distribution systems tracts are not interpreted.

DEPOSITIONAL SEQUENCES – SAVANNAH RIVER AREA

Palaeocene

Lithostratigraphy

Although Harris and Zullo (1988) and Harris *et al.* (1990) recognized four Palaeocene depositional sequences in the Savannah River area, subsequent work indicates that only one Danian and one Thanetian sequence are present. Palaeocene sediments in the Savannah River area are referred to the Ellenton Formation (Fig. 2) of Siple (1967). Dark lignitic clay and coarse sand of the Ellenton Formation overlie dense sticky kaolinite or very coarse kaolinitic sand of Cretaceous Maastrichtian age. The contact with the Cretaceous is often marked by a thin, coarse, poorly sorted gravel up to 2 m thick, which contains lithoclasts of kaolin, booklets of muscovite and some lignitic material.

The Danian sequence consists of lower and upper parts which are gradational over a narrow interval. The lower part includes two or three parasequences that consist of basal, poorly sorted muscovite-rich quartz sand with occasional quartz pebbles and partially lignitized wood fragments, grading upward into dark, organic, sometimes fissile clay that often contains interlaminae of fine quartz sand or silt-sized muscovite flakes. The upper part consists of dark organic clay and claystone with occasional thin quartz silt and sand layers which become more abundant near the top. Diagenetic gypsum is common in the more organic-rich zones. This sequence has a maximum thickness of about 18 m on the Savannah River Site and thins rapidly to the north, west and south. Prowell *et al.* (1985) studied the lower part of the Ellenton Formation and suggested that it was deposited in a marginal-marine deltaic complex. The distribution, thickness and lithology of the sequence also suggest a marginal marine deltaic to perhaps fluvial depositional setting.

One Thanetian sequence is recognized in the area and forms the upper part of the Ellenton Formation. This sequence is generally composed of lower organic-rich quartz sand or glauconitic quartz sand and upper siliceous mudstone, laminated organic claystone or fossiliferous clay or limestone. At some localities two parasequences with the same lithologic successions are present. Fossiliferous clay and limestone become more abundant to the southeast, whereas siliceous mudstone is more abundant on the Savannah River Site. Siliceous mudstone is usually massive, particularly where it contains appreciable amounts of cristobalite, whereas the laminated organic claystone is fissile. The siliceous and organic claystones often contain interbeds or interlaminae of silt to fine sand-sized muscovite-rich quartz sand or partings of silt-sized muscovite flakes in the upper part. This unit, which thins to the northwest and opens to the southeast, is widespread and has a variable thickness that reaches a maximum of about 36 m in Burke County, Georgia. The upper Ellenton Formation was deposited in nearshore shallow marine water above wave base in the lower part and below calm seas wave base in the upper part.

Sequence stratigraphy and correlation

Prowell *et al.* (1985) indicated that the Ellenton Formation is Midwayan in age and suggested that the unit correlated to the Naheola, Porters Creek and Clayton Formations of Alabama. Although Mancini and Tew (1988) suggested that these formations in Alabama represent five depositional sequences, in the Savannah River area only two depositional sequences are recognized. Dinoflagellate cysts, calcareous nannofossils and some invertebrates occur in the Ellenton Formation in cores from the Savannah River area and provide age control for the unit. Samples from the lower sequence of the Ellenton Formation yield dinoflagellates and pollen (Clarke, 1989, pers. comm.) which suggest correlation to the TA1.3 cycle (Fig. 10). Lower poorly sorted quartz sand and fissile clay are assigned to a transgressive systems tract and upper organic clay and claystone to a highstand systems tract (Fig. 10). The upper sequence of the Ellenton Formation contains invertebrates, dinoflagellates and calcareous nannofossils (*Heliolithus riedelli*) assignable to zone NP8 which indicate correlation of the sequence to the TA2.1 cycle (Fig. 10). Lower glauconitic quartz sand is interpreted as a transgressive systems tract; the upper mudstone, claystone and carbonate are interpreted as a highstand systems tract (Fig. 10).

Lower Eocene

Lithostratigraphy

Two lower Eocene depositional sequences are recognized in the Savannah River area (Fig. 2). The lower sequence is represented by an unnamed unit previously referred by Harris and Zullo (1988) to the Williamsburg Formation. The upper lower Eocene sequence was recognized from cores on the Savannah River Site and correlated downdip on the basis of sporomorphs to the lower Eocene Fishburne Formation by Fallaw *et al.* (1989).

The unnamed lower Eocene depositional sequence consists of lower medium to coarse micaceous, kaolinitic quartz sand and upper white to grey to red stiff kaolin. Interbeds of white quartz sand occur at some localities in the upper kaolin. Root zones and palaeosol horizons occasionally occur at the top of this sequence. This sequence reaches a maximum thickness of almost 17 m in the central part of the Savannah River Site and thins rapidly to the west and northwest.

Fallaw *et al.* (1989) indicate that an updip time-rock equivalent of the downdip lower Eocene Fishburne Formation occurs over much of the Savannah River Site. They based correlation of the units on the occurrence of early Eocene sporomorphs and dinoflagellates. Although we have only

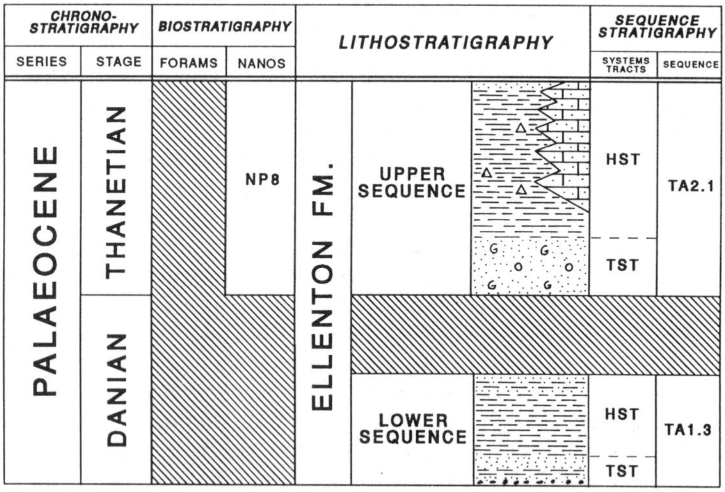

CHRONO-STRATIGRAPHY		BIOSTRATIGRAPHY		LITHOSTRATIGRAPHY		SEQUENCE STRATIGRAPHY	
SERIES	STAGE	FORAMS	NANOS			SYSTEMS TRACTS	SEQUENCE
PALAEOCENE	THANETIAN		NP8	ELLENTON FM.	UPPER SEQUENCE	HST	TA2.1
						TST	
	DANIAN				LOWER SEQUENCE	HST	TA1.3
						TST	

Fig. 10. Generalized biostratigraphy, lithostratigraphy and sequence stratigraphy of Palaeocene sediments in the Savannah River area of South Carolina. Lithologic symbols as in Fig. 4.

recognized this sequence in three core holes in the Savannah River area, Fallaw *et al.* (1989) indicate that it occurs over most of the area. Where present, it consists of quartz sand and grey-green fissile clay that often contains thin laminae of silt and fine to medium quartz sand. The sequence reaches a maximum thickness of about 18 m in the central part of the Savannah River Site.

Sequence stratigraphy and correlation

No age-diagnostic flora or fauna occur in the unnamed lower Eocene depositional sequence. Although Clark (1988, pers. comm.) recognized dinoflagellates at two localities that suggested a Danian to Thanetian age for the sequence, regional correlation and superpositional relationships with over- and underlying sequences indicate a younger age. Therefore, we suggest that the Palaeocene dinoflagellates found in this sequence were reworked from an older unit. We tentatively assign this sequence to the early Eocene TA2.4 Cycle, but recognize that the sequence could also be assigned to either the TA2.2 or TA2.3 Cycle (Fig. 11). On the basis of lithology, the sequence probably represents a fluvial depositional environment. The upper lower Eocene sequence is correlated to the downdip Fishburne Formation on the presence of sporomorphs and dinoflagellates (Fallaw *et al.*, 1989). Harris and Zullo (1991a) assigned the Fishburne Formation to the TA2.5 cycle; therefore, this unnamed updip equivalent is also assigned to the TA2.5 cycle (Fig. 11). Because of the limited occurrence of the unit, we have not determined its distribution or identified systems tracts.

Middle Eocene

Lithostratigraphy

Four middle Eocene depositional sequences are recognized in the Savannah River area. Two are assigned to the Congaree Formation, one to the Warley Hill Formation and one to the McBean Formation (Fig. 2). Nystrom *et al.* (1989) recognized a single lower middle Eocene depositional sequence in the Savannah River area that consists of an updip kaolinitic lithofacies and a downdip sand lithofacies. The updip facies is considered by them to be the Huber Formation and the downdip

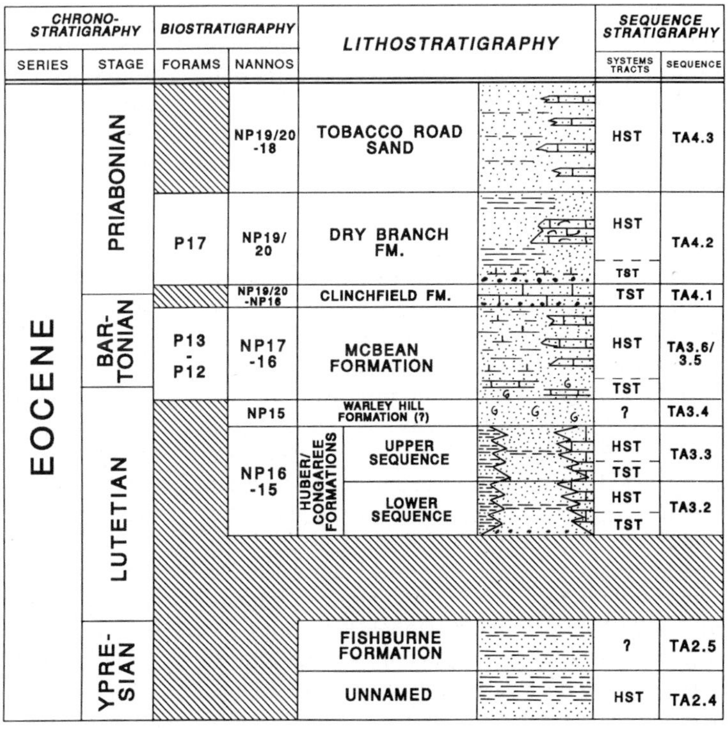

CHRONO-STRATIGRAPHY		BIOSTRATIGRAPHY		LITHOSTRATIGRAPHY	SEQUENCE STRATIGRAPHY	
SERIES	STAGE	FORAMS	NANNOS		SYSTEMS TRACTS	SEQUENCE
EOCENE	PRIABONIAN		NP19/20 -18	TOBACCO ROAD SAND	HST	TA4.3
		P17	NP19/ 20	DRY BRANCH FM.	HST / TST	TA4.2
	BAR-TONIAN		NP19/20 -NP18	CLINCHFIELD FM.	TST	TA4.1
		P13 - P12	NP17 -16	MCBEAN FORMATION	HST / TST	TA3.6/ 3.5
	LUTETIAN		NP15	WARLEY HILL FORMATION (?)	?	TA3.4
			NP16 -15	HUBER/ CONGAREE FORMATIONS: UPPER SEQUENCE	HST / TST	TA3.3
				HUBER/ CONGAREE FORMATIONS: LOWER SEQUENCE	HST / TST	TA3.2
	YPRE-SIAN			FISHBURNE FORMATION	?	TA2.5
				UNNAMED	HST	TA2.4

Fig. 11. Generalized biostratigraphy, lithostratigraphy and sequence stratigraphy of Eocene sediments in the Savannah River area of South Carolina. Lithologic symbols as in Fig. 4.

facies the Congaree Formation. The arbitrary facies line between the two lithostratigraphic units is placed just northwest of the Savannah River Site. Core samples from the Savannah River area indicate two lower middle Eocene depositional sequences which we assign to the Congaree Formation. These two sequences are grouped together as they have similar lithologies and, where only one sequence is present, cannot be assigned to either the lower or the upper sequence. Generally, these two sequences consist of clean, coarse, pebbly, Fe-stained quartz sand with occasional kaolin lithoclasts at the base grading upward into fine-to-medium clean quartz sand with occasional thin green to grey clay stringers. The fine sands are typical of those described by Nystrom *et al.* (1989) as 'sugar sand' that occur in the lower part of the Huber Formation; we have observed the 'sugar sands' in both sequences on the Savannah River Site. Where present, the sequence boundary within the Congaree Formation is sharp, and in some places is overlain by glauconitic quartz sand that grades upward into finer sand with no glauconite. Thin silcrete zones are also present in the upper sequence. Downdip, these two depositional sequences contain a greater abundance of fine siliciclastics and become calcareous. Carbonate occurs principally as thin beds of sandy bryozoan biocalcarenite or sandy molluscan mould grainstone. These sequences are widespread over the area and are only absent to the north and the southwest. A maximum thickness of greater than 50 m is indicated in several places on the Savannah River Site and downdip to the southeast.

Two upper middle Eocene depositional sequences represented primarily by the Warley Hill Formation and the McBean Formation are found in the Savannah River area. Sediments assigned to the Warley Hill Formation, which are recognized at only two locations, consist of glauconitic calcareous quartz sand, and are about 2 m thick. Although the name McBean Formation has been used for a variety of lithologic units of various ages (Nystrom *et al.*, 1989), we follow Cooke and McNeil (1952) and Nystrom *et al.* (1989) and use the name for siliciclastic and calcareous sediments in the inner Coastal Plain that are time-rock equivalent to downdip carbonate sediments containing *Cubitostrea sellaeformis*. In the Savannah River area the McBean Formation generally consists of lower fine-grained glauconitic sand and upper argillaceous, silty and sandy micrite that often contains thin molluscan mould grainstone. The lower glauconitic quartz sand is discontinuous, and in some areas clayey micrite of the upper part of the sequence overlies the underlying sands of the previous depositional sequence. To the southeast, molluscan grainstone increases in abundance throughout the sequence. In some areas, thinly layered to laminated, argillaceous calcarenite or calcareous mudstone commonly forms the entire sequence and is considered to represent the Blue Bluff Member of the Lisbon Formation of Huddlestun and Hetrick (1986). The top of the McBean Formation is a major unconformity that is identified as the 39.5 Ma type 1 boundary. As a result, the upper surface of the unit frequently has high relief. This depositional sequence is widespread in the southwestern part of the study area, but also occurs as outliers on the Savannah River Site. This sequence reaches a maximum thickness of almost 40 m south of the Savannah River Site and thins rapidly to the north.

Sequence stratigraphy and correlation

Nystrom *et al.* (1989) and Harris and Zullo (1991a) correlate the Congaree Formation to the lower middle Eocene Tallahatta Formation of Alabama, which is assigned to the TE2.1 coastal-onlap cycle of Baum (1986) and Mancini and Tew (1988). This cycle probably correlates to the TA3.3 Cycle of Haq *et al.* (1987). In the study area, calcareous nannofossils in the lower and upper sequences of the Congaree Formation are representative of zones NP15–16 (Lauss, 1992); Clark (1988, pers. comm.) also suggested a similar age from study of the dinoflagellates. On the basis of these correlations, the two sequences of the Congaree Formation are assigned to the TA3.2 and TA3.3 cycles. Both sequences contain transgressive and highstand systems tracts and were deposited on a shallow-marine shelf (Fig. 11). Updip, transgressive systems tracts of both sequences consist of basal coarse sand which grades upward into finer sand with minor clay. Downdip, the transgressive systems tract consists of quartz sand, often glauconitic, which becomes more calcareous higher in the section. Highstand systems tracts consist updip of finer quartz sand with minor clay and thin silicified zones and downdip of bryozoan–molluscan biocalcarenite or sandy molluscan mould grainstone.

The Warley Hill Formation is correlated to the TA3.4 Cycle in the Savannah River area by superposition (Fig. 11). Because the sequence is only found at two localities, systems tracts are not identified. The McBean Formation is correlated to the upper Santee Limestone of the middle and outer Coastal Plain of South Carolina, the upper Lisbon Formation of Alabama and sequence 2 of the Castle Hayne Limestone of North Carolina. These units are interpreted by Zullo and Harris (1987) to correlate to the TA3.5/3.6 Cycle (Fig. 11). The occurrence of *Cubitostrea sellaeformis* and calcareous nannofossils indicative of zones NP16–17 in several cores on the Savannah River Site supports this age assignment. Although Haq *et al.* (1987) recognized two cycles in the upper middle Eocene, only one cycle at the top of the middle Eocene has been recognized in North and South Carolina by Zullo and Harris (1987). Lower glauconitic sand and thin sandy molluscan grainstone are interpreted to represent a transgressive systems tract, whereas the upper argillaceous micrite, fine sand and molluscan grainstone are interpreted to represent the highstand systems tract (Fig. 11).

Upper Eocene

Lithostratigraphy

Three upper Eocene depositional sequences are recognized in the Savannah River area which correspond to the Utley Limestone Member of the Clinchfield Formation, the Griffins Landing/Irwinton Sand Members of the Dry Branch Formation and the Tobacco Road Formation. The Utley Limestone, named by Huddlestun and Hetrick (1979, 1986) for sandy, glauconitic, slightly argillaceous, fossiliferous limestone exposed on the upper end of Mallard Pond on the Georgia Power Company's Plant Vogtle Site, Burke County, Georgia, is exposed at Utley Point and Griffins Landing on the Georgia side of the Savannah River and in several cores on the Savannah River Site. On the Savannah River Site the unit seems restricted to two isolated channels that merge and widen to the southeast. Core samples of the unit indicate that it is a sandy molluscan mould grainstone or calcareous quartz sand that has a maximum thickness of 8 to 9 m. Huddlestun and Hetrick (1986) indicate that the unit ranges up to 11 m in Screven County, Georgia. The Utley Limestone was deposited on a shallow shelf that was well washed by wave energy. The remaining two upper Eocene depositional sequences are referred to the undifferentiated Barnwell Group of Huddlestun and Hetrick (1986) and include the Dry Branch Formation and the Tobacco Road Sand. The Dry Branch Formation consists of three distinct lithofacies, a marine clay (Twiggs Clay), bedded sand and sand–clay (Irwinton Sand) and massive calcareous fossiliferous sand (Griffins Landing Member). In the Savannah River area, the Griffins Landing and the Irwinton Sand Members are the principal lithofacies found, although at some localities, Twiggs Clay lithology occurs in minor amounts. In some areas we have been unable to recognize individual members of the Dry Branch Formation or distinguish the Irwinton Sand from the overlying Tobacco Road Sand; therefore, in those areas we have combined all late Eocene depositional sequences above the TA4.1 cycle into the undivided Barnwell Group.

Where the two sequences can be distinguished, the lower consists of lower calcareous sand, often with large disarticulated *Crassostrea* specimens, minor waxy grey-green clay overlain by quartz sand updip (west) or sandy biomicrite downdip (east). In other areas, the entire sequence consists of coarse to medium Fe-stained quartz sand often argillaceous with occasional thin beds of white to tan kaolin. This sequence has a maximum thickness of greater than 40 m in core on Plant Vogtle in Georgia. The upper sequence, which is represented by the Tobacco Road Sand, was named by Huddlestun and Hetrick (1986) for upper Jacksonian deposits that occur in Richmond County, Georgia. The most characteristic lithology of the Tobacco Road Sand is burrowed and bioturbated, massive, moderately to poorly sorted, medium to coarse, pebbly, weathered sand, with flat pebbles occurring at the base of the unit (Huddlestun & Hetrick, 1986). In cores from the Savannah River area, the Tobacco Road Sand consists of medium to coarse, moderately to poorly sorted quartz sand with occasional clay chips, pebbles and thin, green clay beds. Updip the sand is usually highly Fe-stained and occasional zones of silicified molluscs are present. South of the Savannah River Site the Tobacco Road Sand contains thin interbeds of bryozoan-rich carbonate sand. The lower part of this sequence is usually recognized by a thin pebble zone or by a change in sand type from the underlying sequence. In some cases, the sand of this depositional sequence is indistinguishable from the underlying Irwinton Sand of the Dry Branch Formation. The upper contact of the sequence is generally difficult to distinguish and is usually placed where a marked change from original depositional textures to diagenetic textures occurs or where typical 'Upland' unit lithologies occur. The maximum thickness of this sequence is greater than 50 m. Huddlestun and Hetrick (1986) indicate that the Dry Branch Formation and the Tobacco Road Sand were deposited in marginal marine, coastal environments.

Sequence stratigraphy and correlation

The Utley Limestone contains abundant *Periarchus lyelli* in outcrop and in some cores on the Savannah River Site and establishes correlation of the unit to the 'scutella bed' of the lower Moodys Branch Formation of the eastern Gulf Coastal Plain (Huddlestun & Hetrick, 1986). In addition, calcareous nannofossils (*Chiasmolithus oamaruensis*) from a core south of the Savannah River Site indicate zones NP18–19/20. Zullo and Harris (1987) assigned the lower Moodys Branch Formation of the Gulf Coastal Plain to the first eustatic cycle above the 39.5 Ma type 1 unconformity (TA4.1); Baum and Vail (1988) and Mancini and Tew (1988) followed this assignment using a different scheme (TE3.1). Therefore, we assign the Utley Limestone to the TA4.1 cycle and suggest that the sediments represent a transgressive systems tract because of the coarse siliciclastic component and well-worn coarse bioclastics (Fig. 11).

Huddlestun and Hetrick (1986) assigned the Dry Branch Formation a late Eocene age on the basis of planktonic foraminifers and correlated it to the upper Moodys Branch Formation and North Twistwood Creek Clay of Alabama. Calcareous nannofossils indicative of zones NP19–20 occur in this sequence in the Savannah River area (Laws, 1992). Mancini and Tew (1988) did not recognize two parts to the Moodys Branch Formation in Alabama, and placed the Gosport, Moodys Branch and North Twistwood Creek Clay in a single depositional cycle (TE3.1). Baum (1986) separated the lower Moodys Branch from the upper Moodys Branch–North Twistwood Creek and placed the latter two units in the TE3.2 cycle. This latter interpretation is consistent with the correlation made by Zullo and Harris (1987) in which they assigned the same interval to the TA4.2 cycle. The Dry Branch Formation is assigned to the TA4.2 cycle and is interpreted to represent transgressive and highstand systems tracts (Fig. 11). Lower calcareous quartz sand, often with specimens of *Crassostrea*, and waxy grey-green clay (Twiggs lithology) are assigned to a transgressive systems tract, and quartz sand updip grading to sandy biomicrite downdip are assigned to the highstand systems tract.

The Tobacco Road Sand is considered to be latest Jacksonian in age by Huddlestun and Hetrick (1986) and was placed in the TA4.3 cycle by Zullo and Harris (1987). Hazel (1989, pers. comm.) from study of samples in the area suggested correlation of the unit to calcareous nannofossil zones NP18–19/20. Because of the overall lithologic similarity of Tobacco Road sediments, their blanket-like distribution and the absence of normal characteristics of the condensed sections, the Tobacco Road Sand is interpreted as a highstand systems tract (Fig. 11).

Although we have not recognized Oligocene sediments in the area of the Savannah River Site, on the basis of barnacles from four updip localities, Zullo *et al.* (1982) suggested that late Oligocene sediments are present. Recently, Edwards (1992) suggested from study of dinoflagellates that early and late Oligocene sediments occur in the area. Southeast along the Savannah River in the subsurface of Screven County, Georgia, Huddlestun (1988) recognized Oligocene sediments that he assigned to the Suwannee Limestone.

SUMMARY

Concepts of sequence-stratigraphic analysis can be used to interpret the Palaeogene sediments in the Albemarle embayment of North Carolina and in the Southeast Georgia embayment of South Carolina. Through integration of the sequences with their micro- and megafauna, depositional sequences can be correlated to the specific global cycles of coastal onlap recognized by Haq *et al.* (1987) as illustrated in Fig. 12. The following points represent the major conclusions of this study:

1 Palaeocene sediments in the Albemarle embayment and the Santee and Savannah River areas represent two depositional sequences. In the Albemarle embayment the Beaufort Formation is assigned to the TA1.3 and TA2.1 cycles. In the Santee River area the Black Mingo Group is assigned to the TA1.2 and TA2.1 cycles. In the Savannah River area the Ellenton Formation is assigned to the TA1.3 and TA2.1 cycles.

2 One unnamed lower Eocene sequence occurs in the Albemarle embayment and the Savannah River area. In the Albemarle embayment it is assigned to

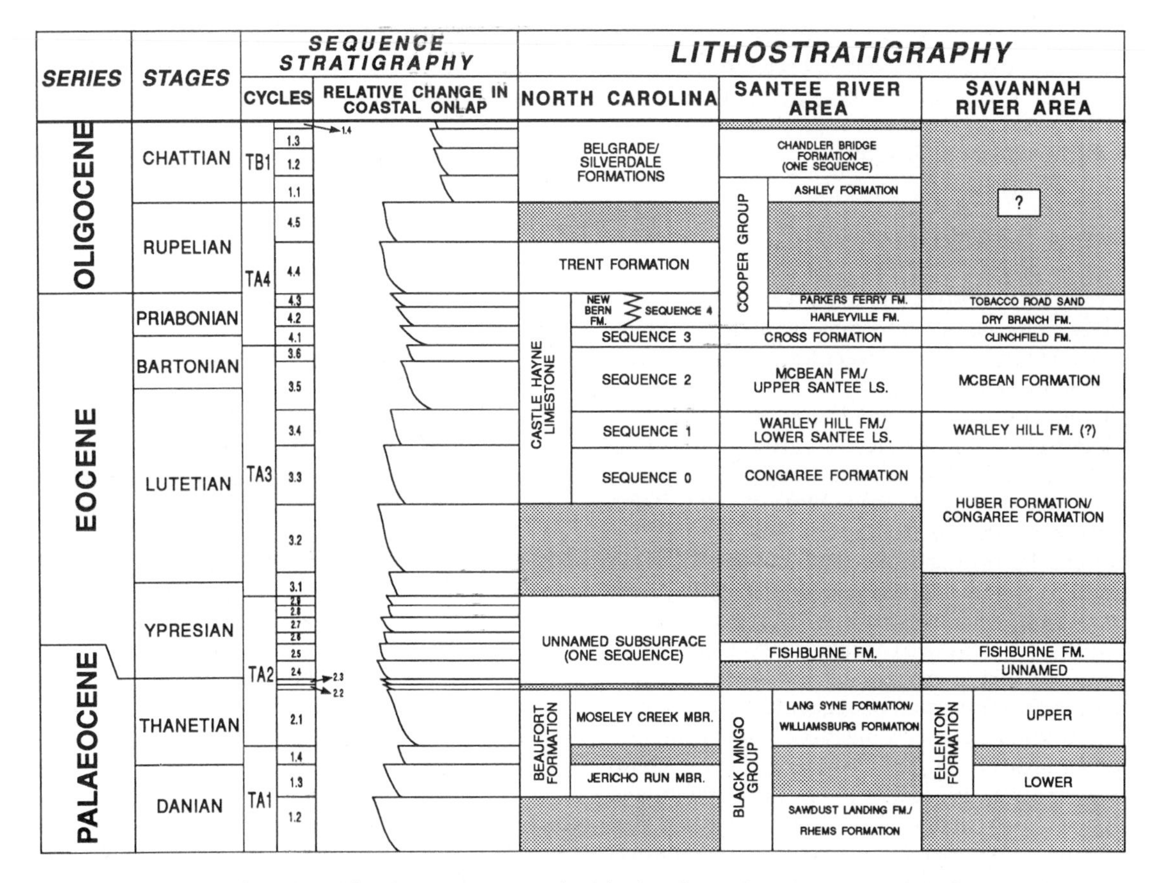

Fig. 12. Correlation of the depositional sequences recognized in the Albemarle embayment and the Santee and Savannah River areas, Southeast Georgia embayment, to the global coastal onlap cycles of Haq *et al.* (1987).

one cycle between the TA2.4 and TA2.9 cycles; in the Savannah River area it is tentatively assigned to the TA2.4 Cycle. In addition, in the Santee and Savannah River areas the lower Eocene Fishburne Formation represents a depositional sequence which is assigned to the TA2.5 Cycle.

3 Five middle and upper Eocene depositional sequences which are represented by the Castle Hayne Limestone and the New Bern Formation are recognized in the Albemarle embayment. They are correlated to the TA3.3 (tentative), TA3.4, TA3.5/3.6, TA4.1 and TA4.2/4.3 cycles. In the Santee River area, six middle and upper Eocene depositional sequences are recognized. They are represented by the Congaree Formation, the lower and upper Santee Limestone, the Cross Formation and the Harleyville and Parkers Ferry Formations of the Cooper Group, and are assigned to the TA3.3, TA3.4, TA3.5/3.6, TA4.1, TA4.2 and TA4.3 cycles, respectively. In the Savannah River area, seven middle and upper Eocene depositional sequences are recognized. They are represented by two sequences in the Congaree Formation, the Warley Hill Formation, the McBean Formation, the Utley Limestone Member of the Clinchfield Formation, the Dry Branch Formation and the Tobacco Road Sand, and are assigned to the TA3.2, TA3.3, TA3.4, TA3.5/3.6, TA4.1, TA4.2 and TA4.3 cycles, respectively.

4 Five Oligocene depositional sequences are recognized in the Albemarle embayment which represent the Trent Formation and the Belgrade and Silverdale Formations. These sequences are assigned to the TA4.4, TB1.1, TB1.2, TB1.3 and the lower part of the TB1.4 cycles, respectively. In the Santee River area, only two Oligocene depositional sequences are recognized representing the Ashley Formation of the Cooper Group and the Chandler

Bridge Formation. They are assigned to the TB1.1 and one of the TB1.2 or TB1.3 cycles, respectively. Although some workers recognize Oligocene age sediments in the Savannah River area, we have not recognized them in this study.

ACKNOWLEDGEMENTS

We thank Van Price and Westinghouse Savannah River Company for providing funding for the research in the Savannah River area, Gerald Grainger of Southern States Company for providing access to cores in Georgia and South Carolina and the South Carolina Division of Water Resources for providing access to cores in the vicinity of the Savannah River Site. We also thank Paul Nystrom and Ralph Willoughby of the South Carolina Division of Geology for their enlightening discussions of South Carolina stratigraphy and the University of North Carolina at Wilmington for providing support for this research. We also thank Henry Posamentier and Colin Summerhaynes for inviting us to participate in the symposium on sequence stratigraphy at the 13th International Congress on Sedimentology in Nottingham, UK, and their invitation to publish this paper. Special thanks to Gerald R. Baum and Bilal U. Haq who reviewed the manuscript and provided many helpful suggestions for improvement.

APPENDIX

Localities

Upper case letters correspond to areas shown in Fig. 1. Specific locations are given in the references cited after each locality.

I North Carolina
- A Albemarle embayment—downdip (see Zarra, 1989 for specific well locations)
- B Albemarle embayment—updip (see Brown *et al.*, 1978 and Harris & Baum, 1977 for specific locations)
- C Cape Fear arch (see Zullo & Harris, 1987 or references cited for specific locations)
  - 1 Martin Marietta New Bern Quarry, Craven County
  - 2 Martin Marietta Belgrade Quarry, Onslow County
  - 3 Jones Silverdale Quarry, Onslow County (Baum *et al.*, 1978)
  - 4 Atlantic Limestone Quarry, Duplin County
  - 5 Billy B. Fussell Quarry, Duplin County
  - 6 East Coast Limestone Quarry, Pender County
  - 7 Martin Marietta Rocky Point Quarry, Pender County
  - 8 Martin Marietta Ideal Quarry, New Hanover County
  - 9 Martin Marietta Castle Hayne Quarry, New Hanover County
  - 10 Haywood Landing, White Oak River, Jones County (Ward *et al.*, 1978)
  - 11 Chinquapin Branch, Jones and Duplin Counties (Zullo, 1984; Worsley & Laws, 1986)
  - 12 Natural Well, Duplin County
  - 13 Maple Hill Lanier Quarry, Pender County (Worsley & Laws, 1986)
  - 14 Carolina Power and Light Company Core, Brunswick County (Harris *et al.*, 1986a)

II South Carolina/Eastern Georgia
- D Santee River area, South Carolina
  - 1 Giant Portland Cement Quarry, Dorchester County (Zullo, 1984)
  - 2 Martin Marietta Georgetown Quarry, Georgetown County (Zullo, 1984)
  - 3 Martin Marietta Berkeley Quarry, Berkeley County (Zullo, 1984)
  - 4 Santee Portland Cement Quarry, Orangeburg County (Zullo, 1984)
  - 5 Wilson's Landing on Santee River, Berkeley County (Powell & Baum, 1982)
  - 6 Santee State Park, southwest shore of Lake Marion, Orangeburg County (Powell & Baum, 1982)
  - 7 Eutaw Springs, Historical Marker, Berkeley County (Powell & Baum, 1982)
  - 8 Roadcut at Warley Hill, US Highway 267, Calhoun County (Nystrom *et al.*, 1989)
  - 9 Roadcut, US Highway 267, Calhoun County (Nystrom *et al.*, 1989)
  - 10 Roadcut on US Highway 21, Calhoun County (Nystrom *et al.*, 1989)
  - 11 Sawdust Landing, southwest bank of Santee River, Calhoun County (Muthig & Colquhoun, 1988)
  - 12 Lang Syne Plantation, Calhoun County (Muthig & Colquhoun, 1988)

13 Lower Bridge, Black River, Williamsburg County (Van Nieuwenhuise & Colquhoun, 1982)
14 Browns Ferry, Black River, Georgetown County (Van Nieuwenhuise & Colquhoun, 1982)
15 Perkins Bluff, Black River, Georgetown County (Van Nieuwenhuise & Colquhoun, 1982)
16 Rhems Landing, Black Mingo Creek, Georgetown County (Van Nieuwenhuise & Colquhoun, 1982)
17 Clubhouse Crossroads corehole #1, Dorchester County (Gohn *et al.*, 1977)
18 Fripp Island corehole, Beaufort County (Colquhoun *et al.*, 1983)

E Savannah River area and Eastern Georgia
1 Savannah River Site, Aiken and Allendale Counties, SC (see Harris & Zullo, 1992 for locations of coreholes)
2 Usserys Bluff, Lower Three Runs Creek, Allendale County, SC (Steele *et al.*, 1986)
3 Shell Bluff, southwest bank of Savannah River, Burke County, GA (Huddlestun & Hetrick, 1986)
4 Griffins Landing, southwest bank of Savannah River, Burke County, GA (Huddlestun & Hetrick, 1986)
5 Blue Bluff, southwest bank of Savannah River, Burke County, GA (Huddlestun & Hetrick, 1986)
6 McBean Creek, Burke County, GA (Huddlestun & Hetrick, 1986)
7 Robinson Creek, Washington County, GA (Zullo, 1984)
8 Riggins Mill Road, Twiggs County, GA (Zullo, 1984)

REFERENCES

BAUM, G.R. (1981) Lithostratigraphy, depositional environments and tectonic framework of the Eocene New Bern Formation and Oligocene Trent Formation. *Southeast. Geol.* **22**, 171–191.

BAUM, G.R. (1986) Sequence stratigraphic concepts as applied to the Eocene carbonates of the Carolinas. In: *SEPM Guidebooks Southeastern United States* (Ed. Textoris, D.A.) Third Annual Midyear Meeting, Raleigh, North Carolina, Soc. Econ. Paleontol. Mineral., pp. 264–269, 413 pp.

BAUM, G.R. & VAIL, P.R. (1988) Sequence stratigraphic concepts applied to Paleogene outcrops, Gulf and Atlantic Basins. In: *Sea-level Changes: An Integrated Approach* (Eds Wilgus, C.K., Posamentier, H., Ross, C.A. & Van Wagoner, J.C.) Spec. Publ. Soc. Econ. Paleontol. Mineral., Tulsa, **42**, 309–327.

BAUM, G.R., COLLINS, J.S., JONES, R.M., MADLINGER, B.A. & POWELL, R.J. (1980) Correlation of Eocene strata of the Carolinas. *SC Geol.* **24**, 19–27.

BAUM, G.R., HARRIS, W.B. & ZULLO, V.A. (1978) Stratigraphic revision of exposed middle Eocene to lower Miocene formations of North Carolina. *Southeast Geol.* **20**, 1–19.

BERGGREN, W.A. (1971) Multiple phylogenetic zonations of the Cenozoic based on planktonic foraminifera. In: *Planktonic Conference 2nd 1970, Proceedings* (Ed. Farinacci, A.) Edizioni Technoscienza, Rome, pp. 41–56.

BLOW, W.H. (1969) Late middle Eocene to Recent planktonic foraminiferal biostratigraphy. In: *Proceedings of the First International Conference on Planktonic Microfossils, Geneva, 1967* vol. 1. (Eds Bronnimann, R. & Renz, H.H.) E.J. Brill, Leiden, pp. 199–422.

BROWN, P.M., BROWN, D.L., SCHUFFLEBARGER, T.E., JR. & SAMPAIR, J.L. (1977) Wrench zone in the North Carolina Coastal Plain. *Spec. Publ. NC Dept. Nat. Econ. Resour., Geol. Resour. Sect.* **5**, 47 pp.

BROWN, P.M., BROWN, D.L., SCHUFFLEBARGER, T.E., JR. & SAMPAIR, J.L. (1978) Wrench zone in the North Carolina Coastal Plain. *Contrib. Basement Tectonics* **3**, 54–73.

CARLSON, T.W. (1976) Opaline claystones in the Paleocene of southeastern North Carolina, MS thesis, University of Minnesota, 73 pp.

COLQUHOUN, D.J., WOOLEEN, I.D., VAN NIEUWENHUISE, D.S. *et al.* (1983) *Surface and Subsurface Stratigraphy, Structure and Aquifers of the South Carolina Coastal Plain.* Report to the Department of Health and Environmental Control, Ground Water Protection Division, Office of the Governor, State of South Carolina, Columbia, 78 pp.

COOKE, C.W. & MACNEIL, F.S. (1952) Tertiary stratigraphy of South Carolina. In: *Shorter Contrib. Gen. Geol. 1952, Prof. Pap. U.S. Geol. Surv.* **243-B**, 19–29.

EDWARDS, L.E. (1992) Dinocysts from the lower Tertiary units in the Savannah River area, South Carolina and Georgia. *Second Bald Head Island Conference on Coastal Plains Geology.* University of North Carolina at Wilmington, pp. 89–90.

FALLAW, W.C., PRICE, V. & THAYER, P.A. (1989) An updip equivalent of the lower Eocene Fishburne Formation in the southwestern Coastal Plain of South Carolina. *Geol. Soc. Am. (Abstr. with Programs)* **21**, 14.

GOHN, G.S., HAZEL, J.E., BYBELL, L.M. & EDWARDS, L.E. (1983) The Fishburne Formation (lower Eocene), a newly defined subsurface unit in the South Carolina Coastal Plain. *Bull. U.S. Geol. Surv.* **1537-C**, C1–C16.

GOHN, G.S., HIGGINS, B.B., SMITH, C.C. & OWENS, J.P. (1977) Lithostratigraphy of the deep corehole (Clubhouse Crossroads corehole #1) near Charleston, South Carolina. In: *Studies Related to the Charleston, South Carolina Earthquake of 1886—A Preliminary Report* (Ed. Rankin, D.W.), Prof. Pap. U.S. Geol. Surv., **1028-E**, 59–70.

HAQ, B.U., HARDENBOL, J. & VAIL, P.R. (1987) Chronology of fluctuating sea levels since the Triassic. *Science* **235**, 1156–1167.

HARRIS, W.B. & BAUM, G.R. (1977) Foraminifera and Rb–Sr glauconite ages of a Paleocene Beaufort Formation outcrop in North Carolina. *Bull. Geol. Soc. Am.* **88**, 869–872.

HARRIS, W.B. & FULLAGAR, P.D. (1989) Comparison of Rb–Sr and K–Ar dates of middle Eocene bentonite and glauconite, southeastern Atlantic Coastal Plain. *Bull. Geol. Soc. Am.* **101**, 573–577.

HARRIS, W.B. & ZULLO, V.A. (1988) Paleogene coastal onlap stratigraphy of the Savannah River region, southeastern Atlantic Coastal Plain. *Geol. Soc. Am. (Abstr. with Programs)* **20**, 269.

HARRIS, W.B. & ZULLO, V.A. (1991a) Eocene and Oligocene geology of the outer Coastal Plain. In: *The Geology of the Carolinas* (Eds Horton, J.W. & Zullo, V.A.) University of Tennessee Press, Knoxville, 251–262.

HARRIS, W.B. & ZULLO, V.A. (1992b) Sequence stratigraphy of Paleocene and Eocene deposits in the Savannah River region. *Second Bald Head Island Conference on Coastal Plains Geology.* University of North Carolina at Wilmington, 122–130.

HARRIS, W.B., THAYER, P.A. & CURRAN, H.A. (1986a) The Cretaceous–Tertiary boundary on the Cape Fear arch, North Carolina, U.S.A. *Cretaceous Res.* **7**, 1–17.

HARRIS, W.B., ZULLO, V.A., LAWS, R.A. & DOCKAL, J.A. (1986b) Coastal onlap stratigraphy of the exposed Oligocene–lower Miocene marine deposits in the North Carolina Coastal Plain. *Geol. Soc. Am. (Abstr. with Programs)* **18**, 629.

HARRIS, W.B., ZULLO, V.A. & PRICE, V. (1990) Coastal onlap stratigraphy and sea-level history of the onshore Palaeogene, southeastern Atlantic Coastal Plain, U.S.A. *13th Int. Sedimentological Congress (Abstr.)* Nottingham, England, p. 212.

HAZEL, J.E., BYBELL, L.M., CHRISTOPHER, R.A. *et al.* (1977) Biostratigraphy of the deep corehole (Clubhouse Crossroads corehole #1) near Charleston, South Carolina. In: *Studies Related to the Charleston, South Carolina Earthquake of 1886—A Preliminary Report* (Ed. Rankin, D.W.), Prof. Pap. U.S. Geol. Surv., **1028-F**, pp. 71–89, 204 pp.

HUDDLESTUN, P.F. (1988) A revision of the lithostratigraphic units of the Coastal Plain of Georgia, the Miocene through the Holocene. *Bull. Georgia Geol. Surv.* **104**, 162 pp.

HUDDLESTUN, P.F. & HETRICK, J.H. (1979) The stratigraphy of the Barnwell Group of Georgia. *Open-File Rep. Georgia Geol. Surv.* **80-1**, 89 pp.

HUDDLESTUN, P.F. & HETRICK, J.H. (1986) Upper Eocene stratigraphy of central and eastern Georgia. *Bull. Georgia Geol. Surv.* **95**, 78 pp.

JONES, G.D. (1983) Foraminiferal biostratigraphy and depositional history of the middle Eocene rocks of the Coastal Plain of North Carolina. *Spec. Publ. NC Dept. Nat. Econ. Resour., Geol. Resour. Sect.* **8**, 80 pp.

LAWS, R.A. (1988) Upper Claibornian–Jacksonian calcareous nannofossils from the Savannah and Santee basins. *Geol. Soc. Am. (Abstr. with Programs)* **20**, 276.

LAWS, R.A. (1990) Correlation of Cenozoic continental margin deposits in North and South Carolina to standard calcareous nannofossil and diatom zonations. *Second Bald Head Island Conference on Coastal Plains Geology.* University of North Carolina at Wilmington, pp. 100–106.

LAWS, R.A. & WORSLEY, T.R. (1986) Onshore-offshore Oligocene calcareous nannofossils from southeastern North Carolina. *Geol. Soc. Am. (Abstr. with Programs)* **18**, 251.

MANCINI, E.A. & TEW, B.H. (1988) Paleogene stratigraphy and biostratigraphy of southern Alabama. *Field Trip Guidebook for the GCAGS-GCS/SEPM*, 38th Annual Convention, New Orleans, Louisiana, 63 pp.

MARTINI, E. (1971) Standard Tertiary and Quaternary calcareous nanoplankton zonation. In: *Proceedings of the Second Planktonic Conference* (Ed. Farinacci, A.). Edizioni Technoscienza, Rome, pp. 739–785.

MUTHIG, M.G. & COLQUHOUN, D.J. (1988) Formal recognition of two members within the Rhems Formation in Calhoun County, South Carolina. *S.C. Geol.* **32**, 11–19.

NYSTROM, P.G., JR, WILLOUGHBY, R.H. & PRICE, L.K. (1989) The Cretaceous and Tertiary stratigraphy of the upper Coastal Plain of South Carolina. In: *Upper Cretaceous and Cenozoic Geology of the Southeastern Atlantic Coastal Plain.* Field Trip Guidebook T172, 28th International Geological Congress, American Geophysical Union, Washington, D.C., pp. 23–42.

OKADA, H. & BUKRY, D. (1980) Supplemental modification and introduction of code numbers to low-latitude coccolith biostratigraphic zonation (Bukry 1973, 1975). *Mar. Micropaleontol.* **5**, 321–325.

PARKER, W. & LAWS, R.A. (1991) Calcareous nanoplankton biostratigraphy of the exposed and subsurface Oligocene and lower Miocene strata in southeastern North Carolina. *Geol. Soc. Am. (Abstr. with Programs)* **23**, 113.

POWELL, R.J. (1984) Lithostratigraphy, depositional environment and sequence framework of the middle Eocene Santee Limestone, South Carolina Coastal Plain. *Southeast. Geol.* **25**, 79–100.

POWELL, R.J. & BAUM, G.R. (1982) Eocene biostratigraphy of South Carolina and its relationship to Gulf Coastal Plain zonations and global changes of coastal onlap. *Bull. Geol. Soc. Am.* **93**, 1099–1108.

PROWELL, D.C., EDWARDS, L.E. & FREDERIKSEN, L.O. (1985) The Ellenton Formation in South Carolina—a revised age designation from Cretaceous to Paleocene. *Bull. U.S. Geol. Surv.* **1605-A**, A63–A69.

RIGGS, S.R., SNYDER, S.W.P., HINE, A.C., SNYDER, S.W., ELLINGTON, M.D. & MALLETTE, P.M. (1985) Geologic framework of phosphate resources in Onslow Bay, North Carolina Continental Shelf. *Econ. Geology* **80**, 716–738.

ROSSBACH, T.J. & CARTER, J.G. (1991) Molluscan biostratigraphy of the Lower River Bend Formation at the Martin Marietta Quarry, New Bern, North Carolina. *J. Paleontol.* **65**, 80–118.

SANDERS, A.E., WEEMS, R.E. & LEMON, E.M., JR. (1982) Chandler Bridge Formation — a new Oligocene stratigraphic unit in the lower Coastal Plain of South Carolina. In: *Stratigraphic Notes, 1980–1982, Bull. U.S. Geol. Surv.* **1529-H**, H105–H124.

SIPLE, G.E. (1967) Geology and ground water of the Savannah River Plant and vicinity, South Carolina. *Water-Supply Pap. U.S. Geol. Surv.* **1841**, 113 pp.

SLOAN, E. (1908) Catalogue of the mineral localities of South Carolina. *S.C. Geol. Surv. Bull.* 2. The State Company Printers, Columbia, South Carolina, 505 pp.

STEEL, K.B., ZULLO, V.A. & WILLOUGHBY, R.H. (1986) Recognition of the Eocene (Jacksonian) Dry Branch Formation at Usserys Bluff, Allendale County, South Carolina. *S.C. Geol.* **30**, 71–78.

Thayer, P.A., Harris, W.B. & Zullo, V.A. (1986) Bentonite from the Castle Hayne Limestone. In: *SEPM Field Guidebooks, Southeastern United States* (Ed. Textoris, D.A.), Third Annual Meeting, Raleigh, North Carolina. Soc. Econ. Paleontol. Mineral., pp. 307–310, 403 pp.

Toulmin, L.D. (1977) Stratigraphic distribution of Paleocene and Eocene fossils in the eastern Gulf Coast region. *Monogr. Geol. Surv. Alabama* **13**, 602 pp.

Turco, K., Sekel, D. & Harris, W.B. (1979) Stratigraphic reconnaissance of the calcareous nannofossils from the North Carolina Coastal Plain, Part II — lower to mid-Cenozoic. *Geol. Soc. Am. (Abstr. with Programs)* **11**, 216.

Van Nieuwenhuise, D.S. & Colquhoun, D.J. (1982) The Paleocene–lower Eocene Black Mingo Group of the east-central Coastal Plain of South Carolina. *S.C. Geol.* **26**, 47–67.

Ward, L.W., Blackwelder, B.W., Gohn, G.S. & Poore, R.Z. (1979) Stratigraphic revision of Eocene, Oligocene and lower Miocene formations of South Carolina. *Geol. Notes, S.C. Geol. Surv.* **23**, 2–23.

Ward, L.W., Lawrence, D.R. & Blackwelder, B. (1978) Stratigraphic revision of the middle Eocene, Oligocene and lower Miocene—Atlantic Coastal Plain of North Carolina. *Bull. U.S. Geol. Surv.* **1457-F**, 23 pp.

Worsley, T.R. & Laws, R.A. (1986) Calcareous nannofossil biostratigraphy of the Castle Hayne Limestone. In: *SEPM Field Guidebooks, Southeastern United States* (Ed. Textoris, D.A.) Third Annual Meeting, Raleigh, North Carolina. Soc. Econ. Paleontol. Mineral. pp. 289–296. 403 pp.

Worsley, T.R. & Turco, K.P. (1979) Calcareous nannofossils from the lower Tertiary of North Carolina. In: *Structural and Stratigraphic Framework for the Coastal Plain of North Carolina* (Eds Baum, G.R., Harris, W.B. & Zullo, V.A.) Carolina Geological Society and Atlantic Coastal Plain Geological Association, Field Trip Guidebook, pp. 65–72, 101 pp.

Zarra, L. (1989) Sequence stratigraphy and foraminiferal biostratigraphy for selected wells in the Albemarle embayment, North Carolina. *Open-File Rep. North Carolina Geological Survey, 89-5,* Depart. Environ. Health, Nat. Resour., Raleigh, 48 pp.

Zullo, V.A. (1984) Cirriped assemblage zones of the Eocene Claibornian and Jacksonian Stages, southeastern Atlantic and Gulf Coastal Plains. *Palaeogeogr., Palaeoclimatol., Palaeoecol.* **47**, 167–193.

Zullo, V.A. & Harris, W.B. (1986) Introduction: sequence stratigraphy, lithostratigraphy and biostratigraphy of the North Carolina Eocene carbonates. In: *SEPM Field Guidebooks, Southeastern United States* (Ed. Textoris, D.A.) Third Annual Meeting, Raleigh, North Carolina, Soc. Econ. Paleontol. Mineral., Tulsa, pp. 257–263, 403 pp.

Zullo, V.A. & Harris, W.B. (1987) Sequence stratigraphy, biostratigraphy and correlation of Eocene through lower Miocene strata in North Carolina. In: *Timing and Depositional History of Eustatic Sequences: Constraints on Seismic Stratigraphy* (Ed. Ross, C.A. & Haman, D.) Spec. Publ. Cushman Found. Foraminiferal Res., **24**, pp. 197–214, 228 pp.

Zullo, V.A., Willoughby, R.H. & Nystrom, P.G. (1982) A late Oligocene or early Miocene age for the Dry Branch Formation and Tobacco Road Sand in Aiken County, South Carolina? In: *Geological Investigations Related to the Stratigraphy in the Kaolin Mining District, Aiken County, South Carolina* (Eds Nystrom, P.G. & Willoughby, R.H.), Carolina Geol. Soc. Guidebook, pp. 34–45, 183 pp.

Spec. Publs Int. Ass. Sediment. (1993) **18**, 563–577

The use of sequence stratigraphy to gain new insights into stratigraphic relationships in the Upper Cretaceous of the US Gulf Coast

A.D. DONOVAN

Exxon Production Research Co., Box 2189, Houston, TX 77252–2189, USA

ABSTRACT

Traditional lithostratigraphic correlations utilize regional coarse-grained lags, flooding (transgressive) surfaces and/or facies changes to define formation boundaries. Since these features can be time transgressive and may have no relationship to major stratal hiatuses, local as well as regional anomalies between lithostratigraphic correlations and biostratigraphic zonations can occur. These anomalies, however, are often reconciled and new insights into regional stratigraphic relationships are attained by subdividing the stratigraphic record into unconformity-bounded stratigraphic units. This sequence-stratigraphic approach also offers a methodology for creating meaningful palaeogeographic reconstructions that can be used to explain and predict the distribution of reservoir, source and seal within a chronostratigraphic framework.

The Upper Cretaceous (Maastrichtian) Prairie Bluff Chalk of central Alabama and the Providence Sand of eastern Alabama and western Georgia (USA) overlie strata mapped as the Ripley Formation and are considered coeval. In central and eastern Alabama, the Ripley–Prairie Bluff/Providence contact is carried along a regional lag and flooding surface that separate sandstones (Ripley) from mudstones (Prairie Bluff/Providence). In western Georgia, however, this surface is carried beneath the facies change from mudstones (Ripley) to sandstones (Providence).

Ripley strata throughout central and eastern Alabama contain elements of the molluscan assemblage zone *Flemingostrea subspatulata,* while the overlying Providence and Prairie Bluff Formations contain elements of a younger molluscan assemblage zone (*Haustator bilira*). Along the Chattahoochee River (Alabama–Georgia border), however, lenticular sandstone bodies, beneath the traditional Ripley–Providence contact, contain elements of the younger *Haustator bilira* assemblage zone. Updip from the Chattahoochee River exposures in easternmost Alabama and western Georgia, coarsening-upward marine sandstones, traditionally included within the basal lithostratigraphic Providence Sand, contain the older *Flemingostrea subspatulata* fauna.

A sequence-stratigraphic analysis of these strata offers a solution to the anomalies observed between the lithostratigraphy and biostratigraphy. A regional unconformity, interpreted as the 68 Ma sequence boundary, can be mapped throughout the study area. This unconformity separates older strata, containing *Flemingostrea subspatulata* fauna, from younger strata, containing *Haustator bilira* assemblage fauna. In central and eastern Alabama, the unconformity typically coincides with the regional flooding (transgressive) surface mapped as the Ripley–Prairie Bluff/Providence boundary. Locally, however, the sequence boundary and flooding surface diverge and bound incised-valley fills. These incised valleys also contain *Haustator bilira* assemblage fauna. Updip, the sequence boundary and transgressive surface can be traced into the traditional lithostratigraphic Providence Sand. This explains why the older *Flemingostrea subspatulata* assemblage fauna is locally observed within the basal updip portions of the classic lithostratigraphic Providence Sand.

INTRODUCTION

Traditional lithostratigraphic correlations utilize regional pebble lags, major flooding (transgressive) surfaces and/or facies changes to define formation boundaries (Fig. 1(a)). While these lithologic criteria provide mappable boundaries, they are often time transgressive and may have no relationship to major stratal hiatuses within a basin. As a consequence, local and regional lithostratigraphic and biostratigraphic anomalies may sometimes result. These anomalies may often be reconciled and new insights into chronostratigraphic relationships attained by subdividing the stratigraphic record into depositional sequences (Fig. 1(b)).

A depositional sequence (Fig. 2) is defined as a relatively conformable, genetically related succession of strata bounded by unconformities or their correlative conformities (Mitchum, 1977; Van Wagoner *et al.*, 1990). Sequence boundaries have chronostratigraphic significance (Mitchum *et al.*, 1977) in that all the strata above a sequence boundary are everywhere younger than all the strata below the sequence boundary. It is this unique relationship that presents stratigraphers with a methodology to subdivide the stratigraphic record into chronostratigraphic units.

Unconformity recognition is an important step in the development of a sequence-stratigraphic framework. Sequence boundaries are best delineated through regional stratal correlations that can reveal surfaces (Fig. 3) associated with stratal terminations (onlap, truncation). At individual well or outcrop locations (Fig. 3), however, sequence boundaries may be more subtly expressed as: (i) a basinward shift in facies; (ii) changes in parasequence stacking; (iii) evidence of abnormal occurence of subaerial exposure; (iv) major biostratigraphic gaps and/or

(a)

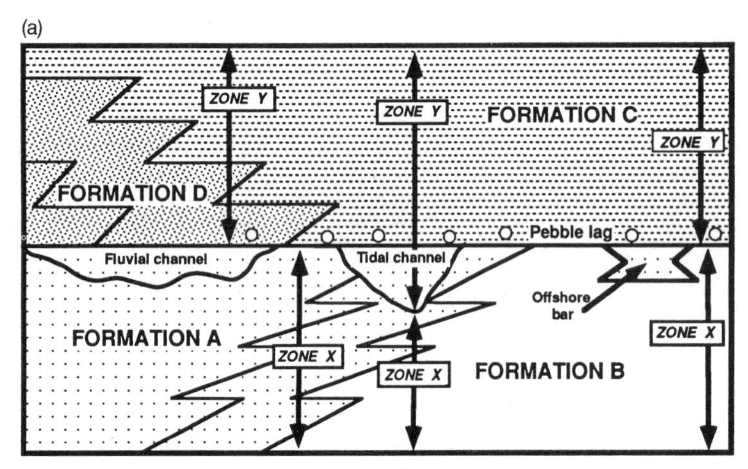

(b)

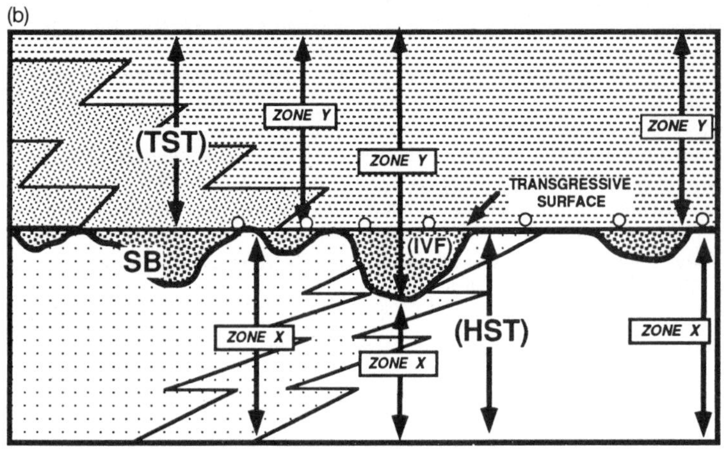

Fig. 1. (a) Traditional lithostratigraphic correlations utilize regional lags, major flooding surfaces or facies changes as the criteria for formation boundaries. These criteria, however, are often time transgressive and do not emphasize major stratal breaks. This methodology may result in regional lithostratigraphic and biostratigraphic anomalies. (b) Sequence-stratigraphic correlations emphasize regional unconformities and their correlative conformities. This methodology offers an additional way to subdivide the stratigraphic record, based on the chronostratigraphic significance of sequence boundaries, and may offer new insights into stratigraphic relationships. SB, sequence boundary; HST, highstand systems tract; TST, transgressive systems tract; IVF, incised-valley fill.

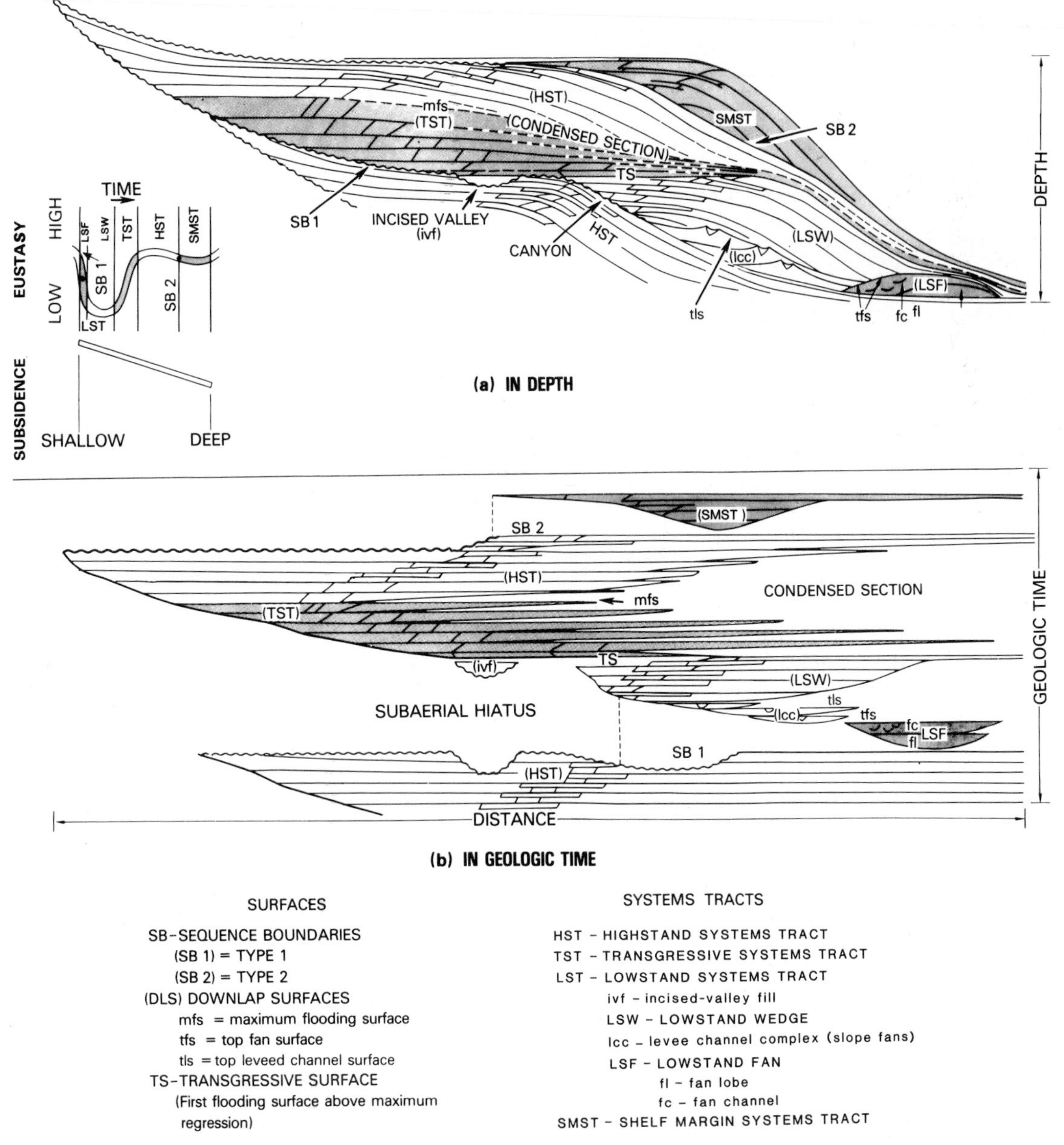

Fig. 2. (a) Stratigraphic cross-section of one complete depositional sequence. (b) Time–distance plot of the same stratigraphic succession shown in (a). Both sequence boundary 1 (SB 1) and sequence boundary 2 (SB 2) separate older strata below from younger strata above and are chronostratigraphically significant surfaces. Note abbreviations for surfaces and systems tracts for subsequent figures. (After Donovan *et al.*, 1988.)

macroinvertebrate assemblage changes; (v) a change in interpreted coastal geomorphology; and/or (vi) a change in interpreted fluvial geomorphology. However, it is only through detailed regional correlations that the differentiation between local depositional and regional unconformable hiatuses can be clearly delineated.

Sequence-stratigraphic correlations indicate that the pebble lags and facies changes associated with regional flooding (transgressive) surfaces do not

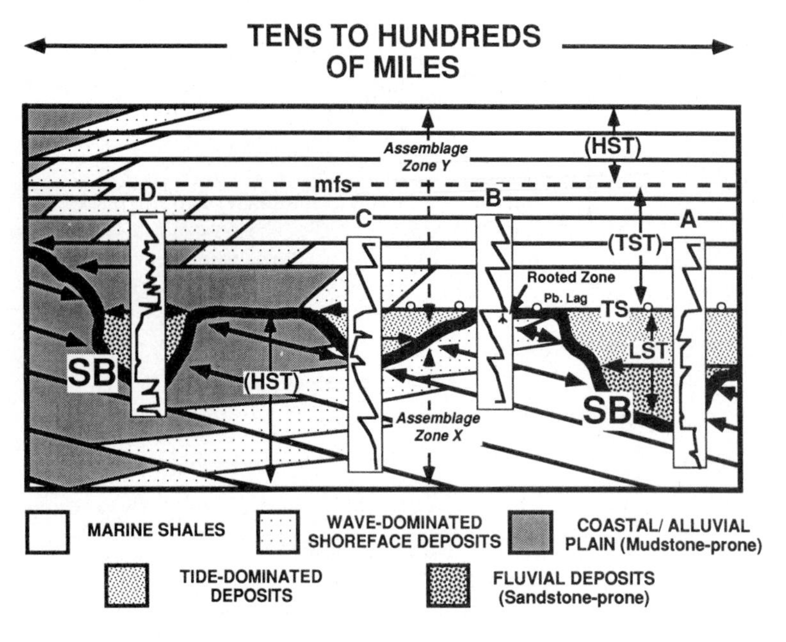

Fig. 3. Sequence boundaries may be recognized by detailed stratal (parasequence) correlations that reveal regional surfaces with onlap, truncation and angular discordance. At individual well or outcrop locations, however, sequence boundaries may be manifested by: (i) abrupt basinward shift in facies (A); (ii) change in parasequence stacking (B, C); (iii) evidence of abnormal occurrence of subaerial exposure (B); (iv) major biostratigraphic breaks (C); (v) evidence for a change in coastal geomorphology: wave-dominated deposits change to tide-dominated deposits (C); and/or (vi) evidence for a change in fluvial geomorphology: meandering fluvial to braided fluvial due to change in base level (D). (After Donovan & Campion, 1991.)

always coincide with regional unconformities (Fig. 4). Sequence boundaries and transgressive surfaces may locally diverge and bound incised valleys updip, as well as other lowstand deposits downdip.

The purpose of this paper is to present a case study in the use of sequence stratigraphy to gain new insights into chronostratigraphic relationships among traditional lithostratigraphic units. The example is typical of many formations as currently mapped and, hopefully, offers insights into the stratigraphic relationships commonly encountered by geologists in both the outcrop and subsurface.

CRETACEOUS LITHOSTRATIGRAPHY OF THE EASTERN GULF COASTAL PLAIN

The exposed Gulf Coastal Plain of eastern Alabama and western Georgia (Fig. 5) consists of a monoclinal succession of Upper Cretaceous and Tertiary strata deposited along a tectonically stable, trailing continental plate margin. The Upper Cretaceous Series consists of approximately 1300 ft (400 m) of Cenomanian- through Maastrichtian-age strata (Reinhardt, 1980). Because the present outcrop belt is oblique to the Cretaceous depositional strike,

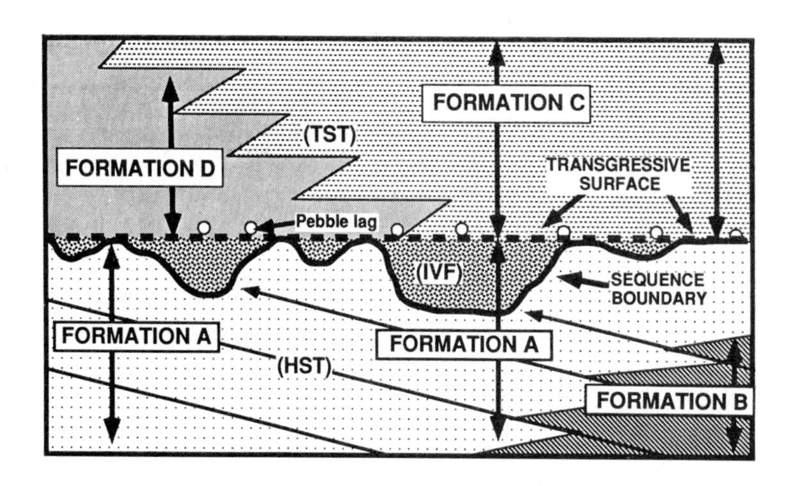

Fig. 4. Pebble lags associated with regional flooding surfaces have been noted by many workers, and are often utilized as formation boundaries. This surface is often referred to as a transgressive surface, ravinement surface, transgressive surface of erosion or transgressive disconformity. The major regional erosional surface or the sequence boundary may coincide with this surface or be stratigraphically beneath it.

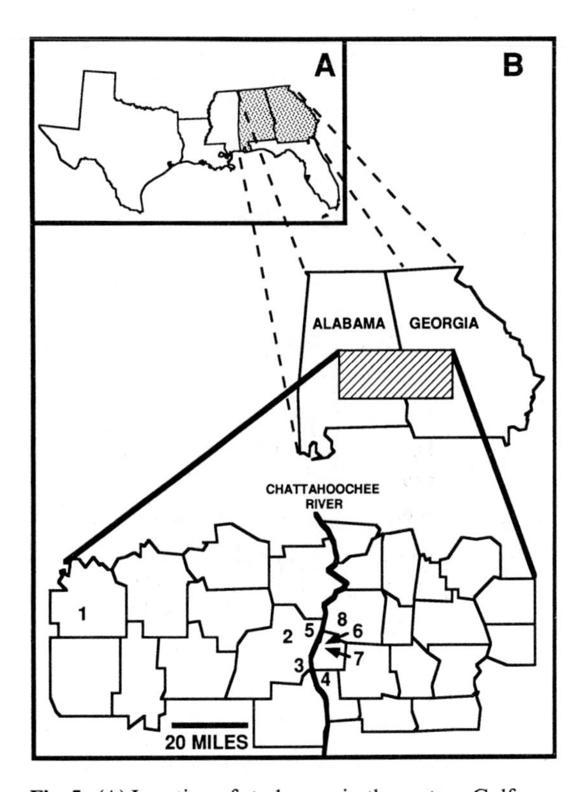

Fig. 5. (A) Location of study area in the eastern Gulf Coastal Plain of Alabama and Georgia. (B) Localities discussed in this paper: 1, Braggs, Lowndes County, Alabama; 2, White Oak, Barbour County, Alabama; 3, Alexander's Landing, Barbour County, Alabama; 4, USGS Ft. Gaines #1, Clay County, Georgia; 5, Eufaula North, Barbour County, Alabama; 6, Georgia Highway 39, Quitman County, Georgia; 7, Firetower Canyon, Quitman County, Georgia; 8, Providence Canyons, Stewart County, Georgia.

coarse-grained, marginal marine deposits crop out to the east, while coeval fine-grained, marine deposits are exposed toward the west. This relationship affords stratigraphers the opportunity to examine updip to downdip stratal and facies relationships along the outcrop belt and in the shallow subsurface.

The lithostratigraphy of Cretaceous strata in this region was developed by Lloyd Stephenson and Watson Monroe of the United States Geological Survey (USGS) (Stephenson & Monroe, 1938; Monroe, 1941). Their lithostratigraphic subdivision (Fig. 6) utilizes regional flooding (transgressive) surfaces, marked by faunal hiatuses, lithologic changes (sandstone to mudstone) and pebble lags, as the formation boundaries. These surfaces were interpreted as unconformities (Stephenson & Monroe, 1938). In central Alabama (Fig. 6), these surfaces were used to define the boundaries of the: (i) Tuscaloosa and Eutaw Formations; (ii) Eutaw Formation and Mooreville Chalk; (iii) Mooreville and Demopolis Chalks; and (iv) Ripley Formation and Prairie Bluff Chalk. The one variation to this methodology was the base of the Ripley Formation. This boundary was placed within a coarsening (shallowing)-upward facies succession from (Demopolis) mudstones to (Ripley) sandstones.

In the Chattahoochee River Valley region (Fig. 6(a)), the interpreted eastward extension of these surfaces was utilized to subdivide the siliciclastic succession in this region into the Tuscaloosa, Eutaw, Blufftown, Cusseta, Ripley and Providence Formations (Stephenson & Monroe, 1938). This subdivision created multi-lithologic, siliciclastic formations bounded by the proposed regional erosional surfaces. The choice of terminology for these formations, however, was unfortunate. Veatch (1909) had previously utilized the terms Blufftown (Marl), Cusseta (Sand), Renfroes (Marl) and Providence (Sand) for distinct (updip) lithologic members within his Ripley Formation (Fig. 7(a)). Stephenson and Monroe (1938) and subsequent workers (Cooke, 1943; Eargle, 1950, 1955; Reinhardt, 1980) all assumed that the updip lithologic Providence and Cusseta Sands of Veatch (1909) were confined within the disconformity-bounded formations bearing the same names (Fig. 7(b)).

Following the work of Stephenson and Monroe (1938) and Monroe (1941), Upper Cretaceous strata were regionally mapped by USGS geologists working in both Alabama (Eargle, 1950; Reinhardt, 1980) and Georgia (Cooke, 1943; Eargle, 1955; Reinhardt, 1980). A correlation chart (Fig. 6(b)) summarizing the resulting lithostratigraphic mapping (Sohl & Smith, 1980) reveals many interesting anomalies between the Cretaceous strata as mapped and the correlations originally proposed by Stephenson and Monroe (Fig. 6(a)). Sohl and Smith's work indicated that the: (i) basal portions of the Blufftown Formation in the Chattahoochee River Valley region are age-equivalent to the uppermost portions of the Eutaw Formation and not the basal Mooreville Chalk; (ii) basal portions of the Ripley Formation in the Chattahoochee River Valley region are age-equivalent to the uppermost Demopolis Chalk to the west; and (iii) basal portions of the Providence Sandstone, as mapped by

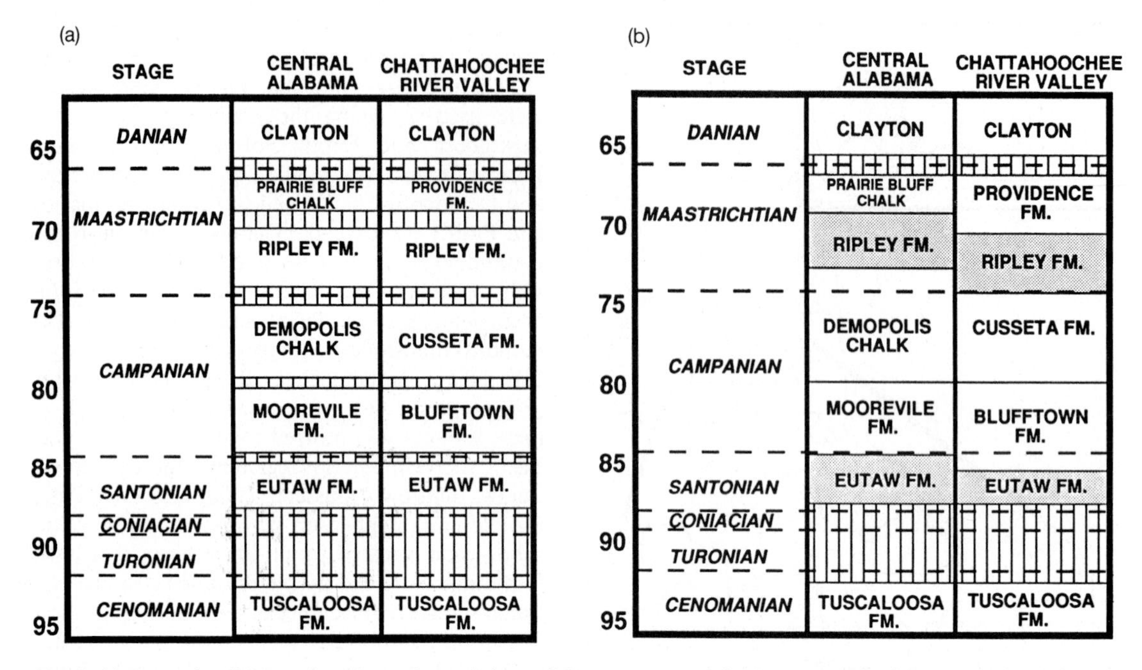

Fig. 6. (a) Formation lithostratigraphy and correlations of Cretaceous strata in the eastern Gulf Coastal Plain. (After Stephenson & Monroe, 1938.) The formation boundaries as defined by Stephenson and Monroe are based on regional surfaces marked by coarse-grained lags, facies changes, major flooding events and biostratigraphic breaks. These surfaces were interpreted as regional unconformities. (b) Regional correlations based on the biostratigraphic work of Sohl and Smith (1980). This work reveals that the base and top of the Ripley Formation, as well as the top of the Eutaw Formation, are not chronostratigraphically significant surfaces regionally. All the strata above these boundaries should be everywhere younger than all the strata below these boundaries.

Eargle (1950, 1955) and Reinhardt (1980) within the Chattahoochee River Valley region, are age-equivalent to the upper portions of the Ripley Formation and not the basal Prairie Bluff Chalk.

MAASTRICHTIAN CHRONOSTRATIGRAPHIC PROBLEMS

In an attempt to understand the local and regional chronostratigraphic variations of the Upper Cretaceous formations in the eastern Gulf of Mexico Coastal Plain, a sequence-stratigraphic analysis of the Ripley and Prairie Bluff/Providence formations was undertaken (Donovan, 1985, 1986). These formations were chosen because of relatively good exposures, excellent biostratigraphic control and the presence of regional chronostratigraphic anomalies (Fig. 8).

Utilizing molluscan assemblage zones that: (i) had widespread geographic distributions (New Jersey to Texas); (ii) could be traced across facies belts (chalks to coarse-grained siliciclastics); and (iii) had chronostratigraphic significance, the Maastrichtian was divided into a younger *Haustator bilira* assemblage zone and an older *Flemingostrea subspatulata* zone (Sohl & Smith, 1980; Sohl & Koch, 1983, 1986; Sohl & Owens, 1991). These assemblage zones (Fig. 8) suggest that: (i) the basal updip portion of the Providence Formation, as mapped by Monroe (1941) and subsequent workers (Cooke, 1943; Eargle, 1950, 1955; Reinhardt, 1980) is age-equivalent to the upper portion of the Ripley Formation in central Alabama; and (ii) the faunal break associated with the Ripley–Prairie Bluff contact in central Alabama does not coincide with the Ripley–Providence contact along the Chattahoochee River as mapped by either Monroe (1941) or Eargle (1950).

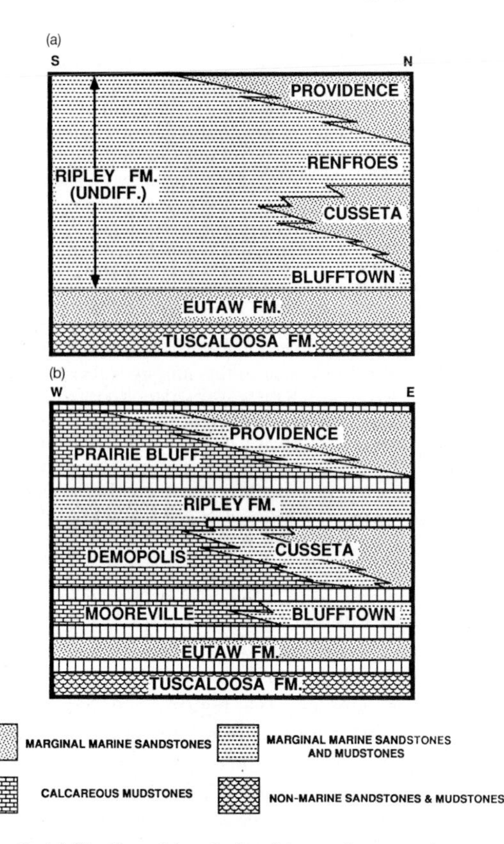

Fig. 7. (a) Stratigraphic relationships and nomenclature proposed by Veatch (1909) within the Chattahoochee River Valley region. (b) Correlations and nomenclature proposed by Stephenson and Monroe (1938) for central Alabama and western Georgia. The unconformity-bounded multi-lithologic formations defined by Stephenson and Monroe in the Chattahoochee River Valley region unfortunately use many of Veatch's (1909) original lithostratigraphic names.

STRATIGRAPHIC RELATIONSHIPS IN CENTRAL ALABAMA

The Ripley–Prairie Bluff contact is exposed near the town of Braggs (NE/NE sec. 8-T12N-R13E) in central Alabama (Figs 5 & 9). At this locality, coarse-grained, fossiliferous sandstones of the Ripley Formation are overlain by calcareous mudstones of the Prairie Bluff Chalk. The contact is overlain by a distinctive lag consisting of: (i) quartz pebbles; (ii) black, rounded phosphatic moulds of body and trace fossils (1 to 5 cm in diameter); (iii) scattered siderite cemented cobbles (10 to 20 cm in diameter); and (iv) scattered worn oyster shells which are highly bored and covered with oyster spat. This surface also coincides with the faunal change (Sohl & Koch, 1983) from the molluscan *Flemingostrea subspatulata* zone (below) to the *Haustator bilira* assemblage zone (above).

According to Baum (Vail *et al.*, 1987), Ripley strata underlying this contact display evidence of subaerial exposure and fresh-water diagenesis by: (i) the development of non-selective microkarst fabrics; (ii) oxidation rinds on oyster shells; (iii) preferential dissolution of aragonite shells with solution enhancement of shell moulds; and (iv) infilling of the solution moulds with overlying Prairie Bluff Chalk. At nearby exposures along the Alabama River, Stephenson and Monroe (1938) also noted angular discordance between Ripley and Prairie Bluff strata along this contact.

The regional stratigraphic relationships observed by Stephenson and Monroe (1938) along the Ripley–Prairie Bluff contact in central Alabama led them to conclude that the boundary was a regional unconformity. The evidence of subaerial exposure and fresh-water diagenesis observed at the top of the Ripley Formation at Braggs, as well as the reported angular discordance along the Alabama River, supports this conclusion. Donovan (1985, 1986) interpreted the Ripley–Prairie Bluff boundary in this region as both a regional subaerial unconformity and transgressive surface (disconformity).

STRATIGRAPHIC RELATIONSHIPS IN EASTERN ALABAMA

The regional surface that marks the Ripley–Prairie Bluff contact of Stephenson and Monroe (1938) in central Alabama was traced to the east by subsequent workers (Monroe, 1941; Eargle, 1950; Donovan, 1985, 1986). Near the White Oak rail junction (Fig. 5) in central Barbour County, Alabama (NE/SE sec. 20-T11N-R27E), an indurated sandstone ledge is overlain by a quartz and phosphate-pebble lag (Fig. 10). This lag coincides with: (i) a flooding surface; (ii) the lithologic change from burrowed sandstones (below) to siltstones and mudstones (above); and (iii) the faunal change from the older *Flemingostrea subspatulata* zone to the younger *Haustator bilira* assemblage zone (Sohl & Koch, 1983). Monroe (1941) placed the Ripley–Providence contact along this surface, and subse-

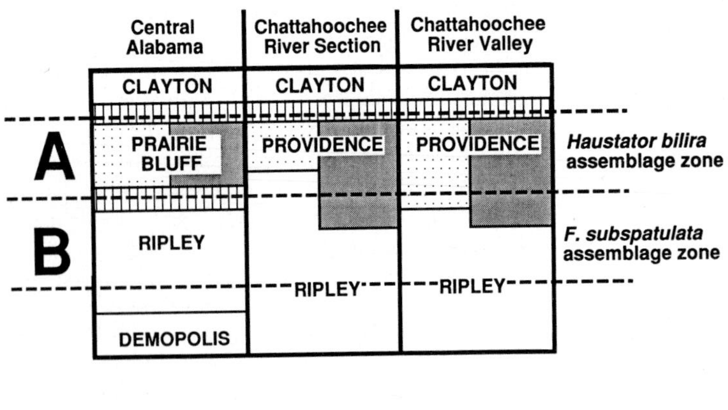

Fig. 8. Maastrichtian lithostratigraphic and biostratigraphic correlations in the eastern Gulf Coastal Plain. In the Chattahoochee River Valley region, the contact between the Ripley and Providence Formations does not coincide with the biostratigraphic break as it does to the west. The stratigraphic placement of this boundary into the Chattahoochee River section also differs among workers. The Ripley–Providence contact, as presently carried in the eastern Gulf Coastal Plain, does not represent a chronostratigraphically significant surface. (Molluscan zones after Sohl & Smith, 1980; Sohl & Koch, 1983, 1986.)

quent workers have agreed with this interpretation. This contact is interpreted as a merged sequence boundary and transgressive surface (Fig. 10).

STRATIGRAPHIC RELATIONSHIPS ALONG THE CHATTAHOOCHEE RIVER

The Providence–Ripley contact was mapped along the Chattahoochee River and its tributaries by Monroe (1941) and Eargle (1950). Monroe (1941) placed the contact approximately 10 miles south of the town of Eufaula, Alabama (Fig. 5) near the old Alexander's Riverboat Landing (sec. 16-T9N-R29E). Following criteria established to the west, Monroe placed the contact at a regional pebble lag that coincided with a marine flooding surface and facies change from coarse-grained sandstones to muddy, fossiliferous sandstones (Fig. 11). Eargle (1950), however, placed the Ripley–Providence contact at a similar lithologic contact exposed at Eufaula Landing (sec. 33-T11N-R29E) approximately 10 miles north and 200 ft stratigraphically lower than the contact Monroe (1941) picked.

Biostratigraphic analysis of the old river localities

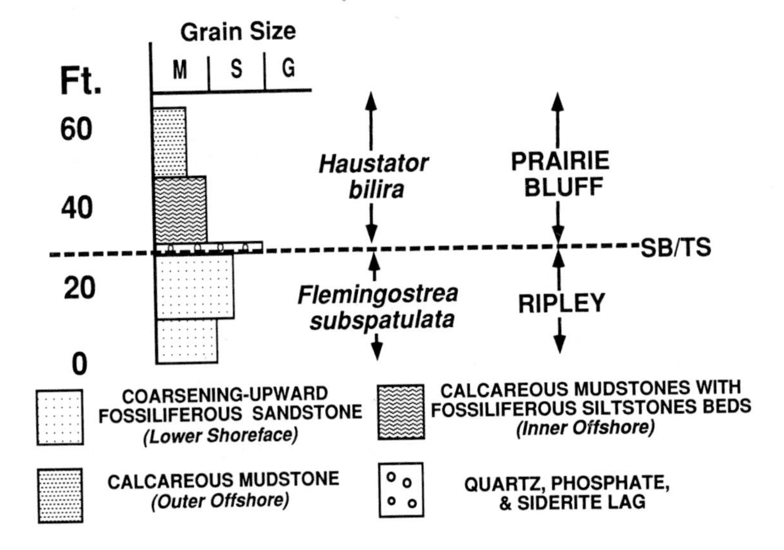

Fig. 9. The Ripley–Prairie Bluff contact near Braggs, Alabama (NE/NE 8-T12N-R13E). A regional lag, flooding surface/facies change and biostratigraphic break mark the contact at this locality. This contact is interpreted as both a sequence boundary and transgressive surface. This middle Maastrichtian unconformity is interpreted as the 68 Ma sequence boundary of the Exxon cycle chart (Haq *et al.*, 1987). (Biostratigraphic data from Sohl & Koch, 1983.)

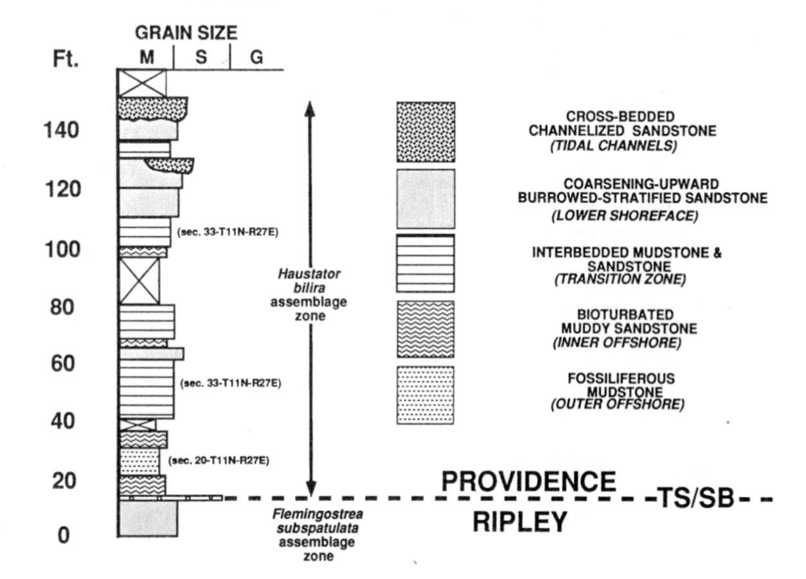

Fig. 10. Composite measured section exposed along Barbour County Road 79 near White Oak, Alabama. At this locality, the Ripley–Providence contact is carried along a pebble lag that coincides with a flooding surface/facies change and biostratigraphic break. This contact is interpreted as a merged unconformity (68 Ma SB) and transgressive surface. (Biostratigraphic data from Sohl & Koch, 1983, 1986.)

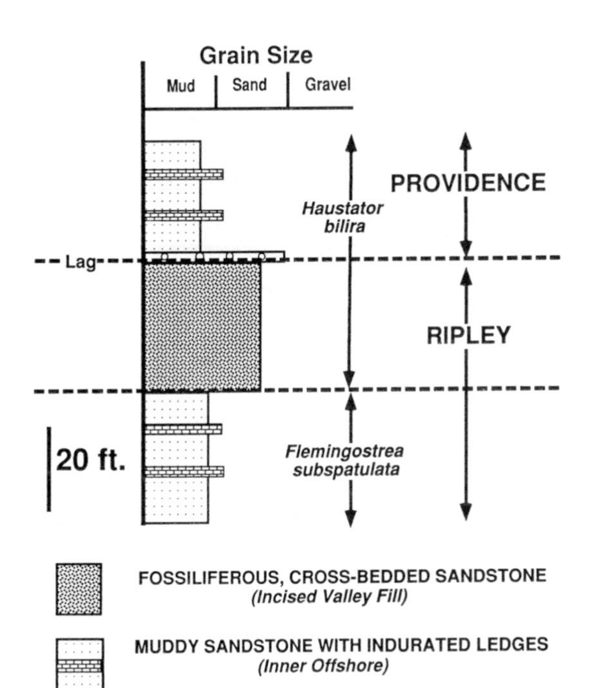

Fig. 11. Measured section near the old Alexander's Landing (SE/NW 9-T9N-R29E) of the Chattahoochee River. (After Monroe, 1941.) The Ripley–Providence boundary is placed at a pebble lag which corresponds to a facies change and flooding surface. The biostratigraphic break, however, occurs below the coarse-grained sandstone which underlies the lag. (Biostratigraphy after Sohl & Koch, 1983.)

(Fig. 8) by Sohl and Koch (1983) reveals that Eargle's (1950) Ripley–Providence boundary occurs within the *Flemingostrea subspatulata* zone while Monroe's (1941) contact occurs within the *Haustator bilira* assemblage zone. Neither surface coincides with the change in the assemblage zones, as is the case to the west (Fig. 8). This biostratigraphic break occurs at the base of the coarse-grained, fossiliferous sandstone that sits directly beneath Monroe's contact (Fig. 11).

Although most of the Maastrichtian localities along the Chattahoochee River were flooded by the damming of the river in the early 1960s, the stratigraphic relationships observed along the old river sections (Fig. 11) also occur in the USGS Ft. Gaines #1 borehole (Fig. 12). In this borehole, a sharp-based, coarse-grained, fossiliferous sandstone with *Haustator bilira* assemblage fauna is present beneath a pebble lag. The faunal break occurs at the base of this sandstone (Sohl, pers. com.).

Subsurface correlations with adjacent wells (Fig. 13) indicate that the coarse-grained sandstone in the USGS borehole is laterally discontinuous and occurs in erosional topography incised into underlying Ripley strata. Donovan (1985, 1986) mapped these laterally discontinuous sandstones in the outcrop and subsurface and observed an orientation at right angles to shoreline trends within the underlying Ripley and overlying Providence Formations. These strata were interpreted as incised-valley fills.

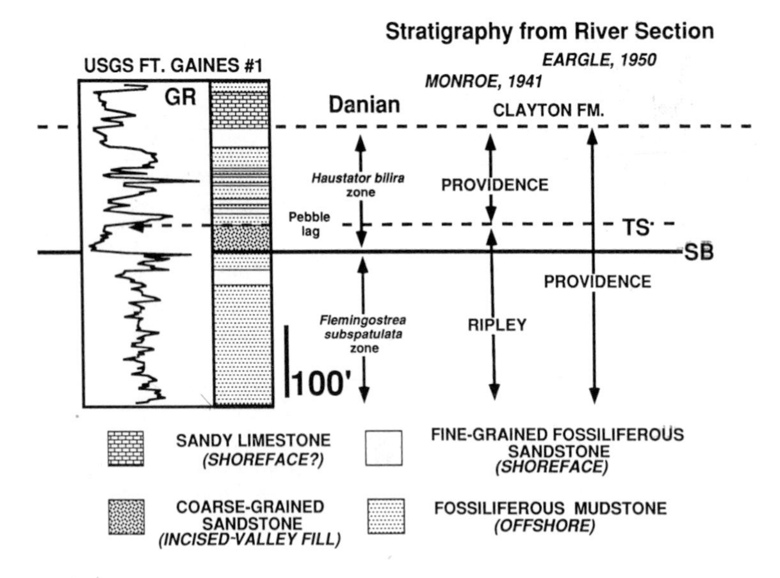

Fig. 12. Facies and stratigraphy of the USGS Ft. Gaines #1 borehole (Lat. 314133/Long. 850508). The stratigraphic relationships present in this well are similar to those described along the flooded river localities (Fig. 11). Monroe's lithostratigraphic contact can be placed along a pebble lag that coincides with a facies change and flooding surface. The biostratigraphic break, however, occurs at the base of the coarse-grained sandstone that underlies the lag. In this area, the regional unconformity (68 Ma SB) and transgressive surface diverge and bound an incised-valley fill. (Biostratigraphic data from Sohl, pers. comm.)

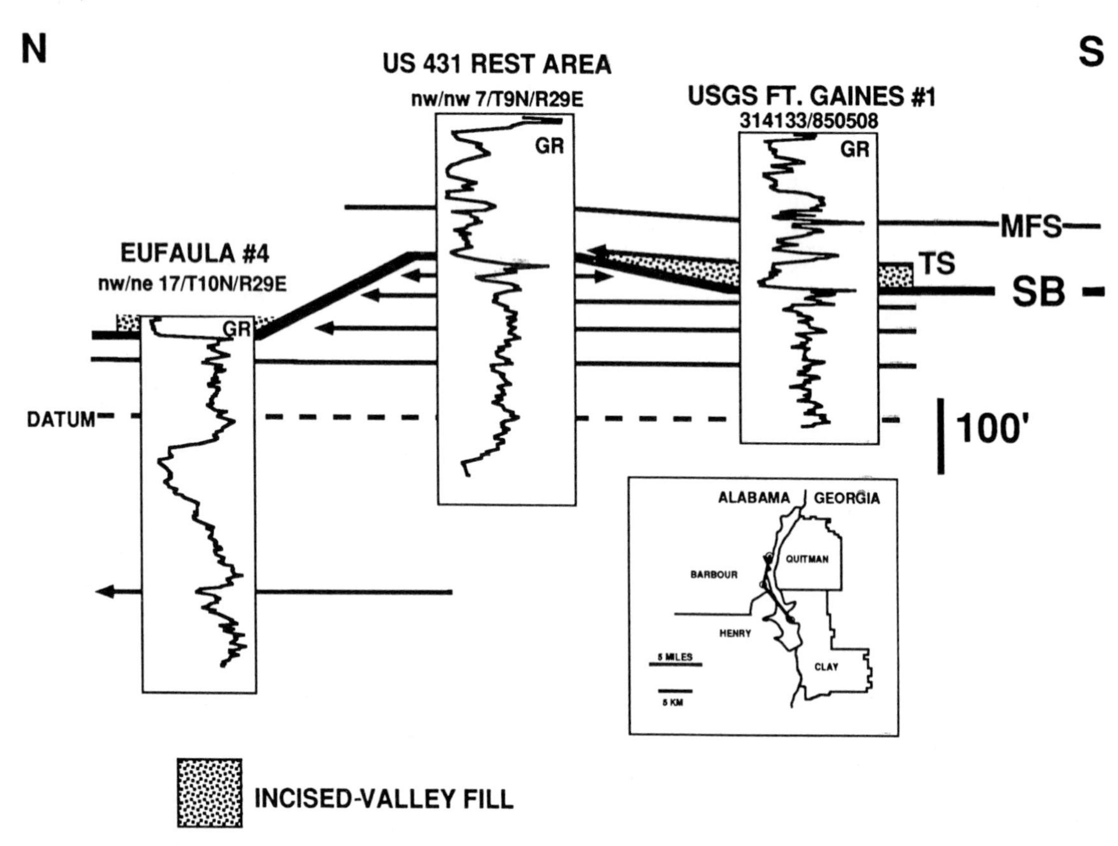

Fig. 13. Subsurface correlations in the Chattahoochee River Valley region. The coarse-grained sandstone within the USGS Ft. Gaines borehole sits erosionally on topography incised into the underlying Ripley Formation, and onlaps toward the adjacent well to the north. These strata are interpreted as part of an incised-valley system related to the 68 Ma sequence boundary.

When these deposits are present the regional unconformity and associated faunal break diverge from Monroe's (1941) lithostratigraphic Ripley–Providence boundary which is carried along the overlying regional flooding (transgressive) surface.

UPDIP STRATIGRAPHIC RELATIONSHIPS

In updip exposures within the Chattahoochee River Valley, all uppermost Cretaceous sandstones are traditionally included within the Providence Formation (Fig. 14). Along Barbour County 97 (approximately 5 miles north of Eufaula, Alabama (Fig. 5)), burrowed to stratified sandstones (Fig. 14) were included within the Providence Formation by Eargle (1950) and Reinhardt (1980). However, biostratigraphic work by Sohl (pers. comm.) indicates that these strata contain imprints of *Flemingostrea subspatulata* assemblage zone fauna and are coeval to Ripley deposits to the west. Based on the absence of a pebble lag, major flooding surface/facies change and the biostratigraphic break, Donovan (1985, 1986) concluded that these strata were stratigraphically below the regional unconformity and transgressive surface mapped as the Ripley–Providence contact to the west.

Along Georgia Highway 39 in Quitman County, Georgia (Fig. 5), a coarse-grained, cross-bedded, sandstone with a sharp erosional base overlies burrowed to stratified, fine-grained sandstones similar to those observed along Barbour County 97 (Fig. 14). These strata were included within the Providence Formation by Cooke (1943) and Eargle (1955). A distinctive quartz-pebble lag is present at the top of the cross-bedded sandstone, while clay clasts up to 20 cm in diameter are present at its base. The clay clasts are fractured and highly weathered, display well-developed clay skins and contain scattered root tubules, indicating that the clasts originated from a soil zone. Donovan (1985, 1986) interpreted this cross-bedded sandstone as an outcrop example of an incised-valley fill, bounded at its base by a sequence boundary and at its top by a transgressive surface. The top of the incised valley as well as the overlying transgressive surface are also exposed at Firetower Canyon (Fig. 14).

Providence Canyons State Park in Stewart County, Georgia is the type section for the Providence Sand (Fig. 5). At this locality, Veatch (1909) identified the Providence Sand and Renfroes Marl as distinct lithologic members of his Ripley Formation, and interpreted the contact as gradational and conformable (Veatch, 1909). Stephenson and Monroe (1938), however, correlated the regional unconformity between their Ripley and Providence Formations beneath Veatch's lithologic Providence Sands. Subsequent workers (Cooke, 1942; Eargle, 1955; Reinhardt, 1980) have also included the sandstones exposed at Providence Canyons within the Providence Formation. Cooke (1943) and Eargle (1955), however, noted no clearly defined unconformity within this succession. Donovan (1985, 1986) identified a root zone within the Providence Sand (Fig. 14). This horizon is located above the strata which contain the *Flemingostrea subspatulata* assemblage zone fauna, and is correlated to the regional unconformity identified to the south and west.

Correlations within the updip Providence Sands (Fig. 14) indicate that the decision by Stephenson and Monroe (1938) to include all of Veatch's (1909) lithologic Providence Sand within their disconformity-bounded Providence Formation was incorrect. The lowermost portions of the updip lithologic Providence Sand (Fig. 15) are chronostratigraphically equivalent to the Ripley Formation to the south and west.

SUMMARY

Sequence-stratigraphic analysis of Maastrichtian strata in the eastern Gulf of Mexico Coastal Plain provides a chronostratigraphic framework to resolve lithostratigraphic and biostratigraphic anomalies within this region. In central Alabama, a regional unconformity and merged transgressive surface separate Ripley sandstones from Prairie Bluff calcareous mudstones. A distinct pebble lag, a facies change, a flooding surface, a molluscan assemblage zone break, evidence of subaerial exposure and local angular discordance denote the contact in this region. Towards eastern Alabama, the Ripley–Providence contact also coincides with a pebble lag, a flooding surface, a facies change and the same molluscan assemblage zone change.

Along the Chattahoochee River, however, the lithological contact (pebble lag, facies change and flooding surface) between the Ripley and Providence Formations (Monroe, 1941) and the molluscan assemblage zone change do not coincide. The faunal change occurs at the base of a laterally

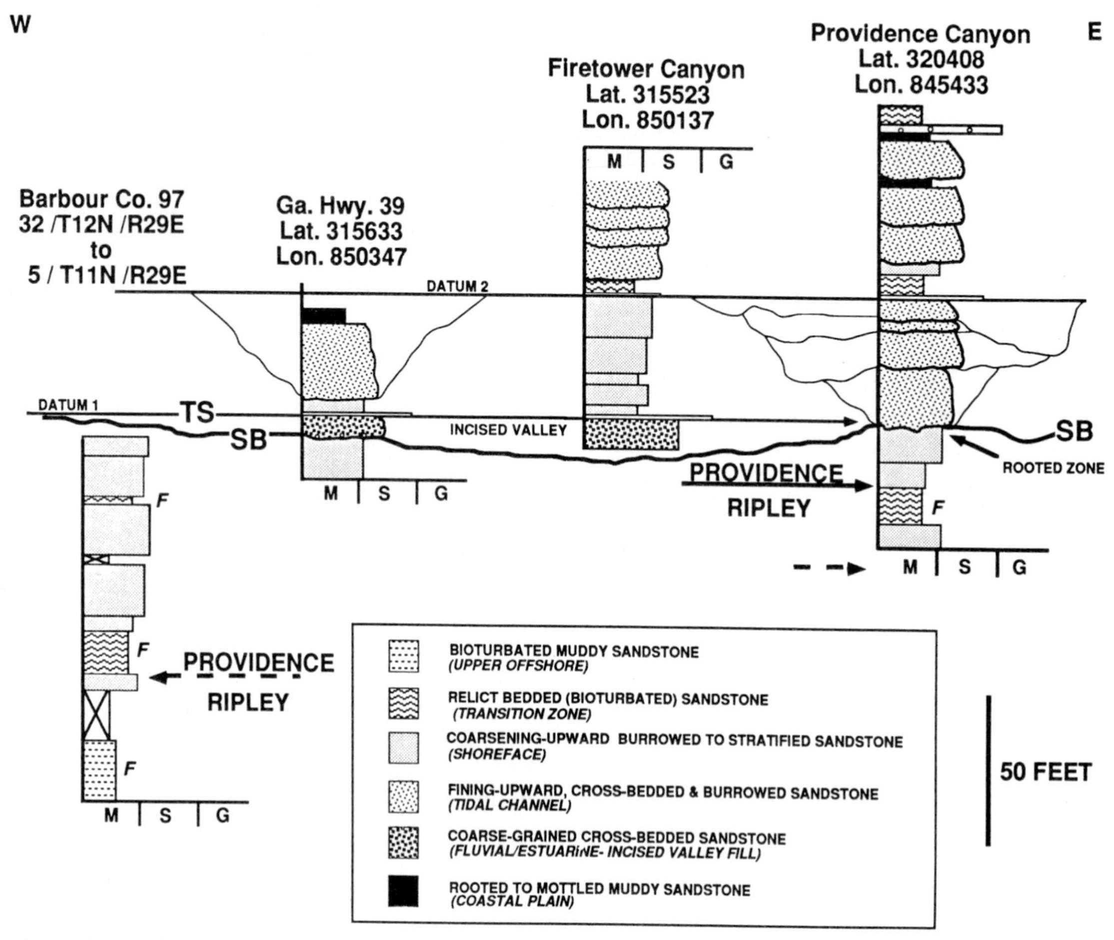

Fig. 14. Facies, biostratigraphy and sequence stratigraphy of the traditional updip lithostratigraphic Providence Sand. Along Barbour County 97, Reinhardt (1980) placed the Ripley–Providence contact at a flooding surface stratigraphically below the thick 'Providence' sandstones exposed at this locality. At Providence Canyons, Stephenson and Monroe (1938), as well as Cooke (1943), placed the Providence–Ripley contact at Veatch's (1909) lithostratigraphic (Renfroes Marl–Providence Sand) boundary. Eargle (1955) and Reinhardt (1980) included all of the Cretaceous section presently exposed within the canyons as the Providence Formation. Sequence-stratigraphic correlations by Donovan (1985, 1986) indicate that the regional unconformity (68 Ma SB), as well as the overlying transgressive surface which marks the Ripley–Providence/Prairie Bluff contact to the south and west, can be traced into the updip lithostratigraphic Providence Sand. F, *Flemingostrea subspatulata* assemblage zone. (After Sohl, pers. comm.).

discontinuous sandstone body that directly underlies the traditional lithologic contact. Regional correlations and biostratigraphic data indicate that the regional unconformity and transgressive surface diverge and bound an incised-valley system. The traditional lithologic boundary occurs at the transgressive surface, but the faunal break coincides with the sequence boundary. These stratigraphic relationships may also be observed in outcrops to the north. This interpretation reveals that regional flooding surfaces, with their associated pebble lags and facies changes, do not always coincide with regional unconformities. This is a basic sequence-stratigraphic observation (Fig. 1) with implications far beyond the Maastrichtian of the eastern Gulf of Mexico Coastal Plain.

In updip regions within the Chattahoochee River Valley, all of the uppermost Cretaceous sandstones

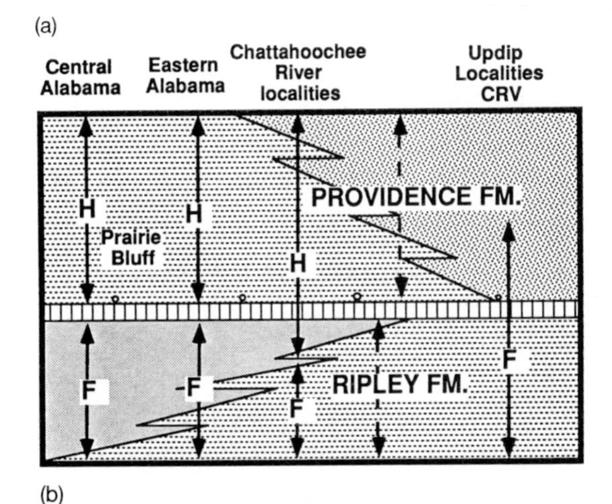

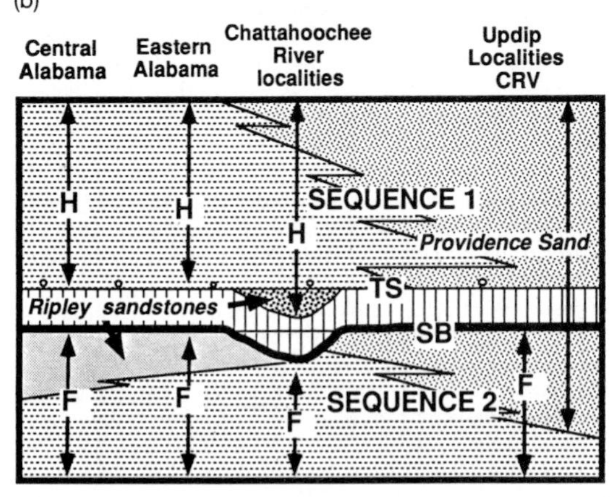

Fig. 15. (a) Classic Maastrichtian lithostratigraphic and chronostratigraphic correlations in the eastern Gulf Coastal Plain. (After Eargle, 1941; Cooke, 1943.) (Biostratigraphic zonation from Sohl & Koch, 1980, 1983, 1986; Sohl, pers. comm.) H, *Haustator bilira* assemblage zone; F, *Flemingostrea subspatulata* zone. (b) Lithostratigraphic, biostratigraphic and chronostratigraphic relationships based on sequence-stratigraphic analysis; SB, sequence boundary; TS, transgressive surface. (After Donovan, 1985, 1986.)

have traditionally been included within the Providence Formation. However, the regional unconformity mapped from the south and west can be traced into the traditional lithostratigraphic Providence Sand in this region. This explains why the basal updip portions of the Providence Sand (Fig. 9) in easternmost Alabama and westernmost Georgia were correlated (Sohl & Smith, 1980) with uppermost Ripley strata to the south and west.

The regional unconformity mapped within the Maastrichtian section of eastern Alabama and western Georgia occurs within the *Lithraphidites quadratus* zone (Sohl & Smith, 1980). Donovan *et al.* (1988) interpreted this unconformity as the 68 Ma sequence boundary of the Exxon cycle chart (Haq *et al.*, 1987).

RECOMMENDATIONS

Sequence-stratigraphic correlations of the Maastrichtian strata in eastern Alabama and western Georgia reveal that the traditional downdip Ripley–Providence contact (a flooding surface/pebble lag/facies change): (i) does not always correspond to the regional unconformity (sequence boundary); and (ii) can be traced into the traditional updip Providence Sand. This strongly suggests that in future stratigraphic work in Georgia and Alabama, the Ripley, Prairie Bluff and Providence Formations, as well as older units, need to be mapped using a single consistent criterion. The formations should be defined by either: (i) regional time-transgressive facies change; or (ii) correlating interpreted coeval regional flooding surfaces. The fact that Stephenson and Monroe (1938) defined the formations based on regional flooding surfaces, interpreted these surfaces as unconformities (downdip) and then correlated the flooding surfaces (updip) into Veatch's (1909) previously defined lithostratigraphic units, has led to much confusion over the lithostratigraphic, biostratigraphic and chronostratigraphic interrelationships among the Upper Cretaceous formations. Simple lithologic criteria, such as those originally proposed by Veatch (1909), would probably be the easiest to use in the field. Utilizing regional flooding (transgressive) surfaces, with associated pebble lags and facies change, to define the formations is also possible. However, this methodology has proven perilous in the past, as is demonstrated by Eargle's (1950) and Reinhardt's (1980) incorrect correlation of the Ripley–Providence contact into the Chattahoochee River section, as well as Stephenson and Monroe (1938) and subsequent workers' (Cooke, 1943; Eargle, 1950, 1955; Reinhardt, 1980) attempts to trace the Ripley–Providence contact updip into more proximal settings. If this second methodology is chosen, the regional flooding surfaces should not be assumed to always correspond to unconformities (sequence boundaries) and/or major biostratigraphic breaks.

ACKNOWLEDGEMENTS

The author would like particularly to thank Norm Sohl of the US Geological Survey for his help and guidance in unravelling the biostratigraphic and lithostratigraphic anomalies within this region. Without his detailed biostratigraphic zonation and extensive collection from Maastrichtian localities throughout the region, the sequence-stratigraphic analysis presented in this paper would have been impossible. A very large debt of gratitude is also owed to Juergen Reinhardt, who inspired and fostered this research at all times.

Discussions with Peter Vail, Gerald Baum, Robert Weimer and John Horne played an important role in helping the author formulate the facies and sequence framework presented in this paper. I am also indebted to many colleagues at the Exxon Production Research Company and Exxon affiliates who have examined and discussed the stratigraphic relationships in this region. Reviews by Keith Shanley, Cindy Yeilding, Fred Wehr, Gerald Baum, Elijah White and Ernest Mancini helped to clarify both the concepts and contents of this paper. The research was part of the author's PhD dissertation at the Colorado School of Mines, which was done in co-operation with the United States Geological Survey (USGS). Researchers at the USGS Water Resource Division offices in Atlanta, Georgia, Montgomery, Alabama and Tuscaloosa, Alabama kindly provided much of the subsurface data used for the original study. Research since 1985 has been supported by the Exxon Production Research Company.

REFERENCES

Cooke, C.W. (1943) Geology of the Coastal Plain of Georgia. *U.S. Geol. Surv. Bull.*, **941**, 121 pp.

Donovan, A.D. (1985) Stratigraphy and sedimentology of the Upper Cretaceous Providence Formation (western Georgia and eastern Alabama). Unpublished PhD thesis, Colorado School of Mines, Golden, Colorado, 236 pp.

Donovan, A.D. (1986) Sedimentology of the Providence Formation: In: *Stratigraphy and Sedimentology of Continental, Nearshore, and Marine Cretaceous Sediments of the Eastern Gulf Coastal Plain* (Ed Reinhardt, J.). Field Trip 3, 1986 Am. Assoc. Petrol. Geol. Meet.—Atlanta, 29–44.

Donovan, A.D. & Campion, K.M. (1991) Sequence stratigraphic analysis—field excursion guidebook to the middle and upper Eocene of the Hampshire Basin, England. *Geological Society of London Conference Field Trip: Exploration Britain into the Next Decade*, 43 pp.

Donovan, A.D., Baum, G.R., Blechschmidt, G.L., Loutit, T.S., Pflum, C.E. & Vail, P.R. (1988) Sequence stratigraphic setting of the Cretaceous–Tertiary boundary in central Alabama. In: *Sea-Level Changes—An Integrated Approach* (Eds Wilgus, C.K., Hastings, B.S., Kendall, C.G.St.C., Posamentier, H.W., Ross, C.A. & Van Wagoner, J.C.) Spec. Publ. Soc. Econ. Paleontol. Mineral. 42, 299–307.

Eargle, D.H. (1950) Geologic map of the Selma Group in eastern Alabama. *U.S. Geol. Surv. Oil and Gas Investigation Preliminary Map*, **105**.

Eargle, D.H. (1955) Stratigraphy of the outcropping Cretaceous rocks of Georgia. *U.S. Geol. Surv. Bull.* **1014**, 101 pp.

Haq, B.U., Hardenbol, J. & Vail, P.R. (1987) Chronology of fluctuating sea levels since the Triassic. *Science* **235**, 1156–1167.

Mitchum, R.M. (1977) Seismic stratigraphy and global changes of sea level, Part 11: glossary of terms used in seismic stratigraphy. In: *Seismic Stratigraphy—Applications to Hydrocarbon Exploration* (Ed. Payton, C.E.). Mem. Am. Assoc. Petrol. Geol. 26, 205–212.

Mitchum, R.M., Vail, P.R. & Thompson, S. (1977) Seismic stratigraphy and global changes of sea level, Part 2: the depositional sequence as a basic unit of stratigraphic analysis. In: *Seismic Stratigraphy—Applications to Hydrocarbon Exploration* (Ed. Payton, C.E.). Mem. Am. Assoc. Petrol. Geol. 26, 53–62.

Monroe, W.H. (1941) Notes on deposits of Selma and Ripley age in Alabama. *Alabama Geol. Surv. Bull.* **48**, 150 pp.

Reinhardt, J. (1980) Upper Cretaceous stratigraphy and depositional environments. In: *Excursions in Southeastern Geology v. II* (Ed. Frey, R.W.). 1980 Geol. Soc. Am. Ann. Meet. Field Trip Guidebook, pp. 385–392.

Sohl, N.F. & Koch, C.F. (1983) Upper Cretaceous (Maastrichtian) mollusca from the *Haustator bilira* assemblage zone in the East Gulf Coastal Plain. *U.S. Geol. Surv. Open File Report*, **83–451**, 204 pp.

Sohl, N.F. & Koch, C.F. (1986) Molluscan biostratigraphy and biofacies of the *Haustator bilira* assemblage zone (Maastrichtian) of the east Gulf Coastal Plain. In: *Stratigraphy and Sedimentology of Continental, Nearshore, and Marine Cretaceous Sediments of the Eastern Gulf Coastal Plain* (Ed. Reinhardt, J.) Field Trip 3, 1986, Am. Assoc. Petrol. Geol. Ann. Meet. pp. 45–56.

Sohl, N.F. & Owens, J.P. (1991) Cretaceous stratigraphy of the Carolina Coastal Plain. In: *The Geology of the Carolinas: 50th Anniversary Volume of the Carolina Geological Society* (Eds Horton, J.R. & Zullo, V.A.) The University of Tennessee Press, Knoxville, pp. 191–220.

Sohl, N.F. & Smith, C.C. (1980) Notes on Cretaceous biostratigraphy. In: *Excursions in Southeastern Geology v.II* (Ed. Frey, R.W.). 1980 Geol. Surv. Am. Ann. Meet. Field Trip Guidebook, pp. 392–402.

Stephenson, L.W. & Monroe, W.H. (1938) Stratigraphy of Upper Cretaceous Series in Mississippi and Alabama. *Bull. Am. Assoc. Petrol. Geol.* **22**, 1639–1657.

Vail, P.R., Baum, G.R., Loutit, T.S., Donovan, A.D., Mancini, E.A. & Tew, B.H. (1987) *Sequence Stratigraphy of the Uppermost Cretaceous and Paleogene of the*

Alabama/Western Georgia Coastal Plain. Lafayette Geol. Soc. 1987 Field Trip Guidebook.
VAN WAGONER, J.C., MITCHUM, R.M., CAMPION, K.M. & RAHMANIAN, V.D. (1990) Siliciclastic sequence stratigraphy in well logs, cores, and outcrops. *Am. Assoc. Petrol. Geol. Meth. Expl. Ser.* **7**, 55 pp.
VEATCH, J.O. (1909) Second report on the clay deposits of Georgia. *Georgia Geol. Soc. Bull.* **18**, 453 pp.

AUSTRALIA

Spec. Publs Int. Ass. Sediment. (1993) **18**, 581–603

Seismic stratigraphy and passive-margin evolution of the southern Exmouth Plateau

R. BOYD*, P. WILLIAMSON† *and* B.U. HAQ‡

**Center for Marine Geology, Dalhousie University, Halifax, Nova Scotia, Canada B3H 3J5;*
†Bureau of Mineral Resources, Canberra, ACT, 2601, Australia; and
‡National Science Foundation, Washington, DC 20550, USA

ABSTRACT

Permian/Carboniferous to Neocomian rifting along northeast Gondwanaland transformed an intracratonic basin fronting the eastern Tethyan continental margin to a new passive margin along northwest Australia fronting the newly created Indian Ocean. Subsequent sedimentation has been relatively thin, resulting in a starved passive margin and an ideal opportunity to utilize high-quality industry shallow seismic data together with ODP drilling methods to investigate the entire history of passive-margin evolution and the resulting sedimentary successions. The two leg 122 sites 762 and 763 combined with results from 11 exploration wells on the southern Exmouth Plateau enable precise dating and geological confirmation of an extensive seismic data base.

A wide range of seismic reflection data sources and over 40,000 line kilometres of data have been used to define eight seismic stratigraphic packages on the southern Exmouth Plateau which overall contain three major clastic depositional wedges and a carbonate blanket deposit. The analysis of these eight packages has provided: (i) an updated interpretation of the regional seismic stratigraphy in the light of new data, research results and advances in interpretation methods; and (ii) a tectono-stratigraphic interpretation of passive-margin evolution. This margin evolution documents a history of intracratonic sedimentation in the Norian–Rhaetian, rift onset and initial breakup from Hettangian–Callovian, a second rift and final breakup between Callovian–Hauterivian, a post-breakup and rift to drift transition from Hauterivian–Cenomanian and ends with a mature ocean phase in transition to a collision margin from Turonian–Holocene.

INTRODUCTION

Tectonic and stratigraphic evolution of the southern Exmouth Plateau

In the early Mesozoic, the present northwest Australian margin (location shown in Fig. 1) was part of a continental rift zone on northeast Gondwanaland that bordered the Tethys Sea to the north (von Rad & Exon, 1983; Fullerton *et al.*, 1989). Northwest Australia was involved in a major late Palaeozoic (possibly Permian or older) rifting event which allowed separation of a basement/lower crust segment and its westward translation to form the Exmouth Plateau (Williamson & Falvey, 1988; Williamson *et al.*, in press). This was followed by subsidence and progradation of a thick Triassic depositional wedge along the western Australian margin (Boote & Kirk, 1989) and termed the Mungaroo Sand/Locker Shale Facies in the Barrow–Dampier Sub-basin (summarized in Hocking *et al.*, 1988). A stratigraphic summary of the southern Exmouth Plateau is shown in Fig. 2. The stratigraphic equivalent of this Triassic depositional wedge extends across the central Exmouth Plateau (Vos & McHattie, 1981) where it is over 3000 m thick and reaches as far as the Wombat Plateau (Exon *et al.*, 1982; von Rad *et al.*, in press). This progradation was terminated by a regional transgression, depositing the largely paralic Brigadier Formation and calcareous lower Dingo Claystone

*Present address: Department of Geology, University of Newcastle, NSW, 2308, Australia.

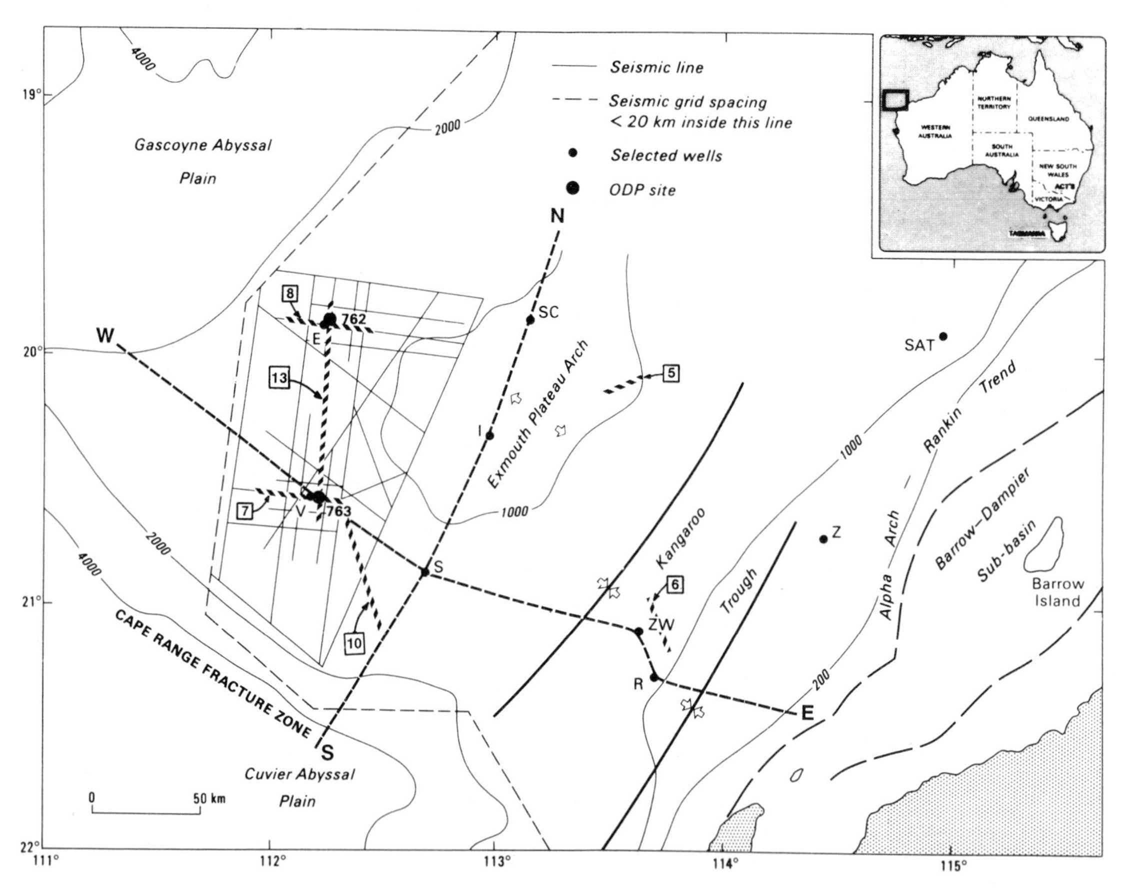

Fig. 1. The study area is located on the southern Exmouth Plateau off northwest Australia. Seismic data were analysed from all regions of the plateau west of the Alpha Arch–Rankin Trend. Intensive seismic correlation was conducted inside the boxed area on the lines shown. Bathymetric contours are in metres below sea level and letters refer to well locations used in the study as follows: SC, Scarborough; SAT, Saturn; Z, Zeepard; ZW, Zeewulf, R, Resolution; S, Sirius; I, Investigator, V, Vinck; E, Eendracht. The Cuvier Rift lies off the figure, to the south of the Kangaroo Trough. E–W is the location of the cross-section of Fig. 3; N–S is the cross-section of Fig. 4. Also shown in numbered boxes are the locations of Figs 5, 6, 7, 8, 10 and 13.

between the Norian and Pliensbachian (Crostella & Barter, 1980; Hocking *et al.*, 1988) in the Barrow–Dampier Sub-basin, on the Rankin Trend and over much of the Exmouth Plateau. Further north on the Wombat Plateau, an uppermost Triassic marginal carbonate facies with reef development occurred in the Rhaetian (Williamson *et al.*, 1989).

A second extended phase of rifting began in the latest Triassic–early Jurassic (Fig. 2) and continued until final breakup occurred in the early Cretaceous (Audley-Charles, 1988; Boote & Kirk, 1989). Rifting began along the margin north from the Exmouth Plateau and resulted in the ?Callovian–Berriasian (see von Rad *et al.*, 1989) breakup of a continental fragment in latest Jurassic–earliest Cretaceous times. Prior to and concurrent with this second rifting event, a thick Jurassic syn-rift sedimentary succession accumulated in restricted rift-graben locations in the Barrow–Dampier Sub-basin (Veenstra, 1985) and south into the Perth Basin (Boote & Kirk, 1989). Barber (1982) documented mid-Jurassic erosion over much of the central Exmouth Plateau in response to the rifting and this region began to accumulate thin holomarine muds only after the Callovian breakup event. Wright and Wheatley (1979) interpreted the main angular unconformity seen on seismic over much of the plateau to be mid-Jurassic in age. Subsequently, a

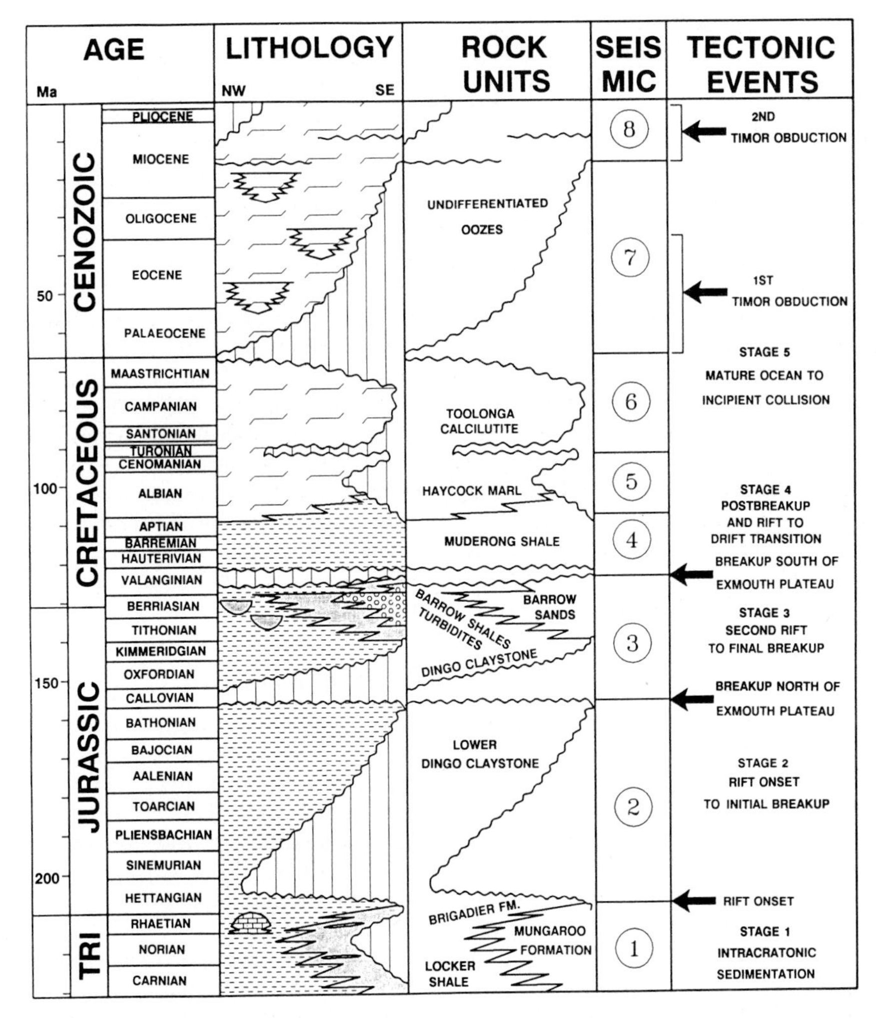

Fig. 2. Stratigraphic summary of the southern Exmouth Plateau from Triassic to Holocene. (Chronostratigraphy is from Haq *et al.*, 1987.) Dashed lithological pattern indicates mudstone; dots, sandstone and conglomerate; bent bars, marl and ooze; and brick, limestone and marl. Vertical lines indicate hiatuses. The seismic column contains correlated seismic packages discussed in the text. Corresponding seismic reflectors for each package lie between that package and the overlying package.

thick mid-Callovian to Tithonian marine sequence filled the Barrow–Dampier Rift and marine shales accumulated up to 2000 m thick in the southern Kangaroo Trough (Barber, 1988).

Subsidence continued accompanied by thin marine shale deposition into the early Cretaceous, except in the southern region where a major clastic source resulted in progradation of the Barrow Group sediments (Fig. 2) northwards into the Barrow Sub-basin and across the southern Exmouth Plateau (Exon & Willcox, 1978; Wright & Wheatley, 1979). In the Barrow Sub-basin, Barrow Group sedimentation occupied the time interval from Tithonian (Hocking *et al.*, 1988) to Valanginian (Kopsen & McGann, 1985). The Barrow Group is also considered to owe its origin to tectonic uplift prior to (Boote & Kirk, 1989) or associated with (Veevers & Powell, 1979) the final (third) phase of continental breakup and the onset of sea-floor spreading.

Palaeomagnetic data from the Cuvier and Gascoyne Abyssal Plains (Fullerton *et al.*, 1989) identify the first magnetic anomaly and hence the final breakup event in both areas to be M10 of early Hauterivian age (Haq *et al.*, 1987). This is supported by the Valanginian age of rift valley sediments directly above basaltic intrusions at site 766 on the Gascoyne Abyssal Plain (Gradstein *et al.*, 1989). Subsidence following breakup enabled a regional transgression to occur along the northwest Australian margin, depositing the Winning Group in the Barrow Sub-basin (Hocking *et al.*, 1988) and a widespread unit interpreted as an equivalent of the Muderong Shale on the Exmouth Plateau (Exon & Willcox, 1978). Due to the lack of previous sampling over much of the Exmouth Plateau from Neocomian to Holocene and the lack of any prospective reservoir units in this interval, little detailed information has previously been available for the area during this time (Barber, 1988). Following breakup, the Exmouth Plateau was considered to have slowly subsided, continuing to deposit a diminishing fine-grained clastic supply equivalent to the Muderong Shale into the middle Cretaceous (Barber, 1982), experiencing an Upper Cretaceous erosional event (Wright & Wheatley, 1979) followed by a transformation to carbonate deposition (Fig. 2). The basin subsided to bathyal depths between the early Cenozoic (Barber, 1982) and the Pliocene–Pleistocene (Exon & Willcox, 1980). Structural reactivation and basin inversion occurred in the late Tertiary in response to obduction of the northern Australian plate margin in the Java Trench (Kopsen & McGann, 1985; Howell, 1988), with major structural growth occurring during the Miocene (Barber, 1988).

Previous seismic investigations

Seismic investigations began on the southern Exmouth Plateau in the 1960s and a range of data were collected up to 1976 by the Australian Bureau of Mineral Resources (BMR), Esso Australia Ltd., Gulf Research and Development Company, Shell Development (Australia) Pty Ltd, and summarized in Exon and Willcox (1980). Major regional seismic surveys were conducted in 1976 by GSI International (summarized in Wright & Wheatley, 1979) and from 1978–81 by Esso Australia Ltd. Surveys on the central Exmouth Plateau were conducted in the late 1970s by the Phillips Group. More recent seismic investigations have been undertaken by the BMR and partners along the northern, western and southern plateau margins and have been documented in Williamson and Falvey (1988) and Exon and Williamson (1988). Some deep seismic results of these surveys were used by Mutter *et al.* (1989) and Williamson *et al.* (1990) to provide evidence for rifting deformation of the Exmouth Plateau. Over 40,000 line km of reflection seismic data make up the southern Exmouth Plateau data base (Fig. 1). Eleven exploration wells were drilled on the Exmouth Plateau between 1979 and 1981. Nearly all of these wells were terminated in Mesozoic sediments, commonly of Neocomian to Triassic age, and the results were discussed by Barber (1982, 1988). Erskine and Vail (1988) used seismic sections from the southern Exmouth Plateau to discuss the sequence stratigraphy and global sea-level correlation of the Neocomian Barrow Group. In addition to these studies on the Exmouth Plateau, a large literature base has accumulated on the seismically and stratigraphically similar sediments found in the adjacent Barrow–Dampier Sub-basin (Kirk, 1985; Veenstra, 1985) and summarized in Purcell and Purcell (1988).

Objectives and methods

The objective of this study is to provide an updated interpretation of the seismic stratigraphy of the southern Exmouth Plateau based upon new data and research which has accumulated since the last regional studies a decade ago (Exon & Willcox, 1978, 1980; Wright & Wheatley, 1979) and using more advanced interpretation methods (Vail *et al.*, 1977; Vail, 1987).

The primary data sources analysed (see Fig. 1 for data coverage) were multi-channel seismic reflection lines from Exon and Willcox (1980), from the unpublished Esso 1978 and 1979 surveys, from the unpublished GSI 1976 Group Shoot, ODP site survey lines used to position sites 762 and 763 and a tie line between the two sites (von Rad *et al.*, 1989) and *Rig Seismic* and *Conrad* seismic lines shot during a joint experiment in 1986 (Exon & Williamson, 1988; Williamson & Falvey, 1988). Record lengths were at least 6 seconds (s) and mostly unmigrated, with Esso data recorded 48-fold, BMR and GSI data 24-fold and ODP data single-fold.

Data analysis employed a seismic/sequence-stratigraphy approach (Vail *et al.*, 1977; Vail, 1987), tied to the Haq *et al.* (1987) time scale. Seismic sequences were defined after Vail *et al.* (1977) as stratigraphic units composed of a relatively conformable succession of genetically related strata bounded at top and bottom by unconformities or their correlative conformities and identified on seismic sections by reflection terminations. Because of the large number found on the Exmouth Plateau, sequences were grouped into eight seismic packages separated by major regional unconformities. These eight packages appear to form natural tectono-stratigraphic units and form the basis of our interpretation.

STRUCTURE

The major tectonic elements of the Exmouth Plateau (Figs 1 & 3) are the western buttress of the Rankin Trend, the Kangaroo Trough which widens to the north and separates the Rankin Trend from the Exmouth Submarine Plateau, the Exmouth Plateau Arch, the rifted western plateau margin and the southern faulted anticline margin (Fig. 4) of the Cape Range fracture zone (Exon & Willcox, 1978; Wright & Wheatley, 1979; Barber, 1982, 1988). Many of these original Mesozoic features have undergone inversion and renewed growth associated with broad folding during the later Cenozoic.

High-angle normal faulting on this megacrustal block dominates the rift onset unconformity and the pre- and syn-rift sediments. Faults follow a broad northeast–southwest or north–south trend paralleling the major structural elements of the plateau and reflecting the pre-existing Precambrian structural trend of the adjacent west Australian cratonic blocks. In contrast, near the southern margin of the plateau, faults follow a northwest–southeast trend (Exon & Willcox, 1980). Faults form an extensive network of horsts and grabens on the south-central Exmouth Plateau. In this region the fault blocks have relatively minor throw which averages less than 200 milliseconds (ms). Closer to the plateau margins and to the Kangaroo Trough, faulting styles change and are characterized by greater throws and increased rotation (Fig. 3). Throw on individual fault blocks is up to 400 ms. Mutter *et al.* (1989) recognize two mid-crustal reflectors which are not penetrated by normal faulting, occur at a depth of around 5 s and are interpreted as detachment surfaces associated with Jurassic rifting. Along the southern plateau margin, the faults swing to a more northwest–southeast trend and exhibit a different style in which seismic packages are difficult to correlate between adjacent faulted blocks and fault surfaces are obscured by numerous diffractions originating from beyond the plane of the section. The location near the Cape Range fracture zone and the faulting style suggest that these are transform faults.

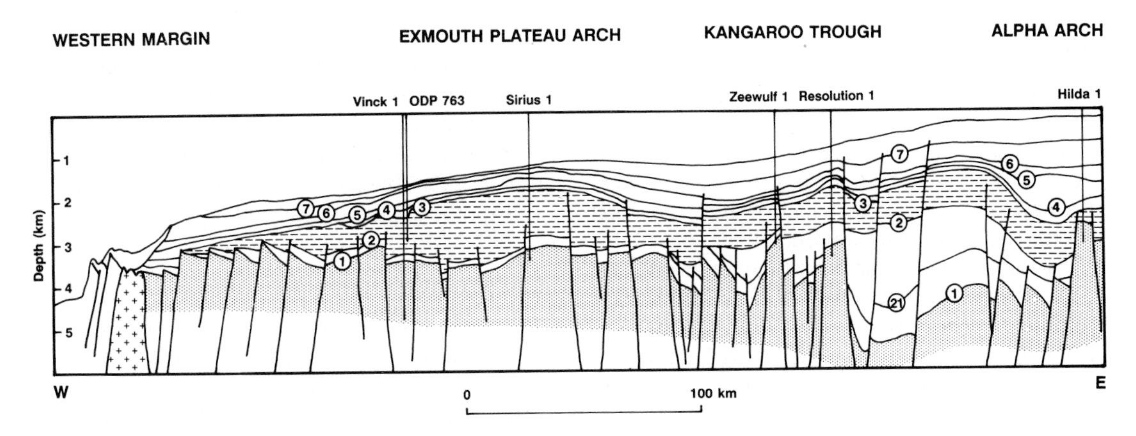

Fig. 3. This E–W cross-section is based on a synthesis of seismic results from this study and those from Exon and Willcox (1980), Barber (1982, 1988) and Mutter *et al.* (1989). The broad tectonic features of the Exmouth Plateau High, the Kangaroo Trough and the Rankin Trend Alpha Arch are clearly seen as well as the normal faulting style across the Exmouth Plateau megacrustal block. Note the thick Jurassic syn-rift sequence in the eastern Kangaroo Trough and its division into late and early Jurassic components separated by the Callovian breakup unconformity. The Barrow–Dampier Sub-basin lies east of the Alpha Arch. Seismic packages discussed in the text are separated by seismic reflectors numbered 1 to 7. Each reflector occurs at the top of its respective package.

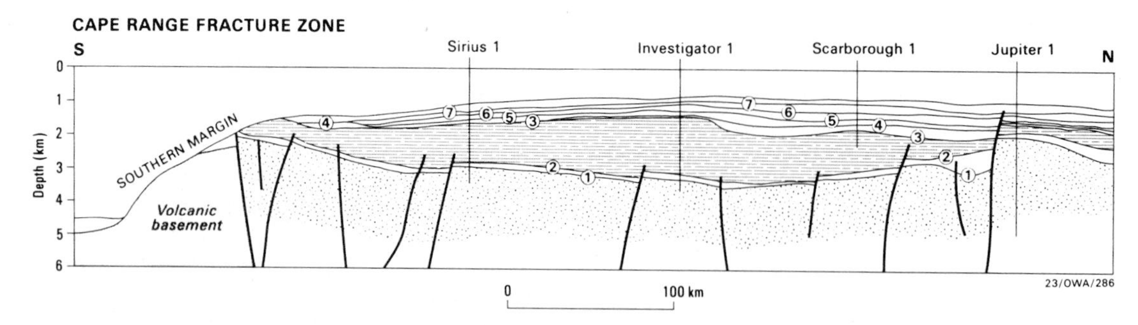

Fig. 4. This N–S cross-section is based on a synthesis of seismic results from this study and those from Exon and Willcox (1980), Barber (1982, 1988) and Erskine and Vail (1988). This N–S cross-section shows the progradation of package 3 away from the southern transform margin and the subsequent uplift and erosion during Hauterivian continental breakup. Packages 3 to 5 thin northward reflecting diminishing clastic supply while packages 6 to 8 thin and onlap southward reflecting the shallower water depths which persisted along the southern margin into the Tertiary. Seismic reflectors are numbered as in Fig. 3.

SEISMIC STRATIGRAPHY

Seismic reflections have been divided into sequences separated by unconformities and their correlative conformities. Several of these unconformities are regional events on the southern Exmouth Plateau and exhibit high amplitude and/or erosional truncation, often associated with significant tectonic events. The seismic sequences have been grouped into packages, separated by the regional unconformities. Packages and their intervening reflectors have been numbered 1 to 8, beginning at the base (e.g. Fig. 3).

Package 1

This package lies below reflector 1 and consistently occurs at or near the top of the major rifted fault blocks (Figs 3 & 4). The base of the package is undefined and extends to the base of the records in most locations. Reflector 1 is often an erosional unconformity that truncates underlying reflections. Where the sequence boundary at the top of package 1 can be clearly identified it is associated with a prominent negative-amplitude anomaly which coincides with a source of numerous hyperbolic diffractions. At other locations, the upper surface of rotated fault blocks has been extensively eroded giving reflector 1 variable amplitude and low continuity. Further west on the plateau, reflector 1 is more conformable with the overlying package 2.

The reflections within package 1 can be grouped into numerous individual seismic sequences. Reflections have highly variable amplitude and frequency together with moderate continuity. Where unobscured by faulting, the sequences in package 1 often alternate between containing high-amplitude–low-frequency seismic facies and low-amplitude–high-frequency seismic facies. Seismic configurations are frequently parallel or slightly divergent and occasionally show low relief, shingled clinoforms (Fig. 5). Occasional reflections exhibit very high amplitudes. Other locations within package 1 exhibit chaotic, low-amplitude, low-continuity configurations of limited lateral extent. Towards the west and northwest the uppermost units of package 1 undergo a change from alternating high-/low-amplitude events, shingled clinoforms and chaotic channelled facies to a more uniform wedge-shaped, low-amplitude sequence. This sequence thickens to the north where it contains occasional high-continuity, high-amplitude reflections. It is the strong negative impedance at the top of this upper sequence that produces the characteristic negative-amplitude anomaly at the upper sequence boundary of package 1.

Interpretation

The parallel/divergent seismic reflection configuration characterized by alternating high/low amplitudes, moderate to good continuity and shingled clinoforms is characteristic of a coastal-plain setting depositing deltaic sediments. The chaotic low-amplitude facies are interpreted as fluvial channels. Particularly high-amplitude intervals are interpreted as coal seams (or at the 5 s two-way travel time (TWT) as possible mylonite zones within the detachment surfaces of Mutter *et al.*, 1989). The

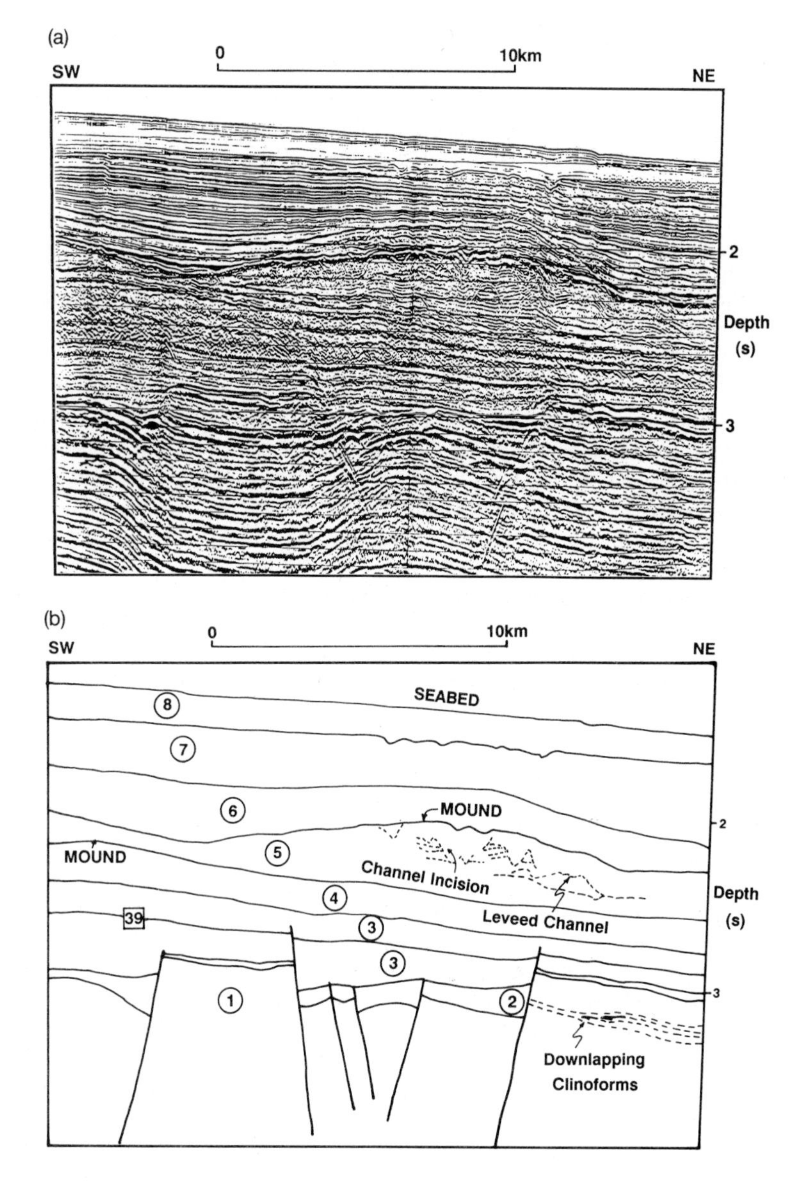

Fig. 5. Horst and graben structures characterize packages 1 and 2 over most of the Exmouth Plateau. Here, downlapping clinoforms can be seen within package 1 (a, uninterpreted; b, interpreted). Thick mounds in package 5 contain extensive channelling and internal bidirectional downlap. Package 8 displays a typical sheet geometry and continuous low-amplitude reflections. The upper unit of package 3 (see Fig. 9) lies above reflector 39. This upper unit is thin to absent over the western Exmouth Plateau (cf. Fig. 7).

upper, northward-thickening transition from alternating high-/low-amplitude events, shingled clinoforms and chaotic channelled facies to a wedge-shaped, more uniform low-amplitude sequence is interpreted to represent the transition from fluvial–deltaic to marine facies.

Package 1 was not penetrated at sites 762 or 763 but it has been penetrated at nearly all other exploration wells on the Exmouth Plateau (Vos & McHattie, 1981; Barber, 1988) and it consistently corresponds to the Mungaroo Formation equivalents and Brigadier Formation of Scythian/Anisian (Crostella & Barter, 1980) to Rhaetian age (Hocking *et al.*, 1988). Dip on shingled clinoforms indicates a southeast source from the Australian continent. The uppermost part of the Mungaroo wedge is a transgressive unit correlative to the Brigadier Formation. Sampling in Vinck 1 (Esso Australia, unpublished well completion report) indicates this upper unit to be a transgressive Rhaetian–Norian marine mudstone and seismic interpretation indicates this grades westward into interbedded marine sandstones and mudstones. The underlying sequences are interpreted from well data as fluvial-deltaic in nature. Northward at the Eendracht 1 site, over 370 m of Norian–Rhaetian marine sediments accumulated in the upper section of package 1. The extension of this transgressive marine unit possibly also correlates to the marine and shelf-edge reef development seen in Rhaetian-age sediments drilled on the Wombat Plateau (Williamson *et al.*, 1989; Shipboard Scientific Party, 1990).

Package 2

This package is thin over most of the southern plateau but thickens dramatically in the southeast corner (Figs 3 & 6). Only one or two seismic sequences can normally be distinguished in package 2 except in the southeast where multiple seismic sequences are developed. Reflector 2 (the upper boundary) and internal reflections of package 2 are often sequence-boundary unconformities or interference composites and have a broad cycle breadth. This, combined with the thinness of the package (often less than 100 ms), usually inhibits detailed seismic-stratigraphic analysis of package 2. The lower sequence is widely distributed on the southern plateau and is commonly only one cycle thick but west of site 763 the transition in package 2 from a single cycle to multiple cycles can be seen (Fig. 7). With this increase in thickness, the lower part of package 2 lies conformably on package 1 and is involved in the block faulting event (e.g. Fig. 6). The upper part of package 2 usually increases in thickness towards the downthrown side of half-grabens and, while involved, appears to post-date the major block faulting event. Internal reflections are rarely discernible in package 2, but where present appear to be discontinuous and of low amplitude. The upper sequence of package 2 thins out of the half-grabens and in places onlaps the elevated fault blocks (e.g. Fig. 8).

In the southeast corner of the Exmouth Plateau, package 1 has undergone significant subsidence in the Kangaroo Trough (Barber, 1988) as a consequence of extensional fault movement (Fig. 6). In this area, reflector 1 is located at depths in excess of 4.5 s and occurs in a depositional trough over 2 s thick (Fig. 3) and filled by several westerly-prograded wedges in package 2. The sequences within these wedges are often characterized by thick seismic facies of low-amplitude reflections, or variable-amplitude low-continuity facies with a clinoform configuration. Sedimentation was confined by uplifted fault blocks on the west side of the Kangaroo Trough, and little of package 2 was able to prograde out of the trough and onto the central plateau.

Interpretation

The complex nature of package 2 suggests syndepositional tectonic activity. The lack of package 2 and erosion of package 1 over the Rankin Trend and fault block crests to the west of the Kangaroo Trough can be attributed to uplift associated with early Jurassic rift shoulders (rift onset unconformity in the sense of Falvey (1974)) and subsequent mid-Jurassic breakup (breakup unconformity or 'main unconformity' of the Barrow–Dampier Sub-basin (Veenstra, 1985)). In contrast, away from the tectonism on the southwest of the Exmouth Plateau in the Investigator site 763 to site 762 region, deposition was thin but continuous during much of the Jurassic (Esso Australia, unpublished well completion reports, Investigator 1, Vinck 1, Eendracht 1), ranging in age from Rhaetian to Tithonian and is time-equivalent to the Dingo Claystone. This indicates that uplift associated with subsequent rifting on the western Exmouth Plateau margin had not yet begun by the late Jurassic. The lower conformable part of package 2 is closely involved with block faulting and is probably pre-Callovian in age. The upper part of package 2 thickens into half-grabens and onlaps emergent fault blocks, suggesting that it is Callovian to Tithonian in age.

Subsidence accompanying deposition of package 2 east of the uplifted Rankin Trend created a large depositional trough to trap Jurassic sediments sourced from the Australian continent to the east (Dingo Claystone equivalents). The eastern Jurassic shelf edge prograded westward, infilling the Kangaroo Trough with over 800 m of late Jurassic sediment above the Callovian breakup unconformity at the Resolution 1 well site.

Package 3

The external geometry of package 3 is a broad tabular wedge (seen in Figs 9 & 10 (Fig. 10 foldout facing p. 594)). Analysis of progradation direction derived from measurements of the orientation of the shelf break (Fig. 9) identifies a source region located to the southeast of the Exmouth Plateau,

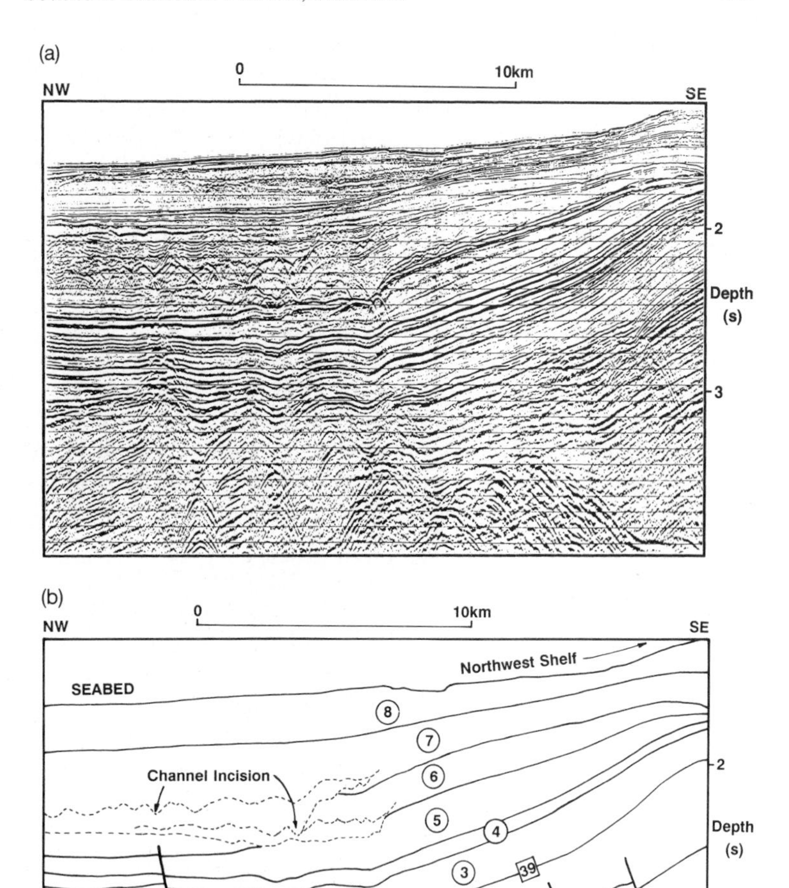

Fig. 6. This NW–SE line from the southeastern Exmouth Plateau is near the Zeewulf well site and shows the steeply rotated Triassic fault blocks of package 1 (a, uninterpreted; b, interpreted). In this region of the Kangaroo Trough, a thick Jurassic graben fill in package 2 has prograded from the east and accumulated against the topographically high fault blocks of package 1. Channels oriented north–south are a prominent feature in packages 6 and 7 indicating that erosion along the base of the prograding northwest shelf and slope (present position shown in upper right-hand corner) has occurred since the late Cretaceous.

coinciding with the intersection of the northwest–southeast trending Cape Range fracture zone and the northwest–southeast trending Cuvier Rift (located south of the Kangaroo Trough). Northwest progradation took place along a broad depositional front, extending the shelf break 150 km northward (Fig. 4) and transporting sediments over 250 km to onlap, as distal marine strata, the uplifted western and northern margins of the present-day plateau and the Australian craton to the east. Successive sequences prograded northeast and northwest away from the sediment source, eventually reaching the Investigator well site and reaching almost as far as site 763 (Figs 7 & 9). A younger, upper sequence was confined to the adjacent Barrow Sub-basin (Tait, 1985) and to the southeast Exmouth Plateau (Figs 5, 6 & 9). Package 3 is thickest in a 50 to 100 km-wide tongue trending northeast from the southern margin in the direction of the Investigator and Scarborough well sites (Fig. 9). Maximum thicknesses of 1.14 s TWT (or 1750 m) occur near the Investigator well site. A depocentre containing over 1.0 s TWT of sediment also occurs in the SE Exmouth Plateau. Following deposition, uplift

(a)

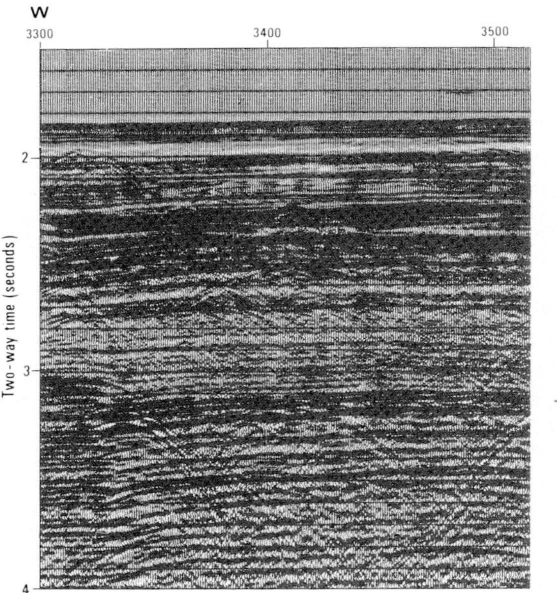
W
3300
3400
3500
2
3
4
Two-way time (seconds)

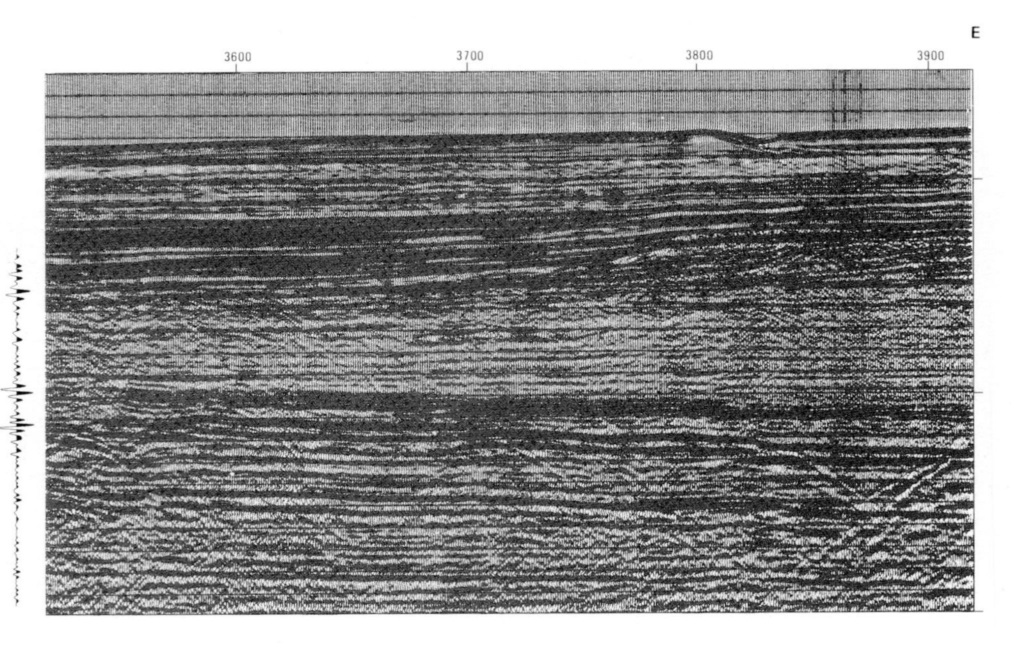
E
3600
3700
3800
3900

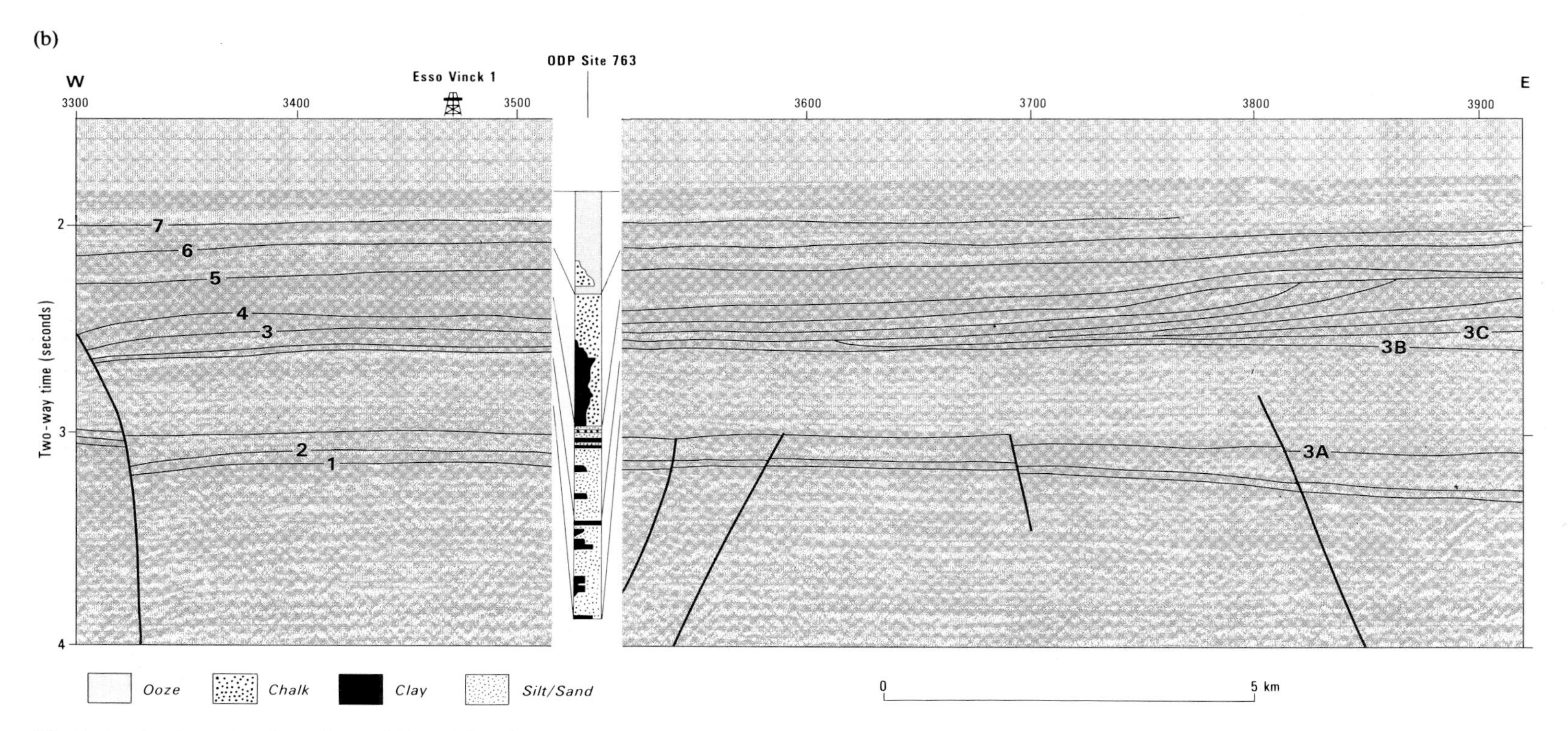

Fig. 7. A seismic section through site 763 and the Vinck 1 well (a, uninterpreted; b, interpreted). All eight seismic packages are present here and separated by reflectors numbered 1 to 7. A simplified lithologic column from site 763 and a synthetic seismogram from the Vinck 1 well site are also presented for comparison. Normal faulting of the rift onset unconformity of reflector 1 (here Rhaetian in age) and thickening of package 2 into the resulting grabens (sps 3500 to 3325) are shown. The age of package 2 here is late Jurassic and post-dates the Callovian breakup unconformity. The upper sequence of package 1 increases in thickness and contains more reflections towards the west. The final progradation of package 3 into this region and its termination less than 10 km east of site 763 (at sp 3790) can also be seen. Packages 4 to 7 thin eastwards onto this depositional platform. Note sea-bed erosion at sp 3825 and similar subsurface features in packages 7 and 8 (e.g. at sp 3310).

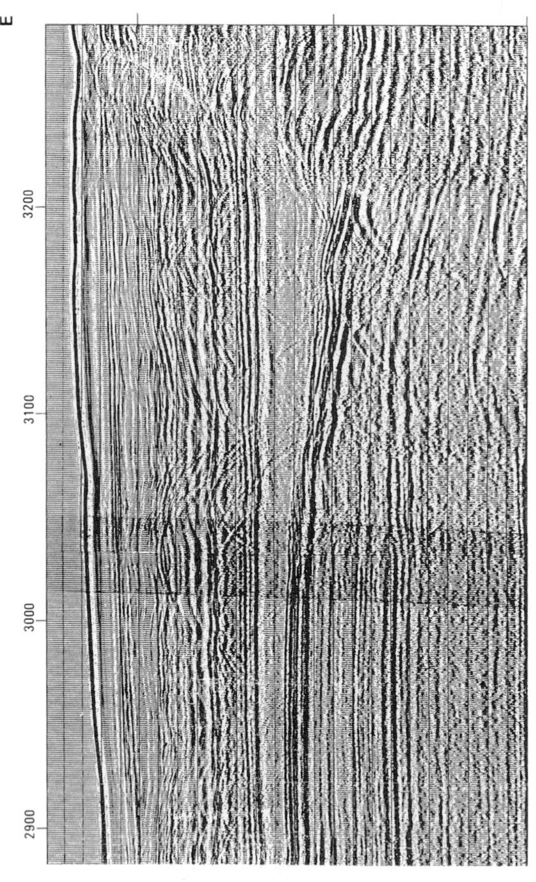
E
3200
3100
3000
2900

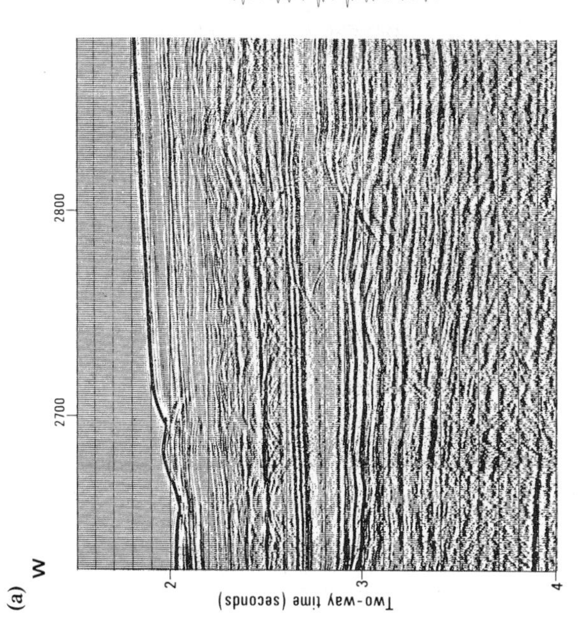
2800
2700
W
(a)
2
3
4
Two-way time (seconds)

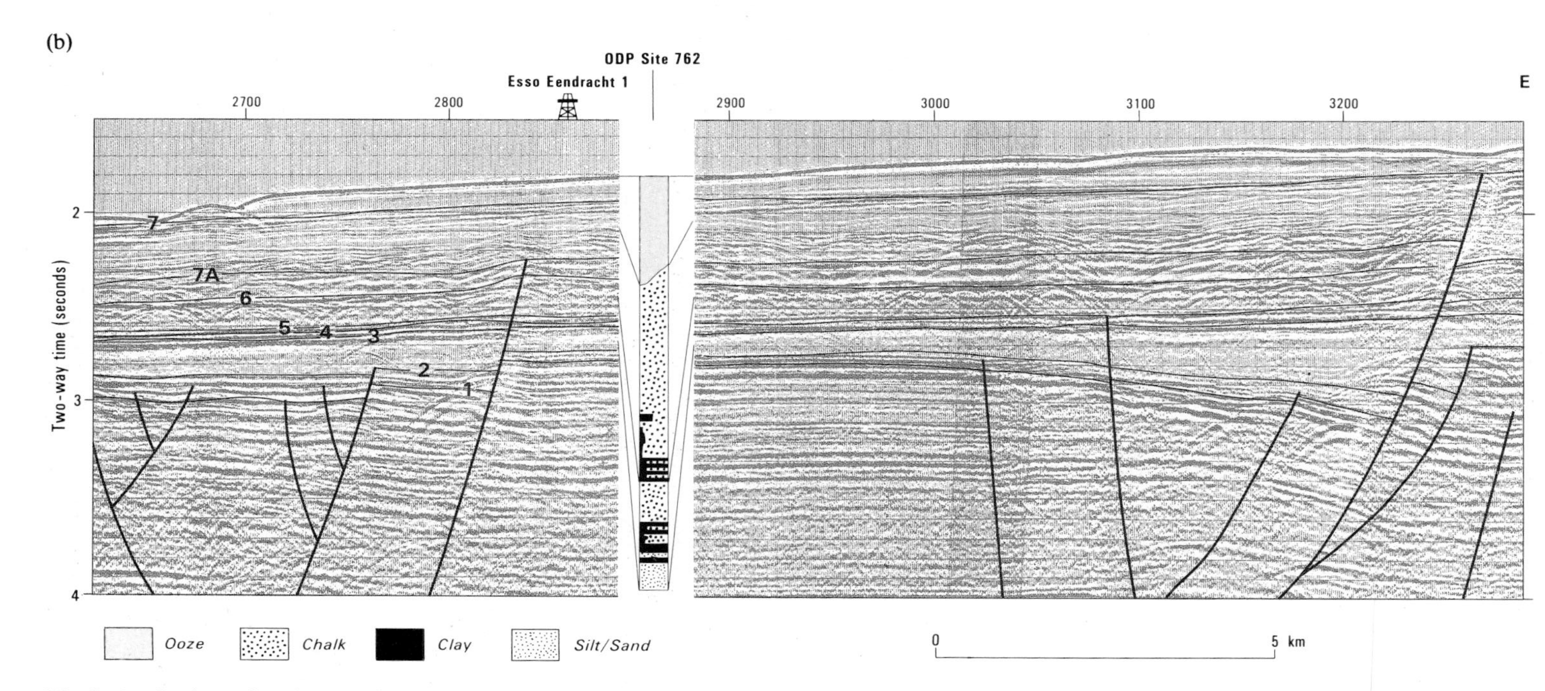

Fig. 8. A seismic section through site 762 and the Eendracht 1 well (a, uninterpreted; b, interpreted). All eight seismic packages are also present here. A simplified lithologic column from site 762 and a synthetic seismogram from the Eendracht 1 well site are presented for comparison. Compared to site 763, throws on normal faults are much greater at site 762 and extend into the Tertiary section (above reflector 6 — K/T boundary). Both the upper sequence of package 1 and all of package 2 are thicker here and contain more reflections than at site 763. Packages 3, 4 and 5 in this distal position are much thinner than at site 763 while package 7 here is much thicker, reflecting higher rates of deep-water Tertiary carbonate accumulation. Reflector 7A here is at the Palaeocene–Eocene boundary and represents an upward transition to gravity-dominated resedimentation shown by the numerous channels in package 7 and the sea-floor erosion in package 8.

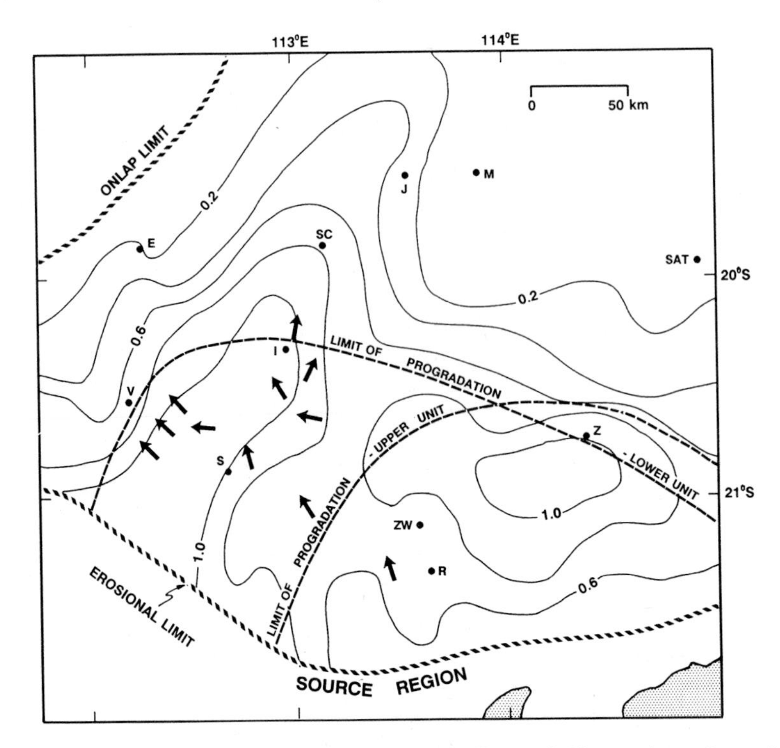

Fig. 9. Isopach of the package 3 Neocomian sediment wedge on the Exmouth Plateau shows that sediment thicknesses of over 1.2 s one-way travel time or over 1700 m occur on the south-central and southeast corner of the plateau. Arrows illustrating the direction of progradation of the continental slope indicate a southern source region centred between the Australian continent and the Cape Range fracture zone extending south into the Gascoyne Sub-basin and not the southern transform margin as suggested by Veevers and Powell (1979) or the adjacent Australian continent as suggested by Boote and Kirk (1989). The Neocomian wedge laps out onto the western margin of the plateau and has been eroded after deposition by uplift along the southern margin during the Hauterivian breakup event. Variations in sediment thickness result from pre-existing topography derived from an earlier Callovian rifting event. The limits of shoreline progradation for the lower unit of package 3 (includes episodes 1 to 4) and the upper unit (episode 5) are shown as dashed lines.

along the southern plateau margin resulted in increasing southward erosion of package 3 from the Airlie 1 well site in the Barrow Sub-basin (Boote & Kirk, 1989) to the western plateau (Exon & Willcox, 1980).

The internal geometry of package 3 shows at least five progradational sequences characterized by an oblique to sigmoidal offlap style with downlap onto the basal boundary and toplap on the upper boundary (Fig. 10). Several of these sequences contain basin-restricted onlapping wedges (as shown also in Mitchum, 1985; Erskine & Vail, 1988) and major incised channels landward. Several onlapping wedge tops are high-amplitude reflections of good lateral continuity basinward of the depositional shelf break and often display a mounded geometry in both depositional strike and dip directions. The uppermost sequence is thickest in the southeast (e.g. Figs 5 & 6) but occupies only a single cycle over much of the depositional platform. In the east, the shoreline of this sequence regresses beyond the previous shelf break (Fig. 9) and is associated with a basinward onlapping wedge. A thick lens of low amplitude, low continuity and chaotic seismic reflections extends westward from the wedge, parallel to the older shelf break.

Age determinations in package 3 are based on a dinoflagellate zonation (Helby *et al.*, 1988). Reflector 2 is an unconformity over much of the Exmouth Plateau, separating late Jurassic from Berriasian sediments. However, in the SE corner the stratigraphic relationship may be conformable

(Hocking *et al.*, 1988) and the base of package 3 extends into the Tithonian (*Pseudoceratium iehiense* zone of Helby *et al.*, 1988). The basal hiatus increases in the direction of northwest progradation. The final progradational episode is Valanginian in age and ends in the *Systematophora areolata* zone (Kopsen & McGann, 1985; Brenner, 1992).

Interpretation

Seismic and stratigraphic interpretation and correlation for package 3 were carried out at ODP sites and nearby wells using synthetic seismograms (e.g. Figs 7 & 8), well-log ties (e.g. Fig. 11) and physical properties (e.g. Fig. 12). Package 3 is seismically (Kirk, 1985) and age-equivalent to the Barrow Group (Hocking *et al.*, 1988) of the Barrow Rift Basin. During the Berriasian, thermal and tectonic subsidence and sediment loading (Swift *et al.*, 1988) allowed over 1700 m of sediment to accumulate in less than 6 Ma. Neglecting compaction effects, each progradational event was characterized by a relatively low depositional front with less than 300 m of relief, indicating an average subsidence rate of at least 300 m per Ma — a very rapid rate compared to most basins (Galloway, 1989). The depositional wedge of package 3 contains all the major physiographic provinces of an Atlantic-style continental margin (in the sense of Kennett, 1982) including a coastal plain, continental shelf, slope and rise consisting of coalesced submarine fans. In most locations the coastal plain passes directly onto a continental slope/prodelta without the development of an extensive continental shelf (Fig. 10).

Thickness distribution is mainly controlled by palaeotopography (Fig. 4). Package 3 is thinner over the old Jurassic rift flanks of the Kangaroo Trough, Barrow–Dampier Sub-basin and adjacent to the Rankin Platform. Package 3 also thins against the Neocomian rift flanks of the current plateau margins. Thick depocentres accumulated over the central and southern plateau and in the axis of the Barrow Sub-basin west of Barrow Island. Although numerous examples of reworked Jurassic and Triassic palynomorphs were found in package 3 sediments at both of sites 762 and 763 (Brenner, 1992) no reflectors interpreted as pre-Cretaceous were seen to outcrop on the Exmouth Plateau, implying a source for package 3 beyond the boundaries of the present Exmouth Plateau. Deposition ended in the Valanginian, prior to the Hauterivian breakup event (Fig. 2). Uplift and subsequent erosion of the

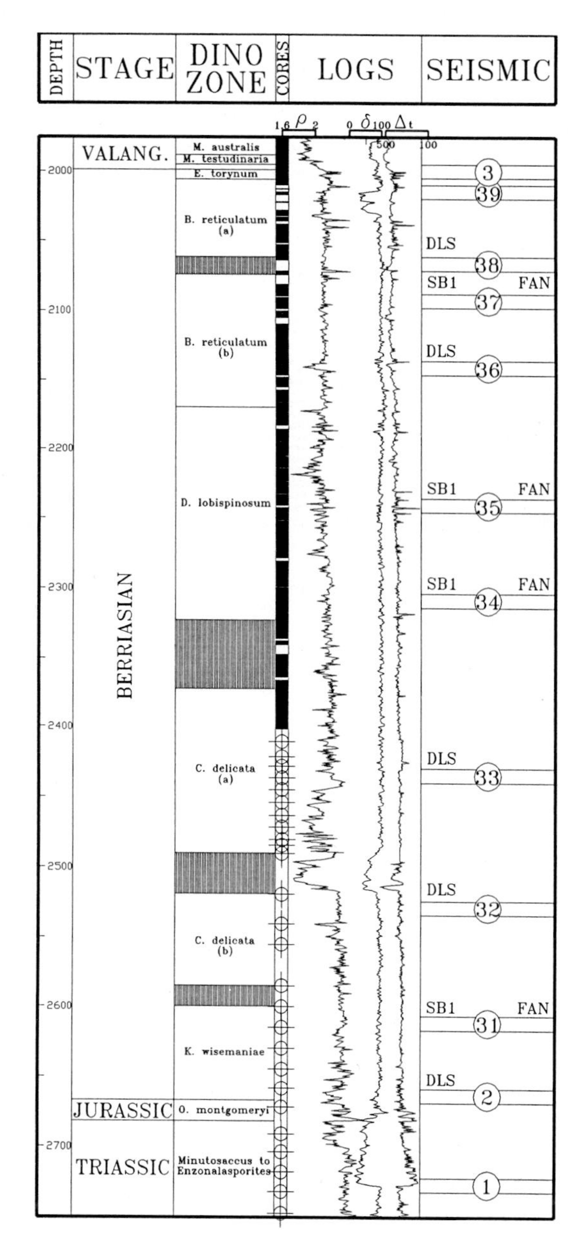

Fig. 11. This figure shows a composite of core and log data from site 763 and the adjacent Vinck 1 well site. Continuous core is shown as a solid block and sidewall cores as crosses. Log data consist of density-log (left), gamma-log (centre) and interval transit time (right). Seismic reflectors correlate with those of Fig. 10. This correlation together with Fig. 12 and others from adjacent site 762 and Investigator 1 well sites enable the seismic stratigraphy from lines such as those shown in Fig. 10 to be verified with borehole lithostratigraphy and chronostratigraphy. SB1, interpreted type 1 sequence boundary; DLS, downlap surface.

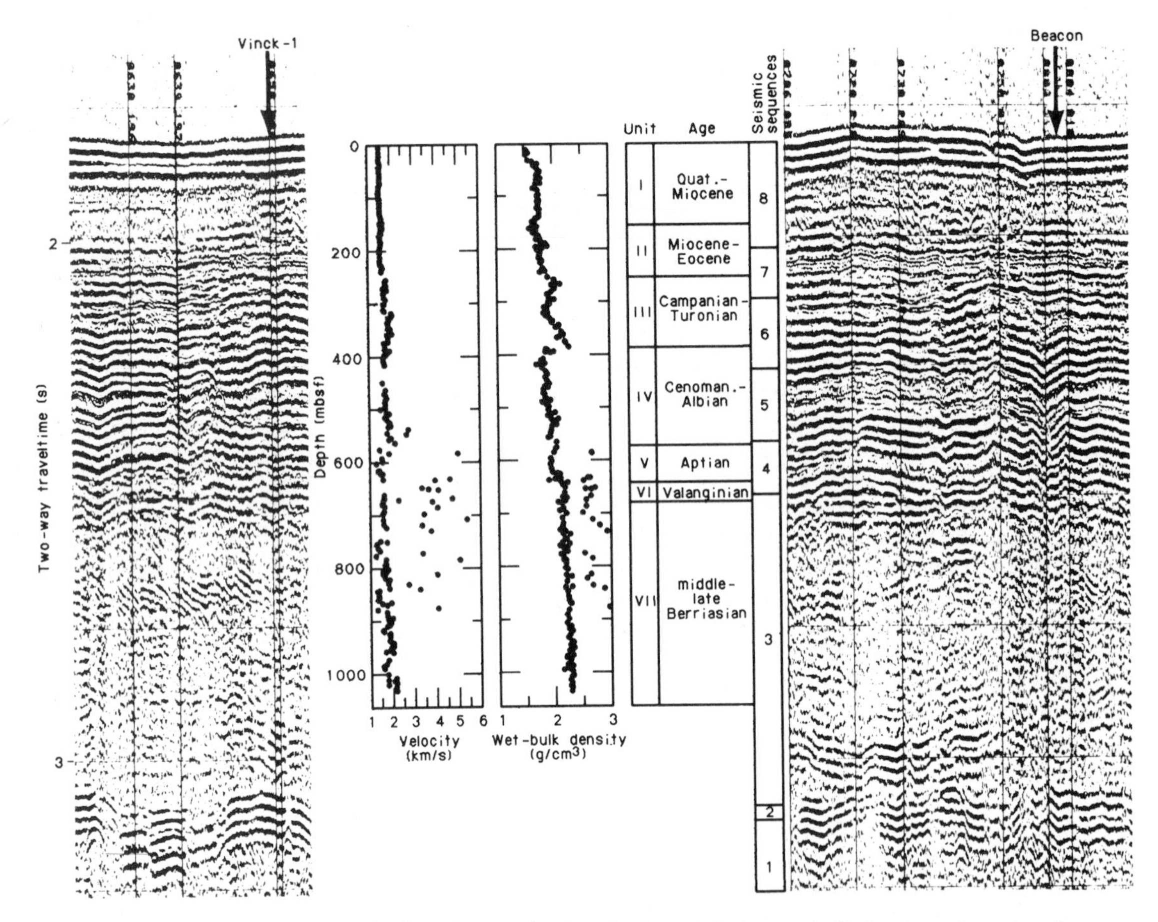

Fig. 12. Correspondence between seismic packages, seismic reflections, velocity, wet-bulk density and stratigraphy at site 763. Also shown is the location of the adjacent Vinck 1 well site. This figure clearly shows the separation of Aptian–Berriasian clastics (variable velocity and density) and the overlying Albian to Quaternary marls and oozes. Seismic packages are not depth-corrected and therefore do not exactly match the well-derived physical properties and lithology. (See Fig. 7 for synthetic seismogram correlation.)

southern margin as India slid past Australia during and after breakup, as suggested by Veevers and Powell (1979), was not the source of the majority of package 3 sediments.

Packages 4 and 5

Packages 4 and 5 are of a relatively similar nature and distribution over the Exmouth Plateau and are discussed here together. They have a broad blanket or sheet geometry 100 to 300 ms thick and are present over most of the southern Exmouth Plateau (Figs 3 & 4). They are thickest in the east where package 5 expands to over 900 ms 5 thickness near the Zeepard 1 well and occurs as a mounded wedge (Fig. 5), prograded northwestward from near Barrow Island. Internally, this wedge contains large overlapping mounds whose bases show bidirectional downlap. These mounds extend laterally for at least 30 km and contain over 800 ms of sediment thickness. Packages 4 and 5 onlap the southern and western margins of the plateau and are locally removed by slumping along the western margin. Both packages are thin over the depositional platform of package 3. In the Barrow–Dampier Subbasin the basal part of package 4 displays an onlapping relationship to package 3 (Kirk, 1985; Kopsen & McGann, 1985) and this relationship is duplicated over much of the Exmouth Plateau.

Reflector 5 at the top of package 5 is an erosional

unconformity sequence boundary of high amplitude and widespread occurrence over all of the southern Exmouth Plateau (e.g. Fig. 5). It has a low-frequency high-amplitude character, particularly on the outer plateau where it is often the most prominent reflection on the post-rift section. It frequently truncates underlying reflections, often with 100 to 150 m of relief over a distance of 5 to 10 km — sufficient occasionally to remove all sediment down to the top of package 3.

The internal reflection configurations of packages 4 and 5 vary widely. Where the sequences are thinner, the internal reflections are commonly of low frequency, high amplitude and high continuity in a sheet geometry. Internal reflections in both packages are disrupted by frequent small-scale diffractions. Some channels are present in the distal mound of package 5 (Fig. 5), but reflections are typically low amplitude or alternating low/high amplitude, variable frequency and low–medium continuity. In the proximal location, internal reflections in the mounded accumulation of package 5 are persistent and of alternating high to low amplitude with little disruption.

Interpretation

Precise dating of packages 4 and 5 from sites 762 and 763 (Shipboard Scientific Party, 1990) and correlation to other commercial wells indicate that they occupy the intervals from Hauterivian–Aptian and Albian–Cenomanian, respectively. Package 4 is thus equivalent to the Muderong Shale and package 5 is equivalent to the Haycock Marl lithological units of the Carnarvon Basin (Hocking *et al.*, 1988).

The distribution of packages 4 and 5 indicates a dominant eastern source during all of this interval except in the southwest where, away from the package 3 depositional platform, package 4 forms a small, but well-defined, depositional wedge 150 ms thick, 15 to 20 km wide, with a flat top and internal clinoforms. This wedge prograded from the southern margin after uplift and truncation of package 3 and is considered coeval with the Hauterivian westward movement of Greater India along the Cape Range fracture zone. This confirms the pre-breakup age of deposition for package 3, the synchronous occurrence of package 3 uplift and erosion with breakup and the concurrent deposition of package 4.

The primary eastern source for packages 4 and 5 suggests a distinct change of sediment source during and after breakup from the eastern end of the southern transform margin (package 3) back to the Australian craton by package 5. Deposition across the highs of the Rankin Trend indicates submergence by the mid-Cretaceous of all antecedent relief derived from the Jurassic rifting event. Onlap against the southern and western plateau margins indicates that these areas remained topographically elevated through the middle Cretaceous.

Parallel, continuous reflection configuration of alternating amplitude in packages 4 and 5 suggests deposition in a low-energy outer-shelf setting over much of the plateau and this is confirmed by the identification of a Cenomanian shelf edge located along the southeast margin of the plateau 20 km west of the Zeepard 1 well and 25 km east of the Zeewulf 1 site. Sediment deposition was thin and blanket-like in package 4 and was controlled by the uplifted western and southern plateau margins and the antecedent platform topography inherited from package 3. Package 5 is dominated by northeast progradation of a lobe-like continental margin onto the east-central Exmouth Plateau along a corridor between the Zeewulf 1 and Saturn 1 sites and centred around the Zeepard 1 site (see Fig. 1 for well locations).

In such a distal setting it is difficult to interpret relative sea-level response but deposition during the basal section of package 4 clearly occupies a rise in sea level indicated by coastal onlap and transgression over the entire Exmouth Plateau in late Valanginian to early–middle Hauterivian time.

Reflector 5 occurs at the Cenomanian–Turonian (C/T) boundary and is marked by black-shale deposition (Shipboard Scientific Party, 1990). Hence it correlates with the worldwide C/T boundary event presumed to mark an oxygen minimum greater than 1000 m deep in the world oceans and related to increased surface productivity (Arthur *et al.*, 1987). On the Australian northwest shelf the C/T event is also considered to represent a significant carbonate dissolution event during a relative rise rather than a lowstand of sea level (Apthorp, 1979). However, the removal of up to 150 m of section in many locations on the Exmouth Plateau and the presence within the black shale of reworked palynomorphs of various ages from Cenomanian to Albian (Brenner, 1992) indicate that a period of physical erosion related to enhanced current activity in deeper water accompanied the chemical event. The time represented by reflector 5 is coeval with separation of Antarctica from Australia and the erosional unconformity may represent submarine erosion resulting from enhanced oceanic circulation at that time.

Packages 6, 7 and 8

Like packages 4 and 5, packages 6 to 8 (inclusive) are of a relatively similar nature and distribution over the Exmouth Plateau, and are discussed here together. Packages 6 to 8 are found over virtually all of the Exmouth Plateau and most often display a conformable drape relationship with reflector 5 and with each other (Figs 2 & 3). They are thickest along the eastern margin of the plateau in a belt which follows the trend of the Kangaroo Trough along the margin of the northwest shelf (Fig. 6). Here these packages are over 1.3 s thick and make up the majority of the section above package 1. Packages 6 to 8 thin in the central Exmouth Plateau over the arch of package 3 and the mound of package 5. They also thin on the margins of the plateau as a result of slumping and mass movement. Along the eastern plateau margin, packages 6 to 8 display a thick prograding wedge geometry with sigmoid-to-oblique offlap and growth to within 200 m of present sea level (distal margin seen in Fig. 6). Downlapping clinoforms of package 6 occur on the eastern plateau across the older package 5 mound. Extensive channelling in package 7 (Fig. 6) occurred in topographic lows controlled by the persisting positive relief of package 5.

Package 6 is often composed of high-amplitude continuous reflections with a parallel configuration. Package 7 is composed of alternating high-/low-amplitude, high-/low-frequency seismic facies which are frequently disrupted by high-amplitude reflections that truncate the underlying reflections of package 6 (Fig. 6). Truncation and erosion here is often extensive and severe, extending as deep as reflector 5 and occasionally incising it. Channelling seems to be intense in specific zones such as at the base of the northwest shelf where the channel fill is up to 1 s thick and individual channels are up to 200 ms deep. Channel orientation seems to parallel the present shelf edge and may connect to large submarine canyons such as Swan Canyon along the plateau margins. Package 8 is characterized by low-amplitude high-frequency reflections that have variable continuity (Fig. 5). Reflector 8 is a prominent reflector between packages 7 and 8 where it marks a change from erosion and channelling below to mostly parallel reflections above. Package 8 has fewer large channels but there is evidence of large-scale slumping, particularly along the plateau margins both within the package and on the present sea floor.

Interpretation

The low-amplitude variable frequency/continuity of the reflections, their presence seaward of the shelf break and the blanket-like geometry of packages 6 and 8 indicate a low-energy deep-water depositional environment probably with a fine-grained lithology. Internal reflector character of packages 6 and 8 on the central plateau is very continuous and, apart from episodes of submarine channelling, reflects deposition by suspension fallout. Correlation of seismic reflectors to sites 762 and 763 indicates that package 6 is Turonian to Maastrichtian in age, package 7 is Palaeocene to middle Miocene in age and package 8 is Miocene to Holocene. Lithologies in all these packages at the ODP sites are dominated by deep-water carbonate chalks and oozes (Shipboard Scientific Party, 1990). The cessation of significant terrigenous input to the Exmouth Plateau and the transformation to a deepening biogenic-dominated depositional environment occurs above the Cenomanian/Turonian erosional event at the boundary between packages 5 and 6, as documented elsewhere throughout the region at this time (Apthorp, 1979). Sediments deposited after this event were sourced mainly from the eastern shelf edge which continued to be located close to its position during package 5 deposition. A second source came from pelagic carbonates generated over the plateau itself. Reflector 6 corresponds to an erosional event around the Cretaceous–Tertiary boundary, and reflector 7 to another unconformity or hiatus of mid-Miocene age. Both these events have been previously recognized in the region (Apthorp, 1979, 1988; Quilty, 1980).

Several locations on the present sea bed, particularly on the plateau margins, have experienced mid–late Cenozoic erosion and mass movement. This pattern may reflect a trend of increasing submarine slopes in response to broad folding on the Exmouth Plateau Arch and associated post-Miocene movement. The topographic relief on package 5 and hence available accommodation indicates that water depths over most of the central Exmouth Plateau at the beginning of the Turonian were at least 300 m. Lapout relationships of packages 6 to 8 on the Exmouth Plateau margins suggest that the western margin subsided during the Tertiary and was progressively covered by deep-water carbonates. In contrast, the southern margin remained high and slow subsidence, possibly as a result of volcanic buttressing, resulted in gradual onlap with thin deposition only extending to the margin crest after the Miocene in package 8.

The small diffractions which disrupt the otherwise continuous reflections in package 6 (Figs 7 & 8) correlate with a zone of high gas concentration in cores at both ODP sites and the diffractions are considered to originate from gas-saturated sediments sourced via faults from the underlying section. This interpretation is supported by frequent direct hydrocarbon indicators such as flat spots in packages 1 and 3, and by the presence of gas chimneys in the region (also reported in Wright & Wheatley, 1979).

DISCUSSION

The deposition of seismic packages 1 to 8 on the southern Exmouth Plateau of the passive northwest Australian margin records a tectonic history consisting of initial Triassic–Jurassic deposition and rifting, early Cretaceous rifting and continental breakup, rift to drift transition, mature ocean-phase sedimentation and a final transition towards collision tectonics. On the Exmouth Plateau, packages 1 to 8 consist of three major clastic depositional wedges and a carbonate blanket deposit. Analysis of the wealth of seismic detail in packages 1 to 8, constrained by excellent chronology provided by continuous core from two ODP sites and broader comparison with 11 industry exploration wells, enables a useful model of passive-margin evolution to be constructed for this region.

Stage 1: intracratonic sedimentation

This model begins with eastern Gondwanaland providing a southerly source of sediments into the western Tethys around palaeolatitude 45 to 50° south (wedge number 1 of Boote & Kirk, 1989). Beginning prior to the Norian and continuing to the Rhaetian, a cratonic downwarp accumulated over 3000 m of sediments, represented on the southern Exmouth Plateau by package 1, over a previously rifted and thinned slab of continental crust (Williamson & Falvey, 1988; Williamson *et al.*, 1990). The early phase of this thick accumulation was fluvial-deltaic, northwestward prograding and arenaceous in nature, but by late Norian to Rhaetian had become primarily a transgressive marine mudstone (base of wedge 2 of Boote & Kirk, 1989). To the north on the Wombat Plateau, this transition began earlier and by Rhaetian time had begun to deposit limestones in reef and back reef settings (Shipboard Scientific Party, 1990).

Stage 2: rift onset and initial breakup

The end of the Triassic coincided with the end of the first depositional wedge and the beginning of a period of sub-crustal heating, isostatic uplift and erosion or non-deposition over most of the early Jurassic on the Exmouth Plateau, leading to normal faulting and the development of a rift onset unconformity. Because of erosion or non-deposition over much of the Exmouth Plateau, the rift onset unconformity is typically developed on a Rhaetian surface. Off northwest Australia narrow, north–south-oriented zones of extensional faulting and subsidence developed the rift valleys of the Barrow–Dampier Sub-basin and the Kangaroo Trough, separated by a structural high over the Rankin Platform. A thick sequence of syn-rift clastics (wedge 2 of Boote & Kirk, 1989) infilled the Kangaroo Trough from an eastern source and makes up the basal fill of package 2 in this area. The rift onset unconformity is best developed over the flanks of the rift valleys and their northward extension into the offshore Canning Basin, for example, on the Wombat Plateau. The period of early–middle Jurassic extension culminated in a Bathonian–Callovian rifting event and a Callovian breakup event along the present northern margin of the Exmouth Plateau, when south Tibet and Burma broke away from northwest Australia and formed Tethys III (Audley-Charles, 1988). Localized subsidence over the eastern Exmouth Plateau in the former rift valley sites produced a second, eastern-sourced clastic infill event (wedge 3 of Boote & Kirk, 1989) of late Jurassic age in package 2 of the Barrow–Dampier Sub-basin and the Kangaroo Trough. This second infill phase in package 2 is separated from the first by a breakup unconformity (Veenstra, 1985; Barber, 1988). The remainder of the Exmouth Plateau remained high throughout the Jurassic deposition of package 2 and either underwent erosion of Triassic strata (e.g. on the eastern rift valley flanks), accumulated an intermittent Jurassic sequence punctuated by non-deposition or erosion (the

Saturn 1 and Mercury 1 well sites (Barber, 1988)) or accumulated a condensed but continuous section in shallow basins such as at the Eendracht 1 well site. Although only a thin section accumulated in such sites as Eendracht 1, both rift onset and breakup unconformities can be seen in package 2, the former conformably involved in block faulting, the latter infilling grabens and onlapping the flanks of uplifted blocks.

Stage 3: second rift to final breakup

The Callovian breakup event was unable to propagate southward along the Australian margin and terminated at the Wombat Plateau, in a location coinciding with the offshore extension of the southern boundary of the Canning Basin. Like many other examples on passive margins (e.g. the Jean d'Arc and Whale Basins of the Canadian Grand Banks in Grant & McAlpine (1990)) where a failed breakup event results in the filling of the inshore rift valley and a subsequent shift to a new seaward rifting location, the Barrow–Dampier Sub-basin/Kangaroo Trough was infilled and abandoned. Heating and extension then shifted to the Gascoyne and Cuvier Rifts at the western margin of the Exmouth Plateau and southward towards the Perth Basin. Major uplift occurred southeast of the Exmouth Plateau in the Gascoyne Sub-basin of the Carnarvon Basin (where Hocking *et al.*, 1988, noted an extended period of erosion) and not on the Australian craton to the southeast (Boote & Kirk, 1989) or the southern Exmouth Plateau transform margin (Veevers & Powell, 1979). A major depositional wedge (cf. number 4 of Boote & Kirk, 1989) prograded from the Gascoyne Sub-basin source northeast into the Barrow–Dampier Sub-basin and northwest across the Exmouth Plateau. Both areas were actively subsiding and accumulated 1000 to 1750 m of clastic infill in package 3 between the Tithonian and Valanginian.

Stage 4: post-breakup and rift to drift transition

The southern clastic source was terminated by Hauterivian continental breakup of Greater India along the Gascoyne Rift to the west and the Cuvier Rift to the south. The oldest magnetic anomalies in both basins are of M10 (early Hauterivian) age (Fullerton *et al.*, 1989), the earliest sediments deposited at site 766 (Gradstein *et al.*, 1989) in the Gascoyne Abyssal Plain are syn-rift Valanginian clastics and the earliest sediments at DSDP site 263 (Veevers *et al.*, 1974) in the Cuvier Abyssal Plain are post-rift Aptian–Albian neritic claystones. At sites 762 and 763, a hiatus or unconformity of two dinoflagellate zones occurs at the Valanginian–Hauterivian boundary in a stratigraphic position comparable to the breakup unconformity of Falvey (1974). Along the southern margin, package 3 sediments are disrupted by transform faults and volcanic intrusions, indicating their pre-breakup age. The two divergent margins of the Cuvier and West Exmouth Rifts were connected by the 400 km-long Cape Range fracture zone. Valanginian–Hauterivian uplift along this fracture zone resulted in extensive erosion of earlier Neocomian sediments in package 3 and deposited the small 100 to 200 m-thick clastic wedge of package 4, 15 km northward across the southwestern Exmouth Plateau. As the spreading centre moved away and the two plates uncoupled, widespread subsidence began elsewhere and provided a location for deposition of a blanket of post-breakup, fine-grained clastics in packages 4 and 5, again sourced mainly from eastern cratonic areas. The dominance of clastic sediment supply waned from Hauterivian to Aptian times following an M4 (Hauterivian–Barremian) ridge jump in the Cuvier Abyssal Plain (Fullerton *et al.*, 1989) which occurred at the same time as plate uncoupling along the Cape Range transform and signals the rift to drift transition. Continued post-breakup subsidence and westward sediment supply in package 5 enabled progradation of a 900 m-thick sediment wedge (cf. number 5 of Boote & Kirk, 1989) onto the eastern Exmouth Plateau from Aptian to Cenomanian. The remainder of the Exmouth Plateau accumulated a sheet drape of claystones, chalks and marls several hundred metres thick, primarily from the eastern continental or pelagic source but also from a secondary source along the southern margin. During this time both the southern and western plateau margins remained high and were onlapped by the sediments of package 5.

Stage 5: mature ocean to incipient collision

The Cenomanian/Turonian anoxic and dissolution event brought an end to clastic deposition on the southern Exmouth Plateau. The southern margin source for clastic supply, shown by downlapping seismic reflections in package 5, ceased and shifted to a pattern of nannofossil/foraminiferal ooze ac-

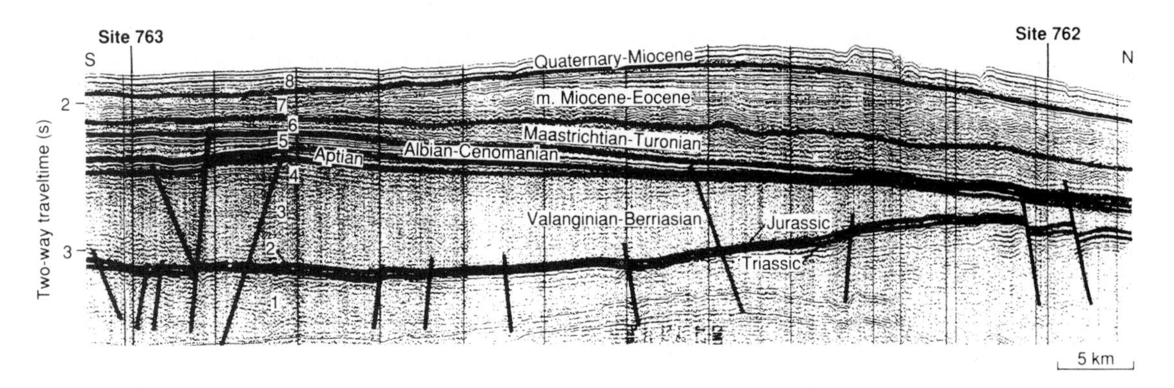

Fig. 13. Seismic correlation between sites 763 and 762 showing the northward thinning of mainly clastic sediments in packages 3 to 5 and the northward thickening of deep-water carbonate sediments in sequences 6 to 8. Southward onlap can be seen at the base of packages 6 and 7. Channelling is common above the base of package 7 and slumping at the sea bed can be seen south from site 762.

cumulation and progressive southward onlap of seismic reflections in packages 6 and 7 (e.g. Figs 4 & 13). The eastern margin also ceased to supply clastic sediments to the Exmouth Plateau and instead switched to carbonate progradation which slowly spread a carbonate wedge (post-wedge 5 of Boote & Kirk, 1989) westward from the Rankin Platform onto the Exmouth Plateau from Turonian to Holocene times. This latest wedge reached over 1000 m thick on the eastern plateau margins beneath which the Kangaroo Trough deeply subsided. Elsewhere, deeper water carbonates accumulated in a blanket deposit, onlapping and finally covering the southern and western margins by Miocene times. Beginning in the early Cenozoic with subduction in the Timor Trench, the passive northwest Australian margin began to collide with the Indonesian crustal collage. This has continued to the present day with additional obduction in the Timor area and collision of the Australian plate in the area of the Java Trench occurring in middle Miocene and again in the Pliocene (Kopsen & McGann, 1985; Howell, 1988). The major structural effect of this collision 1000 km to the north has been to reactivate earlier faults with possible extensional wrenching. A second effect has been to initiate broad folding, particularly during the Miocene which resulted in the elevation of the Exmouth Plateau Arch and depression of the Kangaroo Trough. In the deeper water of the Exmouth Plateau submarine slopes steepened as folding continued which in turn produced resedimentation of water-saturated, low shear-strength oozes resulting in extensive submarine channelling and slumping, seen especially in early Eocene sediments but also continuing up to the present sea floor (Fig. 13). The Miocene collision event correlates with a depositional hiatus across the Exmouth Plateau, the Wombat Plateau and the northwest shelf.

CONCLUSIONS

1 A tectonic–stratigraphic evolution of the southern Exmouth Plateau has been determined from the analysis of eight packages of seismic data, correlated with ODP drilling results and regional correlations. This evolution contains: (i) intracratonic sedimentation (Norian–Rhaetian); (ii) rift onset and initial breakup (Hettangian–Callovian); (iii) second rift to final breakup (Callovian–Hauterivian); (iv) post-breakup and rift to drift transition (Hauterivian–Cenomanian); and ending with a phase of (v) mature ocean to incipient collision (Turonian–Holocene).

2 The source of Tithonian–Valanginian-age sediments of package 3 is clearly indicated to be from the Gascoyne Sub-basin. Progradation of package 3 was contemporaneous with rifting and pre-breakup, and did not owe its origin to uplift along the Cape Range fracture zone during sea-floor spreading. Package 3 was dominated by Berriasian sediments on the southwest Exmouth Plateau and, unlike the Barrow Sub-basin, accumulated few Valanginian-age sediments (cf. Erskine & Vail, 1988), apart from a band of westward-transported contourites. The overall geometry of package 3 is that of a thin, but complete, prograding coastal plain to basin floor physiography with episodes of gravity-fed sedimentation.

3 The clear age of breakup on the southern Exmouth Plateau is Hauterivian and corresponds with a period of intense uplift and the progradation of a previously undescribed sediment wedge in package 4.

4 Sediment supply shifted from a northward-prograding clastic source to a carbonate-dominated southward-onlapping blanket around the Cenomanian/Turonian boundary.

5 Folding related to collision of the Australian and Indonesian crustal collages further north increased slopes on the southern Exmouth Plateau beginning in the Eocene and produced widespread submarine erosion and resedimentation in the water-saturated Cenozoic oozes.

ACKNOWLEDGEMENTS

We would like to acknowledge the assistance given in data collection while at sea on the *Joides Resolution* and the *Rig Seismic.* We would like to thank Esso Australia for access to proprietary seismic and well data and Rod Erskine at Exxon Production Research Company and Henry Posamentier and David James at Esso Canada for provision of seismic data and useful discussions. Two useful reviews were provided by M. Cucci and J. Lindsay. David Boote and Henry Posamentier provided many helpful suggestions. Robin Helby, Wolfram Brenner and Tim Bralower provided data and clarification of biostratigraphic problems. We would also like to acknowledge the Bureau of Mineral Resources for drafting assistance, funding provided by a Canadian NSERC CSP grant to R. Boyd, an Alexander von Humboldt Fellowship provided to R. Boyd and facilities provided by Geomar Institute, Kiel, Germany while R. Boyd was on sabbatical leave.

REFERENCES

Apthorp, M.C. (1979) Depositional history and palaeogeography of the Upper Cretaceous of the Northwest Shelf based on foraminifera. *APEA J.* **19**, 74–89.

Apthorp, M.C. (1988) Cainozoic depositional history of the Northwest Shelf. In: *The Northwest Shelf, Australia* (Ed. Purcell, P.G. & Purcell, R.R.) Proc. Petrol. Explor. Soc. Austral. Symp. 55–84.

Arthur, M.A., Schlanger, S.O. & Jenkyns, H.C. (1987) The Cenomanian–Turonian oceanic anoxic event, II. Paleoceanographic controls on organic matter production and preservation. In: *Marine Petroleum Source Rocks* (Eds Brooks, J & Fleet, A.J.) Spec. Publ. Geol. Soc. London, 26, 401–420.

Audley-Charles, M.G. (1988) Evolution of the southern margin of Tethys (north Australian region) from early Permian to late Cretaceous. In: *Gondwana and Tethys* (Eds Audley-Charles, M.G. & Hallam, A.) Spec. Publ. Geol. Soc. 37, 79–100.

Barber, P.M. (1982) Palaeotectonic evolution of the central Exmouth Plateau. *APEA J.* **22** (1), 131–144.

Barber, P.M. (1988) The Exmouth Plateau deep water frontier: a case history. In: *The Northwest Shelf, Australia* (Eds Purcell, P.G. & Purcell, R.R.) Proc. Petrol. Explor. Soc. Austral. Symp. 173–187.

Boote, D.R.D. & Kirk, R.B. (1989) Depositional wedge cycles on an evolving plate margin, western and northwestern Australia. *Bull. Am. Assoc. Petrol. Geol.* **73** (2), 216–243.

Brenner, W. (1992) Palynological analysis of the early Cretaceous sequence in Site 762 and 763, Exmouth Plateau, NW Australia. In: *Proc. ODP Init. Repts* (Eds Haq, B.U., von Rad, U., O'Connell, S., *et al.*) **122**, College Station, TX (Ocean Drilling Program), 511–528.

Crostella, A. & Barter, T.P. (1980) Triassic–Jurassic depositional history of the Dampier and Beagle Subbasins, Northwest Shelf of Australia. *APEA J.* **20** (1), 25–33.

Erskine, R.D. & Vail, P.R. (1988) Seismic stratigraphy of the Exmouth Plateau. In: *Atlas of Seismic Stratigraphy* (Ed. Bally, A.W.) Am. Assoc. Petrol. Geol. Stud. Geol. 27(2), 163–173.

Exon, N.F. & Willcox, J.B. (1978) Geology and petroleum potential of the Exmouth Plateau area off Western Australia. *Am. Assoc. Petrol. Geol. Bull.* **62** (1), 40–72.

Exon, N.F. & Willcox, J.B. (1980) The Exmouth Plateau: stratigraphy, structure and petroleum potential. *BMR Austral. Bull.* **199**, 52 pp.

Exon, N.F. & Williamson, P.E. (1988) Sedimentary framework of the northern and western Exmouth Plateau. *Bur. Mineral Resour. Geol. Geophys. Rec.* 1988/30.

Exon, N.F., Von Rad, U. & Von Stackelberg, U. (1982) The geological development of the passive margins of the Exmouth Plateau off northwest Australia. *Mar. Geol.* **47**, 131–152.

Falvey, D.A. (1974) The development of continental margins in plate tectonic theory. *APEA J.* **14** (2), 95–106.

Fullerton, L.G., Sager, W.W. & Handschumacher, D.W. (1989) Late Jurassic-early Cretaceous evolution of the eastern Indian Ocean adjacent to northwest Australia. *J. Geophys. Res.* **94**, B3, 2937–2953.

Galloway, W.E. (1989) Genetic stratigraphic sequences in basin analysis I: architecture and genesis of flooding-surface bounded depositional units. *Am. Assoc. Petrol. Geol. Bull.* **73** (2), 125–142.

Gradstein, F., Ludden, J. & Leg 123 Shipboard Scientific Party (1989) The birth of the Indian Ocean. *Nature* **337**, 506–507.

Grant, A.C. & McAlpine, K.D. (1990) The continental

margin around Newfoundland. In: *Geology of the Continental Margin of Eastern Canada*, Chapt 6 (Eds Keen, M.J. & Williams, G.L.) Geol. Surv. Can. Geol. Can. 2, 239–292 (also Geol. Soc. Am. Geol. North Am. vI-1.

Haq, B.U., Hardenbol, J. & Vail, P.R. (1987) The chronology of fluctuating sea level since the Triassic. *Science* **235**, 1156–1167.

Helby, R., Morgan, R. & Partridge, A.D. (1988) A palynological zonation of the Australian Mesozoic. In: *Studies in Australian Mesozoic Palynology* (Ed. Jell, P.A.) Mem. Austral. Assoc. Palaeontol. 4, 1–85.

Hocking, R.M., Moors, H.T. & Van de Graaf, W.G.E. (1988) The geology of the Carnarvon Basin, Western Australia. *W. Austral. Geol. Surv. Bull.* **133**.

Howell, E.A. (1988) The Harriet oilfield. In: *The Northwest Shelf, Australia* (Eds Purcell, P.G. & Purcell, R.R.) Proc. Petrol. Explor. Soc. Austral. Symp. 391–401.

Kennett, J.P. (1982) *Marine Geology.* Prentice-Hall, Englewood Cliffs, 813pp.

Kirk, R.B. (1985) A seismic stratigraphic case study in the Eastern Barrow Subbasin, Northwest Shelf, Australia. In: *Seismic Stratigraphy II* (Eds Berg, O.R. & Wolverton, D.G.) Mem. Am. Assoc. Petrol. Geol. 39, 183–207.

Kopsen, E. & McGann, G. (1985) A review of the hydrocarbon habitat of the eastern and central Barrow-Dampier sub-basin, Western Australia. *APEA J.* **25** (1), 154–176.

Mitchum, R.M. (1985) Seismic stratigraphic expression of submarine fans. In: *Seismic Stratigraphy II* (Eds Berg, O.R. & Wolverton, D.G.) Mem. Am. Assoc. Petrol. Geol. 39, 117–138.

Mutter, J.C., Larson, R.L. & NW Australia Study Group (1989) Extension of the Exmouth Plateau: deep seismic reflection/refraction evidence of simple and pure shear mechanisms. *Geology* **17** (1), 15–18.

Purcell, P.G. & Purcell, R.R. (Eds) (1988) *The Northwest Shelf, Australia.* Proc. Petrol. Explor. Soc. Austral. Symp.

Quilty, P.G. (1980) Sedimentation cycles in the Cretaceous and Cenozoic of Western Australia. *Tectonophysics* **63**, 349–366.

Shipboard Scientific Party (1990) In: *Proc. ODP Init. Repts.* (Eds Haq, B.U., von Rad, U., O'Connell, S., *et al.*) **122**, College Station, TX (Ocean Drilling Program), 826pp.

Swift, M.G., Stagg, H.M.J. & Falvey, D.A. (1988) Heatflow regime and implications for oil maturation and migration in the offshore northern Carnarvon Basin. In: *The Northwest Shelf, Australia* (Eds Purcell, P.G. & Purcell, R.R.) Proc. Petrol. Explor. Soc. Austral. Symp., 539–552.

Tait, A.M. (1985) A depositional model for the Dupuy Member and the Barrow Group in the Barrow Subbasin, Northwestern Australia. *APEA J.* **25** (1), 282–290.

Vail, P.R. (1987) Seismic stratigraphic interpretation procedure. In: *Atlas of Seismic Stratigraphy* (Ed. Bally, A.W.) Am. Assoc. Petrol. Geol. Stud. Geol. 27(2), 1–10.

Vail, P.R. Mitchum, R.M., Jr, Todd, R.G. *et al.* (1977) Seismic stratigraphy and global changes of sea level. In: *Seismic Stratigraphy—Applications to Hydrocarbon Exploration* (Ed. Payton, C.E.) Mem. Am. Assoc. Petrol. Geol. 26, 49–212.

Veenstra, E. (1985) Rift and drift in the Dampier Subbasin, a seismic and structural interpretation. *APEA J.* **25** (1), 177–189.

Veevers, J.J. & Powell, C.M. (1979) Sedimentary wedge progradation from transform-faulted continental rim: southern Exmouth Plateau, Western Australia. *Bull. Am. Assoc. Petrol. Geol.* **63** (11), 2088–2104.

Veevers, J.J., Heirtzler, J.R., *et al.* (1974) *Initial Reports of the Deep Sea Drilling Project,* 27, Washington (U.S. Government Printing Office), 1060pp.

Von Rad, U. & Exon, N.F. (1983) Mesozoic–Cenozoic sedimentary and volcanic evolution of the starved passive margin off northwest Australia. Mem. Am. Assoc. Petrol. Geol. **34**, 253–281.

Von Rad, U. & ODP Leg 122 and 123 Shipboard Scientific Parties (1989) Triassic to Cenozoic evolution of the NW Australian continental margin and the birth of the Indian Ocean (preliminary results of ODP legs 122 and 123). *Sonderdruck Geol. Rundschau* **78** (3), 1189–1210.

Von Rad, U., Schott, M., Exon, N.F., Mutterlose, J., Quilty, P.G. & Thurow, J. (1990) Mesozoic sedimentary and volcanic rocks dredged from the northern Exmouth Plateau: petrography and microfacies. *Bur. Min. Resourc. J. Austral. Geol. Geophys.* **11** (4) 449–476.

Vos, R.G. & McHattie, C.M. (1981) Upper Triassic depositional environments, Central Exmouth Plateau (Permit WA-84-P) Northwestern Australia. Paper presented at the *Fifth Australian Geological Convention*, 17–21 August, 1981.

Williamson, P.E. & Falvey, D.A. (1988) Preliminary post cruise report—Rig Seismic cruises 7 and 8: deep seismic structure of the Exmouth Plateau. *Bur. Mineral Resourc. Rec.* 88/31.

Williamson, P.E., Exon, N.F., Haq, B.U., Von Rad, U., O'Connell, S. & ODP Leg 122 Shipboard Scientific Party (1989) A Northwest Shelf Triassic reef play: results from ODP Leg 122. *APEA J.* **29** (1), 328–334.

Williamson, P.E., Swift, M.G., Kravis, S.P., Falvey, D.A. & Brassil, F. (1990) Permo-Carboniferous rifting of the Exmouth Plateau region, Australia: an intermediate plate model. In: Potential of Deep Seismic Profiling for Hydrocarbon Exploration (Eds Pinot, B. & Bois, C.) Editions Technip, Paris, 237–248.

Wright, A.J. & Wheatley, T.J. (1979) Trapping mechanisms and hydrocarbon potential of the Exmouth Plateau, Western Australia. *APEA J.* **19** (1), 19–29.

Spec. Publs Int. Ass. Sediment. (1993) **18**, 605–631

Application of sequence stratigraphy in an intracratonic setting, Amadeus Basin, central Australia

J.F. LINDSAY, J.M. KENNARD *and* P.N. SOUTHGATE

Australian Geological Survey Organization, GPO Box 378, Canberra, ACT 2601, Australia

ABSTRACT

On the Australian continent, seismic data within broad, shallow intracratonic basins are not always suited to conventional seismic-stratigraphic analysis. Sequence geometries are compressed (vertically), stratal terminations are commonly not evident and many sequences are at or below seismic resolution. Consequently, a new understanding is necessary to apply sequence concepts to sedimentary successions within intracratonic basins.

The Amadeus Basin is a broad intracratonic depression that comprises a thick succession of late Proterozoic and early Palaeozoic deposits, and is typical of a number of basins that developed on the Australian craton at that time. Using a combination of seismic interpretation, outcrop and well-log facies analysis and biostratigraphy, a sequence framework was established for this basin. This framework indicates that the regional depositional patterns were primarily controlled by major tectonic events (two episodes of crustal extension at about 900 and 600 Ma) and that the effects of eustatic sea-level change were superimposed upon these broad regional controls to produce depositional sequences and determine the distribution of facies.

Five sequences of latest Proterozoic–Cambrian age are studied in detail to demonstrate the character and composition of sequences in an intracratonic setting and to assess the variables controlling their development. Slow subsidence rates, low depositional slopes and shallow water depths collectively result in diminished sediment accommodation. Thus, sequences are vertically compressed and erosional unconformities generally subdued. Because depositional slopes are low, seismic terminations are rarely evident and deposits commonly lack progradational geometries. In consequence, sequence boundaries and internal systems tracts can generally only be identified by facies discontinuities, biostratigraphic hiatuses and regional (subaerial or marine) erosion surfaces. Lowstand deposits are poorly developed, incised-valley deposits are areally restricted and many sequences comprise stacked transgressive/highstand deposits that are separated by near-planar unconformities or paraconformities.

Subsidence and eustasy are the principal controlling variables that determined the distribution and geometry of sequences in intracratonic basins. Eustasy by its definition is common to all basins, whilst subsidence occurred in response to tectonic events of cratonic proportions and resulted in near-synchronous development of several basins across central Australia. As a consequence, a predictable depositional pattern occurs within these ancient intracratonic basins.

INTRODUCTION

Recent advances in our understanding of sequence stratigraphy have changed the way we analyse sedimentary basins (Sloss, 1988). The underlying premise of sequence analysis is that sediments are deposited as units (sequences) of genetically related lithofacies bounded by unconformities (sequence boundaries) (Vail *et al.*, 1977a, b, 1984; Vail, 1987; Van Wagoner *et al.*, 1988, 1990). The technique provides a means for subdividing an apparently complex sedimentary succession into packages (sequences) that correspond to chronostratigraphically constrained depositional intervals. Sequence analysis can be used to develop a basin framework to constrain depositional models and assess basin models. However, the techniques used in sequence interpretation have for the most part been developed on passive-margin settings in relatively young sedimentary successions. Here large prograding sequences can be readily discerned so that geometry, the distribution of stratal terminations, sequence boundaries and systems tracts can readily be identified.

On the Australian craton we are faced with a very different problem. A number of broad, shallow intracratonic basins developed across the continent in the late Proterozoic and early Palaeozoic. Seismic data gathered in these basins appear almost featureless and suggest layer-cake stratigraphy. Clearly, the conventional passive-margin seismic models are difficult to apply in these settings. In this paper we explore the application of sequence concepts to intracratonic settings and apply them to the late Proterozoic to early Palaeozoic succession of the Amadeus Basin in central Australia.

SEQUENCE STRATIGRAPHY IN INTRACRATONIC SETTINGS

Whilst seismic data quality in the Amadeus Basin is generally high, seismic sections are not suited to conventional stratigraphic analysis: stratal terminations are seldom visible, seismically defined units commonly have a simple layer-cake geometry and the present limits of the basin are structural and erosional rather than depositional. The first two features reflect the intracratonic character of the Amadeus Basin where marked changes in depositional slope (such as the shelf–slope break or the slope–basin-floor break of continental margins) are only locally developed through time. Furthermore, the basin margins, where stratal terminations might be expected to be most prominent, have been destroyed by overthrusting and erosion. Consequently, an alternative approach is necessary to apply sequence concepts in this, and other, intracratonic basins.

The problem faced in intracratonic settings is not that sequences do not exist or that eustasy does not affect sedimentation. The problem is the geometrical expression of the sequences. Intracratonic basins are extremely broad and shallow — the Amadeus Basin, for example, is 500 × 800 km (Fig. 1) and probably seldom had water depths of more than a few tens of metres. Sedimentation rates were close to subsidence rates and sediments were generally

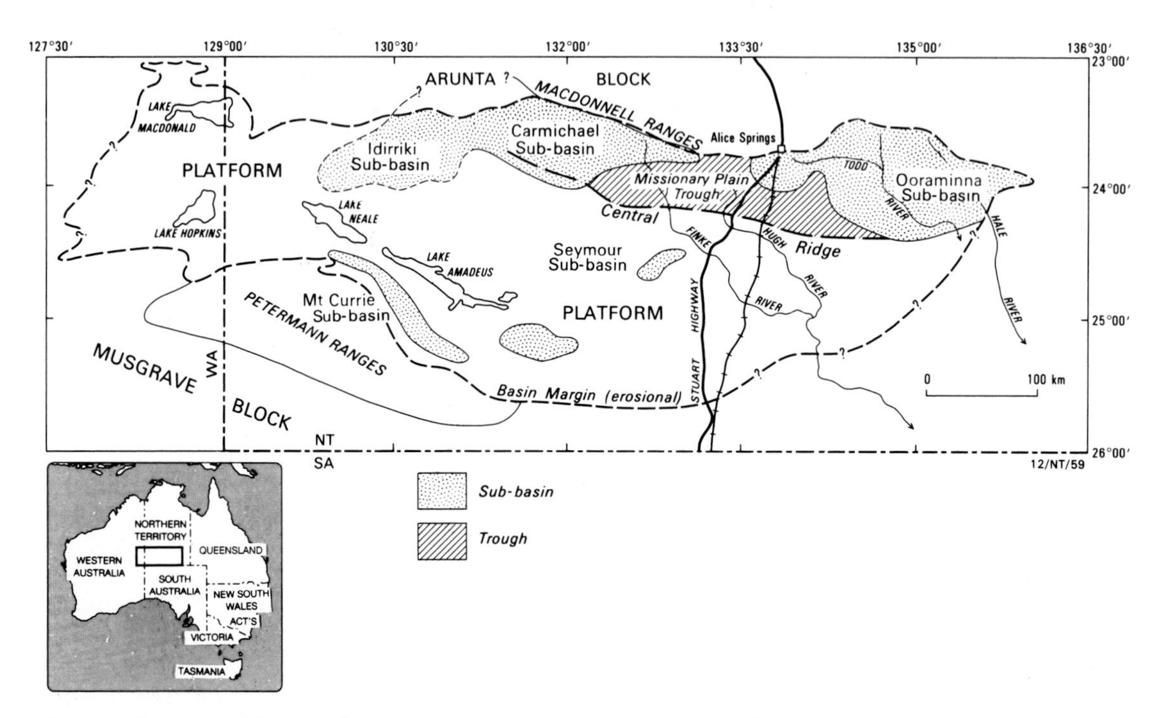

Fig. 1. Main structural features of the Amadeus Basin (sub-basins, trough, platforms and ridges). (After Lindsay & Korsch, 1989, 1991.)

deposited in a shallow-marine or fluvial setting. The resultant sequences are thus extremes of a ramp setting; they form extensive thin sheets that seldom show evidence of progradation and are frequently at or close to seismic resolution. Seismic geometry is thus difficult to use in defining sequences.

In spite of the differences between passive-margin and intracratonic settings, we might expect the lessons learnt on passive margins to be generally applicable. We would expect the various systems tracts to be present within intracratonic sequences, but their geometry (e.g. relative proportions and overall dimensions) may be different. For example, in the absence of a shelf–slope break (continental margin), is it possible to distinguish between type 1 and type 2 sequences? What criteria are needed to recognize lowstand deposits or shelf-margin wedges (if present)? If lowstand deposits are absent, do flooding surfaces coincide with sequence boundaries? All of these require resolution if sequence concepts are applicable in an intracratonic setting. Once these geometrical concepts are understood, they can be used as a guide to identify the sequence and systems tract boundaries. Although throughgoing surfaces are generally prominent on seismic sections, geometric criteria are frequently lacking that permit the sequence boundaries to be separated from flooding and downlap surfaces.

IDENTIFICATION OF SEQUENCE BOUNDARIES

Sequence and systems tract boundaries can be identified in several ways:

1 Stratal terminations. The terminations of cross-cutting strata are the most obvious means of seismically identifying sequence boundaries and other major surfaces. They are visible locally on seismic sections in the Amadeus Basin, especially in the terminal Proterozoic and earliest Cambrian sequences where rapid differential subsidence created steeper depositional slopes along the margins of sub-basins.

2 Discontinuities in facies. Sequence boundaries develop when there is a downward shift in relative sea level followed by a rise in relative sea level. The surfaces are erosional in their upslope extremes and the whole facies package is shifted basinward. In passive-margin settings, basinward shifts of facies are recorded by lowstand deposits such that shallow-water or even non-marine sediments commonly rest directly over the previous highstand systems tract. In contrast, lowstand deposits are not as well developed in intracratonic settings and the sequence boundary is generally marked by a facies discontinuity across a marine flooding surface. Thus, in both settings sequence boundaries can be identified from discontinuities in facies (e.g. Grotzinger, 1986; Lindsay, 1987a, 1989; von der Borch *et al.*, 1988; Lindsay & Korsch, 1989, 1991). The value of facies interpretation is that it can be applied to outcrop and well data, to deformed successions where the seismic method is unsuitable and to ancient successions where biostratigraphy lacks sufficient resolution or is not available. Within any one depositional sequence, sediments can be related to a specific set of depositional processes that operated within well-constrained time limits. That is, the distribution of facies within a single sequence is predictable. In contrast, facies changes across sequences are commonly (but not always) abrupt. These abrupt facies discontinuities can thus be used as an adjunct to seismic interpretation because they provide a means of defining sequence boundaries and help identify key reflectors in the seismic section.

3 Hiatuses indicated by biostratigraphy. Time breaks in the stratigraphic record may be identified by biostratigraphy and may also lead to the recognition of sequence boundaries. It should be noted, however, that not all sequence boundaries record significant time breaks (many hiatuses are below the resolution of the fauna/flora used, particularly in the Palaeozoic), and this method obviously cannot be applied in Proterozoic successions where fossils are absent.

4 Regional subaerial or marine erosion surfaces. These surfaces may separate similar or contrasting facies associations and are generally readily identified in the field. Such surfaces are of particular relevance to intracratonic basins where similar depositional systems from successive sequences (typically highstand systems tracts) are commonly stacked on top of one another and can be differentiated only by the recognition of intervening local erosion surfaces.

5 Systematic changes in parasequence stacking may also help elucidate systems tracts and sequence boundaries.

These five criteria may also help identify downlap and toplap surfaces which bound depositional systems tracts within sequences.

Time (ma)	SEQUENCE STRATIGRAPHY	SYSTEMS TRACTS AND ONLAP CURVES	BASIN DYNAMICS		SEDIMENTOLOGY
	BREWER CONGLOMERATE HERMANNSBURG SANDSTONE PARKE SILTSTONE		STAGE 3 COMPRESSION	*Foreland basin. Advance of major thrust sheets. Shortening of basin.*	*Major coarsening upward syntectonic sequences. Mostly terrestrial detrital sediments.*
400	MEREENIE SANDSTONE CARMICHAEL SANDSTONE				
	STOKES SILTSTONE, STAIRWAY SANDSTONE — FIVE SEQUENCES		STAGE 2 EXTENSION	*Rapidly decreasing subsidence rates. Broad regional subsidence continues and expands.*	*Shallow marine generally tidal, clastic sedimentation. Depositional sequences thinning abruptly below seismic resolution.*
	HORN VALLEY SILTSTONE	HST			
	PACOOTA SANDSTONE 4	LST			
	PACOOTA SANDSTONE 3	HST			
500	PACOOTA SANDSTONE 2	HST			
	PACOOTA SANDSTONE 1	HST			
	UPPER GOYDER FORMATION				
	PETERMANN SST, L. GOYDER FM, JAY CK LST, CLELAND SST, DECEPTION FM, HUGH RIVER SHALE, U SHANNON FM	HST		*Broad regional subsidence beyond sub-basins. Growth on Central Ridge.*	*Shallow marine, carbonates in east and fluvial and deltaic clastics in west.*
	ILLARA SST, L SHANNON FM	LST			
	TEMPE FM, HUGH RIVER SHALE, GILES CREEK DOLOMITE	HST			*Major evaporite/carbonate sequence.*
	CHANDLER FORMATION	LST			
	NAMATJIRA FORMATION, TODD RIVER DOLOMITE	HST		*Extension followed by rapid subsidence to form major sub-basins along northern margin of basin.*	*Major shallowing upward clastic sequences develop in sub-basins. Non-deposition or erosion elsewhere.*
	ARUMBERA SANDSTONE 2	LST			
	ARUMBERA SANDSTONE 1	HST / LST			
600	JULIE FORMATION PERTATATAKA FORMATION	HST	STAGE 1 EXTENSION	*Sub-basins become clearly defined.*	*Major clastic depositional sequences relating in part to sea-level effects due to continental glaciation.*
	PIONEER SANDSTONE, OLYMPIC FORMATION, UPPER ININDIA BEDS	LST			
700	ARALKA FORMATION (LOWER ININDIA BEDS)	HST			
	AREYONGA FORMATION (LOWER ININDIA BEDS)	LST			
800					
	BITTER SPRINGS FORMATION — LOVES CREEK MEMBER	HST		*Increased subsidence with earliest evidence of sub-basins.*	*Major depositional evaporite and carbonate sequence. Minor mafic volcanics.*
	BITTER SPRINGS FORMATION — GILLEN MEMBER	LST			
	HEAVITREE QUARTZITE (SEVEN SEQUENCES)			*Relatively slow regional thermal subsidence.*	*Shallow-marine tidal clastic sequences.*
	ARUNTA COMPLEX; BLOODS RANGE BEDS; MT HARRIS BASALT			*Extension, bimodal volcanism.*	*Rift sequence? Fluvial clastic and bimodal volcanics.*

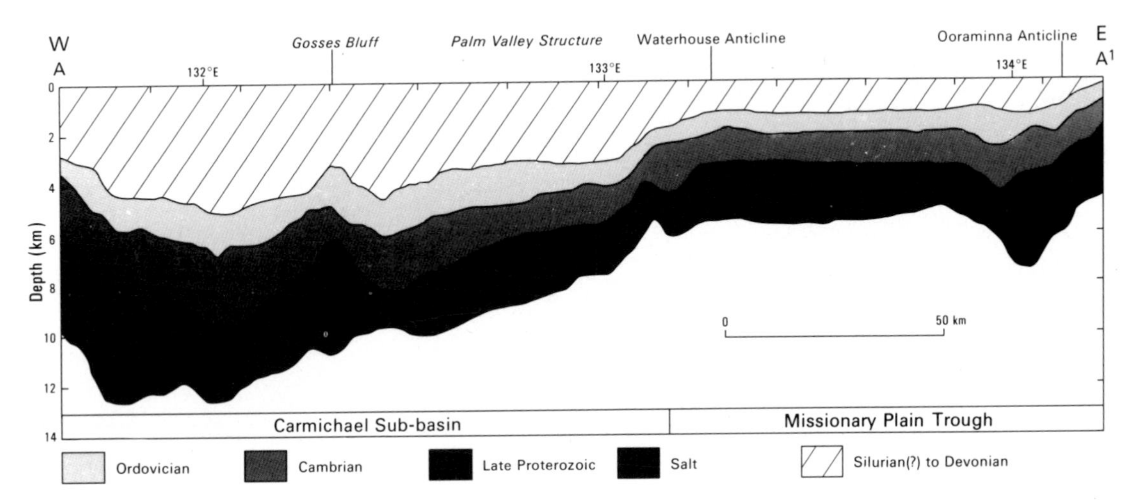

Fig. 3. East–west cross-section through the Carmichael Sub-basin and along the axis of the Missionary Plain Trough. Note the thickening of the salt beneath the major structures.

THE AMADEUS BASIN

A number of broad, shallow intracratonic depressions formed on the Australian craton during the late Proterozoic and early Palaeozoic. The development of these basins almost certainly relates to the breakup of a Proterozoic supercontinent, and basin dynamics appears to be largely tied to this global tectonic event (Lindsay *et al.*, 1987). The Amadeus Basin is typical of these intracratonic basins (Fig. 1) and contains a varied late Proterozoic to mid-Palaeozoic succession. The lithostratigraphy is summarized by Wells *et al.* (1970). More recent detailed sequence studies can be found in Lindsay (1987a, 1989), Lindsay and Korsch (1989, 1991), Gorter (1991), Kennard and Lindsay (1991) and Southgate (1991). The stratigraphy shown in Fig. 2 is simplified but includes the major units discussed in this paper.

The basin can be subdivided into several distinct morphological features (Fig. 1). Major sub-basins along the northern margin of the basin (Ooraminna, Carmichael and Idirriki Sub-basins) are connected by a shallow trough or saddle (Missionary Plain Trough), and are separated from a much larger platform area to the south and west by a low ridge (the Central Ridge). At times this ridge acted as a barrier to sedimentation. The sub-basins contain up to 14 km of late Proterozoic to mid-Palaeozoic rocks, whereas the platform areas typically have less than 5 km of sedimentary section (Fig. 3).

The basin evolved in three major stages (Fig. 4; Lindsay & Korsch, 1989, 1991). Stage 1 began at about 900 Ma with extensional thinning of the crust and the formation of half-grabens. Thermal subsidence was followed by a long post-thermal phase of minimal subsidence during which sedimentation was probably controlled by small-scale tectonic events (Lindsay & Korsch, 1991; Shaw, 1991). Subsequently, a major compressional event (the latest Proterozoic Petermann Ranges Orogeny) affected the southern margin of the basin, and a less intense phase of crustal extension (stage 2) occurred in the northern portion of the basin at about 600 Ma. Stage 2 subsidence was followed by a major compressional event beginning at about 450 Ma (stage 3) in which major southward-directed thrust sheets caused progressive downward flexing of the northern margin of the basin, and sediment was shed from the thrust sheets into a foreland basin (the late Devonian–early Carboniferous Alice Springs Orogeny) (Jones, 1972; Korsch & Lindsay,

---

Fig. 2. (*opposite*) Simplified sequence-stratigraphic chart and onlap curve for the late Proterozoic to Devonian sequences in the Amadeus Basin. (After Lindsay & Korsch, 1991.) Onlap curve based on onlap of the southern margin of the basin. LST, lowstand systems tract; TST, transgressive systems tract; HST, highstand systems tract; SMST, shelf-margin systems tract.

(a)

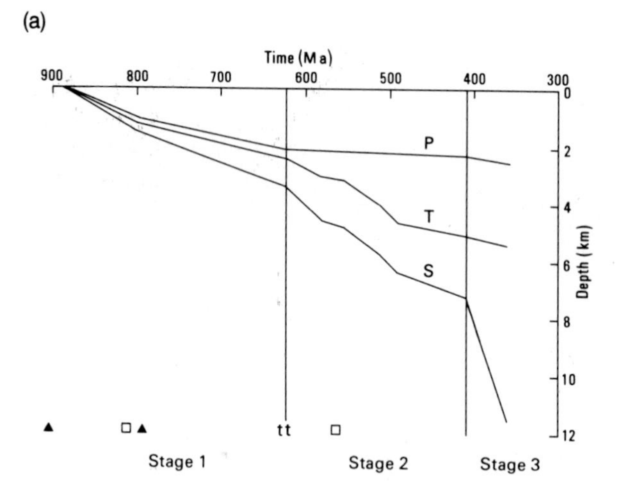

(b)

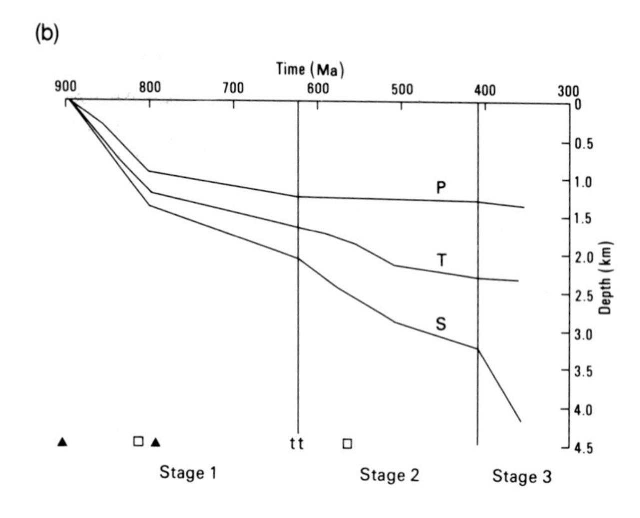

Fig. 4. Subsidence curves for the three major structural regions of the Amadeus Basin. P, platform; T, trough; and S, sub-basin. (a) Total subsidence. (b) Tectonic subsidence. Solid triangles, vulcanism; open squares, evaporite units; t, turbidites.

1989). This event shortened the basin by 50 to 100 km and effectively terminated sedimentation at approximately 360 Ma.

To establish a sequence stratigraphy for the Amadeus Basin, all available data sources (outcrop, wells, seismic) were used. Facies discontinuities, sequences and systems tracts were identified in field outcrops and well logs, and were integrated with more than 6000 km of seismic data. By these means at least 10 sequences have been identified in the late Proterozoic succession, five in the latest Proterozoic–Cambrian and at least nine in the Ordovician (Fig. 2). Isopach maps of these sequences are shown in Kennard and Lindsay (1991) and Lindsay and Korsch (1991). The present discussion is focused on sequences of late Proterozoic to Cambrian age as a means of illustrating the integrated approach necessary to establish a sequence stratigraphy in an intracratonic setting.

Late Proterozoic–Cambrian sequences

Five major depositional sequences are recognized in the latest Proterozoic–Cambrian Pertaoorrta Group of the Amadeus Basin (Figs 2 & 5). These sequences and their component systems tracts are well exposed in the northeast (Figs 6 & 7) in the Ooraminna Sub-basin. In the central and western portions of the basin, however, correlation of sequences and systems tracts is less certain (Fig. 8).

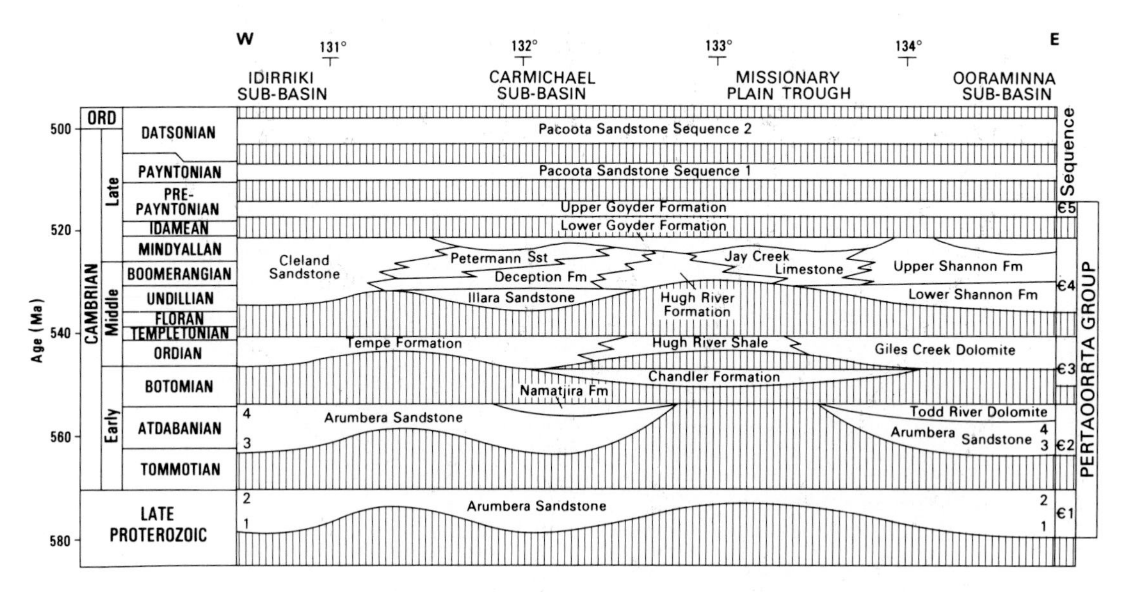

Fig. 5. Interpreted time–space relationship of lithostratigraphic units and depositional sequences of the late Proterozoic–Cambrian Pertaoorrta Group.

Nevertheless, sequence-stratigraphic concepts and models have been used to subdivide these predominantly siliciclastic units into sequences, and to suggest tentative correlations with the better known carbonate-rich sequences in the east (Kennard & Lindsay, 1991).

The base of the Pertaoorrta Group (base Arumbera Sandstone) represents a major sequence boundary, and similarly the top of the group is marked by a major sequence boundary at the base of the overlying Pacoota Sandstone. Between these bounding surfaces five depositional sequences have been identified, but sequence boundaries within the Pertaoorrta Group do not necessarily coincide with formation (i.e. lithostratigraphic) boundaries (Fig. 7).

Sequence 1

This sequence is widely distributed across the northern portion of the basin, and is thin or locally absent across the Central Ridge (Fig. 8). It has a maximum thickness of about 700 m in the Ooraminna Sub-basin and 800 m in the Carmichael Sub-basin, and is 200 to 300 m thick in the Missionary Plain Trough.

The lower contact with shallow-marine oolitic carbonates of the underlying sequence (Julie Formation) is commonly paraconformable, but slight angular discordance and valley incision of at least 100 m are evident between these units at the southern margin of the Carmichael Sub-basin (Fig. 9) (Wells *et al.*, 1965). Near the margins of the Ooraminna Sub-basin and on the flanks of the Central Ridge, sequence 1 disconformably onlaps older Proterozoic units including the much older Bitter Springs Formation. Unconformable relationships can also be identified seismically over localized highs related to the growth of Proterozoic salt structures (Bitter Springs Formation) (Lindsay, 1987a, b).

Lowstand deposits are widespread within the sub-basins (Figs 10 & 11 (Fig. 11 foldout facing p. 616) and form a large-scale, coarsening- and shallowing-upward cycle (Fig. 10). Basal pro-delta or basinal red silty shale is followed by thin-bedded, fine arkosic sandstones which progressively increase in abundance upwards. These sandstones are parallel-laminated and locally culminate in climbing ripples; they are probably turbidites deposited within a lower delta slope environment. Thick arkosic sandstones abruptly appear higher in the sequence, and ultimately become the dominant lithology within a cyclic shale–sandstone interval. These thicker sandstone beds are fine-grained and weakly laminated, have sharp lower contacts with flute and load casts, commonly have poorly developed hummocky cross-stratification and exhibit

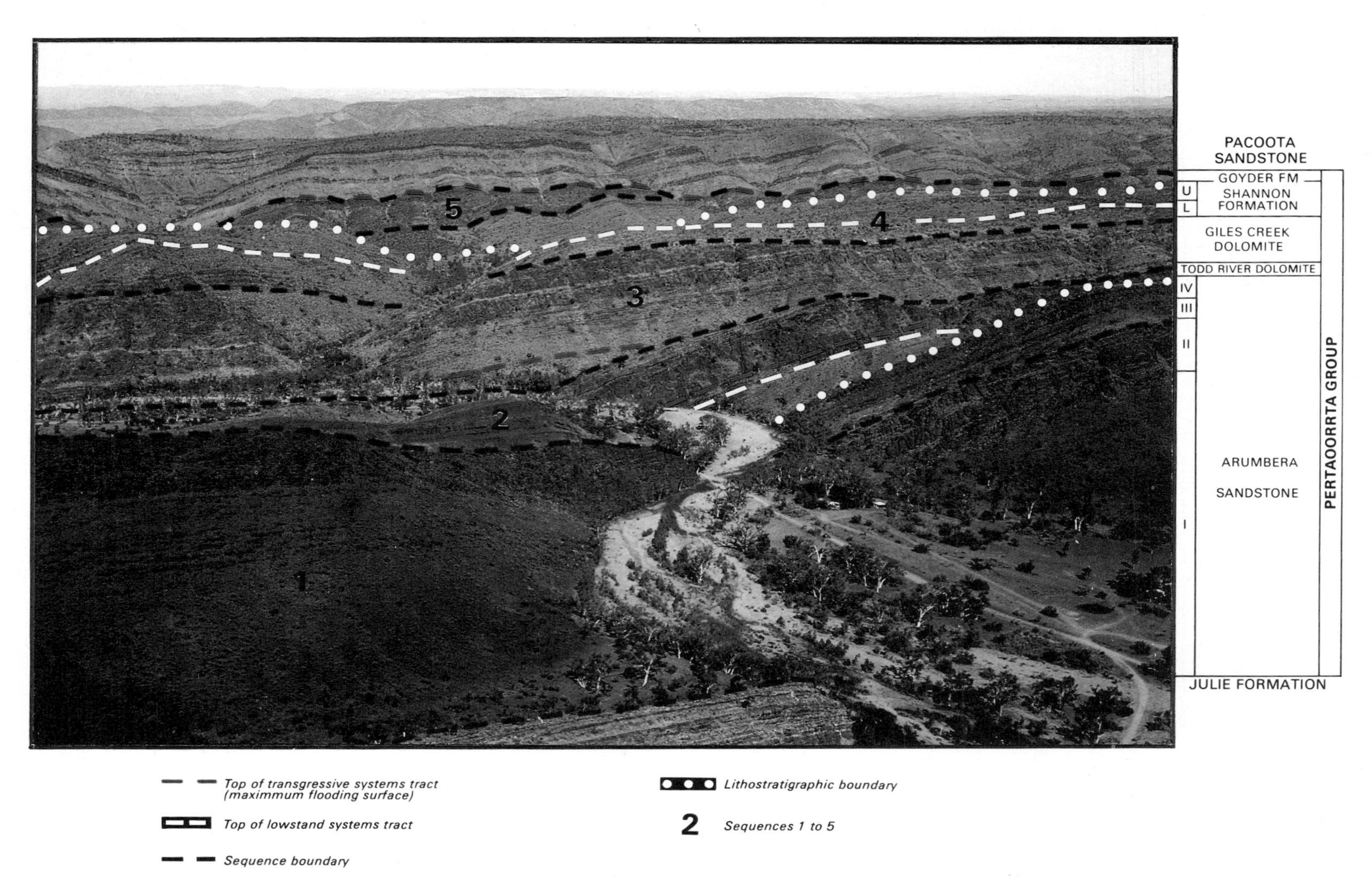

Fig. 6. Oblique aerial view of the late Proterozoic–Cambrian Pertaoorrta Group in the northeast Amadeus Basin (Ross River Gorge), showing sequences and lithostratigraphy.

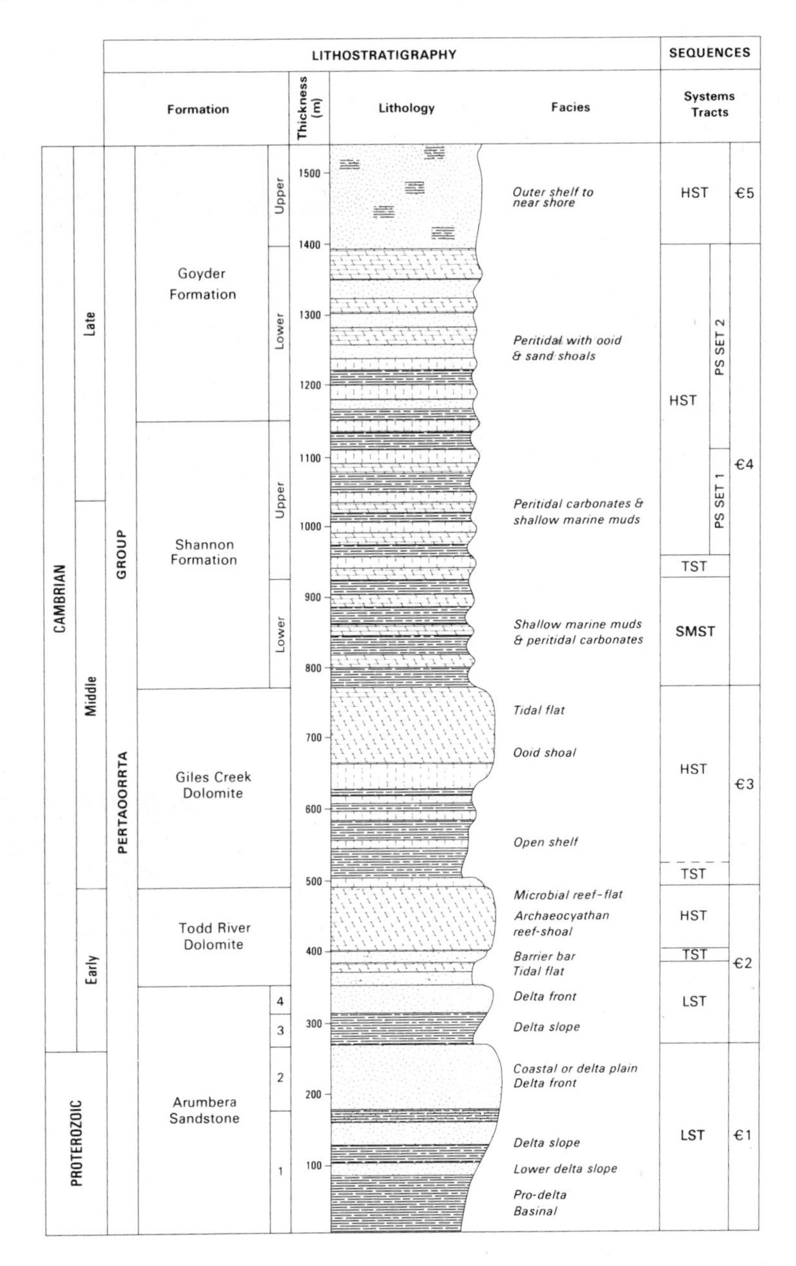

Fig. 7. Lithostratigraphy and sequence stratigraphy of late Proterozoic to Cambrian rocks at Ross River Gorge, northeast Amadeus Basin, showing facies and component systems tracts. LST, lowstand systems tract; SMST, shelf margin systems tract; TST, transgressive systems tract; HST, highstand systems tract.

numerous water-escape structures. These sands were probably generated by storm and flood events on the mid- to upper portions of the delta slope.

The upper part of the lowstand forms a prominent strike ridge of massive fine to medium arkosic sandstone with numerous mud clasts and local large channels. This unit is dominated by fluvial and stream-mouth bar deposits within a delta-plain or coastal-plain environment. Where the sequence onlaps the Central Ridge, these sandstones interfinger with conglomeratic braided stream deposits (Lindsay, 1987a).

Transgressive sands of sequence 1 appear to be restricted to the margins of the sub-basins. They are 20 to 40 m thick on the southwest margin of the Ooraminna Sub-basin (Fig. 12), and about 10 m

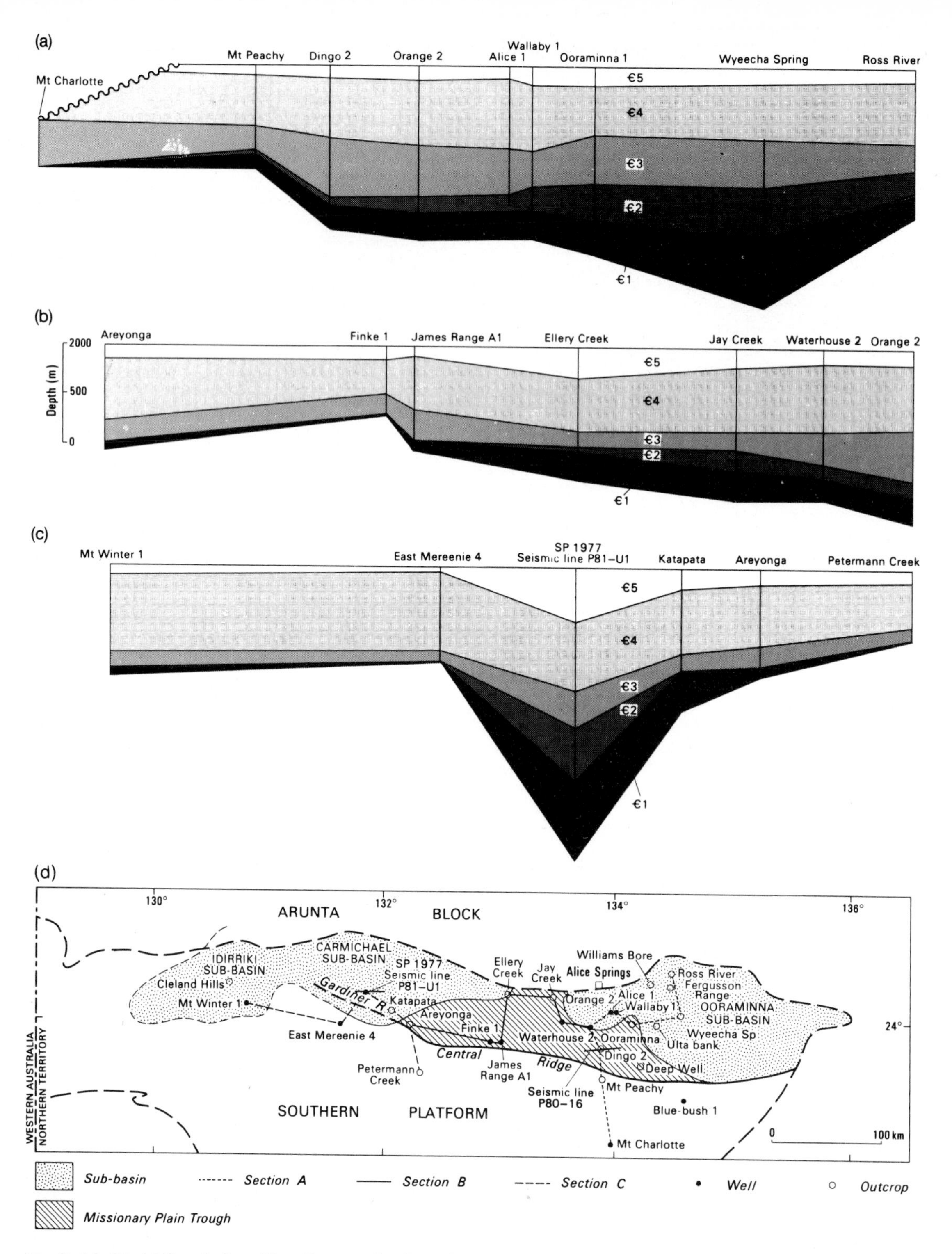

Fig. 8. (a), (b), (c) Correlation of late Proterozoic–Cambrian sequences recognized in outcrop, well logs and seismic sections in the northern Amadeus Basin. (d) Location of cross-sections, field sections and wells.

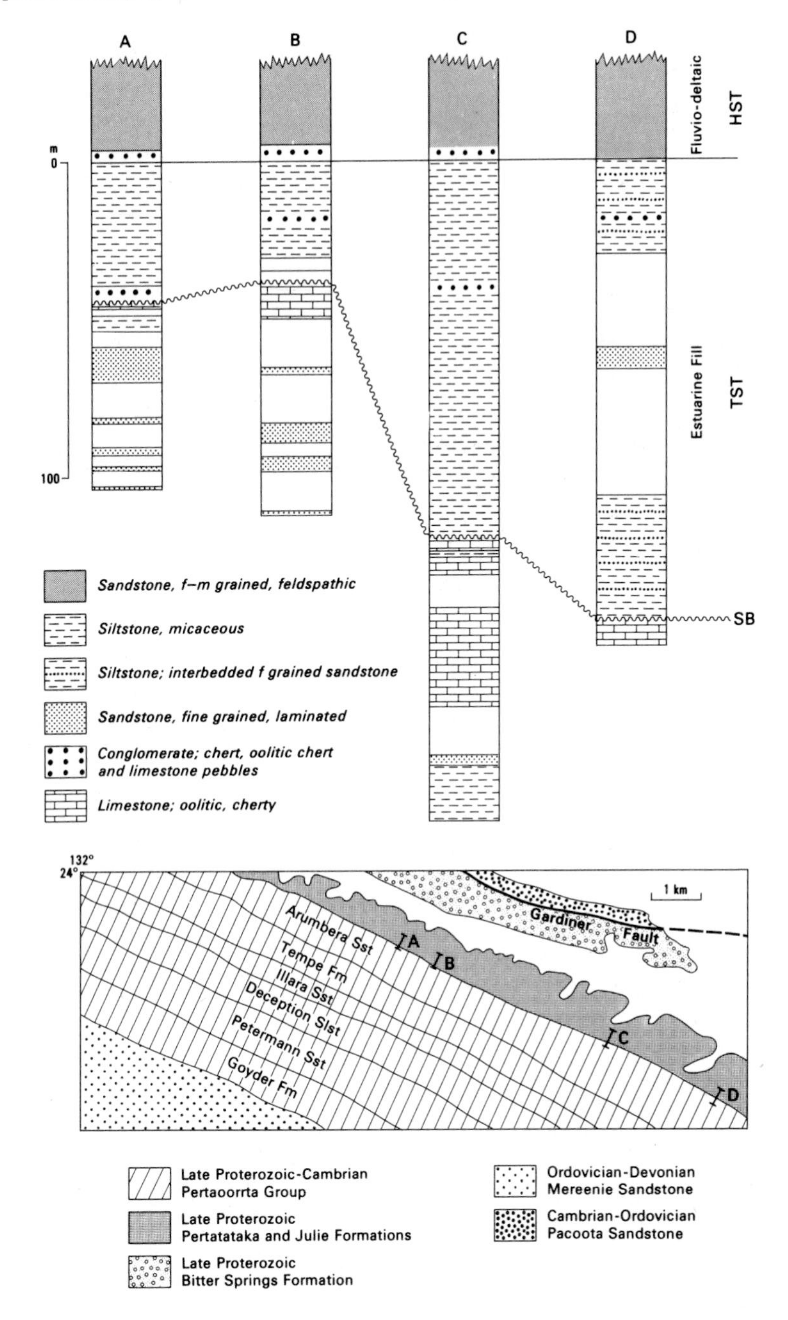

Fig. 9. Lithologic sections showing transgressive estuarine fill of incised valley at base of sequence 1, Gardiner Range, southern margin of Carmichael Sub-basin. (Sequence interpretation based on sections measured by Wells *et al.*, 1965.)

thick on the southern margin of the Carmichael Sub-basin. These transgressive deposits form the reservoir at the Dingo Gas Field, and here they consist of three backstepping progradational shale–sandstone parasequences which have prominent gamma-ray and sonic-log signatures (Fig. 12) (Kennard & Lindsay, 1991; Lindsay & Gorter, 1993). At Ellery Creek the upper part of the highstand is well exposed as a series of thin, parallel sandstone ridges which form the tops of shallowing-upward parasequence sets formed in a shallow-marine coastal-plain setting. Locally, the base of the

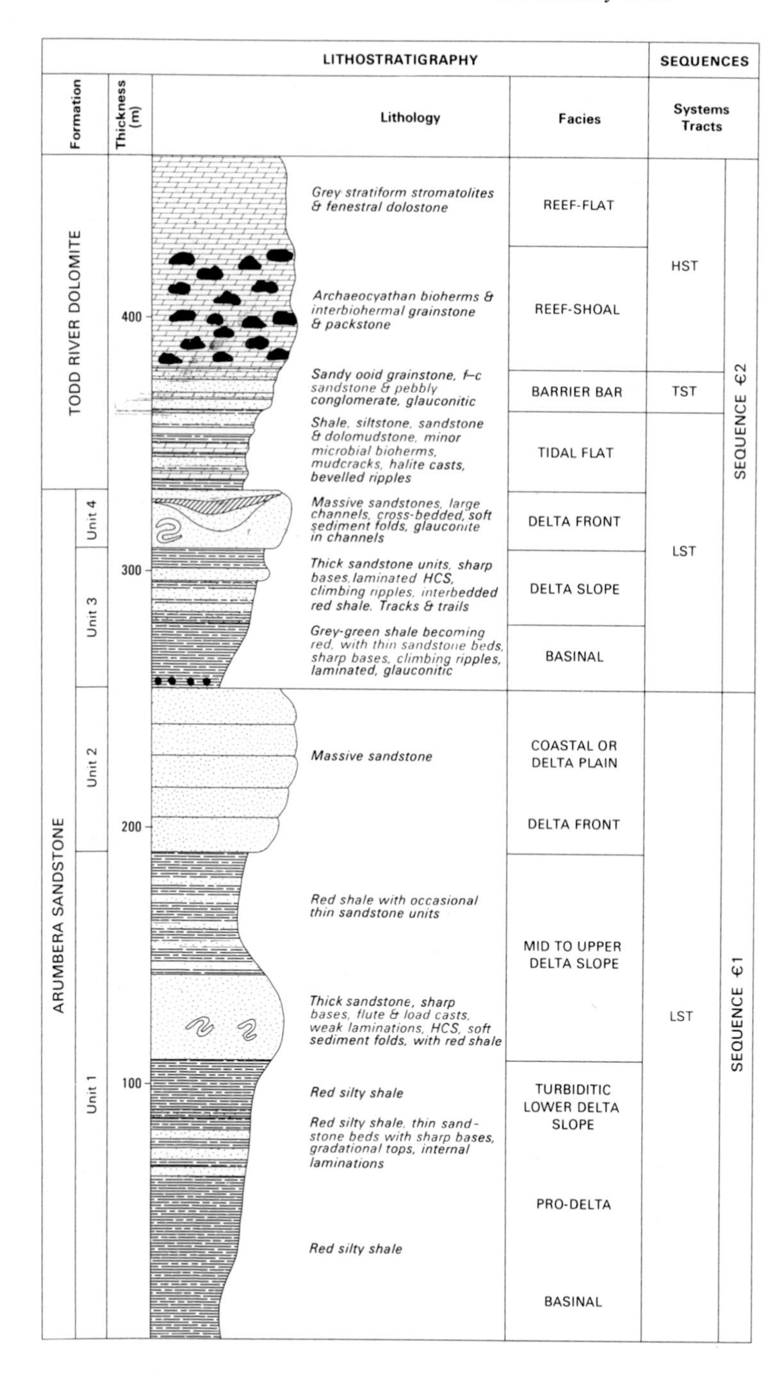

Fig. 10. Lithological section showing facies and systems tracts of late Proterozoic–Cambrian sequences 1 and 2 at Ross River Gorge, northeast Amadeus Basin. LST, lowstand systems tract; TST, transgressive systems tract; HST, highstand systems tract; HCS, hummocky cross-strata.

highstand is exposed and can be seen to be massive red shales with occasional thin turbidites which consist of the latter stages of Bouma sequences. There is thus considerable evidence of rapid upward shallowing.

Sequence 1 displays reciprocal lowstand/highstand sedimentation — thick lowstand deposits occur within the sub-basins and highstand deposits are relatively thinner and restricted to the Missionary Plain Trough and platform areas. Relative

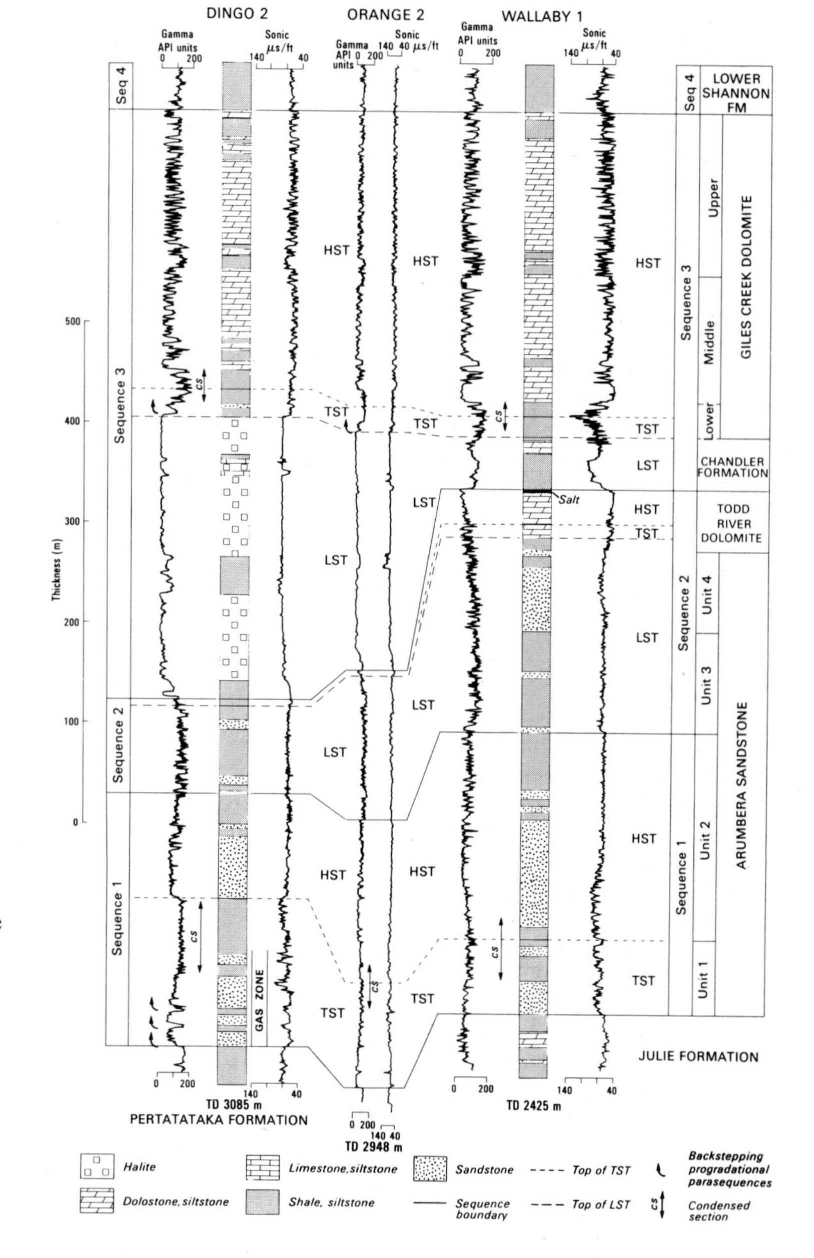

Fig. 12. Well-log correlation of late Proterozoic–Cambrian sequences 1, 2 and 3 in the eastern Amadeus Basin (western margin of the Ooraminna Sub-basin). Backstepping progradational transgressive parasequences at the base of sequence 1 form the reservoir at the Dingo Gas Field. LST, lowstand systems tract; SMST, shelf margin systems tract; TST, transgressive systems tract; HST, highstand systems tract.

sea-level rise is recorded by basal onlap onto the Central Ridge, at the margins of the Missionary Plain Trough to the north, and onto salt growth structures within the sub-basins. Lowstand progradation is clearly evident on seismic profiles near oversteepened margins of the Carmichael Sub-basin (Fig. 11), but elsewhere sequence 1 is largely aggradational (Fig. 13, foldout facing p. 618); if present, highstand progradation was on a scale too small to be resolved seismically. Locally in the Carmichael Sub-basin a poorly defined mounded seismic facies occurs at the base of sequence 1 (Lindsay, 1987a). These mounds probably represent lowstand fans. This facies was probably deposited at the time of local valley incision on the margin of the sub-basin (Fig. 9) but, since it lies below economic depth, it has not been drilled. Transgressive deposits are thin and are restricted to the margins of the sub-basins.

Impressions of metazoan body fossils (see Elphinstone & Walter, in Shergold *et al.*, 1991a) indicate a late Proterozoic (Ediacarian) age for this sequence.

Sequence 2

This sequence consists of a shallowing-upward shale–sandstone succession (the upper part of the Arumbera Sandstone) capped by shallow-marine carbonates (Todd River Dolomite) (Figs 10 & 12). It is largely restricted to the northern sub-basins, and pinches out across the Missionary Plain Trough and against the southern margin of the Carmichael Sub-basin (Gardiner Range) and northern flank of the Central Ridge (Fig. 8). It has a maximum thickness of about 500 m near the centre of the Ooraminna Sub-basin, and has a similar thickness within the Carmichael Sub-basin.

The lower sequence boundary appears to be largely conformable throughout the Ooraminna and Carmichael Sub-basins. At Ross River on the northern flank of the Ooraminna Sub-basin, however, Jenkins and Walter (pers. comm., 1991) have mapped a pebbly erosion surface at the contact of Arumbera Sandstone units 2 and 3 (Fig. 10) which we interpret as a sequence boundary. Jenkins and Walter were able to trace this pebble horizon to the wets where it occurs near the base of the Arumbera Sandstone unit 2, that is, the basal sequence boundary of sequence 2 lies at or near the contact between Arumbera units 2 and 3. Within the Missionary Plain Trough, the basal sequence boundary is commonly erosional, with the underlying sequence exposed and eroded prior to and/or during deposition of sequence 2 (Fig. 13 (foldout facing p. 618)). Onlap to sequence 1 is observed in the Carmichael Sub-basin (Fig. 11) and locally within the Ooraminna Sub-basin.

The shale–sandstone succession that makes up most of this sequence (upper part of the Arumbera Sandstone) represents a prograding deltaic depositional system formed during a relative sea-level lowstand. The coarser grain size and abundance of channels and slump folds within this systems tract probably reflect higher sediment supply and sedimentation rates than the lowstand delta systems of sequence 1; these features are characteristic of lowstand delta systems (Vail & Sangree, 1988). Along the southwest margins of the Ooraminna and Carmichael Sub-basins (Fig. 11) prograding clinoforms appear to represent major delta complexes building from the southwest (Lindsay, 1987a). Thin carbonate siliciclastic tidal flat deposits (basal Todd River Dolomite) (Fig. 10) locally overlie the deltas and represent late lowstand deposits.

A thin glauconitic and intraclastic conglomerate marks a flooding surface at the top of the lowstand deltaic and tidal flat deposits. This flooding surface is overlain by a backstepping progradational quartz–ooid barrier bar complex (Fig. 14) deposited during a rapid relative sea-level rise. The sea-level rise recorded by these transgressive deposits was apparently of limited magnitude since these deposits (and the overlying highstand deposits described below) are largely restricted to the sub-basins.

Transgressive barrier bar deposits are abruptly overlain by archaeocyathan dolostones, and this contact is commonly marked by a marine erosion surface (Kennard, 1991). These archaeocyathan dolostones represent a highstand deposit. Near the northern margin of the Ooraminna Sub-basin, archaeocyaths constructed a high-energy, intermittently emergent reef shoal, the thickest portion of which is capped by an intertidal–supratidal stromatolitic reef flat (Fig. 10). Within the sub-basin to the south, archaeocyathan bioherms and patch reefs represent a somewhat deeper, lower energy, open-shelf environment (Kennard, 1991). Equivalent highstand carbonate deposits have not been recognized beyond the Ooraminna Sub-basin.

Abundant and diverse trace fossils indicate an early Cambrian (Tommotian to early Atdabanian) age for the lowstand deposits of this sequence (Walter *et al.*, 1989; Elphinstone & Walter, in Shergold *et al.*, 1991a), although definitive evidence of a Tommotian age is lacking. Archaeocyaths and phosphatic faunas indicate a late Atdabanian or early Botomian age (Kennard, 1991; Laurie, in Shergold *et al.*, 1991a) for the overlying highstand deposits. The lower boundary of this sequence thus approximates the Proterozoic–Cambrian boundary.

Sequence 3

This is the most widespread Cambrian sequence and records a major eustatic cycle. It consists of evaporites (Chandler Formation) and shallow-marine carbonates (Giles Creek Dolomite) in the east (Ooraminna Sub-basin, Missionary Plain Trough and eastern portions of the Central Ridge and Southern Platform), and more open-marine mudrocks and minor carbonates (Tempe Formation) in the west (Carmichael and Idirriki Sub-

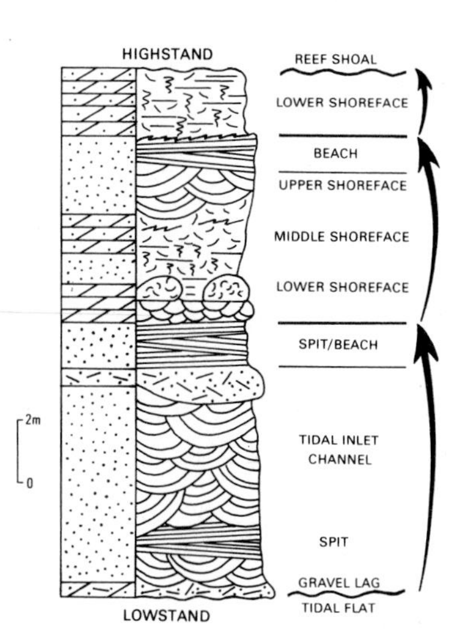

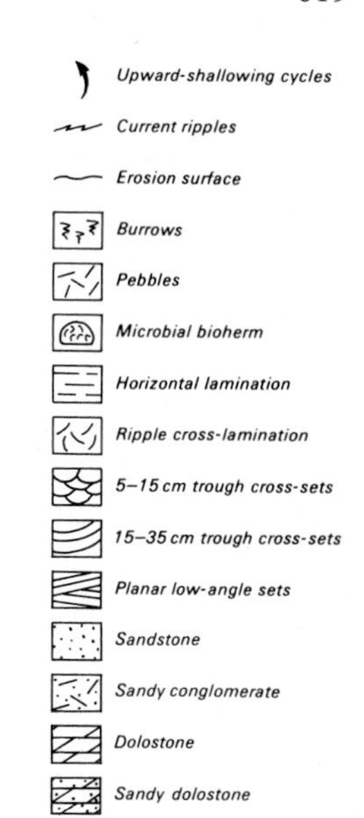

Fig. 14. Upward-shoaling barrier bar cycles (facies 3, Todd River Dolomite) (Kennard, 1991) which form the transgressive systems tract of sequence 2 at Ross River Gorge, NE margin of the Ooraminna Sub-basin.

basins and western portions of the Central Ridge and Southern Platform) (see Figs 5 & 8). It is 250 to 600 m thick in the eastern portion of the basin (much of this variation is due to the presence or absence of the Chandler salt), 300 to 350 m thick in the Carmichael Sub-basin and 150 to 300 m thick in remaining western areas (Idirriki Sub-basin and western portions of the Central Ridge and Southern Platform).

Lowstand deposits consist of basinal evaporites (halite 200 to 400 m thick) and thin red mudrocks and fetid carbonates which locally separate lower and upper salt units (Fig. 12) (Bradshaw, 1991a). These evaporites appear to define two depocentres — one in the eastern portion of the Missionary Plain Trough, and the other extends from the southern portion of the Ooraminna Sub-basin across the Central Ridge and onto the Southern Platform. The salt appears to have extended westward across the Missionary Plain Trough (Bradshaw, 1988, 1991a), but its distribution has been greatly modified by salt flowage, dissolution and tectonic disruption. Salt was apparently thin or absent within the central Ooraminna Sub-basin where equivalent deposits consist of red mudrocks and minor carbonates.

The basal contact of the evaporite unit is unconformable (Bradshaw, 1988, 1991a). It overlies sequence 1 in the Missionary Plain Trough, sequence 2 in the Ooraminna and Carmichael Sub-basins and older Proterozoic units south of the Central Ridge. A palaeokarst surface is locally well exposed between sequences 2 and 3 in the central portion of the Ooraminna Sub-basin (Kennard, 1991). Meteoric diagenesis associated with this exposure surface has resulted in considerable solution-enhanced porosity within the highstand reefal carbonates of the underlying sequence.

There is a major change in the locus of Chandler evaporite sedimentation in comparison to both earlier sequences and overlying units of sequence 3. These relatively short-lived, evaporitic sub-basins may have formed in response to salt flowage in the deeply buried Proterozoic Bitter Springs Formation, and appear to be localized between transfer fault zones (Lindsay & Korsch, 1991). The halite is thought to have precipitated in deeply desiccated sub-basins (Bradshaw, 1988, 1991a) during a sea-level lowstand, and was probably contemporaneous with the subaerial exposure, dolomitization and erosion of carbonates at the top of sequence 2 (Todd River Dolomite) (Kennard, 1991). The local development of lower and upper salt units probably records two cycles of localized subsidence, influx of less saline marine waters and desiccation.

On the western margin of the Ooraminna Sub-basin and adjacent Missionary Plain Trough, lowstand evaporitic deposits of sequence 3 are overlain by a thin transgressive unit at the base of the highstand (Giles Creek Dolomite). This upward-coarsening, shale–siltstone–sandstone unit is 10 to 12 m thick, and forms a prominent backstepping parasequence set on gamma-ray and sonic logs (Fig. 12). The overlying highstand deposits are 300 to 400 m thick. Throughout most of the Ooraminna Sub-basin they consist of interbedded shallow-marine mudrock and peritidal dolostone (Giles Creek Dolomite) (Deckelman, 1985) which form metre- to decimetre-scale, shale–carbonate parasequences. In the subsurface, highstand deposits have been subdivided into three units: a lower shale-rich unit, a middle dolostone-rich unit and an upper unit with subequal amounts of shale and dolostone (Fig. 12). The lower shaly unit is characterized by a high gamma-ray signature and is interpreted as a condensed section at the peak of the marine transgression (period of maximum flooding). The overlying units record the progradation of an inner relatively siliciclastic-rich facies belt (upper unit) across an outer carbonate belt (middle unit) during the later stages of the highstand.

Markedly different facies (also referred to as the Giles Creek Dolomite) are present at the northern margin of the Ooraminna Sub-basin. Here the sequence consists of a basal 5–10 m-thick phosphatic unit overlain by a large-scale shale-to-carbonate cycle (Fig. 15) (Keith, 1974). The basal unit consists of bioturbated, bioclastic wackestone–packstone, dolo-mudstone and minor grainstone, locally with a thin, sandy, dolostone conglomerate containing abundant detrital dolomite grains (Kennard, 1991). This unit contains black chert nodules and irregular phosphatic patches, and is capped by a series of phosphatic hardgrounds (Fig. 15(b)) which represent a condensed section associated with the period of maximum marine flooding. This basal unit is correlated with the transgressive parasequence beneath the Giles Creek Dolomite in the western Ooraminna Sub-basin (Fig. 12). The overlying highstand cycle (Fig. 15(a)) consists of, in ascending order: (i) green shale; (ii) nodular-bedded fetid bioclastic wackestone; (iii) dolo-mudstone; (iv) fenestral dolo-mudstone, intraclast-peloid grainstone and ooid grainstone; and finally (v) metre-scale cycles (parasequences) of thin- to medium-bedded intraclast-peloid grainstone, stratiform and domal stromatolites and fenestral dolo-mudstone. These parasequences commonly culminate in tepee structures and exposure surfaces. This large-scale cycle represents a northeastward prograding, carbonate ramp consisting of a condensed section (shales), open-shelf, ooid-shoal and tidal-flat facies.

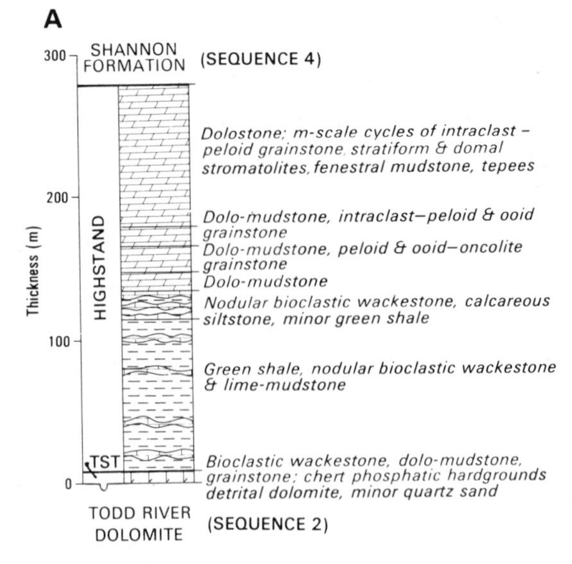

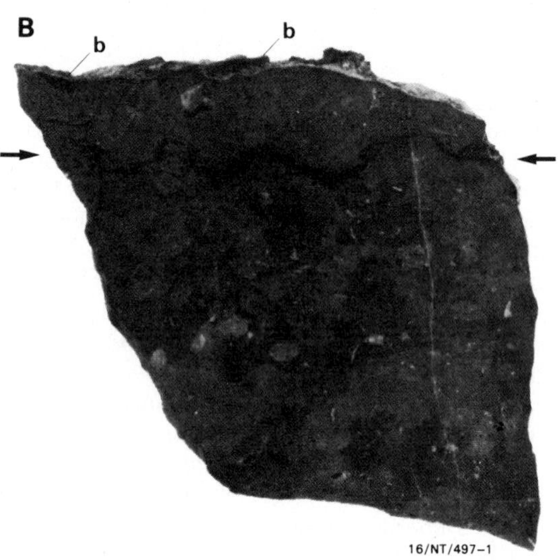

Fig. 15. (A) Upward-shoaling shale–carbonate cycle (Giles Creek Dolomite) comprising basal transgressive unit (TST) and thick highstand deposit, sequence 3, Ross River Gorge, NE margin of Ooraminna Sub-basin. (B) Phosphatic hardground (arrows) and phosphatic burrows (b) within basal transgressive wackestone. Note brachiopod shells (B).

The highstand deposits of the Giles Creek Dolomite extend beyond the Ooraminna Sub-basin southward onto the Southern Platform and westward with no appreciable thinning or facies change until the middle of the Missionary Plain Trough.

Thus the Ooraminna Sub-basin became less distinct at this time, and the highstand deposits of sequence 3 record the inception of a low-gradient carbonate–siliciclastic ramp (Missionary Plain—Ooraminna Ramp) across the eastern Amadeus Basin. These changes reflect the progression from relatively rapid and more localized mechanical subsidence during sequences 1 and 2 to broader thermal subsidence following extension.

This carbonate ramp passed westward into a siliciclastic facies belt, the outer (eastern) portion of which was dominated by shallow-marine muds (lower Hugh River Shale), and the inner portion consisted of shallow-marine silts (Tempe Formation; see Fig. 5). Thickness variations within these latter units suggest that the Carmichael and Idirriki Sub-basins were also less distinct and broader at this time (Fig. 8). In the Carmichael and Idirriki Sub-basins, sequence 3 unconformably overlies sequence 2 and lowstand evaporites are either thin or absent (Fig. 5). Along the margins of the Carmichael and Idirriki Sub-basins and flanks of the Central Ridge, it unconformably onlaps sequence 1 or older Proterozoic units. Transgressive deposits consist of glauconitic shoreface sandstones (basal sandstone of the Tempe Formation) (Bradshaw, 1991b). The overlying siltstones, shales and skeletal limestones (Tempe Formation) record an upward-shallowing cycle of shallow-shelf, shoreface or lagoonal facies, with deeper water facies becoming dominant to the northwest (Bradshaw, 1991b). These deposits also contain abundant glauconite and represent highstand deposits equivalent to the Giles Creek Dolomite in the east.

Sequence 3 extends at least to the western limits of the Idirriki Sub-basin and delineates the westernmost limit of marine deposits within the Pertaoorrta Group.

The eastern highstand deposits of sequence 3 (Giles Creek Dolomite) contain a fauna of hyoliths, brachiopods, gastropods, echinoderm debris and trilobites (Shergold, 1986). The fauna suggests an Ordian to early Templetonian (early middle Cambrian) age for sequence 3, but the initial lowstand evaporite deposits may be of late early Cambrian age. Fossils have also been recorded in the western portion of sequence 3 (brachiopods, trilobites, hyoliths, gastropods, small phosphatic shells and acritarchs); this fauna and flora are consistent with an Ordian (early middle Cambrian) age (Shergold, 1986).

In the Georgina Basin rocks of similar age host economic phosphate deposits interpreted by Southgate and Shergold (1991) as occurring in the transgressive systems tract of the same sequence.

Sequence 4

This sequence exhibits a complex east–west facies mosaic. It consists of mixed clastics and carbonates in the Missionary Plain—Ooraminna Ramp (Shannon Formation, upper part of the Hugh River Shale, Jay Creek Limestone and lower Goyder Formation) (Fig. 5). The sequence is well developed but poorly known in the western portion of the basin where it forms a thick clastic succession. In the Carmichael Sub-basin it appears to be represented by the Illara Sandstone, Deception Formation, Petermann Sandstone and lower Goyder Formation (Fig. 11), and by the Cleland Sandstone in the Idirriki Sub-basin (Fig. 5). It is 600 to 700 m thick in the eastern portion of the basin, and gradually thins westward to 450 to 500 m across the Missionary Plain—Ooraminna Ramp (Fig. 8). It thickens to 700 m in the Carmichael Sub-basin and 750 to 800 m in the Idirriki Sub-basin, and thins southward across the Central Ridge to 500 to 600 m on the Southern Platform.

In the eastern portion of the basin (Missionary Plain—Ooraminna Ramp), this sequence paraconformably overlies highstand deposits of sequence 3. This contact lacks evidence of erosion or obvious depositional hiatus. It is marked by a somewhat subtle change in the relative proportion of shallow-marine siliciclastic mudrocks and peritidal dolostones; shale–carbonate parasequences of sequence 4 (lower Shannon Formation) have thicker shale half-cycles and thinner upper carbonate half-cycles than those in the underlying sequence (Giles Creek Dolomite) (Kennard & Lindsay, 1991). This contact appears to represent a change from largely progradational to aggradational parasequence sets, and correlates with a prominent seismic reflector at the top of the Giles Creek Dolomite (Fig. 13). This contact is thus interpreted as a type 2 sequence boundary and the lower Shannon Formation is accordingly interpreted as a shelf-margin systems tract (*sensu* Van Wagoner *et al.*, 1987, 1988). The seismic expression of the lower boundary of this sequence is poorly developed in the northern portion of the Missionary Plain Trough. In these areas the boundary is thought to occur within the Hugh River Shale, thereby subdividing this unit into upper and lower sequences (Fig. 5).

A marked change in the shale–carbonate parasequences also occurs within sequence 4, and this change has been used to subdivide the Shannon Formation into a lower shale-rich member (140 to 240 m thick) and an upper carbonate-rich member (250 to 400 m thick). In the lower member, shaly half-cycles are typically 1 to 8 m thick and carbonate half-cycles 0.2 to 1 m thick. The carbonate half-cycles comprise dolostone, are dominated by stromatolites of low synoptic relief and typically contain a thin, basal subtidal facies overlain by relatively thicker intertidal facies (Fig. 16) (Kennard, in press). In contrast, parasequences in the upper Shannon Member comprise slightly thinner shaly half-cycles (typically 1 to 5 m) and relatively thicker carbonate half-cycles (typically 0.3 to 3 m). Furthermore, these carbonate half-cycles consist of subequal limestone and dolostone, are characterized by thrombolites (clotted microbialites) (Kennard & James, 1986) of relatively high synoptic relief and typically comprise relatively thick, basal subtidal facies overlain by thinner intertidal facies (Fig. 16).

Parasequences in the lower Shannon Member show no discernible-upward change in either thickness or character. They constitute an aggradational parasequence set, bounded below by a type 2 sequence boundary. Thus, the lower Shannon Member is probably the on-shelf portion of a 'shelf-margin' systems tract deposited during a minor lowering of relative sea level.

In the upper Shannon Member, at least three

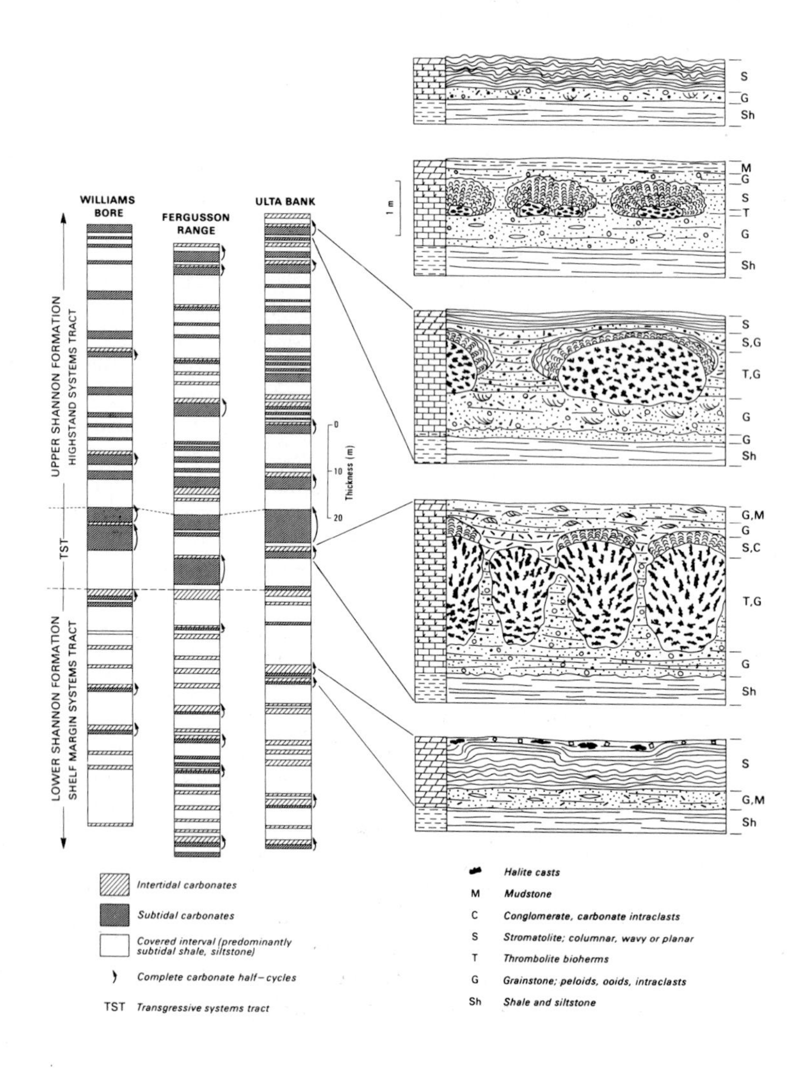

Fig. 16. Pattern of shale–carbonate parasequences in three measured sections across the contact between the lower and upper Shannon Formation, Ooraminna Sub-basin. The lower member (shelf margin systems tract) consists of aggradational parasequences characterized by relatively thin, intertidal-dominated, stromatolitic-carbonate half-cycles. Two or three relatively thick parasequences with thick subtidal-based thrombolitic-carbonate half-cycles occur in the transgressive systems tract at the base of the upper Shannon Formation. Highstand parasequences (upper three cycles right-hand side) have an increasing proportion of intertidal stromatolites to subtidal thrombolites. Section localities shown in Fig. 8(d).

parasequence sets are evident in outcrop and well logs, one in the transgressive systems tract and two in the highstand systems tract (Figs 16 & 17). The basal set is 15 to 25 m thick and comprises three or four relatively thick parasequences dominated by subtidal carbonates (Fig. 16), most notably ribbon and nodular-bedded wackestone and thick thrombolites. These parasequences form distinct marker beds in outcrop and represent a backstepping transgressive systems tract. Parasequences within the overlying parasequence set (approximately 160 m thick) comprise a subequal thickness of shale and carbonate half-cycles and a subequal proportion of subtidal thrombolites and intertidal stromatolites. The third (uppermost) parasequence set is 170 to 250 m thick, and is distinguished by thinner shale half-cycles (and hence a greater proportion of carbonates), a greater proportion of intertidal

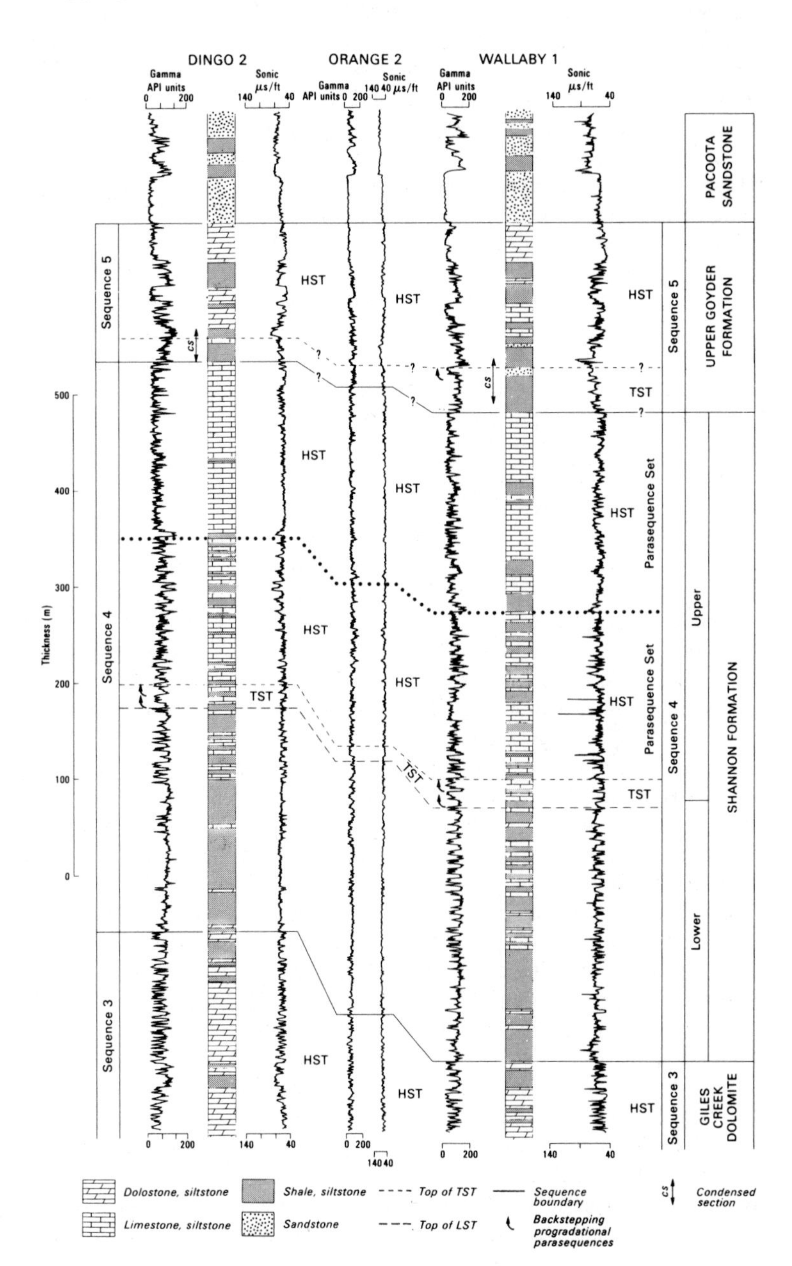

Fig. 17. Well-log correlation of sequences 4 and 5 in the eastern Amadeus Basin (western margin of the Ooraminna Sub-basin). Backstepping progradational parasequences are evident on gamma-ray logs at the base of the transgressive systems tract of sequence 4 (lower Shannon Formation). SMST, shelf margin systems tract; TST, transgressive systems tract; HST, highstand systems tract.

stromatolites than subtidal thrombolites, a decrease in the synoptic relief of both thrombolites and stromatolites and an increase in the proportion of peritidal dolostone, ooid grainstone and quartz sandstone. The upper portion of this set comprises sandstones and dolostones of the lower Goyder Formation. The upper two parasequence sets represent progradational highstand deposits.

The change from aggradational stromatolitic parasequences to backstepping thrombolitic parasequences ('shelf margin' to transgressive depositional systems) can be traced westward to the contact between the upper part of the Hugh River Shale and the basal portion of the Jay Creek Limestone in the Missionary Plain Trough (Fig. 5). Highstand deposits in this area are represented by progradational carbonates and clastics (Jay Creek Limestone) and finally sandstones and dolostones (lower Goyder Formation).

In the central Carmichael Sub-basin an eastward-prograding, fluvial wedge (Illara Sandstone) (Figs 5 & 11) was deposited during the relative sea-level lowstand that generated the lower Shannon Formation 'shelf margin' deposit system in the east. This fluvial succession is overlain by siltstones and sandstones (Deception Formation and Petermann Sandstone). Although these latter units are poorly known, some horizons in the lower portion are bioturbated, and thus probably record the maximum westward expansion of marine sediments that correlate with transgressive deposits on the ramp to the east. In seismic section (Fig. 11), this upper clastic succession appears to represent an eastward-prograding fluvial-deltaic, highstand systems tract.

Across the Central Ridge and Southern Platform, sequence 4 is dominated by fluvial and overbank facies (Deception Formation and Petermann Sandstone) (Sweet, unpublished data). In the Idirriki Sub-basin it consists of gravelly braided stream deposits (Cleland Sandstone).

Late highstand mixed carbonate–siliciclastic marine deposits (lower member of the Goyder Formation) are widely distributed across both the eastern and western portions of the Amadeus Basin (Fig. 5). They consist of fine and medium quartz sandstone, oolitic and minor stromatolitic carbonates and siltstone. Peritidal and nearshore facies dominate in the east, but the deposits are poorly known in the west. Their thickness varies reciprocally with the thickness of the immediately underlying unit; this relationship suggests diachronous lateral changes between carbonate–shale, carbonate–sandstone and sandstone facies. The progressive onlap of these late highstand marine deposits across marine carbonates in the east and siliciclastic deltaic and fluvial facies to the west is thought to have resulted from a decrease in the supply of siliciclastic sediments derived from the west or southwest.

The age of sequence 4 is poorly constrained. Brachiopods, molluscs, hyoliths, gastropods, monoplacophorans and trilobites have been recorded within highstand deposits in the east. This fauna indicates a late Cambrian (late Mindyallan) age (Shergold, 1986; Shergold *et al.*, 1991a). 'Shelf-margin' depositional systems, and all siliciclastic facies in the west, are undated.

Sequence 5

This sequence is widespread across the basin (Figs 5 & 8), but is relatively poorly known. In the northern Ooraminna Sub-basin and Missionary Plain Trough areas it consists of silty, commonly hummocky and planar cross-bedded, fine-grained sandstone (upper member of the Goyder Formation) (Fig. 7). In these areas it is 140 to 280 m thick, and one or more prominent manganiferous horizons commonly occur in the lower half of the unit. This sequence is interpreted as a highstand, nearshore to outer-shelf facies, and the lower manganiferous portion probably represents a condensed section.

In the Carmichael Sub-basin, sequence 5 is considerably thicker (approximately 500 m thick) (Fig. 11). Here it consists of a thick lowstand deposit and an overlying highstand deposit which correlates with outcropping upper Goyder Formation at the margins of the sub-basin. Locally, sequence 5 is unconformably overlain by estuarine sandstones (basal Pacoota Sandstone) which fill an incised valley with at least 55 m of palaeotopographic relief (Zaitlin, unpublished measured section).
Numerous channel-like features are also evident at this horizon on seismic data (Gorter, 1991) and are similarly interpreted as incised-valley deposits.

Sequence 5 contains late Cambrian (post-Mindyallan, pre-Payntonian) trilobites. This suggests a significant time break between it and sequence 4 which contains a Mindyallan fauna (Shergold, 1986; Shergold *et al.*, 1991a). On the basis of its lithology, facies and widespread distribution, this sequence is more closely related to the overlying siliciclastics of the Ordovician Larapinta Group than the underlying Cambrian Pertaoorrta facies which display a marked east to west, carbonate to siliciclastic facies transition (Fig. 5).

Ordovician sequences

Nine thin depositional sequences can be identified within the Ordovician succession above Cambrian sequence 5. These sequences mark a significant change, both in basin dynamics and the pattern of sedimentation within the basin. The sequences progressively onlap the southern margin of the basin (Fig. 18) such that, by early Ordovician time, shallow-marine conditions extended over the entire Amadeus Basin and clastic sedimentation prevailed. Sedimentation rates decreased abruptly from Cambrian sequence 5 into the Ordovician succession and sequences thin to the point where they are close to seismic resolution. During this time, sediment supply kept pace with available depositional space such that, with one exception, water depth remained shallow and commonly in the tidal range. Water depth appears to have increased abruptly for a short period during deposition of the sequence containing the Horn Valley Siltstone (a highstand deposit which has sourced commercial hydrocarbon fields in the basin), although total water depth was not great.

Identification of sequences

Basin dynamics allowed a full range of expression of sequence geometries during late Proterozoic–Cambrian time in the Amadeus Basin. Sequences 1 and 2 were deposited immediately following the second major extensional event when subsidence rates were rapid. However, subsidence (and hence depositional space) was largely confined to the northern sub-basins. This resulted in the development of steep margins on the sub-basins and thus an equivalent of the shelf–slope break of the passive-margin setting. Consequently, these two sequences are relatively thick and prograde into the sub-basins. In this situation seismic data (reflector truncations) can be used directly to identify sequence boundaries and systems tracts. Seismic interpretation was augmented by well-log and outcrop facies analysis which proved of considerable value in identifying: (i) incised-valley deposits at the base of sequence 1 on the margin of the Carmichael Sub-basin; (ii) transgressive reservoir sands along the margin of the Ooraminna Sub-basin; (iii) lowstand deltaic sediments above a cryptic, pebble-

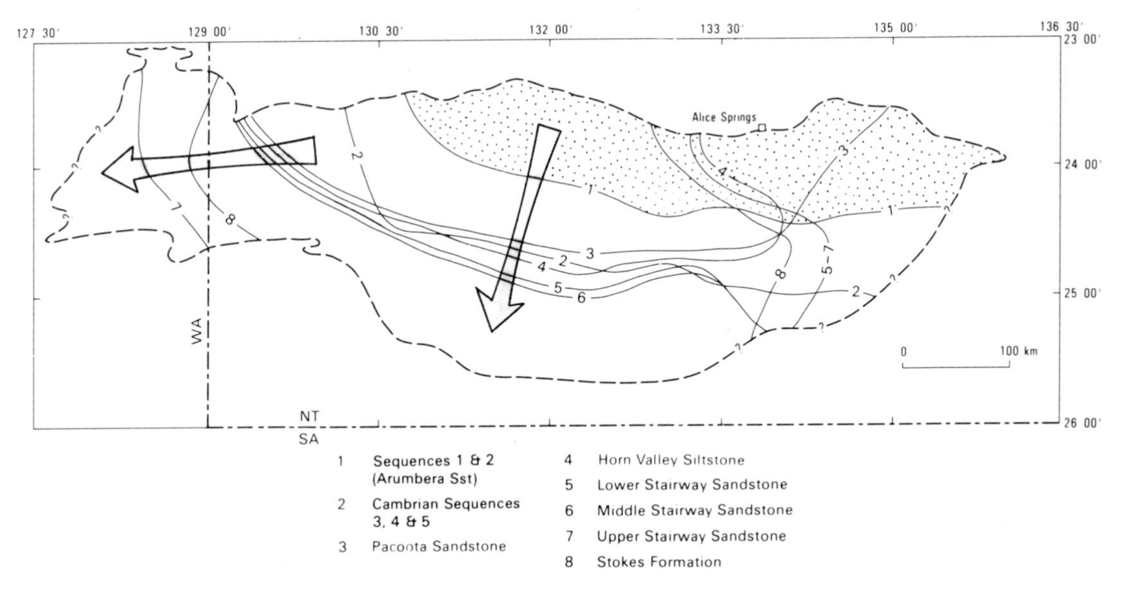

Fig. 18. Growth of the Amadeus Basin during stage 2, showing general onlapping and offlapping relationships of the major depositional units. Arrows show the direction of movement of the limits of sedimentation through time which, along the southern margin of the basin, equates with onlap. Note the abrupt expansion of the basin during late Stairway deposition (lines 7 and 8).

strewn, erosion surface at the base of sequence 2; and (iv) transgressive barrier bars and highstand reefal carbonates in sequence 2.

Sequences 3 and 4 are geometrically intermediate between the thicker, more passive-margin-like, earlier sequences and the overlying more typical layer-cake intracratonic sequences. These sequences are relatively thick but evidence of progradation from seismic sections is only occasionally discernible, and recognition of the sequences depended upon facies studies. Outcrop and well facies provided diagnostic evidence of: (i) local karst erosion on the lower boundary of sequence 3; (ii) tectonically controlled, lowstand evaporite deposits of sequence 3; (iii) marine phosphatic hardgrounds within a condensed interval at the base of prograding highstand carbonate deposits of sequence 3; and (iv) shale–carbonate parasequences and parasequence stacking patterns which define a type 2 sequence boundary at the base of sequence 4, and overlying aggradational 'shelf margin', backstepping transgressive and progradational highstand systems tracts. These sequences were deposited when basin subsidence was still relatively rapid but thermal effects were beginning to decrease and a major global sea-level rise was beginning to take effect.

Sequence 5 and the overlying Ordovician sequences are all at or close to seismic resolution. They appear on seismic sections as a series of well-defined parallel reflectors. The closely spaced nature of these reflectors is distinctive and in itself offers clues to the fact that accommodation was minimal at this time and that thermal subsidence had significantly decreased. These thin sequences can only be understood through detailed facies studies in well and outcrop sections (Gorter, 1991), and can commonly be defined by biostratigraphic hiatuses (Shergold *et al.*, 1991a, b). Seismic data do, however, provide local evidence of valley incision at sequence boundaries.

Duration of sequences

Estimates of the duration of the late Proterozoic–Cambrian sequences can be made on the basis of the correlation of Australian and international Cambrian biochronological scales and international isotopic age determinations (Shergold, 1989). These correlations suggest that the sequences have a duration of about 5 to 15 m.y. (see Fig. 5). This falls within the range of third-order sea-level cycles (Grand cycles) recognized for Cambro-Ordovician passive-margin successions in the United States and northern Canadian Appalachians (James *et al.*, 1989; Read, 1989), but is significantly greater than the 1 to 5 m.y. average duration of third-order eustatic cycles documented by Haq *et al.* (1988) for Mesozoic and Cenozoic strata. This discrepancy may be real, or could simply reflect poorly constrained radiometric age determinations for the Cambrian period and epochs. For example, Odin *et al.* (1983), Jenkins (1984) and Conway Morris (1988) propose that the Proterozoic–Cambrian boundary may be as young as 530 Ma. This younger age would imply that the five sequences identified collectively span a period of about 20 to 30 m.y., and each have a duration of about 5 m.y. Even this figure is probably an overestimate of the period of apparent third-order sea-level cycles due to the likelihood of an incomplete stratigraphic record of eustatic events within this intracratonic setting.

If this scenario is correct, the stratigraphic sequences described in this paper may be analogous to the second-order cycles (supersequences) of Haq *et al.* (1988), and the systems tracts and parasequence sets described herein, many of which are bounded by erosion surfaces, may represent third-order cycles.

DEPOSITIONAL CONTROLS AND AN IDEALIZED SEDIMENTARY SUCCESSION

The depositional successions in all of the early Australian intracratonic basins have broad similarities suggesting common controls on sedimentation (Lindsay & Korsch, 1989). Since all of the basins appear to be the product of tectonic events of global dimensions and share similar subsidence histories (Lindsay *et al.*, 1987), it would seem reasonable to suggest that some simplifying generalizations about depositional controls might be made.

The shape of the relative sea-level/onlap curve (Fig. 2) in the Amadeus Basin was governed over the longer term by basin dynamics, but includes a superimposed eustatic sea-level component. The broad features of the depositional succession can be related to the onset of thermal subsidence and possibly the development of a peripheral bulge. Using a combination of (i) the extensional model and (ii) the observed sedimentary facies in the Amadeus Basin, it is possible to construct an idealized sedimentary succession for a single stage of basin development following an episode of crustal extension (Fig. 19).

In the Amadeus Basin each extensional stage passes through two developmental phases. First, rapid mechanical subsidence produces a rift phase, and second, thermal subsidence begins as the lithosphere thickens and cools. Within the thermal subsidence phase, subsidence is rapid at first then declines gradually. There may be a third, long post-thermal subsidence phase where subsidence is minimal and sedimentation is largely determined by small-scale tectonic events (Shaw, 1991).

These major developmental phases provide the broad controls for sedimentation. Extension initially resulted in the deposition of clastic sediments (Fig. 19). Local fluvial sediments with some bimodal volcanics are assigned to the rift phase during the first extensional stage in the Amadeus Basin (Lindsay & Korsch, 1989). Rift phase sediments do not appear to have been deposited at the start of stage 2 of basin development which supports a lower crustal detachment ramp model as proposed by Shaw (1991).

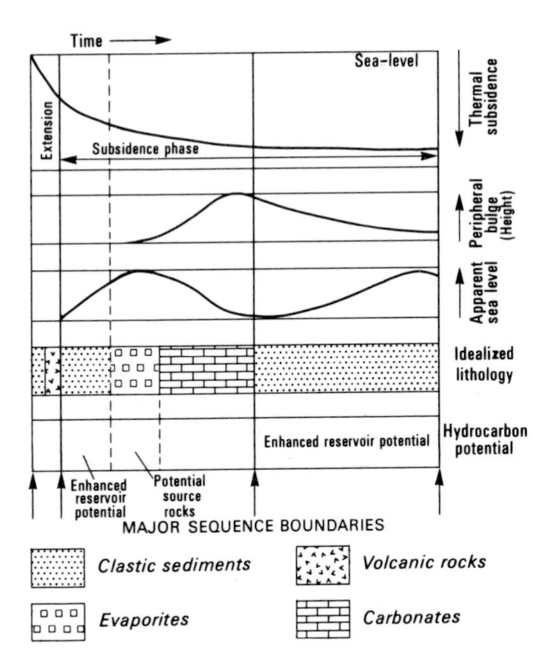

Fig. 19. An idealized sedimentary succession for a single stage of rifting and thermal subsidence showing the relationship between lithology, apparent relative sea level, major sequence boundaries and basin mechanics in a situation where the continents maintain latitudinal stability. (After Lindsay & Korsch, 1989.) Apparent relative sea level is controlled to a large degree by basin dynamics (i.e. extension, thermal subsidence and the growth of the peripheral bulge due to sediment loading).

Following each extensional episode, thermal subsidence took effect, the basin was open to the ocean and clastic sediments derived from an extensive hinterland were deposited in a shallow-marine setting. As sedimentation and thermal subsidence proceeded, sediment loading resulted in crustal flexure and the development of a peripheral bulge. The basin was shallow, so that eventually the bulge progressively restricted the supply of clastic sediments from the hinterland (Korsch & Lindsay, 1989). This, coupled with climatic conditions, resulted in a shift from the deposition of clastic sediments to the deposition of evaporites and carbonates (Heavitree Quartzite to Bitter Springs, stage 1; sequences 1, 2 and 3, stage 2).

Although sequence subdivision in the Proterozoic is incomplete, the gradual change from terrigenous clastics of the Heavitree Quartzite to mixed carbonates, siliciclastics and evaporites of the Gillen Member can be seen to result from rapid subsidence (early extension and thermal subsidence) with a superposed rapid fall in sea level (basinal evaporites of the Gillen Member). This fall was followed by a rapid rise in relative sea level permitting deposition of the Loves Creek Member. Here the transgressive surface is overlain by up to 4 m of cross-bedded quartz sandstone containing lithoclasts eroded from the underlying Gillen Member (Fig. 20). Above this basal unit is a stromatolite bioherm that shows evidence of progressive deepening. In the lower parts of the bioherm, stratiform and small columnar stromatolites are interbedded with thin sandstone beds and have quartz sand infilling intercolumnar areas. Above this interval the columnar stromatolites gradually increase in diameter and eventually pass into domal stromatolites. Throughout this transition siliciclastic components gradually decrease in abundance until they are eventually absent, and synoptic relief gradually increases until large domes up to 10 m across are formed with synoptic relief of up to 1.5 m. This upward-deepening transgressive systems tract is overlain by a condensed interval of thinly bedded and stromatolitic dolostones that contain scattered glauconite grains. Above the condensed section lies a 200 m-thick interval of stromatolitic dolostone arranged into parasequences 2 to 12 m thick (Fig. 20). Within the stromatolite-dominant unit accommodation is greatest in the lower parts of the parase-

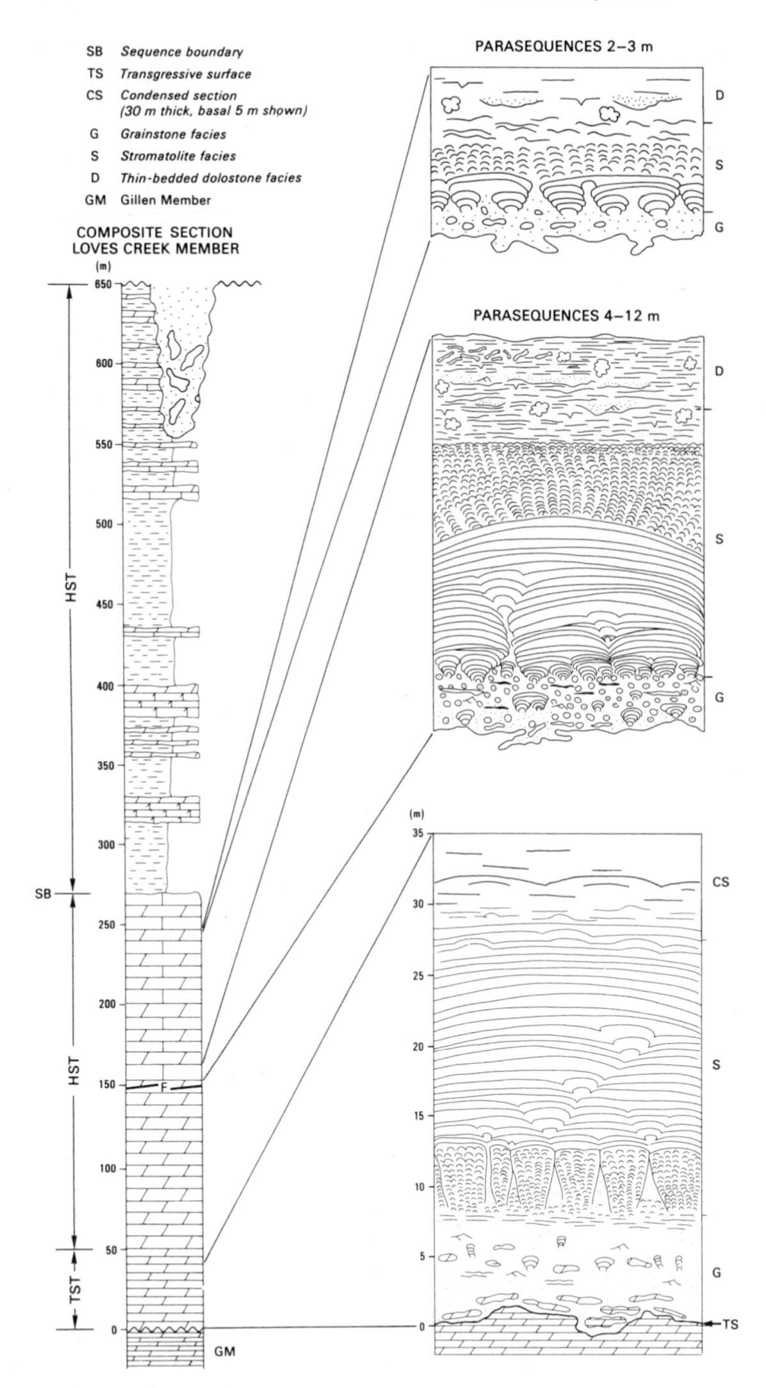

Fig. 20. Composite stratigraphic column for the Loves Creek Member, Bitter Springs Formation. Basal deepening-upward stromatolite biostrome of the transgressive systems tract (TST) is overlain by progressively thinning-upward stromatolite-dominated parasequences of the highstand systems tract (HST). A sequence boundary occurs at or near the top of the stromatolite-dominant highstand and is overlain by an HST composed of lacustrine carbonates and continental redbeds. A major erosion surface with up to 100 m of incision forms a sequence boundary at the top of the Loves Creek Member.

quence set. Here are found the thickest cycles composed of well-developed subtidal bases, but thin and often poorly developed intertidal caps. These cycles contrast with those in the upper parts of the parasequence set where less accommodation produces thinner cycles with less well-developed subtidal bases, but thicker, more evaporitic intertidal caps (Fig. 20). A sequence boundary occurs at or near the contact of the stromatolite-dominant unit and the overlying unit composed of redbeds and lacustrine carbonate rocks. In some places parasequence stacking patterns suggest the sequence boundary lies within the uppermost parts of the marine stromatolite-dominant unit. In these cases a gradational contact separates marine stromatolites from the non-marine redbeds. Elsewhere a sharp erosion surface separates the two units and the sequence boundary is placed at this surface. Sequence subdivision has not been attempted in the non-marine unit of redbeds and lacustrine carbonate rocks. However, the similarity in carbonate facies and evaporite mineral assemblage throughout this interval suggests that marine influences were absent throughout this interval of stratigraphy (Southgate, 1991). The top of the Bitter Springs Formation is a regional erosion surface with up to 100 m of local incision (Fig. 20).

The same sequence of events appears to have occurred in stage 2 of basin development where rapid subsidence (early extension and thermal subsidence) resulted in widespread clastics (sequences 1 and 2) and the local tectonically controlled deposition of evaporites (lowstand deposits of sequence 3), followed by more extensive highstand carbonate and mixed carbonate–clastic units of sequence 3. Basin subsidence, in combination with flexural effects and long-term global sea-level rise, led to the onlapping of these units and the overlying sequences 4 and 5 onto the basin margin and their rapid spread beyond the area of initial extension. During this phase, cyclic carbonate deposition resulted in parasequence stacking patterns similar to those observed in the equivalent phase of stage 1 (compare Figs 16 & 20). This phase of sedimentation ends in a major erosional sequence boundary upon which incised valleys developed (basal sequence of the Pacoota Sandstone) which may relate to flexural effects.

With the diminishing of flexural effects, a broader hinterland once again became available to provide a source for clastic sediments. Relief of the source area was reduced by this time, thermal subsidence effects had also diminished and the sediments were progressively reworked in a shallow peritidal setting. Ultimately, subsidence became negligible and sedimentation ceased.

CONCLUSIONS

Sequence-stratigraphic concepts developed from analysis of Mesozoic–Cenozoic passive-margin successions have been shown to be broadly applicable to latest Proterozoic and Cambrian strata deposited within an intracratonic basin. This analysis is based largely on the recognition of facies discontinuities and biostratigraphic hiatuses in field outcrops and well logs, and is integrated with seismic data. The key to using sequence stratigraphy in an intracratonic setting is understanding the geometry of sequences and the factors controlling them. Once the geometry is understood we can then recognize the sequence boundaries and identify the facies of the various systems tracts.

In an intracratonic setting such as the Amadeus Basin, stratigraphic sequences are generally thin and poorly differentiated compared to their passive-margin equivalents. Slow subsidence rates, low depositional slopes and shallow water depths collectively result in diminished sediment accommodation within these basins. The sequences are extensive and thin, few have recognizable progradational geometries and erosional unconformities generally have minimal topographic relief. During relative sea-level lowstands, little or no accommodation space may be available for sediment accumulation and lowstand deposits may be poorly developed and areally restricted. Thus intracratonic successions commonly comprise stacked transgressive–highstand deposits separated by near-planar unconformities or paraconformities; that is, flooding surfaces commonly coincide with sequence boundaries. Analysis of parasequence stacking patterns suggests that depositional systems comparable to the shelf-margin systems tracts of passive-margin settings also occur within intracratonic basins. That is, despite the lack of a shelf–slope break, type 2 sequences can be locally recognized in intracratonic successions. An understanding of the origin and distribution of these sequences and systems tracts provides a guide to petroleum exploration in the Amadeus Basin.

Our studies suggest that basin dynamics created depositional space and determined the location of major depocentres, whilst eustasy determined the timing of sequence boundaries and the distribution of facies associations. Since some of the major tectonic events appear to be controlled by global events, depositional patterns within the Amadeus and associated Australian intracratonic basins are, to a large degree, synchronous. This suggests that an understanding of major variables associated with basin dynamics and their relationship to depositional sequences may allow the development of generalized depositional models on a basinal scale, and that sequence stratigraphy can be used as a basis for inter-regional correlation; a possibility that has considerable significance in unfossiliferous Archaean and Proterozoic basins. Ultimately this will allow a refined correlation to be developed across the continent.

ACKNOWLEDGEMENTS

We thank Magellan Petroleum Australia Ltd and the Amadeus Joint Venture parties for access to seismic and well data. Discussions with J.H. Shergold assisted greatly with biostratigraphy. Application of sequence concepts benefited greatly from reviews of the original manuscript by N. Christie-Blick and J. Cosgrove and discussions with P.R. Vail and J.F. Sarg.

REFERENCES

Bradshaw, J. (1988) The depositional, diagenetic and structural history of the Chandler Formation and related units, Amadeus Basin, central Australia. Unpublished PhD thesis, University of New South Wales, Sydney.

Bradshaw, J. (1991a) Description and depositional model of the Chandler Formation: a Lower Cambrian evaporite and carbonate sequence, Amadeus Basin, central Australia. In: *Geological and Geophysical Studies in the Amadeus Basin, Central Australia* (Eds Korsch, R.J. & Kennard, J.M.) Bureau of Mineral Resources, Australia, Bull. 236, pp. 227–244.

Bradshaw, J. (1991b) The Tempe Formation: an early Middle Cambrian, open marine, clastic and carbonate sequence, central Amadeus Basin. In: *Geological and Geophysical Studies in the Amadeus Basin, Central Australia* (Eds Korsch, R.J. & Kennard, J.M.) Bureau of Mineral Resources, Australia, Bull. 236, pp. 245–252.

Conway Morris, S. (1988) Radiometric dating of the Precambrian–Cambrian boundary in the Avalon Zone. In: *Trace Fossils, Small Shelly Fossils and the Precambrian-Cambrian Boundary* (Eds Landing, E., Narbonne, G.M. & Myrow, P.) 1987, Memorial University, Proceedings, New York State Geol. Surv. Bull. 463, 53–58.

Deckelman, J.A. (1985) The petrology of the early Middle Cambrian Giles Creek and upper Chandler Formations, northeastern Amadeus Basin, central Australia. Unpublished MS thesis, Utah State University, Logan, Utah.

Gorter, J.D. (1991) Palaeogeography of late Cambrian to early Ordovician sediments in the Amadeus Basin, central Australia. In: *Geological and Geophysical Studies in the Amadeus Basin, Central Australia* (Eds Korsch, R.J. & Kennard, J.M.) Bureau of Mineral Resources, Australia, Bull. 236, pp. 253–275.

Grotzinger, J.P. (1986) Cyclicity and paleoenvironmental dynamics, Rocknest Platform, northwest Canada. *Bull. Geol. Soc. Am.* **97**, 1208–1231.

Haq, B.U., Hardenbol, J. & Vail, P.R. (1988) Mesozoic and Cenozoic chronostratigraphy and eustatic cycles. In: *Sea-level Changes: an Integrated Approach* (Eds Wilgus, C.K., Hastings, B.S., Kendall, C.G.St.C., Posamentier, H.W., Ross, C.A. & Van Wagoner, J.C.) Spec. Publ. Soc. Econ. Paleontol. Mineral. 42, 71–108.

James, N.P., Stevens, R.S., Barnes, C.R. & Knight, I. (1989) Evolution of a Lower Paleozoic continental-margin carbonate platform, northern Canadian Appalachians. In: *Controls on Carbonate Platform and Basin Development* (Eds Crevello, P.D., Wilson, J.L., Sarg, J.F. & Read, J.F.) Spec. Publ. Soc. Econ. Paleontol. Mineral. 44, 123–146.

Jenkins, R.J.F. (1984) Ediacaran events: boundary relationships and correlation of key sections, especially in 'Armorica'. *Geol. Magazine* **121**, 635–643.

Jones, B.G. (1972) Upper Devonian to Lower Carboniferous stratigraphy of the Pertnjara Group, Amadeus Basin, central Australia. *J. Geol. Soc. Austral.* **19**, 229–249.

Keith, J.W. (1974) Carbonate sediments of the Cambrian Pertaoorrta Group, eastern Amadeus Basin, Northern Territory, Australia. Magellan Petroleum Australia Ltd (unpublished).

Kennard, J.M. (1991) Lower Cambrian archaeocyathan buildups, Todd River Dolomite, northeast Amadeus Basin, central Australia: sedimentology and diagenesis. In: *Geological and Geophysical Studies in the Amadeus Basin, Central Australia* (Eds Korsch, R.J. & Kennard, J.M.) Bureau of Mineral Resources, Australia, Bull. 236, pp. 195–225.

Kennard, J.M. (in press) Thrombolites and stromatolites within shale–carbonate cycles, middle-late Cambrian Shannon Formation, Amadeus Basin, central Australia. In: *Phanerozoic Stromatolites II* (Eds Sarfati, J. & Monty, C.L.V.) Springer-Verlag, Berlin.

Kennard, J.M. & James, N.P. (1986) Thrombolites and stromatolites: two distinct types of microbial structures. *Palaios* **1**, 492–503.

Kennard, J.M. & Lindsay, J.F. (1991) Sequence stratigraphy of the latest Proterozoic — Cambrian Pertaoorrta Group, northern Amadeus Basin, central Australia. In: *Geological and Geophysical Studies in the Amadeus Basin, Central Australia* (Eds Korsch, R.J. & Kennard, J.M.) Bureau of Mineral Resources, Australia, Bull. 236, pp. 171–194.

Korsch, R.J. & Lindsay, J.F. (1989) Relationships between deformation and basin evolution in the intracratonic Amadeus Basin, central Australia. *Tectonophysics* **158**, 5–22.

Lindsay, J.F. (1987a) Sequence stratigraphy and depositional controls in late Proterozoic–early Cambrian sediments of the Amadeus Basin, central Australia. *Bull. Am. Assoc. Petrol. Geol.* **71**, 1387–1403.

Lindsay, J.F. (1987b) Upper Proterozoic evaporites in the Amadeus Basin, central Australia and their role in basin tectonics. *Bull. Geol. Soc. Am.* **99**, 852–865.

Lindsay, J.F. (1989) Depositional controls on glacial facies associations in a basinal setting, late Proterozoic, Amadeus Basin, central Australia. *Palaeogeogr. Palaeoclim., Palaeoecol.* **73**, 205–232.

Lindsay, J.F. & Gorter, J.D. (1993) Clastic petroleum reservoirs of the late Proterozoic and early Paleozoic Amadeus Basin, Central Australia. In: *Marine Clastic Reservoirs: Examples and Analogs* (Eds Rhodes, E.G. & Moslow, T.) pp. 39–74. Springer-Verlag, New York.

Lindsay, J.F. & Korsch, R.J. (1989) Interplay of tectonics and sea-level changes in basin evolution: an example from the intracratonic Amadeus Basin, central Australia. *Basin Res.* **2**, 3–25.

Lindsay, J.F. & Korsch, R.J. (1991) The evolution of the Amadeus Basin, central Australia. In: *Geological and Geophysical Studies in the Amadeus Basin, Central Australia* (Eds Korsch, R.J. & Kennard, J.M.) Bureau of Mineral Resources, Australia, Bull. 236, pp. 7–32.

Lindsay, J.F., Korsch, R.J. & Wilford, J.R. (1987) Timing the breakup of a Proterozoic supercontinent: evidence from Australian intracratonic basins. *Geology* **15**, 1061–1064.

Odin, G.S., Gale, N.H., Auvray, B., *et al.* (1983) Numerical dating of the Precambrian–Cambrian boundary. *Nature* **301**, 21–23.

Read, J.F. (1989) Controls on evolution of Cambro-Ordovician passive-margin, U.S. Appalachians. In: *Controls on Carbonate Platform and Basin Development* (Eds Crevello, P.D., Wilson, J.L., Sarg, J.F. & Read, J.F.) Spec. Publ. Soc. Econ. Paleontol. Mineral. 44, 147–166.

Shaw, R.D. (1991) The tectonic development of the Amadeus Basin, central Australia. In: *Geological and Geophysical Studies in the Amadeus Basin, Central Australia* (Eds Korsch, R.J. & Kennard, J.M.) Bureau of Mineral Resources, Australia, Bull. 236, pp. 429–461.

Shergold, J.H. (1986) *Review of the Cambrian and Ordovician palaeontology of the Amadeus Basin, Central Australia.* Bureau of Mineral Resources, Australia, Report, 276, 21 pp.

Shergold, J.H. (Compiler) (1989) *Australian Phanerozoic Time Scales: 1. Cambrian Biostratigraphic Chart and Explanatory Notes.* Bureau of Mineral Resources, Australia, Record, 1989/31, 25 pp.

Shergold, J.H., Elphinstone, R., Laurie, J.R., *et al.* (1991a) Late Proterozoic and early Palaeozoic palaeontology and biostratigraphy of the Amadeus Basin. In: *Geological and Geophysical Studies in the Amadeus Basin, Central Australia* (Eds Korsch, R.J. & Kennard, J.M.) Bureau of Mineral Resources, Australia, Bull. 236, pp. 97–111.

Shergold, J.H., Gorter, J.D., Nicoll, R.S. & Haines, P.W. (1991b) *Stratigraphy of the Pacoota Sandstone (Cambrian–Ordovician), Amadeus Basin, N.T.* Bureau of Mineral Resources, Australia, Bull. 237.

Sloss, L.L. (1988) Forty years of sequence stratigraphy. *Bull. Geol. Soc. Am.* **100**, 1661–1665.

Southgate, P.N. (1991) A sedimentological model for the Loves Creek Member of the Bitter Springs Formation, northern Amadeus Basin. In: *Geological and Geophysical Studies in the Amadeus Basin, Central Australia* (Eds Korsch, R.J. & Kennard, J.M.) Bureau of Mineral Resources, Australia, Bull. 236, pp. 113–126.

Southgate, P.N. & Shergold, J.H. (1991) Application of sequence stratigraphic concepts to Middle Cambrian phosphogenesis, Georgina Basin, Australia. *Bur. Min. Res. J. Austral. Geol. Geophys.* **12**, 119–144.

Vail, P.R. (1987) Seismic stratigraphy interpretation using sequence stratigraphy, Part 1: seismic stratigraphy interpretation procedure. In: *Atlas of Seismic Stratigraphy* (Ed. Bally, A.W.) Am. Assoc. Petrol. Geol. Stud. Geol. 27 (1), 1–10.

Vail, P.R. & Sangree, J.B. (Compilers and Eds) (1988) Sequence Stratigraphy Workbook. Earth Resources Foundation, University of Sydney (unpublished).

Vail, P.R., Hardenbol, J. & Todd, R.G. (1984) Jurassic unconformities, chronostratigraphy and sea-level changes from seismic stratigraphy and biostratigraphy. In: *Interregional Unconformities and Hydrocarbon Accumulation* (Ed. Schlee, J.S.) Mem. Am. Assoc. Petrol. Geol. 36, 129–144.

Vail, P.R., Mitchum, R.M. & Thompson, S., III (1977a) Seismic stratigraphy and global changes of sea level, Part 3: relative changes of sea level from coastal onlap. In: *Seismic Stratigraphy — Applications to Hydrocarbon Exploration* (Ed. Payton, C.W.) Mem. Am. Assoc. Petrol. Geol. 26, 63–81.

Vail, P.R., Mitchum, R.M. & Thompson, S., III (1977b) Seismic stratigraphy and global changes of sea level, Part 4: global cycles of relative changes of sea level. In: *Seismic Stratigraphy — Applications to Hydrocarbon Exploration* (Ed. Payton, C.W.) Mem. Am. Assoc. Petrol. Geol. 26, 83–97.

Van Wagoner, J.C., Mitchum, R.M., Campion, K.M. & Rahmanian, V.D. (1990) Siliciclastic sequence stratigraphy in well logs, cores, and outcrops: concepts for high-resolution correlation of time and facies. *Am. Assoc. Petrol. Geol. Meth. Expl. Ser.* **7**, 55 pp.

Van Wagoner, J.C., Mitchum, R.M., Jr, Posamentier, H.W. & Vail, P.R. (1987) Seismic stratigraphy interpretation using sequence stratigraphy — part 2: key definitions of sequence stratigraphy. In: *Atlas of Seismic Stratigraphy* (Ed. Bally, A.W.) Am. Assoc. Petrol. Geol. Stud. Geol. 27, 11–14.

Van Wagoner, J.C., Posamentier, H.W., Mitchum, R.M., Jr., *et al.* (1988) An overview of the fundamentals of sequence stratigraphy and key definitions. In: *Sea-level Changes: An Integrated Approach* (Eds Wilgus, C.K., Hastings, B.S., Kendall, C.G.St.C., Posamentier, H.W., Ross, C.A. & Van Wagoner, J.C.) Spec. Publ. Soc. Econ. Paleontol. Mineral. 42, 39–45.

Von Der Borch, C.C., Christie-Blick, N. & Grady, A.E. (1988) Depositional sequence analysis applied to late Proterozoic Wilpena Group, Adelaide Geosyncline, South Australia. *Austral. J. Earth Sci.* **35**, 59–72.

Walter, M.R., Elphinstone, R. & Heys, G.R. (1989) Proterozoic and Early Cambrian trace fossils from the Amadeus and Georgina Basins, central Australia. *Alcheringa,* **13**, 209–256.

Wells, A.T., Forman, D.J. & Ranford, L.C. (1965) The geology of the north-western part of the Amadeus Basin, Northern Territory. *Bur. Mineral Resourc. Austral. Report,* **85**, 45pp.

Wells, A.T., Forman, D.J., Ranford, L.C. & Cook, P.J. (1970) Geology of the Amadeus Basin, central Australia. *Bur. Mineral Resourc. Austral. Bull.* **100**, 222 pp.

Index

Page numbers in *italic* refer to figures and tables.